47th Annual Loss Prevention Symposium 2013

(LPS)

Topical Conference at the 2013 AIChE Annual Meeting and 9th Global Congress on Process Safety

San Antonio, Texas, USA
28 April – 2 May 2013

Volume 1 of 2

ISBN: 978-1-62748-447-3

Printed from e-media with permission by:

Curran Associates, Inc.
57 Morehouse Lane
Red Hook, NY 12571

Printed by Curran Associates, Inc. (2013)

For permission requests, please contact AIChE
at the address below.

AIChE
3 Park Avenue
New York, NY 10016-5991

Phone: (203) 702-7660
Fax: (203) 775-5177

www.aiche.org

Additional copies of this publication are available from:

Curran Associates, Inc.
57 Morehouse Lane
Red Hook, NY 12571 USA
Phone: 845-758-0400
Fax: 845-758-2634
Email: curran@proceedings.com
Web: www.proceedings.com

TABLE OF CONTENTS

9TH GLOBAL CONGRESS ON PROCESS SAFETY

Enhanced NOAA Chemical Reactivity Worksheet for Determining Chemical Compatibility

Dave Gorman
The Dow Chemical Company
1897 Building – F Bay
Midland, MI 48667
dbgorman@dow.com

Rob Bellair
The Dow Chemical Company
B-1463/2
Freeport, TX 77541
RJBellair@dow.com

Jim Farr
National Oceanic and Atmospheric
Administration (NOAA)
Emergency Response Division
Western Regional Center
7600 Sand Point Way NE
Seattle, WA 98115
Jim.Farr@noaa.gov

Wade Freeman
University of Illinois – Chicago (Emeritus Professor)
1314 W. Arthington St.
Chicago IL 60607-4103
wfreeman@uic.edu

Dave Frurip
The Dow Chemical Company
1897 Building – F Bay
Midland, MI 48667
djfrurip@dow.com

Al Hielscher
Genwest Systems, Inc.
P.O. Box 397
Edmonds, WA 98020
alh@genwest.com

Harold Johnstone
The Dow Chemical Company (Retired)
205 Peppermint Dr
Lake Jackson, TX 77566

hjohnstone7@gmail.com

Myriam Linke
The Dow Chemical Company
ADD Building 446/1
Terneuzen, ZE (Netherlands) 4530AA
MLinke2@dow.com

Pat Murphy
The Dow Chemical Company
1897 Building – F Bay
Midland, MI 48667
PGMurphy@dow.com

Min Sheng
The Dow Chemical Company
1897 Building – F Bay
Midland, MI 48667
MSheng@dow.com

Klaas Van Gelder
The Dow Chemical Company
ADD Building 446/1
Terneuzen, ZE (Netherlands) 4530AA
kbvangelder@dow.com

Dalina Viveros
National Oceanic and Atmospheric
Administration (NOAA)

Emergency Response Division
Western Regional Center
7600 Sand Point Way NE
Seattle, WA 98115
Dalina.Viveros@noaa.gov

Prepared for Presentation at
American Institute of Chemical Engineers
2013 Spring Meeting
9th Global Congress on Process Safety
San Antonio, Texas
April 28 – May 1, 2013

UNPUBLISHED

Enhanced NOAA Chemical Reactivity Worksheet for Determining Chemical Compatibility

Dave Gorman
The Dow Chemical Company
1897 Building – F Bay
Midland, MI 48667
dbgorman@dow.com

Rob Bellair
The Dow Chemical Company

Jim Farr
National Oceanic and Atmospheric Administration

Wade Freeman (Emeritus Professor)
University of Illinois - Chicago

Dave Frurip
The Dow Chemical Company

Al Hielscher
Genwest

Harold Johnstone (Retired)
The Dow Chemical Company

Myriam Linke
The Dow Chemical Company

Pat Murphy
The Dow Chemical Company

Min Sheng
The Dow Chemical Company

Klaas Van Gelder
The Dow Chemical Company

Dalina Viveros
National Oceanic and Atmospheric Administration

Keywords: NOAA, chemical compatibility, compatibility chart, chemical reactivity worksheet

Abstract

Chemical compatibility is a key consideration throughout the chemical industry wherever two or more chemicals have the potential to mix, either inadvertently or by design. Although several tools exist in the public domain to assist with chemical compatibility decisions, all suffer from various limitations. The development of a single, universally accepted chemical compatibility tool has eluded the chemical industry and has been an ongoing, strategic need. One of the most comprehensive and widely used tools available, the NOAA Chemical Reactivity Worksheet (CRW), has been criticized in the past for being too conservative in evaluating chemical compatibility and being limited in the scope of chemicals that could be evaluated. As a result of a two-year joint effort between the National Oceanic and Atmospheric Administration (NOAA) and The Dow Chemical Company, an enhanced version of this tool now is available. In addition to addressing previously identified limitations, several enhancements have been incorporated into the tool to help with the generation and presentation of chemical compatibility charts. This paper will cover the development and use of this new tool, highlighting the enhancements that position this tool to become the gold standard within the chemical industry for determining chemical compatibility.

1. Introduction

Chemical compatibility is a key consideration throughout the chemical industry wherever two or more chemicals have the potential to mix, either inadvertently or by design. Unfortunately, accidental mixing of incompatible materials continues to be a major cause of reactive chemicals incidents in the global chemical industry and elsewhere. It is critical to have a solid understanding of the potential energy release in the form of heat and pressure from mixing two or more chemicals together. The consequence of misjudging or being unaware of this potential has led in the past to significant loss of property, harm to the environment, and fatalities.[1] Unfortunately, the availability of a single, consistent, universally accepted chemical compatibility tool has eluded the chemical industry and has been an ongoing, strategic need. This need was recognized within the Reactive Chemicals group at The Dow Chemical Company and a team was assembled to address this need. Although the original scope of this work was to develop a standardized chemical compatibility tool for use within The Dow Chemical Company, in the course of pursuing this goal, it was realized that the best path forward was a chemical compatibility tool that could be utilized throughout the chemical industry. As a result of a two-year joint effort between the National Oceanic and Atmospheric Administration (NOAA) and The Dow Chemical Company, an enhanced version of NOAA's Chemical Reactivity Worksheet (CRW) [2] now is available and is described in this paper. The CRW was created originally by NOAA for prediction of chemical hazards that may exist during a chemical spill event involving multiple chemicals (i.e. a train derailment scenario) as an aid to first responders in 1998. It's applicability to laboratory and the industrial chemical process industry became an obvious end result as well. The CRW has been available for download in it's various versions since 1998 (downloaded at a rate of about 1,000 per month to the present).

Enhancements incorporated into the new tool include:

- More accurate (less conservative) compatibility decision logic
- Addition of several new reactivity groups
- Color and letter coding on compatibility charts to provide quick, eye-friendly evaluation of chemical compatibility.
- NFPA codes have been added and automatically appear on compatibility charts when available
- Standard Hill notation for specific chemicals has been added to aid in database searching
- Ability to add custom chemicals
- Ability to override general compatibility decisions to match user experience
- Autosave feature which preserves all user entered data for future use
- The number of chemicals in a mixture has been extended to 100 chemicals
- A section has been added that addresses absorbent incompatibility for chemicals in the NOAA chemical database.
- A new and much improved user software interface

With these enhancements, we believe this tool is positioned to become the gold standard within the chemical industry for determining chemical compatibility.

2. Evaluation of Existing Chemical Compatibility Tools

In the quest to develop a state-of-the-art chemical compatibility tool, the first step was to develop a list of ideal features and capabilities of such a tool. After prioritizing those features and capabilities, the second step was to evaluate existing tools against these criteria. By potentially incorporating existing technology into the new tool, this avoided reinvention of the wheel and allowed for the optimum utilization of limited resources for the project.

2.1 Tools Evaluated

Several different tools were known to exist for determining the chemical compatibility of two materials. Some of these were static charts and others required user input. In all, eight different chemical compatibility tools were evaluated, of which five had been developed and used exclusively within Dow and its subsidiaries. Those tools available in the public domain included the NOAA Chemical Reactivity Worksheet, the EPA Chart [3], and the U.S. Coast Guard Incompatibility Chart [4].

2.2 Evaluation Results

Over 30 criteria were considered for incorporation into a state-of-the-art chemical compatibility tool. These criteria were prioritized based on perceived importance, with thirteen items rated most important. In evaluating the results, it was realized that nearly all of the top-rated criteria for a state-of-the-art chemical compatibility tool could be obtained by merging the strengths of Dow Tool #1 with the NOAA Chemical Reactivity Worksheet (Table 1). Although neither tool had obtained the top overall score in this evaluation, the complimentary features of each tool made this a particularly attractive option to pursue. Based on these results, a NOAA-Dow collaboration to jointly work on an enhanced version of the Chemical Reactivity Worksheet, to be owned and maintained exclusively by NOAA, was initiated in late fall, 2010.

Table 1: Evaluation of NOAA CRW Tool and Dow Tool #1 against Top Criteria

"Should" List for Compatibility Chart Tool	NOAA CRW Tool	Dow Tool #1
Ability to leverage existing compatibility information (no reinventing the wheel for cross-compatibility) - globally shared	0.5	**1**
Searchable database including general functional groups	**1**	0.5
Picklist to generate list of chemicals for chart	**1**	0
One button feature to generate chart from picklist	0.75	0
Appropriate risk evaluation of compatibility (less conservative than NOAA tool); scientifically valid & consistent method for confirming compatibility	0.5	**1**
Easily modified - adding or deleting materials	0.5	0
Template readily accessible	1	1
Tracking of revision history	0.5	0
User friendly - utilizes simple interface, only need basic computing skills, not prone to crashing	1	0.5
Robust - accurate results, broad applicability to various compounds/functionalities	0.75	**1**
Air and nitrogen added to all charts (flammability issues - eg. monomers)	0.5	0
Water added to all charts	0.75	1
Allows input of customized chemicals	0.5	0.5

Key: 1 = strength of tool, 0 = deficiency of tool, other values between 0-1 indicate partial strength of tool. Bolded numbers indicate an exceptional strength of the tool over other existing tools.

For a full summary of all tools evaluated against all criteria, see Appendix A.

3. Key Changes Incorporated into the Enhanced NOAA Chemical Reactivity Worksheet

3.1 Improved Scenario Definition

The scenario-based predictions for the CRW are described elsewhere [2], but were confined to the addition of equal portions of any two materials under "mild" conditions, although we considered ambient conditions to cover heats one may expect in the Southwestern United States. For this version of the CRW, compatible compounds were considered those materials that could not in any conceivable way produce a hazard under these conditions. Compatible combinations are represented by a green box with the letter "Y" within the box. Incompatible combinations were considered as those producing heats of reaction estimated on average to be above 100 calories/gram of reactants, having inherent flammability/explosivity, or producing toxic flammable products. Incompatible combinations are represented by a red box with the letter "N" within the box. Finally, a new category, the cautionary category, was developed to inform the user of certain categorical hazards that exist or where mild reactivity may be present. Caution combinations are represented by a yellow box with the letter "C" within the box. An example of this category can be found in the "Polymerizable Group" that warns the user of the inherent reactivity issues with combining a polymerizable substance with any specific chemical.

3.2 Updated Chemical Compatibility Information

The NOAA Chemical Reactivity Worksheet allows searching by either chemical name for a specific chemical, CAS#, UN #, or by reactivity group according to the molecular structure of a chemical. The database was updated extensively to create the enhanced version of the worksheet. In the course of updating/revising/adding to the NOAA chemical and reactivity group databases, several sources were consulted to ensure consistency of compatibility information. Those sources most widely utilized, in addition to the existing data already in the worksheet, included the following:

- Hazmat Navigator / Bretherick's Reactive Chemical Hazards Database – Version 3.0 [5]
- Fire Protection Guide to Hazardous Materials, 10th Ed. [6]
- "Compatibility of Cargoes," published in 46 CFR Ch. 1 (10-1-93 Edition) (Coast Guard Chart)
- EPA 600/2-80-76, "A Method for Determining the Compatibility of Chemical Mixtures" (EPA Chart)
- Dangerous Properties of Industrial Materials – Sixth Edition (Sax) [7]
- Hazardous Chemicals Desk Reference – Sixth Edition (Lewis) [8]
- Kirk-Othmer Encyclopedia of Chemical Technology [9]
- Ullmann's Encyclopedia of Industrial Chemistry [10]
- National Fire Protection Association, NFPA Publication 491M [11]
- The Merck Index [12]

- Scifinder and Google search engines
- Industrial and academic chemistry experts with extensive experience in the properties of the materials

For compatibility decisions among two reactivity groups, at least three sources generally were consulted, and each cross-compatibility input was double-checked by a second reviewer for completeness and accuracy. A final review of the compatibility decisions was carried out by a chemist at Argonne National Laboratory. In all, over 64 reactivity groups and over 2080 cross-compatibility cells have been evaluated and updated. Of those, 704 combinations were found to be incompatible, 416 were found to be "Cautionary" and 960 were found to be compatible.

3.3 Expanded Reactivity Groups and Special Hazard Codes

Sixty-five Reactivity Groups (Table 2) are available for assignment to chemical compounds. Numbers range from 1-100, with gaps either due to reactivity groups no longer in use or available slots for new reactivity groups.

Table 2 – List of Reactivity Groups

Number	Reactivity Group Name
1	Acids (Strong) Non-oxidizing
2	Acids, (Strong) Oxidizing
3	Carboxylic acid
4	Alcohols and Polyols
5	Aldehyde
6	Amides and imides
7	Amines, Phosphines, Pyridines
8	Azo, diazo, azido, hydrazines, azides
9	Carbamates
10	Base (Strong)
11	Cyanides, inorganic
12	Thiocarbamate Esters and Salt/Dithiocarbamate Esters and Salts
13	Ester, sulfate esters, phosphate and thiophosphate esters, borate esters
14	Ether
16	Aromatic hydrocarbon
17	Halogenated organic compounds
18	Isocyanates and isothiocyanates
19	Ketone
20	Sulfides, organic
21	Metals, Alkali, Very Active
22	Metals, Elemental & Powder, Active
23	Metals, less reactive
26	Nitriles
27	Nitro, nitroso, nitrate, nitrite, organic
28	Hydrocarbons, Aliphatic unsaturated

29	Hydrocarbons Aliphatic saturated
30	Peroxides, organic
31	Phenols, cresols
32	Sulfonates, Phosphates and Thiophosphates, Organic
33	Sulfides, inorganic
34	Epoxide
35	Metal hydrides, metal alkyls, metal aryls, and silanes
37	Anhydride
38	Salts, Acidic Inorganic/Organic
39	Salts, Basic Inorganic/Organic
40	Acid halide, sulfonyl halides, chloroformates
42	Organometallics
44	Inorganic Oxidizing Agents
45	Reducing Agent
46	Inorganic Compounds/Neither Reducing nor Oxidizing
47	Fluorinated compounds
51	Nitrides, phosphides, carbides, silicides
55	Chlorosilanes
58	Siloxanes
59	Halogenating agents
60	Acids (Weak)
61	Base (Weak)
62	Carbonate salt
63	Alkyne (attached hydrogen)
64	Alkyne (no attached hydrogen)
65	Diene
66	Aryl halide
67	Halide salt
68	Anilines
69	Nitrate, nitrite, inorganic salt
70	Acetals, ketals, hemiacetals, hemiketals
71	Acrylates, Acrylic Acids
72	Phenolic salts
73	Quaternary ammonium, phosphonium
74	Sulfite/ thiosulfate salt
75	Oximes
76	Polymerizable
98	Non chemically reactive
99	Unknown
100	Contains Water

In addition, twelve Special Hazard Codes also are available to highlight potentially serious consequences associated with a particular reactivity group (Table 3).

Table 3 – List of Special Hazard Codes

Hazard Code	Description
C	Reaction products may be corrosive
E	Reaction product may be explosive or sensitive to shock or friction
F	Reaction products may be flammable
G	Reaction liberates gaseous products and may cause pressurization
NR	No Known Hazardous Reaction
R1	Exothermic reaction at ambient temperatures (releases heat)
R2	Reaction product may be unstable above ambient temperatures
R3	Reaction may be particularly intense, violent, or explosive
R4	Polymerization reaction may become intense; cause pressurization
T	Reaction products may be toxic
UR	May be hazardous but unknown
X	Possible exposure to radiation

It is recognized that these groups cannot cover all possible molecular permutations, but these groups will accurately encompass a large majority of chemicals. It would be ideal to add several additional functional groups as new Reactivity Groups, but limited resources precluded that from happening in this release of the tool. In many of these cases, a reasonably similar Reactivity Group can be substituted. When substitute groups are used, compatibility decisions must be reviewed carefully, since the groups are only similar, not equivalent. A list of some of these missing groups, with recommended substitute Reactivity Groups, appears below (Table 4).

Table 4 – Missing Reactivity Groups and Recommended Substitute Reactivity Groups

Functional Group	Structure	Similar Reactivity Groups
Alkoxides	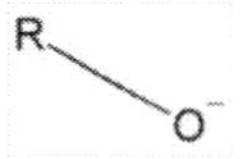	10 Base (Strong)
Beta Lactones	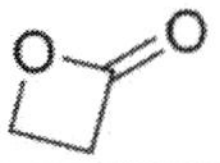	13 Ester, sulfate esters, phosphate and thiophosphate esters, borate esters
Boron Compounds	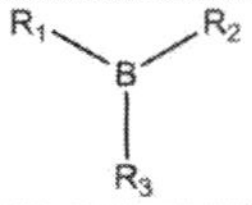	42 Organometallics
C,N heterocycles in general	Various	7 Amines, Phosphines, Pyridines
Cyanohydrins	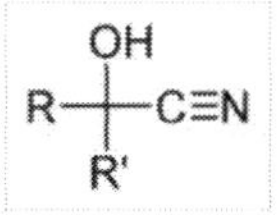	70 Acetals, ketals, hemiacetals, hemiketals

Diazonium salts	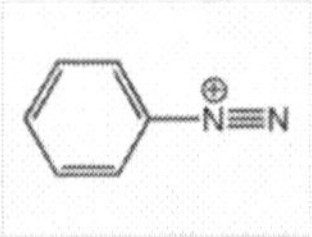	8 Azo, diazo, azido, hydrazines, azides
Hydrazone	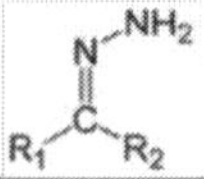	8 Azo, diazo, azido, hydrazines, azides and 19 Ketone
Imidazoles	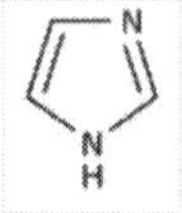 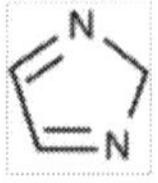	7 Amines, Phosphines, Pyridines
Isoxazoles	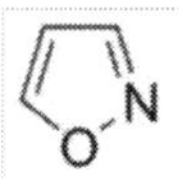	7 Amines, Phosphines, Pyridines
Oxazoles	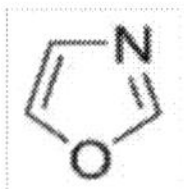	7 Amines, Phosphines, Pyridines and 14 Ether
Phosphazines	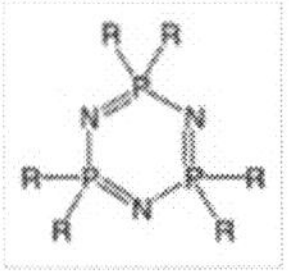	7 Amines, Phosphines, Pyridines
Pyrimidines	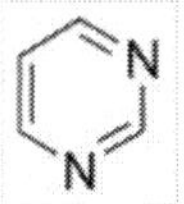	7 Amines, Phosphines, Pyridines
Semicarbazone	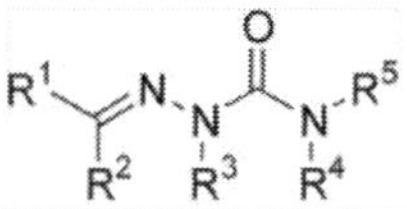	8 Azo, diazo, azido, hydrazines, azides and 9 Carbamates
Thiazines	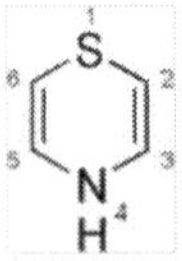	7 Amines, Phosphines, Pyridines And 20 Sulfides, organic
Thiazoles	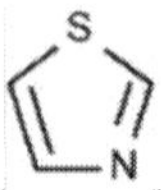	7 Amines, Phosphines, Pyridines And 20 Sulfides, organic
Triazines	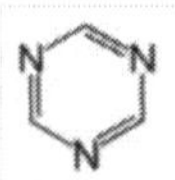	7 Amines, Phosphines, Pyridines
Urea	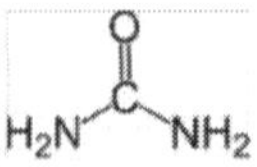	9 Carbamates

3.4 Compatibility Chart Display

In the NOAA Chemical Reactivity Worksheet, the typical output has been a black and white compatibility chart with letter coding corresponding to specific hazards. This can be difficult to read quickly, particularly in an emergency setting. In the enhanced version, a simplified color coded chart displays with three compatibility ratings:

- Y (yes) with a green background indicating the materials are compatible
- N (no) with a red background indicating the materials are not compatible
- C (caution) with a yellow background indicating the materials can be incompatible under certain circumstances, but may be fine to mix under appropriate conditions

The simplified color and letter coding provide a visually appealing chart that is easy to read. By clicking on any cell in the electronic form of the chart, additional information appears in a series of tabs easily navigated by the user. In addition, the chart displays NFPA codes for specific chemicals when the codes are present in the database. The new format is widely used at The Dow Chemical Company and other companies and is entirely consistent with the ASTM Guide for producing these charts.[1,13]

In cases where a compatibility decision predicted by the tool is contrary to what the user knows to be true, the chart may be modified to reflect a different compatibility decision. In those cases, an asterisk will appear next to the letter code to indicate the user has overridden the tool's prediction. At any point, the tool's original decision may be restored, which removes the asterisk from the letter coding.

Modification of any compatibility decision should be done with great care, but it is realized that the tool does not consider the context in which two materials are mixed, which is better appreciated by the user. For instance, the need to modify the tool's prediction of compatibility may arise due to a dilution effect that is not captured by the tool. There are two ways this can occur. One way is from a solvent effect. For example, the temperature rise expected from mixing 50% caustic with 30% hydrochloric acid will be much more severe than mixing 1% caustic with 1% hydrochloric acid. Although the chemistry of neutralization is the same in each case, the potential consequences from inadvertent mixing are not the same in the two cases. A second dilution effect can occur within the molecules themselves. For example, the impact of reacting a short chain acid vs. a long chain acid with an amine will be more severe because the acid comprises a greater weight percentage of the overall molecule. A molecule with a single carboxylic acid group with a MW of 1000, would probably not be a hazard if mixed with a base since the acid-base chemistry, releasing ca. -14 kcal/mol, would not result in a hazardous adiabatic temperature rise. The CRW tool, however, would flag the generally hazardous acid-base exothermic heat release hazard.

In addition, the tool does not predict the kinetics of a given reaction. Two materials may be incompatible, but the rate at which they react in a particular scenario may be readily manageable. Although every reasonable attempt has been made to reflect chemical compatibility under typical processing conditions, exceptions within reactivity groups are possible. Again, care should be exercised whenever a compatibility decision is changed from the tool's prediction.

3.5 Improved Search Functionality

In addition to a much expanded database of chemicals and reactivity groups, new features have been added to increase the versatility of the tool. First, the individual chemicals in the specific chemicals database have had their formulas updated to include the standard Hill notation for each chemical. This ensures consistency when searching for a chemical, which otherwise could be missed if a non-standard formula or the wrong synonym was searched.

3.6 Personalized Chemical Databases

The enhanced tool allows the user to build databases of custom chemicals not contained in the downloaded version of the NOAA database. This database contains about 5100 records of common chemicals that are EPA regulated chemicals, transported in bulk, have been in prior chemical accidents, and other substances that are in commerce (products like gasoline, turpentine, bleaching solutions, etc). In these cases, molecules are added to personalized databases and are characterized based on the reactivity groups, which are related to the functional groups present on the molecule. The tool allows for multiple reactivity groups to be assigned to a single molecule, allowing the tool to handle very complex molecules. Future plans include adding additional reactivity groups, including several heterocycles which currently are not represented specifically among the reactivity groups. In the interim, a list of suggested substitute reactivity groups are recommended for many potential reactivity groups not currently included in the database. As new molecules are added to the user's database, the databases, as well as mixtures used to generate the compatibility charts, are automatically saved by the software as the user interfaces with the program.

3.7 Absorbent Compatibility Information

The enhanced tool includes extensive compatibility information for common absorbents, dividing them into the main classifications of Cellulosic-Based Absorbents, Mineral/Clay-Based, Polymeric-Based, Sand, and Dirt/Earth. For each classification, potential incompatibilities with specific chemicals in the database are noted, as well as incompatible Reactive Groups. This information is important since many materials can and do react with absorbents with potentially negative consequences. A simple problematic example is a strong oxidizing material like sulfuric acid mixed with a cellulosic absorbent like peat moss.

4. Accessing and Using the Enhanced NOAA Chemical Reactivity Worksheet

4.1 Downloading the Software

4.2 Key Software Features with Screen Shots

Upon opening the software, the following screen (Figure 1) appears:

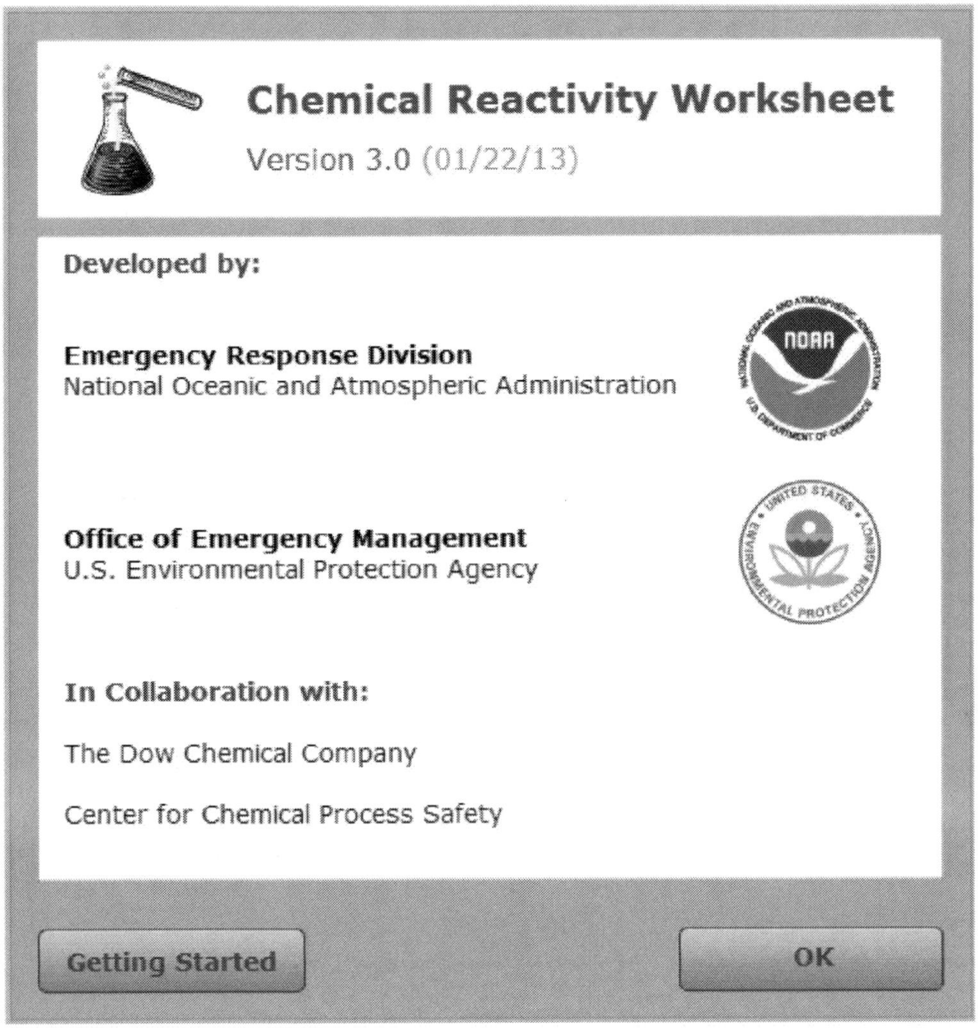

Figure 1: Welcome Screen

First time users should click the Getting Started button to become familiar with the key features of the software. Experienced users would click the OK button, which brings up the main screen (Figure 2).

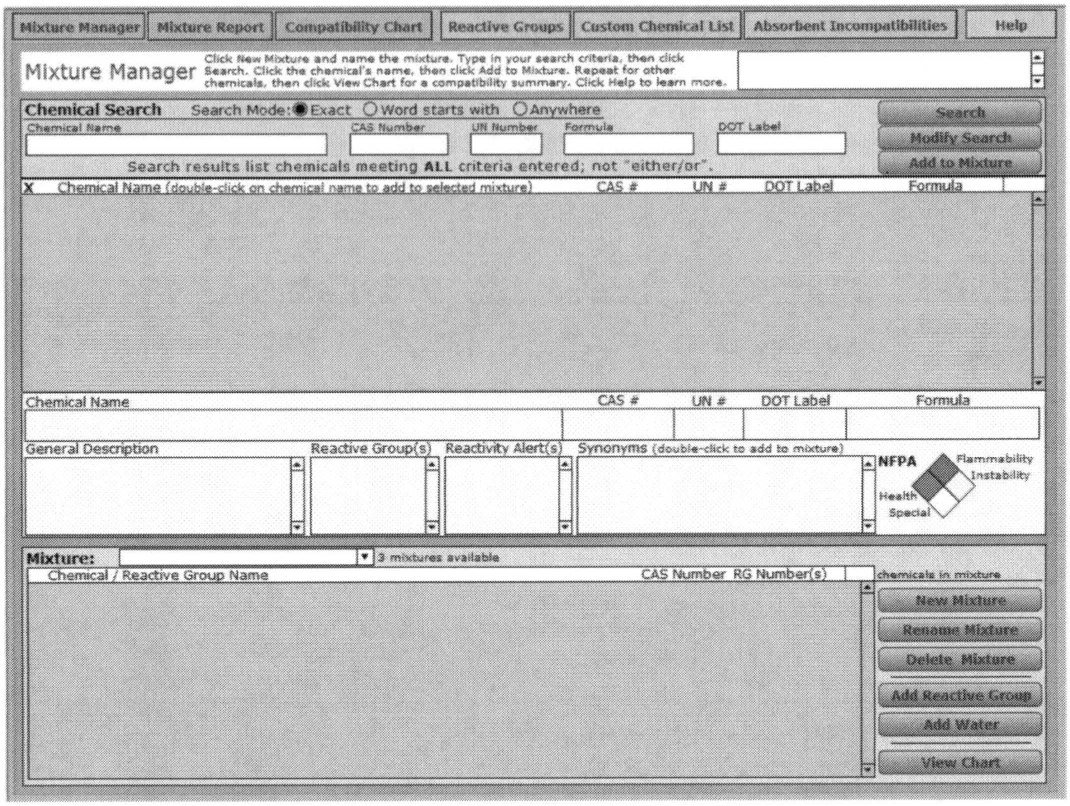

Figure 2: Main Screen

In order to select chemicals to create a chemical compatibility chart or to check cross compatibility, a mixture first must be created. To do this, the new mixture button in the lower right portion of the screen should be selected, at which point a pop up screen (Figure 3) for naming a mixture appears. After creating a mixture name and clicking okay, the mixture name will be added to the pull down menu in the middle/lower left hand side of the screen.

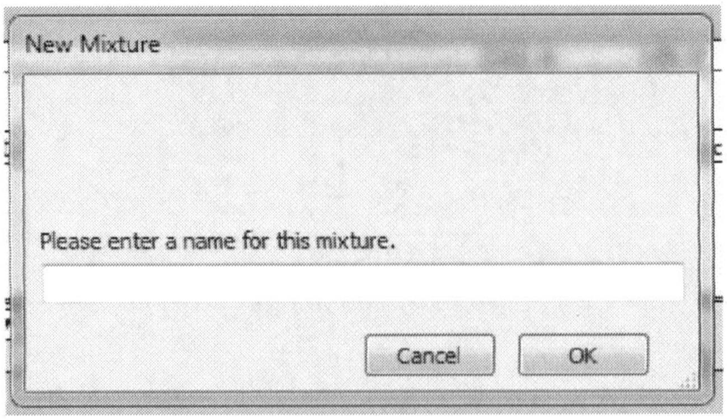

Figure 3: Pop Up Screen for Creating a Mixture

At this point, the upper portion of the screen may be used to search for chemicals to add to the mixture list. Chemicals may be searched by chemical name, CAS number, UN number, chemical formula (including Hill notation), and/or DOT label. Chemicals matching the criteria selected will appear below the search boxes (Figure 4). By highlighting the chemical of interest, additional information about that chemical appears below the list of chemicals. To add the chemical to the mixture, click on the Add to Mixture button below the NFPA code box at the far right where the chemical information is displayed. The chemical also can be added by double-clicking the leftmost box next to the green box in the Chemical Name window. At this point, an X will appear in the Chemical Name box and the chemical will appear in the Mixture Chemical Name list at the bottom of the page.

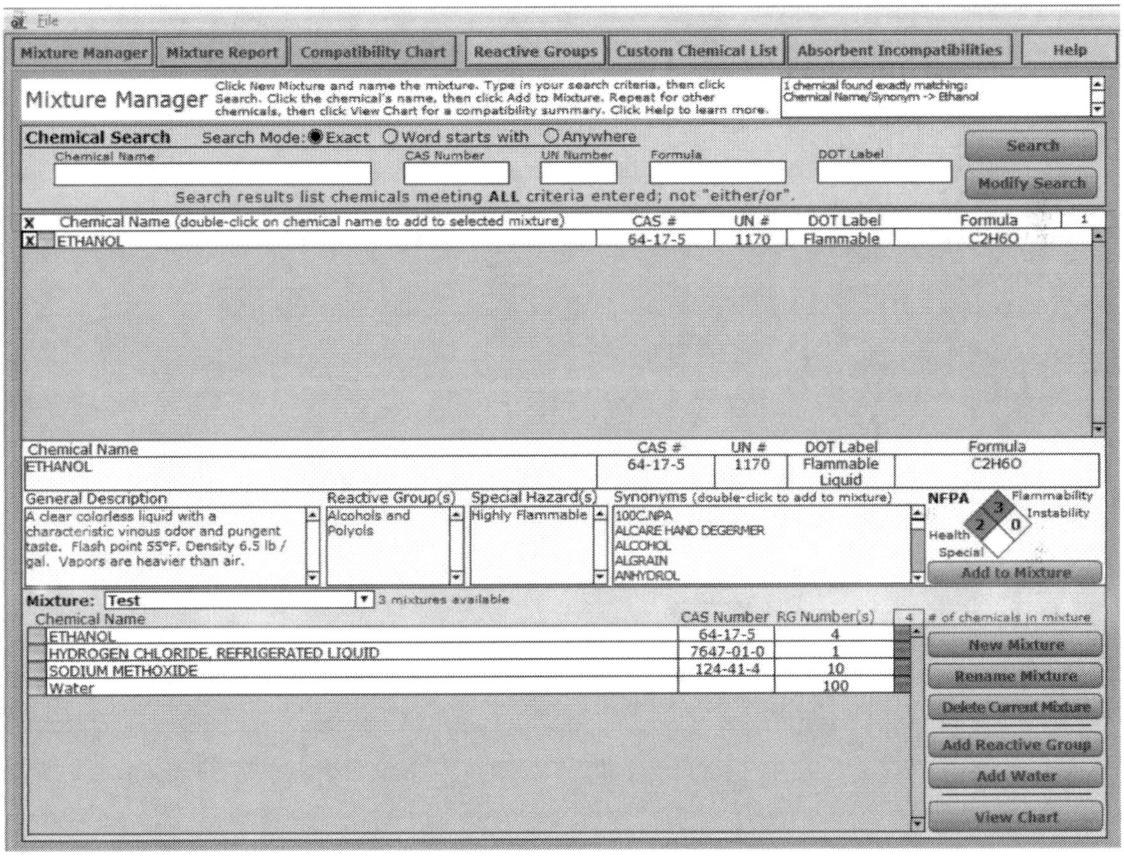

Figure 4: Example Display of Chemical Information

Continuing this process allows the user to generate a list of chemicals for a chemical compatibility chart, which appears at the bottom of the screen when the proper mixture is selected in the Mixture pull down menu box (Figure 5).

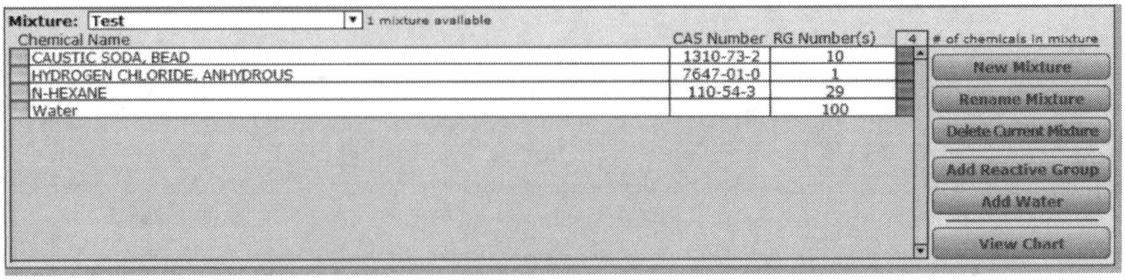

Figure 5: Example Chemical List for a Mixture

A series of buttons in the lower right corner of the screen allow the user to create and modify mixtures (Figure 6). Note that water may be added to mixtures using the Add Water button in this location versus searching for it in the Chemical Search portion of the screen.

Figure 6: Buttons for Creating and Modifying Mixtures

A chemical compatibility chart is generated by clicking on the View Chart button in the lower right corner of the screen. The chart displays a color and letter coded grid for the user, including NFPA codes where available (Figure 7). More detailed information appears below the chart and can be accessed via buttons directly above the white box where the information appears (Figure 8). Clicking on a colored cross-compatibility box in the chart highlights the chemicals being compared and displays additional information below the chart. Clicking on the individual chemicals in the chart does not highlight the box in the chart listing the chemical, but does display chemical information below the chart for that chemical only.

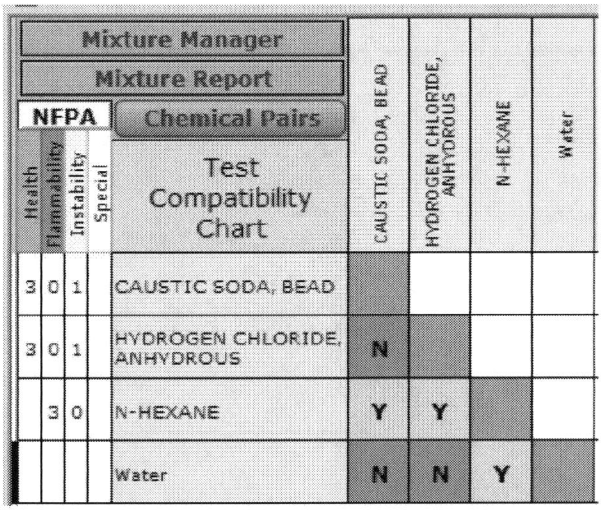

Figure 7: Example Chemical Compatibility Chart

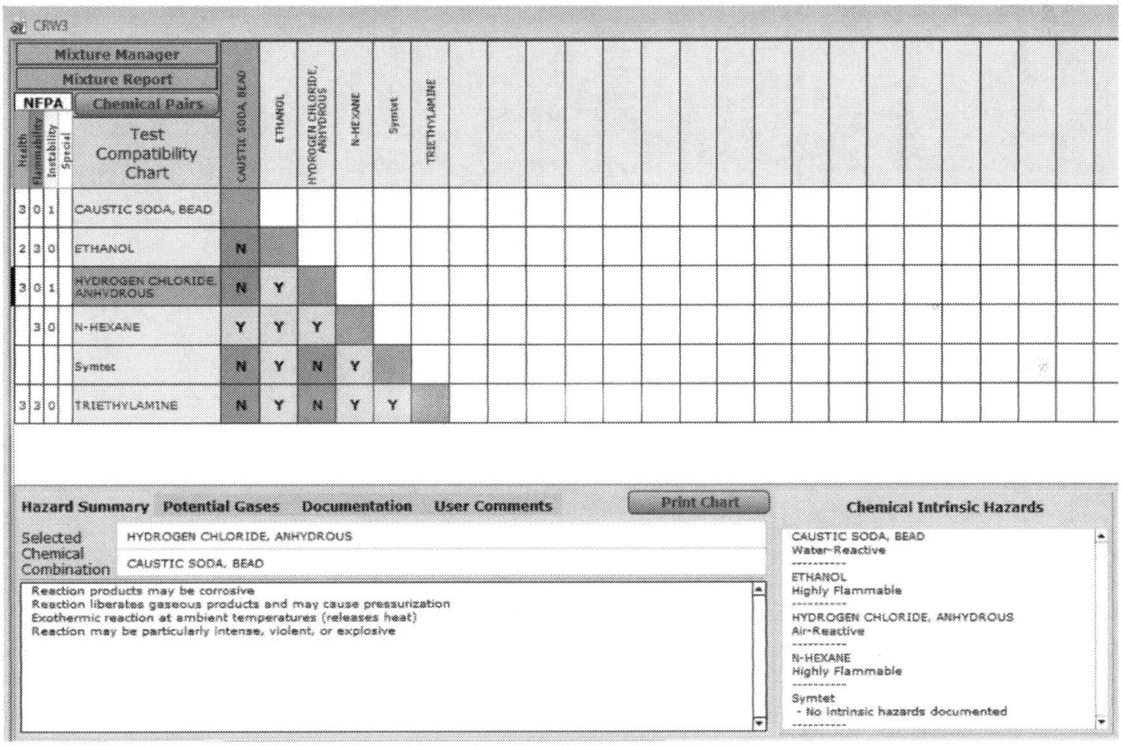

Figure 8: Additional Chemical Compatibility Information displayed for HCl + NaOH

4.3 Creating/Adding Custom Chemicals

To add chemicals which are not located in the Chemical Search database, click on the green Custom Chemical List box, then on the gray Add New Custom Chemical box. This will bring up a pop up box for naming the material to be added (Figure 9). After OK is clicked, a new sheet for entering physical properties will appear (Figure 10).

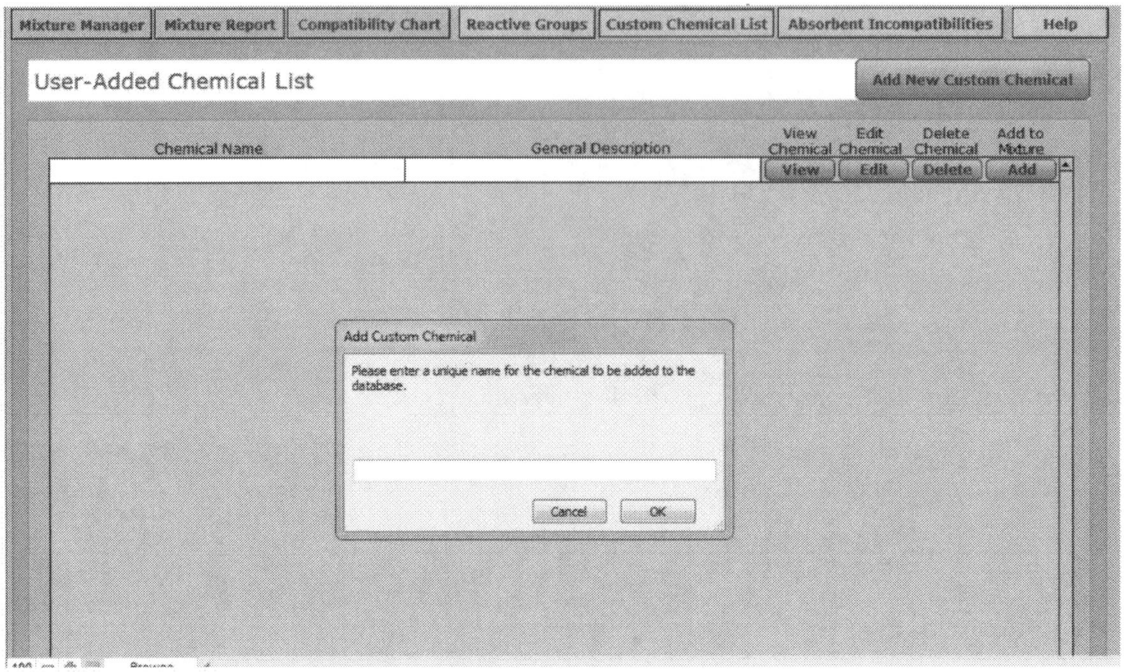

Figure 9: Adding a Custom Chemical

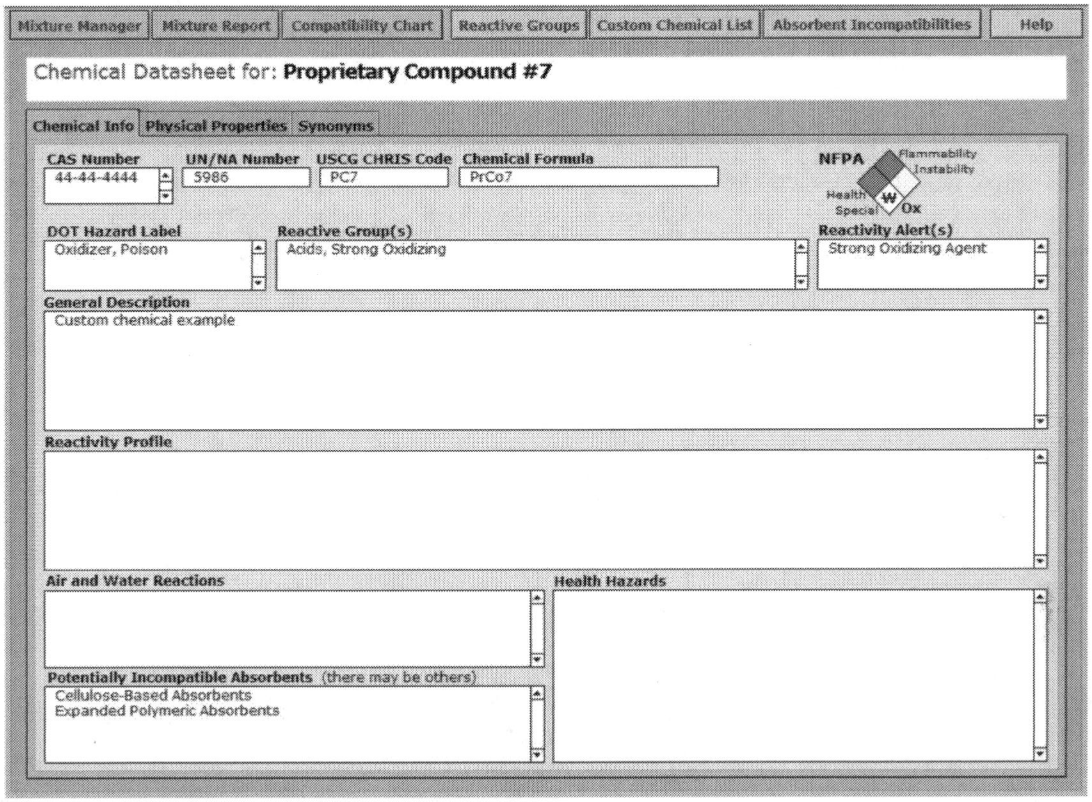

Figure 10: Custom Chemical Information Screen Example

To choose more than one reactivity group to describe the molecule, hold down the Ctrl key on the keyboard until all the Reactivity Groups have been selected, then hit Enter on the keyboard. To delete the entry, highlight the cell, then press the Backspace key. The chemical automatically is added to the Custom Chemical list without needing to press a save button. To add a custom chemical to a selected mixture, clicking again on the green Custom Chemical List button will give the list of custom chemicals (Figure 11). Clicking the far right Add button will add the chemical to the currently selected mixture.

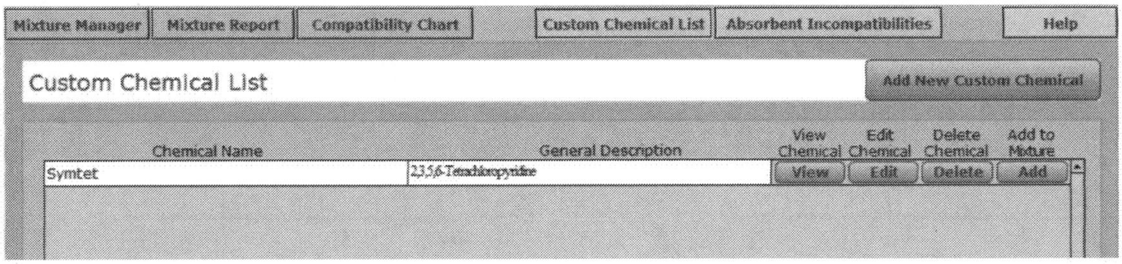

Figure 11: Custom Chemical List Example

4.4 Changing Compatibility Decisions

Another enhanced capability is the ability for the user to input their own compatibility decision if they have solid evidence that the default compatibility decision indicated is either overly conservative and/or in error for their process. This is done from the compatibility chart screen by clicking on the Chemical Pairs button in the upper left area of the chart, and choosing the appropriate combination of chemicals from the pop up menu (Figure 12). In the Selected chemicals compatibility box, control-click of the colored cell brings up a pop up menu (Figure 13) that allows the user to change the color and letter coding, which also changes it on the compatibility chart. When the compatibility decision is overruled by user input, an asterisk appears next to the letter coding in the box and in the compatibility chart (Figure 14). If the user wishes to revert to the original compatibility coding, bringing up the same chemical pairs screen and hitting shift-click in the colored cell brings up a pop up menu that allows the option to revert to the original compatibility code. By reverting to the original setting, this removes the asterisk and restores the box to its original color and letter coding.

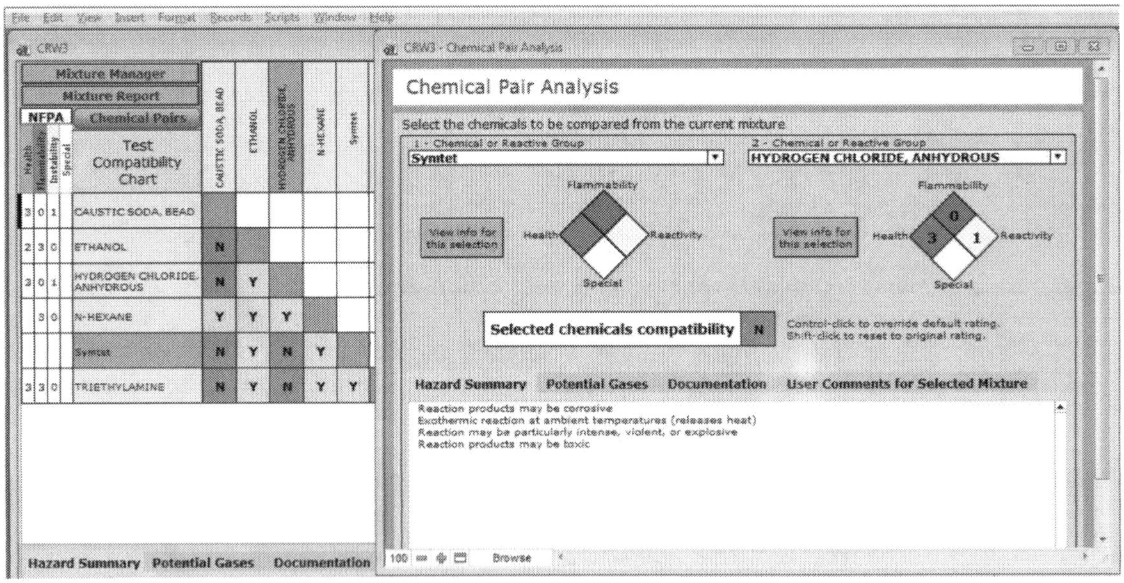

Figure 12: Chemical Pairs Pop Up Menu Example

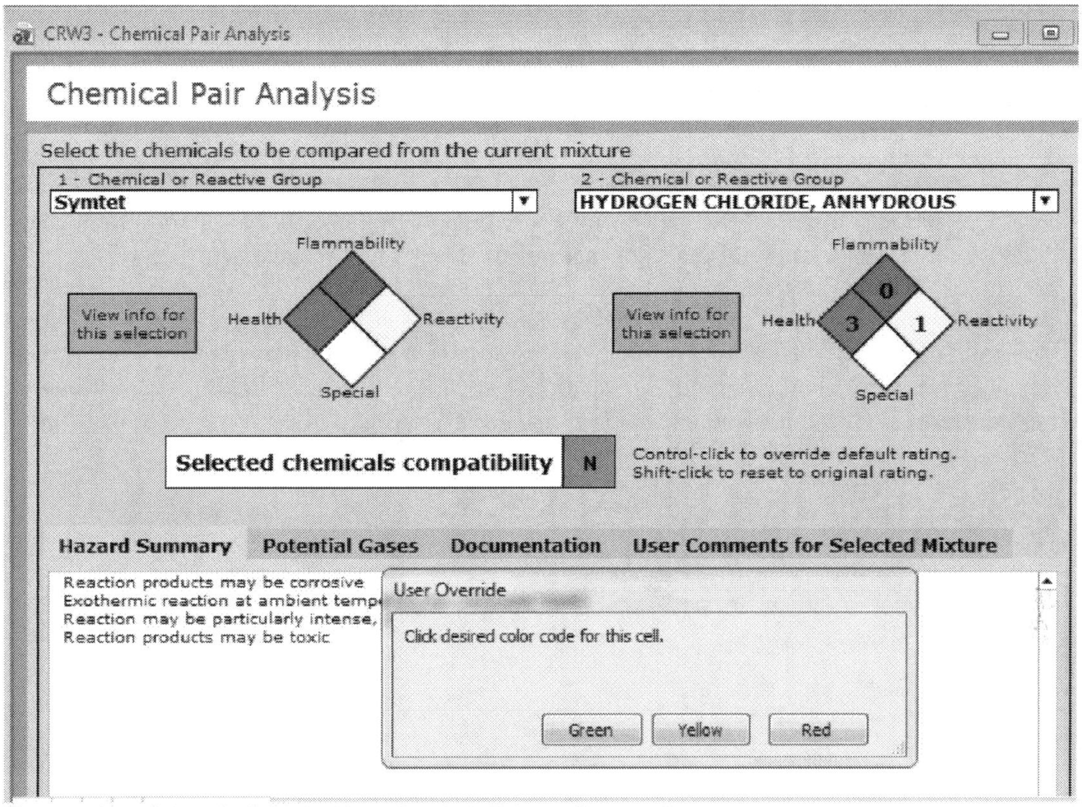

Figure 13: Chemical Pairs Pop Up Sub Menu Example

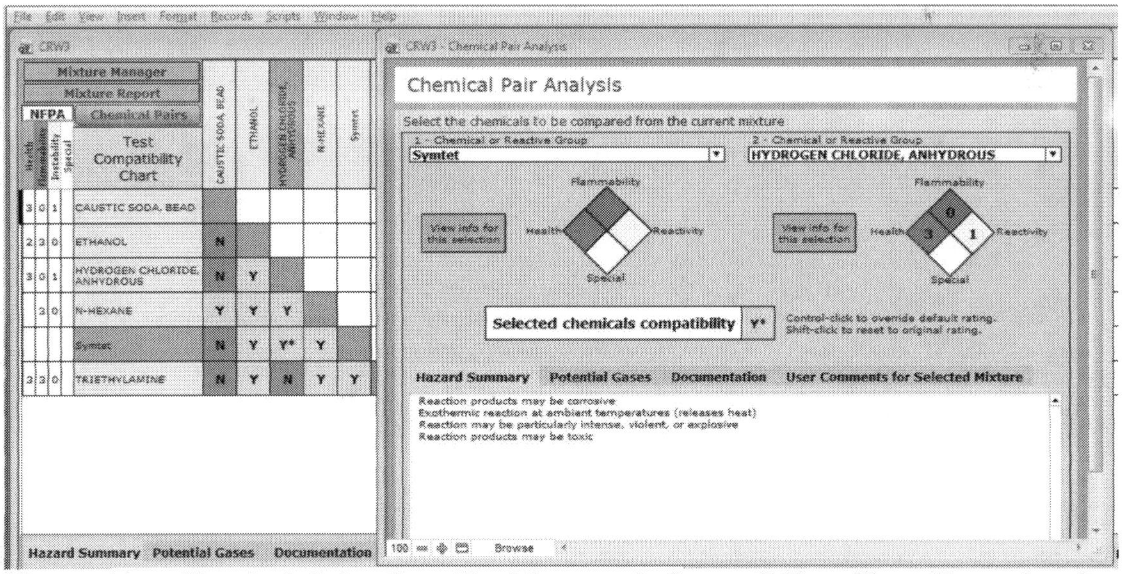

Figure 14: Chemical Pairs Pop Up Menu and Compatibility Chart Example after Compatibility Decision Change

4.5 Printing Charts and Reports

To print a chart, choose print from the File pull down menu at the top of the page where the chart is displayed. A print box will open, allowing the user to adjust settings before the table is printed. The printed form of the table only contains the compatibility chart and not the individual information for each pairing. For that information, a report must be printed.

To print a report, first go to the Mixture Report screen by clicking on the gray Mixture Report button displayed in the top left area of whichever screen currently is displayed. From this screen, the Preview/Print Report button is in the lower right part of the screen. Clicking that button will bring up the report (Figure 15), which can be printed from the File pull down menu at the top left area of the screen.

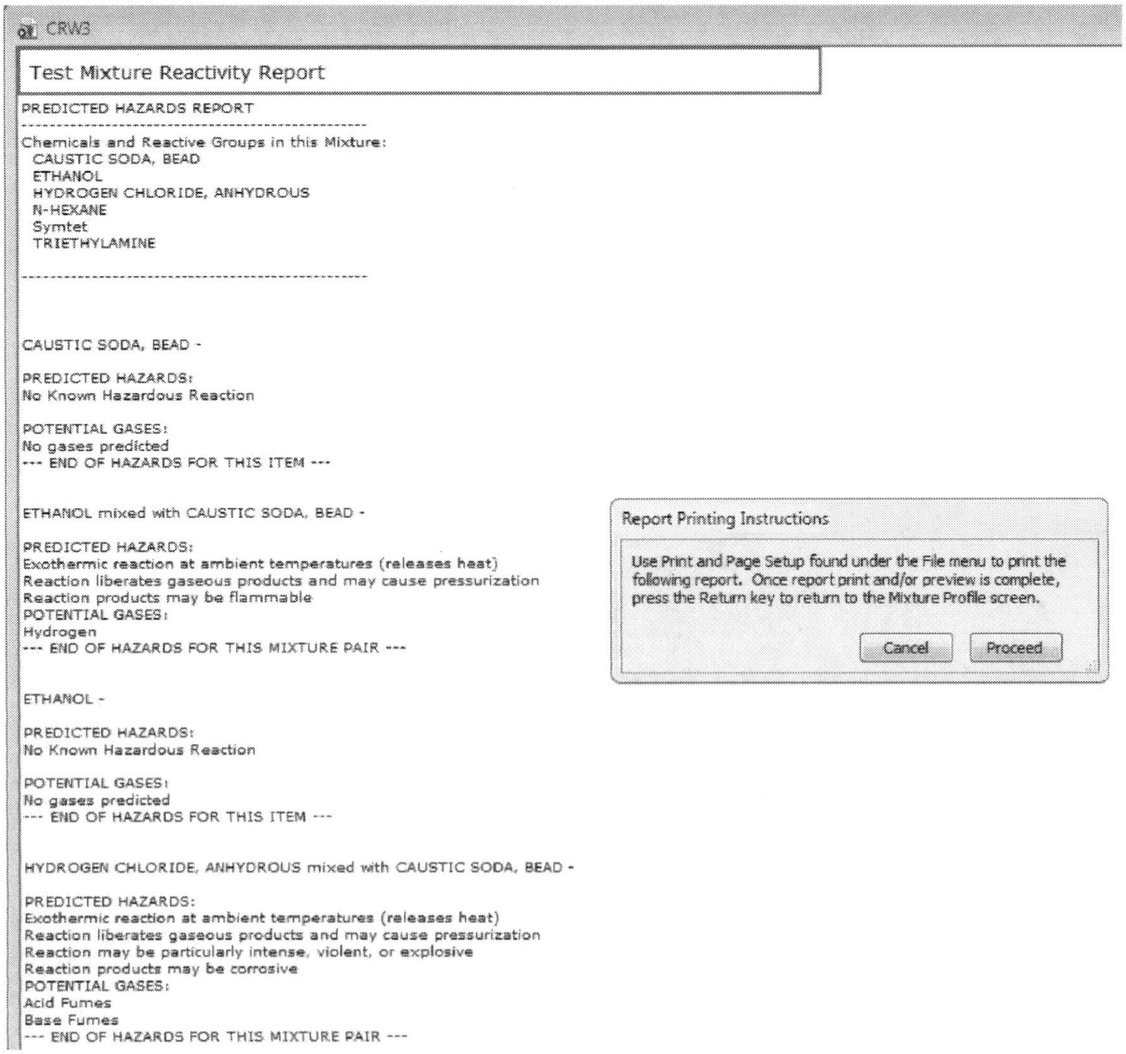

Figure 15: Mixture Reactivity Report Example

4.6 Help Screen

To aid the user, a number of potential aids have been incorporated into the Help screen for the tool. The Help button appears in the upper right hand corner of the Mixture Manager screen. Clicking on this button opens the Help screen (Figure 16).

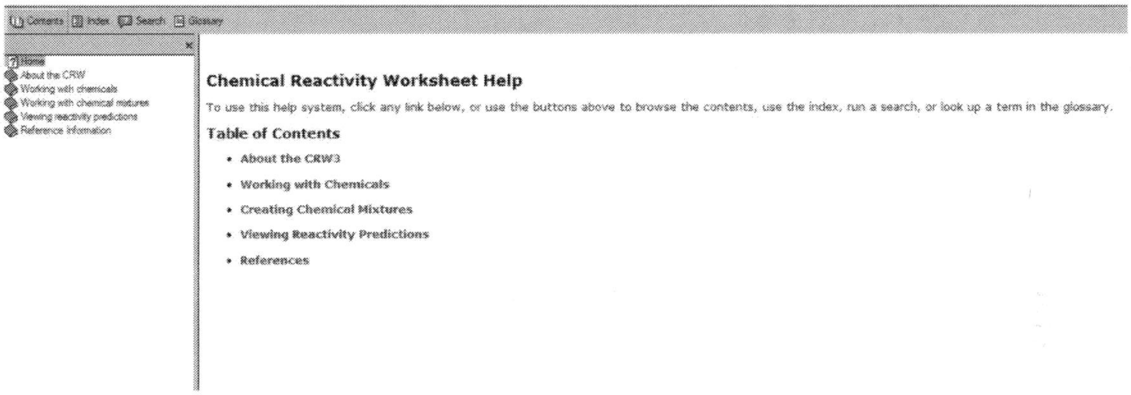

Figure 16: Help Screen

4.7 Reactive Groups Matrix Chart

The tool also comes pre-populated with a reactive groups matrix chart that compares the cross compatibilities of all of the reactive groups (Figure 17). The chart is accessed by choosing Reactive Group Matrix as the mixture within Mixture Manager, then selecting View Chart.

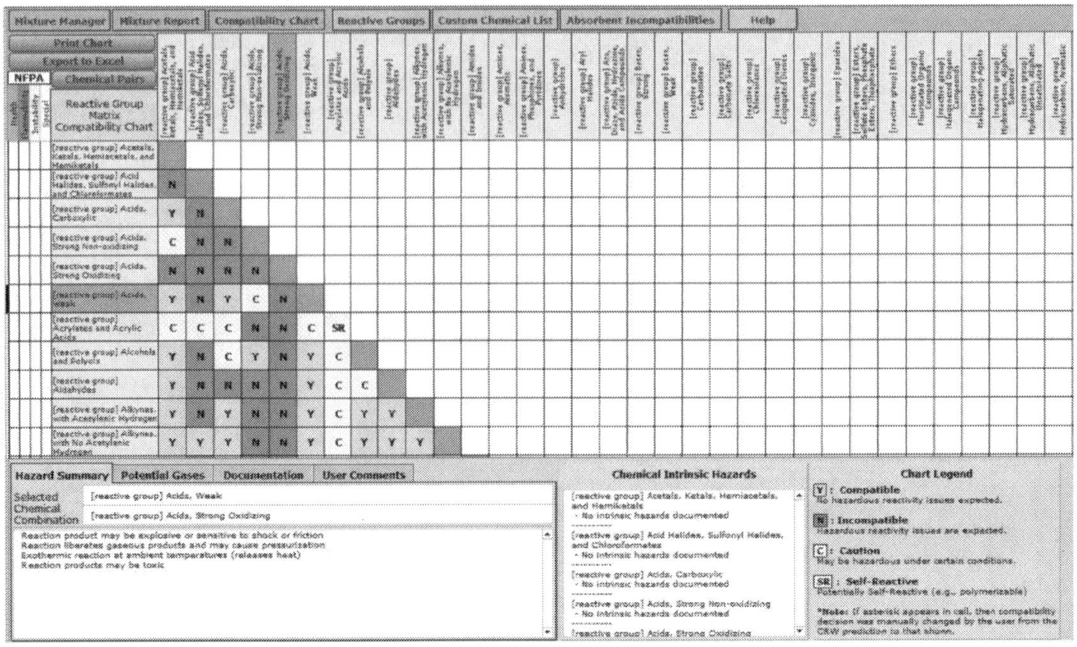

Figure 17: Reactive Groups Matrix Chart (Partial Display)

5. Potential Future Enhancements

Although the enhanced CRW represents a stepwise change in compatibility tools available to the public, it still lacks features that users could find desirable. Rather than delay release of this tool to incorporate these features, the authors believe this tool can aid the chemical industry immediately and incorporating these additional features does not justify delaying release of this tool to the public. Among the future enhancements potentially envisioned for this tool are:

1. Adding Absorbents and Materials of Construction sections to the Chemical Compatibility Charts - this would be helpful for plant settings and already is utilized extensively within The Dow Chemical Company.

2. Adding new Reactivity Groups – see Table 4 for a list of some Reactivity Groups being considered for inclusion in a future release of this worksheet

3. Ability to import data and/or charts into Excel or Word - the current version of the software has limited ability to import information into Excel. Work is underway to determine the longer range feasibility of having this feature be fully operational.

4. Adding a Heat of Reaction Module – this would help the user understand the potential heat/energy release for a given reaction. In its ideal form, it also would contain provision for calculating an adiabatic temperature rise. A working prototype has been developed in the

Reactive Chemicals group within The Dow Chemical Company and is continuing to evolve in scope.

5. Interfacing the popular CHETAH software [1b,14] to the CRW to allow a graphical structure input to be converted into SMILES notation, and then assigned to appropriate reactive groups.

6. Conclusion

As a result of a two-year joint effort between the National Oceanic and Atmospheric Administration (NOAA) and The Dow Chemical Company, an enhanced version of the NOAA Chemical Reactivity Worksheet tool now is available. In addition to addressing previously identified limitations, several enhancements have been incorporated into the tool to help with the generation and presentation of chemical compatibility charts. These enhancements position this tool potentially to become the gold standard within the chemical industry for determining chemical compatibility.

7. References

[1] a) ASTM Standard E2012-06 (2012) , Standard Guide for the Preparation of a Binary Chemical Compatibility Chart; b) **The Role of ASTM E27 Methods in Hazard Assessment: Part I. Thermal Stability, Compatibility, and Energy Release Estimation Methods,** Fruip, Dave, Larry Britton, William Fenlon, John Going, B. Keith Harrison, Jeff Niemer, and Erdem A. Ural, Process Safety Progress, 23, (4), 266-278, (2004).

[2] Process Safety Progress, volume 27, No.3, page 212, CRW 2.0: A Representative-Compound Approach to Functionality-Based Prediction of Reactive Chemical Hazards

[3] EPA 600/2-80-76, "A Method for Determining the Compatibility of Chemical Mixtures," April 1980.

[4] "Compatibility of Cargoes," published in 46 CFR Ch. 1 (10-1-93 Edition) – Part 150 with changes that appeared in the Federal Register dated April 11, 1994 as a Final Rule by Coast Guard (CGD 92-100) for Bulk Hazardous Materials. See also: "Compatibility Guide for Adjacent Loading of Bulk Liquid Cargoes (Final Report)," Report No. USCG-D-156-75.

[5] a) Bretherick L., Bretherick's Handbook of Reactive Chemical Hazards, Fifth Ed., Butterworths, Boston, MA, 1996. b) http://www.hazmatnavigator.com/search/quick

[6] Fire Protection Guide to Hazardous Materials, 10th Ed., National Fire Protection Association, Batterymarch Park, P.O. Box 9101, Quincy, MA 02269-9101, 1991.

[7] Sax, N.I., Dangerous Properties of Industrial Materials, Sixth Ed., Van Nostrand Reinhold Co., New York, NY, 1984. / 12. R. J. Lewis Sr., ed. Sax's Dangerous Properties of Industrial Materials, 8th Edition. Van Nostrand Reinhold, New York, 1992

[8] Lewis, Richard J., Sr., Hazardous Chemicals Desk Reference, Sixth Edition, John Wiley & Sons, Inc., Hoboken, New Jersey, 2008.

[9] Kirk-Othmer Encyclopedia of Chemical Technology, John Wiley & Sons, Inc., 2006.

[10] Ullmann's Encyclopedia of Industrial Chemistry. Wiley-VCH Verlag GmbH & Co. KGaA. 2006.

[11] National Fire Protection Association, NFPA Publication 491M, 1975.

[12] S. Budavari, ed., The Merck Index, 12th Ed. Whitehouse Station NJ: Merck & Co., 1996.

[13] The Determination of Compatibility via Thermal Analysis and Mathematical Modeling, T. C. Hofelich, D. J. Frurip, and J. B. Powers, Proc. AIChE Loss Prev. Symposium (1994), also published in Process Safety Progress, Volume 13, Issue 4, Date: October 1994, Pages: 227-233.

[14] CHETAH version 9.0, ASTM International, Details at http://www.southalabama.edu/engineering/chemical/chetah/index.html; For purchase information concerning CHETAH contact ASTM International.

Appendix – Evaluation Chart of Chemical Compatibility Tools

"Should" List for Compatibility Chart Tool	Importance	Dow Tool #2	Dow Tool #3	NOAA CRW Tool	Dow Tool #5	Dow Tool #1	Dow Tool #4	EPA Chart	Coast Guard Chart
Ability to leverage existing compatibility information (no reinventing the wheel for cross-compatibility) - globally shared	9	1	0.5	0.5	0	**1**	1	1	0.5
Searchable database including general functional groups	9	**1**	0	**1**	0	0.5	0	0	0
Picklist to generate list of chemicals for chart	9	**1**	0.5	**1**	0.5	0	0	0	0
One button feature to generate chart from picklist	9	***1***	**1**	***0.75***	***1***	0	0	0	0
Appropriate risk evaluation of compatibility (less conservative than NOAA tool); scientifically valid & consistent method for confirming compatibility	9	0	1	0.5	0.5	**1**	1	0.25	0.5

Easily modified - adding or deleting materials	9	1	0.5	0.5	1	0	0	0	0
Template readily accessible	9	0.5	1	1	1	1	1	0.5	0
Tracking of revision history	9	0	1	0.5	0	0	0	0	0
User friendly - utilizes simple interface, only need basic computing skills, not prone to crashing	9	0.75	0.5	1	1	0.5	0.5	0.5	0.5
Robust - accurate results, broad applicability to various compounds/functionalities	9	0.5	0.5	0.75	0	1	0.75	0.5	0.75
Air and nitrogen added to all charts (flammability issues - ex. monomers)	9	0.5	0.5	0.5	0.5	0	0	0	0
Water added to all charts	9	1	0.5	0.75	0.5	1	1	1	0.5
Allows input of customized chemicals	9	1	1	0.5	0.5	0.5	0	0	0
Binary vs. gradation for compatibility - prefer some gradation (up to 5 degrees - Yes/No/Caution/Unknown/Desired)	3	0	1	0.5	1	0	0	0	0
Color coding AND letter coding in boxes	3	0.5	1	0.25	1	1	1	1	0.5
Legal disclaimers	3	0	0	0	0	0	0	0	0
NFPA Ratings (GHS or SAP as alternative where appropriate)	3	1	1	0	0.5	0	0	0	0
Can handle multifunctional compounds (one known limitation - amino acids)	3	1	0	1	0	0.5	0.5	0.5	0.5
Multilingual (Spanish, German - legally required, Dutch - legally required, French, others)	3	0	1	0	0	0	0	0	0
Predicts gas release	3	1	0.5	1	0.5	0	0	1	0
Documents rationale behind compatibility designation between two compounds	3	1	0.5	1	0.5	1	1	1	0.5
Predicts heat of dilution effects	3	0.5	0.5	0.25	0.5	0.5	0	0	0
Self-reactivity information where appropriate (ex. monomers)	3	1	0.5	0	0.5	1	0	0	0
Interface to existing search resources	1	0.5	0.5	0.5	0	0	0	0	0

Temperature sensitivity parameter - standardized scenarios depending on geography	1	0	0	0	0	0.5	0.5	0	0
Materials of construction section optional (need best practice resolution - out of scope?)	1	0	**1**	0	0	0	0	0	0
Technical System Support (Alliance)	1	0.5	0.5	0	0	0	0	0	0
Can extract information from MSDSs	1	0	0	0	0	0	0	0	0
Ensure consistency with Dow tool #1 chart	0	0.5	0.5	0.5	0.5	**1**	0.5	0.5	0.5

Additional Screen Shots

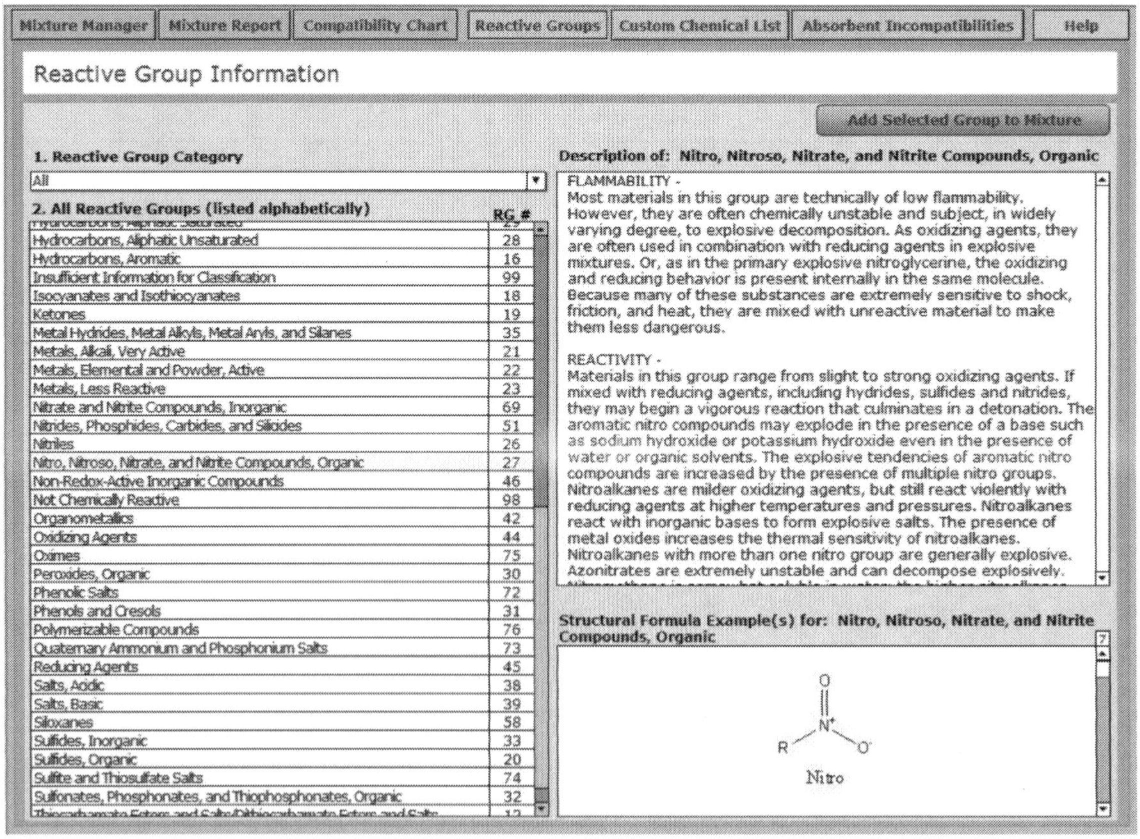

Figure 18: Reactive Group Information Example

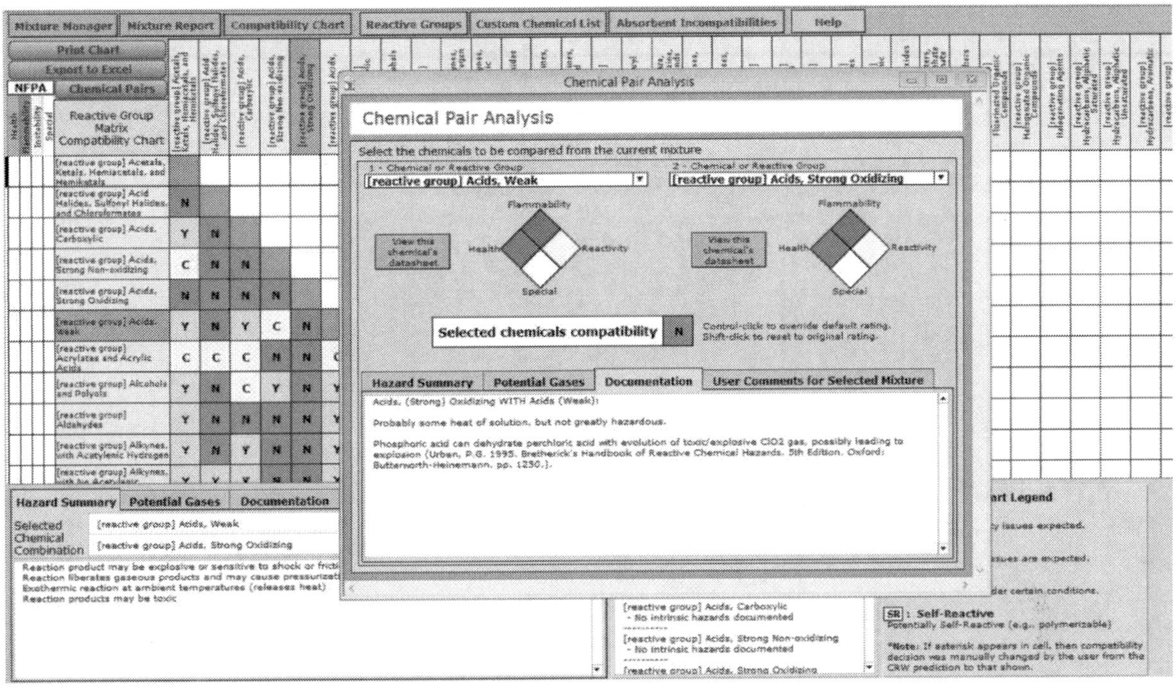

Figure 19: Chemical Pair Analysis Example

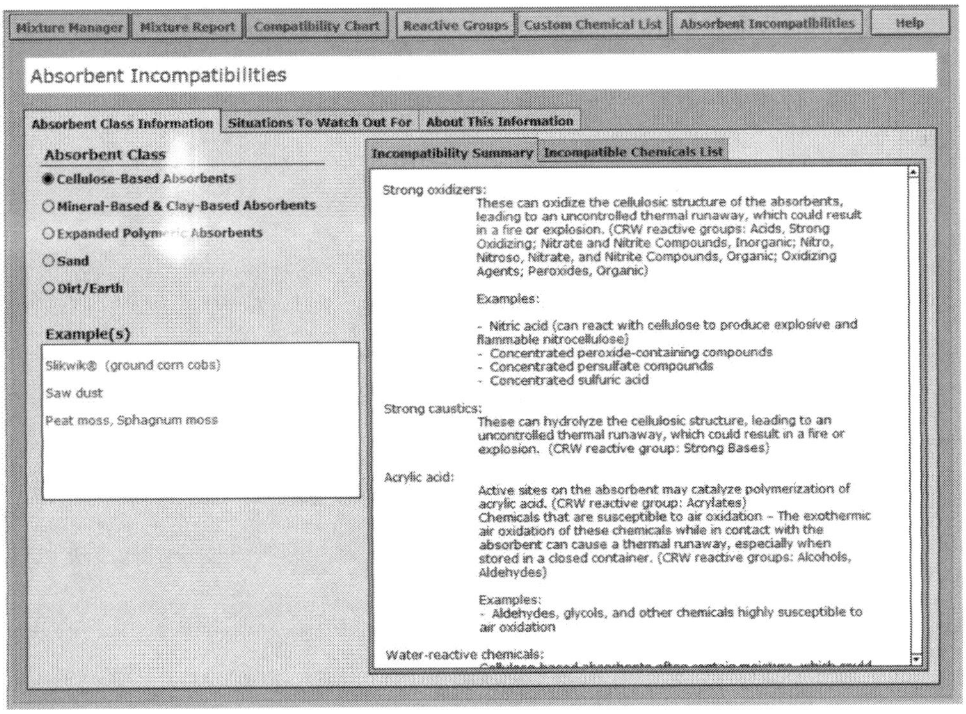

Figure 20: Absorbent Incompatibilities Example

9TH GLOBAL CONGRESS ON PROCESS SAFETY

Effects of the Primary Explosion Site (PES) and Bulk Cloud in VCE Prediction: A Comparison with Historical Accident Data

C. Regis Bauwens
FM Global, Research Division
1151 Boston-Providence Turnpike
carl.bauwens@fmglobal.com

Sergey B. Dorofeev
FM Global, Research Division
1151 Boston-Providence Turnpike
sergey.dorofeev@fmglobal.com

Prepared for Presentation at
American Institute of Chemical Engineers
2013 Spring Meeting
9th Global Congress on Process Safety
San Antonio, Texas
April 28 – May 1, 2013

UNPUBLISHED

Effects of the Primary Explosion Site (PES) and Bulk Cloud in VCE Prediction: A Comparison with Historical Accident

C. Regis Bauwens
FM Global, Research Division
1151 Boston-Providence Turnpike
carl.bauwens@fmglobal.com

Sergey B. Dorofeev
FM Global, Research Division

Keywords: Vapor Cloud Explosions

Abstract

A model for predicting VCE blast loads is described extending a previously developed model. The model considers the contribution of a Primary Explosion Site (PES) as well as the bulk effect of the total flammable cloud. The model results are then compared with historical accidents and other commonly used models considering a single PES. Four historical events are examined: the Flixborough accident of 1974, the Texas City accident of 2005, the Phillips 66 accident of 1989 and the Buncefield incident of 2005. It is found that the combined model produced pressure and impulse values consistent with the damage observed. If the contribution of the total cloud is not considered, the far field pressure, which included overpressures up to 0.35 bar, was under predicted and impulse was significantly under predicted. The other methods, produced, in general, significantly lower overpressures and impulses when compared with the damage indicators seen in the explosion accidents.

1. Introduction

Unconfined vapor cloud explosions (VCEs) can occur when large quantities of flammable mixtures are released in an open area, typically in areas where congestion is present such as chemical process facilities. These explosions can result in significant damage to buildings, injury to personnel and loss of life. In these facilities, loss prevention and risk assessment tasks require accurate estimates of potential blast loads associated with VCEs.

Traditional approaches to estimating the damage potential of VCEs, such as the Baker-Strehlow-Tang (BST) (Baker et al., 1983) methodology or the TNO multi-energy methodology (TNO, 1997), focus on the contribution of a Primary Explosion Site (PES). The PES is a region of congestion and/or confinement within the flammable cloud that is typically responsible for the maximum overpressures that are generated. Most large VCEs, however, span multiple regions of congestion or confinement, all which can contribute to the overall damage to the site. These contributions are either ignored in simple models, or a detailed analysis of the site, defining multiple PESs and explosion zones is performed, such as in (CSB Report, 2005). This type of analysis is highly dependent on the release scenario considered as it is sensitive to the shape of the overall vapor cloud. In addition, realistic estimates should account for the variation in mixture composition within the cloud, which adds to the complexity of the problem. To perform VCE studies over many sites, or to consider several different release scenarios, requires an approximate technique to produce quick, reasonable estimates for VCE damage potential without requiring a detailed site analysis.

In the present study, a simple model for predicting VCE blast loads is described taking into consideration the contribution of a PES as well as the bulk effect of the total cloud (based on averaged properties within the cloud). The results of the combined model are then compared with blast damage indicators from historical accident investigation studies.
Four historical events are examined: the Flixborough accident of 1974, the Texas City accident of 2005, the Phillips 66 accident of 1989 and the Buncefield incident of 2005. In addition, the individual components of the model, as well as the TNT equivalence method and PES-based methods such as the BST methodology, are compared with the historical data. For each accident, the model predictions are compared with pressure and impulse damage indicators gathered by site investigations following the accidents.

2. Model Description

In a VCE accident, the overall peak overpressure generated is typically associated with a small congested region within the cloud of released fuel. This peak overpressure is highly dependent on the size of the congested region, its level of congestion and whether confinement is present. For a large release, however, extending beyond an individual PES, the pressures and impulse generated in the father field are often closely associated with the total energy of the fuel burned. As typical VCE accidents involve large releases of fuel that extend far beyond an individual PES the contribution of this energy must be considered.

When considering the contribution of explosion energy from the PES, a model (Dorofeev, 2007a, Dorofeev, 2007b) considering the geometry and conditions within the PES is used. This method takes into account the flame folding due to interactions between obstacles, the flow produced by the flame and the increase of the burning rate due to turbulence generated in the flow ahead of the flame. Coefficients in this correlation were determined through fitting against a set of experimental data. For a specific fuel, the model results vary depending on the dimensions of PES and the level of congestion and confinement within the PES. For the analysis performed in this study, the PESs are assumed to be filled with a stoichiometric fuel-air mixture.

As emphasized in previous work (Dorofeev, 2007a), the expansion of the combustion products acting as a piston driving the flame is known to be extremely important for flame acceleration. In the cases where the combustion products are completely surrounded by the flame and rigid surfaces, full expansion is available to drive flame acceleration. In vapor clouds with high aspect ratios, however, the combustion products become free to expand into the surrounding air at some point in the flame propagation. In these situations, full expansion is not available to drive flame acceleration.

Effective Expansion Ratio, σ_{eff}:

To account for the loss of expansion, a reduced effective value of the expansion ratio, σ_{eff}, must be used in the PES model instead of the full expansion ratio of the mixture, σ. The value of the effective expansion ratio depends on the relationship between the flame area and the area through which the combustion products vent into air.

To derive a generalized approximation for, σ_{eff}, we take a PES with height, H, width, W, and length, L. We assume that ignition is in the center of the PES and that the resulted flame propagates through the domain. The value of σ_{eff}, must then be estimated at the stage when the flame approaches the edges of the PES.

An effective length, ΔL, may be defined that describes an equivalent area behind the flame front, over which the expansion of the products may be lost. The loss of expansion area, A_{le}, may then be given by the following expression for $L >> \Delta L$:

$$A_{le} = 4\alpha(H + W)\Delta L \ , \qquad \text{[Eq. 1]}$$

where α is the fraction of open area on the lateral surface of the cloud. For this geometry, the flame area can be approximated as:

$$A_f = 2HW \ , \qquad \text{[Eq. 2]}$$

and for the effective expansion ratio, σ_{eff}, one can write:

$$\frac{\sigma}{\sigma_{eff}} = \frac{A_f + A_{le}}{A_f} = 1 + 2\alpha(\frac{\Delta L}{W} + \frac{\Delta L}{H}) \ . \qquad \text{[Eq. 3]}$$

In the extreme case of the flame propagation through a thin layer of combustible mixture H/W approaches 0 and $\sigma_{eff} = \sqrt{\sigma}$, (Feng et al., 1975). If this relationship is applied to (3) for α=1 then one obtains for ΔL:

$$\Delta L = \frac{H}{2}(\sqrt{\sigma} - 1) \ . \qquad \text{[Eq. 4]}$$

Equation (4) shows that ΔL is estimated to be of the order of H for typical fuel-air mixture with $\sigma \approx 8$. In order to extend this approximation for σ_{eff} to the cases where the aspect ratios are close to unity, a term should be added into Eq. (1) that vanishes at $\Delta L \sim L$ and $\Delta L \sim W$, as $\Delta L \approx H$, there should be no loss of expansion in a cloud for aspect ratios equal to unity with central ignition. The following simple expression may give a sufficiently general solution for A_{le} instead of Eq. (1):

$$A_{le} = 4\alpha(H + W)\Delta L(1 - \Delta L / L)^{\theta} \ , \qquad \text{[Eq. 5]}$$

where θ is a constant to be determined, which describes how sharp the loss of expansion vanishes with the aspect ratios approaching unity. Equations (3-5), then, yield:

$$\frac{\sigma}{\sigma_{eff}} = 1 + \alpha(\sqrt{\sigma} - 1)(1 + \frac{H}{W})(1 - \frac{H}{2L}(\sqrt{\sigma} - 1))^{\theta} \; . \qquad \text{[Eq. 6]}$$

Equation (6) gives a description for the effective expansion ratio, σ_{eff}, for a cloud with height, H, width, W, and length, L partially confined on planes LH and LW with an average area blockage of α (planes WH are assumed to be completely open). If such a cloud spreads over the ground surface, this surface should be considered as a plane of symmetry and H equal to the double actual height should be used.

Equation (6) includes one unknown coefficient, θ, which was determined using experimental data. Based on the analysis of a proprietary dataset of large scale partially confined VCE experiments, a value of $\theta = 6.91$ was determined and used for this study.

Total Cloud Model:

To account for the contribution of explosion energy from the total cloud the same model is used. However, instead of using the properties of a PES, the model considers average cloud properties and a "worst case" (Dorofeev, 2007a) non-uniform concentration distribution in the cloud. This "Total Cloud" model uses these average values to calculate the contribution of the overall cloud to the pressures and impulse that develop. The total cloud is considered to be unconfined with an average level of congestion that is assumed to vary with cloud size. In this model, the level of congestion is varied such that the average obstacle size and the characteristic distance between obstacles both increase with the cloud size. Equation (1) is used to approximate the average obstacle size and spacing throughout the cloud:

$$x = 0.35R^{1/2}, \; y = x/3, \qquad \text{[Eq. 7]}$$

where R is the average cloud radius, x is the characteristic spacing between obstructions and y is the characteristic size of the obstructions. Using these approximations, the model results only vary with the total mass of fuel released, for a specific fuel.

The PES model tends to produce higher pressures closer to the PES while the total cloud model higher pressures/impulse at farther distances due to a reduced pressure decay with distance. In this study, the approach taken to combine these two models is to simply take the maximum pressure and impulse estimated by the total cloud and PES models.

The results of these models are also compared with the BST methodology, considering only a single PES. VCloud Version 2.05.00 was used to perform estimates using the BST methodology.

3. Accidents

3.1. Flixborough

3.1.1. Description of accident and model parameters used

On June 1st, 1974 an accidental release of cyclohexane resulted in a large vapor cloud explosion causing 28 deaths and extensive damage to the plant and surrounding area. The explosion source was centered at Section 25A of the plant where a total of 400 tons of cyclohexane was stored in the process unit at a temperature of 155 C and a pressure of 8.8 bar (HSE Report, 1975). The release was the rupture of a 20 in pipe which released the cyclohexane for approximately 60 seconds prior to ignition.

Figure 3.1.1. Aerial image of the Flixborough plant before the explosion, the PES considered is highlighted in red. Original image source (HSE Report, 1975).

As detailed plant diagrams were not available, the entire Section 25A process unit was used as the PES. This area is highlighted in Fig 3.1.1. The dimensions of the process unit were set as 65 x 26 x 10 m^3, and it was considered to have a medium level of congestion throughout with no confinement. For this analysis, a total release of 34 tons of cyclohexane was estimated based on the storage conditions, the ignition delay and leak size reported. This estimate is consistent with the reported estimates of a release between 30-36 tons.

3.1.2. Comparison with observed damage

When the results of the combined model are compared with the pressure and impulse damage indicators found from investigations of the accident site (Sadee, 1976, Roberts et at., 1982) it was found that, in general, the model compares well with the observed damage levels. A plot comparing overpressure as a function of distance from the explosion source is shown in Fig. 3.1.2. From this comparison it is clear that total cloud model performs better than the PES model for overpressures between 0.08 and 0.3 bar (1-4 psi) which occur at a distance between 200-700 meters from the explosion source. Over this range, the overpressures estimates are dominated by the overall cloud size and not the individual PES. Closer than 200m, the PES model produces higher pressures than the total cloud model. The estimates, however, slightly under predict the overpressure in some of the locations. These points correspond to damage occurring southeast of the PES selected and are likely due to an addition PES, or multiple PESs that were not considered in this study.

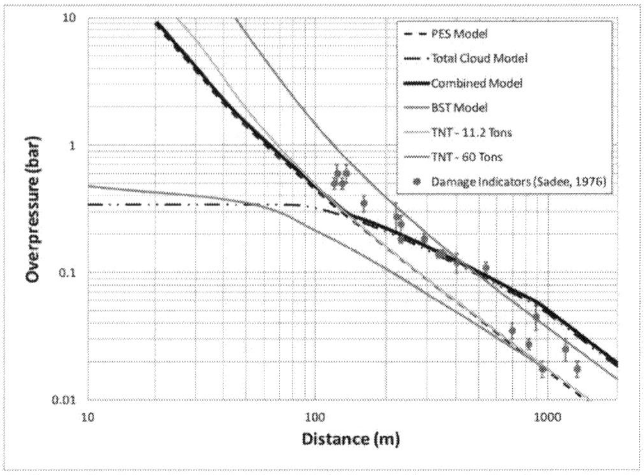

Figure 3.1.2. Comparison of overpressure as a function of distance from explosion source/center for the Flixborough explosion.

When impulse is considered, Fig. 3.1.3, it is clearly seen that the PES model significantly under predicts impulse while the total cloud model, on average, reproduces the impulse indictors.

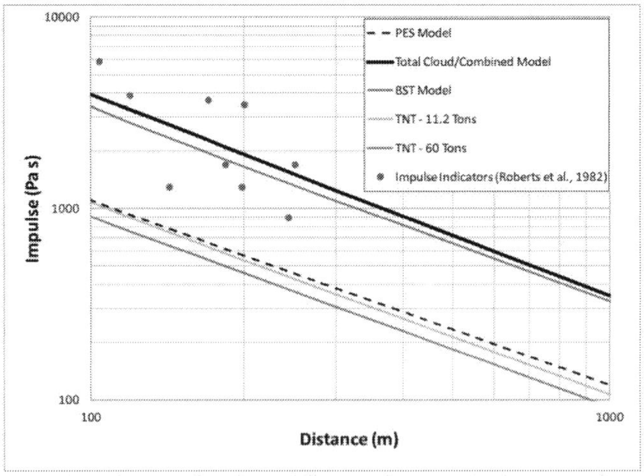

Figure 3.1.3. Comparison of impulse as a function of distance from explosion source/center for the Flixborough explosion.

When the damage observations are compared with the TNT estimates quoted in the literature, it is seen that a 60 ton TNT equivalent (the maximum estimate quoted) compares well with the observed overpressures. In the near field, however, the 60 ton TNT curve produces significantly higher pressures compared to the other models. The lower end of the TNT estimates produces an 11 ton equivalency. This curve closely reproduces the PES only curve and under predicts the pressure as a function of distance as well as the impulse.

3.2. Texas City

3.2.1. Description of accident and model parameters used

On March 23rd, 2005, during the start up of an isomerization unit, a release from the pressure relief system caused an overflow of the atmospheric vent stack on a blowdown drum. This resulted in a large vapor cloud explosion causing 15 deaths and significant damage to the plant.

For this study, the release was approximated as being primarily composed of hexane and a release rate from the vent stack of approximately 260,000 gph (CSB Report, 2005) was used. Over a release time of 107 s this resulted in approximately 40,000 lbs of hexane being released from the process unit. Based on the temperature of the released liquid it is estimated that approximately 25% of the hexane would have vaporized and formed the overall flammable cloud.

Figure 3.2.1. Aerial image of BP Texas City Refinery after the explosion, the PES used is highlighted in red. Original image taken from (CSB Report, 2005)

The PES used for this case study was the region of high congestion shown in Fig. 3.2.1 which corresponded to the region whose damage indicators showed the highest overall overpressure and impulse. Based on images from the scene, the PES dimensions were set to 26 x 9 x 3 m^3 with a high level of congestion and no confinement.

3.2.2. Comparison with observed damage

Although damage surveys of the accident site were performed, individual damage indicators were not published. Instead, overpressure and impulse rings were released (CSB Report, 2005) that roughly correspond to the pressures and impulse derived from damage indicators at the site. These rings were generated using the BST model using multiple PES. In order to achieve a proper description of the pressure and impulse, consistent with damage indicators observed at the site, the analysis included more than 50 individual explosion sources and zones. As this analysis was tuned to match the observed damage levels, these pressure and impulse results will be used to compare with the results of the model.

Figures 3.2.2 and 3.2.3 show the comparison of overpressure and impulse as a function of distance from the explosion source. From the overpressure plot is shown that the PES model does a better job of reproducing the pressures reported in the CSB reports for the higher overpressures in the region of 0.1 bar when compared with the total cloud model. Without taking into account the PES, the total cloud model diverges from the combined model at pressures above 0.05 bar and under predicts the near field overpressure compared with the CSB report. At farther distances, the combined model over predicts the pressure from the CSB report, however, it is unclear whether these points are based on actual damage observations or are simply the product the multi-PES BST. The modeling technique used in the CSB report may have been based exclusively on damage indicators closer to the explosion source which was then extended beyond the observed damage indicators. When compared with the distance that offsite window breakage is observed (CSB Report, 2005), the total cloud/combined models are consistent with the damage observed.

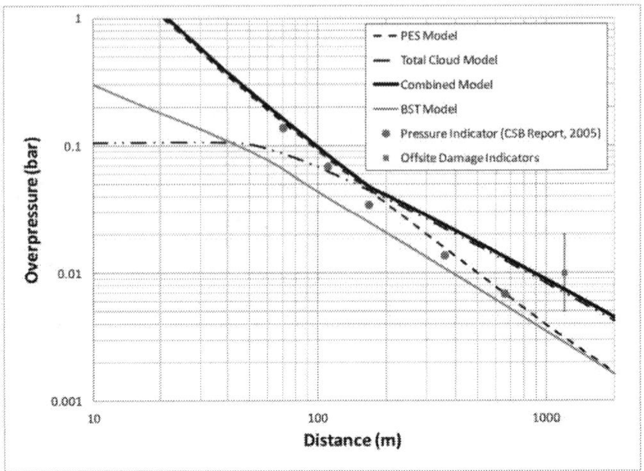

Figure 3.2.2. Comparison of overpressure as a function of distance from explosion source/center for the BP Texas City explosion.

A comparison of impulse as a function of distance shows that the total cloud models closely matches the impulse quoted from the CSB reports while only considering a single PES significantly under predicts the impulse. This is because the CSB results sum up the impulses over each individual PES, to provide results consistent with the observed damage, while the BST model results that are plotted are based on a single PES.

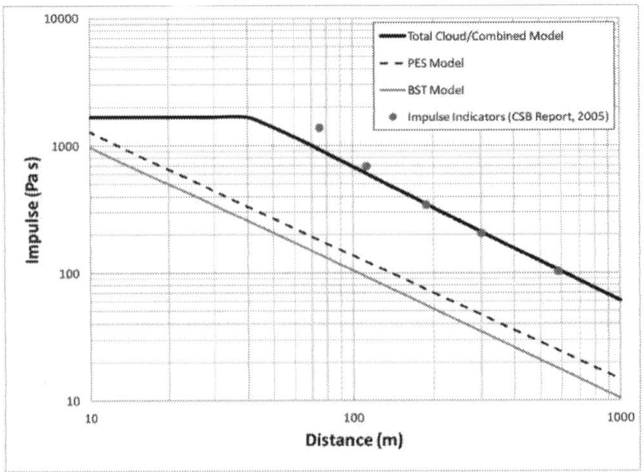

Figure 3.2.3. Comparison of impulse as a function of distance from explosion source/center for the BP Texas City explosion.

3.3. Phillips 66

3.3.1. Description of accident and model parameters used

On October 23rd, 1989 a massive release of process gas occurred during maintenance operations on a polyethylene reactor in the Phillips 66 Houston Chemical Complex. This resulted in a large vapor cloud explosion causing 23 deaths and significant damage to the plant. The release was through an 8 inch pipe and it was assumed that all 75,000 lbs of process gas (a mix of isobutene, ethylene, hexane and hydrogen) was released (USDOL, 1990).

The release material was comprised of primarily isobutane (making up approximately 95% of the process gas). The PES used encompasses reactors 1-6 and is shown in Fig. 3.3.1. This area corresponds to a PES with dimensions of 64 x 38 x 10 m^3. The PES was considered to contain medium congestion obstructions and estimated to be 20% confined.

Figure 3.3.1. Aerial image of the Phillips 66 plant before the explosion, the PES used is highlighted in red. Original image source (USDOL, 1990).

3.3.2. Comparison with observed damage

Due to a lack of published results for pressure or impulse rings for the Phillips 66 case, estimates of damage were made based on the region with the area of major observed damage (Moore, 1989) was seen on the site as well as offsite damage reports such as window breakage.

The plots comparing overpressure and impulse are shown in Fig. 3.3.2 and 3.3.3. The overpressure plot shows similar results between the different models. A single point is added to the plot representing the border of the area of major damage corresponding to overpressures between 2-5 psi (0.14 and 0.34 bar). The range of the uncertainty of this point covers all the different models and from this limited data there isn't any indication of the relative performance of the different models. For this case, however, the spread between the different models is small and all of the models produce results consistent with the damage observed. Also, it is important to note that window breakage was reported as far as 4000 m from the explosion source (USDOL, 1990). This would be more consistent with the results of the total cloud model where the pressure decay with distance is smaller.

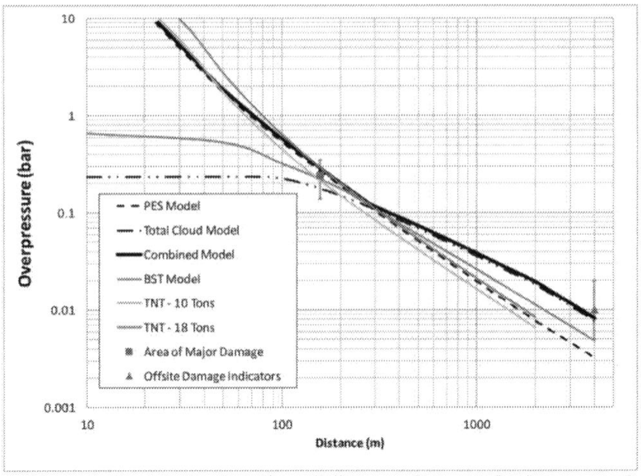

Figure 3.3.2. Comparison of overpressure as a function of distance from explosion source/center for the Phillips 66.

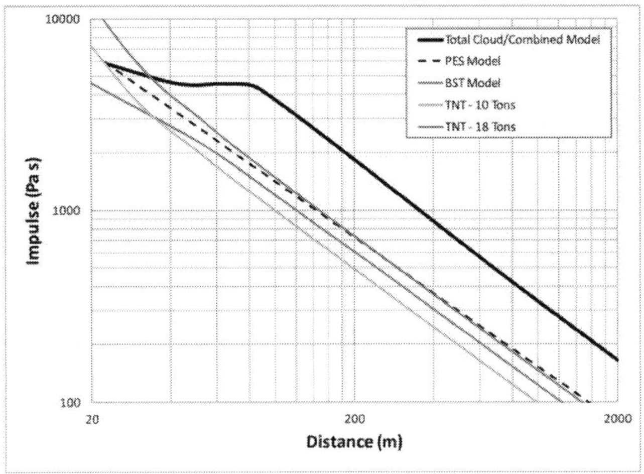

Figure 3.3.3. Comparison of impulse as a function of distance from explosion source/center for the Phillips 66.

3.4. Buncefield

3.4.1. Description of accident and model parameters used

On December 11st, 2005 an accidental release of gasoline resulted in a large vapor cloud explosion and fire causing extensive damage to the plant and surrounding area. The release was caused by a tank overflow releasing approximately 300 tons of gasoline. From the official accident reports (HSE, 2008, Buncefield MIIB, 2008) it is estimated that approximately 10% of this fuel vaporized and formed the explosive cloud.

For this study, the gasoline in vapor form was assumed to be comprised entirely of pentane. While there weren't any process structures in the area that provided significant congestion or confinement, there were large areas of dense vegetation that have been identified (HSE, 2008, Buncefield MIIB, 2008) as providing high congestion. The stretch of vegetation along Cherry Tree Lane is used as the PES for this study.

The dimensions used for the PES was 100 x 6 x 2 m^3 with a high level of congestion and no confinement.

3.4.2. Comparison with observed damage

Good agreement is found between the estimated peak pressures from damage indicators and those predicted by the combined model, particularly considering the uncertainty in both the quantity of fuel in the vapor cloud and in the estimates produced by the damage indicators and witness reports. When comparing pressure as a function of distance, Fig. 3.4.1, it can be clearly seen that the total cloud and combined models perform well when compared with the visual damage indicators. At short distances, the PES model produces high pressures which are consistent with reports that damage indicators inside the cloud reported widespread damage to drums, vehicles and other enclosures with over pressure estimated to require in excess of 200 kPa to produce the damage. Although their positions are not listed, it is assumed that they are in regions close to high congestion vegetation. The total cloud model, however, does not generate pressures above 200 kPa because it does not include the contribution from a PES. It is also seen that BST model with a single PES significantly under predicts the observed overpressures. For impulse it is shown, Fig. 3.4.2, that the combined model predicts much higher impulses compared to the PES alone or BST. For this case study, however, there are no reported impulse indicators and the relative performance of the different models cannot be evaluated.

The farther field damage reported by the HSE report as "zones where multiple cases of light damage… included damage to glazing, door frames, roofing materials, lightweight structures etc." which occurred "at distances of between 1 and 1.5 km" from the explosion. This description of damage is more consistent with the far field pressures estimates produced by the total cloud model than the predictions based on a single PES.

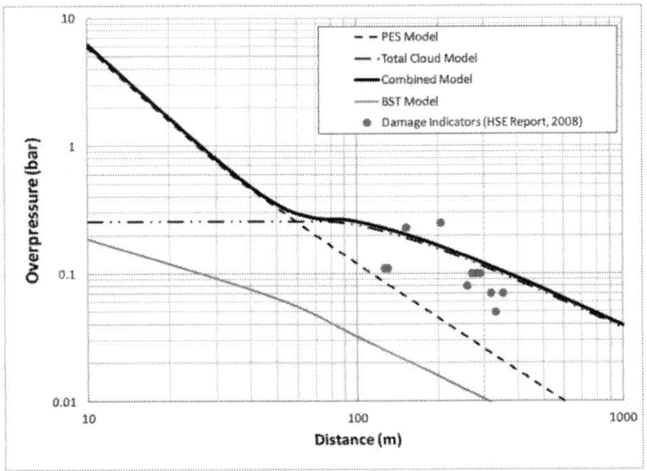

Figure 3.4.1. Comparison of overpressure as a function of distance from explosion source/center for the Buncefield explosion.

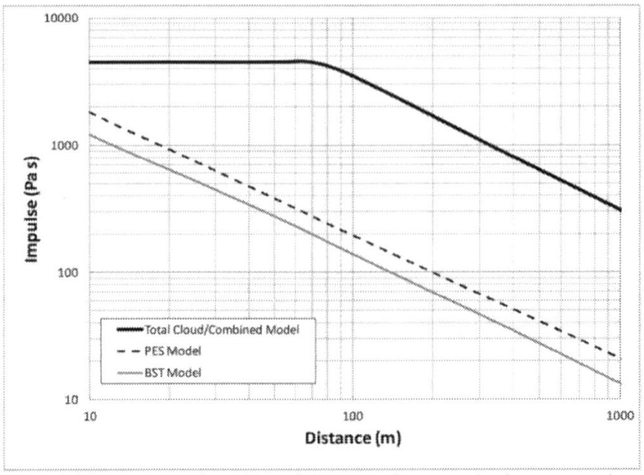

Figure 3.4.2. Comparison of impulse as a function of distance from explosion source/center for the Buncefield explosion.

4. Discussion

Throughout the case studies where pressure and impulse damage indicators are available, the combined model matches the observed damage indicators well. Without considering the total cloud, the PES model under predicts the pressure damage indictors at far field, for pressures typically between 0.07-0.35 bar (1-5 psi), and significantly under predicts the impulse through the cloud. This emphasizes the importance of accounting for the contribution of the entire vapor cloud to the blast damage.

The total cloud model also provided a good match for most of the reported observed damage. It is important to note, however, that the four case studies examined are relatively large explosions with large external clouds. As such, the total cloud model provided a good estimate for far field damage. For smaller releases, where the size of the external cloud is closer to the size of an individual PES, the total cloud model would significantly under predict the overpressure.

Universally, a single PES model predicted lower impulses than all other models, significantly under predicting the cases where impulse indicators were present. This is due to the lack of treatment for the external cloud. In order to produce results matching the pressure and impulse indicators seen in the case studies using a BST approach requires a much more detailed description of the explosion source. This involves defining multiple PESs or explosion zones. In the case of the Texas City explosion, more than 50 explosion zones were used to match the observed damage indicators. To use this method, additional effort would be necessary to evaluate which potential explosion zones may be in the cloud and as well as judgment on how to partition the PESs and calculations of the potential release amount and cloud size would be required. In addition, the choice of which explosion zones to include would significantly influence the damage curves produced.

The combined model, however, only requires the definition of a single PES and the total mass of fuel released. The model produces results consistent with the notion of multiple explosion zones but with fewer required inputs and can be used to produce consistent predictions across multiple different sites. Also, while the notion of multiple explosion zones may reproduce the near field damage in regions where multiple congested areas are present over a large area, this method would not be able to reproduce the damage caused by large vapor cloud explosions without large areas of congestion and confinement. For these scenarios, even a detailed multiple PES analysis of the few areas of congestion would not be able to produce the observed damage contours.

5. Conclusions

A series of case studies were performed comparing a VCE damage prediction models based on a total cloud model as well as models based on a Primary Explosion Site (PES). The models were applied to four historical major explosion accidents.

The study found that, given input conditions based on the available information known about the accident scenarios, a combined model, taking into account both the contribution of a main PES and the total cloud, produced pressure and impulse values consistent with the damage observed across all of the case studies. Without the inclusion of the external cloud, the far field overpressures, typically less than 0.35 bar (5 psi), are under predicted while impulse is significantly under predicted.

It was also found that methodologies using a single PES, such as the BST methodology produced, in general, significantly lower overpressure and impulse compared with the damage indicators seen in the accidental explosion. To properly apply a PES based methodology which does not consider the total cloud, requires the definition of multiple explosion zones to produce results similar to the combined model and the results of the accident investigations.

6. References

[1] Baker WE, Cox PA, Westine PS, Kulesz JJ and Strehlow RA. (1983): Explosion hazards and evaluation. Oxford, New York: Elsevier, Amsterdam.

[2] Buncefield Major Incident Investigation Board (2008): The Buncefield Incident 11 December 2005, The final report of the Major Incident Investigation Board Volume 1.

[3] Chemical Safety and Hazard Investigation Board (2005): Refinery explosion and fire BP, Texas City, Texas, March 23, 2005, Report No. 2005-04-I-TX.

[4] Dorofeev S.B. (2007a): Evaluation of Safety Distances Related to Unconfined Hydrogen Explosions, *Int. J. Hyd. Energy* 32, pp. 2118-2124.

[5] Dorofeev S.B. (2007b): A Flame Speed Correlation for Unconfined Gaseous Explosions, *Process Safety Progress*, 26:2, pp. 140-149.

[6] Feng C.C., Lam S.H. and Glassman I.(1975): Flame Propagation Through Layered Fuel-Air Mixtures, *Comb. Sci. Tech.*, **10**, pp. 59-71.

[7] Health and Safety Executive (1975): The Flixborough Disaster: Report of the Court of Inquiry, HMSO.

[8] Health and Safety Executive (2008): Buncefield Explosion Mechanism Phase 1 Volumes 1 and 2, Report RR718.

[9] Moore L. (1989): Private Communication.

[10] Roberts A.F., and Pritchard D.K. (1982): Blast Effect from Unconfined Vapour Cloud Explosions, *Journal of Occupational Accidents*, 3, pp. 231-247.

[11] Sadee C. (1976): The Characteristics of the Explosion of Cyclohexane at the Nypro (UK) Flixborough Plant on 1st June 1974, *Journal of Occupational Accidents*, 1, pp. 203-235.

[12] The Netherlands Organization for Applied Scientific Research (TNO) (1997): TNO yellow book, 1st ed., TNO, Apeldoorn, The Netherlands.

[13] US Department of Labor (1990): The Phillips 66 Company Houston Chemical Complex Explosion and Fire, A Report to the President, US Department of Labor Occupational Safety and Health Administration.

Determining VCE damage zones using the GAME correlations and explosion regions

Hans Boot MSC
TNO, Dep. Urban Environment & Safety
PO Box 80015, 3508 TA Utrecht, The Netherlands
hans.boot@tno.nl

Martijn (M.M.) van der Voort MSc
TNO, Dep. Explosions, Ballistics and Protection,
PO Box 45, 2280AA, Rijswijk, The Netherlands
martijn.vandervoort@tno.nl

Prepared for Presentation at
American Institute of Chemical Engineers
2013 Spring Meeting
9th Global Congress on Process Safety
San Antonio, Texas
April 28 – May 1, 2013

UNPUBLISHED

Determining VCE damage zones using the GAME correlations and explosion regions

Hans Boot MSC
TNO, Dep. Urban Environment & Safety
PO Box 80015, 3508 TA Utrecht, The Netherlands
hans.boot@tno.nl

Martijn (M.M.) van der Voort MSc
TNO, Dep. Explosions, Ballistics and Protection,
PO Box 45, 2280AA, Rijswijk, The Netherlands
martijn.vandervoort@tno.nl

Keywords: Vapor cloud explosion, GAME, Multi Energy, deflagration

Abstract

Predicting potential consequences of Vapor Cloud Explosions (VCEs) has always been an important issue in safety assessments, because of the devastating damage that this phenomenon can create on (petro chemical) production sites. Although the TNO Multi-Energy method has been recognized as one of the best methods to predict blast overpressures of vapor cloud explosions, its application has been complicated by the required expert knowledge to choose a blast strength category and the congested fraction of the flammable cloud. In the GAME project (Guidance on Application of Multi-Energy, a contract research project for HSE), TNO already developed a quantitative method to derive the blast strength of VCEs, based on well-defined congestion area properties. Recently, this method has been extended and combined with dispersion models to derive a practical cloud coverage fraction, which is used to find the other important influencing parameter: the explosive mass of the congested part of the flammable cloud. This so-called "explosion regions" method can be used to determine the potential damage zones on a realistic site, with various potential sources of releases and multiple congestion areas or "receptors". The paper will give a full description of the calculation procedure used and provide potential benefits and drawbacks of its application.

1. Introduction: VCE blast pressure estimation

The Vapor Cloud Explosion (VCE) is generally seen as one of the most destructive accident scenarios in petrochemical operations. Consequently, much effort is put in analyzing potential damage of VCEs in site safety studies, to fulfill all kinds of regulatory guidelines on the design and location of buildings within refineries and chemical plants. Unfortunately, the exact method to derive VCE overpressure contours is not fully defined yet.
For the prediction of blast effects of vapor cloud explosions, three different methods can be used: TNT equivalency, Blast Curve methods, and numerical simulations [1]. A numerical simulation (often referred as CFD = Computational Fluid Dynamics based) requires a detailed 3 D description of the environment, including all obstructing geometries, which makes the method

very time-consuming and laborious, thus almost impossible to use for a total site analysis. The TNT method uses a "TNT equivalency" factor that has been derived from statistical analysis of the damage observed in a limited number of vapor cloud explosions incidents. However, the TNT equivalency model is a poor model for vapor cloud explosion blast. While a TNT charge produces a shock wave of high amplitude and a short duration, a vapor cloud explosion produces a blast wave, often shocks, of lower amplitude and longer duration. Furthermore, a practical value for the TNT equivalence is an average, based on the wide statistical distribution found in practice. As a consequence, a predictive estimation with TNT-equivalency on the basis of an average has very little statistical reliability. A more deterministic estimate of blast effects is possible if a parameter could be found that correlates with the process of blast generation in vapor cloud explosions. Such a parameter is introduced within blast curve methods, where the (TNO) Multi-Energy method [1,2] is one of the most widely used examples. One of the difficulties in using the Multi-Energy model however is the determination of the appropriate blast strength class, which requires substantial experience. The so-called GAME [3,4] correlations (Guidance on the Application of Multi-Energy) were developed to provide a more quantitative method to correlate blast strength with parameters characterizing the environment in which the vapor cloud is located.
Unfortunately the usage of the GAME correlations requires input parameters that can be difficult to obtain, since they depend on the interaction of a potential flammable cloud with congestion areas located on a chemical site. Because the GAME correlations appeared to provide all the necessary clues to be able to determine potential overpressure levels around obstructed areas, TNO has combined this method with dispersion models into a practical procedure to derive overpressure contours usable for total site analysis.

2. Guidance on the Application of Multi-Energy

2.1 TNO Multi-Energy method

The TNO Multi-Energy (ME) method uses a set of 10 blast curves, describing (dimensionless) side-on overpressure versus scaled distance for 10 typical classes of explosion strength. Additionally, similar curves are provided for blast wave shape and peak dynamic pressure. The availability of information about this wave shape and dynamic pressure (versus distance) allows the usage of building vulnerability criteria, which are often based on a combination of peak overpressure and pressure impulse. Furthermore, the ME method also recognizes that only a portion of the cloud, not the total explosive mass of a vapor cloud, contributes to the blast effect. Thus, usage of the ME method requires to choose two very highly influencing parameters: the blast curve number (or blast strength class), and the percentage of the cloud that is congested. Although the Yellow Book [2] provides suggestions for selecting the blast strength, based on criteria like ignition strength, degree of obstruction and occurrence of parallel plane confinement, such a curve selection includes considerable expert judgment and remains rather qualitative. Apart from that, the determination of the congested fraction of the (potential) flammable cloud requires detailed knowledge of flammable cloud size and explosive mass, which is usually only available through detailed dispersion modeling.

2.2 Basic GAME correlations

The GAME correlations, developed by TNO in a contract research project for HSE [3,4] and aimed at providing a pure quantitative relation to determine the maximum overpressure, basically consist of two formula's:
For an open (3D) environment:

$$\mathbf{P_{max} = 0.84\ (VBR * L_p / D_{avg})^{2.75}\ .\ S_L^{\ 2.7}\ .\ D_{avg}^{\ 0.7}} \qquad \text{[Eq. 1]}$$

For situations with confinement between parallel planes (2D)

$$\mathbf{P_{max} = 3.38\ (VBR * L_p / D_{avg})^{2.25}\ .\ S_L^{\ 2.7}\ .\ D_{avg}^{\ 0.7}} \qquad \text{[Eq.2]}$$

Where:

P_{max} = (dimensionless) maximum overpressure of the VCE
VBR = Volume Blockage Ratio: the fraction of volume occupied by obstacles
L_p = Flame path length; the longest distance for the point of ignition to the outer edge of the congested region
D_{avg} = Average obstacle diameter within the obstructed region
S_L = Laminar burning velocity of the fuel

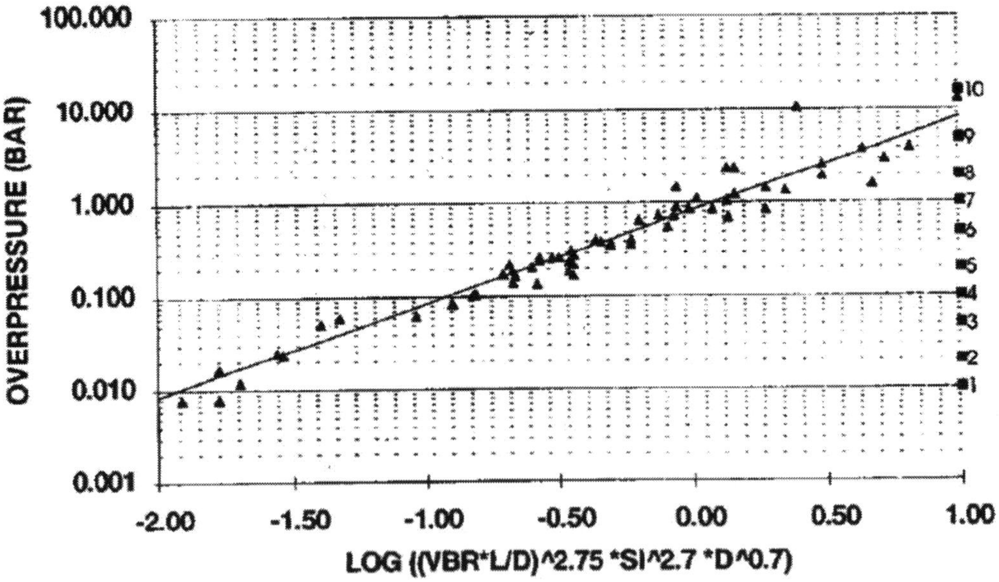

Figure 1: Experimental data (MERGE) versus the GAME correlation [3], [4]

As in the ME curves, the P_{max} overpressure, provided by [Eq1] and [Eq2], is normalized by the ambient pressure to give the initial pressure at r_0 and source strength. This way, the resulting P_{max} can be seen as a replacement for the blast curve number selection procedure, since it determines the starting point of the blast curve.
The VBR and D_{avg} parameters are typical values representative for the congested area defined by the site geometries, and the laminar burning velocity S_L is a chemical dependent property. The flame path length L_p however depends on the overlap of the flammable cloud with a congestion area. This implies that the GAME correlation needs to be combined with an extended dispersion

model, because potential locations and explosive mass within the flammable cloud need to be revealed and combined with congestion area definitions.

2.3 Representative "blast curve" class number

Since the GAME correlation only provides a single P_{max} value, comparable to the initial side-on peak overpressure of the ME method, additional calculations are needed to be able to determine the overpressure at a distance and derive pressure impulse information. In order to relate the GAME result to the original ME blast curve method, a representative blast curve number is calculated using interpolation between the ME classes initial overpressure values. Where the original blast curves have classes 1-10 (value of 10 implying max detonation) , the representative GAME derived value will be decimal: for instance 9.52. This decimal value is subsequently used to construct interpolated "overpressure versus distance", "positive phase duration" and "dynamic pressure versus distance" information, by using interpolation of original surrounding ME class curves. This translation of the single value P_{max} into full graphs thus enables the usage of the GAME correlations in overpressure contour calculation.
When validating this interpolation of ME curves, it was quickly apparent that the calculation examples, provided in [4] could come up with initial pressure well above the maximum side-on overpressure of a class 10 "detonation" curve in the ME charts (which is approx. 16 bar). As the original ME charts are well validated whereas the example is merely used as an illustration of the application of the procedure, it was decided to limit the graphs to this max strength class of 10.

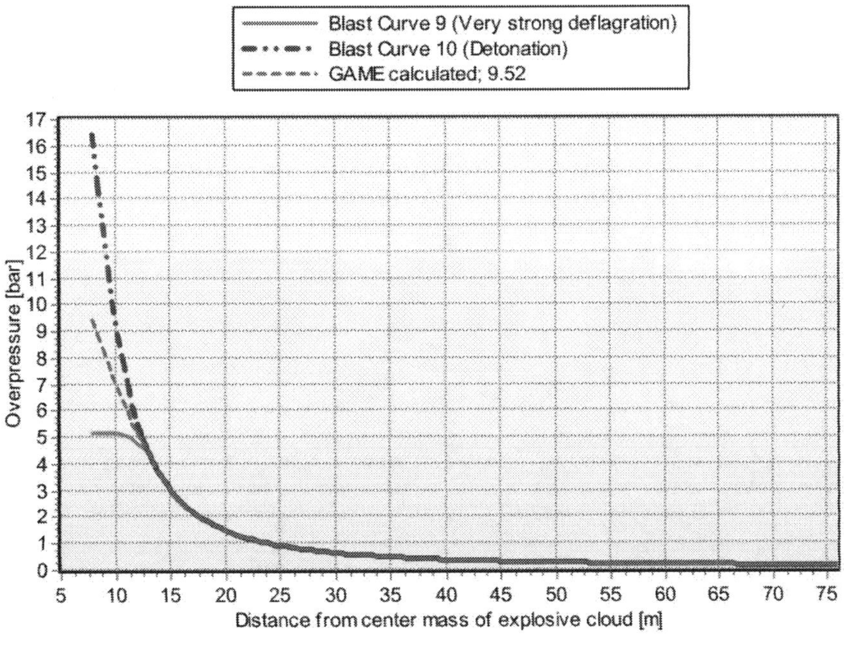

Figure 2: Interpolated overpressure versus distance graphs

3. The explosion regions method

3.1 Extension of dispersion calculation output with "dynamic cloud footprint"

To predict the "spreading" behavior of a flammable cloud, various dispersion models are available, taking into account stability class, wind speed and surface roughness. Since the method is intended for usage in a total site analysis, containing many obstructed regions and multiple potential release locations, thus requiring to evaluate various wind directions and source-receptor combinations, the usage of CFD based dispersion models is (currently) considered as too complex and time consuming. For this application, empirical based dispersion models appear to be more appropriate. Note that these models simplify the influence of congested areas on the dispersion process as an increased roughness length class. The TNO EFFECTS [5] consequence modeling tool, containing both heavy gas (SLAB-based) and neutral gas (Gaussian) dispersion models, is capable of modeling instantaneous, continuous and semi continuous (time limited) source terms. The explosive mass dispersion models available here already provided information about the area of the LFL (Lower Flammability Limit) cloud, and total explosive mass contained within the flammable region of the cloud. For typical instantaneous and semi-continuous releases, the location, size and explosive mass within a flammable cloud are reported as a function of time, as depicted in figure 3.

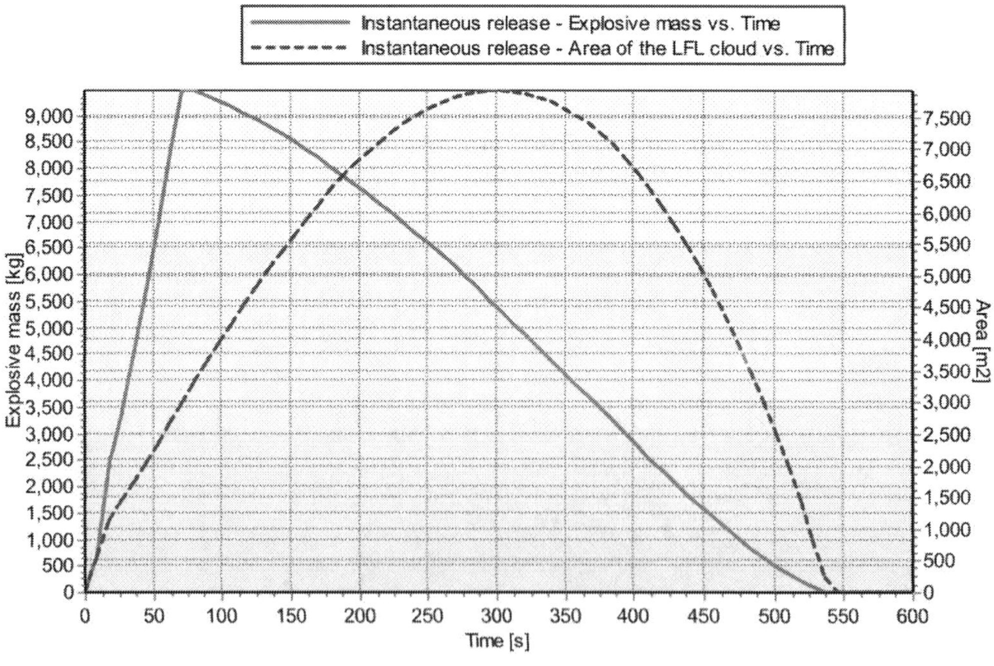

Figure 3: Explosive mass and area of the LFL cloud as function of time

To be able to derive the fraction of the mass that can get enclosed within a potential explosion region, the time depending concentration footprint has been added to the output of the dispersion models. This concentration footprint can be used to derive the outer LFL contour of the cloud, as a function of time.

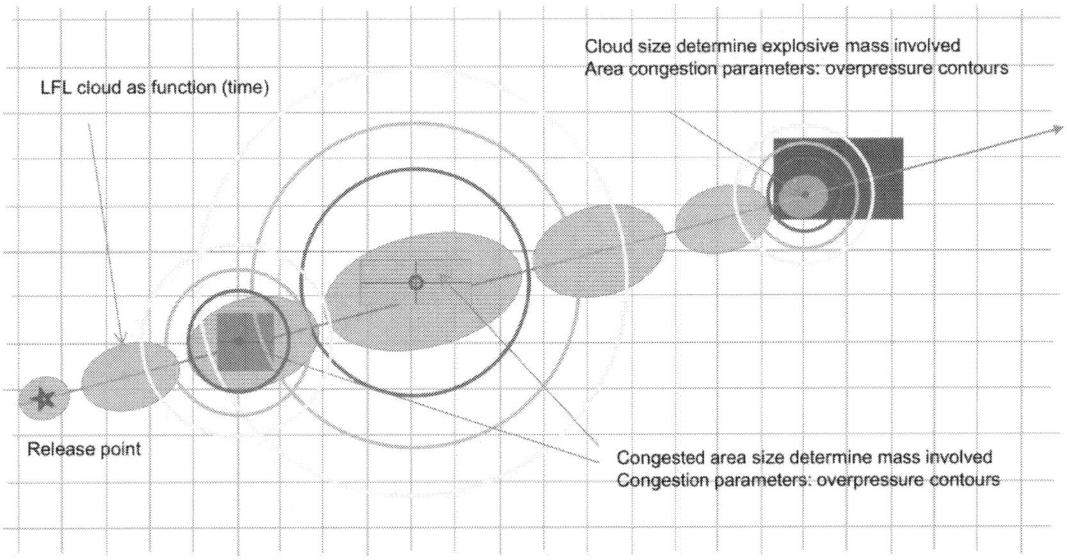

Figure 4: Drifting LFL cloud projected upon congestion areas

3.2 Definition of congestion areas and release locations

Functionality was added to the software to draw congestion areas existing on a chemical site (process equipment; site specific obstructed areas, pinpointed using a background map) as arbitrary polygons, each having its own corresponding "congestion specification". This implies that every defined congestion area will have its own typical congestion properties, which are the VBR and D_{avg} values, and a 2D/3D confinement type selection. These area properties will determine the corresponding GAME formula and provide input parameters for a source-receptor combination involving this area.
After definition of potential release locations, which can be pressure vessels, storage locations, main manifolds etc, (performed on the same site background map), the development of a vapor cloud as a function of time can be plotted on top of these congestion areas, revealing potential overlap situations of the vapor cloud with the areas.

3.3 Overlap of LFL cloud with obstructed area: explosion region

The explosion region method developed uses the (time depending) LFL footprints as provided by the dispersion model, to establish the potential overlap of the cloud with a defined congestion area. This overlap area is called the (potential) "explosion region" and determines both the explosive mass involved in the ME method and the maximum flame path length to use in the GAME relation.

The first approach tested was to use a wind direction (LFL cloud path) aimed at the center of the defined obstructed area, as being the worst case situation. This single cloud path was used to reveal the location of the cloud, having the highest congested explosive mass enclosed within the area.

However, using this approach, it appeared that (when using larger congestion area definitions) this explosion region overlap area would often be located in the center of the area definition, certainly not leading to a worst case overpressure contour. Explosion regions located near the boundaries could potentially lead to higher overpressures at nearby buildings, as compared to an explosion region in the center of the congested area. Apart from that, a congestion area can be L-shaped or have an angle not perpendicular to the release point, creating a situation that outer sides of the obstructed area can even capture more explosive mass.
To overcome this, a search routine was added to determine the worst situation of a maximum enclosed explosive mass within the full boundaries of the obstruction area: iterating both cloud drift angle and distance from release. The resulting worst case explosion region, including the typical congestion parameters VBR and D_{avg} of the obstruction area, will eventually provide the overpressure versus distance information, required to derive overpressure contours.

4. Application of Explosion Region method

The procedure described in the preceding paragraphs has been implemented in the EFFECTS software, combining a dispersion model, an explosion region (maximum mass in overlap) calculation, and the Multi-Energy method that has been extended with the GAME correlations and ME curve interpolation. This chain of models is thus capable of performing an overpressure calculation, based on the definition of a source (location, chemical, storage/working conditions) and congestion areas (VBR, D_{avg}, 2D/3D choice).
Eventually, an important choice relates to the potential location of the determined explosion regions. Because the procedure searches for the worst case situation, basically the highest potential blast wave resulting from an explosion within the defined areas, it does not mean that this location of the explosion area leads to the maximum contour around the region. Since the explosion, with the maximum blast pressures as obtained by the worst case explosion region, can potentially occur anywhere within the congestion area, a practical choice was made to project this overpressure distance as an contour around the full defined congestion areas. The calculated overpressure vs distance profile is used to read out a distance to a threshold level (e.g. 0.1 bar), and this distance is projected around the area as the 0.1 bar contour illustrating the potential damage zone.

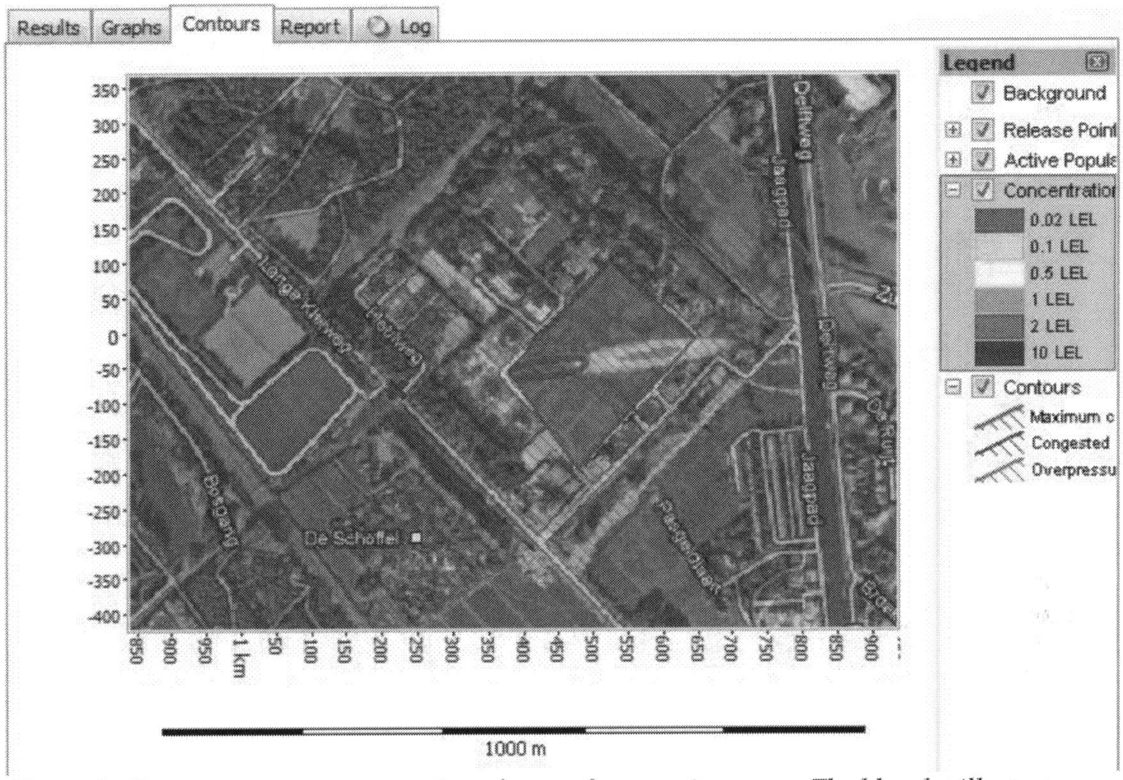

Figure 5: Overpressure contours projected around congestion areas. The blue dot illustrates release point, the red contour shows the (max. mass) explosion region, the colored concentration footprint illustrates occurrence of LFL within congestion area boundaries

Of course this approach neglects the fact that explosions near the boundary of the obstructed region would possibly lead to smaller overpressures, due to venting phenomena or unburned mixture displacement [6] (part of the mass being pushed out of the congestion area during explosion) and potentially smaller flame path length. However, corrections for these "boundary" phenomenon are currently hardly possible, whereas the chosen approach is always more conservative.

When defining larger congestion areas (compared to the size of the LFL cloud itself), the projection of the calculated overpressure distance around the full area would imply explosion regions occurring far beyond the maximum drift of an LFL cloud. This situation is of course highly unrealistic, although congested areas may reduce the dispersion or dilution below LFL. To avoid overpressure contours occurring around boundaries out of reach of the cloud, the obstructed area, as used within the explosion region and contour projection method, is always limited to LFL distance to the source location.

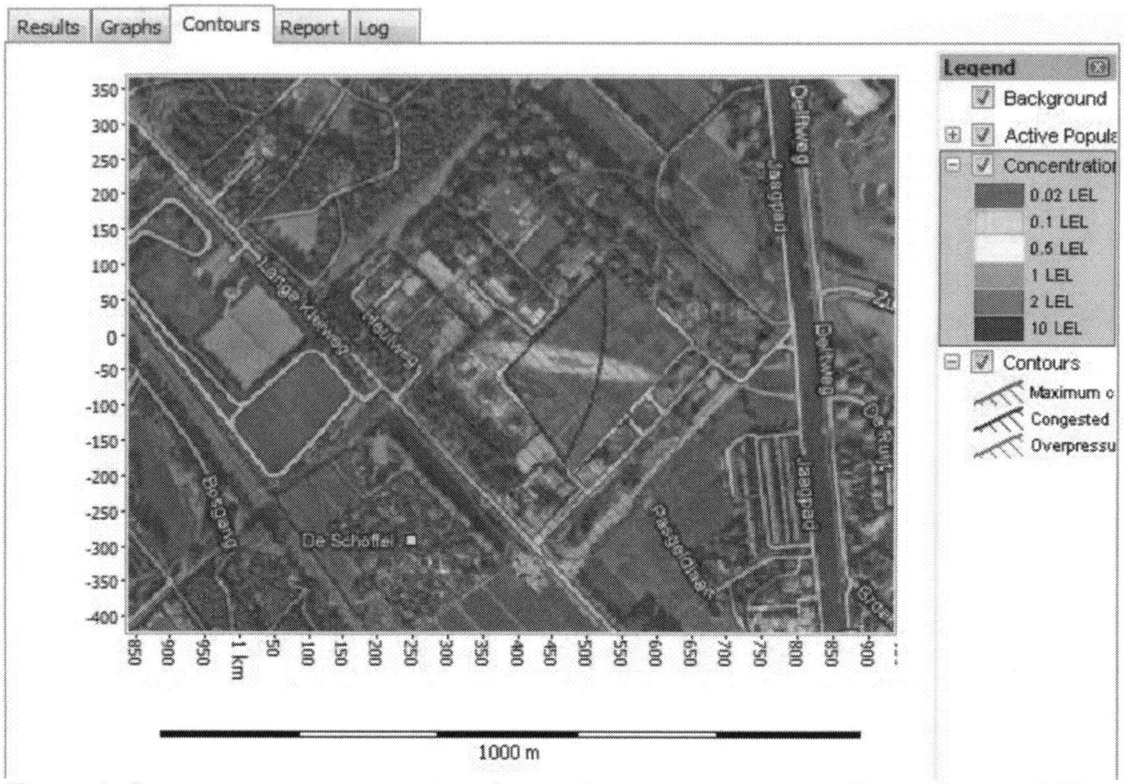

Figure 6: Overpressure contours projected around congestion areas now limited by max. LFL distance

Furthermore, to be on the safe side, it also possible to base explosion region (cloud overlap) and projection method on 50% LFL concentration levels. Note that, although cloud size and maximum distance can be associated with 50% LFL, the explosive mass calculation will always be based upon mass above (100%) LFL.

When evaluating multiple congestion areas, it is also possible to use the guidance on the "Critical Separation Distance" [6] by combining areas into one big area **if** the distance between the congestion areas is less than this separation distance. This feature has not been implemented yet, it involves an important choice with respect to the combination of the potentially different area congestion parameters VBR and D_{avg}.

5. Conclusion

The calculation procedure described, combining a dispersion model with an explosion region (cloud overlap) and multi-energy calculation, proved to be a quick and highly feasible method to present overpressure contours for site analysis purposes. The usage of the GAME correlations avoids the need of expert judgment in selecting blast curves, providing a more transparent and traceable blast pressure calculation. The method described can potentially be extended with damage calculations, using vulnerability criteria for buildings or construction elements and calculated overpressure and pressure impulse at the exact location of critical equipment / structures.

A practical choice was made to use the worst case explosion region (maximum explosive mass) as representative for the contour of the entire congestion area; this eliminates the necessity to perform an explosion calculation for every potential location of an explosion. This solution may be reconsidered in the future if more information about boundary phenomena become available. The explosion region method can be extended with the concept of "Critical Separation Distance", by combining defined congestion areas into one big spatial area. Some practical choices will have to be made for combination of the area congestion parameters.

Although the described procedure can be very practical in its application, some specific disadvantages need to be mentioned as well:
The GAME method, using a VBR and D_{avg} congestion specification, requires detailed knowledge of site geometries and congestion areas, which is not always available.
One of the biggest drawbacks of the method is the fact that it doesn't use a real 3D modeling; the explosion region calculation does not involve any height parameter, and both LFL cloud and congestion areas are 2D projections, although the explosive mass calculation uses 3D volume elements. The height of the release and possible slumping of heavy gas is taken into account by the dispersion model, but concentrations are reported at ground level and height of the congestion area is not included in the overlap calculation.
It is also important to realize that any cloud displacement due to massive buildings/geometries can have big influences on local concentrations, but is not taken into account by the "empirical" dispersion models which simply use an average surface roughness to describe enhanced turbulence due to obstacles. Including this behavior would require a detailed 3D modeling of the surroundings, involving real CFD calculations.

6. References

[1] CCPS, Guidelines for vapor cloud explosion, pressure vessel burst, BLEVE and Flash Fire Hazards, 2nd edition, Center for Chemical Process Safety, 2010.

[2] Committee for the Prevention of Disasters (CPR) publication CPR 14E, The Yellow book, Methods for the calculation of physical effects due to releases of hazardous materials, third edition second revised print 2005, republished as PGS2 in 2006.

[3] Eggen, J.B.M.M (TNO) , GAME: development of guidance for the application of the multi-energy method, HSE contract research report 202, (1998).

[4] Mercx, W.P.M, van den Berg, A.C., van Leeuwen, D, "Application of correlations to quantify the source strength of vapour cloud explsoions in realistic situations" Final report for the project GAMES, Oktober 1998, TNO- report PML 1998-C53 (1998)

[5] EFFECTS Vs 9.012, 2012, Consequence Modeling software by TNO, Utrecht, The Netherlands <www.tno.nl/effects>

[6] van den Berg, A.C., Mos, A.L. Research to improve guidance on separation distance for the multi-energy method (RIGOS), HSE research report 369, (2005)

9TH GLOBAL CONGRESS ON PROCESS SAFETY

NFPA 68 –New Gas Venting Equations

Samuel A. Rodgers
Honeywell
15801 Woods Edge Road, Colonial Heights, VA 23834
samuel.rodgers@honeywell.com

Robert Zalosh
Firexplo
20 Rockland St., Wellesley, MA 02481
bzalosh@firexplo.com

Prepared for Presentation at
American Institute of Chemical Engineers
2013 Spring Meeting
9th Global Congress on Process Safety
San Antonio, Texas
April 28 – May 1, 2013

UNPUBLISHED

NFPA 68 – New Gas Venting Equations

Samuel A. Rodgers
Honeywell
15801 Woods Edge Road, Colonial Heights, VA 23834
samuel.rodgers@honeywell.com

Robert Zalosh
Firexplo
20 Rockland St., Wellesley, MA 02481
bzalosh@firexplo.com

Keywords: NFPA, gas, deflagration venting, vent area, burning velocity, K_G, obstacles, flame enhancement

Abstract

The 2013 edition of NFPA 68 *Standard on Explosion Protection by Deflagration Venting* will replace vent area sizing equations containing the gas Deflagration Index, K_G, for high-pressure applications with equations containing the laminar burning velocity, S_u. Modified Swift-Epstein equations replace both gas venting equations in previous editions of NFPA 68. A scale-dependent flame enhancement factor, λ, has been included in both the low-strength and high-strength methods to account for turbulence and flame instabilities prevalent in many large enclosure deflagrations. The new vent sizing method predicts vent areas that are equal to or larger than the vent areas actually used in the overwhelming majority of tests in unobstructed large enclosures, and in many tests with large-scale obstacles simulating large process equipment. While further work needs to be done to address corrections for enclosures with congested piping arrays, the framework has been set to facilitate this activity going forward. This paper addresses the gas equations implemented in the 2013 edition, modifications to prior vent area corrections and examples of vent sizing.

1. Introduction

The primary purpose of NFPA-68 *Standard on Explosion Protection by Deflagration Venting* [1] is to provide users with suitable vent design methods that can prevent structural failure of enclosures for design basis deflagration scenarios. The 2013 edition of NFPA 68 will have a new gas venting method that, for the first time, has equations and parameters that allow users to specify scenarios that go beyond the particular gas or vapor and the enclosure size, geometry, and pressure resistance. This paper has been written to describe how the new equations have been developed and how they provide a more consistent, as well as more flexible, method over the entire range of enclosure strengths and flammable gases and vapors.

The equations in all editions of NFPA-68 represent the enclosure strength in terms of the design value of maximum reduced pressure, P_{red}, developed in the vented explosion. Since vented gas explosions can have widely varied and complex transient pressures, we start here with a discussion of how P_{red} is related to the transient pressures.

2. Transient and Peak Pressures in Vented Gas Deflagrations

Vented gas deflagration pressure transients have different forms depending on the relative influence of the inherent flammable mixture burning velocity, the vent actuation pressure, turbulent combustion effects, flame instabilities and acoustic waves. Figure 1 is an example of a pressure transient reported by Cooper, et al., [2] in which most of these phenomena occur. There are four labeled pressure peaks associated with the following phenomena. P_1 occurs when the vent is fully open. P_2 occurs at the onset of burned gas venting (and associated increase in vent volume flow rate) and igniting unburned gas outside the vent. P_3 occurs when the enclosure is filled with burned gas such that there is a sharp reduction in the generation rate of gaseous combustion products. P_4, the highest peak pressure in this example, is associated with amplified acoustic oscillations due to a coupling of acoustic waves and residual combustion associated with cellular flames. These amplified pressure oscillations have too high a frequency for most structures to respond. However, low pass filtering of those oscillations to produce smooth pressure at a structural response time scale would probably result in a peak pressure between the P_3 and P_4 values shown in Figure 1. Many tabulations of measured peak pressures from vented gas deflagration tests do not describe the details of which measured or filtered peak pressure is being reported. Therefore, there is some inherent ambiguity associated with the reported peak pressures with regard to their direct applicability to the P_{red} values to be used for vent design.

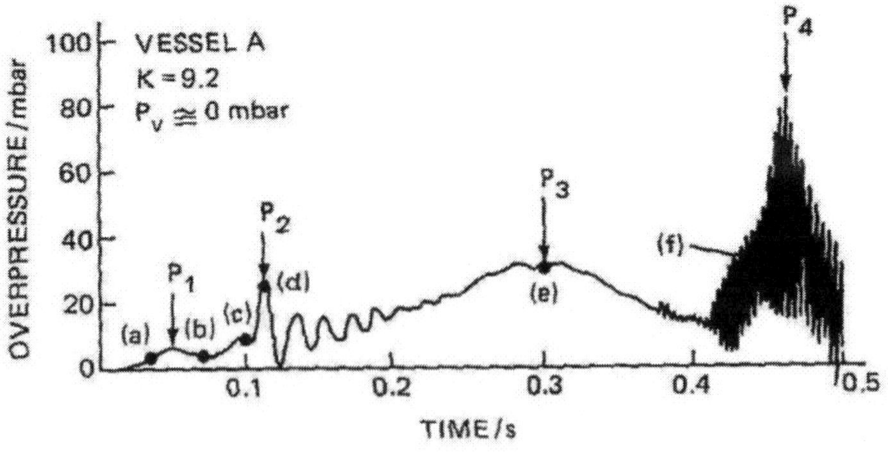

Figure 1: Vented Deflagration Pressure versus Time (from Cooper, et al. 1986)

When flame speeds are high there is yet another issue concerning reported P_{red} values. Whereas the pressure increase is spatially uniform in the enclosure when flame speeds are small in comparison to sound speed, pressure differences start to occur at high flame speeds. This is illustrated in Figure 2, which shows the pressure transients measured at six locations by

Lowensmith, et al., [3] during a methane-hydrogen mixture deflagration in a 69 m^3 enclosure with piping. The measured peak flame speed was 263 m/s, about 75% of the unburned gas-air mixture sound speed. The highest peak pressure in Figure 2 (about 1300 mbar-g) is almost twice as high as the lowest peak pressure measured in the enclosure. The highest peak pressure was measured in a corner of the enclosure where compression waves probably steepen and induce complex interactions with the wall structure.

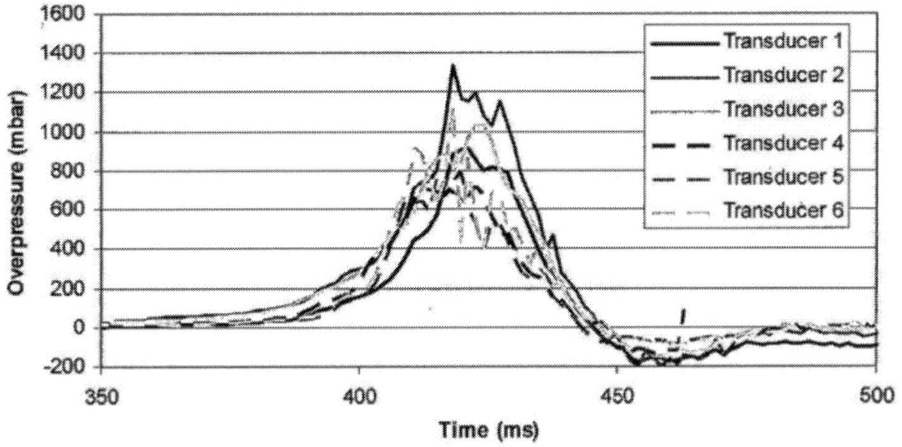

Figure 1: Deflagration Pressures Measured at Six Locations in the Enclosure (Lowensmith, et al. 2010)

In view of the ambiguities and structural significance in the reported peak pressures, it is important for deflagration vent design methods to predict vent areas and P_{red} values that are comfortably larger than the actual reported measured values. The most appropriate vent area design method is not necessarily the one that produces best agreement with reported measured P_{red} values. It is more important to have a design method that can be used readily to determine appropriate required vent areas and vent deployment pressures in applications that go beyond the test conditions. In high flame speed applications where it may be important to determine pressure variations within the enclosure, it could be more appropriate to use Computational Fluid Dynamics methods to estimate local pressures, and to supplement vent designs based on a single P_{red} value.

3. NFPA-68 Gas Venting Equation History

The history of NFPA-68 gas equations is presented to illustrate the various approaches that have been taken and how the prescribed pressures and vent areas compare to various test data. Until the 2013 edition, the NFPA approaches did not explicitly address the effect of internal obstacles on vent area. The constants were, however, chosen to conservatively protect industrial equipment and building enclosures based on available data at the time of a particular edition.

3.1 NFPA-68 1978 Edition

One of the earliest approaches to design of deflagration vents was the vent ratio, that is, the ratio of vent area to enclosure volume. This desired ratio was established based on volumetric scaling of early small-scale test data [4], and was intended to be conservative. In 1978, NFPA-68 [5] presented a model to determine required vent area for gas and dust deflagrations based on the method of Runes [6] as in Equation 1. This model was intended for low-pressure applications, as in buildings. Relying on a limited dataset, the values of the constant, C, were established as in Table 1.

$$A_v = \frac{C \bullet L_1 \bullet L_2}{\sqrt{P}} \qquad \text{[Eq. 1]}$$

where:

A_v = necessary building vent area, ft^2 or m^2

C = constant, characteristic of the fuel gas, dependent upon the units (English or SI) used.
L_1 = smallest dimension of the rectangular building enclosure to be vented, ft or meters.
L_2 = second smallest dimension of the rectangular building enclosure to be vented, ft or meters.
P = maximum internal building pressure which can be withstood by the weakest building member which is desired not to vent or break, lbf/in^2, or kN/m^2

Table 1. C-Factors for Runes Equation

Combustible	C-Factor $(psi)^{1/2}$	C-Factor $(kN/m^2)^{1/2}$
Gases with Flame Speed like Propane	2.6	6.8
Ethylene	4	10.5
Hydrogen	6.4	17
Organic Dusts	2.6	6.8
Organic Mists	2.6	6.8
High Flame Speed Metal Dusts	4	10.5

The relevant C-Factors were separated into groups on the basis of laminar burning velocity, S_u. For a baseline case enclosure with a footprint of 24 ft by 24 ft and a height of 12 ft, L_1 is 12 ft and L_2 is 24 ft. With an enclosure P_{red} of 0.10 bar-g = 1.45 psig and a gas like propane, the required vent area would be 622 ft^2, corresponding to two entire walls and 15% of a third wall area.

3.2 NFPA-68 1988 Edition

In 1988 NFPA-68 [7] adopted two new relations for deflagration venting of gases, one for low-strength enclosures with P_{red} less than 0.1 bar-g and one for high strength enclosures up to a P_{red} of 2 bar-g.. While the Runes equation used the minimum enclosure cross-sectional area, the 1988 equation used the ratio of vent area to enclosure surface area and is a simplification of the Swift-Epstein [8] low pressure derivation. Swift [9] derived C-factors for a number gases and dusts, assuming a burning velocity enhancement factor, λ, of 5 and a discharge coefficient of 0.7. The C-Factors in the 1988 edition of NFPA-68 were developed by the NFPA Explosion Protection Committee enveloping the test data that were available circa 1987. This resulted in the values shown in Table 2. Equation 2 and its C-factors were intended to conservatively envelope the typical situation of initially quiescent gas release in an enclosure with limited or no turbulence-inducing obstacles. Surface area of equipment and contained structures was specifically excluded. The 1988 equation was retained through the 2007 edition of NFPA-68.

$$A_v = \frac{C \bullet A_s}{P_{red}^{\left(\frac{1}{2}\right)}} \quad \text{[Eq. 2]}$$

Where:

A_v = vent area (ft^2 or m^2)

C = venting equation gas/dust constant

A_s = internal surface area of enclosure (ft^2 or m^2)

P_{red} = maximum internal overpressure that can be withstood by the weakest structural element (psi or kPa)

Table 2. C-Factors for NFPA-68 1988

Combustible	C-Factor $(psi)^{1/2}$	C-Factor $(kPa)^{1/2}$
Anhydrous Ammonia	0.05	0.13
Methane	0.14	0.37
Gases with Fundamental Burning Velocity less than 1.3 times that of Propane	0.17	0.45
St-1 Dusts	0.10	0.26
St-2 Dusts	0.12	0.30
St-3 Dusts	0.20	0.51

For the baseline case of an enclosure with a footprint of 24 ft by 24 ft and a height of 12 ft, internal surface area is 2304 ft^2. With an enclosure P_{red} of 0.10 bar-g = 1.45 psig, and a gas like propane, the required vent area would be 325 ft^2, corresponding to one entire wall and 13% of another wall area.

For high-strength (P_{red} > 0.10 bar-g) enclosures, Bartknecht [10] presented a set of nomographs based on mostly unpublished large-scale test data. The nomographs were also eventually reduced to equation form as in Equation 3 and included in NFPA 68-1988 for high strength enclosures. The coefficients for Equation 3 are shown in Table 3.

$$A_v = a \bullet (V)^b \bullet \exp(c \bullet P_{stat}) \bullet (P_{red})^d \quad \text{[Eq. 3]}$$

Where:

A_v = vent area, m^2

V = enclosure volume, m^3

P_{red} = maximum pressure developed during venting, bar-g

P_{stat} = vent closure release pressure, bar-g

Table 3. Coefficients for Bartknecht Relation

Combustible	a	b	c	d
Methane	0.105	0.770	1.230	-0.823
Propane	0.148	0.703	0.942	-0.671
Coke Gas	0.150	0.695	1.380	-0.707
Hydrogen	0.279	0.680	0.755	-0.393

For the baseline case of an enclosure with a footprint of 24 ft by 24 ft and a height of 12 ft, enclosure volume is 196 m^3. The nomographs and equation for a P_{stat} of 0.05 bar-g and a P_{red} of 0.11 bar-g, produce a minimum required vent area of 28 m^2 = 300 ft^2. This required vent area is 92% of the vent area required by the low strength equation at a comparable P_{red} value and the two methods produce similar results at the demarcation.

3.3 NFPA-68 1998 Edition

As a result of the widespread adoption of the Bartknecht nomographs in both Europe and North America, there was a need to have a high-strength enclosure design method that would be applicable to other gases beyond those listed in Table 3. The parameter used by Bartknecht and others to represent a flammable gas or vapor rate of burning and associated rate of enclosure pressure is the gas Deflagration Index, K_G, with the following definition.

$$K_G = \left(\frac{dP}{dt}\right)_{\max} \bullet V^{1/3} \qquad \text{[Eq. 4]}$$

Here $(dP/dt)_{max}$ is the maximum rate of pressure rise measured in a closed vessel deflagration with the worst-case (usually near-stoichiometric) concentration of the flammable gas or vapor. The measured $(dP/dt)_{max}$ is normalized to a vessel volume (V) of 1 m^3. Although simple theoretical considerations, assuming gas burning velocities that are independent of scale, suggest that K_G should be a constant for different size vessels, the closed vessel test data shows that K_G does increase with increasing volume. This effect is discussed in Annex C of NFPA 68-2007 [11].

With the broad acceptance of K_G, Bartknecht [12] published the following equation for gas venting based on K_G. NFPA-68 [13] adopted Equation 5 directly for high-strength enclosures with P_{red} greater than 0.1 bar-g.

$$A_v = \left\{\left[0.127\log_{10}(K_G) - 0.0567\right] \bullet P_{red}^{-0.582} + 0.175 P_{red}^{-0.572}(P_{stat} - 0.1)\right\} \bullet V^{2/3} \quad \text{[Eq. 5]}$$

where:

A_v = Vent area, m^2

K_G = Deflagration Index, ≤550 bar-m/sec

P_{stat} = vent closure release pressure, ≤0.5 bar-g

P_{red} = maximum pressure developed during venting, ≤2 bar-g

V = enclosure volume, ≤1000 m^3

In order to use Equation 5 for different gases, values of K_G need to be "normalized" by the benchmark values of K_G = 100 bar-m/s for propane and 55 bar-m/s for methane. For L/D ≤2, vent area is not adjusted. For L/D from 2 to 5, the vent area, A_v, is increased by adding more vent area, ΔA, calculated from Equation 6, which was based on a review of limited Bartknecht data [12] for elongated enclosures.

$$\Delta A = \frac{A_v K_G \left(\frac{L}{D} - 2\right)^2}{750} \quad \text{[Eq. 6]}$$

For the baseline case of a propane-air deflagration in a 196 m^3 enclosure with a P_{red} of 0.11 bar-g, Equation 5 requires a vent area of 24 m^2 = 259 ft^2, which is somewhat less than the value (28 m^2) determined previously from the Bartknecht nomograph curve fit. The high strength equation vent area was approximately 80% of that using the low-strength equation.

3.4 NFPA-68 2007 Edition

For the 2002 edition some changes were made to account for vent inertia, but the basic gas venting equations remained the same. In the 2007 edition of NFPA-68 [13], the committee developed a correlation for the low-strength Venting Parameter, C, as a function of Fundamental Burning Velocity, shown in Equation 7. This correlation increased the benchmark C-Factor for propane from 0.17 to approximately 0.2 psi $^{1/2}$. Methane also increased from 0.14 to approximately 0.16 psi $^{1/2}$.

For C (psi $^{1/2}$):

$$C = 6.1 \bullet 10^{-5}(S_u)^2 + 6.1 \bullet 10^{-4}(S_u) + 0.0416 \quad \text{[Eq. 7]}$$

When Equation 2 is used with the new C-Factor for propane and the baseline case at a P_{red} of 0.10 bar-g, the result is a required vent area of 380 ft^2 = 35 m^2, which is 46% higher than

Equation 5 produces at a P_{red} of 0.11 bar-g. The change in C-Factor resulted in a larger and more significant discontinuity in NFPA 68 required vent areas near the demarcation between the low-strength and high-strength equations.

3.5 Issues with the NFPA-68 Equations

One fundamental issue with the NFPA-68 2007 edition equations is the use of the laminar burning velocity, S_u, in the low-strength enclosure equation, while the high-strength equation uses K_G. The laminar burning velocity is the more fundamental parameter, and there are numerous reported measured values of S_u for a wide variety of gases and gas mixtures. Table D.1 of NFPA 68-2007 [11] is one such tabulation. Therefore, it is desirable to replace Equation 5 with an equation using S_u.

It is also desirable for the high-strength and low-strength equations to produce the same or much closer vent areas than exists now at the demarcation value of P_{red} for the applicability of the two equations. This is particularly so due to a significant number of enclosures designed near the current demarcation value of 0.1 bar-g.

Since the value of C in Equation 2 is a constant for a given value of S_u, the Swift-Epstein low strength equation in NFPA 68-2007 uses only the enclosure surface area and a fuel factor to determine vent area to produce a given pressure, P_{red}. While attractive in its simplicity, it does not account for the well-known variation of turbulent burning velocities and flame instability manifestations with increasing enclosure scale. Therefore, the use of empirical C values based on a limited range of test volumes may cause calculated vent areas to be under-sized for large volume applications such as large buildings. Conversely, the enveloping of the data to select the C values is probably producing unnecessarily large vent areas for applications with volumes smaller than about 10 m^3. This was the conclusion of Razus and Krause [14] when they compared the pressures calculated from the NFPA 68 equations with those of other vent area equations and to test data that mostly involved enclosure volumes in the range 0.6 m^3 to 1 m^3.

Another major effect not accounted for in the 2007 and earlier editions of NFPA 68 is the influence of internal equipment and structures within the enclosure. The magnitude of the effect of these turbulence-inducing obstacles was investigated by Zalosh in a 2008 NFPA Fire Protection Research Foundation project [15]. Test data on this effect was compiled from eight test programs involving vented gas explosions in enclosures with volumes between 2.5 m^3 and 2688 m^3. This report provided graphs and correlations to show how the ratio of NFPA-68 calculated vent areas to actual vent area varies with the obstacle/enclosure surface area and volume ratios for both small and large obstacles. One of those graphs is shown here as Figure 3. All the vent area ratio data in Figure 3 are less than one when the surface area ratio, $A_{s\_obstacle}/A_{s\_enclosure}$ is greater than 0.46. Some of the data in Figure 3 show vent area ratios less than one at $A_{s\_obstacle}/A_{s\_enclosure}$ ratios as low as 0.21. Thus, a critical value for $A_{s\_obstacle}/A_{s\_enclosure}$, or A_{obs}/A_s, to indicate a need for greater vent area could be expressed as in Equation 8, where the precise value, in the range 0.21 to 0.46, depends on whether the obstacles are small or large compared to the characteristic dimension of the enclosure.

$$0.21 < \left[\frac{A_{obs}}{A_s} \right]_{cr} < 0.46 \qquad \text{[Eq 8]}$$

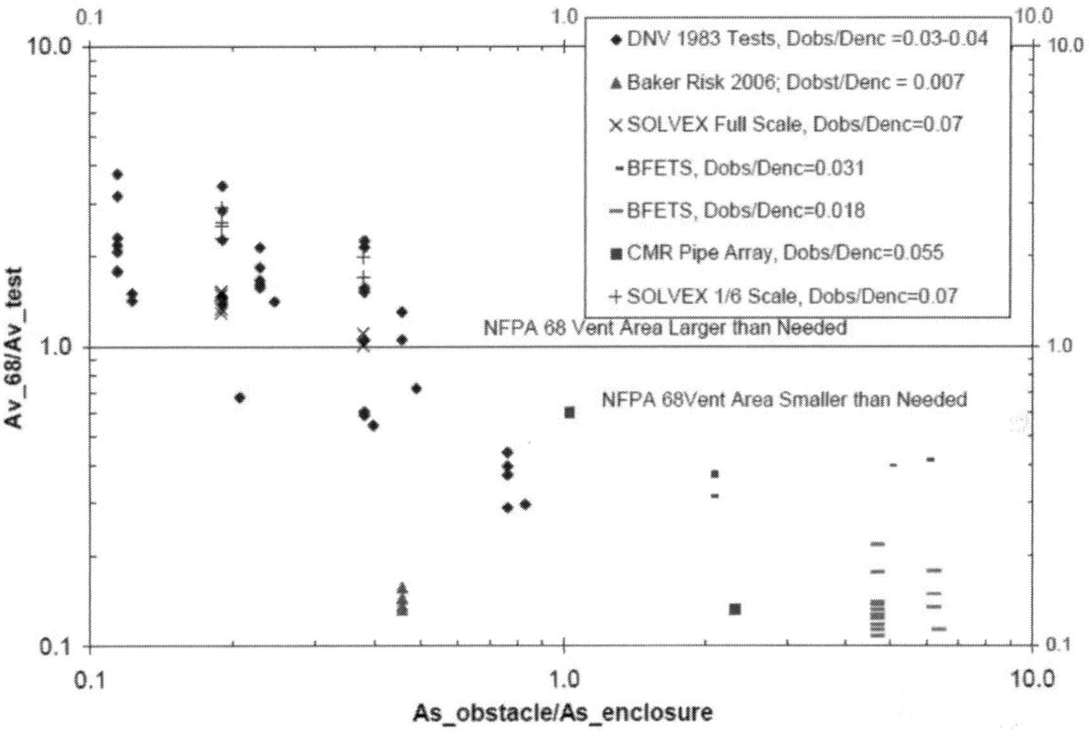

Figure 3. Plot of Calculated to Experimental Vent Area Ratio vs Obstacle/Enclosure Surface Area Ratio [15]

One other important drawback of the 2007 and previous NFPA 68 gas venting methods is the need to design vents for a scenario in which the flammable gas-air mixture is at near-stoichiometric proportions throughout almost the entire enclosure volume. This drawback is particularly problematic for large building applications in which there is only a limited amount of flammable gas or vapor. NFPA 68-2007 does have a partial volume provision in the dust explosion vent design methodology, but not for gas venting.

4. Development of the NFPA-68 2013 Gas Venting Equations

4.1 Development Approach

The latest revision of NFPA-68 is the first step in moving away from the constant C-factor and K_G parameters toward a consistent use of the laminar burning velocity, S_u, and scale-dependent

C values. The goal was to develop a single equation spanning the required strength or P_{red} range, as had been done earlier for dusts. A gas venting task group was charged with compiling test data, evaluating a number of candidate vent area equations, reviewing and incorporating the work of the Fire Protection Research Foundation, and making recommendations on a consistent approach for all P_{red} values.
The NFPA gas venting task group worked on this charge over a period of two years on a volunteer basis. Various gas vent area equations were examined from the standpoint of 1) ease of use; 2) producing vent areas that were at least as large as the vent areas in most large enclosure experiments; and 3) having a fundamental foundation based on the laminar burning velocity and the key enclosure characteristics including volume scaling considerations that can potentially account for flame instabilities, flame stretch, and turbulence inducing flame accelerations. In terms of the key enclosure characteristics, it was considered preferable to use the enclosure internal surface area because that is the maximum flame surface area in the absence of flame wrinkling and stretching. Since it is also the enclosure parameter in the existing NFPA 68 low strength equations, it is already being used in many NFPA 68 applications.

The goal of a single equation was not achieved because fundamentally based vent design equations need to account for the fundamentally different vent flow regimes associated with subsonic and sonic vent flows. The task group recommendation was to retain separate low and high strength relations, both based on the Swift-Epstein formulations described in references 8 and 16, respectively. A cross-over of P_{red} of 0.5 bar-g was determined based on a criterion that the two equations produce similar results for many applications in the P_{red} range of 0.2 bar-g to 0.5 bar-g. However the original constant turbulent flame enhancement factor, λ, in these references has been replaced with a volume scaling prescription based on the empirical equations reported by Rota, et al. [17] based on best fit comparisons to test data available circa 1990. In addition to the Rota turbulent adjustments, there are additional adjustments to include effects of internal obstacles as well as enclosure length to diameter ratio. While not discussed here, a Partial Volume adjustment has also been added, permitting reduced vent areas when the potential quantity of gas released is not sufficient to produce a stoichiometric mixture throughout the enclosure.

4.2 Swift-Epstein Equations

Epstein, Swift and Fauske [16] derived deflagration vent area equations for sonic flow through the vent of a spherical vessel. Subsequently Swift and Epstein [8] developed a similar solution for sub-sonic incompressible flow. Equations 9 and 10 are the related NFPA-68 correlations for sub-sonic flow, applicable to low strength enclosures with $P_{red} \leq 0.5$ bar-g. The form of Equation 9 is identical to the prior NFPA-68 1988 solution with a different expression for C. The Committee modified the original expression for C in Equation 10, making some adjustments that are described in the Postscript section.

$$A_{v0} = \frac{C \bullet A_s}{P_{red}^{(1/2)}} \qquad \text{[Eq. 9]}$$

C is determined from Equation 10, including the burning velocity as the product of fundamental burning velocity and flame enhancement factor.

$$C = \frac{S_u \rho_u}{2G_u} \frac{\lambda}{C_d} \left[\left(\frac{P_{max} + 1}{P_0 + 1} \right)^{\left(\frac{1}{\gamma_b}\right)} - 1 \right] (P_0 + 1)^{\frac{1}{2}} \qquad \text{[Eq. 10]}$$

Where:

A_{v0} = the minimum vent area calculated from Equation 6 (m^2),

A_s = the enclosure internal surface area (m^2),

P_{red} = the maximum pressure developed in a vented enclosure during a vented deflagration (bar-g),

P_{max} =the maximum pressure developed in a contained deflagration by ignition of the same gas-air mixture (bar-g),

P_0 = the enclosure pressure prior to ignition (bar-g),

S_u = fundamental burning velocity of gas-air mixture (m/sec),

ρ_u = mass density of unburned gas-air mixture (kg/m^3) = 1.2 for flammable gases with stoichiometric concentrations less than 5 volume percent, and an initial temperature of 20°C,

G_u = unburned gas-air mixture sonic flow mass flux = 230.1 $kg/m^{2-}sec$ for air at an enclosure initial temperature of 20°C,

λ = ratio of gas-air mixture burning velocity accounting for turbulence and flame instabilities in vented deflagration to the fundamental (laminar) burning velocity,

γ_b = ratio of specific heats for burned gas-air mixture = 1.1 to 1.2 depending on the gas mixture

C_d = vent flow discharge coefficient = 0.70 unless the vent occupies an entire wall of the enclosure, in which case a value of 0.80 can be used.

As with the 2007 low-strength solution, the vent release pressure, P_{stat}, is not explicitly included and continues to be arbitrarily set below P_{red}. When $P_{red} \leq 0.1$ bar-g, P_{stat} must be less than or equal to (P_{red} - 0.024 bar). When $P_{red} > 0.1$ bar-g, P_{stat} must be less than 0.75 P_{red}.

Equations 11 and 12 are the NFPA-68 correlations for sonic flow, applicable to high strength enclosures with $P_{red} \geq 0.5$ bar-g. Contrary to the low-strength correlation, the vent release pressure, P_{stat}, is explicitly included.

$$A_{v0} = A_s \frac{\left[1-\left(\frac{P_{red}+1}{P_{max}+1}\right)^{1/\gamma_b}\right]}{\left[\left(\frac{P_{red}+1}{P_{max}+1}\right)^{1/\gamma_b}-\delta\right]} \frac{S_u\rho_u}{G_u}\frac{\lambda}{C_d}$$ [Eq. 11]

$$\delta = \frac{\left(\frac{(P_{stat}+1)}{(P_0+1)}\right)^{1/\gamma_b}-1}{\left(\frac{(P_{max}+1)}{(P_0+1)}\right)^{1/\gamma_b}-1}$$ [Eq. 12]

Where:

P_{stat} = nominal vent deployment or burst pressure (bar-g)

4.3 Flame Stretch and Instabilities

NFPA 68-2013 evaluates λ in stages with a baseline value, λ_0, then increased for the effect of turbulence-generating obstacles to become λ_1, and finally increased for enclosure length to diameter ratio to become λ.

4.3.1 λ_0 – Baseline Spherical, Unobstructed

Rota, et al [17] developed a similar simplified model for spherical combustion, but continued to determine a value for λ without internal obstructions from the product of two effects, the flame growth inside the enclosure, shown here as φ_1, and the flame flow through the vent, shown here as φ_2. They evaluated three constants, θ, β_1 and β_2, to minimize the error between their simplified approach and a dataset of approximately 160 experiments. The NFPA 68 version simplifies the equations by setting a value of 4000 for the cellular flame Reynolds number originally part of φ_1. Additionally, the original Rota equations use the ratio of flame volume to enclosure volume in the exponent of φ_2 while the NFPA-68 version conservatively evaluates the flame growth effect at the point where the flame volume is equal to the vessel volume. Equations 13 through 18 are required to evaluate λ_0.

$$\lambda_0 = \varphi_1\varphi_2$$ [Eq. 13]

$$\varphi_1 = \begin{Bmatrix} 1, if\ \mathrm{Re}_f < 4000 \\ \left(\frac{\mathrm{Re}_f}{4000}\right)^{\theta} otherwise \end{Bmatrix}$$ [Eq. 14]

$$\mathrm{Re}_f = \frac{\rho_u S_u \left(D_{he}/2 \right)}{\mu_u}$$ [Eq. 15]

$$\varphi_2 = \max\left\{1, \beta_1 \left(\frac{\mathrm{Re}_v}{10^6}\right)^{\left(\beta_2/S_u\right)}\right\}$$ [Eq. 16]

$$\mathrm{Re}_v = \frac{\rho_u u_v \left(D_v/2 \right)}{\mu_u}$$ [Eq. 17]

$$u_v = \left\{ \sqrt{\frac{200000 \bullet (P_{red})}{\rho_u}} if\_P_{red} < 0.9\_bar \right\}, a_u\_otherwise$$ [Eq. 18]

where:

μ_u = the unburned gas-air mixture dynamic viscosity = 1.8x10^{-5} kg/m-sec for gas concentrations less than 5 vol% at ambient temperatures,

D_{he} = the enclosure hydraulic equivalent diameter as determined in Chapter 6 [of NPFA 68-2013] (m),

D_v = the vent diameter as determined through iterative calculation (m),

θ = 0.39,

β_1 = 1.23,

β_2 = 0.0487 m/sec.

Since the vent diameter in Equation 17 is a result of the analysis, the entire evaluation of required vent area becomes iterative. This means not only the basic equation but also corrections due to vent panel inertia and vent ducts are solved iteratively.

Figures 4 and 5 show how the two components of λ_0, φ_1 and φ_2, change with enclosure volume. For both figures, the high-strength Equation 11 was iterated to a small vent area that would produce a P_{red} of 95% of P_{max} with P_{stat} at 90% of P_{max}, essentially a closed volume deflagration. φ_1 increases with volume at a rate proportional to S_u while φ_2 increases at a rate inversely proportional to S_u. As a result of these two effects, λ_0 increases with volume with the magnitude of the increase depending on the gas laminar burning velocity. Very large increases are produced for hydrogen due to flame instability induced increases in surface area, as represented by the φ_1 variation shown in Figure 4. On the other hand, vent flow has more effect on flame enhancement for slower burning gases, as seen by variation in φ_2 shown in Figure 5.

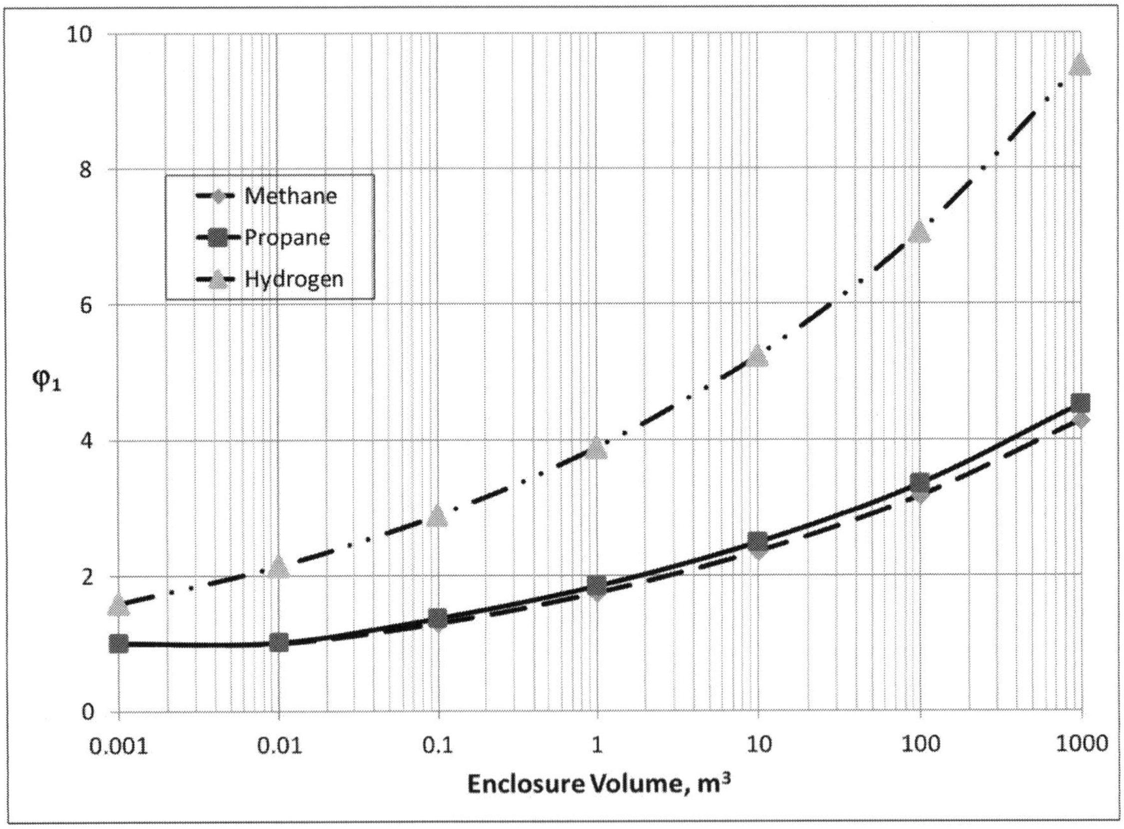

Figure 4: Variation of φ_1 with Spherical Enclosure Volume

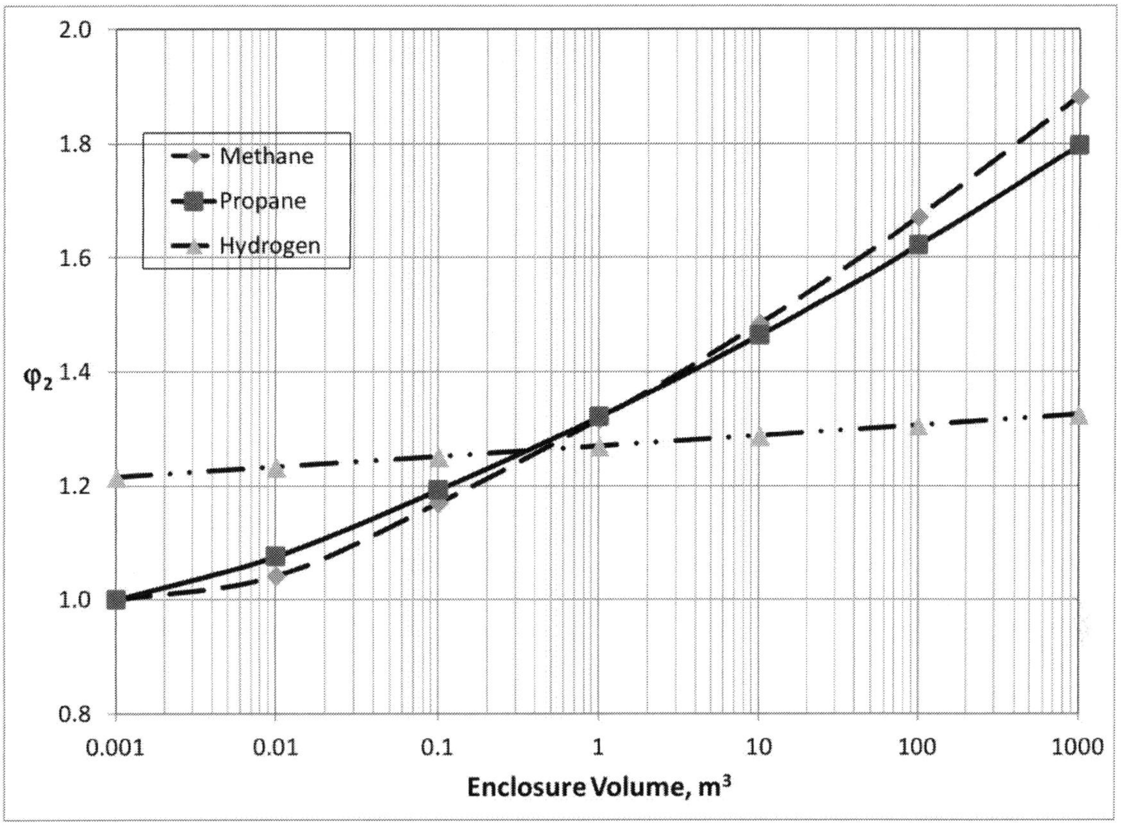

Figure 5: Variation of φ_2 with Spherical Enclosure Volume

4.3.2 λ_1 – Surface Area of Obstructions

In the interest of ease of use, the range of critical internal obstacle surface areas represented by Equation 8 was reduced to a single value, that is, when the surface area of the internal obstacles exceeds 40% of the enclosure surface area. NFPA-68 requires an estimation of the total external surface area, A_{obs}, of equipment and other objects within the enclosure such as:

- Piping – Piping, tubing and conduit with diameters greater than ½ inch.
- Building Supports – Internal structural columns, beams, and joists.
- Stairways – Stairways and railings.
- Smaller Equipment – Equipment with a characteristic dimension in the range 2 inches to 20 inches

As a practical matter, there is conflicting guidance in NFPA-68 concerning idealizing the enclosure surface area while at the same time considering the surface area of internal obstacles. At this time a degree of judgment is needed and the listed elements directly attached to or very near to the enclosure surface would not be "obstacles". For instance, the wall and flat ceiling support beams might not be included while mezzanine steel supports and stairways would be included. A 1 inch thick by 11 inch deep fiberglass composite stair tread with 1-inch square open rectangular construction might be considered a 1 inch by 11 inch rectangle, ignoring all of the surface created by the 1-inch openings.

After estimating the surface area of the obstructions, if this does not exceed 40% of the enclosure surface, no adjustment is made and $\lambda_1 = \lambda_0$. When A_{obs} is $\geq 0.4A_s$, λ is increased linearly as in Equation 19.

$$\lambda_1 = \lambda_0 \bullet \left(0.6 + \frac{A_{obs}}{A_s} \right) \quad \text{for } A_{obs}/A_s \geq 0.40 \qquad \text{[Eq. 19]}$$

4.3.3 λ – Length to Diameter Ratio of Enclosure

The 2007 edition of NFPA-68 effectively limited the low-strength correlation to an L/D of 3 by prescribing the maximum length of a protected enclosure with a single end-mounted vent to be 3 times the hydraulic diameter of the enclosure. At the same time, the high-strength correlation applied a correction factor beginning at L/D = 2. The current formulation compromises and applies a similar correction factor beginning at L/D = 2.5.

If the L/D does not exceed 2.5, no adjustment is made and $\lambda = \lambda_1$. For L/D values from 2.5 to 5, and for P_{red} no higher than 2 bar-g, λ is increased as in Equation 20.

$$\lambda = \lambda_1 \bullet (1 + \left(\frac{L/D}{2.5} - 1 \right)^2) \qquad \text{[Eq. 20]}$$

5. Comparison to Large-Scale Test Data

The task group assembled a database of over 200 tests, mostly in unobstructed enclosures, with volumes of at least 1 m^3. These are in addition to the database of over 150 tests previously compiled by Zalosh [15] for tests in large enclosures with obstructions. A subset of the unobstructed enclosure data pertaining to gas-filled compact enclosures ($L/D \leq 2.5$) was selected for comparison with calculations using the λ-adjusted Swift-Epstein equations described above. The test program enclosures and gases selected for comparison calculations are listed in Table 4.

Table 4. Large-Scale Unobstructed Enclosure Test Programs Used for Vent Area Calculations

Enclosure Geometry	Enclosure Volume (m^3)	Gases Tested	Number of Tests	Reference
Cube, Cylinder	1	Methane	13	Bartknecht 1981 [10]
Cylinder	1	Hydrogen, 29.6%	2	Pasman [18]
Sphere	2	Methane	2	Bartknecht [10]
Cylinder	2.5	Propane	9	DeGood & Chatrathi [19]

Enclosure Geometry	Enclosure Volume (m^3)	Gases Tested	Number of Tests	Reference
Cube	5.2	Propane	1	Van Wingerden et al [20]
Box/Room	11.2	Methane	1	Zalosh [21]
Cube	22	Methane	7	Thorne et al [22]
Cube	22	Propane	5	Thorne et al [22]
Cube	25	Hydrogen 12%	5	Cerchiara&Carcassi,[23]
Cylinder	29	Propane	24	Swift-Epstein [8]
Box/Room	34	Methane	2	Zalosh [21]
Sphere	60	Propane	8	Bartknecht [10]
Box/Room	64	H_2 (18%), C_3H_8, CH_4	10	Bauwens et al [24]
Box/Room	69	Methane	2	Lowensmith et al [3]

After calculating the required vent area from the reported value of P_{red} in each test, the calculated vent area was divided by the actual vent area used in the test. The resulting vent area ratios, $A_{v,calc}/A_{v,exp}$, are shown plotted versus enclosure volume in Figure 6. The calculated vent areas are equal to or greater than the actual test vent areas for all enclosure volumes greater than 1 m^3. The only tests in which calculated vent areas are smaller than actual vent areas are for the 1 m^3 cube data reported by Bartknecht. Comparisons of this particular Bartknecht data with calculations using other vent sizing equations in Table 3 of Razus and Krause [14] indicate that this is one of the few data sets for which most of the vent sizing methods underestimate the reported P_{red} values or vent areas.

Figure 7 is a plot of the ratio of the calculated to actual test vent area against the reported P_{red} values. The plot shows most of the ratios greater than 2 are for P_{red} values less than 0.5 bar-g. This indicates that the low-pressure Swift-Epstein equations (9 and 10) seem to produce larger than required vent areas more often than do the high-pressure versions of the equations (11 and 12). As in Figure 6, the only ratios less than 1 are for the Bartknecht 1 m^3 methane data set.

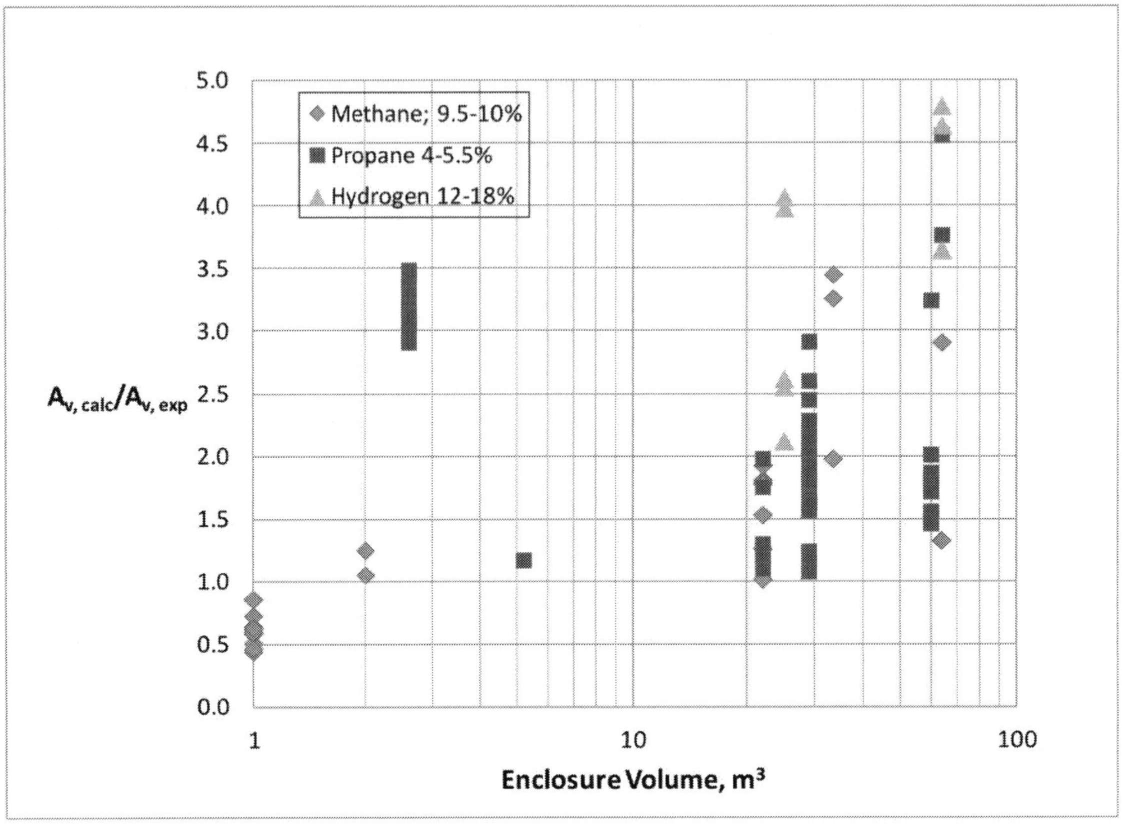

Figure 6: Ratio of Calculated to Actual Vent Area versus Enclosure Volume

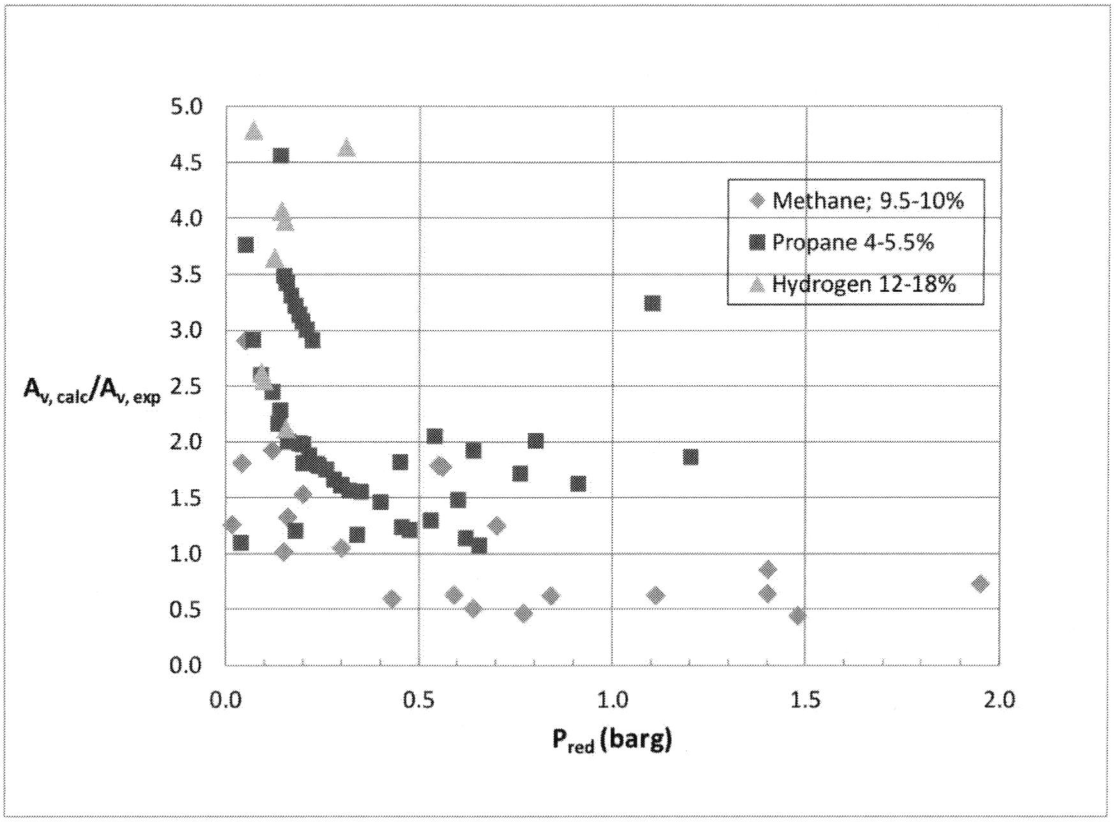

Figure 7: Ratio of Calculated to Actual Vent Area versus Measured P_{red}

The new NFPA 68-2013 Swift-Epstein equations have also been used to calculate vent areas for a subset of the obstructed enclosure data compiled in the Fire Protection Research Foundation report [15]. The five data sets used for the calculations and comparisons are listed in Table 5. Three of the test programs used simulated piping arrays as obstructions, and the other two programs used larger obstructions.

Table 5. Obstructed Enclosure Test Data Used for Vent Area Calculations

Test Program or Organization	Year	Size of Test Enclosure (m^3)	Type of Obstacles	Number of Tests	Reference
Christian Michelson Research Institute	1986	27	Pipe Arrays	12	GexCon Gas Explosion Handbook
Monsanto-Fenwal	1982	82.1	Baffle partitions	8	Howard and Karabinis [25]
Solvex Full-Scale	1993	550	Pipe Arrays	17	Bimson et al. [26]

Test Program or Organization	Year	Size of Test Enclosure (m^3)	Type of Obstacles	Number of Tests	Reference
BFETS	1998	1638, 2688	Low-Density and High-Density Offshore Platform Modules[1]	27	Selby and Burgan [27]
Baker Risk	2006	98	Piping Array	4 reported	Thomas et al [28]

Figure 8 is a plot of $A_{v,calc}/A_{v,exp}$ using the NFPA 68-2013 vent sizing equations versus A_{obs}/A_s, i.e. the analogous plot to Figure 3 for which the calculated vent areas were obtained using the NFPA 68-2007 vent area equations. Unlike Figure 3, most of the vent area ratios in Figure 8 are larger than one even for A_{obs}/A_s values greater than 0.50. The remaining under-sized vent areas in Figure 8 are for the GexCon 1986 and Baker Risk 2006 piping arrays, i.e. for small-scale repeated obstructions. These repeated piping array structures are similar to the simulated piping arrays that produce high flame speeds and pressures in large-scale vapor cloud explosion test programs. In other words, the flame acceleration mechanisms associated with heavily congested piping arrays are not sensitive to the level of confinement associated with varying vent areas. The Baker Risk 2006 tests were, in fact, conducted with all four enclosure walls being vented.

The data shown in Figure 8 suggest that the new NFPA 68-2013 algorithm for including equipment obstructions in the deflagration vent area calculations is a step in the right direction, but does not accurately account for extreme flame accelerations associated with heavily congested piping. Additional work, accounting for the scale and repeated nature of the obstructions is needed.

[1] Low-density BFETS module had compressor and control room mockups, while the high-density module had additional vessels, pipe racks and cable trays.

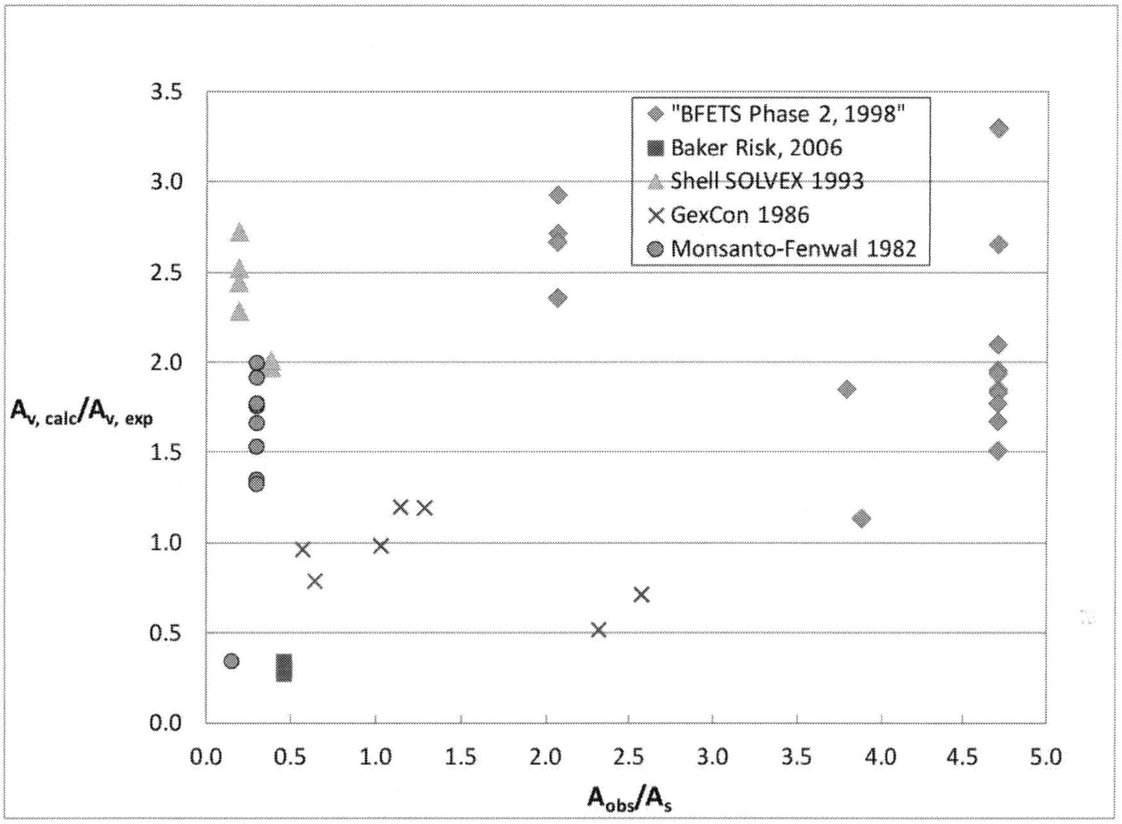

Figure 8. Plot of Calculated to Experimental Vent Area Ratio vs Obstacle/Enclosure Surface Area Ratio – NFPA 68-2013

6. Conclusion

NFPA-68 has developed a new method for required gas deflagration vent area sizing calculations. The 2013 equations are essentially a more transparent casting of the Swift-Epstein-Fauske derivations to provide a framework for including approaches to model flame accelerations and flame wrinkling/stretching in large enclosures with and without internal obstacles. These equations compare favorably with the reference data sets for unobstructed and minimally obstructed enclosures. For a typical medium volume (196 m^3) unobstructed enclosure containing propane, designed for a P_{red} of approximately 0.1 bar-g = 1.45 psig, the new method produces vent area results comparable to prior NFPA vent sizing methods as seen in Table 6.

Table 6. Comparison of Historical Gas Vent Sizing Methods

NFPA 68 Edition	Method	Basic Equation Number	Vent Area, ft^2
1978	Runes	1	622
1988	Swift Simplified (low P_{red})	2	325
1988	Bartknecht Nomograph (high P_{red})	3	300
1998	Bartknecht Correlation (high P_{red})	5	259

NFPA 68 Edition	Method	Basic Equation Number	Vent Area, ft^2
2007	Swift Simplified, $C=f(S_u)$	2 & 7	380
2012	S-E Modified (low P_{red})	9 & 10	389

Calculated vent areas using the new method now also prevent under-sized vent areas for large enclosures with large equipment representative of many petrochemical process structures. However, the new methodology does not yet reproduce large-scale test data with repeated small obstacles simulating highly congested piping arrays and more work on this issue is necessary.

7. PostScript – Background of the NFPA-68 Formulation

The NFPA-68 formulations appear different in two respects from the Swift-Epstein derivations. First, the pressure terms include a factor of 1 and second, the frictionless sonic mass flow, G_u, is shown in both. The first difference is the result of NFPA-68 traditionally using gauge pressures instead of absolute pressures and assuming atmospheric pressure to be 1 bar, as opposed to 1 atm. This is a minor difference for simplicity.

More importantly, the original derivations both are based on a dimensionless group, B, relating the maximum burning rate to the maximum flow through the vent. B is defined slightly differently in each case, including a mass flux term, G', and the vent discharge coefficient, C_d, in the low pressure solution while only including G explicitly in the high pressure solution as in Equations a and b, respectively.

For low pressure (low-strength),

$$B \equiv \frac{\lambda S_u \rho_u A_s}{C_d G' A_v} \qquad \text{[Eq. a]}$$

For high pressure (high-strength),

$$B \equiv \frac{\lambda S_u \rho_u A_s}{G A_v} \qquad \text{[Eq. b]}$$

G' and G are different parameters. By recasting each in terms of sonic velocity, a_u, the NFPA-68 formulation can be explained. G' is described initially as the incompressible mass rate of discharge and defined traditionally as in Equation c.

$$G' = C_d \left[2P\rho_u \left(1 - P_a / P\right) \right]^{1/2} \qquad \text{[Eq. c]}$$

Without changing the symbol, the maximum value of G' is then taken as the frictionless mass flow when $P_a/P << 1$, resulting in Equation d. C_d is then shown explicitly in the dimensionless group B as in Equation a.

$$G' = (2P\rho_u)^{1/2} \qquad \text{[Eq. d]}$$

The factor C_dG'/ρ_u, included in the low pressure group B, can be written in terms of sonic velocity, a_u, for an ideal gas as in Equation e.

$$\frac{C_dG'}{\rho_u} = C_d a_u \sqrt{\frac{2}{\gamma_u}} \qquad \text{[Eq. e]}$$

The description of G in the high pressure derivation is the sonic mass flow per unit area, not the frictionless sonic mass flow, as indicated in the footnote to page 3 of their paper [15]. G/ρ_u can similarly be written in terms of sonic velocity as in Equation f.

$$\frac{G}{\rho_u} = C_d a_u \sqrt{\left(\frac{2}{\gamma_u + 1}\right)^{(\gamma_u+1)/(\gamma_u-1)}} \qquad \text{[Eq. f]}$$

Combining Equations e and f, shows that the ratio of the two original mass flow parameters is a function of γ_u and, equal to approximately 2 when γ_u is 1.36.

$$\frac{C_dG'}{G} = \frac{\sqrt{\frac{2}{\gamma_u}}}{\sqrt{\left(\frac{2}{\gamma_u + 1}\right)^{(\gamma_u+1)/(\gamma_u-1)}}} = 2.09 \approx 2 \qquad \text{[Eq. g]}$$

Finally separating C_d from G clarifies the NFPA-68 formulations.

$$C_dG' \approx 2G = 2C_dG_u \qquad \text{[Eq. h]}$$

8. Acknowledgments

The authors wish to acknowledge the significant assistance of Steve Stuart and Michael Davies as members of the Gas Venting Task Group compiling the test data, as well as Alfonso Ibarreta for investigating and recommending appropriate methods to calculate gas mixture burning velocities.

9. References

[1] NFPA 68, *Standard on Explosion Protection by Deflagration Venting*, 2013 edition, National Fire Protection Association, Quincy, MA.

[2] M.G. Cooper, M. Fairweather, and J.P. Tite, "On the Mechanisms of Pressure Generation in Vented Explosions," *Combustion and Flame* 65 (1986): 1-14.

[3] B. J. Lowesmith, C. Mumby, G. Hankinsom and J. S. Puttock- Vented Confined Explosions Involving Methane/Hydrogen Mixtures- 3rd ICHS, September 16-18, 2009.

[4] E. Cousins and P. Cotton, "The Protection of Closed Vessels Against Internal Explosions," ASME Paper No. 51-RPI-2, 1951.

[5] NFPA 68, *Explosion Venting*, 1978 edition, National Fire Protection Association, Quincy, MA.

[6] E. Runes, "Explosion Venting," *Loss Prevention – Volume 6*, (1972): 68-73, AIChE, New York, NY.

[7] NFPA 68, *Guide for Venting of Deflagrations*, 1988 edition, National Fire Protection Association, Quincy, MA.

[8] I. Swift and M. Epstein, "Performance of Low Pressure Explosion Vents," *Plant/Operation Progress* (Vol. 6, No. 2) 1987: 98-105.

[9] I. Swift, "Designing Explosion Vents Easily and Accurately," *Chemical Engineering* April 11, (1988): 65-68.

[10] W. Bartknecht, *Explosions Course, Prevention, Protection.* Berlin: Springer-Verlag (1981)

[11] NFPA 68, *Standard on Explosion Protection by Deflagration Venting*, 2007 edition, National Fire Protection Association, Quincy, MA.

[12] W. Bartknecht, *Explosions-Schutz: Grundlagen und Anwendung*. Berlin: Springer-Verlag (1993)

[13] NFPA 68, *Guide for Venting of Deflagrations*, 1998 edition, National Fire Protection Association, Quincy, MA.

[14] D. Razus and U. Krause, "Comparison of empirical and semi-empirical calculation methods for venting of gas explosions," *Fire Safety Journal* 36 (2001): 1-23.

[15] R. G. Zalosh, "Explosion Venting Data and Modeling Literature Review," Fire Protection Research Foundation Report, 2008.

[16] M. Epstein, I. Swift and H.K. Fauske "Estimation of Peak Pressure for Sonic-Vented Hydrocarbon Explosions in Spherical Vessels," *Combustion and Flame* 66 (1986): 1-8.

[17] R. Rota, P. Canu, S. Carra and M. Morbidelli, "Vented Gas Deflagration Modeling: A Simplified Approach," *Combustion and Flame* 85 (1991): 319-330.

[18] Pasman H., Groothuisen Th. P., Goojier, P.- Design of pressure relief vents- Proceedings of First International Symposium on Loss Prevention, Delft- Amsterdam: Elsevier, 1974, pp 185-189

[19] R. DeGood, K. Chatrathi, "Comparative analysis of test work studying factors influencing pressures developed in vented deflagrations," *Journal of Loss Prevention* 4 (5) (1991): 297-305.

[20] van Wingerden C. J. M., Zeeuwen J. P.- Venting of Gas Explosions in Large Rooms- 4th International Symposium on Loss Prevention and Safety Promotion in the Process Industries, EFCE Publication Series, No. 33, Vol. 3, F38-F47, 12-16 September (1983)

[21] R. G. Zalosh, "Gas Explosion Tests in Room-like Vented Enclosures" *Loss Prevention*, Vol 13, AIChE, 1979

[22] Thorne P. F., Rogowski Z. W., Field, P.- Performance of Low Inertia Explosion Reliefs Fitted to a 22 m3 Cubical Chamber- 4th International Symposium on Loss Prevention and Safety Promotion in the Process Industries, EFCE Publication Series, No. 33, Vol. 3, F1-F10, 12-16 September (1983)

[23] Cerchiara and Carcassi- "Quantification of the Uncertainty of the Peak Pressure Value in the Vented Deflagrations of Air-Hydrogen Mixtures," 2nd ICHS 2007

[24] Bauwens, C. R., Chaffee, J and Dorofeev, S. B.- Vented Explosion Overpressures from Combustion of Hydrogen and Hydrocarbon Mixtures- 3rd ICHS, September 16-18, 2009

[25] Howard W. B., Karabinis A. H.- Tests of Explosion Venting of Buildings- Plant/Operations Progress, 1, 51-68 (1982)

[26] Bimson, B., Bull, D., Creswell, T., Marks, P., Masters, A., Prothero, A., Puttock, J., Rowson, J. and Samuels, B. (1993). "An Experimental Study of the Physics of Gaseous Deflagrations in a Very Large Vented Enclosure." International Colloquium on the Dynamics of Explosions and Reactive Systems.

[27] Selby, C. and Burgan, B. (1996). "Blast & Fire Engineering for Topside Structures, Phase II, Final Summary Report, The Steel Construction Institute, SCI Publication 253.

[28] Thomas, J. K., Kolbe, M., Goodrich, M., and Salzano, E. (2006). Elevated Internal Pressures in Vented Deflagration Tests. AIChE 40th Annual Loss Prevention Symposium. Orlando, AIChE.

Explosion Hazards Due to Failures of Lithium-Ion Batteries

Vijay Somandepalli
Exponent, Inc.
9 Strathmore Rd., Natick, MA 01760
vsompandepalli@exponent.com

Kevin C. Marr, Ph.D.
Exponent, Inc.
9 Strathmore Rd., Natick, MA 01760
kmarr@exponent.com

Quinn Horn
Exponent, Inc.
9 Strathmore Rd., Natick, MA 01760
qhorn@exponent.com

Prepared for Presentation at
American Institute of Chemical Engineers
2013 Spring Meeting
9th Global Congress on Process Safety
San Antonio, Texas
April 28 – May 1, 2013

UNPUBLISHED

Explosion Hazards Due to Failures of Lithium-Ion Batteries

Kevin C. Marr, Ph.D.
Exponent, Inc
9 Strathmore Rd., Natick, MA 01760
kmarr@exponent.com

Vijay Somandepalli
Exponent, Inc
9 Strathmore Rd., Natick, MA 01760
vsompandepalli@exponent.com

Quinn Horn
Exponent, Inc
9 Strathmore Rd., Natick, MA 01760
qhorn@exponent.com

Keywords: lithium-ion batteries, explosion hazard, battery failure, flammable gas, combustion

Abstract

The failure modes of lithium-ion cells have been extensively studied for applications where the total energy of the battery system is typically below 100 Wh. As lithium-ion cells and systems become larger and more ubiquitous in automotive and vehicular applications, the potential for lithium-ion to be used in energy storage systems in process and refinery facilities is becoming more attractive. However, some failure modes that were rare or non-existent in smaller systems are becoming more common. One potentially hazardous failure mode seen in large lithium-ion systems can occur if flammable gases that are emitted from a lithium-ion cell exceed the lower flammability or explosion limit in a confined space. Flammable gases in spaces such as control rooms or electrical rooms where large back up battery systems may be present would be particularly susceptible to potential ignition by electrical activity or the cells themselves. If this scenario occurs, the confined space can become overpressurized, potentially resulting in an explosion and severe damage to life and property.

With the spread of lithium-ion battery technology into areas like backup power storage, electric and hybrid vehicles and grid stabilization, the safety aspects of using these large energy storage battery packs, especially those related to fires and explosions protection are a significant challenge to address. This paper describes and characterizes the combustion and explosion hazards that can occur when a lithium ion battery pack fails and goes into thermal runaway in an enclosed space. Metrics such as gas composition, maximum overpressure, rate of pressure rise, and flammability limits are described.

This information can be very helpful to first responders, emergency personnel, risk managers and regulatory agencies in developing tactics and strategies to control and prevent explosion hazards from the storage and use of battery packs. Measurements of this type can also be used to provide design guidance for development of adequate ventilation systems, sizing of containment enclosures and explosion venting or blow off panel design that can mitigate the hazard of combustible gases vented during failures in lithium ion cells and large battery packs.

1. Introduction

Rechargeable batteries are ubiquitous in today's electronics, automotive, aerospace, and power storage industries. From smart phones to energy storage systems, the demand for higher capacity batteries in various industry applications has been increasing. Lithium-ion (Li-ion) has become the dominant rechargeable cell chemistry for consumer electronics devices and is poised to become commonplace for industrial, transportation, and power-storage applications. From a technological standpoint, Li-ion technology has enabled entire families of new and smaller devices such as smart phones, electric vehicles etc. because of its high energy density compared to previously popular rechargeable battery types (e.g., nickel metal hydride, nickel cadmium, and lead-acid). However, from a safety and fire protection standpoint, a high energy density coupled with a flammable organic, rather than aqueous, electrolyte has created a number of new challenges with regard to the design, storage and handling of batteries containing Li-ion cells.

There have been numerous incidents of fire and explosion hazards resulting from Li-ion batteries, ranging from consumer electronic devices such as cellular phones [1] and laptops [2] to incidents involving Li-ion battery packs in large vehicles and facilities. Thermal hazards from Li-ion battery failures can be the result of various failure modes. The Li-ion cell chemistry has an inherent high energy density, and a cell failure can generate significant amounts of heat capable of combusting other cell components. A breach in the cell packaging (before or after a thermal event) can allow for flammable electrolyte vapours to permeate and fill any available space, creating a possible explosion risk. Additionally, the degradation and reaction products of a thermal event can create potentially combustible gases [3]. For large power packs used in electric vehicles, and power-storage units, the volume of vented gases may be significant and could pose explosion hazards. Owing to the variability of lithium-ion cell chemical composition due to proprietary manufacture design, the potential explosion severity of the vented gases is not well documented or characterized.

In the present work, the combustion characteristics of vent gases produced due to thermal failure of lithium-ion cells are studied experimentally. The goal of the work is to characterize the explosion characteristics of the vent gases to assess potential explosion hazards associated with venting of lithium-ion batteries during thermal runaway events. The gases vented from small format lithium-ion cells during thermal runaway are sampled and analyzed experimentally. The explosion characteristics including maximum pressure rise, explosion severity and lower explosion limit of the vent gas composition are also measured

2. Experimental Setup

Testing was conducted in at Exponent's laboratories in Natick, MA, using a test chamber specifically designed for sampling and analyzing vent products from Li-ion battery failures. Two different experiments were conducted—one, vent products were sampled and analyzed for composition, and two, vent products were vented into a combustion chamber and ignited to determine combustion characteristics. Lithium-ion cells considered were small format commercially available 2.1 Ah (7.7 Wh nominal) Li-ion batteries. The cells consisted of a negative electrode with graphite active material and a positive electrode with $LiCoO_2$ active material.

2.1 Gas Sampling

Vent products resulting from thermal failure of Li-ion cells were captured in a custom designed 6-in diameter cylindrical stainless steel chamber (Figure 1). The chamber test section is approximately 13-in long and is enclosed on either side with 0.5-in thick stainless steel flanges. On one flange, K-type thermocouples, signal wire pairs to measure battery voltage, and a high voltage power feed thru ports allow for temperature and voltage measurements and power access inside the chamber to monitor the cell health. A pressure transducer, a 50 psig pressure relief valve and a gas sampling and vent port were installed on the other flange to monitor and capture the gases released when these cells were made to fail with thermal runaway conditions. For the Li-ion cell venting experiments, the battery surface temperature, the chamber gas temperature, and the chamber pressure were monitored throughout the duration of each test. Prior to each test, a battery was wrapped with a 100-W, 0.165-in thick heat cable and placed on a thermal brick inside the vent chamber. The vent chamber was sealed, evacuated with a vacuum pump to approximately 0.04 bar, and refilled with argon to atmospheric pressure. The purging process with argon was repeated 3 times before each test.

Catastrophic failure and subsequent thermal runaway of Li-ion cells at 50%, 100%, and 150% state-of-charge (SOC) was induced by applying heat to the cell through the heat cable. Power to the heat cable was terminated once the cell failed and gaseous products vented, and the resultant vent products were allowed to collect in the chamber. The gaseous products were allowed to cool to 60°C and were then sampled through the vent port into an evacuated sample canister. Gas chromatography-mass spectroscopy (GC-MS) was used to analyze the species composition of the gas sample. Prior to analysis, the collected gas samples were heated to 60°C to volatilize any gases that may have condensed in the container after the sampling was completed.

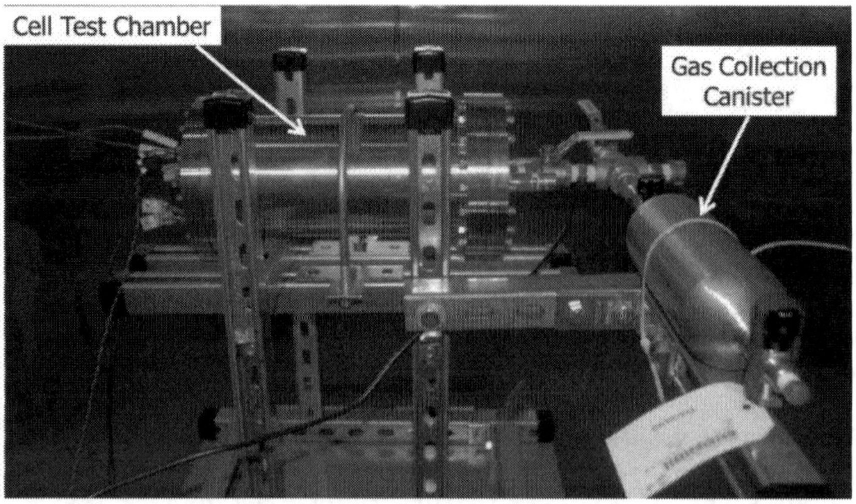

Figure 1. Photograph of battery vent chamber with sample gas canister attached.

2.2 Combustion Chamber Testing

To determine the maximum over pressure, explosion severity, and lower explosion limit (LEL), the battery vent chamber was attached to a 20-L combustion chamber (Figure 2). A description of the 20-L apparatus is detailed in Ref. [4]. The battery vent chamber and combustion chamber were evacuated to approximately 0.15 bar. In this case, the battery vent chamber was not purged with Argon. The valve between the two chambers was closed, and battery failure was induced in the same manner as the gas sampling experiments described above. Once the battery vented, the resultant gases were metered into the evacuated 20-L combustion chamber. After the chamber was filled with a predetermined amount of vent gas, the chamber was filled with air to atmospheric pressure. The resultant mixture of battery vent products and air were ignited using a 100-J chemical igniter. The maximum pressure rise and the explosion severity index were determined from measuring the induced pressure from the resultant deflagration. The upper and lower explosion limits of the vent gases were determined by varying the vent gas-to-air ratio in the combustion chamber.

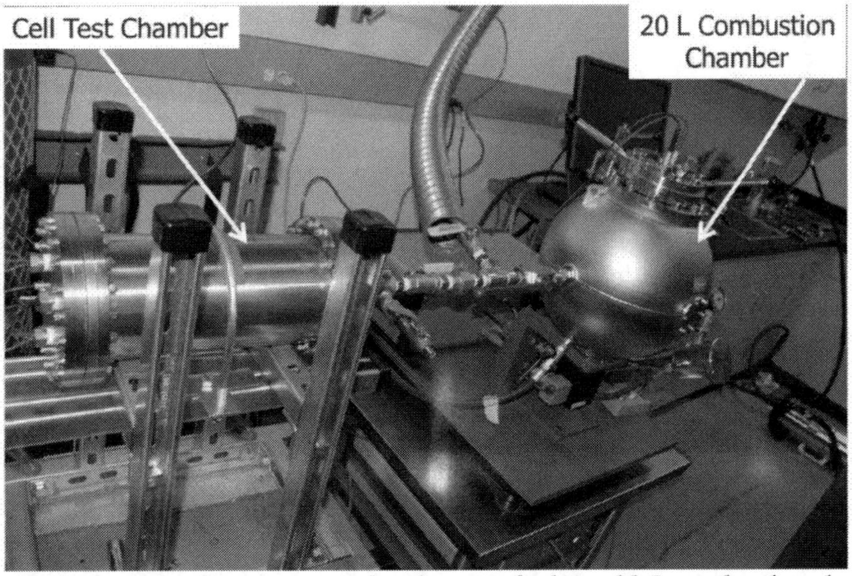

Figure 2. Photograph of battery vent chamber attached to a 20-L combustion chamber.

3. Experimental Results and Discussion

3.1 Volume and Composition of Vent Gases

The chamber gas temperature, chamber pressure, battery surface temperature, and cell voltage was monitored throughout each test. When the battery surface temperature reached approximately 250°C, the cell voltage dropped to zero indicating that the cell had shorted. Immediately following the decrease in voltage, the pressure and temperature of the chamber increased dramatically indicating that the cell had failed and gas had vented. Figure 3 shows an example time trace of chamber pressure, chamber temperature and cell voltage during a 100% SOC test. The total amount of gas produced is estimated from the peak values of the chamber temperature and chamber pressure. At standard pressure and temperature (300K, 1 atm), 2.5 L and 6.0 L of gases, respectively, are released from the failure of cells at 100% SOC and 150% SOC. The 150% SOC cell vents more than twice the volume of gas than the 100% SOC cell. As SOC increases, the available potential energy in the cell increases. As a result, more energy is available for phase transitions of the electrolyte and active materials, and therefore a larger volume of gases are vented. The relationship between SOC and vent gas volume is likely nonlinear. However, further investigation is needed to fully characterize this nonlinearity.

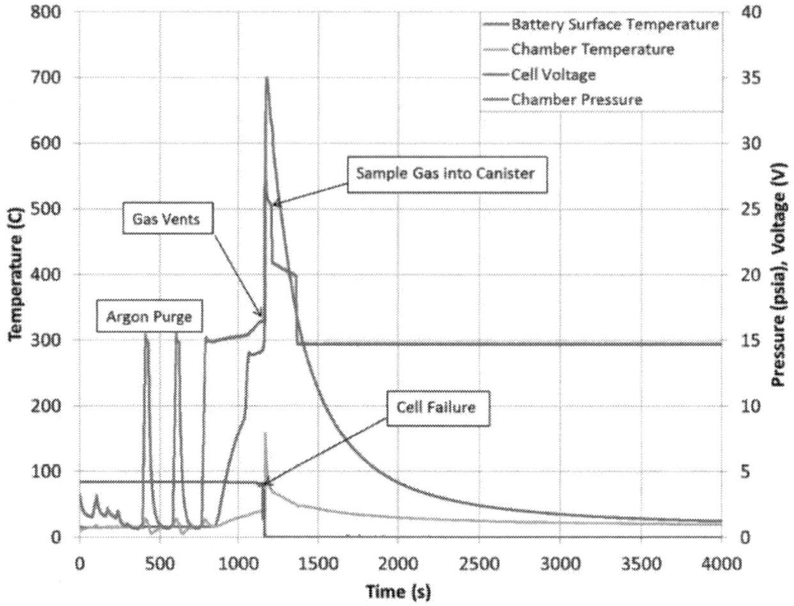

Figure 3. Time trace of cell voltage, chamber pressure, chamber temperature, and battery surface temperature during testing of lithium-ion cells at 100% state-of-charge.

An evacuated sample canister was attached to the sampling port to sample the vent gases from the test chamber. The chamber temperature was allowed to cool to 60°C prior to opening the sampling port valve. The sample gas was reheated to 60°C and then analysed for composition using GC-MS. The results of the GC-MS analysis are summarized in Table 1 for the three states of charge tested. The volume fraction of each species has been normalized to account for the initial argon and trace oxygen and nitrogen in the chamber. Although the batteries were vented into an inert environment, combustion can still take place due to the decomposition of the positive active material, which gives off oxygen. The amount of oxygen produced by the battery is small and is not likely to be sufficient for the fuel to be consumed completely. In all cases tested, significant amounts of hydrogen, hydrocarbons and carbon monoxide were released from the cell. As the SOC increases, carbon dioxide decreases, methane increases and hydrogen remains relatively constant. The total volume fraction of hydrocarbons at the SOC tested range between 20-30%. For 50% SOC, the carbon monoxide produced is much less than both the 100% and 150% SOC cases. It is important to note that Table 1 summarizes the species volume fraction of the vent gases. The absolute volume of each species will depend on the total volume of gas vented, which increases as the SOC increases. Therefore, the total volume of hydrogen released from 150% SOC battery may be significantly more than a 50% SOC battery despite having similar hydrogen volume fractions.

Table 1. Vent gas composition for lithium-ion cells at 50%, 100% and 150% state-of-charge

	50% SOC (%vol)	100% SOC (%vol)	150% SOC (%vol)
Carbon Dioxide	32.3	30.0	20.9
Carbon Monoxide	3.61	22.9	24.5
Hydrogen	31.0	27.7	29.7
Methane	5.78	6.39	8.21
Ethylene	5.57	2.19	10.8
Ethane	2.75	1.16	1.32
Propylene	8.16	4.52	0.013
Propane	0.68	0.26	2.54
Isobutane	0.41	0.20	0.13
n-Butane	0.67	0.56	0.39
Butenes	2.55	1.58	0.60
Isopentane	0.45	0.07	0.036
n-Pentane	1.94	0.73	0.30
Hexanes +	4.94	2.32	0.68
Benzene	0.14	0.11	0.33
Toluene	0.061	0.018	0.052
Ethyl-benzene	0.009	0.002	0.003

Previous studies have reported that the majority of the gas released from a thermally failed Li-ion cell is carbon dioxide [5,6]. Table 2 compares the volume fraction of the major gas species (CO, CO_2, H_2, CH_4) measured in the present study with the measurements by Roth, *et al.* [6]. Roth and co-workers vented Li-ion batteries into an inert argon/helium environment and found that, in comparison to the present study, the vent gas composition consists of significantly less hydrogen and carbon monoxide, and significantly more carbon dioxide. It is possible that the differences in results are due to difference in cell composition and cell geometry. The experimental methodologies are also quite different. Roth *et al.* heated their cells using an Accelerating Rate Calorimeter (ARC) up to 160°C, which most closely simulates thermal exposure at moderately high temperatures over long periods of time. Because the chemical time scale is potentially long in such an exposure, the chemical reaction is likely to be more complete, which would result in a large volume fraction of carbon dioxide and minimal carbon monoxide.

The heating method in the present study is designed to initiate an instantaneous violent thermal runaway event by applying a fast temperature ramp rate (~100°C/min). The large volume fraction of carbon monoxide measured in the current study is evidence that the violent release of vent gases in the current study is likely to have resulted in an incomplete combustion reaction. Therefore, the composition of vent gases due to thermal failure of a Li-ion cell may be highly sensitive to the thermal failure mode. The failure mode considered in the current study is more representative of field failures in individual cells and packs where the transition to thermal runway resulting from mechanical or thermal damage is fast.

Table 2. Comparison of vent gas species volume fractions to previous work.

	50% SOC (%vol)	100% SOC (%vol)	150% SOC (%vol)	*Roth et al. [6] (%vol) Test 1/Test 2
Carbon Dioxide	32.3	30.0	20.9	61.4/75.8
Carbon Monoxide	3.61	22.9	24.5	15.1/6.4
Hydrogen	31.0	27.7	29.7	5.1/5.9
Methane	5.78	6.39	8.21	7.4/1.9

*Data from Roth et al. for two vent test for 100% SOC un-aged cells

3.2 Explosion Characteristics of Vent Gases

For all the tests, the methane and hydrogen volume fractions exceeded the lower explosion limit for their respective species. However, the explosion limits are sensitive to the relative amounts of reactive and inert vent products. The ratio of flammable gases, in particular hydrogen, to carbon dioxide directly affects the explosion limits—a larger ratio will widen the explosibility range, while a smaller ratio will decrease the explosibility range. Depending on cell composition, SOC and other cell characteristics, the vent gas composition, and thus the vent gas explosibility limits, may vary significantly. The species volume fractions summarized in Table 1 is for the vent gas that has not yet mixed with air. For any venting event, the region of explosible gas will be dependent on the mixing associated with the release. The fluid mechanics of the venting event is likely dependent on several parameters including SOC, cell composition, battery packaging type, battery pack design, etc. From an explosions hazard analysis point-of-view, proper mitigation is often designed for the worst case scenario, i.e. maximum overpressure and rate of pressure rise.

Explosion characteristics of the vent gas were determined from the pressure curves obtained from the combustion chamber testing. Figure 4 shows a comparison of the measured pressure curves for different vent gas concentrations. For explosible concentrations, the pressure rises quickly, reaches a maximum pressure, and the decays. The rate of pressure rise, *dP/dt*, is determined by the maximum slope of the pressure curve. When using chemical igniters, a concentration is considered to be explosible if the pressure ratio is greater than one. The pressure ratio, *PR*, is defined as follows [7]:

$$PR = \frac{P_{ex}-P_{ignitor}}{P_{ignition}} \qquad \text{(Eqn. 1)}$$

where P_{ex} is the peak measured pressure, $P_{ignitor}$ is the pressure induced by the ignition source, and $P_{ignition}$ is the intial chamber pressure prior to ignition, which in this study is atmospheric pressure.

Figure 5 shows P_{ex}, and *dP/dt* for the vent gas over a range of concentrations. From the pressure curve, the lower explosible limit (LEL) of the vent gases can be determined. Using the criterion that a gas mixture is explosible if $PR > 1$, the LEL of the vent gas is found to be approximately 6.3% by volume, which is larger than both hydrogen and methane. The larger LEL is due to the presence of carbon dioxide, which partially inerts the mixture. The LEL of the vent gas may vary across different battery manufacturers owing to differences in cell composition. The effect of cell composition on vent gas composition is unknown. However, for any particular cell, the percentage of hydrogen may be sufficiently large such that the LEL of the vent gas could be lower than that of methane.

For confined deflagrations, the measured pressure is effected by the ignition source and cooling from the walls. If the confined volume is sufficiently large, these effects are negligible. For smaller volume chambers, such as the 20-L chamber used in this study, the measured pressure can be adequately corrected. Details of this correction are summarized in Ref [4].

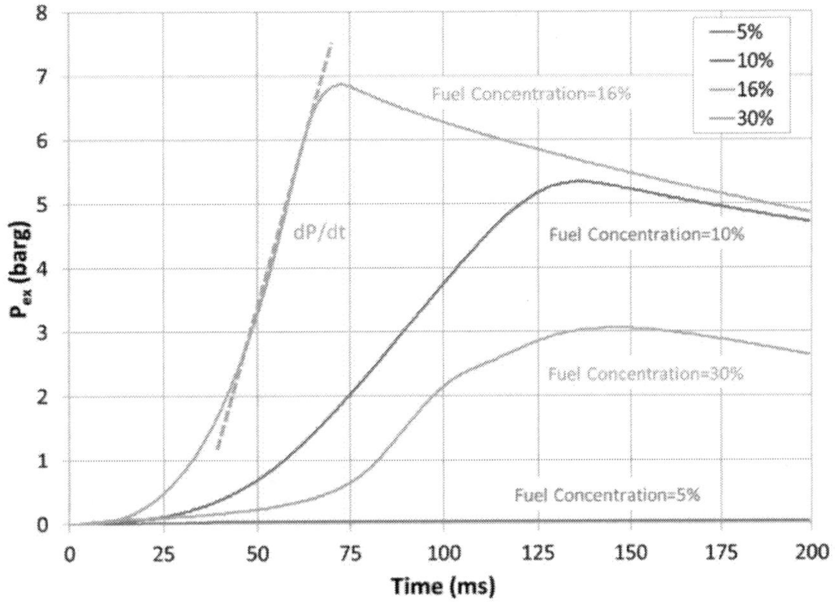

Figure 4. Pressure curves at different concentrations in air of vent gases from a 100% SOC lithium-ion battery obtained from 20-L combustion chamber testing.

The maximum P_{ex} over all concentrations can be corrected to determine the corrected maximum pressure P_m. The maximum dP/dt over all concentrations can also be normalized by chamber volume to determine the explosion severity index, K_g, as follows [7]:

$$K_g = \frac{dP}{dt} V^{1/3} \qquad \text{(Eqn. 2)}$$

where dP/dt is normalized by the cube root of the chamber volume, V. The P_m and K_g values for the Li-ion vent gases were 7.1 barg and 65 m-bar/s, respectively.

Table 3 compares P_m and K_g for the vent gases from a 100% SOC Li-ion battery with several other gas species including methane, propane and hydrogen [8]. The measured values of P_m and K_g vary from study to study. The data from Senecal and Beaulieu was chosen as a comparison because their 22-L test chamber was similar in volume to the 20-L chamber used in this study. The battery vent gas has a P_m similar in magnitude to common hydrocarbons and hydrogen and a K_g value between methane and propane. Despite the large volume fraction of hydrogen, the presence of inert carbon dioxide effectively reduces K_g. However, the similarity to methane and propane may be a coincidence for this particular battery and test condition. At overcharged states, the volume fraction of carbon dioxide decreases, and therefore the vent gas may have a significantly higher K_g. Other parameters, in particular cell composition and battery capacity, are also likely to effect the vent gas composition, which may have a significant effect on K_g. Further parametric tests are needed to fully characterize the variability in explosion characteristics of Li-ion battery vent gases.

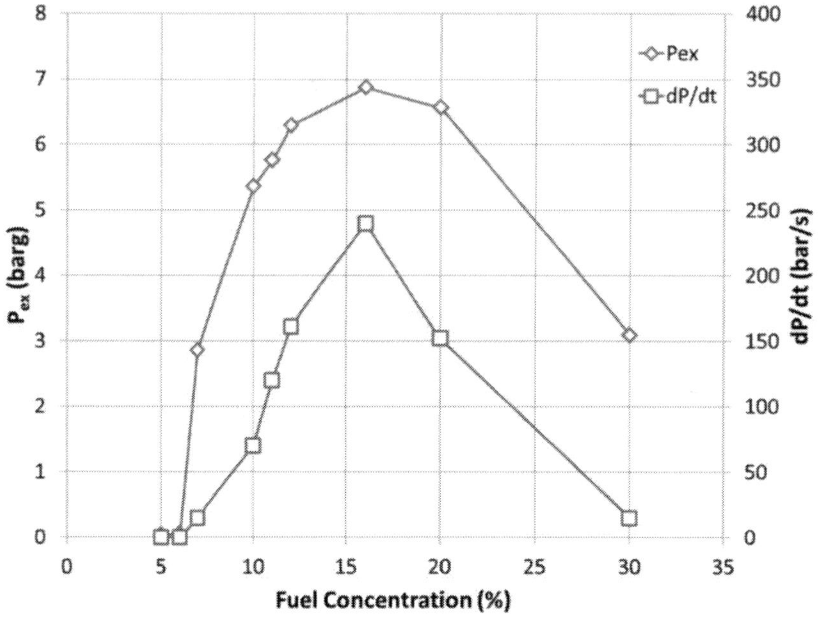

Figure 5. Peak overpressure and rate of pressure rise for different concentrations of vent gases from a 100% SOC lithium-ion battery in air obtained from combustion chamber testing.

Table 3. P_m and K_g values for lithium-ion battery vent gas and other common fuel gases.

Gas	P_m (barg)	K_g (m-bar/s)
Lithium-Ion Vent Gases (100% SOC)	7.1	65
Methane	6.7	46
Propane	7.2	76
Ethane	8.0	171
Hydrogen	6.5	638

*P_m and K_g values for other gas species from Senecal and Beaulieu [8]

5. Conclusion

Characteristics of gas vented from Li-ion batteries during thermal failure have been studied experimentally. The species composition of the vent gas, total volume of gas vented and explosion severity (P_m and K_g) were determined experimentally. Contrary to previous studies, the vent gas composition was found to contain significant amounts of hydrogen, carbon monoxide and hydrocarbons. The differing results are likely due to both differences in chemical composition of the cell and the methodology of inducing thermal runaway. The fast temperature ramp rate used in the current study is more similar to the conditions of field failures. The measured P_m and K_g values for the 100% SOC battery were between those of methane and propane. However, it is likely that the explosion severity is more severe at overcharged states. The vent gas species composition, and hence combustion characteristics, is dependent on the state-of-charge of the cell and may be highly sensitive to the specific chemistry of the Li-ion cell. In addition to state-of-charge and cell composition, cell form factor and capacity may also have a large effect on the vent gas characteristics. To fully characterize the vent gas characteristics, a more complete parametric study is required. In particular, scale up studies for larger format batteries are of interest in light of the increased demand for larger capacity batteries for use in electric and hybrid vehicles and energy storage applications.

Lithium-ion batteries and packs are becoming more and more popular. As Li-ion battery technology grows to meet the demands for higher capacity, proper hazard assessment will be essential in designing battery systems that are safe. Characterizing the explosion severity of Li-ion vent gases is vital in any proper hazard assessment, especially for large capacity battery packs. Although there is currently no comprehensive industry standard that provides guidance for the safety hazards associated with gas vented from Li-ion batteries, there are related standards that provide guidance for combustible dust and flammable gas hazards. *NFPA 68: Standard on Explosion Protection by Deflagration Venting* [9] and *NFPA 69: Standard on Explosion Prevention Systems* [10] provide guidelines for designing mitigation systems for combustible dust and flammable gas hazards. Determining combustion characteristics of the particular fuel is essential in properly designing either deflagration vents or suppression systems. Similarly, proper design of safe Li-ion battery systems will require a detailed understanding of the combustion characteristics of the vent gases.

6. References

1. US Consumer Products Safety Commission, Alert #10-752, "Asurion Recalls Counterfeit BlackBerry®-Branded Batteries Due to Burn and Fire Hazards," August 10, 2010.
2. US Consumer Products Safety Commission, Release #10-240, "HP Expands Recall of Notebook Computer Batteries Due to Fire Hazard," May 21, 2010.
3. Harris SJ, Timmons A, Pitz WJ, "A Combustion Chemistry Analysis of Carbonate Solvents Used in Li-ion Batteries," *Journal of Power Sources*, Vol. 193, 2009, pp. 855-858.
4. Cesana, C, Siwek, R, "50-L Apparatus Manual," http://www.kuhner.com/tl_files/kuhner/product/safety/PDF/B000_071.pdf
5. Crafts C, Borek T, Mowry C, "Safety Testing of 18650-Style Lithium-ion Cells," Sandia National Laboratories, SAND2000-1454C, May 2000.
6. Roth EP, Crafts CC, Doughty DH, McBreen J, "Advanced Technology Development Program for Lithium-Ion Batteries: Thermal Abuse Performance of 18650 Li-Ion Cells,"

Sandia Report: SAND2004-0584, March 2004.
7. ASTM E1226-10: Standard Test Method for Explosibility of Dust Clouds, ASTM International, 2010.
8. Senecal JA, Beaulieu PA, "Kg: New Data and Analysis," *Process Safety Progress*, Vol. 17, 1998, pp. 10-15.
9. "NFPA 68: Standard on Explosion Protection by Deflagration Venting," 2007 ed., National Fire Protection Association, Quincy, MA.
10. "NFPA 69: Standard on Explosion Prevention Systems," 2008 ed., National Fire Protection Association, Quincy, MA.

Unique Failure Mode for Double Mechanical Seals (API Seal Plan 76)

G. Ajith Kumar
Senior Project Manager Maintenance
AYTB- Sipchem Maintenance Project
akumar@sipchem.com

Tariq Alauddin
Staff Process Safety Engineer
Saudi International Petrochemical Company (Sipchem)
talauddin@sipchem.com

Keywords: Seal failure, containment, API 682 Seal Plan 76

Abstract

This article discusses a unique failure mode for double mechanical seals conforming to API standard 682 - Pumps—Shaft Sealing Systems for Centrifugal and Rotary Pumps. This failure mode can result in a catastrophic release of flammable or toxic materials without setting off the seal failure alarm provided as a safeguard against such an event. A case study of a fire caused by seal failure on a butane pump equipped with a cartridge type double mechanical seals is presented. The failure can occur if the set screws holding the seal on the pump shaft slip, changing the clearance on the containment hydrodynamic seal which fails very quickly, followed by the failure of the primary seal. Factors which led to this type of failure and the critical steps in seal maintenance practices will be discussed.

Introduction

The US EPA estimates that an average of 19 tons/yr Volatile Organic Compounds (VOC's) are released as fugitive emissions from pump seals in a typical facility. Besides the fugitive emissions, pump seal leaks can result in catastrophic releases of toxic or flammable materials. Numerous incidents of fires and toxic release due to pump mechanical seal failures have been reported in industry.

To reduce the potential hazard of vapor clouds, fires, toxic releases and fugitive emissions industry has generally adopted double mechanical seals. Various configurations of double mechanical seals are available. One of these, a double mechanical seals with an unpressurized secondary containment chamber, commonly referred to as API Seal Plan 76 (Fig 1a), are used in pumps for Liquefied Petroleum Gas (LPG) and other low boiling hydrocarbons that do not condense at ambient temperatures. This seal plan is considered a robust safeguard in preventing catastrophic releases of toxic or flammable materials.

As stated in API 682, Seal Plan 76 (Fig-1a) is intended to work as follows. Leakage from the inner seal is restricted from escape to atmosphere by the containment seal and goes out through the containment chamber vent. An orifice in the vent line restricts flow, such that high leakage of the inner seal will cause a pressure increase and trigger a pressure switch which provides an alarm in the control room. Seal plan 76 is normally used in combination with a primary seal plan. For clean liquids the primary seal plan is generally seal plan 11(Fig-1b).

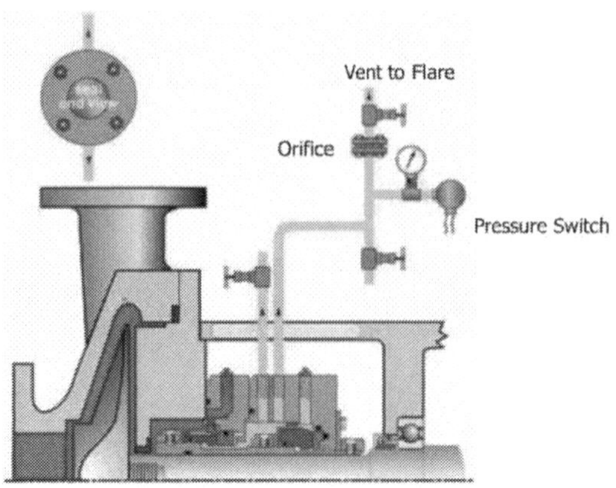

Figure 1a
Seal Plan 76

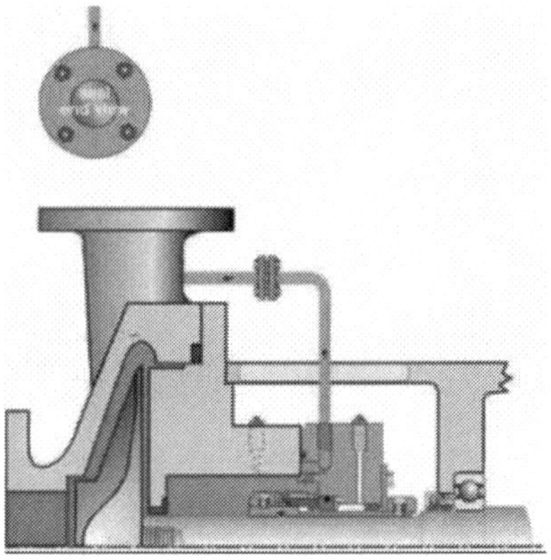

Figure 1b
Seal Plan 11

For ease of installation, both the primary and secondary seals are preassembled in a cartridge assembly. The seal cartridge is installed as a unit to the pump. The fixed part of the seal is bolted to the pump casing, while the rotating part is coupled to the shaft using a numbers of alloy steel grub screws mounted on a drive collar, which are tightened to the pump shaft. Multiple coil springs ensures adequate contact pressure of the rotating face to the stationary mating ring of the seal assembly. To maintain the recommended spring compressions during the assembly and alignment of the cartridge seals, setting plates are provided, which are disengaged after the installation of the cartridge. During the normal operation of the pump the seal face pressure is a combination of spring force and the hydraulic force exerted by the fluid. The primary seal faces are cooled by recycling the pumped fluid, while the secondary seal faces are non-contact gas seals without any buffer gas. See figure 2 for the drive collar and setting plate of the cartridge seal assembly.

Figure 2

International Diol Company (IDC), an affiliate of Saudi International Petrochemical Company (Sipchem), manufactures maleic anhydride (MAn) and butane diol (BDO) as the finished products. A reverse isomerisation unit called Butamer plant produces the n-butane feed stock at the rate of 13.5 tons per hour. A horizontal centrifugal pump manufactured to API 610, type OH-2 is used to transfer n-butane to the maleic anhydride reactor. The operating details of the pump are as given below.

Pump Capacity	27.5 m^3/hr
Discharge Pressure	14.9 bar (g)
Operating Temperature	50 ^{0}C
Fluid Vapor Pressure	6.6 bar (a) @ 50 ^{0}C
Pump RPM	3550
Rated Power	22 kw

Because of the potential for a flammable vapor cloud formation in case of a seal leak, the n-butane feed pump is fitted with a dual unpressurised seal in a tandem arrangement. The primary seal corresponds to seal plan 11. The secondary seal is a dry running containment seal designed as per plan 76, with the containment seal cavity vented to flare through an orifice (Figures 1a and 1b). A pressure switch sets off an alarm in the control room if the pressure in the containment chamber reaches 0.7 bar due to excessive leak from the primary seal.

On November 6, 2008 there was a fire on this pump, which was visually detected by the plant personnel and promptly extinguished with the help of the plant deluge system. The fire occurred an hour after the pump was restarted after an oil seal repair job carried out on the pump bearing housing. Due to the quick identification and fast response by the operations team there was no significant damage to the pump or other equipment. A flame height of approximately one meter was reported by one observer. The photograph of the fire affected area of the pump is shown in Fig-3

Figure 3 Pump after the fire is controlled

Although the incident resulted in minor damage, it could have catastrophic consequences had there been a delay in detection. Investigation of this incident identified a unique systematic failure mode of the double cartridge seal installed on the pump.

Failure Analysis of Mechanical Seal

Observations and Findings

The following observations were made during seal disassembly.

1. The setting plates had been disengaged from the sleeve and found secured in the correct position indicating the seal had been installed correctly.
2. The set screws holding the drive end collar to the shaft were loose and the shaft sleeve could slide on the pump shaft.
3. Deep rubbing marks were found on the drive end collar of the seal and the outboard bearing housing oil seal cover.

Fig 4a Rubbing marks on seal drive end collar

Fig 4b Exploded view of rubbing marks on seal drive end collar

Fig 4c Rubbing marks on oil seal cover

Fig 4d Exploded view of rubbing marks on oil seal cover

4. The set screws holding the drive end collar to the shaft were found flattened and deformed at the edge.

Fig 5 used and new set screws

5. Excessive abrasion and scouring on the seal surfaces of both the inner seal and outer containment seal were observed.

Figure 6

Seal Failure Mechanism

The observations and findings suggest the following mechanism for seal failure:-

1. The set screws (also referred to as grub screws) holding the cartridge seal sleeve in position slipped.

2. The cartridge seal sleeve was pushed towards the bearing housing by the fluid pressure. The force acting on the sleeve under hydraulic pressure can be calculated as:

$$F = \frac{P \times \pi}{4}\left(d_1^2 - d_2^2\right) \times 9.81 \times 10^{-2}$$

Where ,

P = Primary seal chamber pressure in Kg/cm^2
d_1 = Sleeve OD mm
d_2 = Shaft OD mm

Based on the above equation the hydraulic axial force acting on the sleeve would be 2140 Newton, which is countered by the friction/grip of the set screws. This force is acting outwardly on the shaft sleeve.

3. Movement of the cartridge seal sleeve reduced the clearance between the rotating and stationery seal faces resulting in excessive friction and heat.

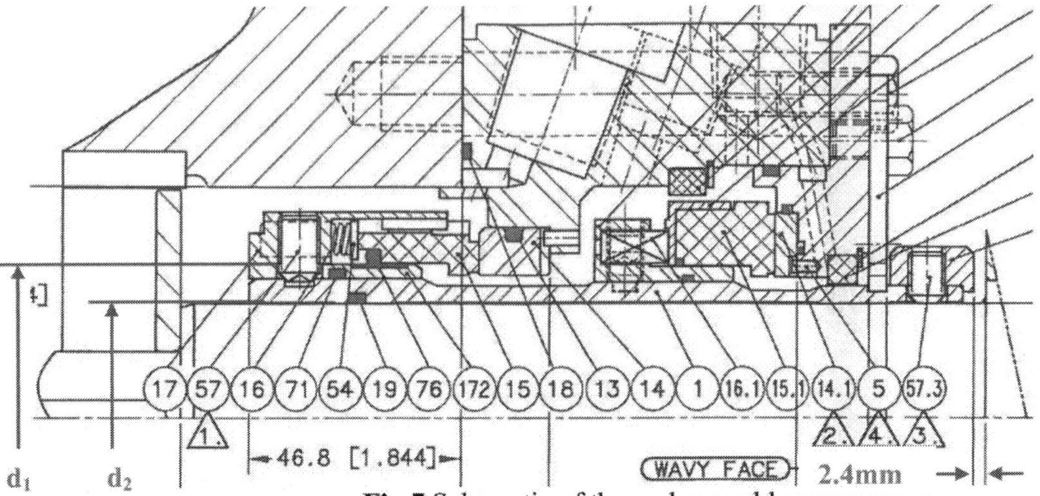

Fig-7 Schematic of the seal assembly

4. The outer containment seal, which is not lubricated or cooled, failed first. Because of this failure the pressure in the containment chamber could not build up to activate the seal failure alarm. This is confirmed by the fact that the seal failure alarm did not sound.
5. The primary seal which is cooled by the product withstood the excessive pressure slightly longer than the secondary containment seal but ultimately failed.
6. Due to the small clearance between the mechanical seal drive end collar and the bearing housing oil seal cover (2.4 mm as shown in Fig 7), axial movement of the shaft sleeve brought the drive end collar and the bearing housing oil seal cover in contact.
7. The rotating oil seal cover pressed against the drive end collar and caused the deep rubbing marks and provided the heat source for the ignition of the leaking butane.

Record of pump amperes reviewed by the investigation team showed significant fluctuations and higher than normal amperes drawn by the pump. This supports the finding of excessive pressure on the seal faces and rubbing of drive end collar with bearing housing

Cause of slippage of Set Screws

The set screws holding the shaft sleeve and the pump shaft used on this pump are called the knurling type grub screws or Knurled Cup Point Set (Grub) Screws similar to DIN 916K. They are designed to deform when tightened to the full torque and are not supposed to be reused. This information is not specifically mentioned in this particular pump or seal manufacturer's documentation and was verbally provided to the investigation team by their engineers.

A review of the maintenance history (Table 1) shows that the seal was installed for the third time with the same original set screws.

Table 1 – Cartridge Seal Installation History for P-6011A	
New mechanical seal Installed	24 June 2006
Removed for pump overhaul due to seizure of thrust bearing (same mechanical seal reused)	28 Oct 2006
Removed to repair bearing housing oil seal – (same mechanical seal reused)	5 Nov-2008

The maintenance procedure conforms to the pump manufacturer's recommendation. As per the normal work practice when seals are replaced all the important steps to ensure seal integrity are followed and usually vendor refurbished cartridge seals are used. The mechanical seal had to be removed for fixing problems with bearings. Since the seal had no performance problem, the same was reinstalled on both occasions. The need for replacing grub screws was not a part of the maintenance instruction and overlooked by all concerned.

It is concluded that the most likely cause of the slippage of the set screws is due to the repeated reuse of the set screws.

Contributing Cause

The design detail of the pump indicates that the seal sleeve could slide all the way to get in contact with the bearing housing. This is due to the fact that the distance between the drive collar and the bearing housing is only 2.4mm. The potential of a physical rub in the event of a grub screw failure always existed.

As seen in the previous calculation, the axial hydraulic force acting on the grub screws in this case is as high as 2140 N. This force is resisted by the frictional grip of grub screw on the shaft. The indentations created by the grub on shaft (see Fig-8a and 8b) provide a shear lock which holds the sleeve in place. The shear lock is only created when the edges of the grub screws are sharp. Repeated use of the same grub screws blunts the edges of the set screws and reduces the frictional force holding the drive end in place.

The design of the set screws, knife edged or knurled is also important. The set screws that were in use on the failed seal were knurled edged type (Fig 5), which has a higher tendency to deform and blunt. Knife edge cup set screws on the other hand have shown less tendency to deform. The indentations formed on the shaft by a new and used knife edge set screw are shown in Fig 8a and 8b. The slight difference in the depth and area of the indentation affects the frictional grip and resistance to the axial force trying to move the shaft sleeve.

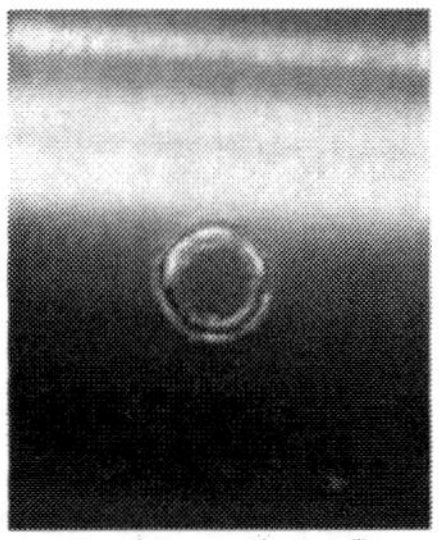

Fig-8a - Indentation formed by a new set screw

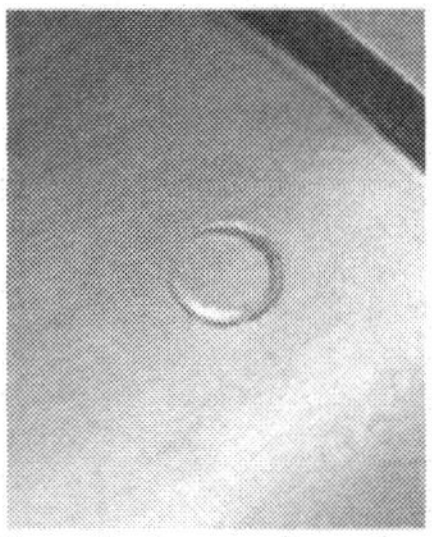

Fig-8b – Indentation formed by a reused set screw

Learning form the Incident

The fire incident on the n-butane feed pump with the potential of escalating into a major catastrophic event was caused by the failure of a minor component in the seal assembly. Several systematic failures (design factors and maintenance practices) led to this incident. Some of these are addressed in the industry standards, while others are best practices based on experience. Increased awareness of the industry in ensuring the following key elements can prevent similar failures.

a) Seal manufacturers should specify the criticality of replacing the set screws and maintenance craftsmen need to be made aware of the importance of this activity. A survey of maintenance instructions for cartridge type pumps indicated that not all manufacturers specify and mention the importance of replacing the set screws whenever the seal is replaced.

b) Clearance between the drive end collar and the bearing housing should be large enough to prevent any chance of the two coming in contact. On some pumps, shaft design incorporates a stepped contour which prevents the free sliding of shaft sleeve beyond a certain point (see fig-9). In such designs, failure of grub screw would not result in a physical rub with the bearing housing.

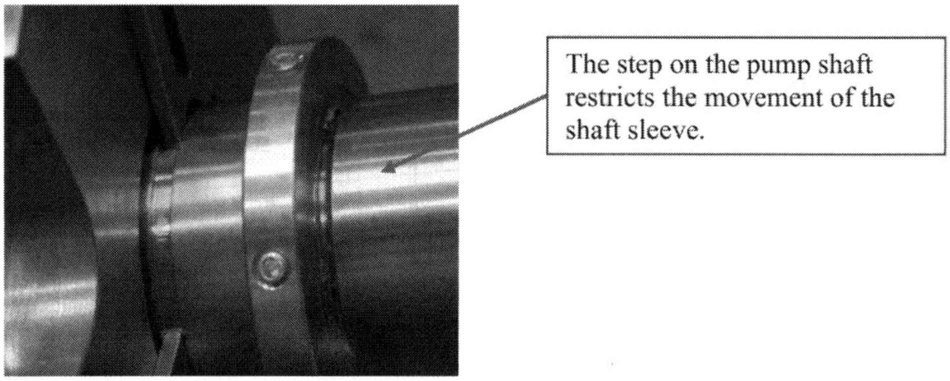

Fig – 9

c) API 682 specifies that the set screws should have sufficient hardness to securely embed in the shaft. The pump manufacturer and seal vendor shall ensure that adequate relative hardness exists between the pump shaft and the drive collar set screws. The hardness of the grub screws should be specified on seal drawings and specification sheets to ensure that proper grub screws are used during repair and maintenance.

d) A survey of maintenance instructions for cartridge type seals indicated that some manufacturers highlight and mention the importance of replacing the set screws whenever the seal is replaced, while others do not. Seal manufacturers should specify this as a critical requirement so that maintenance workers can be made aware of these issues.

e) Management systems meant to address mechanical integrity require that the maintenance practices, procedures and checklist are put in place to prevent dangerous failures of process equipment during operation. Manufacturer maintenance instructions normally do not include situations like the one encountered here when the seal was removed and reinstalled repeatedly. Maintenance procedures for such activities need to be reviewed critically by subject matter specialists to pick up potential issues and failure modes.

Conclusions

Although the seals are designed for effective containment and inherent safety, proper installation and maintenance are essential to ensure they function as intended. The failure of a minor component of a mechanical seal can lead to a loss of containment, resulting in an incident with catastrophic consequences in flammable or toxic services. In order to improve the asset integrity and plant process safety special attention needs to be paid to a) providing good maintenance work instructions by the seal vendors b) development of detailed facility specific maintenance procedures c) implementing a periodic review of the existing procedures d) and effective training of maintenance workers.

VAPOR CLOUD EXPLOSION (VCE) BLAST DAMAGE FAR BEYOND THE COMPANY PROPERTY LINE

Dr. Erdem A. Ural
Loss Prevention Science and Technologies, Inc.
659 Pearl Street
Stoughton, MA 02072, USA
Erdem.Ural@LPSTI.com
Tel. 1-781-818-4114

Prepared for Presentation at
American Institute of Chemical Engineers
2013 Spring Meeting
9th Global Congress on Process Safety
San Antonio, Texas
April 28 – May 1, 2013

UNPUBLISHED

VAPOR CLOUD EXPLOSION (VCE) BLAST DAMAGE FAR BEYOND THE COMPANY PROPERTY LINE

Dr. Erdem A. Ural
Loss Prevention Science and Technologies, Inc.
659 Pearl Street
Stoughton, MA 02072, USA
Erdem.Ural@LPSTI.com
Tel. 1-781-818-4114

Abstract

A massive hydrocarbon leak in 2008 led to a Vapor Cloud Explosion (VCE) in a Texas refinery. This accident inflicted serious injuries, and significant property damage both inside the plant boundaries and far beyond. The explosion was heard at distances as far as 40 miles from the plant. The explosion produced ground motion recorded by seismic stations located approximately 120 to 440 miles away. A large number of houses, located several miles away from the plant, suffered damage. Yet, the analysis shows that the methodologies promulgated by CCPS are incapable of predicting the far field property damage caused by this accident.

This presentation will review the accident, and the observed far field property damage. Simplified analysis including blast propagation, atmospheric inversion, terrain effects, and appropriate damage criteria produces a more accurate representation of the consequences.

1. Introduction

A massive hydrocarbon leak developed in a refinery in Big Spring, TX operated by Alon USA on February 18, 2008. Available information indicates that Propylene splitter system in Alon USA refinery held 162,300 to 198,200 kg of propylene/propane mixture prior to the massive leak. Released vapor cloud was ignited after a delay and resulted in a powerful Vapor Cloud Explosion (VCE), and fire following. It caused catastrophic damage within the plant boundaries, and took the 70,000 barrel-per-day refinery offline for the immediate future. The explosion ripped through Big Spring, the Howard County seat along I-20 with a population of 25,000.

This accident inflicted serious injuries and significant property damage both inside the plant boundaries and outside. The explosion was heard at distances as far as 40 miles from the plant, and produced ground motion recorded by seismic stations located approximately 120 to 440 miles away. Analysis of seismic data from eight stations pinpointed the epicenter to the ALON Yard, Time: 8:15 AM, Magnitude: 2.1, and Depth: 0 km.

The newspapers covered the aftermath in detail and published numerous witness statements such as:

- ★ The shock waves from Monday's oil-refinery explosion in Big Spring rolled as far as San Angelo, 90 miles to the southeast.
- ★ The massive 8:15 a.m. blast injured at least five people - including a woman driving on nearby Interstate 20 - forced the closure of the highway, shattered windows in nearby neighborhoods
- ★ It knocked pictures off walls 45 miles away in Sterling City and led Kat Morgan, an ex-California resident who was playing cards with her husband in Bronte, to wonder whether West Texas had experienced an earthquake. Bronte is about 80 miles by air from Big Spring. "As it rumbled, my house started squeaking," Morgan said. "I could tell that it was moving." The dog, she said, was whining and shaking.
- ★ The blast shattered windows in a nearby school, blew doors open, and cracked and collapsed ceilings. It left Debbie Battle, whose home sits just a quarter-mile from the refinery, wondering whether her attic remains structurally sound. "I thought a plane had crashed into my house," she said via phone. "I've never heard anything that loud."
- ★ "It felt like a bus hit the side of my house" said Howard County Judge Mark Barr. "I thank God nobody was killed."
- ★ Mayor Russ McEwen said he couldn't recall an explosion as big as this one at the refinery. "I was at home at the time and I thought it was thunder" he said. "The windows rattled and all that." "It was extremely scary. You shook you were so scared," Laura McEwen, the wife of Mayor Russ McEwen, who lives about two miles from the refinery, told the Associated Press. "Our walls shook. It jolted your bed. It was like an earthquake."
- ★ John Moseley, managing editor of the Big Spring Herald whose downtown office is also about two miles from the refinery, said, "I thought it would knock the walls down."
- ★ The explosion forced open the doors of the school district's administration building about four miles from the plant, Dunnam said. "Literally pieces of my ceiling came on top of my head," she said.
- ★ Bonnie Duncan, a jailer for the Mitchell County sheriff's office, said the building shook in Colorado City and the office has received reports of the ground shaking in Loraine. Duncan said her sister-in-law in Andrews, 64 miles west of Big Spring, felt the explosion.

There was widespread property damage in Big Spring and beyond. Hundreds of properties sustained damage. ALON contractors quickly replaced many broken windows in exchange for general release of liability. Later, approximately 400 property owners sued ALON. Figure 1 shows a map of the property locations in relation to the Alon USA refinery. The rings centered at the address 200 Refinery Road (Latitude: 32.270N, Longitude: 101.419W) show radii in 1 mile increments. A large number of houses, located several miles away from the refinery, suffered damage.

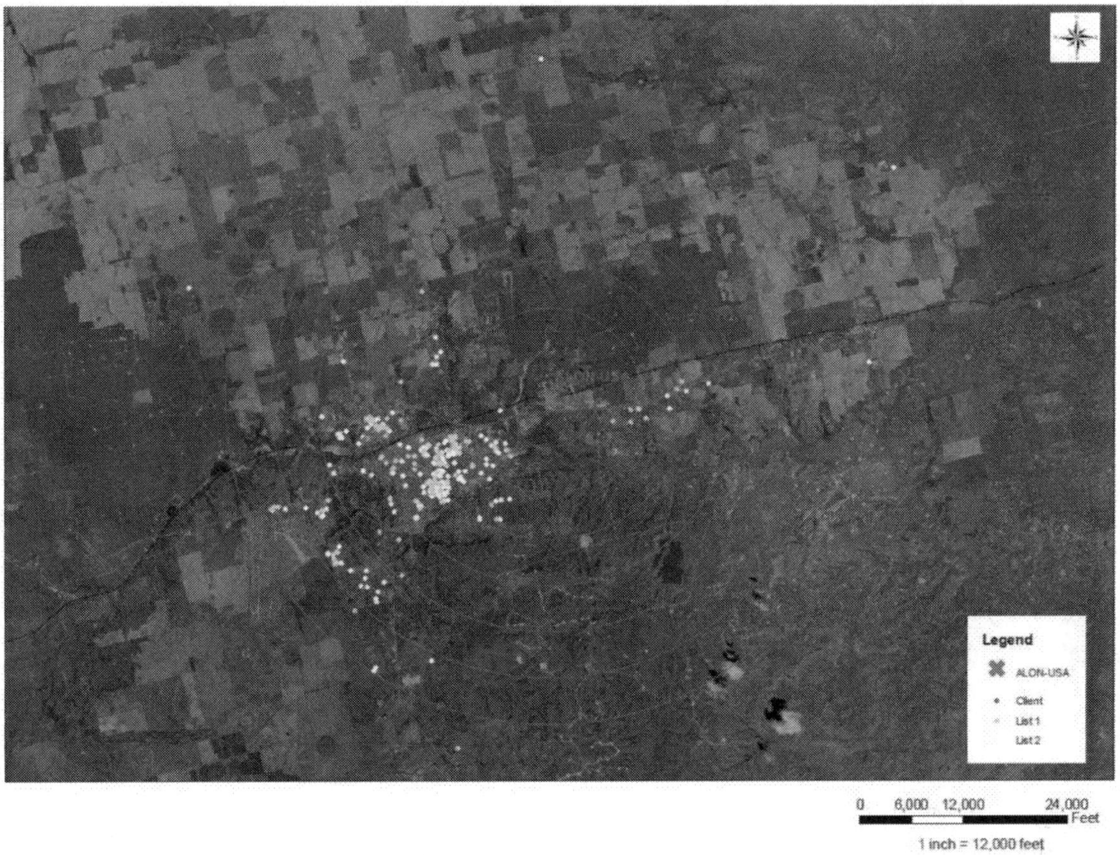

Figure 1. Location of damaged houses in relation to Alon USA

II. Damage Observed Outside the Plant:

Some foundations and floors sustained damage such as pier & beam foundations brought unlevel, plumbing lines under pier & beam structures broken due to movement of structure, fasteners securing floor sheathing above pier & beam foundations caused to recede causing unlevel.

Some windows and doors were hard to open and close, or leaky because of the deformation of the frames. There were cases of exterior doors blown inwards, windows shattered, double pane windows depressurized, garage door knocked off the railing, railing deformed.

Interior damage included buckling of light weight wall paneling, cracks formed on wall and ceiling drywall, wall hangings fell down, wall cabinetry or other objects mounted on the walls dislodged or tilted, and tile damage.

On the exterior walls, there were cases of brick veneer siding cracked, buckling of wood siding, cracks in bricks, mortar damage, external conduit detached from the bracing, CMU (Concrete Masonry Unit) fence (walls) cracked, and wooden fence becoming loose

Roof damage included cases of roof sheathing separated from rafters, damage to metal roofing and siding, localized damage to roof, separation of exterior soffit, separation of exterior fascia board, chimney damage, and ceiling separated from the roof.

III. Known Damage Thresholds:

Vapor Cloud Explosions are rare events. The damage patterns are often well documented within the plant boundaries and in the near vicinity where the mostly industrial structures have experienced moderate and heavy damage. However, this author is unaware of any publicly available damage studies that focused on light damage observed in residential type construction situated in the far field of a VCE.

On the other hand, blasting is a commonly used method in construction and mining. Its effect on residential properties is well documented by many, including the US Government. According to US Bureau of Mines report RI-8485 (1980)[1] entitled "Structure Response and Damage Produced by Airblast from Surface Mining," "There is a consensus that damage becomes improbable below approximately 0.03 psi". This is equivalent to 4.3 psf (lb/ft2). However, the experimental study described in RI-8485 concludes that 0.014 psi (134 dB, 2 psf) is a more appropriate number for the "maximum safe" blast pressure to preclude damage. The 2 psf threshold is also equivalent to the Federal limit (30 CFR 816.67) of 134 dB required to prevent injury to persons, damage to public or private property outside a blasting permit area.

US Army Blast Claims Evaluation Procedures (1994) provides some specific criteria used to judge the legitimacy of public damage claims near firing ranges. A summary of threshold levels for specific kinds of damage are presented in Table 5. The data provided in Table 5 serve as a basis for evaluating claims, but many times the residential component cited is not listed. In those cases, the Army recommends that the item must be compared with a similar item in the table and a subjective judgment made. It should also be kept in mind that overpressure pulses caused by Army firing activities are always concluded in time periods in the order of 0.001 second, whereas, the overpressure pulse caused by the ALON explosion was estimated to last approximately 0.5 second, hence capable of inflicting much more damage at the same overpressure level.

[1] RI-8485, Bureau of Mines Report of Investigations, Siskind et al. "Structure Response and Damage Produced by Airblast From Surface Mining" 1980

Table 5. Overpressure Threshold Criteria for Structural Damage

Interior	PSF	Exterior	PSF
Plaster on wood lath	3.3	Brick (cracks)	18.7
Plaster on Gyplath	7.5	Glass door (loosened)	18.7
Plaster on expanded metal lath	16.0	Mullions (twisted)	8.9
Plaster on concrete block	16.0	Molding (popped)	18.7
Plaster, new	5.4	Stucco (new)	4.9
Plaster, cured	10.4	Light-weight superstructure	209
Nail popping	5.2	Concrete	710
Gypsum board (old-cracks)	4.6	Wood frame wall (fatigue, 80 min.)	6.0
Gypsum board (old-loose paint flaking)	9.6	Roof (fatigue, 20 min.)	7.5
Gypsum board, 1/2-in (nail popping)	10.7	Concrete wall, 8-in thick (fatigue, 10 min.)	18.8
Gypsum board (new-cracks)	16.0		
Bathroom tile (old)	4.4		
Suspended ceiling (new)	3.9		

IV. Estimation of the Peak Pressure Field:

In this accident, as well as in certain other accidents that will be discussed during the presentation, methodologies promulgated by CCPS underpredict the severity of the Vapor Cloud Explosion and damage criteria. In some cases, the actual cloud burning rate or the extent of combustion is larger. In the case of Alon VCE, two additional factors caused the preferential focusing of the blast wave created by the Vapor Cloud Explosion.

1) Temperature inversion:

The shape of the cloud seen in the photographs taken soon after the explosion seen Figure 2 demonstrates that that temperature inversion indeed existed in Big Spring. Figure 3 shows the temperature profile measured during an atmospheric sounding in nearby Midland, TX, approximately 2 hours before the Alon USA vapor cloud release. Figure 3 shows, in fact, there existed two temperature inversion layers, one near the ground, and another at around 2500 m elevation.

The adverse effect of temperature inversion is widely recognized. Strehlow and Baker (1976)[2] wrote "The more significant atmospheric effects which induce non-ideal blast wave behavior are unusual weather conditions which can cause blast focusing at some distance from the source. A low-level temperature inversion can cause an initially hemispherical blast front to refract and focus on the ground in an annular region about the source (Grant et al., 1967). Severe wind shear can cause focusing in the downwind direction. This effect is discussed by Baker (1973) and Reed (1973). Structural damage from accidental explosions has been correlated with these atmospheric inhomogeneities (Siskind, 1973; Siskind and Summers, 1974; Reed, 1968), and complaints of damage from explosive testing were reduced when firings were limited to days when no focusing was predicted (Perkins et al., 1960).

[2] Strehlow and Baker "The Characterization And Evaluation Of Accidental Explosions, Prog. Energy Combust. Sci., Vol. 2, pp. 27-60, 1976

Section 21.5.2 of NFPA 921: Guide for Fire and Explosion Investigations states: "Blast Pressure Front Modification by Refraction and Blast Focusing. Atmospheric inhomogeneities can cause nonideal blast pressure front behavior at times. When a blast pressure front encounters a layer of air at a significantly different temperature, it may cause it to bend, or refract. This occurs because the speed of sound is proportional to the square root of temperature in air. A low-level temperature inversion can cause an initially hemispherical blast front to refract and to focus on the ground around the center of the explosion."

The above-mentioned Bureau of Mines report RI-8485 states: "Air temperatures normally decrease with increasing altitude, with the reverse of this called a "temperature inversion," or warm air layer. The index of refraction of air changes with temperature, so that the normal condition of cooler air at higher altitudes refracts sound away from the ground. Conversely, temperature inversions refract sound downward, leading to higher than normal sound pressure levels at points of focus. Much work had been done on theoretical calculations of airblast focusing from temperature inversions (36-37, 39). Perkins (36) predicted that a single inversion could cause airblast to be 3 to 6 times more intense. Poulter (36) concluded that within a distance of two times the height of the inversion no intensification would occur. Taylor (64) stated that up to a 10-dB increase can occur from inversion-produced refraction. Schomer (50) discussed both low-altitude inversions and jet stream focusing, for propagation distances of 2 to 40 miles and 30 to 300 miles, respectively. The short range case is applicable to mine blasts; a 3-times intensification was the maximum measured and the average was 1.8 times (5.1 dB).

2) Terrain Effects

The valleys around the Alon USA site, seen in Figure 4, can also cause the blast wave to strengthen or focus preferentially. The terrain effects are also widely recognized. Baker et al (1983)[3] states "A ground surface which is irregular can significantly affect the blast wave properties. Gentle upward slopes can cause enhancement, while steep upward slopes will cause formation of Mach waves and consequent strong enhancement. Downward slopes or back surfaces of crests cause expansion and weakening of shocks. These effects are quite localized, however, and "smooth out" quite rapidly behind the irregularities."

Section 21.5.1 of NFPA 921 states: "Blast Pressure Front Modification by Reflection. As a blast pressure front encounters objects in its path, the blast pressure front may amplify due to its reflection. This reflection in some cases will cause the overpressure to increase and will sometimes amplify it as much as eight times at the surface of reflection, depending on the angle of incidence."

The above-mentioned Bureau of Mines report RI-8485 states: "Terrain is another possibly critical factor for airblast propagation. The effect of the bench and blast face on levels and character was treated in the section on airblast characteristics. Wilton (81) discusses experiments of "air bursts" over valleys and the resulting 50-pct increase of intensity compared to flat terrain. He and Wiggins (80) both state that a 300-pct increase is possible. Topographic effects may be responsible for high airblast levels reported in the valleys of the Appalachian Mountain during strip mining."

---

[3] Baker et al (1983), "Explosion Hazards and Evaluation," p. 128

Figure 2. Photographs taken soon after the explosion

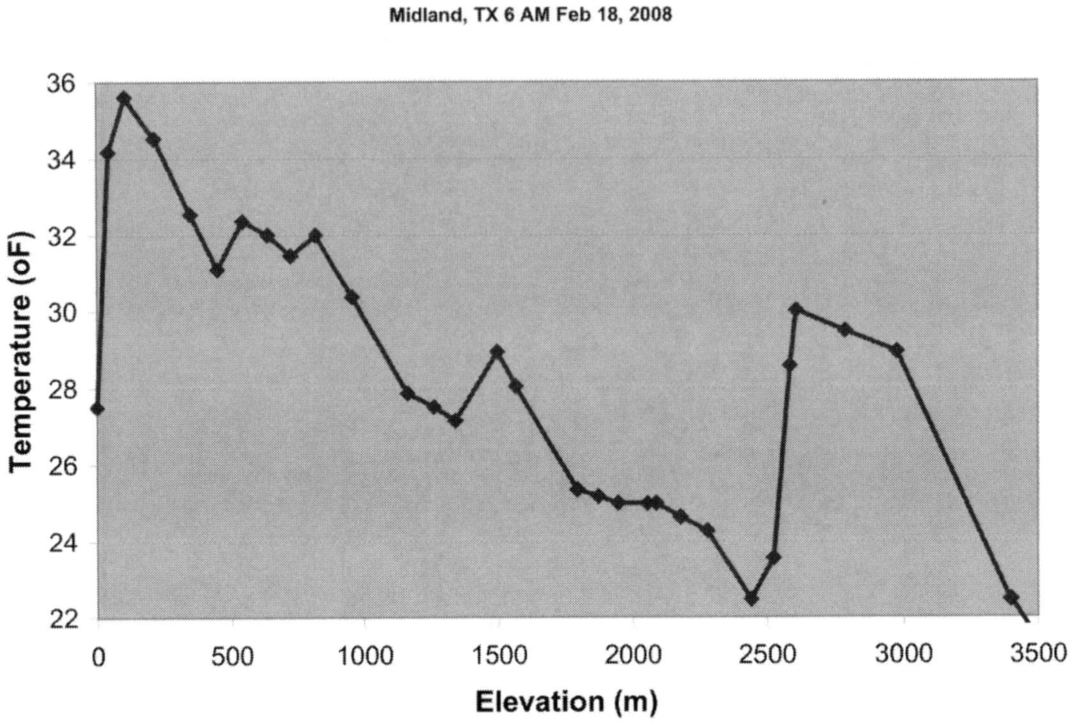

Figure 3. Atmospheric temperature profile approximately 2 hours before the explosion

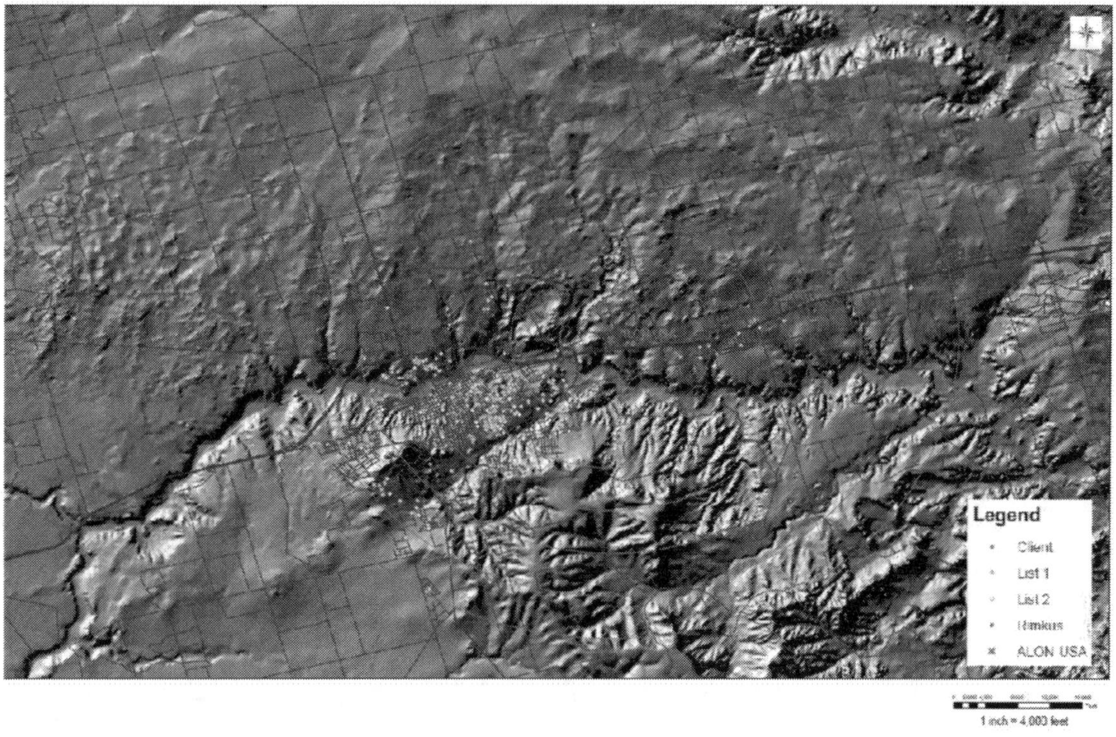

Figure 4. Features of the terrain around Alon USA and damaged houses

Modeling of time-varying dispersion for releases including potential rainout

Henk W.M. Witlox
DNV Software
3 Cathedral Street, London SE1 9DE, UK
henk.witlox@dnv.com

Mike Harper
DNV Software
3 Cathedral Street, London SE1 9DE, UK
michael.harper@dnv.com

Prepared for Presentation at
American Institute of Chemical Engineers
2013 Spring Meeting
9th Global Congress on Process Safety
San Antonio, Texas
April 28 – May 1, 2013

UNPUBLISHED

Modelling of time-varying dispersion for releases including potential rainout

Henk W.M. Witlox
DNV Software
3 Cathedral Street, London SE1 9DE, UK
henk.witlox@dnv.com

Mike Harper
DNV Software

Keywords: dispersion, time-varying, rainout, along-wind diffusion

Abstract

Many commonly-used atmospheric dispersion models cannot accurately simulate time-varying releases of hazardous chemicals into the atmosphere. The current paper discusses a new enhanced dispersion formulation accounting for time-varying effects resulting from a pressure drop in a vessel or pipe and/or from a time-varying pool (following rainout or from direct spill). This new formulation is implemented in the Unified Dispersion Model (UDM), and is planned to be included in a future version of Phast.

The new model includes effects of along-wind diffusion which reduces concentrations in the far-field. This is particularly relevant for toxic releases. The new formulation presumes a number of 'observers' to be released at successive times from the point of discharge or the upwind edge of the pool. The UDM first carries out pseudo steady-state calculations for each observer, where the release data correspond to the time at which the observer is released, and where the observer picks up vapour while travelling over the evaporating pool. Subsequently effects of along-wind diffusion are included by means of Gaussian integration over the downwind distance.

The above new UDM formulation has first been verified against the time-varying HGSYSTEM model HEGADAS-T for the specific case of dispersion directly from a pool. In the near-field the UDM appears to provide more accurate results because of more accurate observer calculations while observers are moving over the pool. Subsequently the UDM has been further tested for more general cases including elevated releases with rainout.

1. Introduction

Many accidents involve releases of hazardous chemicals into the atmosphere. Many commonly used atmospheric dispersion models are limited to continuous or instantaneous releases only. They do not accurately account for time-varying effects following time-varying releases i.e. resulting from a pressure drop in a vessel or pipe, and/or from a time-varying pool (following rainout or from direct spill). These effects include along-wind-diffusion (due to ambient turbulence) at the upwind and downwind edges of the cloud resulting in reduced concentrations while the cloud travels in the downwind direction.

The consequence modelling package Phast includes the Unified Dispersion Model UDM (Witlox and Holt [1], Witlox and Harper [2]) for modelling two-phase outdoor releases including potential rainout (Figure 1). The UDM in the current Phast (versions 6.7 and 7) only accounts for along-wind-diffusion (AWD) for instantaneous releases or finite-duration continuous releases that do not rain out.

Figure 1: discharge, droplet dispersion, rainout and re-evaporation

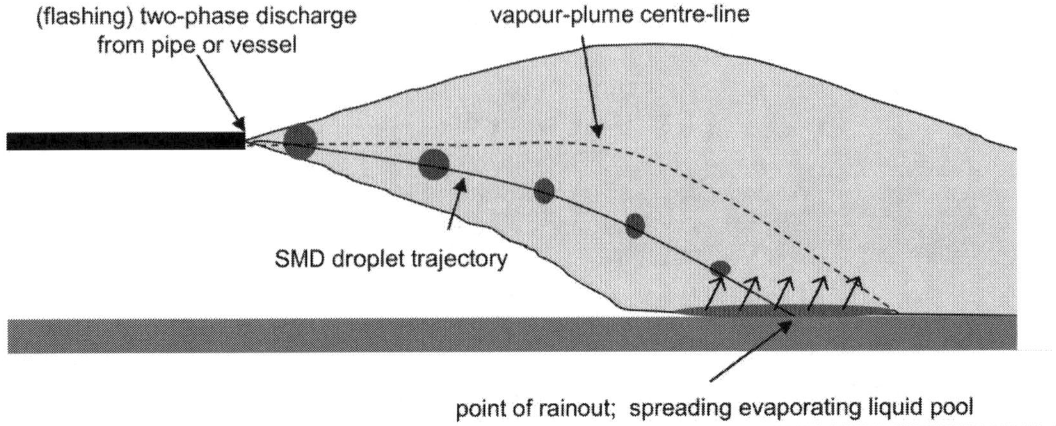

Figure 2: UDM time-varying dispersion – old segment method excluding AWD

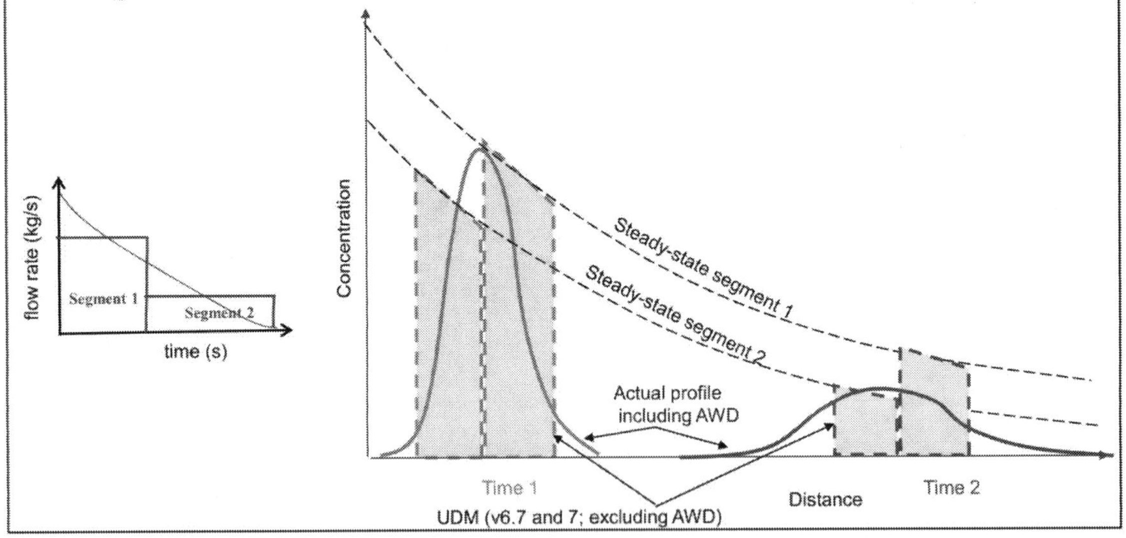

For a time-varying release, Phast divides the calculated discharge mass into a user-specified number of equal-mass segments (Figure 2). Likewise in case of rainout or dispersion directly from an evaporating pool (Figure 1), the evaporated mass from the pool is divided into equal-mass segments. Subsequently the UDM model carries out steady-state dispersion calculations for each segment, and determines for successive times the concentration as a function of distance from these segment data as shown in Figure 2 by the dashed curves. Thus for time-varying releases or for dispersion after rainout, the current UDM does not apply along-wind-diffusion at the upwind and downwind edges of the cloud, or between segments, which may lead to significant over-prediction of concentration and under-prediction of duration in the far-field (see Figure 2 at time 2). This is particularly important for toxic releases, where dispersion calculations are required to be carried out to small concentration levels such as ERPG levels. It is less important for flammable releases with calculations to relatively high concentration levels only, such as LFL or 0.5LFL.

The Shell consequence modelling package HGSYSTEM 3.0 (Post [3]) includes the time-dependent dispersion model HEGADAS-T (Witlox [4]) for modelling the time-dependent dispersion of a heavy-gas cloud moving with the ambient wind. It can be used to model the dispersion downwind of either a time-dependent ground-level source (unpressurised release) or a vertical-plane transition (breakpoint) with a near-source jet model (pressurised release). The transient behaviour of the cloud is approximated by a quasi-steady-state description in which so-called "observers" are released at the source/transition-plane at a series of times. These observers travel with the wind. For each observer, the observed concentration is set from steady-state HEGADAS-S calculations using the observed source/transition data. Thus by calculating the position of each observer at a given time t, the concentration c is set for a number of downwind distances. Subsequently the actual concentration is set from Gaussian integration with respect to the downwind distance x of the above observer calculations. This involves a downwind dispersion coefficient σ_x, which allows along-wind diffusion to be taken into account.

The current paper introduces a new enhanced dispersion formulation (Witlox and Harper [5], [6]) accounting for time-varying effects resulting from a time-varying release. It generalises the above HEGADAS-T formulation to account for along-wind-diffusion both for ground-level unpressurised releases (e.g. evaporating pools) and for elevated two-phase pressurized releases including potential rainout. This method has been implemented in a new version of the UDM model, which is to be included in a future version of Phast.

2. New UDM along-wind diffusion formulation

First Phast discharge calculations are carried out (for release from a hole of a vessel or a pipeline) to determine the time-varying discharge data after expansion to atmospheric pressure and prior to air entrainment [flow rate, velocity, temperature, liquid fraction, droplet size (SMD – Sauter Mean Diameter)].

Figure 3: UDM time-varying dispersion –new observer method including AWD

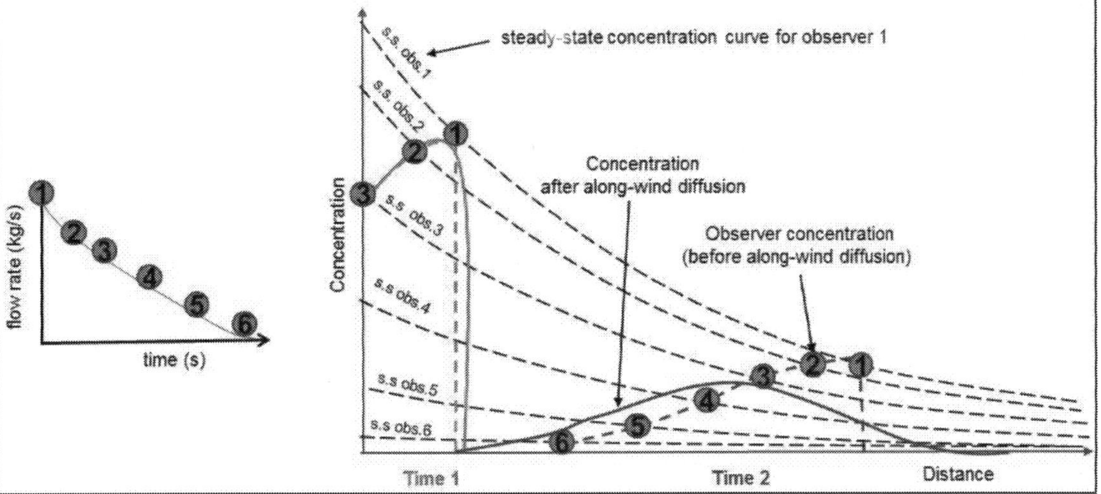

Figure 4: new UDM dispersion phases for time-varying release with rainout

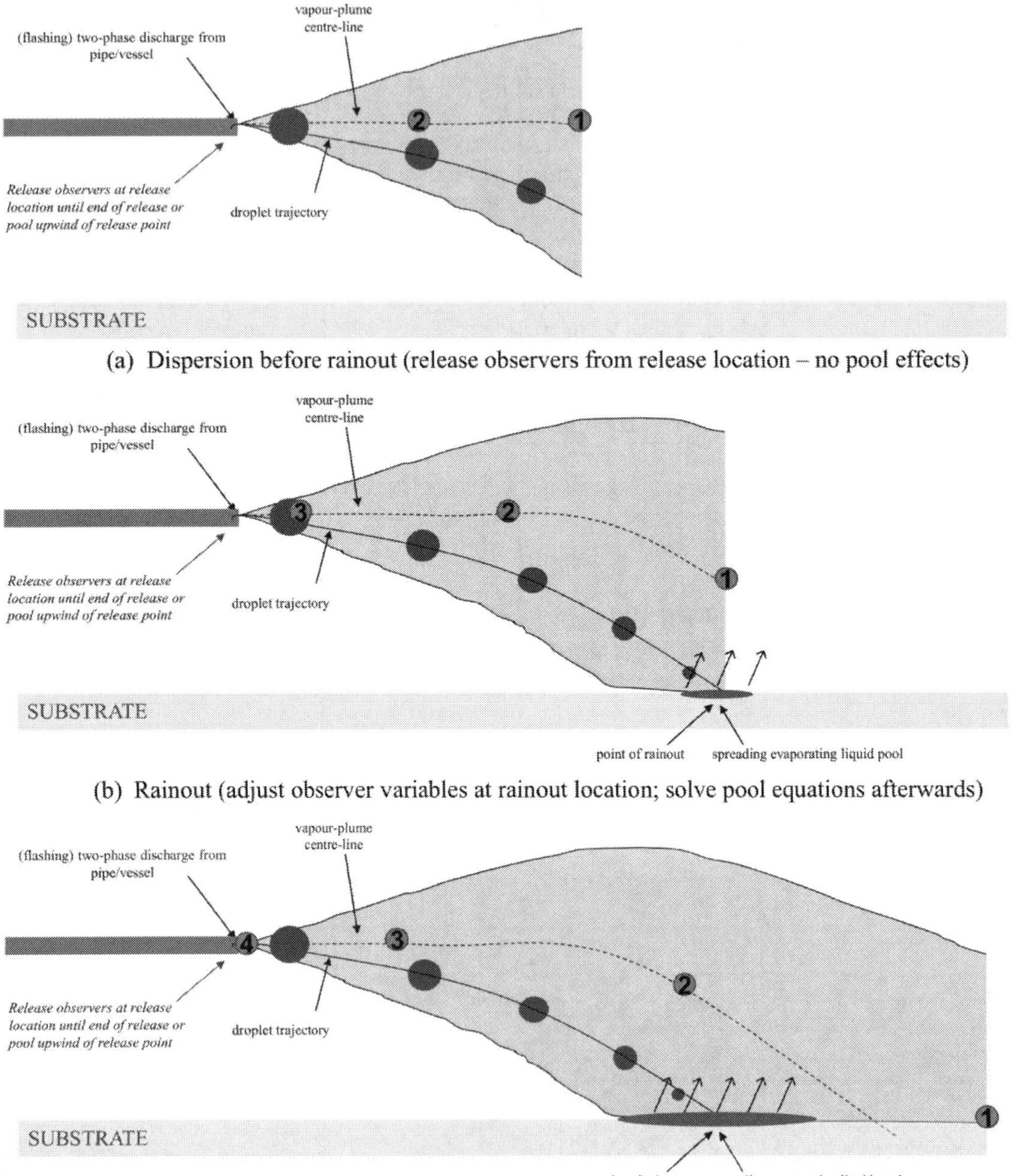

(a) Dispersion before rainout (release observers from release location – no pool effects)

(b) Rainout (adjust observer variables at rainout location; solve pool equations afterwards)

(c) Dispersion after rainout (account for pool vapour pick-up from cloud for observer calculations)

Subsequently 'observers' are released from the release location at intervals which can be chosen to correspond to equal-mass increments as for the old UDM method (using the Phast time-averaging model TVAV; see Figure 3 for the case of 6 release observers). These observers are assumed to move along the cloud centre-line at the UDM cloud centre-line speed. In the case of rainout, additional 'observers' (corresponding to equal pool-mass segments) are released upwind of the pool after the release plume has left the plume behind. This is illustrated by Figure 4, where the first observers (1,2,3,4) start from the release point and subsequent observers (5, 6) start from the upwind edge of the pool.

Figure 4 (continued)

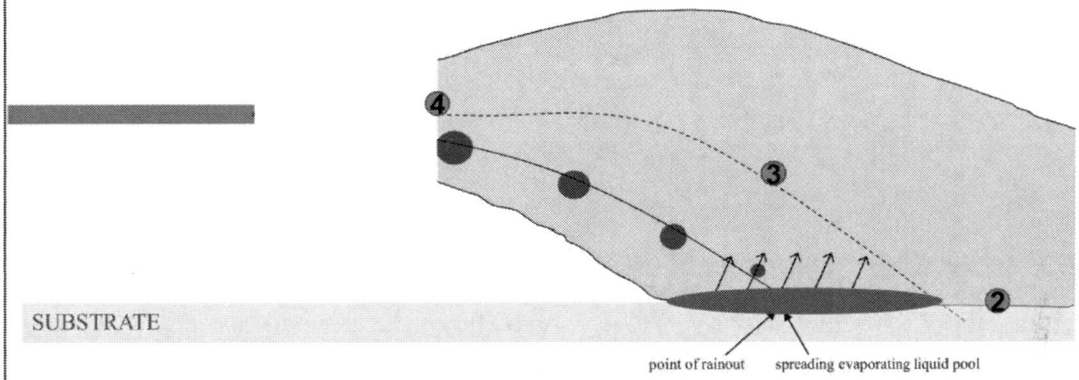

(d) Downwind movement of upwind edge of cloud towards pool (after end of release)

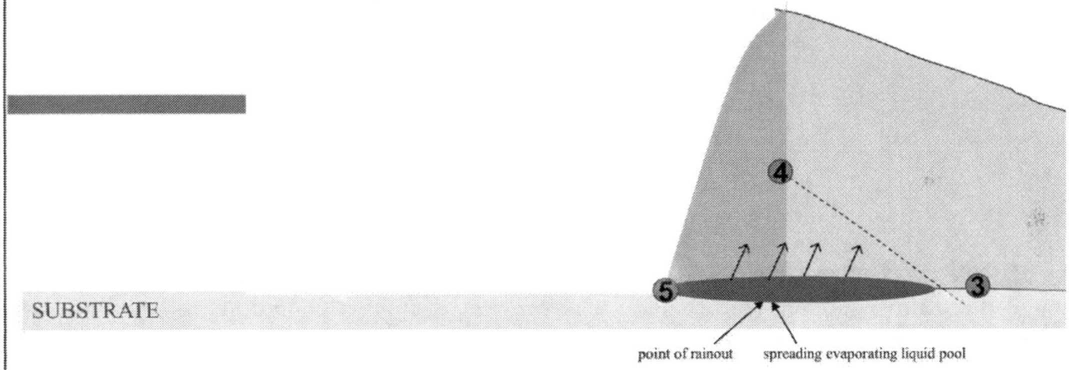

(e) Release observers upwind of pool (after 'release observers' all downwind of upwind pool edge)

(f) Dispersion directly from pool, with residual cloud moving away from pool

For each observer steady-steady calculations based on the observed source data are carried out to evaluate the observer concentration prior to including effects of along-wind diffusion; see Figure 3. While the observer moves above the pool, the observer dispersion equations (conservation of cloud mass and momentum conservation, cloud crosswind gravity spreading, heat transfer from the substrate, etc.) are modified to account for the pool vapour added back to the cloud.

The actual plume concentration c(x,y,z,t) including effects of along-wind-diffusion is a function of time t, distance x downwind of the release location, crosswind distance y, and vertical height z; this function is calculated by means of Gaussian integration of the observer concentration C(ξ,y,z,t),

$$c(x,y,z,t) = \int_0^\infty \frac{C(\xi,y,z,t)}{(2\pi)^{1/2}\,\sigma_x(\xi)} \exp\left\{-\frac{(x-\xi)^2}{2\sigma_x^{\,2}(\xi)}\right\} d\xi \qquad (1)$$

In the above equation ξ is the downwind distance from the release point at time t of an observer travelling with the cloud in the downwind direction. At this position the observer observes the concentration C(ξ,y,z,t). In Equation (1) along-wind diffusion is taken into account by assuming that the concentration C(ξ,y,z,t) spreads out around ξ according to a Gaussian distribution with a downwind dispersion coefficient $\sigma_x = \sigma_x(\xi)$. Figure 3 depicts the pre-AWD observer concentration C and the post-AWD concentration c at a short time after the release (time 1; limited AWD effects), and at a larger time after the release (time 2; larger AWD effects).

The downwind dispersion coefficient σ_x consists of two components,

$$\sigma_x(\xi) = \sqrt{\sigma_{xs}^2 + \sigma_{xt}^2} \qquad (2)$$

where σ_{xs} is the downwind dispersion due to vertical wind shear and σ_{xt} is the downwind dispersion due to turbulent spread caused by downwind velocity fluctuations; σ_{xs} and σ_{yt} are evaluated as function of downwind distance ξ according to the formulation by Ermak [7] [8]. This assumption is fully consistent with the current finite-duration-correction (FDC) formulation adopted in the UDM [2] to apply AWD effects for finite-duration releases..

3. UDM verification against HGSYSTEM for dispersion from pool

The case of an instantaneous spill of 10000 kg of butane liquid is considered with both cases of a bund (radius 10m) and no bund (pool on water; dispersion over land). Ambient data applied are stability D5, temperature 300K and humidity 70%. Furthermore a maximum pool duration of 1 minute is applied.

Figure 5 shows the values predicted by the UDM of butane pool radius and evaporation rate for both cases with and without a bund. It is seen that the immediate butane rainout within the bund results in a virtually steady-state pool of 10 meter radius, 12.3 kg/s evaporation rate and a temperature of 271.8C.

Figure 5: UDM pool predictions (instantaneous 10000kg spill of butane on water)

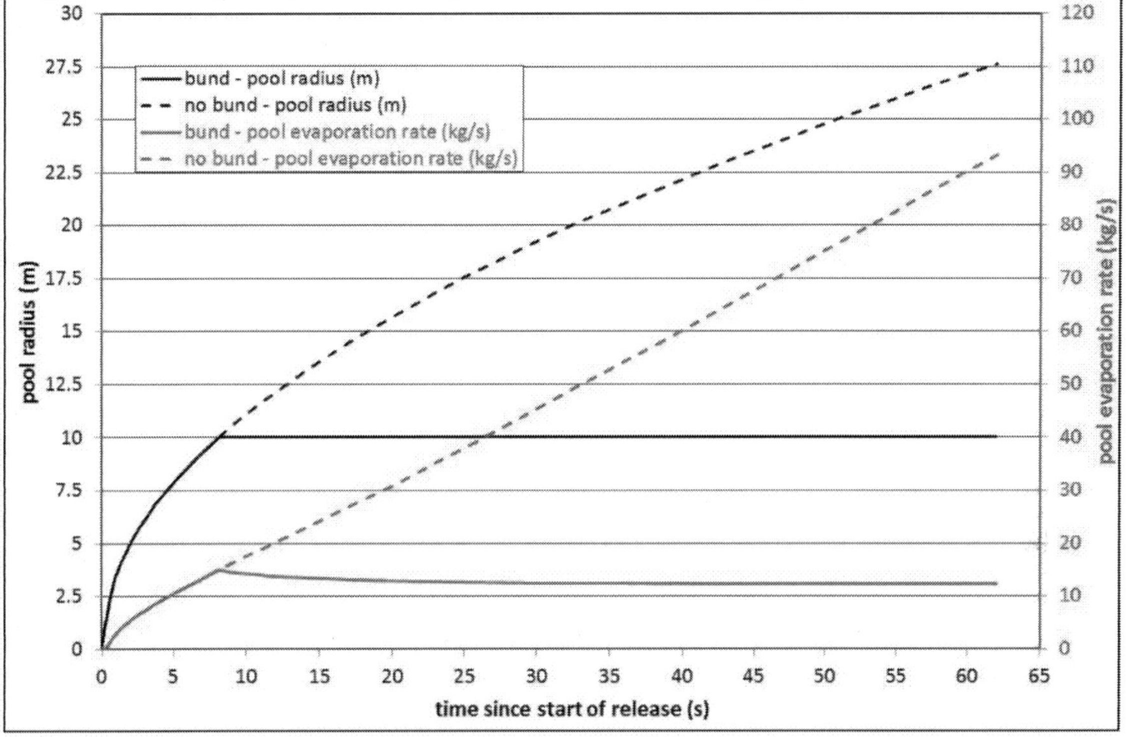

Figure 6 includes results of the following calculations for the case of a bunded pool:

- HGSYSTEM 3.0 calculations (the rectangular HEGADAS source was chosen to be square with the same area as the UDM pool)
  - o steady-state HEGADAS-S calculations (ground-level steady-state heavy-gas dispersion from area source) based on the steady-state pool data mentioned above.
  - o finite-duration correction (using Ermak's formula) to above HEGADAS-S results (60 seconds duration) using the post-processor program POSTHS
  - o HEGADAS-T calculations (ground-level time-varying heavy-gas dispersion from area source). In line with the new UDM model, no gravity-shape correction was applied and the along-wind-diffusion coefficient was chosen to be based on Ermak's formula (non-default HEGADAS-T options)
- UDM calculations
  - o Steady-state UDM calculations based on the above steady-state pool data
  - o FDC correction to the above steady-state UDM results (60 seconds duration)
  - o UDM AWD calculations, with 30 observers released from the upwind edge of the pool

Figure 6: UDM verification against HEGADAS for spill of butane with bund (radius 10m)

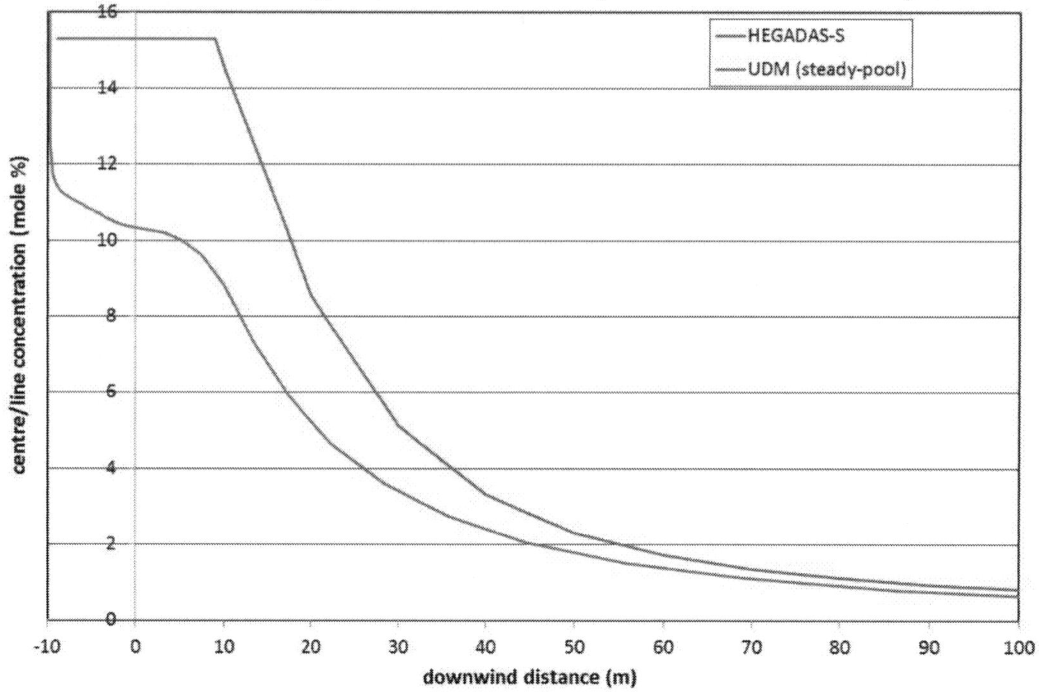

(a) Near-field centre-line concentrations (above pool and downwind of the pool)

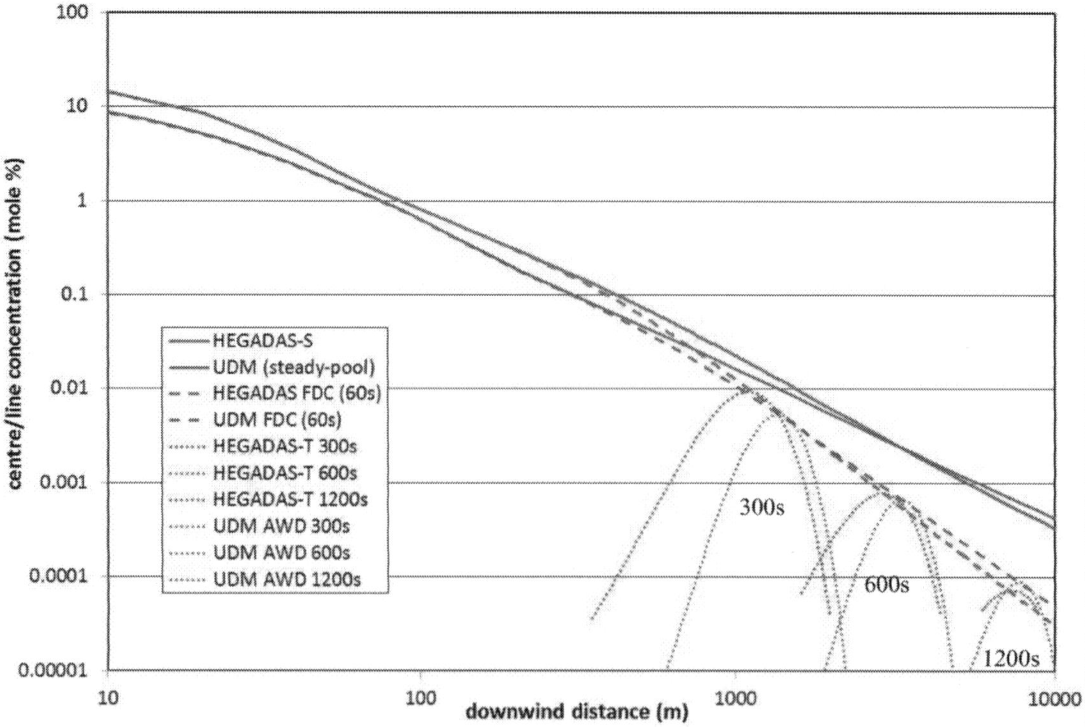

(b) Far-field concentrations

Figure 6a includes the steady-state HEGADAS-S and UDM results of the centre-line ground-level concentrations in the near-field for the case of a steady-state pool. HEGADAS-S assumes a uniform concentration (top-hat profile) above the area source (about 15% mole fraction), which is envisaged to lead to an over-prediction near the source. On the other hand, the UDM model more rigorously solves the dispersion equations across the pool allowing for a variation of the concentration across the pool. The very rapid decrease predicted by the UDM of the concentration near the upwind edge of the pool is a numerical artefact and may not occur in reality. Further downwind from the pool a closer match is observed between the UDM and HEGADAS-S data.

Figure 6b includes results of all the above calculations. It includes results of the time-varying dispersion calculations by UDM AWD and HEGADAS-T at output times 300s, 600s and 1200s. It is seen that the maximum values of these curves correctly touch the maximum centre-line concentration (maximum value over all times) predicted by UDM FDC and HEGADAS-S FDC, respectively. This confirms that the AWD integration is carried out correctly by both programs and is consistent with the FDC formulation. Also close agreement between the UDM and HEGADAS FDC results is observed. It is seen that initially the HEGADAS-T cloud is more heavy (larger concentrations) and therefore moves more slowly than the UDM AWD cloud, but this effect reduces in the far-field when the predictions become closer.

Figure 7: UDM verification against HEGADAS for spill of butane without bund

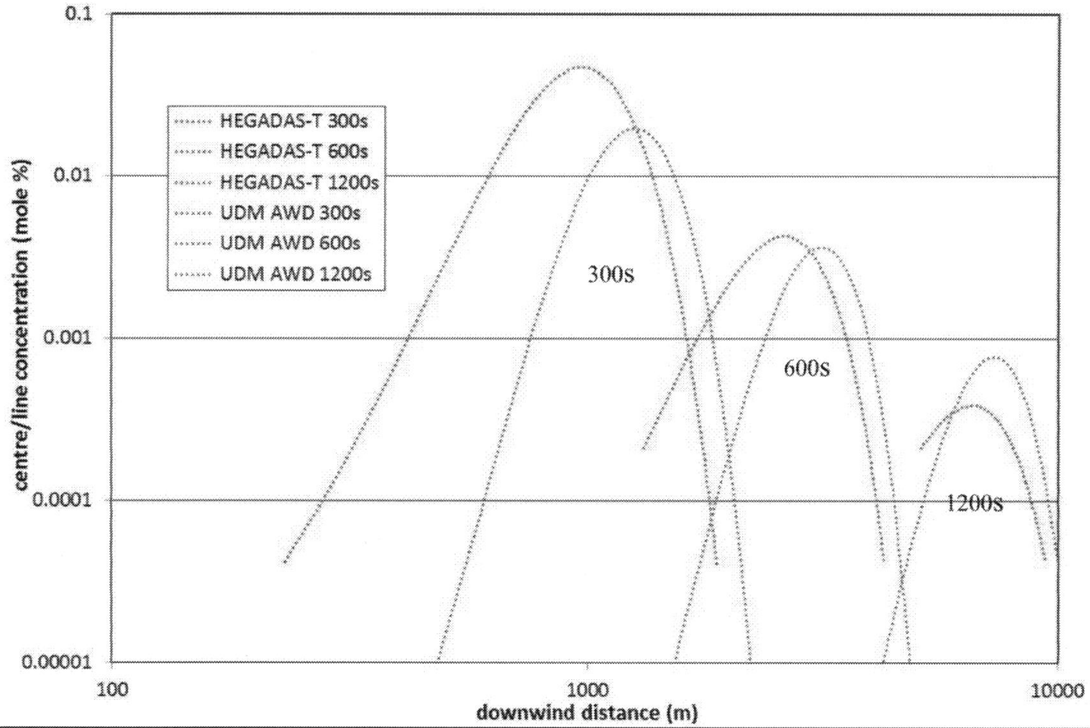

Figure 8: UDM AWD predictions for elevated Cl_2 release with rainout

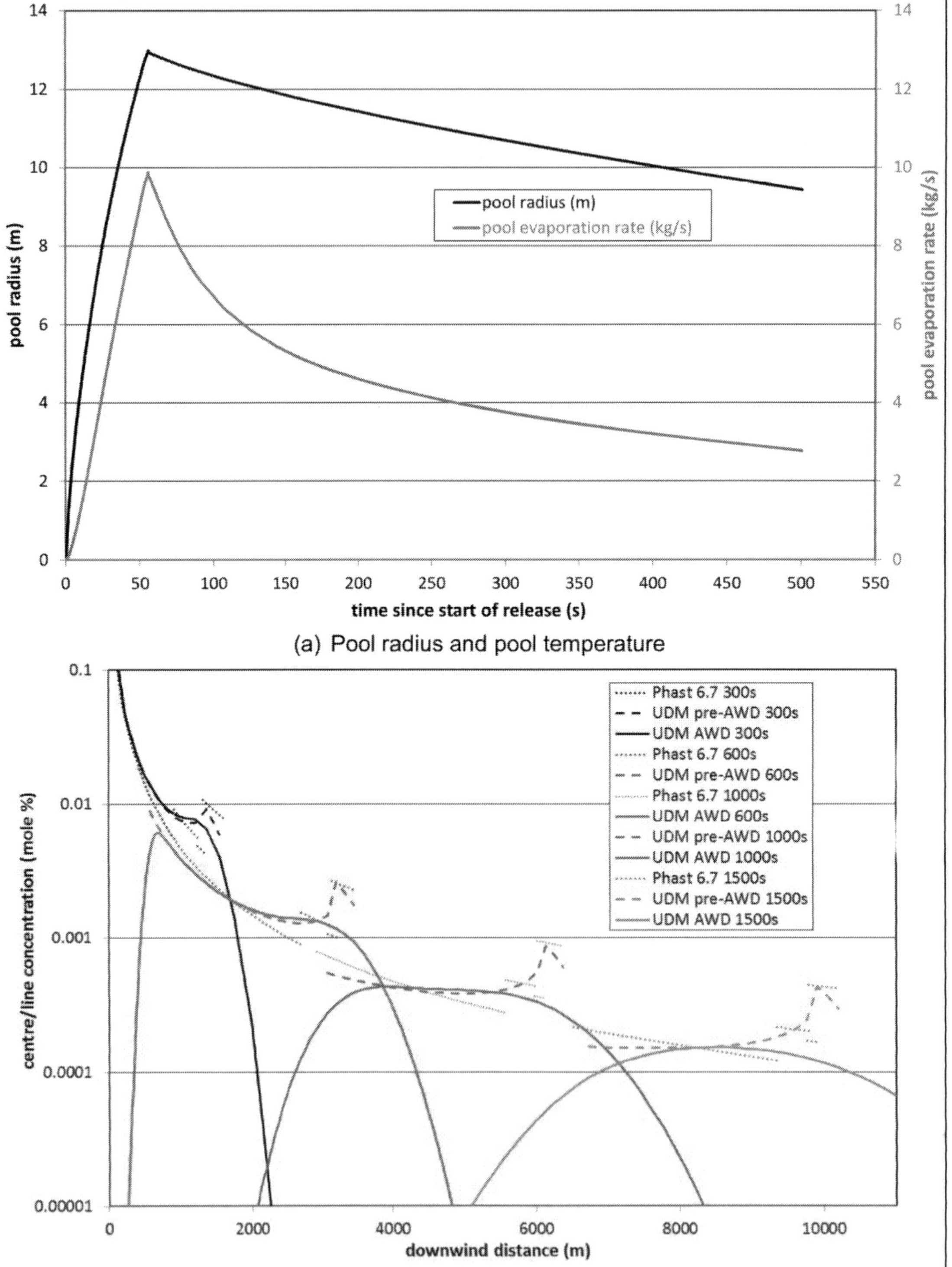

(a) Pool radius and pool temperature

(b) Concentrations at 300, 600, 1000, 1500s: Phast 6.7, UDM pre-AWD and post-AWD

Likewise, Figure 7 includes UDM AWD and HEGADAS-T results at times 300s, 600s and 1200s for the case without the bund. Here the UDM calculated pool data were specified as input to HEGADAS-T with a step size of 2 seconds. The observed differences are confirmed to be very similar to that seen for the case with the bund (compare Figures 6b and 7).

4. UDM results for elevated chlorine release case with rainout

This section includes results for a horizontal chlorine release at elevation height of 1m with a constant release rate of 100kg/s for a duration of 50 seconds. The post-expansion data are chosen to correspond with 100% liquid chlorine at its boiling point (213.9K) with a release velocity of 10m/s and an initially presumed SMD droplet size of 0.001m. Ambient data applied were stability D5, temperature 298K and humidity 70%. Furthermore a maximum pool duration of 500 seconds was applied. Two observers were released from the release point (at start and end of the release) and subsequently 30 observers were released from the pool.

About 90% rainout was predicted at about 5m downwind distance from the release point. Figure 8a shows the values predicted by the UDM of radius and evaporation rate for the un-bunded chlorine pool (pool duration of 500s). The pool radius and evaporation rate are seen to increase rapidly until about 55s at which time the minimum pool thickness is reached. Following this the pool breaks up into bubbles modelled in the UDM by a reducing effective pool radius. The pool evaporation is terminated at the maximum pool duration of 500s.

Figure 8b includes results of both observer concentrations (prior to inclusion of AWD effects; dashed curves) and concentrations after inclusion of AWD effects (solid curves) at output times of 300s, 600s, 1000s and 1500s. Also results are included using the Phast 6.7 UDM segment method (dotted curves). At 300s the release is still on-going, and therefore AWD effects are seen to be only pronounced at the downwind edge of the cloud. With increasing cloud travel times, AWD effects erode concentrations at the upwind and downwind edges of the cloud and increase its cloud length. The pre-AWD observer concentrations predicted by the new UDM observer method are seen to be comparable with the discontinuous segment concentrations predicted by Phast 6.7. The new UDM AWD method predicts reduced concentrations which are considered to be more accurate.

5. Conclusions and future work

The following main conclusions can be drawn:

- The Phast dispersion model UDM has been extended to account for along-wind-diffusion (AWD) effects for time-varying releases and for dispersion after rainout. The new AWD method is based on a generalisation of the HEGADAS-T method for dispersion from area sources. It is fully consistent with the existing UDM finite-duration correction (FDC) method for the case of finite-duration continuous releases.
- In the new model the flow-rate and pool-evaporation rate are no longer divided into discontinuous segments. In the old model the equations were discontinuously modified at the original point of rainout to account for the vapour added back from the entire pool, and no along-wind-diffusion was applied. In the new model the cloud adapts

continuously to the presence of an evaporating pool, without the need for discontinuous segments.
- The new UDM AWD model has been verified against HGSYSTEM for the case of dispersion from a ground-level area source and good agreement was obtained.

The current further work is being planned:

- Extension of UDM instantaneous model to account for along-wind-diffusion after rainout.
- Extension of UDM AWD model to include effects of along-wind gravity spreading, which are particularly relevant in the near-field in the case of small wind speeds
- Validation against experimental data
- Implementation of new UDM AWD model into Phast
- Further refinement of empirical formulas for along-wind-diffusion coefficient σ_x

6. References

[1] Witlox, H.W.M., and Holt, A., "A unified model for jet, heavy and passive dispersion including droplet rainout and re-evaporation", *International Conference and Workshop on Modelling the Consequences of Accidental Releases of Hazardous Materials*, CCPS, San Francisco, California, September 28 – October 1, pp. 315-344, 1999

[2] Witlox, H.W.M and Harper, M.., "UDM theory manual", Phast 6.7 technical documentation, DNV Software, 2011

[3] Post, L. (editor), Section 7C in "HGSYSTEM 3.0 Technical Reference Manual", Report TNER.94.059, Thornton Research Centre, Shell Research, Chester, England, 1994

[4] Witlox, H.W.M., "The HEGADAS model for ground-level heavy-gas dispersion, II. Time-dependent model", *Atmospheric Environment*, Vol. 28, No. 18, pp. 2933-2946, 1994

[5] Witlox, H.W.M., and Harper, M., "Modelling, verification and validation by Phast relating to H2S toxic releases", Contract 96506900/98509900 for TOTAL, December 2008

[6] Witlox, H.W.M., and Harper, M., "Modelling of along-wind diffusion for time-varying dispersion including rainout", Contract 96512500 for TOTAL, RIVM and DNVS, April 2011

[7] Ermak, D.L., "Unpublished notes on downwind spreading formulation in finite-duration release version of SLAB", Lawrence Livermore National Laboratory, California, 1986

[8] Ermak, D.L., "User's Manual for SLAB: an atmospheric dispersion model for denser-than-air releases, Lawrence Livermore National Laboratory, Livermore, California, February 1989

Acknowledgment

Financial support of the work reported in this paper was provided by DNV Software, RIVM (Dutch Government), and TOTAL. The contents of this paper including any opinions and/or conclusions expressed, are those of the authors alone and do not necessarily reflect the policy of these organizations.

New Quantitative Risk Criteria for U.S. LNG Facilities

Ellen H. McInerney, P.E.
Exponent, Inc.
4580 Weaver Parkway, Suite 100, Warrenville, Illinois 60555
emcinerney@exponent.com

Ryan J. Hart, Ph.D., P.E. and Delmar "Trey" Morrison III, Ph.D., P.E.
Exponent, Inc., Warrenville, Illinois 60555

Harri K. Kytömaa, Ph.D., P.E.
Exponent, Inc., Natick, Massachusetts

Prepared for Presentation at
American Institute of Chemical Engineers
2013 Spring Meeting
9th Global Congress on Process Safety
San Antonio, Texas
April 28 – May 1, 2013

UNPUBLISHED

New Quantitative Risk Criteria for U.S. LNG Facilities

Ellen H. McInerney, P.E.
Exponent, Inc.
4580 Weaver Parkway, Suite 100, Warrenville, Illinois 60555
emcinerney@exponent.com

Ryan J. Hart, Ph.D., P.E. and Delmar "Trey" Morrison III, Ph.D., P.E.
Exponent, Inc., Warrenville, Illinois 60555

Harri K. Kytömaa, Ph.D., P.E.
Exponent, Inc., Natick, Massachusetts

Keywords: Quantitative Risk Assessment (QRA), Liquefied Natural Gas (LNG), failure rate

Abstract

The recent increase in natural gas production has expanded the outlook for exporting domestically-produced natural gas. As a result, many existing LNG import facilities in the U.S. are proposing to construct liquefaction facilities collocated with existing import terminals. The Federal Energy Regulatory Commission (FERC) reviews and approves applications for construction. As part of this process, the applicant must perform specific hazardous release consequence analyses to demonstrate flammable vapor dispersion exclusion zones, thermal radiation exclusion zones, and explosion overpressure exclusion zones to meet criteria established in 49 CFR 193 and NFPA 59A.

Over the last few years, the FERC has refined the risk analysis criteria for LNG facility construction projects. While the current regulations do not specify quantitative risk assessment (QRA) as a necessary hazard analysis tool, FERC has recently issued guidance for the selection of leak sizes based upon failure frequencies for piping systems. In particular, scenarios with failure frequencies greater than 3E-05 per year must be considered. The underlying studies that FERC relied upon to derive the failure frequency criterion for single accidental leak source size determination (i.e., 3E-05 per year to 5E-05 per year) are largely based on generic industry data not LNG-specific data. The net result of this criterion has been to require elements of QRA as part of LNG facility permitting process to identify the scenarios that need to be considered and those that don't. This is a shift in the paradigm away from prescriptive leak sizes and single accidental release scenarios that had been used for receiving terminals constructed over the last decade.

This shift in paradigm coincides with and was likely motivated by the North American transition from receiving terminals to liquefaction export terminals. Receiving terminals are considerably simpler in their design than export terminals, which require refrigeration units to liquefy the gas. This paper aims to provide an analysis and discussion of the FERC failure frequency arguments grounded in the strength and weaknesses of the underlying databases. An open question is whether this new criterion provides a reasonable improvement in safety over the prior requirements. The underlying historic databases and studies do not reflect the major differences on the safety records between the LNG industry and general industry. More specifically, they do not necessarily take into consideration the effects of cryogenic processes, management systems, materials of construction, or failure modes on the failure rate data. Given the basis of generic industry failure rates, the question arises as to whether the failure rate thresholds that trigger the requirement to consider certain scenarios are appropriate.

1 Introduction

Many LNG import facilities in the U.S. were proposed in anticipation of the need to import gas; and some were built. The recent increase in domestic natural gas production from non-conventional sources has reversed this trend and become a driver for exporting this domestically-produced natural gas. As a consequence, many existing import facilities are proposing to construct liquefaction facilities collocated with existing import terminals, and new export only projects are also being proposed [1].

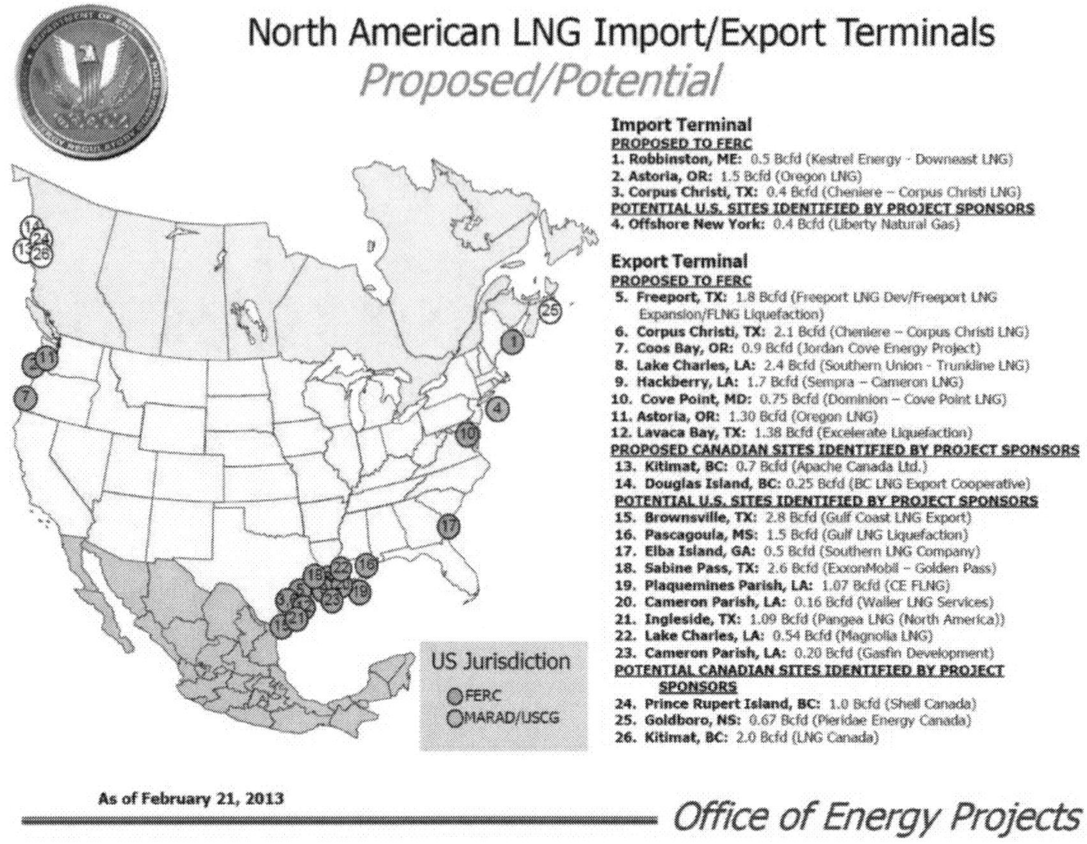

Figure 1. Summary of proposed and potential LNG terminals in North America, provided by the United States Department of Energy Federal Energy Regulatory Commission (FERC) [1].

LNG is processed and conveyed through piping systems as a cryogenic liquid at temperatures of approximately -260°F and moderate to high pressures (e.g., 50 psig up to 1000 psig) depending upon the location in the liquefaction or vaporization process. Flammable pressurized liquid or cryogenic refrigerants may also be present in the liquefaction units at similar temperatures and pressures. Depending upon the type of refrigeration process employed, refrigerants may include ethylene, propane, butane, pentane and higher hydrocarbons, or they may be mixed in unique combinations to produce a mixed refrigerant [2]. A loss of containment (LOC) of these liquids may lead to pressurized jetting and flashing, liquid sprays, and liquid spills. Flammable cloud formation due to evaporation and the subsequent dispersion of the cloud away from the release source pose significant hazards to be considered in the siting of process equipment and buildings at the facility.

The principal hazards of a LOC of refrigerant or LNG relate to flammable vapor cloud formation and dispersion, flammability of the cloud, radiant heat flux from pool fires, flash fires, and damaging overpressures due to ignition of a flammable vapor cloud that exists in a congested or confined portion of the facility. The most hazardous scenarios to consider are those that produce large, high momentum flammable vapor clouds (as opposed to flammable liquid pools in conveyance and impoundment systems, which produce only wind-driven vapor clouds). For LNG and refrigerants, these clouds consist of cold, dense vapor that disperse near the surface of the terrain in the direction of the prevailing wind.

The location of existing or new proposed facilities and the liquefaction, vaporization, storage, and transfer systems on a plant site must be considered in relation to the potential hazards posed by the processes [3]. This evaluation process is termed as facility siting, and the general intent is to limit or reduce the risk to on-site and off-site populations (i.e. on-site workers, neighboring facilities, and the general public) to the hazards if a LOC incident occurs. Existing LNG regulations only address prescribed off-site consequences and will be discussed in detail below. Two alternative methods of evaluating the risk posed by processes at a facility include (1) the use of prescriptive requirements in standards and (2) risk-based analysis [4]. U.S. regulations and safety standards for on-shore LNG terminals impose a prescriptive methodology for defining and evaluating hazard scenarios to satisfy the requirements of the authority having jurisdiction (in this case, FERC as discussed below). In many ways, this is a different approach than traditional risk-based analyses such as Quantitative Risk Assessment (QRA). However, in the current U.S. regulatory environment governing LNG facilities, there has been a gradual shift toward alignment of the traditional prescriptive requirements with risk-based concepts. One reason is likely the significant increase in complexity of liquefaction trains compared to import terminals that were being developed in the mid 2000's. This topic is discussed further in this article by evaluating the current regulatory requirements for hazard scenario selection and comparing it to traditional QRA selection methods.

2 U.S. Regulations for LNG Facilities

Currently, LNG facilities involved in interstate or international commerce must comply with U.S. Federal regulations in regards to facility siting considerations [5, 6]. Capital projects must undergo a review and approval process to receive construction authorization for new operations, facilities, or infrastructure. The process is regulated by the Federal Energy Regulatory Commission (FERC), an agency which ultimately authorizes construction and operation. As part of the process, facilities must demonstrate that they comply with Federal safety standards for LNG facilities that have been promulgated by the Pipeline and Hazardous Materials Administration (PHMSA), part of the U.S. Department of Transportation (DOT), according to Title 49 of the Code of Federal Regulations (CFR) - Part 193. Facility owners will document and communicate the results of site safety analyses through Resource Report 11 and Resource Report 13, which are submitted to FERC as part of the environmental review process. More detail on the substance of these reports is provided in reference [7].

The CFR provides broad requirements for safety and incorporates NFPA 59A (2001) [8], by reference to supplement technical requirements (see [5], part 193.2051). It should be noted that most of the CFR references the 2001 edition of NFPA 59A. Facility siting analyses required by the regulations are focused on the concept of an "exclusion zone." An exclusion zone is defined as "an area surrounding an LNG facility in which an operator or government agency legally controls all activities…" (see [5], part 193.2007). Flammable vapor dispersion distances due to LOC (e.g., spills, ruptures, and leaks) must not yield vapor clouds with a natural gas (typically considered as methane) concentration in air of greater than 2.5% that pass beyond the facility's exclusion zone (see [5], part 193.2059). Natural gas is not pure methane; and assuming that it is usually provides conservatively larger vapor clouds.

NFPA 59A (2001) requires that the 50% LFL cloud from LNG spills be confined to the property in accordance with §2.2.3.3. However, as will be discussed in the following section, liquid spills into the trenches which flow to the main impounding area and the main impounding area itself typically don't cause the worst case flammable vapor clouds. NFPA 59A (2001) provides an additional, more broadly stated criterion for the flammable vapor dispersion distance due to a design spill by stating the following in §2.2.3.4:

> Provisions shall be made to minimize the possibility of a flammable mixture of vapors from a design spill specified in 2.2.3.5, as appropriate, reaching a property line that can be built upon and that would result in a distinct hazard.

This "distinct hazard" is not specific and provides a catchall requirement. Although the regulations prescribe a boundary for the potential hazard - 50% LFL cannot cross the property line that can be built upon - they do not prescribe all of the scenarios that must be considered to generate a flammable vapor source term. In the context of typical risk analyses, accidental release scenarios may be identified by a hazard analysis team or experts according to process hazard analysis (PHA) techniques, standard sets of scenarios for a given type of unit/operation, or through other routes. PHA techniques include Fault Tree Analysis, HAZID, HAZOP, and/or What-if to identify the potential hazard scenarios. These are typical expert- or team-driven hazard analysis techniques that are used to develop scenarios for consideration through formal QRA. In practice, PHA techniques may identify single event failures (e.g., leak from a flange, pipe rupture, or operator error) or they may describe more complex failures including cascading or knock-on type events. The LNG regulations only explicitly consider single event failures, i.e., "single accidental leakage sources." This would seem to simplify the set of analyses that must be considered through the regulatory facility siting process; but that is not the case, as the following sections will discuss.

3 Quantitative Risk Assessment

Quantitative Risk Assessment (QRA) is a valuable technical risk assessment method for understanding the risks associated with industrial processes. QRA is widely used in many industries. International regulations commonly require QRA as a component in the decision-making process for siting of industrial facilities [9]. QRA is not commonly a required element of U.S. regulations; EPA's Risk Management Program (RMP) defines a single worst-case scenario (and an alternative release scenario for RMP Program 2 and 3 sites) used for regulatory permitting [10]. In contrast, OSHA's Process Safety Management (PSM) standard only requires that one of many appropriate process hazard analyses (PHAs) be performed on qualifying processes.

The European Standard "EN-1473: Installation and equipment for liquefied natural gas – Design of onshore installations" provides guidance for performing QRAs at LNG facilities [11], but the standard is not enforceable in the United States. The standard provides details on selecting hazard scenarios, assigning probability, and quantifying the risk associated with various releases. Further, the consequences are considered for both on-site and off-site property and personnel.

NFPA 59A (2013) includes a new chapter on applying risk assessment for LNG plant siting [12]. However, the risk assessment guidelines are provided as an alternative to the existing facility siting requirements. Furthermore, the objective of the NFPA 59A (2013) risk assessment guidance is to assess the risks posed *outside of the LNG facility boundaries* by releases of LNG or other hazardous substances stored or processed on site. The standard cites several references for performing QRAs and provides general QRA guidelines, including an "Example Component Failure Database." However, the standard does not provide a reference for the listed failure probabilities (only references where complementary data may be retrieved).

QRA is an analytical process that systematically evaluates LOC scenarios, their hazardous consequences, and the risk of injury or fatality posed by the scenarios. The risk is calculated as the likelihood of injury on an individual risk (IR) and societal risk (SR) basis. IR is the frequency per year that an individual with continuous potential exposure may be expected to sustain a serious or fatal injury (see [13], p. 408). SR is the cumulative risk to all persons in a population sustaining a serious or fatal injury (see [13], p. 433). IR is typically presented as contours of constant individual risk values, e.g., 1E-05 /yr likelihood of fatality. Figure 2 shows an example of such a rendering that can be used to evaluate the risk to surrounding individuals. SR is typically depicted on an F-N curve, comparing the frequency of fatality (F) to the number of fatalities to be expected (N). A comparison of the tolerance criteria for societal risk is provided in Figure 3. As part of the QRA, each individual scenario outcome would be compared against the tolerance criteria presented by the authority having jurisdiction.

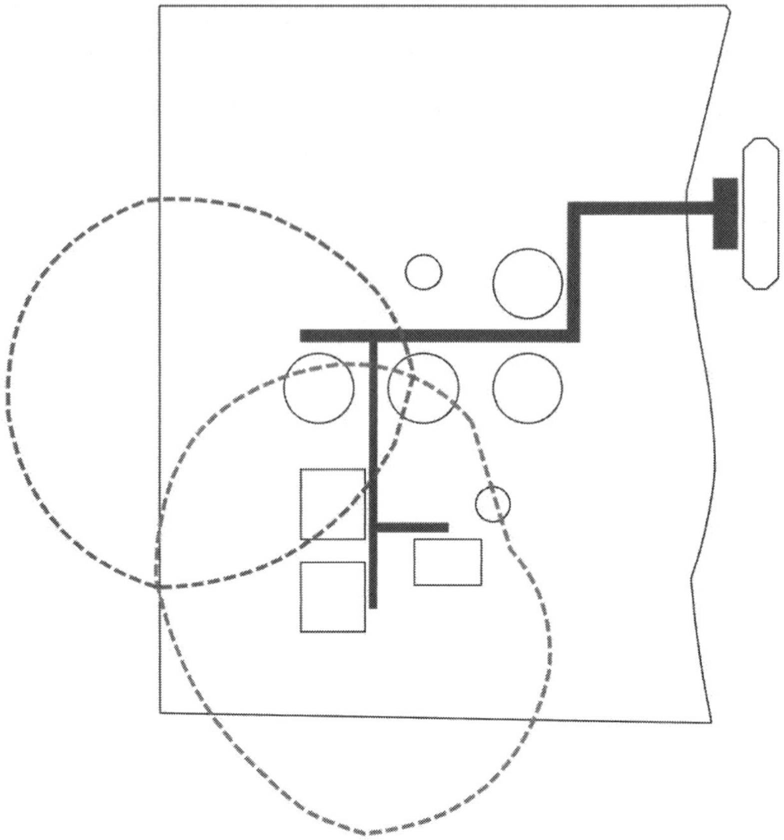

Figure 2. Example of IR contours overlaid on a facility map.

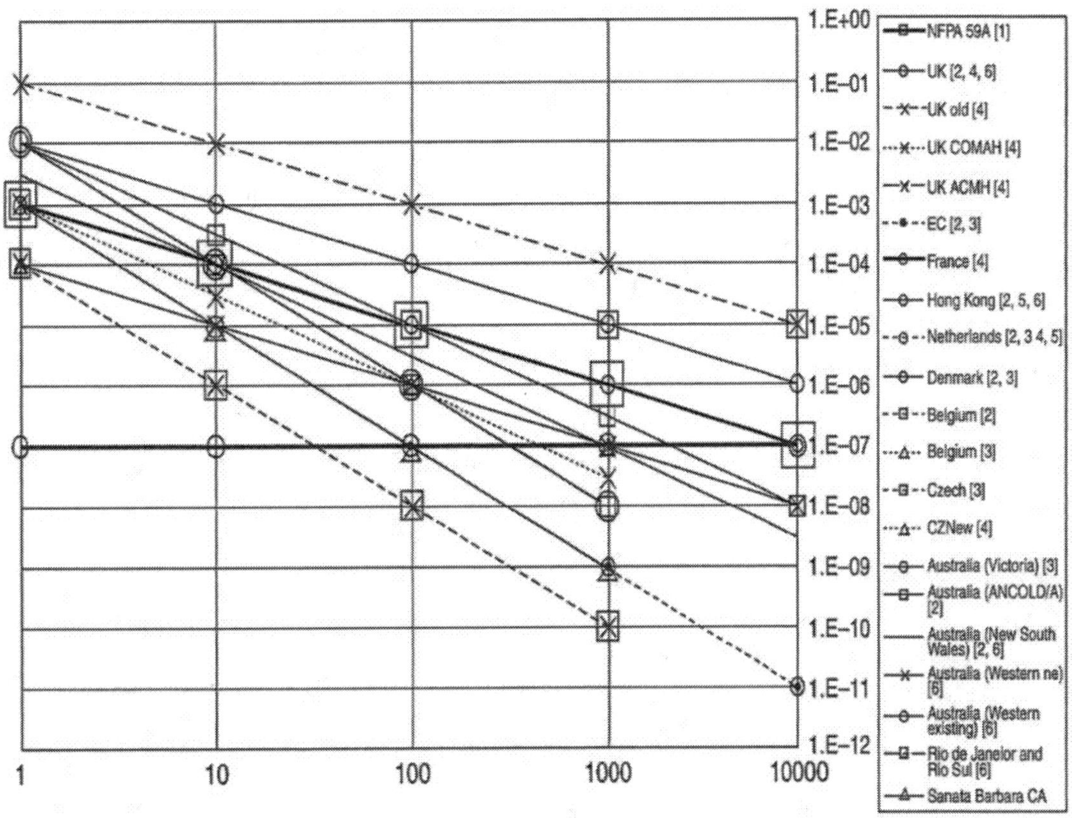

Figure 3. Societal risk tolerance criteria for different international jurisdictions. The frequency of fatality (F) is shown on the vertical axis, and the number of fatalities (N) is shown on the horizontal axis. Reproduced from NFPA59A (2013) [12].

The risk, as determined by QRA, is a result of the individual scenarios that are evaluated. However, it is conceivable that analyzing the risks for any given facility could be a never-ending task if guidance for selecting scenarios isn't provided. The first cut at reducing scenarios to a manageable and representative set is to remove duplicates and low risk processes. For example, a facility may have several identical process units; thus, they do not pose distinctly different hazard scenarios. A facility may also have specific process units that are of much lower hazard in terms of off-site populations, such as an on-site power plant or water treatment plant. These also may not be included in the selection of units to study due to the lower risk posed by them. A company or the authority having jurisdiction may have guidelines for what process units and types of LOC events must be considered that are based upon the nature of the chemical, process conditions, or type of process.

Once the types of LOC scenarios and the process units are decided, the detailed analysis of the release events is performed. The risk, as identified on the F-N or IR contours, is composed of the likelihood of the severe outcome (here taken as fatality) versus the size of the population who are affected. Determination of the scenarios requires a complementary analysis of the frequency (e.g., rate or probability) of the type of equipment failure along with the size/type of the LOC event. NFPA 59A (2013) provides a table illustrating example failure rates, see Table 1 which has been condensed to include only the aboveground piping and components [12]. It should be noted that the annual probability of failure provided for aboveground piping is not reported as a function of pipe length; thus, it is unclear whether such a probability should be applied to an individual section of piping, the piping in a process unit, or all piping at the facility. Additionally, the standard does not reference the exact source of the failure rate data and simply provides other resources containing complementary data. Although there are several industry failure rate databases, none of the databases are specific to the LNG or cryogenic liquid industries. The existing databases will be further explored in subsequent sections.

Table 1. Selected entries in NFPA 59A (2013) table describing failure rate data for various aboveground piping and components in LNG facilities [12].

Aboveground Piping and Component Failure Type	Annual Probability of Failure	Frequency of Failure (Years b/w Failures)
Rupture for Nominal Diameter < 75mm	1E-06	1,000,000
Rupture for 75mm < Nominal Diameter < 150 mm	3E-07	3,333,333
Rupture for Nominal Diameter > 150 mm	1E-07	1,000,000
Pressure Relief Valves – Outflow at Max. Rate	2E-05	50,000
Pumps – Catastrophic Failure	1E-04	10,000
Compressors with Gasket – Catastrophic Failure	1E-04	10,000
Heat Exchanger – Instantaneous Release of Entire Contents from Plate Heat Exchanger	5E-05	20,000
Transfer Equipment – Rupture of Loading/Unloading Arm	3E-08	33,333,333

4 FERC Single Accidental Leakage Source Criteria

49 CFR Part 193 does not explicitly define the set of single accidental leakage scenarios to consider even though it relies upon NFPA 59A (2001), which is a prescriptive standard. NFPA 59A (2001) §2.2.3.5 identifies single accidental leakage scenarios based solely upon the LNG storage tank and not other associated processes. These are referred to as "design spills" in the standard. The design spills are defined as the largest flow rate (of LNG) from any single line that could be pumped when the LNG storage tank's pumps are considered to be delivering the full rated capacity. This could mean that the maximum flow rate for a series of in-tank transfer pumps through an open line should be analyzed. In practice, this yields a liquid spill, which will produce a well-defined vapor cloud based on evaporation in trenches and the main impounding area.

Over the last few years, FERC has developed and refined its requirements for design spills. In 2005, FERC with affirmation from the U.S. Department of Transportation (USDOT) defined the single accidental leakage source (not including the LNG tank fill lines) to include consideration of all small diameter piping connections such as take-off lines, instrumentation taps, flanges, and valve packing [14]. FERC has employed interpretations, letters, and engineering data requests to refine the specified criteria for selecting release scenarios. In 2012, FERC provided a quantitative justification for selection of a consistent set of design spills [15, 6]. Currently, FERC requires selection of release scenarios based upon analysis of the proposed piping sizes, piping lengths, and vessels on-site. FERC has developed a basis for selecting these scenarios that is ostensibly based on LNG plant failure data. The May 7, 2012 memo stated that "based on published failure rates for LNG facilities, the rupture of a storage tank outlet line is on the order of one failure every 20,000 to 30,000 years (5E-05 to 3E-05 failures per year)" [15]. This criterion serves as the basis for selecting single accidental leakage scenarios.

In the memo, FERC provided a table containing piping and equipment failure rates, which is partially reproduced in Table 2 and Table 3. In the case of piping failure rates, the data are based on the likelihood of failure per unit length of piping. This type of failure rate value is commonly used in risk-based analyses for facility siting. It should be noted that the failure rates of piping don't exactly match the cited references (specifically the Failure Rate and Event Data (FRED) provided by the Health and Safety Executive (HSE)) [16]. Apparently, FERC has applied some judgment to create the table.

Table 2. Nominal FERC failure rates for storage tanks and piping [15].

Type of Failure	Description	FERC Nominal Failure Rate
Cryogenic Storage Tanks (General)	Rupture of Storage Tank Outlet/Withdrawal Line	3E-05 /yr (Failure Rate Criterion)
Piping (General)	Rupture at Valve	9E-06 /yr per valve
	Rupture at Expansion Joint	4E-03 /yr per expansion joint
	Failure of Gasket	3E-02 /yr per gasket

Table 3. Nominal FERC failure rates per unit length for piping [15].

Pipe Diameter	Failure Likelihood (Failures/yr/m)			
	Catastrophic Rupture	d_{hole} = 1/3 d_{pipe}	d_{hole} = 10% d_{pipe}, up to 50mm (2-inch)	d_{hole} = 25mm (1-inch)
d_{pipe} < 50mm (2-inch)	10E-07	-	-	50E-07
50mm (2-inch) < d_{pipe} < 149mm (6-inch)	5E-07	-	-	20E-07
150mm (6-inch) < d_{pipe} < 299mm (12-inch)	2E-07	4E-07	-	7E-07
300mm (12-inch) < d_{pipe} < 499mm (20-inch)	0.7E-07	2E-07	4E-07	5E-07
500mm (20-inch) < d_{pipe} < 1000mm (40-inch)	0.2E-07	1E-07	2E-07	4E-07

A flow diagram illustrating the application of the failure rate guidance is provided in Figure 4. The process area piping systems are first catalogued to identify distinct sections of piping and describe their length and diameter. The available guidance does not provide direction on distinguishing the boundaries of a given system of piping. For example, automatic closing valves or check valves may be used to define distinct sections of lines for a given analysis; however, FERC must ultimately approve the selection of system boundaries in the analysis. In past submissions to FERC, applicants have successfully argued that the system boundaries will span sections of piping with constant diameter. For example, a 3-inch diameter pipe may be considered to be a distinct section from its origination at a larger pipe and termination at a vessel.

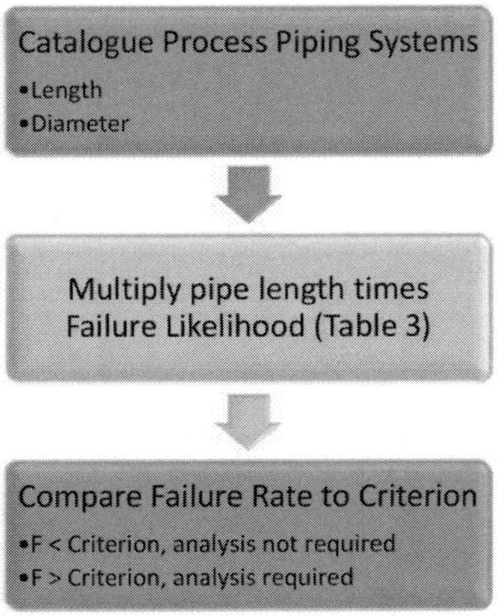

Figure 4. Flowchart for determining if a scenario should be analyzed.

Figure 5 shows an example calculation for a distinct section of LNG piping. In the example, the 328-feet long 8-inch diameter pipe passes the criterion for catastrophic rupture but not for a 1/3 diameter hole or 1-inch hole. Given the nature of the LNG and refrigeration processes, FERC requires analysis for each of the two latter cases because the characteristics of a pressurized release may change from a liquid spill to a two-phase liquid rain-out to a pressurized flashing jet depending upon the hole size and process conditions. Each case may give a different flammable vapor hazard boundary.

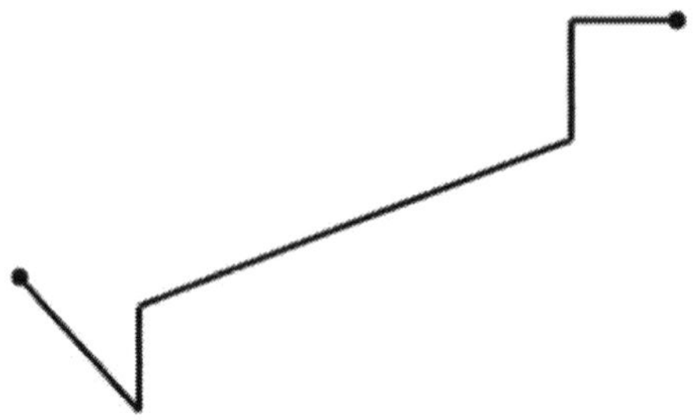

Pipe Size: d_{pipe} = 8-inch
L_{pipe} = 100 m

Catastrophic Rupture:
100 m X 2E-07 = 2E-05 < 3E-05 - **No**

1/3 Diameter Hole:
100 m X 4E-07 = 4E-05 > 3E-05 - **Yes**

25 mm (1-inch) Hole:
100 m X 7E-07 = 7E-05 > 3E-05 - **Yes**

Figure 5. Example failure rate calculation.

Typical QRA approaches provide guidance on release scenarios to be considered and strictly define the allowable consequence. The consequence may apply to both on-site and off-site properties, which intimately links the release scenario to the consequence. For example, when examining the risk of a particular release scenario using the FERC guidelines it may be determined that the ½-LFL flammable cloud approaches the property boundary but does not exceed the boundary. The risk according to FERC is, therefore, acceptable. But when examining the same type of release using a QRA approach, it may be found that although the ½-LFL does stay within the boundary, the consequence from ignition of the flammable cloud may or may not be considered acceptable depending on the nearby structures, their intended use and the intensity of the resulting fire. Additionally, the QRA approach results in consideration of all potential release scenarios given a defined failure – liquid pool, two-phase liquid rainout, or pressurized flashing jet. Each of these releases has a unique consequence associated with ignition of the flammable cloud, dependent on the size and shape of the cloud as well as the fuel characteristics. This relationship between release characteristics and consequence is not captured using the FERC prescription.

There are no explicit on-site or off-site population injury or fatality considerations mentioned in the 49CFR193 or NFPA 59A for ignition of a flammable vapor cloud. In contrast to ignition of liquid spills in impounding areas (see [8] §2.2.3.2), there is no requirement for evaluating the radiant heat flux from a natural gas or refrigerant vapor cloud flash fire. Thus, the regulatory approach does not consider the risks in the same manner as a standard QRA.

5 Discussion of Failure Rates & Databases

FERC provided a table of failure rates to use in identifying single accidental leakage scenarios and a citation list of piping failure rate studies. FERC has stated that the guidance is "based on published failure rates for LNG facilities [15]." However, the underlying studies that FERC relied upon to derive the threshold failure frequency criteria for LNG facilities for single accidental leak source size determination (i.e., 3E-05 per year to 5E-05 per year) are largely based on generic industry data not LNG-specific data [16]. Since historical data of LNG failure events are very rare, there is a lack of available historical data. The cited industry databases provide generic failure rates, which cannot be easily traced back to the original data sources used to create the databases. For the cases where evidence supporting the failure rates can be established, the studies often refer to historical data from other industries in the 1970s – 1980s. These studies and databases typically analyzed a small number of failures, systems, and equipment which may differ from modern LNG facility design. For example, one of the key sources is the U.S. Nuclear Regulatory Commission (USNRC) reactor safety study, which analyzes nuclear industry data. The underlying historic databases and studies do not necessarily reveal the effects of process conditions, management systems, maintenance and inspection practices, construction materials, or failure modes on the failure rate data.

Generic failure rate data may be called upon as the basis for estimating the likelihood of various LOC events at LNG facilities during risk assessments. Generic failure rates used in risk assessment are necessarily derived from various sources of available component failure rates and historical data. When averaging failure frequencies of similar components from other industries, this generic data is not always entirely applicable for LNG facilities while some components of LNG facilities do not have generic counterparts in other industries. For example, stainless steel fully-welded, vacuum insulated piping is not as common as Schedule 80 mild carbon steel piping in the process industries; thus, the question arises as to whether the same failure rates should apply. Since the frequencies are generic and the data is frequently used outside of the context for which it was originally collected, the data is only an approximation for any site-specific application. Even for LNG components with similar generic counterparts in other industries, such as the nuclear and chemical processing industries, the available historical data is often limited to derive failure rates with any degree of confidence. Although such historical system operating data can be used to estimate the frequency of future failures, the limited operating history of LNG facilities and the small number of incidents is insufficient to establish meaningful failure rate information.

In the absence of failure rate data specific to a particular facility or industry, the generic values determined are used as a starting point and can be further modified by expert judgment to account for site-specific and industry-specific factors. Reportedly, generic failure frequencies based on real historical data were blended with earlier sources, whose origins are now practically untraceable and difficult to validate. The current FERC-established failure frequencies for LNG facilities tend to be based on outdated, generic data sources of unknown origin, which further creates uncertainty in application and adaptation of such data to modern LNG facilities. Uncertainty also arises from imprecision of using engineering judgment rather than raw data, which subjectively chooses the component failure rate data.

Table 4 and Table 5 present the failure rate data from FERC-cited sources that FERC appeared to rely upon to derive the threshold failure frequency criteria for LNG facilities. Existing failure rate data appear in a multitude of sources including those cited by FERC and other referenced industry databases [16, 17]. A sampling of relevant data to LNG facilities are summarized below.

Table 4. Summary of selected piping system failure rate data.

Source	Failure Rate (per meter per year)	Pipe Diameter	Definition of Failure
Welker (1979)	3E-05	-	Major and minor failures
Welker (1981)	2E-05	-	Major and minor failures Combined w/Welker (1979)
Pelto (1982)	2E-07	> 3"	Pipe rupture (single component) Section = 100 ft, t = 20 days/yr
USNRC (1975)	1E-07	> 3"	Pipe rupture (single component)
FRED (2010)	0.4E-07	20" < d < 40"	Guillotine/Catastrophic Rupture
TNO Purple Book (2005)	1E-07	> 6"	Full bore rupture
Bevi (2009)	5E-07	N/A for large diameter	Rupture and leakage
COVO (1981)	0.9E-07	> 6"	Catastrophic Rupture
FERC (2012)	0.2E-07	20" < d < 40"	Catastrophic Rupture

The USNRC pipe rupture failure rate assessed is based on nuclear experience for pipes over 3-inches in diameter, where a section is defined as an average length between major discontinuities, such as valves, pumps, etc. (approximately 10 to 100 feet). Given the range for the length of a pipe section, nuclear experience also suggested a range of failure frequency.

Table 5. Summary of failure rate data for nuclear facilities.

Source	Failure Rate (per year)	Definition of Failure
Pelto (1982)	5E-05 (FERC upper bound)	Rupture of tank outlet (fault tree)
	7E-07	Pipe Rupture (excluding expansion joint)
	9E-04	Operating 365 days/yr (fault tree)
	1E-05	Operating 365 days/yr (single component)
	4E-03	Operating 365 days/yr (FERC criteria-component)
Pelto (1984)	3E-05 (FERC lower bound)	Rupture of tank outlet

FERC established a criterion of 3E-05 to 5E-05 failures per year as the tolerable limit for selecting an individual pipe segment for release scenario consequence analysis. This target value range was drawn from two papers by Pelto and Baker in 1982 and 1984 [16]. Pelto and Baker used Fault Tree Analysis to estimate failure rates for an LNG peak shaving facility. Peak shaving facilities are smaller than a typical LNG export facility, and they usually combine liquefaction with re-gasification capabilities. As the name suggests, peak shaving facilities are desired to assist with peak load periods resulting from fluctuating demand; as a result, the demand on a peak shaving facility can fluctuate considerably in a relatively short period of time. Pelto and Baker assumed that a 12-inch diameter storage tank outlet line exited through the bottom of the inner tank and the line only operated 20 days (or 480 hours) per year. Their analysis for a bottom outlet tank also called for an expansion joint and an outlet valve (see Figure 6). Figure 7 depicts the Fault Tree segment, pertinent failure rate assumptions, and calculation results from the 1982 paper. Although the 1982 paper provides the rationalization of the fault tree analysis used to reach the upper bound of FERC's failure criteria (5E-05), the change in failure rate in the 1984 paper is not explained to provide a basis for the lower bound of FERC's failure criteria (3E-05).

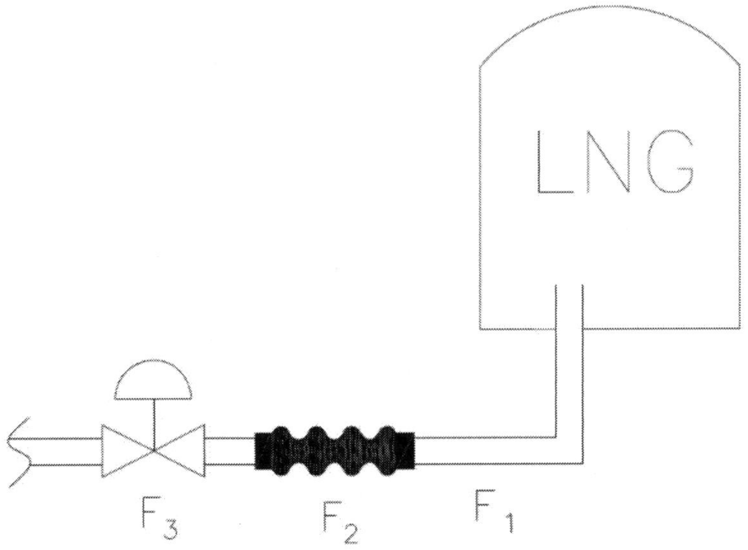

Figure 6. Sketch of system that Pelto & Baker used for the Fault Tree Analysis.

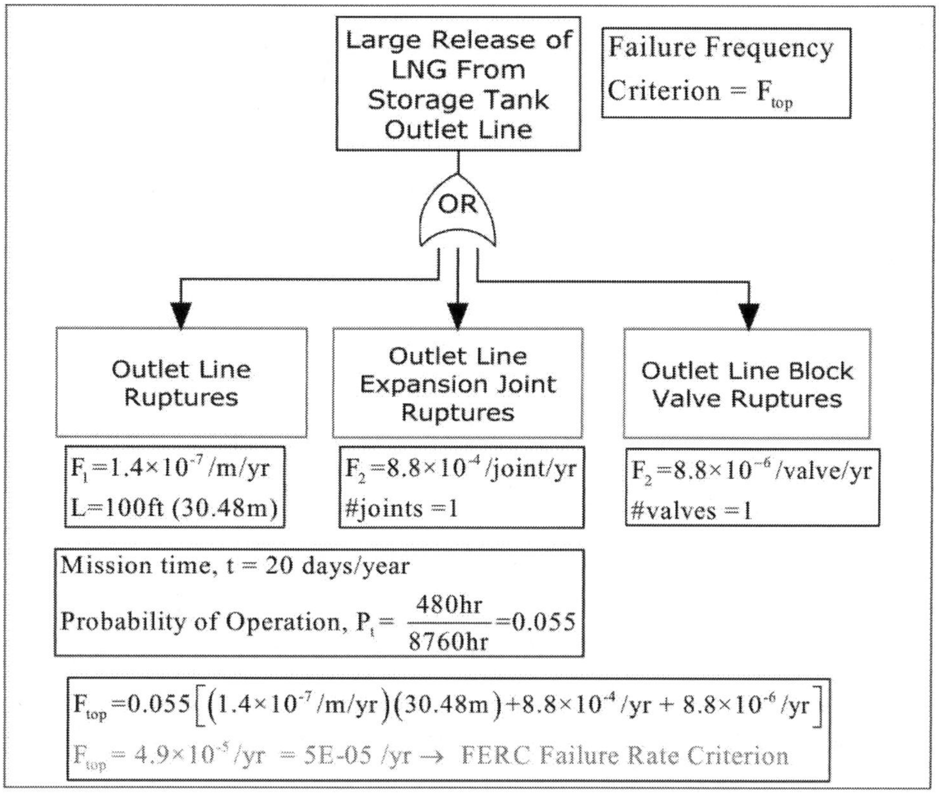

Figure 7. Fault Tree Analysis from Pelto (1982).

Given their assumptions, the authors estimated that the outlet line will rupture about $5.0\text{x}10^{-5}$ times per year, which corresponds to the upper bound of FERC's failure rate criterion. The authors couched their analysis and conclusions with the following reasonable evaluation:

> Unfortunately, the operating history of LNG tanks is too short to provide meaningful failure information. These tanks are unique to the point that generic data for other low-pressure storage tanks is not entirely applicable; however, we were able to make some comparisons with this information. Based on the information available, large releases from LNG storage tanks are estimated to occur about 10^{-5} per year.

The failure frequency is based on fault tree analysis combining the chance of three basic events: outlet line rupture, outlet line expansion joint rupture and outlet line block valve rupture. Of the basic events, the frequency of outlet expansion joint rupture of $4.8\text{x}10^{-5}$ failures per year was dominant in the calculations, which is two orders of magnitude higher than the failure frequency of either outlet line rupture or outlet valve rupture. The failure frequency of a single component outlet line rupture is $2.4\text{x}10^{-7}$ failures per year (per five sections of piping or assuming approximately 100 feet), which converts to $7.9\text{x}10^{-9}$ failures per meter of piping per year. LNG import/export terminal facilities, unlike peak shaving facilities, do not contain the relatively lower reliability expansion joints in their design.

The source database for Pelto and Baker's failure frequency of outlet line rupture was derived from the rupture of a pipe section greater than 3" in diameter according to nuclear industry historical data. The failure frequency of expansion joint rupture was derived in a study by Welker (1976), where the failure rate ranges from $1.0x10^{-5}$ to $1.0x10^{-8}$ failures per hour. This range of failure rates was also derived from the USNRC safety study similar to the failure frequency of outlet line rupture. The concluded failure rate of $1.0x10^{-7}$ failures per hour was based on the authors' engineering judgment and failure rates of similar equipment.

6 Conclusions

Several factors can impact the derivation of failure rates from historical data including inconsistent definitions of "failure", differing criteria for reporting failures, inconsistent contribution of failure modes, the use of unknown sources of historical data, and a lack of convincing support for applied corrective factors and assumptions. Given these uncertainties, it is difficult to compare failure frequency studies and average failure rates. To account for this imprecision, estimates of parameters are often developed on a "worst case" basis. The combination of the imprecision and the use of a worst case approach can result in inappropriately conservative statistics.

It can be expected that the failure rates may vary considerably depending on how the failure rate is defined. Studies of historical data often define "failure" differently. Some studies characterize a "failure" as a shutdown of operations that was not accompanied by a liquid spill or gas leak. Other studies do not clearly distinguish to what extent potential failures were taken into account, including defects found during visual inspections and non-destructive testing. It should be noted that potential failures do not necessarily lead to a release, and if appropriate action is taken to reveal such defects prior to a LOC event, future releases can be prevented.

Failure frequency data is often reported as being independent of differences in inspection and maintenance practices, quality of management, age of construction, operational parameters, and design specifications where historical data is derived from compiling all available reported incidents from various facilities regardless of site conditions and modes of failure. Since these parameters are expected to vary considerably among sites and industries, the failure rates are likewise expected to differ. Furthermore, since the derived failure frequency and studies are outdated and the data does not account for changes in design, maintenance, operation, and management practices, improvements would reduce the likelihood of LOC events for modern LNG facilities.

Studies suggest that a potential risk correction factor can be applied to the base failure rate data for site-specific cases where the magnitude of the correction factor depends on inspection and scheduled maintenance frequency, construction materials, and process conditions. Studies indicate that the failure rate may frequently vary by a factor of 10 or more between individual LNG facilities, which is not an uncommon feature of failure analysis studies. Nevertheless, failure frequency studies available need to provide a better understanding of the actual failure data and the underlying assumptions because site-specific modification factors for release frequencies should not be applied unless the relation is known to the original data to which the modification is made, which is not well-defined.

The accuracy of any QRA is dependent on the quality of the historical data and expert judgment from which the failure rates are derived, and therefore, failure data needs to be up-to-date, reliable and fully justifiable. The derived values represent outdated studies of historical data and expert judgments at a fixed point in time, which have long-term applications regarding current guidelines. The LNG industry operates in a continuously changing and evolving environment with respect to technology and regulations, where only the most recent operating history may be applicable to future considerations. Therefore, these studies need to be updated regularly using current data. The Gas Research Institute (GRI) suggests that the failure rate databases should be periodically updated at about 5-year intervals according to Welker [16].

Updating the failure frequencies would require an extensive investigation into the original data sources to determine the validity of the data and applicability to current practice. Since FERC's failure rate criteria are based on confidential information and several of the sources of failure rates are not available to the public, it is unclear how these failure frequencies were derived, what the source of the data was and what assumptions were made.

The conclusions of our study can be summarized as the following bullets:

- QRA can be a useful tool for siting new LNG export terminals.
- The inherent uncertainty that exists with published failure rate data must be considered when assigning a threshold failure rate for determining release scenarios to be considered.
- Based on analysis of the reliable failure rate data for industrial facilities, the highest degree of certainty that can be assigned to the failure rate data when applied to LNG terminals is one order-of-magnitude.
- The FERC failure rate criterion for single accidental release scenario selection of 3E-05 to 5E-05 /yr is not based on LNG industry data and does not appear to be representative of LNG system failure rates.
- Better LNG industry failure rate databases should be developed to provide useful data for risk assessment.
- Given the weaknesses and uncertainties of the underlying data, an order of magnitude approach for scenario selection criteria, e.g., 1E-04, 1E-05, or 1E-06 /yr should be developed.

7 References

[1] North American LNG Import/Export Terminals – Proposed/Potential. US Department of Energy Federal Energy Regulatory Commission, February 21, 2013. Accessed via <ferc.gov/industries/gas/indus-act/lng.asp>.

[2] Hart, RJ, DR Morrison, AF Ibarreta, HK Kytomaa. "Guidelines for Relative Hazard Ranking of Refrigerants and Siting Considerations for LNG Liquefaction Units." American Institute of Chemical Engineers, 2013 Spring National Meeting, 13th Topical Conference on Gas Utilization, San Antonio, TX, April 28 – May 2, 2013.

[3] Center for Chemical Process Safety (2003). Guidelines for Facility Siting and Layout. Center for Chemical Process Safety/AIChE.

[4] Woodward, John L., Pitblado, Robin M., "LNG Risk Based Safety; Modeling and Consequence Analysis," John Wiley & Sons, Inc., 2010.

[5] "Liquefied Natural Gas Facilities: Federal Safety Standards." Code of Federal Regulations Title 49, Pt. 193, 2011 edition.

[6] Kohout A. U.S. Regulatory Framework and Guidance for Siting Liquefied Natural Gas Facilities – A Lifecycle Approach. Proceedings of Mary Kay O'Connor Process Safety Center, 15th International Symposium, College Station, Texas, October 23-25, 2012.

[7] Draft Guidance for Filing Resource Reports 11 & 13 for LNG Facility Applications, Federal Energy Regulatory Commission, 2005. Resource Report 13: Draft Preferred Submittal Format Guidance, Federal Energy Regulatory Commission, 2006.

[8] NFPA 59A, Standard for the Production, Storage, and Handling of Liquefied Natural Gas (LNG), 2001 Edition.

[9] Guidelines for Quantitative Rise Assessment (TNO Purple Book), Committee for the Prevention of Disasters (CPR), National Institute of Public Health and Environment (RIVM), The Netherlands Organization for Applied Scientific Research (TNO), December 2005.

[10] Chapter 4: Offsite Consequence Analysis, EPA General Risk Management Program Guidance, April 2004. Accessed via <www.epa.gov>.

[11] EN-1473, Installation and equipment for liquefied natural gas – Design of onshore installations, BSI Standards (1997).

[12] NFPA 59A, Standard for the Production, Storage, and Handling of Liquefied Natural Gas (LNG), 2013 Edition.

[13] Center for Chemical Process Safety (2000). Guidelines for Chemical Process Quantitative Risk Analysis, 2nd edition. Center for Chemical Process Safety/AIChE.

[14] Interagency correspondence, FERC Document Accession Nos. 20050615-0175, 20050615-0176, and 20050615-0177, June 15, 2006.

[15] "Meeting Summary, Corpus Christi Project, Docket No. PF12-3-000, LNG Engineering Conference Call, May 7, 2012," FERC Document Accession No. 20120507-4014.

[16] FERC Citations:

- NFPA 59A, Standard for the Production, Storage, and Handling of Liquefied Natural Gas (LNG), 2009 Edition.
- Welker, J.R., Schorr, H.P., LNG Plant Experience Database, AGA Transmission Conference, New Orleans, May 21-23, 1979.
- Welker, J.R., Johnson P.W., Development of an Improved LNG Plant Failure Rate Database, Gas Research Institute, GRI-80/0093, September 1981.
- Pelto, P.J. et al, Analysis of LNG Peakshaving Facility Release Prevention Systems. PNL-4153, Pacific Northwest Laboratory, Richland, Washington, May 1982.
- Pelto, P.J., Baker, E.G., Analysis of Liquefied Natural Gas (LNG) Release Prevention Systems, PNL-SA-12278, August 1984.
- Mniszewski, K.R., Fire Protection Planning for LNG Facilities, AGA Distribution Transmission Conference, San Francisco, California, May 7-9, 1984.
- Mannan, S., Lee's Loss Prevention in the Process Industries, Third Edition, Volume 3, Appendix 14, 2005.
- Guidelines for Quantitative Rise Assessment (TNO Purple Book), Committee for the Prevention of Disasters (CPR), National Institute of Public Health and Environment (RIVM), The Netherlands Organization for Applied Scientific Research (TNO), December 2005.
- Reference Manual BEVI Risk Assessments, Version 3.2, Module C, National Institute of Public Health and Environment, January 2009.
- Failure Rate and Event Data (FRED) for use within Land Use Planning Risk Assessments, United Kingdom (UK) Health and Safety Executive (HSE), 2010.

[17] Industry Databases and References:

- Beerens et al, The Use of Generic Failure Frequencies in QRA: The quality and use of failure frequencies and how to bring them up-to-date, Journal of Hazardous Materials 130 (2006), 265-270.
- Bush, S.H., Statistics of Pressure Vessel and Piping Failures, J. Pressure Vessel Technology, Volume 110, August 1988, 225-233.
- COVO Commission. Risk analysis of six potentially hazardous industrial objects in the Rijnmond area, a pilot study. A report to the Rijnmond public authority. Schiedam: Central Environmental Control Agency Rijnmond, 1981.
- De la Mare, R.F., Y.L. Bakouros and G. Tagaras, "Understanding Pipeline Failures Using Discriminant Analysis: The North Sea Application," Reliability Engineering and System Safety, Vol.39, pp. 71-80, 1993.
- Hurst et al, A classification scheme for pipework failures to include human and sociotechnical errors and their contribution to pipework failure frequencies, Journal of Hazardous Material, Volume 26, 1991, 159-186.

- Hurst et al, Development and Application of a Risk Assessment Tool (RISKAT) in the Health and Safety Executive, Chemical Engineering Research and Design, Volume 67, July 1989.
- Hurst et al, Failure Rate and Incident Databases for Major Hazards, Proceedings of the 7th International Symposium on Loss Prevention and Safety Promotion in the Process Industry, Taormina, Italy, May 1992.
- Keeley et al, Management of the UK HSE failure rate and event data, Journal of Loss Prevention in the Process Industries 24 (2011) 237-241.
- Pasman, Hans J., History of Dutch Process Equipment Failure Frequencies and the Purple Book, Journal of Loss Prevention in the Process Industries 24 (2011) 208-213.
- Phillips, C.A.G., & Warwick, R.G., A survey of defects in pressure vessels built to high standards of construction and its relevance of nuclear primary circuit envelopes, UKAEA AHSB R162, 1968.
- Simpson, Gavin, Review of Failure Rate Data used in Risk Assessment, Department of Civil and Offshore Engineering, Heriot-Watt University, Edinburgh, September 1993.
- Smith, T.A., & Warwick, R.G., The second survey of defects in pressure vessels built to high standards of construction and its relevance to nuclear primary circuits, Safety and Reliability Directorate, SRD R30, 1974.
- Spouge J., New Generic Leak Frequencies for Process Equipment, Process Safety Progress, Vol 24, Issue 4, pp 249-257, December 2005.
- U.S. Nuclear Regulatory Commission (USNRC). Reactor Safety Study. WASH-1400, Appendix III-Failure Data, 1975.
- Welker, J.R. et al, Fire Safety Aboard LNG Vessels, Prepared for the U.S. Coast Guard under Contract No. DOT-CG-42, 355, 1976.

(43a) Learning From High Potential Near Misses in the Refining Sector

Ramesh VM, Research Scholar, Bharathiar University, 641046, India and Dr. Sakthivel R, Management Studies, Govt Arts College, 641018

Abstract: Safety is the first priority for any industry. Though steps are taken to ensure an accident-free environment, "Near Miss" is one phenomenon whose possibility of occurrence cannot be ruled out completely. What matters here is the way in which such instances are tackled preventing accidents. Moreover, these incidents can be used as an effective tool in education and training. Each near miss thus offers a free learning tool and a whole new experience with a scope to preach and practice efficient safety mechanisms. Near misses can be classified into two types- Low potential near miss and High potential near miss. All near misses whether it is of Low or high potential require in-depth analysis leading to identification of the root cause so as to avoid the recurrence of such incidents in future. The outcome of every near miss is to reach down the line so as to spread the useful information to all working in the industry. Many workers do not report "near misses" due to fear of negative actions from the management that can deprive them of their future progressions in the Industry. Steps need to be taken to overcome this fear as capturing of all near misses will definitely help the industry in bringing out innovative ideas to ensure safety. The exercise of accident prevention will bear fruits only if the effort to spread the message of usefulness of near miss reporting is spread down the line. Some of the near misses that occurred in the refinery sector are identified to be of High potential and major mishaps were averted only because of the timely action in the second of need. The ability of the concerned to do the right job at the right hour can be attributed to the training given to every employee on fire fighting which includes the analysis of near miss as well as the analysis of various other accidents happened in the other industries.I wish to share few high potential near misses which can be a learning to all working the refinery sector.

See more of this Session: Electronic Poster Session
See more of this Group/Topical: Global Congress on Process Safety

Felt Leadership:
An Essential Ingredient in Successful PSM Implementation

Brian D. Rains
DuPont Sustainable Solutions
974 Centre Road; CRP-730
Brian.D.Rains@DuPont.com

Prepared for Presentation at
American Institute of Chemical Engineers
2013 Spring Meeting
9th Global Congress on Process Safety
San Antonio, Texas
April 28 – May 1, 2013

UNPUBLISHED

Felt Leadership:
An Essential Ingredient in Successful PSM Implementation

Brian D. Rains
DuPont Sustainable Solutions
974 Centre Road; CRP-730
Brian.D.Rains@DuPont.com

Keywords: Felt Leadership, PSM, High Performing Work Organizations

Abstract

"Felt Leadership" is a term that is frequently mentioned in today's PSM conferences and informal conversations as a quality that must be demonstrated by management if PSM implementation is going to be successful. Interestingly, these two words were first introduced by the DuPont Company over 20 years ago and the context was very different then and not even directly related to safety. Felt Leadership was one of the six attributes of a "High Performing Work System." As one of these 6 attributes, it was deemed essential to achieve outstanding operations and business results through people. This paper will not only describe "felt leadership" in meaningful, practical terms but will also deliver insights as to why it is both appropriate and necessary for effective PSM implementation as well as high performing work organizations.

1. Introduction

One of the clear lessons learned from the Baker Panel Report of the BP-Texas City incident was the role of leadership in successfully implementing and sustaining excellent PSM performance. Of the Panel's ten recommendations, five spoke directly to changes that leadership needed to make. One of the panel members, Dr. Paul V. Tebo, former DuPont Vice President of Safety, Health, and Environment, has actually stated that in his opinion, all ten recommendations speak to leadership. He went on to say that the fundamental, underlying issue at BP-Texas City was "leadership, leadership, leadership."

- **RECOMMENDATION # 1 – PROCESS SAFETY LEADERSHIP ****
- **RECOMMENDATION #2 – INTEGRATED AND COMPREHENSIVE PROCESS SAFETY MANAGEMENT SYSTEM**
- **RECOMMENDATION #3 – PROCESS SAFETY KNOWLEDGE AND EXPERTISE**
- **RECOMMENDATION #4 – PROCESS SAFETY CULTURE ****

- **RECOMMENDATION #5 – CLEARLY DEFINED EXPECTATIONS AND ACCOUNTABILITY FOR PROCESS SAFETY ****

- **RECOMMENDATION #6 – SUPPORT FOR LINE MANAGEMENT ****

- **RECOMMENDATION #7 – LEADING AND LAGGING PERFORMANCE**
- **NDICATORS FOR PROCESS SAFETY**

- **RECOMMENDATION #8 – PROCESS SAFETY AUDITING**

- **RECOMMENDATION #9 – BOARD MONITORING ****

- **RECOMMENDATION #10 – INDUSTRY LEADER**

**** Direct Leadership Responsibility**

As helpful and appropriate as these corporate recommendations were at the time and continue to be, the subject of PSM leadership can, at times, need to be brought down to a very individual and even personal level. In recent presentations at PSM Conferences, the term "felt leadership" has been used to attempt to describe a brand or style of leadership that is also necessary in complex chemical processes and operations. It has been used to describe a quality of leadership that can be sensed or "felt" rather than just heard or seen. It suggests leadership with passion and authenticity and that is genuine and even humble.

2. Origin of Felt Leadership in DuPont

The term "felt leadership" was first used in the DuPont Company in a systemic way in 1991. And the context was very different. In fact, it was not used in connection with safety in any direct way.

A group of Plant Managers, of which I was one, had been formed into a body called the "Plant Managers Board." We had a very specific task and that was to find and identify best practices related to "High Performing Work Systems" or HPWS for short. High Performing Work Systems were described then as "high involvement systems of accomplishing work in which all employees have developed the capability to connect with and drive the quality of business results to be world class."

This board was made up of approximately 10 Plant Managers from all regions of the world. Some were quite young whereas others were more experienced. Some managed very large plants whereas others managed small plants. (I was personally leading a site organization of 20 DuPont employees.) Some spoke English as a second (or third) language whereas others only spoke English. Some were engineers whereas others had educational backgrounds spanning the disciplines. We came together face-to-face at least quarterly for roughly two years. And we almost always visited different operating sites to observe and analyze. Sometimes the facilities were DuPont's whereas in other cases, we visited such diverse facilities as the Honda car assembly plant in Marysville, Ohio, USA.

After a visit to a DuPont facility in rural Illinois, this group met together in a hotel conference room and attempted to synthesize our collective observations into something usable and concise; something that we could attempt to package and leverage across DuPont Operations. We filled tear sheet after tear sheet of paper until we literally covered the four walls in the room. We had seen and observed much, even too much as there was so much data.

Then, the key moment arrived. One of our members, who later rose to become DuPont's Sr. Vice President of Operations (and interestingly spoke English as a second language) took the floor and articulated in simple terms what became to be known as the "Six Attributes of HPWS." The first attribute was "Vision, Mission and Strategic Intent Known by All." The second was "Committed and Dedicated Leadership 'Felt' within the Organization", or simply "Felt Leadership."

Six Attributes of High Performing Work Systems in DuPont

1. Vision, mission, strategic intent known by all.
2. Committed and dedicated leadership "felt" within the organization.
3. Every person knows the paying customer and how their work impacts the customer.
4. Teamwork and people working in teams is part of the structure.
5. Everyone learns, and teaches. "Learn-Teach-Learn".
6. Principle driven.

Phrases that were used at the time to further clarify the meaning of Felt Leadership included:

- Creates vision and energy
- Stretches goals
- Invites positive push back
- Results and output oriented
- "Felt" sense of purpose

Further insight into what was the intended meaning of Felt Leadership was captured in a "Six Attributes of HPWS Self-Assessment Tool." The questions asked were:

- Do the leaders have a clear vision for change based on needs of all stakeholders?
- Are the leaders visible, knowledgeable and open with people?
- Do the leaders value upgrades from anywhere?
- Do the leaders continually set objectives that demand that people develop and grow skills needed to move toward the vision?
- Do the leaders engage people in a process that builds accountability, willingness, and confidence to act in an empowered way?

3. Felt Leadership and PSM

With all this as a backdrop, what really does "Felt Leadership" have to do with Process Safety Management?

A more descriptive term for what process safety is all about might be "Operational Risk Management." Operational Risk Management implies that:

- It is the operation or the materials processing activities that must be controlled.
- It is the reduction of risk to an acceptable level with the ultimate elimination of significant process incidents that is the goal.
- It is the management of these risks that is one of the organization's key responsibilities and challenges.

The degree of success that is achieved in the management of operational risks can be equated with the quality of decision making by and within the operations organization. And leadership is all about influencing and improving the quality of decision making. Thus, Felt Leadership can best be described in the context of PSM as the approach that leaders follow to insure the highest quality of decision making by the organization in the management of the existing operational risks. Decision making links Felt Leadership to PSM Performance. Felt Leadership becomes a necessary ingredient in successful PSM implementation and execution.

4. Demonstrating Felt Leadership

Within the organization of one of our global clients, inspired leaders captured the vision of their roles with respect to PSM and embraced the concept of providing Felt Leadership. But they did not know how to act differently or more effectively in the management of their significant operational risks. They understood the concept of Felt Leadership as it related to PSM but with no role models, they struggled to visualize or project themselves effectively into their/environment or culture in a uniquely additive way. They needed help and they asked for help. They wondered what supportive but probing questions they should even ask to strengthen a culture of PSM Excellence through Felt Leadership. We provided the following as initial guidance:

1. While visiting a control room and speaking to panel board operator, the focus of the discussion could be on safe operating limits for the process:
    - Does he/she know the safe operating limits, and the consequences of operation outside those limits?
    - Is he/she currently operating with any interlocks by-passed? If so, was the by-pass procedure appropriately followed?
    - How long has this situation (interlock bypassed) existed and how long is the by-pass condition expected to last?
    - How many panel alarms are there per shift? Is there any program to reduce the number of panel alarms?

- Is he/she aware of any work going on in the plant, so that he can alert people to any process emergency?

2. While visiting a plant, question the operations supervisor to see what work permits have been prepared for the day. Take time and review them, looking for the quality of the risk assessment and the precautions that have been put in place to mitigate the identified risks.

3. While visiting a plant, a review of a broad range of PSM elements can occur, such as asking operations personnel about:
   - Any incidents that have occurred in other similar operations.
   - The frequency and nature of emergency and disaster mock drills.
   - The number and nature of near misses that have occurred recently and the results of investigations into these and any other incidents that have occurred in the plant.
   - The results of recent PSM audits and what weaknesses were identified.
   - Whether there are any tests or process changes active in the plant.

4. In conversation with operators in the field, discuss the procedure governing the activity underway and how well it is written and followed.

The coaching of these leaders was actually very similar to what had occurred previously inside DuPont as we strived to strengthen our PSM Performance and the quality of Felt Leadership. Simple "Felt Leadership Cards" were printed to help Operations Leaders consider how best to demonstrate their commitment and passion for PSM.

These cards are still in use today and one is duplicated on the next two pages.

Personal "Felt Leadership" – PSM

What is my spoken and unspoken message on PSM expectations to the organization – integral to core values and business success or a "necessary evil"?

What have I done this month to review PSM metrics and performance indicators for the site?

What am I doing to reinforce Operational Discipline?

Is my organization providing adequate "rewards and recognition" for PSM accomplishments?

How is my organization performing versus annual PSM COT's and objectives?

Where do we need to enhance short and long term PSM resources and organizational capabilities?

What is the status of PSM integration improvements at recently acquired sites or businesses?

What are my direct reports doing to reinforce the value, expectations and accountabilities for PSM on an ongoing and consistent basis?

Are we providing enough support and resources to maintain site equipment and infrastructure?

What have I done personally to engage new leaders in my organization that plays key PSM roles to ensure effective MOC- Personnel?

Have I established clear measures on overall PSM program performance (not just a focus on incidents)?

The Goal is Zero

Ten Key PSM Questions for the Line Organization

1) What trends are your PSM tracking systems showing – do you have any overdue PSM action items within the work group?

2) What recent improvements or accomplishments in PSM does your organization feel good about?

3) What is your biggest concern about the area's PSM program right now and what are you doing about it?

4) What additional focus or support do you need, if any, to improve and maintain PSM?

5) What are your first party PSM auditing and incident trend analysis showing across operations?

6) What are you doing to improve your people's knowledge and capability in PSM to eliminate hazards and incidents?

7) How are you driving a consistent "no leak" and "no alarm" operating philosophy in the operating area?

8) What are you personally doing to demonstrate "felt leadership" in PSM on a routine basis and is it effective?

9) How do you ensure all your people training and written operating procedures up to date and complete?

10) Can you show the status of mechanical integrity tests and PSM work orders in the tracking system and are they up to date?

Process Safety and Risk Management Model

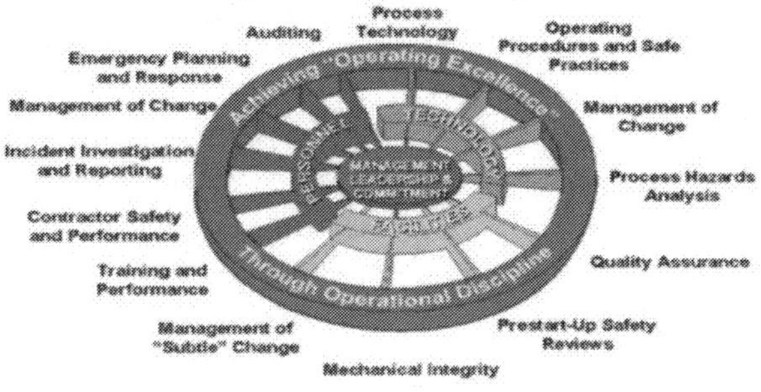

Other words have been used to describe how Felt Leadership applies to PSM Performance. The following ten basic characteristics have been drafted to summarize this thinking.

Ten Characteristics of PSM Felt Leadership

1. Be visible to the organization.
2. Be relentless about time with your people.
3. Recognize your role as a teacher/trainer.
4. Develop your own safety functioning skills and pass them along to the organization.
5. Behave and lead as you desire others to do.
6. Maintain a self-safety focus.
7. Confirm and reconfirm safety as the Number 1 VALUE.
8. Continuous emphasis and clarity around SHE expectations.
9. Show a passion for ZERO injuries, illnesses, incidents.
10. Celebrate and recognize "ZERO" successes.

Some additional guidance for each of these ten characteristics is included in the appendix.

Additional specific actions that management can take to demonstrate Felt Leadership and their commitment to safety include:

- Leading by example in following all safety rules and procedures;
- Showing a passion for safety & a genuine care for individuals;
- Leading safety meetings with direct reports;
- Including safety in all routine meetings and discussions;
- Participating in safety observations regularly;
- Reviewing/critiquing all incident investigations and personally getting involved in high potential and serious investigations;
- Meeting with your organization to express concern and reinforce the learnings from incident investigations; and
- Using company events to recognize/celebrate good safety performance by individuals and groups.

Other attributes of Felt Leadership can be demonstrated by always following this guidance:

- I always practice what I preach with reference to safety.
- I know all potential hazards in my area.
- I audit for hazards.
- I do not permit unsafe shortcuts in order to get a job completed.
- I check on the use of PPE.
- I correct unsafe practices without delay.
- I make sure workers have proper tools.
- I make sure tools are in good condition.

- I encourage reports on unsafe conditions.
- I promptly respond to workers' safety concerns.
- I investigate incidents promptly and fully.
- I encourage safe suggestions.
- I encourage housekeeping for safety.
- I consider my own health and safety.
- I regularly attend safety meetings.
- I set high standards and try to be the example for everyone else.
- I ensure that all workers are properly trained in safety.
- I stop work if necessary to correct unsafe practices.
- I remove unsafe workers from the job.
- I encourage safety communications and follow-up on corrective actions.

5. Conclusion

The successful implementation of robust Process Safety Managing Systems in complex chemical operations requires a high degree of order and high quality decision making. Leadership is fundamentally charged with creating a culture where all members of the organization respond to these needs and perform at a very high level. Although the term Felt Leadership was initially used in DuPont in the context of an essential ingredient of a High Performing Work System, its use as a descriptor of the necessary leadership qualities in Process Safety Management is both appropriate and accurate.

Felt Leadership is essential to the successful implementation and execution of Process Safety Management.

6. Appendix

Ten Characteristics of PSM Felt Leadership

1. **Be visible to the organization:**
    - "Walk the Talk".
    - Provide a positive example to your employees.

2. **Be relentless about time with your people:**
    - Find ways to spend time with employees/contractors.
    - This will encourage you people to do the same.

3. **Recognize your role as a teacher/trainer:**
    - Teach safety through your:
        - Own behavior
        - Interactions with your employees/contractors
    - You are always 'on stage'.

4. **Develop your own safety functioning skills and pass them along to the organization:**
    - Continuously develop and hone skills such as safety observations, incident investigations, improving leadership skills (Felt Leadership), etc.
    - Make sure your people have the time and resources to develop their safety skills.
    - Reinforce and support what they learn when they return from training.

5. **Behave and lead as you desire others to do:**
    - What people see you do is what they believe.
    - When you notice an unsafe situation and do something about it, you are behaving as you want all of your employees to behave.

6. **Maintain a self-safety focus:**
    - Assess yourself in terms of how 'good' or 'bad' a job you are doing.
    - Stop and think about – 'How did that go?'; 'What might I have done differently to get an even better result?'

7. **Confirm and reconfirm safety as the Number 1 VALUE:**
    - Make comments about safety as you walk throughout the work area.
    - Identify and reinforce what is being done right.
    - Identify unsafe conditions and behaviors and correct them before they cause injuries.
    - Learn what safety issues exist on the shop floor.
    - Learn how the issues are being handled.

- Determine how well employees understand and apply the core principles in their work.

8. **Continuous emphasis and clarity around SHE expectations:**
    - Establish and publicize SHE expectations.
    - Include expected safety behavior.
    - Continuously emphasize their importance by:
        - Including them in communications.
        - Weaving them into most, if not all, conversations you have with your organization.

9. **Show a passion for ZERO injuries, illnesses, incidents:**
    - Make clear your expectation that 'zero' is the goal.
    - Put safety upfront on all interactions.
    - Ensure that your safety performance expectations of your contractors is the same as your own employees.
    - Require that everyone work on a steadily improving path in our journey to ZERO.
    - Periodically focus special attention on an aspect of needed safety improvement; be specific.

10. **Celebrate and recognize "ZERO" successes:**
    - Live safety, not just preach it.
    - Recognize and reward improvement and success.
    - Recognize the achievement at all levels and make sure leadership recognizes and shows appreciation.
    - Recognize specific aspects of safety that have improved as a result of targeted efforts.

9TH GLOBAL CONGRESS ON PROCESS SAFETY

Localization of control rooms using a MINLP approach, based on accidental explosion scenarios

Jaime E. Cadena
Chemical Engineering Department, Universidad de los Andes
Cra 1 Este No 19A – 40, Bogotá, Colombia
je.cadena21@uniandes.edu.co

Lina M. Sierra, Sergio E. Rodríguez, Manuel A. Ramos, Jorge M. Gómez, Felipe Muñoz
Chemical Engineering Department, Universidad de los Andes
Cra 1 Este No 19A – 40, Bogotá, Colombia
lm.sierra443@uniandes.edu.co, se.rodriguez68@uniandes.edu.co,
ma.ramos43@uniandes.edu.co, jorgomez@uniandes.edu.co, fmunoz@uniandes.edu.co

Prepared for Presentation at
American Institute of Chemical Engineers
2013 Spring Meeting
9th Global Congress on Process Safety
San Antonio, Texas
April 28 – May 1, 2013

UNPUBLISHED

Localization of control rooms using a MINLP approach, based on accidental explosion scenarios

Jaime E. Cadena
Chemical Engineering Department, Universidad de los Andes
Cra 1 Este No 19A – 40, Bogotá, Colombia
je.cadena21@uniandes.edu.co

Lina M. Sierra, Sergio E. Rodríguez, Manuel A. Ramos, Jorge M. Gómez, Felipe Muñoz
Chemical Engineering Department, Universidad de los Andes

Keywords: facility layout, optimization problem, uncertainty reduction, Monte Carlo simulation, UVCE modeling

Abstract

Facility layout for the process industry plants is an important element in the basic engineering stage, which is fundamental in the design of the process, influencing significantly the future safety performance of the operation and its related costs. A control room serves as an operations centre where the facility can be monitored and controlled, therefore, its location and design is a priority for the safety of the process unit operation. Given that the control room is the 'core' of control and monitoring of the process, any significant damage to it resulting from a process incident might jeopardize the whole process and generate subsequent incidents. This work presents a novel methodology intended to serve as an additional tool for aiding the decision-making process regarding the location of control rooms, taking into account the risk of unconfined vapor cloud explosion (UVCE) and several passive barriers against it. The methodology is based on a mathematical model that bonds the design and localization of a control room into a computational tool, which has been developed taking into account key characteristics of a process or plant layout for explosion scenarios. The tool combines stochastic estimation of atmospheric parameters through raw Monte Carlo simulations with a mixed integer nonlinear programing (MINLP) optimization to determine the control room location. This process also helps to choose the best construction strategy for the control room among a portfolio of strategies fed to the model by the user. The final results of the methodology are the optimal location for a control room in an existing facility and the minimum costs related to the risk level to which it is exposed.

1. Introduction

A process incident such as a fire, explosion or dispersion can cause injuries and/or fatalities, but also a disruption of operations and undesired consequences, business-continuity wise. This makes determining an optimal facility layout and the location of the control room, one of the priorities for safe operations within a production plant. Given that facility layout is one of the elements determined in the early plant design stages, it also represents one of the first safety barriers in a hazard-incident representation such the one presented in Figure 1.

Figure 1: Hazard-Incident representation and location of facility layout as a safety barrier (FLB) and additional safety barriers (SB) [1].

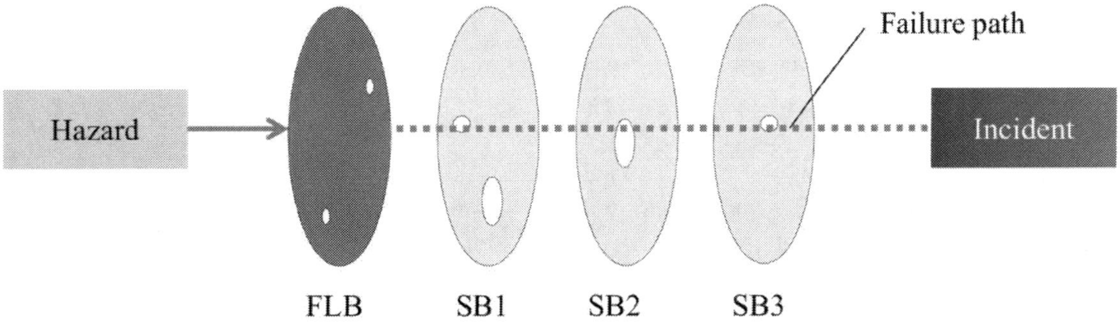

Figure 2 represents the desired function of facility layout as a safety barrier, including the adequate location of the control room. However, this desired representation is not always taken into account by the design philosophy implemented, which leads to process incidents with low probability of occurrence and high consequences, such as Flixborough (1974), Texas City (2005), Buncefield (2005), T2 Labs (2007), Puerto Rico (2009) and Amuay (2012) [2]. Although these incidents involve facility layout issues, Flixborough and Texas City involve serious damage to control rooms and its consequences has made them examples of scenarios and failures that should always be taken into account when determining the final facility layout.

Traditionally the facility layout problem has been approached with heuristic and technical recommendation published by several organizations such as the Center for Chemical Process Safety (CCPS) [3]. However, the integration of such approaches with current edge-cutting engineering tools such as non-linear optimization algorithms and solving platforms allow presenting novel approaches according to characteristics of the problem. In general, three types of problem can be identified: formulation, space treatment, unknown parameters estimation. For each type of problem a different approach is needed and this is summarized in Table 1.

Table 1. Approaches used depending on type of problem

Type of problem	Solution approach	Advantages	Disadvantages
Formulation	Linear [4]	The solution of the problem is usually simple and assures a global optimum under linearity assumptions	Does not allow the modeling of non-linear physic-chemical behavior and/or requires simplifications and strong assumptions
	Non-linear [3][5-7]	Applicable approach to complex systems with a high degree of non-linearity	Obtaining a solution requires more iterations than with a linear approach
Space treatment	Continuous [3][5-7]	Allows determining the exact location of buildings and equipment. Use of different shapes to represent footprints.	Complicated mathematical formulation. Obtaining a solution requires high computational times.
	Discrete [4]	Simple formulation and the solution	Exact locations are not possible. Shapes for

Type of problem	Solution approach	Advantages	Disadvantages
		can be obtained after a few iterations.	footprints are limited to squares o rectangles.
Unknown parameters estimation	Stochastic [3][5-7]	Better representation of real behavior.	Complex procedure that demands large amounts of input data.
	Deterministic [6]	Simple procedure based on worst-case scenario approach.	Derived distribution show few applications.

This work defines describes a methodology which identifies the sets of parameters, variables, functions and restrictions that represent the control room's location problem. Furthermore, the methodology implements a strategy that allows reducing the atmospheric conditions uncertainty when modeling UVCE and dispersion scenarios. The output of the modeling (given by overpressure) allows determining risk values, with consequences on people and infrastructure. Finally a case study is presented and the results analyzed.

2. Methodology

Although no exact and unique solution exists for a highly complex engineering problem such as facility layout, the proposed methodology (Figure 2) intends to support the decision making process in which final layouts are determined and the protection strategies are selected. The methodology is configured as a non-linear optimization of discrete control room's location, having an underlying process of stochastic parameters (atmospheric, risk) estimation. This configuration of the problem requires the following assumptions:

- A square process plant area is selected
- The process plant has only one hazardous process unit, hence only one hazards source
- The process plant area is divided into a uniform computational grid
- A control room can occupy one and only one node of the computational grid
- Historic atmospheric data must be available

The methodology begins by characterizing the hazardous process unit in terms of the possible Loss of Primary Containment (LOPC) scenarios, allowing an estimation of the liberation quantities of hazardous materials (HazMat). From the LOPC, failure trees can be depicted, allowing a classical calculation of the UVCE occurrence probability. Here the liberation of HazMat stops driving the analysis and focus is shifted towards the next driven force: atmospheric conditions. In order to take into account real atmospheric data of the determined location of the plant, it is necessary to acquire the historical atmospheric data, process it, organize it and finally fit it to probability distributions. These distributions are the input for the next step, the Monte Carlo simulation. Since the final aim is to obtain the resulting consequences of the UVCE, both dispersion and an explosion models are implemented within the Monte Carlo simulation that has a fixed number of iterations, *N*. The output of the Monte Carlo simulation is a probability distribution of overpressure-impulse measurements for each node of the grid. This constitutes a risk map, which combined with the calculation of the damages that a determined level of overpressure can cause to humans or infrastructure (i.e. control room) at a specific node, leads to the consequences calculation. The previous steps and costs information on materials, construction and installation of the control room, feed the non-linear optimization algorithm with the required data to run it and obtain an optimum control room's location.

Figure 2: Detailed steps of the proposed methodology

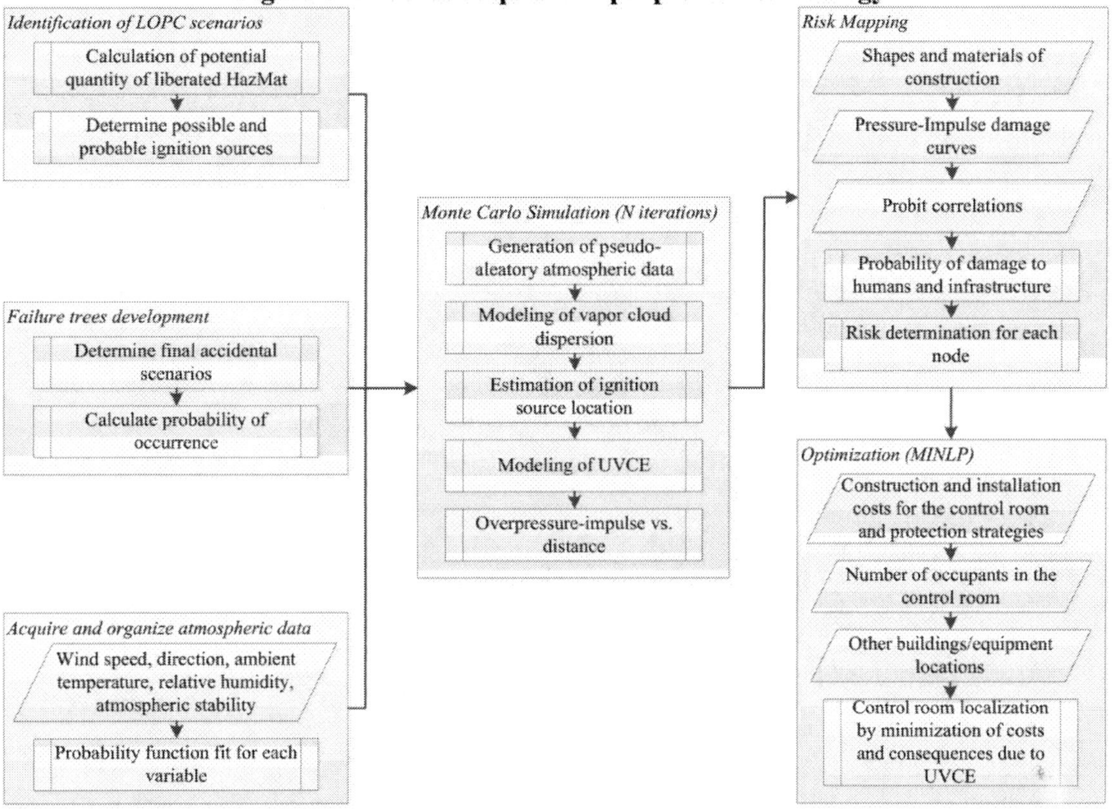

Following the general description of the methodology, it is implemented on a case study and the results are analyzed to establish advantages, disadvantages and improvement opportunities to be implemented in future work.

3. Case study: Hexane-Heptane separation unit

In order to implement the proposed methodology, a facility at Corpus Christi, TX, USA is selected as case study [13]. The hazardous process unit consists of a distillation unit and the following assumptions apply to the subsequent analysis of the system:

- The HazMat consists of 28 tons of flammable material
- The operating temperature of the distillation tower is 130°C and 160°C for top and bottoms, respectively
- The operating pressure is 4 barg
- The vapor transport pipeline has a 0.5 internal diameter and a 25 m length
- The liquid transport pipeline has a 0.15 internal diameter and a 25 m length
- The process plant area is 100 m in length and 100 m wide

- The grid consists of uniform nodes of 10 m by 10 m (see Figure 3)

The previous information and assumptions allow implementing the methodology in 7 sequential steps. The following is a summary of the most important developments in each step.

Figure 3: Representation of the grid and the nodes (labeled) in the process plant area

k01	k02	k03	k04	k05	k06	k07	k08	k09	k10
k11	k12	k13	k14	k15	k16	k17	k18	k19	k20
k21	k22	k23	k24	k25	k26	k27	k28	k29	k30
k31	k32	k33	k34	k35	k36	k37	k38	k39	k40
k41	k42	k43	k44	Hazardous process unit		k47	k48	k49	k50
k51	k52	k53	k54			k57	k58	k59	k60
k61	k62	k63	k64	k65	k66	k67	k68	k69	k70
k71	k72	k73	k74	k75	k76	k77	k78	k79	k80
k81	k82	k83	k84	k85	k86	k87	k88	k89	k90
k91	k92	k93	k94	k95	k96	k97	k98	k99	k100

3.1 Hazards identification

This step aims at identifying a representative group of scenarios that can result in a UVCE accidental event due to the LOPC of the involved HazMat. The found applicable scenarios are catastrophic failure of a vessel within the unit (column, condenser, reboiler), catastrophic failure of the vapor transport pipeline and catastrophic failure of the liquid transport pipeline.

The use of proper source models allow calculating that for both pipeline failures, the total HazMat available would be released in less than 120 seconds in vapor state at averaged atmospheric conditions. Therefore, it is assumed that any of the three mentioned scenarios has a leak/liberation time of 120 seconds. After the characterization of the LOPC, a event tree is developed to obtain the frequency of occurrence of the possible accidental events, using annual occurrence frequencies and event probabilities within the tree, obtaining the results presented in Table 2 [13].

Table 2. Annual occurrence frequencies for the identified accidental events [11]

Scenario	Probability	Frequency (year^{-1})
UVCE	0.34	6.5 x 10^{-6}
Flash Fire	0.34	5.1 x 10^{-6}
BLEVE	0.25	5.1 x 10^{-6}

3.2 Atmospheric data

A 10 year period (2002-2012) of atmospheric data with entries by the hour for the Corpus Christi area was obtained from the National Climatic Data Center (NCDC-NOAA) [19]. Wind speed & direction, ambient temperature, cloud coverage and relative humidity were extracted from the whole data set. This filtered data set allowed determining the Pasquill stability class each entry [8], using the relationship between solar elevation angle (determined an NCDC tool [17]) and cloud coverage as shown in Table 3. The final results of this step allow obtaining 109000 data registries, input for the next step.

Table 3. Relationship between solar elevation angle and cloud coverage

Cloud coverage	Solar elevation angle α		
	60°< α	35°< α<60°	15°< α<35°
Less than 4/8	Strong	Moderate	Minor
Medium altitude clouds	Moderate	Minor	Minor
Low altitude clouds	Minor	Minor	Minor

3.3 Dispersion and explosion models

The SLAB model negative floatability gases, is selected to model the dispersions due to its fit to the case study conditions. Such conditions involve a HazMat (hexane-heptane mixture) with a density higher than that of the air by a 2.5 factor. This model calculates the atmospheric dispersion solving a variant for mass, momentum, energy and species spatial and temporal averages [22]. Additionally, the SLAB model is considered a 'top hat' model, presenting the great advantage of its repetitive use within the Monte Carlo simulation. Once the model is applied, a concentration-distance profile is generated for each data set being evaluated and the ignition point is located as shown in Figure 4. An analysis of the model and the results obtained show that the accuracy of the SLAB model in the near-field is poor [22].
The TNT model is selected for the UVCE intensities calculation (overpressure and impulse) for each node of the computational grid imposed on the process plant area. Even though this model can be considered as the simplest one, its easy implementation in line with the Monte Carlo simulation and its well-known limitations [23] make it an attractive option. This model relies on an explosion efficiency factor (η) that is usually assumed to be between 1% and 10%, however a fixed 1% value is assumed since higher values presented unrealistic results in the near field when compared to typical UVCE overpressure values [4][8][25] .

Figure 4: Assumptions regarding ignition source location

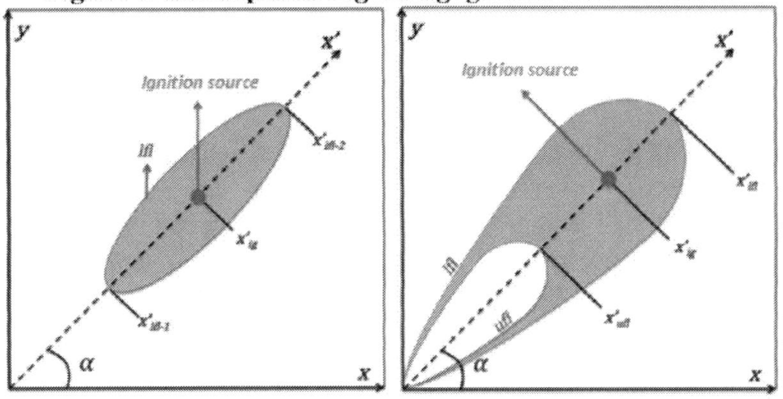

Figure 5: Example of a concentration-distance profile with its ignition source located

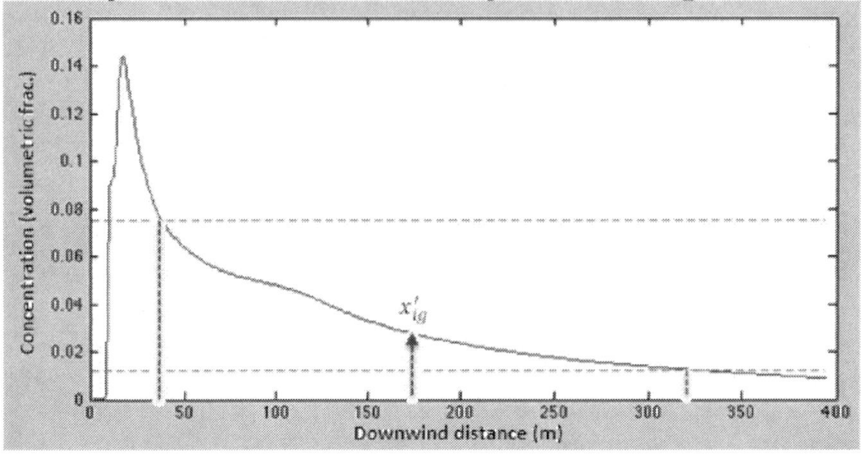

3.4 Monte Carlo simulation and risk mapping

Steps 3.1 and 3.2 are needed inputs for the Monte Carlo simulation and its output is given by the probability distributions of the fitted results of the *N* iterations performed over the dispersion and explosion models. For this case study the number of iterations was taken as 1000 as an initial guess, later proving to be a reasonable number of iterations when comparing computational time against the results' mean behavior. The results (profiles of concentration and overpressure against distance) are organized in form of histograms. From these results the expected values of overpressure (Figure 5) and impulse can be calculated for each node, necessary input to develop the risk maps. It must be noted that the expected value of the wind direction coincides with the non-symmetrical trend in the intensities maps such as Figure 5.

Risk mapping consist on assigning risk values or scores to each node of the grid taking into account structural damage to infrastructures and human affectation. The risk values for infrastructure damage depend on the type of protection strategy selected for the control room, having four different strategies as follows:

1. Steel reinforcement, walls (panels), roof and frame
2. No reinforcement, concrete reinforced walls, concrete roof and steel frame
3. Prefabricated concrete coating, concrete reinforced roof and walls, and steel frame
4. Reinforced masonry walls, light steel roof and no frame

For each construction configuration for the control room, there is an associated risk map calculated from Pressure-Impulse curves (PI) and Probit approach for human affectations. Once this risk maps are established assuming a Risk = Probability x Consequence relationship, the only step left previous to the optimization is the determination of additional parameters such as related costs of installation, construction and the location of buildings and equipment in the process plant area[4][5][12][18].

Figure 5: Overpressure map on the process plant area

35.01	47.41	29.74	29.24	28.53	26.52	24.48	22.61	21.11	20.09
30.49	30.84	30.17	31.36	32.44	31.44	28.44	25.06	22.39	20.70
28.73	29.89	33.48	40.49	48.05	47.33	38.31	29.73	24.43	21.57
28.45	31.39	40.55	64.92	109.75	109.16	63.14	37.52	26.97	22.47
28.60	32.77	48.20	109.78	Hazardous process unit		108.27	45.59	28.86	22.93
28.77	32.17	47.52	109.14			107.84	45.21	28.51	22.59
28.02	29.42	38.43	62.97	108.03	107.67	61.84	36.37	25.94	21.48
27.98	26.05	29.62	37.07	45.07	44.77	36.12	27.84	22.77	20.04
25.63	22.86	23.86	26.08	27.93	27.70	25.36	22.47	20.20	18.80
21.74	20.23	20.29	20.92	21.42	21.29	20.50	19.47	18.66	18.41

3.5 Optimization of the control room's location

The output of the 3.4 step along with some additional parameters, represent the needed input for to carry out the control room's location optimization. This step is performed through a MINLP formulation [24] of the optimization problem and a Simple Branch and Bound (SBB) solving algorithm, implemented on the GAMS software platform [25]. The optimization problem is structured as follows:

$$
\begin{array}{lll}
\min: & z = f(x,y) & \\
\text{s.t.} & gj(x,y) \leq 0 & j \in J \\
 & x \in X \subseteq R^n & \\
 & y \in \{0,1\}^m &
\end{array}
$$

The formulation can be divided into: parameters, sets, variables, constraints and objective function. A brief description of each one is presented as follows:

1. Parameters: the required parameters to solve the optimization problem are comprised of the protection strategies, PI curves, Probit parameters for human affectation, human occupancy, risk maps, construction/interconnection/installation costs, minimum separation distances, existing buildings/equipment & related location and the coordinates of each node within the computational grid (x_k, y_k).

2. Sets: the problem is divided in three sets. The first one (*F*) represents the existing buildings/equipment, the second one (*K*) the node set and the third one (*R*) the protection strategies.
3. Variables: the optimization problem solves two different kind of variables, integer and continuous. The former are given by binary variables that define the location of the control room at a specific node (*k*) and the active protection strategy (*r*). The continuous variables are given by the coordinates of the control room's location (x_c, y_c), but this is merely a formulation requirement since the possible value of these variables is given by the values of the node coordinates (x_k, y_k) specified as parameters.
4. Constraints: the following constraints assure that the obtained solution in the optimization is within physical and logical bounds:
    a. Control room's coordinates: the sum of the node-coordinate by binary variable of location product, of each node allows to determine the location of the control room.
    b. Unique location: the sum of the binary variable of location for each node must equal one.
    c. Protection strategy: the sum of all the binary variables of active protection strategy must equal one.
    d. Separation distances: the control room must be located taking into account the minimum separation distances established. This is a Pythagoras distances' formulation and finds its origins in [26].
5. Objective function: this function comprises the total cost related to the control room's obtained location and the protection strategy selected. Being a cost function, the optimization problem seeks to minimize it.

Once the elements of the optimization problem are defined, it is executed. The results show that the optimum control room's location for the case study is the *k87* node (using the 3rd protection strategy). It is important to note that this location does not correspond to the node with the lowest risk score in the grid, on the contrary, its risk score is 'medium'. The selection of this node balances the cost components involved in the control room's construction, installation and other key parameters evaluated in the objective function.

Regarding the optimum protection strategy, it is found to be the one that resist the highest levels of overpressure even when it also represents the most expensive one. This variable (protection strategy) and its influence demonstrates that it directly affects the decision making process and has a direct relationship with the plant's safety and business continuity.

4. Conclusions

This work proposes a methodology that integrates economic and safety analysis in order to determine the optimum location of a control room through the use of risk scores (risk mapping) and MINLP formulation of an optimization problem. This methodology seeks to aid the decision making process of facility layout.

The proposed methodology was applied to a previously documented case study and the results of it showed that it can effectively locate the control room within a discretized process plant area, while taking into account different protection strategies related to the control room's construction.

Nonetheless, the proposed methodology has diverse and important improvement opportunities in the treatment of the process plant space and the selection of dispersion and explosion models. These models do not allow the modeling of chemical species mixtures and do not take into account the interaction with objects, opposed to a Computational Fluid Dynamics (CFD) approach. The assumption that a building occupies a whole node tends to overestimate the real size of the facilities, inducing uncertainty to the results.

5. Perspectives

Future work should be focused in incorporating risk acceptability within the optimization formulation, as well as the interaction of the dispersion and the blast wave with the facility's buildings and equipment, e.g. through a CFD approach.

A spatially continuous approach should be developed in order to capture in the highest detail the consequences to humans due to structural damages, hence increasing the models sensibility to the number of occupants inside the control room.

Finally, it would be a considerable step forward to include dynamic risk scores related to the location of diverse hazardous process units and the risk represented by these (even if considered fixed).

6. References

[1] Sklet, S. 2006. "Safety barriers: Definition, classification, and performance". Journal of Loss Prevention in the Process Industries, 19, 494-506, http://dx.doi.org/10.1016/j.jlp.2005.12.004.

[2] Vyas, K. & Minaya, E. 2012. "Explosion at Amuay refinery in Venezuela kills 41; fire still burning". Hydrocarbon Processing. Punto Fijo, Venezuela.

[3] C. O. Diaz, R. Vázquez-Román, S. Jung, and M. S. Mannan, "A Comparison of Deterministic and Stochastic Approaches to Solve the Facility Layout Problem with Toxic Releases," in 19th European Symposium on Computer Aided Process Engineering, vol. Volume 26, Elsevier, 2009, pp. 93–98.

[4] S. Jung, D. Ng, C. D. Laird, and M. S. Mannan, "A new approach for facility siting using mapping risks on a plant grid area and optimization," Journal of Loss Prevention in the Process Industries, vol. 23, no. 6, pp. 824–830, Nov. 2010.

[5] S. Jung, D. Ng, J.-H. Lee, R. Vazquez-Roman, and M. S. Mannan, "An approach for risk reduction (methodology) based on optimizing the facility layout and siting in toxic gas release scenarios," Journal of Loss Prevention in the Process Industries, vol. 23, no. 1, pp. 139–148, Jan. 2010.

[6] C. Díaz-Ovalle, R. Vázquez-Román, and M. Sam Mannan, "An approach to solve the facility layout problem based on the worst-case scenario," Journal of Loss Prevention in the Process Industries, vol. 23, no. 3, pp. 385–392, May 2010.

[7] R. Vázquez-Román, J.-H. Lee, S. Jung, and M. S. Mannan, "Optimal facility layout under toxic release in process facilities: A stochastic approach," Computers & Chemical Engineering, vol. 34, no. 1, pp. 122–133, Jan. 2010.

[8] D. A. Crowl and J. F. Louvar, Chemical Process Safety: Fundamentals with Applications, 2nd ed. Prentice Hall, 2001.

[9] C. Cannalire, R. Clarke, B. Mazzarotta, and R. Nicoletti, "Optimizing safety of control rooms in case of flammable releases," Journal of Loss Prevention in the Process Industries, vol. 6, no. 5, pp. 313–317, Oct. 1993.

[10] Q. Baker, Explosion Risk ans Structural Damage Assestment Code (ERASDAC). Arlington, VA: 30th DoD Explosive Safety Seminar, Department of Defense Explosive Safety Board, 2002.

[11] DDESB (Department of Defense Explosive Safety Board), "Technical Paper 14." Approved Methods and Algorithms for DoD Risk-Based Explosives Siting, 2009.

[12] API, "Risk Based Inspection." Base Resource Document. API Publication 581, 2000.

[13] Center for Chemical Process Safety (CCPS), Guidelines for Chemical Process Quantitative Risk Analysis, 2nd ed. Wiley-AIChE, 1999.

[14] S. Mannan, Ed., Lees' Loss Prevention in the Process Industries, Third Edition: Hazard Identification, Assessment and Control, 3rd ed. Butterworth-Heinemann, 2005.

[15] F. Díaz Alonso, E. González Ferradás, J. Francisco Sánchez Pérez, A. Miñana Aznar, J. Ruiz Gimeno, and J. Martínez Alonso, "Characteristic overpressure–impulse–distance curves for the detonation of explosives, pyrotechnics or unstable substances," Journal of Loss Prevention in the Process Industries, vol. 19, no. 6, pp. 724–728, Nov. 2006.

[16] M. Swisdak, Simplified Kingery Airblast Calculations. New Hampshire: Naval Surface Warfare Center, 1980.

[17] Q. Baker and C. Oswald, "Vulnerability model for occupants of Blast Damaged Buildings." 34th Annual Loss Prevention Symposium, Session 3, 2000

[18] Center for Chemical Process Safety (CCPS), Guidelines for Facility Siting and Layout, 1st ed. Wiley-AIChE, 2003.

[19] National Oceanic and Atmospheric Administration, "Climate data online: map and application search," National Climatic Data Center, 24-Oct-2012. [Online]. Available: http://www.ncdc.noaa.gov/cdo-web/. [Accessed: 24-Oct-2012].

[20] National Oceanic and Atmospheric Administration, "Solar Calculation Details," National Climatic Data Center, 2009. [Online]. Available: http://www.esrl.noaa.gov/gmd/grad/solcalc/calcdetails.html. [Accessed: 02-Oct-2012].

[21] Center for Chemical Process Safety (CCPS), Guidelines for Use of Vapor Cloud Dispersion Models, 2nd ed. Wiley-AIChE, 1996.

[22] D. Enmak, User's Manual for SLAB: An Atmospheric Dispersion Model for Denser-than-Air Releases. California: University of California, 1990.

[23] Van der Bosch, C. J. H., Weterings, R. A. P. M. & Stoffen, N. C. P. v. R. d. G. 1997. Methods for the Calculation of Physical Effects: Due to Releases of Hazardous Materials (liquids and Gases): Yellow Book, Directorate-General for Social Affairs and Employment, Committee for the Prevention of Disasters.

[24] J.A. Caballero, I.E. Grossmann. "Una revisión del estado del arte en optimización, Revista Iberoamericana de Automática e Informática Industrial". RIAI, 4 (2007) 5-23.

[25] A. Brooke, D. Kendrick, A. Meeraus, R. Raman. "GAMS User's Guide". Washington DC, USA, 1998.

[26] J.E. Cadena, M.A. Ramos, J.M. Gómez, F. Munoz, "Facility layout optimization detailed research through MILP and MINLP", in AIChE - 2012 AIChE Spring Meeting and 8th Global Congress on Process Safety, Conference Proceedings , 2012.

9TH GLOBAL CONGRESS ON PROCESS SAFETY

Risk analysis for Overpressure from Pressure Vessel Burst

Seungho Jung
Air Products and Chemicals, Inc.
7201 Hamilton Blvd.
Allentown, PA 18195
Jungs2@airproducts.com

Elizabeth M. Lutostansky (lutostem@airproducts.com)
Joan M. Schork (shorkjm@airproducts.com)
Air Products and Chemicals, Inc

Prepared for Presentation at
American Institute of Chemical Engineers
2013 Spring Meeting
9th Global Congress on Process Safety
San Antonio, Texas
April 28 – May 1, 2013

UNPUBLISHED

Risk analysis on overpressure of Pressure Vessel Burst

Seungho Jung
Air Products and Chemicals, Inc.
Allentown, PA 18195
Jungs2@airproducts.com

Elizabeth M. Lutostansky, Joan M. Schork
Air Products and Chemicals, Inc

Keywords: Pressure Vessel Burst (PVB), explosion, Blast curves, geometry factor

Abstract

A pressure vessel burst (PVB) is a type of explosion that involves the burst of a pressure vessel containing gas at elevated pressure. Any vessel that is pressured or has the capacity to build high pressure before bursting can be a candidate for generating a PVB. At burst, a blast wave propagates outward from the source by the sudden expansion of a compressed gas. Fragment hazards are also present when the shell of the vessel along with attached external accessories is thrown. PVBs are possible with vessels containing any gas or mixture including inerts and flammables.

Historically, there have been incidents related to PVBs. Work that has been done to predict the overpressure impacts is covered in Guidelines for Vapor Cloud Explosion, Pressure Vessel Burst, BLEVE and Flash Fire Hazards, 2nd Edition by CCPS [1]. Moreover, as Facility Siting has received significant attention recently, Air Products has included PVBs in Facility Siting studies to safely locate occupied building.

This paper focuses on how to select an appropriate PVB model and correctly use it in calculating vulnerability inside an occupied building. It provides an overview of a simplified method for estimating the characteristics of pressure vessel bursting for practicing engineers.

1. Introduction

Explosions can be classified into two types, oxidations such as Vapor Cloud Explosions (VCE), dust explosions and condensed phase explosion and sudden expansions such as Pressure Vessel Bursting (PVB), Boiling liquid expanding vapor explosions (BLEVE), and rapid phase transitions (RPT). Pressure vessel burst (PVB) is involves a rupture of a vessel containing gas at elevated pressure. The source of the energy released in the blast is the compressed gas. A BLEVE is caused by the rupture of a vessel containing a pressurized liquid above its normal boiling point.

With a BLEVE, the liquid in the vessel undergoes a phase transition from liquid to gas. This flashing contributes to the energy released. Many engineers become confused by the difference between PVBs and BLEVEs because of their similarities. Both liquid and gas may be present in vessels undergoing PVB or BLEVE and blast effect calculations look similar. It is important to remember that in the case of PVB, liquids present in the vessel do not undergo phase transition while for BLEVEs, the primary source of energy is the phase transition. .

The scope of this paper is limited to PVBs. It presents a simplified method for obtaining PVB overpressure results for a building siting study.

2. Methods for predicting blast effects from pressure vessel bursts

A blast curve method is highly recommended because it is more easily implemented than doing numerical simulations. There is an outstanding set of blast curves based on a systematic one-dimensional numerical study by Tang et al. [2]. The Baker-Tang curves assume all of the energy is employed to make a blast effect rather than to be used into the fragments as kinetic energy. These curves are for spherical, free air burst which do not include ground reflection.

Figure 1. Positive overpressure curves for various vessel pressures (courtesy copied from Guidelines for Vapor Cloud Explosion, Pressure Vessel Burst, BLEVE and Flash Fire Hazards, 2nd Edition by CCPS [1])

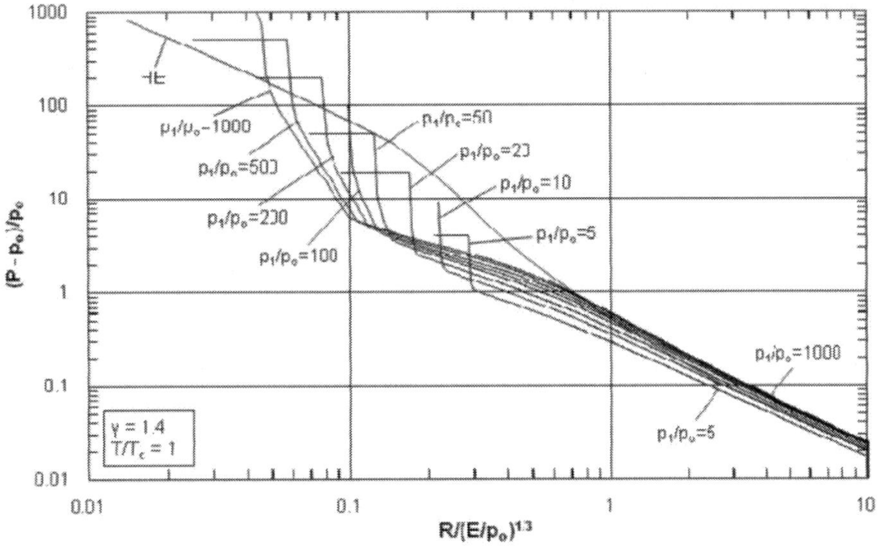

There were not much experimental data for PV burst but Table 1 shows the set of experimental data at relatively lower operating pressure. After early 1980s, there appeared many experimental studies that reported PV burst overpressure measurements but they cover different vessel shapes in different locations such as ground and air [3]. Therefore only data points with spherical free air burst are chosen to compare and validate against the Baker-Tang curves.

Table 1. Summary of literature data

Author	Year	Shape of vessel	Type of burst	Substances	Operating pressure ratio to atmospheric pressure (p1/p0)	Overpressure range
BAKER	1977	Spherical	Free air burst	Air, Argon	6	2.7~50 kPa
BAKER	1977	Spherical	Free air Burst	Freon 12	18 ~ 22	0.73~7.3 kPa

Figure 2 shows good validations against two experimental data sets qualitatively. As the peak overpressure line with the ratio, p1/p0=20 is higher than that with the ratio, p1/p0=5, the experiment data points with Air, Argon are distributed well above the data point with Freon 12.

Figure 2. Validation of Tang's curves on lower pressure cases against experimental data

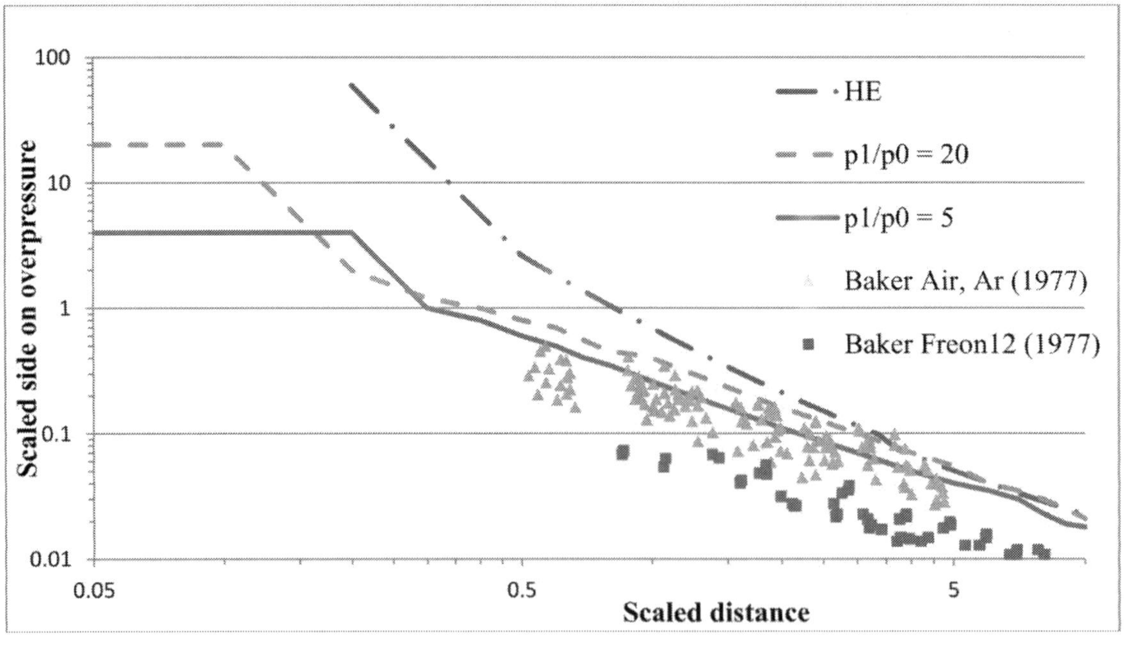

In the far-field, the HE line is close enough to all the other pressure ratio curves as seen in Figure 1. Though a recommendation was made to find the scaled side on overpressure based on p1/p0 ratio, it is quite helpful and faster to use the HE for the far-field further than scaled distance ($\bar{R}$) > 1. 6th order polynomial equation can better fit the HE curve in Figure 1 and suggested as

$$log\overline{P_s} = c_0 + c_1 log\bar{R} + c_2(log\bar{R})^2 + c_3(log\bar{R})^3 + c_4(log\bar{R})^4 + c_5(log\bar{R})^5 + c_6(log\bar{R})^6 \qquad \text{[Eq. 1]}$$

where:

$\{c_0, c_1, c_2, c_3, c_4, c_5, c_6\} = \{-0.4102, -1.8282, 0.3661, 0.1122, -0.0830, -0.0042, 0.0046\}$

$\overline{P_s}$ is the dimensionless overpressure, or the overpressure scaled by ambient pressure (Multiply by 14.696 to convert to psig).

3. General calculating procedure for Pressure Vessel Bursts

3.1 Calculate PV energy

The internal energy U of a system is the sum of the kinetic, potential and intermolecular energies of all the molecules in the system. For an ideal gas it is

$$U = \frac{pV}{\gamma - 1} \qquad \text{[Eq. 2]}$$

where p is the absolute pressure of the gas, V is the volume of the gas, and γ is the ratio of specific heats of the gas in the system, i.e. C_p/C_v. Therefore, the Brode's PV burst energy is

$$E_{ex,brode} = \frac{(p_1 - p_0)V_1}{\gamma_1 - 1} \qquad \text{[Eq. 3]}$$

where the subscript 1 refers to initial state and the subscript 0 refers to the ambient conditions. For ground burst, the effective energy is doubled to account for the ground reflection as

$$E_{eff} = \frac{2(p_1 - p_0)V_1}{\gamma_1 - 1} \qquad \text{[Eq. 4]}$$

3.2 Calculate scaled distance, $\bar{R}$

Once the explosion energy of a PV burst is estimated, the scaled distance, $\bar{R}$ can be obtained from

$$\bar{R} = R\left(\frac{P_0}{E_{eff}}\right)^{1/3} \qquad \text{[Eq. 5]}$$

where R is the distance from the blast at which blast parameters are to be determined.

3.3 Determine scaled positive side on overpressure, $\overline{P_s}$

To use this graphical method, select the overpressure curve in Figure 1 that corresponds to the vessel burst pressure ratio p1/p0. Read the $\overline{P_s}$ = (p1-p0)/p0 value. For simplicity, if $\bar{R}$ is larger than 1, the equation correlated with the Highly Explosive (HE) line can be used

3.4 Adjust scaled positive side on overpressure, $\overline{P_s}$

There are two adjustment factors; one is for temperature, the other is for geometry. Higher temperature increases the overpressure near the vessel but the effect decreases with distance away from the vessel. If the target is far enough from the vessel, this adjustment effect is negligible and the use of ambient temperature in Figure 1 can be adequate.
Geometry adjustment factors are used with of cylindrical vessels and slightly elevated spherical vessels. The multiplier factors shown in Table 2, for$\overline{P_s}$ initially suggested by Baker et al. [4], are widely used.

Table 2. Factors for geometrical effect

Multiplication factors on $\overline{P_s}$ for slightly elevated spherical vessel	
$\bar{R} < 1$	2
$\bar{R} > 1$	1.1
Multiplication factors on $\overline{P_s}$ for cylindrical vessel	
$\bar{R} < 0.3$	4
$0.3 < \bar{R} < 1.6$	1.6
$1.6 < \bar{R} < 3.5$	1.6
$3.5 < \bar{R}$	1.4

However, these factors were made based upon high explosives experiments, and it was found by Geng et al. and written in the CCPS book [1] as the factors are not directly applicable to PVBs. It was suggested using a group of adjustment factor specifically for PVBs with bursting pressure ranges from 10 to 100 for both the cylindrical and elevated spherical PVBs. Note the factors vary continuously with scaled distance, $\bar{R}$. For simplicity, it is recommended that a factor of 1.8 be used for cylindrical vessels and a factor of 2.0 for slightly elevated spherical vessels as these are the maximum values given by Geng.

3.5 Calculate Vulnerability

Given the adjusted $\overline{P_s}$ the vulnerability of people inside a building is next determined. A vulnerability curve can be generated as Figure 3 using information from Guidance for the Location and Design of Occupied Buildings on Chemical Manufacturing Sites, published by the UK's Chemical Industries Association (CIA). The overpressure-vulnerability relationship for type 3, typical domestic building, is presented in Figure 3.

Figure 3. Vulnerability curve against overpressure (courtesy coped from Marx, et al. [5])

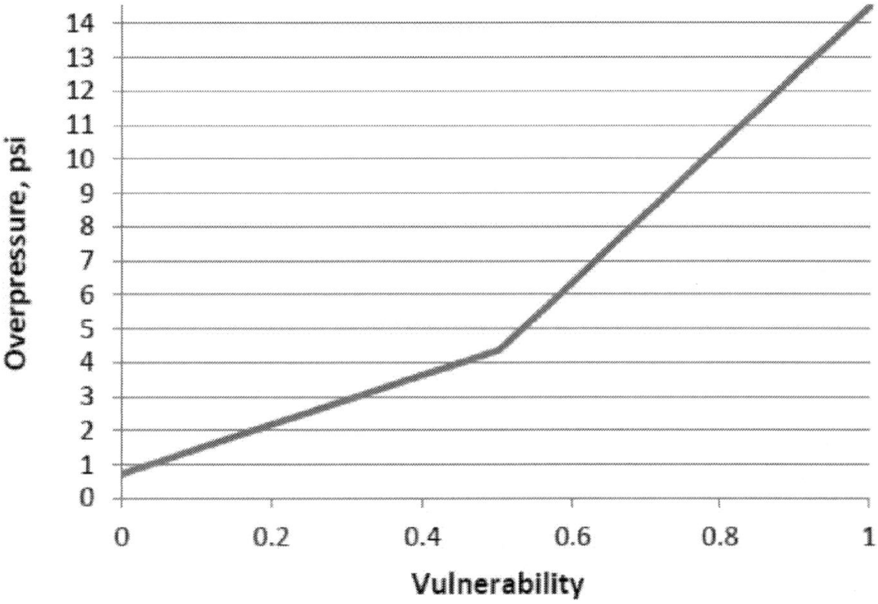

4. Example calculation case study

A $100m^3$-cylindrical pressure vessel at grade was under pressure test with nitrogen gas at 20 barg and it ruptured. There was a control room located 50 me from the vessel. What would be the side on overpressure at the control room? And what would be the vulnerability inside the control room if the building type is CIA type 3.
Table 3 presents the details of PVB calculation and recommendations.

Table 3. PVB calculation flowchart

Procedure	Details of calculation	Note
Collect data	Internal pressure is 21 bara The ratio of specific heats of nitrogen is 1.40 Ground burst, ambient pressure is 1 bara	
Calculate PV burst energy	$E_{ex,brode} = \frac{(p_1 - p_0)V_1}{\gamma_1 - 1} = \frac{(21-1)100}{1.4-1} = 500\ MJ$	
Calculate effective energy	$E_{eff} = \frac{2(p_1 - p_0)V_1}{\gamma_1 - 1} = 1\ GJ$	Multiply 2 if it is ground level explosion
Calculate scaled distance, $\bar{R}$	$\bar{R} = R\left(\frac{P_0}{E_{eff}}\right)^{1/3} = 50\left(\frac{10^5\ Pa}{10^9\ J}\right)^{1/3} = 2.32$	R = 50 m

Read or calculate scaled positive side on overpressure, $\overline{P_s}$	$\overline{P_s} = 0.13$	$\bar{R}$ is larger than 1, equation (1) is proper to use
Adjust scaled positive side on overpressure, $\overline{P_s}$	Multiply 1.8 for a cylindrical vessel geometry factor $\overline{P_s} = 1.8\ \times 0.13 = 0.23\ barg$	
Vulnerability inside building	40 % from Figure 3.	CIA type 3 building
Recommendation	If the consequences are judged to be unacceptable, the blast-pressure effects shall be mitigated by the location of the equipment within suitable shields, walls, or barriers. Or the personnel in the control room should be evacuated when the test is ongoing.	

5. Conclusion

Young engineers are often told that BLEVEs are one of the highest impact phenomena in the process industry, even more so than PVBs. PVBs tend to have lower energy releases involved compared to BLEVEs since PVBs do not involve energy from liquid flashing. However, PVBs cannot be ignored since there can be many pressured gas vessels and pressure tests on a gas industry site. This paper provides an overview of method for estimating the characteristics of pressure vessel bursting for practicing engineers. It has been presented to include advanced modeling technology on PVBs and how to properly apply a simplification in adapting blast curves or equation and geometry effects.

6. References

[1] "Guidelines for Vapor Cloud Explosion, Pressure Vessel Burst, BLEVE and Flash Fire Hazards", 2nd ed., CCPS

[2] Tang, M. J., Cao, C. Y., Baker, Q. A., "Calculation of Blast Effects from Bursting Vessels", 1996 AIChE National Meeting

[3] Chen, Fei, Lutostansky, Elizabeth M., Schork, Joan M., "Improvement of BLEVE Overpressure Method Using Historical Data", 2011 CCPS conference

[4] W.E. Baker, J.J. Kulesz, R.E. Richer, R.L. Bessey, P.S. Westine, V.B. Parr, G.A. Oldham, "Workbook for predicting pressure wave and fragment effects of exploding propellant tanks and gas storage vessels", NASA CR-134906, NASA Scientific and Technical Information Office, Washington, 1975 and 1977

[5] Marx, Jeffery D., "API RP 752 Building Siting Using F-N curves", 2013 MKOPSC international symposium

Effects of Process Variables on Peak Relief Rates Estimated by Dynamic Simulation for a Multiple Distillation Column System

John Wilkins
Staff Process Consultant
Smith & Burgess LLC
5700 Northwest Central Drive Ste. 301
Houston, Texas 77092
John.Wilkins@smithburgess.com

Dustin Smith
Vice President of Operations
Smith & Burgess LLC
Dustin.Smith@smithburgess.com

Prepared for Presentation at
American Institute of Chemical Engineers
2013 Spring Meeting
9th Global Congress on Process Safety
San Antonio, Texas
April 28 – May 1, 2013

UNPUBLISHED

Effects of Process Variables on Peak Relief Rates Estimated by Dynamic Simulation for a Multiple Distillation Column System

John Wilkins
Staff Process Consultant
Smith & Burgess LLC
5700 Northwest Central Drive Ste. 301
Houston, Texas 77092
John.Wilkins@smithburgess.com

Dustin Smith
Vice President of Operations
Smith & Burgess LLC
Dustin.Smith@smithburgess.com

Keywords: dynamic simulation, VMG Sim, distillation columns, sensitivity analysis, pressure relief valves

Abstract

Process engineers and designers have recently turned to using dynamic analysis as a more realistic method for sizing relief systems. The primary goal of using dynamic analysis is to ensure that modifications are truly required instead of making unnecessary and costly upgrades of relief systems based on overly conservative calculations. In work previously presented by the authors, it has been shown that, for certain overpressure scenarios, the dynamic relief rates predicted by dynamic simulation are significantly affected by certain operating conditions [1].

This paper shows the effects of process variables on the relief loads estimated by dynamic simulation for multiple columns. This paper will also show how these effects would impact the flare load for a unit with multiple columns. The multiple column system is based on a system the authors have encountered at a refinery. The sensitivity analysis shows that changes in the column liquid levels significantly affect the initial relief times, relief loads, and relief durations.

1. Introduction

To stay in compliance with industry, regulatory, and typically corporate guidelines, it is important for a plant to conduct and document analyses of their relief systems. It also allows for plant engineers and designers to see the effects of their process on existing or proposed relief systems. Relief system analyses traditionally use steady-state equations to determine the relief rate used to size relief devices. For complex systems, these equations are based on conservative simplifying assumptions. This usually results in a larger relief rate that requires a larger relief device. The drawback to using traditional methods, when analyzing an existing system, is that they can overstate the required relief rate. Sometimes this results in the recommendation of costly relief systems upgrades where only minor modifications are needed.

Dynamic analysis of complex systems involves creating a simulation that calculates changes to the system over time. This analysis can be used to better estimate the relief requirements to protect from overpressure for a specific relief scenario. The primary goal of dynamic relief system analysis is to better understand how the system will respond to the upset scenario. An additional benefit is that the detailed analysis can be used to support making limited changes in lieu of more extensive modifications. The fundamental differences between these two methods are important to understand in order to make sure they are both being used to determine relief rates that are conservative.

Equation 1 shows a simplistic steady-state equation that can be used to determine the relief rate through a relief valve installed on top of a distillation column that has lost cooling, feed, and reflux, but has continued heat input (a typical total power failure scenario).

$$\dot{M}_{relief} = \frac{\dot{Q}_{reb}}{\Delta H_{vap}} \qquad \text{[Eq. 1]}$$

In order to use the above equation, the right hand side of the equation would be evaluated at relief conditions, which may vary from normal operating conditions [2]. Assumptions such as the composition of the fluid on the process side of the reboiler can have effects such as a lower heat of vaporization or a larger reboiler duty that will increase the predicted relief rate.

Equation 2 is an example of how the above equation would be represented in a dynamic environment. It shows the same equation as functions of time as well as initial process conditions such as pressure, temperature, etc.

$$\dot{M}_{relief}(t, P_i, T_i, etc) = \frac{\dot{Q}_{reb}(t, P_i, T_i, etc)}{\Delta H_{vap}(t, P_i, T_i, etc)} \qquad \text{[Eq. 2]}$$

Dynamic simulations calculate the changes in the system based on incremental time steps from an initial starting point. Therefore, the selection of initial conditions is important and will affect the results [1]. These initial conditions are typically based on operating conditions that represent the normal operations of the facility. Operating conditions are controlled in ranges, and depending on the specific item, there may be no such thing as a normal condition. Liquid levels, temperatures, pressures, and reflux rates for distillation columns can vary during operation and can change due to upsets elsewhere in the plant. This is important to consider when global relief scenarios like power failure and cooling failure affect the entire plant, as these variables are all essentially assumptions.

API 521 Section 5.22[3] states that if dynamic simulations are used for column relief system design, the model must be conservative with respect to the maximum relief load. It goes on to state that the assumptions used for the simulation *shall* be checked by sensitivity analyses to assess the impact on the column relief rate. A sensitivity analysis is the study of how the uncertainty of the input variables affects the output of a mathematical model. Therefore, the input variables of a dynamic simulation must be checked in order for the designer to ensure that the simulation is conservative.

2. Methodology

2.1 Setting up the dynamics file

A steady-state model containing 2 debutanizers, a depropanizer, a gasoline splitter, and 12 PSVs was created using VMG Sim. The flow diagram of this system is displayed in Figure 1.

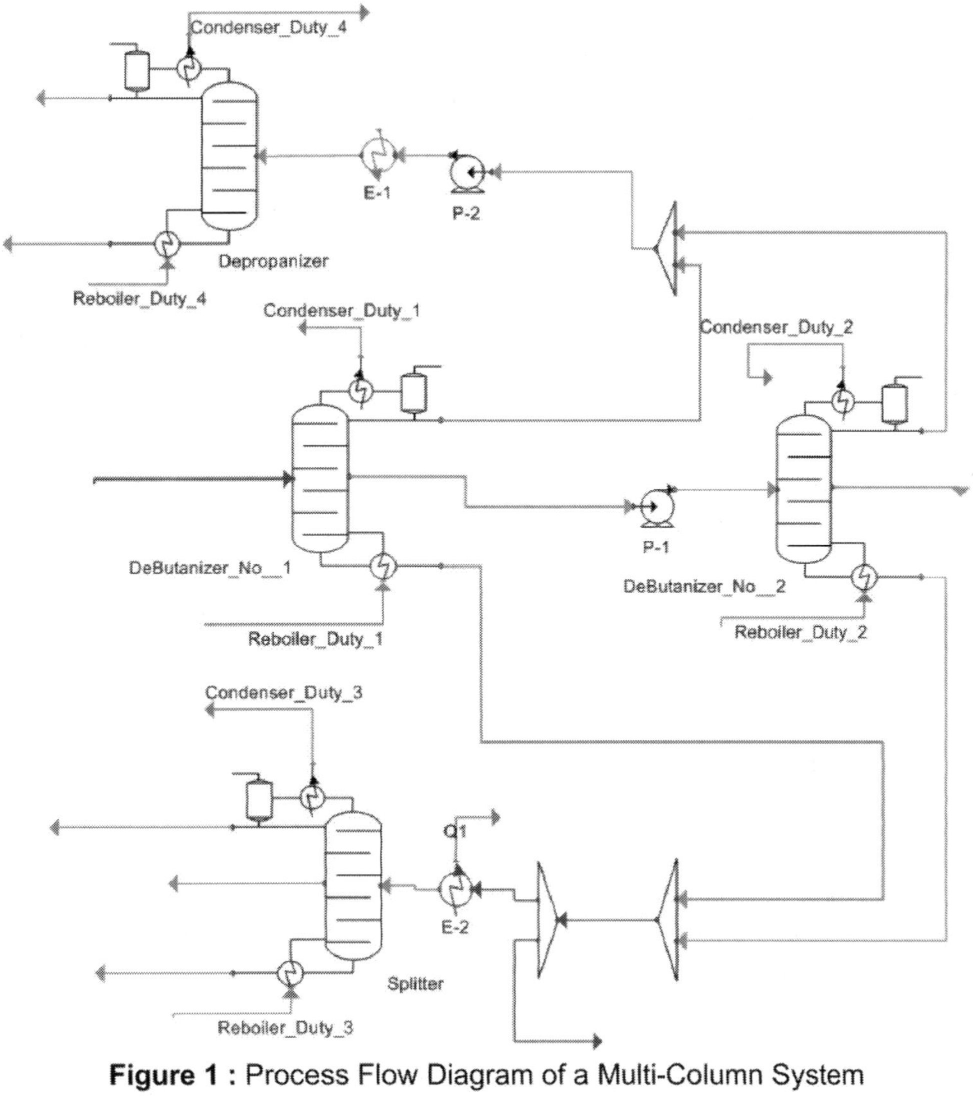

Figure 1 : Process Flow Diagram of a Multi-Column System

The steady-state model was used to create a dynamic model that uses 21 controllers set to match the operating parameters of the equipment. Controllers are used in dynamic simulations to achieve specified flowrates, pressures, temperature, and other physical conditions. The controller attempts to achieve the user's specification by opening or closing a valve located on a material or energy stream. This is different from a steady-state model where flowrates, duties, and physical properties are typically specified or calculated. The action of the controller is based on the tuning parameters provided by the user.

Twelve PSVs were added to the dynamic model to simulate relief. Table 1 displays information on these relief valves.

Table 1: Summary of PSV Information

Name	Location	PRV Size	Set Pressure (psig)
PSV-1	DeButanizer No. 1 vapor overhead	6Q8	184
PSV-2	DeButanizer No. 1 vapor overhead	6Q8	190
PSV-3	DeButanizer No. 1 accumulator	3L4	180
PSV-4	DeButanizer No. 2 vapor overhead	6Q8	190
PSV-5	DeButanizer No. 2 accumulator	1.5H3	190
PSV-6	Splitter vapor overhead	6R8	30
PSV-7	Splitter vapor overhead	6R8	30
PSV-8	Splitter vapor overhead	6R8	30
PSV-9	Splitter vapor overhead	6R8	30
PSV-10	DePropanizer vapor overhead	6Q8	290
PSV-11	DePropanizer accumulator	3K4	274
PSV-12	DePropanizer accumulator	3K4	274

In order to perform the sensitivity analysis, different column liquid levels were used as operating conditions. Setpoints of the respective level controllers for the columns were adjusted to achieve the desired liquid levels. Three simulations were created for a low liquid level of 20%, a medium liquid level of 40%, and a high liquid level of 60%. These percentages represent the fraction of liquid height in the lower section of the column that does not contain trays.

2.2 *Initiating relief*

Cooling water failure was chosen as the overpressure scenario to review as it affects all of the four columns simultaneously. Total loss of cooling water is expected to result in the loss of overhead condenser duties, which may lead to overpressure in the column due to excess vaporization. This was accomplished by specifying duties of zero for all condensers. It was assumed that all of the condensers lost cooling at the same time.

All controllers in the dynamics model were turned off (set to manual) to avoid taking credit for positive controller action. This was done in accordance with API 521 Section 4.2.4[3]. Changing the controller mode from "automatic" to "off" causes the controlled valve to hold its position during relief.

For each liquid level, the integrator (essentially an electronic strip chart recorder in VMG Sim) was turned on and recorded the pressure in the columns as it slowly increased. The overhead PSVs opened at their respective set pressures and flow data was collected. The sum of the mass flow rates relieved by the PSVs was recorded in order to view the total load into the flare.

3. Results

Graphs of dynamic relief for the three liquid levels are displayed below. PSV-5 was not graphed as it does not contribute a significant relief load. The initial time (t = 0), is the time at which the cooling to the tower was stopped and the overpressure scenario began. Note that the change in liquid levels shifts the magnitude of the relief rate and the peak point in time for some of the towers.

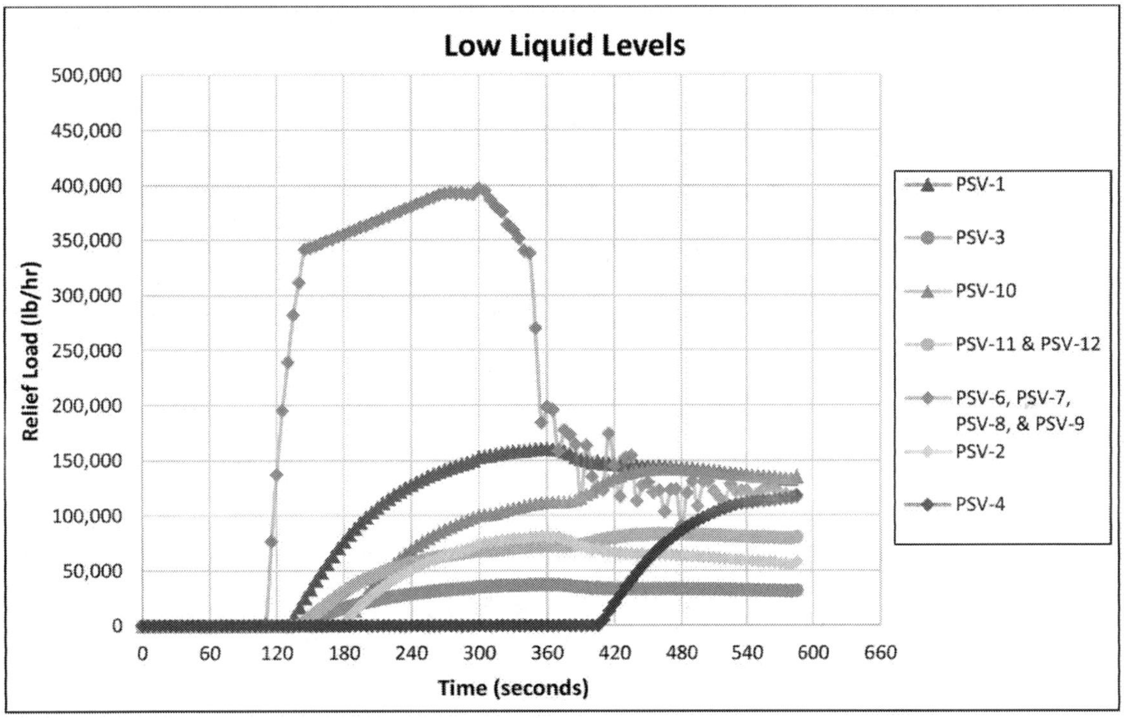

Figure 2: Cooling failure relief rates with low liquid levels

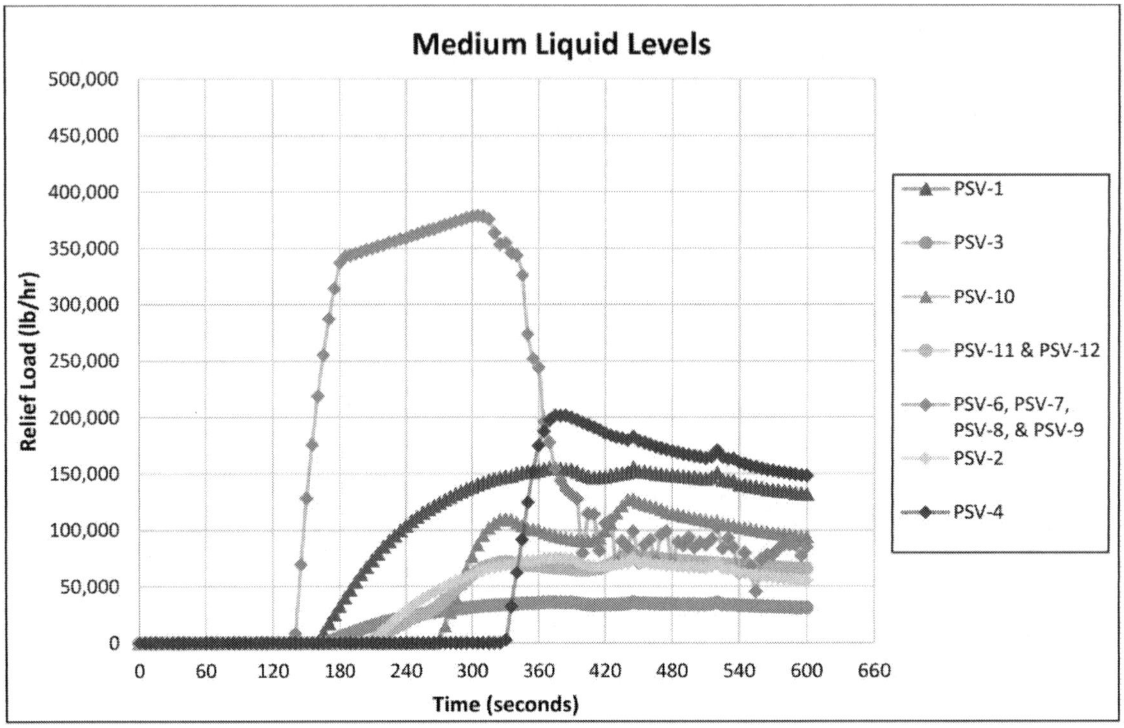

Figure 3: Cooling failure relief rates with medium liquid levels

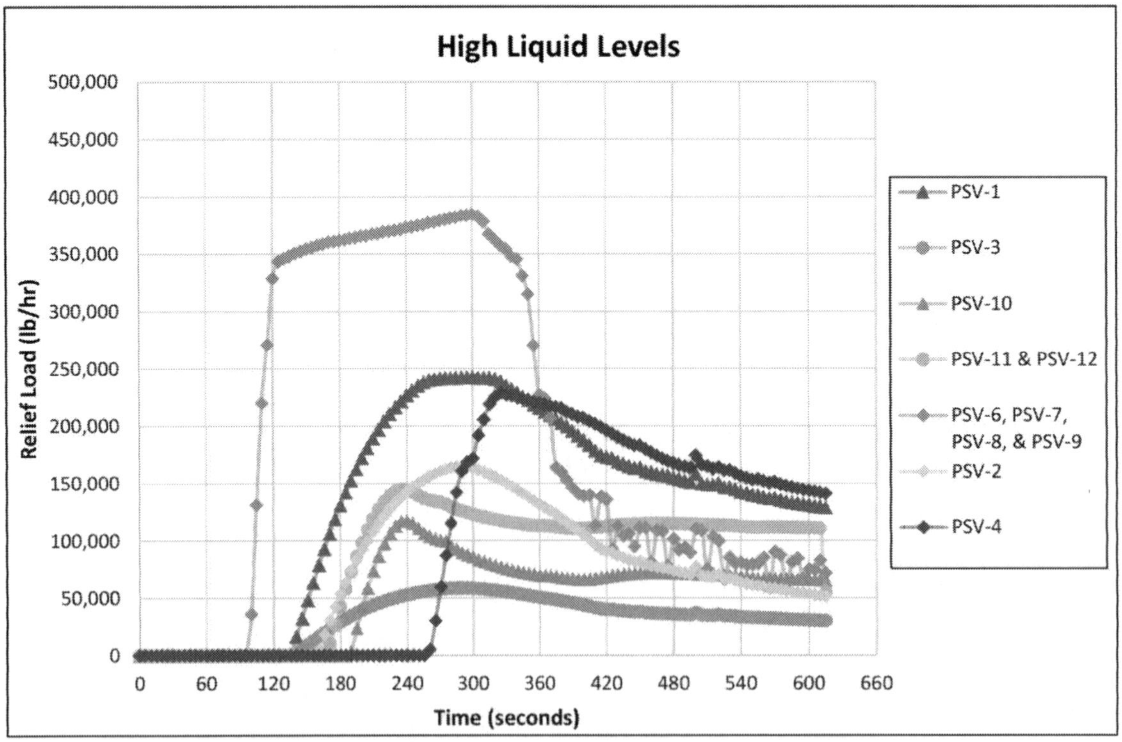

Figure 4: Cooling failure relief rates with high liquid levels

Figure 5 shows the combined flare loads for the three liquid levels. The curves were made by adding the relief loads from the previous graphs over the same time frame. The combined load represents the load that would enter the flare header. The horizontal line represents the steady-state relief load for the four columns estimated by traditional methods.

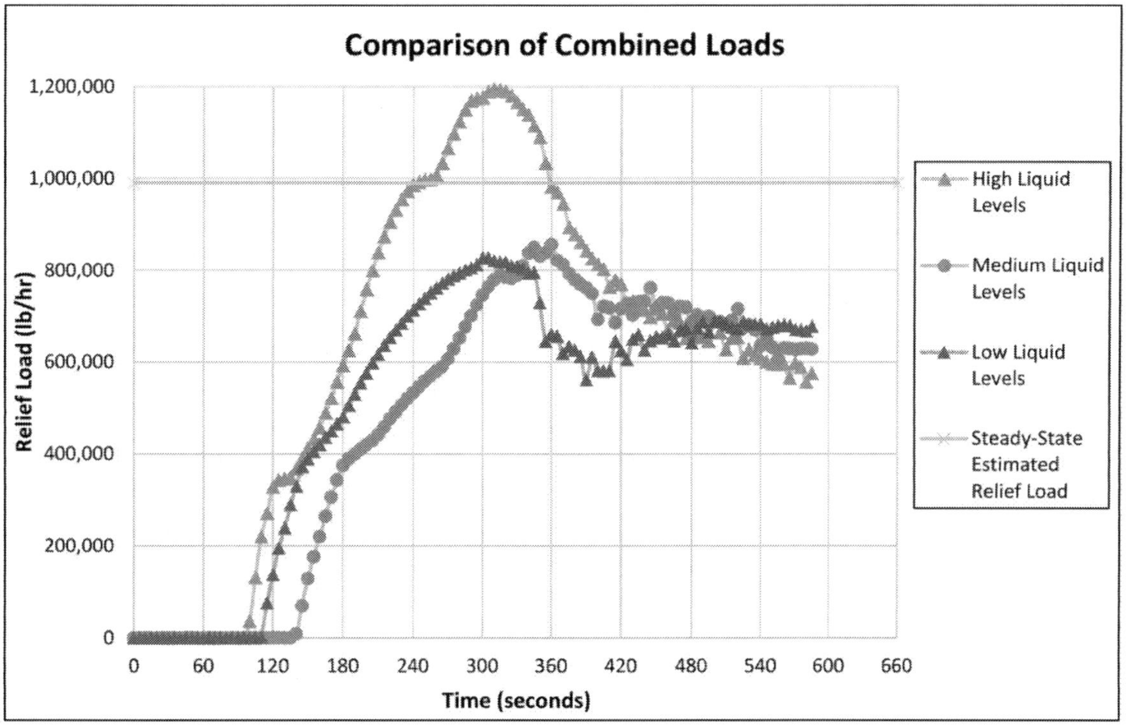

Figure 5: Combined cooling failure loads for different column liquid levels

The steady-state estimation of the combined relief load of the columns assumes that the reboilers have a reduced capacity due to the reduction in the log mean temperature difference (see Reference 1 for an additional explanation). The dynamic model was based on a constant heat input, and thus may over-predict the duties of the reboilers under upset conditions. A further improvement could be made by modeling the reboilers based on their individual UA characteristics.

The three curves in Figure 5 have roughly the same shape as well as time of peak flow. The noticeable difference is the peak relief load for the high liquid level curve. There is a 43% increase in the peak of the combined relief load for all PSVs when the liquid levels are increased by 50%. Lowering the liquid levels by 50% had little effect on the combined peak load.

The increase in the combined peak load appears to be caused by two factors. The peak flows of PSV-1 and PSV-2 located on the 1st debutanizer increased by 57% from normal liquid levels to high liquid levels. The peak time for PSV-4 located on the 2nd debutanizer lowered when the liquid level increased, which resulted in PSV-4 contributing significantly to the combined load peak. Changing the liquid levels in the columns affected the time to initial relief, the time to reach peak load, and the height of the peak for these valves. This behavior is present to a lesser extent in the other relief valves.

4. Conclusions

Dynamics simulation can be a useful tool to show how a system reacts to sudden changes and how the system uses control valves to stabilize itself. Dynamic analysis is very different from hand-calculated methods which are traditionally used to size relief systems. Although dynamic analysis can be used to size relief systems, it must be done knowing how process variations affect the relief rates.

The results show that variations in the liquid levels affect relief times and peaks for a system of multiple columns. Increasing the liquid levels of four columns in a system by 50% increased the combined peak load into the flare by 43%. This was due to increases in the peak flowrate of two PSVs and the shifting of a large peak towards the combined load peak.

This result shows that changes in process variables can have a large impact on the relief rate used to size a flare system. Consideration must be given to the effect process variables can have on the time of initial relief, the time to reach the peak load, and the magnitude of the peak flow.

Because there are multiple variables that can impact the peak relief rate for a flare, a full sensitivity analysis of a system with multiple columns would be time consuming and costly. However, if dynamic analysis is used to size a relief system, a sensitivity analysis must be performed to ensure that the analysis will lead to a safe design.

5. References

[1] Wilkins, John & Dustin Smith, P.E. *Effects of Process Variables on Dynamic Relief Load Estimates for a Depropanizer and Debutanizer*. Oct. 2012. Electronic poster presented at the AIChE Regional Process Technology Conference, Texas City, TX.

[2] Cristea, Nicholas and Dustin Smith, P.E. *Making Relief Load Estimates Match Reality*. 2013

[3] ANSI/API Standard 521 5th Edition, ISO 23251, May 2008

A Hybrid Model for Characterizing the Source of Hazardous Material Release

Shahryar Khajehnajafi, PhD
Anjana Meel, PhD
SAFER Systems LLC
5284 Adolfo Road, Suite 100
Camarillo, CA 93012
shah@safersystem.com

Prepared for Presentation at
American Institute of Chemical Engineers
2013 Spring Meeting
9th Global Congress on Process Safety
San Antonio, Texas
April 28 – May 1, 2013

UNPUBLISHED

A Hybrid Model for Characterizing the Source of Hazardous Material Release

Shahryar Khajehnajafi, PhD
Anjana Meel, PhD
SAFER Systems LLC
5284 Adolfo Road, Suite 100
Camarillo, CA 93012
shah@safersystem.com

Keywords: hazardous chemicals, chemical spill, source of release, source of hazardous material, plume model, Source Area Locator (SAL), source characterization, Gaussian plume model, Bayesian method, Bayesian inference, hybrid model, odor source, source location estimation, burro

Abstract

On Sept 10, 2012, Los Angeles Fire department and 911 operators received many complaints about a strong foul smell. The complaint covered an area of 10,000 square miles. There were many speculations for the source of the release; sewage, chemical spill, pipeline carrying natural gas. However, after many hours of investigation by South Coast Air Quality Management it was discovered that the source of the foul odor was dead fish at the Salton Sea churned up by a thunderstorm that had occurred that evening. The Salton Sea is a 376-square-mile saltwater lake about 150 miles southwest of Los Angeles. The odor remained strong for up to 150 miles downwind of the source, and quite unusual. Fortunately the above case did not pose any health effects.

It is common for industrial sites carrying hazardous chemicals to receive calls from surrounding communities complaining of a chemical presence in the ambient air. Following a complaint, industrial sites must work promptly check and verify if the odor source is related to a chemical release on their property, especially when there are other industrial sites in the vicinity, for mitigating and reducing the extent of the possible hazard. Locating the source and amount of release for predicting the geographical extent of the exposure in a timely manner is, therefore, essential to take an effective response action for evacuation or shelter-in-place.

With only basic knowledge of meteorological information; wind speed, wind direction, and stability along with two or more gas concentration measurements, it is possible to quickly frame the zone of chemical release. This is an inverse problem to the usual forward dispersion calculation in which the source location and release rate is known and the concentrations at sensors positions are predicted. There are many publications in the area of source characterization; location, and amount of chemical being released. The two most cited methods are: Bayesian inference using Markov Chain Monte Carlo (MCMC) sampling methodology, and adjoint advection- diffusion equation. The Bayesian methodology fits well for application to atmospheric dispersion where ambient and gas sensor data are stochastic in nature, however, it takes a long time to converge.

In this paper, we briefly describe a simple least-square based method for source characterization using Gaussian plume model. We apply the Bayesian method to address the same problem and compare the results of the two methodologies for few selected release scenarios as well as a set of field data. Subsequently, we will present a hybrid model using Bayesian inference which utilizes the result of the simple least- square model as a seed for narrowing the search domain and reducing the time for locating source of chemical release. We present the improvement in the run time using the hybrid model which makes it suitable for emergency response and provides a reasonably accurate prediction of source characterization.

1 Introduction

In this paper, the focus is on estimating the source area location of a hazardous release using the sensor measurements obtained from sensors located throughout a facility. If parameters of a system are known, the outcomes of measurements can be predicted. That is known as the 'forward problem'. On the contrary, estimation of the values of the parameters of the system using the results of the measurements is called a 'backward problem'. Identifying source parameters for location, quantity of release using sensor measurements is a backward problem. Tarantola (2005) discussed the inverse problem theory and methods for model parameter estimation in great detail. Usually a forward problem has a unique solution, but the inverse problem can have multiple solutions. Therefore, in an inverse problem, any apriori information should be utilized explicitly and uncertainties in the data should be handled carefully. When nonlinearities are available, it is important to describe uncertainty and apriori information using probability distribution. Bayesian theory based methods can be used to describe uncertainty. When systems are not nonlinear, least-square methods are considered very efficient. Both least-square methods and Bayesian Theory based methods have their own advantages and disadvantages. Least-square method usually minimizes the error between measurements and prediction to calculate the system unknowns while Bayesian theory based methods updates the prior probability distribution of unknown parameters using the sensor measurement data to obtain a posterior distribution of the unknown parameters, which provides the updated information of the system.

Thus the source area location and release rate is indeed a very challenging and nonlinear problem, even for identifying the source area from a single sensor concentration measurement. Even for a single sensor measurement, there can be multiple solutions for release location all of which may lie on a line/curve with varying distance and release rates. For example, a large release rate at a further distance from a sensor may result in the same concentration values observed by a fixed sensor as a small release rate at a location closer to the sensor. When more than one sensor is involved, the problem becomes even more complicated.

The Bayesian theory has been used by several researchers for estimating the source area location. Yee et al. (2007, 2008) developed a Bayesian probabilistic inferential solution for source reconstruction (both location and emission rate) from a limited noise concentrated data and the methods were tested for JU2003 field study in Oklahoma City (urban dispersion) and the ETEX (long-range dispersion on continental scales). Kim et al. (2007) developed a source reconstruction algorithm using dynamic Bayesian inference and applied it on Yonggwang atmospheric tracer experiment in Korea to show that the posterior modes of the release point and the released source rate for this experiment converged to their true values. Keats (2009) developed Bayesian inference technique for source determination in atmospheric environment. Adjoint (backward) dispersion models were used to reduce the computational effort required from calculating one [forward] plume per possible source configuration to calculating one [backward] plume per detector. Markov chain Monte Carlo (MCMC) is used to efficiently sample from the posterior distribution for the source parameters; both the Metropolis- Hastings and hybrid Hamiltonian algorithms are used. These algorithms were applied on a full-scale field experiment (Joint Urban 2003) in Oklahoma City and the results from the algorithms were also compared with the US Environmental Protection Agency's standard CMB model using a test case involving PM data from Fresno, California. Neuman et al. (2005) used an urban puff model [the Defense Science Technology Laboratory's (Dstl) Urban Dispersion Model (UDM)] for prediction plume evolution in complex urban geometries that enabled rapid urban dispersion simulations by combining traditional Gaussian puff modeling with empirically deduced mixing and entrainment approximations. In their work, a preliminary reconstruction of an atmospheric release event was demonstrated using stochastic sampling algorithms and Bayesian inference together with the rapid UDM urban puff model based on point measurements of concentration. This model was applied to a prototype isolated building and for observations and flow conditions taken during the Joint URBAN 2003 field campaign at Oklahoma City. Ideally, the urban puff model is used to approximate the source location and strength. The more accurate CFD model can then be used to refine the solution. Zheng et al. (2011) reviewed inverse calculation approaches for source determination and classified them into probability modeling methods and optimization modeling methods. According to them, probability modeling methods are demonstrated to be insufficient during emergency responses due to their lacking of enough prior information of unknown parameters, while optimization modeling methods are efficient and become a new trend in source determination. Sharan et al. (2011) applied least square data assimilation for identification of point source emissions from limited number of atmospheric concentration measurements. Monache et al. (2008) applied Bayesian inference and Markov Chain Monte Carlo sampling to reconstruct a contaminant source on a continental scale. By running the Bayesian MCMC algorithm on a large parallel cluster the inversion results could be obtained in few hours as required for emergency response to continental-scale releases. Senocak (2008) developed a method for stochastic event reconstruction of atmospheric contaminant dispersion using Bayesian Inference. Cowles and Carlin (1996) wrote a review article describing the markov chain monte carlo convergence diagnostics. Convergence criteria to the right solution and optimum step size for chain movement are important topics in the application of Markov Chain Monte Carlo method. Haario et al. (2001) described the importance of choice of proposal distribution for MCMC methods for convergence of the algorithm. An adaptive metropolis algorithm with Gaussian proposal distribution was introduced.

As shown in the above references, a number of researchers have worked in source location estimation using Bayesian probabilities methods. Since probability based methods alone cannot be sufficient during emergency response due to long convergence time periods, in the approach proposed in this paper, first an initial solution is quickly estimated using a least-square based method based on backward wind corridor. Subsequently, to obtain a more rigorous solution using Bayesian probabilistic method (referred as MCMC), the initial solution based on least square method is provided as a seed solution along with a search area/boundary. Thus, this two-stage method decreases the overall convergence time for probabilistic method by providing an intelligent apriori information. Using this approach, the emergency responders can quickly start responding based on the initial solution and more accurate information using a probabilistic method can be provided to them in due course of time. Thus, a combined least square-Bayesian approach ensures that even a probabilistic method provides a solution in much smaller time frame.

2 Approach

In this section, our hybrid approach for estimation of source area location is described. At first, source area location is estimated using a simple and intelligent technique developed using backward wind corridor and least-square based method (Khajehnajafi and Burla, 2008). The least-square method is an optimization based approach and several key features of this approach are listed in Table. 1. This method is very efficient and gives a good estimate for source location for most of the wind stability conditions, except when the wind is too stable because the slope of wind corridor is very narrow.

Another approach used for source location estimation is a probability based method using Bayesian theory where a probabilistic distribution is assumed for source area location and the sensor measurements and markov chain monte carlo (MCMC) simulations are used to estimate the source area location. Since this is a statistical approach, it takes significantly more time to converge to a value. Key features of this approach are listed in Table 1.

A Hybrid approach is a combination of the two approaches as listed in Table 1. Deterministic approach assumes that the parameters of the system are known completely without any uncertainty while probabilistic approach accounts for uncertainty in the model at the expense of longer time for convergence. Therefore, having a method that can narrow down the initial search zone for the Bayesian approach could be very useful in decreasing the convergence time. Thus, a hybrid approach is suggested in this paper where at first a Source Area Locator (SAL) model is used to narrow down the area and then Bayesian approach (referred as MCMC) is used to zero down on the actual source area location. The idea is that the combination of these two is a great method that accounts for weaknesses of both the models and makes it a very efficient model to narrow down upon a location very close to the actual source for all wind and stability conditions. Each approach on its own can provide reasonable results with some limitation either in terms of a further away solution for certain atmospheric conditions, or really long convergence time. Thus, the hybrid approach is a very valuable approach to estimate the source area location. Next, specific details of each of the methods are provided.

Table 1: Deterministic, Stochastic, and Hybrid Approach to Source Area Location

Deterministic (Least-square based)	Stochastic (Probabilistic)
• Assumes that all the system parameters are known perfectly • Usually faster runtime • Good for emergency response situations • Referred as SAL – Deterministic approach based on least-square based method to narrow down the search area	• Assumes uncertainty in system parameters • Usually longer run-time • Probabilistic approach • Based on Bayesian - Markov Chain Monte Carlo method (MCMC)
Hybrid Approach	
• Initially solved using deterministic approach based on least –square method SAL • Solution of deterministic approach is used to narrow down the search area for Bayesian - Markov Chain Monte Carlo method (MCMC). Thus, reducing the run time and obtaining a more accurate prediction of source parameters.	

2.1 SAL (Least Square Based Model)

The least square based method is referred to as SAL (Source Area Locator) in this paper. In this technique, it was demonstrated that with only the basic knowledge of meteorological information; wind speed, direction, and stability along with two or more gas concentration measurements, it is possible to quickly frame a region for the source of chemical (Khajehnajafi, and Burla 2008). A technique called 'reverse corridor' or 'back trajectory' is used to narrow the search domain for release location as seen in Figure 1 (a). This simple methodology uses the horizontal dispersion obtained from surface data at receptor locations. The construction of the corridor is key to reduce search area, therefore, reducing the computational cost. In Figure 1(a) the yellow region is the search domain if only sensor 1 and 2 were impacted. Introducing sensor 3 value, the search area is further reduced to the meshed zone. Adding additional sensors may narrow the search area; however, it will significantly increase the simulation time and increase the nonlinearity of the problem.

For the simple example shown in Figure 1(a) initially, sensitivity of the model to main parameters for dispersion, e.g. wind meander, sensor concentration fluctuations, and atmospheric stability was studied. A Gaussian model was used for that study. The result of SAL model for different stability classes A through F (1 through 6) showed that the accuracy of the model prediction for source area decreased with the increasing in stability class especially for higher stability classes 5 and 6 and at very low release rates. The results for the source location and the release rates, even when the model was simulated with the noisy sensor data, viz. 40% and 80%, matched very well with the true data when very fine mesh size was use (except for highly stable conditions at low release rates) indicating that the back trajectory method coupled with a simple Gaussian plume model is in fact a very simple yet plausible technique for locating source of pollutant.

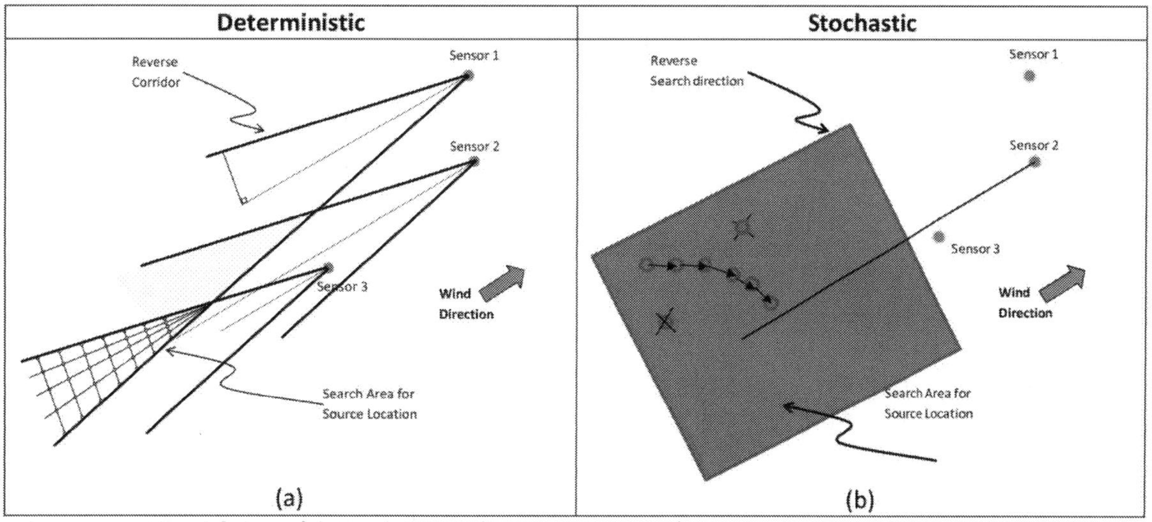

Figure 1: Pictorial view of deterministic and stochastic approach to source location of hazardous material release

2.2 Bayesian Model (Markov Chain Monte Carlo)

As mentioned in the introduction section, MCMC alone takes a lot of time to converge. Herein, based on the solution from the SAL model, an area was outlined as a narrowed search domain for the Bayesian probabilistic method. If the initial guess is chosen at random, the outcome from MCMC may take much longer time to converge and may even fails to converge sometimes because of broader search zone. In Bayesian method, a prior distribution is defined for parameters of interest (For example the location and release rate in this case). The likelihood function is estimated using the predicted concentration and measured sensor concentration. Then both prior distribution and likelihood distribution are used to estimate the posterior distribution. The posterior distribution gives us the latest knowledge of the original parameters of interest. The Bayes's theorem for the source reconstruction can be written as (Monache et al. 2008).

$$p(x,y,q|\,Sensor\ Data) = \frac{p(Sensor\ Data|x,y,q)p(x,y,q)}{p(Sensor\ Data)} \propto p(Sensor\ Data|x,y,q)p(x,y,q) \quad (1)$$

Where p( ) is the probability distribution. The x, y, q are the source coordinates and the release rate. *Sensor data* is the measurements from the different sensors. Bayes theorem relates the posterior distribution, $p(x,y,q|\,Sensor\ Data)$, to the product of the probability of measurements given the source parameters, $p(Sensor\ Data|x,y,q)$, also called the likelihood function, and the probability of the source parameters prior to any knowledge of the measurements, $p(x,y,q)$, also called the prior. For this study, x, y, q are assumed to be unknown parameters. One can include other parameters in the unknown list.

2.2.1 Prior Distribution

Let *m* be the model parameters i.e. all the parameters for which prior need to be defined. x, y, q are the prior parameters, x coordinate of the release location, y coordinate of the release location and the release rate, respectively. The prior distribution can be written as:

$$p(m) = p(x)p(y)p(q) \tag{2}$$

2.2.2 Likelihood Function and Acceptance Condition

The likelihood distribution is function of CP, predicted concentration for a sensor, and CM, measured concentration by an actual sensor. Here, the likelihood function is assumed to be of the following form (Monache et al. 2008):

$$\ln[(CM|S)] = \ln[L(\sigma, CP, CM)] = -\frac{\sum_i^N [log_{10}(CP_i) - log_{10}(CM_i)]^2}{2\sigma^2} \tag{3}$$

Where L is the likelihood function, CP_i is the predicted concentration values at the sensor location *i*, CM_i is the sensor measurements for sensor *i*, σ is an error parameter chosen accordingly to expected errors in the observations and predictions. *N* is the number of sensors. *S* is the state vector formed by all the unknown parameters (*x*, *y*, *q*). If the measured concentration values spread over several order of magnitudes, taking the logarithm of both predicted and measured values prevents large concentrations from dominating in the likelihood computation. The above likelihood is relatively simple, where the *N* prediction errors are assumed independent and expected to be proportional in size to the predicted concentration (as a result of log transforming both the observed and predicted concentration). If large errors are expected in the prediction/or measurements, large values of sigma should be assumed resulting in a broader-flatter posterior distribution. Here a sigma value of 0.8 is assumed. For a more complex scenario, it can also be treated as an unknown parameter and assigned a prior distribution.

The posterior distribution is sampled with an MCMC procedure via a Metropolis-Hasting algorithm.

2.2.3 Concentration Prediction Model

The concentration prediction model is based on a Gaussian plume dispersion model as shown below (Kim et al. 2011).

$$CP_i(x, y, z) = \frac{q}{2\pi U \sigma_y \sigma_z} \exp\left(-\frac{y^2}{2\sigma_y^2}\right) \tag{4}$$

Where, $CP_i(x,y,z)$ is predicted concentration at a particular location (x, y, z). q is the emission rate, U is mean wind speed, H is height of the release, x is distance along the wind, y is distance along the horizontal crosswind direction, z is the distance along the vertical axis, σ_y, σ_z are the standard deviations in the horizontal crosswind and vertical directions that are also called as Gaussian plume dispersion parameters and are defined empirically for different stability conditions. σ_y, σ_z are defined below:

$$\sigma_y = \zeta_1 x\,(1 + 0.0004x)^{-0.5}, \qquad \sigma_z = \zeta_2\, x \qquad (5)$$

2.2.4 Sampling Procedure and Algorithm

Various algorithms exist for the sampling with MCMC. We have assumed uniform prior for the prior parameters. For the MCMC method, x is in the range -10,000 to 10,000 and y is the range 10,000 to 10,000 with (0,0) as the release location. q is uniform prior between 0.1 to 10 kg/s. U is the value given for the experiment of interest. H is assumed at the ground level. Ideally, parameter ranges can be assumed for wind direction, wind speed, ζ_1, ζ_2. However, for this study we have kept these constant. Of course, ζ_1 and ζ_2 will vary based on the stability class.
For the hybrid approach, the release location is first determined using the SAL approach. Limits for x, y are assumed to be (-1000 + x_{sal}, 1000 + x_{sal}) and (-1000 + y_{sal}, 1000+ y_{sal}) where x_{sal} and y_{sal} are release location coordinates obtained using the least-square method (SAL). Range for q is 0.001 to 10 Kg/s. The new value of x, y, q is obtained using adaptive sampling based on the current value and a step size.

2.2.5 Algorithm

An initial value for each parameter is sampled from its prior distribution. System parameters are assumed to be uniform prior with range as described above. The prior distributions are not completely uncertain as a good initial range is provided using the SAL method. Using the likelihood function, these sampled values from the prior distribution are selected to be part of the posterior distribution based on some criteria. A number of iterations involving source term sampling and dispersion simulation are executed until convergence to the posterior distribution is reached. Accepted parameter values collected during all the iterations constitute a Markov Chain which is defined as a set of consecutive values of parameter in which the probability of a value depends only upon the previous sampled value._Each iteration of the solution that becomes part of Markov Chain consists of following steps (Monache et al. 2008):

1. New value for source term is proposed (x_{prop}) according to $x_{prop} = x + dx$. Where x is the current value and dx is displacement from the current value modeled as a random walk sampled from a Gaussian distribution with zero mean and a standard deviation specified from the current step size. The Prior distribution $p\ (m)$ is used as a target distribution to estimate a prior likelihood of x_{prop}. If the prior likelihood of x_{prop} exceeds that of x, the proposed value x_{prop} replaces x. If not, a random (Bernoulli) "coin flip" determines whether the new proposal, even with its lower prior likelihood will be accepted. This ensures that sampling doesn't get stuck at a random point.

2. Repeat step 1 for the y and q parameters considered in the prior distribution. The assumption is that independent sampling of each parameter ensures that they reflect their likelihood better than sampling all of them at once and testing their likelihood
3. The predicted concentration for sensors is simulated using the current values of all the prior parameters using equation 4.
4. Likelihood of the current values of all the parameters which forms the proposed state is calculated using the predicted concentration and observed concentration by sensors. If the proposed likelihood is higher than the likelihood of previous state, it is accepted. If not, a random (Bernoulli) coin flip is used to ensure that the search reconstruction procedures continues to search the complete space of proposed source term parameter states, preventing the procedure from remaining indefinitely within the neighborhood of a local extreme.

After the proposed state is calculated, it is accepted if

$$\ln[L_{new}] - \ln[L] \geq \ln[rnd(0,1)] \quad (6)$$

Where L_{new} is the likelihood value of the proposed state and L is previous likelihood value and rnd(0,1) is a random number generated from a uniform distribution in the internal [0,1].

At a particular time instant, likelihood test for proposed state using measured data from all sensors at a particular time instant determines the acceptance or rejection of this state. Steps 1-4 are repeated to create a new proposed state with respect to all the parameters until the Markov Chain converges for a particular time instant. Note that in case of time varying information; step 1 to 4 can be further repeated for each time instant.

3 Case Studies: Comparison with the Field Data

3.1 BURRO EXPERIMENTS

BURRO experiments from the 1980 LNG (Liquefied Natural Gas) spill tests conducted by Lawrence Livermore National Laboratory (LLNL) and Naval Weapons Center (NWC) were used to compare the predictions from the model with the experiments. BURRO series experiments are one of the most comprehensive experimental analyses available in the literature for sensor measurement data of a chemical release (Koopman et al. 1980, Goldwire et al. 1980). In total nine experiments were conducted. Five experiments from the 1980 LNG (Liquefied Natural Gas) spill tests were used for this analysis. It is very difficult to get access to experimental data in this field and BURRO data are indeed very useful to compare modeling predictions with the experimental data for such a complex set of sensors.

This particular experimental setup employs close to 20-30 sensors. Not all sensors will see concentration values all the time so only the sensor with non-zero concentration reading are analyzed in this situation. This in itself is debatable as non-zero sensors may provide some information about the location which otherwise can be missed in case of uncertainty. Ideally, non-zero sensors should be included as well but for simplicity reasons we have not included those here and can be part of a further study. The next level of complication arises from the dynamic nature of the measured sensor values due to plume meander caused by horizontal turbulent fluctuation in the atmosphere. In ideal situation one expects the sensors to measure concentrations reliably based on the release rates. However, the uncertainty in wind direction, wind speed, measurement errors due to hardware and software and variability in atmospheric stability condition can contribute to a lot of uncertainty in the sensor data. For example, see the BURRO experiments report for concentration values measured by sensors for few experiments. Because of these uncertainties in the values and the simplified nature of our modeling approach, careful and intelligent decisions need to be made about the measured sensor readings that are used for this analysis as only a single measured value is used for each sensor in this analysis. Nonetheless, a more complicated model can be created which can include few/all aforementioned uncertainties at the expense of higher simulation time and resources. These may be included in the future studies of the project after carefully analyzing the tradeoffs involved in various approaches.

In our approach, the spill scenarios conditions and measured sensors values were used to obtain solution from SAL and MCMC approaches and then predictions for source area locations from the two models were compared against the actual experiment release location. In particular, five different spill scenarios from BURRO experimental series i.e. BURRO 2, BURRO 3, BURRO 7, BURRO 8, and BURRO 9 covering a range of atmosphere stability conditions are reported in this study (Table 2).

The release location was denoted as (0, 0) position in a new coordinate system, which is about 45 degree from the true north. The wind conditions were measured along the true north conditions. Therefore, wind directions for the new coordinate system are calculated by adding 45 degrees to wind direction values reported in the BURRO experiments. A plot in the new coordinate system containing actual release location and sensor locations is shown in Figure 2. The sensors were mainly distributed in four different arches as shown in Figure 2. Detailed location of these sensors and other towers can be found in the BURRO report part 1. The wind fields for different experiments are reported in the BURRO report part 2. To run SAL, a constant wind field (i.e. constant wind speed and wind direction) is assumed. For these short duration experiments, wind field does not change a lot so this is a reasonable assumption. Undoubtedly including variability in the wind field would result in a better prediction at the expense of more computational time.

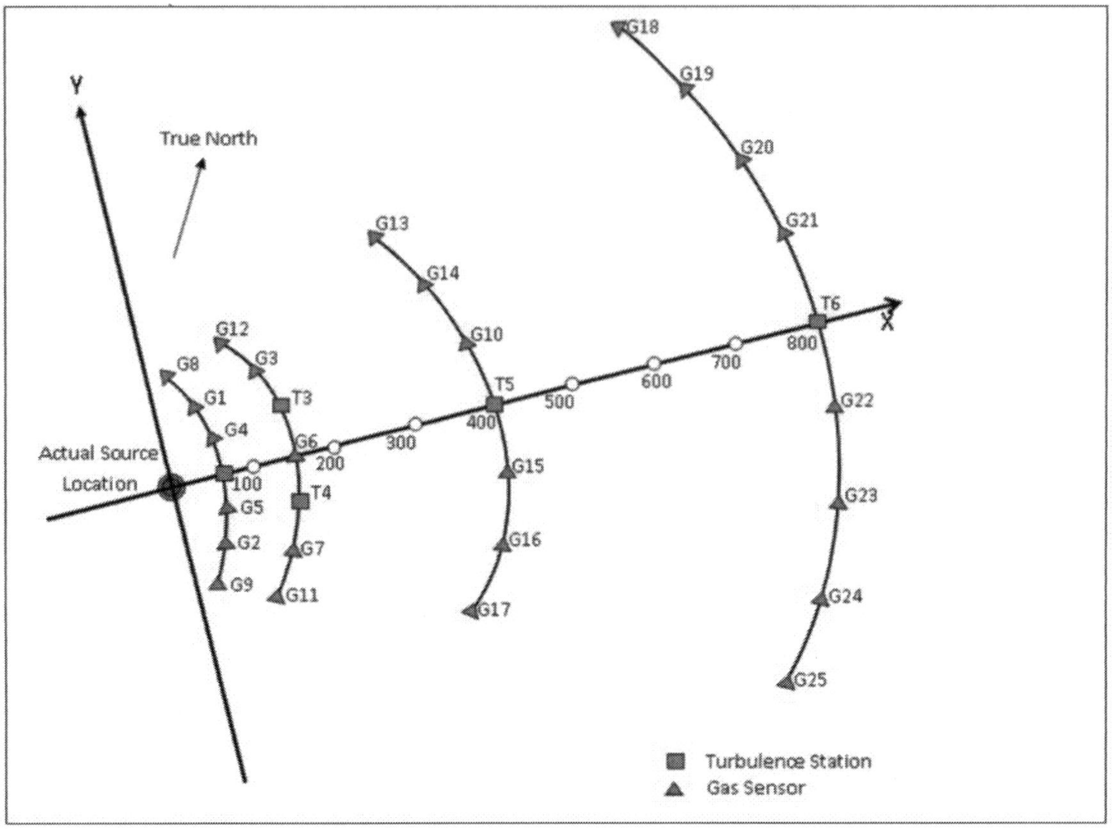

Figure 2: Burro experiment setup of sensors.

Table 2: Spill conditions for BURRO 3, BURRO 7, BURRO 8, and BURRO 9

Experiment Name	Chemical	Wind direction	Stability	Wind speed (m/s)	Spill Rate (m3/min)	Spill duration
BURRO 2	LNG	266	2	5.4	11.9	173
BURRO 3	LNG	270	2	5.4	12.2	166
BURRO 7	LNG	253	3	8.4	13.6	174
BURRO 8	LNG	280	5	1.8	16	107
BURRO 9	LNG	277	4	5.7	18.4	79

3.2 SAL Sensitivity to Different Cost Functions

A reasonably good approximation for source area location using SAL is our first target. To accomplish that goal we had to identify the best cost function. Three cost functions were identified; A simple concentration difference objective function **abs**(c_m-c_p), referred as CF 1, Normalized concentration based objective function [**abs**$(c_m-c_p)/c_m$], referred as CF 2; Distance and concentration normalization based objective function [abs $(c_m-c_p)/c_m$ $*(x_{dist}^2+y_{dist}^2)^{0.5}$], referred as CF 3. Where c_m is the measured concentration and c_p is the predicted concentration.

Only a single measurement value per sensor is used for this analysis. Since it is a dynamic release scenario, various values can be selected from the sensor measurements. A sample of sensors output for Burro 3 is shown in Figure 3. Two set of data was used for SAL study. One was based on the single best sensor value for the entire time zone which reflects the best/average sensor values and the other is based on the maximum value for the entire time. All the sensors with non-zero sensor values were included in the analysis. For SAL simulation, thirteen different values of release rates ranging from 0.01 to 10 are used for source location. A local source point was identified which was average of these thirteen values and was used as an initial point for the next level search. First a coarse grid is used to narrow the search zone. Then a finer mesh grid search was carried out around that zone for various ranges of release rates to identify the source area location.

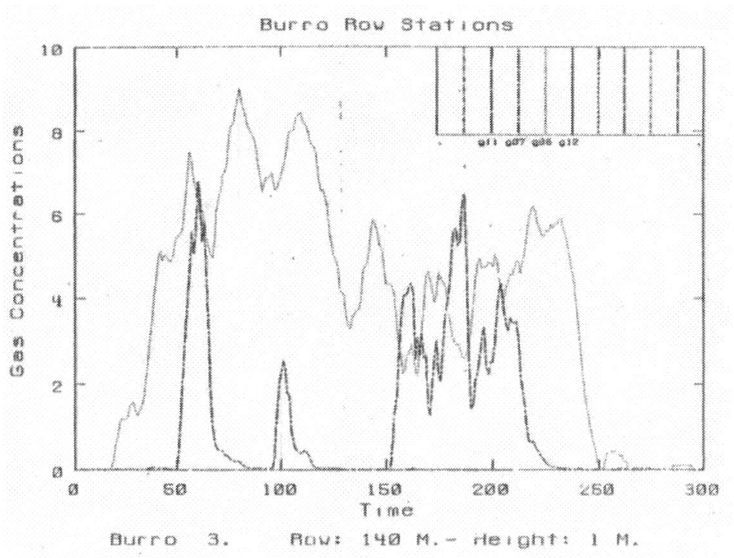

Figure 3: **Gas sensors concentration measurement at 140 m arc.**

The following tables represent the single sensor value obtained from each sensor positioned in the experimental grids for this study along with the source location predicted by SAL using three different cost functions.

3.2.1 Burro 2

Table 3: Sensor's value from BURRO 2 experiment

Sensors	G2	G5	G4	G1	G8	G3	G12	G13
ppm	8	5	1.2	1	2	5	4	0.02

Table 4: Source area location for BURRO 2 using sensor values

WD=266	CF1	CF2	CF3
Stability	2	2	2
X location	-65.1	-57.5	-56.6
Y location	9.2	7.5	8.8

3.2.2 Burro 3

Table 5: Sensor's value from BURRO 3 experiment

Sensors	G2	G5	G4	G1	G7	G6
ppm	9.0	4.0	17	15	4	5

Table 6: Source area location for BURRO 3 using sensor values

WD=270	CF 1	CF 2	CF 3
Stability	2	2	2
X location	-73.4	-117.2	-42.5
Y location	-14.8	40.2	16.3

3.2.3 Burro 7

Table 7: Sensor's value from BURRO 7 experiment

Sensors	G12	G3	T3	G13	G14	G18	G19
ppm	9.0	0.78	0.34	2.8	2.6	0.4	0.05

Table 8: Source area location for BURRO 7 using sensor values

WD = 253	CF1	CF2	CF3
Stability	3	3	3
X location	-111.6	69.5	191.6
Y location	-49.6	59.0	148.6

3.2.4 Burro 8

Table 9: Sensor's value from BURRO 8 experiment

Sensors	G9	G2	T2	G4	G1	G8	G11	G7	T4	G6	T3	G3
ppm	15	6	8	11	5	8	17	12.5	3	6	5	4.8
Sensors	G12	G10	G16	G15	T5	G17	T6					
ppm	2	0.02	0.7	0.7	2	3.8	0.25					

Table 10: Source area location for BURRO 8 using sensor values

WD = 280	CF1	CF2	CF3
Stability	5	5	5
X location	-633.9	-285.4	-355.2
Y location	119.8	70.6	77.9

3.2.5 Burro 9

Table 11: Sensor's value from BURRO 9 experiment

Sensors	G2	T2	G4	G7	T4	G6	T5	G16	T3	T6
ppm	6	1	7	1.2	2.2	6	0.25	0.2	4.5	0.2

Table 12: Source area location for BURRO 9 using sensor values

WD=277	CF1	CF2	CF3
Stability	4	4	4
X location	-70	63.5	66.3
Y location	8.8	-3.9	-5.3

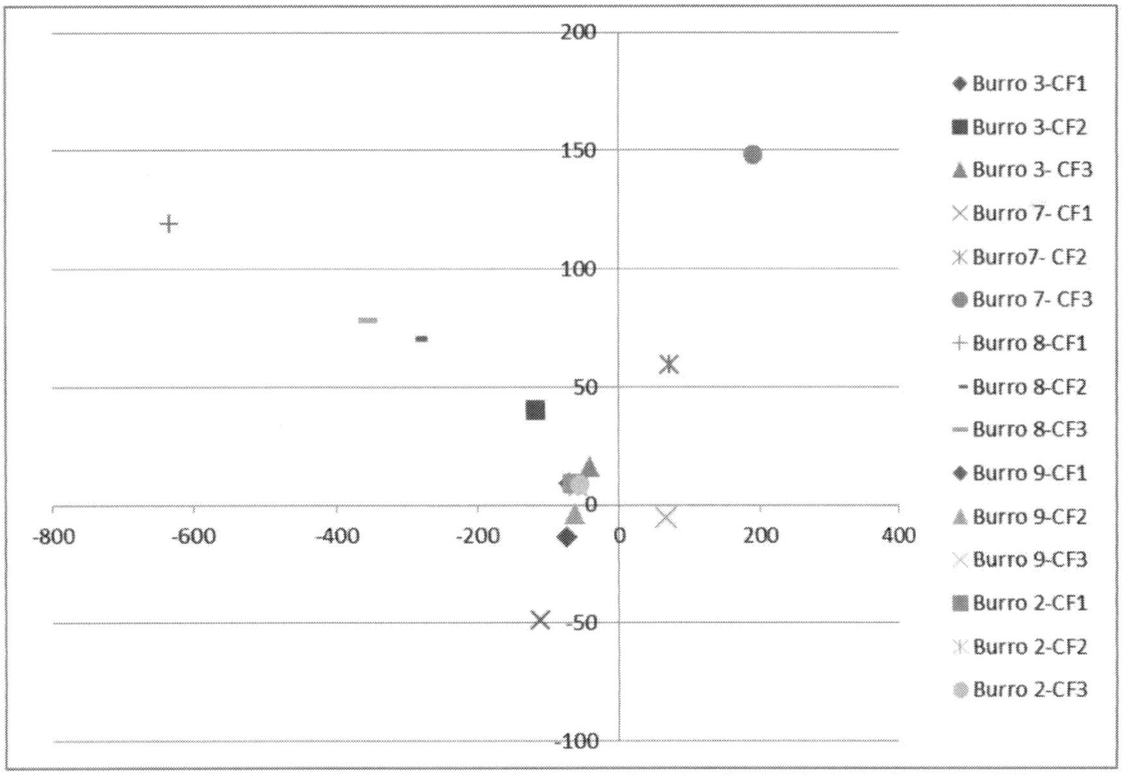

Figure 4: Source location prediction for various Burro experiments using SAL (actual release is at 0,0).

From this analysis we can see that the CF2 give better results out of all three objective functions. It was found that other than stability case 5, in most cases the source area was located within (+/- 70, +/- 70) range of the actual release location. SAL took only few seconds to run for all cases.

3.3 Hybrid Model

Subsequently, the solution from the hybrid approach with an initial guess from SAL to narrow down the search domain using Bayesian theory was obtained. BURRO MCMC refer to the results obtained using the hybrid approach in Figure 5. BURRO SAL refers to results obtained using the least-square based methodology. As evident from Figure 5, BURRO MCMC provides good solution for all stability conditions. Improvements in source location were seen for almost all the cases using the hybrid approach. These simulations were done for a wide variety of stability ranges including unstable to stable atmospheric conditions. Improvements seen in unstable cases using hybrid MCMC approach are more profound than stable cases as shown later in this report. The hybrid model improved the run time by a factor of two to three times.

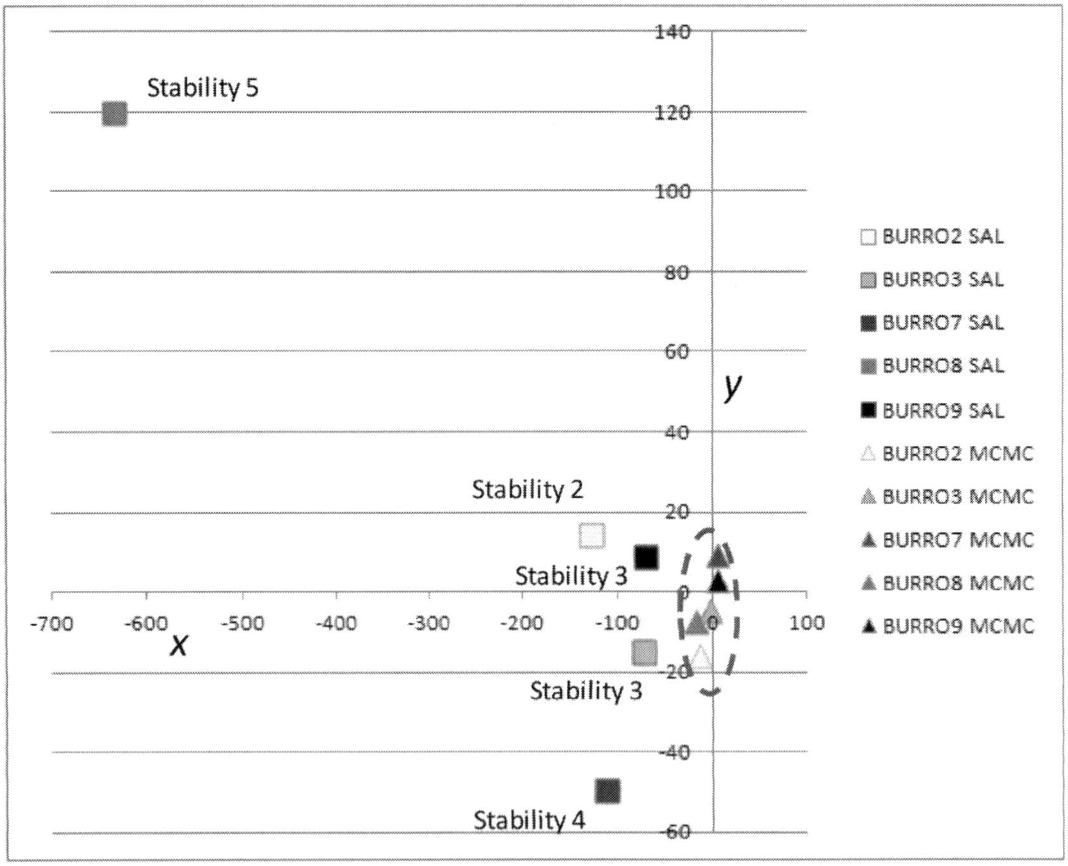

Figure 5: Burro experiment result from SAL and hybrid MCMC approach

Time-varying wind field, uncertainty in wind direction and speed, dynamic sensor readings, release duration, release time may be included in prior parameters for the hybrid approach where the main focus will be on including uncertainties and then identifying the location of the release.

If sensor data is available in dynamic fashion, these hybrid models can be extended to accommodate the dynamic Bayesian model (Johannesson et al. 2004) techniques.

4 Conclusion

The focus of this paper is mainly on estimating release location using the sensor measurements and providing an estimate for the release rate. An important method to estimate source area location using a hybrid approach based on least-square method (SAL) and bayeisan theory (MCMC) was provided. It was shown that the method provides a good solution for BURRO experiment data series for all stability conditions.

5 References

[1] Tarantola A., " Inverse Problem Theory and methods for model parameter estimation", Society for industrial and applied mathematics 2005, Philadelphia.

[2] Johannesson G., Hanley B., Nitao J., Dynamic Bayesian Models via Monte Carlo – An Introduction with Examples -, UCRL-TR-207173, Lawrence Livermore National Laboratory, October 2004.

[3] Cowles M. K. and Carlin B.P. ,Markov Chain Monte Carlo Convergence Diagnostics: A Comparative Review, Journal of American statistical association, vol 91 (434), 1996, pp. 883.

[4] Yee E., Lien F-S, Keats A., D'Amoursc R. , Bayesian inversion of concentration data: Source reconstruction in the adjoint representation of atmospheric diffusion, Journal of Wind Engineering and Industrial Aerodynamics, Vol 96, 2008, pp. 1805–1816.

[5] Yee E., Bayesian Probabilistic Approach for Inverse Source Determination From Limited and Noisy Chemical or Biological Sensor Concentration Measurements, Chemical and Biological Sensing VIII, Proc. of SPIE, Vol. 6554, 65540W, 2007.

[6] Kim J.Y., Jang H.K., and Lee J.K, Source Reconstruction of Unknown Model Parameters in Atmospheric Dispersion Using Dynamic Bayesian Inference, Progress in Nuclear Science and Technology, Vol. 1, 2011, pp. 460-463.

[7] Keats W.A. , Bayesian inference for source determination in the atmospheric environment, PhD thesis, University of Waterloo, Waterloo, Ontario, Canada, 2009.

[8] Neuman S., Glascoe L., Kosovi´c B., Dyer K., Hanley W., Nitao J., Event Reconstruction for atmospheric releases employing urban puff model UDM with stochastic inversion methodology, American Meteorological Society Conference, 2005, Atlanta GA.

[9] Zheng X., Chen Z., Inverse calculation approaches for source determination in hazardous chemical releases, Journal of Loss Prevention in the Process Industries 24 (2011) 293-301.

[10] Sharan M., Singh S. K., and Issartel J.P, Least Square Data Assimilation for Identification of the Point Source Emissions, Pure and Applied Geophysics, August 2011.

[11] Monache L.D. et al., Bayesian Inference and Markov Chain Monte Carlo Sampling to Reconstruct a Contaminant Source on a Continental Scale, Journal of Applied Meteorology and Climatology, pp 2600-2613, vol 47, 2008.

[12] Senocak I., Hengartner N. W., Short M.B., Daniel W.B., Stochastic Event Reconstruction of Atmospheric Contaminant Dispersion using Bayesian Inference, Atmospheric environment, Vol 42 (33), 2008, pp7718-7727.

[13] Koopman R. P., Baker J., Cederwall R. T., Goldwire H. C., Hogan W. J. LLNL/NWC (Lawrence Livermore National Laboratory/Naval Weapons Center) 1980 LNG (Liquefied Natural Gas) Spill Tests. Burro Series Data Report.

[14] Goldiwre H. C., Baker J., Koopman R. P., Cederwall R. T., Hogan W. J. LLNL/NWC (Lawrence Livermore National Laboratory/Naval Weapons Center) 1980 LNG (Liquefied Natural Gas) Spill Tests. Burro Series Data Report: The Appendices.

[15] S. Khajehnajafi, S. Burla, Prediction of Source Location and Rate of a Chemical Release for Emergency Response: Back Trajectory Approach, AIChE Spring Meeting, 2008

[16] Hogan W.R., Cooper G.F., Wallstrom G.L., Wagner M. M. and Depinay J-M, The Bayesian aerosol release detector: An algorithm for detecting and characterizing outbreaks caused by an atmospheric release of *Bacillus anthracis, Statist. Med. 2007;* **26**:5225–5252.

[17] Haario H., Saksman E., Tamminen J., An adaptive metropolis algorithm, Bernoulli, Vol 7 (2), 2011, pp. 223-242.

(43ak) Guidelines for Developing Site-Specific Human Error Rates and Human IPLs for LOPA Using a Safety Climate Approach

Farheen Khan, Human Factors Practice, Exponent, Inc., Chicago, IL, Sunil D. Lakhiani, Human Factors, Exponent, Failure Analysis Associates, Wood Dale, IL and Delmar Morrison, Thermal Sciences, Exponent, Inc., Warrenville, IL

Layer of Protection Analysis (LOPA) is a simple order of magnitude quantitative risk analysis technique commonly applied in the process industries. This powerful tool is especially useful for refining specific hazard scenarios during a process hazard analysis or a risk analysis. In LOPA, a hazard scenario is decomposed into a series of events, which may include the frequency or probability of equipment failures, personnel actions, and physical processes. By one estimate, 80 percent of incidents involve human error; thus, most risk analyses consider the likelihood and effect of human factors on potential scenarios. For LOPA, human error includes events initiated by the human in the work system and events where the human element failed to act as an independent layer of protection against an adverse event.

Although the LOPA method is widely used, there are few guidance sources for refining the human error component. The currently available standard LOPA estimates for the occurrence of human error assume optimal human factors conditions and that work systems are static environments unchanging over time. Work systems are, in fact, dynamic environments, which evolve over time through planned changes, as well as unplanned adaptations. Thus, no work system is completely free of error at any given time. Current LOPA estimates, therefore, fail to account for the dynamic nature of work systems where change from original conditions is not only possible but even necessary from time to time. This paper will examine the deficiencies in the LOPA human error probability estimates and propose a strategy that utilizes a safety climate approach to understanding error causation and incorporating it as a fundamental building block to human error estimation.

See more of this Session: Electronic Poster Session
See more of this Group/Topical: Global Congress on Process Safety

Benchmarking – An Important Milestone in the Journey towards PSM Excellence

Prerna Jain

Reliance Industries Limited

India

prerna.jain@ril.com

[1]CindiePridy, [2]Ravi Ramaswamy, [3]Robert Cravens

[1,3]Phillip Townsend Associates, Texas

[2]Reliance Industries Limited, India

cpridy@ptai.com

ravi.ramaswamy@ril.com

rcravens@ptai.com

Prepared for Presentation at

American Institute of Chemical Engineers

2013 Spring Meeting

9th Global Congress on Process Safety

San Antonio, Texas

April 28 – May 1, 2013

UNPUBLISHED

Benchmarking – An Important Milestone in the Journey towards PSM Excellence

Prerna Jain

Reliance Industries Limited

India

prerna.jain@ril.com

[1]CindiePridy, [2]Ravi Ramaswamy, [3]Robert Cravens

[1,3]Phillip Townsend Associates

[2]Reliance Industries Limited

Keywords: benchmarking, gap analysis, metric, performance gap, process safety management system, risk based process safety element, survey

Abstract

Reliance Industries Limited (RIL) has implemented a Process Safety Management (PSM) system through a structured and systematic approach over the last 6 years. In its PSM journey, RIL realized that there is a continuing need to improve and sustain PSM performance and to develop an understanding of the current status of RIL's existing PSM system, a benchmarking process of PSM practices, with reference to industry peers, is required. This was essential in order to identify and address the gaps to best performers within the peer group, define specific action items to obtain significant performance improvements, and ultimately achieve the best systems in the world in field of Process Safety Management.

To accomplish this goal, RIL conducted a PSM Benchmark Program Study designed by Phillip Townsend Associates (PTAI) and CCPS® at its six largest manufacturing locations in India. By helping to understand the relationship between process safety outcomes and management practices; this benchmarking program provided several key approaches for improving overall PSM performance. The program covered benchmarking of six elements related to a world class process safety management system. They are: process safety culture; compliance with standards, codes, regulations, and laws; hazard identification and risk analysis; asset integrity and reliability; management of change; and measurement and metrics.

The results of the study provided element specific quartile performance of each of the six manufacturing sites and Reliance as a whole in comparison to the industry peers who participated in the PTAI study. The gap analysis report presented by PTAI to the six sites

PSM team and senior management of RIL upon completion of the benchmarking study identified key areas of improvement for implementing and evolving effective Process Safety Management System at RIL.

This paper, presented by RIL and PTAI, discusses the benchmark process, methodology followed for the program, the value proposition of Process Safety Benchmarking, benefits of the program, key applications and lessons learned by RIL and new directions for improving PSM. The paper specifically focuses on the journey to success by realizing the gaps to top performers in the peer group, enhancing PSM knowledge and spreading PSM best practices.

1. Introduction

Chemical production and the associated processing industries dealing with hazardous and toxic materials have a significant commercial and societal responsibility to eliminate and/or mitigate the loss of primary containment or fires related to process materials while also ensuring that adequate safety systems are in place. Process Safety Management systems differ from personnel safety management in that process safety focuses directly on the processes, equipment and technology involved in causing an incident. The broader application of process safety management decision making and leadership behavior needs to address loss of containment events of greater and lesser consequences, safety system challenges and failures that could have led to failure, and operating discipline and management behaviors that ensure adequate layers of protection are being maintained. The focus shifts from the lagging indicators that track incidents to leading metrics that are used in a proactive manner to prevent incidents and continuously improve as shown in Figure 1: Process Safety Indicator Pyramid (API RP 754, 2010).

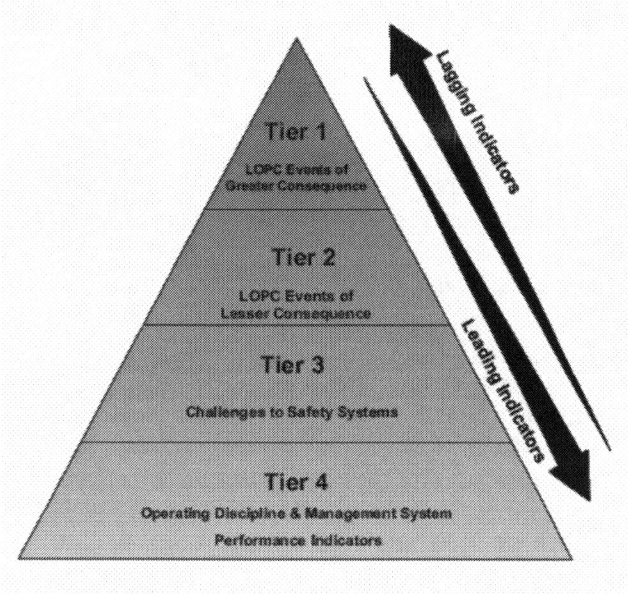

Figure 1: Process Safety Indicator Pyramid

The vision and recommendation of industry-wide process safety originated with the setting up of Centre for Chemical Process Safety (CCPS) by AIChE as a result of the December 3, 1984 Bhopal incident killing more than 3000 people due to toxic Methyl Iso Cyanate gas release. The need for continued focus on process safety got reinforced with the 2005 explosion at the BP refinery in Texas City, Texas. Since then, the Center for Chemical Process Safety (CCPS®) has identified 20 process safety elements to be used as guidelines for a risk based safety process (CCPS®, 2007). In 2009, the CCPS® and Phillip Townsend Associates, Inc. (PTAI) entered into an agreement to develop a benchmarking program for Process Safety Management Systems (PSM).

Viewed from a risk based perspective, the process safety elements can be grouped into four categories: Commitment to Process Safety, Understanding Hazards and Risk, Managing Risk, and Systematic Learning from Experience. PTAI and CCPS® designed the PSM benchmark selecting six of the 20 elements: Process Safety Culture, Compliance with Standards, Hazard Identification and Risk Analysis, Asset Integrity and Reliability, Management of Change, and Measurement and Metrics (CCPS®, 2007). The key goal was to seek an understanding of the relationship between process safety outcomes and management practices.

This paper discusses the details of the PSM benchmark, the value proposition to be gained, and specifically the experience of Reliance Industries Limited (RIL) in developing an understanding of the current status of RIL's existing PSM system, and the practices that will improve and sustain PSM performance. The objective of RIL is to take PSM systems and practices to the top quartile performance through a benchmarking process of industry peers relative to PSM practices. Specifically, the paper focuses on the most important decisions and behavior changes that leadership can make to spread process safety awareness throughout the organization, close the gap to the top performers, develop central and site specific action plans for improvement, and raise the internal level of all organizational units to the that of the top internal performer.

2. Methodology

Benchmarking identifies how well a company is doing compared to its peers in the industry, what the industry top performance is, how large the gap is to the top performers within the industry, and create a stimulus for continuous improvement. Benchmarking programs seek to understand the relationship between outcomes and management practices. The PSM benchmarking program is a systematic study of process safety between companies that will allow sites/companies to compare their management systems with the goal of improving the design and implementation of management practices that impact process safety.

2.1 Value Proposition of Process Safety Benchmarking

PSM benchmarking can elevate a company's process safety performance through four distinct steps:

2.1.1 Identify performance gaps

The identification of a company's performance in relation to Performance Quartiles will highlight key drivers for performance enhancement versus its peers. These quantifiable results replace perceptions with a data-based reality check. In addition, these results identify if the enterprise policies, practices, and procedures are operating as intended and are achieving the desired results.

2.1.2 Create performance enhancement goals

Data driven information provides a measure of the enterprise effectiveness – emphasizing strengths, weaknesses and potential for improvement over time. This helps focus attention on the critical aspects of an enterprise that create risk. It also underscores what improvements should be made – providing quantifiable performance enhancement goals.

2.1.3 Measure the effectiveness of improvement programs

PSM benchmarking can be a starting point for goal setting and performance initiatives aimed at boosting step change improvements in process safety. The clear outcome is that this will add significant shareholder value through reduced risks of catastrophic incidents.

2.1.4 Maintain continuous improvement

The PSM benchmark helps maintain a stimulus for continuous improvement. It provides the necessary motivation to ensure consistent top performance results.

2.2 Benchmarking Process

A peer group is established by PTAI of sites/companies agreeing to compare their process safety programs and performance in an effort to identify best practices and drive improvement. Ensuring consistency and protection of each participant's interests are critical to a successful benchmark. A confidentiality agreement was completed between PTAI and RIL along with execution of a participation agreement. The Data Collection Instrument (DCI), providing clear and consistent definitions for all participating sites/companies was made available. The use of a secure site over the internet ensured security of data and ease of use by participants as they are able to see all the survey questions and select their answers. A kickoff discussion to align key stakeholders at each of the sites is critical. In RIL case, this was done through a series of webinars between PTAI consultants, RIL's Centre for HSE Excellence and stakeholders at each of the participating sites.

RIL chose to benchmark PSM in six of its large Refining and Petrochemical manufacturing sites in India and a team of site focal points, corporate coordinator and PTAI worked in a seamless manner to plan and execute the benchmark program. Following submission of the data, PTAI validates the data on a site and company basis and provides analysis of the results. Validation was enhanced with RIL by individual site visits to each of the six sites by PTAI. This provided the opportunity for discussion and verification of the results reported to the on-line survey. Participation by key RIL personnel during this process added to the effectiveness and accuracy of data reported and provided a seamless voice to the sites between RIL's Center for HSE Excellence and PTAI.

In particular, statistical analysis of the data was performed by PTAI to evaluate various drivers of process safety using a Gap Analysis comparing RIL's performance to the top quartile of the peer group. The quantitative and qualitative results were distributed to participants with numerical scores for each element along with an overall site/company specific score. Strengths, as well as the main opportunities for improvement to close the gap towards the top performers were identified. A discussion of the study results and gap opportunities was held with the executive management in an executive management summary with specific individual gap analysis presentations to the top Line and HSE management of each of the six RIL sites benchmarked.

2.3Benefits of PSM Benchmarking

Management review is an integral part of moving from lessons learned to systematic improvement of PSM. Reviews of the performance gaps can identify potential problems and address them before an incident occurs; as well as, identify opportunities for improvement to move from current levels to the level of top performers. PSM benchmarking helps an organization prioritize the allocation of process safety resources, both financial and human, to focus on the highest identified risk areas.

Among the benefits of PSM benchmarking, there are certain ones that can have a big impact on an organization. There should be a marked increase in staff awareness of process safety issues. This awareness will expand the issues from the Health & Safety group to Engineering to Operations. Based on the organizational level of results discussion, it will potentially increase awareness from site management to impact those that are responsible for setting corporate strategy. A key learning for RIL was the differences between the individual site scores resulting in a changed perception by the Center for HSE Excellence and executive management. PSM benchmarking creates a common understanding and language on process safety matters. Benchmarking can also reduce risks by potentially providing an incident early warning signs.

Sometimes the opportunities for improvement can be very different from the perceived gaps in an organization. Identifying and gaining insight on perceived gaps versus actual gaps is greatly enhanced by an independent benchmark which effectively creates consistency across multiple sites within a company and with the others in the peer group.

3. Key Applications and Lessons Learned

Traditional process safety surveys have identified a list of both quantitative and qualitative issues. Translating the qualitative issues into measurable, actionable steps is a key issue for senior management decision making and allocation of resources. Benchmarking results must be measurable and use a scoring methodology that will not only identify the performance gaps, but also quantify the impact of selected actions. The

PSM benchmarking process designed by CCPS® and PTAI includes a scoring methodology which enables participants to have measurable action plans. Benchmarking is a systematic way to make site-to-site company comparisons and view a company's performance level in relation to that of top performers.

The application of the benchmarking results to identify performance gaps provides meaningful insight for organizations as they develop goals and the associated improvement plans to reach those goals. Repeated benchmarking monitors the rate of improvement. One example of a lesson learned that applies to numerous organizations relates to Process Safety Culture. The findings from the Process Safety Culture section indicate the largest range of results compared to the other sections. The underlying cause of the low scores was a direct outcome of not monitoring the process safety culture through the use of an employee survey. Without a way to assess their work force's values, attitudes and behaviors pertaining to process safety, an organization will have no meaningful basis for making and measuring improvements.

The global process safety incidents have produced tremendous need for the use of PSM benchmark assessment to measure and improve process safety performance. Initially focused on fatality rates, manufacturers quickly learned that having strong PSM system and effective layers of protection were important to prevent any loss of containment scenarios.

At RIL, PSM Benchmark assessment and results helped in identifying following three main interrelated but distinct benefits or observations: (a) Internal Benchmark of various RIL sites, (b) learning through process of understanding the benchmark questionnaire, and (d) need for a detailed analysis of existing systems implementation. We briefly discuss and illustrate examples of each benefit or observation, highlighting how the results could be used.

(a) Internal Benchmark of RIL multiple sites:

PSM Benchmark program conducted for six RIL sites brought to the surface best practices followed at different sites like: Number of indicators tracked and analysis of metrics trend, Quality control reviews of Process Hazard Analysis reports, action plan clearly defined from the Operational Discipline survey results, employee involvement through various PSM campaigns and some other actions related to operational discipline. These practices were beyond the existing standard and system requirements and became good learning for other sites. Also, these best practices will be additional inputs to the group level company documents on PSM. For example, the delta between the site with the highest score and the site with the lowest score in the Compliance with Standards section was 17%; offering a significant opportunity to analyze internal best practices and improve the level of performance to the highest score.

(b) Learning through process of understanding the questionnaire

The benchmarking questionnaire itself generated opportunities for learning new knowledge. Understanding the "why" part of questions called for discussions among various functions at each of the manufacturing sites and reinforced the understanding of PSM. A self-assessment process got started on the compliance to PSM requirement. Following table gives certain examples of learning and new knowledge for the questionnaire itself:

Table 1. Examples of learning and new knowledge from Benchmark questionnaire

Sr. no.	PSM Element	Question	Learning
1	Hazard Identification and Risk Analysis	How are overdue HIRA action items escalated in your organization?	Tracking system to be further strengthened for overdue HIRA action items.
2	Asset integrity and reliability	How is the metal thickness data from the periodic inspections analyzed for trends?	Thickness measurement trends should be analyzed, remaining life projections should be calculated, and adjustments to be made to inspection intervals.
3	Process safety culture	If your organization has utilized a survey tool to assess Process Safety Culture, how many questions were on the survey?	PSM Culture Survey to be established for the organization.

The validation visits from PTAI enhanced the quality of responses and the participation of various functions in the PSM benchmarking program.

(c) Need for a detailed analysis of existing systems implementation

The PSM benchmark results identified areas for further analysis to enhance the systems implementation for improved performance. At RIL, the trend of Process Safety incidents is varying for the last three years and there is no clear trend of reduction in loss of containment events from the metrics, inspite of implementing PSM systems for the last four years. For this, RIL has initiated a group level six sigma project on reduction in loss of containment events with support of an inter-site team. The project follows Define, Measure, Analyze, Improve and Control

(DMAIC) methodology of six sigma. Following is a brief of activities initiated for the above:

- A multisite team has been formed with a six sigma Black Belt guide.
- Literature on Process Safety and Six sigma circulated to team members.
- Project charter was prepared and circulated to team members and other stakeholders for inputs/suggestions and firmed up.
- Cause-Effect Matrix Format developed and Process mapping completed for PSM elements.
- Audit Report Analysis.
- Incident investigation report analysis.

4. New Directions and Next Steps

The PSM benchmark program results gave new insights to the RIL team on the status of PSM within RIL and in relation to the other participants. The gap analysis was converted to actionable recommendations at Group Level and individual site level. Next steps in the PSM excellence journey for RIL have been identified and these are

- Develop a formal system of Process Safety Culture Survey.
- Review of existing PSM metrics and make necessary amendments.
- Review Human Reliability/Human Factors/ Human-Machine Interface requirements, develop best practice requirements and make necessary amendments in Standards.
- Develop program for assessment of compliance with standards/codes/practices for existing equipment.
- Establish system to manage deviations from the industry consensus standards.
- Roll out the Management of Change –Personnel process formally.
- Actual implementation of Standards/Procedures/Practices in field:
    - Meeting target inspection/testing intervals, backlog of overdue inspections
    - Closure of recommendations of PHA, QRA studies
    - HIRA team members to meet qualification requirements
    - Closure of Audit recommendations
    - Implementation of temporary Management of Change
- Review RIL Group PSM Standards and Procedures for gaps and make necessary amendments.

A key benefit of the program for RIL is an Internal Benchmarking among its major manufacturing sites and sharing the best practices within the group. All the sites have developed action plans to close their own gaps in PSM practices.

The collaboration between PTAI and RIL gave new directions for RIL as well as for the Benchmark program itself. A development of two distinct Plan Do Check Act Assess (PDCAA) cycles, one for RIL and the other for PTAI has been the result of this collaboration.

4.1 Overview of PDCAA cycle

PDCAA is a 5-step methodology for process improvement. The stages are shown in Table 2.

Table 2. PDCAA stages

Plan	Establish objectives and how they will be achieved
Do	Put the plan into effect
Check	Verify that the process achieved the desired results
Act	Analyze any differences and their causes. Take action to improve things.
Assess	Analyze and reflect on evidence to make informed and consistent judgments on an ongoing basis for continual improvement.

The PDCAA Cycle for RIL is addressed through the following:

- Planning upgrade of RIL PSM standards to industry best levels and establishing systems to manage deviations, and developing appropriate leading and lagging metrics.
- The do part of the cycle will address the actual implementation of standards and practices in the field at different sites through a process of sharing the best practices by institutionalizing the inter-site network teams for the PSM elements.
- Check part will address strengthen the current PSM metrics tracking and review systems, IT enablement of dashboards for Process Safety at different levels of organization, and reflecting the updated requirements of compliance in the PSM audit protocols.
- A robust system of closure of recommendation and actions arising out of various studies and audits through a system of IT enabled universal work list and linking it to individual performance management system is under development which form the Act step of the cycle.

- Periodic reviews and assessment of PSM will complete the whole cycle of improvement.

4.2 Plan-Do-Check-Act-Assess- RIL Example: (Hazard Identification and Risk Assessment)

Plan

In this step, process needs are analyzed by examining a range of data available from the Benchmark assessment gap analysis relating questions in program to performance results. Data can be analyzed at the individual sites level and the company level. In this process, aspects on HIRA management system, including HIRA methodologies, team composition, criteria for HIRA team leaders/facilitators, level of training, techniques, risk criteria, recommendations and their status were covered.

Do

This step asks what group standards require and how they will be assessed. Forward action plan is made by looking at what is lagging in actual implementation in field ; such as, the timely closure of HIRA recommendations and minimum criteria/qualification for team members. Sites develop individual goals to improve their performance.

Check

In the "check" step, formal and informal assessments take place continually, to strengthen the performance of Metrics on HIRA elements. Sites can make use of comparative data to identify trends to study; for example, overdue HIRA action items.

Act

In this example the "act" step calls for standardization or having a robust system to ensure closure of overdue action items. Results from this cycle become input for the "assess" phase.

Assess

In this step, process of periodic reviews and monitoring of HIRA action items is done and tracked to ensure continual improvement.

4.3 Plan-Do-Check-Act-Assess- PTAI Example: (The benchmarking process and Process Safety Culture)

The PDCAA Cycle has also provided the opportunity for continuous improvement for PTAI and the PMS benchmarking survey.

Plan

Through the validation process the iterative nature of conference calls and site visits proved to be of significant value in providing accurate and consistent understanding of the survey questions. This resulted in comparable and meaningful results among the sites and RIL total scores.

Do

Reviewing the current benchmarking process and seeking feedback from RIL identified that enhanced validation and the creation of a more iterative process could be incorporated into the benchmarking process. A key area of needed improvement supporting the benchmarking process in the Process Safety Culture section is the measurement of Process Safety Culture. Both RIL and PTAI determined the need for a formal system for a Process Safety Culture Survey that provides more consistent measurement and comparable results of the safety culture that can be utilized for RIL internal site to site comparisons and RIL's sites to the external peer group. RIL and PTAI have discussed future collaboration on this project and the possible implementation to all of the RIL sites for internal benchmarking and review.

Check

Informal and formal process assessments are the key to continuous improvement of the PSM Benchmarking Survey and process.

Act

The act step calls for standardization of a formal process assessment tool to measure the benchmarking efforts required by participants and the actual benchmarking survey tool.

Feedback from participants needs to be evaluated and appropriate improvements to the program parameters and the benchmarking process implemented.

Assess

This is the process of continuous improvement and an environment of continuous learning. PTAI will periodically review the process and enhancements with CCPS to ensure an evolving process that grows in value-added elements.

5. Key success factors

The success of the PSM benchmarking survey hinged on four key factors; management support of a systematic approach, the strength of the validation process, the ability to report results with a detailed gap analysis that could be translated into specific action plans at a central and site level, and the seamless voice of RIL's Center for HSE Excellence and PTAI through strong communication and a systematic effort.

The management presentation of results was opened with comments from RIL's President, HSE, emphasizing the strategic importance and focus of PSM as a systematic process and not simply a process safety program. His comments were directed and focused on the areas he felt most important to taking RIL to the next level. This sent a very clear message of the strategic focus placed on PSM at RIL.

The continued dialogue and collaboration between RIL and PTAI has enhanced the double loop, PDCAA cycle. As a critical success factor this creates ongoing and new opportunities for both.

6. Conclusion

Process safety is a mature, but continuously evolving concept with practices and procedures varying widely among the refining, chemical and chemical processing industry. Moving from a compliance model to a risk-based model requires a comprehensive approach to managing process safety. Substantial improvement opportunities exist for even the current top performers. Specifically, no site has made a perfect score on any of the risk-based elements of the PSM benchmark.

The PSM benchmarking program is a unique methodology to allow companies to identify how large the gap is to top performers and to "learn" how to prioritize resource allocations based on the greatest impact opportunity to improve their process safety management system. This shifts the understanding of current practices to the lessons learned that are measurable and actionable. The benchmarking process identifies top quartile management practices while highlighting not only a company's strengths, but also potential development opportunities. The program's scoring system provides a marker to track improvements and measure the gaps within the company sites and against external top performers.

Using a systematic Process Safety Management System benchmark drives lessons learned to strategic and organizational learning.RIL considers periodical assessment of its PSM through repeat benchmarking at regular intervals as a good value proposition to monitor trends of performance and to achieve pacesetter performance compared to industry peers.

7. References

[1] American Petroleum Institute, API RP 754, "Process Safety Performance Indicators for the Refining and Petrochemical Industries, First Edition," 2010.

[2] Center for Chemical Process Safety CCPS®, "Guidelines for Risk Based Process

Safety," John Wiley & Sons, Inc., Hoboken, NJ, ISBN 978-0-470-16569-0: liii-liv, 2007.

9TH GLOBAL CONGRESS ON PROCESS SAFETY

"Process Safety Management - A Practical View"

German Luna-Mejias, P. Eng.
ACM Facility Safety
Suite 700 940 6th Ave SW, Calgary, AB, Canada
gluna@acm.ca
guatire2@gmail.com

Prepared for Presentation at
American Institute of Chemical Engineers
2013 Spring Meeting
9th Global Congress on Process Safety
San Antonio, Texas
April 28 – May 1, 2013

UNPUBLISHED

"Process Safety Management - A Practical View"

German Luna-Mejias, P. Eng
ACM Facility Safety
Suite 700 940 6th Ave SW, Calgary, AB, Canada
gluna@acm.ca
guatire2@gmail.com

Keywords: Process, Safety, Management, Practical

Abstract

Process Safety Management (PSM) was developed over 20 years ago to reduce hazards and risk for the Oil Industry. In the last 3 years serious accidents have happened in the Oil Industry that requires us to re-evaluate PSM programs. It has been observed in many HAZOP studies, that the HAZOP (part of a PSM program) does not reflect a real commitment to safety, but the "need" to meet regulations. This paper's initiative is to call organizations/companies to think and be safe instead of just fulfilling project or government rules. Full involvement of the organization, starting with managers, is required for the PSM program to work. In this paper, practical examples will be drawn from the successful implementation of a PSM program from several years ago in a refinery. Additionally, the reasons for this implementation to be considered successful will be discussed at length.

1. Introduction

1.1 History

On June 1st, 1974, a Cyclo-hexane vapor cloud explosion killed 28 workers and injured 53 others; 1800 houses were damaged in the nearby area in Flixborough, England. [1]

On Dec 3, 1984, an uncontrolled release of methyl-isocyanate (MIC) gas from a pesticide plant in Bhopal, India, resulted in more than 3,800 fatalities and approximately 11,000 injuries with different degrees of severity. [2]

In 1985, the American Institute of Chemical Engineers (AIChE) formed the CCPS (Chemical Center for Process Safety) quickly set out to publish its first process safety guideline book: Guidelines for Hazard Evaluation Procedures.

On July 6, 1988, the offshore platform Piper Alpha, located in the North Sea and owned by Occidental Petroleum, suffered a series of explosions and fires that resulted in the deaths of 165 crew members and the total loss of the platform. [3]

On Oct 23, 1989, an explosion in the Phillips refinery in Pasadena, Texas killed 23 people and injured 314 more. Impact of the explosion was registered as a 3.5 on the Richter scale. [4].

In 1990, the American Petroleum Institute issues the recommended practice 750: "Management of Process Hazards" which can be considered the first step onto the Process Safety Management programs. [5].

In 1992, OSHA formally recognized the vital importance for managing highly hazardous chemical processes within the US. A company that is involved in highly hazardous chemical processes or activities is mandated to implement a process safety management system. [9]

In 1994, CCPS modified the elements of the PSM program recommended in API RP-750 [5] to issue a new "standard" for PSM that includes more elements: "Guidelines for implementing Process Safety Management System"

After a period of "relative calm", big accidents in the oil and gas industry occurred again:

- BP Texas city, 2005 [6]
- BP Macondo, Gulf of Mexico 2010 [7]
- San Bruno, California 2011 [8]

Many disasters have impacted the process industry, from Flixborough to Macondo and elsewhere many more. Through the ashes of these tragedies, strong change has been advocated by the public and their government's regulators to hold companies accountable for their negligence.

1.2 Process Safety Management

Process Safety Management (PSM) is a methodology that was developed after the lessons from several major accidents showed that such events can be difficult to prevent using traditional occupational health and safety approach, which tends to focus on the interface between the individual operator and the equipment or process.

The 3 keywords of this philosophy are:

- PROCESS: related to process issues. i.e. fires, explosions, etc
- SAFETY: Safety regulation, safety concern
- MANAGEMENT: anyone is a manager in some degree, requires employment participation.

PSM is the application of management principles and systems to the identification, understanding and control of process hazards to prevent the release of highly hazardous chemicals such as toxic, reactive, flammable and/or explosive substances (process-related events), that may lead to injuries and incidents.

The goal is to minimize process incidents by evaluating the whole process. The PSM standard (OSHA 29- Part 1910.119) [9] is the regulatory framework for process industries in the US and is quickly becoming the industry "best practices" standard for responsible companies worldwide.

The PSM system originally suggested by the CCPS (Guidelines for implementing Process Safety Management System) consists of 12 elements:

1. Accountability: objectives and goals
2. Process knowledge and documentation
3. Capital project review and design procedures
4. Process risk management
5. Management of change
6. Process and equipment integrity
7. Human factors
8. Training and performance
9. Incident investigation
10. Company standards, codes and regulations
11. Audits and corrective actions
12. Enhancement of process safety knowledge.

The intent of this paper is to present some of the elements of PSM that were considered the most important for a specific company; not to debate the number of elements or which one is more important than the other.

Regulations should not be the sole driving force for the process industry to "think safe". Applying the PSM program accurately and effectively requires more than just the requirement to fulfill government regulation but a whole company culture to be involved in making the workplace safe. This paper's initiative is to encourage organizations/companies to think and be safe instead of simply fulfilling a project or complying with government regulations.

The examples given in this paper correspond to a company where PSM was not mandatory by law, but it was part of the company's standard as a way to make sure the company complied with the International O&G industry standards.

2. The practical Approach

In this section, examples of the approach taken by an O&G company to implement PSM are given for what was considered the most relevant part of the program. Not all elements are discussed as the key elements described below will support all other elements. The practical approach also includes unfortunate situations that happened before the implementation of the program to emphasize the need for it.

Elements analyzed in this document are:

- Management commitment (Element 1)

- Process Safety Information (Element 2)
- Process Hazard Analysis (Elements 4 & 5)
- Operating Procedures (Elements 6,7 & 8)
- Hot work Permit (Element 6)
- Training (Element 8)
- Pre-start up Safety Review (Element 9)
- Emergency Planning and Response (Elements 4 & 12)
- Program Audits (Element 11).

2.1 Management Commitment

Management commitment is fundamental to the PSM program. Exercising and eating well is something that people need to do for their own good and not because it is a mandate or a command to obey. Exactly the same happens with the PSM program: if people are not convinced of the importance of safety, the PSM program won't last. The first safety best practice identified, and one which is essential to any good safety program, is a management commitment to safety.

No company can have a successful PSM program if the management of that company is not committed to the concept of safety. Every company has an obligation to ensure, as much as possible, the safety of its employees. Not only does the company have a moral obligation to ensure the safety of their workforce, it is also good business to do so. Keeping the number of incidents down will also help the company to maintain few lost time days, low work injury indexes and increased profitability.

In a particular refinery, the General Manager wanted to develop a PSM program and designated a top manager to be in charge of the PSM program. This designated manager's first action was to convince all top managers of the importance of the PSM program. Convincing the other managers was not based on: "I said so, and I have been appointed by the GM" but with education and conviction.

The training of managers and cascading this knowledge to all personnel with supervisory responsibilities started with the video forum about the Piper-Alpha accident (one of the objectives of the forum was to recognize that this scenario may happen to that facility). The Piper-Alpha video, in conjunction with API 750 video, was presented to all personnel.

"Safety oriented Management behavior" was a key, as the employees know who the managers are, and they observed their commitment to safety. All personnel understood that they were "living by example" instead of "giving lip service to their own safety rules". The exposure of management risk and demonstration of Safety Concept by action prevented the scenario of: "Well, management doesn't need to operate safely, neither do we."

Additionally, financial incentives to personnel were also created during the earlier phases of the program to promote participation within it. After a period of time, people became conscious about safety and they started to behave safely even when "nobody is looking"; not because of a reward but they were convinced of PSM as a way of life.

Part of the management commitment to safety was the implementation of safety related training programs with the same level of importance as technical training. The program was successful because there was support from the entire management team, and management was involved in the formulation, communication and enforcement of its safety program. All levels of management were visibly demonstrating their commitment to safety.

2.2 Process Safety Information.

Process Safety information (PSI) includes design information, mechanical design and working knowledge. These 3 elements are linked to good quality P&IDS, which was the driving reason the company decided to move in the direction to update the P&IDs as the first step under the PSI element.

This was critical to the facility as it was a 60 year old refinery. In several cases the installation of jump-overs, minor changes and similar modifications were done in the field without any analysis or documentation on the P&IDs. Some minor incidents (spills, spray, etc.) were found to be a consequence of inadequate information on P&IDs.

As a result, a team composed with very experienced operators and young engineers was formed to update all the P&IDs of the facility, with two main purposes: pair young engineers with very experienced personnel and teach them how to update P&IDs and to comply with PSM.

A few years later, when most of the 30+ year operators were retired, these updated P&IDs were successfully used to train the new group of operators in the facility.

Another group, supported by Technical Services, began work to update and maintain the MSDS (Material Safety Data Sheets) for the facility. This action proved to be very effective for the firefighters as they were properly trained in the materials they may have to deal with in case of emergency. Before that they were relying on an operator telling them what to do during an emergency scenario.

Nowadays, this situation (outdated PSI) is a concern for many facilities: In a study performed in 2010 by OSHA IN 2010 [10]; 22.2% of the deviations (not compliance with PSM) were associated to non-updated PSI.

2.3 Process Hazard Analysis (PHA)

The purpose of this analysis is to minimize the likelihood of the occurrence and the consequences of a dangerous substance release by identifying, evaluating, and controlling the events that could lead to releases.

The management team and all personnel were asked to answer the following questions:

- Is my facility safe to operate?

- In the case of an event happening, am I protected?
- Is my facility protected?

This is how the HAZOP concept was introduced to the facility. A group of people were trained in HAZOP facilitation and they were in charge of implementing the PHA program.

Implementation of PHA, as with any other type of new procedure, was a challenge. At that time only the HAZOP was considered for implementation; later, additional PHA methods were added to the portfolio. Manual recording of PHA scenarios was a challenge, but the use of PHAs (manual or automated) continues to be vital to gauge whether a facility is protected or not.

In the earlier stages of the implementation, PHA-HAZOP was used as part of Management of Change element, but later was included in the evaluation cycle of all projects.

2.4 Operating Procedures

In the old facility, procedures were transmitted by oral tradition and there were very few written procedures, but even those written were not clear for non-experienced individuals. They were not developed with consideration for new personnel in that corresponding area.

In several opportunities, lack of written procedures led to accidents when one of the "guru's" was not there to execute a particular procedure or when a step was missed in a complicated procedure.

To prevent this from happening repeatedly, a team was built consisting of very experienced operators with experts in writing procedures and process engineers to start writing the procedures for each unit. Managers defined the most critical and most urgent ones to be done first. It was a medium-term effort that produced great results when a few years later the operations crews were modified to include people with technical background but almost zero operational experience.

2.5 Hot Work Permit

In traditional operation, the Operations Supervisor is also in charge of supervising the maintenance activities in his area. About 25 years ago, a request for a work permit was made to the Operations Supervisor; he erroneously determined that the risk was not too high, as he was busy with the day-to-day operation. He did not check the actual situation in the field. Unfortunately, the line where the work was going to be performed was not checked before and was filled with light naphtha (not properly drained) and this led to a fatality when welding.

After this accident and as a part of PSM program, a new work permit procedure was established:

- A dedicated Operations Supervisor for maintenance activities, being the only authority to sign-off work permits.
- A procedure was developed to include documentation and proper sign-off
- Training was developed for all involved parties about how to perform using the new Hot Working Permit, including contractors.

2.6 Training

The training of all personnel is a key element of Process Safety. The training must be on a continuous basis and at all levels of the organization. In this particular company, the kick-off of the PSM was to present the Piper-Alpha accident video to all levels of the organization from the top Management to the bottom of the organization. Also the API-750 Video was presented to all personnel, the objective: understand the risk, study the failures and realize that this situation was very likely to happen to that facility.

The importance of training is that it must to be continuous, refreshed on regular basis. Training must be developed for new employees and all contractors, with a rule of "nobody in" without safety training. The safety training included an exam; and those who do not pass the exam cannot get into the facility. To make sure PSM works, training is a key element and dedicated personnel to fulfill this target is recommended.

2.7 Pre start-up Safety Review

According with API RP 750 [5] the PSSR (Pre Start-up Safety Review) should be performed for new and modified facilities that are subject to this recommended practice to confirm that:

- Construction is according with specifications
- Procedures are in place (including safety procedures)
- PHA recommendations are closed out.
- Training for all operators is provided

It is fairly common to see confusion between PSSR for commissioning and PSSR during normal S/D for maintenance; as API 750 indicates, it is needed under both scenarios.

Before the implementation of PSSR, the responsibility of checking the facility prior to start up was only for operations. Normally this was done by an Operations Supervisor with other duties in the area and sometimes relying on another operator or on a Maintenance Supervisor. In this facility the need for PSSR was identified in an unfortunate incident:

An Operations Supervisor on his last night shift was doing the plant checking for start-up after he was informed by maintenance about completion of work. He was the only person in charge of checking the facility and provided the team with the "go ahead" for start-up 30 minutes before the end of his shift. The heating of the plant started and about 2 hours later a hot naphtha leak spilled over a hot heat exchanger producing a large fire that required 2 months of additional plant shutdown to repair damages. A good example of what "not to do".

A PSSR exercise takes a longer time to perform and needs a multidisciplinary team but will prevent incidents or accidents.

2.8 Emergency Planning & Response

Firefighter training is very important. In 1984, 100 municipal firefighters died when they were called to fight a fire in a heavy diesel tank in another location [11]. They handled it as a regular fire and high quantities of water were directed to the roof of the tank. No effort was made to cool off the walls of the tank and they were not aware of the presence of water in the bottom of the tank. When the diesel in the bottom of the tank reached the appropriated temperature, a boil over situation occurred.

One of the key elements in an Emergency Response plan is the knowledge of the hazards of a system that may create an emergency. As part of a PSM program, an intensive training on Process for non-technical personnel was given to firefighters (permanent and volunteers). Prior to the implementation of the PSM program, in the event of an emergency, the Firefighting Supervisor communicated directly with the Operations Supervisor on site, to obtain information about the potential hazards to people and facility. The Firefighting Supervisor had only a very limited knowledge of the facilities and associated hazards.

Emergency dry-runs were also implemented weekly according with a program developed on the most critical areas. This practice involved regular firefighters and volunteer firefighters.

Another step that was taken as a part of keeping the firefighters engaged was the annual recognition to all firefighters (rewards, celebrations, etc.).

2.9 Program Auditing

All the PSM elements should be audited on a regular basis to ensure the program effectiveness and performance. Programs like PSM require a lot of involvement from the Auditing team, the PSM leading team will define which parts of the program are most critical and it will be the start-up point for the auditing.

Audit process will always be needed, even when every person in the organization is "in the safety mindset".

3. Conclusions

- Implementation of the PSM program is a time consuming effort.
- Dedicated personnel are required for leading the change.
- Management actively involved in the PSM program is a must.

3.1 Tangible benefits from the implementation of PSM

- In a facility with a record of 4 fires per year, they were able to reduce the number of fires down to 1/year.
- Training of operators improved the operations of the facility reaching production records and reliability records.
- The Safety concept was built-in (be safe when nobody is looking)
- Ongoing, non-prescriptive, performance-based, participation at all levels.
- Facility received a Safety Award as "Safest in class" facility.

4. Recommendations

- PSM is a MUST for all O&G facilities
- It is a way of living! I encourage all O&G companies to follow this "way of living".

5. References

[1] Health and Safety Executive, 'The Flixborough Disaster: Report of the Court of Inquiry', HMSO, ISBN 0113610750, (1975)

[2] Chouhan TR "The Unfolding of Bhopal Disaster". Journal of Loss Prevention in the process industry **18** (4–6): 205–208. (2005).

[3] PatC-Cornell, M. E. "Learning from the Piper Alpha Accident: A Postmortem Analysis of Technical and Organizational Factors". Risk Analysis, Vol. 13, No. 2, 1993

[4] Bethea, R.M. "Explosion and Fire at the Phillips Company – Houston Chemical Complex, Pasadena, TX" Chemical Engineering Department, Texas Tech University, Lubbock, TX 79409-3121 (2007)

[5] API Recommended Practice API RP-750. "Management of Process Hazards". American Petroleum Institute, 1990.

[6] CSB. Investigation Report: Refinery Explosion and Fire (15 killed, 180 injured). Report # 2005-04-I-TX; Chemical Safety Board, Washington, DC. 2007

[7] British Petroleum Oil Company (BP). "Deepwater Horizon Accident Investigation Report" (2010)

[8] NTSB "Pacific Gas And Electric Company Natural Gas Transmission Pipeline Rupture and Fire" San Bruno, California. NTSB Report # PAR-11-01 (2011).

[9] OSHA. "Process safety management of highly hazardous chemicals". OSHA 29- Part 1910.119 (1992)

[10] Jordan Barab., "Learning from Industry Mistakes" presented at NPRA National Safety Conference, May 2010

[11] MARAVEN S.A. "Investigacion del Accidente de Tacoa" (Tacoa's Accident Investigation, In Spanish). Caracas, 1985.

(43an) Auditing the Process Hazard Analysis - Key Characteristics to Scrutinize in a Good-Quality HAZOP Study and Supporting Effective HAZOP Study Updates

Steven T. Maher[1], Michael A. Nour[1] and Kristin D. Norton[2], (1)Process Safety, Risk Management Professionals, Irvine, CA, (2)Process Safety, Risk Management Professionals, Inc., Irvine, CA

Although Process Safety Management (PSM), requiring good-quality Process Hazard Analyses, was promulgated in the United States in 1992, and HAZOP Study techniques date back to the mid-1970's, the HAZOP Study technique is still not uniformly applied in a high quality fashion. The objective of this paper is to provide experience-based tips on Process Hazard Analysis auditing, and some key characteristics that the auditor can scrutinize to assure quality implementation. Analogously, a similar set of characteristics that can facilitate future HAZOP Study updates can also contribute to a quality HAZOP Study that supports the spirit of Safety Management System objectives for the identification of hazards potentially impacting plant personnel, the community, and the environment. This paper will also review tips for structuring the HAZOP Study to facilitate later updates and revalidation.

See more of this Session: Electronic Poster Session
See more of this Group/Topical: Global Congress on Process Safety

9TH GLOBAL CONGRESS ON PROCESS SAFETY

Vapor Cloud Explosion Test of Air-Inflated Fabric Structure

Brad Horn
Baker Engineering and Risk Consultants, Inc.
3330 Oakwell Court, Suite 100
San Antonio, TX 78218-3024
BHorn@BakerRisk.com

Matthew T. Edel, P.E.
Baker Engineering and Risk Consultants, Inc.
3330 Oakwell Court, Suite 100
San Antonio, TX 78218-3024
MEdel@BakerRisk.com

Prepared for Presentation at
American Institute of Chemical Engineers
2013 Spring Meeting
9th Global Congress on Process Safety
San Antonio, Texas
April 28 – May 1, 2013

UNPUBLISHED

Vapor Cloud Explosion Test of Air-Inflated Fabric Structure

Brad Horn
Baker Engineering and Risk Consultants, Inc.
3330 Oakwell Court, Suite 100
San Antonio, TX 78218-3024
BHorn@BakerRisk.com

Matthew T. Edel, P.E.
Baker Engineering and Risk Consultants, Inc.
3330 Oakwell Court, Suite 100
San Antonio, TX 78218-3024
MEdel@BakerRisk.com

Keywords: vapor cloud explosion, explosion test, air-inflated structure, fabric structure, blast load, thermal load, structural response

Abstract

Portable buildings are gaining widespread use at refining and chemical processing facilities, particularly for facility siting requirements. Although trailers and blast resistant modules (BRMs) can be used for many modular applications, in certain situations fabric structures may provide a viable alternative. However, the blast response of fabric structures is neither clearly defined nor well understood throughout the blast design community.

One such modular fabric structure is an air-inflated structure. A manufacturer of this type of structure recently funded the testing of one of their specimens for the effects of a vapor cloud explosion (VCE). The VCE for the test was initiated using propane and an internal matrix of congestion inside a three-walled cubicle, referred to as the Deflagration Load Generator (DLG) test rig. The test involved a deflagration event that imparted a thermal flux and a maximum free-field pressure of 2.9 psi with an average duration of 31 ms at the front wall of the air-inflated structure.

The reduced blast loads and thermal effects inside the structure were compared to those outside. In addition, the attenuation of blast loads behind the structure was also examined, as well as shielding effects. The structural response of the test specimen was captured with high-speed video, and observations on the motion of the structure are provided.

1. Introduction

Baker Engineering and Risk Consultants, Inc. (BakerRisk) has developed a Deflagration Load Generator (DLG) test rig that can supply VCE blast loads to a variety of structures in order to provide structural response test data to manufacturers (i.e., product proof testing). The DLG rig produces a VCE blast load with a blast wave shape, pressure, and duration similar to the blast loads that are generated by accidental VCEs at chemical processing facilities. Dynamic Air Shelters, Ltd. (DAS) contracted a test of its manufactured air-inflated shelter.

The test involved imparting a VCE blast load on a DAS air-inflated shelter. The shelter was erected to the manufacturer's specifications and then subjected to a VCE blast load using the DLG test rig. The target blast loading for this test was a centrally peaked blast wave with a peak free-field blast pressure of 3.5 psi and 50 ms duration at the shelter wall closest to the DLG rig. Data recorded include pressure histories, thermocouple and heat flux measurements, and video.

2. Test Setup

This section describes BakerRisk's DLG test rig, the DAS air beam shelter, and the instrumentation used to measure and record data during the test.

2.1 DLG Test Rig

The DLG test rig is shown in Figure 1. The test rig is a steel box with one open side that is initially covered with a thin plastic sheet to allow a fuel-air mixture to be developed within the test rig. During testing, the test rig is filled with a propane-air mixture at a desired concentration, which is then ignited at the middle of the back wall. An obstacle array within the test rig causes the flame to accelerate and generate a blast wave. The resulting blast wave propagates out toward the target structure.

2.2 Air Beam Shelter

DAS constructed the air-inflated shelter shown in Figure 1, which was located directly in front of the DLG test rig during the test. The shelter system consisted of a ballistic-resistant wall and an air-inflated shelter with overall dimensions of approximately 46 ft. (14 m) wide by 56 ft. (17 m) long and 16 ft. (4.9 m) tall. The blast wall was made of PVC trapezoidal bags that were filled with material, in this case sand, and then hung between supporting wall frames spaced at approximately 6.7 ft. (2.0 m).

Figure 1. Air-Inflated Shelter (on left), Blast Wall, and DLG Test Rig (on right)

The air-inflated members were continuously pressurized and created the structural support for the shelter. Doors were located on two walls parallel to the direction of the blast propagation. The interior air beams are shown in Figure 2. Except for the instrumentation, the interior of the shelter was vacant during the test.

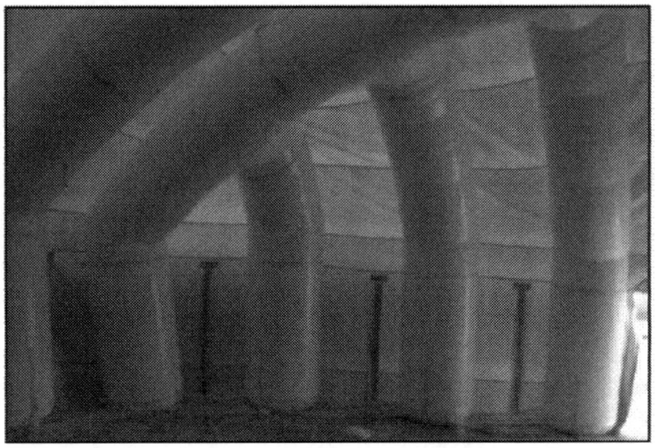

Figure 2. Air-Beam Columns

A schematic showing the location of the test rig, shelter and instrumentation is provided in Figure 3. The standoff distance between the test rig and the shelter was 21 ft. (6.4 m).

Figure 3. Instrumentation Layout Map

2.3 Instrumentation

During the test, the pressure, heat flux, and temperature histories were measured at locations of interest. Both regular and high speed video of the shelter structural response were captured.
A total of 29 dynamic pressure gauges sampled at 10,000 samples per second were used to measure the blast loads. Pressure gauge locations are shown in Figure 3; the gauge locations relative to the front of the test rig are provided in Figure 5. A total of 13 pressure gauges were positioned to measure the blast load at points of interest on the shelter and its components. Six gauges measured blast loads at the ballistic resistant wall, with three on the ground at the base of the wall and three mounted on the wall 4 ft. (1.2 m) above the ground gauges in order to measure the reflected load. One pressure gauge was located on the ground between the ballistic resistant wall and the shelter wall, and one pressure gauge was located at each shelter doorway. Three pressure gauges were located within the shelter interior, and one pressure gauge was located just outside the rear wall of the shelter.

Thermocouple and heat flux sensors were placed at the ballistic resistant wall and inside the shelter, coincident with pressure gauge locations 20 and 23. Each sensor was elevated 4 ft. (1.2 m) above grade and oriented toward the DLG test rig. Each thermocouple was sampled at 20,000 samples per second.

Video recording devices were used to document the test and record the shelter structural response to the blast loading. The high speed video (HSV) camera was positioned in the same plane as the leading edge of the ballistic resistant wall, while two high definition (HD) video cameras, two exterior spy cameras, and one interior spy camera recorded other angles.

3. Test Results

Figure 4 shows the structural response recorded by the high definition video camera, capturing at 30 frames per second.

Figure 4. VCE Loading Applied to Shelter

3.1 Blast Loads

A peak pressure measurement map with pressure values listed in green font is shown in Figure 5. Table 1 provides the blast load summary which includes the pressure gauge location, standoff distance from the test rig, measured peak pressure, positive impulse, and both the measured and calculated durations. The calculated duration assumes a perfectly triangular blast pressure history (i.e., duration is twice the impulse divided by the peak pressure). Table 2 provides a summary of the peak reflected blast pressures measured at the ballistic resistant wall and the corresponding calculated peak free-field (i.e., side-on, incident) blast pressures.

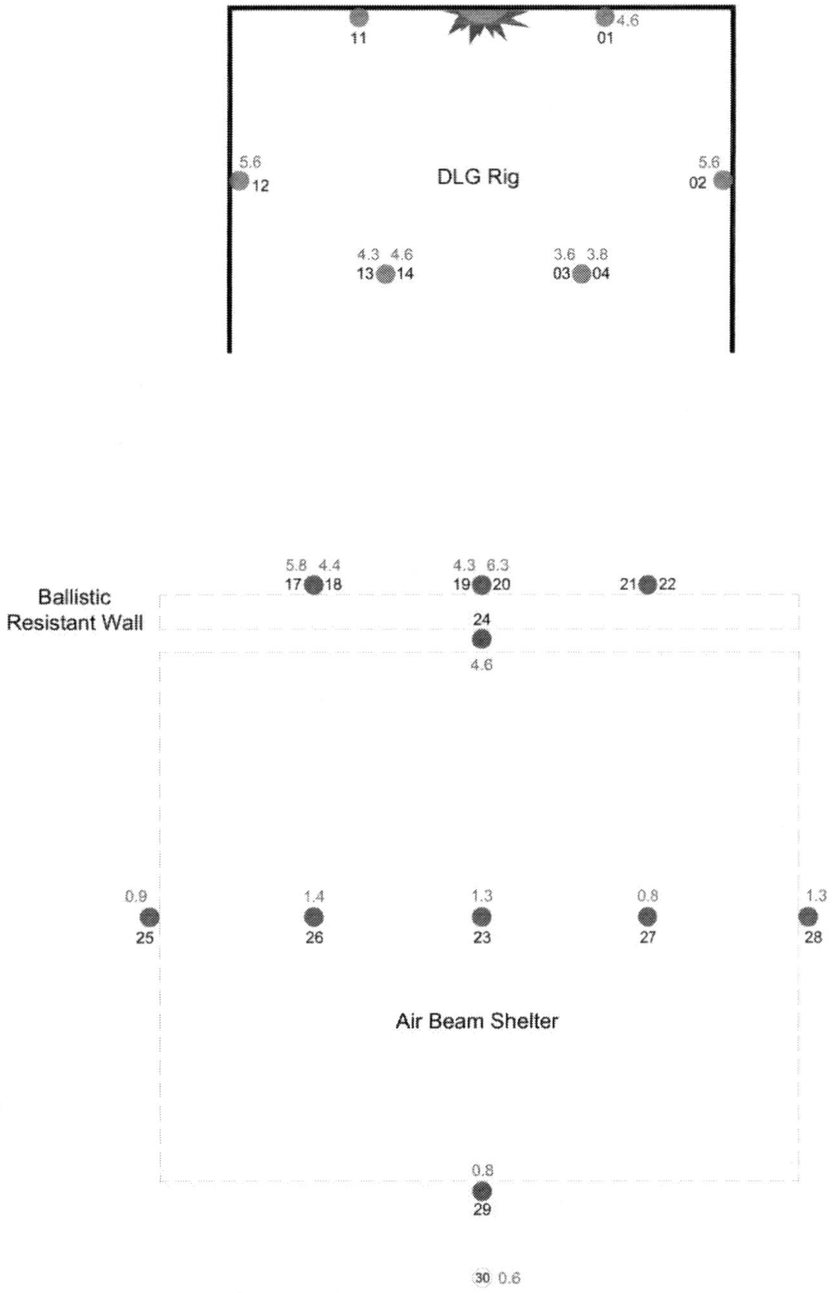

Note: Pressure values directly in front and behind the ballistic-resistant wall are reported as the adjusted peak values.

Figure 5. Peak Pressure Measurement Map (psig)

Table 1. Blast Load Summary

Location	Gauge	Distance From Test Rig (ft.)	Measured Peak Pressure (psig)	Measured Positive Impulse[1] (psi-ms)	Measured Duration (ms)	Calculated Duration (ms)
Rig Interior	2	0	5.6	206	120	74
Ballistic Resistant Wall[1]	17	21	5.4	90	48	31
	18	21	3.9	67	42	31
	19	21	3.9	70	51	32
	20	21	5.8	93	48	30
Front Shelter	24	24	4.6	95	57	41
Shelter Doorway	25	48	0.9	19	57	41
	28	48	1.3	28	57	41
Shelter Interior	23	48	1.3	29	53	45
	26	48	1.4	29	53	41
	27	48	0.8	16	53	41
Rear Shelter	29	71	0.8	16	45	41
Far Field	30	76	0.6	10	30	33

Table 2. Ballistic Resistant Wall Blast Pressure Summary

Location	Gauge	Distance From Test Rig (ft.)	Measured Peak Pressure (psig)	Adjusted Peak Pressure[1] (psig)	Calculated Free Field Pressure[2] (psig)
Ballistic Resistant Wall[1]	17	21	5.4	5.8	2.7
	18	21	3.9	4.4	2.0
	19	21	3.9	4.3	2.0
	20	21	5.8	6.3	2.9

[1] *Pressure values adjusted for pre-event signal ringing*
[2] *Calculated from reflection factor*

The maximum peak pressure in the DLG test rig was 5.6 psig (0.39 bar). Gauges at the ballistic resistant wall were positioned to measure the reflected load on the face of the wall or shelter.

The peak pressures and impulse at the wall were adjusted to account for the signal ringing prior to the arrival of the blast wave and are shown in Figure 7. The maximum adjusted peak pressure was 6.3 psig (0.43 bar). The pressure signals along the centerline of the shelter in the direction of the blast are shown in Figure 6. Two of the six gauges were invalid and were therefore discarded.

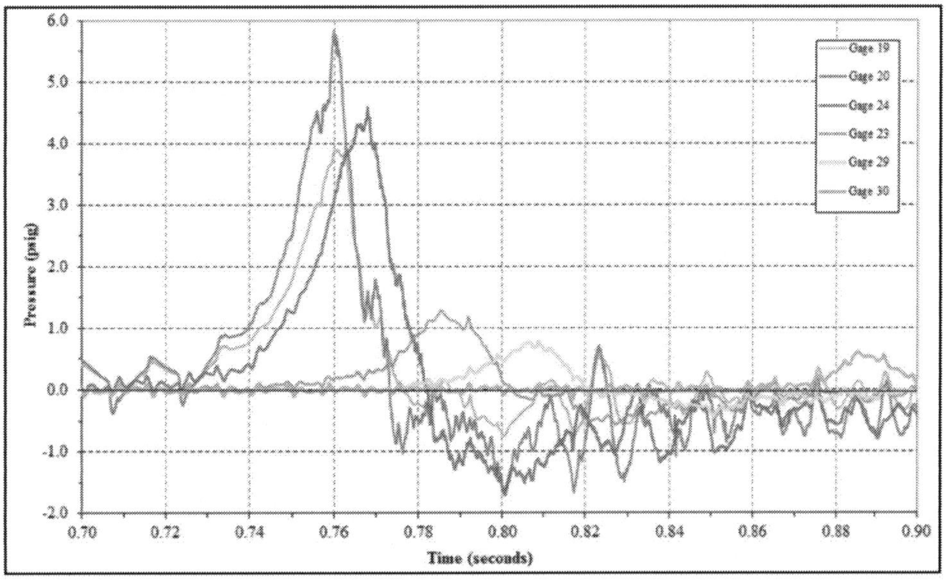

Figure 6. Pressure Histories along the Front Center Gauge Lane

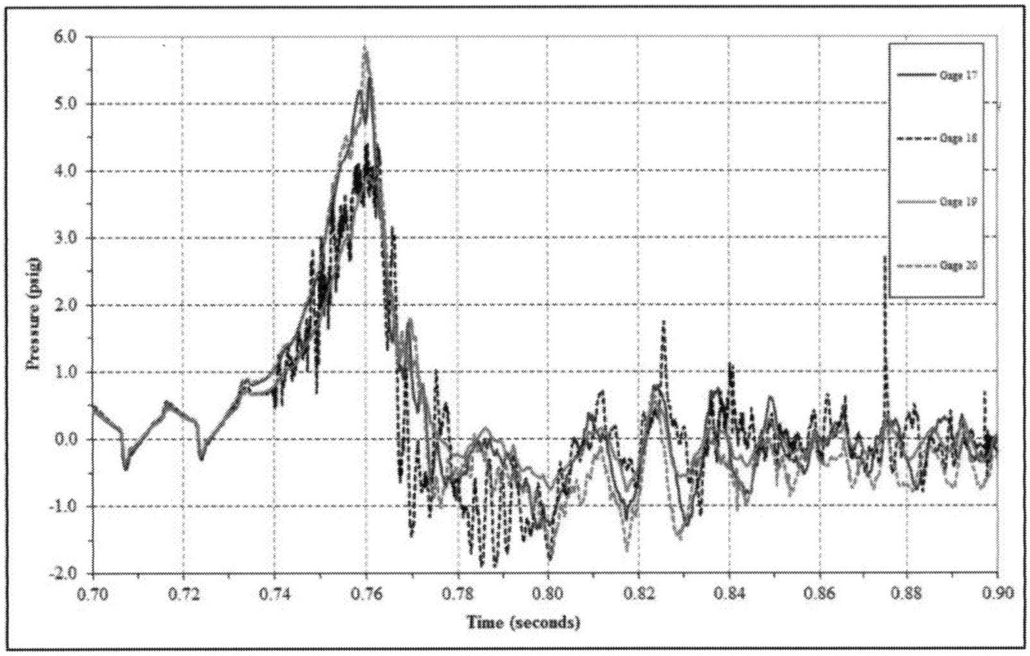

Figure 7. Pressure Histories at the Ballistic-Resistant Wall

The measured peak pressure and impulse between the ballistic resistant wall and air shelter were 4.6 psig (1.4 bar) and 95 psi-ms (6.6 bar-ms), yielding a calculated duration of 41 ms. Pressure histories were measured along the mid-span of shelter. The pressure histories at these mid-span locations are shown in Figure 8. For the doorway exterior locations, the largest measured peak pressure was 1.3 psig (0.09 bar) with an impulse of 28 psi-ms (1.9 bar-ms) and a calculated duration of 41 ms. The largest measured peak pressure among the interior gauges was 1.4 psig (0.1 bar) with an impulse of 29 psi-ms (2.0 bar-ms) and a calculated duration of 41 ms. Behind the shelter, the measured peak pressure was 0.8 psig (0.06 bar) with an impulse of 16 psi-ms (1.1 bar-ms) and a calculated duration of 41 ms.

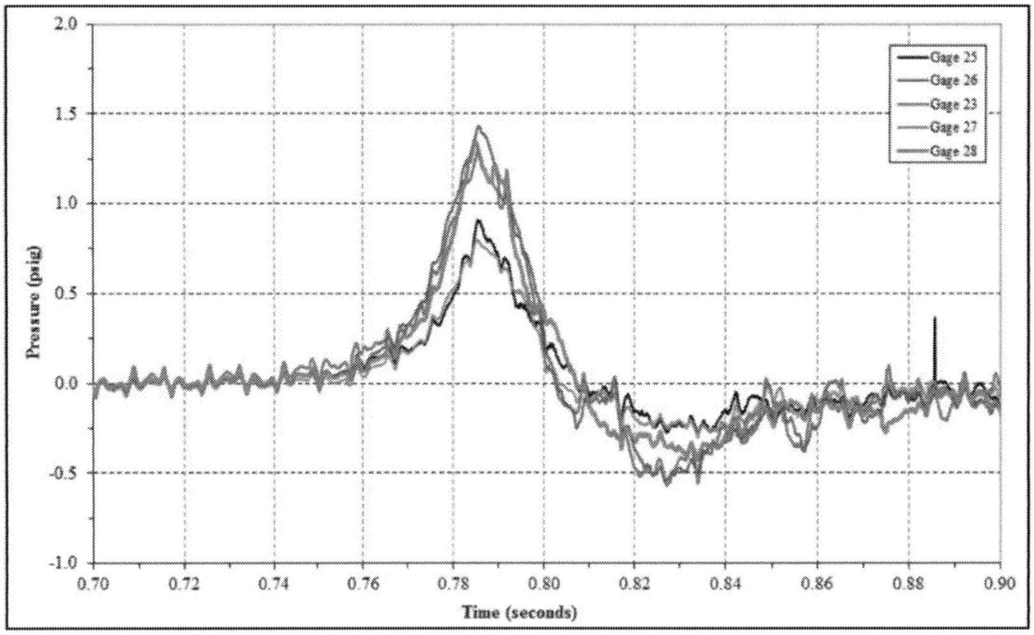

Figure 8. Pressure Histories at Shelter Mid-Span Locations

3.2 Thermal Loads

Figure 9 shows the measured temperature just past the exterior face of the ballistic resistant wall. This peak signal exceeded the range of the signal conditioner and was "clipped" at approximately 1100 °F (590 °C). The estimated actual maximum temperature at this location was just under 1500 °F (815 °C) based on an extrapolation of the signal. The measured signal inside the shelter indicates that the temperature had increased less than 5 °F (3 °C), as shown in Figure 10. That is, while the temperature just outside the shelter increased dramatically due to exposure to the fireball, the temperature within the shelter barely rose.

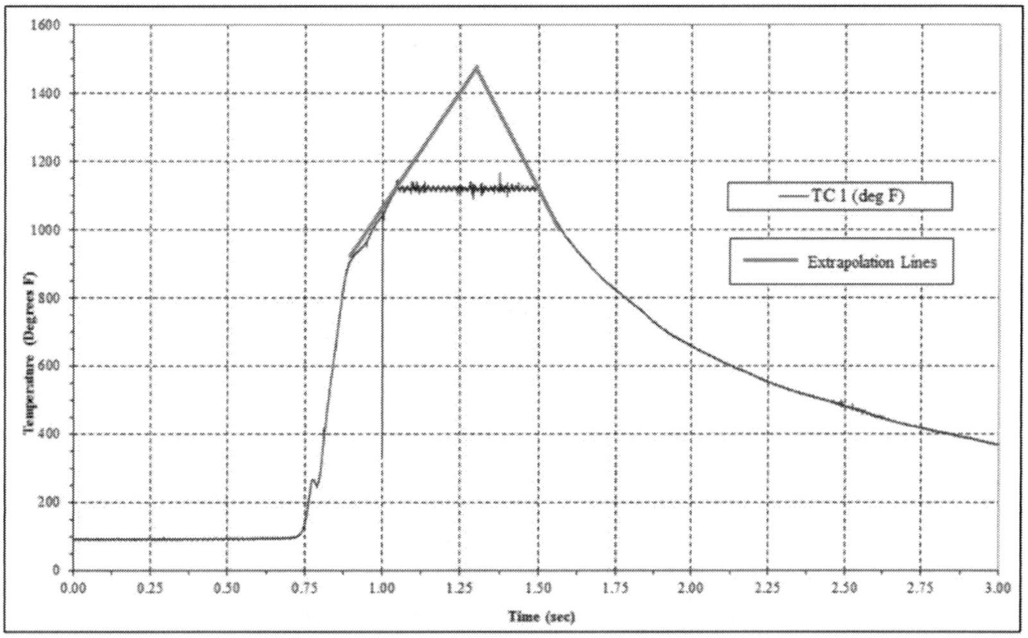

Figure 9. Exterior Temperature History

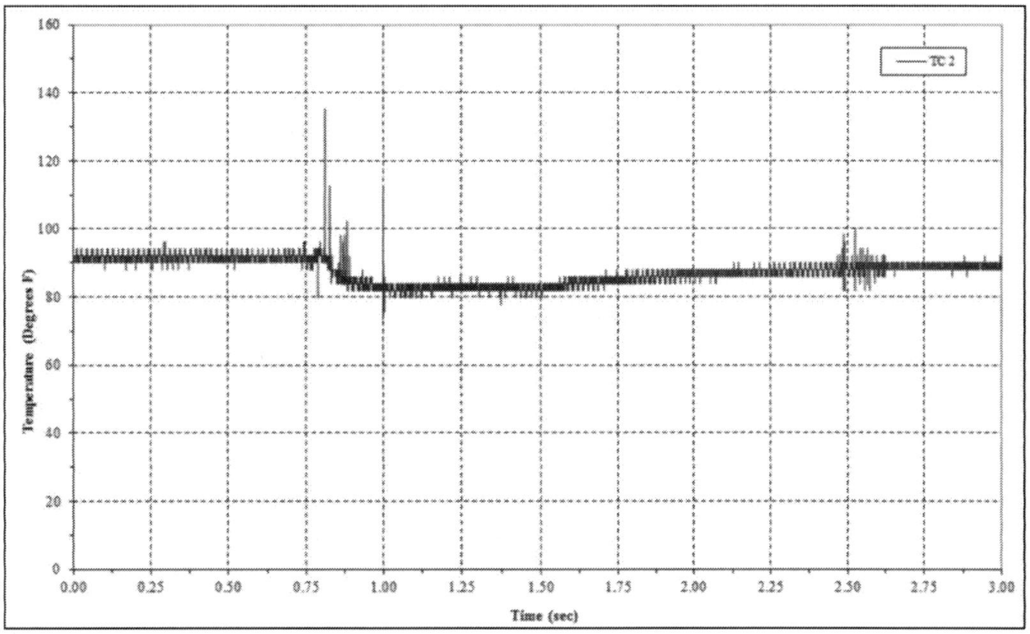

Figure 10. Interior Temperature History

The exterior heat flux signal is shown in Figure 11. This signal was filtered to provide a smooth curve with a maximum heat flux of 6 kW/m^2. The interior heat flux signal was invalid due to the low signal to noise ratio. The virtually zero interior heat flux is consistent with the measured temperature at this location.

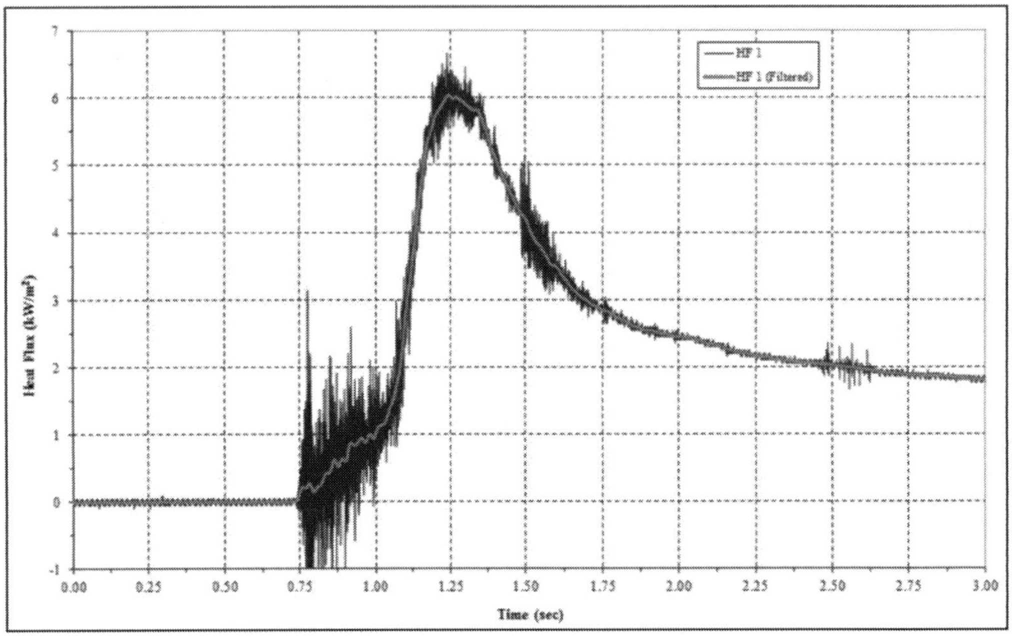

Figure 11. Exterior Heat Flux History

3.3 Structural Response

The air-inflated shelter sustained only minor damage from the VCE blast loading. The air-beams that create the arch section of the shelter experienced rapid temporary deflection as the blast wave propagated across the shelter. The ballistic resistant wall remained functional following the test, allowing safe egress from the shelter after the VCE. Figure 12 shows the wall deflection of about 1 ft. (0.3m) at the top. The yellow grid lines are approximately 1 ft. (0.3 m) apart. The yellow and blue markers indicate the roof deflected approximately 1.5 ft. (0.5 m) downward during the test and returned to the initial position within 2 seconds.

Figure 12. Shelter Deflection Comparison

4. Conclusions

A Dynamic Air Shelters, Ltd. air-inflated structure was subjected to a VCE having a blast load with a peak free-field pressure of 2.9 psi and a 31 ms duration at the wall nearest the DLG rig. The test resulted in a deflagration-type load with a centrally-peaked blast shape. The blast loads inside the structure were reduced from the levels recorded at the exterior ballistic resistant wall by approximately 75% but were comparable to the exterior pressures measured outside of the two doors. Shielding effects of the blast loads were anticipated directly behind the structure, but none were observed at this location relative to the loads recorded closer and further from the rig.

Significant thermal loads were observed on the front face of the structure due to its close proximity to the fireball emanating from the test rig. Yet even at nearly 1500 °F (815 °C) and 6 kW/m^2 at the exterior front face, the interior temperature increased less than 5 °F (3 °C) with virtually no heat flux. The air-inflated structure provided almost complete shielding from the thermal loads with negligible observable thermal damage.

The high speed and high definition video show a maximum of 1 ft. deflection of the ballistic resistant wall and less than 2 ft. deflection of the air-inflated shelter. The structure returned to its initial position within 2 seconds and had no permanent deflection. Internally, the structure had no debris hazards and remained inflated, allowing safe egress from the air beam structure.

9TH GLOBAL CONGRESS ON PROCESS SAFETY

Revisiting the Conditions for a CO2 Tank Explosion

Yu Zhang
Air Products and Chemicals Inc.
7201 Hamilton Blvd, Allentown, PA 18195-1501
zhangy15@airproducts.com

Joan Schork
Keith Ludwig
Air Products and Chemicals Inc.
7201 Hamilton Blvd, Allentown, PA 18195-1501
schorkjm@airproducts.com
ludwigka@airproducts.com

Prepared for Presentation at
American Institute of Chemical Engineers
2013 Spring Meeting
9th Global Congress on Process Safety
San Antonio, Texas
April 28-May 1, 2013

UNPUBLISHED

Revisiting the Conditions for a CO2 Tank Explosion

Yu Zhang
Air Products and Chemicals, Inc.
7201 Hamilton Blvd, Allentown, PA 18195-1501
zhangy15@airproducts.com

Joan Schork
Keith Ludwig
Air Products and Chemicals, Inc.
7201 Hamilton Blvd, Allentown, PA 18195-1501
schorkjm@airproducts.com
ludwigka@airproducts.com

Keywords: Boiling Liquid Expanding Vapor Explosion (BLEVE), Carbon dioxide, Overpressure

Abstract

Carbon dioxide is broadly used in a number of applications including food freezing, beverage production, enhanced oil recovery, supercritical extraction and welding. It is commonly stored as a pressurized liquid (e.g.: -18 °C/0 °F and 2 MPa/300 psia) in carbon steel tanks.

A number of CO2 tank explosions have been reported in the literature over the past 50 years. In order to design appropriate safety systems for liquid CO2 storage tanks, it is important to understand the worst case scenario. Should the energy release be modeled as a rapid vapor phase expansion (PV) or a Boiling Liquid Expanding Vapor Explosion (BLEVE)? The question arises because of a common definition of a BLEVE as an event resulting from the sudden loss of containment of a liquid that is at or above its atmospheric superheat limit temperature. Under the conditions listed above, CO2 is above its atmospheric boiling point but below its superheat limit. In this temperature range, will the vaporization rate be rapid enough to contribute to a blast overpressure in the event of a tank failure?

To answer this question, the authors made a critical review of information available in the literature including laboratory data and incident reports. This paper will describe the analysis conducted and present the results indicating that the potential for rapid vaporization should be included in evaluating the consequences of a CO2 tank rupture below the superheat limit.

1. Introduction

Carbon dioxide (CO2) is broadly used in a number of applications including food freezing, beverage production, enhanced oil recovery, supercritical extraction and welding. Since CO2 at its bubble point becomes a solid at a pressure below 5 bars, CO2 is commonly stored as a pressurized liquid at an elevated pressure of 20 bars absolute in carbon steel tanks.

When a pressurized liquid in an equilibrium state is depressurized instantaneously, it enters a superheated state (a meta-stable liquid above its boiling point) from which the liquid could vaporize in an explosive manner resulting in a blast wave and risk of flying fragments of the vessel. This is called a Boiling Liquid Expanding Vapor Explosion (BLEVE) [1,2]. When the degree of superheat of the liquid is small, the vaporization rate is not large enough to develop a significant blast effect. According to the Superheat Limit Theory developed by Kim and Reid [3], an explosive evaporation occurs when the depressurized liquid enters the spinodal decomposition temperature-pressure range. Based on thermodynamics principles and equation of state of fluids, Kim and Reid [3] predict a so-called BLEVE envelope (Figure. 1) which indicates the range of initial temperatures and pressures which will lead to a BLEVE following instantaneous depressurization. The superheat limit theory is cited in some industrial hazard evaluations as a criterion to determine whether or not a BLEVE should be considered [4,5].

The mechanism of BLEVE based on Kim and Reid's theory is briefly described here. When the boiling point of a liquid is reached, vapor bubbles start to form at active sites such as vessel walls. Nucleation at these sites is called *heterogeneous nucleation*. If bubble nuclei are large (approaching millimeter sizes), a small degree of superheat (defined as heating beyond the boiling point) is needed to make the bubble grow. If the bubble nuclei are small, for example when the vessel is very clean and smooth, they need a large superheat to grow. When there are not many active sites, the bubble nucleation must take place in the bulk of the liquid. Boiling in this case generally takes place at submicron nucleation sites at suspended particles, impurities, or ions. This requires a large degree of superheat to make the bubble grow. This means the liquid normal boiling point can be exceeded without boiling, and the liquid can be significantly superheated. There is, however, a temperature limit at a given pressure above which a phase transition from liquid to vapor has to happen, even though the bubble nulei are small. This superheat limit temperature (SLT) and pressure correspond to the conditions at which a spinodal decomposition of liquid happens. This can produce a very violent phase transition with boiling caused by *homogeneous nucleation* in the bulk of liquid. The superheated liquid state could also be reached by sudden pressure drop. For CO2, the experimental value of SLT at atmospheric pressure is 267 K (-6.2°C) [6].

Although the theory is commonly accepted, some authors question its applicability to a real case. For example, Birk and co-works [7] report that BLEVEs have been observed with propane when the pressure-liquefied gas was at ambient temperature (20 °C), well below its atmospheric superheat limit temperature of 53 °C. In the 2010 edition of the Center for Chemical Process Safety (CCPS) Guidelines to evaluate BLEVE [2], it says "There is no experimental evidence with a real pressure vessel where the phase change was by homogeneous nucleation. Homogeneous nucleation is extremely unlikely in industrial use due to temperature gradients, impurities, surface roughness and other factors that cause localized nucleation."

2. CO2 BLEVE

The end state of liquid CO2 depressurized to 1 atmosphere is a solid and vapor mixture, which makes its depressurization process unique. CO2 may evaporate explosively even when its temperature is below the STL, because CO2 crystallizes at ambient pressure, thus presenting enough nucleation sites for liquid to flash [1].

Several fundamental questions are raised when considering CO2 system. How does the formation of dry ice affect the blast energy? During the decompression process, does the vapor expand isothermally or isentropically? This initial report will only touch on two questions based on various incident reports on large scale CO2 tank energy releases:

1. *Could a CO2 BLEVE happen when the storage conditions (temperature and pressure) are outside the so-called BLEVE envelope predicted by the superheat limit theory?*
2. *If it goes BLEVE, could we use the algorithm presented in the CCPS guide to perform the consequence analysis?*

3. Historical experiences

Table 1 summarizes major CO2 tank incidents we found with documentation in the open literature. The Worms, Haltern, and Repcelak incidents are summarized in CCPS's guide to BLEVE [1-2]. Information about the incident at Yuhang is collected from news reports in Chinese [8-9], and information about Fukushima's incident is translated from reports in Japanese [10-12].

Table 1. Some major CO2 incidents

Date	Location	Cause	Initial Pressure [bar]	Initial Temperature [◦C]
13 Nov. 2008	**Yuhang, China***	Brittle fracture	23	-15
21 Nov. 1988	Worms, Germany	Overpressure	20	-20
2 Sept. 1976	Haltern, Germany	Brittle fracture	23	-15
1 Mar. 1969	Fukushima, Japan	Overpressure	20	-20
2 Jan. 1969	**Repcelak, Hungary***	Overfilling	15	-30

* Overpressure tags are extracted from these incidents

Out of the five incidents, only two overpressure damage data points (tags) are obtained, which are used to validate the algorithm employed in the CCPS guidelines. There were many tags on projectiles, but we find these less reliable than the overpressure tags. One data point is from the Repcelak incident: a warehouse about 15-30 meters away from the explosion center was torn down due to heavy damage. The other data point is from the Yuhang incident: Window glass shattered within a 400 meter radius of the explosion center. Criteria on overpressure damage to buildings are obtained from CCPS's guide to BLEVE [1].

3.1 Large scale CO2 tank energy release

Yuhang, Hangzhou, China, 13 November 2008

A transport ship carrying 130 cu. meter (95% full) of CO2 exploded on November 13, 2008 in Yuhang, Zhejiang, China [8-9]. The tank's content was stored under a pressure of 23 bar at a temperature of -15 °C. At the moment of explosion, the ship was in the dock belonging to Rongsheng Chemicals. It was not in operation and there were few workers around as it was before regular business hours. The explosion completely destroyed the CO2 ship and sank H2SO4 and H2O2 carrying ships nearby. Two workers on the CO2 ship lost their lives instantly and 3 were injured due to projectiles.

Pieces of the ship were scattered over a 500 meter area around the explosion. The weight of these fragments varied from a few grams to several hundred kilograms. The shrapnel caused damage in several adjacent residential buildings. All the windows of a metal machining factory located around 500 meters away shattered. In the Yujiadouzhen, a village 500 meters away from the explosion, windows shattered in many residential buildings.

The cause of this accident is believed to be overloading and brittle failure of the CO2 tank. The tank was designed for on-shore use, but was modified by Rongsheng Chemicals for ship transportation. To lower the transportation cost, the company modified the level indicator, locked the relief valve, and overloaded the tank (95% filling level). The tank material, 16 MnR steel, has a working temperature from -20 °C to 475 °C. It was not suited for use under low temperature conditions.

Worms, Germany, 21 November 1988

This is one of the most catastrophic incidents involving a CO2 tank explosion. The detailed description of Clayton and Griffin is summarized here [13]. The liquid CO2 was stored in a 30 tons capacity tank at 20bar and -20°C. The vessel was shattered into a number of pieces in the explosion. Most of the tank was propelled into the Rhein River about 300 meters away from the epicenter. Only 20% of the tank was recovered originally. The Worms's investigation concluded that the vessel failed catastrophically, most likely due to overpressure from overheating with a frozen relief valve. The failure pressure was estimated to be between 31 and 51 bara. It was speculated that the failure caused a CO2 BLEVE [2].

Haltern, Germany, 2 September 1976

A rail car carrying 231,000 kg of CO2 exploded in Haltern, Germany [1]. The tank's contents were at 7 bar and -15°C. At the moment of the explosion, the car was passing through a railroad shunting yard. As it passed checkpoints, the car was observed to be releasing plumes of CO2 from the safety valve. It then exploded 15 m in front of a group of other rail cars. Parts of the tank were thrown up to 360 m. One person was killed in the explosion. Analysis of a recovered piece of the tank car showed that failure was due to brittle fracture. Brittle metal fracture is consistent with the near instantaneous overpressure created by a BLEVE event.

Fukushima, Japan, 1 March 1969

A CO2 storage tank in a steel mill exploded [10-12]. A slate-roofed factory within a 50 m radius was completely destroyed, and only its pillars remained. Windows of houses located within a 500 m radius were damaged. Before the explosion, the main safety valves were closed during repairs. Liquid remained in the tank (50% filling level) and the heat source of the external evaporator was left on. The vapor phase temperature and pressure in the tank kept rising, and eventually the tank exploded when the pressure exceeded the burst pressure of the tank. Based on heating element power, CO2 usage, and heating time, it is estimated that almost all liquid CO2 was turned into vapor before the explosion. Thus, this incident is determined to be a PV release.

Repcelak, Hungary, 2 January 1969

Two vessels containing liquid CO2 (15 bar, -30°C) in a CO2 production and filling plant exploded in rapid succession [1, 14]. The explosion completely destroyed the tank yard consisting of four liquid CO2 storage vessels and the equipment of the technological laboratory. Large fragments were scattered in a circle of approximately 400 m radius. A 2800 kg shell landed at a distance of 150 m and a fragment weighing 1000 kg landed 250 m away. During filling, the first vessel exploded. Some minutes later, another close-by vessel exploded. The second vessel probably failed as a result of impact from a fragment from the first vessel. The explosions tore the two vessels into pieces, and tore the third vessel off its foundation bolts. The third vessel was shot into the process laboratory like a rocket due to liquid CO2 rapid expansion through a hole on the bottom. The probable cause of the accident was overfilling the first tank due to a level indicator failure. The explosion of the second vessel was speculated to be caused by a CO2 BLEVE.

3.2 Small scale experimental investigation

Most experimental work on BLEVE has focused on flammable substances like LPG and propane [1,2]. Recently, there were several small scale lab tests performed to study CO2 BLEVE [15-18].

Kim and Reid, 1978

An important reason for considering the possible occurrence of BLEVE with CO2 is this paper by Kim and Reid which presents experimental and modeling work to investigate the phenomena of rapid depressurization of hot, high pressure liquids or supercritical fluids (CO2) [3].

The authors applied a thermodynamic model based on the spinodal fluid state, which represents the limit-of-stability for the liquid phase during expansion. For a BLEVE to occur, the substance has to find itself on the spinodal curve between 1 bar and the critical point where the curve ends. When a catastrophic failure occurs there is not time for heat to pass into the system and so the path during failure is adiabatic. In thermodynamic terms the entropy of the system is constant during failure and the conditions for BLEVE to occur means that the initial entropy has to be higher that the entropy of the intersection of the spinodal curve and the 1 bar isobar. As a result, a region of initial conditions that could give a spinodal state can be identified (Figure 1). Kim and Reid performed some simple experiments on depressurization of saturated liquid and supercritical CO2. None led to explosions regardless of the initial conditions relative to the envelope in Figure 1. The authors suggested the reasons for disagreement between theory and experiments were associated with factors like vapor bubbles on the wall of the vessel (sites for heterogeneous nucleation) and shock waves or disturbances from the mechanism that ruptured the burst disk of the test vessel.

Pettersen, 2002

Pettersen performed an experimental study aimed at clarifying the possible occurrence of BLEVE in CO2 systems [17]. A test vessel was built and instrumented in order to measure transient pressures during initial depressurization, and to record possible subsequent overpressure peak(s). The study concluded that none of the experiments, which were conducted over a wide range of initial conditions (3.5, 4.5, 6.0 and 10.0 MPa initial pressures with varying liquid fill level and discharge opening), showed signs of BLEVE, which would be characterized by a sudden repressurisation to a level significantly above the initial pressure.

Wei, 2009
This master thesis presents experimental demonstration of CO2 BLEVE which happened below the superheat limit temperature (outside the BLEVE envelop in Figure 1) [18].

The list below highlights (taken from Wei's thesis) the most significant findings:

- The 'Superheat Limit Temperature' theory for predicting occurrence of a BLEVE was not supported with experiments in this work. A CO2 BLEVE can also occur when the initial temperature is below the superheat limit temperature of CO2 (-13.8 °C). Nevertheless, considering influencing factors including CO2 filling level, potential gas leaking and CO2 mixture with different phase compositions, the theory may still be acceptable.
- Two possible approaches for a safer CO2 storage include using an initial storage pressure of around 2 MPa and developing a safety valve that can further slow down pressure drop when an unexpected vessel opening and depressurization occurs.

3.3 Consequence analysis

An established procedure for calculation of the consequence of a BLEVE event can be found in CCPS's guide to BLEVEs [1,2].

1. Calculate the explosion energy. Ongoing debates on the calculation of the explosion energy from a BLEVE event involve discussions about the BLEVE mechanism and are outside the scope of this study. In this work, we consider three options where the explosion energy is
   a. Internal energy difference between the initial and final states of an isentropic expansion process [1],
   b. Internal energy difference between the initial and final states of an irreversible adiabatic expansion [19],
   c. The PV energy (energy from the vapor phase, excluding that from flashing liquid).

   Please note that the explosion energy will need to be multiplied by a factor of 2 if the burst is a ground level, to obtain the effective working explosion energy.
2. Once the explosion energy of a BLEVE is estimated, overpressure can be determined by employing the overpressure- scaled distance curve by Baker.
3. Once overpressure is calculated, a multiplier has to be applied to account for vessel shape effect.

According to CCPS, the method presented above is based on the similarity of the blast waves of pressure vessel bursts and high explosives. This similarity holds only at some distance from the explosion. In the near field (scaled distance < 2), the peak overpressure and impulse from a pressure vessel burst differ greatly from those of a detonation of a high explosive, except when the pressure vessel is filled with a very hot high-pressure gas. Baker et al. developed a method which is used to solely predict blast pressures in the near field. However, it is based on results of numerical simulations and involves solving an implicit equation. The details of this refinement can be found in CCPS's Guidelines [1]. Note that this refinement for the near field cannot be made for non-ideal gases.

4. Results

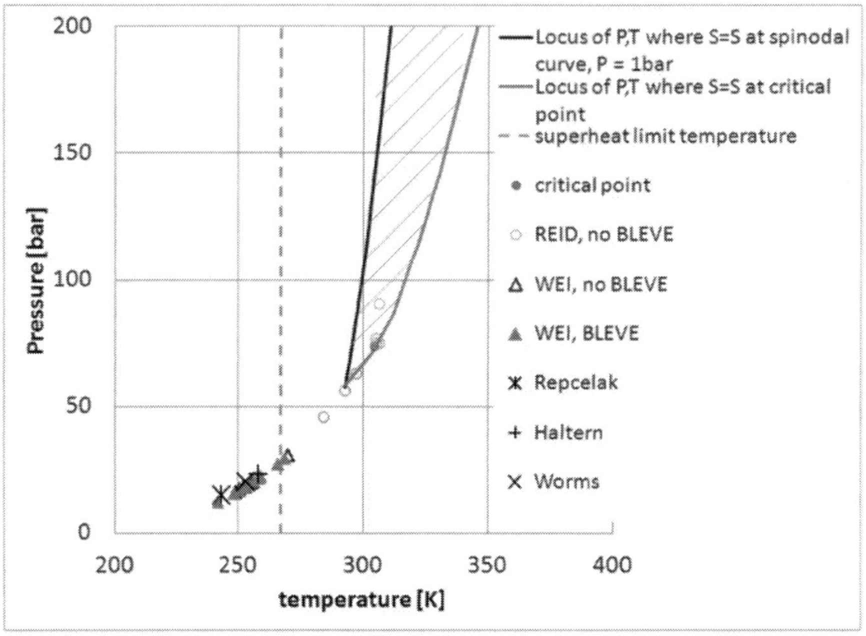

Figure 1. Summary of conditions involving CO2 BLEVE. Empty symbols are experiments with no BLEVE, and solid ones and crosses are experiments/events with BLEVE.

We summarize the literature data discussed in Section 3 in Figure 1. From Figure 1, we see that there is no clear dividing line between a BLEVE and a non-BLEVE event based on the superheat limit temperature, as indicated by experimental results and historical experiences. This might be due to the various processes involved in such an event, including nucleation of bubbles at the vessel wall. It is conclusive that the superheat limit temperature has no special significance to the BLEVE event observed in either past accidents or controlled experiments. Furthermore, Kim and Reid's experimental data along with others failed to produce a BLEVE in the laboratory when initial conditions were within the envelope predicted by the superheat limit theory.

We obtained two overpressure tags, one from Repcelak and the other fromYuhang:

1. The Repcelak: "light structure warehouse was so seriously damaged that it had to be pulled down." [14] The warehouse was 15 meters from the epicenter, and the overpressure was about 0.15-0.25 bar[1];
2. Yuhang: The windows of a metal machining factory located around 500 meters away, all shattered. In the Yujiadouzhen, a village 500 meters away from the explosion, windows shattered in many residential buildings [9]. The overpressure was about 0.035-0.07 bar [1].

Three different explosion energies, as discussed in Section 3.3, for each event are summarized in Table 2. The actual overpressure data *vs.* calculated overpressures using the CCPS's recommended method is shown in Figure 2. If the correct energy is chosen, the points should fall on or near the $y = x$ curve. Clearly, these results suggest that the explosion energy released from the incidents is more consistent with BLEVE type of explosion-energy where energy from the liquid phase contributes significantly.

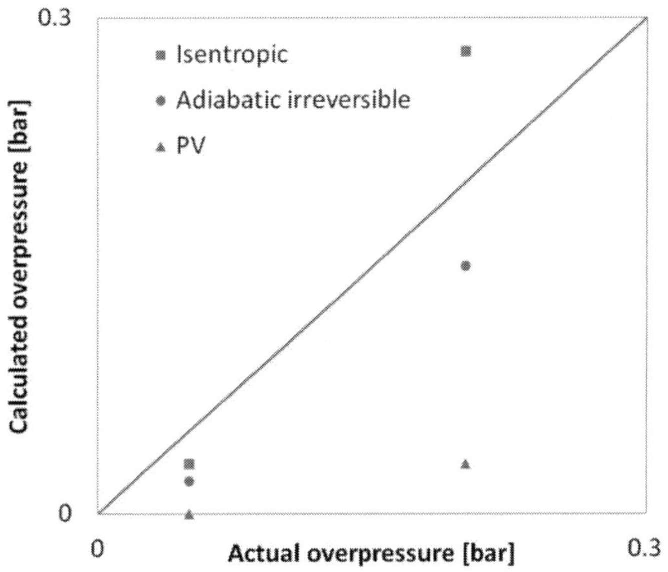

Figure 2. Actual overpressure versus calculated overpressure using different explosion energies for two events.

Table 2. Explosion energy, calculated overpressure (COP), and actual overpressure (AOP) tag for Repcelak and Yuhang incidents

Location	Initial filling level	Isentropic expansion [J]/COP[bar]	Adiabatic irreversible [J]/COP[bar]	PV [J]/COP[bar]	OP tag distance to epicenter R_tag [m]	AOP at R_tag [bar]
Repcelak	80%	3.5E9/0.28	9.9E8/0.15	3.0E7/0.03	15±5	0.20±0.05
Yuhang	95%	1.3E10/0.03	4.9E9/0.02	3.6E7/0.003	400±100	0.05±0.02

5. Conclusion

Through a critical review and analysis of historical and experimental reports of CO2 tank explosion, we conclude that a CO_2 tank rupture could lead to a BLEVE event even when the initial storage temperature is below the superheat temperature limit of CO_2. The consequence of a BLEVE event should be considered when assessing the risk associated with CO2 storage.

6. References

[1] "Guidelines for Evaluating the Characteristics of Vapor Cloud Explosions, Flash Fires, and BLEVEs, Center for Chemical Process Safety", *American Institute of Chemical Engineers*, New York, 1994

[2] "Guidelines for Vapor Cloud Explosions, Pressure Vessel Burst, BLEVE and Flash Fire Hazards, Center for Chemical Process Safety", *American Institute of Chemical Engineers*, New York, 2010

[3] Kim, M.E., Reid, R.C., "The rapid depressurization of hot, high pressure liquids or supercritical fluids", *Chemical Engineering at Supercritical Fluid Conditions*, Ann Arbor Science, 1983

[4] Calculation of BLEVE conditions for carbon dioxide, www.criticalprocesses.com/BLEVE.htm (accessed Jan. 10, 2013)

[5] "Final public report safety study for liquid logistics shipping concept (DNV)", http://cdn.globalccsinstitute.com/sites/default/files/publications/19006/final-public-report-safety-study-liquid-logistics-shipping-concept.pdf (accessed Jan. 10, 2013)

[6] Abbasi, T., Abbasi, S.A., "Accidental risk of superheated liquids and a framework for predicting the superheat limit," *Journal of Loss Prevention in the Process Industries,* 20 (2007) 165-181

[7] Birk, A.M., Cunningham, M.H., "The boiling liquid expanding vapor explosion", *J. Loss Prev. Process Ind.,* 7 (1994) 474–480

[8] Li, W., Di, G., Wang, R., "The cause analysis of a liquid CO2 tank explosion on a ship", *Materials Science Forum,* 689(2011)461-467

[9] Yuhang CO2 cargo ships explosion, http://zjnews.zjol.com.cn/05zjnews/system/2008/11/14/014973390.shtml (in Chinese, accessed Jan. 10, 2013)

[10] "High Pressure Gas Safety Inst. of Japan. Rupture of liquefied carbon dioxide in a low-temperature storage tank", *Accident investigation report (outline),* 1969 (in Japanese)

[11] "High Pressure Gas Safety Inst. of Japan", *Examples of accidents, Collection of examples of high-pressure gas accidents,* 1982 (in Japanese)

[12] Kitagawa, T., "Vapor explosion in liquid carbon dioxide tank", *Analysis of explosion hazard.* 1980 (in Japanese)

[13] Clayton, W.E., Griffin, M.L., "Catastrophic failure of a liquid carbon dioxide storage vessel", *Process Safety Progress* 13 (1994) 202-209.

[14] Voros, M., Honti, Gy, "Explosion of a liquid CO2 storage vessel in a carbon dioxide plant", *Loss Prevention and Safety Promotion in the Process Industries*, 1974

[15] van der Voort, M.M, van den Berg, A.C., Roekaerts, D.J.E.M., Xie, M., de Bruijin, P.C.J., "Blast from explosive evaporation of carbon dioxide: experiment, modeling and physics", *Shock Waves* 22 (2012) 129-140

[16] Bjerketvedt, D., Egeberg, K., Ke, W., Gaathaug, A., Vaagsaether, K., Nilsen, S.H., "Boiling liquid expanding vapour explosion in CO2 small scale experiments", *Energy Procedia* 4 (2011) 2285-2292

[17] Pettersen, J, "Experimental study on boiling liquid expansion in a CO2 vessel", 5th IIR Gustav Lorentzen Conference on Natural Working Fluids, Guangzhou, China, September 2002

[18] Ke, W., "CO2 BLEVE (Boiling Liquid Expanding Vapor Explosion)", Master Thesis, Telemark University College, Porsgrunn, Norway, 2009

[19] Chen, F, Schork, J.M., "Improvement of BLEVE Overpressure Method Using Historical Data. Conference Proceedings", 7th Global Congress on Process Safety. Chicago, IL, 2011.

New Ignition Source "Exploding Wire" for the Determination of Explosion Characteristics of Combustible Dusts in the 20-L-Sphere

Marc Scheid
Federal Institute for Materials Research and Testing (BAM),
Department 2 "Chemical Safety Engineering"
D-12200 Berlin, Germany
marc.scheid@bam.de

Alexander Klippel, Rico Tschirschwitz, Volkmar Schröder, Stefan Zirker,
Christian Kusche
Federal Institute for Materials Research and Testing (BAM), Department 2 "Chemical Safety Engineering"
D-12200 Berlin, Germany

Prepared for Presentation at
American Institute of Chemical Engineers
2013 Spring Meeting
9th Global Congress on Process Safety
San Antonio, Texas
April 28 – May 1, 2013

UNPUBLISHED

New Ignition Source "Exploding Wire" for the Determination of Explosion Characteristics of Combustible Dusts in the 20-L-Sphere

Marc Scheid
Federal Institute for Materials Research and Testing (BAM),
Department 2 "Chemical Safety Engineering"
D-12200 Berlin, Germany
marc.scheid@bam.de

Alexander Klippel, Rico Tschirschwitz, Volkmar Schröder, Stefan Zirker, Christian Kusche
Federal Institute for Materials Research and Testing (BAM),
Department 2 "Chemical Safety Engineering"
D-12200 Berlin, Germany

Keywords: dust explosions, dust explosion testing, exploding wire, explosion characteristics, fuse wire, ignition source

Abstract

According to international standards several safety characteristics of dusts are determined in the 20-L-sphere (also known as SIWEK-Chamber). Dust cloud ignition is carried out using pyrotechnical igniters. Due to various disadvantages of pyrotechnical igniters the need for alternative ignition sources arises again and again. An alternative ignition source could be the so-called "exploding wire" or "fuse wire". The paper presents test results of a comparative study between both ignition sources for the determination of the safety characteristics Maximum Explosion Pressure and Maximum Rate of Explosion Pressure Rise of selected dusts in the 20-L-sphere. In addition to that the flame/arc propagation of both ignition sources was analysed by high speed and IR camera recordings. Tests in a windowed autoclave were performed to get information on the influence of turbulence generated due to dispersion of the dust on the shape of the generated flame and its propagation. Turbulence measurements in the windowed autoclave as well as in the 20-L-sphere with a LDA system allowed investigation if the igniter led to turbulence increase at the moment of ignition in comparison to tests without ignition.

1. Introduction

Safety characteristics are essential for the determination of explosion hazards during handling of combustible dusts and for the design of safety measures. The safety characteristics Maximum Explosion Pressure p_{max}, Maximum Rate of Explosion Pressure Rise $(dp/dt)_{max}$ and Lower Explosion Limit LEL are determined in closed vessels such as the 20-L-sphere (also known as SIWEK-Chamber). Tests are preformed according to international standards, for example EN 14034 part 1 and part 2 [2,3] or ASTM E1226 [1]. In order to determine the safety characteristics the dust samples are dispersed in air and ignited using two pyrotechnical igniters with energy contents of 1000 J (LEL) or 5000 J (p_{max}, $(dp/dt)_{max}$). Since the igniters could have an influence on the test results they are defined in the standards.

Due to various disadvantages of pyrotechnical igniters such as high costs, legal requirements concerning its storage and use as well as high energy input in comparison to most ignition sources in practice, the need for alternative ignition sources arises again and again; see also Krietsch [6]. Such an ignition source should be less expensive, readily available and the operator should be able to use it without a certificate of competence. In addition to that, the ignition energy should be adjustable over a wide range and if possible allow energy amounts less than 1000 J. Compared to the ignition energies described in the standards such energy amounts are more realistic for ignition sources found in practice, such as electrostatic discharges, mechanical sparks or hot surfaces.
An alternative ignition source which fulfils these requirements is the so-called "exploding wire" or "fuse wire". This type of ignition source is used for the determination of explosion limits of gases and is described in the standard EN 1839 [4].
This paper presents test results of a comparative study in order to find out, if the exploding wire is suitable as an alternative ignition source for the determination of explosion characteristics of dusts. For the study Maximum Explosion Pressure and Maximum Rate of Explosion Pressure Rise values from different dusts were determined with both ignition sources in the range of 100 J to 2000 J. The dusts were selected such that different combustion mechanisms were considered. High speed and IR camera recordings allowed comparison of flame front and electrical arc generated from the igniters. Calorimetric and electric measurements obtained information on the ignition energy of the igniters delivered to the mixture. With additional tests in a windowed autoclave it was possible to get information on the influence of pre-ignition turbulence on the shape of the generated flame/arc and its propagation as well as information on the turbulence generated from the ignition source. Turbulence measurements in the windowed autoclave were realised with a LDA system. The same system was used for tests in the 20-L-sphere to get information on the turbulence at the moment of ignition.

2. Test assembly and test settings

2.1 20-L-sphere

Main component of the test assembly was the 20-L-sphere (Cesana and Siwek [5]; EN 14034 1-2 [2,3]). In order to reach almost homogeneous dust dispersion into the test chamber the so called "rebound nozzle" was used.
Tests with pyrotechnical igniters as an ignition source were performed using the standard control unit of the 20-L-sphere. For tests with exploding wires an external ignition unit was necessary. Depending on the ignition energy different systems with isolating transformer, or capacitors were used. An isolation amplifier allowed the determination of the ignition energy by measuring current and voltage.

2.2 Windowed autoclave

A windowed autoclave with cylindrical shape was used for testing in order to get information on flame/arc propagation in a closed vessel as well as on the shape of flame/arc generated from both ignition sources. In this autoclave it was also possible to determine the influence of turbulence on the generated flames. The volume of the autoclave was approximately 11 L. Its diameter was 220 mm. Optical accessibility via the window allowed high speed recordings at different positions and turbulence measurements. A fan which was placed on the bottom of the autoclave enabled average turbulent velocity fluctuations (RMS-value) up to 1 m/s. In addition to that the system allowed measurements of turbulence generated from the ignition source.

2.3 Ignition sources

One of the fundamental differences of the two ignition sources pyrotechnical igniter and exploding wire is the ignition mechanism. While for the exploding wire an electrical arc and plasma is generated the pyrotechnical igniter emits flames and burning solids.

The ignition sources were placed in the middle of the vessel and connected to the lid of the vessel, such as it is described in the test standards. According to the European standards for the determination of the explosion characteristics Maximum Explosion Pressure and Maximum Rate of Explosion Pressure Rise two pyrotechnical igniters with energy of 5000 J shall be used, firing horizontally in opposite directions (EN 14034 1-2).

Instead of the ignition energy described in the standards tests were performed with ignition energies of 100 J, 500 J, 1000 J and 2000 J. In the first tests only a single ignition source with ignition energies of 100 J, 500 J and 1000 J was used. Further tests with a modified lid allowed tests with two igniters with ignition energies of 500 J and 1000 J each. For all tests with one ignition source the igniter was mounted such that it produced a flame or an electrical arc downwards. Electrodes and connected igniter for tests with a single ignition source are shown in Figure 1. In tests with two igniters the ignition source was installed such that the flame propagated in nearly horizontal direction such as it is described in the standard.

Pyrotechnical igniter

All pyrotechnical igniters used for this work have the same composition as the standard igniters used for determination of explosion characteristics in the 20-L-sphere. They contained 40 % by weight zirconium metal, 30 % by weight barium nitrate and 30 % by weight barium peroxide.

Exploding wire

The principle of the exploding wire is the evaporation and ionization of metal particles from a wire due to a sufficient current flow within milliseconds. During that process a metal vapour is created in which an electrical arc is generated between the electrodes because of the electrical conductivity of the plasma.

For all tests a nickeline wire with a diameter of 0.12 mm was used. In tests with a single ignition source the wire was tensed between the ends of two auxiliary electrodes which were connected two the main electrodes of the 20-L-sphere, see Figure 1. For tests with two exploding wires the lid of the 20-L-sphere had to be modified. It was designed such that it contained four electrodes.

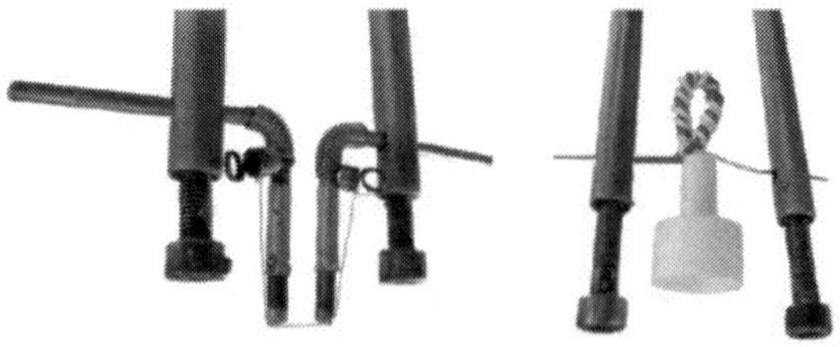

Figure 1. Ignition source for tests with a single igniter: Exploding wire with electrodes (left side) and pyrotechnical igniter (right side)

Ignition unit with isolating transformer

First tests with the exploding wire were performed with ignition energy of 100 J. In these tests the electrical power which is needed for melting the wire and generating the arc was supplied by an isolating transformer similar to that used for the determination of explosion characteristics of gases and vapors according to EN 1839 [4]. This system and its disadvantages such as not reproducible ignition delay are described by Scheid [7].

Ignition unit with capacitors

In tests with ignition energies of 500 J, 1000 J and 2000 J an ignition unit with capacitors was used instead of the isolating transformer. The main advantages of this new type of ignition unit are the higher ignition energies which the unit can produce and that it is able to ensure an exact ignition delay. A choke enabled elongation of the electrical arc during each test.

2.4 Recordings of flame front and electrical arc

In order to record the propagation of flame/arc of both igniters a high speed camera system was used. It allowed recordings of 5000 frames per second. So it was possible to determine maximum flame volume and length as well as the time delay for activating the ignition source and its burning time. In addition to that a fast acting IR camera system was used to estimate the maximum temperature level generated from the igniters. The IR camera allowed recordings with 850 frames per second.

2.5 Measurements of ignition energy (delivered to the mixture)

A combustion calorimeter working with the dry method allowed measurements of the ignition energy of both ignition sources. Dry method means, that the brass bomb of the calorimeter is covered with an insulating styrofoam and not with a water bath. This allows a more sensitive direct measurement of the temperature changes in the metal. Additionally an isolation amplifier was used to measure the burning current in the exploding wire as well as the voltage and enabled the determination of ignition energy generated in the electric arc during each test.

2.6 Turbulence measurements

A one-dimensional laser-Doppler-system (LDA) was used for turbulence measurements in the 20-L-sphere as well as in the windowed autoclave. Due to the big window of the autoclave the turbulence was measured at three different positions around the ignition source. Since the optical accessibility in the 20-L-sphere could only be realized via a windowed flange with a diameter of 40 mm it was only possible to measure the turbulence at one position next to the ignition source.

2.7 Tested dusts

Dust explosions generally show complex reaction mechanism, which depend e.g. on the chemical composition of the dust, the dust concentrations and the flow conditions. Van der Wel [8] differs between different reaction mechanisms depending on whether the reaction takes place in the gas phase due to evaporation or at the solid surface in form of gaseous products, solid or liquid material. The dusts used in this work were chosen such that they represent the different chemical classes and mechanisms as well as the dust explosion classes found in practice. As a result dusts from lignite, maize starch, niacin and anthraquinone were chosen. In addition to that a steel dust which was generated during a shot blasting process was tested.

3. Test results

3.1. Ignition energy measurements

Measurements with the isolation amplifier lead to different test results depending on the boundary conditions of the measurements. Table 1 shows the test results for measurements of ignition energy outside the test vessel without dust and flow, inside the test vessel with flow but without dust and inside the test vessel with flow and with a dust concentration of approximately 1000 g/m³ - 1250 g/m³ maize starch. Ignition unit was the isolating transformer.

Table 1. Ignition energy of igniter determined with the isolation amplifier; ignition source isolating transformer; ignition energy 100 J

Experiment-No.	Determined ignition energies		
	Outside the apparatus [J]	Inside the apparatus without dust [J]	Inside the apparatus with dust [J]
1	93.61	98.42	98.88
2	86.84	94.19	84.14
3	90.92	89.27	88.24
4	89.80	101.59	48.85
5	91.62	82.96	94.13
6	84.46	98.58	83.89
7	90.40	94.74	100.63
8	94.41	91.37	74.36
Ø	**90.26**	**93.89**	**84.14**
SD	**3.30**	**5.97**	**16.67**

Table 1 shows that the ignition energy values for tests without dust (inside and outside the test vessel) were almost similar. As a trend the values with flow (tests inside the vessel) and their standard deviation seem to be slightly higher. Tests with dust led to slightly lower ignition energies and a significant higher standard deviation of the values. Most of the tests led to ignition energies of more than 90 % of the adjusted value (set point adjustment). However, some tests with dust led to lower values. In one test less than 50 % of the adjusted value was measured. As a result the dust particles seem to influence the test results in some tests.

Figure 2 shows the arithmetic mean of the determined ignition energies as a function of dust cloud concentration for maize starch and lignite for adjusted ignition energy of 100 J. From the course of the mean values follows, that there is only a minor influence of the dust concentration on the ignition energy up to concentrations of approximately 1250 g/m³. For higher concentrations the influence is significantly higher.

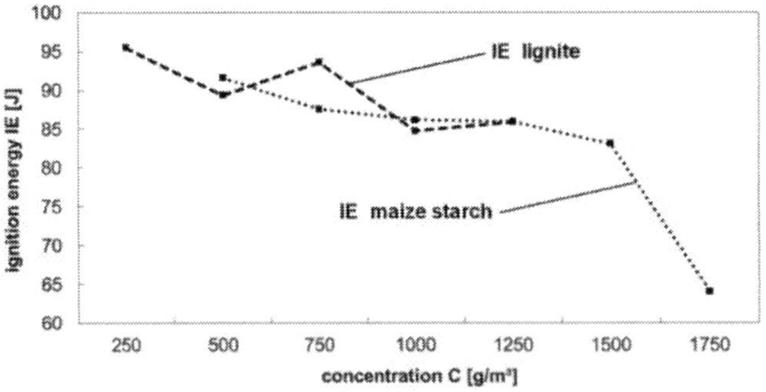

Figure 2. Arithmetic mean of ignition energy of igniter as function of dust cloud concentration for maize starch and lignite; adjusted ignition energy 100 J

Comparable test results were determined for ignition energies of 500 J, 1000 J and 2000 J using an ignition unit with capacitors. But as a trend the influence of the dust cloud on the ignition energy seems to be lower. For all tests the determined ignition energy was at least 80 % of the set point adjustment. Energy measurements with the combustion calorimeter confirmed the test results.

3.2 Propagation of flame/arc of ignition source

Tests in the windowed autoclave with maximum fan speed were performed in order to get information if pre-ignition turbulence has an influence on the flame/arc propagation of a 1000 J pyrotechnical igniter or an exploding wire with the same energy. However, for both ignition sources no visible influence on flame/arc propagation and maximum flame/arc volume was found.
Figure 3 shows the flame propagation for a 1000 J pyrotechnical igniter in comparison to an exploding wire with the same energy for tests in the windowed autoclave (no reaction of dust). The flame recordings show, that the flame of the pyrotechnical igniter propagates 3-4 times faster than the flame of the exploding wire and reaches a much higher maximum volume.

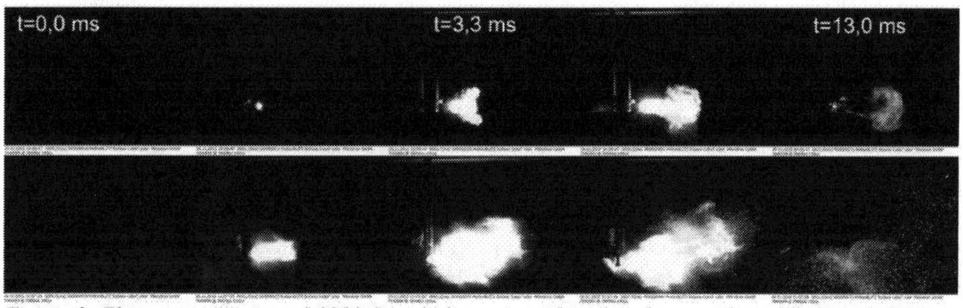

Figure 3. Flame propagation of 1000 J exploding wire (above) and pyrotechnical igniter (below) in a windowed autoclave with a volume of 11 L recorded with high speed camera

Recordings with a fast IR camera allowed an estimation of the maximum temperature level in the generated flame/arc and the maximum volume of flame or arc. Figure 4 shows four sequences of the flame/arc propagation of exploding wire (first and third sequence) and pyrotechnical igniter (second and fourth sequence) recorded with the IR camera. The sequences show a period of approx. 5 ms within the first 15 ms after triggering the igniter. Temperature range in the first two sequences was adjusted such that temperatures between 200 °C and 650 °C were visualized.

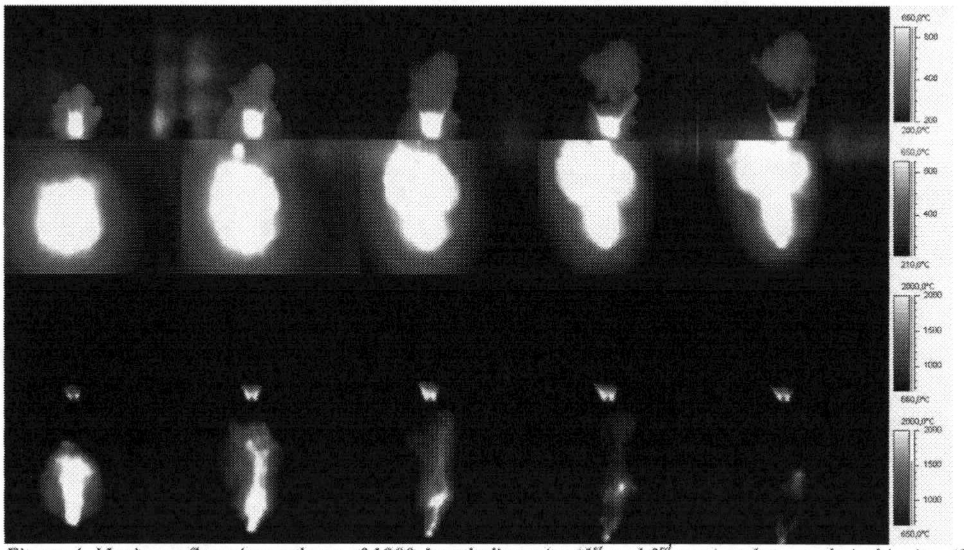

Figure 4. Maximum flame/arc volume of 1000 J exploding wire (1st and 3rd row) and pyrotechnical igniter (2nd and 4th row) recorded with a fast IR camera system

A comparison between sequence one and sequence two shows that the volume with the highest temperatures generated from the pyrotechnical igniter is much larger than the volume with such temperatures generated from the exploding wire. While for the exploding wire the volume with the highest temperatures does not increase with the propagation of the flame, it increases for the pyrotechnical igniter.
Sequence three and four visualize the temperature range between 650 °C and 2000 °C. For the exploding wire the highest temperatures were determined close to the wire. The volume in which such temperatures were determined stays almost constant in the whole sequence. In contrast to that the pyrotechnical igniter seems to generate a large volume with such high temperatures in from of a jet immediately after it was triggered. However, the volume with the highest temperatures seems to decrease significantly within 2 ms.

3.3 Turbulence measurements in 20-L-sphere and windowed autoclave

In tests in the windowed autoclave with operating fan (generated RMS-value of approx. 1 m/s) no differences in the measured RMS-value could be determined between tests without ignition and with a 1000 J exploding wire. Tests with the pyrotechnical igniter resulted in RMS-values of 2.5 m/s - 4.5 m/s depending on the measuring position. In tests in which the fan was turned off the exploding wire led to RMS-values of 0.5 m/s while the pyrotechnical igniter led to values of 2 m/s - 3 m/s. As a result turbulence measurements in the windowed autoclave showed that the turbulence generated from a 1000 J pyrotechnical igniter is stronger than the value generated from an exploding wire with the same ignition energy.
LDA-measurements in the 20-L-sphere showed, that the dispersion of dust with the rebound nozzle generated a flow with a maximum velocity of more than 40 m/s. This flow velocity decreased rapidly with decreasing pressure in the dust storage chamber and with closing the inlet valve. At the moment of ignition the flow velocity reached a value of approximately 10 m/s. The measured RMS-value at the moment of ignition was approximately 15 m/s.
In comparison to the turbulence generated in the windowed autoclave the value determined in the 20-L-sphere is much higher. Hence the test results from the windowed autoclave could not be transferred to the 20-L-sphere. As a result LDA-measurements in the 20-L-sphere were necessary. These measurements showed for tests with 1000 J ignition sources that at the moment of ignition for none of the igniters an increase of the RMS-value could be measured. However, the pyrotechnical igniter showed a not negligible velocity increase in one direction in comparison to the velocity generated due to the dispersion of dust.

3.4 Explosion behavior

Figure 5 presents the explosion pressure of maize starch as a function of the dust cloud concentration determined with both ignition sources with ignition energy of two times 1000 J. Both curves show a comparable maximum of approximately 8.2 bar for the exploding wire and 8.4 bar for the pyrotechnical igniter. For both igniters the maximum was determined for a concentration of 1000 g/m³. Deviations in explosion pressure between the different series were very low for dust explosion testing.
The determined rate of explosion pressure rise values as a function of the dust concentration are found in Figure 6. All test results show a good conformity concerning maximum rate of explosion pressure rise values which were almost similar. However, the reproducibility was not as good as for the explosion pressure. As a result the test values of the different test series differed more for both ignition sources. But this is very often found in dust explosion testing. Comparable test results were found for all tested dusts and ignition energies.

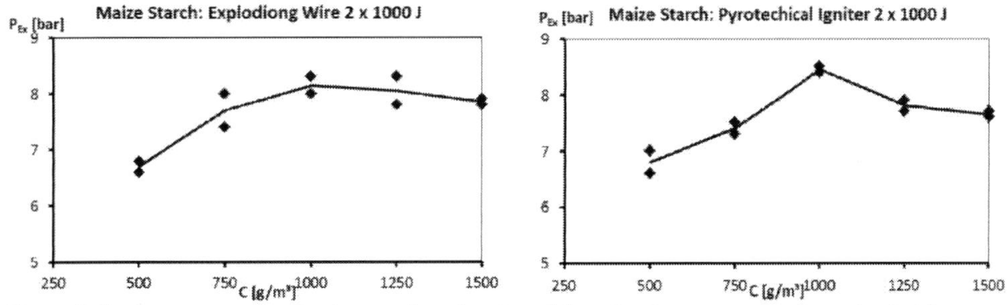

Figure 5. Explosion pressure of maize starch as function of dust cloud concentration determined with pyrotechnical igniters and exploding wires with ignition energies of 2 x 1000 J

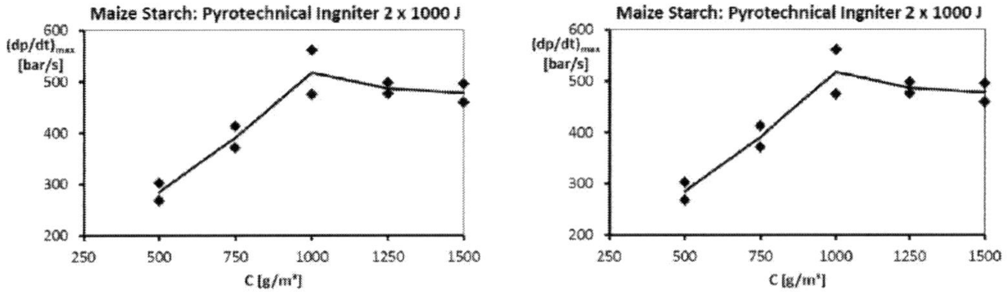

Figure 6. Rate of Explosion pressure rise of maize starch as function of dust cloud concentration determined with pyrotechnical igniters and exploding wires with ignition energies of 2 x 1000 J

Table 2 gives an overview of all determined p_{max} and $(dp/dt)_{max}$ values as well as the K_{St} values which were calculated from the $(dp/dt)_{max}$ values.

A comparison of the results determined with two 500 J ignition sources with one 1000 J ignition source in order to find out if the use of two ignition sources leads to different test results did not come to a clear result. While for lignite tests with two ignition sources led to slightly higher values for the Maximum Explosion Pressure, for maize starch one ignition source led to slightly higher values. The determined rate of explosion pressure rise values were for both dusts slightly higher in tests with two ignition sources. However, for both tested dusts the deviations in explosion pressure and explosion pressure rise between the different test series seem to be slightly lower for tests with two ignition sources.

Table 2 also includes test results which have been determined with two 5000 J igniters, which is the ignition energy described in the test standard for determining p_{max} and K_{St}. The tests showed that the Maximum Explosion Pressures values determined with ignition energies of 100 J, 500 J, 1000 J and 2000 J are less than 10 % lower than those with 10000 J. For steel dust and niacin the values determined with 10000 J were slightly lower than the values determined with the other ignition energies. However, the determined $(dp/dt)_{max}$ values and therefore the K_{St} values differed a lot from the values determined according to the standard procedure, especially for tests with comparatively low ignition energies. Tests with 10000 J led to values which were up to 30 % higher than the values determined with low ignition energies. Only tests with 2000 J ignition sources led to reasonable test results.

Table 2. p_{max}, $(dp/dt)_{max}$ and K_{St} values for all tested dusts determined with exploding wires and pyrotechnical igniters with ignition energies of 1 x100 J, 1 x 500 J, 2 x 500 J, 1x 1000 J 2 x 1000 J and 2 x 5000 J.

Dust		1 x 100 J		1 x 500 J		2 x 500 J		1 x 1000 J		2 x 1000 J		2 x 5 kJ
		Ex. Wire	Pyro. Igniter	Ex. Wire	Pyro. Igniter	Ex. Wire	Pyro. Igniter	Ex. Wire	Pyro. Igniter	Ex. Wire	Pyro. Igniter	Pyro. Igniter
Lignite median 37 µm	p_{max}	8.4	8.1	8.1	7.6	8.1	8.5	7.9	7.7	8.3	7.9	8.4
	$(dp/dt)_{max}$	601	545	681	630	711	619	671	608	723	705	784
	K_{St}	163	148	185	171	193	168	182	165	196	191	213
Maize starch median 14 µm	p_{max}	8.4	8.5	8.4	8.6	8.0	8.2	8.2	8.4	8.2	8.4	8.7
	$(dp/dt)_{max}$	408	470	528	480	500	527	487	468	504	530	614
	K_{St}	111	127	143	130	136	143	132	127	137	144	167
Niacin median 14 µm	p_{max}	-	-	8.3	8.2	-	-	8.2	8.0	-	-	7.9
	$(dp/dt)_{max}$	-	-	953	942	-	-	920	874	-	-	1051
	K_{St}	-	-	259	256	-	-	250	237	-	-	285
Anthraquinone median 3 µm	p_{max}	-	-	-	-	-	-	8.1	8.0	-	-	8.4
	$(dp/dt)_{max}$	-	-	-	-	-	-	1113	999	-	-	1379
	K_{St}	-	-	-	-	-	-	307	271	-	-	374
Steel dust median 24 µm	p_{max}	-	-	-	-	-	-	4.1	4.1	-	-	3.8
	$(dp/dt)_{max}$	-	-	-	-	-	-	398	394	-	-	432
	K_{St}	-	-	-	-	-	-	108	107	-	-	117

4. Conclusion

Measurements of ignition energy delivered to the mixture of the ignition source exploding wire using two different methods showed that the ignition energy generated with the exploding wire was reproducible. For ignition energies of 500 J and higher, the dust cloud did not influence the ignition energy significantly. However, some tests with ignition energy of 100 J and dust concentrations with more than 1000 g/m³ showed that the dust cloud was able to influence the ignition process such that the electric arc extinguished earlier. This resulted in lower ignition energy.

High speed recordings showed that the flame generated from the pyrotechnical igniter propagated faster than the flame/arc generated from the exploding wire and reached a larger volume. Recordings with a fast IR camera system confirmed these test results. In addition to that the recordings showed that as well the volume with a temperature level comparable to the Minimum Ignition Energy of most dusts as the volume in which very high temperatures were measured was much larger for the pyrotechnical igniter. In contrast to that in tests with the pyrotechnical igniter the temperature decreased earlier as in tests with the exploding wire. As a result of high speed and IR recordings as well as energy measurements for the 100 J exploding wire the igniter should have a significant influence on the determined p_{max} and $(dp/dt)_{max}$ values. However, such an effect was not determined during the measurements of p_{max} and $(dp/dt)_{max}$ of five different dusts. All tests resulted in comparable values of p_{max} and $(dp/dt)_{max}$ for both ignition sources.

A comparison of p_{max} values determined with two 500 J ignition sources with one 1000 J ignition source in order to find out, whether the use of two ignition sources will lead to different test results, did not come to a clear result. The determined pressure rise values were for both dusts slightly higher for tests with two ignition sources. However, for both tested dusts deviations in explosion pressure and rate of explosion pressure rise seem to be slightly lower for tests with two ignitions sources.

The p_{max} values determined with ignition energies of 100 J, 500 J, 1000 J and 2000 J were less than 10 % lower than values determined according to the test standard with two igniters of 5000 J each. As a result the influence of the ignition energy of the igniter on the Maximum Explosion Pressure seems to be almost negligible.
In contrast to that $(dp/dt)_{max}$ values determined according to the test standard led to 30 % higher values, especially in tests with ignition energies of 100 J and 500 J. Only tests with 2000 J ignition sources led to reasonable test results. The reason for that could not be solved totally. It is assumed that a combination of different effects has to be considered such as turbulence generated from the ignition source and the flame/arc volume of the ignition source. However, the effect of the flame/arc volume should result in different test results for exploding wire and pyrotechnical igniter. However, this could not be determined. Probably the turbulence generated from the igniters has a strong influence especially on the $(dp/dt)_{max}$ values. Tests in the 20-L-sphere showed for all ignition sources with energies up to 2000 J that the turbulence generated from the igniters seems not to have a significant influence on the turbulence field generated due to dispersion of the dust (pre-ignition turbulence). This is probably different for higher ignition energies.
In a next step tests with ignition energies up to 10000 J with two pyrotechnical igniters and two exploding wires will be performed. In addition to that a validation of the exploding wire with a greater number of dusts is planned.

5. References

[1] ASTM E1226-12, "Standard Test Method for Explosibility of Dust Clouds," ASTM International, US, 2012

[2] EN 14034–1, "Determination of explosion characteristics of dust clouds - Part 1: Determination of the maximum explosion pressure p_{max} of dust clouds," Beuth-Verlag, Berlin, Germany, 2011.

[3] EN 14034–2, "Determination of explosion characteristics of dust clouds - Part 2: Determination of the maximum rate of explosion pressure rise $(dp/dt)_{max}$ of dust clouds," Beuth-Verlag, Berlin, Germany, 2011.

[4] EN 1839, "Determination of explosion limits of gases and vapours," Beuth-Verlag, Berlin, Germany, 2003.

[5] Cesana C., Siwek R., "Manual KSEP 7.0," Kühner AG, <kuhnersafety.com/tl_files/kuhner/product/safety/PDF/B000_071.pdf>, Accessed 2013-01-04

[6] Krietsch A., Scheid M., Lasry M., Suter G., Teske H.-J., Stahmer K. W., Arnold A., Nemcak O., Cesana C., Siwek R., "Validation of alternative igniters for the determination of explosion characteristics of dust clouds in 20-l-sphere and $1 m^3$-vessel," Proceedings 13th Int. Symp. on Loss Prevention and Safety Promotion in the Process Industries 6th – 9th June 2010, Brugge, 2010

[7] Scheid M., Kusche C., Schröder V., Barth U., "Tests on Suitability of the Ignition Source "Exploding Wire" for the Determination of Explosion Characteristics of Combustible Dusts in the 20-L-Sphere," Chemical Engineering Transactions, vol. 31, 2013 (in print)

[8] Van der Wel P. G. J., "Ignition and Propagation of Dust Explosions," Dissertation, Delft University Press, the Netherlands, 1993

Operational Discipline

Improving Process Safety Management Implementation through Behavior and Cultural Change

Charles A. Soczek
Principal Consultant
DuPont Sustainable Solutions
E.I. DuPont De Nemours and Company

Abstract

In 2007, US OSHA initiated a Petroleum Refinery Process Safety Management National Emphasis Program. This was followed in 2009 with a similar program focused on Chemical operations. These programs developed a series of findings, many of which were results of failure to implement OSHA 1910.119 PSM regulatory requirements. Many facilities had well documented programs, but there was poor implementation. In DuPont it was recognized that implementation is an Operational Discipline issue.

Operational Discipline (OD) as defined by DuPont is "the deeply rooted dedication and commitment by every member of an organization to carry out each task, the right way, and each time". OD is recognized as the key to successful implementation of a Process Safety Management program, turning a paper program into reality. Typically, companies have addressed OD improvements through behavioral modifications, but while the gains from programs to modify behavior can be substantial, they are difficult to sustain. Cultural change, while far more difficult to achieve, can produce more sustainable results. When done together, substantial improvements in PSM performance can be both rapid and sustained. This presentation will discuss the key attributes of operational discipline as well as provide an outline of the process to address behavior and culture change is provided.

Introduction

Operational Discipline (OD) as defined by DuPont is "the deeply rooted dedication and commitment by every member of an organization to carry out each task, the right way, and each time". Per the Business Dictionary, OD is defined as assurance that specific activities necessary for successful daily operation of a company's operations are running as they should be. This can include specific goals, missions, and guidelines. Whichever definition is chosen, OD is important to success of all aspects of an operating facility.

Process Safety Management has been in place for over twenty years. Many companies have good PSM systems, but based on the CSM Special Emphasis studies, implementation is poor. It is being increasingly recognized that Operating Discipline is a key factor that drives PSM implementation. Operating Discipline helps turn a paper PSM program into a reality

Improving Operating Discipline

Advances in Social/Psychological theory are showing that operating discipline improvements are influences by two factors, Behavior Change and Culture change. The Dejoy model helps explain this.

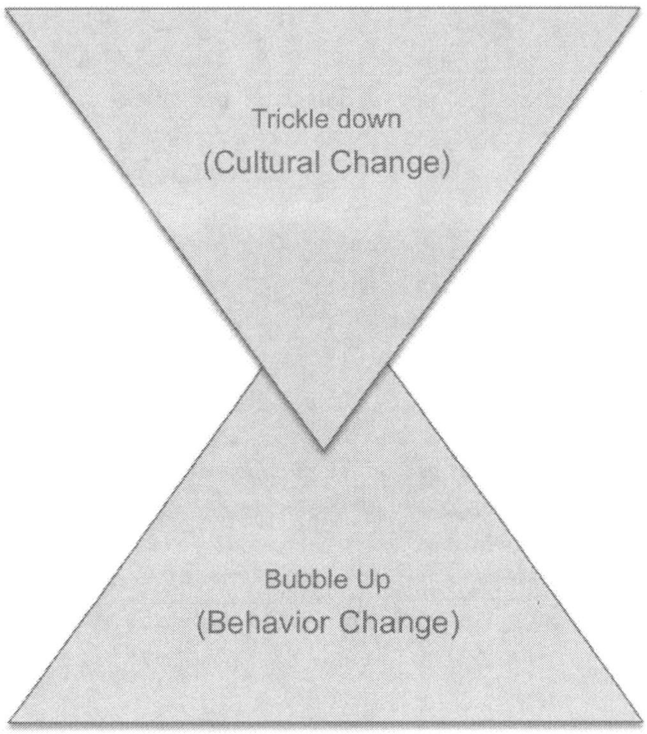

The basis for the Dejoy[1] model is that behavior change can be made quickly, but improvements are not sustainable without a large continuing effort, and that even with this, results will plateau (Bradley Curve). Safety culture change takes much longer, but once in place, superior performance can be both achieved and sustained when coupled with behavior improvement. In the Dejoy model, cultural change is driven by leadership whereas behavior change is driven by supervisors. This is an important distinction and helps focus the role of leadership and the role of supervision.

In a study of some DuPont sites, it has been found that multi-faceted Operating Discipline programs are in place. These programs address both behavior change and cultural change. Programs vary, and can consist of consists of a large number of small efforts or fewer larger programs involving all members of the organization. Either can be successful.

Dejoy[1] did a comparison of a cultural based program and a behavior based program.

Characteristic	Behavior -Based changes approach to safety	Cultural Based changes approach to safety
Background/origin	Operant psychology/ behavior modification	Organizational behavior/ anthropology
Key aspects	Bottom up approach	Top-down
	Analytic/data driven	Intuitive/ethnographic
	Setting Specific	Setting Specific
	continuous process	Self-sustaining
Typical Implementation	Identify and define critical behaviors	Assess aspects of culture (values, beliefs, assumptions, etc.)
	Set performance goals	Devise alternative vision
	Observe/sample behavior	Work with leadership and employees to implement change
	Provide contingent feedback and reinforcement	
Principal Strengths	Specific technology	Emphasizes organizational change
	Objective/empirical	Focuses on basic causes
	Shop floor focus	participatory (often)
	Participatory (usually)	Comprehensive
	Positive	
Principal Weaknesses	Victim-blaming	Diffuse technology
	Minimizes environment	Subjective/intuitive
	Focuses on immediate cause	Indirect

The above chart has highlighted a few of the key differences between the behavioral and cultural approach. In the bottom up, the emphasis is on the operators, maintenance workers, and contractors to follow procedures and rules. This requires good processes and standards and includes items as good communications, housekeeping, documentation, following procedures, and adequate resources. The top-down cultural improvement process is driven by leadership and is reflected in efforts such as leadership by example, building pride in the organization, expression of values, risk sensitivity, and being a learning organization that is innovative, agile, and continuously improves..

Some Definitions

- Operant conditioning psychology (sometimes referred to as instrumental conditioning) is a method of learning that occurs through rewards and punishments for behavior
- Psychological anthropology is an interdisciplinary subfield of anthropology that studies the interaction of cultural and mental processes. The subfield tends to focus on ways in which humans' development and enculturation within a particular cultural group—with its own history, language, practices, and conceptual categories—shape processes of human cognition, emotion, perception, motivation, and mental health. It also examines how the understanding of cognition, emotion, motivation, and similar psychological processes inform or constrain our models of cultural and social processes. Each school within psychological anthropology has its own approach
- Ethnography is a qualitative method aimed to learn and understand cultural phenomena which reflect the knowledge and system of meanings guiding the life of a cultural group

In behavior based approach, performance goals are set, whereas in a cultural based approach leadership works to build a vision of a better place (zero incidents).

Behavior programs are focused on the shop floor, whereas cultural change is focused on everyone in the organization.

Operating Discipline Characteristics

Organizations that have a high level of operating discipline exhibit an identifiable set of characteristics. DuPont has defined these characteristics as follows:

OD Characteristics

» Leadership by example
» Sufficient & capable resources
» Employee involvement
» Active lines of communication
» Strong teamwork
» Shared values
» Up-to-date documentation
» Practice consistent with procedures
» Absence of shortcuts
» Excellent housekeeping
» Pride in the organization

» Risk sensitivity
» Innovation, agility, and continuous improvement
» Justice and fairness

It is recognized that behavior change of people and cultural change are very complex. It is not proposed that the set of characteristics above is comprehensive, and it is expected that over time additional characteristics will be recognized and added to this list.

Implementation of an Operating Discipline Improvement Program

The process developed to implement an operating discipline program starts with an assessment based on 6 Sigma methodology. In the assessment, each characteristic is defined by a set of CTQs (Critical to Quality) and CTXs that define the CTQ.

These steps are

1.Pre-work

- Define scope of the assessment
- Form client OD team
- Develop assessment schedule
    - Capture input from 50% or more of the population
    - Capture current relevant data (see appendix A)
- Assessment team orientation
- Develop current state hypothesis

2. Field data collection (reach 50% of the workforce)

- Focus Groups – operators, mechanics, first line (2 hours)
- One-on-One- managers and leaders (60 minutes)
- Field Tours- employees actively performing work

3. Analysis

- Review and summarize key findings
- Apply a rating to each CTQ per rating guidance

4. Recommendations

- Each low rate CTQ has a recommendation(s) based on best practices. These best practices come from DuPont sites.
- The completed Assessment Model is used as the report to management

5. Action Plan

- Owned by line management based on findings

6. Continuous Improvement

- Conduct reviews of the assessment and periodically update the action plan with additional "best practices"
- Conduct re-assessments no more frequently than once per year

Examples of the Assessment Questions

Below is an example of the assessment tool developed that is used to understand each of the fourteen OD characteristics.

OD Characteristic: Leadership by Example

Leaders are demonstrating unrelenting passion for SHE and continuously model the behavior they expect of others

CTQ-1 Visible leadership presence and involvement across the organization

CTX – Build value in leaders that this is a critical part of their role and hold them accountable with specific expectations

- Is visible leadership evaluated as part of the overall personnel review process for line leaders, and if so, how ? Do Leaders at all levels have COTs relating to Operational Discipline?
- Do line leaders consistently take specific corrective action steps for non-conformance to SHE requirements?
- Line management is actively involved with planning, leading safety meetings & communications, and emphasizing broad participation
- Is visible leadership included in new supervisory/manager training and if so, is it effective?
- Is there consistency in approach and focus over time as leaders change?
- Does leadership review action plans to assure alignment with goals and objectives?
- Does leadership allocate adequate resources to support the safety program?
- Does leadership provide a personal example of following standards and procedures? (Wearing required personal protective equipment when in the field; using handrails, etc.)

CTX – Define and communicate what is "Visible Leadership" throughout the organization

- Does the site have formal/informal guidelines for leadership time spent in the field? How does the site measure this and is it effective
- Has the site clearly defined opportunities for visible leadership and personal leadership participation at all levels?
- Is leadership at all levels routinely in work areas conducting Safety Observations audits or other field evaluations?
- Has site leadership clearly articulated a vision and goals for Operational Discipline?
- Does leadership conduct periodic reviews of activities with all safety related teams?
- Does leadership critique audit and reports and follow-up to assure implementation of recommendations?

This characteristic has three CTQs and six CTXs. This is similar to the other characteristics.

The assessment process generates a large volume of information. The assessment team takes this information and develops action items. These items must be prioritized. Below are some examples of action items for leaders taken from an operating facility that could be generated from the process described above.

1. Increase field time and double safety audits
2. Devote 50% of PM meeting to safety
3. Off hours audits and safety coaching
4. Continue small things matter theme
5. Establish designated time in the field for leadership on safety
6. Revamp weekly safety audit meeting
7. Focus team on elimination of slips trips and falls
8. Review all letters of commendations

Summary

Sustainable improvements in Operating Discipline take time, and require facilities to continuously work on improving the safety culture and the behavior of people. This continuing focus on Operating Discipline is a common characteristic of high performing facilities.

References

(1) "Behavior change versus culture change: Diverging approached to managing workplace safety" David M. Dejoy Safety Science 43 2005.

(43g) Risk Analysis for Pressure Relief Devices – Managing Plant Safety and Reliability

Jose R. Aguilar-Otero Sr., Technology Application, COMIMSA, Saltillo, Mexico

Risk Analysis for Pressure relief devices – Managing Plant Safety and Reliability

Pressure relief devices are one of the most critical assets for plant risk and safety, therefore it is essential to be able to ensure PSV's operation, reliability and availability through the implementation of risk mitigation or prevention actions, either, corrective, preventive or predictive; these actions will allow plant owners to manage the risk associated with the device operation (or failure). This paper is intended to show a risk assessment using Weibull analysis focused on the identification for risk reduction opportunities and the selection for the best chance and dates to implement risk mitigation actions. Risk assessment methodology is based on API RP 581, Risk Based Technology.

Historically, maintenance planning for a PSV was carried out, based on pressure tests and engineering judgment; a pressure (POP) test is usually conducted to evaluate device's availability and functional condition and its frequency is linked to the protected equipment's maintenance program, regardless whatever the failure mode. Nevertheless, now a days, more quantitative and rigorous methodologies are required not just for maintenance planning but for developing a risk management strategy.

The methodology is a risk assessment. We evaluate the probability of failure (POF) times the consequence of failure (COF), for both, a valve failure and a protected equipment failure. Two failure modes are evaluated: fail to open on demand (FOD) and leak (LEAK). POF is calculated analyzing three main issues:

1. Historical inspection records and time in service, using a two parameters Weibull distribution, as follows:

(1)

2. Probability of failure from the protected assets,

(2)

3. Probability of failure from process over pressure scenarios.

(3)

Consequences of failure (FOD, LEAK) are evaluated as a function of consequence area and as a function of its economic impact, for a valve failure and for a protected equipment failure.

FOD consequences are evaluated for four leak scenarios (1/4", 1", 4" and catastrophic rupture) for each protected equipment and using an overpressure as the discharge pressure. This overpressure is a higher pressure than the normal operating conditions. In this approach impact zones are used to evaluate fire, explosion and toxicity effects, in the same way we do it for a traditional quantitative consequence analysis. Meanwhile, for the leak failure mode, consequences are evaluated using a financial approach, no impact area is considered. Consequences categories for the leak mode are as follows:

- Loss of production costs
- Maintenance cost (repair costs)

- Product costs

The leak mode includes both, a small leak and a moderate leak. Moderate leak means valve stuck open, while small leak means leakage past valve (No seal).

Therefore, the global risk is calculated using the two risks approaches,

(4)

Risk calculation involves a straightforward review of the process operating conditions using a layer of protection analysis (LOPA), it also requires analyzing the process conditions and initiating events than can challenge or demand the PSV and the layers affecting the initiating event frequency. The methodology also involves a process for evaluate as many overpressure scenarios as possible; this scenarios will also challenge the pressure relief device.

This risk analysis methodology has the following benefits:

a) Perform a What if...? Scenarios analysis, and evaluate risk mitigation actions

b) Update analysis modeling, using new data

c) Understanding risk drivers

d) Understanding event sequence for risk contributors

e) Set up a risk management strategy

The risk assessment was performed for 100 PSV, located on several offshore facilities. Results show risk contribution for every failure mode, risk ranking for the whole set of PSVs, new maintenance dates as well as reliability and availability estimations. The use of this approach is very useful, because we are able to deal with large data samples and also fits well with small data samples. Results can be expressed using failure mode information or time to failure information.

See more of this Session: Electronic Poster Session
See more of this Group/Topical: Global Congress on Process Safety

(43h) Enterprise Risk Management--Adopting Risk-Based Decision Making in a Chemical Enterprise From the Board Room On Down Through All Functions and All Levels

Jerry L. Jones, PE, Consultant, Menlo Park, CA

Enterprise risk management (ERM) practices have been in place for decades in many businesses. However, following a number of failures of large financial, energy and telecommunications ventures, the interest in broader adoption of effective ERM systems has increased dramatically. During the last decade in response to recommendations and guidance by leading professional societies concerned with financial accounting and corporate governance practices, many more publicly owned non-manufacturing and manufacturing businesses have begun implementing very structured and rigorous ERM systems.

While risk has historically been defined in relation to hazards (including natural perils) as a function of the probability and consequences (harm and financial loss) of an undesired exposure event, under various recent ERM guidelines risk has been more broadly defined in relation to business objectives:

o **Risk isthe effect of uncertainty on meeting objectives** (ISO 31000-2009: Risk Management Principles and Guidelines adopted in the US as ANSI/ASSE Z690.2-2011)

o **Risk isthe possibility that an event will occur and adversely affect the achievement of objectives** (COSO Framework for Integrated Risk Management, (Committee of Sponsoring Organizations for the Treadway Commission, 2004)

A key argument made for adopting ERM principles by both public and private companies and using risk-based decision making processes throughout the enterprise is the opportunity the approach offers to more effectively and efficiently protect current value (preventing losses, the downside of risk) and to create value (the upside of some risk exposures) for owners and other stakeholders. A further incentive is the potential to establish long-term strategic advantage by being able to more effectively and efficiently manage risks than competitors.

Both of the ERM system guidelines cited above recommend rigorous and comprehensive oversight of risk management practices by the Board of Directors and active engagement of the executive leadership team in managing the key risks. Such practices are exactly what have been recommended in relation to chemical process safety by a number of investigation panels or boards in the US and abroad following major chemical process industry accidents during the last decade.

In this presentation the wide range of risk exposures that must typically be managed by a chemical enterprise (showing that process safety is one of many requiring leadership attention) will be briefly reviewed with some specific examples provided for relevant consensus standards or guidelines for managing those risks. The following list of objectives that an ERM should achieve will be discussed along with some the challenges they present.

The ERM system should assure that:

o The company's significant existing risk exposures are identified, assessed and managed in a consistent manner by the various functional organizations and the results of such risk management activities regularly reviewed by the executive leadership team and the Board of Directors,

o The "right" risks are being taken based on consideration of risk/reward tradeoffs and the "risk appetite" of the organization, and resources for risk management are allocated commensurate with the risk exposures and risk control needs,

o Individual at all levels are actively engaged in risk-based decision making and risk management so as to leverage off the knowledge and experience of the entire organization,

o Individual responsibilities and accountabilities for ongoing risk control and monitoring are clearly defined in all functional organizations at all levels, including risk management related reporting requirements with metrics (both leading and lagging indicators) that go to the executive leadership team and to the Board of Directors,

o Internal auditing and self-assessment processes are in place to regularly evaluate the effectiveness of risk management processes and controls so as to keep risk exposures below the agreed upon risk tolerance limits.

A four-step business process for risk identification, analysis, risk-based decision making, and risk controls monitoring/reporting will be described that is consistent with ERM consensus guidelines and others including those for risk-based process safety. This generic business process can be readily adopted by various functional organizations within companies of all sizes to manage a wide range of types of risk exposures and used to support other business processes requiring risk-based decision making (e.g. process hazard analysis, product safety analysis, product quality assurance, management of change, incident investigations, supply chain security, mechanical reliability analysis, and business continuity analysis).

See more of this Session: Electronic Poster Session
See more of this Group/Topical: Global Congress on Process Safety

(43i) The Chemical Safety Board and the Prevention of Combustible Dust Hazards

Manuel R. Gomez, Chemical Safety Board, Washington , DC and Christina Morgan, Office of Recommendations, US Chemical Safety Board, Washington, DC

THE CHEMICAL SAFETY BOARD AND THE PREVENTION OF COMBUSTIBLE DUST HAZARDS

This presentation will:

- Briefly summarize the CSB's rationale for recommending an OSHA general industry combustible dust standard;
- Review OSHA actions up to date;
- Comment on the difficulties OSHA may be facing in the conduct of combustible dust rulemaking; and,
- Describe the CSB's ongoing and planned efforts to advocate successful rulemaking.

See more of this Session: Electronic Poster Session
See more of this Group/Topical: Global Congress on Process Safety

9TH GLOBAL CONGRESS ON PROCESS SAFETY

Performing a HAZOP on Start-up/Shutdown Procedures

Fareed Ebrahim
ACM Automation Inc.
#700, 940 – 6th Avenue SW
Calgary, Alberta
febrahim@acm.ca

German Luna-Mejias, P. Eng
ACM Automation Inc.
gluna@acm.ca

Prepared for Presentation at
American Institute of Chemical Engineers
2013 Spring Meeting
9th Global Congress on Process Safety
San Antonio, Texas
April 28 – May 1, 2013

UNPUBLISHED

Performing a HAZOP on Start-up/Shutdown Procedures

Fareed Ebrahim
ACM Automation Inc.
#700, 940 – 6th Avenue SW
Calgary, Alberta
febrahim@acm.ca

German Luna-Mejias, P. Eng
ACM Automation Inc.

Keywords: Procedure HAZOP, Start-up/Shutdown

Abstract

Statistics show that approximately 65% of process safety related incidents occur during the start-up/shutdown operations of the plant. Yet, the bulk of the risk assessments performed today focus on the normal operation of the plant. A major contributor to the lack of start-up/shutdown risk assessments is the lack of a strongly documented methodology. One method of performing this assessment is through the HAZOP of start-up/shutdown procedures. This HAZOP involves a step by step review of the procedure with consideration given to the sequencing, delay, or missing of steps involved in the procedure. By performing this review, required permissives are identified, contingencies are developed and essential safeguards are identified. This will help ensure the operations team is confident that they are performing the start-up/shutdown in a safe manner. This paper focuses on the framework for guidelines, procedures and templates that are necessary to perform this HAZOP. A practical approach with examples is presented.

1. Introduction

Currently, it is common practice to reserve a deviation in a continuous process HAZOP for start-up and shutdown issues. Unfortunately, during analysis, this deviation if often not explored to its full potential. This abrupt analysis is often due to the lack of available information and expertise, lack specificity within the deviation, and time constraints. Yet, statistics provide a compelling argument to improve the rigour associated with start-up/shutdown analysis. Approximately 65% of process safety related incidents occur during these modes of operations1. To improve the rigour associated with the assessment of start-up/shutdown, the HAZOP technique must be further extended to the procedures for these modes of operation. Many sources are available which explore the fundamentals of performing such a HAZOP1,2, and therefore, this paper will focus on developing a methodology by expanding upon the commonly used HAZOP technique for normal operating modes.

2. Preparing for the HAZOP

In order to ensure that a complete and accurate assessment is performed, compiling the correct documents and team members for the HAZOP is of paramount importance. Just as there are key documents required for a process HAZOP, specific documents are also required for the start-up/shutdown procedure HAZOP.

2.1 Required Documentation

Firstly, the procedure for start-up/shutdown must be available. Just as the P&IDs are the central document during a process HAZOP, the procedure will be the main focus in the start-up/shutdown analysis and therefore, should be the most up-to-date revisions available. This procedure can then be split into "nodes" or subsystems for analysis. Nodes or subsystems may be defined as each individual step in the procedure.

Depending on the complexity of each step, it may be required to split the step into simplified actions, which in turn become the node for analysis. Conversely, procedure steps may be combined to develop a node based on timing requirements and the requirement for actions needing to be performed simultaneously. Timing concerns associated with procedures will be further discussed with the HAZOP deviations. As the implications of each procedure step may not be immediately clear to the HAZOP facilitator, it is of utmost importance that nodes are developed in conjunction with the operations personnel.

Once the nodes have been identified, and the complexity of the procedure understood, a decision must be made on the degree of analysis for the HAZOP to be performed. Two methods of analysis can be employed for the HAZOP – a "What-If?" approach, or a guideword method. The "What-If?" approach follows the same method as that used for a continuous process. If only an extremely high level analysis is required, the "What-If?" methodology may be applied to the entire procedure instead of applying it to nodes. Conversely, the application of the guideword method provides a more structured approach. Generally, as the complexity of the procedure increases, more guidewords should be employed in order to ensure a thorough analysis is completed. PHA software3 contains pre-packaged guidewords to be used for analysis. Explanation and application of the guidewords will be explored in Section 3.

Supporting documents for the procedure HAZOP include the following:

- As-Built P&IDs
- Isolation markup drawings - showing spec blind placement and position as per start-up/shutdown modes
- Shutdown keys - including permissives
- Alarm lists
- Alarm/Trip bypass procedures - including contingency plans

It is important to note that the shutdown keys and alarm lists may not outline which systems may be bypassed during a start-up and shutdown. Therefore, it is critical to obtain the bypass information, as well as the operating window for all instrumentation that will be referenced during the start-up/shutdown operation.

2.2 Required Personnel

In addition to compiling the correct documents for the session, it is of equal importance to ensure the correct personnel are in the room during the HAZOP. Similar to a process HAZOP, representation from the process team, instrumentation, and operation is required. It is suggested to have extra representation from the operations group to provide a more diverse perspective for the brainstorming process. Furthermore, if the HAZOP is being performed at the commissioning phase, it is recommended for the construction manager or equivalent representative to be present.

3. Performing the HAZOP

3.1 Introducing Guidewords

Risk assessment software[3] often includes preloaded libraries of guidewords for the HAZOP of procedures. Table 1 below lists seven guidewords that can be used for the HAZOP. These guidewords are listed in a suggested order for use in the HAZOP. Guidewords may be changed, added or deleted based on the HAZOP team experience, process being analyzed, and corporate guidelines. Note that the performance of a start-up/shutdown procedure HAZOP does not negate the requirement to perform a pre-start-up safety review.

Table 1. Suggested Guidewords and Description for Procedure HAZOP

Guideword[3]	Description
Less	Not enough is done to adequately execute procedure action.
More	Action is performed to a further extent than specified.
No	Action is fully omitted.
Reverse	Actions in procedure are performed out of specified sequence.
Other than[3]	Action is performed that does not align with what is specified.
As well as[3]	The operator does another action, in addition to the specified step being done correctly.[2]
Sooner/Later[3]	Timing of action is incorrect.

All seven guidewords may not be used for each start-up/shutdown procedure being analyzed. Guidewords should be applied based on the complexity of the procedure. An example of a simplified method is the two-guideword approach[2] is contained in Table 2 below.

Table 2. Two-Guideword approach for Procedure HAZOP[2]

Guideword	Description
Omit	Step is not done or part of the step is not done.
Incorrect	Step is not performed as intended.

3.1 Application of Guidewords

In order to further illustrate the application of each guideword, a subsection of a mock, simplified procedure (listed below) for a Reformer will be used. Guidewords from the two word approach will not be further explored as they can be considered a subset of the initial seven. Note that nodes will be assumed to be the individual procedure steps for the purpose of this illustration. Consequences are not to be considered complete for use in industry HAZOPs but for the purpose of example only.

Simplified Excerpt of Reformer Procedure:

1. Once pilots are lit, start lighting main burners (two per cell).
2. After completing the first ten pilots, light another ten but evenly distributed per cell.
3. Wait for 30 minutes or until the heater outlet temperature is stable
4. Continue lighting burners to increase the heater temperature at a rate of 50°C/hour until a temperature of 380°C is reached.
5. Once temperature is stable and maintains at 380°C, start injecting HP Steam and continue lighting burners as required to prevent the heater temperature from dropping below 350°C.

3.2.1 Less

Example Deviation from Step 3: Adequate time is not allotted by operator for heater outlet temperature to stabilize in an attempt to speed up start-up.

Consequence: Potential for a sudden temperature increase in heater leading to refractory damage or flame impingement.

3.2.2 More

Example Deviation from Step 1: More than two burners are lit per cell in order to speed up start-up of unit.

Consequence: Potential for a sudden temperature increase in heater leading to refractory damage or flame impingement.

3.2.3 No

Example Deviation from Step 4: Additional burners are not lit by operator.

Consequence: If steam is injected into the heater when temperature is below 350°C there will be deactivation of catalyst within heater tubes.

3.2.4 Reverse

Example: Deviation from Step 3: Additional burners are lit prior to stabilization of heater outlet temperature.

Consequence: Potential for a sudden temperature increase in heater leading to refractory damage or flame impingement.

3.2.5 As well as

Example: Deviation from Step 2: Steam is injected by operator as well as lighting the additional burners.

Consequence: If steam is injected into the heater when temperature is below 350°C there will be deactivation of catalyst within heater tubes.

3.2.6 Other than

Example: Deviation from Step 5: Natural Gas is injected into the heater instead of Steam.

Consequence: Overheating of tubes in reactor leading to tube damage and potential explosion.

3.2.7 Sooner/Later

Example: Deviation from Step 5: Steam injected to heater prior to temperature reaching 350°C.

Consequence: If steam is injected into the heater when temperature is below 350°C there will be deactivation of catalyst within heater tubes.

3.3 Applying a Risk Ranking

As HAZOPs have developed over time, there has been a continued emphasis to add more semi-quantitative aspects to the study. A major element of this semi-quantitative aspect is the application of a risk matrix. Unfortunately, risk matrices are designed using a per year time frame. This per year concept is of course not applicable to a procedure. Therefore, when performing the procedure HAZOP, two options are available. One, is to avoid the risk ranking and perform a fully qualitative analysis. Second, is to apply the risk matrix by avoiding the yearly references contained in the likelihood, and setting a default likelihood of making an error in the procedure to correspond with highest likelihood available (ie. frequent) on the matrix. This is to quantify the fact that there is a possibility to make a mistake in executing the procedure at any given opportunity for execution. For example, if a procedure step is incorrectly worded, there is an opportunity for error every time that step is executed. The consequence levels outlined in the risk matrix are not time constrained and therefore are easily applied to a start-up/shutdown procedure HAZOP. Safeguards can then be applied in a fashion similar to process HAZOPs by dropping the likelihood by one level.

3.4 Adding Safeguards

When applying safeguards to a procedure, it is of course of utmost importance to adhere to the guidelines set out in IEC-61511 if they will be used as independent protection layers. In addition to this, it is extremely important to ensure that the safeguard will not be bypassed during the start-up or shutdown. In some cases, safeguards through a safety instrumented system may be disabled during start-up or shutdown. It is extremely important to identify these bypasses and ensure safeguards are not credited that will not be active. If a safeguard is bypassed, refer to contingency planning documents in order to identify an alternative. Furthermore, it is of utmost importance to ensure that the safeguards will in fact be within their operating range during start-up and shutdown. One control related safeguard that is far more predominant in start-up/shutdown procedure HAZOPs when compared to process HAZOPs are permissives. It is therefore important to have these permissives identified, optimally on the shutdown key. In addition to instrumentation safeguards, a double check to the operator performing the specified action also acts as a valid safeguard.

4. Conclusion

The application of the HAZOP technique to a start-up/shutdown procedure is an effective way to improve the quality of analysis in these modes of operation. Just as in a normal operating mode HAZOP, specific documentation and team members are required. Furthermore, the level of rigour of the study may be adjusted by using a "What-If?" approach or by varying the number of guidewords to be used. A risk ranking may be applied during a procedure HAZOP, but caution must be applied in order to avoid diluting the risk of a situation by ensuring the procedure is analyzed on a per opportunity (instead of a per year) basis. Safeguards must also be applied with caution due to the potential for bypasses or conditions being out of the operating window of certain instruments during start-up/shutdown. It is hoped that the application of the HAZOP methodology to start-up and shutdown procedures will greatly reduce the number of incidents that occur during these modes of operation.

5. References

[1] Bridges, Bill. "Paper: How to Efficiently Perform the Hazard Evaluation Required for Non-Routine Modes of Operation (Start-up, Shutdown, Online Maintenance)." Proc. of 2011 Global Congress on Process Safety., 2011

[2] CCPS. "9. Extensions and Special Applications." *Guidelines for Hazard Evaluation Procedures.* New York: Center for Chemical Process Safety of the American Institute of Chemical Engineers, 1992. 257-67.

[3] Dyadem International Ltd. *PHA-Pro.* Computer software. Vers. 8.2.2.0.

(43p) Chasing the Protection of Intermediate Bulk Containers, Year 10

John A. Leblanc, Engineering Standards, FM Global, Norwood, MA

The challenges of protecting composite intermediate bulk containers (IBC) storing liquids that burn have been under investigation by a number of groups for more than 10 years. The path to resolving this challenge chosen by FM Global looked at defining the fire hazard created by the liquids and then understanding how the containers reacted to fire. This effort has resulted in a strong understanding of the fire hazards and protection options for several liquid-IBC combinations:

Rack storage of Oil and alcohol in IBCs

Silicone fluids in palletized IBCs

Very high flash point liquids in palletized IBCs

Propylene glycol in palletized IBCs

Storage of Empty IBCs

Defining an FM Approved IBC

This paper will provide an overview of the fire hazards associated with storing liquids that can burn in IBCs and present the new protection options and their associated test programs.

Process Safety Challenges in the Hazardous Waste Industry

Hope A. Luebeck
Veolia Environmental Services, L.L.C.
700 East Butterfield Road, Suite 201
Lombard, IL 60148
Hope.Luebeck@Veoliaes.com

Prepared for Presentation at
American Institute of Chemical Engineers
2013 Spring Meeting
9th Global Congress on Process Safety
San Antonio, Texas
April 28 – May 1, 2013

UNPUBLISHED

Process Safety Challenges in the Hazardous Waste Industry

Hope A. Luebeck
Veolia Environmental Services
700 East Butterfield Road, Suite 201
Lombard, IL 60148
Hope.Luebeck@Veoliaes.com

Keywords: Process Safety, Hazardous Waste Industry, Recyclers' Safety Council

Abstract

The OSHA Process Safety Management (PSM) Standard, 29 CFR 1910.119, was promulgated in 1992 to prevent the release of highly hazardous chemicals (HHCs) following a series of tragic industrial accidents. During the last 20 years, the emphasis of this standard has primarily been focused on refineries, chemical plants, and similar processes. However, companies in the Hazardous Waste Industry (HWI) also handle HHCs and face a special set of challenges as they strive to be PSM-compliant in their operations.

Hazardous Waste Industry facilities are often smaller than typical petrochemical, chemical and pharmaceutical plants. They have fewer resources dedicated to their operations and, as a result, generally make slower progress in advancing their PSM programs. Reasons for these tendencies will be discussed, as will initiatives which are being taken to strengthen their programs.

1. Introduction

History has taught us that the most influential changes result from tragedies. The Process Safety Management (PSM) Standard 29 CFR 1910.119 [1] was established by the Occupational Safety and Health Administration (OSHA) in 1992 after a series of deadly industrial accidents including the methyl isocyanate release at a Union Carbide pesticide plant in Bhopal, India, in 1984 [2] and the vapor cloud explosion at the Phillips 66 chemical complex in Pasadena, TX in 1989 [3]. The 14-element safety standard includes specific requirements for managing risks associated with processes which utilize Highly Hazardous Chemicals (HHCs) such as were involved in the noted catastrophes. More recent events such as the BP Texas City, TX disaster in 2005 [4] have motivated OSHA to renew its scrutiny of process safety practices of PSM-covered organizations through a series of National Emphasis Programs [5].

There have been serious accidents in the Hazardous Waste Industry (HWI) as well, though these have not had the significant impact or notoriety as those in the chemical or petrochemical industries. Such recent HWI events include The Environmental Quality Company fire in 2006 [6] and the Veolia Environmental Services explosion and fire in 2009 [7] (See Table 1). As these sorts of incidents continue in the HWI as well as in other industries, companies must continue to further develop their PSM programs and prepare for potential future NEP inspections.

Table 1. Some Noteworthy Industrial Process Safety-Related Incidents

Industry	Company	Location	Year	Deaths	Injuries
Petrochemical	Union Carbide	Bhopal, India	1984	2000+	Unknown
	Phillips 66	Pasadena, Texas	1989	24	132
	BP	Texas City, TX	2005	15	180
Hazardous Waste	The Environmental Quality Company	Apex, NC	2006	0	30
	Veolia Environmental Services	West Carrollton, OH	2009	0	4

Many of the same issues that challenge the petrochemical industry in their quest to be PSM-compliant are shared by the HWI. However, the size and nature of their operations are often quite different; thus, their approach to PSM compliance can vary significantly.

This paper explains some unique aspects of PSM which apply to companies in the Hazardous Waste Industry (HWI). When the acronym HWI is mentioned, this is intended to apply broadly to all companies having business operations involving hazardous waste. A subset of HWI companies is Hazardous Waste Treatment (HMT) facilities. The HMT acronym is used to emphasize operations related to the treatment (e.g. purification or neutralization), recycling, or destruction of hazardous wastes.

2. Operational Differences by Industry

Companies in the refinery, chemical or pharmaceutical industries typically utilize the same types and quantities of raw materials on a daily basis. They tend to manufacture or store a limited number of products or families of products. Their manufacturing operations occur according to cycles or routine production runs. A given plant has one or more primary technologies which are key to their particular business. These facilities often have recognized that PSM was applicable to their operations long ago and thus have reasonably mature PSM programs.

Companies in the HWI have more variable operations. They utilize an assortment of unit operations to handle, process, recycle or destroy the chemical and biological wastes they receive on a regular or periodic basis. They combine, decant, recycle, fuel blend, distill, incinerate or otherwise dispose of the materials they receive from their customers. These are shipped to their facilities in a variety of container types and material phases. PSM challenges shift on a daily basis since the nature of what they need to process changes frequently and can fluctuate widely. PSM programs in the HWI must be adept at effectively managing these inconsistent hazards.

3. General PSM Challenges for the HWI

Companies in the HWI have to be prepared for a wider variety of process hazards due to the nature and kind of substances they accept from their customers. They refer to waste shipments of a given type they receive as a 'stream'. Even trace amounts of certain chemicals can have a tragic effect if these contaminate incoming streams without the HWI recipients' knowledge. HMT facilities need close ties with their customers to truly understand the nature of the wastes they receive and know how repeat streams may vary in chemical concentrations of major components and may contain different kinds of constituents.

The number and volume of these streams are always changing, even from the same customer. Customers of HWI companies don't intentionally generate waste regularly and the amounts of each are often related to either their production volumes or the number of process upsets they have experienced. So companies in the HWI can't always anticipate how much of what kinds of streams they will process in a business cycle. The generation of customer waste is extremely unpredictable. HWI companies have to be prepared for almost anything and have equipment which is designed for just that. Since waste can be generated at any time, HMT facilities need to be nimble and be able to respond quickly to new business or business for a newly-generated stream of waste from an existing customer. Compliance concerns which may result from new business need to be promptly addressed.

HWI companies are usually comprised of small facilities (with less than 100 employees per site) which have a multitude of regulations which they need to meet, including RCRA, RMP, and numerous others in addition to PSM. Personnel in these more-compact organizations tend to wear many hats and are distracted by the assorted duties cast upon them during the normal workday. They many even include the owner/operator of such a facility. Expediency in addressing PSM compliance issues may be affected by the amount of resources available in such organizations.

HMT facilities tend to have smaller budgets than comparable size facilities in larger companies or those in other industries. As such, monies to upgrade facilities, build infrastructure, or address hazard reduction projects can be more limited. Business cycles have a huge impact on what funding is available to address all compliance needs. And with the HWI downstream of the manufacturing industry, the HWI often goes as manufacturing does.

Companies and operating sites in the HWI have had multiple owners over the years, perhaps more than those in the chemical or petrochemical industries. This has led to mixed safety cultures both at individual sites and across organizations. Dealing with PSM issues may be controversial, since senior personnel or management might have experienced one or more safety cultures and younger personnel others. Companies must apply extra effort to address identified deficiencies since these cultural viewpoints may lead to disagreement regarding the extent of work necessary to effectively resolve compliance issues.

Organizations in the HWI industry previously treated their activities as simply waste-related operations, even if they were recycling or creating renewed end-products through processes such as distilling or fuel blending. But they really are another kind of chemical plant, albeit a more unique sort. These facilities need to be just as focused on maintaining their waste processing equipment as though they are generating a virgin product. Operations can have deadly consequences from improperly processing or having uncontrolled releases of highly hazardous chemicals, which is the reason for the PSM regulation in the first place. HWI companies have improved their performance in this arena in recent years, especially as OSHA continues its Chemical NEP activities [5].

4. Specific Challenges by PSM Element

Employee Participation
Effective Employee Participation can be limited by the number of personnel available at HMT facilities to participate in, lead and drive improvements and PSM-related activities. Companies must allot time and resources for their employees to participate in training, hazard analyses, safety committees, operating procedure reviews, incident investigations, emergency drills, etc. required for their PSM program to be successful. Site management may need to be resourceful in giving personnel time to meet PSM program obligations.

Process Safety Information (PSI)
Efforts must be ongoing to keep PSI current. Personnel must be encouraged to review and update P&IDs, equipment files, chemical and process hazard information, consequences of deviation in procedures and similar documents so they remain accurate and useful. One of the hardest challenges seems to be keeping good records of PSM-covered equipment in easy to access files. Older facilities have limited equipment design or code information available, which can prompt re-inspection or re-certification of equipment to verify these critical details. Sometimes design literature may need to be requested from a manufacturer if it becomes damaged or misplaced. If changes of service (say of storage tanks) have occurred but were poorly documented, records may need to be re-created and verified before inclusion in PSI databases.

Some HMT facilities have difficulty describing their 'process chemistry' per the requirements of the PSM standard. This is a dilemma unique to the HWI. True process chemistry is often not intended in typical HWI operations. If this applies to a given location, the site's process safety documentation must clearly explain this circumstance.

Process Hazard Analysis
A comprehensive Process Hazard Analysis (PHA) program is critical to understanding any site's hazards and ensuring that appropriate controls and acceptable levels of risk are assumed. Many PSM-covered sites in the HWI are older facilities which suffer from process designs which did not adequately cover facility siting concerns at the time of construction. If rebuilding such sites is not possible, locations instead have to focus on administrative and engineering controls to reduce or mitigate risk to their personnel and property from their operations. And facilities with multiple sites operated independently can face inconsistencies in risk tolerances when coming under a corporate PSM umbrella.

Some HWI organizations heavily rely on consultants to lead their PHAs. This can create a double-edge sword: a site gets support for a needed activity, but if they don't define the requirements for the PHA adequately nor hire a suitable consultant, the PHA results yielded may be substandard. The PHAs may not cover enough hazardous aspects of a process if their consultant is not properly informed and knowledgeable. PSI details and reports for these PHAs may be limited in their scope and may omit critical information (background or otherwise) relevant to the study. Sites must be vigilant to ensure that their PHA activities be thoroughly completed as required.

Operating Procedures
HMT facilities typically have very good basic operating procedures due to the nature of their operations. There would either be more accidents or more hazardous spills if these facilities were not being run safely. But, as is sometimes the case in any industry, the written procedures which exist may not be adequately comprehensive in their descriptions of consequences of deviation or emergency actions to take, including explaining what conditions warrant the shutdown of a process. Some procedures may direct the operator to consult with supervision during atypical or emergency operations. While this is not necessarily incorrect, what does an operator do when the process goes haywire on a weekend or off-shift when less supervision or technical support is available? Authors of such procedures need to clearly spell out common out-of-control responses and/or define when emergency shutdown should take place to provide the operators with the best ready guidance at all times.

On the maintenance side of operations, many more procedures often need to be composed to describe inspection, testing and preventive maintenance (ITPM) activities and how repairs and replacement of equipment ought to take place. A significant number of procedures should exist to define ITPM for PSM-critical equipment systems and controls. When predictive and reliability-centered maintenance (RCM) programs become utilized by more advanced organizations, the number of maintenance procedures should increase proportionally to define and address these specialized activities.

Training
Training can be executed in a variety of ways. Some companies depend upon computer-based training for much routine or refresher training, since employees can determine when they have time available to meet the required obligations and can fit in training modules while they are present during normal work hours. When facilities do not have this option available, they may need to be resourceful with training solutions. A familiar pattern is to schedule hourly workers for day long training sessions when they have a day off. Even this becomes an issue for companies who do not have excess staff to backfill work slots for other employees undergoing training. Companies can also organize special 'training day or days' when operations are intentionally curtailed so that particular training or orientation programs may be held for the entire facility at one time. Possibly other creative solutions may be sought to balance staffing with training needs and requirements as situations dictate.

Contractors
Managing contractors can be tricky, depending on the number and type of crafts or specialty groups on which a facility may rely. Overseeing the screening, performance and insurance

requirements, among other factors, can require constant review. Staff dedicated to birddogging records and updates of same can help to keep these appropriately current. Ensuring that contractors have proper desired skills, certifications, and current site orientation may be another story. Some sites find it cumbersome to track down qualifications of contractors and have regular orientation sessions with them. That is why many smaller sites often partner with local/regional safety and health groups which specialize in OSHA and other common safety program training. In the greater Chicago area, one such group is the Three Rivers Manufacturing Association, which has a specialty Safety Center [8]. Similar such organizations exist in other parts of the country. For an annual membership fee, the safety groups can ensure their customers that a variety of contractors serving a region have certain basic safety and health skills, which then make it easier for member companies to quickly orient contractor personnel to their PSM-covered sites when jobs are required.

Sites in the HWI may also have their own employees serve as contractors for customer facilities when they work on special projects or are embedded at customer sites. These HWI personnel may need additional PSM-related training to be appropriately compliant both for their employer as well as for the customer. This is another unique aspect of the HWI business.

Pre-Start Up Safety Reviews (PSSR)

It is less common in the HWI for companies to debottleneck or re-design their processes. They don't generate as many cases where PSSRs are required. As a result, the PSSR process tends to be less well-known and less understood by site personnel. More time needs to be taken whenever PSSR opportunities exist to refresh personnel's memory of the purpose of the review and to ensure that a thorough and proper study is made of the new equipment or system before it is commissioned and HHCs are introduced to the process.

Mechanical Integrity

Maintaining the mechanical integrity of PSM-critical equipment is complicated for any organization. What makes this particularly difficult for smaller companies is establishing and staying on top of inspection, testing and preventive maintenance (ITPM) for what is often older equipment that had historically been run until it was broken. Emphasis is needed on implementing aspects of preventive and predictive maintenance to enhance the reliability and service life of their critical equipment, whether this is done in-house or via contract resources. Maintenance employees need appropriate training and/or certification according to the ITPM skills expected of them. Improving ITPM practices may even involve a type of 'culture change' which shifts focus from reactive to preventive/predictive maintenance practices. Management influence is essential to support this enhanced execution of maintenance activities.

Management of Change (MOC)

The MOC process is not only to be used for new equipment, capital projects, or changes in personnel with PSM responsibilities, but is also for new streams to be processed in an HMT facility. As waste compositions change from allowed norms, reviews need to occur to ensure that new streams or significant variations in existing ones will not result in any new hazards in the anticipated handling of the materials or the associated process operations. Changes in streams should likewise result in re-evaluating the materials of construction of the process equipment which may handle the new or different stream so that unanticipated failures of equipment do not occur. Furthermore, since newer equipment is less common in the HWI, employees need to be diligent in understanding the difference between 'replacement in kind' (RIK) and non-RIK situations so they know when to apply MOC appropriately.

Hot Work

Hot Work operations in an HMT facility can be confusing to personnel. Some sites have difficulty defining hot work areas due to the fact that handling of waste streams may occur in many more locations around the site than at a standard manufacturing plant. Employees may need additional orientation and training to recognize Hot Work situations and zones where Hot Work operations may apply. As employees become more knowledgeable of their site's Hot Work requirements, proper preparation for and permitting of all Hot Work activities should occur as expected.

Incident Investigation

HMT facilities tend to have well-established incident investigation and emergency response planning. Many of their environmental permits already require these to be fairly robust. HWI organizations, like many other organizations, need to place more emphasis on leading indicators rather than lagging indicators. The new focus on standard API RP 754 *Process Safety Performance Indicators for the Refining and Petrochemical Industries* [9] is further emphasizing this trend. One leading indicator that is easy to track is near misses. Good PSM programs already have discipline in identifying incidents or spills to report. Now these need to identify events which may not have had an adverse result, but may have had the potential for the release of a highly hazardous chemical. Understanding the dangerous potential of these near miss situations can help sites to avoid tragic consequences from ignoring such warning signs.

Emergency Planning and Response

There needs to be diligence in conducting enough annual emergency response drills and in holding debriefings to help identify weaknesses in response programs and make necessary changes to improve. HWI organizations focus heavily on spills due to the nature of their environmental permits, but should equally consider drills on equipment failures or releases from pressure relief valves, for example. This 'balanced approach' to emergency drills should result in more comprehensive emergency response procedures and staffs which are better prepared to handle a greater range of adverse events.

Compliance Audits

Sites in the HWI are used to compliance audits from regulators on EPA and similar programs and tend to be well-versed in preparing for such audits. PSM auditing activities likewise need to be ongoing, including self-audits, to support PSM gap analyses which ought to be routinely occurring. Management should regularly review the results of these audits and put in place systems to successfully address identified deficiencies.

Trade Secrets

HMT sites do not tend to have many trade secrets, but, as with other organizations, they need to ensure that these are made available to personnel who need them to conduct PSM compliance activities. These personnel may need to enter into confidentiality agreements regarding access to this critical information to perform their normal duties.

5. Advancing PSM Programs at any PSM-Covered Facility

PSM programs are only as good as their corrective actions and improvement efforts. If you find problems in your systems but don't address them, it is as though you had no knowledge of them. Shortcomings identified in audits but which were not addressed have been contributing factors to major incidents, such as that which occurred in Texas City, TX in 2005 [4]. Management teams in every company need dedication to follow through on the recommendations from hazard analyses, incident investigations and compliance audits, among others, to address deficiencies and to correct issues. They must ensure continual improvement activities and projects exist and are led by appropriate champions. A variety of solutions may be needed to engage the work force to achieve better success with these aspects of PSM.

Sister plants can work together or company divisions can pool resources when there is an inadequate number of technical staff at a given location to lead PSM efforts. Today's internet-based applications and array of electronic gadgetry make these interactive communications easier than ever without excessive travel burdens.

Companies may consider utilizing process safety consultants to supplement their programs. This is a great way to get new ideas and access to industry best practices from trusted sources. Third-party auditors may also be employed to make unbiased suggestions for improvement. However, be sure your company actively partners with these groups for more successful PSM program advancement.

Organizations should get involved with trade organizations, professional societies and other groups that conduct seminars and training on process safety topics. Technical conferences are a great way to network with others in your own industry and learn about current PSM topics and newly identified best practices. The Global Congress on Process Safety (GCPS), which is a part of the Spring Meeting of the American Institute of Chemical Engineers (AIChE), is a fine example of an ongoing conference series dedicated to Process Safety education [10].

Companies in the HWI also have a new organization to supplement their PSM compliance efforts. The Recyclers' Safety Council (RSC) was created in 2011 by a consortium of industrial companies who recycle waste solvents and purify them for reuse. The intent of the RSC is to share best practices related to safety programs and compliance with the PSM standard. No similar group previously existed specifically for companies who recycle waste solvents. The RSC promotes collaboration among organizations who understand the unique set of challenges they face in achieving safety and process safety goals. My company, Veolia Environmental Services, is one of the founding members of this Council. Please contact me, the author of this article, for more information on participating in RSC activities, including an annual safety-themed conference.

6. Conclusions

HWI facilities, and HMTs in particular, require special diligence to effectively manage and progress their PSM programs. The size and nature of their operations require them to seek creative and possibly atypical measures to meet PSM compliance challenges. Though they are sometimes less advanced than their counterparts in other industries, they remain dedicated to addressing process safety gaps and continually improving their PSM systems. The recent formation of the RSC is an example of this commitment to process safety excellence.

PSM program success for any company is a function of the availability and dedication of the employees and the leadership of management. Limited resources, while a challenge, should not be an impediment to compliance. Be proactive and focused on your program and it will be possible to have a thriving process safety culture in your organization, regardless of size or the nature of your company's business.

7. References

[1] Process Safety Management of Highly Hazardous Chemicals, 29 CFR 1910.119, Occupational Health and Safety Administration , Occupational Safety and Health Standards.

[2] Jackson B. Browning, Jackson Browning Report, "Union Carbide Disaster at Bhopal (1993)" Available at www.bhopal.com/reports-and-studies/union-carbide-disaster-at-bhopal.

[3] "OSHA: Phillips 66 Company Houston Chemical Complex Explosion and Fire: A Report to the President, U.S. Department of Labor, Washington, D.C. (1990)".

[4] Baker, James et al 2007 "The Report of the BP U.S. Refineries Independent Safety Review Panel".

[5] OSHA National Emphasis Programs: Available at: http://www.osha.gov/SLTC/processsafetymanagement/index.html

[6] U.S. Chemical Safety Board Case Study 'Fire and Community Evacuation in Apex, N.C.', April 16, 2008. Available at: www.csb.gov/investigations .

[7] U.S. Chemical Safety Board Case Study 'Veolia Environmental Services Flammable Vapor Explosion and Fire', July 21, 2010. Available at: www.csb.gov/investigations .

[8] Three Rivers Manufacturing Association; Available at www.trma.org.

[9] API RP-754 Process Safety Performance Indicators for the Refining and Petrochemical Industries, 2010. Available at: www.api.org .

[10] Global Congress on Process Safety; Available at: www.aiche.org/resources/conferences.

9TH GLOBAL CONGRESS ON PROCESS SAFETY

A Novel Methodology for Maintaining Mechanical Integrity of High Pressure Relief Systems

Sandipan Laskar
Sr. Process Safety Manager
Clean Harbors Inc.
2027 Independence PKWY S
La Porte, TX 77571
laskar.sandipan@cleanharbors.com

Prepared for Presentation at
American Institute of Chemical Engineers
2013 Spring Meeting
9th Global Congress on Process Safety
San Antonio, Texas
April 28 – May 1, 2013

A Novel Methodology for Maintaining Mechanical Integrity of High Pressure Relief Systems

Sandipan Laskar
Sr. Process Safety Manager
Clean Harbors Inc.
2027 Independence PKWY S
La Porte, TX 77571
laskar.sandipan@cleanharbors.com

Keywords: Relief System, Mechanical Integrity (MI)

Abstract

Safe operation and maintenance of emergency systems are of primary importance to the process industries, including upstream oil and gas. The upstream production facilities are usually protected from overpressurization by high integrity pressure protection systems (HIPPS) which shuts in the well when local pressure levels exceed a certain threshold. The pressure relief system is often installed as a last safety barrier, when HIPPS fails. One of the critical elements of emergency system is proper design, documentation and to maintain high reliability of this system.

In particular, the pressure relief system is an integral part of upstream design and operation; prevention of an accidental release of toxic, flammable or noxious materials from a production facility requires correct design and construction, safe operation, and proper documentation of this system.

This paper proposes a basic change in the way relief systems are documented - a requirement in order to comply with OSHA 1910.119 and thereby describes a new methodology and a change in documentation to meet OSHA requirements and other industry standards (such as API 520/521, API-RP 75, 14J). This paper also describes the steps involved in a high pressure relief system evaluation and introduces concept of ever-greening to maintain a high reliability of this system. The paper illustrates how an up-to-date relief system leverages management of change (MOC) and helps in setting up the foundation for a strong overall process safety management (PSM) program. The paper will identify the drivers and the benefits of the proposed methodology for pressure relief documentation as well as explore the interdependency of the proposed program elements.

The paper will assist upstream facility personnel to utilize good engineering practice in design and documentation of safety relief systems, to meet both onshore and offshore process safety requirements, and to develop consistent company best practices for process safety management. In addition the paper will describe a comprehensive equipment based analysis protocol, preparation of documentation sufficient for design, operational, and audit purposes, and discuss maintenance both of the physical system and design data.

1. Introduction

Most oil and gas production, refining and petrochemical organizations have some kind safety management program that stems from the requirements outlined under OSHA's CFR 1910.119. Some have a very extensive program that covers most of the 14 elements of Process Safety Management (PSM) and is managed via an integrated platform while others follow an ad hoc approach that does not span all of the elements or is disjointed.

Maintaining accurate and current Process Safety Information (PSI) is a cumbersome and difficult task to accomplish due to the numerous sub-elements of Safety Management Program (SEMP). One of the key sub-elements of PSI maintenance is the management of the pressure relief documentation. Historically this is an element that has not been given the attention it deserves. However, recent incidents in the industry have elevated the importance of maintaining accurate and current pressure system documentation.

The ultimate goal is to allow all plant personnel to understand the basis of the current plant relief and flare header design and ensure that this current plant relief and flare header design provides adequate overpressure protection for all equipment. Thus, a new methodology for relief system documentation is created that is customized to fit the needs and resources of the operating company and had the ability to relate to the other Process Safety Information elements.

2. Pressure Relief Analysis

Pressure relief systems are applied in various installations in the oil and gas industry. The function of a pressure relieving system is to provide a positive means of protection for plant equipment from overpressure above limits permitted by the relevant codes and to dispose safely of any resultant relief of the process fluids.

The pressure relief analysis provides an efficient way to document the basis for relief system design basis. The major reasons for undertaking the pressure relief analysis by an operating company are listed below.

- Safety Initiatives
- Design/Capital Projects
- Audit/Compliance Regulations
- Corporate Standardization
- Acquisitions

Identification of the causes of overpressure, quantification of relieving rates required complying with codes, sizing of relief devices and the design of the relief disposal system is a complex task. Use of a systematic method for identification of contingencies, together with a set of project specific standards, provides the basis for sound relief system design. Quantification of contingencies using established calculation methods and process simulation is then used to size pressure relief devices and the associated relief header piping network. This approach provides an auditable safe design for a new plant and existing facilities. The different steps involved in a pressure relief analysis are illustrated below.

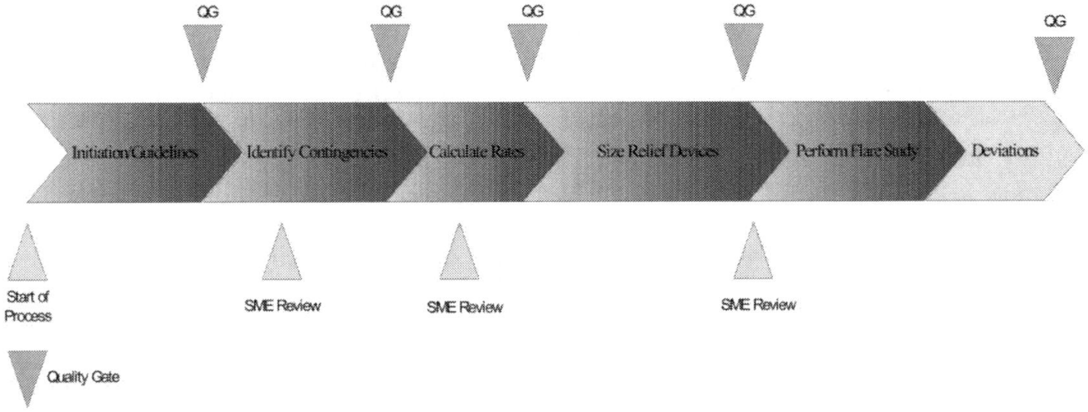

Figure 1 –Typical Phases of Pressure Relief Analysis

Mostly the pressure relief analysis is done as a one-time comprehensive project during the design phase or if there are any changes in the process. The major issues with one time comprehensive studies are listed below.

- Reactive instead of proactive approach
- Based on snap shot
- Underutilized valuable dormant data
- Missed training opportunities
- Out-of-sync with typical MOC programs
- Spend millions in 1-2 years for complete overhaul
- Increase risk of missing safety issues

3. Mechanical Integrity

Mechanical integrity is defined as the program or activities which ensure that equipment in a specified process is designed, constructed, installed and maintained to minimize the risk of accident release of high hazardous chemicals. Mechanical integrity can be achieved through the use of followings.

- Proper design to assure that equipment is suitable for its intended service
- Procedures to assure that we know how to properly operate, inspect and maintain equipment
- Training to assure everyone understands the procedures and the importance of doing things the right way
- Inspection to find flaws or deficiencies that may occur over a period of time or to confirm the status of equipment if nothing has changed
- Repair to maintain integrity
- Quality assurance to make sure that whenever something is done it is the done the right way.

Though there are many codes, standards and recommended practices that are available as a guide to ensure compliance with this element of PSM but still Mechanical Integrity (MI) presents the greatest implementation challenge. Mechanical integrity can be defined as an element of process safety management used to encompass activities that are designed to assure that mechanical equipment in an operating facility has the integrity to perform in the manner and service for which it was designed, locate, characterize, and rectify areas of deficiency that might lead to a condition that would be unsafe and endanger human life, health and/or property, assure that equipment is built and maintained in such a way as to perform it's desired function and ensure the ongoing durability and functionality of equipment achieved through a systematic approach. Although maintenance departments have successfully installed, repaired, and replaced plant equipment for decades, many of these same maintenance departments have struggled with OSHA's PSM requirements. One major challenge is prioritizing resources. Opportunities to improve will always exist; however, it is often difficult to effectively allocate money and manpower. Another challenge is simply getting organized. To meet the requirements of maintaining mechanical integrity, it is necessary to develop a program of testing safety devices and inspecting for both internal and external corrosion and erosion.

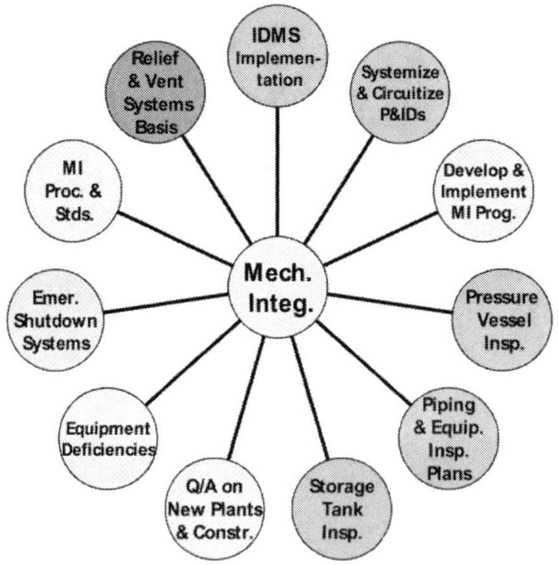

Figure 2 -Elements of Mechanical Integrity (OSHA's 29 CFR 1910.119, Para. j)

Mechanical integrity assures integrity and reliability of pressure equipment which in turn results in fewer safety incidents; reduction of mechanical failure (minimizes downtime and enables better planning), less process upsets and improves maintenance consistency and efficiency.

OSHA 1910.119 was written so that it would cover highly complex and hazardous processes handling a wide variety of hazardous chemicals. Production facilities are some of the simplest processes covered under this regulation. On the other hand, API RP 75 is written specifically for oil and gas production. It addresses offshore operations but much of the information is applicable to onshore operations as well. API helps in interpreting the application of OSHA's requirements to oil and gas facilities.

Many of the mechanical integrity activities are performed as part of a plant's maintenance program. Generally the OSHA guidelines rely on recognized and generally accepted good engineering practice (RAGAGEP), but allows companies to utilize alternate methods as long as they can prove that performance resulting from that method meets the standard. The specifications and techniques that form RAGAGEP are generally understood to mean recommended practices from the American Petroleum Institute (API), codes from the American Society of Mechanical Engineers (ASME), standards from the International Standards Organization (ISO), certification agency standards (DNV, Lloyd's, ABS), and other sources in regular use. The different elements of the PSM and SEMP program, is as shown below.

OSHA's PSM	SEMP
1. Employee Participation 2. Process Safety Information 3. Process Hazard Analysis 4. Operating Procedures 5. Training 6. Contractors 7. Pre-startup Safety Review 8. Mechanical Integrity 9. Hot Work 10. Management of Change 11. Incident Investigation 12. Emergency Planning and Response 13. Compliance Audits 14. Trade Secrets	1. Safety and Environmental Information 2. Hazard Analysis 3. Operating Procedures 4. Training 5. Pre-Startup Review 6. Assurance of Quality and Mechanical Integrity of Equipment 7. Safe Work Practices 8. Management of Change 9. Investigation of Incidents 10. Emergency Response and Control 11. Audit of Safety and Environmental Management Program Elements 12. Records and Documentation

The OSHA PSM guidelines are enforceable throughout the United States and specific US holdings for facilities having a threshold volume of hazardous chemicals, but are not applicable where another federal agency has occupational safety and health jurisdiction, including in federal waters controlled by Bureau of Ocean Management Regulation and Enforcement (BOEMRE, formerly Minerals Management Service (MMS)). Oil and gas drilling and servicing operations and normally unmanned remote facilities are specifically excluded from the standard, but typical facilities in the upstream industry such as production facilities, gas oil separation plants, and gas plants are covered if the threshold chemical volumes, typically hydrocarbons, are present.

In today's economic climate, we are operating with fewer people and demanding more from them than ever before. Therefore, it is imperative that the amount of documentation required be reduced to the minimum necessary to improve communication and increase safety. Excessive documentation places a time requirement that distracts people from their true and safe work practices. Documentation must be kept simple and concise so that it will be easy to use, with essential information readily at hand. The goal of the methodology presented in this paper is to improve safety without creating a large paperwork burden.

4. Methodology

The main goal of the proposed methodology is to develop a program that impacts the culture of the organization, thereby, drives and sustains accurate pressure relief information on a continual basis. The new program must leverage the overall, existing Safety Management Program with focus on the specific task at hand; keep perspective of risk in the current design. The new methodology eradicates the short comings of a one-time comprehensive pressure relief analysis and provides a standardized approach to ensure high reliability of the pressure relief system. The proposed methodology is further sub-divided into six different elements. These elements are Workflow, Resources and Processes, Tools, Training, Key Performance Indicators and Auditing. All these elements are interdependent for an overall success of the proposed methodology. The workflow process starts with a relief system review and generates the PRV Impact checklists, this checklist can be handed over to the MOC program. The actual workflow of the process is shown below.

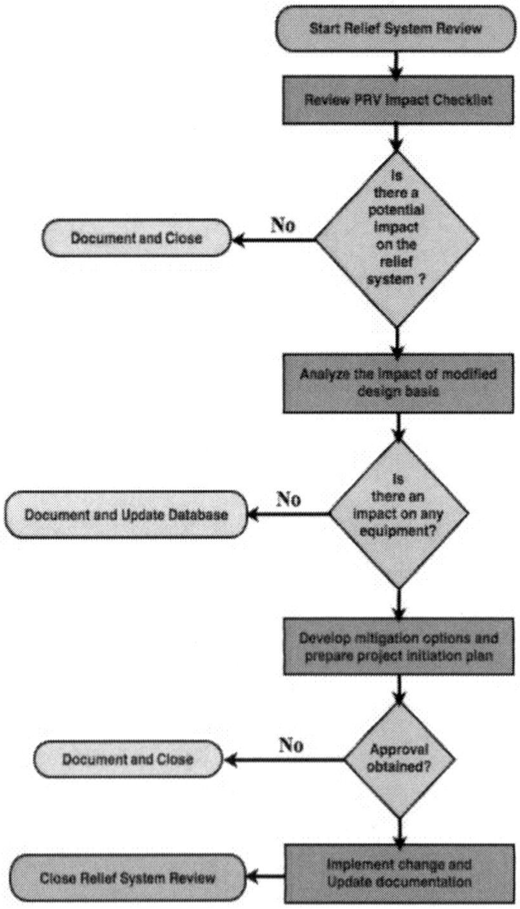

Figure 3 –Pressure Relief Valve Program Workflow

The steps involved in maintaining the data integrity and control of the proposed methodology are mentioned below.

- Develop a Relief System Impact checklist
- Incorporate checklist in the MOC program
- Control access to electronic databases
- Establish within MOC an exception check process for reconciling and validating data changes
- Develop a revalidation process for the plant-wide flare/relief system evaluation
- Complete a revalidation of each study within five years of the previous study
- Perform a complete redo of the study on a cyclical basis
- Manage the resolution of recommendations
- Establish a system of records for data
- Pool data into a central repository

The other elements of the program are to designate an owner for the program and appoint subject matter experts for engineering and information technology. An advisory group both internally at the site and externally at corporate level should be formed to support the program. There are readily available software in the market today to facilitate the actual calculations and sizing of pressure relief valves. Care should be taken to ensure that the tools used meet corporate standard, could be tailor-fitted to compliment internal company practices and also adapt to industry technology and technical practices.

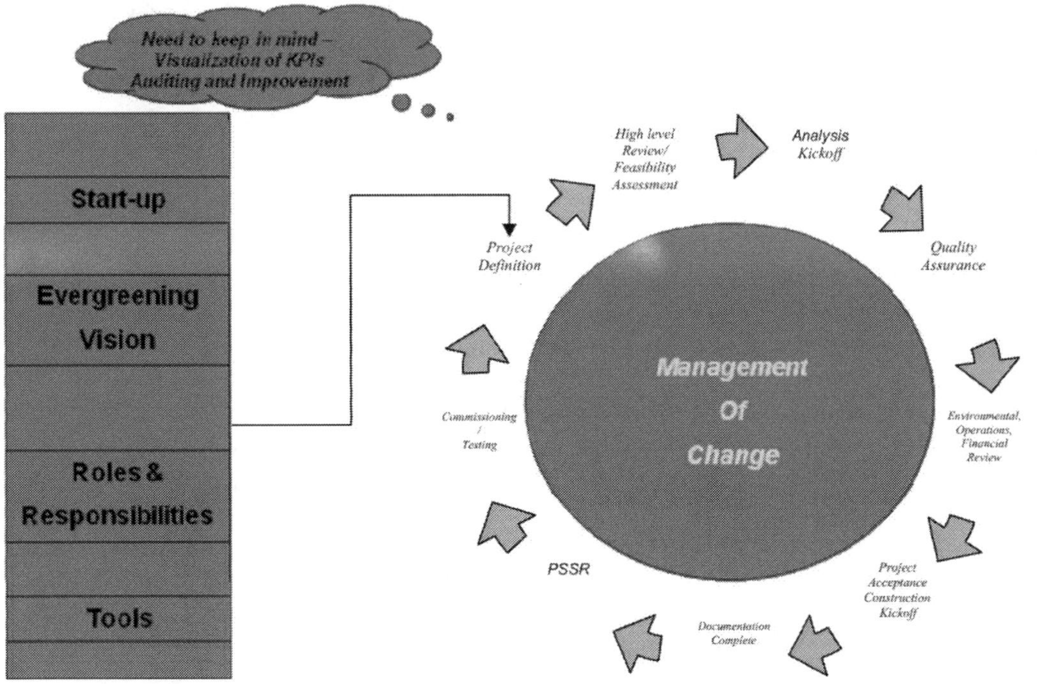

Figure 4 –Pressure Relief Valve Program Leverage Management of Change (MOC)

The above figure enunciates the importance of the proposed methodology in leveraging the existing management of change (MOC) program. The key performance indicators of the program are the data accuracy between different repositories, leverage existing information to measure performance and the program pull dormant data from relief system analysis to measure other PSM metrics. The major benefits of the proposed program are shown below.

- Standard approach
- Continual training
- Defined success factors
- Repeatable process
- Known roles and responsibilities
- Updated and accurate documentation

5. Conclusion

A good program work process starts from the projects to operations handover step on new construction or debottlenecks and runs all the way through the lifecycle of the plant encompassing modifications as a result of small and large projects. It provides a platform to communicate the PSI requirements through all disciplines at the operating facility in a concise manner and holds everyone at a high standard to meet these set requirements by continuously auditioning its effectiveness and providing training and improvement measures.

The cause of overpressure and the amount of fluid that is required to be removed from the system (vessel and piping) is unique for each installation. The proposed methodology ensures process safety without creating a large paperwork burden. The proposed methodology provides operating and maintenance staff, supervisors and foreman with the information, training and motivation they need to conduct their business in a safer manner. It is also intended that the information presented will help in proper documentation and thereby enable a faster response in case of an emergency.

The above mentioned methodology ensures most cost effective process for maintaining the mechanical integrity of high pressure relief systems while contributing up-to-date documentation of the emergency systems and thereby leveraging the existing MOC program. Particularly for high profile vertically integrated companies operating both onshore and offshore facilities, coordination of programs and practices provides an opportunity to streamline internal processes, comply with the new offshore regulatory environment, and operate more safely than ever before.

6. References

[1] "API RP 75, Recommended Practice for Development of a Safety and Environmental Management Program for Offshore Operations and Facilities", third edition. 2004. Washington, DC: API

[2] "Oil and Gas and Sulphur Operations in the Outer Continental Shelf--Safety and Environmental Management Systems", Bureau of Ocean Energy Management, Regulation and Enforcement Implementation (BOEMRE), *Final Rule*, 30 CFR Part 250, Volume 74, Number 199, (October 15, 2010)

[3] "Petroleum Refinery Process Safety Management National Emphasis Program", U.S. Occupational Safety and Health Administration (US OSHA), Directive CPL 03-00-004, (June 7, 2007)

[4] "PSM Covered Chemical Facilities National Emphasis Program", U.S. Occupational Safety and Health Administration (US OSHA), Directive CPL 09-06-002, (July 27, 2009)

[5] "Survey for API RP 75", Bureau of Ocean Energy Management, Regulation and Enforcement Implementation (BOEMRE), (September. 17, 2010)

[6] "API 520 Part 1, Standard for Sizing, Selection, and Installation of Pressure-Relieving Devices in Refineries, eighth edition. 2008. Washington, DC: API

A Framework for Managing Near-Miss Incidents

Presenter: GC Shah/Principal Author: GC Shah
Affiliation: Wood Group Mustang inc.
Address: 16401 Park Ten Place Blvd.; Houston, TX. 77084
email address: ghanshyam.shah@mustangeng.com

[

Prepared for Presentation at
American Institute of Chemical Engineers
2013 Spring Meeting
9th Global Congress on Process Safety
San Antonio, Texas
April 28 – May 1, 2013

UNPUBLISHED

A Framework for Managing Near-Miss Incidents

Presenter: GC Shah/Principal Author: GC Shah
Affiliation: Wood Group Mustang Inc.
Address: 16401 Park Ten Place Blvd.; Houston, TX. 77084
email address: ghanshyam.shah@mustangeng.com

Keywords:

ANSI/API RP-754-2010: Standard that creates "performance indicators" for process safety.
HAZID: Hazard Identification.
HAZOP: Hazard and Operability Study.
Lagging Indicators: Failures which have resulted in loss.
Leading Indicators: Indicators or evaluations that help predict the risk of unsafe events.
Near-miss incidents: Incidents that did not result in any loss, but have a potential for loss.
PHA: Process Hazard Analysis.

Abstract

Today, there is considerable emphasis being placed on learning from near-miss incidents and applying additional safeguards which would minimize possibility of escalating adverse impact of these events in the future. Not all near-miss incidents have potential for major adverse impact. It is important to focus on near-miss incidents which could result into incidents with major adverse impact. This paper applies a modified "Risk Assessment Matrix," that is commonly used for conducting PHA assessments such as HAZID and HAZOP, to estimate potential future risk indication conveyed by near-misses. In a strategic sense, near-miss management spans all work systems of an organization and its culture: IT system, operations, legal, and many others.

1. INTRODUCTION

In the hydrocarbon industry, as well as in many other industries for that matter, the "minor" events occur fairly often. While most of these events probably would end there and not result into major adverse impact, some of the events, if they recur perhaps under different circumstances, could have major impact on safety, environment, asset or a company's reputation. For instance, the event where a worker misses a rung on the ladder and falls and gets bruised may not develop into a major event. On the other hand, an event where there is a pipe leak which creates a cloud of flammable vapor that gets dissipated because of high wind, could under slightly different circumstances, result into a major event (potential fire and explosion). Even small leaks of flammable gases, if allowed to accumulate, could cause a fire and/ or an explosion.

OSHA defines near miss as "*incidents where no property was damaged and no personal injury sustained, but where, given a slight shift in time or position, damage and/or injury could have occurred.*" The definition seems to suggest that, the existing safeguards failed to work; but luckily the incident turned out to be a minor event. However, sooner or later luck may run out and a near miss can turn into a major adverse event. In terms of probabilities, one may say that statistically, there is a greater probability of occurrence of 'small impact" events and much smaller probability of high consequence or high impact events. From cost considerations, high impact events with high frequency simply can't be sustained financially.

Once near-miss events which have a potential to turn into "major" events have been identified, resources can be allocated to either strengthen the existing safeguards or add new safeguards in order to minimize the likelihood of occurrence of a major event.

2. IDENTIFYING POTENTIAL "TIME BOMBS."

Some experts suggest that near-miss is an accident waiting to get you into big trouble. The goal is to let it keep waiting!

How do we identify with a good degree of certainty that a near-miss will grow into a large-impact event? Experience with operations and maintenance could be a big help. You can develop checklists of incidents likely to develop into major events and incidents which are not likely to develop into large-scale events. Before we look at these lists, it is worth considering the meaning of the term "large-scale?" Stated broadly, the events that could cause one or more of the following may be regarded as the events that are likely to develop into "large-scale events."

OSHA recordables
Lost-time
Environmental RQ (Reportable Quantity) events
Environmental or safety violations
Events that could become a legal liability to a company
Public complaints

Some examples of specific events which could potentially turn into events with major adverse impact are:

Pin-hole leaks (e.g., pipes or flanges and fittings handling flammable, explosive or toxic materials; steam or condensate leaks on old boiler systems);
Leaking flanges, leaking mechanical seals;
Leaking tanks with flammables and/or toxic materials;
Spills under Reportable Quantity;
Lack of or poor lubrication of large pumps and/or compressors because pre-startup checks were not made or there was no checklist;
A worker narrowly escapes being asphyxiated in a confined space where nitrogen is being compressed;
Purge cycle on a furnace not working properly;

Welding sparks falling on hydrocarbon containing pipes;
Workers fail to understood a procedure (and therefore don't follow it) as it contained high-level technical jargon;
An isolation valve fails in open position;
Security infraction incidents (including cyber security);
Others.

It may be convenient to categorize the above events into the following groups:

- *Failure to follow procedures (OR poorly written or nonexistent procedures)*
- *Control system (and safety system) failures*
- *Old (older) equipment in challenging service*

Similar other systems of classification may be developed for your company.

Examples of events which will not likely develop into major events are:

Minor bruises or cuts;
Falls with a minor injury;
Potable water leaks;
A worker experiencing headache and taking non-prescription medication;
Other similar examples may be cited.

3. A FRAME WORK FOR NEAR-MISS MANAGEMENT

To be effective, the framework should be efficient. Workers should be encouraged to report ALL near-misses. Near-miss management start NOT after a near-miss, but way before such an event occurs. Guidance on leading indicators (as well as lagging indicators) is developed by API-754-2010. As Figure 1 shows near-miss is not a narrowly defined safety issue of simply reporting events; it spans through the entire organization and systems.

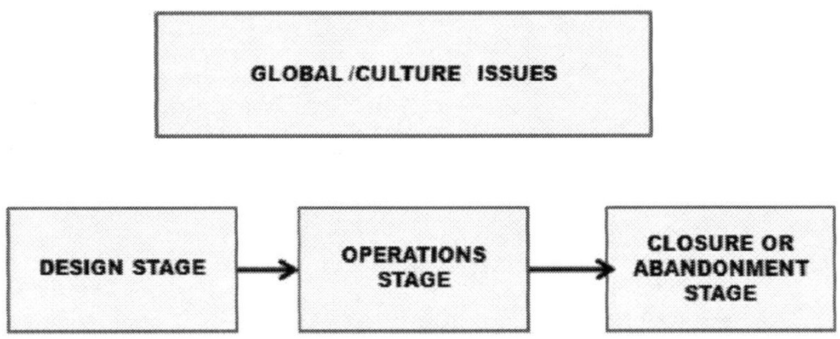

FIGURE 1: NEAR-MISS INCIDENT MANAGEMENT STAGES

3.1 Global and Cultural Issues

- Follow "good neighbors" policy. Establish good rapport with neighbors and regulators.
- Strategically important documents should receive thorough checks so as to avoid mis-interpretation.
- Encourage mutual trust between workers and management.
- Avoid bureaucratic systems in near-miss management.
- Encourage high-tech professionals (e.g., IT networks, advanced controls) to translate technical jargon into lay-person's language. The key is that persons, for whom the procedures (developed by technical professionals) are intended, should understand clearly.
- Ensure there are adequate resources to address near-miss incidents in reasonable time. Do not "dilute," safety.
- Acquisitions and divestitures also should (and typically they do) receive due diligence.

3.2 Design Stage:

- Ensure design is in accord with the applicable industrial standards and regulations.
- As far as practicable, follow the philosophy of fault-tolerant design. Stated differently, design should a mishap occur, there should be appropriate mitigative measures to constrain impact resulting from the mishap.
- Ensure documentation is easy to access and not ambiguous. (Workers referring to the documents in the future should get a clear understanding of the meaning and intent of the design).
- Perform periodic risk assessments; address findings thoroughly, and document.

3.3 Operations Stage:

- Encourage a collaborative rapport between your technical experts or engineers and operations and maintenance workers and contractors.
-
- Introduce New Technology only after a thorough risk assessment.
- Implement frequent training.
- Ensure the procedures and instructions are CLEAR to workers. Seek their input before implementing them.
- Conduct periodic audits of the facility. Some companies term these as "Loss Control Audits."
- Encourage workers to report ALL near-misses.
- Identify near-miss incidents which have potential to develop into major events. You may work out checklists workers can use to identify such incidents or consider the Risk Matrix shown later in this section. Of course checklists should be thorough so that workers won't miss out a near-miss that has a potential to grow into a major event.
- Provide resources to address near-misses within a reasonable time.

3.4 Closure and Abandonment

- Develop closure and abandonment policies. Apply Risk Management criteria to this phase of the organization.

3.5 Risk Matrix for Near Miss Incident Management

Figure 2 below shows a proposed Risk Matrix for near miss incident management. Risk is defined as a product of likelihood and consequence (severity of the outcome of an incident). Risk is grouped into three levels:

- High
- Medium
- Low

All near-miss incidents must be looked into. However, incidents identified as those having a potential for developing into "major" events require further investigation and possibly adding new safeguards or making existing safeguards more reliable or robust. Near miss management should be efficient. A small team of experienced workers can apply Risk Matrix (or checklists).

Figure 2: Risk Matrix for Near Miss Incident Management

		Risk Matrix for "Near-Miss" Incidents		
Consequence	3	3	6	9
	2	2	4	6
	1	1	2	3
		1	2	3
		Likelihood		

		Risk Matrix for "Near-Miss" Incidents		
		Likelihood		
		1	Relatively low likelihood of occurrence once in 10 yrs.	
		2	Relatively moderate likelihood of occurrence once every yr.	
		3	Relatively high likelihood of occurrence several times a year.	
		Consequence		
		1	Relatively "low" or "potentially low" level of impact: This events are NOT likely to develop into recordable or lost time or RQ events or higher.	
		2	Moderate level of impact or potential for medium level of impact such as OSHA recordable or lost time incident or RQ events.	
		3	High impact or potential for "high" impact: For example, large scale environmental or safety impact, or potential for damage to reputation.	
		Suggested Corrective Action		
		BLUE	Document the event. Further resources for corrective action may not be warranted.	
		YELLOW	Investigate and implement corrective action in "reasonable "time.	
		RED	Investigate. Should receive top priority and urgency in implementing corrective action.	

4. CONCLUSION

Near miss incident management starts way before a near miss occurs. In a strategic sense, the near miss management pervades all layers of an organization and its interactions with external systems. Strong safety culture and trusting relationships are important components of an effective near miss management framework.

HAZOP Analysis and Debottleneck for Laboratory Operation in the Semiconductor Industry

Tianxing Cai
Dan F. Smith Department of Chemical Engineering
Lamar University, Beaumont, TX 77710, USA
tcai@lamar.edu

Qiang Xu
Dan F. Smith Department of Chemical Engineering
Lamar University, Beaumont, TX 77710, USA
qiang.xu@lamar.edu

Prepared for Presentation at
American Institute of Chemical Engineers
2013 Spring Meeting
9th Global Congress on Process Safety
San Antonio, Texas
April 28 – May 1, 2013

UNPUBLISHED

HAZOP Analysis and Debottleneck for Laboratory Operation in the Semiconductor Industry †

Tianxing Cai
Dan F. Smith Department of Chemical Engineering
Lamar University, Beaumont, TX 77710, USA
tcai@lamar.edu

Qiang Xu*
Dan F. Smith Department of Chemical Engineering
Lamar University, Beaumont, TX 77710, USA
qiang.xu@lamar.edu

Keywords: HAZOP Analysis, Semiconductor Industry, Laboratory Hazards, Operation Debottleneck

Abstract

Unlike the semiconductor manufacturing process with fully automatic equipment, the laboratory operation in the semiconductor industry will still involve multiple types of manual handling for chemical analysis, reliability test, material characterization and failure analysis. The test operators or lab engineers have different levels of exposure to a lot of chemicals and physical hazards in multiple phases. This paper will describe the straightforward HAZOP (Hazard and Operability Study) process to identify the hazards and assess the risks of laboratory operations. The analysis will be deployed in the ways of manpower, machine, materials, method and environment. The analysis output can be regarded as the aid to safer laboratory design, operator training and real-time fault diagnosis. This will help to debottleneck and achieve operation safety and occupational health in the semiconductor industry.

† For presentation at AICHE 2013 Spring Meeting ,9th Global Congress on Process Safety.
* All correspondence should be addressed to Prof. Qiang Xu (Phone: 409-880-7818; Fax: 409-880-2197; E-mail: Qiang.xu@lamar.edu).

1. Introduction

Semiconductor industry revenue in 2012 is expected to reach $323.2 billion, up a slight 3.3 percent from last year's revenue of $312.8 billion, according to an IHS iSuppli Global Manufacturing Market Tracker report from information and analysis provider IHS.

Based on the professional estimation, the growth in the semiconductor industry from 2013 to 2015 will average between a more encouraging 6.6 to 7.9 percent, as shown in the figure below, with total semiconductor revenue by 2015 rising to some $397.7 billion. In the coming several years, the strong semiconductor growth can be seen.

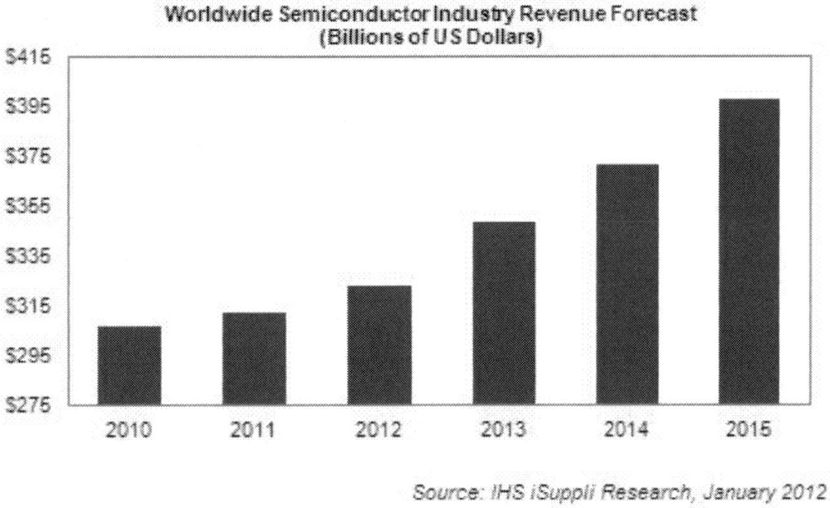

Figure 1: Worldwide Semiconductor Industry Revenue Forecast

Coupled with the fast development, the semiconductor employment has globally stretched. In USA, recently on Dec 11th, 2012, the Semiconductor Industry Association (SIA) announced that U.S. semiconductor industry jobs stretch across almost every region of the country and into the majority of states, according to an analysis of government data. California leads all states with 47,100 semiconductor jobs, followed by Texas (28,800), Oregon (23,400), Arizona (18,800), Massachusetts (10,100) and New York (7,600). According to BLS, the U.S. semiconductor industry's workforce grew by 3.7 percent over the previous year. In comparison, jobs throughout the broader U.S. economy increased by 1.2 percent over the same time period.

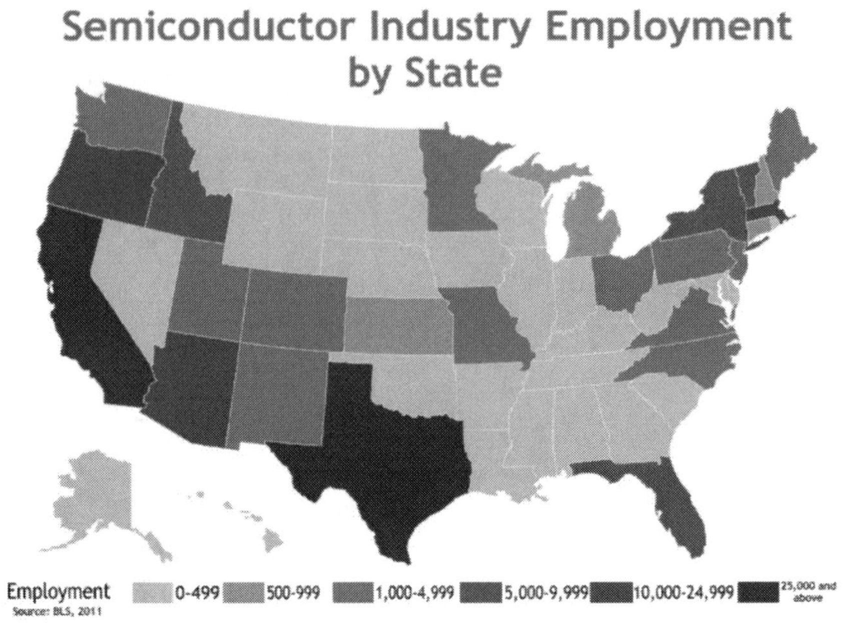

Figure 2: Semiconductor Industry Employment by State

Ever since the invention of transistor by the Bell Telephone Laboratories in 1948, the semiconductors industry has expanded greatly. The continuous fast development and innovations have expedited the increase of the semiconductor industry employment, which has further invoked an unavoidable concern of occupational safety to the semiconductor manufacturing operation.

Unlike the semiconductor manufacturing process with fully automatic equipment, the laboratory operation in the semiconductor industry will still involve multiple types of manual handling for chemical analysis, reliability test, material characterization and failure analysis. These operations will involve the handling of advanced and complex equipment or instruments such as Gas Chromatography (GC), Gas Chromatography-Mass Spectrometry (GC-MS), Fourier Transformed Infra Red Spectroscopy (FT-IR), Ultraviolet Visible Spectroscopy (UV/Visible), Liquid Chromatography (LC), Thermal and Elemental Analysis (TEA). Furthermore, due to rapid changes in this industry, manufacturing processes and their associated hazards may change completely every few years. These changes make hazard assessments more difficult to complete and require that they be conducted more often. Common hazards may include exposure to solvents, acid and caustic solutions, toxic metals, and radiation. Below Table 1 has listed the typical hazardous chemicals with their roles in the manufacturing process and potential health problems linked to the exposure. The test operators or lab engineers have different levels exposure to a lot of chemicals and physical hazards in multiple phases.

Table 1:Hazardous Chemicals in the Semiconductor Industry

Chemical name	Role in manufacturing process	Health problems linked to exposure
Acetone	Chemical-mechanical polishing of silicon wafers	Nose, throat, lung, and eye irritation, damage to the skin, confusion, unconsciousness, possible coma
Arsenic	Increases conductivity of semiconductor material	Nausea, delirium, vomiting, dyspepsia, diarrhea, decrease in erythrocyte and leukocyte production, abnormal heart rhythm, blood vessel damage, extensive tissue damage to nerves, stomach, intestine, and skin, known human carcinogen for lung cancer
Arsine	Chemical vapor deposition	Headache, malaise, weakness, vertigo, dyspnea, nausea, abdominal and back pain, jaundice, peripheral neuropathy, anemia
Benzene	Photoelectrochemical etching	Damage to bone marrow, anemia, excessive bleeding, immune system effects, increased chance of infection, reproductive effects, known human carcinogen for leukemia
Cadmium	Creates "holes" in silicon lattice to create effect of positive charge	Damage to lungs, renal dysfunction, immediate hepatic injury, bone defects, hypertension, reproductive toxicity, teratogenicity, known human carcinogen for lung and prostate cancer
Hydrochloric acid	Photoelectrochemical etching	Highly corrosive, severe eye and skin burns, conjunctivitis, dermatitis, respiratory irritation
Lead	Electroplated soldering	Damage to renal, reproductive, and immune systems, spontaneous abortion, premature birth, low birth weight, learning deficits in children, anemia, memory effects, dementia, decreased reaction time, decreased mental ability
Methyl chloroform	Washing	Headache, central nervous system depression, poor equilibrium, eye, nose, throat, and skin irritation, cardiac arrhythmia
Toluene	Chemical vapor deposition	Weakness, confusion, memory loss, nausea, permanent damage to brain, speech, vision, and hearing problems, loss of muscle control, poor balance, neurological problems and retardation of growth in children, suspected human carcinogen for lung and liver cancer
Trichloroethylene	Washing	Irritation of skin, eyes, and respiratory tract, dizziness, drowsiness, speech and hearing impairment, kidney disease, blood disorders, stroke, diabetes, suspected human carcinogen for renal cancer

In order to mitigate the potential risk during the chemical process operation, HAZOP study is always conducted. A hazard and operability study (HAZOP) is a structured and systematic examination of a planned or existing process or operation in order to identify and evaluate problems that may represent risks to personnel or equipment, or prevent efficient operation. The

HAZOP technique was initially developed to analyze chemical process systems, but has later been extended to other types of systems and also to complex operations and to software systems. A HAZOP is a qualitative technique based on guide-words and is carried out by a multi-disciplinary team (HAZOP team) during a set of meetings.

This paper will describe the straightforward HAZOP process to identify the hazards and assess the risks of laboratory operations. The analysis will be deployed in the ways of manpower, machine, materials, method and environment. The analysis output can be regarded as the aid to safer laboratory design, operator training and real-time fault diagnosis. This will help to debottleneck and achieve operation safety and occupational health in the semiconductor industry.

2. General Methodology

Several tools and methodologies are adopted to deploy the HAZOP Analysis and debottleneck for laboratory operation in the semiconductor industry. This has included Laboratory Operation Flow Diagram (LOFD), HAZOP Meeting, Parameters and Guide Words

2.1 Laboratory Operation Flow Diagram (LOFD)

The method applies to existing or planned laboratory operation for which design information is available. Similar with the commonly used process flow diagram, this commonly includes a laboratory operation flow diagram, which is examined in small sections, such as individual items of laboratory test instruments, glassware or chemicals between them. For each of these, a operation design intention is specified. For example, in a failure analysis laboratory, a standard solution of aqua regia (the combination of HNO_3 and HCl with the volume ratio of 1:3) may have been used. The intention to remove gold layer in the integrated circuit so that the further visual inspection can be conducted to check the failure mode. The HAZOP team then determines what are the possible significant deviations from each intention, feasible causes and likely consequences. It can then be decided whether existing, designed safeguards are sufficient, or whether additional actions are necessary to reduce risk to an acceptable level. Besides the operation safety, HAZOP can also give the detailed instruction on the opinion whether the potential deviation will affect the quality of experiment or test analysis result.

2.2 HAZOP Meeting

When HAZOP meetings were proceeded by the involved laboratory engineers, technicians and operators, they were generally scheduled for one to two hours per day. For a medium-sized laboratory where the total number of test items to be considered is 600 (items of equipment and pipes or other transfers between them), about 20 such meetings would be needed.

2.3 Parameters and Guide Words

The key feature is to select appropriate parameters which will apply to the design intention. The general words used in the laboratory operation are usually solution concentration, solution temperature, sampling volume, and instrument detection limit. In the above example, it can be seen that variations in these parameters could constitute deviations from the design intention of experiments. In order to identify deviations, a set of guide words is applied to each step of the experiment process. The current standard Guide Words are as follows:

Table 2: Guide Words for HAZOP Study of Laboratory Operation in the Semiconductor Industry

Guide Word	Meaning
NO OR NOT	Complete negation of the design intent
MORE	Quantitative increase
LESS	Quantitative decrease
AS WELL AS	Qualitative modification/increase
PART OF	Qualitative modification/decrease
REVERSE	Logical opposite of the design intent
OTHER THAN	Complete substitution
EARLY	Relative to the clock time
LATE	Relative to the clock time
BEFORE	Relating to order or sequence
AFTER	Relating to order or sequence

2.4 HAZOP Software

Various software programs are now available to assist in HAZOP meetings. Even though they are always used in the chemical industry, the principles will be quite applicable to laboratory operation in the semiconductor industry.

2.5 HAZOP Team

HAZOP is normally carried out by a team of people with roles as follows

Table 3: Team Members for HAZOP Study of Laboratory Operation in the Semiconductor Industry

Name	Position	Role
HAZOP Group Leader	Lab Manager	someone experienced in HAZOP and laboratory operation, to ensure that the method is followed carefully
Recorder	Lab Engineer 1	someone to ensure that problems are documented and recommendations passed on
Designer	Lab Engineer 2	someone to explain any design details or provide further information
User	Lab Operator	someone to consider it in use and question its operability, and the effect of deviations
Specialist	Lab Engineer 3	someone with relevant technical knowledge
Maintainer	Lab Technician	someone concerned with maintenance of the process.

3. HAZOP Management System

For the purpose of effective implementation of the HAZOP management policy and solid application of introduced general methodology, the laboratory HAZOP system should be set up. The system has coupled general management policy (operation specification and work instructions), environment control (temperature and humidity), operation safety (chemical storage, hazards assessment, chemical identification and MSDS), equipment management (PM, calibration, MSA), chemical control (data analysis and shelf life time control), training and certification (orientation, OJT, regular qualification). The whole system has covered multiple layers of analysis factors. The first layer will include manpower, machine, materials, method and environment.

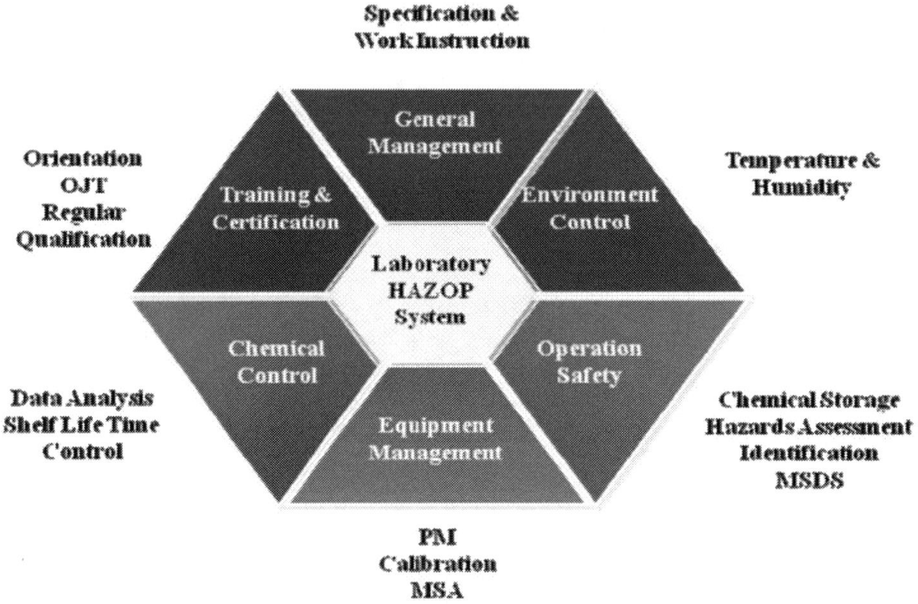

Figure 3: HAZOP Management System for Laboratory Operation in the Semiconductor Industry

These factors can be expanded with below fishbone diagrams so that the second layer of factors will be included into HAZOP matrix system: for example, sample collection and sample preparation will be considered in the material part; training, parallax, reproducibility, practice and ergonomics will be considered in the manpower part; samples, test methods, workmanship and standards will be considered in the method part; discriminations, repeatability, bias, linearity and calibration will be considered in the instrument part; vibration, lighting, temperature and humidity will be considered in the method part.

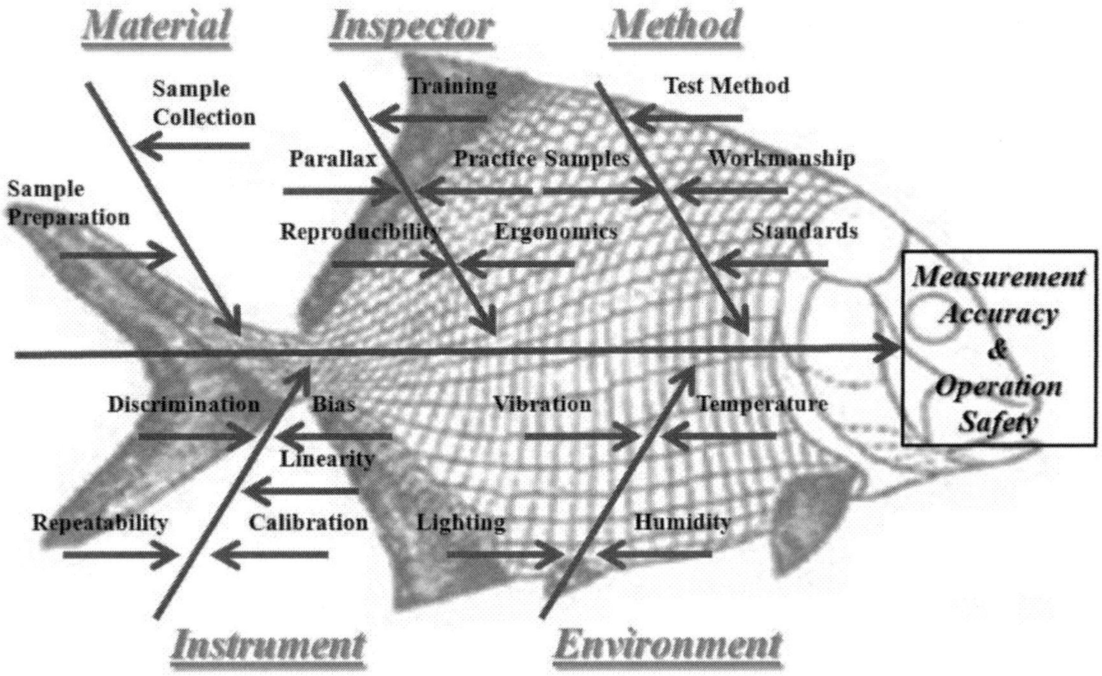

Figure 4: Fish Bone Diagram for HAZAP Study for Laboratory Operation in the Semiconductor Industry

Next, these practices can form a closed PDCA (Plan, Do, Check and Act) loop for continuous improvement of HAZOP activity. The close loop include four phases: plan, do, check and act. In the first phase of planning, the procedures and schedules to conduct the study should be identified and the input data should be collected. Then, the deviation should be identified by the application of key words and the causes and consequences should also be determined in the second phase. Next, the improvement actions need to be followed up. It is better to restudy the necessity of current HAZOP system. After the verification of the above items, the study report needs to be generated. Then the activity of HAZOP study should be evaluated to confirm whether the current scope, objective and responsibility need to be updated. Furthermore, the team should decide whether they need more team members to cover a larger scope.

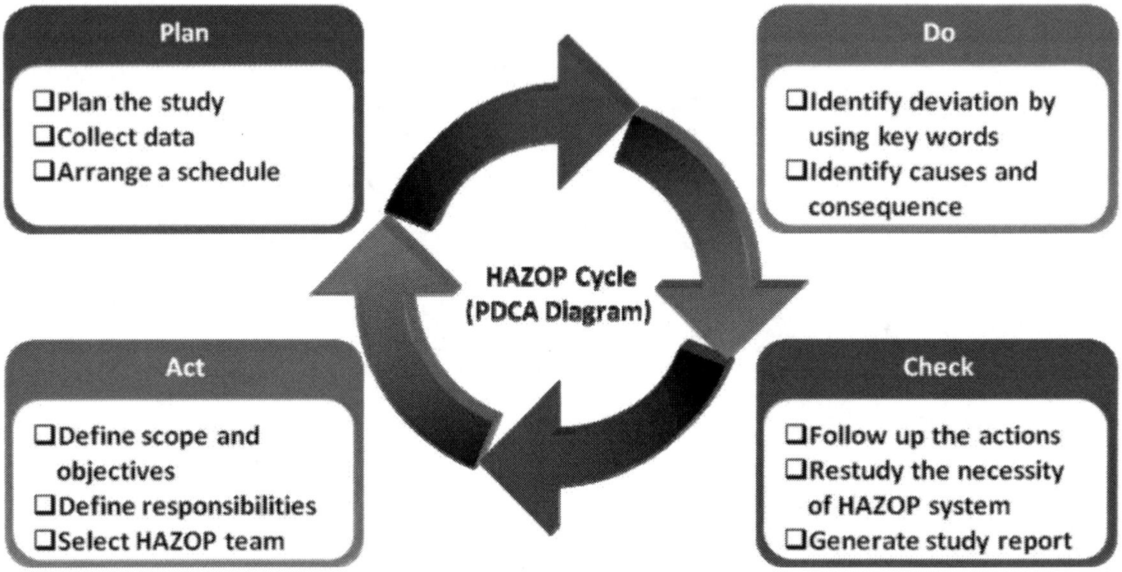

Figure 5: PDCA (Plan, Do, Check and Act) Diagram for HAZAOP Study for Laboratory Operation in the Semiconductor Industry

The analysis output can be regarded as the aid to safer laboratory design, operator training and real-time fault diagnosis. This will help to debottleneck and achieve operation safety and occupational health in the semiconductor industry.

4. Application Illustration

To demonstrate the efficacy of the developed systematic methodology, one case study has been conducted as the illustration of the methodology application in the real practice of plating bath monitoring of wafer bumping in the semiconductor industry. Wafer bumping is an advanced packaging technique that attaches solder spheres or gold bumps to the bond pads on a selected die. Gold bumping is generally used for Tape Automated Bond (TAB) assembly or Anisotropic Conductive Film (ACF) flip-chip assembly. Solder bumping is primarily used for reflow flip-chip assembly which unites the components face down, directly with the substrate or board through conductive bumps on the bond pads. These bumps deliver the electrical, mechanical and thermal interconnection which provides direct contact between the chip package and the device. Another process is called stud-bumping. Unlike gold and solder bumping, this process can be done at the die level. Wafer bumping also known as flip chip has traditionally been used for high-end niche applications. However in the last few years it has gone mainstream to include a wider range of consumer applications geared towards both the performance driven market and the form factor driven market.

Wafer bumping provides the electrical and mechanical connection between a die and substrate, which is one of the most critical elements of any flip chip package structure. Predominantly lead-tin and lead-free solders at the present time, these connections - or bumps - must exhibit superior adhesion to the die, minimal resistance, and result in high assembly yields. Solder bumps are formed by using either thin film metal deposition or ball loading techniques. Currently the advanced semiconductor assembly and test company such as Amkor offers state-of-the-art capability in electroplated solder technologies including the electroplating capability to support fine pitch bumping and copper pillar structures.

Wafer bumping utilizes thin film deposition, photolithography and plating techniques to place solder or gold onto IC (Integrated Circuit) lead pads. The wafer solder bump process is to produce the solder bumping of the wafer pad. Then it is going to melt the bumping by heat in assembly. This technology can substantially reduce IC size and have the advantages such as high density, low sensitivity, low cost and effective heat dissipation. This bumping process is suitable for flip chip assembly for LCD, memories, microprocessors, and microwave RF ICs applications. The process includes sputter UBM (under Bump Metallurgy), Photolithography, Plating and Etching etc., This bumping is suitable to apply on TAB, Flip Chip. The detailed process flow chart has been illustrated in below flow chart.

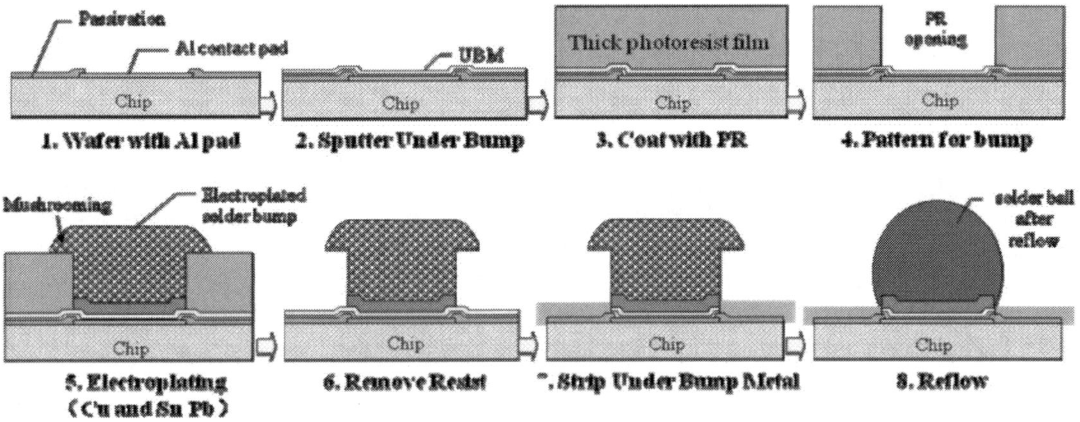

Figure 6: Process Flow Chart for Wafer Bumping Process

One of the key test item for wafer bumping operation is the electroplating bath monitoring and composition control of solder materials and additives. Plating solutions must be routinely analyzed in order to maintain the recommended bath formulation and to preempt the occurrence of problems related to improper levels of bath constituents. Contaminant levels in the solutions must also be monitored. Manufacturers of plating systems establish optimum specifications to ensure maximum solution efficiency and uniformity of deposits. The various factors that cause the concentrations of bath constituents to deviate from their optimum values are as follows: 1. drag-out; 2. solution evaporation; 3. chemical decomposition; and 4. unequal anode and cathode efficiencies. Regular analysis of plating solution becomes the responsibility of process engineering and quality control group.

The techniques employed for the quantitative analysis of plating solutions are classified as volumetric (titrimetric), gravimetric, and instrumental. Volumetric and gravimetric methods are also known as "wet" methods. The analyst must select the method that is best suited and most cost effective for a particular application. The wet methods outlined here are simple, accurate, and rapid enough for practically all plating process control. They require only the common analytical equipment found in the laboratory, and the instructions are sufficiently detailed for an average technician to follow without any difficulty. The determination of small amounts of

impurities and uncommon metals should be referred to a competent laboratory, as a high degree of skill and chemical knowledge are required for the determination of these constituents. Instrumental methods differ from wet methods in that they measure a physical property related to the composition of a substance, whereas wet methods rely on chemical reactions. The selection of an instrument for the analysis of plating solutions is a difficult task. Analysts must decide if the cost is justified and if the analytical instrument is capable of analyzing for the required substances with a high degree of accuracy and precision. Instruments coupled to computers can automatically sample, analyze, and record results. Mathematical errors are minimized and sample measurements are more reproducible than with wet methods. Instrumental methods are also extremely rapid when compared with wet methods. Analytical instruments frequently used in the analysis of plating solutions can be categorized as spectroscopic, photometric, chromatographic, and electro-analytical. Spectroscopic methods (flame photometry, emission spectrometry, X-ray fluorescence, mass spectrometry, and inductively coupled plasma) are based on the emission of light. Photometric methods (spectrophotometry, colorimetry, and atomic absorption) are based on the absorption of light. Chromatographic methods (ion chromatography) involve the separation of substances for subsequent identification. Electro-analytical methods (potentiometry, conductometry, polarography, amperometry, and electrogravimetry) involve an electric current in the course of the analysis. The instrumental methods, comprehensively reviewed below, are most applicable to plating environments.
Here, the manual titration procedures to determine copper concentration in the sample plating bath is used as illustration because the manual titration will have more potential safety concerns compared with the instrument analysis methods. Below is the procedure of the test.

PROCEDURE:

1. Preparation of EDTA (Ethylenediaminetetraacetic Acid) (0.01M). Dry EDTA for 2 hours at 80°C. GMW = 372.2., or check the reagent bottle. Dissolve needed grams in a 500 ml vol. flask. Store in a conditioned polyethylene container. EDTA solutions attack glass and deteriorate after 1 or 2 days.
2. Preparation of known copper solution. Place 0.2 g to 0.25 g of pure Cu. wire (or shot) (weighed on an analytical balance) in a 100 ml beaker. Add 15 ml conc. HNO3 and 1 ml conc. H2SO4. Place a watch glass over the beaker. Boil until dense white fumes of SO3 cease. Transfer carefully to a 200 ml vol. flask. Remember to rinse the watch glass. Dilute to the mark.
3. Preparation of unknown copper solution. Weigh 0.5 to 0.6 grams of unknown into a 200 ml vol. flask. Add 30 ml of 1:5 nitric acid. Heat until all particles are dissolved. Dilute to the mark.
4. Titration with EDTA - murexide indicator. Do this step for 4 samples of known and 4 samples of unknown.

KNOWN: Place 10 ml of Cu known solution (step 2) in 250 ml Erlenmeyer flask. Add pH buffer drop-wise until the solution turns completely blue and the precipitate first formed dissolves. Add about 100 ml. Add more buffer if solution becomes cloudy. Add 1 tablet of murexide indicator and titrate with EDTA until color changes from yellow (green) to deep violet. If the volume of EDTA is less than 10 ml, repeat using 25 ml liquid of Soln. 2.

UNKNOWN: Place 25 ml of Cu unknown solution (step 3) in 250 ml Erlenmeyer flask. Add pH buffer drop-wise until the solution turns completely blue and the precipitate first formed dissolves. Add 1 tablet of murexide indicator and titrate with EDTA until color changes from yellow (green) to deep violet. If the vol. of EDTA solution is less than 10 ml. use a larger liquid of Soln. 3.
Remarks: The copper-murexide complex is pure yellow, but the color at the starting point of the titration is greenish depending on the amount of copper present. The green is the mixture of yellow (indicator complex) and blue (copper-tetramine complex). As the titration proceeds the tetramine complex is transformed to the less intensely colored EDTA complex and the yellow of the indicator complex becomes permanent.
CALCULATION: Calculate the molarity of the EDTA. Calculate percent copper in the sample using the value of the Molarity of the EDTA found. Include the spread sheet for the calculation.

Based on the available procedures to determine the copper concentration in the sample solution of plating bath, HAZOP factors can be expanded so that the second layer of factors will be included into HAZOP matrix system: for example, sample collection and sample preparation will be considered in the material part; training, parallax, reproducibility, practice and ergonomics will be considered in the manpower part; samples, test methods, workmanship and standards will be considered in the method part; discriminations, repeatability, bias, linearity and calibration will be considered in the instrument part; lighting, temperature and humidity will be considered in the method part. These factors are listed as the column. Then, the HAZOP guide words will be listed as the row. The related information of the deviation stored in LOD(Laboratory Operation Database) and HPD(HAZOP Project Database) can be selected into the blank spaces from the drop-down list box in the above expanded table. Next, the data record of regular monitor results will be proceeded into data analysis to validate the impact of the above deviation . The data analysis result will couple with the expanded HAZOP table to quantitatively identify the potential deviation of operation safety and test accuracy.

5. Conclusion

The laboratory operation in the semiconductor industry will have more possibilities of manual handling for chemical analysis, reliability test, material characterization and failure analysis. Therefore higher attention should be paid to the laboratory operations. As illustrated in this paper, the methodology of HAZOP study procedure can be applied to identify the potential deviations in both experiment accuracy and operation hazards. Multiple layers of factor consideration by the application of 4M1E (Man, Machine, Material, Method and Environment) will help the brainstorm to give the aid to safer laboratory design, operator training and real-time fault diagnosis. This will help to debottleneck and achieve operation safety and occupational health in the laboratory operation semiconductor industry.

6. Reference

[1] Chepesiuk, R., "Where the Chips Fall: Environmental Health in the Semiconductor Industry." Environmental Health Perspectives, Volume 107, pp.452-457, 1999

[2] Swann, C. D., & Preston, M. L., "Twenty-five years of HAZOPs", Journal of Loss Prevention in the Process Industries, Volume 8, No 6, pp.349-353, 1995

[3] Amkor Technology Wafer Bumping Service Introduction, http://www.amkor.com /go/turnkey-services/wafer-bumping

(accessed Jan. 5, 2013)

[4] Nolan, D.P., "Application of HAZOP and What-If Safety Reviews to the Petroleum, Petrochemical and Chemical Industries". William Andrew Publishing/Noyes, 1994

[5] Lawley, H. G., "Operability studies and hazard analysis", Chemical Engineering Progress, Volume 70, No 4, pp.45, AIChE,1974

[6] Chemical Industries Association., "A Guide to Hazard and Operability Studies", 1977

[7] IHS iSuppli Research., "Semiconductor Industry Revenue to Ensure Slow Growth in 2012",http://www.isuppli.com/Semiconductor-Value-Chain/News/Pages/Semiconductor-Industry-Revenue-to-Endure-Slow-Growth-in-2012.aspx

(accessed Jan. 5, 2013)

[8] U.S. Bureau of Labor Statistics (BLS) Data Report,2011

[9] Semiconductor Industry Association, http://www.sia-online.org/

(accessed Jan. 5, 2013)

[10] Bolmen, Richard A., "Semiconductor Safety Handbook: Safety and Health in the Semiconductor Industry", Noyes Publications, 1998

9TH GLOBAL CONGRESS ON PROCESS SAFETY

Best Practice Management of Change & PSSR Process

Michael Bearrow
Optimized Systems and Solutions Inc.
8303 McHard Road, Suite 200, Houston, TX 77053
Michael.E.Bearrow@o-sys.com

Prepared for Presentation at
American Institute of Chemical Engineers
2013 Spring Meeting
9th Global Congress on Process Safety
San Antonio, Texas
April 28 – May 1, 2013

UNPUBLISHED

Best Practice Management of Change & PSSR Process

Michael Bearrow
Optimized Systems and Solutions Inc.
8303 McHard Road, Suite 200, Houston, TX 77053
Michael.E.Bearrow@o-sys.com

Keywords: MOC, process safety code, recommendation,

Management of Change (MOC) and Pre Start-up Safety Review (PSSR) are still the most challenging elements of OSHA's Process Safety Management (PSM) standard, the EPA's Risk Management Program (RMP) rule, and now, the US Department of Interior's Bureau of Safety and Environmental Enforcement's Safety and Environmental Management Systems (SEMS). Even though the PSM standard has been around since 1992 and the industry has been managing change for several decades, we still can get it wrong, sometimes with disastrous results. The mindful or diligent efforts of many working in concert are necessary to ensure that change is identified, analyzed and executed in a quality way. Just when we get it right on paper and get the workforce up-to-speed, we have employee turnover, neglect and sometimes regulatory change. New actors and a constantly changing script make it hard to manage change efficiently and effectively. This paper discusses how the chemical process industry has defined the MOC and PSSR best management practices and how they should be automated.

Management of Change (MOC) and Pre Start-up Safety Review (PSSR) are still the most challenging elements of OSHA's Process Safety Management (PSM) standard, the EPA's Risk Management Program (RMP) rule, and now, the US Department of Interior's Bureau of Safety and Environmental Enforcement's Safety and Environmental Management Systems (SEMS). Even though the PSM standard has been around since 1992 and the industry has been managing change for several decades, we still can get it wrong, sometimes with disastrous results. The mindful or diligent efforts of many working in concert are necessary to ensure that change is identified, analyzed and executed in a quality way. Just when we get it right on paper and get the workforce up-to-speed, we have employee turnover, neglect and sometimes regulatory change. New actors and a constantly changing script make it hard to manage change efficiently and effectively.

Many high-profile companies have addressed this challenge by creating and automating standard processes across their manufacturing or production facilities. MOC and PSSR are two of these major processes that have been successfully automated. Technology has been used to automate each step of the MOC workflow, enabling the collection of change ideas, formalizing the analysis of the ideas, and documenting the approval and execution of the change. Reports and dashboards can provide visibility of the health or current state of your MOC and PSSR processes. The use of software can enable the standardization of best practices. This paper discusses how the chemical process industry has defined the MOC and PSSR best management practices and how they should be automated.

What is a best practice?

"Best Practice" is an overused buzzword, like so many, and they are hard to define. Similar terms include Industry Standard, Best of Breed, Lesson Learned, Core competency, and Customer-centric, Downsizing, Drinking the Kool-Aid, Eating your own dogfood, Granular, Herding cats, Holistic, Low Hanging Fruit, Mindshare, Mission Critical, Management Visibility, Pain point, Return on Investment, Seamless, Touch point, Value-add and Visibility. Fortunately, there is a working definition of a best practice.

A **best practice** is a method, process or technique that consistently shows results superior to those achieved by other means. It is hard to prove that you have developed a standard process--like MOC--that enables your company, facility or organization to achieve repeatable superior performance, but it has been done.

Some other terms synonymous with best practices are good operating practice, good agricultural practices, good manufacturing practice, good laboratory practice, good clinical practice, good distribution practice, and **R**ecognized **A**nd **G**enerally **A**ccepted **G**ood **E**ngineering **P**ractices (REGAGEP). In the chemical process industry, the PSM standard relies on REGAGEP to ensure facilities adopt best practices. REGAGEP is specified in federal statutes from OSHA and EPA, and is also mentioned in the American Chemistry Council (ACC) Responsible Care® Process Safety Code.

How do you identify a Best Practice?

Everyone would like to be known as the best practice company. Even if you did not create the best practice, identifying them and using them can achieve the same results. If you can tick the box next to these five requirements, then you may have a best practice:

1. Gives a company or facility a tangible competitive advantage.
2. Takes advantage of technology.
3. Improves EHS performance, product quality, and also lowers cost.
4. Gives management more visibility, control and influence over outcomes.
5. Must be a working system; i.e. not a theoretical best practice

The benefits of having such a best practice might separate your company from your competition or keep you from costly business disruption, fines and/or law suits. Many of the best practices have common themes, including: ensuring consistency; standardization; simplifying processes; zero redundancies; leveraging technology and web-based software solutions; and single source for all data (one source of truth). It may be difficult for you to prove you have adopted/created a best practice. However, it is easy to identify a best practice when a confluence of large and small companies adopt the same technique or process and are successful.

In the context of this paper, we are focused on best practice MOC and PSSR automation. Now that we know what a best practice is, let's define change and discuss how to manage it.

What is Change & Why Do We Manage it?

"Change" means to transform or cause to be different. In the chemical process industry, change is omnipresent and required to stay competitive. William Edwards Deming, a famous American statistician, professor, author, lecturer and consultant, once said, "It is not necessary to change. Survival is not mandatory." In order to survive, a company must learn, change and continuously improve to survive. Successful change requires you to learn from your experience and the experience of your peers and make changes that will improve your competitiveness.

Change is a normal part of business, but uncontrolled or poorly managed change has resulted in some of the worst disasters in the chemical process, oil and gas and refining businesses. According to Ian Sutton, a noted PSM expert, *"The root cause of all accidents is uncontrolled change. Leaving aside sabotage and other malicious acts, all industrial facilities are designed and operated to be safe, clean and profitable - yet incidents continue to occur. In every case, the fundamental cause of the incident is that someone, somewhere lost control of the operation, i.e., they allowed operating conditions to deviate beyond their safe range. Hence, the proper management of change is the foundation of all safety and accident prevention programs; an effective Management of Change (MOC) creates an atmosphere of "no surprises".*

Whether it is Bhopal, Piper Alpha or Deepwater Horizon, time after time the difficulty of managing change confounds some of the most proven processes of the most successful companies in the world. As a consequence of these disasters and hundreds more, we have rules, standards, regulations and laws to help ensure that companies manage change better. Guidance documents on managing change include the following sources:

- *US EPA's Risk Management Programs for Chemical Accident Prevention (40 CFR PART 68) or EPA RMP*
- *US OSHA's Process Safety Management of Highly Hazardous Chemicals standard (29 CFR OSHA 1910.119) or PSM,*
- *US Department of Interior's Bureau or Safety and Environmental Enforcement (BSSE) - 30 CFR Part 250 Subpart S - Safety and Environmental Management Systems (SEMS),*
- *OHSAS 18001, 2007 revision - ANSI/AIHA Z10-2005.*
- *European Union Law Seveso II Directive (see Annex III of EU Council Directive 96/82/EC).*
- *API-691 - Risk Based Machinery Management (RBMM) (in development)*

What kinds of change are there?

All ideas for improvement require change. No one ever suggests that we should do exactly what we are doing today, tomorrow. This fact is lost on a segment of the chemical process industry who do not think of recommendations as changes. Recommendations should be treated as "in-kind" changes because they are. They do not have to follow a formal and rigorous MOC and PSSR process, but they are certainly recommendation for change.

All change can be separated into three categories: corrective actions; preventive actions; and continuous improvements. Corrective actions are meant to fix something that is broken, or at least to improve efficiency and functionality. These ideas for change originate from accident investigations, incidents and negative operating experiences. Preventive actions can also come from incidents, but are more commonly created from near misses, safety observations, audits, hazard reviews, inspections, emergency response critiques, and/or general surveillance or observations. These ideas for change help ensure that we address upstream issues or causes before a negative outcome (incident, fine, loss) results. The final kind of change is continuous improvement. These are generally ideas for increasing production, reducing manufacturing costs or increasing product quality.

No matter what kinds of changes you have, they all should be managed. No matter how good an idea is, it must be analyzed for impact (positive and negative) and executed expertly. Ideas can also be rejected or denied if the risk or cost of implementation exceeds the benefits or value of the change. If the change affects the design and/or operation of a facility, it is likely going to require formal change management or MOC. The diagram below illustrates three kinds of change in the idea universe.

Figure 1.0

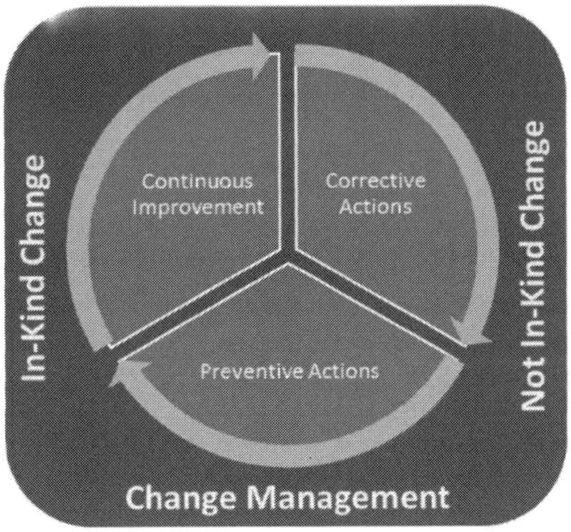

OSHA'S Two Categories of Change

The common way to refer to an idea that requires formal change management as defined by OSHA PSM or EPA RMP is an MOC or Management of Change (Not in Kind). MOCs are managed for changes to equipment, materials, technology, personnel, processes, procedures or anything that deviates from the original design, processes, operating procedures, or maintenance program.

In the chemical process industry, we refer to two categories of change. These are "Replacement In-Kind" and "Replacement Not In-Kind". These odd-sounding categories can have ambiguous meanings, but are explained by Ian Sutton, a noted process management expert, as follows:

"If an equipment item is to be replaced with one that is functionally identical, i.e., if the new item is built to the same specification as the old one, then the change is "in-kind". Otherwise it is "not-in-kind", and the MOC process has to be followed before the change can be implemented."

... The most challenging aspect of managing change is identifying that the proposed modification is in fact a change.

1. *Management of Change is needed if the change is Not-In-Kind.*
2. *A Not-In-Kind change is one where Management of Change is needed.*

In other words, the terms "Management of Change" and "Not-in-Kind" tend to be defined in terms of one another.

The second difficulty to do with the In-Kind/Not-In-Kind decision, noted above, is that all changes are, when analyzed deeply enough, not-in-kind."

What is the role of PSSR in the MOC process?

Pre-Startup Safety Review, or PSSR, is not actually part of the MOC process per se. It is intended to be a redundant check or review to ensure that an MOC has been implemented in a quality and safe manner. The PSM standard states:

The employer shall perform a pre-startup safety review for new facilities and for modified facilities when the modification is significant enough to require a change in the process safety information. Process safety information includes all information about the process. That covers about everything. The pre-startup safety review shall confirm that prior to the introduction of highly hazardous chemicals to a process:

- *Construction and equipment is in accordance with design specifications;*
- *Safety, operating, maintenance, and emergency procedures are in place and are adequate;*

- *For new facilities, a process hazard analysis has been performed and recommendations have been resolved or implemented before startup; and modified facilities meet the requirements contained in management of change, paragraph (l);*
- *Training of each employee involved in operating a process has been completed.*

When is Management of Change (MOC) Required?

OSHA requires the employer to establish and implement written procedures to manage changes (except for "replacements in kind") to process chemicals, technology, equipment, and procedures, as well as changes to facilities that affect a covered process. The MOC procedures shall assure that the following considerations are addressed prior to any change:

- The technical basis for the proposed change;
- Impact of change on safety and health;
- Modifications to operating procedures;
- Necessary time period for the change;
- Authorization requirements for the proposed change;
- Employees involved in operating a process and maintenance and contract employees whose job tasks will be affected by a change in the process shall be informed of, and trained in, the change prior to start-up of the process or affected part of the process;
- Updates to process safety information and operating procedures.

The industry has been practicing process safety for over 20 years. The definition of MOC and PSSR are well understood. Some companies have developed great procedures for addressing both of these elements in an integrated fashion using Internet-based web applications and powerful database engines. These software systems simplify the complex MOC process and enable standardization, ensuring a higher level rigor which is difficult to impose with paper-based execution. While the best practices for MOC and PSSR work practices continue to evolve, the following sections reveal what is currently accepted as best practice.

Best Practice MOC Part 1 – Collecting Change Ideas from the Idea Generators

As established above, continuous improvement is impossible without implementing change ideas that can come from anywhere. The problem is how can organizations empower their workforce to initiate ideas, capture and evaluate them then act on the ones that will improve the business? Can you imagine how successful your business would be if you could leverage all of the grey matter at your disposal to solve production problems?

In order to efficiently and effectively capture these ideas, we need to leverage technology. Using Internet–based software can make it easy for ideas to be collected at the source. The easier we make it for staff members to pass on their ideas, the more of these ideas we will receive with time to act on them. One best practice that has surfaced to address this need is to design an electronic suggestion or idea box. Web-based quick entry screens can be accessed using a browser on any company computer to easily suggest a change or improvement. Once these ideas are entered, they are automatically communicated to idea reviewers (supervisors) for vetting, categorization (in-kind or not in-kind) and prioritization. No matter what happens to these ideas, the reviewer must provide feedback to the idea initiators on what happened to their ideas. If an idea results in tangible savings or competitive advantage, a tangible reward will serve to prime the idea generator for the future.

You can visualize the best practice MOC process using a series of funnels (below). Changes begin as ideas and they can be entered by anyone in the company into a single quick entry screen and collected in the main funnel. The supervisor is automatically notified that a new idea has been entered for their unit or area. Each idea is meticulously reviewed and categorized as a recommendation or a MOC. Using electronic submission, this centralized collection funneling system is infinitely more efficient and effective than word-of-mouth, a paper system and/or multiple systems that collect and manage ideas. One common system for collecting, vetting, prioritizing and categorizing ideas is a best practice.

Figure 2.0

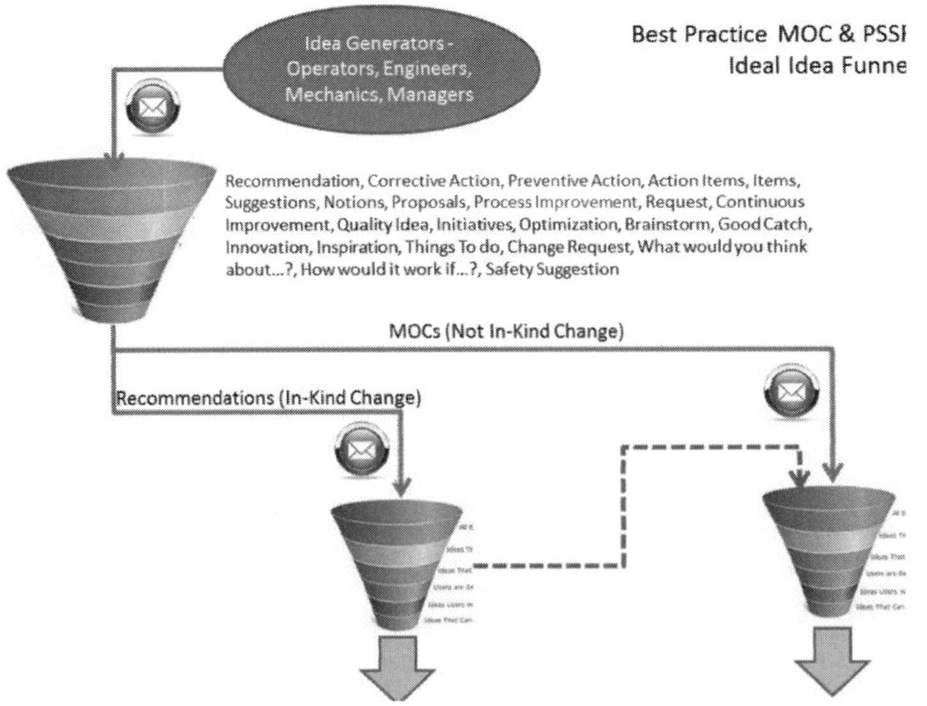

Best Practice MOC Part 2 – Managing the Recommendations (In-Kind Change)

Recommendations can come from audits, incidents, risk assessments, process hazard analysis, emergency response critiques, inspections, reviews, and surveillances. These recommendations are generally ideas for continuous improvement aimed at improving safety, environmental performance, quality or efficiency. Automating and standardizing the recommendation management process makes it repeatable, measurable and more readily understood. Utilizing software that illustrates the correct process and visually progresses users through the workflow is key to the acceptance and the eventual success of the MOC program.

Once the idea has been reviewed, assigned and categorized as a **recommendation**, or in-kind change, a simple workflow can be leveraged to ensure the idea is addressed in an efficient manner. A simple database can be used to document the steps in this process, but a best practice is to use software that illustrates the work process on screen as it is executed. The use of a visual workflow to navigate from step to step, showing progress with check marks or color-coding and enabling steps only when precursors (predecessor steps) are satisfied is a best practice. In addition, email notifications should be used to alert persons assigned to manage, approve and execute actions along the workflow. The more proactive the communication and visual the workflow is, the more intuitive it is to the end-user. The workflow presented below is used for hundreds of thousands of recommendations per year. The entire work process is automated from beginning to end.

Figure 3.0

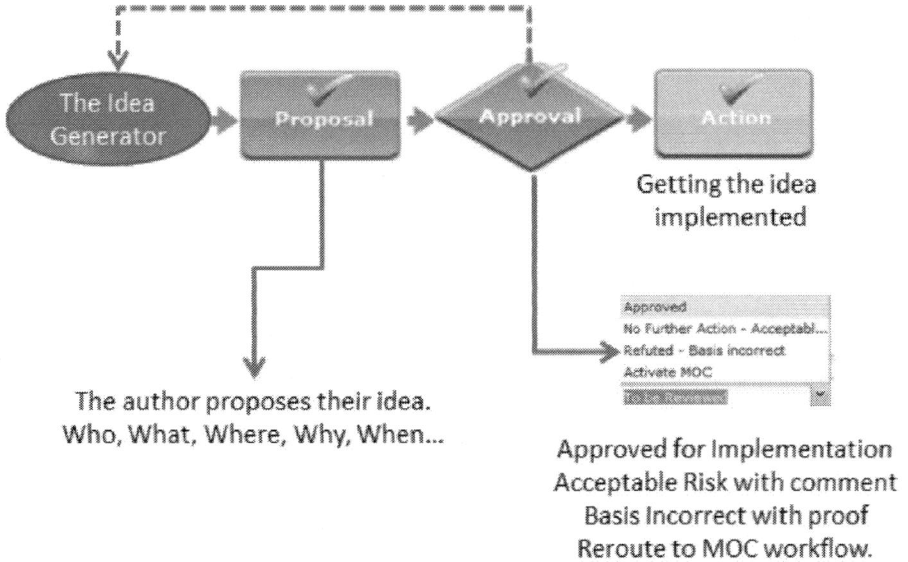

The workflow begins in the proposal step or in the "idea generator's" mind. The recommendation proposal must contain a title, a description (what needs to be done), a justification (why it is important) and a deadline (when it needs to be done) for getting this idea accomplished and a responsible person (RP). The reviewers of the proposal will ensure the accuracy and the resources are available to address the idea. The next step is to accept the idea into the system and automatically notify the RP. The acceptance is illustrated by a check or tick mark in the proposal icon.

After a review of the recommendation, the RP moves to the Approval step. It is in this step that they must decide whether to approve the idea, refute the idea, close the idea as an acceptable risk or decide it needs to be treated as an MOC. Approval of the idea will enable the creation of action items. Refuting a recommendation or addressing it as an acceptable risk should require proof or explanation. These comments must be documented for future reporting and analysis and automatically communicated to the idea generator.

Best Practice MOC Part 3 – Managing MOCs (Not In-Kind Change)

If the idea needs to be managed as a "Not in-Kind" change or MOC, a more complex workflow is required. Similar to the "In-Kind" change, it still requires a proposal, acceptance and assignment to a RP, but additionally will need a formal hazard review and planning. In almost all cases, a MOC also requires the pre-startup safety review (PSSR) step, a Startup Approval step and some post startup steps. Because of its complexity and the need to enforce a rigorous process, a "live" workflow showing what steps are necessary, which ones have been accomplished and which ones are left to do is essential. The workflow below is considered the best practice MOC and PSSR. There are no steps left out and no superfluous steps either.

Figure 4.0

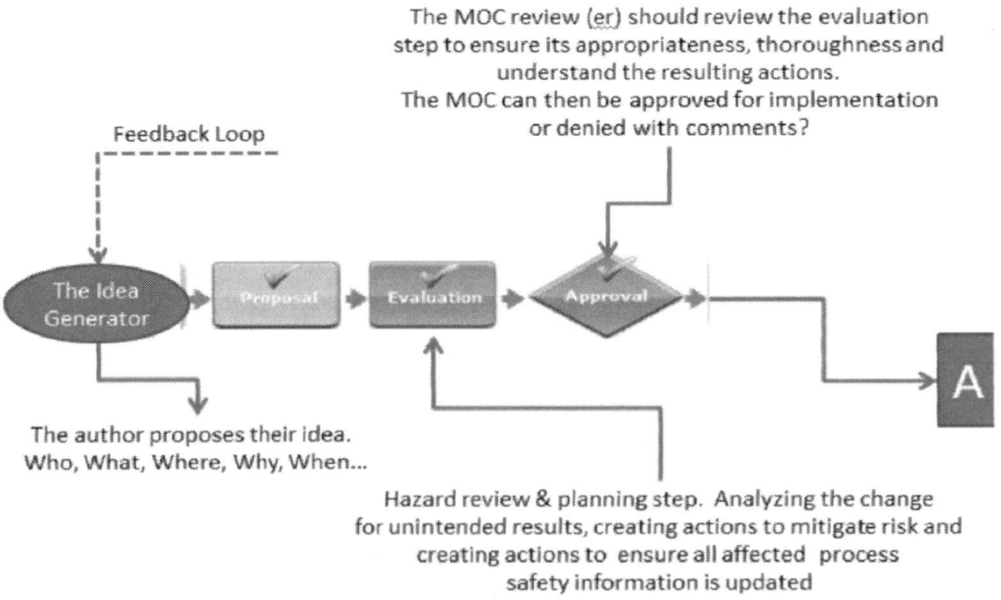

After the acceptance of the idea into the system, the RP receives a notification that they have been assigned a MOC to manage. The next step in the workflow is to "Evaluate" the MOC. This evaluation process is meant to ensure that the hazards of the change are reviewed, understood and mitigated, or at least a plan for mitigation is created. It is also meant to reveal the process safety information that will be affected and the training that will be necessary. Of vital importance is the creation of an action plan to make sure hazards are addressed and that appropriate training is done and all process safety information is completed. Just identifying the need falls very short of hitting the target. In fact, if you identify the need for action and then do not follow through, you have created the proverbial "smoking gun" and a red flag for auditors.

A smart checklist of questions should be presented to the RP in this step which are based on the category of the change. This eliminates the need to answer questions that have nothing to do with the change. Automation gives you the ability to tailor the evaluation process so that it is more effective and efficient to use. Asking evaluation questions unrelated to the change is very inefficient and can also make finding the right questions difficult. Even more automation can be leveraged to pre-assign reviewers to questions that only they should be answering and configuring automatic/predictable actions with associated questions. Here in this step we can make the comments and closure of all questions mandatory for moving to the next step. Until all questions are closed, we cannot move to the next step, the Approval step.

The Approval (to implement) step can be done by a person or groups of people. Their job is to look back at the MOC proposal, review the evaluation that was undertaken and judge whether it was done effectively by the right people. They should also take notice of the mitigating actions that resulted from that review. If more evaluation is required, the reviewer can loop back in the process and create more evaluation questions to satisfy. They should also ensure that actions have been tagged appropriately as pre- or post-startup. If the resulting actions or time have otherwise made this MOC a bad idea to execute, the MOC can be denied at this point with closure comments. Those comments are sent to the idea generator and should be documented in the system. If they are satisfied, then approval can be selected and we can begin implementing the MOC.

The MOC Action Step

Now the real work begins--updating procedures, marking up drawings, making sure all affected personnel are trained, writing work orders, getting an environmental permit, and all of the other actions required for completion to do a quality job. Once all actions (pre-startup) are completed, then you get a check mark over the action icon and we move on to the PSSR step.

Figure 5.0

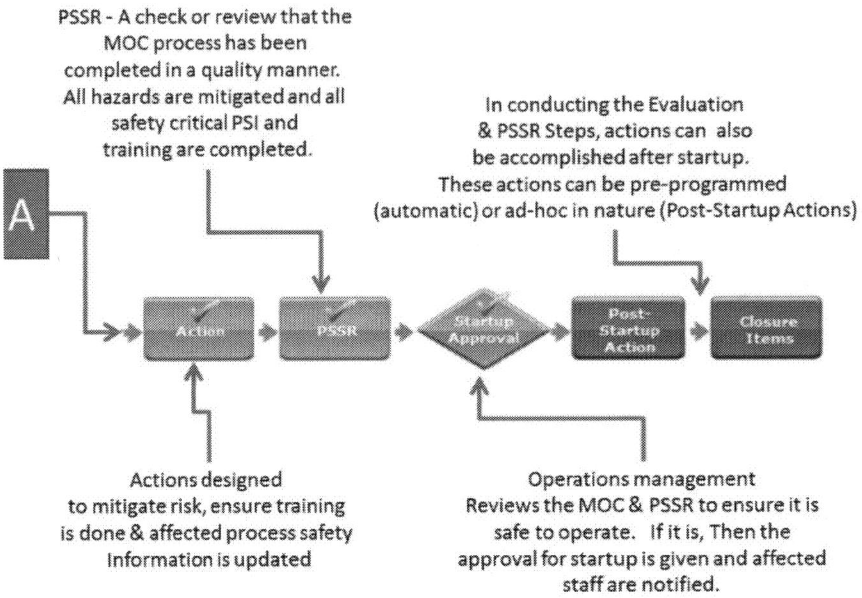

The PSSR Step

As previously stated, the PSSR is a redundant check to ensure the MOC has been properly executed. Like the Evaluation step, once we have chosen a category of change, automation can be used to ensure that the appropriate PSSR checklists/questions are presented to the RP. This increases the efficiency and effectiveness of this step as well. Once the PSSR is completed, we receive our check mark in that workflow step, and an automatic email is sent to the appropriate manager asking for permission to Startup (Startup Approval).

Startup approval is the formal acceptance of the change and the acceptance of the risk associated with it by operations or production. The person who completes this step should review all of the previous work steps in the workflow as well as physically review the work in the field.

The last two steps in the MOC workflow are closure items and post-startup actions. Although these are not referred to in any of the governing regulations, there will be non-safety related activities that are not required to be completed by startup and still need to be managed. This best practice will ensure these actions are accomplished and documented. Past Due email notifications are sent appropriately if the MOC becomes past due.

There are some distinct advantages to adopting the best practice MOC and PSRRs workflows discussed herein. As the best practice evolves, so must the users of the technology. Refresher training, upgrades and constant tuning and tweaking of the workflow are all part of continuous improvement of an MOC and PSSR program. This minimal investment is a small price to pay to avoid poorly managed changes or the opportunity cost of not collecting ideas from your idea generators.

How does Automation help?

Automation of the MOC and PSSR process helps in several ways. First, we can collect ideas at the source of the risk, whether it is the plant or on the platform. This is where the risk is and where continuous improvement is accomplished. Furthermore, it is also where most of the practical/achievable ideas for continuous improvement come from. The best practices must be useful and used where the money is made. Technology makes that possible.

Today's technology is perfectly suited for collecting, analyzing and sharing ideas. Accomplishing these actions without automation is nearly impossible. The days of paper-based MOCs are almost non-existent because automation makes these tasks more efficient and effective as well as less likely to be overlooked.

Reporting metrics and automatic alerts when MOC and PSSR actions are assigned, closed and past due ensure diligence and accountability. Technology can spread the news quickly to those who can create and execute a plan of action and do something with new ideas. Morse code, paper reports and snail mail do not get the job done like email communication and automatic reminders.

A central corrective and preventive action tracking software can help ensure that identified issues are dealt with in a safe and timely manner. Idea or recommendation management and action tracking can be accomplished with one database engine worldwide. That means one system to learn, maintain and continuously improve. It also means you have one global change management system to get things done, communicated and for generating reports. Managers can get the visibility they need to measure performance without bothering the people making the money.

An automated best practice MOC and PSSR system helps ensure that once you get it "right", you keep it "right". It also makes sure that change is appropriately identified, evaluated, approved and executed. It also makes poor change management easy to identify and correct through management surveillance.

Automation is not the goal; best practice is. Like reporting and management visibility and lots of other byproducts of this best practice, the goal is managing change efficiently and effectively.

Words of Caution

A cardinal sin of MOC & PSSR process automation is using a finance or maintenance technology that you already have to automate your best practice, because you already have it. This will sub-optimize a best practice faster than anything you can do, and it will likely defeat the purpose of the automation. Remember that 95% of your users are engineers, operators and mangers--not accountants, IT "geeks" or maintenance planners.

And finally, you can develop your own best practice MOC and PSSR, but to be successful as best practice software, you must have a full-time MOC best practice expert leading the development. Otherwise there will be several months of trial and error. Make no mistake; software development is difficult enough with unlimited resources, expert developers and patient end users. You cannot automate a business process you do not understand. Remember that managing change is a complicated management system for operations or production personnel. You cannot write a requirements document so thoroughly and clearly that a software development team can automate your MOC best practice workflow without constant oversight from the MOC experts.

Unifying Consequence Models for Natural Gas Pipelines in Petrobras Transpetro

Ana Paula Rodrigues Bahiense
PETROBRAS TRANSPORTE - TRANSPETRO
Av Presidente Vargas 328
Centro CEP 20091060
Natural Gas Division
Rio de Janeiro, Brazil
anapaulabahiense@petrobras.com.br

Maria Beatriz Winter Caputo
PETROBRAS TRANSPORTE - TRANSPETRO
Av Presidente Vargas 328
Centro CEP 20091060
Natural Gas Division
Rio de Janeiro, Brazil
Email: beatriz.caputo@petrobras.com.br

Prepared for Presentation at
American Institute of Chemical Engineers
2013 Spring Meeting
9th Global Congress on Process Safety
San Antonio, Texas
April 28 – May 1, 2013

UNPUBLISHED

Unifying Consequence Models for Natural Gas Pipelines in Petrobras Transpetro

Ana Paula Rodrigues Bahiense
PETROBRAS TRANSPORTE - TRANSPETRO
Av Presidente Vargas 328
Centro CEP 20091060
Natural Gas Division
Rio de Janeiro, Brazil
anapaulabahiense@petrobras.com.br

Maria Beatriz Winter Caputo
PETROBRAS TRANSPORTE - TRANSPETRO
Av Presidente Vargas 328
Centro CEP 20091060
Natural Gas Division
Rio de Janeiro, Brazil
Email: beatriz.caputo@petrobras.com.br

Keywords: pipeline, natural gas, consequence model

Abstract

Petrobras Transpetro has been contracting quantitative risk analysis for natural gas pipelines and their associated installations as a legal and internal requirement. These studies have been carried out by a series of consultants that estimate the magnitude of possible damage using different software available in the market. This fact led to a variety of results based on the premises adopted by consultants (most of them depending on the risk analyst) and differences regarding the peculiarities of mathematical models.

Petrobras Transpetro team, mainly field personnel, must use these data to carry out preventive actions inside vulnerable areas, and also to be aware of these ranges in case of a real emergency, thus providing evacuation and escaping actions, as necessary.

Based on this scenario, Petrobras Transpetro's Natural Gas Division had invested in purchasing a consequence-analysis software in order to model the effects of high-consequence loss events as an effort to unify consequence models for natural gas transportation pipelines at Petrobras Transpetro.

As per the above description, this paper aims to present results and benefits of the practical usage of unified consequence modeling information in risk management at Petrobras Transpetro's Natural Gas Division.

1. Introduction

Transpetro, a Petrobras wholly owned subsidiary, is the Brazilian leading logistics company dealing with transportation and storage of oil and byproducts, biofuels and natural gas. It operates 48 terminals (land & water), a fleet of 60 cargo ships, and a network of 15,000 kilometers of oil and gas pipelines interconnecting these facilities.) In addition, Petrobras Transpetro provides services to several distributing companies and to the petrochemical industry.
Although the contribution of natural gas to the country's energy matrix is still comparatively low, it is increasing in accelerated rate. On the other hand, the extent of the gas transmission pipeline network has doubled in the last 4 years. Nowadays, Petrobras Transpetro's Natural Gas Division is responsible for operating and maintaining 7.327km of onshore natural gas pipelines, 12 Compression Stations and 129 City Gates, all of them remotely operated by the National Operation Control Center (CNCO), located in Rio de Janeiro.
In Brazil, State & Federal Environmental Protection Agencies require every installation to carry out a Quantitative Risk Analysis including consequence and vulnerability analysis, frequency estimation and societal and individual risk calculation. These Agencies establish risk tolerability criteria and have, therefore, the responsibility of issuing operational licenses based on the risk results.
Beyond that, ANP (Agência Nacional do Petróleo, Gás Natural e Biocombustíveis), the Federal Agency responsible for regulating the oil & gas industry in Brazil, also requires risk assessments to be implemented for onshore pipelines, which may be either qualitative or quantitative, depending on the installation's complexity.
Based on governmental requirements and on internal procedures, Petrobras Transpetro has carried out Quantitative Risk Analysis (QRA) of natural gas pipelines, conducted by different Consultants, in different periods of time. The first QRA is often carried out when the Company requests an Environmental License, and then it's updated based on a regular basis. The review cycle for these studies depends on internal procedures and the License validity.
The results of consequence and vulnerability analysis, which are part of QRA studies, are the basis for pipelines' Emergency Plans. These results are used basically to assess whether there is population inside the vulnerable area and if it's necessary to evacuate people.
In case of onshore pipelines, that cross several communities and points of interest, such as schools, churches, hospitals and others, the information related to vulnerable areas is vital to adopt more adequate emergency response procedures.

2. Methodology

2.1 Software

Petrobras Transpetro uses PHAST - Process Hazard Analysis Software Tools, by DNV Risk Management Software (DNV RMS), to perform consequence and vulnerability analysis. PHAST contains discharge, dispersion and effects models to assess major hazards, foreseeing complex consequences from possible releases of hazardous materials.
The algorithm for evaluation of jet fire scenarios considers product characteristics (liquid, gas or two-phase), release rate and duration, height of release, orifice diameter and angle of the formed

jet (0° for horizontal jets and 90° for vertical jets). The software estimates flame dimensions and intensity.
For fireball scenarios, the model calculates its radius and duration. The critical parameters to be defined for a fireball model are: product mass involved in the fireball; strength of the thermal radiation source; and the fraction of the total heat that is released as thermal radiation.
The model simulates flash fires due to ignition of a vapor cloud, on an integrated basis, to the algorithm of atmospheric dispersion. The flammable cloud's size is used to model a transient front of flames that travels through the cloud. Thermal radiation in the receptor is calculated as a transient function due to variation of the flames' sizes inside the cloud.
This algorithm considers the speed, width and height of the flame, and concentration of the flammable vapors of the cloud during calculation time. The distance which corresponds to the Lower Flammable Limit (LFL) is assumed as the flame length. Result is calculated in terms of the distance reached by the flash fire.

2.2 *Premises*

2.2.1 Accidental Scenarios

Failure of a high-pressure natural gas pipeline can lead to various outcomes, some of which can pose a significant threat to people and property in the immediate vicinity of the failure location. For a given pipeline, the type of hazard, and the damage or injury potential associated with the hazard, will depend on the mode of line failure (i. e., leak vs. rupture), nature of the gas discharge (i.e., vertical vs. inclined jet, obstructed vs. unobstructed jet) and the time to ignition (i.e., immediate vs. delayed). The various possible outcomes are summarized in Figure 1. In mathematical simulations, methane was used as a representative for natural gas.

Figure 1: Possible outcomes from failure of a high-pressure natural gas pipeline

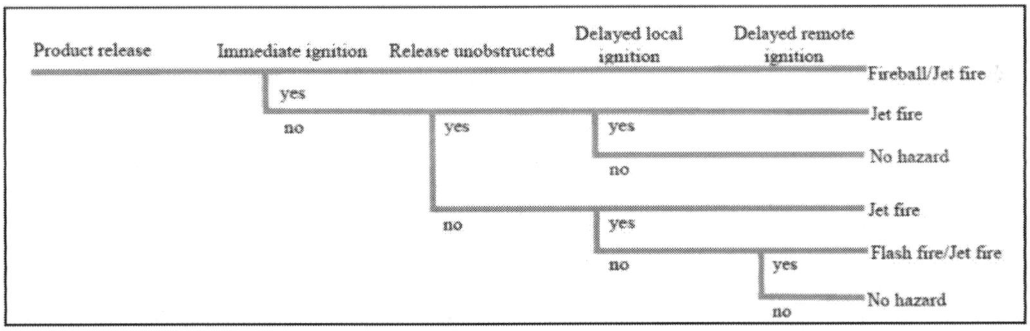

Although there is a great variety of outcomes that can result from failure of a high-pressure natural gas pipeline, the present study aims to analyze thermal effects due to jet fire.

2.2.2 Typical Pipelines

A set of 11 typical natural gas transportation pipelines was established to estimate physical effects due to natural gas possible scenarios. Typical pipelines considered in this study are listed on Table 1.

Table 1. Typical pipelines considered in this study

Pipeline	Pipeline Diameter	Operational Pressure
Pipe 1	8 in	70 bar
Pipe 2	10 in	100 bar
Pipe 3	12 in	100 bar
Pipe 4	14 in	100 bar
Pipe 5	18 in	100 bar
Pipe 6	20 in	120 bar
Pipe 7	22 in	74 bar
Pipe 8	26 in	100 bar
Pipe 9	28 in	100 bar
Pipe 10	38 in	100 bar

2.2.3 Meteorological data

The meteorological data considered in the mathematical simulations are presented in Table 2.

Table 2. Meteorological Data

Parameter	Day Condition	Night Condition
Ambient Temperature	25°C	20°C
Wind Speed	3 m/s	2 m/s
Atmospheric Stability classification*	C	E
Humidity	80%	80%
Soil Temperature	30°C	20°C

* Pasquill-Gifford Stability Categories

2.2.4 Physical Effects Considered

The Eisenberg Vulnerability Model, developed by the United States Coast Guard, was used for calculation of the exposed community vulnerability. This model uses probabilistic functions of the PROBIT type. It is a widely used approach to relate the phenomena intensity (like fire heat radiation) to their consequences.

Probit is a number related to the probability of damages, injuries or fatalities, expressed through the percentage of the population or physical installations affected.

It is relevant to emphasize that, conservatively, the vulnerable area corresponds to the area that could be affected in case of an accident scenario. The area predominantly affected will be influenced by local climatic and geographical conditions at the instant of the leak, such as interference by wind direction and speed, and may vary from person to person and equipment to equipment.

The following levels of interest were considered in this study:

1. Thermal radiation level of $4kW/m^2$ is frequently suggested in specialized literature as the pain limit after exposure of 30 seconds;
2. Thermal radiation level of $12.5kW/m^2$, which represents the limit to cause deaths - 1% fatality probability for an exposure time of 30 seconds;

The effects of thermal radiation arising from eventual jet fires were estimated for a height of one meter, understanding that this is a representative height for evaluating physical effects on people.

2.2.5 CETESB Requirements

CETESB is a State Environmental Protection Agency, from São Paulo State, publisher of the Quantitative Risk Analysis Manual: "Accidental risk from technological origin. Method for decision and reference terms", extensively used in Brazil, which establishes most of the premises adopted in this case study.

Although CETESB is a "local" Agency, it's considered as a benchmark all over the country. So, most of State Environmental Protection Agencies base their rules on CETESB requirements, including IBAMA, the National Environmental Protection Agency.

Premises based on CETESB Manual and used in this study are listed below:

- Simulations were carried out considering only pipeline buried stretches, calculating the effects of angular (45°) jet fires;
- Simulations also considered pipe rupture, considering the leak hole as 100% of pipe diameter.

3. Results

Results are presented in Table 3.

Table 3. Reaches (m) for Jet Fires Scenarios

Pipeline	Pipeline Diameter	Operational Pressure	Jet Fire Reaches (m)			
			Daily (kW/m^2)		Nightly (kW/m^2)	
			12,5	4	12,5	4
Pipe 1	8 in	70 bar	154	245	151	252
Pipe 2	10 in	100 bar	228	360	227	369
Pipe 3	12 in	100 bar	271	425	271	436
Pipe 4	14 in	100 bar	312	490	315	503
Pipe 5	18 in	100 bar	395	616	400	634
Pipe 6	20 in	120 bar	433	675	440	695
Pipe 7	22 in	74 bar	477	743	486	766
Pipe 8	26 in	100 bar	554	860	565	887
Pipe 9	28 in	100 bar	592	919	606	949
Pipe 10	38 in	100 bar	782	1209	803	1251

4. Conclusion

The results presented in item 3 show that the higher operational pressure and hole diameter, the higher would be the reaches of an accidental scenario of jet fire. The reaches shown in Table 3 could not be considered as an absolute value, instead it should be considered as a reference for the decision making process.

All models, including consequence models, present uncertainties. Input parameters that are modeled using the consequence analysis software may, and probably will, be different in an eventual emergency real case. Therefore, the results presented should be used as guidance to the emergency response team, so that adequate response is provided considering possible effects of

a jet fire on the pipe neighborhood, such as community evacuation, roads blocking, among others.
This study may be completed by simulating other possible outcomes, such as flash fires and fireballs, and by calculating thermal radiation levels.
As originally intended, the results present unified consequence modeling for jet fire scenarios involving natural gas pipelines, providing a quick reference to response preparedness and field personnel use in decision making process, in case of eventual emergencies.

5. References

[1] Manual de Orientação para a Elaboração de Estudos de Análise de Riscos, P4.261, CETESB, May 2003.

Additional references not cited:

[2] 8th Report of the European Gas pipeline Incident data Group - EGIG, December 2011.

[3] AMERICAN INSTITUTE OF CHEMICAL ENGINEERS (AICHE), 2000, "Guidelines for Chemical Process Quantitative Risk Analysis", New York, USA, 2nd Edition.

[4] "A Model for Sizing High Consequence Areas Associated with Natural Gas Pipelines", Gas Research Institute (GRI), 2000.

9TH GLOBAL CONGRESS ON PROCESS SAFETY

Automation Strategies: Avoiding Accidents During the Process of Draining Water From LPG Spheres

Carlos André Vaz Junior
Departamento de Engenharia Química, Escola de Química, Universidade Federal do Rio de Janeiro
Centro de Tecnologia, Bloco E, Sala E-209, Ilha do Fundão
cavazjunior@eq.ufrj.br

Maurício Bezerra de Souza Junior
Departamento de Engenharia Química, Escola de Química, Universidade Federal do Rio de Janeiro
Centro de Tecnologia, Bloco E, Sala E-209, Ilha do Fundão
mbsj@eq.ufrj.br

Diego da Silva Lione
Escola de Química, Universidade Federal do Rio de Janeiro
Centro de Tecnologia, Bloco E, Sala E-209, Ilha do Fundão
diegolione@ig.com.br

João Marcio Almeida Villaça
Escola de Química, Universidade Federal do Rio de Janeiro
Centro de Tecnologia, Bloco E, Sala E-209, Ilha do Fundão
joaovillaca@hotmail.com

Leonardo Salim Damian Aguiar
Escola de Química, Universidade Federal do Rio de Janeiro
Centro de Tecnologia, Bloco E, Sala E-209, Ilha do Fundão
leonardosdaguiar@gmail.com

Prepared for Presentation at
American Institute of Chemical Engineers
2013 Spring Meeting
9th Global Congress on Process Safety
San Antonio, Texas
April 28 – May 1, 2013

UNPUBLISHED

Automation Strategies: Avoiding Accidents During the Process of Draining Water From LPG Spheres

Carlos André Vaz Junior
Departamento de Engenharia Química, Escola de Química, Universidade Federal do Rio de Janeiro
Centro de Tecnologia, Bloco E, Sala E-209, Ilha do Fundão
cavazjunior@eq.ufrj.br

Maurício Bezerra de Souza Junior
Departamento de Engenharia Química, Escola de Química, UFRJ

Diego da Silva Lione
Escola de Química, UFRJ

João Marcio Almeida Villaça
Escola de Química, UFRJ

Leonardo Salim Damian Aguiar
Escola de Química, UFRJ

Keywords: LPG spheres, automation strategies, process of draining water

Abstract

The present paper has as its main objective to evaluate different strategies of automation, focusing on the reduction of risks on the water drainage procedures in spheres of liquefied petroleum gas (LPG). The condensation phenomenon of the fraction of the water found in gases during the storing process is largely observed in chemical plants. The condensed water will be accumulated in the bottom of the storing sphere and needs to be removed through a drainage procedure. Although being considered an operational routine process, such event was the main cause of at least two severe accidents: in 1966, in France, and in 1972, at the Refinaria Duque de Caxias (Duque de Caxias Refinery), in Rio de Janeiro, Brazil. One of the most critical decisions viewing to avoid that a new accident was to interrupt the drainage before the water could get completely drained. When the drainage continuous after all the water has been removed, the LPG begins to leak. The gas leaking phenomenon promotes the formation of a toxic and inflammable cloud, *par et passu* with a strong reduction of the temperature. In order to analyze the system, mathematical simulations of the drainage were carried out using the UniSim@ Design (Honeywell) simulator. Variations of the different properties (parameters) of the system were studied along the drainage. Understanding the dynamic of the process it is possible to develop automation systems which can preview and anticipate the best moment for the interruption of the water draining process. Traditional control architectures were suggested based on the PID control architecture. The "monitoring processes", that had been proposed by Seborg et al. [2], were tested as well. The monitoring system was considered more efficient, avoiding that natural fluctuations would anticipate and/or retard the action of the controller.

1. Introduction

The present study has as its objective to propose and to evaluate different controlling strategies for water draining systems in LPG (Liquefied Petroleum Gas) spheres. This analysis obtained its stimulus from the study of the accident that occurred in the Refinaria Duque de Caxias/Duque de Caxias Refinery, in Rio de Janeiro, in 1972. Such accident caused the explosion of a LPG sphere, resulting in several fatal victims and a large number of wounded people. The sphere's water draining system consisted, at the time, of a manual opening of the valve localized in the lower part of the reservoir. The employee in charge of such operation should be waiting for the estimated period of time so that practically all the water would have been removed. After this time, the employee should close the valve, interrupting the drainage and avoiding the gas leak. During the whole time of the drainage, the employee should stay in the area, following the process. In the accident of 1972, the employee left the place, not returning for the closing of the valve, causing the LPG leak which formed an inflammable cloud and the freezing of the equipment nearby. The low temperature generated by the liquefied gas leak promoted the freezing of the drainage valve, not allowing its closing. After the accident, the sphere's drainage system received several improvements, such as the inclusion of the blockage double system and safety instruments, even though the drainage procedures continued to be carried out essentially in a manual way.

Proposing to offer a more efficient and safe solutions for the drainage procedures of the water, this study used dynamic simulation. Through the simulator were made several drainage simulations, evaluating the behavior of several parameters of the process. Therefore, the automation strategy was based on the fact that the dynamics of the variables of the process can anticipate under a safe and exact way the ideal moment in which the end of the drainage should happen.

2. The Accident

The description made here is based on statements of the professionals that were working in the location at the time of the accident. The accident occurred in 1972, in the park of the LPG spheres of the Duque de Caxias Refinery, in Rio de Janeiro, Brazil. The storage of the LPG occurs under elevated pressure, allowing the maintenance of the gas in its liquefied state. Nevertheless, the elevated pressure also generates the formation of a condensed water layer in the bottom of the sphere. The valve and pipes localized in the lower part of the sphere are used for the water drainage. To remove the water is fundamental to ensure that the final product will meet the required specifications when in use. At the time of the accident, the space where the drainage pipes sat had a sole manual valve.

The conventional procedure to be carried out in 1972 by the employee on duty was to connect the hose to the drainage pipes. This hose was then positioned in a nearby ditch. The server would return to the place in order to open the valve. The drained water would be routed this way to the ditch. As the amount of water formed during the compression is known, as well as the flow of drained water, it is possible to estimate the necessary time for the water to be removed from the sphere.

According to instructions being followed at the time, it was expected that the employee on duty would stay in the location, following the process. Nevertheless, it is said that on the day of the accident the operator was absent at the exact moment of the end of the drainage, therefore not closing the valve as expected. Due to this fault, not only the water, but the gas as well, was sent to the draining pipes. Besides the formation of an inflammable cloud, the quick decompression of the liquefied gas promoted a strong reduction of the temperature of the line, causing the freezing of the valve. From this moment on, the operator became incapable to stop the leak. The emergency brigade of the company was called, but did not succeed in stopping the gas leak.

It was reported that a dense and large cloud of LPG was formed over the whole park of the spheres of the Refinery. The presence of an ignition source, never identified, detonated the cloud. This detonation, although causing light burning wounds in a few employees, was the main cause of the improvements that followed the accident. The detonation created a jet fire, thus forming a flame that touched the sphere, heating it.

Due to the seriousness of the accident and its duration, the fire was battled not only by the emergency brigade of the refinery, but also with the help of the employees of the companies of the neighborhood. Moreover, the first emergency teams were helped by the next shift of workers. This fact contributed to the increase of the number of victims of the accident.

After almost two hours of fire, the sphere that was leaking exploded in a phenomenon known as BLEVE. The explosion sent fragments over the other spheres, thus propagating the destruction. A sequence of explosions totally destructed the park of spheres of the refinery. Pieces of the reservoirs were found 800m away from their original place.

2.1 Implemented improvements

Similar to the accident reported, a few others occurred in other industrial plants in the same period, as the one in Feyzin, in France, in 1966 [1]. Although the task to remove the water formed in storage spheres of LPG continues to exist until today, several improvements were implemented. The ground of the spheres is now concreted including a slope, avoiding the accumulation of LPG under the sphere in case of a leak (pool fire). Other important changes were implemented in the drainage piping, which now present a larger number of valves, as shown in Figure 1. The opening-and-closing of the valve nearest to the sphere became remote, driven from the control room. The functioning from a distance has as its goal to allow that, even in case of fire, the valve will be controlled. The valve counts on a thermo-sensitive fuse, which promotes its closing in case the external temperature increases. Finally, the length of the drainage piping was also extended. The physical distance between the drainage point and the valves is set so that–even in a leaking scenario– the low temperatures would allow the freezing of the equipment. A larger number of valves in the system also contribute for a safer process. A project of "closed drainage" is being tested in Brazil, nowadays. This system should eliminate the gas release to the atmosphere, in case a leak happened. If a gas leak occurred during the drainage process, the system would send the gas to flare.

Figure 1: Drainage piping with a larger number of valves

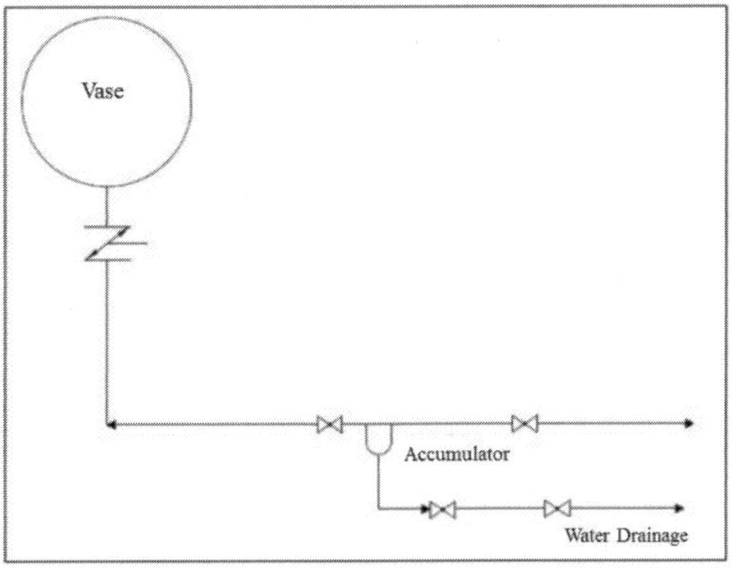

3. Simulator

Through the use of a dynamic simulator, it was possible to carry out several simulations of the drainage system. These results permitted the evaluation of the dynamic behavior of different variables of the process. From the behavior of the process parameters, different control strategies were proposed. The automation would allow the interruption of the drainage process in its ideal moment, thus avoiding the LPG leak.

In this paper we applied the software UniSim® Design. The simulation considers: the storage sphere, the valve directly connected to the bottom of the sphere, and the last blocking valve of the drainage line, as seen in Figure 2. For the simulation it was applied the Peng-Robinson's state equation. The LPG composition was presumed as the mixture formed by 70% of propane and 30% of butane. The amount of the condensed water in the bottom of the sphere, before the beginning of the drainage, was estimated in approximately 5% of the total volume. The diameter of the simulated sphere was 18.25 m. The diameter of the valve nearest to the sphere (VLV-100) was estimated to be 10 in, while the diameter of the most remote valve (VLV-101) was 1 in. The operating pressure of the sphere was presumed to be 8.5 kgf/cm^2 and its initial temperature, 29°C (environmental temperature). In order to simulate the drainage, the outlet flow (flow 5) pressure was adjusted for 1atm (atmospheric pressure). Finally, at the start of the drainage process the initial opening of both valves was presumed to be 100%.

Figure 2: System without PID control

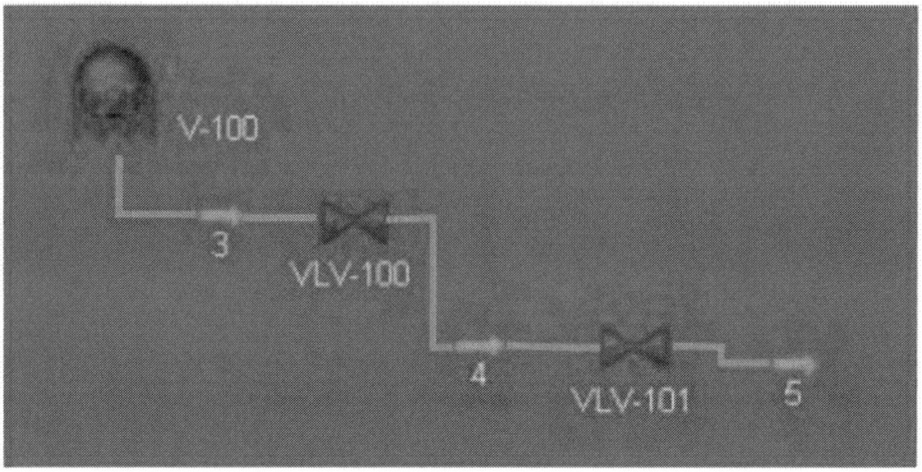

The system represented in Figure 2 does not have the PID controller to interrupt the process. This way, once both valves open, all the condensed water will be removed. After that, the gas release will occur in a situation similar to that observed in the 1972 accident.

After the simulations of the system represented in Figure 2, a PID controller was implemented (Figure 3). The control system sees, as its main goal, to allow that the drainage procedure be carried out safely and efficiently. To achieve this, it is necessary that the controller interrupts automatically the drainage before a leak of the LPG occurs.

Figure 3: System with PID control

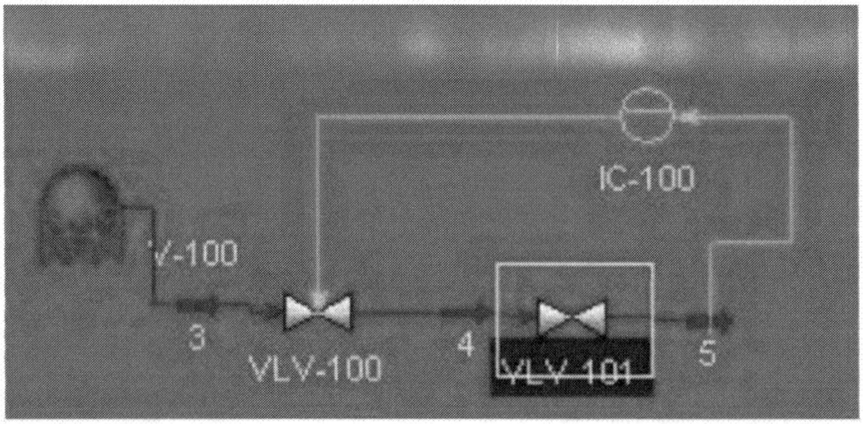

From the simulations study of the flow-diagram without PID controller (Figure 2), the controlled and manipulated variables were defined. The out-flow temperature (flow 5) was defined as being the controlled variable. Such choice was due to the fact that such process parameter presented high sensitivity to the beginning of the LPG leak, varying abruptly. The manipulated variable was defined as being the out-flow drainage of the tank, through the opening control of the valve (VLV-100).

4. Results

4.1 Process simulation without PID controller

First, the simulation of the system was done without the controller (Figure 2). In this simulation, the two valves were maintained opened during all the simulated time (1h40m). This time window was sufficient to simulate not only the water drainage, but the LPG leak as well.

The fractions of the phase water and LPG at flow 3 (closest to the sphere) are illustrated in Figure 4. One may notice that initially there is only a way-out for the water from the spherical tank. This occurs because the water is denser than the LPG, being accumulating so far at the bottom of the sphere. At instant t=1852 seconds, i.e., approximately 30 min. after the beginning of the simulation, the water fraction begins to decrease, initiating the way-out of the liquid LPG from the sphere. From this moment on, the water fraction of flow 3 decreases.

Figure 4: Fraction of the Phase x Time

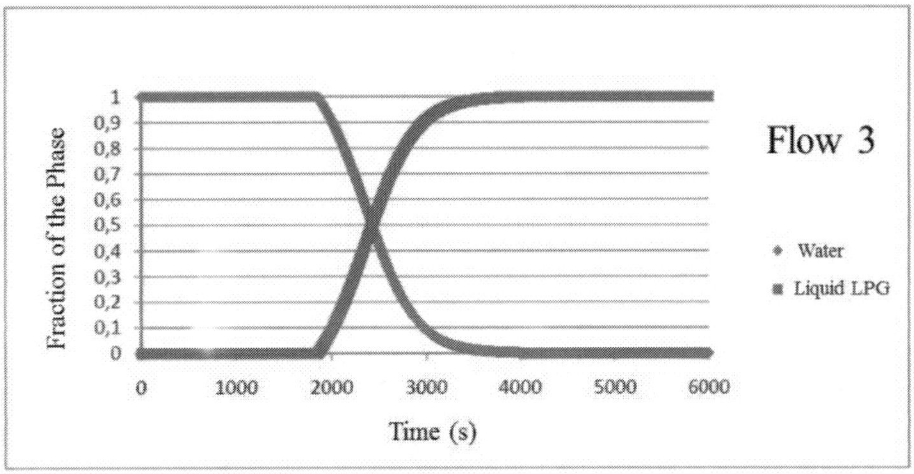

On the other hand, the fractions of flow 5 are represented in Figure 5. In a similar way to flow 3, one may notice that during the first moments of the simulation, the water fraction in the flow equals 1. From instant t=1855 seconds, the water fraction decreases, while the LPG fraction under a vapor state increases. The LPG presence under a vapor state occurs because this flow is being discharged on the environment, with an atmospheric pressure and an initial temperature of 29°C. Under such conditions, the LPG is found in a vapor phase. Therefore, the liquid LPG originated from the tank will vaporize after valve VLV-101.

As observed in the 1972 accident, the vaporization promotes a strong reduction of the line temperature, reaching -30°. C, as we can see in Figure 6. From instant t= 2170 seconds, the flow 5 temperature is already registering -23°C, and the LPG starts to leak in its liquid form. We shall highlight that, in a real scenario, this condition should be sufficient to promote the freezing of the nearest valve.

Figure 5: Fraction of the Phase x Time

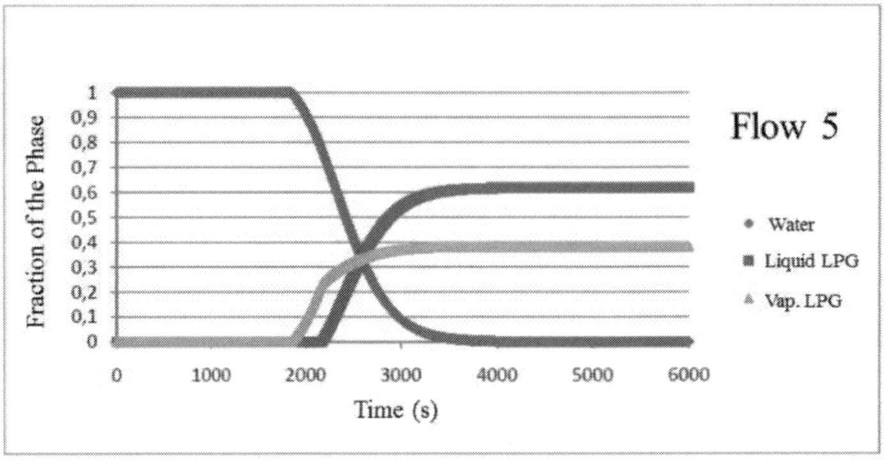

Figure 6: Temperature x Time

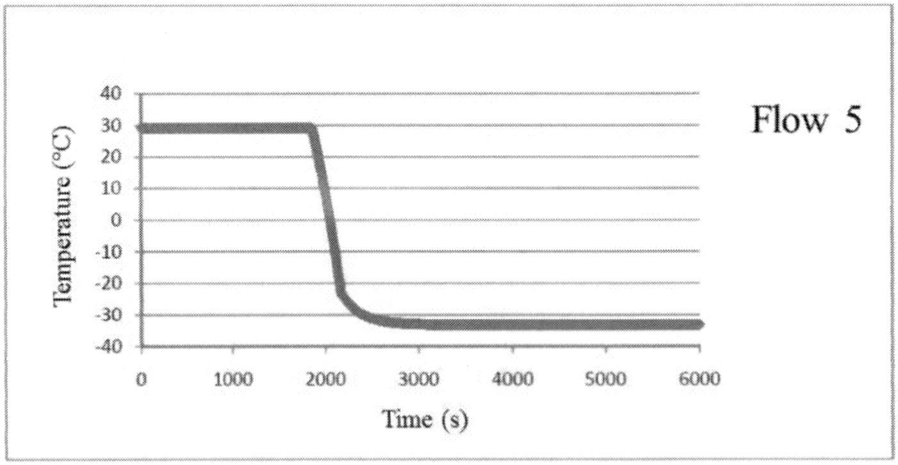

Through Figure 6 we can observe that the out-flow temperature (flow 5) is stable, approximately at the environmental temperature of 29° C, during the first 1855 seconds. In this period of time, only the water of the reservoir leaves it (Figure 5); from then on, the temperature decreases abruptly. At 1915 seconds, i.e., 1 minute after the beginning of the LPG leaking, the temperature was 20°.C, a fall rate of 10 degrees per minute. After 10 minutes, the temperature of the flow reached -30°C. This strong change of the temperature occurs due to the fast change of the phase suffered by the LPG.

From the simulation it was possible to observe that within the variables analyzed, the most sensitive to the beginning of the LPG leaking was the flow 5' temperature. This way, as mentioned, the control system was based on the temperature measures of flow 5.

4.2 Simulation of the process with different tunings of the PID controller

Intending to propose and test different automation strategies for the sphere's water drainage, the use of the PID methodology was initially investigated. Viewing to maintain the controller as simple as possible in a first approach, we opted for the use of a controller known as "P" (proportional). The tuning of the controller had, as its main goal, to maximize the withdrawing of the water from the spheres, thus avoiding the occurrence of a significant LPG leak. The chosen *modus operandi* for the controller was "direct": if the temperature starts to decrease the controller should be activated in order to reduce the flow by closing the valve.

Several simulations were made varying the gain (Kc) of the proportional controller. The set-point was adjusted for 19°.C, 10 degrees lower than the environmental temperature (29°C). In each simulation the following variables were evaluated: opening of the valve, the water fraction at flow 5 and the temperature in flow 5. The results are shown in Figure 7.

Figure 7: Water fraction in the flow 5 x Time

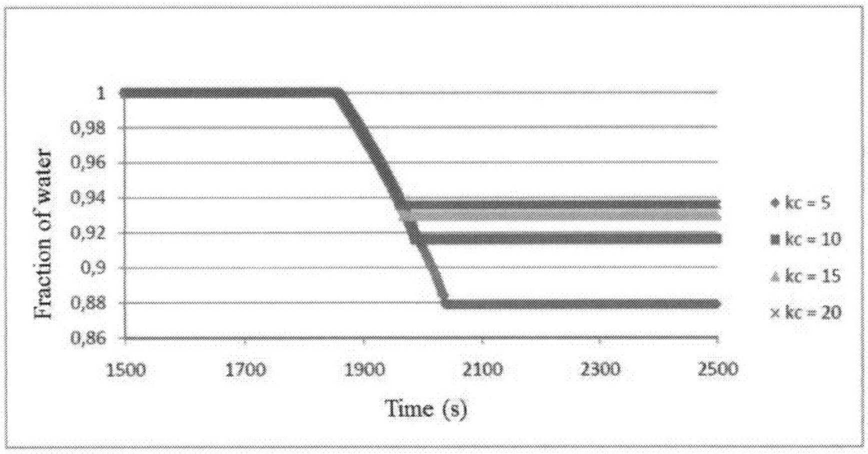

Figure 7 illustrates the behavior of the water fraction in flow 5 as time goes by, for different tunings of the control network. First of all, one may observe that only the water is being drained. From then on, a fall of this fraction occurs, indicating the beginning of the LPG leaking. As the water fraction decreases, the amount of LPG in the flow increases. As a consequence, the temperature decreases severely, according to Figure 8. It is possible to notice that the behavior of the temperature in the outflow follows the same profile of the water fraction in this flow.

Figure 8: Temperature of flow 5 x Time

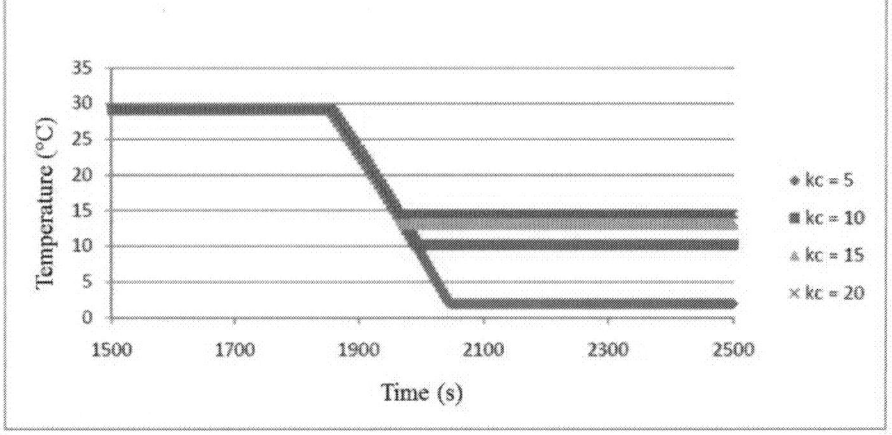

Figure 9 shows the percentage of the valve VLV-100 opening, as a function of time. Initially, while only the water is being removed from the sphere, the valve is kept in a totally open position (100%). As the simulation is carried out, such percentage decreases at the same time that the water content and the temperature decrease. The valve movement proceeds until the valve is totally closed (0% opening). The closing of the valve indicates that the controller is acting expecting the decrease of the leak, since temperature of flow 5 is coming down.

Figure 9: Opening of the valve x Time

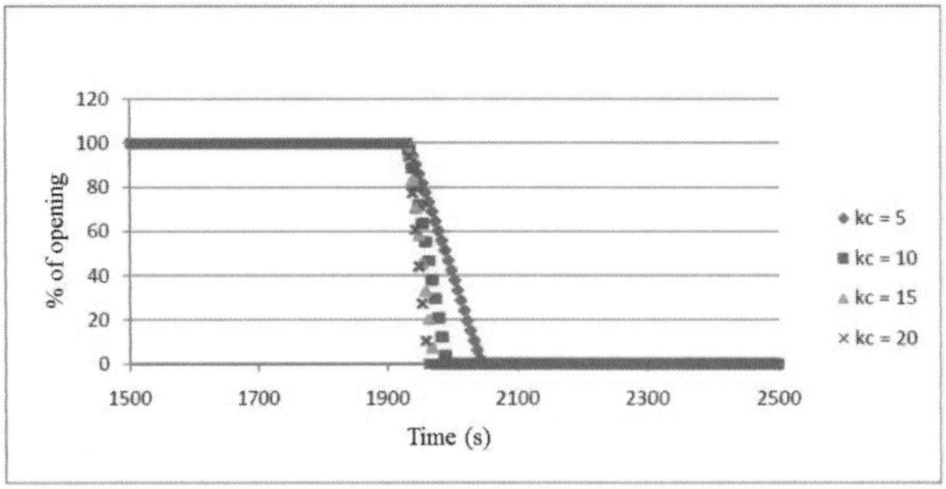

Comparing the behavior of the analyzed variables in relation to the values of the proportional controller gains (Kc), one may observe that the larger the gain, the quicker the valve closes and, therefore, the temperature and the water fraction are higher in flow 5 (Table 1). High gains minimize the amount of the LPG leaked, reducing the risks of frozen valves. Nevertheless, it is worth to highlight the fact that the closing time of the valve is still a function of the valve dynamics, which was not considered in this paper.

Therefore, for a given set-point, the increase of the gain of the controller promotes a quicker closing of the valve and drainage with a smaller LPG leak. It is interesting to notice that for a determined value of gain an increase on the value of the set-point would cause the same results.

Finally, we shall say that very elevated gains of the controller take the valve to act practically as a valve on-off. In this case, there are only two possible positions for the valve: totally closed or totally opened. Other types of control action (integral - I or derivative - D) were not tested, once the results based on the PID methodology took us to choose the on-off action.

Table 1. Dynamic behavior of the parameters of the process

Kc	Final Temperature - Flow 5 (°C)	Final Water Fraction - Flow 5
5	1.95	0.879
10	10.21	0.916
15	13.09	0.929
20	14.49	0.938
100	18.04	0.952
1000	18.74	0.955

4.3 Processes monitoring

In industrial processes, the maintenance of the operational conditions in general uses strategies of feedback or feed forward controller. A different approach was proposed by Seborg et al [2] and denominated "processes monitoring". This approach can be employed with the purpose of following the dynamical behavior of the variables. The monitoring, when associated to tools of supervision and control, guarantees that the variables stay within specific limits and rapidly detect abnormal conditions of operation, implementing corrective actions, etc. This way, the monitoring contributes to the reduction of the variability of the process. In given scenarios, this strategy may present advantages on the use of traditional approaches of control.

In the present study, the variable measured is the temperature of the out flow. The methodology of "process monitoring" is based on the realization of a series of measures of the temperature of the out flow carried out in the beginning of the draining. During the initial phase, the parameter stays practically stable, its average value being used as reference. In a real situation this initial value would correspond approximately to the environmental temperature. As the drainage proceeds and there is a space for a small fraction of LPG, the temperature decreases. The values measured are then compared in real time with the reference value and the differences are then added. When the sum overcomes a specific value, the control action will occur; in this case, the closing of the valve. The strategy of processes monitoring is known as "Cumulative Sum" [2].

It is important to highlight the fact that the temperature of flow 5 – before and at the beginning of the draining process – is subjected to natural fluctuations, such as alterations of the environmental temperature. This normal variation can reduce the efficiency of the traditional approach of control based in a fixed set-point.

In this study, through a simulation the average of the temperature of the out flow was measured during the first 10 minutes of drainage, thus obtaining the average value of 29.17ºC. Through previous simulations it was observed that the decrease of the temperature, caused by the beginning of the leak, was of approximately 0.5º.C This was the maximum value of the cumulative sum adopted to determine the closing of the valve.

Using the cumulative sum, two seconds after the beginning of the LPG leak, the cumulative term became larger than the value stipulated as maximum, this way determining the closing of the valve. The quick closing of the valve, immediately after the beginning of the LPG leak, prevents the freezing of the valves. This approach seems to be much more stable and efficient than the proportional controller.

So, it is important to highlight the fact that in real operations the existence of fluctuations noises (interference) and fluctuations of temperature, or the composition of the LPG yet, could well difficult these parameters. The adoption of higher values as the maximum allowed at cumulative sum could appear adequate.

5. Conclusion

The large accidents of the past reveal the catastrophic potential of the events involving large LPG storage spheres. The draining procedure of the water accumulated in the bottom of the spheres was the main objective of this paper. Focusing on the automation of the procedure, we tried to simulate such process, based on dimensions, parameters and typical industrial conditions. We searched for the identification of a variable that could be indicative of gas leaking, being able to be applied as reference to indicate the best time to interrupt the draining. The temperature of the out flow seemed to be the most adequate parameter.

Different tunings of the proportional controller were tested. It was noticed that the larger the gain value (Kc), the smaller was the amount the LPG liberated at the end of the drainage. This way, the results converged to an action of control of the type on-off. This approach seemed to be strongly influenced by noise (interference) on the variable measured and by the environmental temperature observed.

The use of the strategy of process monitoring was also tested, based on the cumulative sum method. The results obtained were higher than those found for the proportional controller. The drainage was immediately interrupted after the occurrence of a small leak of LPG through the out flow. A "cumulative sum" of only 0.5º C, was adopted as the best. A quick interruption of the drainage process would have reduced the risks of freezing in the line and the emission of LPG to the atmosphere. Therefore, it seems to be an interesting option to increase the safety of the procedure.

6. References

[1] Kletz, Trevor. *Learning from Accidents.* 3rd edition, Gulf Professional Publishing, Woburn, MA, 2001.

[2] Seborg, Dale; Edgar, Thomas; Mellichamp, Duncan, *Process Dynamics and Control*, New York, 2nd edition, Wiley, New York, 2003.

(44aa) Complex BMS Made Simple

Dave Sheppard, Emerson Process Management, Hazelwood, MO and Nikki Bishop, Industry Solutions Group, Emerson Process Management, Roswell, GA

Complex BMS systems with multiple fuels and multiple burners have long presented a struggle for engineers and operators alike. This presentation will cover the methods used to simplify the implementation of a complex BMS in DeltaV SIS. The approach to design and use of standardized sequence modules and graphics greatly increases operator awareness of the operating conditions of the BMS. Included in this presentation will be a description of the design philosophy, the configuration standards, and the approach to operator graphics.

A three step approach is implemented. Break the BMS sequences down in to incremental parts that are easy to understand. Develop standard modules that implement these parts individually. Use a graphical interface the represents these individual parts in an easy to understand way for the operator. This approach is highly scalable to meet various complex BMS implementations.

With this approach it is easy to take a complex BMS and break it down in to small and simple to understand pieces. This allows greater alignment and understanding between the safety engineer, configurator, and operator. Ease of use and troubleshooting for the operator is also greatly increased through a standard set of graphics that show actions needed and first outs for trips.

Through this approach it is possible to take the design of how a complex BMS needs to function and actually implement that design cleanly and simply. An accurate system to capture the first out for a trip or loss of a permissive and also indication of the current state and actions needed to progress forward is needed to increase operator knowledge of what is going on in the BMS at all times.

See more of this Session: Poster Session
See more of this Group/Topical: Global Congress on Process Safety

9TH GLOBAL CONGRESS ON PROCESS SAFETY

"Using ESD Valves as Safeguards, Myth or Reality?"

German Luna-Mejias, P. Eng.
ACM Facility Safety
Suite 700 940 6th Ave SW, Calgary, AB, Canada
gluna@acm.ca
guatire2@gmail.com

Prepared for Presentation at
American Institute of Chemical Engineers
2013 Spring Meeting
9th Global Congress on Process Safety
San Antonio, Texas
April 28 – May 1, 2013

UNPUBLISHED

"Using ESD Valves as Safeguards, Myth or Reality?"

German Luna-Mejias, P. Eng.
ACM Facility Safety
Suite 700 940 6th Ave SW, Calgary, AB, Canada
gluna@acm.ca
guatire2@gmail.com

Keywords: Process, Safety, Emergency, Shutdown, valve, bypass

Abstract

An Emergency ShutDown (ESD) valve is a safety device to be actuated during an emergency situation to prevent hazardous consequences. The ESD valves are to be designed not to create additional hazards when actuated. Sometimes, due to operational constraints ESD valves are bypassed which contradicts best practices and creates an incremental risk for the facility, people and environment. Lack of regulations sometimes drives ESD valves' users to overlook or ignore best practices with the associated incremental risk. This paper presents experiences during HAZOPs/field observations where installed ESD valves are claimed as safeguards but are not (can be bypassed). The intent of this paper is to call the attention of designers, operators and managers of the potential hazardous consequences they may face if ESD valves are not used as per recommended practices.

1. Introduction

Our police force has to be right every time to prevent worst case scenarios, the "bad guys" only have to be right once for a major impact to occur. Exactly the same happens with safety; safety devices have to work when required; only one failure is enough to cause a major impact.

The use, absence or abuse of safety devices –e.g. emergency isolation valves- have been the cause of several accidents in the industry:

- A.P. Green Refractories Company – Explosion at Brick Kiln during start-up due to a main gas isolation valve bypass line leaking with 2 injured. Georgia, USA 1985. [1]
- Oil & Gas, Sonat Exploration Company - Explosion and fire due to misalignment of manual valves when feeding a low pressure system with HP Gas with 4 fatalities. Louisiana, USA. March 1998.[2]

- Ammonia Plant: Fire and explosion with substantial damage to a reformer-furnace due to bypass of control valve -with shutdown function- inadvertently open. No injuries, India, May 1998.[3]
- Esso-Australia Gas Plant Explosion due to the improper use of bypass valve around safety valve as the safety valve was not operable with 2 fatalities and substantial public welfare impact. Victoria, Australia. Sept 1998.[4]

As per these four examples, problems can happen anywhere in the world. In the last 3 years, when performing Process Hazards Analyses (PHA), the PHA teams indicated no concerns about bypasses around ESD valves, because there are no regulations that prevent this. Common responses were: "this is not regulated, why worry about it" or even worse: "recommended practices are recommendations, they are not mandatory". These practices are not only recommended practices but common sense. Additionally, as AIChE members (including professional denominations), the Code of Ethics indicates that welfare of the public and environment is paramount.

2. Definitions

Emergency isolation valves, emergency block valves, Emergency Shutdown valves or Automated Shutdown valve are different names for the same type of function valve.

A paper presented to GCPS in 2012 [5] indicates that the Emergency Block Valves (EBV) are namely different as per different codes, regulations and companies.

2.1 Additional definitions:

Emergency Block valves are used to control a hazardous incident. These are valves for emergency isolation and are designed to stop the uncontrolled release of flammable or toxic materials [6]

The ASV (Automated Shutdown Valve) is most commonly use in the Pipeline Safety Act 2011 [7]. The concept of the ASV is to mitigate the release of flammable or toxic materials, not the prevention of it. Pipeline Safety concern is H&SE not Process Safety.

2.1 ESD valve:

A valve with the following characteristics:

- Actuated valve
- Actuate upon a detection of a dangerous event defined by designers/operators and confirmed during PHA study.
- Provide protection against possible harm to people, assets or environment.
- Not used for control

An Emergency Shutdown valve –for the scope of this paper- is a valve that will prevent a hazardous situation from occurring (i.e. closing to prevent the overpressure of the vessel downstream of the valve) or mitigating the impact after the situation occurred.

2.2 Bypass valve:

Definition of bypass:

- go around something
- channel carrying a fluid around a part and back to the main stream

In the oil and gas industry is sometimes common to find bypasses around control valves; this situation is commonly used when it is expected that the valve may need maintenance during the operation of the plant. The plant can continue to run for a short period of time with a manual controlled function using a globe valve. This is a common accepted practice.

In this paper, the word "bypass" is a key element of the study, and follows the second definition: "channel carrying a fluid around a part and back to the main stream"; more specific: "around the ESD valve".

2.3 Norms and Standards

Depending on the system under review, there are several norms, recommended practices, regulations, etc. associated with the operation and installation of ESD valves. Nevertheless, none of these documents indicates or dictates providing a bypass around an ESD valve. A list of these documents is provided in the appendix A.

3. Installation and Use of ESD valves

In Viviano's work [5], the use of the "5W" (What, Who, When, Where and Why) helps to identify the need for emergency isolation valves. In this paper, it is considered that all 5W are being taken in consideration and the evaluation is moving onto the next questions: Is it correctly installed? It is a valid layer of protection? Does it protect against a hazardous event? Does it mitigate the impact?

Is my safeguard device reliable? In the O&G business a spurious trip is also a problem, but does not affect public welfare. Once the ESD valve is included in the detail design, how do we ensure it is going to work in the different scenarios that led to the installation of the ESD valve?

The first phase of the analysis it to perform the risk evaluation or risk assessment.

3.1 Risk Analysis

During a PHA, analysis will be done to ensure the different ESD valves are valid safeguards to prevent the hazardous event from happening. In this phase of the study a qualitative analysis is performed and team members will agree, or disagree, on the protection being used for prevention. A good facilitator should challenge the team about the effectiveness of the safeguard (e.g. is there enough Process Safety Time, Is the actuation of the ESD valve creating another hazardous condition(s)?).

On the other hand, since the PHA is a qualitative analysis, each safeguard is considered to be of the same "weight" or of the "same importance". If required, an additional safeguards evaluation may be carried out: LOPA.

LOPA (Layers of Protection Analysis) is a semi-quantitative analysis to determine the PFD (probability of failure on-demand) of the safeguards to meet the tolerable frequency requirements of each company.

Based on LOPA an Emergency Shutdown Valve may require being a SIL rated device to achieve the tolerable frequency for a specified event. SIL rated means that the associated elements of the safety device has to be designed, constructed and maintain to a minimum level of "probability of failure when demanded to operate" [8]

Previously it was mentioned that ESD valves should not be used to perform a control function as the reliability or SIL rating of the ESD valve may be compromised. The PFD of a typical BPCS (Basic Process Control System) is 0.1 (1 failure every ten times it is demanded) and the ESD valve may require a higher reliability (e.g. SIL 2).

3.2 Use of ESD for Accident/Incident Prevention

When referring to Emergency Block Valves or Emergency Shutdown valves, two functions (or purposes) should be studied: 1. Use of ESD valves for preventing the hazardous situation from happening (preventive barrier). 2. Mitigating the consequence (acting after the hazardous event happens).

3.2.1 Accident/Incident Prevention

Valves in this category are mainly used in industries (e.g. Oil & Gas) to prevent the hazardous event from happening.

In several scenarios briefly described in the introductory section of this paper, ESD valves were installed to prevent a hazardous incident from occurring as per the original plant design concept.

3.2.2 Accident/Incident Mitigation

Valves in this category are mainly used to prevent public exposure to a hazardous event, after the initiating event occurs. These scenarios are largely described and analyzed in documents such as Pipeline Safety Act 2011 [7].

The concerns of the regulator are focused more on spurious trips that may create a public impact due to the effect of cutting the gas supply to communities. In this case, the regulator may recommend a manually activated Isolation Valve. For this scenario, a LOPA is recommended to ensure that the Isolation valve will have a low PFD (probability of failure on demand).

4. ESD Valve Configuration

Sometimes for "ease the operation" the bypassing of ESD valves are permitted as there are no regulations or norms against it but common sense. These "ease operation" scenarios may remove the protection intent of the ESD valve.

4.1 Creating additional hazards.

Installing an ESD valve to reduce the risk is a good idea unless this addition creates additional hazards. If the ESD valve closes, it will isolate two systems: one at high pressure and other at lower pressure. What happen if the high pressure system cannot withstand the pressure? This analysis should be part of the PHA analysis.

Sometimes, a risk analysis is improperly done just to complete a "check mark" in a paper that means the activity was performed. Therefore a proper evaluation is not accomplished and the operator may believe the facility is safe to operate and it is not.

Unfortunately in many projects the PHA is just a "check mark" to achieve the project approval and the PSM is not strong enough to ensure a proper PHA is accomplished. In an evaluation done by OSHA [9] it has been found that improper or inaccurate hazard assessment is one of the main deviations from the safety programs.

In many PHAs studies it has been found that as no regulations "mandate" the owner to avoid bypassing the ESD valves they are frequently bypassed as it makes the start-up much easier and faster.

If the ESD valve is designed to "Provide protection against possible harm to people, assets or environment" or "actuated upon a detection of a dangerous event" it should not be bypassed. A general rule of Safety indicates that: "the important thing to do is to be safe when nobody is watching"

4.2 Bypassing ESD valve for start-up.

After the PHA and LOPA team agrees to install an ESD valve with specific SIL rating and confirms that no added risk will be created if valve is actuated, what is next?

Even if no regulation "mandates it" common sense indicates that if a bypass is required around an ESD valve the bypass has to be equipped with similar protection or equivalent to prevent the inadvertent operation of bypass or minimize the possibility of leaking.

What are the problems associated with improper bypassing of ESD valves?

- Operator error may remove protection
- Valve leakage may remove protection
- Any of these scenarios will invalidate the PHA previously performed and "due diligence" may be difficult to justify if an accident happens.

Filling the lines with products before start-up is a common practice, occasionally done by opening the bypass around the ESD valve. The filling of lines is not a problem but the increased pressure on the downstream side of the ESD valve may be a problem. That is what happened in the Esso accident (Victoria, Australia) as a valve misalignment created an overpressure of the receiving vessel.

Use of CSC (car seal closed) valves in the bypass around ESD valves is a partial protection as the isolation valve may leak overtime without the operator knowing it. Pressure equalization is a bigger issue than flow.

In the incident that occurred in the Brick Kiln the "bypass valve" around the main isolation valve was "somehow inadvertently open"; it means that to start-up the facility the protection was removed.

5. Conclusions

Potential damages to public and environment can be serious, including fatalities, as it has been shown in several accidents around the world.

In the Pipeline Safety Act 2011 [7], the major concern is the amount of gas release and the impact on public after the incident happens. It does not direct the users to provide preventive safeguards.

As there are no specific regulations or codes, the bypassing of safety devices is a concern that is very likely to be overlooked.

6. Recommendations

6.1 General

- Accident prevention requires ACTIVE leadership by management of safety issues.
- Effective implementation of the right controls to manage, mitigate or eliminate hazards and reduce risk.
- Management systems and attitudes towards safety go hand-in-hand in creating robust defenses.

This paper is written with the intention that the reader perform "due diligence" to make sure the area under his/her responsibility is safe.

6.2 Safe Design

- As a general rule, ESD valves should be installed based on PHA/LOPA
- The actuation of ESD valves must not create additional hazards.
- The required reliability of an ESD valve (PFD) must be evaluated by a LOPA or similar analysis.
- Bypass around ESD valve should only be permitted if an additional ESD valve with the same functionality is installed in the bypass.

6.3 Operating practices

In the accidents reviewed at the beginning of this paper, in most of the cases the inadvertent opening or closing of a valve was a key element to the problem. It is highly recommended to implement a PSSR (Pre Start-up Safety Review) before any start-up.

7. Final Remarks

- Bypass of a safety device may save time for operator, but may cost lives!
- Follow PSM or equivalent because you want to be safe, not because a regulation requires it!
- A general rule of Safety indicates that: "the important thing to do is to be safe when nobody is watching"

8. References

[1] OSHA. Accident Report: Explosion in Brick Kiln. Accident # 14542070, report ID 0418100. Washington, DC. 1985

[2] CSB. Investigation Report: Catastrophic Vessel Overpressurization (4 deaths). Report # 1998-002-I-LA; Chemical Safety Board, Washington, DC. 2000

[3] R.M. Chopde & K.M. Patel ., "Ammonia Plants: Past accidents and lessons learnt", Krishak Bharati Cooperative LTD, India. Presented at International Fertilizer Industry Association (IFA) Technical Conference, Louisiana, 2000.

[4] A. Hopkins, PhD ., "Lessons from Esso's Gas Plant Explosion at Longford"; Australian National University, 2000.

[5] J. Viviano ., "Emergency Block Valves"; presented at VIII Global Congress on Process Safety, 2012.

[6] " API Recommended Practice API RP-553:"Refinery valves and accessories for Control and Safety Instrumented Systems", Chapter 7. American Petroleum Institute, 1998.

[7] "Pipeline Safety Act 2011" U.S. Congress, Law 112-90 – Jan 3, 2012.

[8] CCPS., "Layer of Protection Analysis: Simplified Process Risk Assessment", Center for Chemical Process Safety. New York, NY. 2001.

[9] Jordan Barab., "Learning from Industry Mistakes" presented at NPRA National Safety Conference, May 2010

APPENDIX A

Regulations, codes and recommended practices for the installation of emergency block valves.

- API RP-553 "Refinery valves and accessories for Control and Safety Instrumented Systems" American Petroleum Institute, 2007.
- NFPA 30 "Flammable and Combustible liquid Code", National Fire Protection Association, 2012
- NFPA 497 "Recommended Practice for the Classification of Flammable Liquids, Gases, or Vapors and of Hazardous (Classified) Locations for Electrical Installations in Chemical Process Areas" National Fire Protection Association, 2012
- API RP-500 "Electrical Area Classification" American Petroleum Institute, 2000
- NFPA-58 "Liquefied Petroleum Gas Code" National Fire Protection Association, 2011
- API RP-2510 "Design and Construction of LP-Gas Installations at Marine and Pipeline Terminals, Natural Gas Processing Plants, Refineries, Petrochemical Plants, and Tank Farms", 4th Edition, 1978. American Petroleum Institute; recommended practice.
- NFPA 85 "Boiler and Combustion System Hazards Code", National Fire Protection Association, 2011
- API RP-560 "Fired heaters for General Refinery Services" American Petroleum Institute, 2001
- API RP-556 "Instrumentation, Control and Protective Systems for Gas Fired Heaters", American Petroleum Institute, 2000.
- API RP-520 "Part I: "Sizing, selection, and installation of pressure-relieving devices in refineries". Part II – "Installation" American Petroleum Institute, 2000
- API RP-521 "Guide for pressure-relieving and de-pressuring systems", American Petroleum Institute, 1998
- Pipeline Safety Act 2011: U.S. Congress, Law 112-90 – Jan 3, 2012.

(44ac) Contours When Pipelines Meets Facilities

Karthick Deivasagayam, Loss Prevention, SNC Lavalin Inc, Calgary, AB, Canada

Pipelines involved in oil & gas sector carries risk of fires and explosion along with the actual pipelines. This risk of fire and explosion are vulnerable in the places it could affect and become even more vulnerable in places of proximity of facilities where the risk could turn out into initiating of a major catastrophic events.

This should be addressed in the way of doing things.

QRA done for the most of the projects are specific to their scope alone. Boundary limits of the QRA Study is the same as that as the contractual scope of work. Whereas the risk and hazard boundaries are not contractual but as long as it could affect.

This paper will discuss about few pipeline accidents their consequences and the probable effect if they occurred in a different location (close to the facilities) and from discussing about the potential of what a pipeline accident can do and the focus will shift to the way these risks are approached the way what we do.

This paper will also discuss the way the risks of pipeline projects are captured and also drawbacks and grey areas of the existing approach. This paper also recommends the proposed approach that could be followed for the future risk assessments.

See more of this Session: Poster Session
See more of this Group/Topical: Global Congress on Process Safety

(44ae) Ozone Flammability and Its Affect On the Lower Flammable Limit of Simple Hydrocarbons

Paul Osterberg, Thomas Johnson and Thomas Frawley, Fauske & Associates, Burr Ridge, IL

As a result of ozone's strong oxidizing capabilities and ease of generation on an as-needed basis, ozone applications in process operations are gaining momentum. Previous work has demonstrated that ozone is flammable at high concentrations with a lower flammable limit of 8.4 vol.% in oxygen. However, there is limited data regarding the effect of ozone on the lower flammable limit of other gases known to be flammable in air. This work investigated the lower flammable limit of methane, ethane, and propane in various concentrations of ozone at atmospheric pressure and temperature. The tests were conducted in a 2-L vessel using a fuse wire ignition source at 1 percent ozone intervals spanning from no ozone to the LFL of ozone in the vessel. It is anticipated that the addition of ozone will decrease the LFL of each flammable gas as the concentration of ozone increases.

See more of this Session: Poster Session
See more of this Group/Topical: Global Congress on Process Safety

Evaluate Plant-Wide Safety Of Your Interlock System

Principal Author: Mohammed Naved Khan
Ingenero Technologies India Pvt Ltd.
Mumbai, India
nkhan@ingenero.com

Presenter: Swapnil Pathak
Ingenero Technologies India Pvt Ltd.
Mumbai, India
spathak@ingenero.com

Co-author: Jim Brigman
Ingenero Technologies Inc.
Houston, Texas, USA
jbrigman@ingenero.com

Prepared for Presentation at
American Institute of Chemical Engineers
2013 Spring Meeting
9th Global Congress on Process Safety
San Antonio, Texas
April 28 – May 2, 2013

UNPUBLISHED

Evaluate Plant-Wide Safety Of Your Interlock System

Principal Author: Mohammed Naved Khan
Ingenero Technologies India Pvt Ltd.
Mumbai, India
nkhan@ingenero.com

Presenter: Swapnil Pathak
Ingenero Technologies India Pvt Ltd.
Mumbai, India
spathak@ingenero.com

Co-author: Jim Brigman
Ingenero Technologies Inc.
Houston, Texas, USA
jbrigman@ingenero.com

Keywords: IRM, interlock relationship, interlock-process interference, holistic, isolated information, PHA, mitigation

Abstract

Interlocks serve as important safety systems in industrial settings, where they protect equipment from damage and employees from toxic and harmful releases from that equipment arising out of unsafe conditions. The same safety systems can lead to damage of process systems, if not analyzed properly. In some of the cases we encountered, initiation of interlocks in equipment caused severe damage to other equipment (due to change in composition) or initiation of interlock caused damage to the same equipment it was intended to protect (due to incorrect sequencing of interlocks).

The information available with the interlock design documents (like cause and effect diagrams) is limited to the vicinity of that particular interlock. It does not give any information about the relationship of the interlock, under study, with other process equipment or even with other interlocks (except for initiation). The relationship between safety interlocks and process system arises out of the behavior of the process system (e.g. change in composition) to the actions taken by the interlock. This paper proposes how a holistic view for these behavioral responses can be obtained by developing the IRM (Interlock Relationship Matrix). IRM is a single process document (excluding instrumentation and logic details) that provides information on the effects of all interlocks on each element of process system. Developing IRM is crucial in evaluating the interlock-process interference (behavioral responses) and is beneficial in managerial decision making to assess the likely damage to equipment and provide mitigation thereof.

1. Introduction

Interlocks form the most critical and independent layer of safety in industrial settings. They are primarily designed for:

- Equipment - protect the equipment from damage due to unsafe conditions (eg. over speed) and the resulting economic loss arising out of replacement and downtime.
- Personnel - protect the personnel from exposure to harmful and toxic materials that may get released from equipment damage (e.g. Seal leaks from rotating equipment).
- Plant - maintain the integrity of the plant (e.g. flare mitigation methods employed to limit the flaring during global utility failures and avoid exposure of un-burnt hydrocarbons to the general public).

The design intention of the approaches employed for interlock systems are two-fold; first is the isolated view to protect the equipment at hand and second is the global view to maintain integrity of the plant. Although the approach seems to be looking fair, they (even the global interlock system) lack in considering the interlock-process systems relationships in designing interlock systems.

The very definition of an independent protection layer provided by interlock systems (whether a single or global interlock) has an inherent element of relationship between interlock actions and the behavior of process systems i.e. the actions of the interlock system are taken independent of the responses of process system. This relationship may or may not be relevant as far as the functions of that particular interlock is concerned, but in no case it should cause upset in other process units that could jeopardize their integrity and cause irreparable damage. In the following sections, we will look into some of the incidences that lead to irrevocable equipment damage as a direct result of the interlock-process interference and we will suggest an approach as to how a holistic view for these behavioral responses of the process systems to interlock actions can be obtained for effective mitigation. Also the major points of differences between global interlock systems and holistic view of individual interlock systems will be highlighted.

2. Interlock Design Information

The design of interlock systems starts with the development of interlock diagrams or narratives. They provide key information as to the analysis and identification of scenarios that can cause damage to particular equipment (initiating causes) and the actions that are required to be taken for the protection of that equipment. They are of the following three types:

2.1 Narratives

Narratives provide descriptive statements for the various elements of the interlock system. Narratives are a short description stating the primary function of the interlock, events that can lead to unsafe conditions (initiating causes) and finally actions required by the interlock to mitigate these events. They are written in a simple but precise manner without making use of detail identification tags.

2.2 Cause and Effect (C&E) Diagram

C&E diagrams are the next level of detail information about the interlock. C&E diagrams are presented in a tabular format, detailing the initiating causes and the actions taken in the form of individual tag numbers, usually initiating causes are activation of suitable switches due to unsafe conditions and corresponding actions taken (opening or closing a valve). Actions taken are segregated for each initiating cause. They also provide information about time delays. C&E diagrams provide detail accounting of each element of the interlock.

2.3 Logic and Ladder Diagram

The information provided in logic and ladder diagrams is specific to the instrumentation part of interlock system. They provide the circuit diagrams of the relay logic hardware used for the interlock system under consideration.

3. Limitations of Design Information

Interlock design information is limited to the vicinity of the particular interlock under consideration. The actions taken (or required) are bound by the isolated view of the equipment to be protected. They are limited in their ability to evaluate and analyze the effect of actions taken by the interlock systems on other process systems.

The structure of the C&E and ladder-logic diagrams is incompatible with identifying and analyzing such kind of relationships. Interlock narratives can incorporate this information in a piece-meal fashion (similar to the way information is presented in a narrative - one interlock at a time), however the utility of such disintegrated information is questionable due to the following facts –

- Language - The information presented in narratives is written in easy to understand language and hence lacks a structured and objective approach.
- Lack of specific information - They are prepared usually at a stage where system specific detail information is not available and lot of re-work might be required to make them consistent and compatible to receiving detail information.

4. Need for Holistic View

We will discuss two incidences where interlock activation in an equipment led to severe damage in the same or other equipment.

4.1 Incident one – Activation of Interlock in Acetylene Hydrogenation Reactor of Ethylene Unit

Ethylene cracker unit was operating with an acetylene hydrogenation reactor located between fourth and fifth stages of cracked gas compressor after Deethanizer and before the chilling train that produces tail gas (front end configuration). Tail gas, usually, is a methane-rich stream that is used as fuel in cracking furnaces. Fuel gas system for crackers also has backup from imported LPG to handle contingencies.

For this case, interlock was designed for shutting down, boxing up and depressurizing acetylene reactor to purge out reactive hydrocarbons from the catalyst bed and prevent runaway reactions that might cause a meltdown of the reactor. However, shutting down the reactor also meant stopping of tail gas flow to furnaces. Thus LPG from backup system was lined up, which changed the calorific value of fuel to the furnaces. The cracking furnace temperature controller was working without duty correction to account for change in composition and, therefore, was sluggish in taking appropriate corrective actions and closing the valves. This time gap led to increased heat input to the furnaces which resulted in a spike in the tube metal temperatures of the coils and subsequent coil damage. Thus to protect acetylene reactor, interlock actions lead to damage of cracking furnace coils due to inappropriate handling of fuel gas composition change.

In this case, the behavioral response of process system to the actions of interlock was change in fuel gas composition to the cracking furnaces. This information was not apparently or directly visible from interlock design information.

4.2 Incident two - Activation of Total Shutdown interlock of Cracking furnace of Ethylene Unit

Cracking furnaces in Ethylene units usually have two levels of interlock shutdown in case of upset, first is steam standby with isolation of hydrocarbon feed and operating the firebox in hot condition and second is the total shutdown where firebox is cooled.

Usually, first interlock is designed for faster restart after a partial shutdown by sustaining close to operating conditions inside the firebox and maintaining minimum inventory of BFW in the steam drum. The second interlock is designed to maintain integrity of the furnace and cool the firebox, which might be operating close to 1100 deg C, to a safe condition. Both interlocks have different functional requirements and are designed in a way so as to be independent of each other. However, from process point of view, initiation of first interlock also provides the necessary time lag for controlled rate heat release from the firebox, by maintaining minimum firing, before the cooling down of the firebox is initiated by total shutdown interlock.

For the case under consideration, in order to maintain minimum BFW inventory for first shutdown, the trigger was set at significantly low-level point. This resulted in activation of both interlocks with minimum time delay resulting in rapid cooling of firebox. The coke layer deposited inside the coils has very different coefficient of expansion than coil metallurgy leading to unequal contraction and subsequent mechanical failure of coils.

The incident gives the interlock interference or dependence that may be sometimes necessary due to process behavior. Design information of both interlocks, when observed in isolation, was sufficient to perform their designated tasks. However without a holistic view of looking at them, integrity of the equipment was jeopardized.

With the above-discussed instances, it is evident that sole reliance on interlock design information to provide the hindsight necessary to avoid damage to other equipment is insufficient. It requires a holistic view, encompassing all individual interlock systems, to study the interlock-process relationship for each element of the process system to better understand these relationships rather than piece-meal and isolated interlock design information.

One can argue that global interlock systems responsible for maintaining the integrity of the entire plant can take care of these relationships. We think that such a system is practically infeasible. The reasoning starts with the very definition of an interlock which states that it is the actuation of the initiating causes that triggers the interlock responses irrespective of the process. For global interlock systems to take care of interlock-process relationships, all the elements of the process system must be allowed to trigger them (along with the upsets in global utility networks).

Another approach could be to trigger specific interlock systems put in place to protect these relationships. The practicality of such an interlock system is bound to fail because of two reasons. First the investment required in putting up these extra interlock systems or triggers is huge as the permutations and combinations of initiating causes from all process elements is gigantic. Second it can lead to loss of earnings and operational prudence due to the higher cost of production for increased number of instances of unsteady plant operation due to spurious activations.

5. IRM

We propose to develop IRM as a tool to give the necessary holistic view for understanding the interlock-process relationship. IRM is an acronym for Interlock Relationship Matrix. It is the final outcome of the study for evaluation of interlock actions and is a single document that shows the effect of actions of each interlock systems on all elements of the process system.

It is prepared in a tabular format with interlock systems arranged in rows and process system elements arranged in columns. A typical IRM is shown in Table 1 for reference layout.

Table 1. Typical IRM Layout

Elements of Process System	Element 1	Element 2	Element 3	Element 4	...
Interlock Systems					
Interlock 1					
Interlock 2					
Interlock 3					
...					

The effect of each interlock action is documented in the corresponding column for process elements. For example, if the actions of Interlock 2 change the feed composition of Element 3, then "change in feed composition" is typed in cell corresponding to Interlock 2 and Element 3.

IRM should include only those effects of interlock actions that have the potential to cause significant damage to process elements. This will avoid ambiguity and redundancy of information. Populating IRM with all effects of interlock actions reduces the utility of the tool. Existing safeguards should also be considered for evaluating the potential damage to the equipment. If damage is possible after considering existing safeguards, only then IRM should be updated with the relevant effect. For earlier example, it can be analyzed whether change in feed composition can cause damage to the equipment if it is not corrected (by the absence of density or composition meter) for changed composition.

However, reference of existing safeguards, for effects that have potential for damage, can be made in IRM to ensure and document that the analysis of the interlock actions has been conducted.

6. Why IRM

There are generic approaches available for evaluating hazards for process system. They can be studied under two categories: Software Evaluation and PHA.

6.1 Software evaluation techniques

These techniques include Software Systems Hazards Analysis like Fault/Event Tree Analysis, Failure Mode and Effects Analysis, Software Failure Mode, Effects, and Criticality Analysis, etc. They are deemed to analyze and identify whether software system components' operation or failure (functioning or lack of functionality) could result in hazards for process systems. It begins when the components are designed sufficiently, and is updated as their design matures to take care of the potential hazards.

Functional difference between these analyses and IRM is that they are more concerned with the effects of software (interlock) systems on process elements in case the software goes rogue and malfunctions (lack of functionality). Moreover, the analysis of components' operation (functionality) on process systems is unable to understand interlock-process relationship as it is concerned only for segregating and avoiding intermixing interlock (safety actions) and process (control actions) systems.

Whereas, in the case of IRM, it analyzes the effects of interlock actions on process element considering that software (interlock) system has fulfilled its functionality and there is no sharing of functions from either side. It is not intended to find any design fault in software (interlock) systems but to evaluate and understand the probable responses of process system to software (interlock) system actions.

6.2 PHA

PHA techniques include operational and process hazards analysis like HAZOP, LOPA, HAZAN, etc. These techniques use guidewords in combination with process parameters to evaluate deviation scenarios from normal operation and verify the utility of existing safeguards. If the existing safeguards are not sufficient, new safeguards are suggested. Safeguards can include interlocks, process control, alarms, relief valves, operating procedures, etc.

Key difference between PHA and IRM is that PHA considers only the process parameters as the potential causes for deviation and subsequent consequences. To mitigate such consequences, PHA provides the scope for defining actions required by a safety system. However, it does not evaluate the actions of a safety system like an interlock as probable causes of deviation.

IRM fills this gap of analyses of actions of an interlock as potential causes of deviation for the process elements and gives insight to unearth latent harmful events of interlock actions.

7. Developing IRM

IRM can be developed in a structured way such that the aftereffects of interlock actions can be analyzed under the considerations of core parameters critical to the industrial setting. This approach will avoid ambiguity and subjectivity.

Simple IRM layout discussed in Table 1 can be modified to include critical parameters as shown in Table 2.

Table 2. Typical IRM Layout with critical parameters (cp)

Elements of Process System	Element 1	Element 2	Element 3	Element 4	...
Interlock Systems					
Interlock 1	*cp1:*	*cp1:*	*cp1:*	*cp1:*	...
	cp2:	*cp2:*	*cp2:*	*cp2:*	...
	cp3:	*cp3:*	*cp3:*	*cp3:*	...
	...	...	...	...	...
Interlock 2	*cp1:*	*cp1:*	*cp1:*	*cp1:*	...
	cp2:	*cp2:*	*cp2:*	*cp2:*	...
	cp3:	*cp3:*	*cp3:*	*cp3:*	...
	...	...	...	...	...
Interlock 3	*cp1:*	*cp1:*	*cp1:*	*cp1:*	...
	cp2:	*cp2:*	*cp2:*	*cp2:*	...
	cp3:	*cp3:*	*cp3:*	*cp3:*	...
	...	...	...	...	...
...	...	...	...	...	...

For the refining and petrochemical industry, critical parameters are pressure, temperature, flow and composition. Other parameters like chemical reaction, liquid level, etc. can be evaluated from these four parameters. Each action of an interlock system should be evaluated with respect to above four parameters for understanding the relationship with each process element.

We will prepare IRM for the two incidents discussed in Section 4 as case study (Table 3 and 4). These tables will form a subset of the overall IRM for the plant.

Table 3. IRM for interlocks and equipment involved in Incident 1

Elements of Process System	Cold box	Cracking Furnace
Interlock Systems		
Front End Acetylene Reactor shutdown interlock	*Pressure:* no change, cold box is boxed up	*Pressure:* no change, OSBL pressure controller is designed for backup
	Temperature: no effect, temperature will rise gradually due to ambient heating	*Temperature:* no effect
	Flow: no flow to cold box resulting in loss of tail gas and H2	*Flow:* flow of tail gas will be replaced by LPG

Elements of Process System	Cold box	Cracking Furnace
Interlock Systems		
	Composition: no effect on cold box streams composition	*Composition:* methane-rich tail gas will be replaced by propane-butane rich LPG increasingly the volumetric calorific value, no heat value correction available. Likely to increase firing duty due to sluggish temperature control of furnace coils.

Table 4. IRM for interlocks and equipment involved in Incident 2

Elements of Process System	Cracking Furnace
Interlock Systems	
First shutdown interlock of Cracking Furnace	*Pressure:* no change in firebox and furnace coil pressure
	Temperature: firebox temperature is maintained at lower level by minimum firing.
	Flow: less flue gas flow due to low firing
	Composition: NA
Second shutdown interlock of Cracking Furnace	*Pressure:* firebox pressure will rise to ambient due to unavailability of ID fan
	Temperature: firebox temperature will reduce rapidly to ambient due to fuel gas shutdown. Controlled rate heat release is required to avoid rapid cooling by maintaining minimum firing conditions (first shutdown interlock) for some duration or stopping flue gas instantly by closing the damper. However, neither the duration of firebox operation under first shutdown is sufficient, nor the damper will close to allow controlled rate heat release and avoid thermal shock. Likely damage to coils.
	Flow: no flue gas flow
	Composition: NA

8. Integration of IRM with PHA

A structured IRM described in Section 7 provides compatibility to integration with PHA techniques such as HAZOP, LOPA, etc. PHA technique like HAZOP analyzes each and every parametric deviation on process system that is applicable for an industrial setting. For refining and petrochemical industry, the parameters include pressure, temperature, level, flow, composition, chemical reaction, mixing, contaminants, special procedures, etc.

Development of IRM along with PHA study starts by identification of a subset of critical or primary parameters out of the PHA parametric set on which other parameters depend (like pressure, temperature, flow and composition). It then incorporates and analyzes actions of a safety system (interlock) as potential sources of deviation (in addition to the parametric deviations) for each process element (equipment or node) where the deviations of interlock actions are evaluated under the guidance of critical parameter subset.

9. Conclusion

Interlock design information, in the form of narratives, cause and effect diagrams or ladder/logic diagrams, is limited to the vicinity of the interlock system. It is unable to evaluate and analyze the effect of actions taken by the interlock systems on other process systems. This inability might result in irrevocable damage of equipment or potential personnel exposure to toxic materials as evident from the two case studies presented in Section 4. These damages can be avoided by understanding the interlock-process relationship to gauge probable process responses to interlock actions.

The interlock-process relationship is not visible explicitly in evaluation techniques like Software Systems Hazards Analysis (Fault/Event Tree Analysis, Failure Mode and Effects Analysis) and PHA (HAZOP, LOPA, HAZAN). These techniques intend to eliminate operational hazards caused by either malfunctioning of the software (interlock) system or deviations in process parameters. However, they do not take into consideration the probable causes that can occur in process elements after the actions of a software (interlock) system have successfully executed.

IRM (Interlock Relationship Matrix) can overcome the isolated nature of interlock design information by providing a holistic view of all interlock systems and process elements in the plant (in the form of a matrix) and evaluate critical interlock-process relationship under the guidance of critical parameters applicable for the industrial setting. It can also fill the gap in other evaluation techniques by analyzing actions of a safety system (interlock) as potential causes of deviation for the process elements. It provides efficient incorporation of new interlock systems or functions for existing process systems and vice-a-versa by giving a ready platform for analyzing the interlock actions on the process system. IRM is a managerial decision making instrument that provides valuable information to assess the likely damage to equipment and provide mitigation thereof.

Understanding interlock-process relationship is crucial to avoid jeopardizing mechanical integrity of the process system by the very interlock system installed to protect them and IRM makes it happen.

10. References

1. "Guidelines for Hazard Evaluation Procedures", 3rd edition, Center for Chemical Process Safety (CCPS), John Wiley & Sons, 2008.
2. "Lees' Loss Prevention in the Process Industries, Hazard Identification, Assessment and Control", Volume 1, 3rd edition, Sam Mannan, Elsevier Butterworth-Heinemann, 2005.

Additional references not cited:

1. "Instrumentation and Control systems", Princeton Plasma Physics Laboratory, ES&HD 5008 Section 2, Chapter 10, Revision 6, 2005.
2. "Code Walk-Through", 1984, Dunn, Robert, Software Defect Removal, McGraw-Hill, Inc.
3. "Event Trees and their Treatment on PC Computers", Limnious, N. and J.P. Jeannette, Reliability Engineering, Vol. 18, No. 3, 1987.
4. "A Guide to Hazard and Operability Studies", Chemical Industry Safety and Health Council of the Chemical Industries Association, Alembic House, London, UK.
5. "HAZOP and HAZAN", Klutz, T.V., Institution of Chemical Engineers, UK, 1986.
6. "Nuclear Surety Design Certification for Nuclear Weapon System Software and Firmware", Air Force Regulation 122-4, Department of the Air Force, 24 August 1987.
7. "Petri Net Theory and Modeling of Systems", Peterson, J.L., Prentice Hall, 1981.
8. "Procedures for Performing A Failure Mode and Effect Analysis", MIL-STD-1629A, Department of Defense, 24 Nov 1980.
9. "Generic Techniques in Reliability Assessment", Fussel, J., Noordhoff Publishing Co., Leyden, Holland, 1976.

Virtual Reality Applied to Training and Process Safety: A Semi-Industrial Boiler Case Study

Carlos André Vaz Junior
Departamento de Engenharia Química, Escola de Química, Universidade Federal do Rio de Janeiro
Centro de Tecnologia, Bloco E, Sala E-209, Ilha do Fundão
cavazjunior@eq.ufrj.br

Thiago Souza de Oliveira
Escola de Química, Universidade Federal do Rio de Janeiro
Centro de Tecnologia, Bloco E, Sala E-209, Ilha do Fundão
thiagowsousa@gmail.com

Maurício Bezerra de Souza Junior
Departamento de Engenharia Química, Escola de Química, Universidade Federal do Rio de Janeiro
Centro de Tecnologia, Bloco E, Sala E-209, Ilha do Fundão
mbsj@eq.ufrj.br

Prepared for Presentation at
American Institute of Chemical Engineers
2013 Spring Meeting
9th Global Congress on Process Safety
San Antonio, Texas
April 28 – May 1, 2013

UNPUBLISHED

Virtual Reality Applied to Training and Process Safety: A Semi-Industrial Boiler Case Study

Carlos André Vaz Junior
Departamento de Engenharia Química, Escola de Química, Universidade Federal do Rio de Janeiro
Centro de Tecnologia, Bloco E, Sala E-209, Ilha do Fundão
cavazjunior@eq.ufrj.br

Thiago Souza de Oliveira
Escola de Química, Universidade Federal do Rio de Janeiro

Maurício Bezerra de Souza Junior
Departamento de Engenharia Química, Escola de Química, Universidade Federal do Rio de Janeiro

Keywords: virtual reality, process simulation, training, 3D plant model

Abstract

The short-term time given to learn a particular task associated with the demands for higher levels of production to increase profits are potential reasons for the generation of accidents. Seeking to overcome this situation in training a large number of operators in a similar environment in which they will work, virtual applications were developed. A 3D reality, integrating phenomenological simulation developed in MATLAB, and Virtual Reality (VR) using Plant Design Management System© (PDMS), were applied in this work. Process simulators using virtual reality are applied in several areas, like flight simulators and firefighters' training. Additionally, the developed 3D plant model based tool allows different safety tasks, such as the optimum location of fire extinguishers, fire hoses and escape routes, among others. As a case-study, a semi-industrial boiler, located in Universidade Federal do Rio de Janeiro, Rio de Janeiro, Brazil, was chosen. The interaction between the user and 3D model was addressed by developing a process data interpreter, so that the model may been modified on the basis of actions taken by the operator and by the phenomenological simulation. As the developed tool improves the operator's understanding of the process, it may minimize rework, accidents and damage. Additionally, this methodology may be replicated in other industrial plants.

1. Introduction

Combine safety process, higher levels of production and environment cares is a permanent challenge to be overcome. Face this fact becomes much harder in the current days due to the increasing complexity of plant process, much ambitious quality range and the short-term time given to learn a particular task. In particular, the complexity of the chemical industry also affects in process safety.

Modern chemical processes use several electronic controls and layers of protection. However, even with all the complex components and systems, the human error still are the responsible by an expressive number of incidents in the industry [1]. It is understood as essential to the operator the full knowledge of the system in which he works. But with the growing demand for increased productivity, professionals have with this feature becomes increasingly rare.

For this scenario, industries are increasingly opting for the use of Virtual Reality (VR) training, where the VR can be defined as a computer-generated three-dimensional environment, updated in real time, allowing human interaction across multiple input-output devices [2]. Virtual reality is used in several studies that allow a greater understanding of safety and process control in many areas of engineering, such as analyzes of alarms allocation, analysis of electrical area classification, determination of routes of leakage, among others.

This work has as objective the development of a simulation tool of operational dynamics of industrial equipment, engaging in 3D graphical representation through an electronic model. To improve the study and the application, this tool was implemented to a plant containing a semi-industrial boiler, as a case study.

2. Material and Methods

In this work, the integration between process simulation and 3D graphical representation was done to provide the interaction with the user. Two specific tools were used: MATLAB and AVEVA Plant Design Management System (PDMS). The MATLAB was used to develop the phenomenological model of the boiler studied in the plant. The PDMS was used to develop the 3D model of the plant. PDMS also work as a bridge between data generated from process simulator (MATLAB) and the 3D model itself in PDMS. The system was used to train operators and safety projects.

2.1 virtual reality

The PDMS is a software engineering that has the capability of modeling three-dimensional CAD (Computer-Aided Design), allowing multiple concurrent users on a single model. The software can be used in several areas. He has a complete interactivity with the environment detailing dimensional, possessing style similar to that adopted by the software of the Windows environment.

The software has a number of benefits that among the most important, is the ability to extract a list of quantitative component containing its data of materials, weights, and other information. Importantly, such a list, depending on the size of the unit may have more than a thousand pages, which greatly increases the chance of inconsistencies, considered the possibility of creating this document manually. Thus, the software allows an important gain in terms of time, eliminating delays, the automated generation of various documents and reports, plans and two-dimensional views and isometric piping, and other documents. The user can check the information contained in the 3D model and compare them to equivalent 2D designs.

2.2 phenomenological simulation

The boiler operation was simulated in MATLAB software. To provide this, was used the model created by ÅSTRÖM & BELL [3], who presented a dynamic model for a boiler, based on basic principles of conservation of mass and energy. Justifies the adoption of this model here for its simplicity and because it has been validated with real data by the authors [3].

2.3 case study: boiler studied

The usability of the tool developed in this work has been tested and applied in a small boiler, model CV-500-VDM. This boiler is located in a pilot plant, which in turn is in the Utilities Central Laboratory of Chemical Engineering, Escola de Química, Universidade Federal do Rio de Janeiro, Rio de Janeiro, Brazil. In the Figure 1 will be shown a picture taken from the plant. All the related equipment was identified in it.

Figure 1: Studied pilot plant containing the boiler simulated

3. Results and discussion

3.1 process simulation

Using the MATLAB and the developed program was possible integrate "process simulation" and "virtual reality". To verify if the data is trustful was produced some graphics and analyze their behavior through steps given in some input variables. The Figure 2 represents the temperature behavior after some different steps given in heat flow, Q.

Figure 2: Temperature behavior after steps in heat, Q

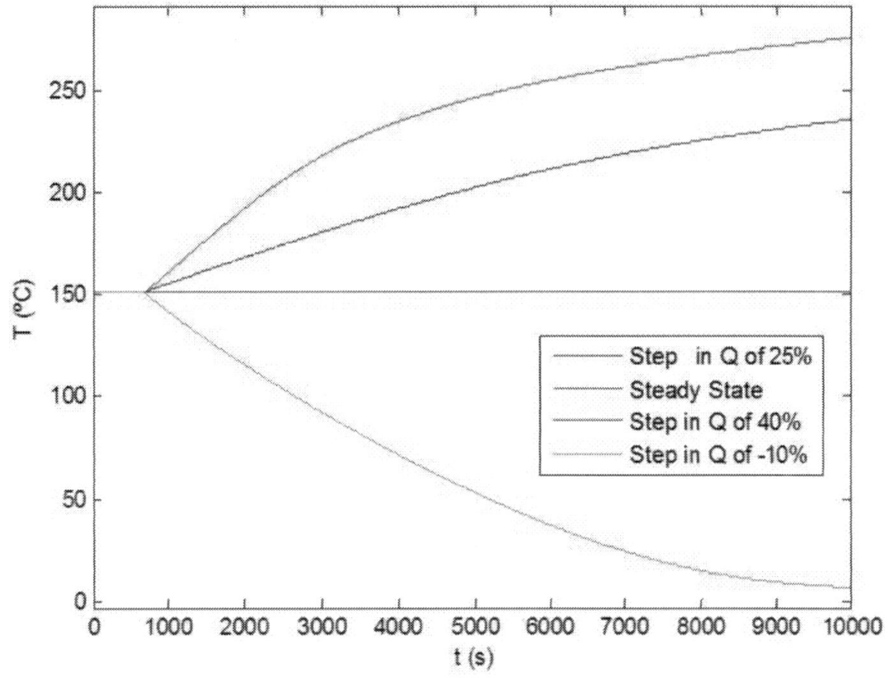

It is possible to see that if the heat is increased, the temperature consequently increase as well. By other side, if the heat is decreased, the temperature also decreases, following the dynamics of the boiler from energy and global mass balance. In the Figure 3, a similar result is obtained, where a step is given in Q and the pressure is observed after it.

Figure 3: Pressure behavior after steps in heat, Q

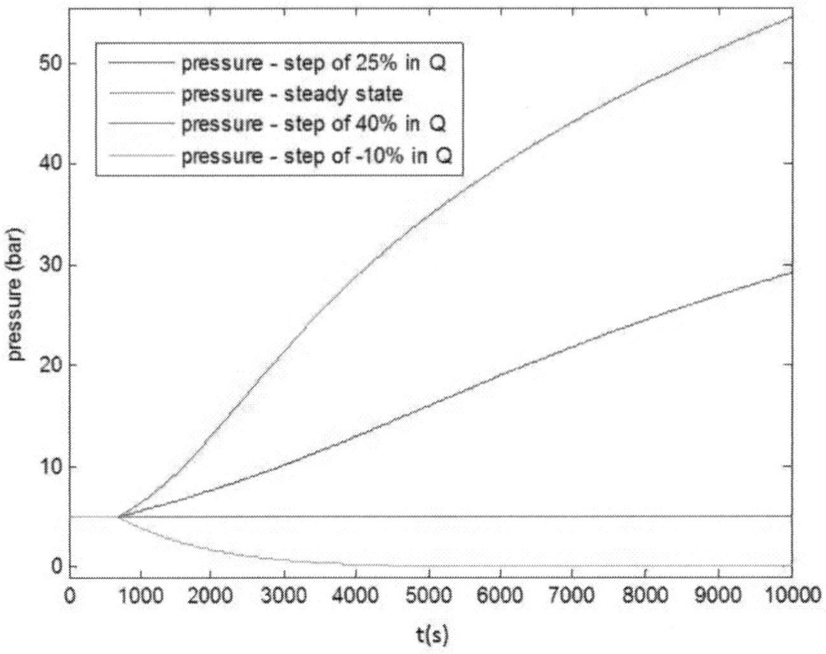

3.2 virtual reality tool

The developed tool has in summary three main functions as results. The first one is provide the integration between the simulations taken in MATLAB. The 3D model shows to the operator if the plant is operating in normal situation or if there is some equipment in alarm status. The model can shows data tables too. In the Figure 4, is possible to see some equipment analyzed by the integration system and showing if they are in normal operation (green), in maintenance (gray), in stand by (yellow) or in alarm (red). Beyond that, it is possible to see relevant data process, like pressure, temperature and level, directly in the model.

The second function is providing the training of the operator. In this work, the training was applied to "boiler starting". During the training the operator can see instructions and complementary texts. The 3D model change through the actions taken, interacting with the operator. The user has the free movement possibility during the training. In the Figure 5 is possible to see the virtual operator during the training. The water level and pump status are shown in the Figure 5 too.

Figure 4: Equipment analyzed by the integration tool

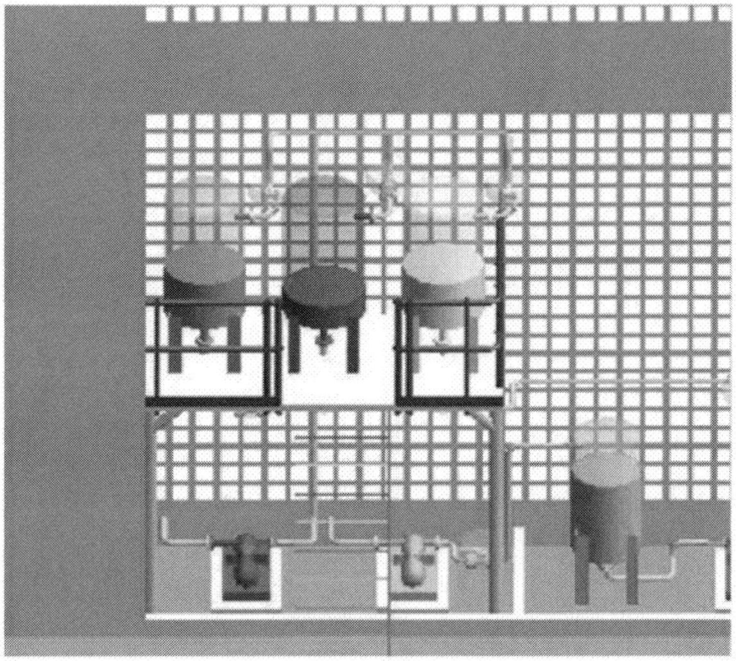

Figure 5: Model during the training

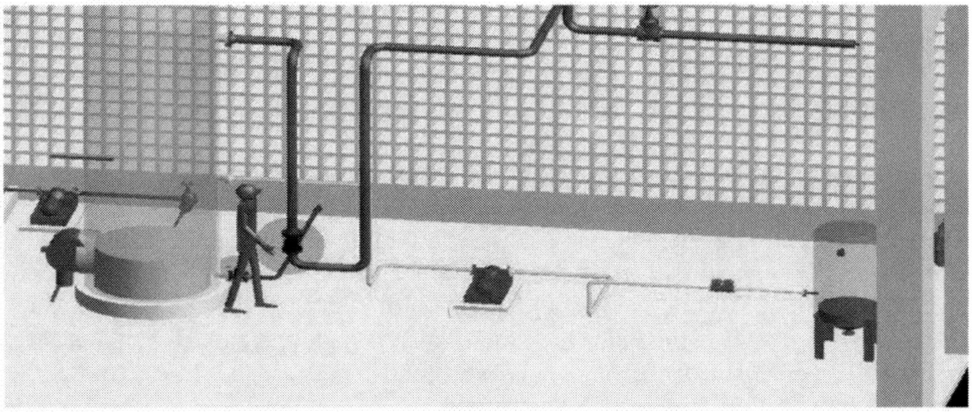

The third function is work as a "support tool", helping the safety analyzes in the plant. The safety analyzes can be applied during project of new process. The virtual reality can be used to design escape routes and emergency signaling. An example of the escape route chooses is showed in the Figure 6.

Figure 6: Virtual tool used to know the best exit to be taken in a case of emergency

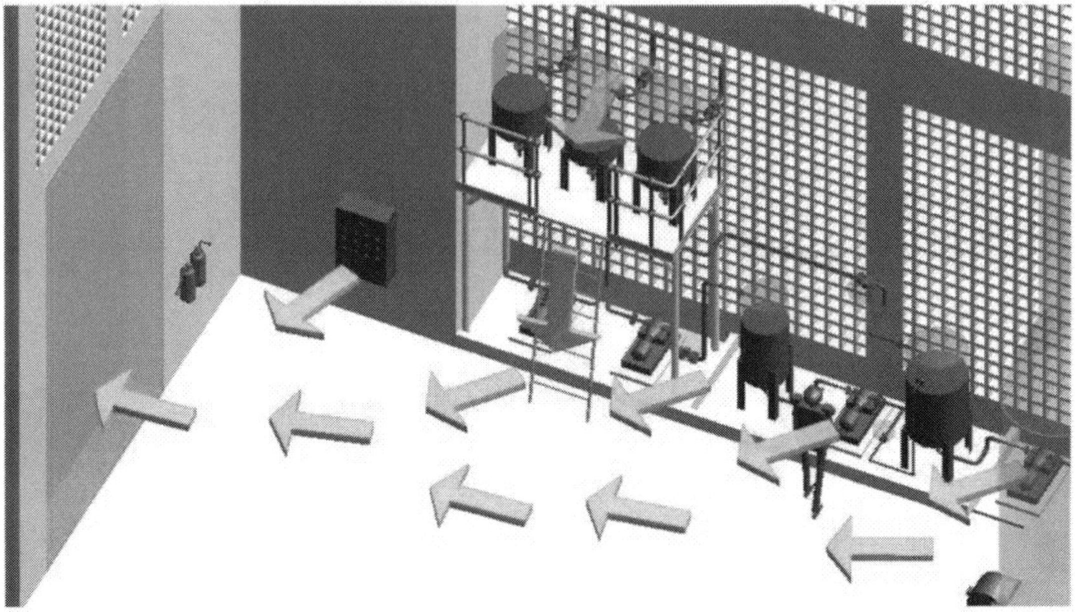

4. Conclusion

The integration of virtual reality and process simulation allow plant operators acquire knowledge about real process. This solution offers low cost and it's safer than use real equipment during operation training. In other words, performing training in the virtual reality tool developed, the user can get information about the plant without actually going to it.

About the development of a three-dimensional supervisory system prototype, it is concluded that it is valid and interesting for training purposes. With the training tool, it is possible to obtain many advantages, due to the operator interaction with the virtual three-dimensional model. So, it is concluded the development of a similar system for all industrial plants can facilitate the training of new operators, facilitating the determination of operations and/or maintenance actions in the plan. Through this, it is possible study the best way to access the site without the need to perform a complete analyzes of the activity at the site to be developed. Furthermore, the safety analyzes can be very useful for the operation team.

About study in general, it can be concluded the use of virtual reality, primarily for training purposes and in addition as a support tool to the operation unit, has much contributing to healthy plant operability. The implementation of virtual reality is becoming an increasingly viable option in view of the development of systems software and hardware capable of supporting the tools developed so that the cost of such equipment is not high as it would at the last decade of previous millennium.

5. References

[1] Lasala, K.P; Filgueiras, L.V.L; Gigley, H.; Ullman, R., *Designing Systems for Reliable Human Performance*, IEEE Tutorial Video, 1997.

[2] Pamplona Jr, A. R. S., *Hidrogeradora Virtual: Utilização de Técnicas de Realidade Virtual Desktop para o Estudo de uma Unidade Hidrelétrica de Energia*, Dissertação de Mestrado, Programa de Pós Graduação em Engenharia Elétrica, Centro Tecnológico, Universidade Federal do Pará, 2006.

[3] Åström, K. J.; Bell, R.D., *Drum Boiler Dynamics*. Automatica, v.36, n.3, p. 363-378, 2000.

9TH GLOBAL CONGRESS ON PROCESS SAFETY

Investigations on the influence of different inert gases on the laminar flame propagation in lycopodium/air mixtures

Hannes Kern
Montanuniversität Leoben
Franz-Josef Strasse 18, 8700 Leoben
hannes.kern@unileoben.ac.at

Bruno Wimmer, Harald Raupenstrauch
Montanuniversität Leoben
Franz-Josef Strasse 18, 8700 Leoben
bruno.wimmer@stud.unileoben.ac.at, harald.raupenstrauch@unileoben.ac.at

Prepared for Presentation at
American Institute of Chemical Engineers
2013 Spring Meeting
9th Global Congress on Process Safety
San Antonio, Texas
April 28 – May 1, 2013

UNPUBLISHED

Investigations on the influence of different inert gases on the laminar flame propagation in lycopodium/air mixtures

Hannes Kern
Montanuniversität Leoben
Franz-Josef Strasse 18, 8700 Leoben
hannes.kern@unileoben.ac.at

Bruno Wimmer, Harald Raupenstrauch
Montanuniversität Leoben
Franz-Josef Strasse 18, 8700 Leoben
bruno.wimmer@stud.unileoben.ac.at, harald.raupenstrauch@unileoben.ac.at

Keywords: dust explosion, flame velocity, limiting oxygen concentration

Abstract

The laminar flame velocity is a key parameter in the description of explosions. The determination of the flame velocity (laminar or turbulent) of dust/air flames is relatively complex due to the characteristics of dusts in general. In context of investigations on the limiting oxygen concentration of different powders, the influence of the oxygen concentration and different inert gases on the laminar flame propagation in lycopodium/air mixtures was investigated. In the course of this study, a new apparatus for the determination of the flame velocity was developed. The apparatus mainly consists of a vertical duct with a diameter of 140mm that is split in three main sections. The dust input system lies in the head section of the apparatus and is realized by a small conveyor screw. The bottom section of the apparatus houses the electrical spark ignition system and a gas inlet system. The gas inlet creates a steady flow of a chosen gas. The flame propagation is detected by IR photodiodes along the 2m long main section of the apparatus. The dust concentration is measured by light extinction. The determination of the flame velocity was carried out at dust concentrations of 100g/m^3, 200g/m^3 and 300g/m^3 as well as around the lower explosion level of lycopodium. The oxygen concentration was varied in steps of 1Vol.-% from 21Vol.-% down to the limit where no ignition could be achieved. To investigate the influence of the inert gas on the flame propagation carbon dioxide and nitrogen were used as inert gases. The results of the recent studies are used for further studies on the limiting oxygen concentration of different dusts and the validation of a new CFD model for the simulation of dust explosions in large vessels.

1. Introduction

Dust explosion research mostly relies on sum parameters like the minimun ignition energy (MIE), the maximum explosion pressure or the limiting oxygen concentration (LOC) for the description of explosion processes and material characterization. For mathematical modeling these sum parameters show a lot of uncertainties due to the influence of experimental design. Therefore parameters like the laminar flame velocity are commonly used for the description of the combustion process itself. For combustible gases the determination of the flame velocity is relatively simple compared to combustible dusts. The biggest challenge concerning dusts is the creation of a homogeneous dust cloud with low turbulence.

In the course of this experimental study about the influence of the oxygen concentration on the flame velocity of lycopodium/air flames a new testing device was designed. With this apparatus it is possible to measure the flame speed within a 2m combustion tube, a new method for the calculation of the laminar flame speed from the results is yet to be developed. For current research we relied on present methods. Even if the laminar flame velocity can't yet be measured directly with this device, the influence of changing atmospheric conditions can be observed very well.

2. Experimental investigations

2.1 Method

Various apparatus for the measurement of the flame speed of dust/air mixtures are found in literature [1]. The testing apparatus used for current research consists of three major parts. As shown in Figure 1 the main sections of the apparatus are the top section, the main explosion tube and the bottom section. All elements are built out of a Plexiglas (PMMA) tube with a diameter of 140mm and a wall thickness of 5mm.

The head element houses the dust input system and is equipped with a flame arrestor (Figure 2). The dust input is carried out by a screw conveyor that can be regulated by the operating voltage. The input system allows the generation of dust concentrations from 50g/m³ to 500g/m³ (Lycopodium).

The bottom section contains a gas inlet system and the ignition source. The ignition source is an electric spark igniter (~200mJ), consisting of two 3mm steel electrodes and a high voltage generator. The gas inlet is designed as a circular tube with multiple nozzles to generate a homogeneous gas flow along the main section of the apparatus. The 2000mm long main section of the apparatus is equipped with 5 photodiodes to measure the flame propagation along the tube. The first diode is mounted 10cm above the ignition point, further diodes are placed every 40cm.

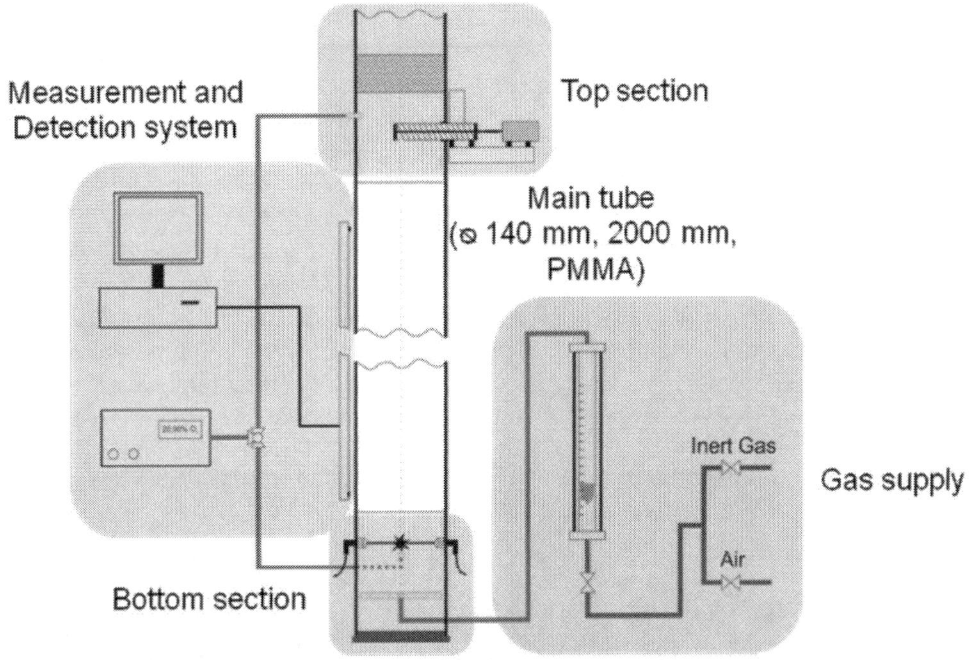

Figure 1: Experimental setup for the determination of the flame velocity

Two of the photodiodes are used to monitor the dust concentration in the tube before ignition by the extinction of IR light emitted by IR diodes. The emitter diodes are turned off shortly before the ignition by the electric spark.

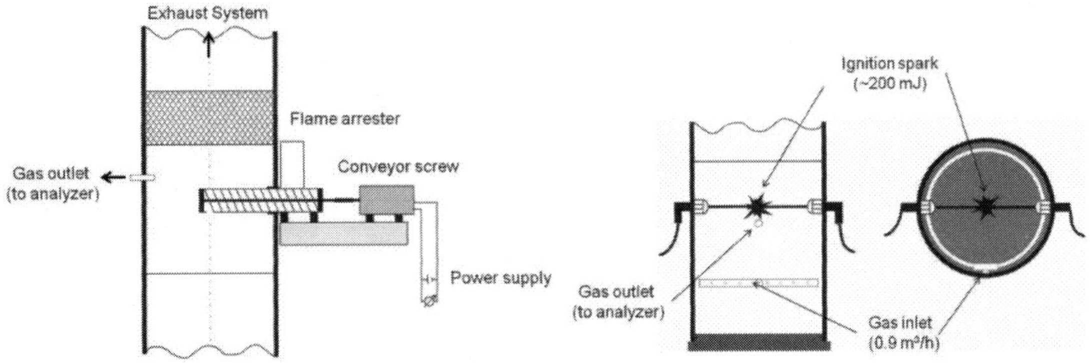

Figure 2: Top section (left) and bottom section (right) of the experimental setup

In the measuring mode a gas flow of 0,015m³/min is provided through the bottom to realize different atmospheric compositions. The dust concentration is observed throughout IR extinction. The oxygen concentration inside the tube is measured by an oxygen monitor at the head section of the apparatus. When the extinction reaches a constant level along the tube, the emitter diodes are switched off and the ignition source is activated.

2.2 Analysis of flame propagation

For dust explosion research in general, laminar flame velocity is a relevant parameter. Different methods for the determination of the laminar flame velocity are known, like using burners to create laminar dust flames or tube methods like described above. The data sets produced from all of these methods are more or less deviating from laminar conditions due to the experimental conditions. Figure 3 shows a sequence of pictures taken during the measurement of the flame velocity at a dust concentration of 300g/m³ and an oxygen content of about 15Vol.-% using nitrogen as inert gas. The picture shows a more or less turbulent combustion close to the ignition point and a relatively closed flame shape at around 800mm. At lower concentrations the flame tends to be more closed throughout the whole measuring distance. The flame therefore shows a more laminar character at lower concentrations. The turbulence effects may have occurred due to the falling particles themselves.

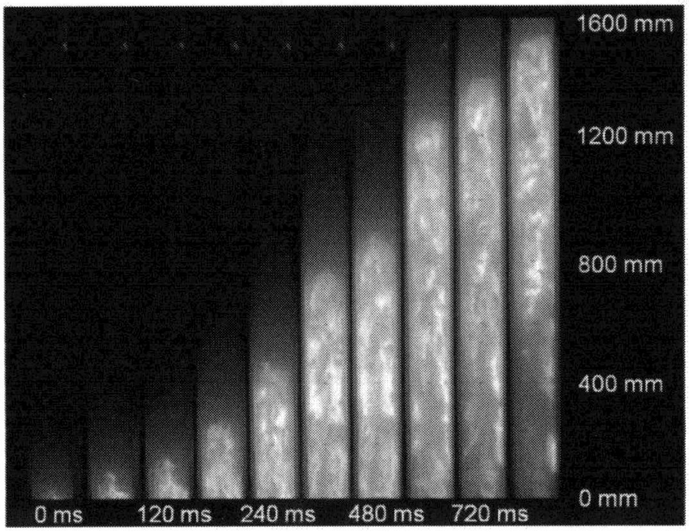

Figure 3: Flame propagation within the testing device

Andrews and Bradley [2] described a method for the calculation of the laminar flame velocity out of relations regarding shape and surface of the flame. Andrews and Bradley's approach (1) was taken into account for further investigations. A video analysis system was installed to determine the flame geometry.

$$S_L = \frac{A'}{A_{fl}} \cdot \left(S_{fl} - u\right) \qquad (1)$$

For a lycopodium concentration of 50g/m³, the flame surface (A_{fl}) and the area cross section (A') were interpolated and calculated by a Matlab code. The measured flame velocity (S_{fl}) was 0.62m/s. Calculations with Andrews and Brandly's relation resulted in a laminar flame velocity (S_L) of around 0.20m/s.

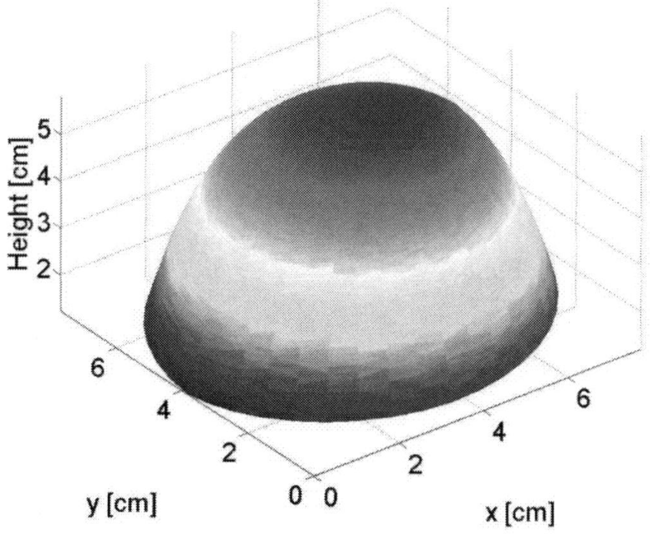

Figure 4: Flame shape at 50g/m³, surface area around 150cm²

Figure 4 shows an example for the flame shape of a 50g/m³ lycopodium/air flame. Compared to the results described by Eckhoff [1], the measured flame velocity is well in the range of velocities measured in other apparatus. For concentrations of around 50g/m³ the flame became very unstable in some measurements, so that higher concentrations were chosen for further research on the influence of the oxygen concentration. Due to the clear connection between actual and laminar burning velocity, only the actual flame speed was measured further on, to avoid deviations from the flame shape analysis.

2.3 Results

Measurements of the flame velocity (S_{fl}) have been carried out at three different dust concentrations. In order to keep the flame stable, concentrations of 100g/m³, 200g/m² and 300g/m³ were investigated. The oxygen concentration was lowered by nitrogen and carbon dioxide as inert gases. Measuring sequences started at ambient conditions (~21Vol.-% oxygen) and were lowered in steps of 1Vol.-% down to the point where no ignition could be reached.

Table 1: Flammability limits

	N_2	CO_2
Dust conc. [g/m³]	cO_2 [Vol.-%]	
100	12.0	15.0
200	12.0	15.0
300	13.0	15.5

As described in Table 1, no ignition could be achieved below 12Vol.-% during nitrogen inerting and 15Vol.-% when carbon dioxide was used. Bartknecht [3] described a relation between the oxygen concentration and the required ignition energy for lycopodium/air mixtures. Regarding the observed ignition limits during the experiments, the results show very good accordance with Bartknecht's values. For an oxygen content of around 12Vol.-% Bartknecht estimated an ignition energy between 150 and 250mJ which is well in the range of the ignition source used for the experiments. Figure 5 shows the development of the flame velocity for nitrogen inerting.

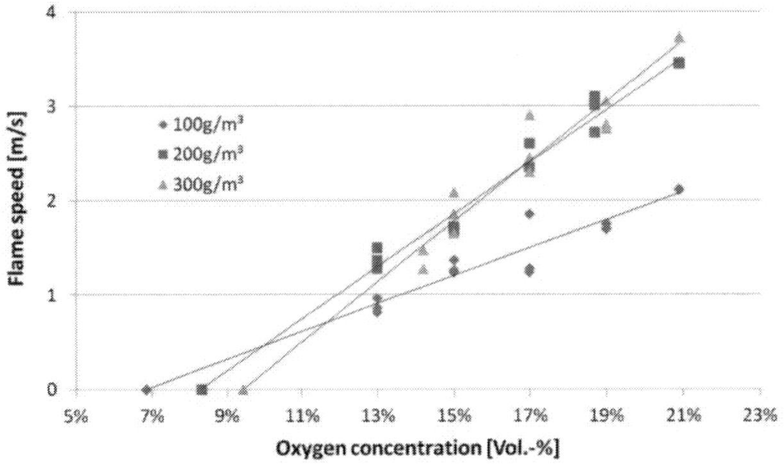

Figure 5: Development of flame speed during nitrogen inerting

At ambient conditions the flame velocity increases with the dust concentration. The increase of the flame velocity in the range from 100g/m³ to 200g/m³ is significantly bigger than from 200g/m³ to 300g/m³. This effect can be explained by stoichiometry [4].

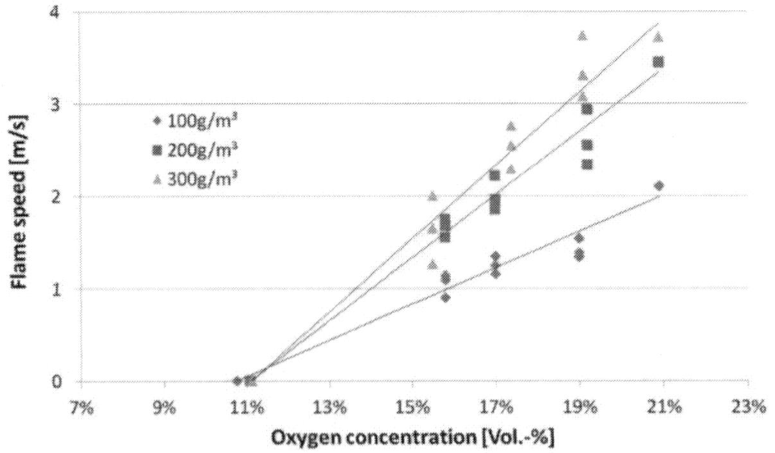

Figure 6: Development of flame speed during carbon dioxide inerting

At lower dust concentrations the combustible material is the limiting component for combustion while at higher dust concentrations oxygen becomes the limiting factor, which is why the flame velocity nearly stagnates. For all dust concentrations the flame velocity decreases with lower oxygen concentrations. The flame velocity drops faster at higher concentrations than at 100g/m³. That is also true for carbon dioxide inerting, as it can be seen in Figure 6. At Figure 5 the lines for 300g/m³ and 200g/m³ intersect at around 17Vol.-%. This is presumably due to problems with the gas measuring system and could not be observed at the carbon dioxide experiments. The values for zero flame speed are estimated.

As mentioned above values for zero flame speed were estimated to be able to compare possible results with the limiting oxygen concentrations of lycopodium. The extrapolated values are given in Table 2.

Table 2: Estimates limiting oxygen concentrations

	N_2	CO_2
Dust conc. [g/m³]	cO_2 [Vol.-%]	
100	6.8	10.8
200	8.4	11.1
300	9.4	11.1
average	8.2	11.0

Siwek [5] found a relation between the limiting oxygen concentrations (LOC) measured during nitrogen inerting and the LOC values for carbon dioxide inerting (2).

$$LOC_{CO2} = 1.36 \cdot LOC_{N2} \quad (2)$$

Due to the described problems with measurement attempts were made to recalculate the LOC for nitrogen inerting out of the measured values for carbon dioxide inerting. The estimation resulted in a value of 8Vol.-% for nitrogen inerting which is highly consistent with the 7.5Vol.-% published in BIA Report 12/97 [6] and the average value that was measured.

3. Conclusions

During the investigations on the flame speed of lycopodium/air mixtures a new apparatus for the determination of the flame velocity was developed. The results from the method regarding laminar flame speed showed good accordance with published values at ambient conditions. With the present experimental setup it was also possible to determine the influence of varying atmospheric conditions on the flame propagation in a vertical duct. Carbon dioxide and nitrogen were used as inert gases. Due to the relatively low ignition energy provided by the spark igniter, the flame propagation around the known limiting oxygen concentrations for lycopodium could not be reached. Extrapolations on the limiting oxygen concentrations based on the measured data have shown good accordance with values measured in the ISO 20L sphere. In addition to the the influence of inert gases on the flame propagation other deviations from atmospheric conditions will be investigated in future research. The results from the current experimental work are used for the validation of a new CFD model for the calculation of dust explosions in complex geometries.

4. References

[1] Eckhoff R. K. (3rd edition) (2003) Dust Explosions in the Process Industries. Gulf Professional Publishing

[2] G.E. Andrews, D. Bradley (1972), Determination of burning velocities: A critical review, Combustion and Flame, Vol. 18/1, Elsevier 1972

[3] Bartknecht W. (1993), Explosionsschutz – Grundlagen und Anwendung. Springer – Verlag

[4] Crowl D. A. (2003), Understanding Explosions, Center for Chemical Process Safety (CCPS) of the American Institute of Chemical Engineers, 1st edition, Wiley-AIChE, New York 2003

[5] Siwek R. (1989), Zuverlässige Bestimmung explosionsschutztechnischer Kenngrößen in der 20l-Laborapparatur, VDI Berichte 701, S215-262

[6] BIA – Report 12/97, Brenn- und Explosionskenngrößen von Stäuben, Hauptverband der gewerblichen Berufsgenossenschaften, November 1997

(44al) Fire Protection and Life Safety Issues in Semiconductor and Related Facilities

Jonathan Eisenberg, Rolf Jensen & Associates, Framingham, MA

Semiconductor and related process facilities contain numerous fire and explosion hazards that impact life safety and operations. This presentation discusses the major fire protection and life safety design aspects for such facilities, including specific requirements contained in the International Building & Fire Codes, as well as applicable NFPA codes and standards. Project examples will be used to illustrate how these aspects are addressed in facility and process design.

The presentation will include the following:

Location and design of Use Group H-5 occupancies – Major IBC Chapter 4, IFC Chapter 18, and NFPA 318 fire protection and life safety requirements, including vertical and horizontal location, occupancy separation, code approach for associated mechanical spaces including interstitial levels, electrical classification, ventilation, fire suppression (overhead and local), fire and gas/vapor detection, and emergency alarms.

Location and design of Use Group H-2, H-3 and H-4 storage areas – Building and fire code requirements for vertical and horizontal location, occupancy separation, electrical classification, ventilation, fire suppression, fire and gas/vapor detection, and spill control/secondary containment. Approaches and ideas on the number, location and function of flammable liquid, flammable and pyrophoric gas, and highly toxic gas storage areas, including piping and delivery philosophy.

Special suppression and detection – Fire protection system approaches for hazards found in a clean room/fab area, eg. wet bench chemistry involving flammable liquids, and processing that uses hydrogen and/or silane mixtures. Discussion of local suppression options, as well as specialty gas and flame detection. We will give our thoughts and experiences on how these systems interact with the rest of the building and the responding fire department.

Industrial Issues – Hydrogen detection; combustible dust hazards; NEC Article 500 electrical classification; explosion protection; and several examples from our industrial/chemical project experience

Licensing and permitting – Discussion of issues faced by owners of fab and clean room facilities that store and use hazardous materials, where the hazards are not clearly understood outside of the semi-conductor industry.

See more of this Session: Poster Session
See more of this Group/Topical: Global Congress on Process Safety

Lower Flammability Limits of Mixtures of Hydrogen and Methane at Atmospheric and Sub-atmospheric Pressures

Hai Le
Mary Kay O'Connor Process Safety Center, Artie McFerrin Department of Chemical Engineering, Texas A&M University
Room 200, Jack E. Brown Building, Texas A&M University, 3122 TAMU College Station, TX 77843-3122
haile@tamu.edu

Yi Liu, M. Sam Mannan
Mary Kay O'Connor Process Safety Center, Texas A&M University
Room 200, Jack E. Brown Building, Texas A&M University, 3122 TAMU
College Station, TX 77843-3122
mannan@tamu.edu, liuyi.chemical@mail.che.tamu.edu

Prepared for Presentation at
American Institute of Chemical Engineers
2013 Spring Meeting
9th Global Congress on Process Safety
San Antonio, Texas
April 28 – May 1, 2013

UNPUBLISHED

Lower Flammability Limits of Mixtures of Hydrogen and Methane at Atmospheric and Sub-atmospheric Pressures

Hai Le
Mary Kay O'Connor Process Safety Center, Artie McFerrin Department of Chemical Engineering, Texas A&M University
Room 200, Jack E. Brown Building, Texas A&M University, 3122 TAMU College Station, TX 77843-3122
haile@tamu.edu

Yi Liu, M. Sam Mannan
Mary Kay O'Connor Process Safety Center, Texas A&M University

Keywords: lower flammability limit, hydrogen, methane, pressure

Abstract

The knowledge of the LFLs of hydrogen mixtures with hydrocarbons is essential since hydrogen is present in various mixtures with hydrocarbons in many processes and applications. The lower flammability limits of various mixtures of hydrogen and methane in air were determined experimentally in a closed cylindrical stainless steel vessel (inner diameter 10.22 cm, length 100 cm) with upward flame propagation at room temperature (20 °C) and initial pressures of 1.0 atm, 0.5 atm, and 0.1 atm. It was found that the LFLs of mixtures of the same hydrogen fuel fractions did not change much when the initial pressure decreased from 1.0 atm to 0.1 atm. And pressure has more impact on the LFLs when there is more hydrogen in the fuel mixture. The application of Le Chatelier formula to predict the LFLs of the mixtures was also evaluated. Overall, Le Chatelier formula can predict the LFLs of mixtures of hydrogen and methane very well at all initial pressures.

1. Introduction

Lower flammability limit (LFL) is an important parameter used in the prevention of fire/explosion when handling combustible gas mixtures. By definition, LFL is the minimum concentration of gas in air capable of propagating flame upon ignition;[1] in other words, if the concentration of a gas is smaller than its LFL, it is no longer flammable. Therefore, LFL is widely used in the calculation of the level of inerting needed to prevent the formation of flammable mixtures.[2] In these calculations and other fire risk assessments, it is advised to select and use the proper value of LFL for the operating condition since LFL is affected by pressure, temperature, direction of flame propagation and other conditions.[1]

The literature has a vast number of LFLs of various gases determined with different apparatuses at both atmospheric and non-atmospheric conditions, e.g. high temperatures, pressures, etc. LFLs of pure gases are more readily available than LFLs of mixtures of gases. Whenever LFLs of mixtures of gases are not available, methods have been developed to predict them. One of the widely used methods is Le Chatelier formula. It is an empirical mixing rule to predict the LFL of a mixture of gases as below: [3]

$$LFL_{mix} = \frac{1}{\sum_i^n \frac{y_i}{LFL_i}}$$

where LFL_{mix} is the LFL of the mixture, y_i is the mole fraction and LFL_i is the LFL of the ith component. For mixtures at non-atmospheric condition (e.g. high pressure, temperature) the LFLs of the components at the same condition should be used to accurately estimate the LFLs of the mixtures.

The knowledge of the LFLs of hydrogen mixtures with hydrocarbons is important since hydrogen is present in various mixtures with hydrocarbons in many processes and applications. Examples of such mixtures are refinery gas streams which contain various percentages of hydrogen (up to 90%) and hydrocarbons (methane, ethane, C_3, C_{4+}.).[4] Hydrogen is also added to natural gas, consisting mostly methane, to reduce CO, CO_2 and NO_x emissions[5-7], improve flame stability,[6, 7] and extend the lean operating limit of spark ignited engines.[8] It is critical to understanding the LFLs of these hydrogen mixtures in order to handle them safely.

Reference to the literature shows that studies on the LFLs of mixtures of hydrogen and hydrocarbons at non-atmospheric conditions, particularly sub-atmospheric pressure conditions, are very limited. Therefore, to enhance our understanding about the LFLs of hydrogen mixtures, this study focuses on investigating the LFLs of various mixtures of hydrogen and methane at atmospheric and sub-atmospheric pressures. Specifically, the LFLs of mixtures of hydrogen and methane were determined experimentally at room temperature (20 °C) and initial pressures of 1.0 atm, 0.5 atm and 0.1 atm. All experiments were carried out in a closed cylindrical vessel with upward flame propagation. The application of Le Chatelier formula to predict the LFLs of mixtures of hydrogen and methane at atmospheric and sub-atmospheric pressures was also verified using the LFLs of pure hydrogen and methane generated by the same experimental method and conditions.

2. Experimental setup

The experimental apparatus and procedure is illustrated schematically in Figure 1 and briefly described below.

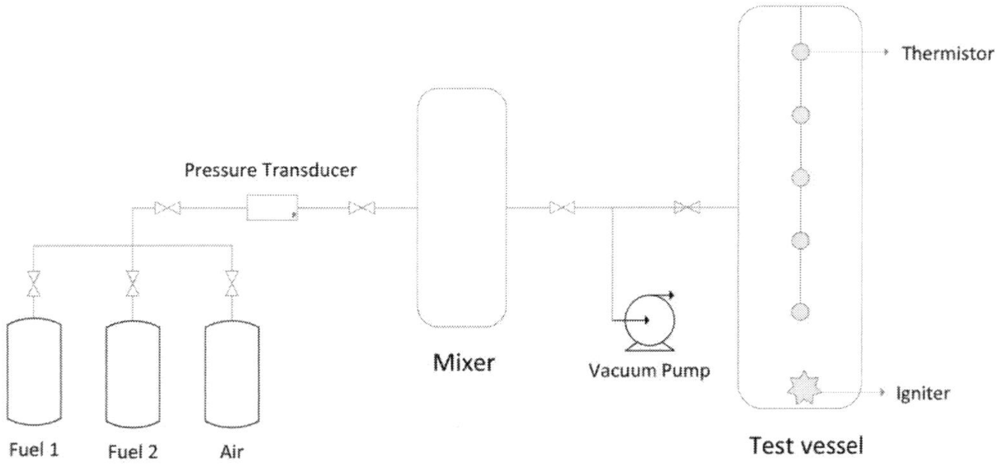

Figure 1: Experimental Setup

The LFLs of mixtures of hydrogen and methane were determined in a test vessel which was a closed stainless steel cylindrical vessel whose design was similar to that developed and used by the U.S. Bureau of Mines.[1] The vessel was 100 cm long with the inner diameter of 10.12 cm and outer diameter of 11.43cm. The igniter was installed at the bottom of the vessel and was capable of providing 10 J of energy consistently. The pressure inside the test vessel was measured by a pressure transducer placed at the top of the vessel. A set of 5 thermistors was positioned along the center axis of the test vessel to detect flame propagation. The distance between the top thermistor and the igniter was 75 cm. Further details about the configuration of the test vessel and its components can be found in our previous published work.[9]

High purity fuels and air from pressurized cylinders were loaded into the mixing vessel. The amount of the fuel and air were determined based on the partial pressure basis. The fuels and air were thoroughly mixed in a mixer which was a stainless steel vessel containing a Teflon block which slid along the vessel when it rotated lengthwise to enhance mixing. After the mixing, the test mixture was permitted to flow to the test vessel at the desired initial pressure measured by the pressure transducer. Then the mixture was ignited from the bottom, and the flame propagation was detected by the 5 thermistors inside the test vessel. Signals from the thermistors were obtained by a Keithley data acquisition card (Keithley® KPCI-3102, 225 signals per second at 0.05% accuracy) installed in a desktop computer, and the experimental process was controlled by a LabView® program. All experiments were carried out at room temperature (20 ^{o}C) and initial pressure of 1.0 atm, 0.5 atm and 0.1 atm. After each experiment, combustion products were purged from the test vessel which was then flushed with nitrogen and vacuumed to be ready for the next experiment.

A mixture is considered flammable if it can propagate flame to the top thermistor upon ignition;[9, 10] this behavior was defined as continuous flame propagation.[9, 10] For each mixture composition, 10 experiments were repeated and the probability of continuous flame propagation was recorded. Between two concentration values different by one experimental step size, one has the probability of continuous flame propagation equal or less than 50%, the other has the probability more than 50%. The former was chosen as the LFL, unless the former's probability is 0%, the later was chosen as the LFL. [9, 10]

3. Experimental Results and Discussion

3.1 Lower flammability limits (LFLs) of mixtures of hydrogen and methane at atmospheric and sub-atmospheric pressures

The experimentally obtained LFLs of various mixtures of hydrogen with methane at room temperature (20 °C) and initial pressure of 1.0 atm, 0.5 atm and 0.1 atm are shown in Figure 2. Overall, the LFLs of the mixtures of the same hydrogen fuel fractions did not change much when the initial pressure decreased from 1.0 atm to 0.1 atm. This is understandable since the LFLs of pure hydrogen and methane also changed very little with the initial pressure. When the fraction of hydrogen in the fuel mixture increased from 0% to 100%, the LFLs of the mixtures showed larger deviation from each other under the change of pressure from 1.0 atm to 0.1 atm. This suggests that pressure has more impact on mixtures of hydrogen and methane when there is more hydrogen in the mixture. However the impact is not significant since the degree of changes of the LFLs is small. It is worth to note that the LFLs of the mixtures was the lowest at 0.5 atm which means that there is an increased risk of fire/explosion when the initial pressure is reduced from 1.0 atm to 0.5 atm. Thus, more attention and precautions are recommended when handling mixtures of hydrogen and methane at pressures lower than 1.0 atm.

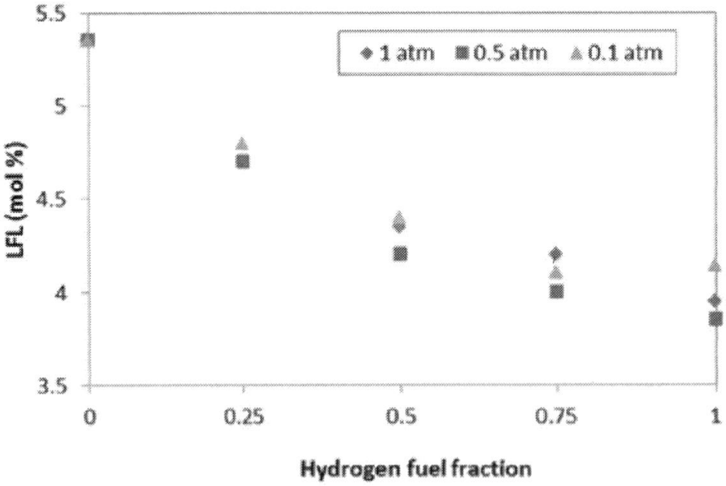

Figure 2: LFLs of mixtures of hydrogen and methane at atmospheric and sub-atmospheric pressures and room temperature (20 °C)

3.2 The application of Le Chatelier formula to the LFLs of mixtures of hydrogen and methane at atmospheric and sub-atmospheric pressures

The LFLs of mixtures of hydrogen and methane were calculated based on Le Chatelier formula using the LFLs of pure hydrogen and methane determined experimentally with the same experimental setup and conditions. Comparisons between the calculated LFLs and experimental LFLs are presented in Figure 3, Figure 4, and Figure 5 for 1.0 atm, 0.5 atm and 0.1 atm respectively. In general, Le Chatelier formula can predict the LFLs of the mixtures very well at all initial pressure conditions; in other words, the prediction of Le Chatelier formula for hydrogen and methane mixtures is not affected by changes in the pressure from 1.0 atm to 0.1 atm. The maximum deviation between experimental and calculated LFLs is 7.0 %. And the deviation is the largest when the fraction of hydrogen in the fuel mixture is 0.5.

The application of Le Chatelier formula to the upper flammability limits (UFLs) of the mixtures of hydrogen and methane will be evaluated in the future when we complete the determination of the UFLs of the mixtures experimentally.

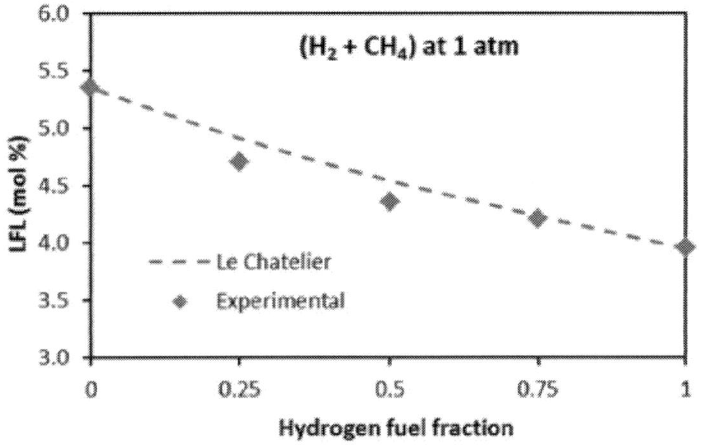

Figure 3: LFLs of mixtures of hydrogen and methane at room temperature and 1.0 atm (experimental and calculated by Le Chatelier formula)

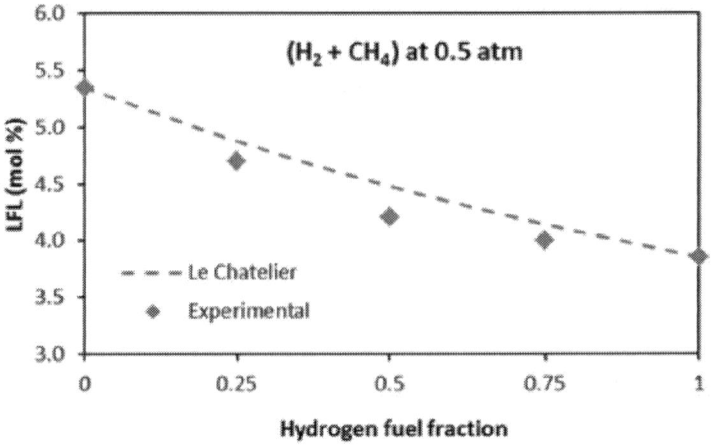

Figure 4: LFLs of mixtures of hydrogen and methane at room temperature and 0.5 atm (experimental and calculated by Le Chatelier formula)

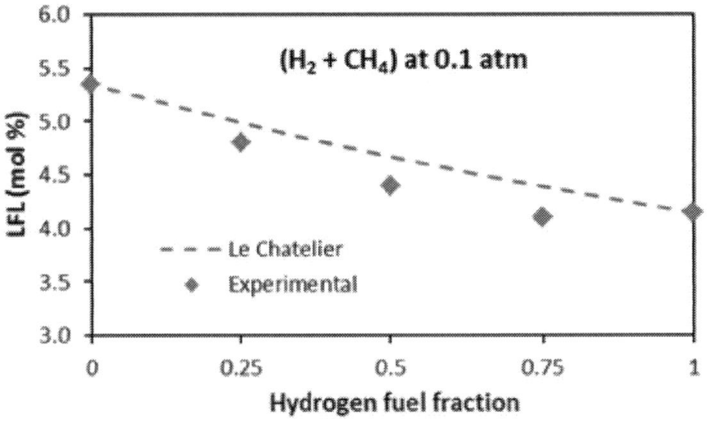

Figure 5: LFLs of mixtures of hydrogen and methane at room temperature and 0.1 atm (experimental and calculated by Le Chatelier formula)

4. Conclusion

The lower flammability limits (LFLs) of mixtures of hydrogen and methane in air were determined experimentally at room temperature and initial pressure of 1.0 atm, 0.5 atm and 0.1 atm. The application of Le Chatelier formula to predict the LFLs of the mixtures was also evaluated. It was found that:

- The LFLs of mixtures of the same hydrogen fuel fractions did not change much when the initial pressure decreased from 1.0 atm to 0.1 atm. Pressure has more impact on the LFLs of the mixtures when there is more hydrogen in the fuel mixture. The LFLs of the mixtures were the lowest at 0.5 atm which implies an increased risk of fire/explosion when the initial pressure decreases from 1.0 atm to 0.5 atm.

- Le Chatelier formula can predict the LFLs of mixtures of hydrogen and methane very well, and the prediction of Le Chatelier formula is not affected by changes in the pressure from 1.0 atm to 0.1 atm.

5. Acknowledgement

This research was sponsored by the Mary Kay O'Connor Process Safety Center, Artie McFerrin Department of Chemical Engineering, Texas A&M University.

6. References

[1] Coward, H.F. and G.W. Jones, *Limits of flammability of gases and vapors*, in *Bureau of Mines Bulletin*, 1952.

[2] Crowl, D.A. and J.F. Louvar, *Chemical Process Safety: Fundamentals with Applications, Second Edition*. 2001: Prentice Hall.

[3] Mashuga, C.V., *Determination of the combustion behavior for pure components and mixtures using a 20L sphere*, in *Chemical Engineering*, 1999: Michigan.

[4] Kaldis, S.P., G.C. Kapantaidakis, and G.P. Sakellaropoulos, *Simulation of multi-component gas separation in a hollow fiber membrane by orthogonal collocation — hydrogen recovery from refinery gases.* Journal of Membrane Science, 2000. **173**: p. 61-71.

[5] TerMaath, C.Y., et al., *Emissions reduction benefits from hydrogen addition to midsize gas turbine feedstocks.* Int J Hydrogen Energy, 2006. **31**(9): p. 1147-1158.

[6] Guo, H., et al., *The effect of hydrogen addition on flammability limit and NOx emission in ultra-lean counterflow CH4/air premixed flames.* Proc Combust Inst, 2005. **30**(1): p. 303-311.

[7] Hawkes, E.R. and J.H. Chen, *Direct numerical simulation of hydrogen-enriched lean premixed methane–air flames.* Combust Flame, 2004. **138**(3): p. 242-258.

[8] Bell, S.R. and M. Gupta, *Extension of the lean operating limit for natural gas fueling of a spark ignited engine using hydrogen blending.* Combust Sci Technol, 1997. **123**(1-6): p. 23-48.

[9] Le, H., S. Nayak, and M.S. Mannan, *Upper Flammability Limits of Hydrogen and Light Hydrocarbons In Air at Sub-atmospheric Pressures.* Ind. Eng. Chem. Res, 2012. **51**: p. 9396-9402.

[10] Wong, W.K., *Measurement of flammability in a closed cylindrical vessel with thermal criterion*, in *Artie McFerrin Department of Chemical Engineering, Texas A&M University,* 2006.

Informatics Rich Integrated Chemical Safety Risk Assessment (ICSRA) Platform

Muthukumarasamy Karthikeyan1, Renu Vyas2, HV Prasad3

1 Centre of Excellence in Scientific Computing and Digital Information Resource Centre, CSIR-National Chemical Laboratory, Pune, India, CSIR-National Chemical Laboratory, Pune, India
2 Chemical Engineering and Process Development, CSIR-National Chemical Laboratory, Pune, India, CSIR-National Chemical Laboratory, Pune, India
3 Chemical Engineering and Process Development, CSIR-Indian Institute of Chemical Technology, Hyderabad, Habsiguda, Hyderabad-500 607, India India

Abstract

Inherently Safer Practices (ISP) is a systematic innovative scientific methodology integrated with the process, design and operations for facilitating the improvement of process safety at all stages of process life cycle. ISP constitutes intensification (minimization), substitution, attenuation, limitation of effects and simplification as procedural options for implementation in process plants.

For this study we conducted literature data search focused on recent domestic and international incidents where the primary cause was related to chemical reactivity and runaway reactions; Several parameters were taken into consideration including chemical manufacturing (i.e., raw material storage, chemical processing, and product storage) and other industrial activities involving bulk chemicals, such as storage/distribution, waste processing, and petroleum refining.

For purposes of the incident search, only reactive incidents that caused serious consequences would be examined. Source of data: Four decades of data collection involve searching over 40 data sources, focusing on incidents where the primary cause was related to chemical reactivity. For the purposes of the investigation and analysis, an "incident" was defined as a sudden event involving an uncontrolled chemical reaction—with significant increases in temperature, pressure, and/or gas evolution etc.,

The proposed experimental and simulation platform will serve as a repository for plant-specific information, and facilitates the classification and ranking of Indian chemical industry - for the first time - through a Risk index.

Introduction

Computational and Simulation of Industrial chemical reactions to predict runaway reactions and conditions as safety measures. To our knowledge, no systematic studies on inherent safety are yet reported in India. The studies comprise important research areas e.g., process evaluation, estimation of reactivity, hazard identification and evaluation, assessment and prevention of thermal runaways, early fault diagnosis etc. The computational risk assessment platform intends to collect new chemical process safety and risk data, and plant-specific incident data from participating industries across the world for model building and simulation.

To integrate and include a dynamic interactive information platform, viz., chemical safety risk ranking and incident reporting system, by the application of High performance Computing (Cloud, Grid and Distributed systems) for simulation, modeling and response for use by Govt., Industry, Emergency Responders, Research and academic communities, and the public. Upgradation of Thermo chemical Laboratories to assess the proposed/existing industry from the context of implementing ISP, i.e., towards safer process chemistry including selection of solvents, safer process routes, safer process operating conditions for preventing Runaways, early detection of runaways, evaluation/selection of process and designs to handle lesser inventories etc.

Methodology

The methodology devised comprises recognition of the hazards and risks posed by the process, and efforts to reduce or control them to the lowest practical levels, keeping in view the principle objectives of the business. Here we studied data based on literature to identify and collect chemically significant industrially important data pertaining to physico-chemical, biological and toxicological profiles of chemicals (reactants, reagents, catalysts, solvents, products, side products) for their suitability and reusability in simulation and modelling. This includes development of suitable algorithms, chemical reaction component descriptors and methods to simulate experimental conditions and validation through experiments to improve the predictions of risk assessment and prevention of disaster in industrial conditions (Lab-scale, pilot plant and large scale manufacturing).

This process also includes mapping known literatures (Publications, patents and reports) to the existing knowledge for interactive risk assessment of chemicals and incidents in industrial environment. The computational approaches include classical methods, semi-empirical methods, and quantum methods to determine properties at the molecular level. We are developing a computational problem-solving platform and these tools, data and methods to build ICSRA a cloud-computing infrastructure for integrated chemical safety risk assessment of chemicals manufactured, imported and exported from world chemical sources. Proper documentation and accurate predictive models on chemicals of Indian origin would facilitate compliance with other global compliance on chemicals being exported especially in REACH, EPA, FDA and other global federal regulation context.

Development of QSPR (Quantitative Structure Property Relationships) models to relate macroscopic behaviour to molecular properties with detailed molecular computations to obtain thermo chemical data for reactive materials and to predict calorimetric data based on molecular properties. These open source based tools are integrated to this platform (ChemInfoCloud) includes identification and assessment of hazards due to chemical reactivity and risk assessment.

For this systematic study we considered major potential chemical group (CG) which are frequently encountered in industrial reagents like Organic peroxides, Organic Nitrites, Unsaturated Hydrocarbons, Epoxides, Hydrides and Hydrogen, Metal acetalides etc. Efforts to study and model major Chemical Reaction Types (CRT) include Decomposition, Polymerization, Pyrophoric, Peroxide Former, Water sensitive and Oxidation reactions. Molecular computing would help to predict heat of formation, electronic energy, core-core repulsion, dipole, no. of filled levels, ionization potential, molecular weight, topological, electronic and three dimensional descriptors for critical components of reactants, products, solvents.

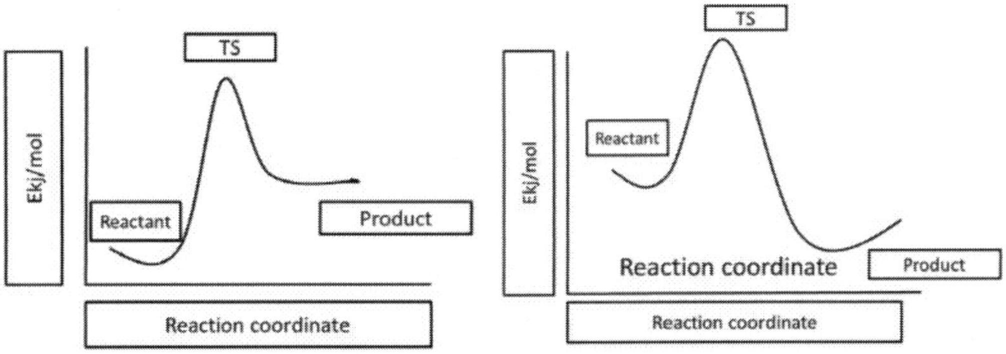

Figure-1: Computational Modeling generic types of chemical reactions along the reaction co-ordinates

Theoretical study of transition states of chemical reactions, computation of activation energies and experimental validation to predict that one reaction pathway is favoured over another reaction pathway. In order to predict the type of polymerization and runaway reactions enthalpy of reactions and kinetics was calculated using the semi-empirical quantum methods, molecular modelling and dynamics simulations of reactants under simulated diverse experimental conditions of Industrial Chemical Processes (ICP), including homogeneous and heterogeneous catalysis.

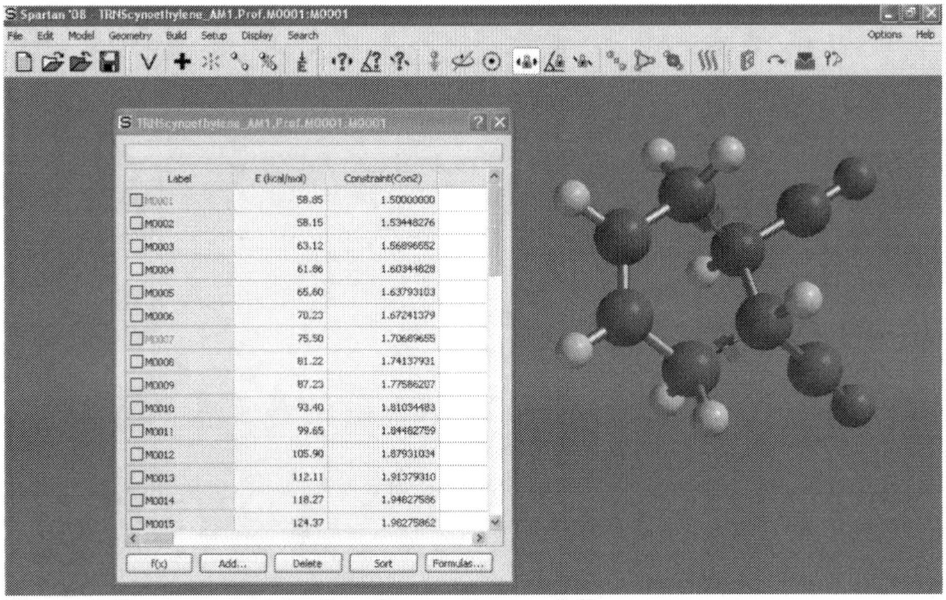

Figure 2: Spartan user interface for reaction modeling

Reactants **Transition State** **Product**

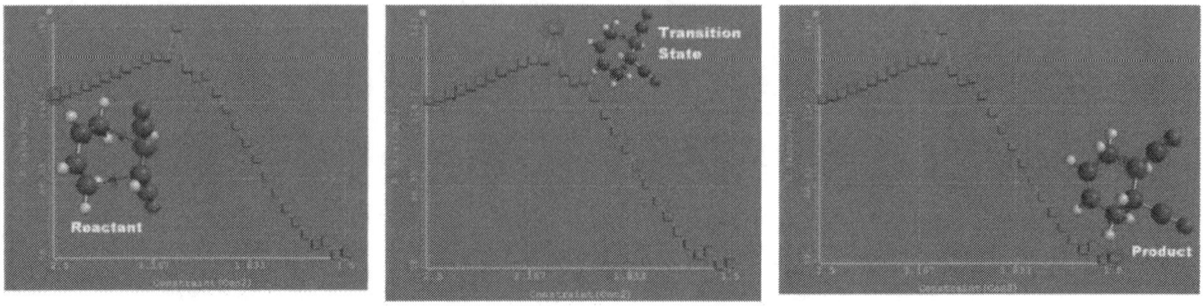

Figure: 3 Computational modelling of organic reactions (Exothermic)

Reactants	Transition State	Product
Heat of Formation: **489.651** kJ/mol Solvation Energy: SM5.4/A -9.308	Heat of Formation: **595.583** kJ/mol Solvation Energy : SM5.4/A -21.665	Heat of Formation: **246.229** kJ/mol

Table:1 Sample data of predicted Heat of formation for reactants, transition state and product

E (TS-Reactant)	E (Reactant-Product) ** heat generated!
E1=595.583-489.65 =105.933 kJ/mol =25.32 kcal/mol	E2=489.65-246.229 =243.421 kJ/mol =58.19 kcal/mol

Table:2 Computed energy difference

Digital mapping of incidental data concerning process safety and their availability to decision makers would help effective risk assessment and rapid response during unexpected events. During the presentation the detailed roadmap, current scenario and future direction of the chemical safety risk assessment platform will be discussed and the input from delegates for this meeting would be considered for integration into the tool if required.

References

1) Organic Reactions Series. John Wiley & Sons
2) http://downloads.wavefun.com/Spartan10Manual.pdf

IT Enablement For Effective Implementation And Monitoring of Process Safety Management

Ajay Kumar/ Pawan Mittal
Reliance Industries Limited
Jamnagar
Ajay.kumar@ril.com/ Pawan.mittal@ril.com

IT Enablement for various PSM elements

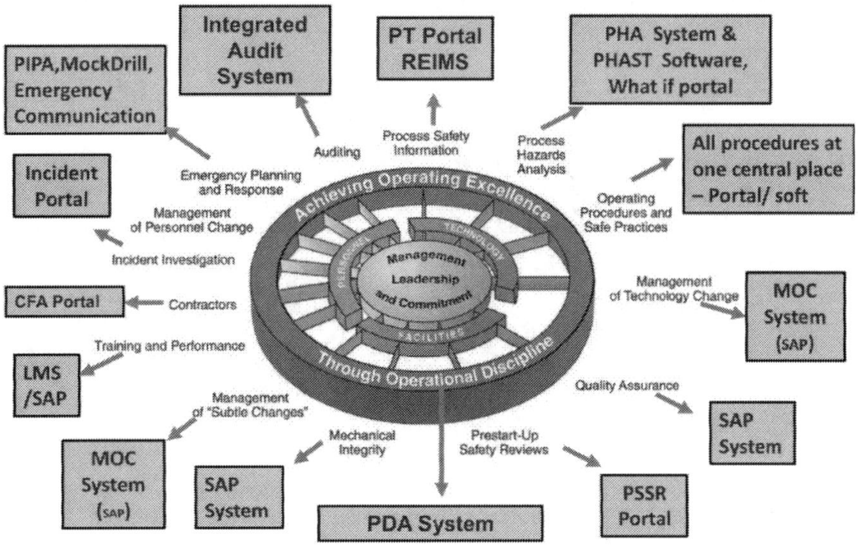

Governance and Control : Corrective Action Tracking System

Advantages of IT Enablement of PSM Elements

- **Process Technology (PT)**
  - ✓ Single repository for storage of all PT related documents
  - ✓ Easy access to PT documents to all concerned.
  - ✓ Easy updation of PT documents

- **Process Hazard Analysis (PHA)**
  - ✓ Automation of all 11 steps of PHA
  - ✓ Linkage of all relevant information required for PHA
  - ✓ Easy Tracking of PHA study progress
  - ✓ Consequence Modelling using PHAST software
  - ✓ Easy recordkeeping of PHA study reports & effective networking
  - ✓ Recommendation management in CAT system

Advantages of IT Enablement of PSM Elements

- **Management of Change Technology & Subtle (MOC – T & S)**
  - ✓ Single repository for recordkeeping of all changes
  - ✓ Controls provided for Hazard Analysis (Pre Screening PHA Checklist)
  - ✓ Tracking & recordkeeping of approval process
  - ✓ Tracking of PT document updating while MOC close out

- **Pre Start up Safety Review (PSSR)**
  - ✓ Automation of PSSR Checklist
  - ✓ Control provided for steps of PSSR e.g. Self Assessment , Level II & Level I
  - ✓ Pending checklist point tracking in CAT system
  - ✓ Permanent records of PSSR points & its close out status

Advantages of IT Enablement of PSM Elements

- **Learning Management System (LMS)**
  - ✓ Online nomination for the training program is available
  - ✓ Online training modules & validation sheets are available
  - ✓ Easy retrieval & tracking of employees' training records
  - ✓ Tracking of the pending training modules for planning training programs
- **Incident Reporting Investigation & Learning**
  - ✓ Reporting of incident in iRIL system.
  - ✓ Online team formation.
  - ✓ RCA tool interface
  - ✓ Uploading of investigation report in system.
  - ✓ Interface with CAT system for Recommendation till closure
  - ✓ LFI sharing to all concerned
  - ✓ Incident Investigation Quality
  - ✓ SAP based incident management system is under development.

Advantages of IT Enablement of PSM Elements

- **PIPA (Pre Incident Planning & Assessment)**
  - ✓ PIPA software used for management of emergency scenarios
  - ✓ Each plant selected credible scenarios & developed plan with the help of PIPA software
  - ✓ Rehearsal of PIPA scenarios & plans is ensured through emergency mock drills
- **Contractor Field Audits (CFA)**
  - ✓ Online scheduling & recordkeeping of field audits for various contractor agencies
  - ✓ Effective records of audit observations e.g. % Good Citizens observed during audit round, Score achieved by contractor agencies
  - ✓ Overall Tracking of Severity Index (Plant/Site wise)
  - ✓ Tracking of Contractor Field Audits e.g. Scope Vs. Completed

Advantages of IT Enablement of PSM Elements

- **Progressive Disciplinary Action System (PDA)**
  - ✓ Developed for recordkeeping of violation / negligence of safety rules & procedures
  - ✓ Easy retrieval of records
  - ✓ Uniformity in record maintenance of PDA at all RIL sites.
- **Corrective Action Tracking (CAT)**
  - ✓ Interface with iRIL, Audit Management, PSSR, Mock Drill
  - ✓ Tracking of Recommendation at one location
  - ✓ Mail Alert for Pending and Overdue Recommendation

Thanks

(44at) Sulphur Dust Explosions an Overview

Karthick Deivasagayam, Loss Prevention, SNC Lavalin Inc, Calgary, AB, Canada

This paper will focus on the sulphur dust, its explosion energy, available models and capabilities of modeling its explosion energy, the equipment susceptible for sulphur dust explosion, various prevention& mitigation strategies of sulphur dust explosion, and industrial applied technology to sulphur dust explosion mitigation. This paper will begin with increase of solid sulphur production from the refineries, its present forms of products and case studies.

The technologies and methodologies for sulphur dust explosion, frequency data, consequence modeling, prevention and mitigation of sulphur dust as applied in a present day sulphur plant.

Discussion for future developments will be discussed in this paper as well.

Simplified Uncertainty Analysis of LOPA Results

Raymond "Randy" Freeman
S&PP Consulting
12303 Lake Shore Ridge
Houston, TX 77041

rafree@yahoo.com

ABSTRACT

Layer of Protection Analysis (LOPA) is a semi-quantitative risk evaluation tool. The methodology is widely used in setting Safety Integrity Level (SIL) targets for Safety Instrumented Systems (SIS). The methodology is often used to evaluate the adequacy of existing protective systems against corporate risk targets. The LOPA method is based on the use of conservative estimates of the failure probability of protective systems.

Freeman previously presented a method for the quantification of uncertainty in a LOPA study based on the application of variance contribution analysis (VCA) techniques. The use of VCA to evaluate the uncertainty in a LOPA study requires the evaluation of the sensitivity of the LOPA equation to changes in each of the variables. Freeman used direct analytical computation of the partial derivatives of the LOPA equation for each of the uncertain variables. While mathematically correct, this use of a partial derivative analysis during an actual LOPA review is unlikely. Faster and easier methods are needed if LOPA uncertainty analysis is to be applied by practicing engineers.

This paper explores the use of approximations to compute the required uncertainty information without the use of the partial derivative computations. Numerical approximations (perturbation methods) are found to allow for the estimate of the partial derivatives with a maximum error of 16%. When evaluating simple safety systems designed per IEC 61511, the error is found to be no more than 10% for SIL-1 or SIL-2 capable systems. Compared to direct Monte Carlo simulation methods, the numerical approximation method is much more efficient. Monte Carlo simulation takes approximately 10,000 function evaluations to arrive at a result. For the base line problem studied, the numerical approximation method will take only 5 function evaluations.

BACKGROUND

Layer of Protection Analysis (LOPA) [Ref. 1] is a widely used method for the evaluation of the frequency of undesired events. LOPA is one of the methods proposed in the ANSI/ISA 84.00.01 standard [Ref. 2] for the determination of Safety Integrity Level (SIL) targets. The SIL target [Ref. 3] is used by process and instrument engineers for specification of equipment to reduce the frequency of undesired events such as overpressures, spills from tanks, releases to flare systems, fires, etc.

Freeman [Ref. 4, 5] developed a method for evaluating the uncertainty in a LOPA study using a mathematical method called Variance Contribution Analysis. The VCA method outlined in Freeman's paper require the evaluation of the partial derivative of the basic LOPA equation for each of the uncertain variables. There are several problems with the application of this method in an actual LOPA study, including:

1. Most engineers would not be comfortable evaluating a series of partial derivative functions while leading a LOPA study.
2. The LOPA team leader must develop the uncertainty parameters for each initiating event and each IPL during the analysis. A set of pre-calculated tabulations of the uncertainty parameters would be needed to speed the analysis.
3. The application of Monte Carlo simulation as used in Freeman's paper is not a topic familiar to most process engineers - nor is Monte Carlo simulation a simple methodology to implement for a real LOPA scenario.

The fundamental issue with the use of VCA in an actual LOPA study is the evaluation of the variance contribution from each uncertain parameter. This paper is directed at rectifying these deficiencies in the application of VCA to the evaluation of uncertainty in a LOPA study.

REVIEW OF LOPA METHODOLOGY

A LOPA [Ref. 1] review is a semi-quantitative risk evaluation of undesired events in a process. The LOPA begins with the definition of the set of undesired events (fires, explosions, toxic chemical releases, etc.). Next, potential safeguards (such as pressure relief valves, high/low level or pressure safety shutdown systems, emergency shutdown systems, etc) are reviewed to determine if there are independent protection layers (IPLs) that will prevent the undesired event from occurring. The frequency of the undesired event is then computed by combining the frequency of initiating event with the combined probability of failure on demand (PFD) of the IPLs as:

$$\text{Frequency of undesired event} = (\text{Initiating event freq}) \times \prod_{i=1}^{n} \text{PFD of the IPLs} \qquad \text{(Eqn. 1)}$$

Or in more compact form:

$$\text{Freq} = (\text{IE}) \times \prod_{i=1}^{n} P_i \qquad \text{(Eqn. 2)}$$

Where:

Freq = Frequency of the undesired event being studied by the LOPA method
IE = Initiating event frequency or the starting event frequency for the scenario under study

P_i = Probability of Failure on Demand (PFD) of an Independent Protection Layer (IPL) counted as reducing the frequency of the undesired event.
i = Subscript indicating the ith IPL
n = Total number of IPLs counted in the LOPA study as satisfying the criteria for an IPL

Equation 1 is referred to as the basic LOPA equation. The basic LOPA equation of Eqn. 2 can be restated as:

$$Freq = (IE)\ P_1\ P_2\ P_3\ \ldots\ P_n \qquad \text{(Eqn. 3)}$$

All of the variables in Eqn. 3 are random variables with inherent uncertainty. The resulting calculated frequency of the LOPA scenario will therefore be uncertain.

APPLICATION OF VARIANCE CONTRIBUTION ANALYSIS TO THE BASIC LOPA EQUATION

Using variance contribution analysis, the mean and variance for the LOPA equation are found directly per the work of Freeman [Ref. 4, 5]. For the expression:

$$Y = F(x_i) \qquad \text{(Eqn. 4)}$$

Where:

Y = dependent random variable computed from x_i
$F(x_i)$ = function that converts (maps) random variable x_i into Y

The mean of Y can be estimated using the following approximation:

$$E(Y) = F[E(x_i)] \qquad \text{(Eqn. 5)}$$

Where:

E(Y) = expected value of random variable Y = mean of Y
$E(x_i)$ = expected value of random variable x_i = mean of x_i

The variance of Y can likewise be estimated as:

$$V(Y) = \sum_{i=1}^{n}\left[\frac{\partial Y}{\partial x_i}\right]^2 V(x_i) \qquad \text{(Eqn. 6)}$$

Where:

$V(Y)$ = variance of random variable Y
$V(x_i)$ = variance of random variable x_i

The application of the VCA method to the evaluation of the uncertainty in a LOPA calculation requires the evaluation of the partial derivative term in Eqn. 6. Previously, Freeman computed the partial derivative directly using the methods of calculus and partial differential equations. However, the direct analytical evaluation of each of the terms in a LOPA expression is time-consuming and prone to error.

The basic LOPA equation of Eqn. 3 is restated as:

$$Freq = (IE)\ P_1\ P_2\ P_3\ \ldots\ P_n \qquad \text{(Eqn. 3)}$$

Note that Freq is first order with respect to each random variable. Therefore, the partial derivative of the Frequency (Freq) with respect to a random variable will become a function of the form:

$$\frac{\partial Freq}{\partial P_1} = (IE)\ P_2\ P_3\ \ldots\ P_n \qquad \text{(Eqn. 7)}$$

The partial derivative may be evaluated numerically rather than analytically. The numerical approximation to the partial derivative may be found by simple evaluation of the Freq near the mean values of each random variable. For all cases to be reviewed in this paper, a 10 percent perturbation about the expected value of the random variable will be used to complete the numerical approximation calculations for the partial derivative term in a VCA study.

For the simple case where all of the IPLs are independent and share no common inspection or repair times, the numerical evaluation of the partial derivative of Freq with respect to the variable P_1 in Eqn. 3 may be estimated as:

$$\frac{\partial Freq}{\partial P_1} \approx \frac{\Delta Freq}{\Delta P_1} = \{[E(IE)\ 1.1E(P_1)\ E\ (P_2)\ldots E(P_n)] - [E(IE)\ E(P_1)\ E(P_2)\ldots E(P_n)]\}/0.1E(P_1)$$
(Eqn. 8)

Where:
$E(P_i)$ = expected value of random variable P_i = mean of P_i

The numerical estimate reduces to:

$$\frac{\partial Freq}{\partial P_1} \approx \frac{\Delta Freq}{\Delta P_1} = \left(\frac{1.1-1}{0.1}\right)\ [E(IE)\ E(P_2)\ E(P_3)\ \ldots\ E(P_n)] \qquad \text{(Eqn. 9)}$$

$$\frac{\partial Freq}{\partial P_1} \approx \frac{\Delta Freq}{\Delta P_1} = [E(IE)\ E(P_2) \ldots E(P_n)] \quad \text{(Eqn. 10)}$$

Equation 10 is exactly the same as the analytical expression presented in Eqn 7. Thus, for simple LOPA expressions involving multiple independent protection layers, the result obtained by a numerical estimate will be the same as that obtained by the analytic expression obtained from the partial differentiation of the scenario basic LOPA equation. No error will be introduced by using the numerical approximation method to evaluate the partial derivative.

There are occasions in a LOPA review where the resulting basic LOPA equation may become non-linear. For example, if two or more IPLs are implemented in a safety instrumented system, the resulting LOPA equation becomes non-linear with respect to inspection frequency. For example a SIL-2 interlock design might involve two redundant instrumentation systems installed in parallel to reduce the frequency of an undesired event.

Appendix A to this paper evaluates the cases up to, and including, a SIL-4 capable design with 4 safety interlocks all tested on the same test period. The results of the analysis of Appendix A are summarized in Table 1. For most practical LOPA problems, a numerical analysis of the partial derivative term (Eqn. 6) will produce a result within 16% of that obtained using analytical methods.

Table 1.
Error Introduced in LOPA Uncertainty Calculations by the Use of Numerical Estimates of the Partial Derivative Term

Case	Type of System	Error In LOPA Uncertainty* %
n = 1	SIL-1 Design or BPCS interlock	0
n = 2	SIL-2 Design or two SIL-1 interlocks	5
n = 3	SIL-3 Design or three SIL-1 interlocks	10
n = 4	SIL-4 Design or four SIL-1 interlocks	16

* Error = (Numerical Approximation Estimate – Analytical)/Analytical

The second problem with the application of the VCA method to LOPA uncertainty analysis is the need for pre-calculated tabulations of the mean and variance of the initiating event frequencies and IPL PFDs. Appendix B of this paper presents the method used as well as the results of calculating the needed uncertainty parameters using the data presented in the CCPS LOPA book. These tabulations may be used directly in a LOPA uncertainty study. The application of these simplifications is best illustrated using an example.

EXAMPLE PROBLEM

Previously, Freeman [Ref. 4, 5] presented an example problem to demonstrate the direct application of VCA to the uncertainty in the LOPA calculations. The same example is used for comparison in this work.

Consider the process system shown in Figure 1. The bottom sump level in separation tower (T-200) is controlled by a level controller (LIC-101) that regulates a control valve LCV-102. The tower is normally operating at a pressure of 500 psig and is designed for a maximum pressure of 750 psig. The liquid flow from T-200 goes to an intermediate vessel (Hold Drum D-205) prior to additional processing. D-205 is designed for a maximum pressure of 100 psig and normally operates at 75 psig.

Of concern is the failure of the level control system in T-200 to maintain the liquid seal at the bottom of the tower. If the liquid level in the T-200 sump is lost due to an instrumentation failure, the Hold Drum could be exposed to pressures as great as 500 psig. A pressure of 500 psig is a factor of 5 above the design pressure of the Hold Drum. Catastrophic failure of the Hold Drum is probable.

This problem is generically referred to as a high-low pressure interface problem or as a liquid blow-through case. Two protective systems are in place to prevent the Hold Drum from being over pressured.

1. The Separation Tower T-200 is provided with a SIL-1 low level interlock to prevent the loss of the liquid seal in the tower sump.
2. The Hold Drum D-205 is provided with a relief valve (PSV-206) sized to prevent overpressure if the tower liquid seal is lost.

LOPA EVALUATION OF THE EXAMPLE

For the example problem, the initiating event of concern is a basic process control system (BPCS) failure, resulting in loss of level control in the Separation Tower. From the CCPS LOPA book [Ref. 1], the initiating event frequency of a BPCS failure is 1 event in 10 years.

The low level protective interlock system was designed and installed in accordance with the ANSI/ISA 84.00.01 standard [Ref. 3] as a SIL 1 protective system. By definition, the probability of failure on demand (PFD) of a SIL-1 system is between 0.01 and 0.1. The LOPA book recommends using 0.1 as the PFD of a SIL-1 system.

The relief valve on the Hold Drum D-205 is designed for this hypothetical event. A value of 0.01 is suggested by the LOPA book as an estimate of the PFD for the relief valve of not opening on demand. However, for this analysis, a more conservative PFD value of 0.016 is used based on a review of the work by Bukowskia [Ref 6].

The LOPA formulation of the frequency of the Hold Drum being overpressured is:

$$\text{Freq} = \text{IE x SIS failure x PSV failure} \quad \text{(Eqn 11)}$$

$$\text{Freq} = (0.1 \text{ event/yr}) \times (0.1) \times (0.016) = 1.6 \times 10^{-4} \text{ events/yr} \quad \text{(Eqn 12)}$$

In a LOPA, the frequency of 1.6×10^{-4} events per year would typically be compared with applicable risk acceptance criteria to determine if additional safeguards are needed. Many LOPA teams might recommend the addition of a critical high level alarm outside of the basic process control system, with adequate time for operator intervention, to reduce the risk of this liquid blow-through scenario.

Due to conflicts in turnaround schedules, the site management is interested in extending the test interval on the SIS system by 3 months, from 2 years to2.25 years. The management asks the LOPA team:

> "Will extending the SIS test interval by 3 months require the provision of additional IPLs to protect the system?"

At this point in the analysis, the engineer would not be able to explain to management the magnitude of the underlying uncertainty in the computed frequency of the Hold Drum overpressure. The next section of this paper attempts to answer this, and related, questions.

LOPA UNCERTAINTY

The basis for the evaluation of the LOPA uncertainty comes from two sources:

1. The LOPA model presented in Equation 11
2. The uncertainty in the parameters of the LOPA model

Rewriting Equation 11 in shorter terms, we obtain:

Freq = (IE) (SIS) (PSV) (Eqn 13)

Where:

Freq is the frequency of the Hold Drum D-205 being over pressured to the point of failure
IE is the initiating event frequency of the basic process control system failing in a dangerous state
SIS is the PFD of the SIL-1 low level interlock on the Separation Tower T-200
PSV is the PFD of the relief valve (PSV-206) opening and relieving the overpressure event

The variables in Equation 13 are all random variables. The variable SIS is actually the result of a simple calculation involving the equipment failure rate and test interval. For a negative exponentially distributed failure rate for a non-repairable device, the PFD is approximately:

$SIS = \lambda\ T/2$ (Eqn 14)

Where:
λ is the failure rate of the SIL-1 low level interlock on the Separation Tower T-200
T is the inspection or test interval for validation that the interlock is working and is functional
Equation 14 is an approximation of the negative exponential distribution. For λ T/2 < 0.2, the approximation is very good (error of 10%) and will be used throughout this paper.

Substituting the SIS failure rate and test interval into Equation 13 yields the basic LOPA equation for this scenario as:

$$Freq = (IE)(\lambda)(T/2)(PSV) \quad \text{(Eqn 15)}$$

The initiating event frequency in the basic LOPA analysis was taken directly from the LOPA book as 0.1 events per year. Discussions with site operations group indicates that the IE frequency is uncertain and might be as low as 0.05 events per year or as high as 0.5 events per year.

Likewise, the failure rate of the SIS interlock (λ) has significant uncertainty. The original plant design targeted this interlock as SIL-1 with a risk reduction factor of 50 based on a two-year testing interval, or a SIS failure rate of 0.02 failures per year. However, the failure rate may be as low as 0.01 events/year or as high as 0.1 events/year. From Appendix B (Table B-6) for a SIL-1 interlock designed with an RRF=50, the mean PFD is 4.3E-2 and the variance is 4.1E-4.

The test interval (T) may be targeted as once every 2 years. However, it might be as short as once every 1.5 years if the site takes advantage of an unexpected shutdown to test the system. It might be as long as 2.5 years, if the production run of the system is extended.

Appendix B has a discussion of relief valve PFDs. The Bukowskia [Ref 6] Process Safety Valve (PSV) PFD of 0.016 is higher than the CCPS LOPA book value and was used in the above LOPA calculation of the frequency of overpressure of the Hold Drum. For this uncertainty analysis, a mean PSV PFD of 0.016 with a standard deviation of 4.7×10^{-3} was used as the best estimate of the probability of failure of the relief valve to open on demand. The PSV variance would be $[4.7 \times 10^{-3}]^2$ or 2.2×10^{-5}.

These data may be used to establish the probability distribution for each of the uncertain variables in Eqn. 15. Assuming a triangular probability distribution (Ref. 7), the mean and variance may be calculated as:

$$Mean = (A + B + C)/3 \quad \text{(Eqn 16)}$$
$$Variance = (A^2 + B^2 + C^2 - AB - AC - BC)/18 \quad \text{(Eqn 17)}$$

Where:
A = Minimum Value
B = Mode
C = Maximum Value

The mean and variance for the initiating event (IE), the failure rate of the SIS (λ) and the inspection interval are calculated and presented in Table 2.

Table 2
Triangular Probability Distribution Parameters

Model Variable	Units	Minimum A	Mode B	Maximum C	Mean by Eqn 16	Variance by Eqn 17
Initiating Event (IE)	Events per Year	0.05	0.1	0.5	2.167E-01	1.014E-02
λ of SIS (From Table B-6 with RRF=50)	Failure per Year	0.01	0.02	0.1	4.3E-02	4.1E-04
Inspection Interval (T)	Years	1.5	2	2.5	2.000E+00	4.167E-02

The variance contribution for each uncertain (random) variable may now be computed using the numerical approximation for the partial derivative term presented above and derived in detail in Appendix A to this paper. Restating Eqn 15 as the basic LOPA equation for this scenario:

Freq = (IE) (λ) (T/2) (PSV) (Eqn 15)

Using a 10% perturbation about the means of each random variable, the partial derivative term may be estimated. For example the partial derivative portion of the contribution from the uncertainty in the initiating event may be estimated as:

$$\frac{\partial Freq}{\partial IE} \approx \frac{\Delta Freq}{\Delta IE} = \{[1.1\ E(IE)]\ E(\lambda)\ E(T/2)\ E(PSV) - E(IE)\ E(\lambda)\ E(T/2)\ E(PSV)\}/(0.1\ E(IE)) \quad \text{(Eqn 16)}$$

$$\frac{\partial Freq}{\partial IE} \approx \frac{\Delta Freq}{\Delta IE} = [1.1]\ E(\lambda)\ E(T/2)\ E(PSV) - E(\lambda)\ E(T/2)\ E(PSV)/(0.1) \quad \text{(Eqn 17)}$$

$$\frac{\partial Freq}{\partial IE} \approx \frac{\Delta Freq}{\Delta IE} = [(1.1 - 1)/0.1]\ E(\lambda)\ E(T/2)\ E(PSV) \quad \text{(Eqn 18)}$$

$$\frac{\partial Freq}{\partial IE} \approx \frac{\Delta Freq}{\Delta IE} = E(\lambda)\ E(T/2)\ E(PSV) \quad \text{(Eqn 19)}$$

Likewise the partial derivative contribution to the variance of the other random variables may be found as:

$$\frac{\partial Freq}{\partial \lambda} \approx \frac{\Delta Freq}{\Delta \lambda} = E(IE)\ E(T/2)\ E(PSV) \quad \text{(Eqn 20)}$$

$$\frac{\partial Freq}{\partial T} \approx \frac{\Delta Freq}{\Delta T} == E(IE)\, E(\lambda)\, (1/2)\, E(PSV) \qquad \text{(Eqn 21)}$$

$$\frac{\partial Freq}{\partial PSV} \approx \frac{\Delta Freq}{\Delta PSV} = E(IE)\, E(\lambda)\, E(T/2) \qquad \text{(Eqn 22)}$$

It should be noted that these results are identical to those obtained by taking the partial derivative of the basic LOPA equation for this scenario (Eqn 15) with respect to each of the random variables. With these estimates for the partial derivatives, the variance contribution from each of the uncertain random variables may now be computed (see Table 3). Note that the results in Table 3 are identical with those previously obtained by Freeman [Ref. 4, 5] using more formal mathematical methods. The same result could be obtained by simple numerical substitution. Equations 19 through 22 and those in Table 3 are presented to allow the reader to understand why the numerical results of a numerical approximation are the same as obtained using analytical methods.

Table 3.
Contribution of Each Model Variable to Frequency of Drum Overpressure Variance

Model Variable	**Units**	**Equation for Contribution**	**Variance Contribution**	**Percent of Total Variance**
Initiating Event (IE)	Events per Year	$[E(\lambda)\, E(T/2)\, E(PSV)]^2\, V(IE)$	4.874E-09	41%
λ of SIS	Failure per Year	$[E(IE)\, E(T/2)\, E(PSV)]^2\, V(\lambda)$	4.874E-09	41%
Inspection Interval (T)	Years	$[E(IE)\, E(\lambda)\, (1/2)\, E(PSV)]^2\, V(T)$	2.351E-10	2%
Relief Valve (PSV)	none	$[E(IE)\, E(\lambda)\, E(T/2)]^2\, V(PSV)$	1.947E-09	16%
		Totals	**1.193E-08**	**100%**

Since the variance contribution and total variance are the same whether computed via an analytical method or by a numerical approximation method, the accuracy of the resulting will be the same as previously reported by Freeman [Ref 4, 5]. Table 4 presents a comparison of the computed results with those previously obtained by Freeman using both an analytical and a Monte Carlo method for estimating the frequency of overpressure of the Hold Drum in the example problem.

Table 4
Comparison of Results via Numerical Approximation,
Using Theoretical and Direct Monte Carlo Simulation of Hold Drum Overpressure

Method	Mean event/yr	Standard Deviation event/yr	Value of 70% Upper Limit event/yr	Value of 95% Upper Limit event/yr
Theoretical	1.50E-04	1.09E-04	2.07E-4	3.29E-04
Numerical Approximation	1.50E-04	1.09E-04	2.07E-4	3.29E-04
Monte Carlo	1.51E-04	1.19E-04	1.78E-4	3.90E-04
Basic LOPA	1.6E-04	Not Computed	Not Computed	Not Computed

With the results of the uncertainty analysis now available, we can answer the managements question in the example problem on the impact of extending the inspection and test interval by 3 months. As seen in Table 3, the variance contribution of the inspection interval is only 2% of the total. Increasing the inspection interval by 3 months will not appreciably impact the computed 95% upper limit on the frequency of the Hold Drum overpressure.

SIMPLIFIED METHOD FOR QUANTIFYING LOPA UNCERTAINTY

Based on the results obtained using the numerical approximations outlined in this paper, a simplification of the LOPA uncertainty method outlined by Freeman [Ref. 4, 5] is possible. With the exception of the method used to compute the partial derivatives for the variance contribution, the revised procedure will be the same as that which was previously presented. The change to the previously presented method is indicated in *italic font* below.

1. Complete the LOPA study using the methods outlined in the CCPS LOPA book [Ref. 1].

2. Write the basic LOPA equation for the scenario of interest in the form of the product of initiating event frequency and the PFD of the IPLs. An example of the basic LOPA equation is given in Equation 15.

3. Define the uncertainty in the terms of the expanded LOPA equation. The uncertainty can be given as the upper and lower range of the possible values (uniform probability distribution), as the upper, lower and recommended values (triangular distribution), or as a mean and standard deviation (normal distribution). See the example above for guidance in the evaluation of safety instrumented system interlocks.

4. Compute the expected value of each variable in the LOPA equation. The equations for the mean and variance for the uniform, triangular and normal probability distributions are presented for various probability distributions in Vose [Ref. 7]. *For commonly used LOPA initiating events and IPLs, the mean and variance are tabulated in Appendix B to this paper and may be used directly.*

5. Compute the expected value or mean of the LOPA equation using the mean value of each of the variables in the basic LOPA equation.

6. *Compute the sensitivity of the result from the LOPA equation by estimating the partial derivative of the basic LOPA equation with respect to each of the variables using a 10% perturbation about the mean or expected value of each random variable. See the problem worked above as an example of this step. The theory for this simplification is presented in Appendix A to this paper*

7. Compute the variance of the scenario frequency of the basic LOPA equation by use of the variance contribution of Equation 6. This entails multiplying the variance of each of the uncertain variables in the basic LOPA equation by the square of its sensitivity (obtained in step 6), as evaluated at the mean. Sum the resulting terms to obtain the overall variance of the result of the basic LOPA term.

8. Compute the 95% upper confidence limit on the computed frequency by use of the standard normal factor, Z, [Ref. 8] as:

$$Z = [x_i - E(x)]/\sigma \qquad \text{(Eqn 23)}$$

Where:

σ = standard deviation of the frequency of the scenario of interest from the basic LOPA equation obtained from step 7. Note that the variance of a random variable is the square of the standard deviation of the random variable.

$E(x)$ = the expected value of the frequency of the scenario of interest from the basic LOPA equation obtained from step 5

For the 95% upper limit, Z = 1.64. Rearranging Eqn. 10 allows for the direct calculation of the corresponding value of the 95% upper confidence limit as:

$$X_{95\%} = 1.64\,\sigma + E(x) \qquad \text{(Eqn 24)}$$

Where

$X_{95\%}$ = is the upper 95% limit on the computed frequency of the scenario of interest from the basic LOPA equation.

9. Compare the 95% upper confidence limit on the frequency of the scenario of concern with that obtained using the basic LOPA method (step 1). If the difference is greater than one order of magnitude, revise the LOPA study accordingly to incorporate better information obtained in the uncertainty analysis. Improve the integrity of the LOPA IPLs or identify additional IPLs to drive the risk to a tolerable level.

RESULTS AND CONCLUSIONS

A numerical approximation can be used to estimate the variance contribution of each uncertain variable in a basic LOPA. Use of a 10% perturbation about the mean or expected value of each random variable was found to be adequate to define the variance. The numerical approximation is exact for cases where all of the IPLs are independent. In the case where 4 IPLs share a common test or maintenance frequency, the resulting error in the uncertainty estimate will be less than 16%.

A simplified method has been developed to evaluate the uncertainty in the basic LOPA equation. Using available data, the simplified numerical approximation method estimates the mean and variance of the basic LOPA equation and then computes the uncertainty range in the resulting frequency of the scenario of concern. For the example studied with 4 random variables, a total of 5 evaluations of the basic LOPA equation were all that were required to characterize the uncertainty. The results are found to be in excellent agreement with those obtained using analytical methods. When compared to a Monte Carlo simulation of 10,000 function evaluations, the calculation speed improvement is approximately a factor of 2,000.

A set of tables for the mean and variance of initiating events (IE) and independent protection layers (IPLs) are presented in Appendix B of this paper. These can be used directly by the analyst to estimate the uncertainty in an IE or IPL.

The methodology presented in this paper is simple and direct. It may be programmed into any standard LOPA software package to allow the users the ability to estimate the uncertainty in their calculated result.

ACKNOWLEDGMENTS

This paper is an expanded version of a paper entitled ''Uncertainty in LOPA; Part 2 – Simplified Methods Applied to Safety Systems'' presented at the 2012 Mary Kay O'Connor Process Safety Symposium, October 23-25, 2012, College Station, TX. The author wishes to thank Kathy Kas, Angela Summers, Robert Johnson, Dan Crowl and other individuals who gave comments and suggestions for improving this article.

LITERATURE CITED

1. CCPS, *Layer of Protection Analysis – Simplified Process Risk Assessment*, American Institute of Chemical Engineers, New York, 2001
2. Instrumentation, Systems and Automation Society, Functional Safety: Safety Instrumented Systems for the Process Sector—Parts 1, 2, and 3, Standard ANSI/ISA 84.00.01, 2004. USA implementation of IEC 61511.
3. R.A. Freeman, "Using Layer of Protection Analysis to Define Safety Integrity Level Requirements, pp 185 – 194, *Process Safety Progress*, Vol. 26, No. 3, 2007
4. R.A. Freeman, "Uncertainty in LOPA Studies," pages 623-634, Proceedings of Mary Kay O'Connor Process Safety Center International Symposium held October 25-27, 2011 in College Station, Texas
5. R.A. Freeman, "Quantifying LOPA Uncertainty," pp 240 – 247, *Process Safety Progress*, Vol. 31, No. 3, 2012. Published in Wiley Online Library (wileyonlinelibrary.com), DOI 10.1002/prs.11493
6. Julia V. Bukowskia and William M. Goble, "Analysis of Pressure Relief Valve Proof Test Data," pp 24-29, *Process Safety Progress*, Vol 28, No.1, 2009
7. David Vose, *Risk Analysis – A Quantitative Guide, 3rd Edition*, John Wiley, 2008
8. Paul L. Meyer, *Introductory Probability and Statistical Applications*, John Wiley, New York, 1972

Appendix A – Numerical Approximation to Partial Derivatives

The basic LOPA equation of Eqn. 2 can be restated as:

$$\text{Freq} = (\text{IE})\ P_1\ P_2\ P_3\ \ldots\ P_n \qquad \text{(Eqn. A-1)}$$

Where:
Freq = Frequency of the undesired event being studied by the LOPA method
IE = Initiating event frequency or the starting event frequency for the scenario under study
P_i = Probability of Failure on Demand of the Independent Protection Layer (IPL) counted as reducing the frequency of the undesired event.
n = Total number of IPLs counted in the LOPA study as satisfying the criteria for an IPL

The probabilities of failure on demand of the IPLs may be restated in terms of a failure rate and a time between tests as:

$$\text{Freq} = (\text{IE})\ (\lambda_1)\ (T_1/2)\ (\lambda_2)\ (T_2/2)\ \ldots\ (\lambda_n)\ (T_n/2) \qquad \text{(Eqn A-2)}$$

Where:
λ_1 is the failure rate of the safety instrumented system, i

T is the inspection or test interval for validation that the interlock is working and is functional

i is a subscript indicating the ith IPL

If all the IPLs are tested or inspected at the same test interval

$$T_1 = T_2 = T_3 = T_n \qquad \text{(Eqn A-3)}$$

Forming a new parameter K as:

$$K = (\text{IE})\ (\lambda_1)\ (\lambda_2)\ \ldots\ (\lambda_n) \qquad \text{(Eqn A-4)}$$

The basic LOPA equation may now be re-written as:

$$\text{Freq} = K\ (T/2)^n \qquad \text{(Eqn A-5)}$$

When evaluated at the expected value of the random variables, the parameter K will not change and can be treated as a constant. How good is the numerical approximation for the partial derivative of Freq with respect to time as n increases from 1 to n?

The numerical approximation to a derivative requires the selection of a step size for the change in the variable of interest. In the limit as the step size goes to zero, the resulting answer will become the exact analytical expression for the derivative. However, from a practical point of view the selection of the numerical step size is constrained by two problems:

1. Too large and the function may change too much and the resulting numerical approximation will be very bad.
2. Too small and the numerical approximation will be limited by round-off error of the computer.

The step size of 0.1 T was chosen as the balance point between the two numerical problems. For the following cases (1 through 4), a step size of 0.1T was used as the basis for estimating the partial derivatives required for computations in the variance contribution analysis. Note that the step size for the evaluation of the derivative is arbitrary. A smaller step size could have been selected (say 0.01 T) and the resulting errors would decrease. However, the potential for numerical calculation errors would increase.

Case 1, n=1

For n = 1, restating the LOPA frequency becomes:

$$Freq = K\ (T/2) \qquad \text{(Eqn A-6)}$$

$$\frac{\partial Freq}{\partial T} = 0.5\ K \qquad \text{(Eqn A-7)}$$

The numerical approximation would become:

$$\frac{\Delta Freq}{\Delta T} = K\ (\frac{1.1^1 - 1}{0.1T})\ (T/2)^1 \qquad \text{(Eqn A-8)}$$

$$\frac{\Delta Freq}{\Delta T} = K\ (\frac{0.1}{0.1})\ 1/2 \qquad \text{(Eqn A-9)}$$

$$\frac{\Delta Freq}{\Delta T} = 0.5\ K \qquad \text{(Eqn A-10)}$$

This numerical approximation is exactly the same as the analytical result. For n =1, there will be no error in the computation of the partial derivative.

Case 2, n = 2

For n = 2, restating the LOPA frequency becomes:

$$Freq = K\ (T/2)^2 \qquad \text{(Eqn A-11)}$$

From basic calculus, the analytical expression for the partial derivative of equation A-5 when n=2 is:

$$\frac{\partial Freq}{\partial T} = 0.5\ K\ T \qquad \text{(Eqn A-12)}$$

The numerical approximation would become:

$$\frac{\Delta Freq}{\Delta T} = (\frac{1.1^2 - 1}{0.1T})\ K\ (T/2)^2 \qquad \text{(Eqn A-13)}$$

$$\frac{\Delta Freq}{\Delta T} = (\frac{0.21}{0.1})\ (1/2)^2\ K\ T \qquad \text{(Eqn A-14)}$$

$$\frac{\Delta Freq}{\Delta T} = (\frac{2.1}{4})\ K\ T \qquad \text{(Eqn A-15)}$$

$$\frac{\Delta Freq}{\Delta T} = (0.525)\ K\ T \qquad \text{(Eqn A-16)}$$

The error in the numerical approximation is:

Error = (Estimate - Actual)/Actual x 100 (Eqn A-17)

Error = (0.525 -0.5)/0.5 x 100 = 5% (Eqn A-18)

Thus, for a SIL-2 type of design, we would expect that no more than a 5% error in the LOPA uncertainty estimate would be introduced by the use of the numerical approximation for the partial derivatives.

Case 3, n =3

For n = 3, restating the LOPA frequency becomes:

$$Freq = K\ (T/2)^3 \qquad \text{(Eqn A-19)}$$

From basic calculus, the analytical expression for the partial derivative of equation A-5 when n=3 is:

$$\frac{\partial Freq}{\partial T} = K\ (3/8)\ T^2 \qquad \text{(Eqn A-20)}$$

The numerical approximation would become:

$$\frac{\Delta Freq}{\Delta P_1} = (\frac{1.1^3 - 1}{0.1T})\ K\ (T/2)^3 \qquad \text{(Eqn A-21)}$$

$$\frac{\Delta Freq}{\Delta P_1} = (\frac{0.331}{0.1})\ (1/2)^3\ K\ T^2 \qquad \text{(Eqn A-22)}$$

$$\frac{\Delta Freq}{\Delta P_1} = (\frac{3.31}{8})\ K\ T^2 \qquad \text{(Eqn A-23)}$$

$$\frac{\Delta Freq}{\Delta P_1} = (0.3875)\ \mathrm{K}\ \mathrm{T}^2 \qquad \text{(Eqn A-24)}$$

The error in the numerical approximation is:

Error = (Estimate - Actual)/Actual x 100 (Eqn A-25)

Error = (3.31/8-3/8)/(3/8) x 100 (Eqn A-26)

Error = (3.31-3)/(3) x 100 = 10% (Eqn A-27)

Thus, for a SIL-3 type of design, we would expect no more than a 10% error in the LOPA uncertainty estimate to be introduced by the use of the numerical approximation for the partial derivatives.

Case 4, n =4

For n = 4, restating the LOPA frequency becomes:

$\mathrm{Freq} = \mathrm{K}\ (\mathrm{T}/2)^4$ (Eqn A-28)

From basic calculus, the analytical expression for the partial derivative of equation A-5 when n=4 is:

$$\frac{\partial Freq}{\partial T} = (1/4)\ \mathrm{K}\ \mathrm{T}^3 \qquad \text{(Eqn A-29)}$$

The numerical approximation would become:

$$\frac{\Delta Freq}{\Delta P_1} = \left(\frac{1.1^4 - 1}{0.1T}\right) \mathrm{K}\ (\mathrm{T}/2)^4 \qquad \text{(Eqn A-30)}$$

$$\frac{\Delta Freq}{\Delta P_1} = \left(\frac{0.464}{0.1}\right) (1/2)^4\ \mathrm{K}\ \mathrm{T}^3 \qquad \text{(Eqn A-31)}$$

$$\frac{\Delta Freq}{\Delta P_1} = \left(\frac{4.64}{16}\right) \mathrm{K}\ \mathrm{T}^3 \qquad \text{(Eqn A-32)}$$

$$\frac{\Delta Freq}{\Delta P_1} = (0.290)\ \mathrm{K}\ \mathrm{T}^3 \qquad \text{(Eqn A-33)}$$

The error in the numerical approximation is:

Error = (Estimate - Actual)/Actual x 100 (Eqn A-34)

Error = (0.29 – 0.25)/(0.25) x 100 = 16% (Eqn A-35)

Thus, for a SIL-4 type of design, we would expect no more than a 16% error in the LOPA uncertainty estimate to be introduced by the use of the numerical approximation for the partial derivatives.

Table 1 presents a summary of the results of these calculations. In summary, the use of a numerical approximation for estimating uncertainty in a LOPA analysis will introduce no greater than a 16% error in the results. A 16% error in the LOPA uncertainty is well within the accuracy of the Variance Contribution Analysis method for estimating the LOPA uncertainty. This result eliminates the need for time-consuming evaluation of partial derivatives during a LOPA uncertainty analysis. Instead, a simple numerical analysis using a 10% perturbation about the mean of the inspection frequency is adequate to evaluate the partial derivatives needed in a LOPA uncertainty study. The results of this appendix are general and will apply to any basic LOPA equation formulated up to a 4th degree in one or more random variables.

Appendix B – Derivation of Tables for LOPA Uncertainty Analysis

The use of Variance Contribution Analysis to estimate the uncertainty in an estimate of the frequency of an accident requires a source for estimating the statistical parameters of the initiating events (IEs) and independent protection layers (IPLs). This appendix presents a methodology for estimating the required statistical parameters using available published data.

Initiating Events

The initiating event frequencies commonly used in a LOPA study are presented in Table 5.1 CCPS LOPA book [Ref. 1]. Included in Table 5.1 are ranges for the frequencies of each initiating event as determined from a literature search during development of the book. The uncertainty in the frequency data is expressed as a range (minimum to maximum), with a particular value assigned for use during a LOPA review. This type of uncertainty can be described by the triangular distribution [Ref. 7]. The triangular distribution has three parameters:

A = Minimum value
B = Mode
C = Maximum value

For a triangular distribution, the mean and variance are:

$$\text{Mean} = (A + B + C)/3 \qquad \text{(Eqn. B-1)}$$
$$\text{Variance} = A^2 + B^2 + C^2 - AB - BC - AC \qquad \text{(Eqn. B-2)}$$

For the data in Table 5.1, the minimum value becomes A, the maximum value becomes C, and the value used in LOPA becomes the mode, B. The mean and variance of the initiating event frequencies of the CCPS LOPA book Table 5.1 are calculated using Equations B-1 and B-2 and are presented in Table B-1 below. Note that the triangular distribution parameters (A, C and B) are defined in Table B-1 as the Min A, Max C, and Used in LOPA B respectively.

Table B-1. Mean and Variance of LOPA Initiating Events

Initiating Event	Units	Min A	Max C	Used in LOPA B	Mean per unit	Variance per unit^2
Pressure Vessel residual failure	per year	1E-07	1E-05	1E-06	3.7E-06	5.0E-12
Piping residual failure - 100 meter - full breach	per year	1E-06	1E-05	1E-05	7.0E-06	4.5E-12
Piping Leak - 10% section - 100 meter	per year	1E-04	1E-03	1E-03	7.0E-04	4.5E-08
Atmospheric Tank Failure	per year	1E-05	1E-03	1E-03	6.7E-04	5.4E-08
Gasket Packing Blowout	per year	1E-04	1E-02	1E-02	6.7E-03	5.4E-06
Turbine / diesel engine overspeed with casing breach	per year	1E-04	1E-03	1E-04	4.0E-04	4.5E-08
External impact by backhoe, vehicle, etc.	per year	1E-04	1E-02	1E-02	6.7E-03	5.4E-06
Crane Drop	per lift	1E-04	1E-03	1E-04	4.0E-04	4.5E-08
Lightning strike	per year	1E-04	1E-03	1E-03	7.0E-04	4.5E-08
Safety valve operates spuriously	per year	1E-04	1E-02	1E-02	6.7E-03	5.4E-06
Cooling water failure	per year	1.E+00	1E-02	1E-01	3.7E-01	5.0E-02
Pump seal failure	per year	1E-02	1E-01	1E-01	7.0E-02	4.5E-04
Unloading / loading hose failure	per year	1E-02	1E-01	1E-01	7.0E-02	4.5E-04
BPCS failure	per year	1E-02	1E-01	1E-01	7.0E-02	4.5E-04
Regulator Failure	per year	1E-01	1E+00	1E-01	4.0E-01	4.5E-02

The CCPS LOPA book Tables 6.3, 6.4, and 6.5 present the probability of failure on demand of passive, active mechanical and human operator IPLs. Using the same of data as for the initiating events allows for the uncertainty of IPLs to be calculated using Equations B-1 and B-2. The results of these calculations are summarized in Tables B-2, B-3, and B-4 below.

Table B-2. Mean and Variance of LOPA Passive IPLs

IPL Description	Min A	Max C	Used in LOPA B	Mean per unit	Variance per unit^2
Dike	1E-03	1E-02	1E-02	7.0E-03	4.5E-06
Underground Drainage System	1E-03	1E-02	1E-02	7.0E-03	4.5E-06
Open Vent (no valve)	1E-03	1E-02	1E-02	7.0E-03	4.5E-06
Fireproofing	1E-03	1E-02	1E-02	7.0E-03	4.5E-06
Blast Wall / Bunker	1E-03	1E-02	1E-03	4.0E-03	4.5E-06
Flame /Detonation Arrestors	1E-03	1E-01	1E-02	3.7E-02	5.0E-04

Table B-3. Mean and Variance of LOPA Active Mechanical IPLs

IPL Description	Min A	Max C	Used in LOPA B	Mean	Variance
Relief Valve*	1E-05	1E-01	1E-02	3.7E-02	5.1E-04
Rupture Disk	1E-05	1E-01	1E-02	3.7E-02	5.1E-04
BPCS	1E-02	1E-01	1E-01	7.0E-02	4.5E-04

Note the asterisk (*) in Table B-3 for relief valves The issue of relief valves not opening when needed is s topic that has been discussed extensively by Bukowskia [Ref. 6]. Bukowskia found that there is a significant probability that a relief device is not working when installed. A PFD value of 0.016 was suggested as an estimate for the PSV not opening on demand due to initial installation problems. Table 1 in the paper by Bukowski presents an estimate for the uncertainty in the PFD of the relief valve in terms of the standard deviation. For the data sets evaluated, the mean PFD of the PSV may be taken as 0.016 with a standard deviation of 4.7×10^{-3}. The corresponding PSV variance would be $[4.7 \times 10^{-3}]^2$ or 2.2×10^{-5}. The data of Bukowski indicate a PFD of about half that of the mean based on the range of the CCPS book Table 6.4 with a smaller variance. The LOPA value of 0.01 is lower than the value obtained using Bukowski's paper. The analyst completing a LOPA uncertainty study should carefully evaluate which data set to use for representing the uncertainty in the relief valve PFD.

Table B-4. Mean and Variance of LOPA Human Operator IPLs

IPL Description	Min A	Max C	Used in LOPA B	Mean	Variance
Human action with 10 minutes response time	1E-01	1E+00	1E-01	4.0E-01	4.5E-02
Human action with 40 minutes response time	1E-02	1E-01	1E-01	7.0E-02	4.5E-04

Safety Instrumented Systems
Safety instrumented systems are often provided in process design to control undesirable events such as overpressure, runaway reactions, or loss of level interface control. Note for the purposes of this paper the use of the phrase "safety interlock" means:

> "A Safety Instrumented Function (SIF) implemented in a Safety Instrumented System (SIS) to achieve a target Safety Integrity Level (SIL) designed and installed in accordance with the requirements of ANSI/ISA S84 – 2004 [Ref. 2]"

The S84 standard defines a range of acceptable performance operation of the interlocks in terms of a safety integrity level (SIL). The various SIL levels are defined by the probability of failure on demand (PFD) of the interlock, where the RRF is related to the PFD as:

RRF = 1/PFD (Eqn. B-3)

For example, an interlock designed for SIL-1 must have a PFD between 0.1 and 0.01, or by Equation B-3, an RRF between 10 and 100. Table B-5 presents the defined range of PFDs and RRFs for various SIL classes.

Table B-5. Probability of Interlock Failure on Demand of Various SIL Classes

IPL Description	Min PFD	Max PFD	Used in LOPA	Min RRF	Max RRF
SIL-1	1E-02	1E-01	1E-01	10	100
SIL-2	1E-03	1E-02	1E-02	100	1,000
SIL-3	1E-04	1E-03	1E-03	1,000	10,000
SIL-4	1E-05	1E-04	1E-04	10,000	100,000

There are two strategies available for the interlock design engineer to use:

1. Complete the design using rules of thumb, evaluate the RRF, and accept the design if the RRF falls in the desired range (as specified by the standard).
2. Complete the design to a desired risk reduction factor (RRF)

These two different design strategies will result in different probability distributions for the resulting uncertainty in the performance of the interlock. In the first case, any design that falls in the desired RRF range will be acceptable. The resulting uncertainty will be uniformly distributed from the minimum to maximum acceptable RRF. The uniform probability distribution may be characterized by two parameters, A and C:

A = Minimum value
C = Maximum value

The mean and variance of the uniform distribution are:

Mean = (A + C)/2 (Eqn. B-4)

Variance = $((C-A)^2)/12$ (Eqn. B-5)

In the second case, the design is targeted to a particular value of the RRF. This RRF becomes the mode (parameter B) of the triangular probability distribution of outcomes. Equations 1 and 2 may be used to compute the mean and variance of the resulting distributions.

Since a range of RRFs are available for each selected interlock SIL, a set of three possible targets was selected to represent the SIL range. These are:

Target RRF = 25 percent of allowed range (Eqn. B-6)

Target RRF = 50 percent of allowed range (Eqn. B-7)

Target RRF = 75 percent of allowed range (Eqn. B-8)

The uncertainty in the interlock performance was calculated for both design strategies for SIL-1, SIL-2, SIL-3 and SIL-4 at each of the three targeted SIL levels (Eqns. B-6, 7, and 8). The results of these calculations are presented in Tables B-6, B-7, B-8, and B-8.

For design strategy 1, Table B-6 presents the uncertainty parameters for each interlock SIL class when no target RRF is assigned prior to design. For example, such a case could occur if the design engineer were told to simply provide a SIL 2 interlock with no specification of the corresponding RRF.

Table B-6. Uncertainty Parameters for Interlock PFD for Various SIL Classes
(No target RRF set prior to design)

IPL Description	Mean PFD	Variance of PFD
SIL-1	5.5E-02	6.8E-04
SIL-2	5.5E-03	6.8E-06
SIL-3	5.5E-04	6.8E-08
SIL-4	5.5E-05	6.8E-10

For design strategy 2, the interlock RRF is selected prior to the start of design. This selected RRF then determines the design target PFD, which becomes the mode (parameter C, Equations B-1 and 2) for the triangular distribution. Tables B-7, B-8 and B-9 present the results for the computation of the mean and variance of the PFD of interlocks designed for various RRFs.

Table B-7. Uncertainty Parameters for Interlock PFD for Various SIL Classes with Target RRF at 25% of Allowed Range

IPL Description	Min PFD A	Max PFD C	Used in LOPA	Design RRF	Mode B	Mean PFD	Variance of PFD
SIL-1	1E-02	1E-01	1E-01	25	4.0E-02	5.0E-02	3.5E-04
SIL-2	1E-03	1E-02	1E-02	250	4.0E-03	5.0E-03	3.5E-06
SIL-3	1E-04	1E-03	1E-03	2,500	4.0E-04	5.0E-04	3.5E-08
SIL-4	1E-05	1E-04	1E-04	25,000	4.0E-05	5.0E-05	3.5E-10

Table B-8. Uncertainty Parameters for Interlock PFD for Various SIL Classes with Target RRF at 50% of Allowed Range

IPL Description	Min PFD A	Max PFD C	Used in LOPA	Design RRF	Mode B	Mean PFD	Variance of PFD
SIL-1	1E-02	1E-01	1E-01	50	2.0E-02	4.3E-02	4.1E-04
SIL-2	1E-03	1E-02	1E-02	500	2.0E-03	4.3E-03	4.1E-06
SIL-3	1E-04	1E-03	1E-03	5,000	2.0E-04	4.3E-04	4.1E-08
SIL-4	1E-05	1E-04	1E-04	50,000	2.0E-05	4.3E-05	4.1E-10

Table B-9. Uncertainty Parameters for Interlock PFD for Various SIL Classes with Target RRF at 75% of Allowed Range

IPL Description	Min PFD A	Max PFD C	Used in LOPA	Design RRF	Mode B	Mean PFD	Variance of PFD
SIL-1	1E-02	1E-01	1E-01	75	1.3E-02	4.1E-02	4.3E-04
SIL-2	1E-03	1E-02	1E-02	750	1.3E-03	4.1E-03	4.3E-06
SIL-3	1E-04	1E-03	1E-03	7,500	1.3E-04	4.1E-04	4.3E-08
SIL-4	1E-05	1E-04	1E-04	75,000	1.3E-05	4.1E-05	4.3E-10

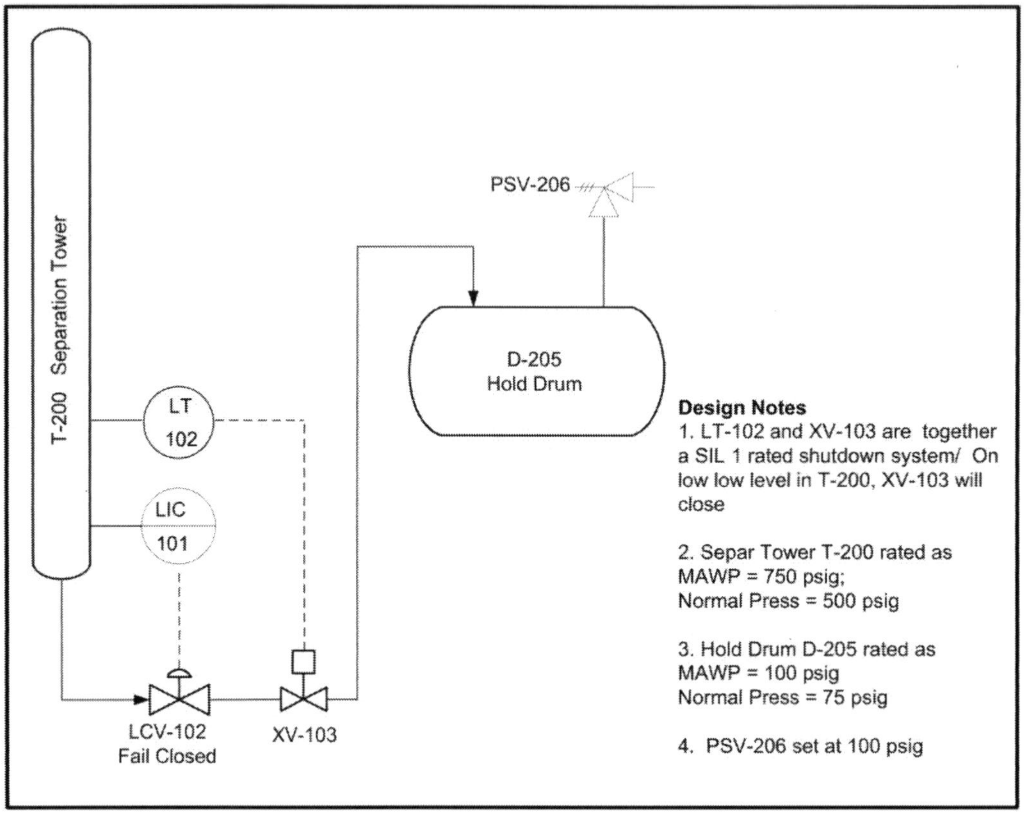

Figure 1. Piping and Instrument Diagram for Example

The Impact of Ozone on the Lower Flammable Limit of Hydrogen in Vessels Containing Savannah River Site High Level Waste

Paul Osterberg
Fauske & Associates, LLC
16W070 83rd Street
Burr Ridge, IL 60527
Osterberg@fauske.com

Carol Sherburne
Savannah River Remediation, LLC
Savannah River Site, Aiken, SC
Carol.Sherburne@srs.gov

Tom Johnson, Thomas Frawely
Fauske & Associates, LLC
16W070 83rd Street
Burr Ridge, IL 60527
Johnson@fauske.com, Frawely@fauske.com

Prepared for Presentation at
American Institute of Chemical Engineers
2013 Spring Meeting
9th Global Congress on Process Safety
San Antonio, Texas
April 28 – May 1, 2013

UNPUBLISHED

The Impact of Ozone on the Lower Flammable Limit of Hydrogen in Vessels Containing Savannah River Site High Level Waste

Paul Osterberg
Fauske & Associates, LLC
16W070 83rd Street
Burr Ridge, IL 60527
Osterberg@fauske.com

Carol Sherburne
Savannah River Remediation, LLC
Savannah River Site, Aiken, SC
Carol.Sherburne@srs.gov

Tom Johnson
Fauske & Associates, LLC
16W070 83rd Street
Burr Ridge, IL 60527
Johnson@fauske.com

Thomas Frawely
Fauske & Associates, LLC
16W070 83rd Street
Burr Ridge, IL 60527
Frawely@fauske.com

Keywords: hydrogen, ozone, flammability, lower flammability limit

Abstract

The Savannah River Site, in conjunction with AREVA Federal services, has designed a process to treat dissolved radioactive waste solids with ozone. It is known that in this radioactive waste process, radionuclides radiolytically break down water into gaseous hydrogen and oxygen, which presents a well defined flammability hazard. Flammability limits have been established for both ozone and hydrogen separately; however, there is little information on mixtures of hydrogen and ozone. Therefore, testing was designed to provide critical flammability information necessary to support safety related considerations for the development of ozone treatment and potential scale-up to the commercial level. Since information was lacking on flammability issues at low levels of hydrogen and ozone, a testing program was developed to focus on filling this portion of the information gap. A 2-L vessel was used to conduct flammability tests at atmospheric pressure and temperature using a fuse wire ignition source at 1 percent ozone intervals spanning from no ozone to the Lower Flammable Limit (LFL) of ozone in the vessel, determined as 8.4%(v/v) ozone. An ozone generator and ozone detector were used to generate and measure the ozone concentration within the vessel *in situ*, since ozone decomposes rapidly on standing. The lower flammability limit of hydrogen in an ozone-oxygen mixture was found to decrease from the LFL of hydrogen in air, determined as 4.2 % (v/v) in this vessel. From the results of this testing, Savannah River was able to develop safety procedures and operating parameters to effectively minimize the formation of a flammable atmosphere.

1. Introduction

Currently, in support of the Federal Facilities Agreement Act, tanks at the Savannah River Site (SRS) used to collect high level nuclear waste are being emptied, cleaned and grouted; and the waste permanently dispositioned in a series of specialized treatment processes. The majority of this waste removal is accomplished by a Bulk Waste Removal (BWR) campaign which empties the tank of easily accessible liquid supernatant and other liquid, followed by salt dissolution/removal and, finally, heel removal. At the end of the BWR campaign, fewer than 20,000 liters of sludge (less than 2 inches depth) remains in the tank. Currently, this final residual material is cleaned from the tank during a process called Bulk Oxalic Acid Cleaning (BOAC), whereby heated, 6 weight percent oxalic acid is added to the tank to a sufficient level to enable mixing. While BOAC has been effective in sludge dissolution and removal, the metal oxalates formed during oxalic acid dissolution cause downstream complications during evaporation (e.g. foaming) in addition to requiring excessive sludge washing (thereby greatly increasing the overall volume of waste to be dispositioned) to meet feed qualifications of the Defense Waste Processing Facility. This prompted research, testing and acceptance of a treatment process that utilized the proven effectiveness of oxalic acid for sludge dissolution and mobilization, yet removed the offending metal oxalates from the solution without adding liquid volume or contaminants. The Enhanced Chemical Cleaning (ECC) process was thus developed to replace BOAC stage of waste removal.

The Enhanced Chemical Cleaning (ECC) process uses ozone to promote the oxidation of metal oxalates produced during the dissolution of sludge in the Savannah River Site (SRS) waste tanks. The process is held at 70°C to improve decomposition kinetics and 10 psig to improve ozone solubility. A simplified depiction of the ECC process is shown in Figure 1, below.

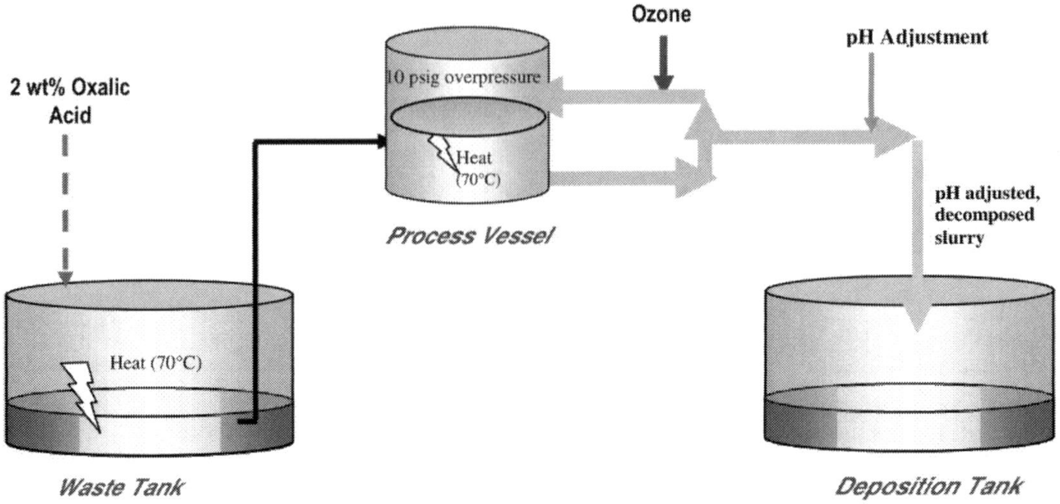

Figure 1: Schematic of the Enhanced Chemical Cleaning Process

In this process, the ozone reacts with the metal oxalates to form metal oxide and hydroxide precipitants, and the CO_2, O_2, H_2O and any unreacted O_3 gases are discharged into the vapor space. The major chemistry employed in the ECC process vessel is the reaction of ozone with the metal oxalates to form metal oxides, metal hydroxides, CO_2 and H_2O by the reactions shown below:

$$O_3 + H_2O \rightarrow O_2 + 2OH\bullet$$
$$Fe_2(C_2O_4)_3 + 6\ OH\bullet \rightarrow 6CO_2 + Fe_2O_3 + 3H_2O$$
$$H_2C_2O_4 + O_3 \rightarrow 2CO_2 + H_2O + O_2$$

While only the Iron oxalate reaction is listed above, formation of the powerfully reactive hydroxyl radical initiates many oxidizing reactions, effectively putting most waste constituents into their highest oxidation state.

In addition to the non-radioactive metals in the waste, however, the SRS radioactive waste also contains a variety of radionuclides, hence, hydrogen gas is also present in the vapor space of the ECC system. In order to prevent a buildup of hydrogen, and, hence, a flammability/explosion concern, waste tanks are purged and the hydrogen concentration monitored to ensure that the vapor space of each waste tank and all related processing equipment remains below the Lower Flammability Limit (LFL) of hydrogen. Furthermore, a comprehensive flammability control program is utilized to ensure that the LFL of the vapor space of a waste tank will not be attained for at least 7 days without forced ventilation. Because the use of ozone/oxygen in the processing equipment may also introduce ozone/oxygen into the vapor space of the waste tanks, the impact of this gas stream on the LFL of hydrogen must be understood for all possible operating scenarios of both normal and off-normal situations, with particular emphasis at the elevated temperatures and pressures of the typical process operating conditions. Oxygen is a known accelerant in combustion reactions, but while there are data associated with the behavior of hydrogen/oxygen environments, relevant studies addressing the effect of ozone on the flammability limit of hydrogen proved scarce. Further, discussions with industry experts verified the absence of data in this area and indicated that laboratory testing, specific to defined operating parameters was needed to comprehensively address the issue. Testing was thus designed and commissioned to provide the data necessary to support safety related considerations for the ozone treatment process.

2. Experimental

2.1 Flammability Apparatus

In order to ensure consistency between this testing and other documented flammability testing, the test methodology utilizes the guidelines established in ASTM E-918-83 (2005), "Standard Practices for Determining Limits of Flammability of Chemicals at Elevated Temperature and Pressure" [1]. Using this standard as the testing basis, a test setup was developed and is given in Figure 2.

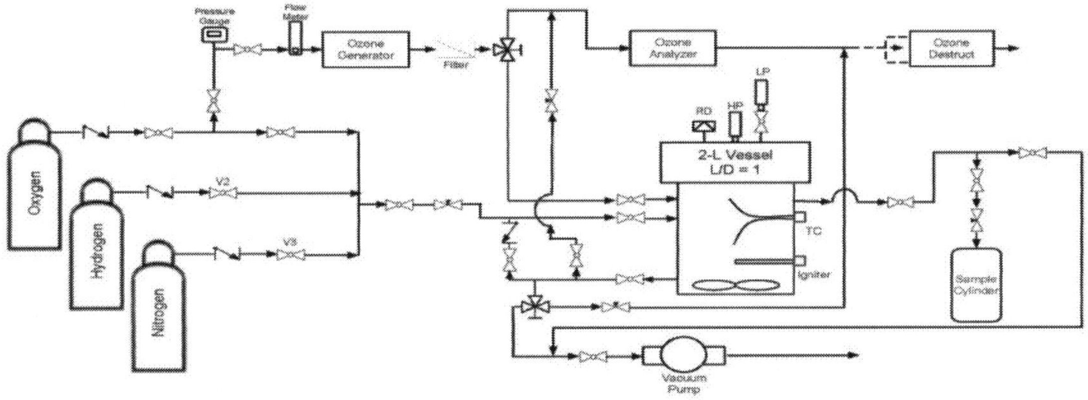

Figure 2: Test Apparatus for Testing the Lower Flammability Limit of Specified Gas Mixtures

The testing was conducted in a 2-L vessel rated to 3,000 psi at 500°C with an L/D ratio of one. There are two pressure transducers mounted on the vessel: a high accuracy (0.05% full scale) low range pressure transducer (0-25 psia) used for preparing the gas mixtures by partial pressure and a high accuracy (0.05% full scale) high range pressure transducer (0-200 psia) used to monitor the pressure after the mixture is ignited. High purity research grade gases were used for the preparation of the gas mixtures: nitrogen (99.9999%), oxygen (99.999%), and hydrogen (99.9999%). Two thermocouples were placed in the vessel to measure the temperature at the center and at the upper area of the vessel. A stir bar was placed at the bottom of the vessel to provide mixing of the gases. The gas mixture was ignited using an exploding fuse wire positioned a third of the way from the bottom. The energy provided by the fuse wire was approximately 10 joules. An ignition energy of 10 J was chosen because it produces very little wire spatter and a cylindrical shock wave is formed upon the vaporization of the wire because it is above the 5 J/cm threshold [2].

Ozone is generated on-site using a high ozone concentration generator. The ozone concentration of the ozone/oxygen stream is analyzed with an in-line UV absorption unit designed to enable accurate measurements of ozone from 0.01 wt. % ozone in oxygen to 20 weight percent.

In order to minimize undesired ozone decomposition, the materials used for this testing were carefully chosen. The test vessel, tubing, and fittings are all made of 316 stainless steel and passivated per ASTM A967 [3]. The packing material and/or seals for the valves were made of PTFE or fluorocarbon FKM.

2.2 Experimental Method

Prior to testing, the thermocouples, pressure transducers, ozone analyzer, and data acquisition instrumentation are checked to ensure they are functioning properly. The 2-L vessel is opened and wiped down with acetone. A 10mm 40 gauge tinned copper wire is wrapped around the igniter electrodes. The vessel is sealed and checked for leaks through pressurizing the vessel. This is followed by a vacuum and pressure purging procedure with a diluent gas (N_2 or O_2) depending on the test conditions. The gas manifold and piping are vacuumed and pressure purged with the respective gas prior to addition into the vessel. The purging helps minimize trace contamination of other gases as well as to remove residual moisture.

By assuming that these gases behave ideally, the gas concentration is determined based on partial pressure addition. Prior to gas filling, the vessel is evacuated down to approximately 0.2 psia. The magnetic stirrer is turned on to provide mixing during the gas filling step. Each gas is added slowly to the desired concentration based on the partial pressure calculation. During addition, there is a slight temperature rise due to the compression of the gases; therefore, after each gas addition, the mixture is allowed to reach equilibrium. The gas manifold and piping are purged with the subsequent gas prior to being added to the mixture.

For the ozone tests, oxygen is fed into a corona discharge ozone generator. A potentiometer on the unit controls the power to the elements. Thus, by adjusting the oxygen flowrate through the generator and tuning the potentiometer, the precise ozone/oxygen ratio specified in the test matrix can be produced. The ozone concentration is measured with an in-line ozone analyzer that is placed on the outlet line of the vessel.

Early scoping tests performed to optimize the test setup indicated that if ozone was added to the test vessel while the vessel was under a vacuum, the ozone decomposed very rapidly, making it difficult to establish the desired test conditions. By feeding the ozone/oxygen mixture through the process vessel at atmospheric pressure, then monitoring the ozone concentration at the discharge, an accurate determination of the ozone concentration in the test vessel, just prior to the addition of hydrogen, may be made. The test vessel can then be evacuated just enough to add the prescribed volume of hydrogen for the test, minimizing loss of ozone.

Once the mixture has been prepared, the magnetic stirrer is turned off. The mixture is allowed to relax prior to ignition so that a quiescent mixture is tested rather than a turbulent mixture. The low pressure transducer valve is closed and the data collection is started. The mixture is manually ignited using the fuse wire providing a nominal 10 joules. The data acquisition records the pressure and temperature time traces in the vessel. A mixture is considered to be flammable if the pressure rise after the ignition event is greater than or equal to 7% in accordance with ASTM E918. The procedure is repeated until there are a minimum of two demonstrations conducted at the lowest fuel concentration to be considered flammable and three at the highest fuel concentration considered non-flammable.

3. Results and Discussion

3.1 Hydrogen Lower Flammable Limit

The lower flammable limit for initially quiescent hydrogen-air mixtures was determined in order to validate the test equipment setup and the procedure. The LFL for hydrogen in air was determined as 4.2 volume percent based on the ASTM 7% pressure rise criteria. A comparison of this result to data obtained by other researchers is summarized in Table 1.

Table 1: Hydrogen LFL in Air, this testing compared to others

Source	Hydrogen Concentration (vol. %)	Comments
LFL using 5% pressure rise criteria		
SAFEKINEX [4]	4.1	2.8-L vessel
	3.9	6-L vessel
	4.1	14-L vessel
	4.7	40-L vessel
	5.8	1250-L vessel
	4.1	2000-L vessel
LFL using 7% pressure rise criteria		
Present Work	4.2 ± 0.1	2-L vessel (L/D = 1)
Crowl and Jo [5]	5.7	20-L spherical vessel
Cashdollar et al. [6]	5.0 ± 0.5	8-L chamber
	6.0 ± 0.5	20-L chamber
	6.5 ± 0.5	120-L spherical vessel
LFL using a visual criterion		
Coward and Jones [7]	4.0	Flammability Tube

The experimentally measured LFL is in close agreement with the accepted lower flammable limit of 4 vol. % which was obtained through visual criterion [7]. Many other researchers have shown that it is difficult to obtain this value using the 7% pressure rise criteria. This has been explained through hydrogen's high diffusivity as compared to oxygen and nitrogen [6].

The majority of tests using the pressure rise criteria are performed with the ignition source in the center of the vessel. Central ignition is used for the determination of the explosion severity of flammable mixtures because it gives the maximum rate of pressure rise and the maximum overpressure. However, central ignition may not be useful when determining the limits of flammability. It has been shown that ignition source location has an impact on the flammability limits of methane [8]. As the location of the ignition source is lowered, a larger percentage of the combustible mixture participates in the reaction with minimal heat loss to the vessel wall. This results in a higher explosion overpressure, thereby widening the flammable region. The present work uses an ignition source placed approximately one-third from the bottom of the vessel.

3.2 Ozone Lower Flammable Limit

Initially, tests were conducted to determine the extent of ozone decomposition over time inside the combustion vessel at ambient conditions. This was accomplished by preparing an ozone-oxygen mixture and holding the mixture in the combustion vessel for approximately 30 minutes while stirring. After the specified hold time the ozone concentration was measured. It was found that there were negligible amounts of ozone decomposition that occurred in the vessel.

The LFL of ozone in oxygen was determined as 8.4 volume percent. Previous studies on the flammability of ozone in oxygen are presented in Table 2.

Table 2: Ozone LFL in Oxygen, this testing compared to others

Source	Ozone Concentration (vol. %)	Comments
Present Work	8.40 ± 0.17	2-L vessel (L/D = 1)
Mizutani et al. [9]	10.5	1-L cylindrical vessel, Electric spark
Koike et al. [10]	10.0	1-L cylindrical vessel, 1 J electric spark
Cook et al. [11]	9.5 ± 0.5	10 cm i.d. glass bulb, 0.47 J electric spark

The present work reveals that the explosion limit of ozone in oxygen is consistent with previously reported values; however, it is lower than other literature data which can be attributed to the differences in the experimental method and procedure. The lower explosion limit obtained in this work is most likely a result of a combination of factors: the type of ignition source, location of the ignition source, and vessel size and geometry. In these tests, the ignition source energy was setup to give approximately 10J which is much greater than that used in previous work. Flammability limits are a function of ignition energy, thus, by increasing the strength of the ignition source the flammable limits tend to widen [12]. There also tends to be some variation in the flammable limits which is influenced by the type of ignition source: a fuse wire as compared to an electric spark [13]. This may be an effect of the energy density provided by each type of ignition source and the amount of energy actually being transferred to the mixture. As discussed previously, the ignition source location may be playing a part in the lower value obtained.

The differences in the size and dimensions of the explosion vessels could have affected the lower flammable limit [13]. As the vessel size is increased, the heat losses to the wall become negligible. Through minimizing heat losses to the vessel wall, more heat is transferred to the combustion reaction, therefore, promoting flame propagation. This will result in a widening of the flammable region and combustion can occur at lower temperatures.

3.3 Lower Flammable Limit of Hydrogen in Ozone/Oxygen Mixtures

The experimental results for the lower flammable limits of hydrogen in the ozone/oxygen are shown in Figure 3. The solid circles represent a flammable mixture and the open circles represent a non-flammable mixture.

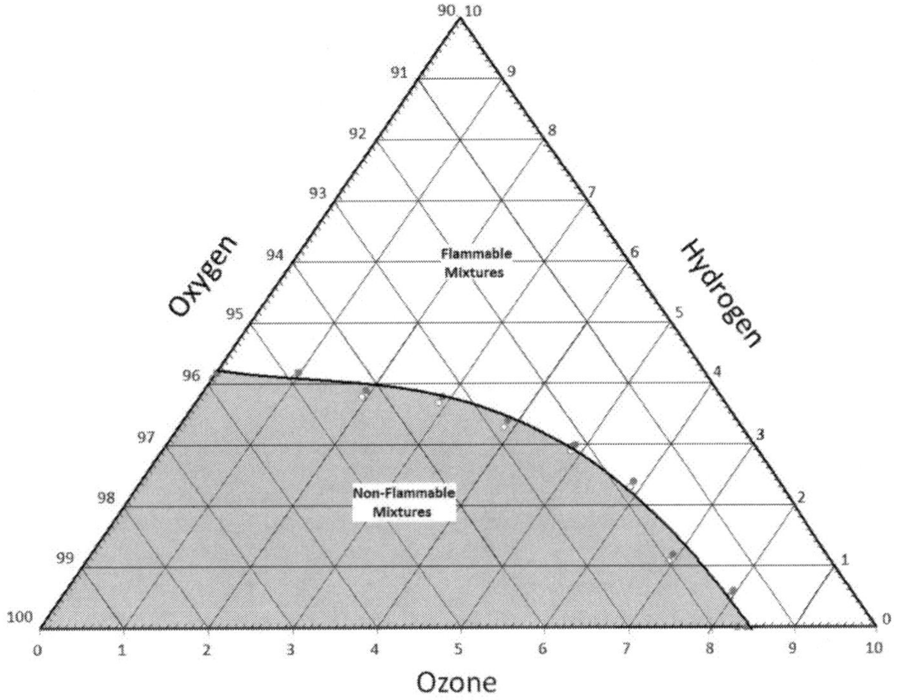

Figure 3: Flammability envelope for hydrogen/ozone/oxygen mixtures (all values are in volume percent).

Typically, it is of interest to estimate the flammability limits of multi-component mixtures. In the absence of experimental data, estimation methods can be useful if the flammable limits of pure components are known. Le Chatelier [14] developed an empirical equation for predicting the flammable limits of fuel mixtures,

$$LFL_{MIX} = \left(\sum_{i=1}^{n} \frac{y_i}{LFL_i} \right)^{-1} \qquad \text{Eq. [1]}$$

where y_i is the mole fraction of the ith component on a combustible basis and LFL_i is the lower flammable limit of the ith component in volume percent.

A comparison of the LFL predictions to the experimentally determined LFL of hydrogen in various ozone-oxygen mixtures is shown in Figure 4. It was found that the LFL of hydrogen decreases as the ozone concentration in the mixture increases. This result is consistent with the values approximated through Le Chatelier's rule; however, the experimental results are higher than the predicted calculations. Previous studies have shown that Le Chatelier's rule provides close approximations for many simple hydrocarbons; however, it is less accurate for mixtures with hydrogen and generally not applicable for substances that undergo decomposition [15]. However, from a safety stand point, Le Chatelier's rule provides a conservative estimate of the LFL.

The experimental data was further compared to flammability predictions from the calculated adiabatic flame temperature (CAFT) method. The flame temperature criteria used to assess whether a mixture is flammable or not was established by first calculating the adiabatic flame temperature at the determined LFL of hydrogen in air and ozone in oxygen. Three flame temperatures were used – the flame temperature of 8.4% ozone in oxygen, the flame temperature of 4.2% hydrogen in air and an average of these two flame temperatures. The LFL of hydrogen in various concentrations of ozone/oxygen was then calculated from these flame temperatures.

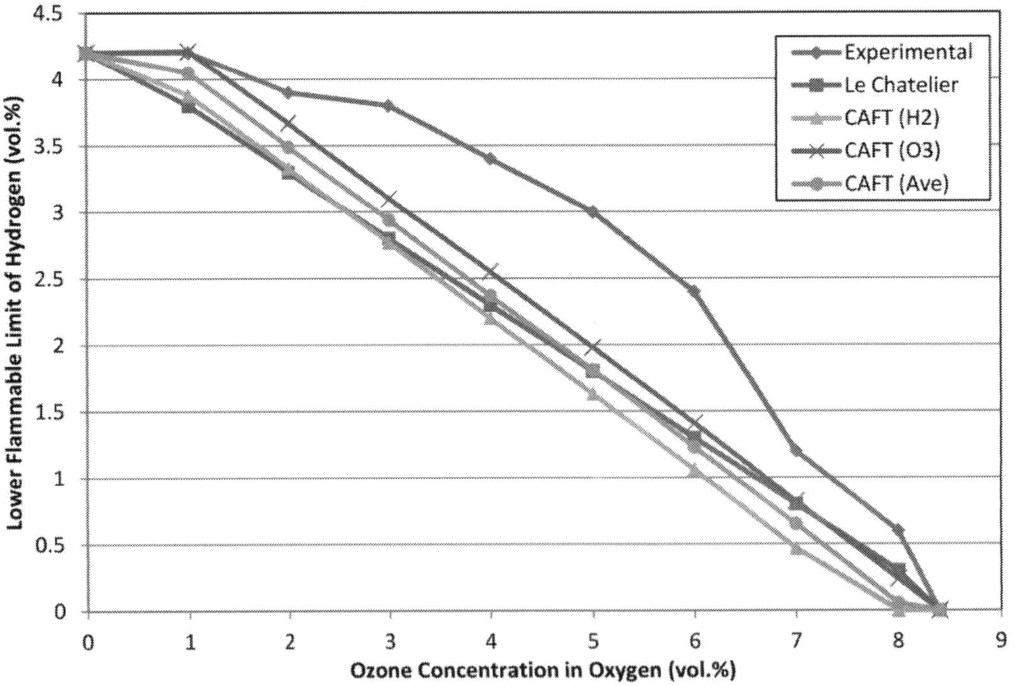

Figure 4: Comparison of lower flammability limit of hydrogen in ozone/oxygen to calculated adiabatic flame temperature predictions.

The calculated adiabatic flame temperature (CAFT) predictions are very similar to the predictions given by Le Chatelier's empirical equation. The deviations of the experimental data from the estimations in the flammability limits of a hydrogen/ozone/oxygen mixture lie within the unpredictability of ozone. This unpredictability comes from the instability of ozone, which tends to readily decompose to form oxygen. Furthermore, there are many different reactions that could be taking part in these experiments: H_2-O_2, O_3 decomposition, O_3-O_2, and H_2-O_3. The data seems to suggest that as the ozone concentration is increased, the reaction mechanism that takes precedence changes.

The deviation between the predicted and the experimental values may be two-fold. First, the reaction mechanism taking preference during the combustion reaction changes as the ozone concentration is increased. Secondly, the concentration of ozone participating in these reactions may be changing as well. Tests conducted on ozone decomposition in oxygen have shown that only 5.3% of the ozone decomposes for a gas mixture containing 6.7 mole % ozone and as high as 90% decomposition when the ozone concentration was 11.2% in the mixture [11]. In any case, additional work is needed to fully understand the nature of these experiments.

3.4 Temperature and Pressure Effect on the Lower Flammable Limit of Hydrogen-Ozone-Oxygen Mixtures

It has been reported previously that the ozone decomposition rate increases at the temperature of its environment increases [16]. Therefore, stability tests were performed again to determine the extent of ozone decomposition at the elevated test conditions. This time it was found that the ozone decomposition rate could not be considered negligible. With initial mixtures containing 5 vol.% ozone in oxygen at approximately 75°C, the ozone decomposition rate was approximately 0.1 vol.% ozone every 10 minutes.

The results from the tests performed at elevated temperatures and pressures are shown in Figure 5. It was found that as the temperature is increased the LFL decreases and the opposite was observed when the pressure was increased.

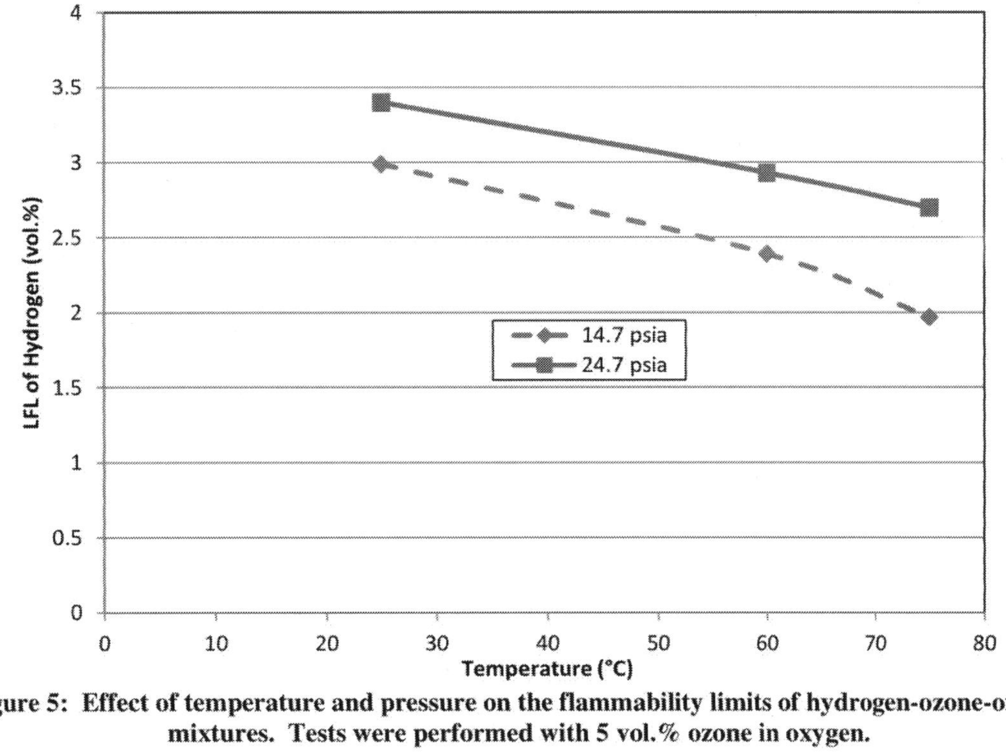

Figure 5: Effect of temperature and pressure on the flammability limits of hydrogen-ozone-oxygen mixtures. Tests were performed with 5 vol.% ozone in oxygen.

The increase in the LFL as a result of the increase in pressure is generally not characteristically observed for flammable gases; however, this is a characteristic of hydrogen flammability [17]. This result leads to the conclusion that the reaction mechanism responsible for the increased LFL as the pressure is increased for hydrogen is still the dominate mechanism when ozone is added. Furthermore, it is noticed that as the temperature is increased the decrease in the LFL is fairly linear throughout the range of temperatures tested. The LFL results from these tests show that the LFL is more susceptible to elevated temperatures when compared to results of hydrogen-air or hydrogen-oxygen. One explanation for this is that at the increase temperatures more of the ozone participates in the reactions that occur in the vessel. Previous research on ozone decomposition has found that only a part of the ozone decomposes when the ozone concentration is below its LFL. At the higher temperatures, the reaction rates increase as well as the instability in the ozone molecule.

3.5 Effect of Relative Humidity on the Lower Flammable Limit of Hydrogen-Ozone-Oxygen Mixtures at Elevated Temperatures and Pressure.

Tests conducted with the addition of water vapor have shown a slight increase in the LFL as shown in Figure 6. Since there is an additional component (water vapor) added to the mixture, the results should be compared on a total combustibles (hydrogen plus ozone) basis rather than just on the concentration of hydrogen in the mixture. This is because as more water vapor is added to the mixture, the concentration of ozone decreases; therefore, even if more hydrogen was needed to be added to the mixture, the total amount of combustibles may have remained constant to produce a flammable mixture.

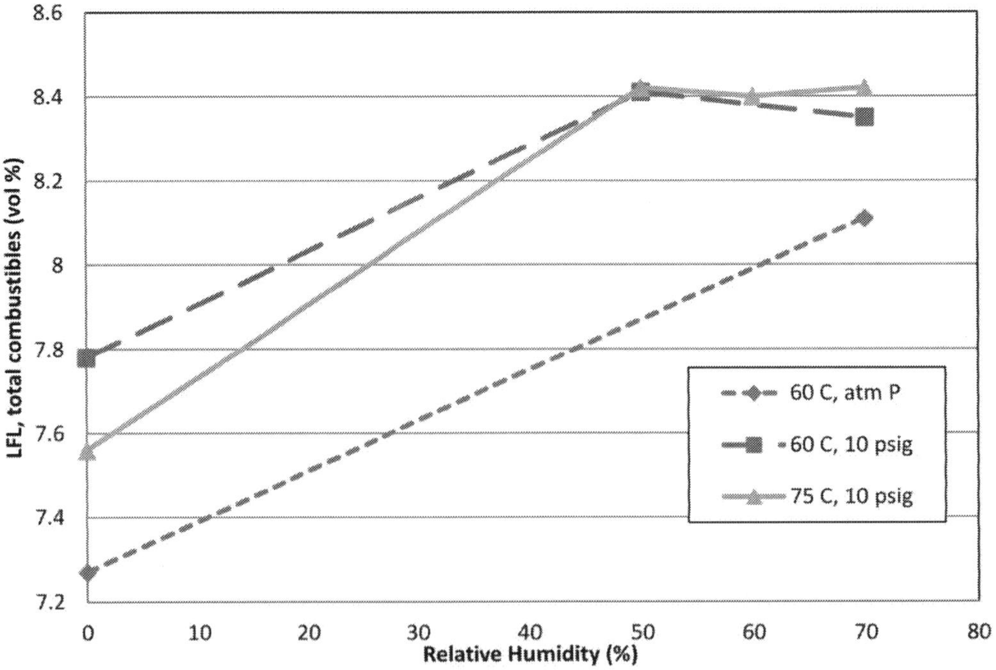

Figure 6: Effect of relative humidity on the LFL of hydrogen-ozone-oxygen mixtures. Tests were performed with 5 vol.% ozone in oxygen.

The slight increase in the LFL could be a result of either the addition of the water vapor or decrease in ozone concentration. The addition of water vapor to the mixture helps extract heat from the reactions occurring in the vessel (combustion or decomposition), thereby, quenching the flame propagation by lowering the reaction rate. Furthermore, the observed increase in the LFL could be a result of a decrease in the ozone concentration. It is known that both hydrogen and water act as a catalyst for ozone decomposition; therefore, the actual concentration inside the vessel prior to igniting the mixture is lower than the reported concentrations.

4. Impact on Operation

This testing was designed to provide data under the conditions considered most optimal to produce a deflagration. The geometry and materials of construction of the testing vessel; the location of the fuse wire; the magnitude of the supplied energy; the careful minimization of diluents and other contaminants; and meticulous procedural detail to maintain integrity of the ozone to the maximum extent practical, result in data that reflect not the expected process conditions, but those that enhance the possibility of flame propagation. For this reason, there is believed to be considerable conservatism in the indicated results. Per the vendor, the maximum possible ozone concentration producible by the planned ECC Ozone generator is 8 volume percent (typical maximum expected concentration is 6.8 vol%), less than the 8.4 minimum volume % concentration shown to be flammable in a pure O_2 environment at the optimally conservative conditions established in this testing. As both water vapor and nitrogen, together constituting approximately 6.5 % of the actual feed stream, are powerful diluents, it is believed not credible that deflagration can occur at this maximum feed concentration condition. In addition, once the ozone stream contacts the waste stream, the many simultaneous oxidizing reactions will rapidly decompose available ozone to well below flammable levels. Further, because the radiolytically generated hydrogen quantity is negligible compared to the supplied ozone/oxygen stream (0.0004 moles per minute H_2 vs 76 moles per minute ozone/oxygen), even the total H_2/O_3 mixture, without crediting decomposition reactions, does not approach flammable concentrations. Finally, even at the "end" of the ECC batch cycle, when most of the metal oxalates have been decomposed, testing has indicated that the ozone concentration in the vapor space of the ECC process vessel reaches a concentration of no more than 3 vol%, remaining well below concentrations of concern.

The major issue for the ECC operation established by this testing is the impact of the data when applied to off normal conditions. While it is possible to discontinue ozone addition to the reaction vessel at any time, the radiolytic hydrogen generation rate continues, varying slightly as ambient pressures and temperatures change. Relative to the data generated and analyzed in this testing, the ECC hazards analysis team will re-evaluate off normal conditions (e.g. those during which process exhaust ventilation is lost) such that issues involving mixtures of hydrogen and ozone in the vapor space can be appropriately controlled.

5. Conclusion

The LFL of hydrogen in air was determined and is in good agreement with the literature data. Ozone-oxygen mixtures were found to be flammable at concentrations above 8.3 vol.% based on the ASTM E918 7% pressure rise criteria for flame propagation. This result is lower than previously reported values which can be explained through the variations in the test setup and procedure. Tests performed with hydrogen in various concentrations of ozone in oxygen have shown that the LFL of hydrogen decreases as the concentration of ozone in the mixture increases.

The estimations of the LFL for various mixtures of hydrogen and ozone in oxygen was performed using Le Chatelier's rule as well as the CAFT method. It was found that both methods underestimate the LFL for mixtures of hydrogen and ozone.

The effects of temperature, pressure, and relative humidity were found for hydrogen-ozone mixtures. The trends observed through these tests were consistent with the general trends in reported results for hydrogen. Overall, the difficulty with some of these tests ultimately lies with the unpredictability of ozone.

6. References

[1] ASTM. (2005). ASTM E 918-83 Standard Practice for Determining Limits of Flammability of Chemicals at Elevated Temperature and Pressure. Philadelphia, PA: American Society for Testing and Materials.

[2] Mashuga, C. V. (1999). *Determination of the Combustion Behavior for Pure Components and Mixtures using a 20-Liter Sphere.* Phd Thesis. Michigan Technological University. Ann Arbor: UMI Company.

[3] ASTM. (2005). ASTM A967-05, Standard Specification for Chemical Passivation Treatments for Stainless Steel Parts. Philadelphia, Pa: American Society for Testing and Materials.

[4] Project SAFEKINEX, Contract No. EVG1-CT-2002-00072, Deliverable No.8, *Report on experimentally determined explosion limits, explosion pressures and rates of explosion pressure rise – Part 1: methane, hydrogen, and propene,* March 2006.

[5] Crowl, D. A., & Jo, Y.-d. (2009). A Method for Determining the Flammable Limits of Gases in a Spherical Vessel. *Process Safety Progress , 28* (3), 227-236.

[6] Cashdollar, K. L., Zlochower, I. A., Green, G. M., Thomas, R. A., & Hertzberg, M. (2000). Flammability of methane, propane, and hydrogen gases. *Journal of Loss Prevention in the Process Industries , 13*, 327-340.

[7] Coward, H. F., & Jones, G. W. (1952). *Limits of Flammability of Gases and Vapors.* BuMines, 503.

[8] Van den Schoor, F., Norman, F., & Verplaetsen, F. (2006). Influence of the ignition source location on the determination of the explosion pressure at elevated initial pressure. *Journal of Loss Prevention in the Process Industries* , 459-462.

[9] Mizutani, T., Matsui, H., Sanui, H., & Yonekura, M. (2001). Decomposing detonation and deflagration properties of ozone/oxygen mixtures. *Journal of Loss Prevention in the Process Industries*, *14*, 559-565.

[10] Koike, K., Nifuku, M., Izumi, K., Nakamura, S., Fujiwara, S., & Horiguchi, S. (2005). Explosion properties of highly concentrated ozone gas. *Journal of Loss Prevention in the Process Industries, 18*, 465-468.

[11] Cook, G. A., Spadinger, E., Kiffer, A. D., & Klumpp, C. V. (1956). Explosion Limits of Ozone-Oxygen Mixtures. *Industrial and Engineering Chemistry*, *48* (4), 736-741.

[12] Bartknecht, W. (1981). *Explosions, Course Prevention Protection.* New York: Springer-Verlag.

[13] Project SAFEKINEX, Contract No. EVG1-CT-2002-00072, Deliverable No. 2, *Report on the experimental factors influencing explosion indices determination*, September 2003.

[14] Le Chatelier, H. (1891). Note on the Determination of Firedamp by Flammability Limits. *Ann. Mines*, *19*, 388-395.

[15] Babrauskas, V. (2003). *Ignition Handbook.* Issaquah: Fire Science Publishers.

[16] Koike, K., Inoue, G., Suzuki, M., Saida, J., & Komatsu, I. (1998). Decomposition Characteristics of Concentrated Ozone. *Journal of Chemical Engineering of Japan*, *31* (2), 195-200.

[17] Schroeder, V., & Holtappels, K. (2005). Explosion Characteristics of Hydrogen-Air and Hydrogen-Oxygen Mixtures at Elevated Pressures. *International Conference on Hydrogen Safety.* Pisa.

9TH
GLOBAL
CONGRESS
ON PROCESS
SAFETY

Process Safety Considerations for New DOE Depleted Uranium Hexafluoride Conversion Plants

Peter M. Burban, Ph.D.
Department of Energy
Portsmouth Paducah Project Office
Peter.Burban@lex.doe.gov

Prepared for Presentation at
American Institute of Chemical Engineers
2013 Spring Meeting
9th Global Congress on Process Safety
San Antonio, Texas
April 28 – May 1, 2013

UNPUBLISHED

Process Safety Considerations for New DOE Depleted Uranium Hexafluoride Conversion Plants

Peter M. Burban, Ph.D.
Department of Energy
Portsmouth Paducah Project Office
Peter.Burban@lex.doe.gov

Keywords: depleted uranium hexafluoride, dry conversion, fluidized bed, uranium oxide, hydrofluoric acid, layers of protection, independent safety system, defense in depth, basic process control system, HF detectors, passive safety systems, Department of Energy

Abstract

The Department of Energy (DOE) OH & KY depleted uranium hexafluoride (DUF6) facilities are first of a kind operating plants designed to convert DUF6 into a safer form for reuse and/or disposal. Safety played a lead role in process design with HAZOP results guiding implementation of layers of protection to reduce risks associated with several materials including UF_6, H_2 and HF. A stand-alone independent safety control system functions to provide independent monitoring and control capability for those processes that have significant safety considerations.

1. Introduction

The Department of Energy Piketon, OH and Paducah, KY DUF6 facilities, Figure 1, are new fluid bed operating plants designed to convert approximately 700,000 metric tons of depleted uranium hexafluoride into a safer form for reuse and/or disposal. Depleted uranium hexafluoride has been generated in the United States since the mid-1940's as a byproduct of enriching uranium for both civilian and military applications at three gaseous diffusion plants located near Oak Ridge, Tennessee; Paducah, Kentucky; and Piketon, Ohio. This paper will highlight important process safety design considerations implemented that are critical to ongoing operations. Key aspects of the process hazards analysis and some lessons learned will be shared regarding chemical hazards associated with HF.

Figure 1. Portmouth (top) and Paducah DUF6 facilities

2. DUF6 Conversion Process and Safety Considerations

2.1 Process Description

The source DUF6 material, stored at both plants in approximately 63,000 large steel cylinders, is fed direct from the storage cylinders. A DUF6 feed cylinder, heated in a specially designed autoclave, vaporizes DUF6 which is fed with steam and hydrogen, blended with nitrogen, to a high temperature fluidized bed reactor for conversion to uranium oxide powder and hydrofluoric acid vapor. Some elements of the chemical process are described by Urza [1] and a general schematic of the process is given on the internet [2]. The uranium oxide powder transfer system loads powder into modified feed cylinders. The HF recovery system condenses the water and HF vapors in the process offgas, producing aqueous HF acid which is temporarily stored in tanks and later loaded for shipment to HF vendor facilities. The environmental green aspects of this approach include reuse of the feed DUF6 cylinders, regeneration of KOH used in scrubber operation and recovery of aqueous HF for commercial sale.

2.2 Overall Process Safety

A major safety advantage of the conversion process is low pressure reactor operation. The 'dry' conversion process uses superheated steam and nitrogen diluent to ensure safe reactor hydrogen concentrations. Administrative control of the feed cylinders ensures only depleted UF_6 is processed, eliminating any nuclear criticality event.

Hazard evaluations were performed to develop accident scenarios with the potential for exposure to HF, the greatest chemical hazard. Unmitigated consequences and risks were estimated for the credible events and compared to DOE risk guidelines. This paper focuses how safety significant engineered features and administrative controls are integrated in layers of protection to prevent or mitigate accidents associated with the major HF risks [3].

An Integrated Control System (ICS) controls and monitors the facility automated systems and associated components during normal operations. The system also mitigates accidents involving the conversion process and any of the systems or activities that have an interface with the process. The ICS consists of BPCS (Basic Process Control System) and ISS (Independent Safety System) subsystems. The purpose of the BPCS is to provide monitoring and control for normal operations from the Control Room and/or local operator workstations. The purpose of the ISS is to provide independent monitoring and control capability for those processes that have significant safety considerations. The ISS does not rely on the ISS human machine interface or BPCS to perform interlock actions.

3. Safety Features

3.1 Feed System Safety Features

A major component of the feed system is the autoclave, which serves to vaporize DUF6. The autoclave also serves as a pressure containment boundary (at several multiples of the supply pressure) to mitigate the consequence of an accidental release of DUF6 from a breached feed cylinder/piping in the autoclave. Since UF_6 reacts with atmospheric moisture to generate HF vapor, both HF and UF_6 releases have similar hazards classifications.

In the event of an over pressurized autoclave leading to activation of the pressure relief, the vent gases are directed to a controlled plenum. Automatic interlocks shut off DUF6 flow and/or electrical power to the cylinder heating system for an autoclave high pressure condition, as well as the following abnormal operating conditions: cylinder over-temperature, DUF6 header high-pressure, and an open autoclave door. The ISS monitors UF_6 and HF detectors located in the autoclave room which, upon alarm, terminate DUF6 flow to the downstream piping via an autoclave isolation valve. Additional administrative controls lead to shutdown of the reactor system. These layers of protection provide 'defense in depth' to reduce worker exposure from an abnormal UF_6 feed cylinder/system release.

3.2 Reaction System Safety Features

Each reactor serves as a primary containment boundary for reactants and products including DUF6 and HF. The ISS monitors UF_6 and HF detectors located in the reactor system room and adjoining HF condenser room which, upon alarm, terminate DUF6 flow to the downstream piping via the autoclave isolation valve. The BPCS in-line process oxygen monitors alarm on detection of oxygen in the process off gas. Additional administrative controls lead to shutdown of the reactor system.

Upon significant UF6 or HF leak detection, specific administrative controls instruct control room operations to terminate hydrogen flow via an automatic hydrogen feed isolation valve (a redundant manual isolation valve is also closed). Additional administrative controls lead to shutdown of the reactor system. These layers of protection provide 'defense in depth' to reduce worker exposure resulting from breach of the reactor system.

3.3 HF Recovery System Safety Features

Off gases from each reactor pass through individual partial condensers which produce an aqueous HF stream collected in individual primary receivers located in the HF recovery room. Off gases continue flowing to first a water and then to a KOH scrubber. All scrubber outlet streams merge and flow to a final backup KOH scrubber prior to venting out the facility stack. Multiple scrubbers reduce the likelihood of stack HF emissions. A secondary overflow HF receiver for each primary mitigates primary receiver level control problems. The HF recovery system units are skid mounted and reside in a common secondary containment dike mitigating the effect of HF leaks.

The ISS monitors several HF detectors located in the HF recovery room which, upon alarm, terminate DUF6 flow to the specific reactor via the autoclave isolation valve. The ISS also monitors HF detectors located in the HF recovery area for HF leaks in the HF transfer line which, upon alarm, terminates flow from the HF receiver tanks. Upon significant HF leak detection, specific administrative controls instruct control room operations to terminate DUF6 and hydrogen flow via an automatic hydrogen feed isolation valve (a redundant manual isolation valve is also closed). Additional administrative controls lead to shutdown of the reactor system. These layers of protection provide 'defense in depth' to reduce worker exposure resulting from breaches in the HF recovery system.

An important series of lessons learned include proper maintenance of HF process lines to eliminate chronic leaks. In addition, care was taken for stress relief of all piping flange connections, and modifications were made to address piping areas experiencing vibration. The use of HF vapor highly sensitive paint was helpful to identify the onset of leaks. Additional isolation valves were installed to improve the safety and ease of corrective maintenance.

3.4 HF Storage Tank Farm Safety Features

The HF storage area includes several passive, active and administrative safety controls. Each HF storage tank is situated in a secondary liquid dike that can hold the entire capacity of the 10,000 gallon tank in case of a leak. The tanker truck and rail loadout areas have a gravity drain equipped with a pumping system to transfer spilled HF to either a containment dike or a storage tank. This minimizes the spill area to reduce the potential on and off site consequences of an HF leak.

The ISS also monitors HF detectors located in the HF storage area which, upon alarm, terminates flow via the HF storage tank automated isolation valve. In addition, the ISS terminates HF flow from the HF receiver tanks. Specific administrative controls instruct an operator to be in required attendance for all transfers. This operator observes each HF storage tank to mobile tanker transfer and can terminate the transfer via an emergency button which stops HF transfer flow even before HF detection by the ISS. Upon alarm, following administrative control, the operators in the control room isolate the HF recovery system and then shutdown the reactor system.

These layers of protection provide 'defense in depth' to reduce worker exposure resulting from leak accidents at the HF storage tank system.

3.5 Additional Safety Features

Several fire suppression systems minimize the likelihood of a fire causing a cylinder breach and release of HF and/or DUF6. Strategically placed vehicle barriers add "defense in depth" to reduce consequences associated with collision with storage tanks and staged feed cylinders. Safety design elements were also incorporated for natural event accident initiators so that process buildings and equipment meet design criteria for forces such as seismic and high winds. Safety management requires the use of appropriate personal protection equipment for all work activity with potential HF exposure. These layers of protection provide 'defense in depth' to reduce worker exposure resulting from HF leak accidents in various systems described above.

4. Discussion

The process safety approach at the DUF6 facilities blends active, passive and specific administrative controls. Considering the other chemical and radiation hazards at the facility, the ability to maintain a safe operation depends on reliable machines and human performance of required operator actions. A rigorous preventative maintenance program is administratively controlled to maintain the reliability of all active safety systems at the facility in line with DOE safety requirements. The Integrated Safety Management Program at the DUF6 facility includes systematic approach to training for Operations. Training includes preparing each individual with the toolbox needed to address all specific administration controls, responses to alarms and recovery from abnormal operating conditions. Monitoring of human performance is an essential element for maintaining safe operations and continuous improvement.

Major facility start up challenges included accurate procedure development and updates, initial training and continuous training as the Operations knowledge base grew. Unique lessons were learned for operations in the four season climates of Ohio and Kentucky, necessitating adaptation of procedures, and emphasis on minor equipment changes to minimize the need for corrective maintenance. Both sites have had a good start to their multiple decade operations campaigns built on the cornerstone of a defense in depth process safety approach.

5. References

[1] Urza, Inaky J., "Conversion of uranium hexafluoride to uranium dioxide" US Patent 4,830,841

[2] "The DUF6 Process," www.bwconversionservices.com, accessed on Feb. 11, 2013

[3] DOE, Final Environmental Impact Statement for Construction and Operation of a Depleted Uranium Hexafluoride Conversion Facility at the Portsmouth, Ohio, Site, DOE/EIS-0360, prepared by Argonne National Laboratory for U.S. Department of Energy- Oak Ridge Operations, Office of Environmental Management, Oak Ridge, Tenn. June 2004. energy.gov/nepa/downloads/eis-0360-final-environmental-impact-statement

(44b) Minimum Ignition Energy Study of Flowing Heat Transfer Fluid Aerosols: Experimental and Theorectical Approach

Szu-ying Huang[1], Xinrui Li[1] and M. Sam Mannan[2], (1)Mary Kay O'Connor Process Safety Center, Artie McFerrin Department of Chemical Engineering, Texas A&M Univerisity, College Station, TX, (2)Mary Kay O'Connor Process Safety Center, Artie McFerrin Department of Chemical Engineering, Texas A&M University, College Station, TX

Liquid-gas aerosol has been a significant concern in process industries for its complexity and unpredictability of formation, ignition probability, combustion mechanisms, and flame propagation with impact on onshore/offshore facilities. Heat transfer fluid, specifically, has been proved to have the possibility to form aerosol from pressurized leaking, spraying, and potential liquid impingment from surfaces. Having more combustible matter and higher energy density, the high-flash point aerosol may induce more hazardous fires and explosions even operated or stored under the safety limits. Among the indicators describing flammability, minimum ignition energy (MIE) is one of the most widely used and studied ones. It is capable to reflect the internal chemical properties of the material, as well as to make connection to analyze the possible facility ignition sources. This paper demonstrates MIE experimental result of commercial heat transfer fluid aerosols, and develops COMSOL-Multiphysics models as theoretical approach on MIE prediction of similar chemical species. The influence of aerosol droplet sizes, initial spray velocity, fuel-air equivalence ratio, and fluid temperature are analyzed. The characteristic ignition temperature, which is the "flash point" of aerosol system, is also developed to compare with single-phase material. Results of the study may be used for industry liquid handling, working fluid selection, mitigation design, and ignition elimination to prevent fire and explosion of aerosols.

See more of this Session: Poster Session
See more of this Group/Topical: Global Congress on Process Safety

Domino Effect Model Applied to the Hertfordshire Oil Storage Terminal (Buncefield Oil Depot)

Manuscript ID: 295031

Esteban Bernechea
Centre for Technological Risk Studies (CERTEC). Department of Chemical Engineering.
ETSEIB. Universitat Politècnica de Catalunya.
Av. Diagonal, 647. 08028-Barcelona (Spain).
esteban.bernechea@estudiant.upc.edu

Xavier Seguí
Centre for Technological Risk Studies (CERTEC). Department of Chemical Engineering.
ETSEIB. Universitat Politècnica de Catalunya.
Av. Diagonal, 647. 08028-Barcelona (Spain).
xavier.segui@upc.edu

Josep Arnaldos
Centre for Technological Risk Studies (CERTEC). Department of Chemical Engineering.
ETSEIB. Universitat Politècnica de Catalunya.
Av. Diagonal, 647. 08028-Barcelona (Spain).
josep.arnaldos@upc.edu

1. Introduction

On December 11, 2005, a series of explosions occurred at the Buncefield Oil Storage Terminal, located in Hemel Hempstead, Hertfordshire, England. The explosions resulted in large fires which spread throughout the installation, causing more than 40 injuries, the destruction of most of the installation and the emission of large clouds of black smoke to the atmosphere; the accident also caused significant damage to residential and commercial properties in the vicinity of the terminal, the evacuation of population centers near the terminal, and an overall economic loss estimated at close to £ 1 billion [1].

The Buncefield disaster is a perfect example of domino effect in the oil industry; the initial explosion, caused due to the overflow of one of the tanks, caused fires that engulfed more than 20 storage tanks, and resulted in the destruction of a significant part of the terminal. The domino effect is defined as a cascade of events in which the consequences of previous accidents are increased both spatially and temporally by following ones, thus leading to a major accident [2]. This definition describes perfectly the situation that occurred in the Buncefield terminal.

One of the characteristics that made this accident the object of many researches was that the overpressure wave ensuing from the first explosion was much higher than would be expected from a Vapor Cloud Explosion (VCE). However, the aim of this manuscript is not to model the exact explosion, or to determine the reasons that caused the overpressure resulting from it to be higher than expected; much works have been dedicated to this endeavor, producing satisfactory conclusions to the phenomenon. The objective of this work is rather to use the Buncefield Terminal layout to test a domino effect model proposed by the authors [3]; to achieve this, possible accident sequences after an explosion will be simulated, to ascertain if they are similar to the chain of accidents that took place during the real event. Also, Quantitative Risk Assessments (QRAs) will be performed using the layout: one in a traditional fashion [4], without accounting for domino effect, and another taking domino effect into account through the application of the proposed model. This is done to analyze the change in the iso-risk curves obtained for both QRAs.

First, a more detailed description of the accident is presented, followed by an explanation of the proposed domino effect model. After this, the way in which the model has been applied to the layout of the installation is described and the results obtained are presented and discussed. Finally, the conclusions reached by the end of the study are shown.

2. Description of the accident

A brief abstract of the incident as described in Volume 1 of The Final Report of the Major Incident Investigation Board [1] is presented in this section.

On Saturday 10, December 2005, a delivery of unleaded petrol started to arrive at Tank 912 in bund A. At about 05:30 am on 11 December, the safety systems that prevented the overfilling of the tank failed to operate. Petrol cascaded down the side of the tank and was collected in Bund A. As the released continued, a vapor cloud mixture of fuel and vapor was formed and started dispersing westwards. Up to 300 tonnes of petrol were released, and about 10% turned to vapor and mixed with air, eventually reaching dangerous concentrations. At 6:01 am on Sunday 11, December 2005, the first, in a chain of explosions, occurred. The explosions caused a huge fire which engulfed more than 20 tanks and burned for 5 days, destroying great part of the terminal.

3. Domino effect model

The model used in this work [3] was developed to assess the impact that the domino effect could have on the frequencies of the accidents used to perform QRAs on storage installations; however, it can also be used to produce stand-alone accident sequences.

The model is an algorithm that has been programmed in MATLAB; it follows the steps used to perform a QRA in a cyclical manner to produce accident sequences which, in the end, will produce frequencies of accidents in all the tanks that are part of the installation.

The first step to obtain the final frequencies of accidents is to input some initial data that describes the installation; these are: the number of tanks in the installation, the substances stored and their properties, the mass contained in each tank, the sizes of the containment dikes, the types of tanks (atmospheric, pressurized), the initial frequencies of accidents per tank and the distances between tanks.

After the characteristics of the installation have been defined, the initial Loss of Containment events (LOCs) that will occur in each of the units, along with their frequencies have to be specified. Using these LOCs, and knowing the substance stored, event trees can be used to propose sets of possible accidents, together with their probability of occurrence; when these probabilities are multiplied by the frequency of the LOC, the initial frequency of accident is obtained. The initial LOCs and the event tree used in this work are those proposed in "The Purple Book" [4]. The LOCs for atmospheric storage tanks and their frequencies are presented in Table 1.

Table 1: Initial LOCs in atmospheric storage units

Code	Definition	Frequency (y^{-1})
G.1	Instantaneous release of the complete inventory.	5×10^{-6}
G.2	Continuous release of the complete inventory in 10 min at a constant rate of release.	5×10^{-6}
G.3	Continuous release from a hole with an effective diameter of 10 mm.	1×10^{-4}

The steps followed up to this point are the same as for a regular QRA. To produce the accident sequences, one of the tanks will be selected to suffer the initial LOC, and the effects and consequences of the accidents in this unit will be calculated for all other tanks. Depending on the consequences, a level of damage, corresponding to a new LOC will be assigned to each of the affected units. This was done using the threshold values presented in Table 2.

Table 2: Threshold values applied in the model

Accident	Effect	Threshold Values		Consequence
		Pressurized Tanks	Atmospheric Tanks	
BLEVE	Thermal radiation (kW/m2)	≥ 100	≥ 100 [5]	G.3 type release
		< 100	< 100 [5]	No consequence
Explosion	Overpressure wave (kPa)	≥ 100	≥ 25 [6]	G.1 type release
		$53 < \Delta P < 100$ [7]	$20 < \Delta P < 25$ [7]	G.2 type release
		$30 \leq \Delta P < 53$ [7]	$14 \leq \Delta P < 20$ [7]	G.3 type release
		$\Delta P < 30$ [5]	$\Delta P < 14$ [8]	No consequence
Jet fire	Jet length (m) Heat Radiation (kW/m2)	$dj \geq d$ [5]	$dj \geq d$ [5]	G.1 type release
		$dj < d$ & $I \geq 40$ [5]	$dj < d$ & $I \geq 10$ [5]	G.3 type release
		$dj < d$ & $I < 40$ [5]	$dj < d$ & $I < 10$ [5]	No consequence
Flash fire	-	-	-	No consequence
Pool fire	Thermal radiation (kW/m2) Pool duration (min)	$dp \geq d$ [5]	$dp \geq d$ [5]	G.1 type release
		$60 \leq I$ & $tp \geq 10$ & $dj < d$ [9]	$15 \leq I$ & $tp \geq 10$ & $dj < d$ [9]	G.2 type release
		$I \geq 8$ & $dj < d$	$I \geq 8$ & $dj < d$	G.3 type release
		$I < 8$ & $dj < d$	$I < 8$ & $dj < d$	No consequence
dj = jet length; dp = pool diameter; d = distance between tanks; tj = duration of jet fire; tp = duration of pool fire				

Once new LOCs resulting from an initial accident have been defined in the affected units, the effects and consequences of subsequent events on surrounding remaining units will be estimated; this will lead to other LOCs in other tanks, producing more and more branches of possible events. For each of the levels of propagation, the frequencies of accidents will be calculated, using the frequency of the previous accident as the initiator and the same event tree for the probabilities of possible accidents. Once one of the accidents in a branch reaches a frequency lower than 1×10^{-9}, the sequence will stop, as it is deemed that this frequency is too low to be taken into account in the QRA.

Figure 1. Domino effect model algorithm

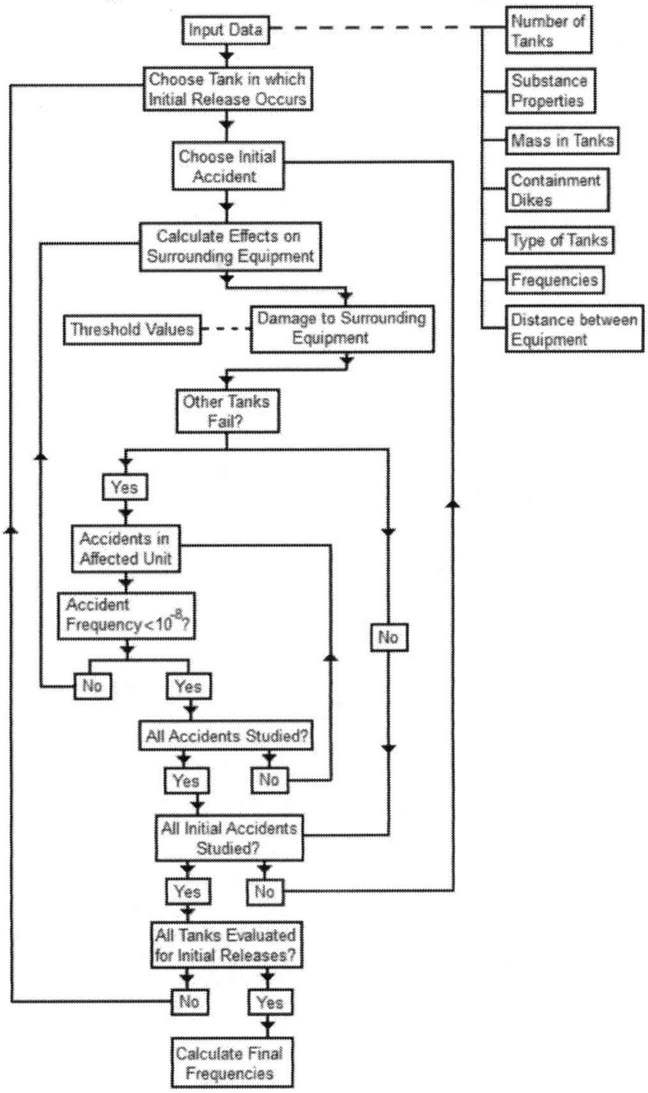

The process has to be repeated in loops, until all sequences have been explored, and all the tanks have been selected as initiators (suffering the initial LOC). In each step, the frequencies with which the different accidents occur in the different tanks are stored. Finally, these frequencies are added, by accident and by tanks, (for example, adding all the times that "accident 1" possibly occurs in "Tank 1"), whether due to an initial LOC event or due to the effect of an accident in another tank; these frequencies are referred to as final, and are the ones used in the QRA. A diagram of the algorithm that is followed in the methodology is presented in Figure 1.

The results obtained using the model can be compared against those obtained from a traditional QRA in order to assess the impact of domino effect on the risk associated to the installation. The model can also be used to produce a singular accident sequence, resulting from a specific accident in one of the units.

Figure 2. State of the terminal and proposed layout

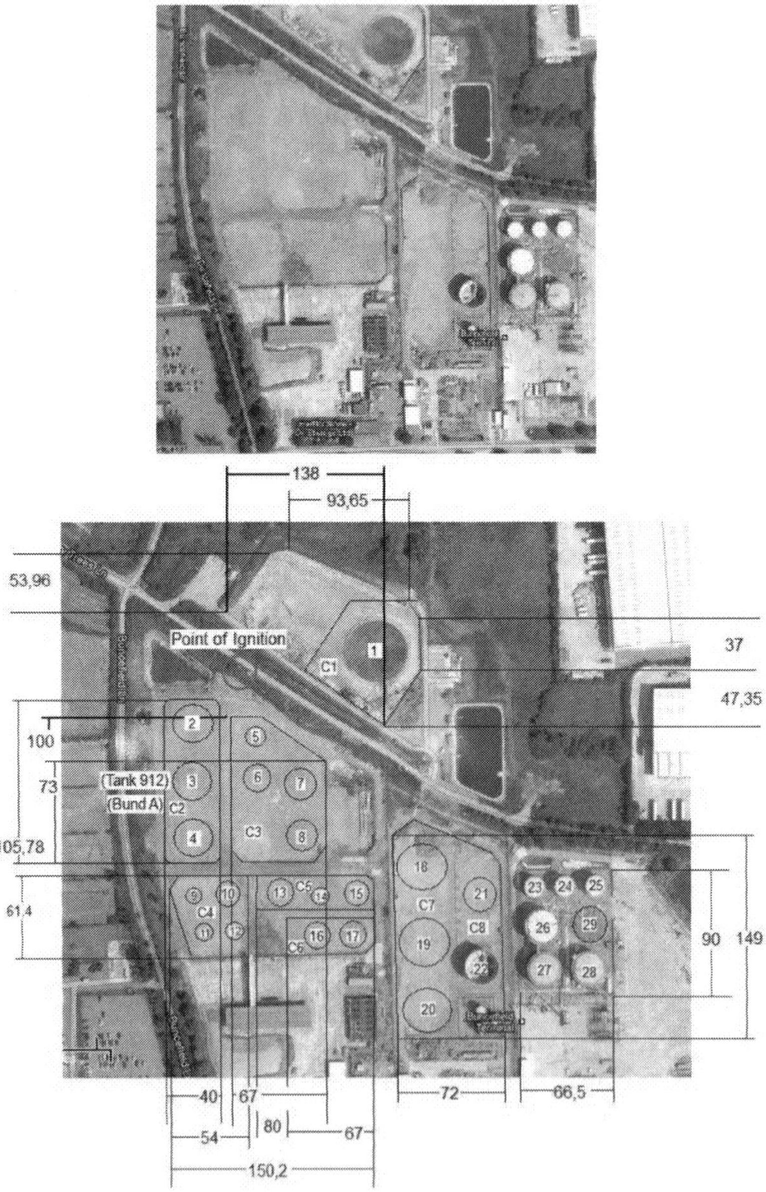

4. Applying the model to the Buncefield terminal

In this section the domino effect model will be applied to the terminal. First, to produce an accident sequence in a situation similar to the real accident and later, to produce iso-risk curves in two different QRAs.

4.1 Modeling the installation

To model the installation, the "Google Maps" tool was used to view the current state of the terminal, and obtain measures similar to those of the tanks that were destroyed and of the tanks that are still standing, as well as the distances between the tanks and the dimensions of the containment bunds. Figure 2 shows, on top, the latest image found of the state of the installation and on the bottom, the distances and dimensions used to model the installation. After estimating the dimensions of the installation, the tanks and bunds were renumbered for the application of the model; for example, Tank 912 becomes Tank 3, and Bund A is called C2.

4.2 Approaches to Accident Modeling

The vapor cloud that resulted in the explosion in the Buncefield Terminal is thought to have ignited at the emergency pump house located close to the containment bund A (in which Tank 912 was located), as shown in Figure 2. Since the domino effect model works by going through a fixed set of possible accidents in each of the tanks, the approach followed to produce the accident sequences was to define a fictional tank "Number 30" in the area in which the ignition of the flammable cloud is thought to have occurred; this tank is modeled containing a mass close to 150,000 kg of pentane.

This approach was followed to produce the accident sequence, but not to develop the QRAs, which were completed using the layout proposed in Figure 2, with all the tanks filled at full capacity.

4.3 Accident Sequence

Figure 3 shows the most probable incidents in the sequence produced when the domino effect method is applied to the case of an explosion derived from a G.2 type release in Tank 30 (the fictional tank created to model the initial explosion). The tanks are represented inside circles, while the possible accidents are shown inside diamonds (E stands for explosion, P for pool fire). The lines generated from the accident diamonds represent the level of damage caused by an event; these lines later divide, reaching different tanks, which means that the tank suffers the associated type of release after the previous incident. All the lines shown in the figure represent possible accidents identified by the methodology. The continuous lines represent the most probable accidents, which derive in the destruction or heavy damage of the affected unit, while the dotted lines represent accidents that do not cause catastrophic damages, or that are less probable; however, they are presented to show other possible scenarios produced when applying the algorithm. Once a tank is reached by a continuous line, it means that it has suffered a level of damage that will result in its destruction.

When an explosion (from a G.2 LOC) occurs in tank 30, the overpressure wave affects tanks 2, 3, 5, 6 and 7 in a catastrophic way; tank 1 suffers damages that lead to a G.2 release and tanks 4 and 8 are affected in a minor way. This is the first sequence of accidents, and it is understood that the failures and later releases occur almost simultaneously. After the mentioned tanks fail due to the explosion, other accidents will generate in each of them, specifically, explosions, flash fires or pool fires; since these last type of accident will be the most probable one (according to the event trees used to produce the sequences), it will be shown for each of the affected tanks.

Figure 3. Accident sequence developed using the domino effect model

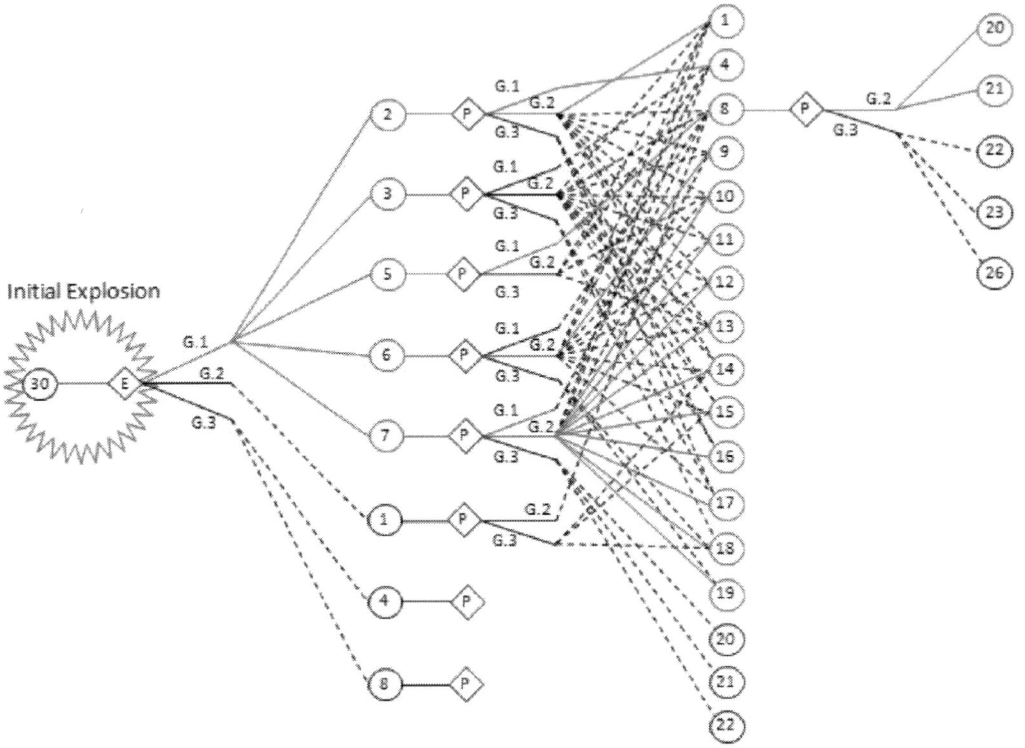

In the case that a pool fire occurs in Tank 2, Tank 4 will fail catastrophically due to the thermal radiation load it receives, while Tank 1 will also be affected in a major way, and also fail. There are other affected tanks, but these possible paths are not explored in detail here. It has to be noted that by the time Tank 2 fails, Tank 3 is also failing. A catastrophic failure in Tank 3 would also result in a pool fire that would further the chain of accidents, but none of these occurrences is explored. Pool fires in tanks 5, 6 and 7 would, according to some paths of the sequence obtained, result in the failure, and grievous damage, of tanks 8 through 19. If the sequence is furthered following a pool fire in Tank 8, tanks 20 and 21 would be damaged beyond repair, while tanks 22, 23 and 26 would suffer minor damages. It has to be noted that in the sequence explored, tanks 24, 25, 27, 28 and 29 do not suffer damages.

The sequence of accidents explored, which is the one that presented higher frequencies in each of its events, results in the destruction of 21 out of the 29 tanks that are part of the installation; this is in accordance to the level of damage in the real Buncefield terminal after the situations resulting from the explosion were controlled. It has to be said that this is one of many possible sequences produced by the algorithm, several of which are different from the real occurrence; however, branches of the possible sequences could be trimmed, in order to choose occurrences that are deemed to be more probable.

4.4 Application to QRA

Two QRAs were performed using the layout presented in Figure 2, the first using the normal procedure and the regular frequencies obtained from the original LOCs and the event trees; the second, using the frequencies of accident obtained from applying the domino effect model. The iso-risk curves obtained for the studies are presented in Figures 4 and 5.

Figure 4. Regular iso-risk curves

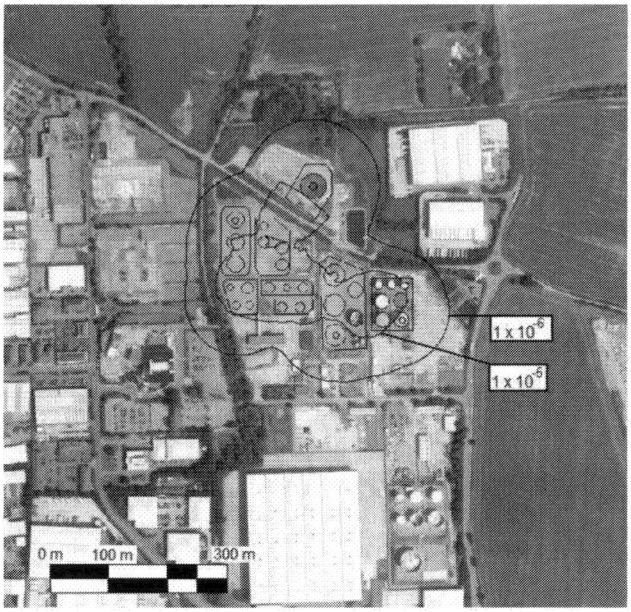

It is clear when observing the figures presented above that the individual risk associated to the installation increases when the domino effect methodology is applied. When regular frequencies are used, only 10^{-5} and 10^{-6} curves appear, while a 10^{-4} curve appears when using the modified frequencies; it encompasses the complete layout of the installation, and in some points, extends some 80 m from the limits of the terminal. The 10^{-4} increases it size so much because many lesser accidents with low frequencies appear in the accident sequences generated by the algorithm. The 10^{-5} curve also increases its area significantly, growing to the point of covering the complete evaluated installation, and reaching close to a 100 m from the limit of the terminal on its furthest point. The 10^{-6} curve is the one that presents a less appreciable change; to the north of the installation the curve suffers almost no alteration, to the south, it increases its length some 30 m and to the east and west, it increases some 20 m in each direction.

Figure 5. Iso-risk curves accounting for domino effect

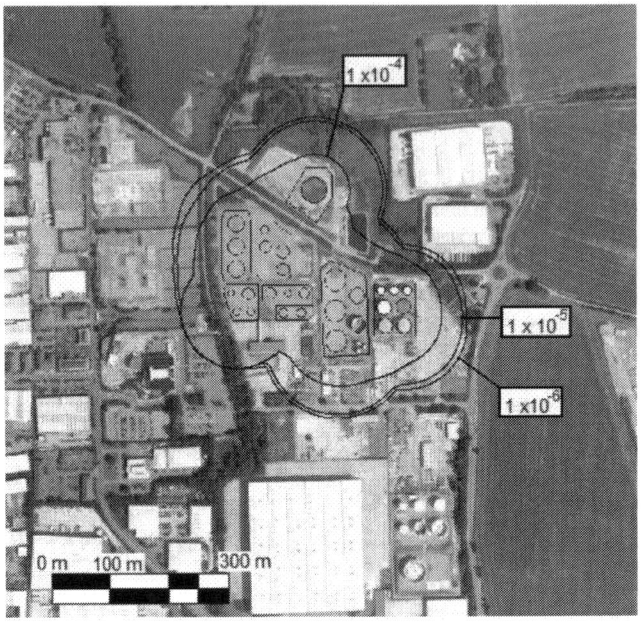

It is interesting to note that when performing a QRA of a storage installation composed of atmospheric tanks, the accidents with highest frequencies of occurrence are the pool fires, followed by flash fires and explosions. This means that the iso-risk curves related to the type of installation studied will not extend over great distances, as occurs when evaluating installations that store pressurized flammable liquids or toxic gases. Therefore, although the increase in the 10^{-6} curve, which, in many European countries is the most representative at the moment of evaluating external affectation, seems marginal, it is important, due to the type of installation that has been evaluated.

5. Conclusions

The domino effect model applied to a layout similar to that of the old Buncefield terminal produced many accident sequences; one of them is thought to be similar to the real accident that occurred in the facility in 2005. This means that the model is indeed suitable to be applied on real layouts, and that it can be used to help assess the risk related to a projected or existing installation, taking into account the possibility of domino effect occurrence.

The model can be applied with a minimum cost of time, and the frequencies obtained can be introduced in commercial software like "Riskcurves" without additional cost; QRAs could be performed taking domino effect into account, without the need to spend more time and money in the development of the study.

An increase in the risk associated to the installation was observed when the domino effect frequencies were used. The 10^{-6} curve does not suffer a significant increase in size due to the nature of the installation and its equipment.

Domino effect should be taken into account when performing QRA of industrial installations, as it can have an appreciable effect on the risk associated to the plant. This could lead to an impact on the Land Use Planning of the zone in which the facility is located.

References

[1] The Buncefield Incident 11 December 2005. The Final Report of the Major Incident Investigation Board. Volume 1, 2008. Buncefield Major Incident Investigation Board.

[2] Delvosalle, C, "Domino Effects Phenomena: Definition, Overview and Classification", First European Seminar on Domino Effects, Leuven, 1996.

[3] Bernechea, E.J., Vílchez, J.A., Arnaldos, J., 2012. A model for estimating the impact of the domino effect on accident frequencies in quantitative risk assessments of storage facilities. Process Safety and Environmental Protection, http://dx.doi.org/10.1016/j.psep.2012.09.004

[4] CPR 18E, 2005. Guidelines for Quantitative Risk Assessment (Purple Book). Directorate General for Social Affairs and Employment, The Hague, Netherlands.

[5] G. Antonioni, G. Spadoni, V. Cozzani, Application of domino effect quantitative risk assessment to an extended industrial area, Journal of Loss Prevention in the Process Industries, 22, 614-624, 2009.

[6] F. I. Khan, S. A. Abassi, Estimation of probabilities and likely consequences of a chain of accidents (domino effect) in Manali Industrial Complex, Journal of Cleaner Production, 9, 493-508, 2001.

[7] V. Cozzani, E. Salzano, The quantitative assessment of domino effects caused by overpressure: Part I. Probit models, Journal of Hazardous Materials, 107, 67-80, 2004.

[8] V. Cozzani, E. Salzano, Threshold Values for domino effects caused by blast wave interaction with process equipment, Journal of Loss Prevention in the Process Industries, 17, 437-447, 2004.

[9] G. Landucci, G. Gubinelli, G. Antonioni, V. Cozzani, The assessment of the damage probability of storage tanks in domino events triggered by fire, Accident Analysis & Prevention, 41, 1206-1215, 2009.

Proven Approach to Investigating Near Misses

Greg Smith, Senior Engineer
Process Improvement Institute, Inc. (PII)
7502 Chad Road
Harrison, TN 37341
Phone: (423) 413-0038
Fax: (865) 671-4944
e-mail: gsmith@piii.com

Gerald D. Burch, Senior Engineer
Process Improvement Institute, Inc. (PII)
e-mail: gburch@piii.com

William G. Bridges, President
Process Improvement Institute, Inc. (PII)
e-mail: wbridges@piii.com

Prepared for Presentation at
9th Global Congress on Process Safety
San Antonio, TX
April 28-May 1, 2013

UNPUBLISHED

9TH
GLOBAL
CONGRESS
ON PROCESS
SAFETY

Proven Approach to Investigating Near Misses

Steven (Greg) Smith, Senior Engineer
Process Improvement Institute, Inc. (PII)
7502 Chad Road
Harrison, TN 37341
Phone: (423) 413-0038
Fax: (865) 671-4944
e-mail: gsmith@piii.com

Gerald D. Burch, Senior Engineer
Process Improvement Institute, Inc. (PII)
e-mail: gburch@piii.com

William G. Bridges, President
Process Improvement Institute, Inc. (PII)
e-mail: wbridges@piii.com

Keywords: investigation, near miss, accident, incident, root cause, RCA

Abstract

Getting near misses reported is a major hurdle for most companies, one reason being is that the investigation system is not ready for the large influx of reports. Workers and management respond by being reluctant to potentially overload the investigation system (and overload themselves). This paper explains approaches to manage efficiency and effectiveness in near miss investigations that have been successful for handling near misses reporting within both large and small companies and to ensure high value from the investment of reporting and analysis. This paper builds on the updated paper presented in 2012 at GCPS on "Gains from Getting Near Misses Reported."

1. Introduction

Much attention has been given to the detection and correction of early warning signals as a means of improving safety performance.

Process safety events often have entirely different causal factors (human error, equipment malfunction, natural causes) and root causes (management system weaknesses) from those found in more traditional personal injuries.[1,2] Process safety incidents are more technical and complex in nature, usually involving multiple failures of protective barriers. The human actions and errors which cause them are often several steps removed from the incident itself. The human factors involved are not as obvious to investigators; the causes are more systemic than individual in their nature.

Even though personal injury rates have declined, significant injuries and financial losses have continued in a series of high profile, catastrophic (multiple injuries) process safety events. Because of inconsistencies in reporting standards and definitions, it is difficult to tell if these events are actually increasing in frequency or if their visibility has simply increased due to media attention and heightened public concerns. Either way, improved understanding and enhanced tools are required to improve our process safety performance.

In the process industries one such tool is near miss reporting, the early warning system equivalent to unsafe behaviors for process safety. Companies which have successfully developed the ability to recognize, report, investigate, and learn from near misses have shown dramatic reductions in loss events, losses, and injuries due to process incidents. In some cases, loss reductions of more than 90% have been achieved.[2]

2. Definitions

The following definitions are used throughout this article:

> *An incident is either an accident or a near miss.*
>
> *An accident is a sequence of unplanned events and conditions that result in harm to people, environment, process, product or image.*
>
> *A near miss is an unplanned sequence of events that could have caused harm if conditions were different or is allowed to progress, but did not in this instance.*
>
> *A causal factor is a human error (typically an error by the at-risk employee performing a task/job in the process) or a component fault/failure. Note that these human errors and component failures are probably caused by other humans making mistakes, and all errors are controlled by management systems. An incident typically has multiple causal factors. Natural phenomena can also be a causal factor.*
>
> *A root cause is a management system weakness that results in a causal factor. A casual factor typically has multiple root causes.*

The definitions above are the same as used in *Guidelines for Investigating Chemical Process Incidents, Second Edition* (CCPS, 2003).

3. Importance of reporting and learning from near misses

Investigating near misses is critical to preventing accidents, because near misses share the causes and root causes of accidents; they are one or two barriers away from the loss/accident. We are very likely preventing many apparently unrelated accidents when we prevent the ones that are obviously related. Figure 1 illustrates a hypothetical relationship between causal factors and root causes of accidents and those for near misses. From our experience, this relationship appears to be valid; see the case study section of this paper to see some of the proofs of this concept.

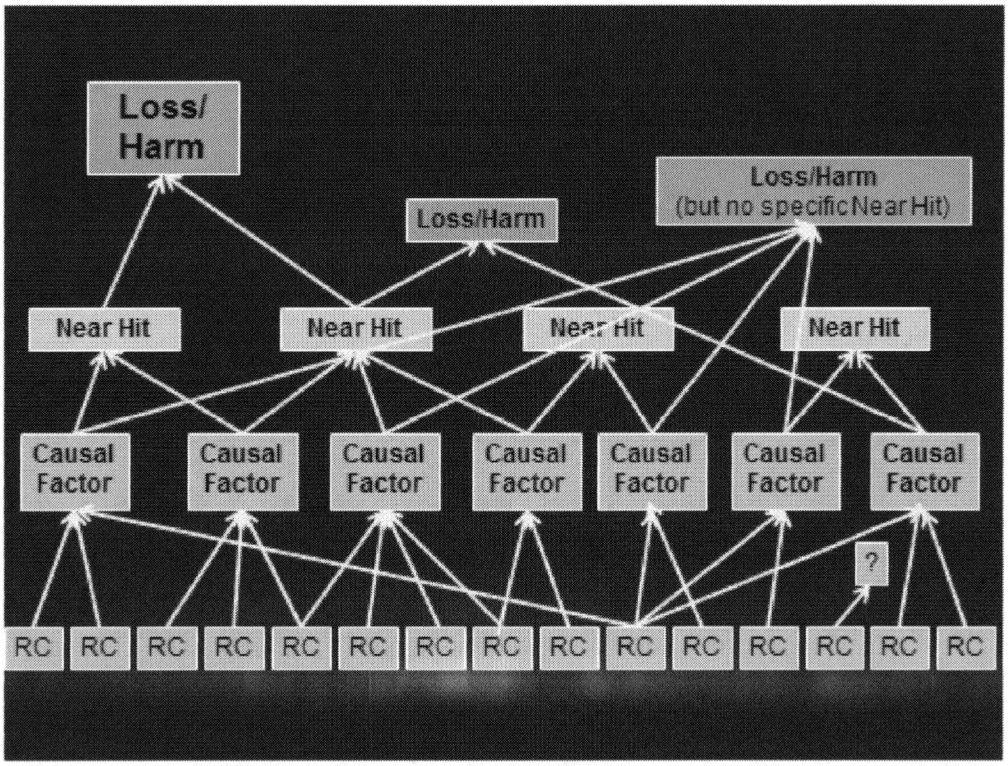

Figure 1 Interrelationship between the causes of Accidents (Losses) and the causes of Near Misses

As a brief explanation of Figure 1, root causes (management system weaknesses) make it more likely for a causal factor to occur. Combinations of causal factors (or in rare cases, perhaps a single causal factor) result in near misses. Additional causal factors result in losses (accidents) in increasing severity. A root cause can increase the likelihood of seemingly unrelated causal factors.

Formal and informal organizational surveys[3] detail the difficulties of implementing an effective near miss reporting system. In this paper, a number of barriers are outlined and potential solutions are proposed. The identified barriers are:

1. *Fear of disciplinary action*

2. *Fear of teasing by peers (embarrassment)*
3. *Lack of understanding of what constitutes a near miss versus a non-incident*
4. ***Lack of management commitment (no training provided on investigation techniques and procedures) and lack of follow-through once a near miss is reported (time is not allocated to investigate near misses, or corrective actions not implemented)***
5. ***An apparently high level of effort is required to report and to investigate near misses compared to low return on this investment***
6. ***There is No Way to investigate the thousands of near misses per month or year***
7. *Disincentives for reporting near misses (e.g., reporting near misses hurts the department's safety performance)*
8. ***Not knowing which accident investigation system to use (or confusing reporting system)***
9. *Company discourages near-miss reporting due to fear of legal liability if these are misused by outsiders*

4. Primary Barriers to Effective Investigation of Near Misses

Four of the barriers (4, 5, 6, and 8) tie directly to the organization's ability to effectively and efficiently investigate and resolve near miss incidents.

Of the four barriers related to getting near misses reported, the two that are most related to effective near miss investigations are **Barriers 6 and 8.**

Overcoming these barriers is described below by looking at the experience of organizations as they implement these concepts.

4.1 Barrier 6. There is No Way to investigate all of the thousands of near misses per month or year!

Normally, when the discussion of having a huge number of near misses reported, such as four per worker per month, the reaction will be shock and then a statement such as Barrier 6. This barrier is closely related to Barrier 5*: An apparently high level of effort is required to investigate near misses compared to the small gain perceived*. If a site has 500 staff as operators and maintenance craftsmen, then likely 25,000 near misses could be reported. At first glance, it can appear impossible to cope with, let alone investigate that number of near misses (incidents).

Maybe the most important solution to Barrier 5 and 6 is to first decide which near misses and losses/accidents need to be investigated. The best solution is normally stated as:

> ***Let front line foremen or supervisors decide if a near miss or accident needs to be investigated to root causes; the decision is made on the apparent Learning Value of the incident. How do you decide it a high or low learning value? The foremen or supervisors will easily decide based on their hands-on experience.*** *For instance, an additive pump failure may directly result in a*

minor and temporary impact to quality or production rates but does not appear to have the potential to cause a large loss. So the foreman of the shift decides to designate this as a low learning value incident and therefore not do an immediate RCA. But, then other similar pumps fail over a six month period. A query shows that after 12 months, there have been 3 failures of the same additive pump and 8 failures of 5 other similar pumps in different services. Based on the number of failures and cumulative losses, the company performs an RCA of the 11 failures, resulting in findings which when applied to all similar pumps and services, greatly reduces the production/quality impacts. The RCA of the grouping of incidents proved very cost effective.

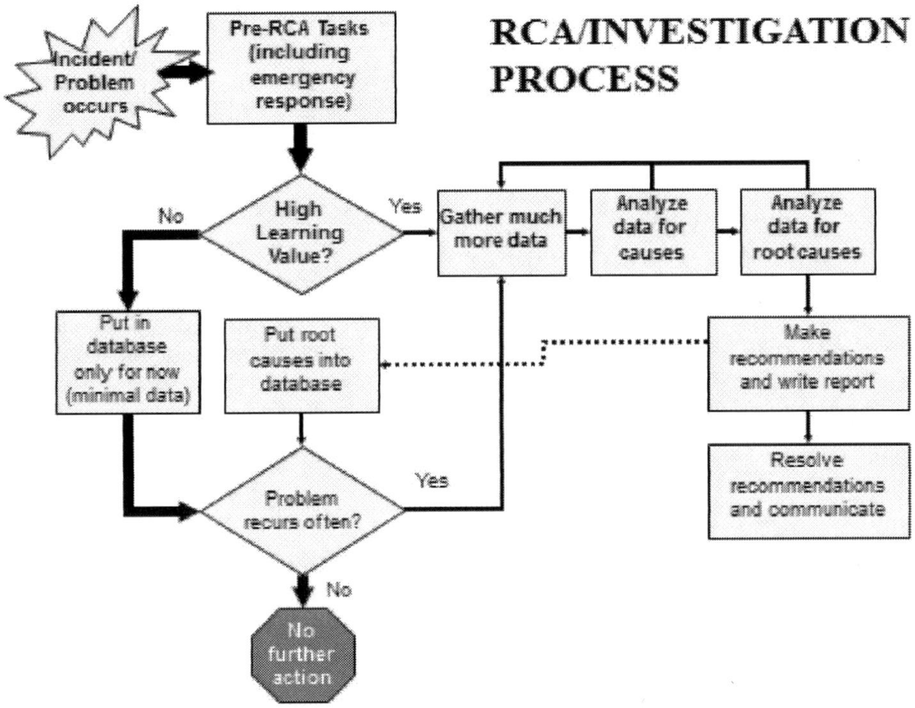

Figure 2: RCA/Investigation Flow Process

Figure 2 illustrates the process flow for an investigation system than can handle a large volume of near misses and losses/accidents. For this process to work:

- Be prepared for investigations by having enough staff trained in root cause analysis methods (or to help in the analysis, such as being able to interview peers).
- When the near miss, etc., is first noticed or reported by staff, let the frontline supervisor or foreman decide if it has high learning value.
- For high learning value incidents, investigate now. For low learning value incidents, put in the database now, along with the little data you have and any obvious causes. Do not investigate yet!

- Query the database every one to six months and perform Pareto or similar analysis to help decide which recurring events need to be analyzed in more detail.
- Take the root causes from investigations/RCAs and put in the database as well.
- Query the database every one to six month

4.2 Barrier 8. Not knowing which accident investigation system to use

One consideration that is not related to any of the barriers mentioned above, except marginally to Barrier 3 (lack of understanding of what a near miss is), is the scope of the investigation program. Some companies have one investigation system for occupational safety incidents, another one for process safety incidents, another for environmental releases, another for reliability issues, and yet another for quality and customer services issues. We have found that the same investigation approach and investigator training works well for incidents in any facet of a business. We believe there is merit in combining the systems and, in particular, in combining the incident databases. Combining the incident systems will require more work on defining near misses and in determining success in reporting near misses.

Consider having ONE incident reporting system with ONE approach for teaching employees the definition of a near miss and with ONE approach for doing incident investigations (including one approach for root cause analysis).

5. Effective and Efficient Incident Investigation and Root Cause Analysis

The incident investigation and root cause analysis (II/RCA) process should enhance organizational learning, reduce accidental losses, invest wisely in protective and mitigative systems, better protect people and equipment from harm, and continuously improve EHSS performance. An efficient and effective process ensures that scarce organizational resources are used appropriately in activities that matter.

The overall II/RCA process consists of the following steps:

- Data collection
- Causal factor analysis
- Root cause analysis
- Develop recommendations
- Reporting and follow through

5.1 Data collection

Every investigation begins by collecting, organizing, and preserving all relevant data as quickly as possible after the incident has been stabilized. Data comes in the form of personnel interviews, physical evidence, and documentation. All data is fragile and susceptible to loss or damage if it is not properly collected and preserved in a timely manner. All front line supervisors should be trained in the II/RCA process with particular emphasis on initiating the investigation, interviewing people, and collecting the physical evidence. Provide front line supervisors with an investigation tool kit containing a camera, barrier tape, collection bags and bottles, and appropriate forms and checklists.

5.2 Causal factor analysis

Once collected, data must be organized and analyzed to determine what happened – the chain of events (causal factors) that led to an undesirable occurrence. The most effective and proven methods for analyzing data are Event and Causal Factor Charting (ECFC) supplemented, if needed, with a Simplified Fault Tree Analysis for complex, chronic, or acute events. These are only effective when the rules are followed for their use – but with training and practice most people can be taught to identify causal factors using these basic tools for most of the incidents they will encounter.

Event and Causal Factor Charting: When an investigation team begins a root cause analysis (investigation), the first step is to prepare an Event and Causal Factors Chart (ECFC).

- Arranges event building blocks (sticky notes) to graphically depict cause-effect relationships between known events
- Allows for testing of "necessity" and "sufficiency"
- Best method when timing of events is important

Events make up the backbone of the ECFC. Event statements describe specific occurrences (e.g., "4-12 shift operator filled Tank 123," or "Control room operator acknowledged level alarm for Tank 123"). Conditions usually provide descriptive information (e.g., "Pressure was 1000 psig" or "The operator did not have hands-on training in this task") as opposed to stating action (e.g., "Operator placed Valve ABC into open position"). Primary events form the basic sequence in the diagram. Secondary events are actions that impact the primary events, but which are not directly involved in the situation. Figure 3[3] provides the steps for developing a casual factor chart.

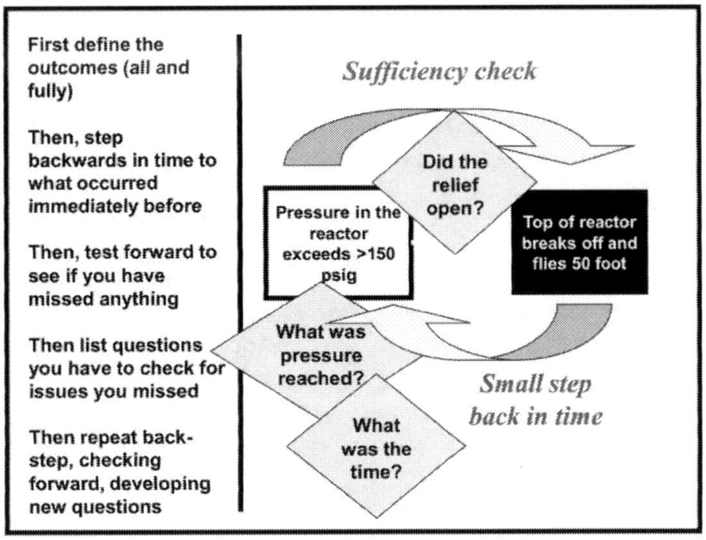

Figure 3: Steps for Casual Factor Charting

If applied correctly, ECFC will aid the investigators in generating a great many new questions. These will lead to more data gathering. Above are a couple of examples of how we show these questions on an ECFC. The answers to the questions sometimes become new events or conditions on the ECFC, as illustrated in Figure 4[3]:

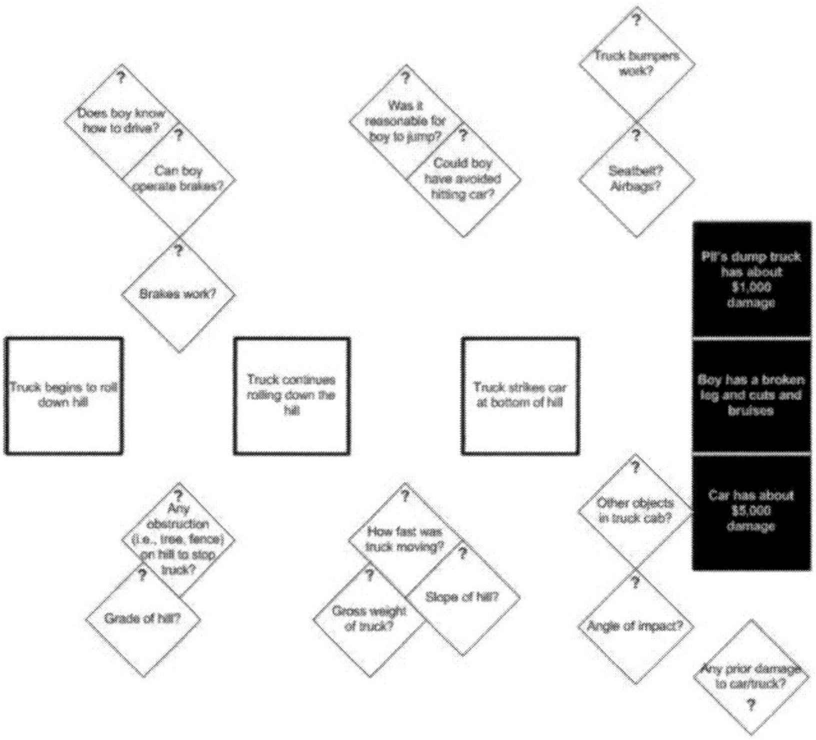

Figure 4: Example of a Partial Casual Factor Chart

The investigators will go back and forth between doing the ECFC, generating new questions, getting new data to answer the questions, and taking more steps backwards in time and sequence. This process continues until all of the data is exhausted and/or the investigators fully understand the sequence of the events and all of the mistakes and failures that led to the final ending condition or loss.

After the ECFC has been completed, the investigators are in a good position to identify the negative factors that influenced the course of events. These components are labeled <u>causal factors</u>. Causal factors, which may be in the form of events or conditions, are those items that are considered to be major contributors to the incident and as defined earlier, are either component failures or human errors or severe environmental problems.

The logic rules for constructing an ECFC are critical to success. These govern how small of step backwards the investigators take backwards in time and how many questions they generate as they test to ensure they understand each pair of building blocks.

5.3 Root cause analysis

When all the causal factors have been identified, the investigation team should begin asking "Why?" each causal factor occurred. Using a combination of the "5 Whys" and a Root Cause Chart™ (derived from the management oversight and risk tree process developed for US DOE's Savannah River Laboratory) shown in Figure 5, step down each of the root cause paths by:

- Select a causal factor from the causal factor chart or fault tree
- Work through the map for each causal factor
- Ask Why to help gathering the right data to get to the lowest level (root causes)
- Step down paths, identifying:
    - causal factor type (equipment, personnel, other)
    - root cause category
    - near root cause
    - root cause
- Record results on forms, at each step (see next page)

The Root Cause Chart™ structures the reasoning process for identifying root causes, identifies detailed root causes for each causal factor, facilitates consistency across all investigations, supports trending of "root causes" and "categories", and gives more thorough coverage of root causes by reminding investigator of all possible root causes.

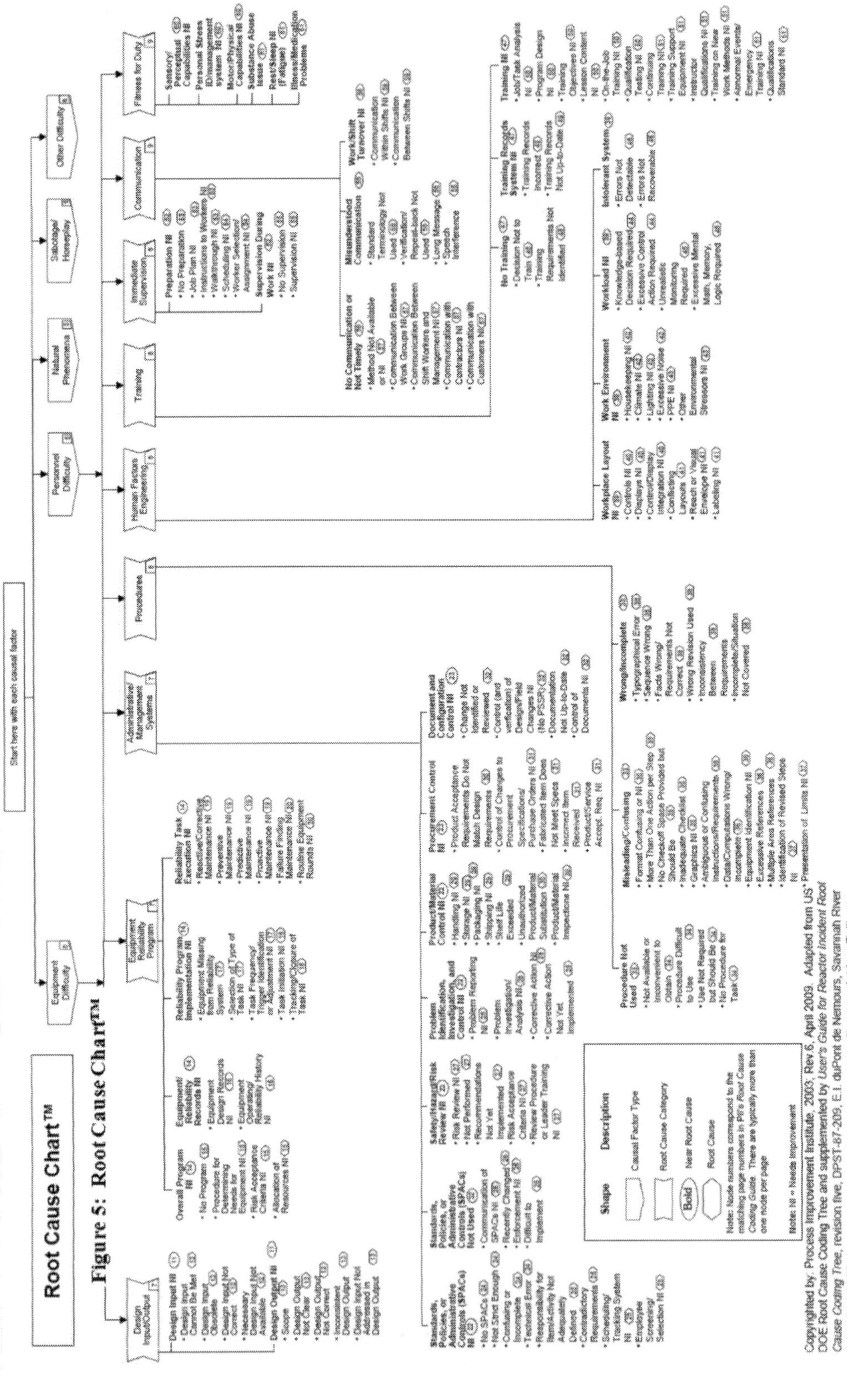

Figure 5: Root Cause Chart™

5.4 Develop recommendations

The investigation team should develop recommendations to remedy each root cause of each causal factor. Recommendations should describe the general objective to be accomplished and the consequences if not completed. Recommendations can also include possible solutions considered by the team and an example of how this could be accomplished.

5.5 Reporting and Follow Through

5.5.1 Simplify the reporting of the investigation/RCA results to the bare minimum needed

Part of the reason for the belief that it is impossible to investigate large numbers of near misses stems from the large reports currently required by the company for investigating incidents. Some companies insist on producing what they call "professional" reports of accidents, and these grow to 50 or 100 pages (half of the pages are attachments). Why produce such a large report? What is the use of that large of a report? What makes "size" of a report equivalent to "professionalism" of a report? Think about every aspect of the report and make sure it is needed. Normally, all that is needed is a:

- Cover sheet that includes the date, time, location, one or two sentence description of the near miss or incident, and a title that summarizes the incident at a glance. The cover sheet should also list the team members.
- Forms that have the causal factors filled in and the root causes filled in, with perhaps one or two sentences that explain the root cause. These forms can also contain the recommendations necessary to correct the root causes.
- That's It!

So, most near miss investigation results will be two pages or so and most loss/accident reports will be four pages or so.

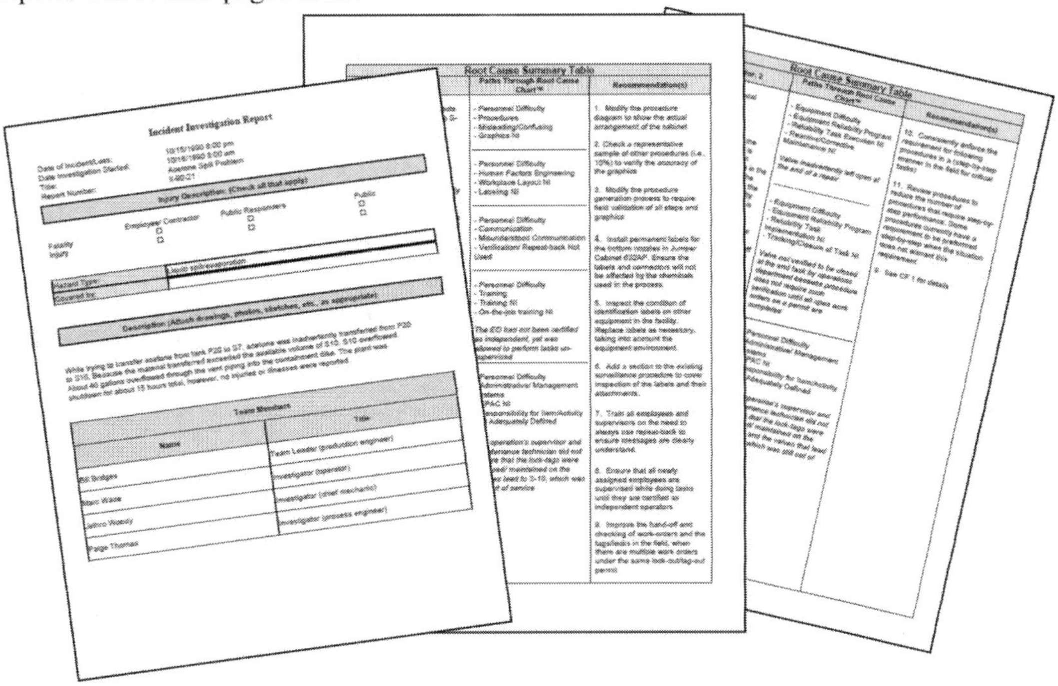

The complexity of results reporting has grown from the legacy of only investigating losses/accidents. When an organization gets a large ratio of near misses report and therefore a large number of investigations going on, the reports must shrink. This is a good thing. However, if you still have a major accident (which you won't have if you get a large number of near misses reported and investigated), then add more documentation to meet the needs related to litigation, regulatory interface, etc.

5.5.2 Deciding which incidents should be investigated

All near miss incidents should be reported and cataloged with basic information and apparent causes, but not all near miss incidents will warrant a full investigation. Only those incidents with either high potential impact or high learning value should be investigated using the II/RCA process. This is an excellent opportunity to engage staff and give them control over the process by granting front line supervisors the authority to decide which incidents have high learning value. For example, establish a shift target of investigating 15-20% of the near miss incidents reported and then let them decide which ones they want to investigate to meet their goal.

6. CASE STUDY: One Company's Experience with II/RCA

A large pharmaceutical company (more than 50,000 employees) initiated a drive to report and analyze large numbers of near misses. Before this initiative, they had about 3 near misses reported per loss event. Their loss categories included quality impacts, reliability issues, safety mishaps, and environmental issues. Also before the initiative, they had five incident reporting systems, investigation protocols, and databases. One of the first steps the company took towards the stated goal was to consolidate the five reporting and investigation systems into one system. They tried to use one of the existing database systems for this consolidation, but decided that all were deficient for this purpose. So, they modified the reliability incident database to make it suitable for all types of incidents. This took several months of effort (not full time though).

In parallel, they adopted a new investigation approach, which is the one described in this paper. This involved changing the database structure to provide pull-down menus for each branch on the Root Cause Chart. The chart was not modified. Next, a training course for lead investigators was customized to meet their new needs and to use their examples in the exercises and case studies. Then, this course was presented seven times in the USA and UK, thereby training an initial group of 200 investigators but more importantly training 7 instructors within the company. The eventual goal was to train 15% of the operating staff how to perform root cause analysis (RCA) of near misses and losses. The new database and consolidation of reporting and analysis systems (developed in parallel with the development of the customized training materials), was rolled out at the same time as the initial training courses. The reporting system was incorporated into their Lotus Notes e-mail system. The reporting forms were very simple and any worker could report a near miss or incident. Initial reports were routed automatically by the e-mail system to the worker's supervisor and viewable by any qualified investigator. A lead investigator on the same shift rotation and from the site (but not from the same group) was assigned to each incident; investigations typically began within one hour and on average were completed within the next hour.

These initial efforts were extremely successful and within about one year, the company achieved a ratio of more than 100 misses being reported per loss event. All near misses and loss events were captured in the database of incidents and about 20% were high value and immediately investigated. Typical reports were 2-3 pages. After the first year, the sites performed queries of the databases and based on Pareto analysis, performed RCA on the highest number of repeated "similar" incidents. In subsequent years, the queries were performed every three months.

This new system was much more productive in reducing losses than the previous collection of five systems. Overall, production and quality losses were reduced about 95% with little or no increase in time spent in RCAs (because the number and magnitude of losses decreased dramatically and so there was much less time spent on fire-fighting and major investigations).

7. Conclusions

Near miss reporting and investigation has been very effective in reducing losses due to process safety incidents. However, implementation has proven difficult for many companies due to several implementation barriers. The proven methods described will improve both the efficiency and the effectiveness of the II/RCA process. Consistently applying these methods will enable near miss reporting, providing early detection and correction of process safety warning signals, thereby reducing losses and improving process safety performance.

8. References

1. Busch, Carsten, *Heinrich's Common Cause Hypothesis: A tool for creating safety*, July 30, 2012, http://www.predictivesolutions.com/safetycary/heinrichs-common-cause-hypothesis-a-tool-for-creating-safety/
2. Wright, Linda and van der Schaaf, Tjerk, *Accident versus near miss causation: a critical review of the literature, an empirical test in the UK railway domain, and their implications for other sectors*, Journal of Hazardous Materials 111 (2004) 105–110.
3. Bridges, William, *Gains from Getting Near Misses Reported,* AIChE 8th Global Congress on Process Safety, 2012.
4. *Incident Investigation and Root Cause Analysis Course Materials*, Process Improvement Institute, Inc., 2003-2013.

9TH GLOBAL CONGRESS ON PROCESS SAFETY

Key Strategies for Implementing a Corporate Fatigue Risk Management System

David E. Cummings
E.I. DuPont de Nemours & Co.,Inc.
1007 Market Street, Wilmington DE 19898
david.e.cummings-1@dupont.com

Prepared for Presentation at
American Institute of Chemical Engineers
2013 Spring Meeting
9th Global Congress on Process Safety
San Antonio, Texas
April 28 – May 1, 2013

UNPUBLISHED

Key Strategies for Implementing a Corporate Fatigue Risk Management System

David E. Cummings
E.I. DuPont de Nemours & Co.,Inc.
1007 Market Street, Wilmington DE 19898
david.e.cummings-1@dupont.com

Keywords: fatigue, fatigue management, ANSI/API 755, human factors

Abstract

The First Edition of ANSI/API Recommended Practice 755 "Fatigue Risk Management Systems for Personnel in the Refining and Petrochemical Industries" was published in April 2010 as a result of the BP Texas City catastrophe of March 2005. In the last two years, many companies in the petroleum and chemical processing sectors have been working to implement the basic tenants of this new practice. An effective fatigue management program involves a multi-disciplined team approach to engage various functions of an organization, including PSM/EHS, HR, Medical, and Operations. This paper will explore the process, associated elements, and key issues integral to the design, development, and implementation of a formal fatigue management system across DuPont as a global manufacturing organization with a diversity of high hazard processes. This includes internal policies, training and development, standardized procedures and tools, and internal collaboration strategies.

1. Introduction

ANSI/API Recommended Practice 755 [1] was issued in 2011 based on a recommendation issued to the American Petroleum Institute by the U.S. Chemical Safety Board as an outcome of the BP Texas City 2005 incident investigation report. Prior to that time, there was little industry guidance or standards on the subject of fatigue management (FM). There is and has been substantial literature published and technical resources developed on the general issues and concerns with fatigue management in terms of circadian rhythms and the effects on human performance variables, and many industries such as aviation and commercial transportation have adopted specific standards with accompanying regulatory requirements.

E.I. du Pont de Nemours & Co. is a global science-based products and services company with over 200 manufacturing units across 70 countries involving nine business segments. It is engaged in technology solutions and innovation across a range of disciplines, including agriculture and industrial biotechnology, chemistry, biology, materials science and manufacturing, offering a wide range of products and services for markets including agriculture and food, building and construction, electronics and communications, general industrial, and transportation. Manufacturing technologies and site operations involves a myriad of process hazards, including flammable liquids and gases, highly toxic materials, reactives, combustible dusts, corrosives, asphyxiants, and thermal/mechanical energies. DuPont's philosophy and current practice is to apply process safety management systems and requirements to all manufacturing operations and locations using one global set of internal elements and technical requirements. Based on the company's current risk profile, process safety systems that help maintain a high degree of operational discipline and reduce the potential for human error, including those that could be related to employee fatigue, are key elements of an integrated risk approach.

DuPont's process safety management model and system has included a long standing element and requirements for "Training and Performance" where the performance portion of the element requires a specific focus on physical capabilities (fitness for duty) as well as diminished capacity (mental and psychological capabilities for process safety sensitive jobs, including the effects of fatigue) for operating and maintenance personnel working at all manufacturing locations. DuPont updated the "Training and Performance" element to specifically include an internal fatigue management policy as an integral part of the global process safety management program in January 2011. Initial implementation of the new technical requirements has been progressing since that time. The program embraces the following key elements of fatigue management as described in ANSI/API 755:

- Maximum hours of service
- Training for employees and line management
- Staff workload balance
- Work environment evaluations
- Medical Evaluations (personal risk factors)

Implementing a fatigue management system was integrated into corporate process safety management systems; however, active involvement of other key functions within the organization is important in this area. Within DuPont, the process safety competency serves as the corporate "owner" on FM, while other key functions such as Human Relations (HR), Integrated Health Services, and Occupational Health (EHS) all have played a meaningful and contributing role in developing and implementing portions of the FM process in the initial design stages, during implementation, and in an ongoing role as part of the "run and maintain" phase of this work. This collaboration has contributed greatly to the overall results.

2. Foundational Activities for FM Implementation

Building an internal fatigue management system based on the elements in the API standard requires a number of key strategic decisions and initiative planning prior to formal roll-out, as well as ongoing support, training, and tools during the active implementation phase. The following provides examples of many questions that were posed as part of DuPont's initial design strategy for fatigue management:

- Should the fatigue management program be a global or U.S. based only focus?
- What is the current state at all sites (does anyone already have an effective FM system)?
- What will be the major impacts on Operations at each site (resources, costs, tracking tools)?
- Who are the key stakeholders who should be involved in the design of the system?
- Which sites will be required to implement a local fatigue management system?
- How will fatigue management requirements be "codified" in corporate policies?
- What elements are consider mandatory versus suggested practices/guidance?
- What site roles and positions will be included in FM (and not included)?
- What is the required internal timing for implementation and how will it be tracked?
- When and how will the PSM auditing process incorporate FM elements during site reviews?
- How will interpretations and guidance on FM be captured and distributed?
- Will the program be required for contractors, and if so, which ones?
- Are there leveraged solutions and tools that can be developed to assist sites for effectiveness and efficiency?
- Who has primary responsibility to ensure compliance with Table 1 (employees or line managers)?

2.1 Site Surveys to Define Current State FM Practices

The first step that was taken to develop a fatigue management program across DuPont was to conduct a detailed survey across all manufacturing locations to accurately determine the present state relative to the main elements of ANSI/API 755. This included the following self-assessment parameters:

- Existence of site level fatigue management written programs and policies
- Existence of local policy limits on overtime and maximum consecutive work hours
- Past and present fatigue management training
- Existence of tools or tracking systems to measure work hours and overtime by work group and employees
- Local site systems and capabilities to integrate fatigue management into existing medical programs
- Forecasted requirements and needs with local labor organizations and contracts for fatigue management

Once the site survey was completed, the data was collated and utilized as inputs for the development of corporate policies, timing and implementation plans, management training, and development of internal FM training and guidance documents. The survey was collected and analyzed in a simple excel® tool similar to that depicted in Figure 1.

Site Fatigue Management Survey – Current State

What is your site's current level of implementation of Fatigue Management requirements? List all numbers that apply. 1. Leadership is aware. 2. The site is tracking hours. 3. Employees have received general awareness training. 4. The site is actively managing hours for Shutdowns/TurnArounds, Special Circumstances, or Normal Operation. 5. A Fatigue Management policy has been written. 6. An Exception Management policy has been developed.	If you listed No. 2 above, how is the site tracking hours? Is the tracking historical or real time? What system or tools are being used?			If your answered No. 4 above, when and how is the site managing hours?		Have you considered which employees would be included in a Fatigue Management program? If yes, how many at your site? Does this number include employees in LHO's? If yes, how many of the total are in LHO's?	How are your shifts set up, ie. 8 hour, 10 hour, 12 hour, rotating, 5 days, 7 days, etc.?	If your site has changed any policies, procedures, tracking, people's work times, etc. to move toward compliance, can you describe what you have done?	Site Changes
Level of Implementation	Tracking Hours	Tracking Hrs.?	How Tracked ?	Managing Hours	Mgt. Hrs.?	Which Employ.	Shifts	Changes	Changed Schedule/ Pay?

Figure 1: Sample Site Survey Baseline Questions and Spreadsheet

2.2 Corporate PSM Standard Policies and Requirements

DuPont drafted and included a new fatigue management section in the global process safety management standard as part of the" Training and Development" element as depicted in Figure 2. Specific new content in the corporate standard was intended to incorporate the minimum mandatory requirements involving fatigue management elements described in ANSI/API 755, and not a be a re-statement of the guideline's full technical content. It was also defined with the standard scope that new internal fatigue management requirements would be applied to vital PSM roles within DuPont sites as well as to specific types of contractors, but only at higher hazard site operations (DuPont classifies all global manufacturing locations as higher hazard processes or lower hazard operations). The standard changes went through a broad internal peer review and was published in January 2011.

Figure 2: DuPont's Process Safety Management Model

2.3 Education and technical guidance to drive consistency

Internal webinars were developed and delivered which were targeted to several different key audiences, including site PSM leaders, operations leadership, site HR leaders, and contract administrators in order to develop foundational knowledge of the new FM program elements across the organization. The initial ANSI/API 755 standard as published is a performance based standard with specific prescriptive FM requirements such as Table 1 maximum hours of service limits, however, it became apparent early in the FM implementation process that numerous technical and interpretive questions existed both on the standard, as well as within DuPont's local site implementation efforts. In order to attain as much consistency as possible, a series of Frequently Asked Questions (FAQ's) were collected and formally documented with specific guidance. Once vetted and resolved, these FAQ's were distributed to sites and posted on the internal PSM website for retention and routine access. Additional questions and needs for interpretations have continued to surface from sites and functions and been added to the FAQ "evergreen" document even after almost two years of implementation progress. There is currently no effective platform within industry to collect and address these types of interpretations in a consistent manner.

As another step to implement consistency across all locations, DuPont developed and distributed a standard site based written fatigue management program or template which can be applied by each site as a base document and modified to add in local responsibilities, procedures, and line management practices.

There are also unique situations that may develop for plant operations that do not fit neatly into the existing definition and guidance in ANSI/API 755 involving "outages" as it applied to extended consecutive work shifts in Table 1. The existing definition for outages includes the following:

> *"Planned or unplanned interruption in the normal operations of a unit or plant, including mobilizing or de-mobilizing. Outages include but are not limited to such things as turnarounds, unit shutdowns, operational responses, etc."*

This definition does not cover all situations where an extended work set and shift schedules may be applicable and necessary to minimize fatigue, including commissioning new plants or substantive MOC projects where construction, PSSR's, start-ups, and lining out the process may be necessary involving operations, technical and R&D resources, response to natural disasters (e.g., hurricanes, earthquakes, etc.), mothballing or dismantlement activities, new product trials in pilot plants, etc. Each company should define when and how these types of situations should be managed in terms of unique conditions and the application of extended work hours and shift schedules with potential fatigue factors.

DuPont adopted an internal policy at the site level to develop a formal exception process for situations where the Hours of Service Limits (ANSI/API Table 1) may not be met due to specific circumstances. An exception process is discussed in the standard in Section 4.8.5 with general criteria, however; the specific procedures and details should be developed by each company. For DuPont, the exception approval process has been designed to be executed at the local level for a case-by-case exception by site or unit management. For longer term, permanent, or "standing" exceptions, off-site leadership approvals are required. In both cases, exceptions will document the situation, the rationale, and any relevant alternative controls or methods to be taken to mitigate the potential effects of fatigue. This includes use of a standardized form and line management approvals for each exception situation.

3. Shift Schedules and Maximum Hours of Service

API 755 includes specific requirements for maximum hours of service provided in Table 1 of the standard. The hours of service element in FM has received the most internal discussion, review and implementation work. A detailed analysis of existing shift rotation schedules at all U.S. site operations was conducted by DuPont HR to ascertain if any inherent conflicts existed in the current schedule structures that were not be fully compatible with Table 1. This included a specific emphasis on the number of consecutive shifts and the total hours of recuperative time for both day and night shift rotations. This review revealed that DuPont had over one hundred (100) different shift rotation schedules in use at all U.S sites for 8, 9, 10, and 12-hour operations and maintenance shifts. In a very limited number of cases, an inherent conflict in the site shift structure existed which did not fully meet the maximum "hours of service" in Table 1 (mostly involving the transition from day to night 12 hour shift schedules). Figure 3 describes the shift schedule analysis results at all US sites, and figure 4 provides an example of the detailed shift schedules which were generated and reviewed at a typical multi-unit site.

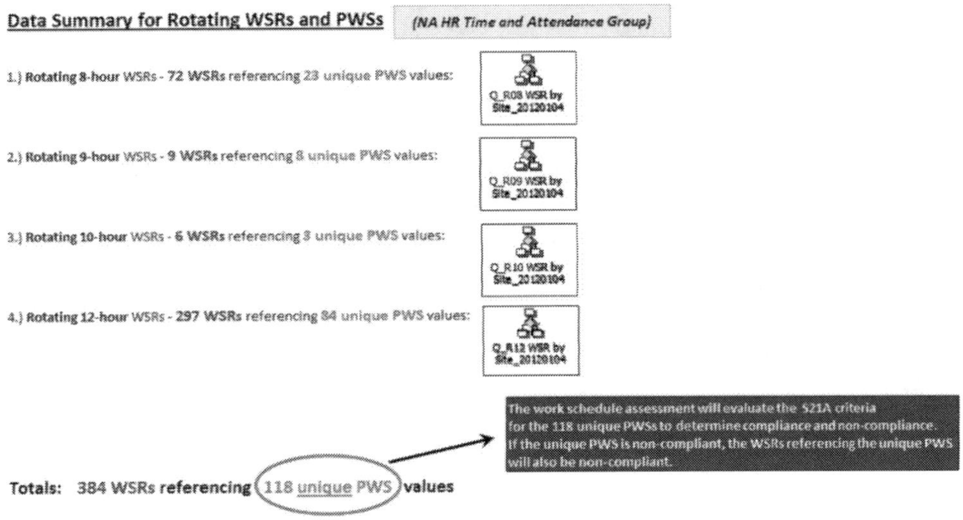

Figure 3: Data Summary for Plant Rotating Work Schedules

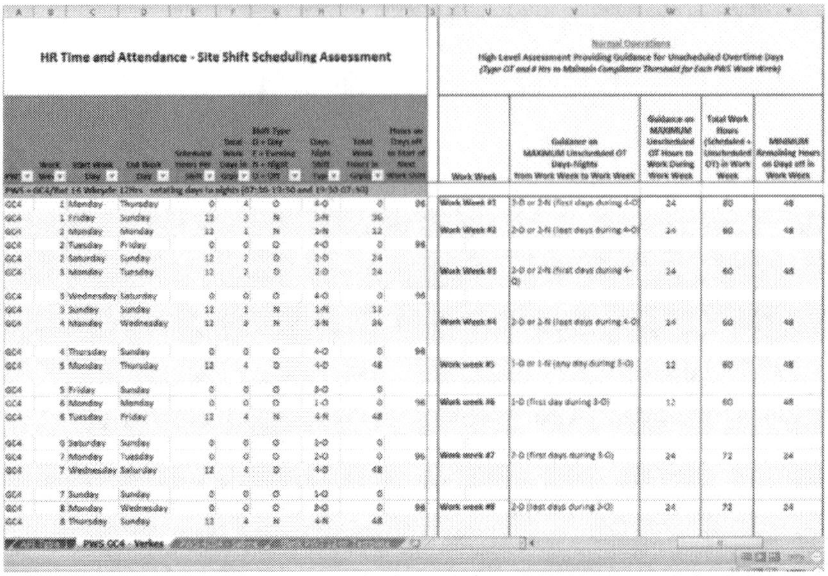

Figure 4: Sample Site Analysis – Time and Attendance Schedules

A significant part of DuPont's initial evaluation of existing shift schedules and site data survey questions focused on the types of tracking and software (scheduling) tools that were either already in use or could require development to track work hours, including scheduled and unscheduled overtime call-ins. In most cases, site systems tracking overtime rates for work groups but not specific employees other than annual salary statistics. Site level tracking systems which were cataloged that could be modified and used to monitor maximum hours of service and off site recuperation times included the use of an HR work hours and pay status, web based tools, security gate electronic logs, and access data bases. None of these existing systems as structured were deemed an acceptable and reliable means to proactively schedule, manage, and track individual employee work hours. A six sigma black belt resource and project was initiated in order to review each of the work hour and overtime tracking system attributes, and develop a standardized solution that could be used at all sites to help line management ensure that the maximum hours of work in ANSI/API 755 Table 1 maximum hours of work. As part of the six sigma project, the main attributes of an overtime tracking system were captured as illustrated in figure 5.

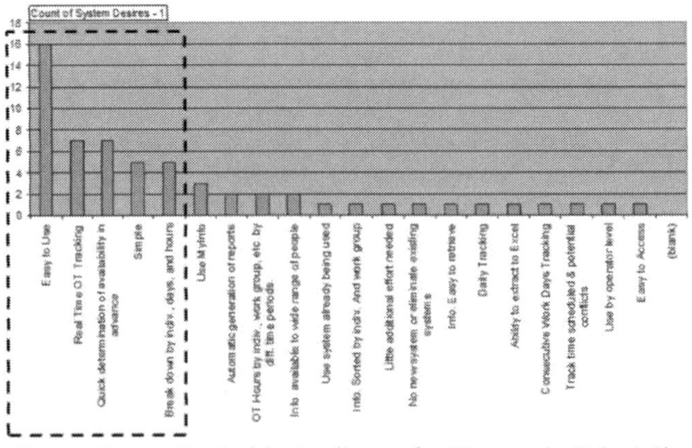

Figure 5 : Site Survey Data – Desirable Attributes for Electronic Scheduling and Tracking

4. Training

DuPont developed a standardized and leveraged training package which could be applied at each site in order to provide initial and refresher training on fatigue management. Two packages were developed ; one to target employees who were subject of FM (operators and mechanics), and one for line managers who supervise these employees. The guidance and criteria in ANSI/API 755 was used to create the technical content of each package. Initial training was designed to be one hour of information and cover the following topics:

- Basics of sleep and fatigue including causes, symptoms, and prevention strategies
- Understanding risk of on the job fatigue and impacts on safety
- Overview of the site fatigue management system including maximum hours of service criteria

DuPont occupational health resources have also previously partnered with external groups such as the Porcupine Health Unit and the Sudbury & District Health Unit in Canada to develop additional sources of fatigue management training such as the Shiftwork like Clockwork (SWLCW) training modules illustrated in Figures 6 and 7 . These materials provide significant details over nineteen (19) modules that can be used for both initial training of new employees and refresher training for existing employees. DuPont has adopted an internal FM refresher training frequency of three years.

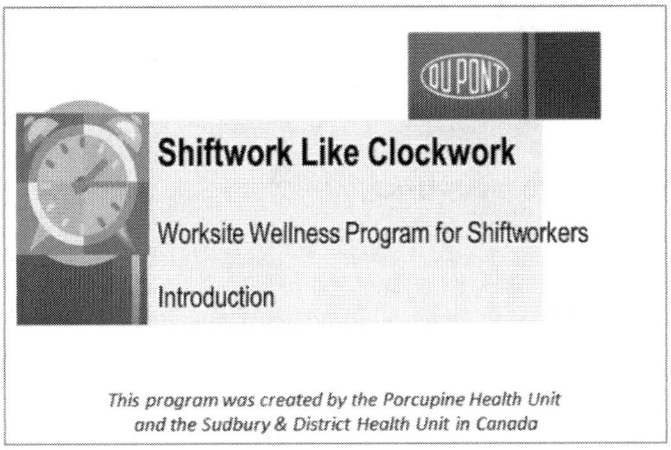

Figure 6: "Shiftwork Like Clockwork" Training Program

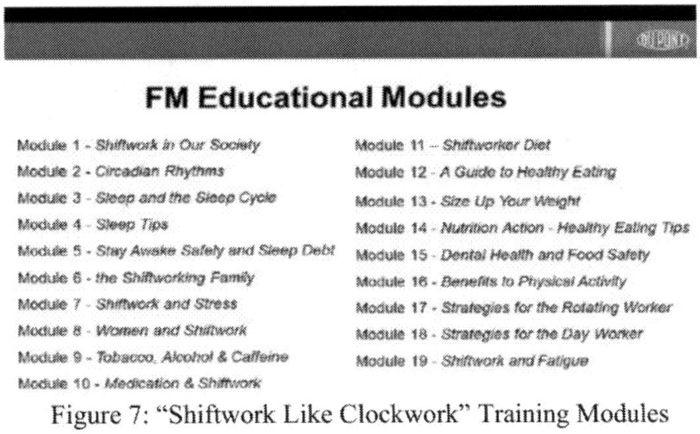

Figure 7: "Shiftwork Like Clockwork" Training Modules

5. Staff Workload Balance

Another key tenant of the ANSI/API 755 standard includes a focus to help ensure a staff workload balance is achieved and maintained within each operating area at each site. This should include an analysis of work activities on each shift, and a focus on considerations and impacts related to employee turnover, long term disability, extended time off due to personal reasons (vacation, leave of absence), staffing vacancies, and other sources of long term absenteeism. DuPont has adopted a policy that each site will conduct and document an annual review of staff workload balance within each higher hazard operating unit to periodically evaluate the status and performance in this area. A standardized set of evaluation criteria for the annual review was developed and distributed to all sites as a leveraged tool to help ensure consistency across all locations.

6. Work Environment

DuPont has utilized a series of PHA related tools and checklists to conduct baseline, cyclical and MOC related evaluations for over the past twenty years. A human factors checklist has been one of these historical tools. DuPont's strategy to address FM work environmental factors involved adding content to these existing tools in order to help ensure that fatigue management workplace factors were considered and evaluated during all cyclic and project related PHA's. Integrating this review into existing PHA processes has proven to be a more sustainable and effective practice rather than creating a "stand alone" fatigue management work environment assessment process.

7. Medical Evaluations (Personal Risk Factors)

DuPont Integrated Health Services, in coordination with the process safety competency, developed a fatigue management medical screening approach and screening form which could be used as a voluntary tool during periodic health exams for shift employees who may be subject to fatigue risks. The health services group also provided targeted FM training to site medical professionals (nurses, doctors, and NP's) in order to consistently apply FM medical aspects, including both the proactive (voluntary) approaches as well as "for cause" evaluations where potential fatigue has been observed by line management in the work place.

DuPont Medical also worked with the corporate HR, labor relations, and legal functions to develop and communicate consistent medical guidance and case management methods where employee fatigue concerns have been identified. This is an area where emerging issues, specific interpretations, and case law should be considered by each entity such as the following (examples only):

1) Where additional FM medical testing is prescribed based on an employee's physical status and personal risk factors (obesity, concerns about sleep apnea) which could relate to fatigue in the workplace, should the costs for further medical evaluations be borne by the employee as part of personal health care or as part of the occupational medical program for a given company?
2) How should an employee be managed if they refuse to participate in additional medical screening or specific diagnostic tests that focus on fatigue management?
3) Are certain medical conditions that can cause fatigue in the work place subject to the 1990 Americans with Disabilities Act (ADA) in the United States and if so, which ones, and what may be the expectations for reasonable accommodations?

8. Contractors

DuPont has also adopted internal process safety requirements for a fatigue management program applied to selective contractors who work in higher hazard process areas which typically is limited to resident maintenance service providers, test and inspection consultants, and specialized services resources who may be on-site only during major turnarounds.

Contractors are required to develop their own internal programs with the basic elements of fatigue management as described in ANSI/API 755. DuPont will typically audit their systems and performance at the site level on a periodic basis to verify the program is adequately implemented. Many major contractors have embraced an enhanced focus to manage fatigue risks. In one example, a resident maintenance contractor developed a modified approach for their employee work scheduling during an annual shutdown with a DuPont site in Protection Technologies in order to improve fatigue management practices. The shutdown typically lasts 15 days involving 12 hour shifts and the contractor reduced the shift duration to 11 hours and inserted a day off for each employee during the process. Feedback from the employees was very positive, and this modifications were managed so that there were no additional costs or delays in the shutdown schedule. This unique approach serves as a good example where the owner/operator (DuPont) worked in collaboration with contactor partners to achieve a common objective which also supported the business and operations.

9. Incident Investigation Processes

DuPont maintains a global internal incident investigation reporting, classification and investigation policy for both EHS and process safety incidents which addresses communications, investigation teams, root cause failure analysis (RCFA), recommendations tracking, and overall documentation. The existing policy does not specifically include or consider fatigue management components as they may relate to contributing factors and human error as part of the overall RCFA process. At some point in the future, the policy will be updated to include a focused consideration of fatigue during all formal investigations, particularly when human error has been identified. This may include a standardized review of employee work shift schedules and overtime for at least the previous two weeks prior to the incident date.

10. Conclusion

Based on the "lessons learned" from the BP Texas City incident in 2005 and the "new" ANSI/API 755 Recommended Practice, internal company initiatives to integrate fatigue management into existing risk management systems is a prudent and necessary step to help minimize the potential for human error and the associated episodic incidents in high hazard process operations. The process safety function should be the logical "process owner", however, the initial systems design and implementation activities involving fatigue management necessitate a collaborative effort between other key company and site level resources and functions, including operations, human relations, health services and occupational health, in order to develop an integrated system and achieve maximum risk management benefits. Since the ANSI/API RP 755 standard includes primarily performance based content, implementation also may involve a substantial number of internal corporate or local policy decisions, and potential issues which require focused discussions and interpretations to promote consistency with internal applications at each site.

The initial FM implementation process may span two to three years in order to plan, educate, develop, implement and monitor site level organizations and accompanying results. Organizations that include fatigue management requirements in existing PSM policies should recognize that implementation progress may vary at each site depending on size, complexity, shift schedules, the existence of labor unions with collective bargaining agreements, and other factors. These potential variations on the rate of progress require a simplified process to track implementation across all applicable operating sites (see figure for an example of tracking based on the six main program elements). Integration of fatigue management evaluation criteria into existing PSM audit protocols and systems is typically the last step in a long term FM implementation and performance measurement process.

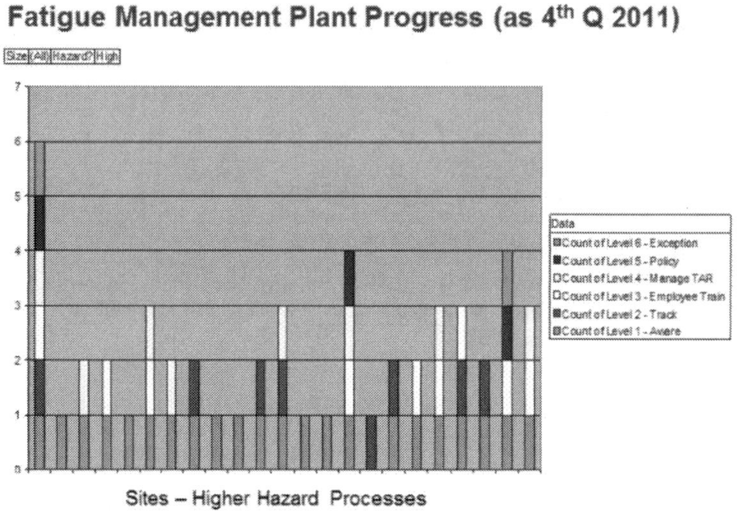

Figure 8: Example of Visual Management Board – Site Progress

11. References

(1) American National Standards Institute (ANSI)/American Petroleum Institute (API) Fatigue Risk Management Systems for Personnel in the Refining and Petrochemical Industries - RP 755 (First Edition, April 2010)

9TH GLOBAL CONGRESS ON PROCESS SAFETY

Consequences calculation for boilover and storage tank fires

Jaime E. Cadena
Chemical Engineering Department, Universidad de los Andes
Cra 1 Este No 19A – 40, Bogotá, Colombia
je.cadena21@uniandes.edu.co

Alejandra Cano[1], Juan Carlos Díaz[2], Alexis Vignes[3], Felipe Muñoz Giraldo[4]
[1,2,4]Chemical Engineering Department, Universidad de los Andes
Cra 1 Este No 19A – 40, Bogotá, Colombia
[3]INERIS
Technologique ALATA, 60550, Verneuil-en-Halatte, France
[1]na.cano744@uniandes.edu.co, [2]jc.diaz111@uniandes.edu.co, [3]Alexis.VIGNES@ineris.fr,
[4]fmunoz@uniandes.edu.co

Prepared for Presentation at
American Institute of Chemical Engineers
2013 Spring Meeting
9th Global Congress on Process Safety
San Antonio, Texas
April 28 – May 1, 2013

UNPUBLISHED

Consequences calculation for boilover and storage tank fires

Jaime E. Cadena
Chemical Engineering Department, Universidad de los Andes
Cra 1 Este No 19A – 40, Bogotá, Colombia
je.cadena21@uniandes.edu.co

Alejandra Cano[1], Juan Carlos Díaz[1], Alexis Vignes[2], Felipe Muñoz Giraldo[1]
[1]Chemical Engineering Department, Universidad de los Andes
Cra 1 Este No 19A – 40, Bogotá, Colombia
[2]INERIS
Technologique ALATA, 60550, Verneuil-en-Halatte, France

Keywords: Storage tank fires, boilover modeling, process safety, emergency response

Abstract

During the past decade, high loss incidents have been recorded due to fires in hydrocarbon storage facilities worldwide. An example is the storage tanks fire at a refinery in Amuay, Venezuela (August 25 2012, under investigation), which led to the loss of 650000 barrels of naphtha, around 41 fatalities and 80 injured. These incidents have high magnitude consequences, therefore an a-priori consequences analysis of these events become a key element for the decision-making processes to prevent, control and mitigate fire risks. The risk evaluation and mitigation for fire scenarios usually follows a prescriptive approach, therefore this work presents a tool for obtaining quick and conservative calculation of the thermal risks due to fire scenarios of hydrocarbons storage tanks facilities. These intensities are translated into consequences, emergency response requirements and key information for Contingency Plans development. To complement the fire scenarios, the tool also includes the calculation of effected distances for boilover phenomena. The results obtained by the fire tool were within an acceptable margin of error but it still conservative in comparison with real data. For the boilover tool different models were implemented among them the only one conservative for all the validation cases was the one proposed by Casal et. al. [1] for the feedback heat and the one proposed in the book Risk analysis in industrial facilities [2] for the boilover time.

1. Introduction

Storage tanks are a key element of any Oil & Gas operation since they constitute intermediate steps in the upstream operation e.g. in a refinery or a pipeline operation, and also in the downstream operations in which they are the source of final products to be delivered to the clients and customers. Nevertheless, even when hydrocarbons are stored without physico-chemical changes, risks are still present such as fire and boilover risks [3]. A fire in a storage tank is usually assumed as a pool fire in the surface of its contents, while a boilover is the ejection of the stored HazMat due to the fire and posterior boiling of a liquid with lower boiling point and higher density than the HazMat [2]. Storage tank fires and boilover scenarios represent low probability but high consequences events, making it necessary to apply available knowledge and strategies to analyze, assess, control and mitigate the related risks in order to avoid high capital and human loss as illustrated in Table 1 [4], which highlight that regardless of location or date, such events continue to occur. This makes it a key topic of study to avoid major hazard events and irreparable loss, especially for a region with a high industrial development such as Latin America [5, 6], in which the deployment of new Oil & Gas operations not only present emerging risks but also modify the existing ones [7].

Table 1. Storage tank fires and/or boilovers and related loss

Date	Location	Event	Related loss
13/12/1982	Bogotá, Colombia	Esso oil storage facility fire [8]	• 8000 people evacuated • Loss of 3 storage tanks
19/12/1982	Caracas, Venezuela	Explosion and subsequent oil tank fire [9]	• 153 fatalities • 7 missing • Severe damage to the facility
24/02/1986	Thessaloniki, Greece	Fire at an oil storage tank facility [10]	• 10 out of 12 tanks destroyed • US$330 M
11/12/2005	Buncefield, UK	UVCE and fire at hydrocarbon storage facility [11, 12]	• 43 injured • Over 2000 people evacuated • Low and moderate damage to surrounding buildings • Total destruction of most of the complex • Possible Detonation to Deflagration Transition (DDT) [13]
26/10/2009	Bayamon, PR, USA	Caribbean Petroleum Company fire (currently under investigation) [14]	• 1 injured • Costs around US$6.4 million • Around 5.6 million of product loss • Around 600 people evacuated • Damage to buildings in a radius over 1 mile (1.6 km)
25/08/2012	Amuay, Venezuela	PDVSA Amuay refinery explosion and fire (currently under invetsigation) [15]	• 48 fatalities (latest information) • Over 80 injured • Stop of the refinery with 640000 bpd capacity • Damage to at least two contiguous storage tanks, generating fire • 18 national guard and 15 civilian fatalities • Damage to around 200 residential buildings and 11 businesses • 33 families displaced • Damage to a military base

Although the subject has been addressed through different approaches [16, 17], in the field of process safety it is fundamental to understand the phenomena underlying the events of interest. For tank fires and boilover events this has been carried out for several decades, and a wide variety of empirical, semi-empirical and CFD models have become available to model such events [1, 18-23]. This work aims to develop a methodology to apply a set of the available models to calculate the consequences of the events of interest through its implementation in a user-friendly computational tool that allows obtaining quick and conservative results for either a preliminary risk analysis stage in which critical scenarios are identified or during actual emergencies in which knowledge of the possible outcomes is needed to establish mitigation strategies.

2. Methodology

In order to achieve the objectives stated above, a methodology (Figure 1) was proposed for the development and validation of the tool. First step is defining the scenario to evaluate, the tool use variables that influence tank fire and a boilover like some specific hydrocarbon properties such as viscosity, initial and final boiling point temperatures and combustion heat as input variables, next the tool implements models for the calculation of heat released by the fire the feedback heat and boilover and duration of fire times. Based on the last calculations the tool use models to determine the safe distances for both phenomena (boilover and fire) and in the case of the fire the tool also returns as output variables some mitigation strategies like refrigeration rings.

Figure 1. Methodology

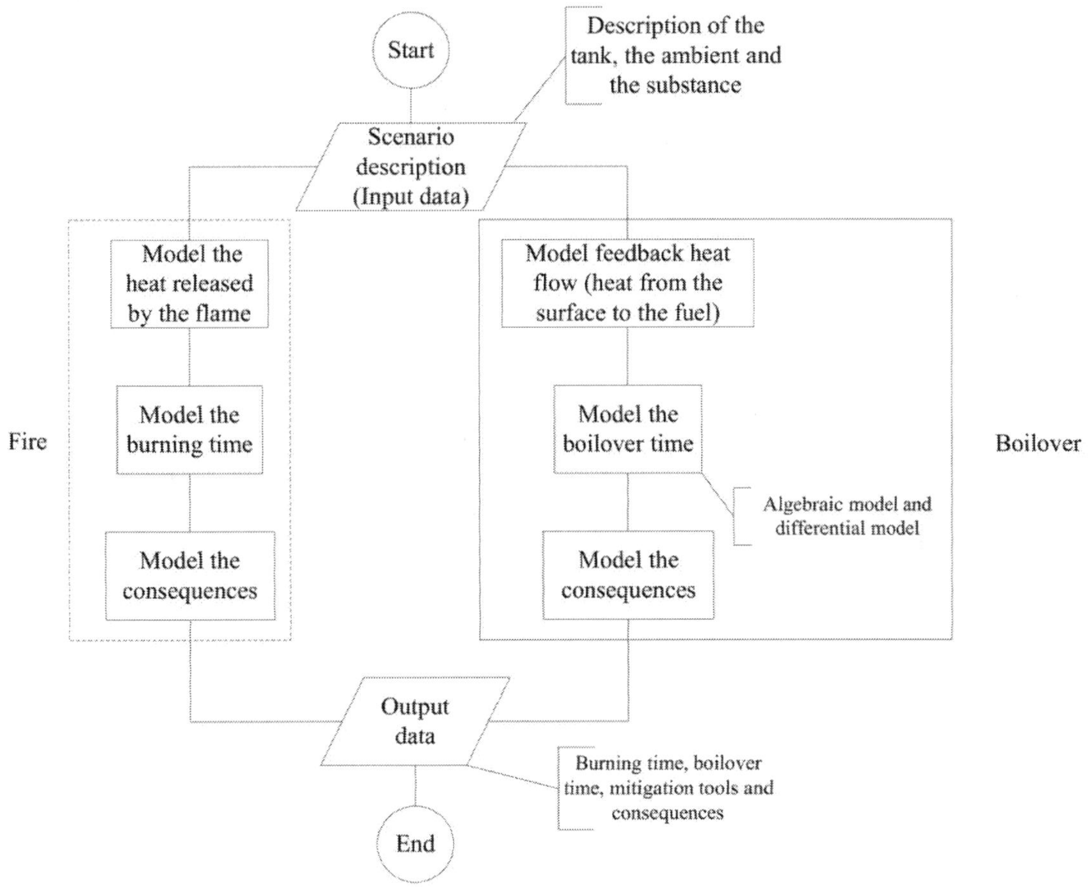

3. Tool implementation

The tank fire tool was developed in excel (visual basic) using macros for both, the heat released by the fire and the burning time. For the boilover there are multiple approximations, for the feedback heat there is an average value reported [2] and models for calculating it using a fraction of the total heat released by the fire, the difference among the last ones is the calculation of the total heat released that can be done using models with combustion heat and experimental corrections depending on the fuel [24] or heat balances between the tank and the surroundings [1], for the boilover time there is an algebraic approximation derived from a heat balance [2] and differential approximations; varying among them the initial and boundary conditions [1] . Some of them were implemented in excel and the ones involving differential approximations were implemented in matlab as this software allows multiples ways for solving differential equations in a short and easy way. Both Excel and Matlab have a user-friendly interface. The models selected for the two tools are presented on Table 2.

Table 2. Models used for the tool implementation.

		Model	Assumptions	Ref
Fire	Time calculation	$t_b=4V/\pi D^2 v$ $v=m''/\rho$	-	[24]
	Heat calculation	$Q = m''\Delta H_{c,eff}(1-e^{-\kappa\beta D})A_{dike}$ $D=(4A_{dike}/\pi)^{1/2}$	- Experimental correction depending on the fuel	[24]
Boilover	Time calculation	1) $t_{boilover}=\rho c_p h_{HC}(T_{hw}-T_a)/(Q_f-m''(H_v+c_p(T_{0av}-T_a)))$ $T_{0av}=(T_{0min}T_{0max})^{1/2}$		[2]
		2) $\delta^2T/\delta y^2=(1/\alpha)\delta T/\delta t$	-Unidimentional	[21]
	Heat calculation	1) $Q_f \approx 60$ kW	-	[2]
		2) $Q_f=(4\chi/\pi)\rho_\infty c_{p\infty}(T_\infty g(T_f-T_\infty)D)^{1/2}$ $\chi=H_v/H_c$	- A heat conduction mechanism is assumed, no conduction in the fuel layer -The radiation is fully absorbed at the surface	[21]

The values for T_f in the heat calculation for the boilover, change depending on the assumptions used for the estimation, some authors use the adiabatic flame temperature as an approximation and others recommend approximated values like 1100 K [1] or values ranging from 1600 to 1900 K depending on the substance.

To solve the differential equation for time of boilover the following boundary conditions were established.

- Initial condition. [1]

$$\mathbf{T(y,0) = T_\infty} \quad \text{[Eq. 1]}$$

- Boundary condition 1.[1]

$$T(y(t),t) = T_s \quad \text{[Eq. 2]}$$

Where **y(t)**is defined by the following equation.

$$\delta y(t)/\delta t = m'' / \rho_f \quad \text{[Eq. 3]}$$

- Boundary condition 2.

$$-k_f * \delta T / \delta y \,|_{fuel} = -k_w * \delta T / \delta y \,|_{water} \quad \text{[Eq. 4]}$$

To achieve boundary condition 2, the modeling of the water layer is needed and it can be solved by the same equation for the fuel reported on Table 2. The first boundary condition for the water layer is stated on Equation 4 and the second boundary condition stated in Equation 5. This boundary condition implies assuming adiabatic conditions in the limit of the water layer with the bottom of the tank.

$$\delta T / \delta t \,|_{y=0} = 0 \quad \text{[Eq. 5]}$$

3.1 Input and output variables

For the fire tank tool is needed from the tank the diameter, the height, the type of roof and the capacity; the tool have all the characteristics for the fuel, it only need to be specified with the volume present in the tank. In addition if the studied facility has more storage tanks near to the one in fire (storage farm tanks) the distances from the one in fire to the others must be known. The output variables are the duration of the fire, the area involved on the fire, the zone classification (hot, warm, cold) and the mitigation and prevention actions [24] like if the refrigeration ring should be activated.
For the boilover the input values are the diameter of the tank the ambient conditions, specifying the fuel involved in the fire, the initial height of the fuel in the tank before the ignition the temperature of the heat wave when the boilover occurs and for one of the models the flame temperature. The output data consists of the boilover time and the distances at which lethally and irreversible consequences occur.

3.2 Consequence estimation

The consequences for both tank fire and boilover were calculated using Excel and both have dependence from the heat irradiated from the fire in the case of the tank fire and for the resulting fireball in the case of the boilover.
The consequence of the fire depends on the heat radiation determining the hot, warm and cold zone. In the hot zone the equipment will fail and the heat release will be lethal for human beings. The warm zone can be accessed with protection equipment, the firemen attack the fire there. Finally the cold zone is the one where the equipment is displayed, there is no need of protection equipment in this zone. The hot zone is determined by values where the heat is over 10 kW/m^2, the warm zone is the one with heat values between 2,5 kW/m^2 and 10 kW/m^2 and the cold zone is the one with values lower than 2,5 kW/m^2 [24]. The consequences of a boilover event are fireballs and in the ejection of ignited fuel around the tank. According to the radiation heat

release of the fire there can be determined lethal and irreversible (serious burns) effect distances corresponding to radiation heats of 1000 kW $m^{-2}$and 600 kW m^{-2} each one, they are estimated using the conservative assumption that the entire fuel in the tank participates to the fireball. The models used for the safe distances and zones are reported on Table 3 [2].

Table 3. Models used for consequence calculations.

	Consequence	Model	Ref
Fire	Irradiated heat	$d_{hotzone}=(QXr/4\pi\ q''_{hotzone})^{1/2}$ $d_{hotzone}=(QXr/4\pi\ q''_{warmzone})^{1/2}$	[24]
Boilover	Fireball	$d_{lethality}=k_{lethality}W^{0.45}$ $d_{irreversible}=k_{irreversible}W^{0.45}$	[2]

4. Results and analysis

To validate the tool it was compared with selected cases for each event (fire and boilover), the case selection was made using two criteria, the first one is searching for cases that are not affected by other events after the fire starts and the second one is searching for cases with complete information about the outputs and inputs for the tool mentioned above. The cases and the comparison with the tool results are described in this section with the comparison among the boilover time and the burning time.

4.1 Tank fire tool

For the tank fire tool the case selected was an accident that occurred at ConocoPhillips Company's Glenpool South tank farm, Glenpool, Oklahoma on April 7 2003. The event involved the tank 11 of the facility during the transport of Diesel from another tank [25] . The results obtained by the tool are reported on Table 4 and Table 5.

Table 4. Results related to the tank and the fire variables

Burning Tank Name:	11
Type of Roof:	CONE
Product:	Diesel
Nominal Volume (m3):	8420,00
Contents (m3):	1338,67
Height (m):	14,63
Diameter (m):	33,22
Area of Fire (m2):	1526,84
Time of Fire (hr):	8,39
Time of Fire (day):	0,35
Refrigeration Range (m):	71,89

Table 5. Results related to the consequences of the fire

	From (m)	To (m)
Hot Zone	0	64,298
Warm Zone	64,298	128,596
Cold Zone	128,596	N/A

The variable to compare among the case and the tool was the burning time because that is the variable reported in the few cases that also have all the input parameters. The percentage of error among the real case and the results reported by the tool is lower than 5%. In the case of the burning time the percentage of error reported may be due to the model, because the variables involved in its calculation (volume of the fuel and dimensions of the tank) don't require any assumptions and are directly specified by the case to evaluate. The possible source of error for this validation case may be directly related to it, as the tank was being filled at the moment of the fire the initial volume of fuel is not an accurate value.

4.2 Boilover tool

For the boilover tool different models were used, a comparison among them is also included in this section.

Case 1:

For the boilover, a large scale experiment was used for the study of boilover in large tanks using crude oil [26]. The results obtained by the tool with different models are presented on Table 6 and the distance effects are reported on Table 7. They are independent from the model used for heat and time of boilover so there is just one value for $d_{lethality}$ and one for $d_{irreversible}$ reported for the case.

Table 6. Case 1 results for the boilover time with the models proposed on Table 2.

	Tflame (K)		**tbo (s)**		**treal (s)**	**%error**	
Heat model 1	NA	NA	4067,67		4728	13,96637056	
Heat model 2	1100	1900	1127,07357	692,031935	4728	76,1617264	85,3631147
Time model 2	NA	NA	60604		4728	NA	

Table 7. Safe distances for the boilover consequences for case 1

dletal (m)	14,7314678
dirrev (m)	20,0281754

Case 2:

A fire on tank 3 containing crude at the Czechowice-Dziedzice Refinery on June 26, 1971 at 19:50 occurred due to a lightning hit [27]. For this case the results obtained by the tool are reported on Table 8.

Table 8. Case 2 results for the boilover time with the models proposed on Table 2.

	Tllama (K)		tbo (hr)		treal (hr)	%error	
Heat model 1	NA	NA	29,77614629		5,3	461,8140809	
Heat model 2	1100	1900	2,4275687	1,64560344	5,3	54,196817	68,9508784
Time model 2	NA	NA	1068,095		5,3	NA	

Table 7. Safe distances for the boilover consequences for case 2

dletal (m)	350,936098
dirrev (m)	477,115369

For case 1 the most accurate value reported by the tool was the one calculated by using an approximated value of 60 kWm^{-2} for the feedback heat, while the most accurate values in case 2 were the ones calculated by using the heat model 2, for this case the value using the heat model 1 is not conservative. This can be explained by the fact that the heat proposed by literature is very small compared with the one calculated by the second heat model, it worked well for a tank with an small amount of fuel but with great amounts of fuel that heat is not enough and will take more time to increase the temperatures in fuel layer and to take the water to its boiling point, making the boilover time bigger. So the second model may be better for modeling large tanks with big amounts of fuel, keeping the time in the same order of magnitude and conservative compared to the real time.

In the case of the differential model mentioned before for both cases the boilover time calculated with the tool is four orders of magnitude or more above the real time, this can be explained by the assumption around the model, first the model is one dimensional assuming that all the heat is transferred from the top to the bottom of the tank when it exist also a radial heat transference, this can also be seen by comparing the results from case 1 with the results from case 2 where the difference among the real time and the calculated time is proportional to the amount of fuel and the tank dimensions. Also the thermal diffusivities along with the rest of the fluid properties were assumed constant during the fire, despite the changes of temperatures in the fuel and water layers. One of the assumptions of the model implies assuming only a heat conduction mechanism, no convection inside the layer occurs, this last phenomena implies the change of density in different heights of the fuel layer and this difference in densities implies that the fuel in the top of the tank will eventually move to the bottom and the fuel in the bottom will eventually move to the top making the heat transference quicker and the boilover time smaller.

4.3 Comparison among boilover time and burning time

The case 2 described for the validation of the boilover tool have the complete information to use also the tank fire tool and can be used to compare the time of burning if the boilover doesn't occur with the time at which boilover occurs. Giving a little notion of what fraction of the total burning time takes the boilover to occur.

The results obtained by tool for the fire are presented on Table 8 and the consequences on Table 9.

Table 8. Results related to the tank and the fire variables for case 2

Burning tank name:	1
Type of roof:	cone
Product:	Crude Oil

Nominal volume (m3):	80000,00
Content (m3):	10142,89
Height (m):	14,7
Diameter (m):	33,00
Area of fire (m2):	1523,99
Time of fire (hr):	93,33
Time of fire (day):	3,89
Refrigeration rate (m):	57,11

Table 9. Results related to the consequences of the fire for case 2

	From (m)	To(m)
Hot zone	0,000	51,083
Warm zone	51,083	102,167
Cold zone	102,167	N/A

The boilover results are reported in the above section, the comparison was made using the real value reported in literature for the case and the only conservative value reported by the tool that was calculated using the heat model 2. The percentages reported on Table 10 are the fraction that takes the boilover to occur with respect to the total time of fire.

Table 10. Percentage boilover time/ burning time

	%
Real Value	5,67877424
Heat model 2	2,59294975

The boilover time is a little fraction of the total burning time, showing how important is to estimate the first one, modeling only the fire will give more time for mitigation actions but if a boilover occurs this time is reduced on a 95 % and being conservative on a 97%, the consequences are also greater for the boilover where the safe distances are superior than 477,11 m while the cold zone for the fire is above 102,167 m, about a fifth times less the boilover distance. Making the boilover a more dangerous event than the fire itself and a matter to be worried about.

4. Conclusions

The fire tool's validation results are within an acceptable error margin but still conservative, which implies that the used model describes the event and can be used for real calculations and being also simple to manipulate for the user because there is no need of complicated interfaces or calculations. In the case of the boilover the recommended model based on the results reported on this work is the one reported by Casal et. al. [1] for the feedback heat and the algebraic approximation for time because with them there were obtained for both cases used for validation a conservative result. This implies that some of the proposed models can actually be implemented on tools that can be used in real life cases helping to generate and design mitigation strategies.

5. Perspectives

For future work the differential solution for time will be studied calculating parameters like the thermal diffusivities depending on the temperature profiles and solving a two-dimensional model. Also validating the tool against more cases including cases that involve alternative fuels.

Nomenclature

Q: Pool fire heat release rate (kW)
m": Mass burning rate of fuel per unit surface area (kg $m^{-2}sec^{-1}$)
$\Delta H_{c,}$: Effective heat of combustion of fuel (kJ/kg)
$A_f = A_{dike}$:Surface area of pool fire (area involved in vaporization)
κβ: Empirical constant (m^{-1})
D: Diameter of the pool fire (m^2)
t_b: Burning duration of pool fire (sec)
V: Volume for liquid (m^3)
D: Pool diameter (m)
v: Regression rate (m sec^{-1})
ρ:Liquid fuel density (kg m^{-3})
c_p: Specific heat of the fuel at T_a (kJ $kg^{-1}K^{-1}$)
h_{HC} : Height of the fuel in the tank after ignition (m)
ΔH_v : Vaporization rate atT_{0max} (kJ kg^{-1})
T_a : Ambient temperature (K)
T_{hw} : Heat wave temperature when the boilover occurs (K)
T_{0av} : Average boiling temperature of the fuel (K)
Q_f: Thermal flow entering the fuel from its surface (kW m^{-2})
T_{0min}:
T_{0max}:
α: Crude thermal diffusivity
ρ_∞ : Air density (kg m^{-3})
$c_{p\infty}$: Specific heat of the air (kJ $kg^{-1}K^{-1}$)
T_∞: Ambient temperature (K)
T_f: Average flame temperature (K)
Xr:
$q''_{hotzone}$: Heat per unit area that humans and equipment can't tolerate (10 kW/m^2)
$q''_{warmzone}$: Heat per unit area that can be tolerated with protection equipment (2,5 kW/m^2)
W: Mass of the fuel at the beginning of the fire.
$k_{lethality}$: Lethality constant depending on the fuel.
$k_{irreversible}$: Irreversible consequences constant, depends on the fuel.

N. References

[1] J. Hristov, E. Planas-Cuchi, J. Arnaldos, and J. Casal, "Accidental burning of a fuel layer on a waterbed: a scale analysis of the models predicting the pre-boilover time and tests to published data," *International Journal of Thermal Sciences,* vol. 43, pp. 221-239, 2004.
[2] J. Casal, *Análisis del riesgo en instalaciones industriales*: Edicions UPC, 2001.
[3] K. Trevor, "Chapter 5 - Storage Tanks," in *What Went Wrong? (Fifth Edition)*, ed Boston: Butterworth-Heinemann, 2009, pp. 97-121.
[4] B. Zheng and G.-h. Chen, "Storage tank fire accidents," *Process Safety Progress,* vol. 30, pp. 291-293, 2011.

[5] E. Corportation, "2012 The Outlook for Energy: A view to 2040," 2012.
[6] CIA. (2011, 11/01/2012). *The World Factbook :: Oil Production*. Available: https://www.cia.gov/library/publications/the-world-factbook/fields/2173.html
[7] G. Gómez, S. Rodríguez, J. Cadena, and F. Muñoz, "Kletz's legacy for developing countries: Simple systems based on inherently safer design," *Journal of Loss Prevention in the Process Industries,* vol. 25, pp. 843-847, 2012.
[8] AP, "8000 flee Colombia gas fire," in *The Telegraph*, ed, 1982, p. 1.
[9] S. Mannan, F. P. Lees, and Knovel (Firm). (2005). *Lee's loss prevention in the process industries hazard identification, assessment, and control (3rd ed.).* Available: http://www.knovel.com/knovel2/Toc.jsp?BookID=1470
[10] J. I. Chang and C.-C. Lin, "A study of storage tank accidents," *Journal of Loss Prevention in the Process Industries,* vol. 19, pp. 51-59, 2006.
[11] D. Bradley, G. A. Chamberlain, and D. D. Drysdale, "Large vapour cloud explosions, with particular reference to that at Buncefield," *Philosophical Transactions of the Royal Society A: Mathematical, Physical and Engineering Sciences,* vol. 370, pp. 544-566, February 13, 2012 2012.
[12] B. Investigation Board, "The Buncefield Incident," Health and Safety Commision, HSC2005.
[13] B. J. Lowesmith, G. Hankinson, and D. M. Johnson, "Vapour cloud explosions in a long congested region involving methane/hydrogen mixtures," *Process Safety and Environmental Protection,* vol. 89, pp. 234-247, 2011.
[14] U. S. C. S. a. H. I. B. CSB, "CSB Conducting Full Investigation of Massive Tank Fire at Caribbean Petroleum Refining; Investigative Team Plans to Thoroughly Examine Facility Safety Practices," in *News Room* vol. 2012, ed. Washington DC, USA, 2009.
[15] K. Vyas and E. Minaya. (2012) Explosion at Amuay refinery in Venezuela kills 41; fire still burning. *Hydrocarbon Processing.* Available: http://www.hydrocarbonprocessing.com/Article/3080380/Environment-and-Safety/Explosion-at-Amuay-refinery-in-Venezuela-kills-41-fire-still-burning.html
[16] C. Crippa, L. Fiorentini, V. Rossini, R. Stefanelli, S. Tafaro, and M. Marchi, "Fire risk management system for safe operation of large atmospheric storage tanks," *Journal of Loss Prevention in the Process Industries,* vol. 22, pp. 574-581, 2009.
[17] C. D. Argyropoulos, M. N. Christolis, Z. Nivolianitou, and N. C. Markatos, "A hazards assessment methodology for large liquid hydrocarbon fuel tanks," *Journal of Loss Prevention in the Process Industries,* vol. 25, pp. 329-335, 2012.
[18] C.-I. Lin and H.-F. Wang, "Research on Safety and Security Distance of Flammable Liquid Storage Tank," *Procedia Engineering,* vol. 11, pp. 51-60, 2011.
[19] M. Arai, K. Saito, and R. A. Altenkirch, "A Study of Boilover in Liquid Pool Fires Supported on Water Part I: Effects of a Water Sublayer on Pool Fires," *Combustion Science and Technology,* vol. 71, pp. 25-40, 1990/05/01 1990.
[20] T. Inamura, K. Saito, and K. A. Tagavi, "A Study of Boilover in Liquid Pool Fires Supported on Water. Part II: Effects of In-depth Radiation Absorption," *Combustion Science and Technology,* vol. 86, pp. 105-119, 1992/11/01 1992.
[21] I. M. Shaluf and S. A. Abdullah, "Floating roof storage tank boilover," *Journal of Loss Prevention in the Process Industries,* vol. 24, pp. 1-7, 2011.
[22] M. Muñoz, J. Arnaldos, J. Casal, and E. Planas, "Analysis of the geometric and radiative characteristics of hydrocarbon pool fires," *Combustion and Flame,* vol. 139, pp. 263-277, 2004.
[23] M. Muñoz, E. Planas, F. Ferrero, and J. Casal, "Predicting the emissive power of hydrocarbon pool fires," *Journal of Hazardous Materials,* vol. 144, pp. 725-729, 2007.
[24] NFPA, *SFPE handbook of fire protection*, 3rth ed., 2002.

[25] "Storage Tank Explosion and Fire in Glenpool, Oklahoma," National Transportation Safety Board, Washington DC2003.

[26] H. Koseki, Y. Natsume, Y. Iwata, T. Takahashi, and T. Hirano, "Large-scale boilover experiments using crude oil," *Fire Safety Journal,* vol. 41, pp. 529-535, 2006.

[27] H. Persson and A. Lönnermark, "Tank Fires Review of fire incidents 1951–2003," SP Swedish National Testing and Research Institute.

9TH GLOBAL CONGRESS ON PROCESS SAFETY

A practical view of risk reduction management of hardware and software for safety critical applications

Luis Duran
Product Marketing Manager Safety Systems
ABB
3700 West Sam Houston Parkway South suite 600
luis.m.duran@us.abb.com

Mike Krywonos
Safety Execution Center Manager
ABB
3700 West Sam Houston Parkway South suite 600
mike.krywonos@us.abb.com

Prepared for Presentation at
American Institute of Chemical Engineers
2013 Spring Meeting
9th Global Congress on Process Safety
San Antonio, Texas
April 28 – May 1, 2013

UNPUBLISHED

A practical view of risk reduction management of hardware and software for safety critical applications

Luis Duran
ABB
3700 West Sam Houston Parkway South suite 600
luis.m.duran@us.abb.com

Mike Krywonos
ABB
3700 West Sam Houston Parkway South suite 600
mike.krywonos@us.abb.com

Keywords: Safety, Risk Reduction, Safety Systems, Independent Protection Layers, Common Cause Failures, Functional Safety Management System, IEC61508, IEC61511, TÜV, third party certification, Human Factors

Abstract

Best in Class companies (both National and Global) continue to invest in safety systems; this in most cases is because safety is at the core of their production process and among their key performance metrics, closely linked to their productivity.

Many unfortunate industrial accidents such as the recent explosion of the Deep Water Horizon with the largest marine oil spill in history are evidence of the negative consequences they have on people, environment, production assets and corporations.

This paper discusses the accepted definition of safety and risk reduction in the industry and will discuss the best practices applied by engineers in the design of products for mission critical functions and the implementation of those products in projects.

Both in product design and implementation, this session will dedicate special attention to methodologies or work processes used to reduce potential common cause failures that might impair the operation of the safety system and ultimately expose the facility to a higher risk; and the human factor considerations recommended to ensure proper operation upon a critical condition or an emergency in the industrial process.

1. Introduction

Best in Class companies worldwide, defined as such for their high percentage of Overall Equipment Effectiveness (OEE) and low injury frequency rate; continue to show strong initiatives intended to expand the safety of their operation. From executive sponsored programs at the corporate level to the definition of proactive risk management strategies, these companies are investing in safety systems and processes; simply because in most cases, safety is identified among the core values of those companies, at the center of their production process and among their key performance metrics, closely linked to their productivity.

There are numerous publications and records of Industrial Accidents, a large number of them with terrible consequences, including not only loss of human life or environmental impact but also affecting the production assets in the site. As an example, in researching for this paper, the authors found record of at least 20 "notable" oil and gas offshore blowouts between 1980 and 2010.

As an example for this paper, the authors compared two serious incidents in the Oil & Gas sector; Piper Alpha operated by Occidental Petroleum and Deep Water Horizon leased to BP. Although both incidents are 22 years apart from each other, occurred in different geographies and have many differences in causes and consequences. They illustrate the risk and the potentially catastrophic dimension they can take in this industry and the impact on people, environment and corporations, which might cease to exist after these incidents.

2. Defining Safety and Risk

Safety is defined in the industry as a reduction of existing risk to a tolerable or manageable level while risk is a combination of the probability of a harmful incident and magnitude of the harm. Traditional design practices allocate the risk reduction across different and independent protection layers, the rationale behind it is simple "any system that can fail will fail" so the engineering best practice is to distribute the risk reduction tasks across multiple independent functions or systems one of these systems is a Safety Instrumented System.

2.1 Safety Instrumented Systems one of many Independent Protection Layers to Reduce Risk

A Safety Instrumented Systems (SIS) is a mission critical system designed following international design practices such as IEC61508 [1] Functional Safety Standard to reduce risk to the people in and around the production environment, the environment, the production asset and the business. In many cases the Safety Instrumented Systems are the last resource to prevent disaster.

SIS performance is measured by Safety Integrity Level (SIL) SIL 1 low, SIL 3 high o Risk Reduction Factor.

As mentioned earlier, the Safety Instrumented System is one of many functionally independent systems each intended to perform a task, as shown on Table 1 typically referred to as Independent Protection layers [2]

Table 1: Expected Functionality of Independent Protection Layers

Layer of Protection	Expected Functionality
BPCS:	Keeping the process under control
Alarms:	Alerting the operator of abnormal conditions and providing guidelines for appropriate operator response
SIS:	Automatically taking the process to a safe condition in case the abnormal condition goes out of control and the operators can't take any corrective action in time
Other layers:	Intended to mitigate the consequence of the hazard

The concept of Independent protection layers (IPL) can also be illustrated in Figure 1

3. Common Cause Failures

The idea of functional independent systems is an attempt to avoid Common Cause Failures (CCF).

Common Cause Failures [3] are failures that might affect more than one of these protection layers at the same time. As engineers, the tendency will be to focus on elements such as Heat, Humidity, Shock, and radio interference among others. However this paper also considers with special interest the human elements related to the design of safety critical systems which can contribute to additional failures.

As with other systems, some SIS problems are related to the Commercial Off the Shelf (COTS) products designed for a specific function (i.e. hardware limitations or poor documentation) while other problems are related to the use of those COTS product such as misapplication, user application programming or poor maintenance practices. Both areas can be addressed by implementing appropriate design best practices to reduce risk.

The industry has conceived these "best practices" as a series of steps that must be performed before putting one of these systems in operation (design, installation and commissioning). This paper will describe the two areas:

- Product Design and Implementation of the Commercial Off-the-Shelf (COTS) Product
- Application Design

Figure 1: Independent Protection Layers

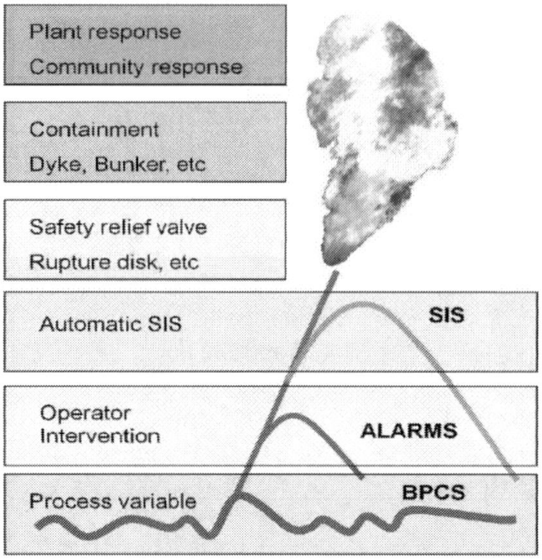

3.1 Reducing Common Cause Failures by Design

Traditionally or historically, Systems designed for SIS functions have relied on physical separation and redundancy [4] to reduce common cause failures. The authors identified technology changes across multiple generations of SIS over a period of 30 years. However over the same time frame the industry applied lessons learned from incidents to develop best practices in design and implementation of systems, the international industry standards used more often are IEC61508 and IEC61511 [5] (Figure 2)

3.2. Hardware Fault Tolerance (1st Generation SIS)

1st Generation systems rely on Hardware Fault Tolerance or redundancy to achieve reliability and availability as required for these applications, this practice is rooted in the late 1970s and early 1980s technology. As the majority of these systems were designed prior to the release of the Functional Safety standards, they do not follow the best practices found with regards to design, documentation and testing found in those standards.

Figure 2: Evolution of National Prescriptive Standard to International Performance Base Standards

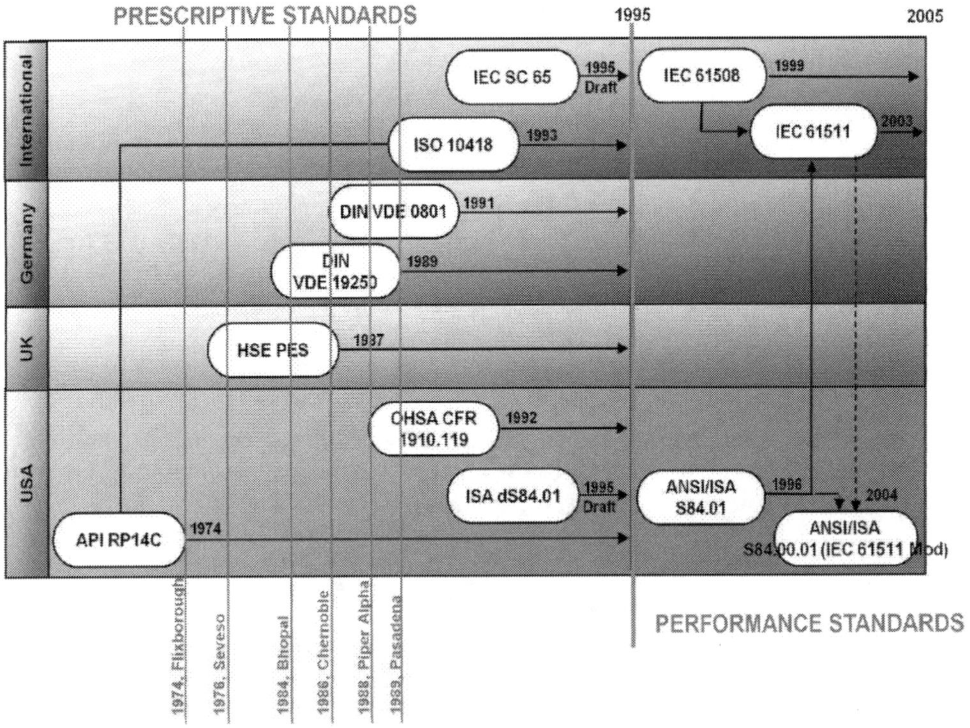

3.3. 2nd Generation Systems

As the standards became available (mid 1990s), vendors started to apply the design best practices and to pursue certification to those standards. This generation of systems was produced having third party assessment and certification but continued to rely on Hardware Fault Tolerance to satisfy the performance requirements. The authors found that a large portion of the systems applied as SIS in the market today, although designed to satisfy Functional safety standards haven't change dramatically in their use of hardware fault tolerance and basic software and hardware diagnostics.

3.4. Diverse Architecture and Implementation a 3rd Generation Systems

A more recent and different type of system, uses diverse redundancy, diverse implementation and active diagnostics [6] and can deliver not only reliability and availability required for the application while minimizing common cause failures found by the use of the same technology but also introduce additional protection to systematic failures normally related to human factors

3.4.1 Hardware Design – Diverse Hardware

This newer generation of systems use diverse processing hardware (multiple technologies) such as diverse operating systems and diverse hardware and diverse redundancy, both enhanced with diverse implementation (different implementation teams).

3.4.2 Application Execution in a SIL Compliant Environment

The use of diverse operating systems (use of different technologies) including the use of third party certified COTS and different execution path with different compiler rules extends the risk reduction found in the hardware design to the firmware/software environment in the system, including the use of limited variability software functions and provision of systematic capabilities as specified in the Functional Safety standards for critical applications.

3.4.3. Systematic Capabilities and Human Error

The concept of systematic capabilities and its associated performance measurement was introduced in the latest revision of IEC61508 standard (didn't exist as such in 1st Generation systems and was emerging in 2nd generation systems).

3.5. Safety and Network Security

1st and 2nd generation Safety Systems were not designed to reside on a networked plant, for the most part these systems allowed some communication interface or gateway but were not intended for integration, therefore in general are not equipped to counter security threats as those the industry has experienced in recent years.

A brief analysis of the definitions will show a clear differences Safety issues are random in nature and statistic analysis is possible security intentional in nature and statistical analysis will not measure performance.

The authors concur with industry standards and experts in the field. Network security may affect system performance and the safety of the installation. A system can't be safe without it being secure.

3.6. Design Best Practices and V-Model

As indicated earlier, industry standards in Functional Safety introduced the concept of Functional Safety Management System (FSMS) which includes a series of steps in the design, documentation and testing of the system, and should include Network Security as well.

Figure 3: Product Development V-Model

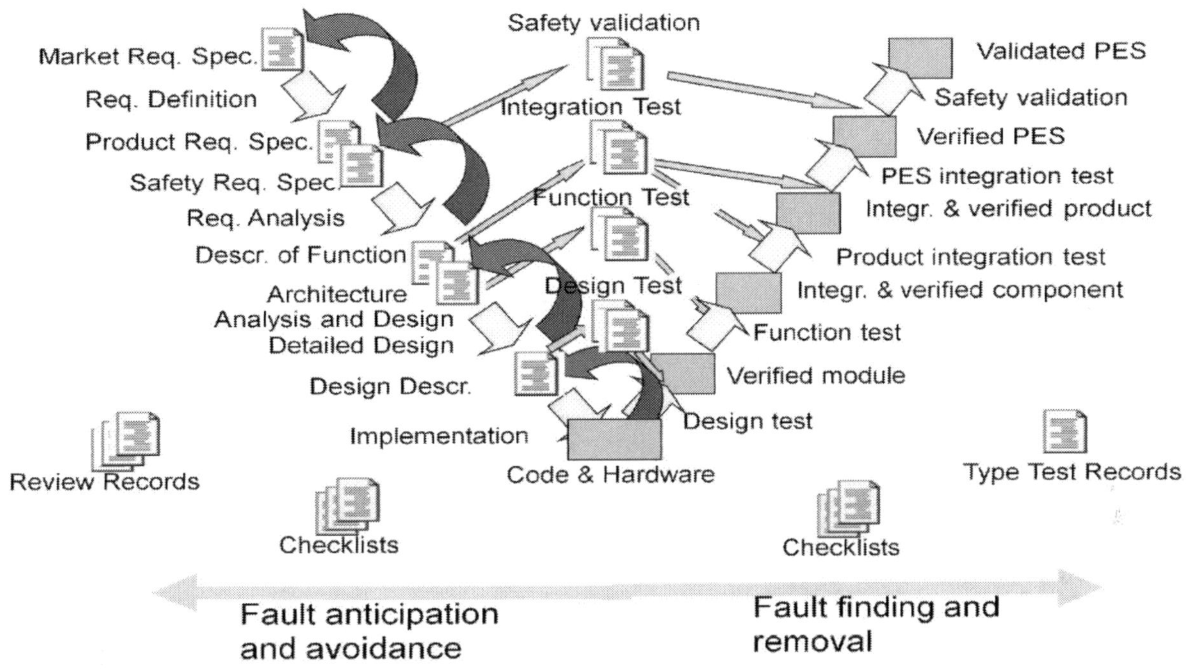

Today product development or design organizations responsible for releasing COTS products (hardware and software) intended for safety applications use design best practices as the V-Model (Figure 3) and are evaluated by a third party organization (i.e. TUV) and their FSMS is approved prior to the assessment of their products.

4. Engineering best practices to reduce risk in application design

As indicated previously, best practices in design apply both in the design of a Commercial Off the Shelf (COTS) Product and the Application Design or the application of a COTS tailored to reduce a particular risk in an industrial installation.

Best Practices in Application Design will apply a model similar to the V-Model used in the development of a product (Figure 3) and reinforced by a structured Quality Management system appropriate for Functional Safety applications.

Application Design starts with Hardware Design, which starts with a design that follows the COTS Product's Safety Manual and considers the particularities of the application as described on the Application Safety Requirements Specification to design a system that is less susceptible to product failures by considering aspects such as

- Power feeds
- Power Supplies
- CPUs
- I/O Modules Hardware
- Application Software
- Networks

4.1 Hardware

It's clear that hardware is not perfect, it's subject to fails, and the application design must consider fail recovery modes or additional risk reduction methodologies or options including diverse methods of performing shutdowns as:

- Hardwired wired pushbutton,
- Hardwired logic systems (with appropriate SIL rating)
- Solid State relays
- Etc.

4.2 Software

One critical aspect that has gained visibility in recent years is the software design of the application. Efforts in this area include the development of a well documented Software Functional Design, appropriate selection of software tools and libraries that meets the Safety Integrity Level (SIL) required for the application, Code Review practices including Independent Validation and Verification depending on the SIL, competence of the programmer and other elements described later in section 5.1. The features available in the COTS product to address the systematic capabilities requirements are a valuable asset for the programmer and contribute to reducing potential errors that will later affect the safety of the installation.

4.3 Functional Testing and Periodic Proof Testing

Another critical aspect is Functional Testing and Periodic Proof Testing, particularly because Not ALL System Faults are Self-Revealing and Covert Faults that may inhibit SIS action on Demand can only be detected by testing the entire system, this requires not only a Full Functional Test prior to commissioning, typically conducted as an integration test in the vendor or System Integrator facility and later in the final instrumentation room in the plant; but Periodical Functional Tests using a documented procedure to Detect Covert Faults and covering the entire SIS.

Functional Testing should Record and Analyze activation of SIS functions, and spurious activation of an Emergency Shutdown Valve due to a Process Shut Down, but this does not test the Entire Function of the same valve during an ESD action.

5. Automation can't check for human intervention

Once the design is complete, all of these systems will have different degrees of interaction with users, and different type of users with different levels of competence depending to their role (i.e. operation, maintenance personnel and engineers)

The most sophisticated automation can't prevent human error, industrial plants are designed with the highest accuracy in mind, with several separate safety loops often checking the integrity of process systems. This apparatus however can't check for human intervention and a small human error could cause an enormous catastrophe. 70% of reported incidents in the oil and gas industry worldwide are attributable to human error and account for in excess of 90% of the financial loss to the industry.

Results from research conducted by the Health and Safety Executive and published in the book Out of control: Why control systems go wrong and how to prevent failure [7] show the impact of the human element in industrial incidents throughout project lifecycle. This is shown on Figure 4.

International Functional Safety Standards (IEC 61508 and IEC61511) and previously ISA 84 introduced the concept of the Safety Lifecycle, describing the phases that should take place from concept to design to implementation and operation of a Safety Instrumented System. The Safety Lifecycle is a step in the direction to reduce the impact of human factors by establishing the proper design best practices, documentation reviews and validation and verification steps in the execution of a safety project. Additionally the standards introduce two important elements: Competence of Personnel mentioned briefly in Section 4.2 and Functional Safety Management System (FSMS).

Recent changes to International Functional Safety Standard IEC61508 have turned the requirements on FSMS and Competence of Personnel into normative clauses of mandatory compliance instead of a recommendation as was presented in the previous version of the same standard.

5.1 Functional Safety Management System (FSMS) and Competence Requirements

A Functional Safety Management System (FSMS) consists of proven and validated procedures, methodologies, templates and report outlines, covering those essential elements of hazard and risk assessment; SIL determination; SIS design and SIL capable hardware, SIS installation, SIS commissioning and validation and SIS operations and maintenance. A FSMS is used in the design of a COTS product and during the Application Design. It's typically an extension of the existing Quality Assurance/Quality Management process and lives from the conception to decommission of the SIS and beyond to become part of the organization best engineering practices and work processes.

Figure 4: Root Cause of Control System Failures

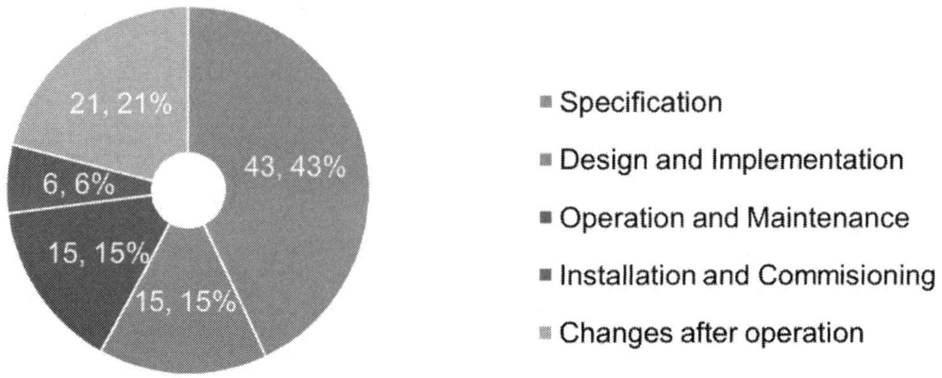

As indicated, the safety standards call for competence of personnel. This requirement shouldn't be interpreted as a need for safety experts in every aspect of the design, commissioning or operation and maintenance of a SIS, but instead as the need for an evaluation of the appropriate expertise and experience required to perform a particular job function in any phase of the safety lifecycle. In practical term those involved with activities related to a safety system should have demonstrated "competence" either by formal training, certification or accreditation or on-the-job relevant experience.

The reader might conclude that the standards today require competent and independent review groups because system designers don't trust product designers or other application designers and therefore we would rather be safe than sorry.

6. Summary

In summary

- Best in Class companies link safety to their success and invest in programs and systems to reduce their risk
- Engineers have attempted to reduce risk by minimizing the potential for common cause failures

- Common cause failures can occur in products (hardware or software) or the implementation of the application
- The industry has conceived best practices (i.e. FSMS) to minimize the impact of human error
- Human factors can't be ignored in the design, particularly in the application design
- Enforcement of these best practices via formal FSMS has proven to be a way to reduce the risk introduced by engineers… it's vital!

References

[1] IEC 61508 Functional safety of electrical/electronic/programmable electronic safety-related systems International Standard 2010

[2]Integrating Control and Safety: Where to draw the line.
Robin McCrea-Steele, TÜV FSExpert, Invensys-Premier Consulting Services

[3] Common Cause and Common Sense, Designing Failure Out of Your SIS
Angela E. Summers, Ph.D. and Glenn Raney, Oct 2000

[4] Integrated but separate, Advances in integrated and safety control,
Roger Prew, ABB , June 2009

[5] IEC 61511 Functional safety - Safety instrumented systems for the process industry sector
International Standard 2003

[6] Providing Independent Layers of Protection with Integrated Safety Systems
Luis M. Duran, ABB and Ron Johnson Dow Chemical, Oct 2009

[7] Out of control: Why control systems go wrong and how to prevent failure

HSE Books ISBN 0-7176-2192-8

9TH GLOBAL CONGRESS ON PROCESS SAFETY

A practical view of risk reduction management of hardware and software for safety critical applications

Luis Duran
Product Marketing Manager Safety Systems
ABB
3700 West Sam Houston Parkway South suite 600
luis.m.duran@us.abb.com

Mike Krywonos
Safety Execution Center Manager
ABB
3700 West Sam Houston Parkway South suite 600
mike.krywonos@us.abb.com

Prepared for Presentation at
American Institute of Chemical Engineers
2013 Spring Meeting
9th Global Congress on Process Safety
San Antonio, Texas
April 28 – May 1, 2013

UNPUBLISHED

A practical view of risk reduction management of hardware and software for safety critical applications

Luis Duran
ABB
3700 West Sam Houston Parkway South suite 600
luis.m.duran@us.abb.com

Mike Krywonos
ABB
3700 West Sam Houston Parkway South suite 600
mike.krywonos@us.abb.com

Keywords: Safety, Risk Reduction, Safety Systems, Independent Protection Layers, Common Cause Failures, Functional Safety Management System, IEC61508, IEC61511, TÜV, third party certification, Human Factors

Abstract

Best in Class companies (both National and Global) continue to invest in safety systems; this in most cases is because safety is at the core of their production process and among their key performance metrics, closely linked to their productivity.

Many unfortunate industrial accidents such as the recent explosion of the Deep Water Horizon with the largest marine oil spill in history are evidence of the negative consequences they have on people, environment, production assets and corporations.

This paper discusses the accepted definition of safety and risk reduction in the industry and will discuss the best practices applied by engineers in the design of products for mission critical functions and the implementation of those products in projects.

Both in product design and implementation, this session will dedicate special attention to methodologies or work processes used to reduce potential common cause failures that might impair the operation of the safety system and ultimately expose the facility to a higher risk; and the human factor considerations recommended to ensure proper operation upon a critical condition or an emergency in the industrial process.

1. Introduction

Best in Class companies worldwide, defined as such for their high percentage of Overall Equipment Effectiveness (OEE) and low injury frequency rate; continue to show strong initiatives intended to expand the safety of their operation. From executive sponsored programs at the corporate level to the definition of proactive risk management strategies, these companies are investing in safety systems and processes; simply because in most cases, safety is identified among the core values of those companies, at the center of their production process and among their key performance metrics, closely linked to their productivity.

There are numerous publications and records of Industrial Accidents, a large number of them with terrible consequences, including not only loss of human life or environmental impact but also affecting the production assets in the site. As an example, in researching for this paper, the authors found record of at least 20 "notable" oil and gas offshore blowouts between 1980 and 2010.

As an example for this paper, the authors compared two serious incidents in the Oil & Gas sector; Piper Alpha operated by Occidental Petroleum and Deep Water Horizon leased to BP. Although both incidents are 22 years apart from each other, occurred in different geographies and have many differences in causes and consequences. They illustrate the risk and the potentially catastrophic dimension they can take in this industry and the impact on people, environment and corporations, which might cease to exist after these incidents.

2. Defining Safety and Risk

Safety is defined in the industry as a reduction of existing risk to a tolerable or manageable level while risk is a combination of the probability of a harmful incident and magnitude of the harm. Traditional design practices allocate the risk reduction across different and independent protection layers, the rationale behind it is simple "any system that can fail will fail" so the engineering best practice is to distribute the risk reduction tasks across multiple independent functions or systems one of these systems is a Safety Instrumented System.

2.1 Safety Instrumented Systems one of many Independent Protection Layers to Reduce Risk

A Safety Instrumented Systems (SIS) is a mission critical system designed following international design practices such as IEC61508 [1] Functional Safety Standard to reduce risk to the people in and around the production environment, the environment, the production asset and the business. In many cases the Safety Instrumented Systems are the last resource to prevent disaster.

SIS performance is measured by Safety Integrity Level (SIL) SIL 1 low, SIL 3 high o Risk Reduction Factor.

As mentioned earlier, the Safety Instrumented System is one of many functionally independent systems each intended to perform a task, as shown on Table 1 typically referred to as Independent Protection layers [2]

Table 1: Expected Functionality of Independent Protection Layers

Layer of Protection	Expected Functionality
BPCS:	Keeping the process under control
Alarms:	Alerting the operator of abnormal conditions and providing guidelines for appropriate operator response
SIS:	Automatically taking the process to a safe condition in case the abnormal condition goes out of control and the operators can't take any corrective action in time
Other layers:	Intended to mitigate the consequence of the hazard

The concept of Independent protection layers (IPL) can also be illustrated in Figure 1

3. Common Cause Failures

The idea of functional independent systems is an attempt to avoid Common Cause Failures (CCF).

Common Cause Failures [3] are failures that might affect more than one of these protection layers at the same time. As engineers, the tendency will be to focus on elements such as Heat, Humidity, Shock, and radio interference among others. However this paper also considers with special interest the human elements related to the design of safety critical systems which can contribute to additional failures.

As with other systems, some SIS problems are related to the Commercial Off the Shelf (COTS) products designed for a specific function (i.e. hardware limitations or poor documentation) while other problems are related to the use of those COTS product such as misapplication, user application programming or poor maintenance practices. Both areas can be addressed by implementing appropriate design best practices to reduce risk.

The industry has conceived these "best practices" as a series of steps that must be performed before putting one of these systems in operation (design, installation and commissioning). This paper will describe the two areas:

- Product Design and Implementation of the Commercial Off-the-Shelf (COTS) Product
- Application Design

Figure 1: Independent Protection Layers

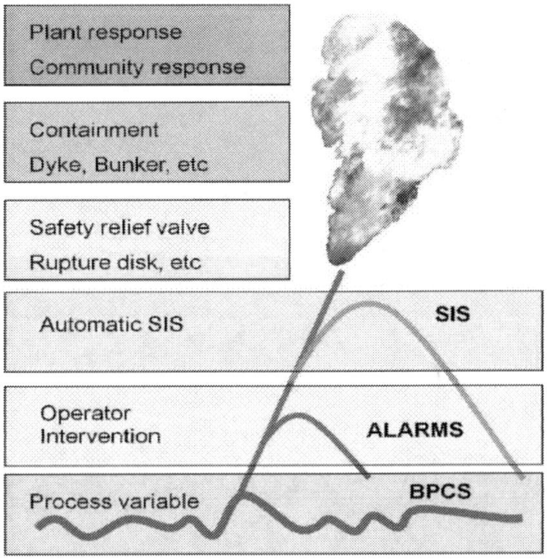

3.1 Reducing Common Cause Failures by Design

Traditionally or historically, Systems designed for SIS functions have relied on physical separation and redundancy [4] to reduce common cause failures. The authors identified technology changes across multiple generations of SIS over a period of 30 years. However over the same time frame the industry applied lessons learned from incidents to develop best practices in design and implementation of systems, the international industry standards used more often are IEC61508 and IEC61511 [5] (Figure 2)

3.2. Hardware Fault Tolerance (1st Generation SIS)

1st Generation systems rely on Hardware Fault Tolerance or redundancy to achieve reliability and availability as required for these applications, this practice is rooted in the late 1970s and early 1980s technology. As the majority of these systems were designed prior to the release of the Functional Safety standards, they do not follow the best practices found with regards to design, documentation and testing found in those standards.

Figure 2: Evolution of National Prescriptive Standard to International Performance Base Standards

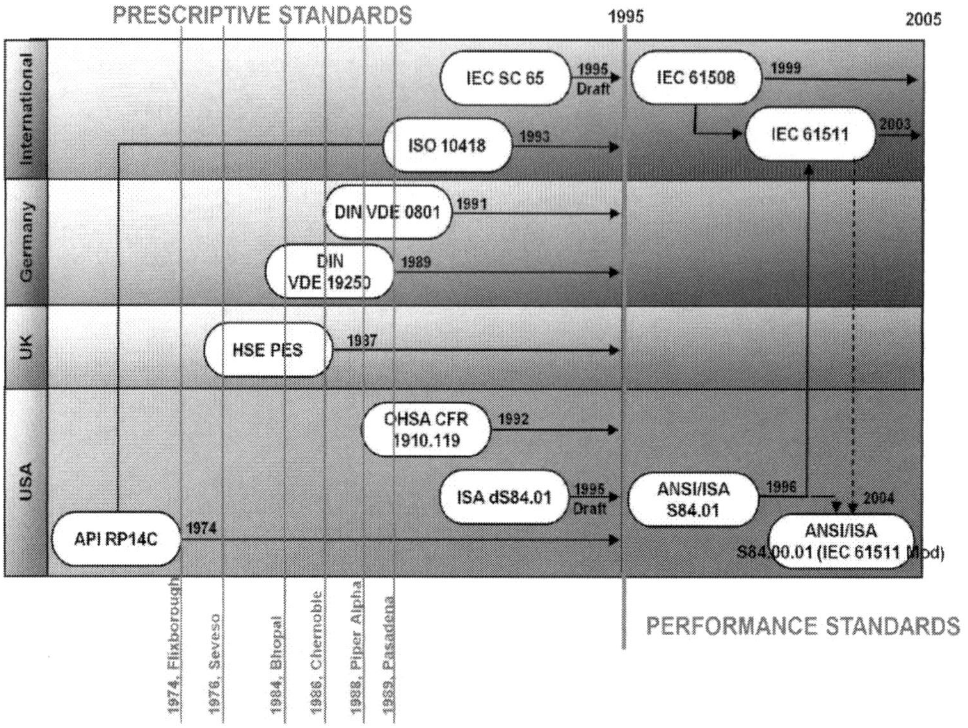

3.3. 2nd Generation Systems

As the standards became available (mid 1990s), vendors started to apply the design best practices and to pursue certification to those standards. This generation of systems was produced having third party assessment and certification but continued to rely on Hardware Fault Tolerance to satisfy the performance requirements. The authors found that a large portion of the systems applied as SIS in the market today, although designed to satisfy Functional safety standards haven't change dramatically in their use of hardware fault tolerance and basic software and hardware diagnostics.

3.4. Diverse Architecture and Implementation a 3rd Generation Systems

A more recent and different type of system, uses diverse redundancy, diverse implementation and active diagnostics [6] and can deliver not only reliability and availability required for the application while minimizing common cause failures found by the use of the same technology but also introduce additional protection to systematic failures normally related to human factors

3.4.1 Hardware Design – Diverse Hardware

This newer generation of systems use diverse processing hardware (multiple technologies) such as diverse operating systems and diverse hardware and diverse redundancy, both enhanced with diverse implementation (different implementation teams).

3.4.2 Application Execution in a SIL Compliant Environment

The use of diverse operating systems (use of different technologies) including the use of third party certified COTS and different execution path with different compiler rules extends the risk reduction found in the hardware design to the firmware/software environment in the system, including the use of limited variability software functions and provision of systematic capabilities as specified in the Functional Safety standards for critical applications.

3.4.3. Systematic Capabilities and Human Error

The concept of systematic capabilities and its associated performance measurement was introduced in the latest revision of IEC61508 standard (didn't exist as such in 1st Generation systems and was emerging in 2nd generation systems).

3.5. Safety and Network Security

1st and 2nd generation Safety Systems were not designed to reside on a networked plant, for the most part these systems allowed some communication interface or gateway but were not intended for integration, therefore in general are not equipped to counter security threats as those the industry has experienced in recent years.

A brief analysis of the definitions will show a clear differences Safety issues are random in nature and statistic analysis is possible security intentional in nature and statistical analysis will not measure performance.

The authors concur with industry standards and experts in the field. Network security may affect system performance and the safety of the installation. A system can't be safe without it being secure.

3.6. Design Best Practices and V-Model

As indicated earlier, industry standards in Functional Safety introduced the concept of Functional Safety Management System (FSMS) which includes a series of steps in the design, documentation and testing of the system, and should include Network Security as well.

Figure 3: Product Development V-Model

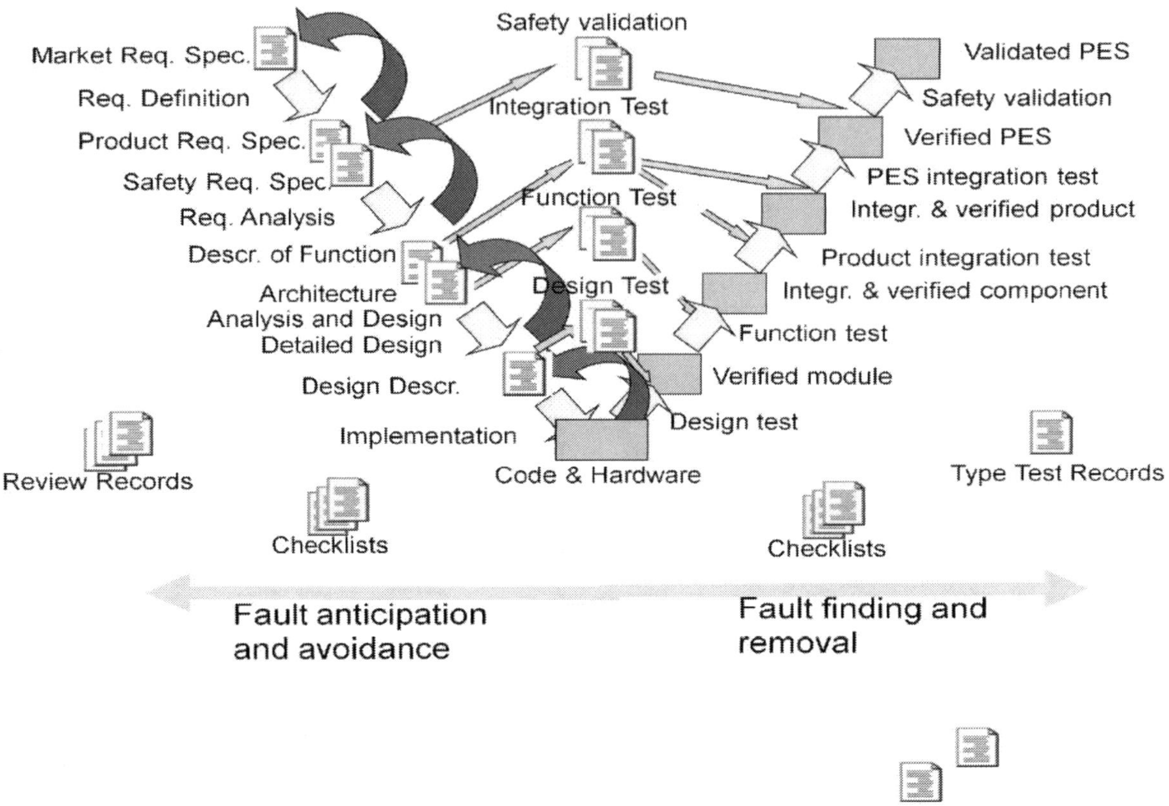

Today product development or design organizations responsible for releasing COTS products (hardware and software) intended for safety applications use design best practices as the V-Model (Figure 3) and are evaluated by a third party organization (i.e. TUV) and their FSMS is approved prior to the assessment of their products.

4. Engineering best practices to reduce risk in application design

As indicated previously, best practices in design apply both in the design of a Commercial Off the Shelf (COTS) Product and the Application Design or the application of a COTS tailored to reduce a particular risk in an industrial installation.

Best Practices in Application Design will apply a model similar to the V-Model used in the development of a product (Figure 3) and reinforced by a structured Quality Management system appropriate for Functional Safety applications.

Application Design starts with Hardware Design, which starts with a design that follows the COTS Product's Safety Manual and considers the particularities of the application as described on the Application Safety Requirements Specification to design a system that is less susceptible to product failures by considering aspects such as

- Power feeds
- Power Supplies
- CPUs
- I/O Modules Hardware
- Application Software
- Networks

4.1 Hardware

It's clear that hardware is not perfect, it's subject to fails, and the application design must consider fail recovery modes or additional risk reduction methodologies or options including diverse methods of performing shutdowns as:

- Hardwired wired pushbutton,
- Hardwired logic systems (with appropriate SIL rating)
- Solid State relays
- Etc.

4.2 Software

One critical aspect that has gained visibility in recent years is the software design of the application. Efforts in this area include the development of a well documented Software Functional Design, appropriate selection of software tools and libraries that meets the Safety Integrity Level (SIL) required for the application, Code Review practices including Independent Validation and Verification depending on the SIL, competence of the programmer and other elements described later in section 5.1. The features available in the COTS product to address the systematic capabilities requirements are a valuable asset for the programmer and contribute to reducing potential errors that will later affect the safety of the installation.

4.3 Functional Testing and Periodic Proof Testing

Another critical aspect is Functional Testing and Periodic Proof Testing, particularly because Not ALL System Faults are Self-Revealing and Covert Faults that may inhibit SIS action on Demand can only be detected by testing the entire system, this requires not only a Full Functional Test prior to commissioning, typically conducted as an integration test in the vendor or System Integrator facility and later in the final instrumentation room in the plant; but Periodical Functional Tests using a documented procedure to Detect Covert Faults and covering the entire SIS.

Functional Testing should Record and Analyze activation of SIS functions, and spurious activation of an Emergency Shutdown Valve due to a Process Shut Down, but this does not test the Entire Function of the same valve during an ESD action.

5. Automation can't check for human intervention

Once the design is complete, all of these systems will have different degrees of interaction with users, and different type of users with different levels of competence depending to their role (i.e. operation, maintenance personnel and engineers)

The most sophisticated automation can't prevent human error, industrial plants are designed with the highest accuracy in mind, with several separate safety loops often checking the integrity of process systems. This apparatus however can't check for human intervention and a small human error could cause an enormous catastrophe. 70% of reported incidents in the oil and gas industry worldwide are attributable to human error and account for in excess of 90% of the financial loss to the industry.

Results from research conducted by the Health and Safety Executive and published in the book Out of control: Why control systems go wrong and how to prevent failure [7] show the impact of the human element in industrial incidents throughout project lifecycle. This is shown on Figure 4.

International Functional Safety Standards (IEC 61508 and IEC61511) and previously ISA 84 introduced the concept of the Safety Lifecycle, describing the phases that should take place from concept to design to implementation and operation of a Safety Instrumented System. The Safety Lifecycle is a step in the direction to reduce the impact of human factors by establishing the proper design best practices, documentation reviews and validation and verification steps in the execution of a safety project. Additionally the standards introduce two important elements: Competence of Personnel mentioned briefly in Section 4.2 and Functional Safety Management System (FSMS).

Recent changes to International Functional Safety Standard IEC61508 have turned the requirements on FSMS and Competence of Personnel into normative clauses of mandatory compliance instead of a recommendation as was presented in the previous version of the same standard.

5.1 Functional Safety Management System (FSMS) and Competence Requirements

A Functional Safety Management System (FSMS) consists of proven and validated procedures, methodologies, templates and report outlines, covering those essential elements of hazard and risk assessment; SIL determination; SIS design and SIL capable hardware, SIS installation, SIS commissioning and validation and SIS operations and maintenance. A FSMS is used in the design of a COTS product and during the Application Design. It's typically an extension of the existing Quality Assurance/Quality Management process and lives from the conception to decommission of the SIS and beyond to become part of the organization best engineering practices and work processes.

Figure 4: Root Cause of Control System Failures

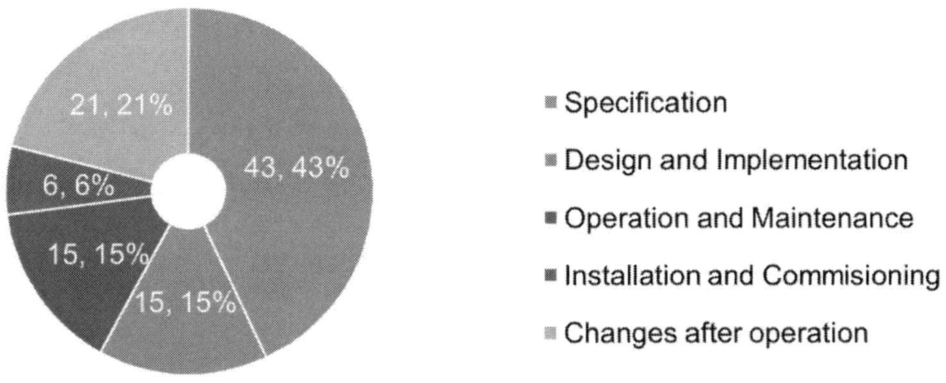

As indicated, the safety standards call for competence of personnel. This requirement shouldn't be interpreted as a need for safety experts in every aspect of the design, commissioning or operation and maintenance of a SIS, but instead as the need for an evaluation of the appropriate expertise and experience required to perform a particular job function in any phase of the safety lifecycle. In practical term those involved with activities related to a safety system should have demonstrated "competence" either by formal training, certification or accreditation or on-the-job relevant experience.

The reader might conclude that the standards today require competent and independent review groups because system designers don't trust product designers or other application designers and therefore we would rather be safe than sorry.

6. Summary

In summary

- Best in Class companies link safety to their success and invest in programs and systems to reduce their risk
- Engineers have attempted to reduce risk by minimizing the potential for common cause failures

- Common cause failures can occur in products (hardware or software) or the implementation of the application
- The industry has conceived best practices (i.e. FSMS) to minimize the impact of human error
- Human factors can't be ignored in the design, particularly in the application design
- Enforcement of these best practices via formal FSMS has proven to be a way to reduce the risk introduced by engineers… it's vital!

References

[1] IEC 61508 Functional safety of electrical/electronic/programmable electronic safety-related systems International Standard 2010

[2]Integrating Control and Safety: Where to draw the line.
Robin McCrea-Steele, TÜV FSExpert, Invensys-Premier Consulting Services

[3] Common Cause and Common Sense, Designing Failure Out of Your SIS
Angela E. Summers, Ph.D. and Glenn Raney, Oct 2000

[4] Integrated but separate, Advances in integrated and safety control,
Roger Prew, ABB , June 2009

[5] IEC 61511 Functional safety - Safety instrumented systems for the process industry sector International Standard 2003

[6] Providing Independent Layers of Protection with Integrated Safety Systems
Luis M. Duran, ABB and Ron Johnson Dow Chemical, Oct 2009

[7] Out of control: Why control systems go wrong and how to prevent failure

HSE Books ISBN 0-7176-2192-8

(44bj) Establishing a 3 Credit Course in PSM for Undergraduate Engineers

Deborah L Grubbe, Executive Mgmt, Operations and Safety Solutions, LLC, Chadds Ford, PA

Considering academic direction and industrial needs, the Purdue Chemical Engineering faculty and an industrial alumni team jointly sponsored, developed, executed and audited a 3 credit course in Process Safety Management for undergraduates in Engineering. This "team taught" course is now a required in the Purdue ChE curriculum, and has received enthusiastically high marks from the students. Originally conceived in 2010, the course went from concept to required in less than 24 months.

One of the alumni team members, with over 30 years in the process industries, will describe each step of the process: creating the support, drafting the syllabus, designing the course, executing the lectures, planning the ancillary support activities, and auditing the results of the 15 week, 45 lecture series. The final exam was the simulation of an emergency situation, which was modeled after a real life process safety incident. Actual student feedback will be shared. An eBook text has been developed for this course, and the speaker will conduct a short demonstration of this product as part of the overall presentation.

See more of this Session: Poster Session
See more of this Group/Topical: Global Congress on Process Safety

9TH GLOBAL CONGRESS ON PROCESS SAFETY

Investigations on the self heating potential of garage waste as basis for the hazard evaluation of bulk storage facilities

Hannes Kern
Montanuniversität Leoben
Franz-Josef Strasse 18, 8700 Leoben
hannes.kern@unileoben.ac.at

Christoph Spijker, Harald Raupenstrauch
Montanuniversität Leoben
Franz-Josef Strasse 18, 8700 Leoben
christoph.spijker@unileoben.ac.at, harald.raupenstrauch@unileoben.ac.at

Prepared for Presentation at
American Institute of Chemical Engineers
2013 Spring Meeting
9th Global Congress on Process Safety
San Antonio, Texas
April 28 – May 1, 2013

UNPUBLISHED

Investigations on the self heating potential of garage waste as basis for the hazard evaluation of bulk storage facilities

Hannes Kern
Montanuniversität Leoben
Franz-Josef Strasse 18, 8700 Leoben
hannes.kern@unileoben.ac.at

Christoph Spijker, Harald Raupenstrauch
Montanuniversität Leoben
Franz-Josef Strasse 18, 8700 Leoben
christoph.spijker@unileoben.ac.at, harald.raupenstrauch@unileoben.ac.at

Keywords: self heating, garage waste, differential reactor

Abstract

The self heating of bulk material is a major point to be considered in hazard analysis especially for waste management and waste treatment companies. The self heating potential for bulk material mainly depends on the composition of the material itself and on systemic influences like the geometry of the storage area, weather conditions or the storage temperature. Besides microbiological activity, chemical oxidation at low temperatures is the main reason for the self heating and auto ignition of bulk material. Data from the measurement of the chemical and biological oxidation kinetics can be used as basis for the simulation of storage areas by computational fluid dynamics. The measurement of the low temperature oxidation kinetics is carried out in a differential reactor, which is characterized by a small amount of sample material, isothermal conditions and very low turnover rates. Due to the fact that chemical oxidations reactions at low temperatures show very low turnover rates themselves, the experimental setup is laid out as a cycle process. The oxidation kinetics are measured by the analysis of the carbon monoxide (CO) and carbon dioxide (CO2) production rate at different temperatures. Usually a temperature range from 40°C up to 120°C is investigated. Waste materials from workshops contain significant amounts of grease and oil as well as residuals of metal dusts or chips dispersed on the surface of paper or cloth material used for cleaning purposes. The large surface combined with the fine dispersion of grease and oil accelerates the chemical oxidation significantly. To determine the influence of the oil and grease content on the oxidation kinetics, measurements of paper and cloth samples with different contents of oil were carried out in a differential cycle reactor. For the investigations biological cutting oil and oil with a petrochemical origin was investigated. Furthermore the influence of metal dusts on the oxidation characteristics of the different materials was investigated. The results of the investigations were compared with the oxidation kinetics of garage waste material with a mean diameter of 0- 40mm.

1. Introduction

The self heating of bulk material is a major point to be considered in hazard analysis especially for waste management and waste treatment companies. Besides safety issues the auto ignition of bulk storage areas is also a serious economical threat and occurs quite frequently. In industrial branches where material with an inhomogeneous composition (like waste treatment facilities) are processed the problem of self heating is even more present. Especially at residual derived fuel storage areas fires are also a serious ecological threat, due to composition of the materials themselves. The measurement of the low temperature oxidation kinetics combined with CFD [1] modeling used for the prediction of the self heating potential of bulk storage areas is an important element of hazard analysis and loss prevention. In the present study the self heating potential of garage waste was investigated. Garage waste shows a high tendency of self heating and therefore auto ignition is a major hazard. In the course of this study different influencing parameters on the self heating potential of garage waste were investigated, to be able to evaluate the risk of auto ignition. Self heating in general is a complex process that mainly depends on the parameters presented below [2]:

Material specific parameters
- composition
- chemical kinetics
- particle size distribution
- bed porosity
- heat conductivity

Apparatus specific boundary conditions
- geometry of the storage area
- size surroundings

Operating conditions
- temperature
- gas velocity

The chemical oxidation at low temperatures is besides microbiological activity the main reason for the self heating and auto ignition of bulk material. Basicly two main oxidation reactions occur at low temperatures:

$$a_{Fuel(s)} + b_{O_2(g)} \rightarrow c_{CO(g)} + d_{\mathrm{Rest}(s)} \quad (1)$$

$$e_{Fuel(s)} + f_{O_2(g)} \rightarrow g_{CO_2(g)} + h_{\mathrm{Rest}(s)} \quad (2)$$

The reaction of the solid fuel with oxygen gives carbon monoxide (1) and carbon dioxide (2). In a simple kinetic approach it is assumed that the solid fuel can be seen as bulk with a constant concentration compared to the oxygen concentration. After some simplifications the relations (3) and (4) can be derived from the Arrhenius relation for chemical reaction kinetics:

$$r_{CO} = k_0 \cdot \exp\left(\frac{-E_A}{R \cdot T}\right) \cdot y_{O_2}^n \qquad (3)$$

$$r_{CO} = k_0 \cdot \exp\left(\frac{-E_A}{R \cdot T}\right) \cdot y_{O_2}^n \qquad (4)$$

For further research it is assumed that the postulated reactions undergo only chemical limitation, because chemical limitation exceeds transport limitation by several orders of magnitude. The determination of the kinetic parameters is carried out by the observation of the CO and CO_2 production rates.

2. Experimental investigations

2.1 Sample Material

Waste materials from workshops contain significant amounts of grease and oil as well as residuals of metal dusts or chips dispersed on the surface of paper or cloth material. The large surface combined with the fine dispersion of grease and oil accelerates the chemical oxidation significantly [3]. To determine the influence of the oil and grease content on the oxidation kinetics, paper and cotton cloth samples with different contents of oil were chosen as model substances for further investigations. The calorific values as well as the chemical composition of the investigated materials can be seen in Table1. The oil content of the garage waste sample material was determined by Soxhlet extraction with cyclohexane as solvent. The sample had an oil content of around 15% and a particle size of up to 30mm.

Table 1: Composition of the sample material

	Carbon [%]	Hydrogen [%]	Nitrogen [%]	Calorific Value [kJ/kg]
Paper	40.41	6.70	0.61	16277
Cotton cloth	41.81	6.80	0.35	16484
Oil (arecal)	78.7	12.1	0.01	35911
Oil (WD 40)	85.7	13.3	0.04	42258

The product "arecal" is a cutting oil based on rape oil while "WD40" is a cutting oil derived from fossil hydrocarbons. Pre tests in the differential reactor described in 2.2 showed that arecal is far more reactive than WD40. That may be explained by the high content of unsaturated carbon bonds in rape oil, which are more reactive in terms of oxidation processes. Further investigations on the oxidation behavior have been carried out only with the more reactive arecal oil.

2.2 Method

The measurement of the low temperature oxidation kinetics is performed in a differential reactor (Figure 1), which is characterized by a small amount of sample material, nearly isothermal conditions and very low turnover rates. Due to the fact that chemical oxidations reactions at low temperatures show very low turnover rates themselves, the experimental setup is laid out as a cycle process. The oxidation kinetics is measured by the analysis of the carbon monoxide (CO) and carbon dioxide (CO_2) production rate at different temperatures. Usually a temperature range from 40°C up to 120°C is investigated.

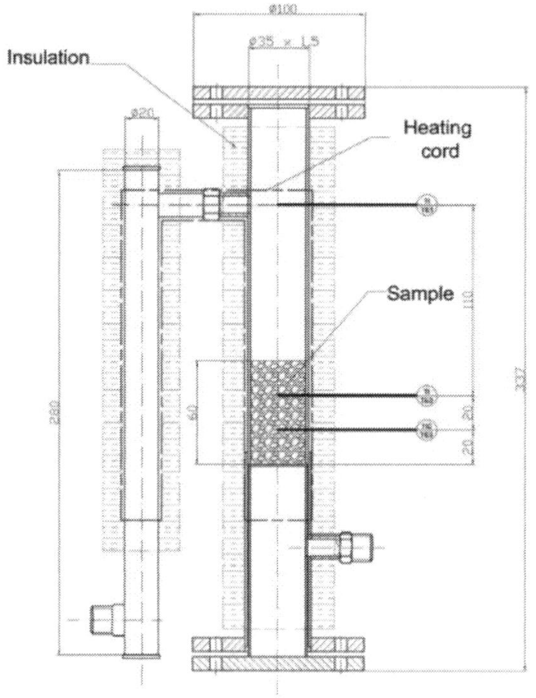

Figure 1: Principal layout of the differential reactor [2]

Figure 2 shows a typical of temperature development and gas production (e.g. carbon dioxide). At the beginning of the measurements the sample material inside the reactor is heated up above the highest planned measuring temperature to desorb carbon dioxide and carbon monoxide that may be adsorbed at the surface of the sample material. After the desorption process the sample material is cooled to the first measuring temperature and constantly purged with synthetic air (80%N_2, 20%O_2). When the sample reaches a constant temperature value the measurement process is started and the measuring cycle is closed. The areas of constant CO_2 increase in the concentration graph of Figure 2 represent these measurement cycles at different temperatures. At higher temperatures an increase of carbon dioxide production can be observed, which is directly linked to oxidation kinetics. The calculation of the kinetic parameters is carried out by the Arrhenius relation after the measuring procedure. A measuring cycle normally takes about 30min for one temperature.

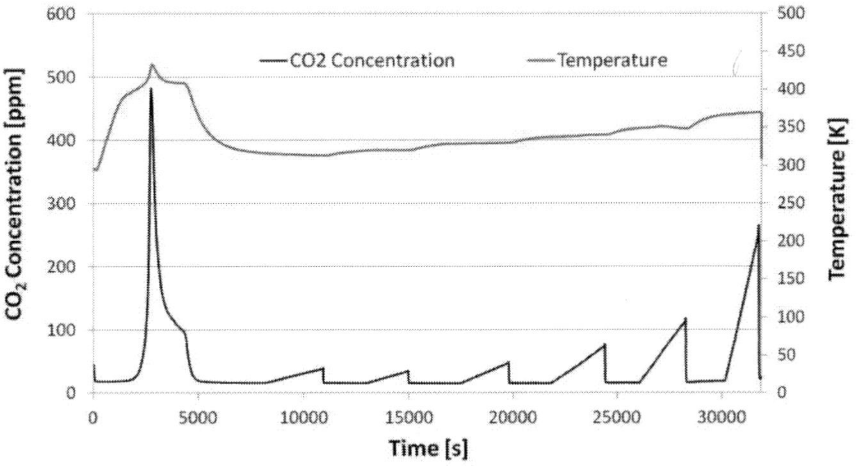

Figure 2: Principal layout of the differential reactor

2.3. Paper

Figure 3 below shows the CO production rates of paper with different oil contents. The "0% Oil content" line represents paper material not contaminated with arecal oil. The CO production rate increases with the oil content of the sample. The progress of the CO production rate with an oil content of 6.9% is widely comparable to the results of garage waste (~15% oil content).

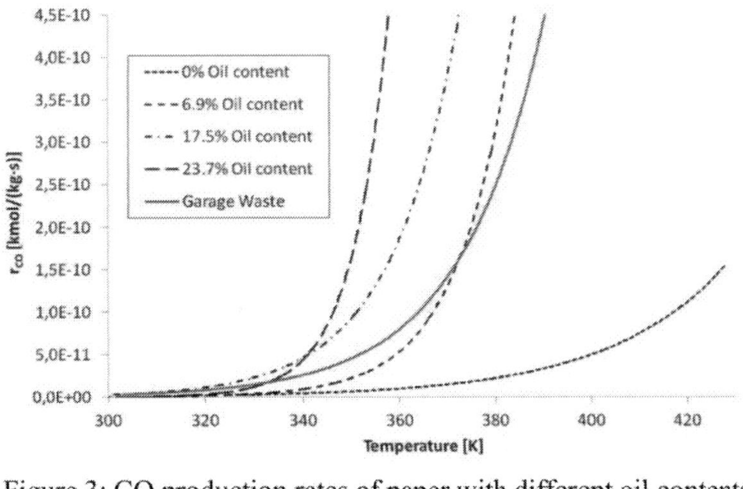

Figure 3: CO production rates of paper with different oil contents

For CO_2 production, as shown in Figure 4, the results were a bit different. The CO_2 production rate of garage waste was higher than all other rates measured. Interestingly the CO_2 production rate for the paper with an oil content of 17.5% was higher than for 23.7%. That may be explained by kinetic effects and is maybe linked to the significant amounts of CO that were produced by this paper sample.

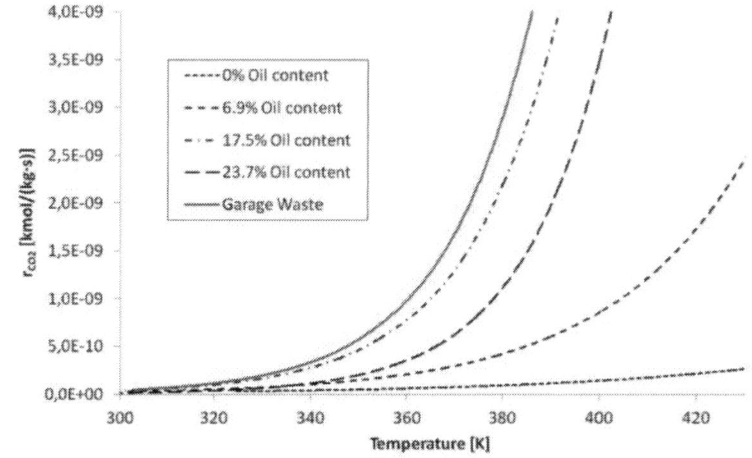

Figure 4: CO_2 production rates of paper with different oil contents

The kinetic data resulting from the calculations based on the Arrhenius relation are presented in Table 2.

Table 2: Kinetic data of the paper sample material

	CO		CO_2	
	E_a [kJ/mol]	K_0 Kmol/kg*s	E_a [kJ/mol]	K_0 [Kmol/kg*s]
Paper 0.0% Oil cont.	4.89E+04	3.25E-09	2.52E+04	9.74E-12
Paper 6.9% Oil cont.	8.70E+04	7.45E-03	3.45E+04	6.63E-10
Paper 17.5% Oil cont.	6.69E+04	3.17E-05	4.94E+04	3.84E-07
Paper 23.7% Oil cont.	1.02E+05	1.91E-00	5.48E+04	1.04E-06
Garage waste	5.59E+04	3.35E-07	5.22E+04	1.14E-06

2.4 Cotton cloth

During the preparation of the cotton sample material oil contents that were similar to the paper samples have been tried to be reached. Although there are smaller deviations in the oil content, the results are comparable.

Figure 5 shows the results for the investigation of the cotton cloth material. The results do not show good accordance with the garage waste material. But again it can be observed, that the sample materials with a high oil contents also show higher reactivity than the sample with lower contents.

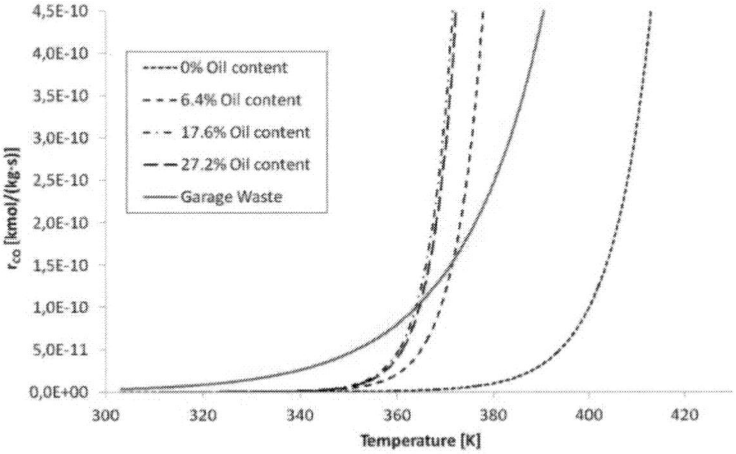

Figure 5: CO production rates of cotton cloth with different oil contents

The production rates from Figure 6 do not show a big deviation regarding the oil content, but the samples contaminated with oil clearly show a higher reactivity than the cotton material itself. In general all CO_2 production rates are comparable with the original garage waste sample material, which again indicates a strong link between oil content and reactivity.

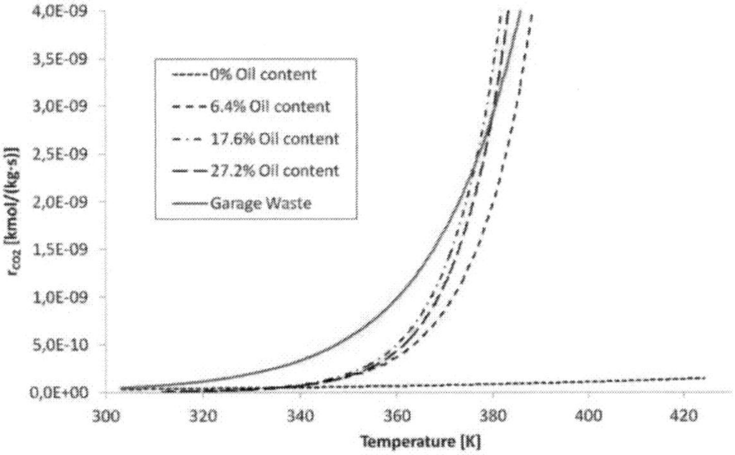

Figure 6: CO_2 production rates of cotton cloth with different oil contents

The kinetic data of the cotton cloth samples is presented in Table 3.

Table 3: Kinetic data of the cotton cloth sample material

	CO		CO_2	
	E_a [kJ/mol]	K_o Kmol/kg*s	E_a [kJ/mol]	K_o [Kmol/kg*s]
Cotton 0.0% Oil cont.	3.68E+04	1.66E-10	1.34E+04	1.99E-13
Cotton 6.4% Oil cont.	1.90E+05	2.90E+12	9.01E+04	1.52E-01
Cotton 17.6% Oil cont.	1.27E+05	3.14E+04	9.96E+04	4.82E+00
Cotton 27.2% Oil cont.	1.02E+05	1.80E+00	5.48E+04	9.78E-07
Garage waste	5.59E+04	3.35E-07	5.22E+04	1.14E-06

2.5 Influence of iron dusts

Due to normal workshop operations garage wastes always contain significant amounts of metal residues. To investigate the influence of metal residues on the oxidation kinetics small amounts of iron dust were added to a cotton cloth sample with an oil content of around 11%.

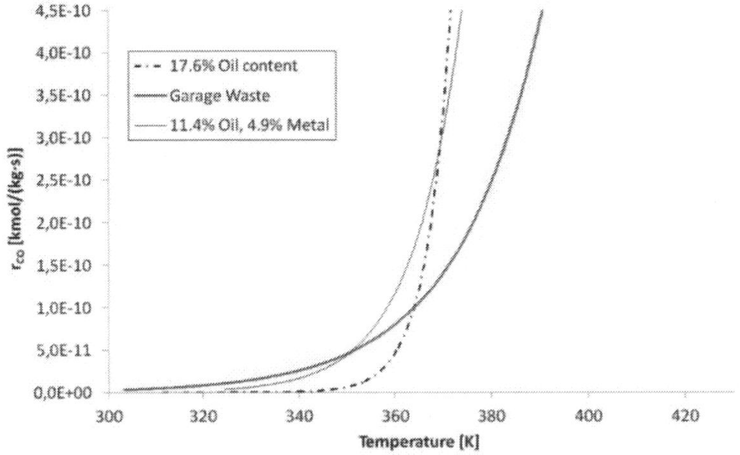

Figure 7: CO production rates of cotton cloth contaminated with oil and iron dust

Figure 7 shows the production rates for the cotton/oil/iron dust sample and also the results for pure cotton cloth with 17.6% oil content. The results show that the iron dust has an influence on the reactivity of the material. Both materials show comparable CO production rates even though the sample containing iron dust contains less oil. The same effect can be observed for the CO_2 production rate in Figure 8. The CO_2 production rate in general shows good accordance to the results of the garage waste sample which indicates the representativeness of the chosen sample materials.

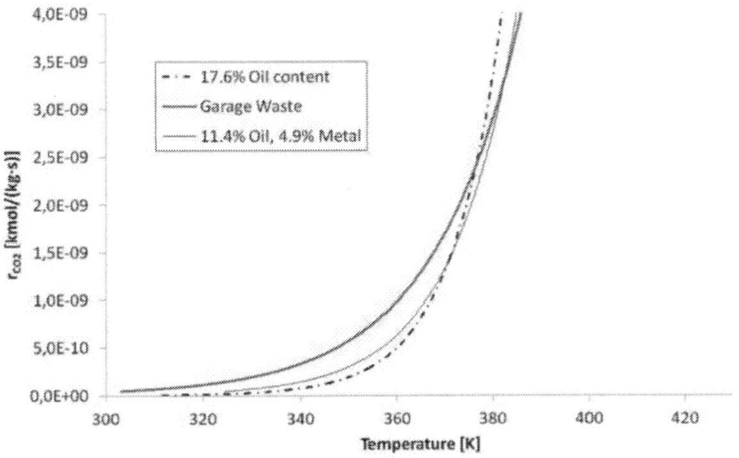

Figure 8: CO_2 production rates of cotton cloth contaminated with oil and iron dust

3. Conclusions

The results of the investigations of garage waste and different reference materials show a clear connection between the oil content of garage waste and its reactivity. Garage waste in general should be considered as highly reactive. Field investigations show, that a significant temperature rise can be observed within days from the beginning of the storage period. High contents of oil and metal residues lead to a significant fire risk. To determine the permissible storage conditions for garage waste CFD models could be applied. The results from measured kinetics can be used as the basis for CFD modeling in combination with the layout of the storage area and the further surrounding conditions [4].

4. References

[1] Wallner, S. (2004). Ein Modell zur Untersuchung des Selbsterwärmungsverhaltens von Schüttgütern. Dissertation, Technische Universität Graz

[2] Walkner, R. (2005) Untersuchung des Selbsterwärmungsverhaltens eines Lagers für aufbereiteten Substitut Brennstoff (ASB). Diplomarbeit, Technische Universität Graz

[3] Wallner, S., Walkner, R., Raupenstrauch, H. (2004). Untersuchung des Selbsterwärmungsverhaltens gelagerter Kunststoffmaterialien mittels numerischer Simulation. Endbericht für das FFF- Projekt „Entwicklung neuer Brennstoffqualitäten aus Abfall". Technische Universität Graz

[4] Raupenstrauch, H. (1997) Gasdurchströmte, chemisch reagierende Schüttschichten. Habilitationsschrift, Technische Universität Graz

Estimate Vibration Risk for Relief and Process Piping

Georges A. Melhem, Ph.D.
ioMosaic Corporation
93 Stiles Road | Salem, NH 03079
melhem@iomosaic.com

Prepared for Presentation at
American Institute of Chemical Engineers
2013 Spring Meeting
9th Global Congress on Process Safety
San Antonio, Texas
April 28 – May 1, 2013

UNPUBLISHED

Estimate Vibration Risk for Relief and Process Piping

Georges A. Melhem, Ph.D.
ioMosaic Corporation
93 Stiles Road | Salem, NH 03079
melhem@iomosaic.com

Keywords: Acoustic induced vibration (AIV), flow induced vibration (FIV), vibration risk, resonance, singing relief valve.

Abstract

Current API, AIChE/CCPS, and AIChE/DIERS pressure relief and flare systems guidelines and standards do not formally address vibration risk. They do not offer specific guidance on velocity limitations other than backpressure calculations and they do not offer guidance on acoustic induced or flow induced piping vibration fatigue failure. This paper provides a summary of experience based methods for the estimation of vibration risk in relief and process piping. In addition, this paper extends the applicability of the experience based methods to two-phase flow.

1. Introduction

Fatigue failure of relief and/or process piping caused by vibration can develop due to the conversion of flow mechanical energy to noise. Factors that have led to an increasing incidence of noise vibration related fatigue failures in piping systems include but are not limited to (a) increasing flow rates as a result of debottlenecking which contributes to higher flow velocities with a correspondingly greater level of turbulent energy, (b) frequent use of thin-walled piping which results in higher stress concentrations, particularly at small bore and branch connections, (c) design of process piping systems on the basis of a static analysis with little attention paid to vibration induced fatigue, (e) and lack of emphasis of the issue of vibration in piping design codes. Piping vibration is often considered on an ad-hoc or reactive basis. According to the UK Health and Safety Executive (HSE), 21 % of all piping failures offshore are caused by fatigue/vibration. Typical systems at risk include large compressor recycle systems and high capacity pressure relief depressuring systems. For relief and flare piping, flow induced turbulence and high frequency acoustic excitations are key concerns.

2. Flow Induced Vibration (FIV)

Fluid flow in pipes generates turbulent energy (pressure fluctuations). Dominant sources of turbulence are associated with flow discontinuities in the piping systems (e.g., partially closed valves, short radius, mitered bends, tees or expanders). The level of turbulence intensity is a function of pipe size, fluid density, viscosity, velocity, and structural support. High noise levels are generated by high velocity fluid impingement on the pipe wall, turbulent mixing, and if the flow is choked, shock waves discontinuities downstream of flow restriction. This can lead to high frequency excitation / vibration risk.

3. Acoustic Induced Vibration (AIV)

High frequency acoustic energy is often generated by a pressure reducing device such as a relief valve, control valve, or orifice plate. Acoustic induced piping failure is of a particular concern for safety related systems (e.g. relief and blowdown/depressuring). The severity of high frequency acoustic excitation is primarily a function of the pressure upstream and downstream of the pressure reducing device, pipe diameter, and the fluid volumetric flow. Acoustically induced piping failures are known to occur at non-asymmetric discontinuities in the downstream piping such as small bore and branch connections and welded supports.

4. Additional Causes of Vibration

Additional causes of piping vibration include mechanical excitation, pulsation, vortex shedding, surge or momentum changes associated with valves, cavitation, and flashing. Mechanical excitations are often associated with pipes connected with reciprocating compressors, pumps, or rotating machinery. Such connection machines cause vibration of the pipe and its support structure. Thermowells are intrusive fittings and are subject to static and dynamic fluid forces. Vortex shedding is the dominant concern as it is capable of forcing the thermowell into flow-induced resonance and fatigue failure. This is particularly important at high fluid velocities.

5. Relief and Depressuring Systems

Depressuring systems are often subjected to acoustic energy (rapidly fluctuating pressure forces) generated by flow turbulence which is accentuated by flow restricting devices within the flow path. The magnitude of pressure fluctuations depends on the mass flow rate, speed of sound, and density. For choked flow, intense noise due to large shock discontinuities and pressure fluctuations is generated. The generated noise is non-periodic due to the randomness of the pressure fluctuations. Choked flow typically leads to a wide frequency spectrum with peak values than can exceed 1000 Hz. Vibrations associated with small fittings and branch connections are of special concern because they introduce discontinuities and stress concentration points.

In many situations resonance can onset which can lead to magnification of static piping loads up to a factor of 50 times. The presence of discontinuities such as tees and welded pipe supports can further increase these loads.

6. Noise Generation

A pressure reducing device or relief device controls flow by converting internal energy into kinetic energy. Some energy is converted to heat through friction (viscous forces) by intense turbulence and shock formation. Some of the energy is also transferred to the pipe wall as vibration, and a portion of this is radiated as noise. The primary noise generating mechanism is the confined jet of fluid formed between the upstream and downstream locations. Flow noise can be modeled as noise of a confined jet. As a result, the noise-generation mechanisms are turbulent mixing, turbulence boundary interaction, shock, shock/turbulence interaction and flow separation.

Since the noise is generated downstream of a flow restriction, most of the acoustic energy is radiated to the downstream piping, which becomes the transmitting medium. As the noise travels downstream along the inside of the pipe in the fluid it radiates through the pipe wall along its entire length.

7. How to Calculate Sound Power Level

We can calculate the sound power level in a fundamental way by assigning an efficiency factor to provide the fraction of flow mechanical energy that is converted to noise:

$$W = \eta \frac{1}{2} \dot{m} u^2 \tag{1}$$

$$L_W = 10\log_{10}\left[\frac{W}{10^{-12}}\right] = 10\log_{10} W + 120 = 10\log_{10}\left[\eta \frac{1}{2}\dot{m}u^2\right] + 120 \tag{2}$$

where W is the flow mechanical power or energy, L_W is the sound power level in dB, and η is defined as the acoustical efficiency factor. If the flow is choked, then W becomes:

$$W = \eta \frac{1}{2} \dot{m} u_{sonic}^2 \tag{3}$$

If a value of η can be estimated for single and multi-phase flow, then the sound power level can be easily calculated not only for pressure reducing devices but also for pipe flow. Attenuation due to friction and temperature changes can be calculated from pipe flow equations in a more detailed manner. Computer codes such as SuperChems can then calculate the sound power level at every axial location for flow piping for single and multi-phase flow. For incompressible flow, the value of flow velocity for an ideal nozzle can be calculated from the mechanical energy balance:

$$u = \sqrt{\frac{2}{\rho_l}\Delta P} \tag{4}$$

Substituting the above equation for u in Equation 1 yields the following Equation for W for liquid flow through an ideal nozzle:

$$W = \eta \dot{m} \frac{\Delta P}{\rho_l} \tag{5}$$

For liquid flow, a typical value of η is approximately 10^{-8}.

All gas flow acoustic efficiencies have been reported to approach 1 % of the total flow mechanical energy for rockets[i]. This is illustrated in Figure 1. When the acoustic efficiency is plotted against the flow mechanical power Figure 1 is obtained. The measured curve indicates that the acoustic efficiency falls off as the flow mechanical energy gets larger while the calculated curve[ii] indicates increasing acoustic efficiencies.

Figure 1. Acoustic efficiency trends

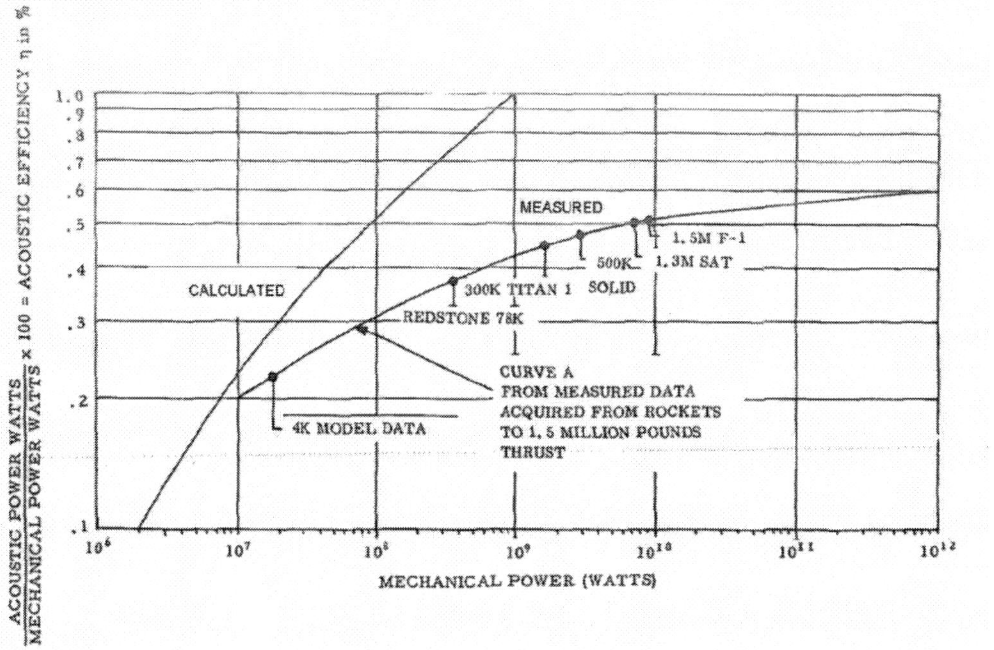

Two-phase flow sound power level can also be calculated using Equation 1. Recent work by Singh et al.[iii, iv] demonstrated that more attenuation of noise is exhibited by two-phase flow. Therefore, using the gas acoustic efficiency values will over predict sound power and noise levels for two-phase flow.

8. Existing Methods and Guidance

Current API, AIChE/CCPS, and AIChE/DIERS pressure relief and flare systems guidelines and standards do not formally address vibration risk. They do not offer specific guidance on velocity limitations other than backpressure calculations and they do not offer guidance on acoustic induced or turbulence induced piping vibration fatigue failure. The Marine Technology Directorate Limited (MTD) has published in 1999 "Guidelines for the Avoidance of Vibration Induced Fatigue in Process Pipework"[v]. A second edition of these guidelines were published in 2008 by the Energy Institute[vi]. The methods outlined in these guides have been incorporated in the SuperChems Expert ioVIPER modules. The MTD/Energy Institute guidelines provide qualitative and quantitative methods for the assessment of piping vibration failure risk and depending on the calculated risk level they provide generic guidance for the mitigation of vibration risk.

Many operating companies have established their own internal guidance for evaluating and minimizing piping vibration risk. Although these criteria vary from company to company, they all in general include a limit of flow velocity in some form.

A common criteria used in relief systems is to limit the value of flow Mach number to a value ranging from 0.3 to 0.9:

$$0.3 \le M = \frac{u}{u_{sonic}} \le 0.9 \tag{6}$$

Another common criteria often encountered is to limit the dynamic pressure component or kinetic energy of a flow stream to a value of 100,000 Pascals for gas flow:

$$\rho u^2 \le 1 \times 10^5 \ \text{kg/m/s}^2 \tag{7}$$

and 50,000 for two-phase flow:

$$\rho u^2 \le \frac{1}{2} \times 10^5 \ \text{kg/m/s}^2 \tag{8}$$

The 2006 fifth edition of the NORSOK Process Design Standard P-001 limits the flow velocities for all flare lines to $\rho u^2 \le 200{,}000 \ \text{kg/m/s}^2$ for single and multiphase flow. The piping for flare headers and sub-headers are designed for a maximum Mach number of 0.6 and lines downstream of pressure relief valves to the first sub-header are designed for a maximum Mach number of 0.7. For process lines the maximum design velocity for gas pipes is limited to 60 m/s or $u \le 175\frac{1}{\rho^{0.43}}$ whichever is less. The maximum velocity for two-phase lines is limited to $u \le 183\frac{1}{\rho^{0.5}}$ or:

$$\rho u^2 \le 183^2 \le 33{,}489 \ \text{kg/m/s}^2 \tag{9}$$

9. The Method of Carucci and Mueller

Carucci and Mueller have published guidance for the estimation of sound power levels for control valves and pressure reducing devices[vii, viii].

$$L_W = 10 \times \log_{10}\left[\left(\frac{P_1 - P_2}{P_1}\right)^{3.6} \times \dot{m}^2 \times \left(\frac{T_1}{M_w}\right)^{1.2}\right] + 126.1 \tag{10}$$

$$= 10 \times \log_{10}\left[4 \times \left(\frac{P_1 - P_2}{P_1}\right)^{3.6} \times \dot{m}^2 \times \left(\frac{T_1}{M_w}\right)^{1.2}\right] + 120 \quad (11)$$

$$= 36\log_{10}\left(\frac{P_1 - P_2}{P_1}\right) + 20\log_{10}(\dot{m}) + 12\log_{10}\left(\frac{T_1}{M_w}\right) + 126.1 \quad (12)$$

where P_1 is the upstream pressure or source pressure, P_2 is the downstream pressure, T_1 is the source temperature, $\dot{m}$ is the mass flow rate, and M_w is the gas molecular weight.

Attenuation of noise due to friction and heat conduction losses is estimated from the following equation:

$$L_{W,At} = 0.06\frac{L}{D_i} \quad (13)$$

where L is the pipe length and D_i is the internal pipe diameter. For an L/D of 50 the attenuation loss is 3 dB. Abrupt changes in flow area (expansion) in the piping can also lead to attenuation. The decrease in sound power level can be estimated from the following equation:

$$L_{W,Ex} = 2\left(\frac{D_2}{D_1} - 1\right) \quad (14)$$

where $D_2 > D_1$. A 3 dB reduction is typically applied to the flow leaving a tee in each direction or entering into a header or a large drum or vessel. The Carucci and Mueller Equation 12 is used by the Energy Institute Guidelines for the assessment of failure likelihood for high frequency acoustic excitation [vi]. Note that the Carucci and Mueller equation cannot be easily applied to complex piping systems, multi-phase systems, and relief piping with multiple chokes. Some companies also add 6 dB to the sound power level estimate when sonic flow exists at a branch connection to account for amplified dynamic strain response.

9.1 Analysis of the Carucci and Mueller Data Set

Careful analysis of the Carucci and Mueller data set (see Table 1) indicates that the source pressures were high enough to produce choked (sonic) flow through a flow limiting orifice or valve upstream of the failure point further downstream, often at a branch or line connection. This important point was missed by Eisinger who used the downstream (non-choked) flow velocity to establish his Mach number based failure criterion. We were also able to reproduce Eisinger's estimates of the original Carucci and Mueller data set based on his published paper.

Carucci and Mueller provided enough data in their original paper to allow the calculation of the upstream flow limiting flow area and choke point conditions (this is possible because they reported the actual flow rates) as well as the conditions downstream of the flow limiting device in the discharge piping. Note that the sound power level estimate upstream of the choke points should be based on the pressure difference (or pressure ratio) of the source pressure and the choke point. The sound pressure level downstream of the choke point (the primary cause of acoustic induced fatigue failure in downstream piping of the choked point) should be based on the difference (or pressure ratio) of the choke point and the shock discontinuity pressure downstream of the choke point. The original equation proposed by Carucci and Mueller used the pressure difference (or ratio) between the source pressure and the exit pressure downstream of the choke point. This yields an overestimate of the sound pressure. This overestimate is somewhat tempered by the fact that the pressure difference ratio to the source pressure is raised to the 3.6 power.

The following information was provided by Carucci and Mueller regarding the five failures and two high vibrations cases observed in Table 1.

A. Failure occurred during startup. Sonic velocity is achieved at the 6 inch branch connection of a 24 inch pipe run downstream of the recycle valve. The sound power level estimate considers the combined acoustic energy generated by the letdown valve and the sonic condition at the branch connection. The upstream pressure at the branch connection was estimated to be 98.8 psia.

B1/B2. This system consisted of a control valve letting down pressure to a safety valve / flare header system. The failure occurred after five to ten hours of operation as a crack at a 10 inch branch connection to 28 inch header. Sonic velocity was achieved at both the control valve and the 10 inch branch connection to the 28 inch flare header. As such, two pertinent data points are established:

1. Acoustic energy within the 10 inch line as generated by valve (not the failure point), and
2. Combined acoustic energy within the 28 in header as generated by valve and sonic condition at the 10 in branch connection (the failure point).

C. High vibrations, no failures.

D. Failure during startup. Cracks were observed in a 24 inch line downstream of the compressor recycle valve after twelve hours of operation. The cracks were near the branch connections of a 6 inch line and a 3/4 inch drain valve and at an I-beam support welded to the pipe at the elbow immediately downstream of the control valve.

E. High vibration, no failures.

F. Failure at severely undercut weld on a 300 mm (12 in.) line made to fasten conduit support clip. Points of high stress concentration were later eliminated.

G. This system consisted of six, parallel, high pressure steam letdown valves, each with a downstream, three pass contra-flow attemporator and an in-line silencer. Failure of the 18 inch attemporator shells occurred after four hundred hours of operation.

H. This system included four parallel desuperheaters located downstream of control valves. Several cracks developed in this system after two to three months of operation. Two to three of the four letdown valves were discharging steam through a 10 inch connection to a 20 inch header which swaged up to a 30 inch diameter. Longitudinal cracking occurred at the bottom of the 20 inch header at a transverse guide 0.5 meters downstream of the fourth 10 inch branch connection. In addition, a 1 inch bypass line for a block valve downstream of the third letdown valve cracked and the 20 inch header cracked around a pressure tap downstream of the transverse guide.

Table 1 compares the acoustic efficiencies implied in the Carucci and Mueller method (at the upstream choke point) vs. acoustic efficiencies established using IEC methods described later on in this paper. The acoustic efficiencies implied by the Carucci and Mueller method at the upstream choke point ranged from 0.1 to 5 percent while the IEC acoustic efficiency ranged from 0.67 to 1 percent. The Carucci and Mueller method is based on acoustic energy theories encompassing both jet and choked flow noise as well as test data from work performed by Exxon research and engineering. As discussed in later sections, the experience based failure criteria originally developed by Carucci and Mueller can only be used with sound power levels calculated by Equation 12.

The implied acoustic efficiency by Equation 12 is proportional to the mass flow rate and inversely proportional to u^2:

$$\eta = 8\left(1-\frac{P_2}{P_1}\right)^{3.6}\left(\frac{T_1}{M_w}\right)^{1.2}\left(\frac{\dot{m}}{u^2}\right) \tag{15}$$

If we consider the case of an ideal gas undergoing isentropic choked (sonic) flow through a restriction orifice, we can establish the following expressions for variables of interest at the choke point:

$$T_2 = 2\frac{T_1}{\gamma+1} \tag{16}$$

$$P_2 = P_1\left(\frac{2}{\gamma+1}\right)^{\frac{\gamma}{\gamma-1}} \tag{17}$$

$$\rho_2 = \frac{P_2 M_w}{R_g T_2} \tag{18}$$

$$u_2 = \sqrt{\frac{\gamma R_g T_2}{M_w}} \tag{19}$$

$$\dot{m} = \rho_2 u_2 A_o \tag{20}$$

Substituting the values of mass flow rate $\dot{m}$, choked flow velocity u_2, and choke pressure P_2 in Equation 15, we obtain the following simplified expression for the Carucci and Mueller acoustic efficiency after some algebraic manipulations:

Table 1. Vibration risk data set used by Carucci and Mueller and reproduced by Melhem

CASE	SERVICE	T_1, K	P_1, bara	T_2, K	P_2, bara	D_2, m	u2, m/s	Exit Mach Number, M_2	Mass Flow Rate, kg/s	Heat Capacity Ratio, C_p/C_v	Restriction Orifice / Valve Flow Speed, m/s	Carucci and Mueller Acoustic Efficiency, %	IEC Acoustic Efficiency, %	PWL_cm, dB	PWL_mel, dB	PWL_cm - PWL_mel, dB	PWL_mel (IEC), dB
A	Propylene Compressor Recycle	319.26	20.55	238.91	1.59	0.610	32.50	0.137	29.74	1.13	224.30	0.65	0.85	168.5	165.0	3.5	161.5
B1	Natural Gas Pressure	318.71	42.75	226.00	6.27	0.254	125.60	0.399	48.26	1.22	322.99	2.54	0.91	170.9	158.1	2.8	163.6
B2	Natural Gas Letdown to Flare	318.71	5.17	270.38	2.07	0.711	58.20	0.169	48.26	1.22	357.00	0.41	0.69	171.4	158.9	2.5	163.3
C	Natural Gas Letdown to Flare	308.15	47.57	195.68	4.14	0.406	64.40	0.218	47.63	1.23	311.29	4.17	0.95	171.8	169.8	2.0	163.4
D	Nitrogen Compressor Recycle	310.93	6.96	280.31	3.72	0.610	93.80	0.297	119.70	1.20	324.76	1.05	0.79	168.3	168.2	0.1	167.0
E	Nitrogen Compressor Recycle	310.93	16.82	267.59	7.03	0.457	60.10	0.194	84.80	1.21	328.35	0.79	0.66	168.8	165.6	3.3	164.8
F	Natural Gas Compressor Recycle	330.37	51.09	220.53	13.86	0.305	29.80	0.093	35.28	1.45	272.52	0.59	0.82	164.9	159.5	5.4	160.3
G	Desuperheater Steam	802.59	99.97	480.32	10.34	0.457	49.80	0.093	36.89	1.29	619.65	2.51	0.94	175.5	173.1	2.4	169.2
H	Desuperheater Steam	658.15	43.09	476.11	10.34	0.508	49.60	0.093	45.99	1.29	586.00	0.36	0.83	163.9	168.3	-4.4	168.9
1	Process Gas Compressor Recycle	308.15	13.17	204.98	1.03	0.406	122.80	0.461	27.09	1.19	256.81	2.22	0.95	165.9	164.2	1.7	160.6
2	Process Gas Compressor Recycle	311.48	38.75	265.88	10.00	0.305	14.20	0.048	13.86	1.13	258.54	0.32	0.83	156.6	151.8	4.8	155.9
3	Propylene Compressor Recycle	316.48	20.55	262.14	3.24	0.406	15.70	0.065	12.60	1.11	209.52	0.31	0.91	156.0	153.5	2.5	154.0
4	Process Gas Compressor Recycle	300.37	2.48	257.91	1.03	0.508	145.60	0.450	34.78	1.21	325.00	0.26	0.77	161.3	156.7	4.6	161.5
5	Process Gas Compressor Recycle	310.93	15.86	215.12	2.21	0.406	68.70	0.226	25.33	1.23	357.01	1.21	0.91	165.2	152.9	2.3	161.7
6	Process Gas Compressor Recycle	300.93	39.44	250.67	15.51	0.254	26.10	0.075	20.66	1.24	330.10	0.20	0.78	158.3	153.6	4.7	159.4
7	Propylene Compressor Recycle	310.93	16.69	233.88	1.38	0.356	131.30	0.575	38.05	1.13	216.35	3.52	0.95	166.6	165.4	1.2	159.3
8	Propylene Compressor Recycle	310.93	16.69	234.72	1.38	0.508	59.20	0.259	35.03	1.13	216.35	3.51	0.95	165.9	164.7	1.2	158.9
9	Waste Steam Dump	403.71	2.76	370.82	1.03	0.406	245.10	0.566	18.90	1.10	436.60	0.15	0.77	160.5	154.4	6.1	161.4
10	Process Gas Compressor Recycle	302.59	10.69	207.90	1.03	0.406	95.80	0.357	20.92	1.19	295.15	1.49	0.94	163.2	161.3	1.9	159.3
11	Process Gas Compressor Recycle	308.15	54.81	263.43	27.51	0.305	7.20	0.016	9.15	1.29	441.77	0.10	0.79	150.5	149.7	0.8	158.5
12	Natural Gas Compressor Recycle	308.15	3.79	282.54	2.48	0.914	28.40	0.075	39.69	1.26	315.99	0.19	0.73	155.4	155.7	-0.3	161.6
13	Natural Gas Compressor Recycle	308.15	6.48	279.42	4.00	0.914	17.30	0.046	39.44	1.25	334.77	0.24	0.75	157.2	157.2	0.0	162.2
14	Natural Gas Compressor Recycle	308.15	11.45	282.82	6.76	0.762	14.40	0.039	38.93	1.19	346.05	0.28	0.76	158.1	158.1	0.0	162.5
15	Natural Gas Compressor Recycle	308.15	21.75	266.72	11.65	0.660	10.10	0.027	36.04	1.30	359.11	0.33	0.78	159.4	158.8	0.6	162.6
16	Natural Gas Compressor Recycle	308.15	40.75	272.56	21.99	0.610	6.30	0.017	35.91	1.25	344.77	0.34	0.78	159.2	158.6	0.6	162.2
17	Natural Gas Compressor Recycle	308.15	80.12	271.44	40.89	0.508	4.80	0.013	35.41	1.23	324.89	0.40	0.78	160.1	158.8	1.3	161.6
18	Natural Gas Compressor Recycle	308.15	172.99	240.53	79.98	0.406	3.80	0.010	35.41	1.47	326.41	0.71	0.82	161.5	161.3	0.2	161.9
19	Natural Gas Compressor Recycle	308.15	3.86	280.10	2.34	0.914	27.60	0.074	37.17	1.24	340.74	0.24	0.75	157.3	157.1	0.2	162.1
20	Natural Gas Compressor Recycle	308.15	6.21	281.91	3.93	0.914	16.50	0.044	36.79	1.24	325.37	0.20	0.74	155.9	155.9	0.0	161.6
21	Natural Gas Compressor Recycle	308.15	10.62	274.09	6.27	0.762	14.30	0.038	36.04	1.29	345.37	0.26	0.76	157.4	157.4	0.0	162.1
22	Natural Gas Compressor Recycle	308.15	19.72	266.35	10.62	0.660	10.10	0.027	33.01	1.31	359.23	0.30	0.78	158.4	158.0	0.4	162.2
23	Natural Gas Compressor Recycle	308.15	36.82	263.14	19.79	0.610	6.40	0.017	32.89	1.34	346.54	0.31	0.78	158.4	157.9	0.5	161.9
24	Natural Gas Compressor Recycle	308.15	64.47	270.22	36.96	0.508	4.90	0.013	32.82	1.31	319.94	0.32	0.77	157.3	157.2	0.1	161.1
25	Natural Gas Compressor Recycle	308.15	139.48	320.30	64.54	0.406	4.20	0.012	32.76	0.95	311.67	0.56	0.80	160.7	159.5	1.2	161.1
26	Propane Compressor Recycle	308.15	12.07	234.24	0.97	0.457	143.50	0.644	50.40	1.12	217.75	4.62	0.96	168.9	167.6	1.3	160.6
27	Steam Desuperheater	558.15	43.09	522.78	10.34	0.762	22.00	0.041	45.99	1.19	554.55	0.51	0.83	161.2	168.3	-7.1	167.9

$$\eta = (4.7950\gamma - 2.5882)\left(\frac{M_w}{T_1}\right)^{0.3} P_1 A_o \tag{21}$$

where A_o is the effective restriction orifice flow area in m^2, P_1 is the upstream pressure in bara, T_1 is the upstream temperature in Kelvin, and η is the calculated Carucci and Mueller acoustic efficiency in percent. This expression shows that the Carucci and Mueller acoustic efficiency is directly proportional to upstream pressure, and the restriction orifice flow area (flow rate). It can be shown that the Carucci and Mueller acoustic efficiency produces unrealistic values for high pressure systems, large flow area, and/or mixtures with high molecular weights.

If we consider the case of methane discharging through a restriction with an effective flow area of $0.1\ \mathrm{m}^2$, upstream temperature of $373.15\ \mathrm{K}$, and an upstream pressure of $100\ \mathrm{bara}$, the calculated acoustic efficiency is:

$$\eta = (4.7950 \times 1.282 - 2.5882)\left(\frac{16}{373.15}\right)^{0.3} \times 100 \times 0.1 \tag{22}$$

$$= 3.559 \times 0.388 \times 100 \times 0.1 = 13.8\% \tag{23}$$

The same value can also be calculated by calculating $\dot{m}$, u_2, P_2, and η directly from the above ideal gas flow equations and Equation 15:

$$T_2 = 2\frac{T_1}{\gamma+1} = 2\frac{373.15}{1.282+1} = 327.05\ \mathrm{K} \tag{24}$$

$$P_2 = P_1\left(\frac{2}{\gamma+1}\right)^{\frac{\gamma}{\gamma-1}} = 100\left(\frac{2}{1.282+1}\right)^{\frac{1.282}{1.282-1}} = 54.9\ \mathrm{bara} \tag{25}$$

$$\rho_2 = \frac{P_2 M_w}{R_g T_2} = \frac{54.9 \times 100000 \times 16}{8314 \times 327.03} = 32.306\ \mathrm{kg/m^3} \tag{26}$$

$$u_2 = \sqrt{\frac{\gamma R_g T_2}{M_w}} = \sqrt{\frac{1.282 \times 8314 \times 327.03}{16}} = 466.76\ \mathrm{m/s} \tag{27}$$

$$\dot{m} = \rho_2 u_2 A_o = 32.306 \times 466.76 \times 0.1 = 1507.63\ \mathrm{kg/s} \tag{28}$$

$$\eta = 8\left(1 - \frac{P_2}{P_1}\right)^{3.6}\left(\frac{T_1}{M_w}\right)^{1.2}\left(\frac{\dot{m}}{u^2}\right) \tag{29}$$

$$= 8 \times (1 - 0.549)^{3.6} \times \left(\frac{373.15}{16}\right)^{1.2} \times \left(\frac{1507.63}{466.76^2}\right) \tag{30}$$

$$= 8 \times 0.0568 \times 43.78 \times 69.2 \times 10^{-4} \tag{31}$$

$$= 0.1376 \text{ or } 13.76\% \tag{32}$$

A similar equation to 21 can be derived for the acoustic efficiency downstream of the choke point:

$$\eta = (68.63 - 4.836\gamma)\left[1 - (1.0397 + 0.6096\gamma)\frac{P_b}{P_1}\right]^{3.6}\left(\frac{M_w}{T_1}\right)^{0.3} P_1 A_o \tag{33}$$

where P_b is the superimposed backpressure downstream of the choke point. Applying Equation 33 to the same example above we calculate an acoustic efficiency of 227 percent at a backpressure of 1 bara:

$$\begin{aligned}\eta &= (68.63 - 4.836\times 1.282)\left[1 - (1.0397 + 0.6096\times 1.282)\frac{1}{100}\right]^{3.6}\left(\frac{16}{375}\right)^{0.3} \times 100 \times 0.1 \\ &= 62.430 \times 0.9359 \times 0.388 \times 100 \times 0.1 \\ &= 226.70\%\end{aligned} \tag{34}$$

It is evident from the actual data reported by Carucci and Mueller and from the above theoretical proof that the acoustic efficiency used by Carucci and Mueller in their equation can produce unrealistic values, well in excess of 1 percent.

10. Failure Criteria

Several methods are now available for screening and analyzing the potential failure risk of piping caused by vibration. These methods include:

- Experience Based - D/t method or D Method
- Experience Based - Mach number method (Not recommended)
- MTD methods
- Detailed structural dynamics methods

The experience based methods center around correlating likelihood of failure based on actual experience and/or test data. The most widely cited reference is that of Carucci and Mueller (see vii and viii) for steel pipe.

Figure 2 illustrates the failure criteria developed originally by Carucci and Mueller and later modified by Eisinger. This failure criteria establishes a design sound power level vs. the ratio of pipe diameter to thickness. This criteria is fundamentally better than the other criteria based on pipe diameter only since thicker wall pipes are stronger than thinner wall pipes.

The original work by Carucci and Mueller suggests a limit provided by the following equation (Figure 2):

$$L_{W,limit} = 184.6 - 0.215\frac{2D_i}{D_o - D_i} = 184.6 - 0.215\frac{D_i}{\delta} \tag{35}$$

where δ is the pipe thickness.

The lower allowable limit developed by Eisinger also shown in Figure 2 is given by the following equation:

$$L_{W,limit} = 173.6 - 0.125\frac{2D_i}{D_o - D_i} = 173.6 - 0.125\frac{D_i}{\delta} \quad (36)$$

The NORSOK standard published guidance in 2006 using the same equation for $L_{W,limit}$ as proposed by Eisinger above.

Note that the above limits are based on sound power levels calculated using the source pressure and the downstream exit pressure. We have recalculated the sound power levels based on the correct pressure ratios downstream of the choke point. This data is shown in Figures 3, 4, and 5.

Another design limit criterion that was originally proposed by Carucci and Mueller is based on pipe diameter only for steel pipe. This criterion applies to pipe diameters ranging from 10 inch to 36 inch (approximately between 200 mm to 800 mm) and with wall thicknesses ranging from 0.219 in to 0.439 in (approximately 5.5 mm to 10 mm). The design limit is shown in Figure 3 for the corrected data and can be approximated by:

$$L_{W,limit} = 192.8 - 9.8\ln D \quad (37)$$

where D is the nominal pipe diameter in inches and L_W is the sound power limit in dB.

Recent theoretical and experimental work by Chiyoda Engineering[ix] indicates that AIV risk will increase for smaller pipe diameters at the same diameter to thickness ratio. As a result, the limiting criteria established above should be decreased by 6 dB for an 8 inch pipe, or by 20 $\log_{10}$ (16/D).

The design limit correlations developed in this paper can be used with computer programs such as SuperChems Expert which calculates the sound power level at every axial locations for single and multiphase flow to decide if the steel piping inside diameter to thickness ratio exceed the allowable limits as shown by Figure 6.

11. Other Methods for Calculating η

The simplest method for calculating η is to assume the same efficiency as an expanded jet (see Figure 7) and apply it equally to single and multi-phase flow. The efficiency shown in Figure 7 is the same as what is currently used by API for the estimation of flare noise.

The flow regime types are defined by varying jet shapes in the area upstream and downstream of the vena contracta. These jets change their shape when certain differential pressure ratios are exceeded. In regimes II to IV, higher Mach numbers arise downstream of the vena contracta. Yet, M at the vena contracta itself remains unchanged at 1. The IEC method produces equivalent values of efficiency to the data shown in Figure 7 (as $F_L \rightarrow 1.0$) as shown in Figure 9. Note that Figure 9 uses $x = 1 - P_2 / P_1$ as the X axis while Figure 7 uses $P_1 / P_2 = \frac{1}{1-x}$.

Figure 2. Piping failure limit as originally proposed by Carucci and Mueller and by Eisinger for steel pipe

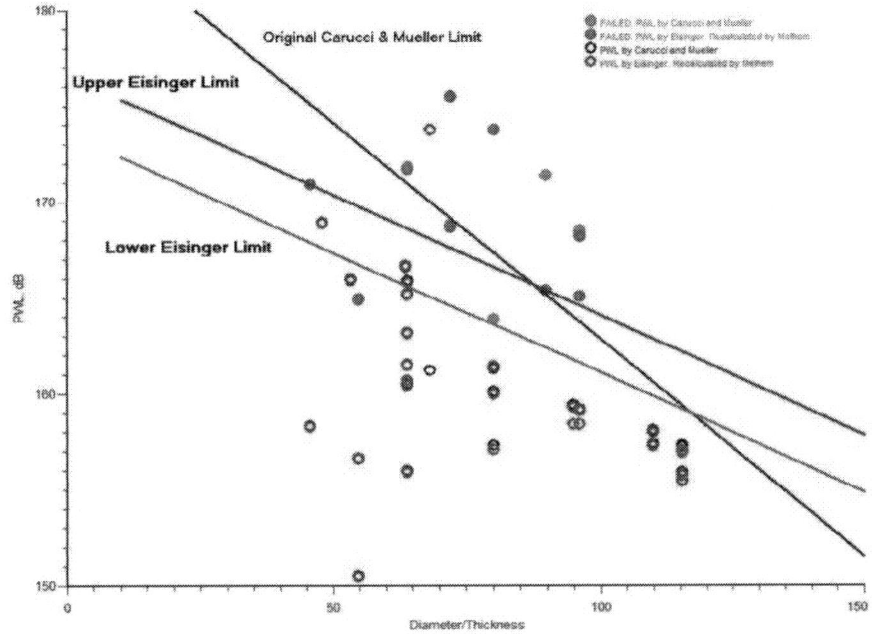

The efficiency is related to the flow pressure ratio and the flow regime as well. IEC calculates the efficiency based on five different flow regimes (see Figure 8) depending on the value of the downstream pressure, P_2:

> **Regime I** The flow is sub-sonic. The sound generation has the character of a dipole jet. The highest Mach number is reached at the vena contracta, not exceeding Mach 1 at the maximum. Downstream of the vena contracta, the jet expands, leading to partial pressure recovery (F_L factor).
>
> **Regime II** Sonic and supersonic flows exist together, which means that strongly turbulent flow and shock cell structure dominate. Pressure recovery drops until the top limit of regime II is reached.
>
> **Regime III** The rise in pressure is non-isentropic. The flow is supersonic and shear turbulence predominates.
>
> **Regime IV** The shock cells disappear and a Mach disk forms. The dominant mechanism is the interaction between shock cells and turbulence.
>
> **Regime V** The acoustical efficiency is constant.

Figure 3. Piping failure correlation developed by Melhem using Carucci Mueller acoustic efficiency and corrected sound power level estimates for steel pipe. PWL vs. D

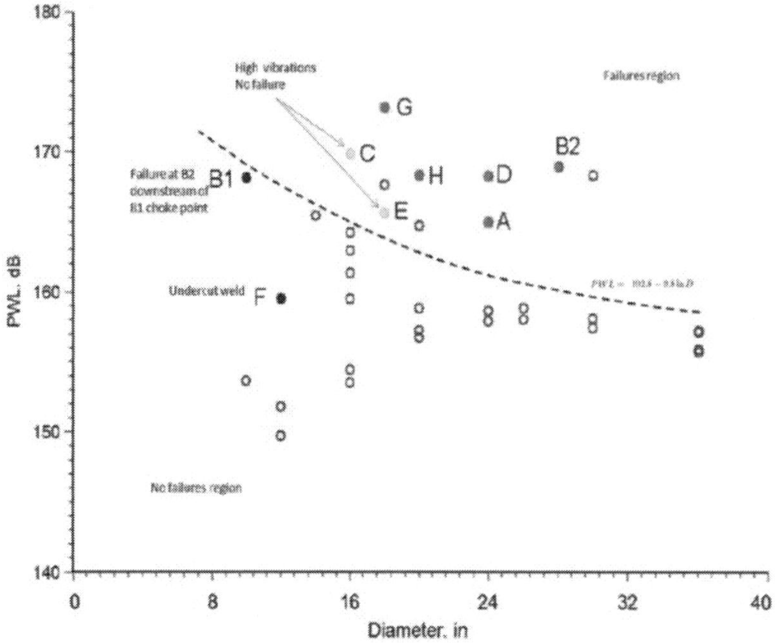

Figure 4. Piping failure correlation developed by Melhem using Carucci Mueller acoustic efficiency and corrected sound power level estimates for steel pipe. PWL vs. D/t

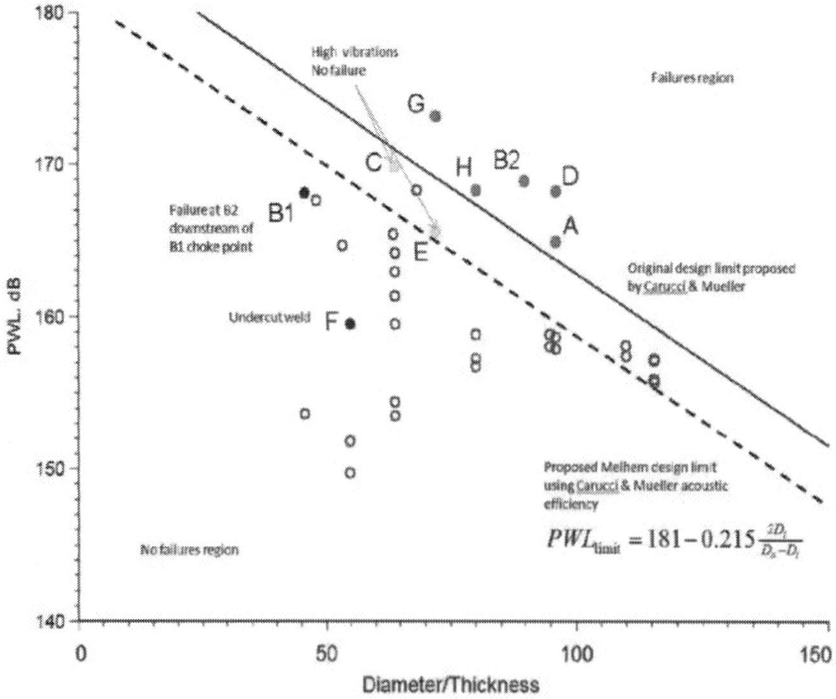

12. Vibration Frequencies

As a fluid moves through piping components there is typically a separation of the fluid from the constraining wall as the fluid changes flow direction. As a result a vortex is formed and then swept into the main stream. This vortex shedding occurs at fairly well defined dimensionless frequencies. The strength of the vortex varies but does not need to be very strong to cause damage especially if the shedding frequency couples with the natural frequency of the piping system. The shedding peak frequency for a vortex for subsonic and sonic flows (Regimes I and II up to a Mach number of 1.4) is given by:

$$f_p = \frac{N_{Str} u}{D} = 0.2 \frac{u}{D} \tag{38}$$

Figure 5. Piping failure correlation developed by Melhem using IEC acoustic efficiency and corrected sound power level estimates for steel pipe. PWL vs. D/t

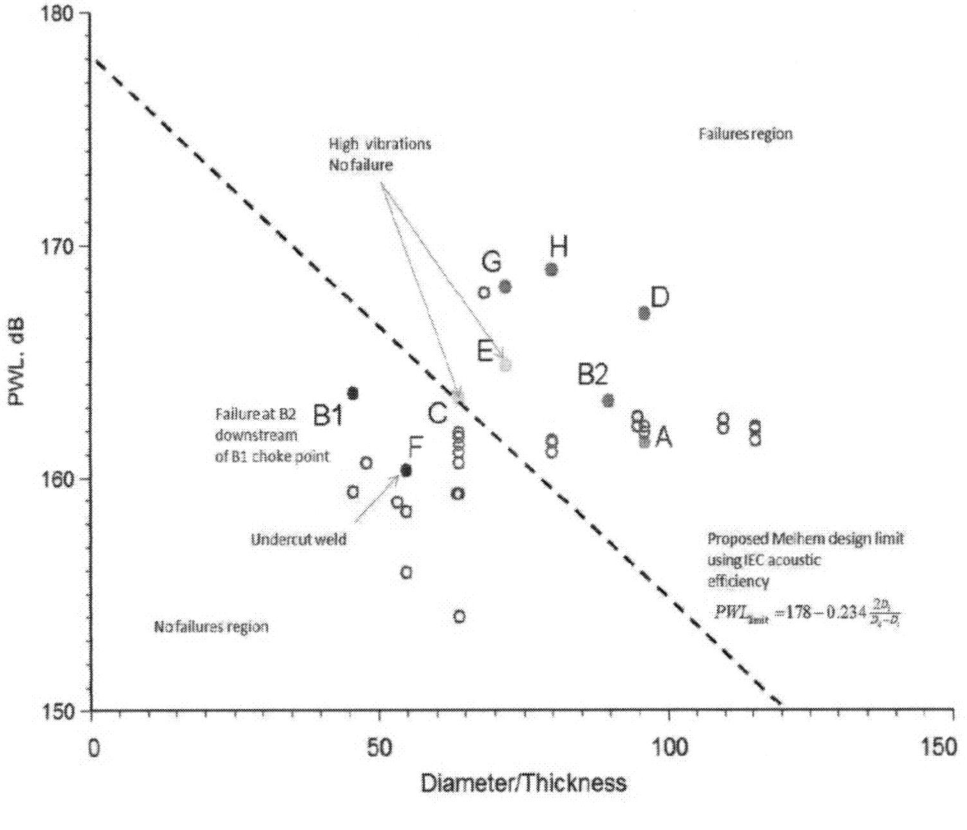

where u is the flow velocity in m/s, D is a characteristic flow dimension (perpendicular to flow), f_p is the peak frequency in Hz and N_{Str} is the Strouhal Number which varies depending upon the geometry causing the separation of the boundary layer. For a circular cylinder its value is 0.2 over a wide range of Reynolds numbers. It is usually between 0.1 and 0.3. Vortex shedding frequencies are generally more than 30 Hz [upper limit for most piping system natural frequencies].

For flows with Mach numbers larger than 1.4 ($M > 1.4$, Regimes III to V), the peak frequency f_p is given by:

$$f_p = \frac{0.4u_{sonic}}{1.25D\sqrt{M^2 - 1}} \tag{39}$$

For pipe flow, D is equal to the inside flow diameter. For a control valve, D is given by:

$$D = 0.0046\sqrt{\frac{C_v F_l}{N_o}} \tag{40}$$

Figure 6. Sample piping sound power level calculated by SuperChems Expert v6.4mp vs. experience based allowable limit for steel pipe

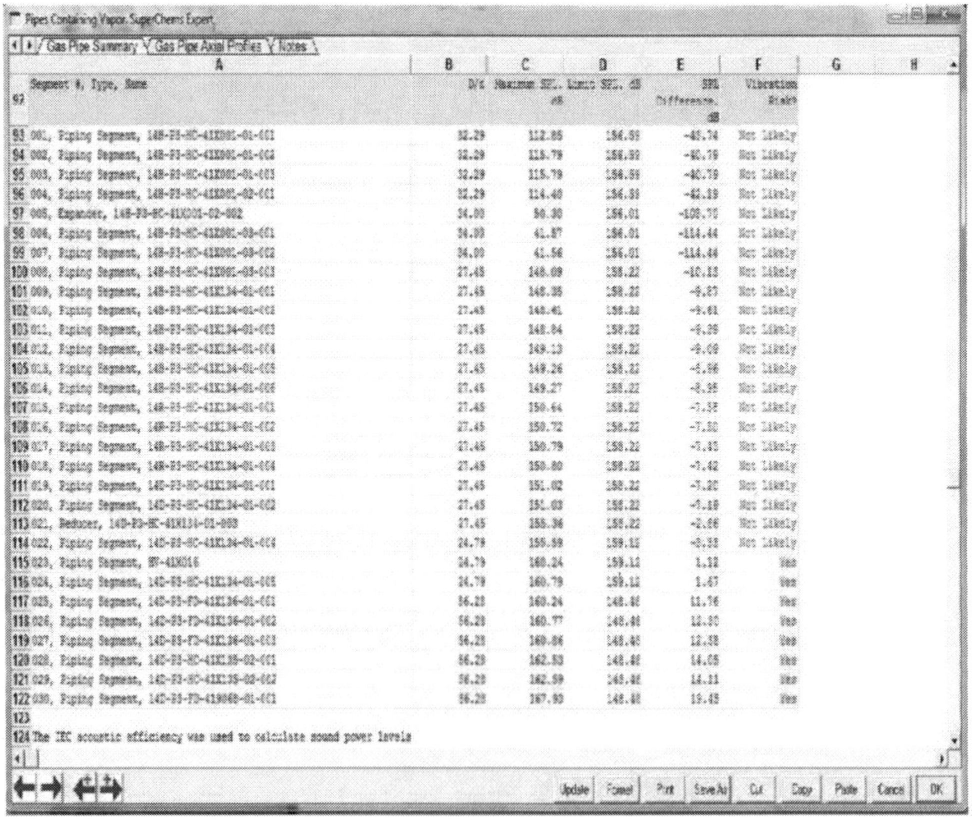

where D is in meters, N_o is the number of separate flow passages, C_v is the valve flow coefficient, $\frac{C_v}{N_o}$ is the channel flow coefficient, and F_l is the valve recovery factor. The pressure recovery factor is defined as:

$$F_l = \sqrt{\frac{P_1 - P_2}{P_1 - P_{vc}}} \tag{41}$$

where $P_1 - P_2$ is the pressure differential across the valve and P_{vc} is the pressure at the vena contracta (note that P_2 is larger than P_{vc} since the pressure at the exit of the valve recovers):

$$P_{vc} = P_1 - \frac{P_1 - P_2}{F_l^2} \tag{42}$$

For liquid flow, P_{vc} can reach the vapor pressure of the liquid and cause choked flow.

$$P_{vc} = F_f P_v \approx \left[0.96 - 0.25\sqrt{\frac{P_v}{P_c}}\right] P_v \tag{43}$$

Figure 7. Acoustic efficiency of shock noise generated by choked jets, η vs. jet pressure ratio P_1 / P_2x

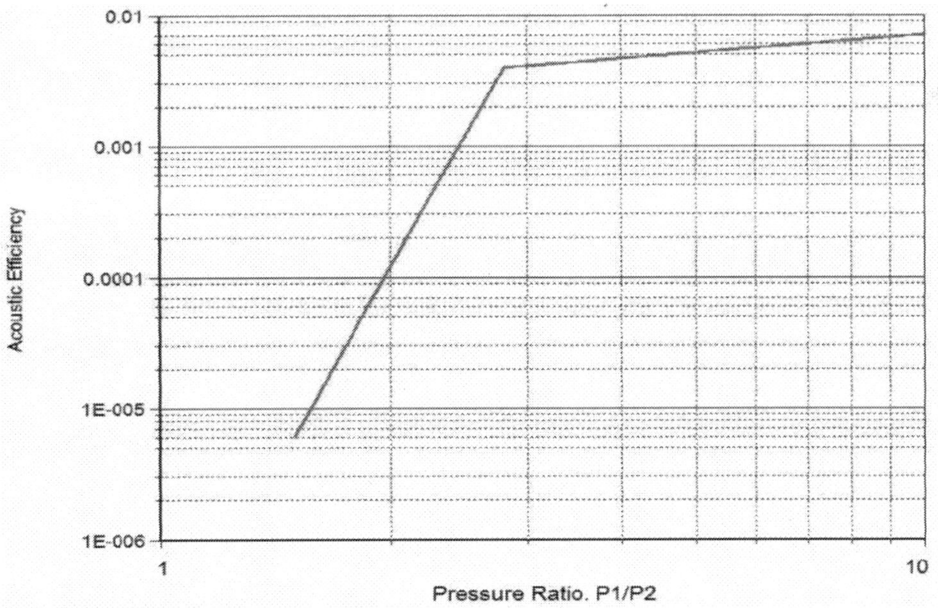

where F_f is the liquid critical pressure factor, P_v is the liquid vapor pressure, and P_c is the liquid critical pressure.

The lowest possible value of P_{vc} in a valve flowing liquid would be vacuum or 0. For this particular limiting case, Equation 41 can be solved for the pressure ratio P_1 / P_2:

$$P_d = \frac{P_1}{P_2} = \frac{1}{1-F_l^2} \tag{44}$$

where P_d is the damaging pressure ratio. If the valve is operated at a pressure ratio exceeding P_d, the flow will be choked, noisy, and subject to excessive vibration. Table 2 shows typical valve values for F_l and P_d.

Within every flowing pipe there will also be a standing wave moving axially back and forth in the pipe. The frequency of this wave depends on the effective acoustic length of the pipe and the sonic velocity of the fluid in the pipe. The effective acoustic length of the pipe is the distance between obstructions or acoustic barriers. Examples of obstructions would be valves, pumps, and orifices. An acoustic barrier would be an opening into a larger pipe, a reservoir, the end of a pipe run such as a 'T' intersection where the branch of interest requires a right angle turn. Piping components such as expanders or reducers could be an obstruction. An assessment of vibration risk should consider both the frequencies with and without the presence of expanders as obstructions.

The frequency of the standing wave can be calculated as shown below and then compared to the natural frequencies of valve components and the piping system to determine if there is a potential for vibration and/or resonance.

Closed End Pipe $f = \frac{i*u_{ac}}{4L}$

Open End Pipe $f = \frac{i*u_{ac}}{2L}$

where $i = 1, 2, ...$ is the wave number, L is the length between acoustic barriers, and u_{ac} is a characteristic acoustic speed throughout the pipe contents defined as follows:

$$u_{ac} = \frac{u_{sonic}}{\sqrt{1+\frac{dK}{\delta E}}} \tag{45}$$

where u_{sonic} is the speed of sound in the fluid, K is the bulk modulus of elasticity in the fluid, d is the pipe diameter, δ is the pipe thickness and E is the pipe material of construction modulus of elasticity.

The above equation is derived from a more general form that depends on the elastic properties of the pipe:

$$u_{ac} = \frac{u_{sonic}}{\sqrt{1+\frac{K}{E}\psi}} \tag{46}$$

where ψ is a function of the elastic properties of the pipe. Typical expressions of ψ are shown in Table 4. Typical material properties for ψ are shown in Table 3. Note that materials properties change with temperature as shown in Figure 10.

The speed of sound will change as a function of pressure, temperature, and composition. The presence of dissolved gas nuclei (such as air or other gases) in compressed liquids can significantly reduce the sonic velocity following a pressure drop which leads to the formation of gas bubbles. Up to 40 % reduction in sonic velocity has been observed[xi].

To control the vibration caused by a standing wave it is necessary to change the magnitude and/or the frequency of the standing wave or to change the natural frequency of the pipe or components being excited by the wave. The best approach is to address the magnitude of the standing wave. The magnitude is related to the fluid turbulent energy that is enforcing the wave. The most dominant source of this turbulence is the kinetic energy generated by the fluid jet exiting the valve trim. Thus a valve change with a trim that reduces this jet energy will eliminate this wave influence. Trying to change the frequencies is usually not beneficial. There is such a wide range of frequencies present in the turbulent flow that excitation can continue to establish a strong wave at the new frequency and continue the piping vibration.

Figure 8. Flow regimes considered for the estimation of acoustical efficiency

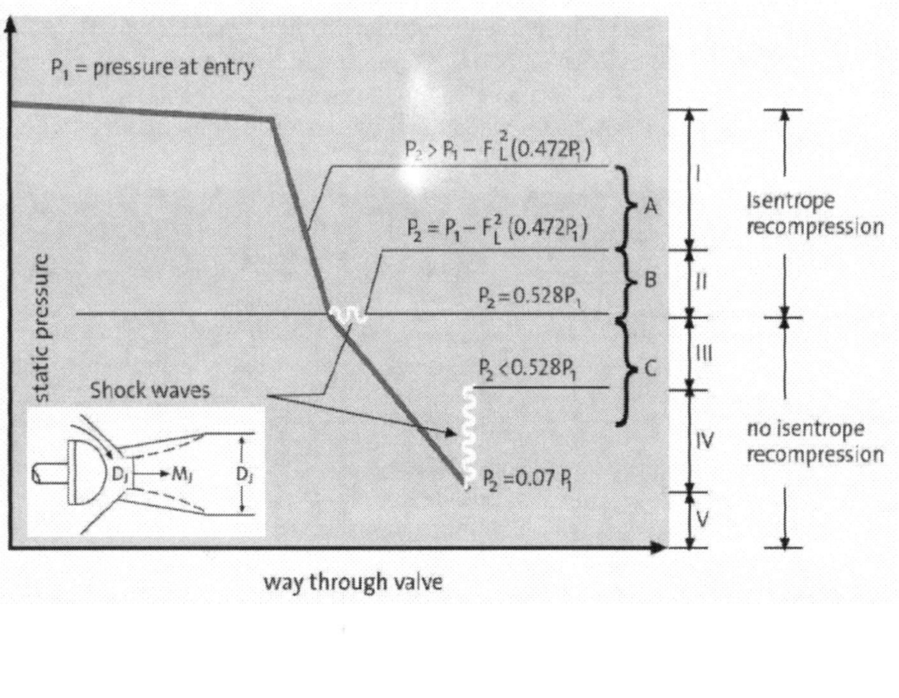

Figure 9. IEC flow acoustical efficiency as a function of F_L and differential pressure ratio

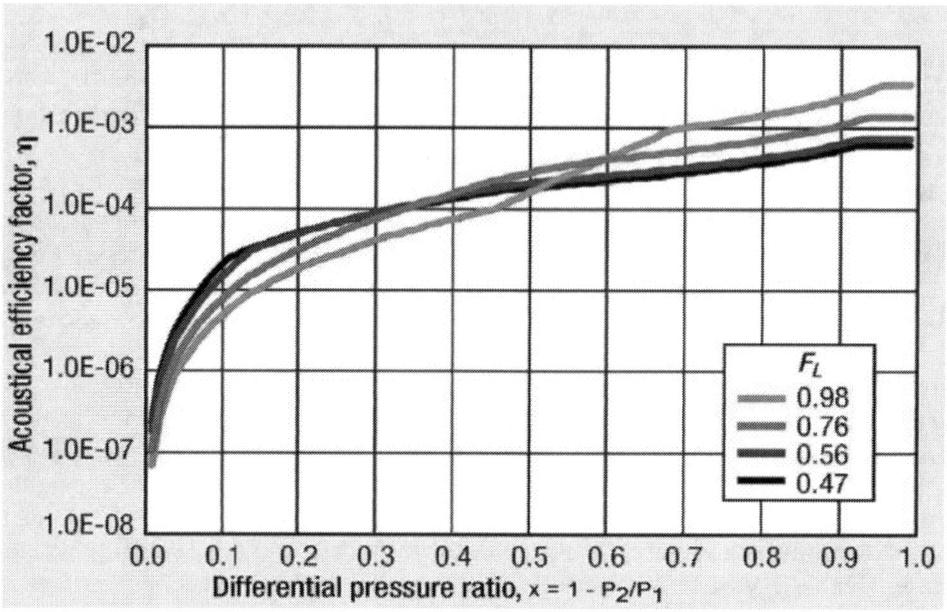

13. The Singing Safety Relief Valve Problem

As a result of increasing steam flow rates, several boiling water reactor (BWR) nuclear power plants have recently experienced the excitation of acoustic standing waves in closed side branches, e.g., safety relief valves (SRVs), due to vortex shedding generated by steam flow in the main stream lines. Flow past a valve entrance cavity excites a standing wave, resulting in noise and vibration[xii]. A similar tone is produced when air is blown across the mouth of a glass bottle.

Table 2. Typical valve values for F_l and P_d

Body	Trim	Flow Direction	F_l	$P_d = \frac{P_1}{P_2}$
Single Seat Globe	Cage	Open	0.90	5.3
	Cage	Closed	0.80	2.8
	V Plug	Open	0.90	5.3
	V Plug	Closed	0.90	5.3
	Contoured	Open	0.90	5.3
	Contoured	Closed	0.80	2.8
Double Seat Globe	V Plug		0.90	5.3
	Contoured		0.85	3.6
	Standard Bore		0.55	1.4
	Characterized		0.57	1.5
Angle	Cage	Open	0.85	3.6
	Cage	Closed	0.80	2.8

Ball	0.8 dia. Orifice		0.55	1.4
Butterfly	60°	Open	0.68	1.8
Butterfly	90°	Open	0.55	1.4

The amplitude of the acoustic pressure waves can be several times higher than the dynamic pressure present in the system (see Figure 12). The acoustic waves propagate in the steam lines, eventually reaching sensitive components such as steam dryers and turbine stop valves. In addition, the acoustic waves generated in the side branches may generate vibration problems locally and may lead to complications such as valve-seat wear. Therefore, the structural components are subjected to high-cycle fatigue loads, which over time may severely impact those components' functionality and safety.

Resonance occurs when the vortex shedding frequency coincides with the acoustic frequency of the standpipe or the valve components. The natural frequency of the standpipe/valve combination for a closed end pipe is given by the following equation:

$$f_a = \left(\frac{2n-1}{4}\right)\frac{u_{ac}}{L+L_e} = \left(\frac{2n-1}{4}\right)\frac{u_{ac}}{L+0.425d} \tag{47}$$

where n is the mode number (1 for 1st mode, 2 for 3rd mode, etc.), u_{ac} is the acoustic speed through the pipe contents as defined earlier, and L_e is an end correction corresponding to Rayleigh's upper limit.

The frequency of pressure oscillations (sound) created by vortex shedding, the energy source for the standing waves, is given by the following equation:

$$f_s = N_{str}\frac{u}{d+r} \approx 0.33(n-0.25)\frac{u}{d+r} \tag{48}$$

Typically peak oscillations occur at a Strouhal Number around 0.4 as shown in Figure 12. Note that the root mean square pressure amplitude shown in Figure 12 is the ratio of pressure oscillations divided by dynamic pressure ($\frac{1}{2}\rho u^2$). RMS begins increasing at a specific onset Strouhal Number and flow velocity depending on acoustic speed, pipe diameter, and pipe length, reaches a peak value and then decreases.

There are many similar installations of pressure relief valves in the process industries where the valves are mounted on large process lines such as overhead lines for distillation columns. In order to avoid fatigue failure from resonance caused by the coupling of normal flow vortex shedding frequency and the acoustic frequency of the standpipe ($f_s = f_a$), the normal flow velocity in the main line has to be limited to less than this critical value:

$$u_{max} < \frac{f_s}{N_{st}}(d+r) < \frac{1}{4}\left(\frac{u_{ac}}{L+L_e}\right)\left(\frac{d+r}{N_{str}}\right) \tag{49}$$

$$< \frac{1}{4}\left(\frac{u_{ac}}{L+0.425d}\right)\left(\frac{d+r}{0.6}\right)$$

As shown in Figure 12 the pressure fluctuations start to increase at a Strouhal number of 0.6 and then decrease after they reach a peak value around a Strouhal number of 0.4. The same approach can be applied to the flow through the inlet and/or discharge line of a pressure relief valve:

$$u_{max} < \frac{1}{2}\frac{u_{ac}}{L}\frac{D}{0.6} \tag{50}$$

where L is the acoustic length of the inlet or discharge line. Resonance can also be checked by comparing an open pipe/contents frequency with the natural frequency of the pressure relief valve, f_n:

$$f_n = \frac{1}{\tau_n} = \frac{\omega_n}{2\pi} = \frac{1}{2\pi}\sqrt{\frac{K_s}{m_D}} \tag{51}$$

$$\tau_n = \frac{2\pi}{\omega_n} = 2\pi\sqrt{\frac{m_D}{K_s}} \tag{52}$$

where τ_n is the undamped natural period in s, and f_n is the undamped natural frequency in Hz where one Hz equals 1 cycle/second, K_s is the spring constant in N/m, m_D is the mass of the valve disc and moving parts in kg, and ω_n is the undamped circular natural frequency in radians/s.

14. Conclusions

A general method for the estimation of vibration risk for piping systems for single and multiphase flow is developed and implemented. An experience based vibration risk failure criteria is outlined using the work of Carucci and Mueller for gas flow. Because the Carucci and Mueller sound power level equation can produce unrealistic values of acoustic efficiency, well in excess of 1 percent for high pressure systems and/or systems with large mechanical flow energy, the revised experience based failure criteria developed in this paper (see Figure 5) based on the IEC acoustic efficiency is recommended for multiphase flow.

Figure 10. Temperature effects on material of construction

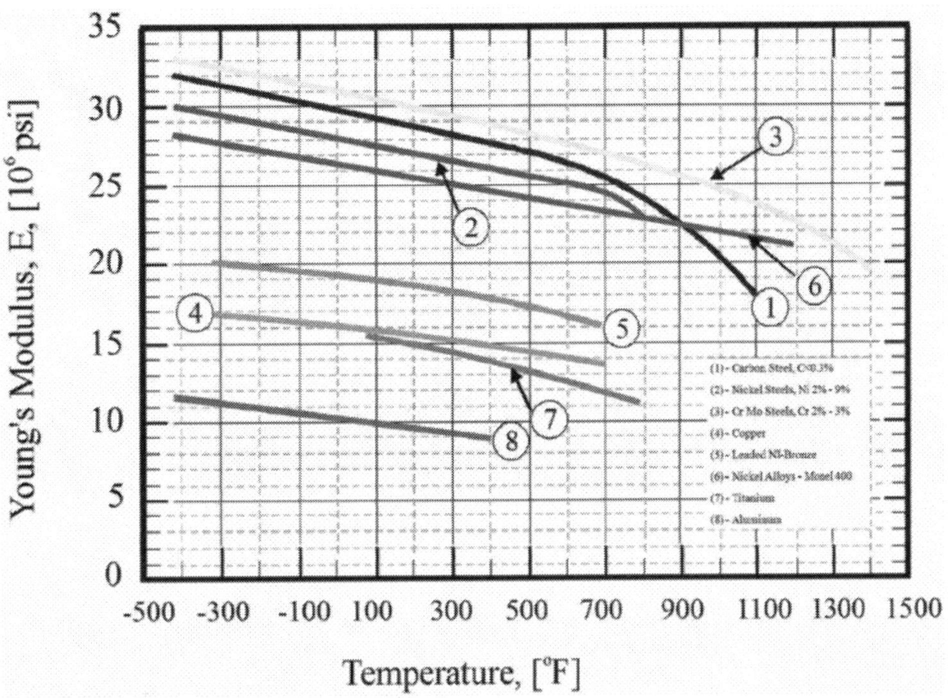

Figure 11. The singing safety relief valve[xiii]

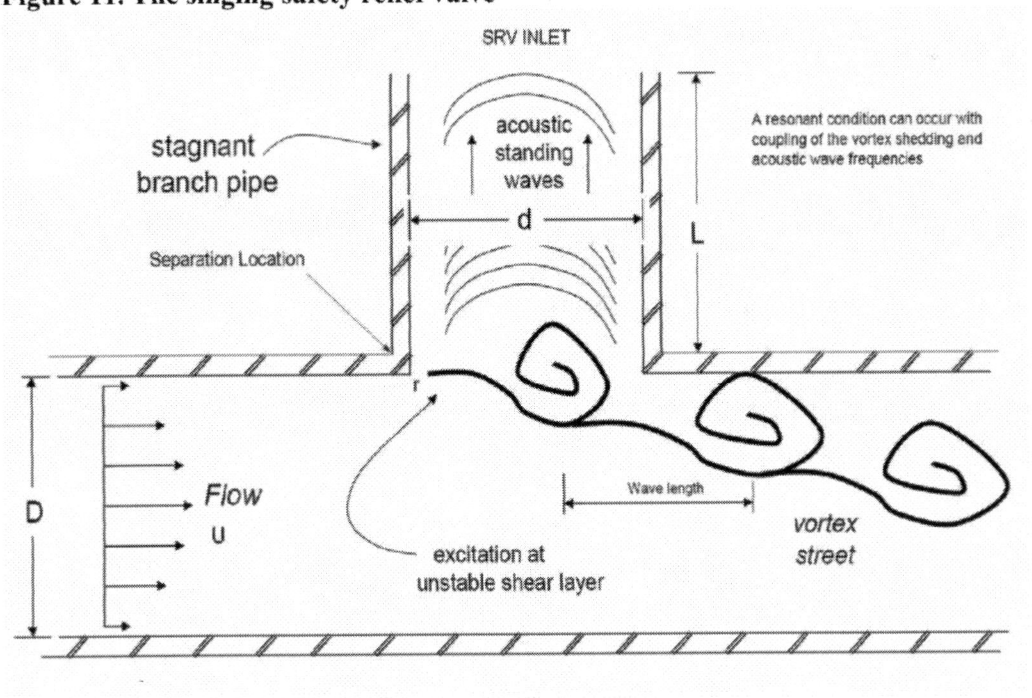

Figure 12. Comparison of calculated and measured pressure fluctuations as a function of Strouhal number[xiv]

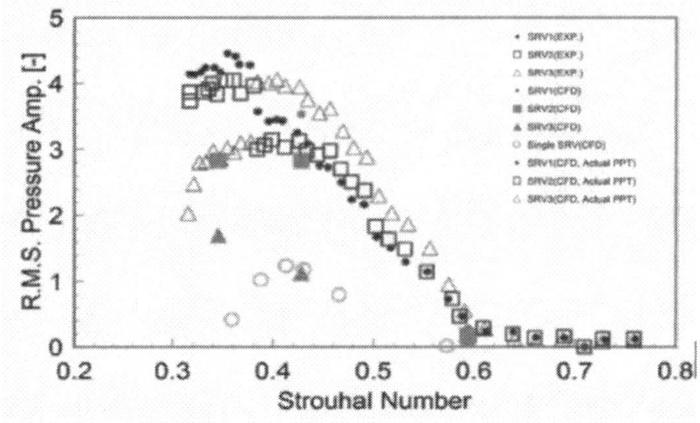

E is typically referred to as Young's modulus of elasticity

G is typically referred to as modulus of torsion, $G = \frac{1}{2}\frac{E}{1+\nu}$

Table 3. Typical data used in the estimation of sonic velocity in pipelines

Material	E (GPa)	Poisson's Ratio ν	$K = \frac{1}{\kappa}$ (GPa)	ρ in kg / m3
Aluminum	69	0.33		
Brass	78-110	0.36		
Carbon steel	202	0.303		
Cast iron	90-160	0.25		
Concrete	20-30	0.15		
Copper	117	0.36		
Ductile iron	172	0.30		
Fibre cement	24	0.17		
High carbon steel	210	0.295		
Inconel	214	0.29		
Mild steel	200-212	0.27		
Nickel steel	213	0.31		
Plastic / Perspex	6.0	0.33		
Plastic / Polyethylene	0.8	0.46		
Plastic / PVC rigid	2.4-2.75			
Stainless steel 18-8	201	0.30		
Water - fresh			2.19	999 at 20 C
Water - sea			2.27	1025 at 15 C

Table 4. Typical expressions for ψ

Pipe condition	ψ
Rigid	0
Anchored against longitudinal movement through its length	$\frac{d}{\delta}\left(1-\nu^2\right)$
Anchored against longitudinal movement at the upper end	$\frac{d}{\delta}\left(1.25-\nu\right)$
Frequent expansion joints present	$\frac{d}{\delta}$

References

[i] W.A. Skipwith. Acoustics technology. Survey SP-5093, NASA National Aeronautics and Space Administration, 1970.

[ii] C.M. Harris. *Handbook of Noise Control*. cGraw-Hill Book Co., Inc., 1957.

[iii] N.R. Miller, G.M. Singh, E. Rodarte and P.S. Hrnjak. Modification of standard aeroacoustic valve noise model to account for friction and two-phase flow. In *ACRC TR-162*, pages 1-11. Mechanical and Industrial Engineering Department, University of Illinois, 2000.

[iv] N.R. Miller, G.M, Singh, E. Rodarte and P.S. Hrnjak. Prediction of noise generated by expansion devices throttling refrigerant. In *ACRC TR-163*, pages 1-13. Mechanical and Industrial Engineering Department, University of Illinois, 2000.

[v] MTD. *Guidelines for the Avoidance of Vibration Induced Fatigue in Process Pipework*. Marine Technology Directorate Limited (MTD), 1999.

[vi] Energy Institute. *Guidelines for the Avoidance of Vibration Induced Fatigue in Process Pipework*. The Energy Institute, London, England, 2008.

[vii] V.A. Carucci and R.T. Mueller. Acoustically induced piping vibration in high capacity pressure reducing systems. In *82-/WAPVP-8*, pages 1-13. American Society of Mechanical Engineers, ASME, 1982.

[viii] F.L. Eisinger. Designing piping systems against acoustically-induced structural fatigue. In *PVP-VOL 328, Flow-Induced Vibration*, pages 397-404. American Society of Mechanical Engineers, ASME, 1982.

[ix] Hisao Izuchi, *Investigation of Pipe Size Effect Against Vibration Risk*, Presentation to ioMosaic Corporation, March 2013, Houston.

[x] I. Heitner. How to estimate plant noises. *Hydrocarbon Processing*, 47, 67-74, 1968.

[xi] V.L. Streeter and E.B. Wylie. Water hammer and surge control. *Annual Reviews in Fluid Mechanics*, pages 57-74, 1974.

[xii] T.M. Mulcahy, S.A. Hambric and V.N. Shah. Flow-induced vibration effects on nuclear power plant components due to main stream line valve singing. In *Ninth NRC/ASME Symposium on Valves, Pumps, and Inservice Testing, NUREG/CP-0152,* volume 6, pages 3B:49-3B:69. NCR/ASME, 2006.

[xiii] P. Sekerak. Potential adverse flow effects at nuclear power plants. In *16th International Conference on Nuclear Engineering*. ICONE16-48900, 2008.

[xiv] K. Okuyama, F. Inada, Y. Ogawa, R. Morita, S. Takahashi and K. Yoshikawa. Evaluation of acoustic and flow induced vibration of the BWR main stream lines and dryer. *Journal of Nuclear Science and Technology,* 48 (5): 759-776, 2011.

AUTHOR INDEX

AUTHOR INDEX

47th Annual Loss Prevention Symposium 2013

(LPS)

Topical Conference at the 2013 AIChE Annual Meeting and 9th Global Congress on Process Safety

San Antonio, Texas, USA
28 April – 2 May 2013

Volume 2 of 2

ISBN: 978-1-62748-447-3

Printed from e-media with permission by:

Curran Associates, Inc.
57 Morehouse Lane
Red Hook, NY 12571

Printed by Curran Associates, Inc. (2013)

For permission requests, please contact AIChE
at the address below.

AIChE
3 Park Avenue
New York, NY 10016-5991

Phone: (203) 702-7660
Fax: (203) 775-5177

www.aiche.org

Additional copies of this publication are available from:

Curran Associates, Inc.
57 Morehouse Lane
Red Hook, NY 12571 USA
Phone: 845-758-0400
Fax: 845-758-2634
Email: curran@proceedings.com
Web: www.proceedings.com

TABLE OF CONTENTS

(44bt) Performance Metrics Development Management and Reporting Reaching Sustainability by Design

Robert C. Smith, ABB, Inc., Houston, TX

Overview

The saying is, "You can't manage, what you can't measure". This presentation will discuss various performance management measurement frame-works; How to use organizational goals and objectives to develop appropriate performance measures; Design effective performance measures to implement your organization's goals and strategies; Link performance measures to a variety of management processes; Develop a performance management framework that drives results. Presentation goals

Create a clear understanding of setting and managing key performance metrics surrounding highly hazardous environments.

- Create a practical base knowledge of leading and lagging metrics and how to practically apply them into your environment.
- Learn about the most common errors people make when selecting metrics and you'll gain new ideas around how metrics can be visualized better.
- This session will give you an in-depth look at the often confusing and misunderstood world of performance metrics. You'll learn the truth about metrics, and how an effective performance measurement system can help you make better decisions.
- Recognize management's role in leading and implementing performance metrics.
- Have a foundation to build metrics and scorecards by using the tips, techniques, and tools acquired in the presentation.

Who should attend?

This presentation is intended for any executive, manager or employee with responsibility for implementing or maintaining score carding performance management systems, and is interested in learning more about how to establish and measure metrics. What you'll learn:

- Detailed knowledge of leading and lagging metrics
- Measuring program success
- Key points continuous improvement and sustainability of measures

- Understand the purpose of process management
- How to analyze an existing metric system
- How to create well designed metrics that support strategy
- How to steer away from some of the classic pitfalls of metric development

See more of this Session: Poster Session
See more of this Group/Topical: Global Congress on Process Safety

(44bu) Alarm Management - the Human Factors Challenge

Robert C. Smith, ABB, Inc., Houston, TX

Refinery alarm systems are supposed to support the operator, yet in several notable accidents they have proven to be ineffective. Part of the problem is that these systems are not being designed or managed with human needs in mind. Do alarm systems help the operator or are they a hindrance? This paper is a short review of how human factors contributed to several incidents including:

- July 1994 explosion at Texaco, Milford Haven, UK
- September 1998 explosion at Esso, Longford, Australia
- March 2005 explosion at BP, Texas City, US.

Human operators are needed to ensure plants and processes continue to operate safely. The control room and control system interfaces must therefore be built with human needs in mind. Includes a review of how implementation of ISA 18.2 and/or EEMUA 191 can improve alarm management practice.

See more of this Session: Poster Session
See more of this Group/Topical: Global Congress on Process Safety

Dust Flame Propagation in Industrial Scale Piping
Part 2: CFD Study of a Conveying Vessel-Pipeline System

Jim Vingerhoets
Fike Europe
Toekomstlaan 52, 2200 Herentals, Belgium
jim.vingerhoets@fike.com

Thomas M. Farrell
Fike Corporation, Industrial Protection Group
704 SW 10th Street, Blue Springs, MO 64015, USA
tom.farrell@fike.com

Jef Snoeys
Fike Europe
Toekomstlaan 52, 2200 Herentals, Belgium
jef.snoeys@fike.com

Prepared for Presentation at
American Institute of Chemical Engineers
2013 Spring Meeting
9th Global Congress on Process Safety
San Antonio, Texas
April 28 – May 1, 2013

UNPUBLISHED

Dust Flame Propagation in Industrial Scale Piping
Part 2: CFD Study of a Conveying Vessel-Pipeline System

Jim Vingerhoets
Fike Europe
Toekomstlaan 52, 2200 Herentals, Belgium
jim.vingerhoets@fike.com

Thomas M. Farrell
Fike Corporation Industrial Protection Group
704 SW 10th Street, Blue Springs, MO 64015, USA
tom.farrell@fike.com

Jef Snoeys
Fike Europe
Toekomstlaan 52, 2200 Herentals, Belgium
jef.snoeys@fike.com

Keywords: Dust explosion, flame propagation, vented vessel, pipelines, CFD

Abstract

The use of pipes to connect vessels and to transport hazardous materials, such as flammable gases or dusts, is a common practice in the process industry. The potential of fires and explosions has been recognized for decades and is addressed in standards such as NFPA 68 [1] and NFPA 69 [2]. Unlike for gases, the experimental evidence on dust flame propagation is scarce, and this study is particularly aimed at expanding the knowledge in this field. The purposes of this paper are to present the results of a parametric computational fluid dynamics (CFD) study validated by experimental evidence and to further develop a detailed understanding of combustion propagation in industrial scale piping.

The results of an ongoing research program which involves a piping system connected to a 5 m3 vented vessel are presented in the accompanying paper [3]. CFD was used to broaden the field of application beyond the tested configurations, and to study the influence of vessel volume and pressure, fuel characteristics and flow conditions.

Upstream and downstream explosion propagation behavior is studied as a function of fuel reactivity and type, vessel fuel concentration, pipeline fuel concentration, flow velocity, vessel volume, vent area, vent location, ignition location, pipe diameter and length.

1. Introduction

The dust explosion hazard continues to represent a constant threat to process industries that manufacture, use and/or handle powders and dust of combustible materials. In the process industry, the majority of process equipment is connected via pipes or ducts, and accidents have shown that the propagation of the explosion through the connection pipe work worsens the consequences of the explosion. Understanding the mechanisms of explosion flame propagation is a prerequisite for designing effective explosion protection solutions.

The problem of dust explosions in large-scale connected equipment has been much less investigated than isolated equipment, mainly due to cost and complexity to conduct experiments. The information in the public domain on dust explosion propagation is scarce; refer to the bibliography section at the end of the paper for relevant contributions. An alternative to large scale testing is the use of mathematical modeling and simulating the problem using CFD - computational fluid dynamics. CFD uses numerical methods and algorithms to simulate the combustion phenomena defined by boundary conditions. Although the use of CFD in gas and dust explosion applications is relatively new, this technique is irrefutably gaining interest.

The present research investigates, by experiment and CFD, flame propagation from a vented vessel through connected pipes. The aim is to examine the mechanisms that govern the velocity and acceleration of flame through pipelines, thus providing valuable information for the design of explosion protection solutions and systems. Once sufficient correlation between experimental and numerical results was demonstrated, the scope was extended with a numerical sensitivity study to quantify the influence of several application parameters on the flame propagation behavior. These parameters include:

- fuel type and reactivity,
- fuel concentration in vessel and pipelines,
- flow velocity and direction,
- vessel volume, pipe diameter and length,
- vent area, location,
- ignition location.

2. Simulation code

The numerical explosion simulations have been performed using DESC - Dust Explosion Simulation Code. DESC resulted from a research project supported by the European Commission under the Fifth Framework Program and was reviewed by Skjold in 2007 [4]. The main purpose of the project was to develop a simulation tool based on computational fluid dynamics (CFD) that could predict the potential consequences of industrial dust explosions in complex geometries. DESC is presently based on the existing Flame Acceleration Simulator (FLACS) code for gas explosions. The dust particles are modeled as gaseous components that follow the flow of the other gaseous components in the system. The combustion model used in FLACS and adopted by DESC is the so-called β flame model developed by Arntzen [5]. The turbulent burning velocity used by the β model is a function of the root mean square of the turbulent velocity fluctuations and the Karlovitz stretch factor. More specifically, the turbulent burning velocity is formulated as:

$$S_T = 15.1 S_L^{0.784} u'^{0.412}_{rms} l_I^{0.196}$$

with S_T the turbulent burning velocity, S_L the laminar burning velocity, u'_{rms} the root mean square of the turbulent velocity fluctuation, also called the turbulence intensity and l_I the turbulence integral length scale. The inputs for the combustion model are laminar burning velocities extracted from the measured K_{st} and P_{max} values in standardized 20 L sphere tests. The code has been validated against large scale tests reported in the public domain. It was found that DESC is sensitive to the numerical grid resolution, the presence of bends and interconnected vessels in general. The grid resolution in the current research has been chosen based on recommendations from this validation, and bends were excluded from the study.

3. Experimental setup

A test rig was built at the test site of Fike Corporation consisting of an inlet pipeline, a vessel, and an outlet pipeline, arranged as shown in the following image. The inlet and outlet pipelines were oriented 90° to each other, with the outlet pipeline aligned with the major axis of the vessel. All explosion venting was constrained to a single panel on the vessel, on the wall opposite the outlet pipeline interconnection. The system was equipped with a fan, enabling adjustable pipe flow velocities. By using separate constant mass flow rate dust injectors on the pipelines in addition to traditional dust injectors on the vessel, the vessel dust concentration and pipeline dust concentration could be varied independently of each other.

Figure 1: Photo of experimental setup

In order to cover several industry branches, different fuel types were used to conduct the research. The properties of the fuels used in this program are listed in Table 1.

Table 1: Closed vessel explosion characteristics of the dust fuels used

Dust type	Dust concentration (g/m^3)	20 L sphere	
		K_{st} (bar.m/s)	P_{max} (barg)
Wood flour	250	50	5.7
	500	108	7.0
	750	109	7.2
Phenolic dust	250	203	7.3
	500	217	6.9
	750	200	6.6
Corn Starch	250	93	6.6
	500	199	8.7
	750	205	8.6

Many tests were conducted to determine explosion pressures and flame arrival times as a function of venting area, vessel fuel concentration, pipeline fuel concentration and conveyance velocity. More information can be found in the accompanying paper [3].

4. Simulation setup

Figure 2 shows the basic setup of the explosion simulations. It represents as much as possible the experiment setup. The vessel itself consists of two identical cubes of 2.55 m^3 each, attached to each other without a partitioning wall in between. The cube assembly is equipped with dust jet injectors, 4 vent panels, a 10 kJ igniter and flanges to connect two pipelines. Each of the 4 vent panels can be closed off (to vary the total vent area) or opened at a controlled burst pressure. In some simulation scenarios, the location of the vent panels was moved to another wall of the vessel.

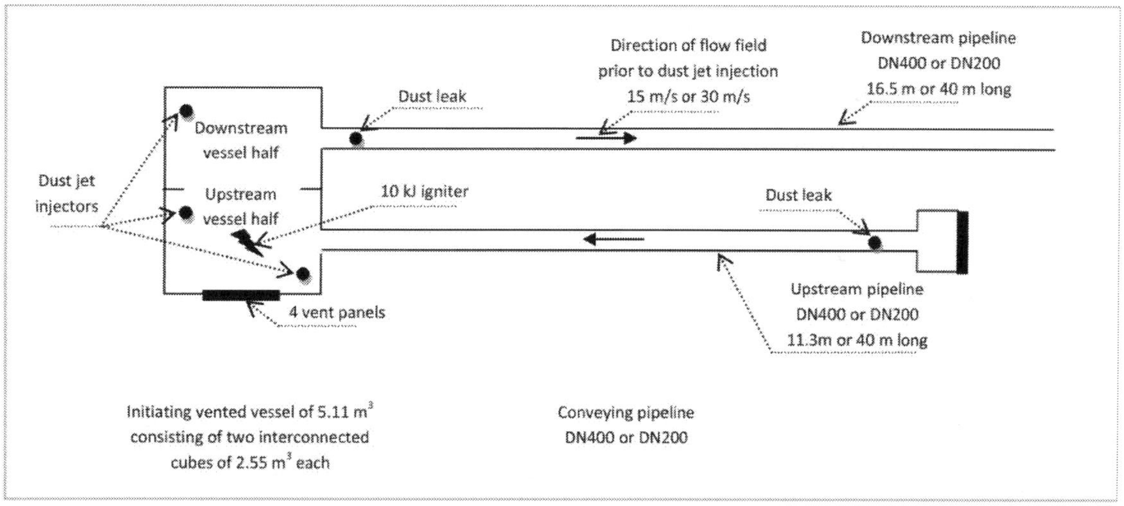

Figure 2: Sketch of basic setup of the explosion simulations.

Figure 3: Three dimensional CAD drawing picture of the simulation setup

The default location of the igniter is in the centre of the upstream vessel half, but in several simulation scenarios the igniter was moved to the centre of the downstream vessel half or to the interconnection with one of the pipelines. The delay time between start of dust jet injection and ignition was identical to the experimental program. The pipelines were either 400 mm or 200 mm in diameter and 11.3 m, 16.5 m or 40 m long. In order to restrict the simulation domain and associated computer processing time, the downstream pipeline was rotated 90° with respect to the experimental setup. Flow of 15 m/s or 30 m/s was established in the pipelines, depending on the simulation scenario. Dust leaks were programmed to fill the pipelines with a dust/air mixture of known concentration. Other simulations were done in a 62 m^3 vessel with proportionally scaled vent panels.

All simulation scenarios have been numerically solved on cubical grid cells with optimized size conforming to DESC guidelines.

5. Comparison between experiments and numerical simulations

Figure 4 to Figure 6 display comparative plots of vessel pressure and flame position as a function of time for three example scenarios including each of the three dusts of the program. Flames propagating towards the blower are represented by negative pipeline distances.

Table 2: Comparison between experiments and simulations giving the main characteristic values of the explosion propagation

Comparison of main characteristic values		Example 1		Example 2		Example 3	
		Exp	Sim	Exp	Sim	Exp	Sim
Vessel peak pressure		3.2 psig	2.0 psig	4.2 psig	6.7 psig	8.6 psig	2.3 psig
Time to reach 1,5 psig in vessel		64 ms	161 ms	46 ms	56 ms	54 ms	110 ms
Average flame speed in first 10 meters of pipe	Downstream	90 m/s	36 m/s	90 m/s	116 m/s	154 m/s	116 m/s
	Upstream	79 m/s	55 m/s	116 m/s	143 m/s	221 m/s	112 m/s
Time to reach 2,7 meters in the pipe	Downstream	110 ms	382 ms	114 ms	117 ms	88 ms	214 ms
	Upstream	127 ms	233 ms	106 ms	95 ms	85 ms	135 ms

The simulations under-predict the explosion peak pressure and pressure rise time in the vessel for two out of the three scenarios. The modeled explosion peak pressure for example 2 is higher (6.7 psig) than its experimental equivalent (4.2 psig), with similar pressure rise time (1.5 psig @ 56 ms and 1.5 psig @ 46 ms). Similar observations are made for the flame front pipeline propagation: the modeled explosions for example 1 and 3 propagate at lower speed (36 m/s downstream, 55 m/s upstream and 116 m/s downstream, 112 m/s upstream respectively), whereas the modeled dust explosion of example 2 propagates faster. Finally, the modeled explosions of example 1 and 3 enter the pipes later whereas the modeled dust explosion pipe entrance time of example 2 is similar to the experiment.

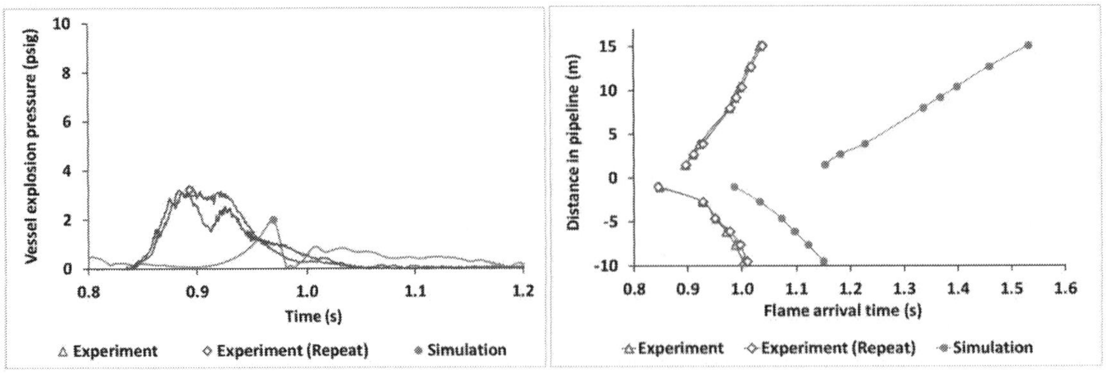

Figure 4: Comparison experiment and simulation: example 1

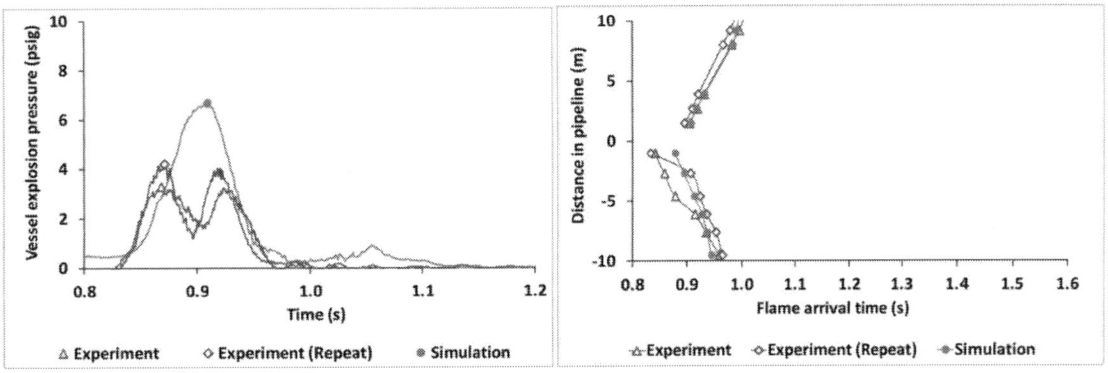

Figure 5: Comparison experimental and simulation: example 2

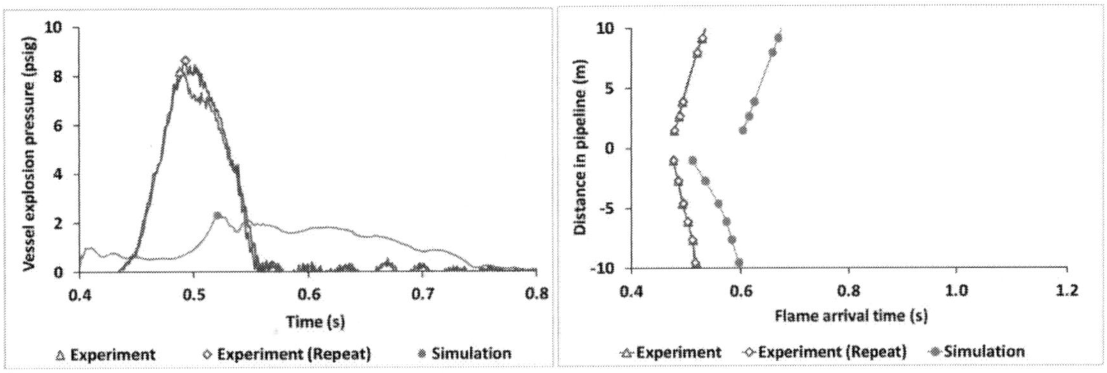

Figure 6: Comparison experimental and simulation: example 3

A likely explanation for the divergence between experimental and simulated results is that the modeled dust jet injectors generate different turbulence conditions than the injectors used in the experiments. This is a plausible explanation since the dust particles are modeled as gaseous phase fuel. Figure 7 shows the turbulence intensity in the centre of the upstream vessel half during simulated dust injection for different vessel dust concentrations and for a pipe flow of 15 m/s. The simulated dust jet injectors create an average turbulence intensity level between 2.0 and 3.4 m/s in the first 200 ms after ignition. 200 ms is the typical duration of an explosion in a 5 m^3 vessel. Although slightly higher than the reported turbulence in 20 L and 1 m^3 tests, the numbers are quite reasonable. Unfortunately, the turbulence could not be measured during the experiments.

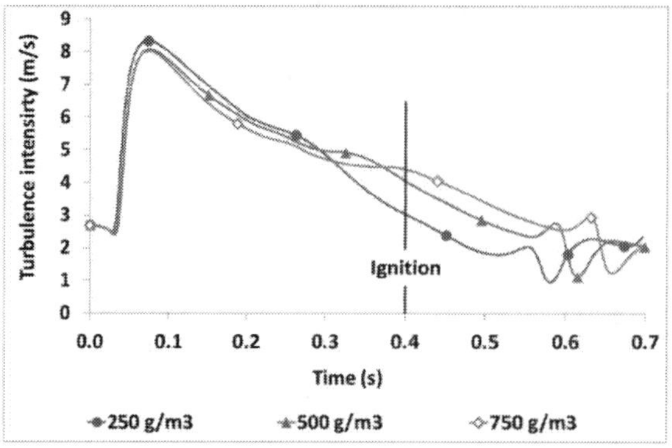

Figure 7: Turbulence intensity in the centre of the upstream vessel half during simulated dust injection for different vessel dust concentrations; initial pipe flow = 15 m/s

Another possible reason is that the 20 L flame data used by DESC do not scale up well to larger volumes. Several researchers postulated scaling problems related to turbulence and the powerful igniter used in the closed vessel 20 L and 1 m^3 standardized tests. This problem was investigated by additional explosion experiments in a closed 1m^3 vessel with varying ignition delay [3]: the K_{st} of two of the three dusts were more or less independent of volume, although strongly affected by ignition delay. The other dust showed an opposite behavior and is relatively insensitive to decreased ignition delay.

Figure 8 shows the effect of increasing the laminar burning velocities which was used as a surrogate for boosting the turbulence. The agreement becomes much better which indicates that the prediction potential of DESC is improved, if both the fundamental combustion parameters (the laminar burning velocity derived from K_{st} and P_{max}) and the initial process turbulence are estimated correctly.

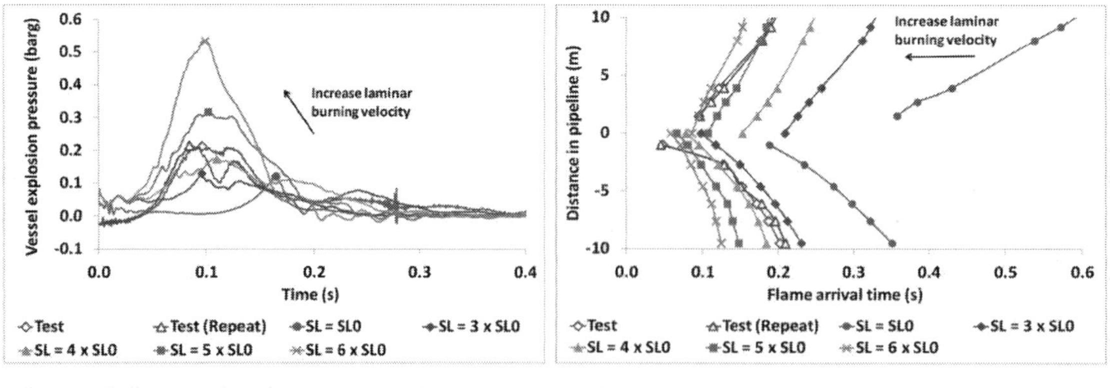

Figure 8: Sensitivity of vessel explosion pressure and flame propagation to the modeled laminar burning velocity and comparison with experiment (example 1)

In order to deal with the uncertainties in turbulence and combustion parameters for different vessel volumes and injector assemblies, simulations were performed in the reference test volume with all vent panels and connections to the pipelines closed off, and the K_{st} and P_{max} values were recorded. The results obtained by DESC should be referenced to K_{st} and P_{max} as obtained by DESC for the closed vessel. By following this approach, together with additional specific considerations as pointed out further, it is believed that simulations can represent industry reality with similar level of confidence as experiments.

6. Mechanisms of flame propagation

A thorough analysis of 94 explosion simulations for different fuel types, vessel and pipeline fuel concentrations, vent areas, flow velocities, significantly contributes to our understanding of the fundamental mechanisms which govern the propagation of dust flames through pipelines which are connected to a vented vessel. For all these scenarios the vessel and pipeline explosion pressures have been recorded as well as the flame arrival times at several distances along the pipelines. Besides pressure and flame, other physical variables, such as flow velocity and dust concentration have also been recorded. In addition to dust fuels listed in Table 1, also other dust fuels were used in the simulation study, so covering a wider range of K_{st} values, from 28 bar.m/s up to 281 bar.m/s.

In all scenarios the flame propagated with high speed up until the end of both pipelines, including the ejection of a large fireball, even against strong initial flows of 30 m/s. The highest recorded flame speed was 434 m/s. This was at the end of the downstream pipeline for the dust and concentration with highest reported K_{st} value and with the smallest venting area programmed. The lowest recorded flame speed was 34 m/s, for the dust and concentration with lowest K_{st} value and the largest vent area programmed.

It was confirmed that flame enters both pipes and propagates through a dust cloud that is pushed from the vessel into the pipes, as shown in Figure 9 below. In this scenario, the pipelines were not filled with dust fuel prior to ignition.

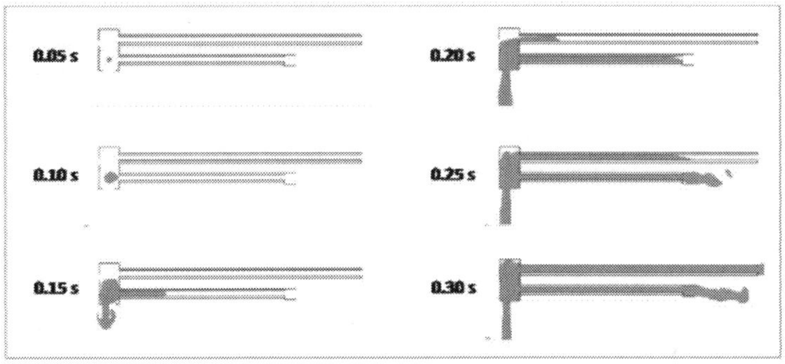

Figure 9: Flame propagation shown at subsequent time steps during an example explosion. No flow through pipe, no dust in pipe. 2 vent panels open. Vessel dust concentration = 500 g/m^3.

Figure 9 also illustrates that flame accelerates as it enters the pipe: It takes 0.1 seconds for the flame to reach the pipe and in only another 0.1 seconds it manages to travel the full pipeline distance.

In the vessel the freely expanding flame front propagates at a moderate speed of about 5 to 10 m/s. When the explosion flame enters the connected pipes, it will accelerate because the flame front will travel along with the combustion gases flowing at high speed out of the vessel:

$$S_{\mathrm{f,\,pipe}} = S_{\mathrm{g}} + S_{\mathrm{T}}.E$$

The pipe gas flow velocity induced by the explosion pressure in the vessel (S_g) is often a lot larger than the speed of turbulent burning and associated local expansion ($S_T.E$) and can reach up to 300 m/s. For low to moderate explosive reactive materials the vessel pressure induced gas flow will dominate the pipe flame speed.

Figure 10 shows the correlation between gas flow velocity in the pipelines and flame front velocity, for three different fuels and for three different dust concentrations. The more reactive the fuel and the larger the gas flow velocity, the more the flame velocity exceeds the gas velocity. This is because the turbulent burning velocity increases with fuel reactivity and with increased flow turbulence. The flow turbulence is higher at higher gas velocities.

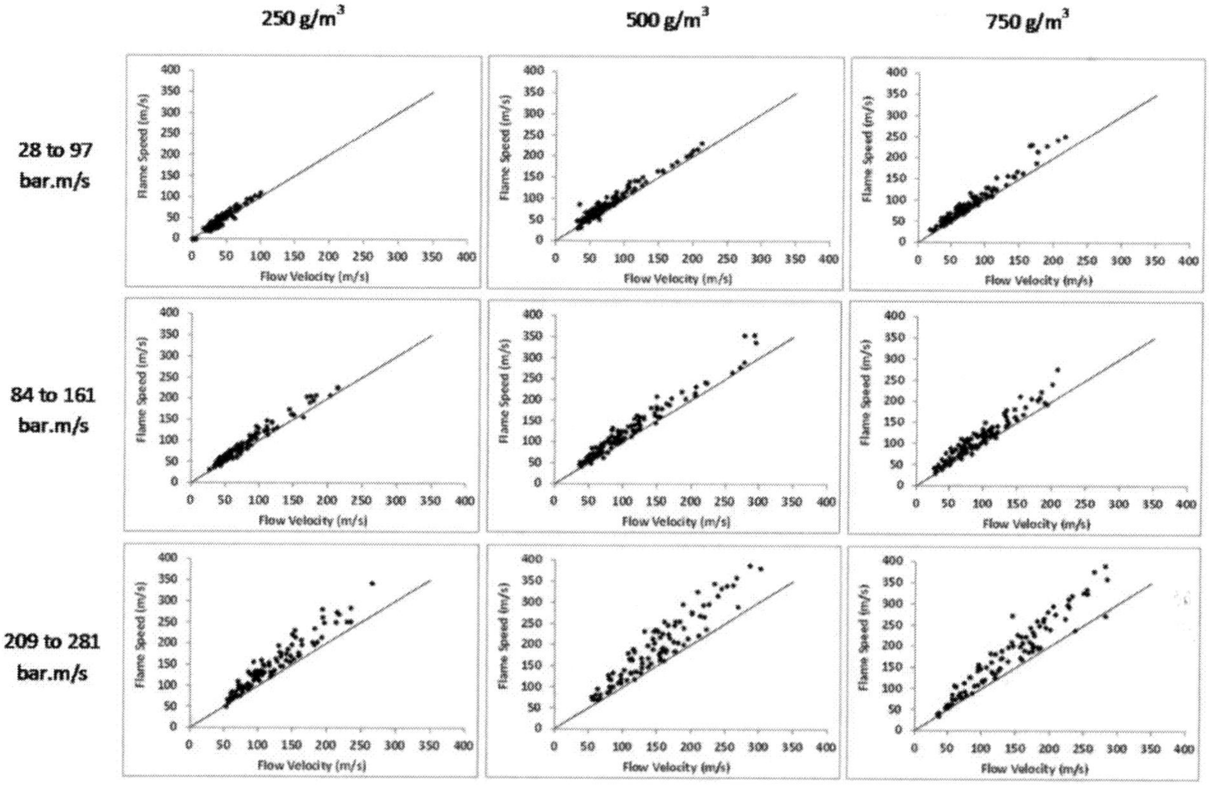

Figure 10: Correlation between flame speed and explosion induced gas velocity in the pipelines for different fuel types and dust concentrations

7. Sensitivity study

To enable extrapolation of the findings to real industry applications, 172 additional scenarios have been simulated in which the flow direction, vessel volume, vent area, vent location, ignition location, pipe diameter and length were varied.

7.1. Effect of vent area

In all cases, the flame speed decreases with larger vent areas. Figure 11 shows example simulation results for two dust types. This finding was consistently confirmed by experiments [3] and is attributed to lower vessel reduced pressures associated with larger venting areas. The decrease in flame speed is less pronounced in the upstream pipe, because the ignition was located closer to the upstream pipe, so that the flame could enter the upstream pipe before the vent panel opened.

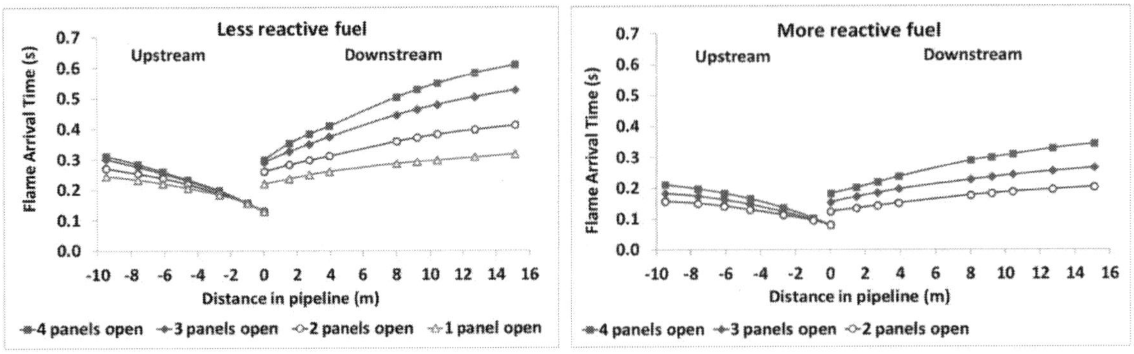

Figure 11: Effect of vent area on flame propagation. Flow = 15 m/s; Vessel and pipeline dust concentration = 750 g/m^3; Ignition location = centre of upstream vessel half.

7.2. Effect of pipeline dust concentration

Many different types of dust transport exist in the industry, including dust aspiration and dust extraction and lean to dense phase transport. Some dusts will stick to the pipe wall and build layers that may become airborne by the explosion propagating through the pipeline, resulting in combustible dust-air cloud generation in transport systems that typically operate at or below the minimum explosible concentration. It is therefore interesting to study the effect of pipeline dust concentration on the explosion propagation. Figure 12 shows examples for two dust types with the pipeline dust concentration changed from 500 g/m^3 to 50 g/m^3. The vessel dust concentration was kept constant at 500 g/m^3. Although the fuels are more reactive at the higher dust concentration, flames are slower at high pipeline dust concentrations. This effect is attributed to the change of the inertia of the fluids in the pipelines. Also this finding was confirmed by the experiments [3].

Figure 13 demonstrates that the combustion in the first meters of the pipeline is fueled by dust ejected from the exploding vessel and not so much by the pipeline dust.

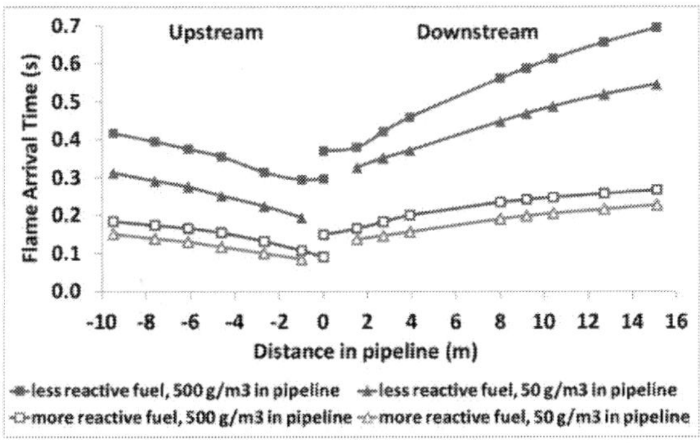

Figure 12: Effect of pipeline dust concentration on flame propagation for two dust fuels; Vessel dust concentration = 500 g/m^3.

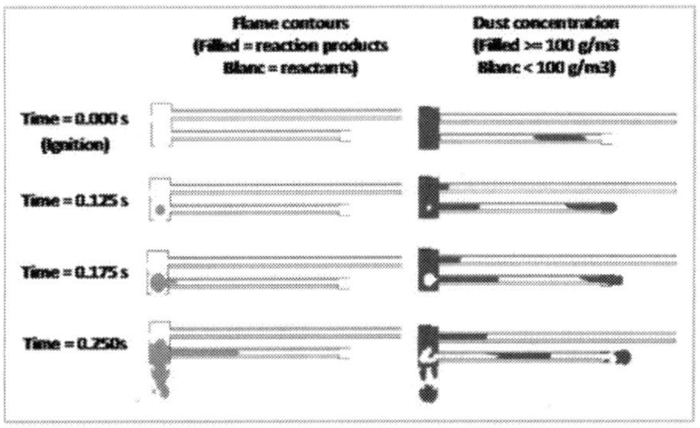

Figure 13: Illustration of how dust is ejected out of the initiating vessel before the flame front arrives for an example explosion; Vessel dust concentration = 500 g/m3; Pipeline dust concentration = 50 g/m^3.

7.3. Effect of ignition location

Scenarios testing the effect of ignition location were run with the igniters located in the three positions indicated in Figure 14 below, including ignition near the vent (1), at the interconnection with the upstream pipeline (2), and in the center of the downstream vessel half, far from the vent (3)

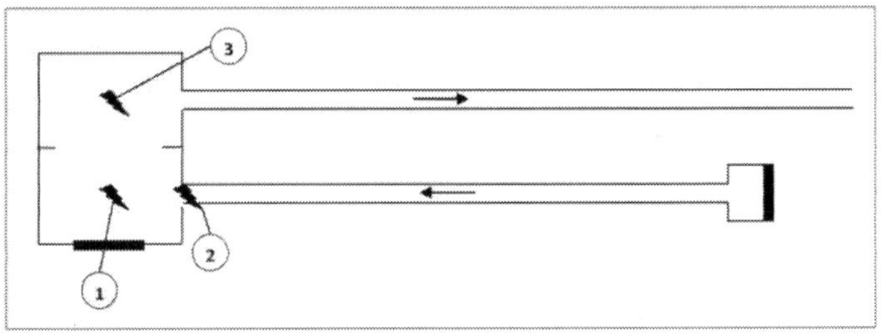

Figure 14: Ignition locations

The effect of the ignition location is twofold: the closer the ignition point is located towards the pipe entry, the sooner the explosion flame will enter the pipeline; and the further the ignition point is located from the pressure venting device, the higher will the explosion pressure in the vessel be, with faster flames as a result. These phenomena are illustrated in Figure 15, where for examples with two dust types, the flame arrival times are compared for three explosion scenarios with varying ignition location. It is clearly shown that flames enter the pipeline earlier for pipeline ignition (2), and that ignition far away from the vent panel (3) generates significantly faster flames

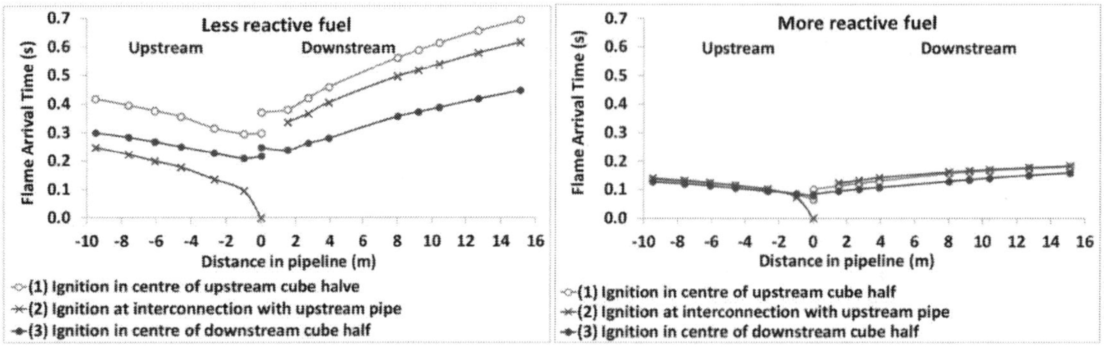

Figure 15: Effect of ignition location on flame propagation. Examples for two dust fuels.

7.4. Effect of vent location

Ideally, the explosion vent panel should be positioned as close as practically possible to the expected ignition location. This will provide an early escape of the combustion gases and will limit the resulting explosion overpressure in the vessel and consequently the flame speed in the pipeline. This is illustrated in Figure 16 and Figure 17: the further the vent panel is moved away from the ignition position (– with vent location alternative 3 further than alternative 2, and alternative 2 further than alternative 1), the faster the explosion flame propagates in both pipelines. The driving factor is the increased explosion intensity in the initiating vessel.

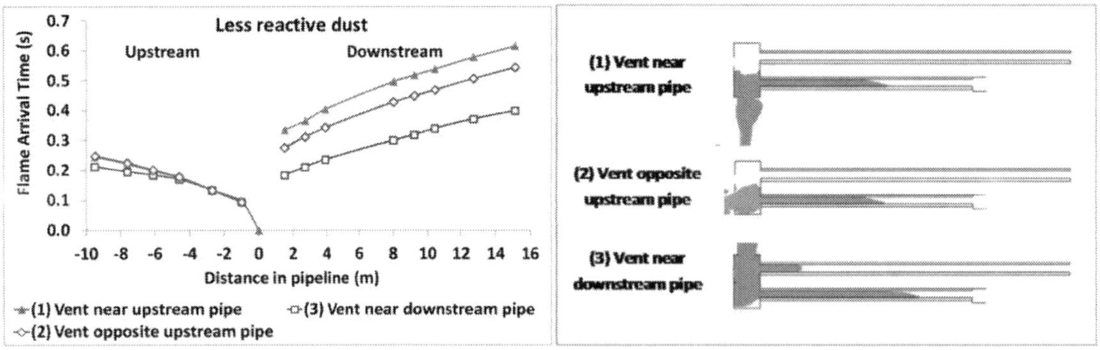

Figure 16: Effect of vent location on flame propagation for an example explosion scenario. Ignition location = interconnection of vessel with upstream pipeline. Left: flame arrival times. Right: flame contours at 0.2 s after ignition.

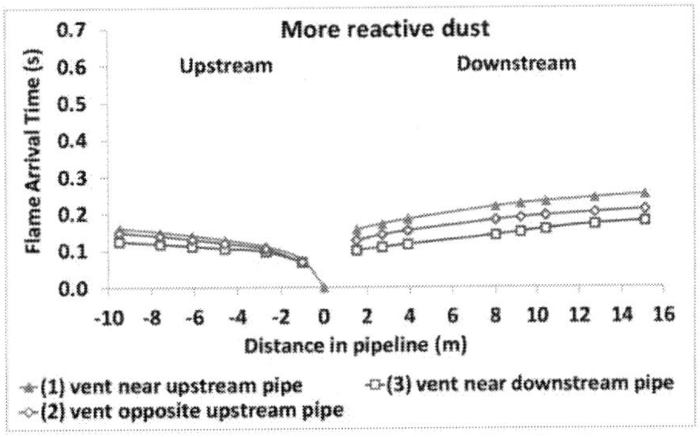

Figure 17: Effect of vent location on flame propagation for another example explosion scenario. Ignition location = interconnection of vessel with upstream pipeline.

7.5. Effect of pipe length

Longer pipes will increase the flow resistance and gas column inertia, resulting in a less aggressively accelerating flame along the pipeline. The effect is most pronounced for cases with low vessel explosion pressure, where flow resistance and gas column inertia play a relatively stronger role, and for cases with low K_{st} value, where the flame velocity is stronger linked with the gas flow velocity. Figure 18 shows the simulated flame arrival times for three example explosions, where the length of the piping was varied from the 11.3/16.5 m of the basic configuration to 40 meters.

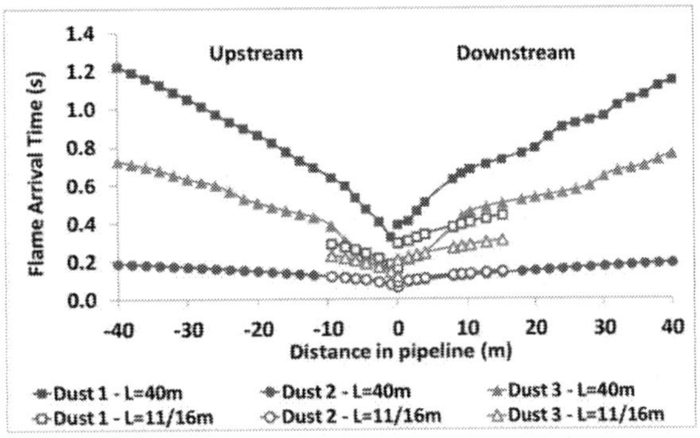

Figure 18: Effect of pipe length on flame propagation for three example explosions.

7.6. Effect of pipe diameter

As with the pipe length, changing the pipe diameter affects the flow resistance. But, in addition to this, the pipe diameter also changes the Reynolds number and thus the gas turbulence in the pipeline. Flames will accelerate faster in smaller pipes as a result of higher turbulence. Note that heat losses of the hot burning gases to the cold pipe wall material have not been modeled. Figure 19 shows the effect on flame propagation when the pipe diameter is decreased from 0.4 m to 0.2

m. For the example with lowest fuel reactivity and generating relatively low vessel explosion pressures, the flame propagates slower in the smaller diameter pipes than in the larger diameter pipes. For the example with highest fuel reactivity and generating relatively high vessel explosion pressures, the flame propagates faster in the smaller diameter pipes than in the larger diameter pipes. For the example with medium fuel reactivity, it is seen that in the initial phase of the explosion, the flame propagates slower in the smaller diameter pipe geometry, but later on, when the explosion pressure develops and the fluid turbulence enhances, the flame will propagate faster.

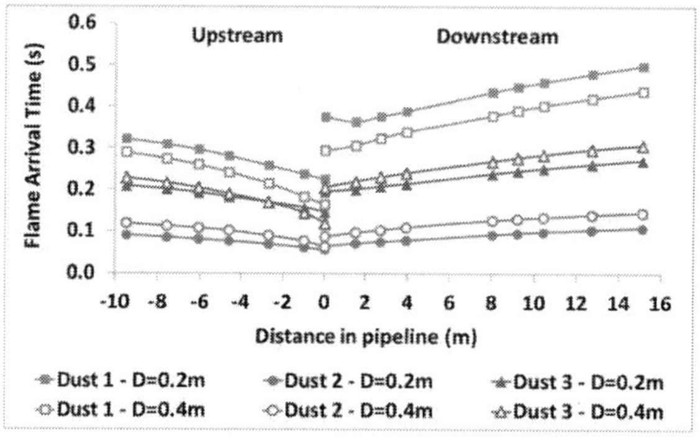

Figure 19: Effect of pipe diameter on flame propagation for three example explosions.

7.7. Effect of vessel volume

When the volume of the initiating vessel is increased, delayed pipeline flame propagation can be expected, because the explosion pressure rise in the vessel will be slower. This is clearly illustrated in Figure 20, where for three example explosions the flame propagation is compared for the cases when initiated in a 5 m^3 vessel and in a 62 m^3 vessel, all other parameters either kept identical or scaled with the volume. The ignition was located in the centre of the 5 m^3 upstream vessel half, which is 0.68 m away from the interconnection with the upstream pipeline, which is then of course out of the centre for the 62 m^3 vessel half. To make the comparison explosion pressure independent, the vent area on the 62 m^3 vessel was increased by factor of 4, which approaches the factor $\left(\frac{62}{5}\right)^{2/3} = 5.36$ which would be required theoretically. Initially the pressure rise is slower which leads to a delayed flame entry into the pipes, but as soon as the vessel pressures in the 62 m^3 exceed the ones recorded in the 5 m3, the flame acceleration rate for the 62 m^3 vessel cases becomes higher.

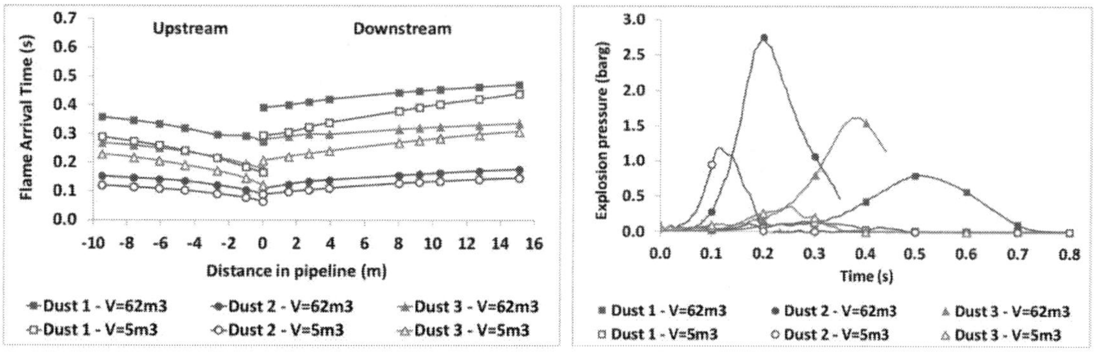

Figure 20: Effect of vessel volume on flame propagation for three example explosions. Left: flame arrival times in pipelines. Right: vessel explosion pressures.

8. Summary

DESC, a CFD code with an integrated combustion model, has been used to study the behavior of propagation of dust explosions through pipelines connected to vented vessels. A 5 m^3 vented vessel in which explosions were initiated, was connected to two DN400 pipelines, respectively 11.3 m and 16.5 meter long. The system was equipped with a fan to establish a flow field prior to dust injection and explosion ignition. Ignition always occurred in the vessel; neither the present study nor the accompanying empirical program [3] considered ignition in the pipeline. Input for the combustion model are laminar burning velocities extracted from K_{st} and P_{max} values determined by standardized 20 L tests.

Comparison between experimental and simulation results revealed that for two out of the three dust fuels, the experimental dust explosions were more intense than the simulated equivalents. A method was devised to improve the confidence level of DESC simulations. By following this approach, and by understanding the underlying physics, it is believed that simulations can represent industry reality with similar level of confidence as experiments.

A total of 94 explosion simulations have been performed in the basic configuration, where the fuel type, vessel and pipeline fuel concentration, vent area and flow velocity were varied. The effects of the parameter variations on flame arrival times in the pipelines were fully in line with the experimental results [3]. In all cases the flame propagated along the full length of the pipes, both upstream and downstream. A high degree of correlation was revealed between explosion induced expansion flow and flame speed.

172 additional simulations were performed in which flow direction, vessel volume, vent location, ignition location, pipe diameter and length were varied. This extended scope enabled to confirm the pronounced effect of the vessel reduced explosion pressure and pipeline geometry.

9. Conclusion

Understanding the mechanisms of explosion flame propagation is paramount for designing effective explosion protection solutions for interconnected process equipment that handle flammable substances. The present research has demonstrated that dust explosions do propagate through pipelines, even against strong flows and that many different application parameters besides K_{st} and P_{max} considerably influence the speed of dust flame propagation. When these parameters are not taken into account, the protection system may fail to protect against the explosion hazard. It has been irrefutably demonstrated that controlling the explosion pressure in the initiating vessel is very effective in controlling the speed of flame propagation. The explosion protection design should therefore always consider the explosion protection problem in its totality: a combination of vessel explosion pressure protection and pipeline explosion propagation protection.

10. References

[1] NFPA 68, Standard on Explosion Protection by Deflagration Venting, 2007

[2] NFPA 69, Standard on Explosion Prevention Systems, 2008

[3] T.M. Farrell, J. Vingerhoets, J. Snoeys, J.E. Going, "Dust Flame Propagation in Industrial Scale Piping - Part 1: Experimental Study in a Conveying Vessel-Pipeline System", Submitted for 9th Global Congress on Process Safety, San Antonio, April 28 - May 1, 2013

[4] T. Skjold, "Review of the DESC project", *Journal of Loss Prevention in the Process Industries*, Volume 20, p291-302, 2007

[5] Arntzen, B.J., "Modeling of turbulence and combustion for simulation of gas explosions in complex geometries", Dr. Ing Thesis, NTNU, Trondheim, Norway, 1998

Additional references not cited:

[6] A.E. Dahoe, J.F. Zevenbergen, S.M. Lemkowitz, B. Scarlett, "Dust explosions in spherical vessels: the role of flame thickness in the validity of the cube-root law", *Journal of Loss Prevention in the Process Industries*, Volume 9, p33-44, 1996

[7] A.E. Dahoe, R.S. Cant. and B. Scarlett, "On the decay of turbulence in the 20-litre explosion sphere", *Turbulence and combustion*, Volume 67, p159-184, 2001

[8] A.E. Dahoe, K. van der Nat, M. Braithwaite and B. Scarlett, "On the sensitivity of the maximum explosion pressure of a dust deflagration to turbulence", *KONA*, Volume 19, p178-195, 2001

[9] G.A. Lunn, "Dust explosions in totally enclosed interconnected vessel systems", *Journal of Loss Prevention in the Process Industries*, Volume 9, p45-58, 1996

[10] P. Holbrow, S. Andrews, G.A. Lunn, "Dust explosions in interconnected vented vessels", *Journal of Loss Prevention in the Process Industries*, Volume 9, p91-103, 1996

[11] M. Roser, A. Vogl, S. Radandt, W. Malalasekara, R. Parkin, "Investigations of flame front propagation between interconnected process vessels. Development of a new flame front propagation time prediction model", *Journal of Loss Prevention in the Process Industries*, Volume 12, p421-436, 1999

[12] A. Vogl, "Explosionsübertragung aus Behältern in Rohrleitungen", *VDI Berichte*, Volume 1272, p215-236, 1996

[13] J. P. Pineau J.P., G. Ronchail, "Propagation of dust explosions in ducts", Symposium on the control and prevention of dust explosions, Bâle, 1982

[14] A. Vogl, S. Radandt, "Explosionsübertragung durch dünne Rohrleitungen", FSA Report F-05-9903, 2005

[15] U. Bielert, M.Sichel, "Numerische Simulation von Staubexplosionen in pneumatischen Saug-Flug-Förderanlagen", *VDI Berichte*, Volume 1601, p449-472, 2001

[16] W. Jian, L. Xinguang, Z. Shengjun, S. Randandt, W. Fuli, R. Chunli, "Flame propagation through potato starch/air mixtures in pipe of interconnected vessels", *Measurement and Control of Granular Materials*, Volume 8, p376-379, 2009

(44c) The Use of CFD to Evaluate the Interactions Between Multiple-Leak Sources and to Assess Effectiveness of Integral Modeling Techniques in a Multiple-Leak Scenario

Joshua Richardson, Ha Nguyen and M. Sam Mannan, Mary Kay O'Connor Process Safety Center - Texas A&M University, College Station, TX

In many practical applications within the chemical industry, it is of vital importance to know the concentration profile that may occur from the dispersion of a material. For example, effect footprints for the release of a toxic material are needed to properly assess risk to both a facility and its surrounding communities and are useful for emergency response planning. Likewise, knowing the concentration gradient of a flammable gas can affect, among many other things, the layout of a facility to prevent or lessen the consequence of explosion and fire scenarios or for the placement of detection systems.

There are two main tools for the modeling of dispersion scenarios. Computational fluid dynamics (CFD) is a useful tool for accurately assessing the dispersion of materials, whether toxic, flammable, pollutant, or otherwise. It has found extensive use in academic as well as industrial settings when scenarios and geometries are well-defined because it takes into account the surroundings in the release environment. However, it is relatively slow and computationally expensive to use.

Also used extensively in the chemical industry are integral consequence modeling tools (such as Phast) which are generally more user-friendly and have the advantage of delivering results in a more time-efficient manner, but cannot take into account complex geometry considerations. A particularly interesting problem to consider in practice is that of multiple simultaneous release sources. This could be caused by a common failure of two units causing concurrent release, a domino effect where an event occurs that causes another event leading to multiple releases, or even an intentional event such as a terrorist attack. While CFD has the capability to resolve the possible interactions between the multiple release sources, integral modeling does not.

A simple current approach to the problem of multiple-source scenarios is to model them independently and superimpose the results upon one another. This may be a good approximation in some cases, but it does not take into account the complex interactions that may take place between the sources. The practical question then becomes whether it is acceptable in certain cases to model a certain set of simultaneous releases as an independent set of sources or whether it could be possible to model them as a single combined source.

This study aims to utilize CFD modeling of a set of varying scenarios to classify multiple-release dispersion scenarios into one of three different categories:

- Releases that are independent and may be modeled as such
- Releases that may be combined and modeled as one source
- Releases that must be considered together due to inherent interactions between them

Based upon criteria such as material, distance between sources, dispersion rate and duration, weather conditions, and geometry considerations, a general methodology for classifying multiple-release scenarios into these three different categories will be proposed so that these releases can ultimately be modeled in the appropriate manner.

See more of this Session: Poster Session
See more of this Group/Topical: Global Congress on Process Safety

(44ca) Understanding, Managing, and Communicating High Consequence Risk

Charles A. Soczek, DuPont Sustainable Solutions, Wilmington, DE

This presentation discusses the relationship between the tools used to identify, understand, mitigate and communicate high consequence risks. The understanding and reporting of high consequence risks, along with providing management, senior leadership, and the Board of Directors with the information necessary to ensure successful risk management, are fundamental outputs to a risk management process. Tools such as the Risk Matrix, Bow Tie Analysis, and Risk Register will be discussed along with key issues such as risk mitigation and risk information flow from the operating level to senior leadership to the Board of Directors. Fundamental to this discussion is the concept that all hazardous events are managed to an acceptable risk through engineering and administrative controls, and a robust Process Safety Management system.

See more of this Session: Poster Session
See more of this Group/Topical: Global Congress on Process Safety

Frequencies of Accidents as Functions of Design Variables

Manuscript ID: 295029

Esteban Bernechea
Centre for Technological Risk Studies (CERTEC). Department of Chemical Engineering.
ETSEIB. Universitat Politècnica de Catalunya.
Av. Diagonal, 647. 08028-Barcelona (Spain).
esteban.bernechea@estudiant.upc.edu

Cristian Massaguer
Department of Chemical Engineering. ETSEIB. Universitat Politècnica de Catalunya.
Av. Diagonal, 647. 08028-Barcelona (Spain).
cristian.massaguer@gmail.com

Josep Arnaldos
Centre for Technological Risk Studies (CERTEC). Department of Chemical Engineering.
ETSEIB. Universitat Politècnica de Catalunya.
Av. Diagonal, 647. 08028-Barcelona (Spain).
josep.arnaldos@upc.edu

1. Introduction

The mainstream definition of Risk is that it is a measure of the damage caused on humans, the environment or material property, in terms of the probability of occurrence of an accident and the magnitude of the damage it causes. This can be expressed mathematically as follows:

$$Risk = Frequency \cdot Consequences \quad \text{[Eq. 1]}$$

Risk is then, a function of the frequency of occurrence and consequences of a possible accident.

Often, a Quantitative Risk Assessment (QRA) of a specific equipment or installation has to be performed; in Europe the Seveso Directive makes it compulsory for installations that work with a certain amount of dangerous materials. In order to perform a QRA it is necessary to define initial Loss of Containment Events (LOCs) in the studied equipment, to later estimate the frequencies of the accidents that may result from these releases, and the consequences of said events. These values can finally be combined in order to obtain a measure of the risk that the installation poses over individuals or communities.

There is a wide variety of techniques that can be applied to estimate the frequency of an accident, like historical analysis, failure trees or event trees, subjective estimation, Bayesian analysis or Monte Carlo simulations. In Catalonia and Spain it is common to use the values of frequency provided by "The Purple Book" (CPR18E) [1] for the possible LOCs that can occur in different equipment. Later, event trees which depend on the probabilities of occurrence of certain events (like the immediate or delayed ignition of a flammable release) are used to obtain the final frequencies of the set of accidents that will be studied in the QRA.

However, the base frequencies presented in The Purple Book do not explicitly include the possibility that the LOCs will occur due to a domino effect; this phenomenon is defined as a cascade of events in which the consequences of previous accidents are increased both spatially and temporally by following ones, thus leading to a major accident [2]. The complexity of domino effect makes it difficult to model, and more so to include in the calculations performed in a QRA. This leads to the problem of how to take into account a phenomenon that can have a significant impact on the risk associated to the installation, but that is virtually impossible to include (without significantly increasing the number of scenarios to be considered). In some cases in which there are various similar equipment, the common criteria is to select a representative scenario, and multiply its frequency by the number of units and a factor of 2. In this way, domino effect may be included in the calculations, but not in a rigorous way, which could lead to the under or over-estimation of risk.

In this work, a methodology proposed by the authors to produce accident sequences in a domino effect has been exploited to obtain equations that describe frequencies of accidents in a specific installation. The studied plant stores pressurized liquefied propane in horizontal cylinders. The equations will depend on the number of units and the mass that is stored in the installation; they will also depend on the layout, which in turn depends on the size of the tanks, and therefore on the mass and number of equipment. Distances between tanks were defined by using the NFPA58 standard [3].

By finding equations that can be used to estimate the frequency with which a certain accident will occur in the installation, the time required to include domino effect into a QRA could be decreased importantly; only the frequency obtained through the equations, applied to a representative scenario would be necessary to take the phenomenon into account.

The main objective of this work is to use a mathematical model to find equations that relate the frequency of accidents in an installation to some of its design variables. This could be useful not only at the moment of performing QRA, but also at the moment of designing the installation; also, it can help understanding how the frequency of the diverse accidents is affected by the design of the plant.

In the next sections, the model that allows developing the accident sequences is briefly explained, after this, the design of the experiment, which is how the mentioned model was exploited, is presented; later, the results obtained are shown, and finally, some conclusions are reached.

2. Model for the development of accident sequences

The model used to develop the domino effect sequences [4] is based on the use of event trees in a sequential way; it evaluates the damage suffered by one tank in the case that an accident occurs in another, and establishes the possible accidents that may occur afterwards. The LOCs and event trees used to develop the sequences are those presented in the Purple Book [1], and are shown in Table 1 and Figures 1 and 2.

Table 1: LOCs used to define initial accidents

Code	Definition
G.1	Instantaneous release of the complete inventory.
G.2	Continuous release of the complete inventory in 10 min at a constant rate of release.
G.3	Continuous release from a hole with an effective diameter of 10 mm.

The first step to produce the accident sequences is to select one of the units, choose an initial LOC and define the possible accidents that may occur. After this, one of the accidents must be selected, and its effects must be calculated on all surrounding units and compared against the threshold values presented in Table 2.

Depending on the effects of the previous accident each of the affected tanks will suffer a specific LOC, which will in turn, generate new possible accidents; the sequence will continue to unfold until all the tanks have suffered catastrophic events, or the frequencies of the accidents become lower than 10^{-9} y^{-1}. This procedure has to be repeated for each of the possible accidents and for every one of the initial LOCs, starting the sequence in each of the tanks. Finally, the frequencies of all the occurrences of a specific accident in all the sequences, (for example, a flash fire derived from an instantaneous release), has to be summed, in order to know the number of occurrences of the incident.

The frequencies of the accidents down the sequence are calculated using the event trees and probabilities of accidents; each time an accident occurs caused by a previous incident, its initial frequency will be that of the event that caused its failure, multiplied by its probability of occurrence.

Figure 1. Event Tree for the Instantaneous Release of a Pressurized Flammable Gas

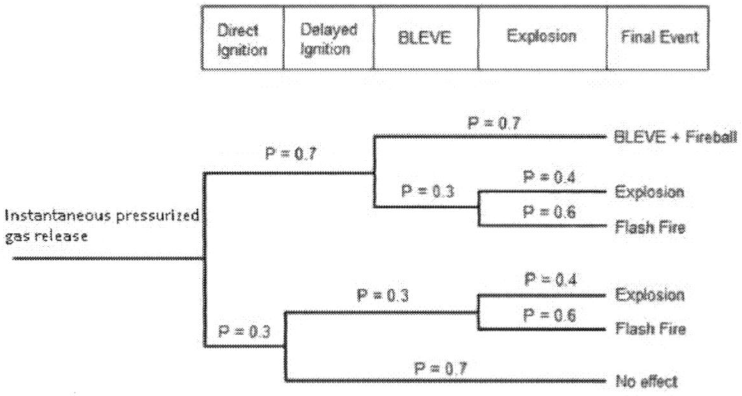

Figure 2. Event Tree for the Continuous Release of a Pressurized Flammable Gas

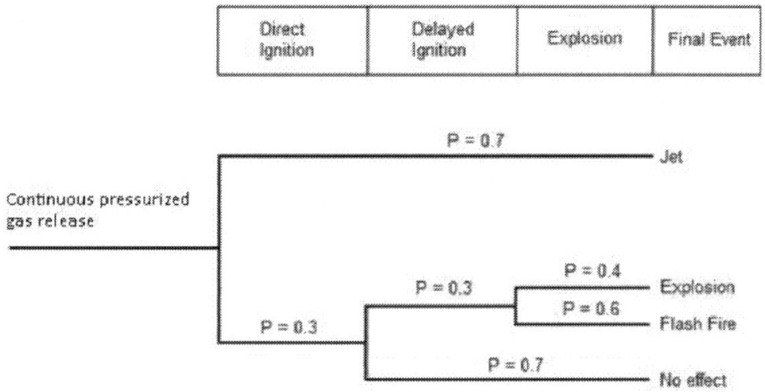

3. Experimental design

The equations that describe the frequencies of the different accidents will be developed for storage terminals of pressurized liquefied propane that use horizontal cylinders. The equations will be functions of the number of tanks used and the total mass of propane stored in the terminal. The number of tanks will be varied from 1 to 4, while 5 values of stored mass will be used: 5,000 kg, 10,000 kg, 20,000 kg, 50,000 kg and 100,000 kg; these possibilities for number of equipment and stored mass will be used to develop different layouts (for example 3 tanks which store 50,000 kg of propane) in which the total mass stored will be equally divided amongst all the tanks (all of which will have the same dimensions). The propane will be said to be stored at the conditions presented in Table 3.

Table 2: Threshold values for tank damage and associated LOCs

Accident	Effect	Threshold Values		Consequence
		Pressurized Tanks	Atmospheric Tanks	
BLEVE	Thermal radiation (kW/m2)	≥ 100	≥ 100 [5]	G.3 type release
		< 100	< 100 [5]	No consequence
Explosion	Overpressure wave (kPa)	≥ 100	≥ 25 [6]	G.1 type release
		$53 < \Delta P < 100$ [7]	$20 < \Delta P < 25$ [7]	G.2 type release
		$30 \leq \Delta P < 53$ [7]	$14 \leq \Delta P < 20$ [7]	G.3 type release
		$\Delta P < 30$ [5]	$\Delta P < 14$ [8]	No consequence
Jet fire	Jet length (m) Heat Radiation (kW/m2)	$dj \geq d$ [5]	$dj \geq d$ [5]	G.1 type release
		$dj < d$ & $I \geq 40$ [5]	$dj < d$ & $I \geq 10$ [5]	G.3 type release
		$dj < d$ & $I < 40$ [5]	$dj < d$ & $I < 10$ [5]	No consequence
Flash fire	-	-	-	No consequence
Pool fire	Thermal radiation (kW/m2) Pool duration (min)	$dp \geq d$ [5]	$dp \geq d$ [5]	G.1 type release
		$60 \leq I$ & $tp \geq 10$ & $dj < d$ [9]	$15 \leq I$ & $tp \geq 10$ & $dj < d$ [9]	G.2 type release
		$I \geq 8$ & $dj < d$	$I \geq 8$ & $dj < d$	G.3 type release
		$I < 8$ & $dj < d$	$I < 8$ & $dj < d$	No consequence

dj = jet length; dp = pool diameter; d = distance between tanks; tj = duration of jet fire; tp = duration of pool fire

As the frequency of the accidents will depend, not only on the number of tanks and the mass stored, but also on the distances between the units, this parameter has also been taken into account; this was done by using the distances proposed in the NFPA58 [3], which depend on the dimensions of the tank.

Figure 3 shows an example of the used layouts. In it, L is the length of the cylinder, while ϕ stands for the diameter; B represents the sum of the diameters of the tanks and the distances between them (d). The complete width of the facility will come from multiplying B by a factor of 1.1, and its length, from multiplying L by an equal factor.

Table 3: Atmospheric and Storage Conditions Used for the Application of the Model

Parameter	Value
Ambient Temperature (K)	298.15
Relative Humidity (%)	70
Atmospheric Stability Class	D
Wind velocity (m/s)	5
Ground Roughness Coefficient (cm)	10
Tank Pressure (bar)	9.51
Tank Temperature (K)	298.15

Since the substance stored in the facility is a liquefied pressurized flammable gas, the LOCs presented in Table 1, as defined in The Purple Book are used. The base frequencies of the different accidents for each tank are obtained by using the event trees presented in Figures 1 and 2 for the instantaneous and continuous releases.

Figure 3. Layout example

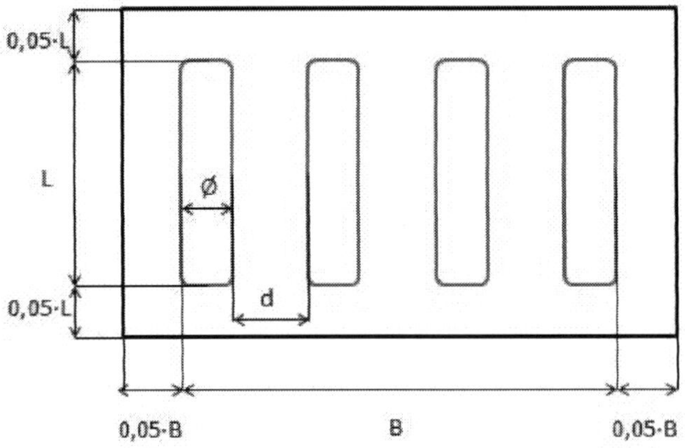

4. Results

It was necessary to use the model a total of 180 times to develop accident sequences for every possible initial accident, combining the number of tanks and values of stored mass. However, since all the proposed layouts are symmetrical, it was only necessary to perform 90 simulations.

Figure 4. G.1 accident sequences for two tanks storing 50,000 kg each

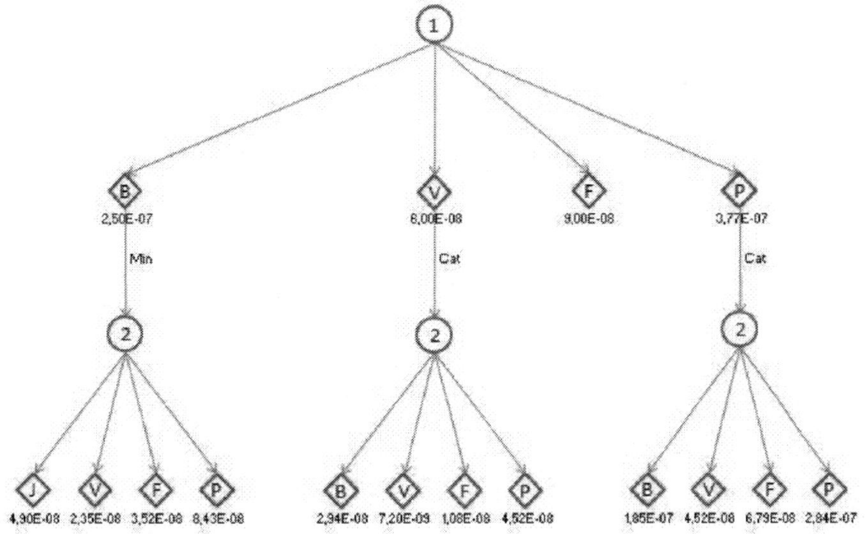

Figure 4 presents the possible accident sequences obtained by applying the model for the case of an instantaneous release from one of two tanks, each storing 50,000 kg of propane. In it, it can be seen that four possible accidents can occur after the G.1 release, a BLEVE (B), a VCE (V), a flash fire (F) or a pool fire (P); the frequencies with which these accidents occur is presented, as well as the level of damage they cause on the second tank, and the possible accidents that may occur in it, due to the initial ones.

The final values of frequency obtained for the case of BLEVE, depending on the mass and number of tanks are presented in Table 4. They serve as examples of the complete results obtained when the model is applied.

After the final frequencies for all possible combinations of design variables have been obtained, they can be plotted in 3D graphs, (in which one of the axis represents the number of tanks, another, the stored mass, and the last one, the frequency of occurrence). The plots obtained for all the studied accidents are presented in Figures 5 through 7.

Table 4: Final frequencies for the BLEVE case

Mass (kg)	Number of Tanks			
	1	2	3	4
5,000	$1.75x10^{-7}$	$1.90x10^{-6}$	$5.77x10^{-6}$	$1.04x10^{-5}$
10,000	$1.75x10^{-7}$	$1.90x10^{-6}$	$4.12x10^{-6}$	$6.89x10^{-6}$
20,000	$2.45x10^{-7}$	$3.44x10^{-6}$	$5.13x10^{-6}$	$6.87x10^{-6}$
50,000	$2.45x10^{-7}$	$3.44x10^{-6}$	$9.95x10^{-6}$	$2.21x10^{-5}$
100,000	$2.45x10^{-7}$	$3.44x10^{-6}$	$1.18x10^{-5}$	$2.10x10^{-5}$

Figure 5. Frequency vs. Number of tanks and Stored Mass for G.1 accidents

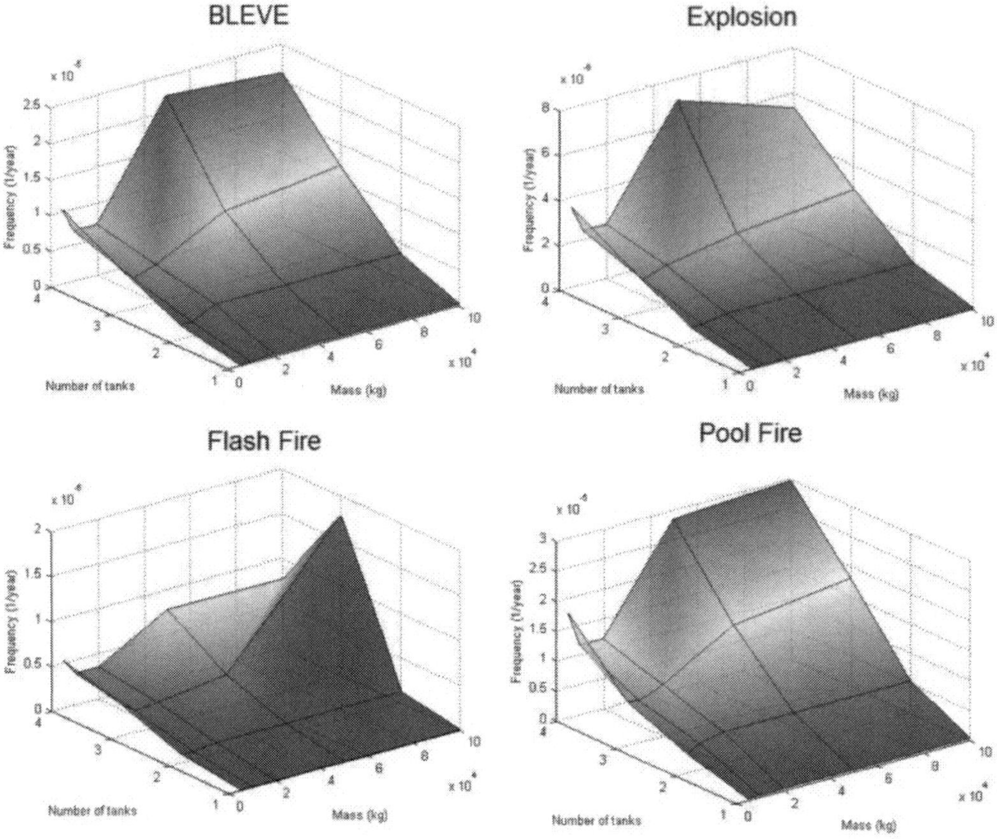

After the plots were obtained, they were fitted using different types of equations, like power, exponential or logarithmic among others. Finally, a Gaussian equation of the form presented in Eq. 2 proved to be the best fitting.

$$f = A \cdot exp\left(\frac{(n-B)^2}{C} + \frac{(m-D)^2}{E}\right) \quad \text{[Eq. 2]}$$

Where n is the number of tanks, m the stored mass in kg and A, B, C, D and E are dimensionless coefficients; these coefficients for all the studied accidents, as well as the values of R^2 obtained for the fitting are presented in Table 5, and the plots obtained when using the Gaussian equation with the presented coefficients are shown in Figures 8 through 10.

Figure 6. Frequency vs. Number of tanks and Stored Mass for G.2 accidents

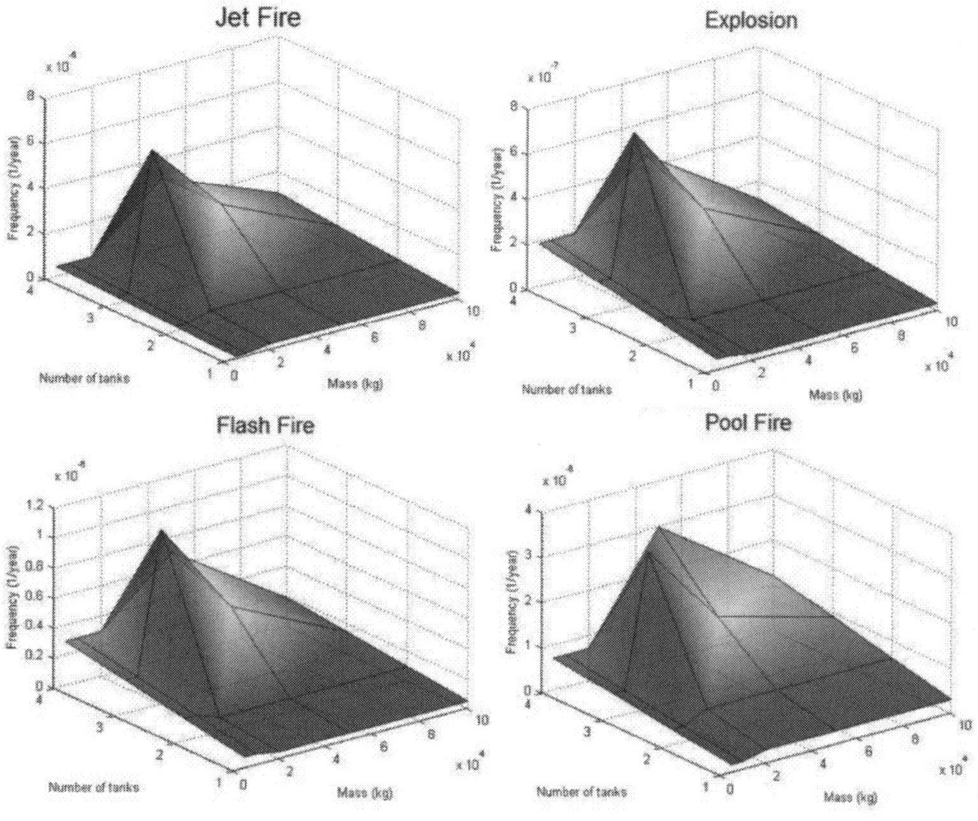

Table 5: Coefficients for the Gaussian equation for all accidents

Accident	Coefficients					R2
	A	B	C	D	E	
BLEVE	2.8x10-5	4.79	-4.15	7.7x104	-4x109	**0.94**
G.1 Explosion	9.9x10-6	5.37	-5.10	7.0x104	-4x109	**0.91**
G.1 Flash Fire	2.8x10-2	3.33	-0.8	1.0x102	-1.2x1011	**0.85**
G.1 Pool Fire	3.4x10-5	4.57	-3.96	8.3x104	-6.3x109	**0.95**
G.2 Jet Fire	2.2x10-5	3.14	-0.56	3.4x104	-1.5x108	**0.84**
G.2 Explosion	1.1x10-6	3.28	-0.82	3.3x104	-3.9x108	**0.77**
G.2 Flash Fire	1.6x10-6	3.27	-0.82	3.3x104	-3.9x108	**0.77**
G.2 Pool Fire	2.7x10-6	3.53	-2.17	5.5x104	-2.3x109	**0.57**
G.3 Jet Fire	1.5x10-4	6.13	-0.15	2.2x106	-1.8x1012	**0.98**
G.3 Explosion	7.8x10-5	6.04	-0.15	2.3x106	-1.9x1012	**0.99**
G.3 Flash Fire	9.2x10-5	6.10	-0.15	2.1x106	-1.7x1012	**0.98**
G.3 Pool Fire	2.2x10-4	6.10	-0.15	2.1x106	-1.7x1012	**0.99**

Figure 7. Frequency vs. Number of tanks and Stored Mass for G.3 accidents

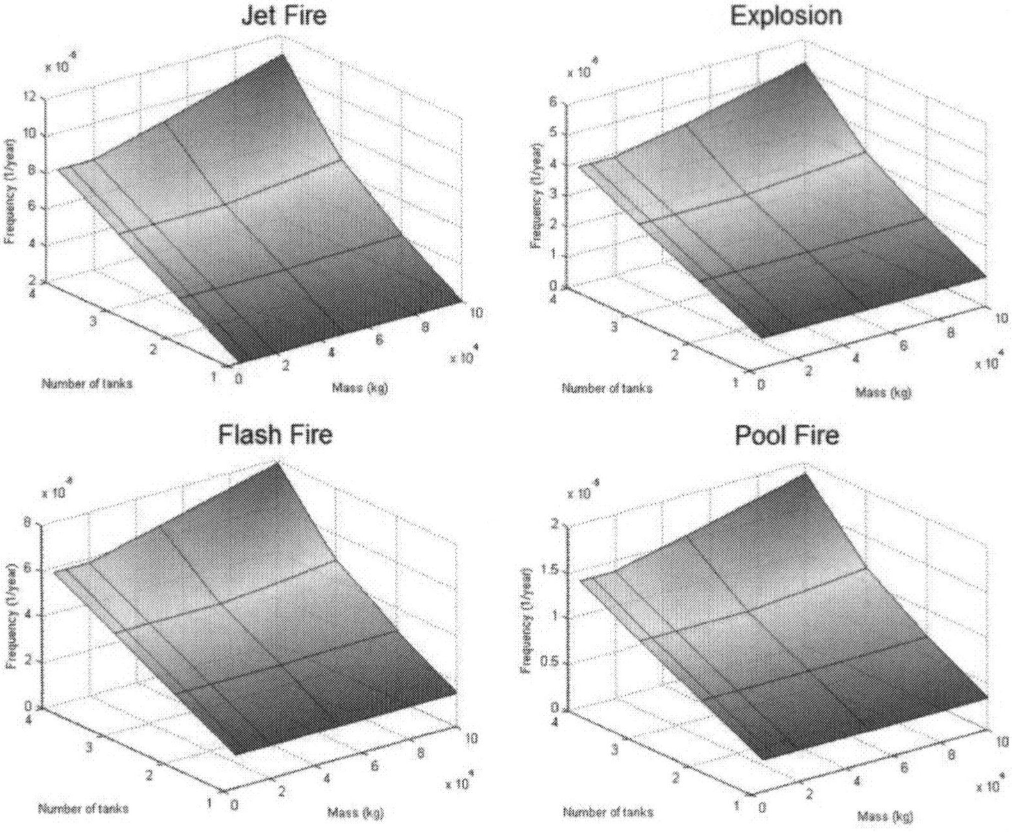

5. Conclusions

Equations that depend on the number of tanks and mass stored in the installation have been found, which, in many of the cases, present good tendencies in replicating the data obtained in the experiment.

The equations obtained could be used to take domino effect into account when performing a QRA on a storage installation of characteristics similar to those presented in the experiment, without increasing significantly the time required to complete the study.

Also, equations of this kind could be used during the design stages of a storage installation to help decide the number of equipment and quantities of substance stored; they could be used to know how risk behaves depending on different designs, and which design is inherently safer.

Figure 8. Frequency vs. Number of tanks and Stored Mass for G.1 accidents using the Gaussian Equation

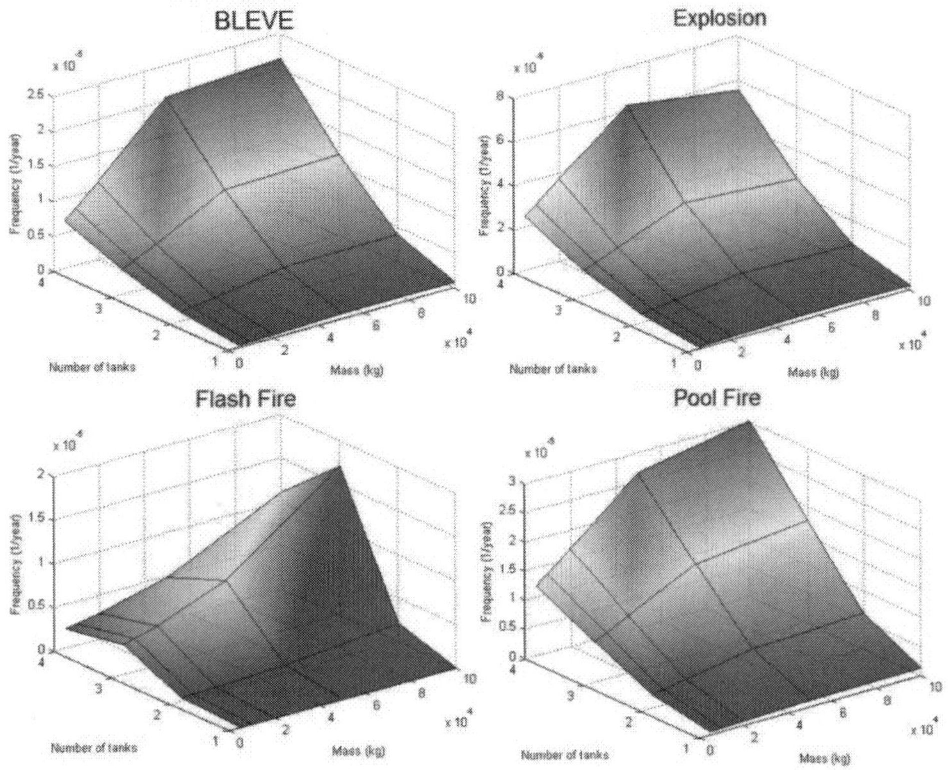

Figure 9. Frequency vs. Number of tanks and Stored Mass for G.2 accidents using the Gaussian Equation

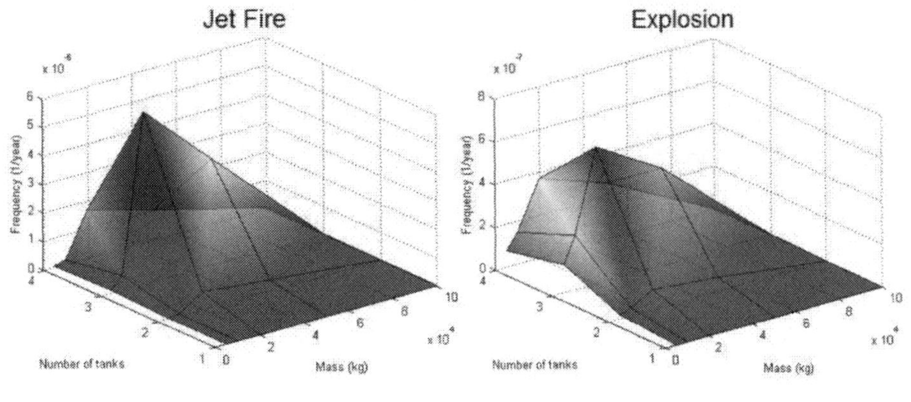

Figure 9. Frequency vs. Number of tanks and Stored Mass for G.2 accidents using the Gaussian Equation (Cont.)

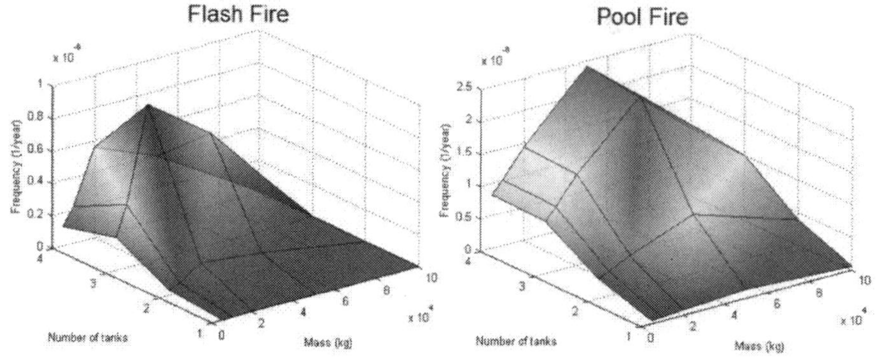

Figure 10. Frequency vs. Number of tanks and Stored Mass for G.3 accidents using the Gaussian Equation

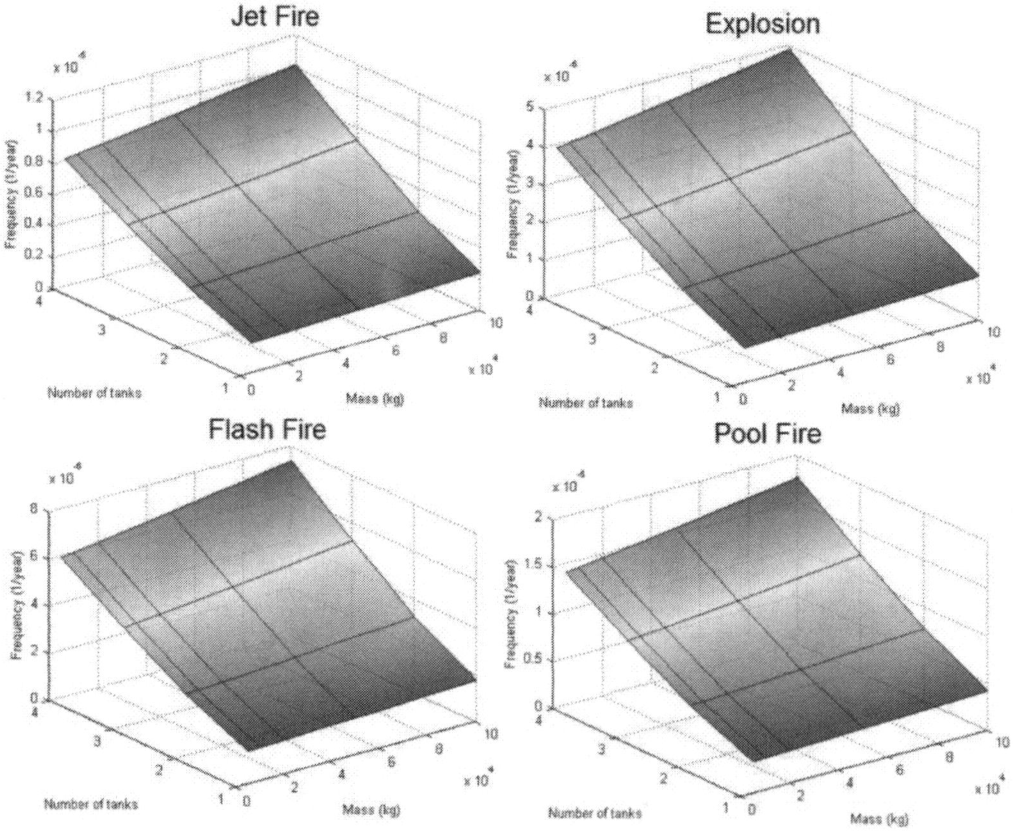

References

[1] CPR 18E, 2005. Guidelines for Quantitative Risk Assessment (Purple Book). Directorate General for Social Affairs and Employment, The Hague, Netherlands.

[2] Delvosalle, C, "Domino Effects Phenomena: Definition, Overview and Classification", First European Seminar on Domino Effects, Leuven, 1996.

[3] "Liquified Petroleum Gas Code". National Fire Protection Association (EEUU), 2001.

[4] Bernechea, E.J., Vílchez, J.A., Arnaldos, J., 2012. A model for estimating the impact of the domino effect on accident frequencies in quantitative risk assessments of storage facilities. Process Safety and Environmental Protection, http://dx.doi.org/10.1016/j.psep.2012.09.004

[5] G. Antonioni, G. Spadoni, V. Cozzani, Application of domino effect quantitative risk assessment to an extended industrial area, Journal of Loss Prevention in the Process Industries, 22, 614-624, 2009.

[6] F. I. Khan, S. A. Abassi, Estimation of probabilities and likely consequences of a chain of accidents (domino effect) in Manali Industrial Complex, Journal of Cleaner Production, 9, 493-508, 2001.

[7] V. Cozzani, E. Salzano, The quantitative assessment of domino effects caused by overpressure: Part I. Probit models, Journal of Hazardous Materials, 107, 67-80, 2004.

[8] V. Cozzani, E. Salzano, Threshold Values for domino effects caused by blast wave interaction with process equipment, Journal of Loss Prevention in the Process Industries, 17, 437-447, 2004.

[9] G. Landucci, G. Gubinelli, G. Antonioni, V. Cozzani, The assessment of the damage probability of storage tanks in domino events triggered by fire, Accident Analysis & Prevention, 41, 1206-1215, 2009.

(44ce) Routes Optimization for Hazardous Materials Transportation in Bogotá, Colombia

Jorge Victoria, Universidad de los Andes, Bogota, Colombia, Nubia Velasco, Universidad de los Andes, Eliecer Gutierrez, Industrial Engineer, Universidad de los Andes, Bogota, Colombia and **Felipe Munoz**, Universidad de los Andes, Bogota D. C., Colombia

The present work presents an optimization approach to design the set of routes for the transport of hazardous materials (HazMats) at Bogotá. HazMats transportation importance lies in the social, environmental and economic impact that might happen when an accident, during the transportation occurs. This paper is focused in developing a decision support system supported on optimization tools with the purpose of find the routes, -from gas distributor to gas stations-, that minimize: the total travel distance of heterogeneous tank trucks, the social risk and the infrastructure risk around the routes created.

The problem is modeled as a vehicle routing problem and it is solved using the principle of Route first, Cluster second. In a first stage the algorithm generates a big tour where all clients are visited, ignoring the vehicle capacities. Secondly, thanks to a Split procedure the tour is segmented considering the capacities. Then routes are improved with a MultiObjective Metaheuristic called MultiObjective Variable Neighborhood Search (MO-VNS). The algorithm is tested on two set of instances, the first one randomly generated and the second from Bogotá Geographic Information System (GIS). Several approaches to generated solution are proposed.

See more of this Session: Poster Session
See more of this Group/Topical: Global Congress on Process Safety

(44cf) Mejores Prácticas En Los Procesos De Gestión Del Cambio Para Seguridad De Procesos y La Evaluación De Seguridad Previa Al Inicio (PSSR)

J. Luis Aguirre, KMS, Optimized Systems and Solutions, Houston, TX

La Gestión del Cambio continúa siendo el elemento más retador/complicado de la Gestión de Seguridad de Procesos (PSM), del Programa de Gestión de Riesgos (RMP) y de los requerimientos regulatorios de los Sistemas de Gestión de Medio Ambiente y Seguridad (SEMS). A pesar de que el estándar PSM ha existido desde 1992 y que la industria ha gestionando cambios desde hace mucho tiempo, aún es común no obtener el resultado deseado. Los esfuerzos conscientes de muchas personas trabajando en conjunto son necesarios para garantizar que el cambio sea identificado, analizado y ejecutado con calidad. Justo cuando logramos capturarlo en papel y mantener informado al equipo de trabajo involucrado, se presentan cambios, negligencias o las regulaciones cambian. Nuevos involucrado y responsabilidades dificultan la gestión del cambio como sabemos que debe hacerse.

Sin embargo, la tecnología ayuda en todas las etapas del flujo de la Gestión del Cambio (MOC) al permitir la recopilación de ideas, formalizando el análisis de las mismas, su aprobación y ejecución del cambio. También puede ayudar a proporcionar visibilidad del estado actual y análisis de tendencias, así como la identificación de peligros a los que están expuestos los diferentes equipos, sitios, y bienes. Correctamente aplicada, la tecnología podría ayudar a predecir el futuro. Además, la tecnología facilita y ayuda a implementar sus estándares resultando en Mejores Prácticas de Gestión del Cambio, Evaluación de Seguridad Previa al Inicio (PSSR) y el seguimiento al cumplimiento de acciones, notificaciones de alerta y el cierre.

See more of this Session: Poster Session
See more of this Group/Topical: Global Congress on Process Safety

(44cg) Lesha – Multi-Layer Progressive Stress & Impact Assessment On Health & Behavior

Salvador Ávila Filho, PEI TECLIM, UFBA, Salvador, Brazil

The chemical process industry seeks, constantly, to achieve the demands social and environmental (mandatory and voluntary) within the international economic chain. The Integrated Management Systems meet issues in the areas of quality, environment, safety, health, social responsibility, now including risk management (ISO31000). This excess of valuation in integrated management system just prioritizes management functions in place of the functions related to core activity. By integrating management activities into the routine of work, the task will be performed in a more committed form avoiding the domain of management system in place of production skills.

Mandatory Standards for work in the areas of health, hygiene and safety are concerned in dealing the most damage, searching through punishment, correct managerial failure, or correct the ignorance. The preventive tools are not fully developed leading to leaders to be worried about maximum allowable values in the areas of safety, health, hygiene, and also the environment.

The Voluntary activities are not developed due to difficulties initiating deployment of management systems that were born in a TOP-DOWN philosophy. The human and operational reliability was borned in a shape BOTTOM-UP avoiding the occupation of operator, most part of time, doing management systems reports. Operational risks are analyzed including the role of man, the role of group in the planning and execution of tasks. The techniques involve subjectivities as effect of cultures fight in the body of work(social - mass culture, regional culture, organizational, and security). The analysis of dominant cultures involves identifying the various vectors impacting cultural influences in the workplace. Psychological stress causes physical and behavioral changes (affecting the decision) and in the body, both, can cause the loss of quality of life.

Various techniques were developed and presented in GCPS to assess the hazard and the risk of social conflicts and that can cause human error. This error initiates or increases the process of failure in equipment or processes through the intervention performed on the steps of the task. Thus, social conflict in new business (APRSH), cognitive traps and human error in critical task (Social HAZOP), causality analysis of cognition / decision, task and process losses (COGNOSP), analysis of organizational and cultural factors as root cause for accidents, decision analysis operation in progressive stress (LODA), and analysis of cognitive failure of operator in the control of equipment and components (FMEAH).

At the present moment we present a technique that try to study the effect of progressive psychological and physical stress on the body and quality of life, here dubbed as LESHA (Stress Analysis & Progressive Effect on Human Health and Behavior). This study highlights the importance of analyzing the life cycle of faults that impact on health by studying the trend of biological indicators resulting from psychological and physical stress. We also emphasize the search for signs of failure before occupational diseases being installed, showing the most suitable installation of barriers against failure, bringing the barrier closer to the root cause.

The LESHA technique therefore intends to monitor the level of physical and psychological stress of environmental factors, and stress from events related to task, which can turn a pathogen resident in symptom (Holnagel and Reason) and hence in occupational disease.

The events that initiate failures, and the stressed environments can cause stress progressive state on individuals, activating, or causing the trigger of failure through the stages: pathogen resident; uncontrolled organic process; symptom; disease; replication of the disease; and fatality or permanent labor shortage. Each level of stress is analyzed for signs and / or indicators of failure that impact over worker health permitting taking preventive or corrective action (biological safety barriers).

The following levels of analysis are proposed: Level 1 - Identification of Human Type, and the respective pathogen resident or signs that may cause the future disease in the body and in the behavior. Level 2 - initiators of failure in the task, or, bad social relationship that can cause the lack/ fault of organic metabolism, generating growth signals, variables still toward the disease. Signs of lost productivity interlinked with organic symptoms ... Level 3 - Shares of the social environment (society) or technical (technical knowledge), and secondary causal events and their impact on the body by increasing the metabolism of organic lack/ fault of symptom onset with fixed initial stage of the disease. Signs of lost productivity interlinked with organic symptoms...

Level 4 - Emergence of social ills (irritation, and communication difficulties), emergence of occupational diseases (slips related to the task and can cause accidents), and biological diseases like ulcer, high blood pressure and other diseases related to workplace using indirect biological ways (early onset of diabetes or other). Level 5 - Replication of the disease by adopting bad habits in the workplace where colleagues, "introject" behavior and even biological processes of "friends." Level 6 - fatality or permanent labor shortage causing low morale in the team, and developing or activating resident pathogen not active or nonexistent.

Understand the signs of this biological process is challenge of safety, hygiene and occupational medicine in conjunction with studies of human reliability. Different disciplines come together to prevent the disease from the body (regardless of the root cause) had been evident. The diseases are originate from labor, directly, or resulting from the influence of the job, and, from society influence. One should avoid firing the trigger of failure and the development of the pathogen resident.

The result of this analysis allows build the programs to avoid diseases. Prevention and correction programs depend on the intensity of the interference signals and the stress level reached in this process. It is proposed for each level and intensity of the biological and behavior signals following through risk mapping:
* Intensity Level X frequency.

Building Design for Resilience Beyond Design Basis Blast Loads

Michael A. Polcyn
Baker Engineering and Risk Consultants, Inc.
3330 Oakwell Court, Suite 100
San Antonio, TX 78218
MPolcyn@BakerRisk.com

Raymond H. Bennett, Ph.D., P.E.
Baker Engineering and Risk Consultants, Inc.
3330 Oakwell Court, Suite 100
San Antonio, TX 78218
RBennett@BakerRisk.com

Prepared for Presentation at
American Institute of Chemical Engineers
2013 Spring Meeting
9th Global Congress on Process Safety
San Antonio, Texas
April 28 – May 1, 2013

UNPUBLISHED

Building Design for Resilience Beyond Design Basis Blast Loads

Michael A. Polcyn
Baker Engineering and Risk Consultants, Inc.
3330 Oakwell Court, Suite 100
San Antonio, TX 78218
MPolcyn@BakerRisk.com

Raymond H. Bennett, Ph.D., P.E.
Baker Engineering and Risk Consultants, Inc.

Keywords: blast resistant design, consequence-based design, risk-based design

Abstract

One of the guiding principles of API RP-752 is that buildings intended for occupancy are to be designed, constructed, installed, modified, and maintained to protect occupants against explosion, fire, and toxic material releases. Owners/operators may use either a consequence-based approach or a risk-based approach to evaluate the siting of both new and existing buildings for blast loads. While a consequence-based approach should be based on a maximum credible event (MCE) for each building and type of hazard considered, a risk-based approach may consider numerous scenarios, accounting for both the consequences and the frequencies of the scenarios. However, when designing either new buildings or upgrades to existing buildings, designers are typically provided only a single set of blast loads and a single response criteria (for example, medium response per ASCE "Design of Blast-Resistant Buildings in Petrochemical Facilities"). A lower response level implies lower occupant vulnerability, while higher response levels imply higher occupant vulnerability. A design for higher magnitude loads to low response criteria would result in very conservative design. On the other hand, a design for lower magnitude loads to higher response limits would result in a building without capacity to protect occupants in the event of a more severe explosion.

A better approach would be to consider multiple levels of blast loads when developing a design. Lower magnitude loads (usually from higher probability events) should require designs to meet the lower response limits, while higher magnitude loads (usually from lower probability events) should require designs to meet the higher response limits. This paper demonstrates this approach by evaluating typically designed building systems for a range of blast loads commonly encountered at refining and chemical plants. Recommendations are provided for the design of new buildings or upgrades to existing buildings. By including the concept of resilience in the design process, the designer may provide protection from future yet-to-be-determined hazards or scenarios.

1 Introduction

The structural design of buildings for conventional loads, including wind, snow, and seismic loads, is a well-defined process that is followed by engineers on a routine basis. Loads are specified by various codes or design manuals (e.g., International Building Code [1] or ASCE 7-05 [2]). Similarly, the design approach for determining the capacity of various types of structures is defined in corresponding design manuals. For example, the American Concrete Institute publishes "Building Code Requirements for Structural Concrete" [3], and the American Institute of Steel Construction publishes the "Manual for Steel Construction" [4].

In contrast, the design of buildings subjected to blast loads is not as formalized. Design case blast loads are not specified in any guidance documents and are left for the owner to define appropriate loads based on the processes and materials in place at their facilities. There are several guidance documents that describe approaches for designing and analyzing structures subjected to blast loads, which include dynamic analysis methodology, material properties, and structural response limits. Response or deformation limits, such as those defined by the American Society of Civil Engineers [6], are given ranges of response, where Low Response refers to a component with no more than slight visible permanent damage, Medium Response corresponds to a component with some permanent deflection, but is generally repairable, and High Response indicates that the component has not failed, but has undergone significant permanent deflections and is likely unrepairable following the design case loads. It is up to the owner to select the response level, and thus the level of protection to building occupants, for their building design.

This paper provides an overview of the design process, highlighting the differences between a consequence-based design approach and a risk-based design approach. The discussion continues by presenting features to the design that will enhance the structure's ability to be usable following an explosion event and to provide some degree of protection to occupants for events exceeding the selected design load cases. It should be noted that only explosion hazards are addressed in this paper; building and upgrade designs should also consider hazards from fire and toxics.

2 Dynamic Analysis Method

When designing buildings or building upgrades for blast loads, structural components are usually assessed using nonlinear dynamic Single-Degree-of-Freedom (SDOF) analytical methodology. The SDOF methodology summarized in Biggs [6] is a widely accepted approach for estimating damage to structural components subjected to time-varying blast loading. The SDOF approach is a government and industry accepted methodology that has been described in many guidelines such as UFC 3-340-01 [7], UFC 3-340-02 [8], and ASCE [5].

The dynamic response of structural components from prescribed blast loads is determined by modeling the components as an SDOF system, such as the equivalent spring-mass system shown in Figure 1. Structural components such as walls, windows, beams, doors, and panels will deform and vibrate when loaded with a shock wave. The SDOF model for each component is constructed using the component's physical structural properties so that the model will exhibit the same displacement history as the point of maximum deflection in the component. The displacement history of the SDOF model is obtained with numerical integration techniques using a time-stepping computer program to solve the equation of motion of the equivalent system at discrete time steps.

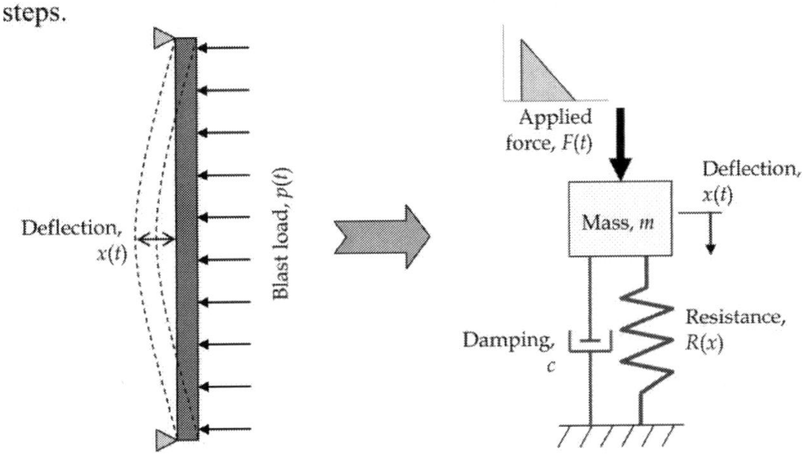

Figure 1. Equivalent Spring-Mass SDOF System

The calculated peak deflection (see Figure 2 for example) is used to obtain the support rotation and ductility ratio, which represent the deformation limit criteria most commonly used in blast design. As shown in Figure 2, the support rotation is the angle between the original un-deflected shape of a component and a straight-line segment between the point of maximum deflection and the support. The ductility ratio expresses the maximum deflection in terms of the minimum deflection of the component when it reaches its ultimate capacity. Therefore, ductility ratios that are greater than unity (1.0) indicate that permanent deformations have been sustained.

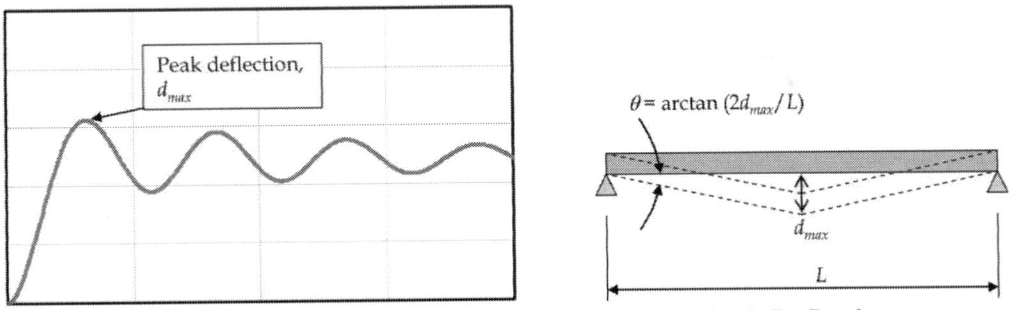

Figure 2. Calculation of Support Rotation from Peak Deflection

The design process for blast loads is an iterative process, where the pertinent design blast loads and response criteria are defined, "trial" structural members are selected based on the building geometry, and the structural components are analyzed using the dynamic analysis methods discussed above. If the response of the component does not meet the required response criteria, a heavier section is evaluated or the building geometry is modified until the analysis shows that the response criteria are met. Connections are typically designed to resist the maximum end reactions predicted from the dynamic analysis.

Pressure-impulse (P-i) diagrams are commonly used to assess building or component response to multiple sets of blast loads. P-i diagrams are plots of constant damage levels for various combinations of blast overpressure and impulse. Each individual curve represents the suite of all pressure and impulse pairs that will result in the same structural damage to a given component. The combination of the three individual curves forms a complete component P-i diagram to define Low, Medium, and High Damage Levels. An example of a P-i diagram is provided in Figure 3. As shown, the region above the upper curve would imply component failure.

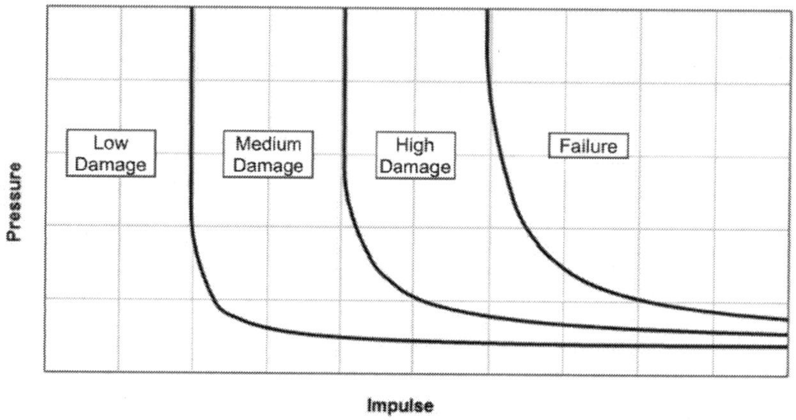

Figure 3. Example of a Pressure-Impulse Diagram

The use of P-i diagrams allows for quick consideration of a variety of blast loading situations. If the blast load of interest falls below or to the left of a particular damage curve, then the given component is predicted to receive damage less than or equal to the level represented by that P-i curve. In addition, component curves can be combined to develop a set of building P-i damage curves, or BDL curves.

3 Consequence-Based vs. Risk-Based Design

Traditionally, design of buildings or upgrade to buildings has been performed using a consequence-based approach where the frequency of events is not considered. According to API RP-752, a consequence-based approach is to be based on maximum credible events (MCEs) for each building. For example, a precast concrete wall for a control room could be designed to meet the ASCE Medium Response criteria when subjected to the applied blast loads from the MCE determined for that building. The resulting P-i curves for the wall would be similar to those shown in Figure 4.

The figure also demonstrates the concept of resiliency in the design. Four pressure-impulse pairs are shown, corresponding to two Design Load Cases and two Potential Overload Cases. The potential overload is 1 psi greater than the design pressure for both cases, while the impulse is held constant. In both Design Cases, the component just meets the Medium Response criteria. However, the curves show that for Design Load 1, the component has the capacity to resist the overload, albeit with larger deformations, but without failing (the component undergoes a high level of response). For low probability events, such behavior might be acceptable. However, the implied resiliency of the component cannot be determined solely by setting the response limits. For example, in Design Case 2, the corresponding overload will result in the failure of the component, even though the Design Case load is farther below the boundary between medium and high response. Thus the true resiliency of a design has to be evaluated in light of the potential for an overload.

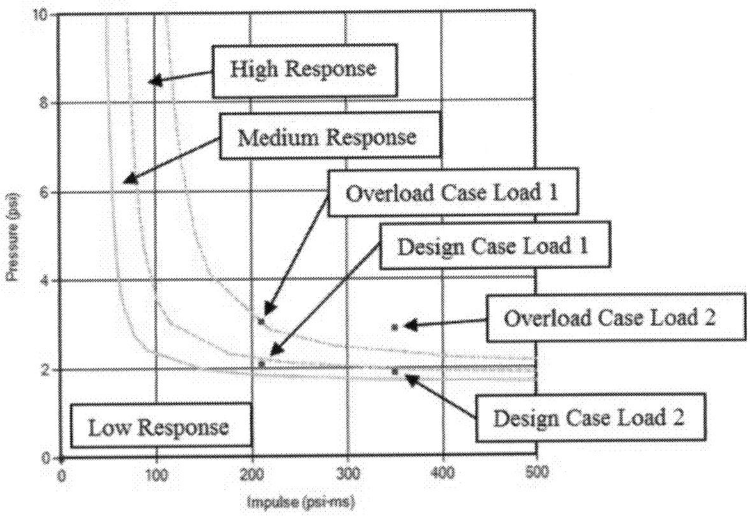

Figure 4. Component P-i Curves

In more recent years, companies have conducted quantitative risk assessments (QRAs) in support of facility siting study efforts. Such studies include estimates of the expected frequency and consequences of potential accidents, typically resulting in an estimate of risk, either as an aggregate risk, an individual risk, or a relationship of the expected frequency to the consequence (e.g., an F/N curve). This approach, which considers a large number of possible explosion scenarios, can be extended to a risk-based design approach for buildings or building upgrades.

When using a risk-based approach, it is necessary to consider the overall building damage rather than the response of the individual components independently. By doing so, the predicted building damage for each scenario can be related to the expected occupant vulnerability [9], which is used to perform the risk calculations. For example, the Building Damage Level (BDL) curves in Figure 5 show that the building is expected to undergo moderate damage, or BDL 2a, when the components are designed to meet the ASCE Medium Response criteria. Note that the inflection points in the BDL curves are caused by a change in the governing building component. For example, the upper left portion of the BDL 3-4 curve may be set by roof purlin response and the lower left portion may be set by a roof girder response. If the design case loads are defined by the MCE for all of the postulated explosion scenarios, the BDL plot in Figure 6 can be generated showing the loads from "all" postulated scenarios with the BDL curves. As shown, none of the scenarios will result in greater than moderate damage (BDL of 2a), and the resulting risk will be very low. It can be concluded that such a design has significant resiliency to withstand loads greater than those postulated.

Alternatively, the building could be designed for higher response limits under the design load cases, possibly allowing partial building failure under the most severe load cases. The BDL curves in Figure 7 demonstrate a case where the components were designed to meet the ASCE High Response limits for the MCE and ASCE Medium Response Limits for the majority of the load cases. As shown, the expected damage for the MCE approaches higher BDLs, but building collapse (BDL of 4) is not expected. In this case, the expected risk to building occupants would still be relatively low, although somewhat higher than the risk demonstrated in Figure 6. However, there is less resiliency in the design should there be an overload case. Note particularly that if the scenario shown in Figure 5 were 0.5 psi greater, the BDL for the building portrayed in Figure 6 would still be 2A, while the BDL for the design represented in Figure 7 would be increased to BDL of 3.

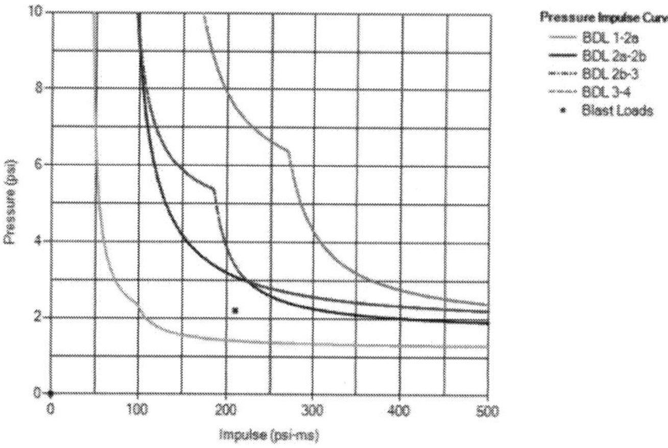

Figure 5. Building Damage Level (BDL) Curves

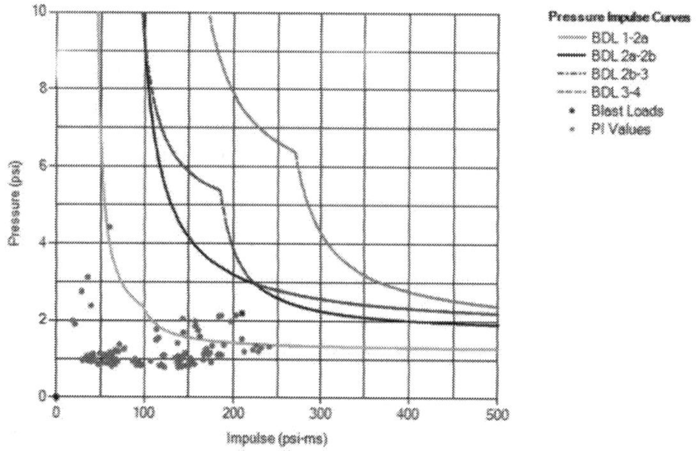

Figure 6. BDL Curves with Loads from "All" Postulated Scenarios

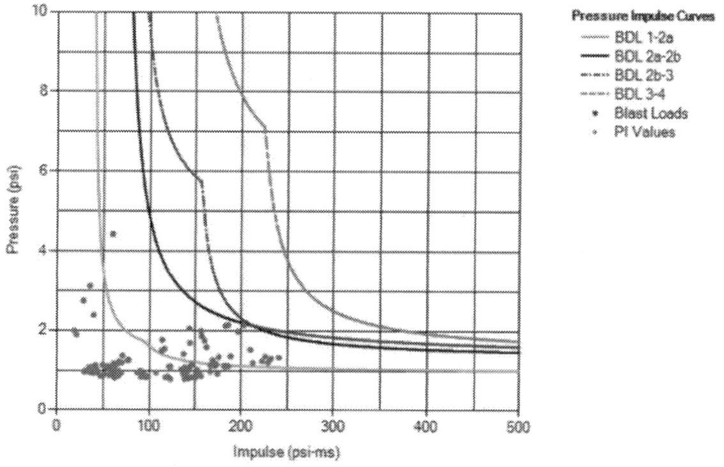

Figure 7. BDL Curves for "Weaker" Building

4 Resiliency in Design of Occupied Buildings

The examples discussed in the previous section demonstrate three approaches to providing a resilient structural design:

1. Structural systems that can develop a high degree of ductility are greatly preferred over structural systems that undergo a more brittle response. Most materials commonly used for blast resistant structures (steel and concrete) can undergo significant plastic deformations, typically well beyond those associated with published response limits, before failing. This plastic deformation allows the building to absorb the energy applied by the blast load. In addition, structural components designed to undergo large plastic deformations will have lower support reactions than those designed to behave elastically or with small plastic deformations, thus resulting in lower demands for connections and the building lateral load-carrying systems.

2. When selecting response limits, the likelihood of the selected design loads should be considered. Typically, high probability scenarios result in lower magnitude loads, while more severe load cases are the result of lower probability scenarios. This would imply that higher response limits could reasonably be used when designing for low probability, high magnitude load cases. Similarly, when designing for higher probability / lower magnitude loads, or loads with considerable uncertainty associated with their prediction, lower response limits should be considered.

3. Since one of the objectives of a blast resistant building is to protect building occupants, the layout of the building and the location of occupants relative to the sources of the potential explosions should be considered. By placing unoccupied areas of the building on sides facing the explosion source, localized failure of non-loadbearing components may be acceptable.

5 Conclusion

The design of blast resistant buildings does not follow an approach that is as prescriptive as the approach used for conventional design. Blast loads must be defined based on the conditions at the facility. Appropriate response limits must be selected for the components comprising the building. In addition, process conditions and building use may change to meet future requirements. Due to the uncertainty that is inherent in this design process, it is important to provide a building design that protects building occupants with a degree of resiliency. This can be developed by selecting the building structural system, building layout, and required response limits in a manner that is consistent with the postulated loads and the expected potential for these loads. In addition, the potential for more severe loads should also be considered in the design process.

6 References

[1] 2012 International Building Code, published by the International Code Council, Inc., May 2011.

[2] ASCE/SEI 7-05, "Minimum Design Loads for Buildings and Other Structures," published by the American Society of Civil Engineers, 2006.

[3] ACI 318-11, "Building Code Requirements for Structural Concrete and Commentary," reported by ACI Committee 318, published by the American Concrete Institute, 2011.

[4] "Steel Construction Manual," 14th Edition, Second Printing, published by the American Institute of Steel Construction.

[5] "Design of Blast Resistant Buildings in Petrochemical Facilities, Second Edition," Task Committee on Blast Resistant Design, American Society of Civil Engineers, 2010.

[6] Biggs, J., Introduction to Structural Dynamics, McGraw-Hill, Inc. New York, NY, 1964.

[7] UFC 3-340-01, "Design and Analysis of Hardened Structures to Conventional Weapon Effects," June 2002. (Formerly Technical Manual for Army TM 5-855-1, Air Force AFPAM 32-1147(I), Navy NAVFAC P-1080, and Defense Special Weapons Agency DAHSCWEMAN-97, August 1998.)

[8] "Structures to Resist the Effects of Accidental Explosions," UFC 3-340-02, 5 December 2008.

[9] Oswald, Charles J. and Baker, Quentin A., "Vulnerability Model for Occupants of Blast Damaged Buildings," presented at the 34th Annual Loss Prevention Symposium, March 6-8, 2000, Session 3.

(44g) Reliability Analysis of Man-Machine Response to an Emergency Situation Based On a Detailed-Action Plan

Najmeh Vaez, North America Risk, Environmental Resources Management (ERM), Houston, TX; Environmental Science and Safety Department, University of Houston Clear Lake, Houston, TX, Farshad Nourai, Aftab Imen Parto Consulting Engineers Co., Tehran, Iran, Yaneira Saud, North America Risk, ERM Southwest, Houston, TX and Suresh Yelisetty, Americas Risk Practice, Environmental Resource Management (ERM), Houston, TX

The primary emergency response actions in oil and gas industries are the functions of the process safety control system and the actions completed by the control room operators. The performance of the safety control system and the control room operating crew has a significant role in proper emergency response. To overcome each critical scenario, an emergency response operating procedure is provided with specific instructions, that in this research paper it is named a detailed action plan (DAP). This term is recommended for this purpose by the Center for Chemical Process Safety (CCPS). The reason of using a DAP is to reduce the probability of error in the emergency response of control room operators by providing scenario-specific written instructions. However, most of the DAP's are developed during design phase when there is no opportunity to test them through emergency drills. Therefore, a framework is proposed to analyze the reliability of the DAP's. This framework should be able to determine reliability of every single human and equipment action as well as the combined human-machine reliability, modeled by means of a Reliability Block Diagram to calculate the reliability of integrated automatic-operator emergency actions. The framework focuses on incorporating the operational and cognitive operator's errors in the process/equipment reliability analysis. For this purpose, a human reliability analysis method named SPAR-H was utilized. The proposed framework was applied for analyzing DAPs of an emergency well control operation in order to check and improve the applicability of this framework. The presented framework facilitates the DAP assessment process by performing task analysis modeling the detailed tasks, and completing a human-machine reliability analysis. Besides, the application of the presented framework, its findings and recommended actions enables a considerable improvement in the reliability of DAPs and therefore proves its efficiency and effectiveness.

(44i) Hazards of Gasoline Contamination – Case Study of a Tanker Truck Explosion Resulting From Switch Loading

Scott G. Davis, GexCon US, Bethesda, MD and Kees Van Wingerden, GexCon AS, Bergen, Norway

When a tanker truck is drained of a Class I liquid (e.g., gasoline) and refilled with a similar liquid, there is no flammable mixture on the surface of the rising fluid because Class I fluids create mixtures at their surface that are too fuel rich to burn. Similarly, if a tanker truck is drained of Class II liquid (e.g., diesel) and refilled with a Class II liquid, there is no flammable mixture created because Class II liquids do not produce ignitable mixtures at ordinary temperatures. An extremely hazardous condition occurs, however, if the liquids are *switch loaded*, which is when a Class II or III liquid is loaded into a tank vehicle that previously contained a Class I liquid. Under these conditions, the atmosphere above the fluid surface can become ignitable. An explosion can occur if a spark occurs from the fluid surface to another object, igniting the flammable mixture in the headspace of the tank.

This paper describes a recent incident involving an explosion during the filling of a tank of a transport truck with diesel fuel. This particular tank was previously filled with gasoline and was not completely emptied of the gasoline prior to the switch loading. After only filling one-quarter of the programmed load, a spark ignited the flammable mixture within the headspace. The resulting explosion and overpressure caused significant damage to the truck and neighboring tanks. The study will address issues and lessons learned regarding spark promotion with the tank, reduction of static generation by the fluid being filled into the tank, and elimination of the flammable mixture within the tank prior to switch loading.

(44j) Is Your MOC Process Doing All That It Should Be?

William Fink and Carolina Herrera, Sage Environmental Consulting, L.P., Austin, TX

OSHA 1910.119 Process Safety Management of Highly Hazardous Chemicals was the driver for refining and chemical industry to implement Management of Change (MOC) programs some 20+ years ago. However, companies have now recognized the benefits of implementing and maintaining an effective MOC program beyond regulatory compliance. Effective MOC programs enhance a company's ability to respond to changing "market" or processing needs with up to date process safety information. Management of Change has even been incorporated into staffing levels and personnel expertise by some Companies.

Management of Change programs typically focus on changes that are "not in kind" changes. Significant effort is placed on proper evaluation of these changes and the potential impacts on the operation. This presentation emphasizes both the typical and non-typical areas of MOC that have been found to be overlooked. MOC programs rigorously evaluate the potential to introduce new hazards of the process such as fire, explosion, toxicity, and environmental impacts. Is that far enough to ensure that you are meeting all of your performance needs?

Focus on environmental impacts such as water contamination, air pollutants, and hazardous waste generation may not be going far enough. Gradual increases in highly hazardous material inventories may cross the lines of "threshold quantities" and introduce new OSHA or EPA requirements. Does your MOC program look at this potential? In addition, management of "in kind" changes may also not be keeping pace with the changes required by the EPA Leak Detection and Repair (LDAR) requirements or the Benzene Waste Operations NESHAP (BWON) requirements. These requirements as applied to MOC are further explained in this presentation. Several "gaps" in MOC programs will also be examined and suggestions for improvements provided.

(44I) Beyond DSC, Improved Thermal Hazards Screening Methodology

Peter J. Ralbovsky IV, Netzsch Instruments North America, LLC, Burlington, MA

Thermal Hazards Screening is a primary part of any chemical process safety program. Differential Scanning Calorimetry (DSC) is used extensively in studying the thermal properties of materials and is often used to screen for thermal hazards. This method is quick and inherently safe as it can be done with a small amount of material (typically milligrams). These two benefits cannot be understated as there may be hundreds of screening test which a lab may have to do in a given year and sometimes little or nothing is known about its thermal behavior, its toxicity, or the toxicity of the decomposition products. DSC also has some short comings as it is unable to measure pressure, which can be a significant hazard, even when the heat release is relatively low. Also it can be difficult to handle multiphase samples or samples which react together at room temperature.

The Multiple Mode Calorimeter (MMC) is an excellent tool to be used in conjunction with the DSC to provide a more comprehensive look at thermal hazards. The MMC uses samples sizes up to about 3 grams so it is possible to do injection and stirring. In the MMC temperature and pressure are recorded as a function of time. Sample energies measurement of both endotherms and exotherms is possible. Heat capacity can be calculated as a function of temperature. This method offers a quick and inexpensive way to perform isothermal tests which can be used to help design larger scale tests done in 2L+ reaction calorimeters or pilot plant operations.

The Multiple Mode Calorimeter consists of an electronic base and a detachable calorimeter assembly. There are multiple types of calorimeters available which can be operated by the electronic base. This allows the user to select the right calorimeter for the type of experiment that is needed. The calorimeters can also be tethered to the electronic base with an electronic cable. This allow for the calorimeter to be placed remotely from the electronics. The calorimeter can then be easily placed in the hood and/or behind explosion barriers or even placed inside a freezer for sub-ambient operations. This instrument has been described in detail elsewhere.

There are a number of calorimeters available; the best two for Thermal Hazards Screening is shown in Figure 1 & 2. The first calorimeter (Figure 3) is capable of operating as a low phi adiabatic calorimeter but more importantly, for thermal hazards screening, can run in constant power or constant temperature rate mode. The MMC external heater module does not run in adiabatic mode but can be operated in scanning mode. Both MMC modules incorporate the VariPhi technology and can run true isothermal calorimetry tests. These MMC modules can measure endotherms and exotherms as well as calculating heat capacity during those periods where no transitions or reactions are occurring. The MMC measures sample pressure as a function of time/temperature which is critical in screening for thermal hazards.

DSCs, like shown in Figures 3, have been used routinely as an initial step in the thermal hazards screening process. They have some unique advantages which make them suitable for this task. Scanning tests can be done very rapidly and with an auto-sampler they can be used to screen many samples efficiently. They sample size is small which makes them intrinsically safe in case of explosion or toxic gas production. These systems are small, inexpensive and have low operating costs. There are numerous DSCs available on the marketplace and they are used in many different applications. Two features are important from the thermal screening application. The first is an auto-sampler as discussed earlier. The second is the material of construction used in the DSC. It is likely that a DSC system used for thermal hazard screening will see explosions and the release of corrosive materials. DSC systems that can run hundreds of polymer melt tests without fail may have a limited life in the thermal hazards screening field. This is often unlooked. Figure 5 shows the cut-away of a robust DSC system.

Their continues to be a need to be able to safely and effectively screen thermal hazards for organizations involved in the manufacture, processing, storage or transportation of chemicals. Numerous screening tools are available in the marketplace, perhaps the most widely used is the Differential Scanning Calorimeter (DSC). This method has many advantages but the largest disadvantage is that the sample pressure is not measured. Thermal hazards are normally associated with pressure increases and these pressure increases are what ultimately provide the hazard to equipment and personnel. A storage drum does not melt due to a thermal runaway it ruptures due to the pressure increase because of the thermal runaway. Screening tools should be sensitive enough to see small exotherms that in near adiabatic conditions can fuel a runaway, they should be easy to use, provide quick results, and be as flexible as possible to handle different scenarios.

[1] Chippett, A New, Low Φ Factor, Scanning, Adiabatic Calorimeter, 18th Annual CCPS International Conference, 2003

[2]Chippett, US Patent Number 7,021,820 B2, April 4, 2006

Using Explicit Finite Element Analysis to Simulate the Effects of External Chemical Explosions on Single and Double-Walled Storage Tanks

Phillip E. Prueter
The Equity Engineering Group Inc., Shaker Heights, Ohio
peprueter@equityeng.com

David J. Dewees, P.E.
The Equity Engineering Group Inc., Shaker Heights, Ohio
djdewees@equityeng.com

Prepared for Presentation at
American Institute of Chemical Engineers
2013 Spring Meeting
9th Global Congress on Process Safety
San Antonio, Texas
April 28 – May 1, 2013

UNPUBLISHED

Using Explicit Finite Element Analysis to Simulate the Effects of External Chemical Explosions on Single and Double-Walled Storage Tanks

Phillip E. Prueter
David J. Dewees, P.E.
The Equity Engineering Group Inc., Shaker Heights, Ohio
pepprueter@equityeng.com

Keywords: Explicit Finite Element Analysis, Blast Loading, CONWEP, Accidental Explosions, Storage Tanks

Abstract

Accurately simulating the overpressure or shock wave associated with a given far-field chemical explosion is extremely valuable in assessing the structural response and possible failure modes of critical process equipment such as storage tanks, piping, or pressure vessels. Furthermore, assessing the potential damage from an overpressure wave can provide valuable information about protecting structures from external explosions and improving designs that advocate blast damage mitigation. Such analyses become all the more important should the contents of the tank or vessel exposed to such a potentially catastrophic event pose a risk to humans or the surrounding environment. This paper discusses the underlying theory and examines the practical application of multiple finite element based explicit computational techniques for simulating the load acting on a structure due to an external blast. Explicit three-dimensional blast analysis of single and double-walled storage tanks that carry an extremely high consequence of failure is performed. The structural response of the tanks due to postulated accidental explosions is investigated and likely failure modes are discussed.

The two explicit computational blast loading methods discussed in this paper are the incident wave loading model and the Conventional Weapons Effects Blast Loading Model or CONWEP. The CONWEP model is appropriate for detonations of conventional explosions, and in this case automates many crucial features of analysis, such as developing the overpressure time history (positive and negative phases), defining overpressure spatial decay, and accounting for spatially varying reflection effects. Large vapor cloud explosions are known to behave very differently from conventional explosives, and for these cases, the more manual, but also general, incident wave loading approach permits the user to define their own time-varying overpressure amplitude. Additionally, this method also allows for coupled structural-acoustic analysis (should reflection effects need to be considered in detail). While it is not necessary to model the fluid medium using acoustic elements when using the incident wave loading model, it does become essential if reflection effects dominate the response of the given problem. If the fluid medium is modeled (air in this case), the acoustic elements can be coupled to the structural elements and the loading surface becomes the boundary of the acoustic mesh closest to the blast source. While this approach can be accomplished, it is extremely resource and time intensive, and is generally not necessary (which will be discussed later in this paper). Three-dimensional examples of the above blast loading approaches are presented in this paper; incident wave, and CONWEP. Comparisons are made between these methods and the corresponding computational results.

Furthermore, commentary on the structural response and likely failure modes of the storage tanks is provided and a discussion regarding design features that supplement blast damage mitigation and process safety is rendered. The advanced computational methods discussed herein permit realistic evaluation of complex, three-dimensional process equipment subjected to accidental explosions. Simulating and understanding the possible failure modes due to blast loading of critical process equipment in the petrochemical and related industries is very beneficial in not only improving process safety, but also providing plant operators and safety personnel with crucial information regarding the possibility of equipment failure should a catastrophic explosion occur.

1. Introduction

An explosion in air creates a blast wave as a result of the atmosphere surrounding the explosion being pushed back. In general, the pressure of the compressed air at the blast wavefront decays as it moves away from the blast source. As discussed in references [1-6], a typical blast wave, as observed at a location removed from the center of a chemical explosion, reaches a peak value of overpressure and begins to decay exponentially (as shown in Figure 1), eventually decreasing below atmospheric pressure. This peak overpressure decreases as the distance from the explosion source increases. The amount of time it takes the wavefront to reach a given point is known as the arrival time. The overpressure profile can be divided into the positive pressure portion (positive phase) and negative pressure portion (negative phase). Furthermore, the amount of time it takes the peak overpressure to decay below atmospheric pressure is known as the positive phase duration and is closely related to the damage capabilities of a given blast wave. Similarly, the amount of time it takes for the negative pressure to return to atmospheric pressure is the negative phase duration [1]. The area under the curve during the positive phase duration is the positive blast wave impulse.

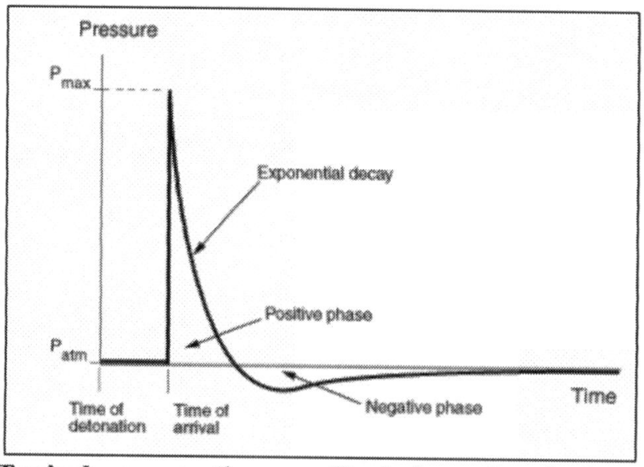

Figure 1: Typical pressure-time amplitude for a chemical explosion [7].

In some cases, the positive phase impulse is calculated and an equivalent triangular impulse with the same peak overpressure is used to approximate the realistic overpressure amplitude shown in Figure 1. In many cases, this approximation is appropriate because the positive phase portion of the incident pressure wave is typically the most damaging. Generally, structural response is more sensitive to capturing the correct incident wave reflection. Accurately simulating a realistic blast wavefront on a structure for a given explosive scenario is challenging because the peak overpressure, positive phase duration, amount of explosives, and distance from the explosion source all affect the overpressure amplitude. Furthermore, accounting for reflection effects on a loading surface adds further complication to the blast loading model.

Two different explicit finite element techniques used to simulate blast loading due to external chemical explosions in air will be discussed herein; incident wave loading and the Conventional Weapons loading model (CONWEP) [8]. CONWEP is well suited for simulating the overpressure wave associated with the detonation of conventional explosives while incident wave loading permits reasonable analysis of vapor cloud explosions. The underlying methodology of each approach will be discussed as well as the advantages and applications of each. Furthermore, both computational methods will be compared for the explicit dynamic simulations of realistic explosive overpressure waves (from a postulated concentrated explosion source and a vapor cloud) impacting single and double-walled storage tanks that carry an extremely high consequence of failure.

2. Blast Curve Generation

In order to develop the blast parameters required to generate realistic overpressure curves for conventional explosives, the equivalent TNT method is most commonly used. TNT equivalency of an explosive is simply the ratio of the mass of TNT to the mass of explosive such that both yield equal pressure and impulse [9]. In order to compare shock wave parameters to available published data, a scaled distance from an explosion source to the loaded structure may be used. As discussed in [10], the most common form of blast scaling is the Hopkinson-Cranz or cube-root scaling method following

$$Z = R/W^{(1/3)}$$

where Z is the scaled distance, R is distance from the explosion source to the loading point, and W is the weight of TNT. This formulation implies that self-similar blast waves are produced at identical scaled distances when two different sized explosive charges (of similar geometry and chemical composition) are detonated in the same atmosphere. Using published empirical data, this scaled distance from the explosion source to a loading surface can be used to determine specific blast parameters. The CONWEP blast loading model discussed herein is based on a TNT explosion (assumed surface blast). Thus, when using CONWEP, it is necessary to convert the mass of explosives in question to an equivalent amount of TNT.

As outlined in [11], a major limitation associated with the equivalent TNT method is that the blast yield is a function of the amount of explosives and not the combustion mode; making this method not preferred when analyzing vapor cloud explosions. As discussed in References [12-

16], vapor cloud explosions differ from conventional explosive detonations in that flame accelerations, which can be exacerbated by confinement due to intervening structures, define the overpressure associated with a given explosion more so than simply the amount of fuel ignited. In order to quantify blast curves associated with vapor cloud explosions several widely accepted methods have been developed; TNO [15] blast curves and Baker-Strehlow [17] blast curves among others. In this study, triangular impulses of varying peak overpressures are used for simplicity when analyzing the structural response of a double-walled storage tank subjected to a postulated vapor cloud explosion (discussed below). This also permits investigation into the sensitivity of structural response to varying blast wave parameters.

3. CONWEP and Incident Wave Loading

A significant challenge associated with computational blast modeling is accounting for reflection effects. The peak blast pressure that a structure is subjected to is greater than the peak overpressure and is known as the reflected pressure. This pressure is a function of the peak overpressure and the angle of incidence (the angle between the loading direction and the surface normal of a point on the loading surface). Neglecting incident wave reflection can yield non-conservative results, particularly for high peak overpressures, while assuming the maximum reflection effect at all points on a (cylindrical) structure can potentially be overly conservative.

The CONWEP blast loading model is based on a collection of conventional weapons effects calculations from the equations and curves of TM 5-855-1 [8] and is implemented into the commercial finite element program, Abaqus. The main advantage of using CONWEP over the incident wave loading approach is that realistic overpressure amplitudes (including positive and negative phase) and other blast wave parameters are inherently calculated based on a defined amount of TNT at a given distance from the explosion source. Furthermore, the CONWEP blast loading model does not require modeling the fluid medium (air) to account for reflection effects.

Incident wave loading is an explicit finite element technique in which an incident wave overpressure at a point can be defined as a function of space and time. This pressure amplitude is defined at a standoff point, located between the detonation point and the loaded structure. When simulating blast loading, realistic overpressure amplitudes need to be defined by the user. An incident wave can be defined as planar or spherical. A schematic showing the incident wave loading model in Abaqus is given in Figure 2 below. For a more rigorous overview of the underlying theory behind both the CONWEP blast loading model and the incident wave loading as well as a benchmark of the overpressure reflection calculations of CONWEP, refer to Reference [18].

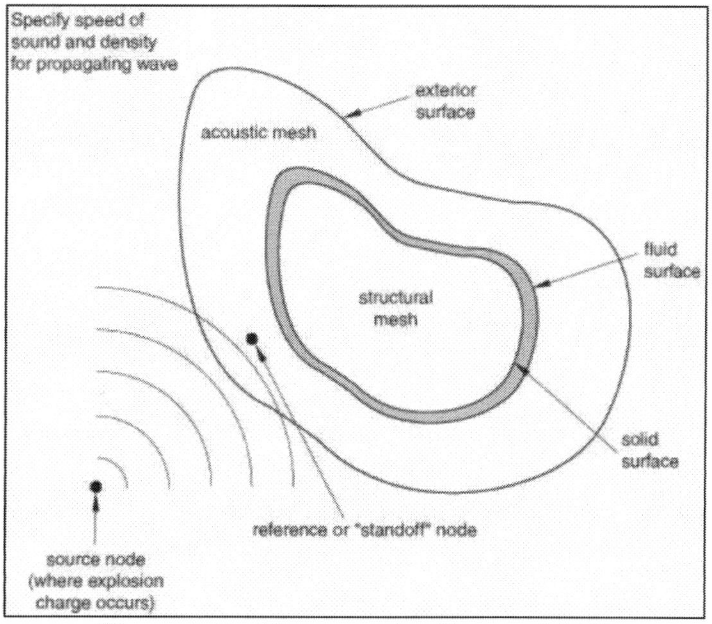

Figure 2: Incident wave loading model [7].

4. Explicit Analysis of a Storage Tank Subjected to a Vapor Cloud Explosion

As discussed in [19, 20] the consequences of storage tank failure in chemical or petrochemical facilities can be dire. Furthermore, it is estimated that 85 percent of storage tank accidents are due to fire and explosion [21]. Analyzing the structural response of storage tanks subjected to potential blast loading from conventional detonations or vapor cloud explosions is of great interest to the process safety community. The explicit finite element analyses discussed below provide real-world examples of the usefulness that computational blast analysis provides in the petrochemical and related industries.

The first computational blast analysis examined herein relates to a double-walled storage tank subjected to a postulated external vapor cloud explosion. The rough size of the outer tank is a 140 foot diameter with a shell height of approximately 125 feet. A complete, elastic-plastic, three dimensional finite element model was constructed using linear shell elements. Figure 3 shows the half-symmetry model with detailed geometric features. The baseline finite element model includes all major structural features, including the suspended deck, bracket supports, and the lower support straps. Quadratic beam elements are used to model the suspended deck hanger rods and the strap supports, and both the inner and outer tanks are modeled for completeness. Contact is defined between the suspended deck and the shell wall.

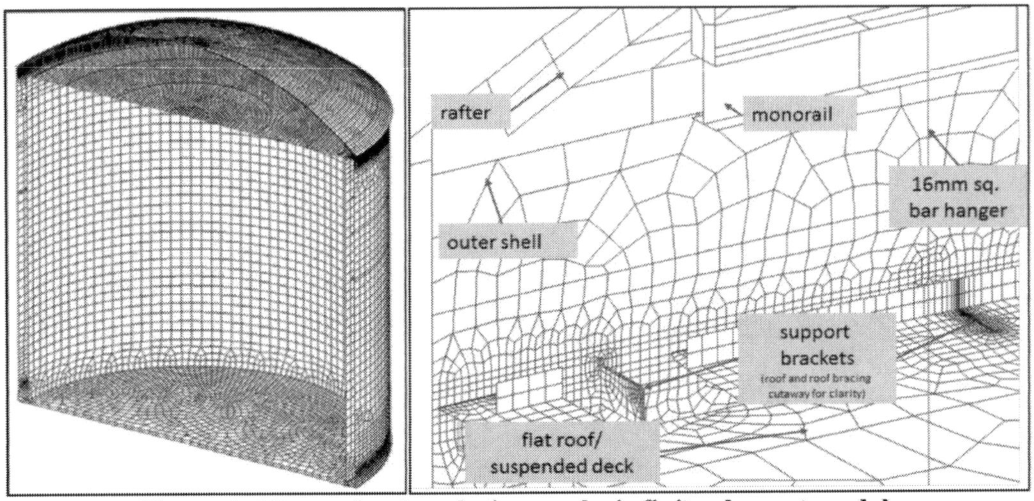

Figure 3: Vapor cloud explosion analysis finite element model.

A simplified, lumped-mass model is also analyzed. In this model, the mass associated with the suspended deck is lumped into the domed roof and the mass of the inner tank shell is lumped into the outer tank shell. The mesh is coarsened in the simplified model to aid in computational efficiency. Incident wave loading is used to analyze a range of peak overpressures (assumed to be peak reflected pressure because acoustic elements are not used) applied as planar triangular impulses. The effects of gravity (self-weight) are included.

In this case, because of the range of potential vapor cloud explosive scenarios, iterations are performed to determine the maximum permissible overpressures and loading durations at the tank to ensure containment and safe operability. That is, the focus is on establishing a sort of blast rating for the structure independent of a specific explosion scenario. This provides an essential characteristic of the structure that can be referred to for any future scenarios or concerns that might arise, and is similar in some sense to the maximum allowed working pressure (MAWP) of a pressure vessel. Peak reflected pressures of 5, 8, and 11 psi are applied for an impulse duration of 50 milliseconds to the lumped model. Figure 4 below compares maximum deflections and stresses in the lumped model from the domed roof (opposite impact side) for peak applied overpressures of 5, 8 and 11 psi. This figure demonstrates the relatively constant nature of the largest displacements. In the 5 psi case, maximum deflections are generally less than 12 inches with relatively little accumulation in plastic strain.

These results indicate that a peak reflected pressure of 5 psi is near the limit to achieve tank containment following blast loading. Furthermore, for the 8 and 11 psi overpressure cases, results indicate that plastic collapse of the domed roof and tank shell would be likely and containment of the vessel contents would be jeopardized. Figure 4 also shows an example transient deformation corresponding to the full model results for 5 psi. In general, there is reasonable agreement between the simplified, lumped parameter model and the complete model with the suspended deck across cases. However, the suspended deck is a critical feature of the tank, and is also an indicator of safe operability. As Figure 4 illustrates, the compete model indicates that the suspended deck is one of the weakest points of the tank internals; results from the complete model series of analyses show that a maximum peak reflected pressure of 2 psi is advisable to ensure safe tank operability (again, impulse duration assumed to be 50 milliseconds). This corresponds to a maximum side-on overpressure of 1 psi (uniform/constant reflection coefficient of 2).

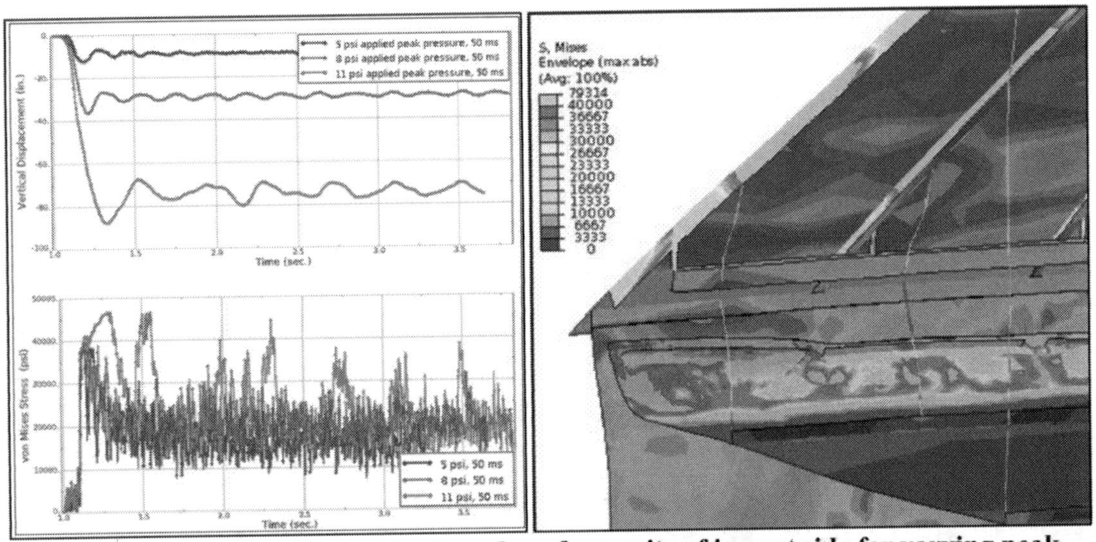

Figure 4: Deflection/stresses in domed roof opposite of impact side for varying peak overpressures and suspended deck deformation.

Additionally, in order to compare the effect of varying blast wave impulse durations, a simulation at 11 psi peak applied overpressure with an impulse duration of 10 milliseconds is carried out. Figure 5 compares maximum domed roof deflections and stresses for the 50 and 10 millisecond loading durations. As expected the duration of the impulse has a significant effect on the overall structural damage. It is noted that much lower peak overpressures acting at longer durations could also be damaging.

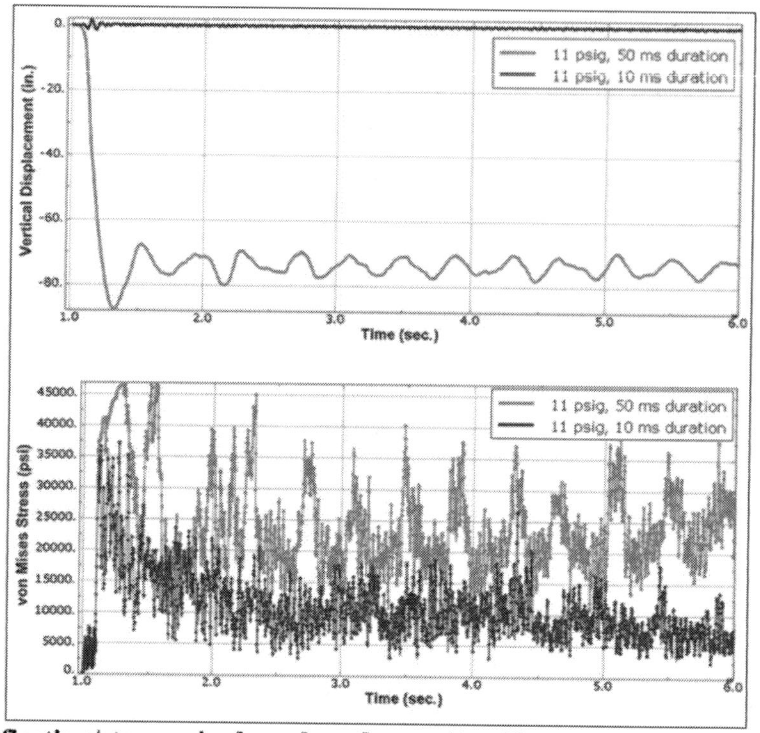

Figure 5: Deflection/stresses in domed roof opposite of impact side for varying impulse duration.

5. Explicit Analysis of a Storage Tank Subjected to Conventional Explosion

A different double-walled storage tank is analyzed; this time the postulated blast loading is due to accidental detonation of a known amount of concentrated, combustible fuel. The approximate outer tank diameter is 100 feet, and the outer tank height is roughly 75 feet. Again a complete, elastic-plastic, three dimensional finite element model is constructed using linear shell elements. Figure 6 shows a cutaway view of the model with contours of shell element thickness. The inner tank has four stiffening rings that protrude into the annular space between the inner and outer tank. The outer tank roof support structure consisting of channels and angles is also modeled in detail and included in the finite element model. Contact is modeled in between the tank structures and the effects of self-weight are included.

An initial analysis is performed on just the outer tank to investigate the plastic strains and wall deflections due to the applied blast wave. Both the CONWEP and incident wave loading models are compared (the equivalent TNT method is used to determine the overpressure profile in accordance with CONWEP). Due to the size of the finite element model and the potentially high computational cost, the fluid medium is not explicitly modeled for the incident wave simulation. An overpressure amplitude that accounts for reflection effects is used (peak reflected pressure applied in the incident wave model). Furthermore, in order to mimic the behavior of the CONWEP simulation as much as possible, the overpressure profile defined for the incident wave simulation as a function of time does, in fact, consist of positive and negative phases and decays exponentially once it reaches its peak value. The side-on reflection coefficient for the overpressure magnitude in this case is 2.

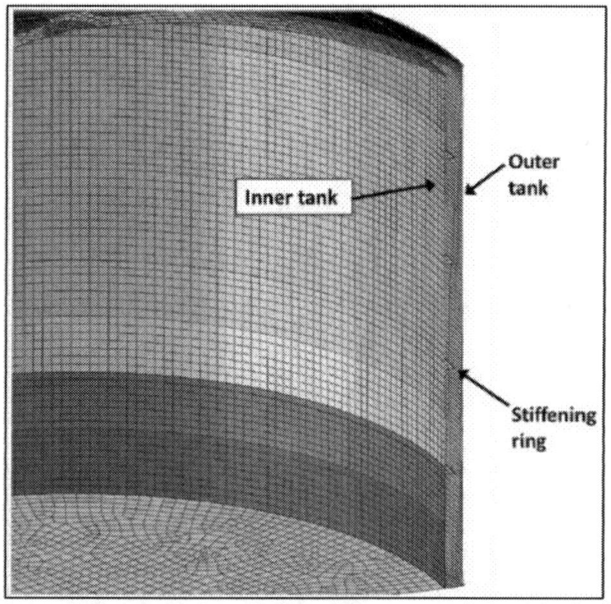

Figure 6: Cutaway view of the complete finite element model with shell thickness contours shown.

Even for the relatively small peak reflected overpressure of roughly 1.9 psi, the outer tank wall deflections prove to be significant, to the point of local collapse. A static post-buckling analysis (using the Riks solver), with surface pressure applied using an analytical field definition to match surface pressure in the dynamic simulations confirms that the outer tank wall buckles before the peak reflected pressure is reached. Riks post-buckling analyses are discussed in more detail below.

For comparison, Figure 7 shows contours of outer tank plastic deformations (identical scale) due to the overpressure wave for both methods. As expected, the incident wave approach produces higher deflections than CONWEP because the applied peak reflected overpressure does not decrease as angle of incidence increases. However, the general agreement between methods is actually quite good. This suggests that the simplification of a single maximum reflection coefficient in this case for the incident wave approach is not excessively conservative, and that reasonable results were obtained without modeling the entire acoustic domain (in which case reflection effects would be rigorously accounted for).

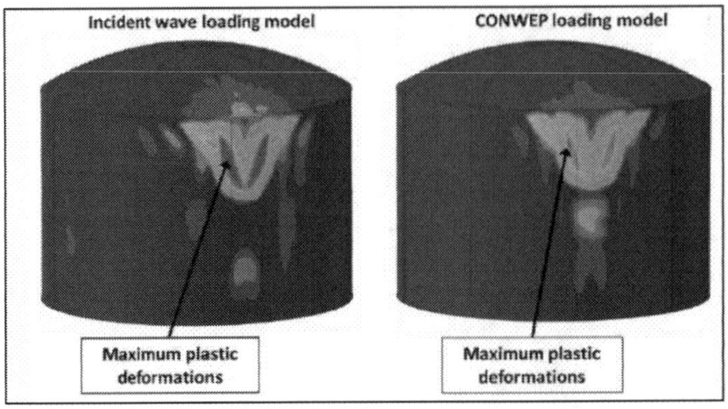

Figure 7: Outer tank plastic deformations due to blast loading using the CONWEP and incident wave loading models (contour scales identical).

For the blast analysis of the complete finite element model that includes both the inner and outer tanks with contact defined between shells, the outer wall deflections are much less significant because the outer tank impacts the inner tank stiffeners. Essentially, the inner tank stiffening rings absorb the impact from the outer tank wall and plastically deform. Furthermore, the deflections in the inner tank wall are minimal. This is demonstrated in Figure 8, which shows the maximum deflection during blast loading from the CONWEP simulation of the inner tank and the outer tank. The contour scale for the inner tank corresponds to 10 percent of the outer tank contour scale. Additionally, the maximum inner tank wall deflection is less than 10 percent of the maximum outer tank wall deflection, and the maximum outer tank deflections occur in the regions in between the inner tank stiffening rings. The outer tank wall never directly impacts the inner tank wall. The minimal inner tank deflections suggest that the sloshing effects of the liquid contained by the inner tank should also be minimal.

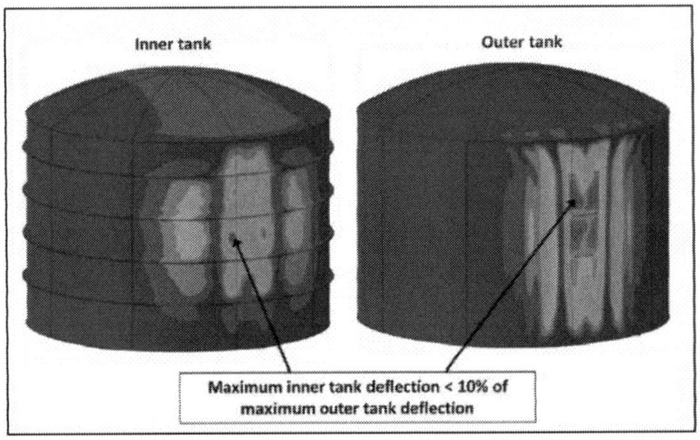

Figure 8: Maximum deflection of the inner and outer tanks due to blast loading from the CONWEP model (inner tank contour scale corresponds to 10 percent of outer tank scale).

These results indicate that from a blast damage mitigation perspective, the four inner tank stiffening rings prove to be very beneficial in reducing the severity of the impact between the inner and outer tanks. While both explicit computational techniques used to analyze this complete, three dimensional model give reasonably similar results, from a convenience standpoint, CONWEP proves to be advantageous for conventional detonations because the appropriate overpressure and spatial decay are automatically calculated for the given equivalent amount of TNT. Additionally, due to more rigorous accounting for reflection effects, some conservatism is removed in the CONWEP case, making it the natural choice for well-defined, conventional explosive detonation scenarios.

6. Explicit Analysis of a Cluster of Single-Walled Storage Tanks

The postulated blast source in this case is the same source as discussed in the previous example (conventional detonation). However, this cluster of single-walled tanks is located much closer to the blast source, meaning peak overpressures are much higher than the above example. All three tanks are roughly 30 feet in diameter and 35 feet high. The tanks also have self-supporting roofs and are anchored to their respective foundations via hold down clips and hold down bolts. As is the case for the aforementioned storage tanks in this study, these tanks carry an extremely high consequence of failure due to their process contents.

Three dimensional, elastic-plastic finite element models of each tank are constructed using linear shell elements and explicitly analyzed using CONWEP. Peak reflected overpressures at each tank as a function of time are provided in Figure 9 below. Again, the effects of gravity are included. A node at the center of each hold-down clip is fixed in all degrees of freedom and then coupled to the surrounding bolt hole to simulate the restraint of the bolts as shown in Figure 10. A rigid surface representing the tank foundation is constructed and included in the model, and contact between this rigid surface and the tank bottom is defined. This ensures that the tank will not unrealistically sag due to gravity alone (essentially penetrate the foundation due to self-weight and induce non-realistic stresses in the tank floor/walls) while allowing the tank bottom to physically lift off its foundation when the blast load is applied. In order to account for the weight of the process fluid, an equivalent pressure is applied to the tank bottom that corresponds to the weight of fluid up to the maximum fill height. This pressure is applied throughout the dynamic simulation (sloshing effects of the fluid are not considered; this assumption is discussed below).

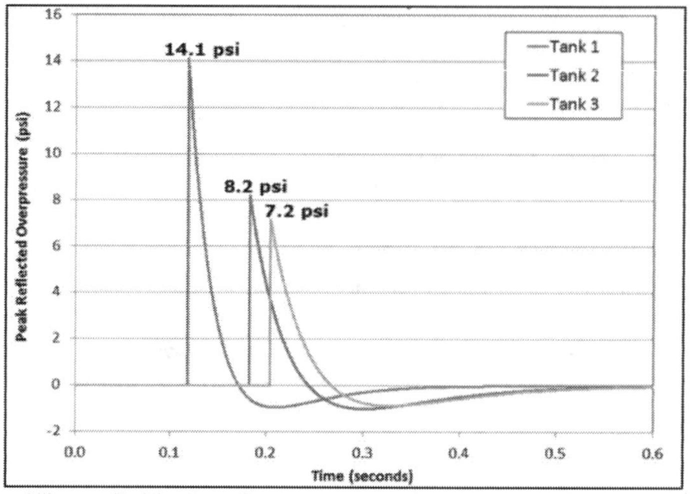

Figure 9: Peak reflected overpressures at each tank.

After the onset of the blast wave, the wall and the roof of each tank significantly deform. In order to better visualize the tank behavior following the onset of the blast wave impact, Figure 11 shows contours of Tank 1 deflection as time elapses (time scale given is seconds after detonation). These large deflections are more indicative of plastic collapse like behavior than a dynamic or vibratory failure. Furthermore, the stresses observed in the tank shortly after blast wave impact are significant.

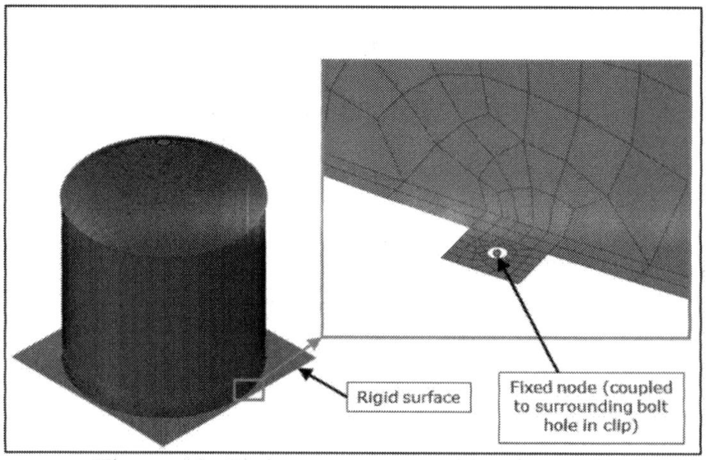

Figure 10: Finite element model of Tank 1.

Additionally, the stresses in the tank at the hold-down clips and reaction forces at the assumed bolt locations predict likely failure of either the bolts themselves or the hold-down clips. Figure 12 shows equivalent plastic strains in Tank 1 after the impact of the blast wave (contour scale capped at 5%). The hold-down clips and portions of the tank wall near the hold-down clips exceed 100% equivalent plastic strain. Results for Tanks 2 and 3 are similar.

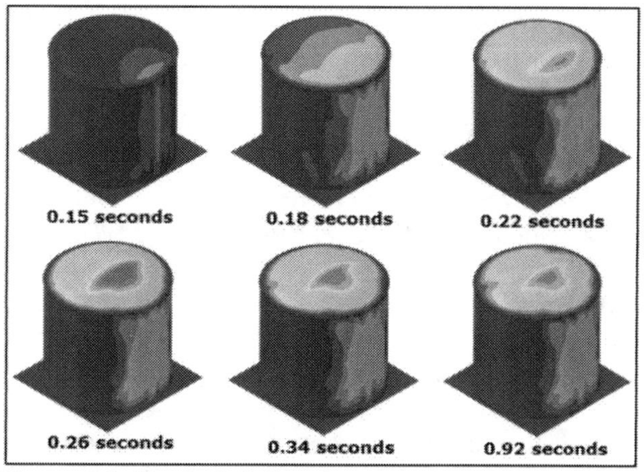

Figure 11: Tank 1 deflections during blast loading.

In order to further investigate the possibility of plastic collapse or buckling, static Riks post-buckling analyses are performed on each tank (a static pressure distribution is applied to mimic peak blast loading pressures). A Riks analysis is useful for simulating non-linear instability problems, and includes the effects of large deformations and elastic-plastic material behavior. It can also be used to look at post-buckling or collapse shapes. The applied load in a Riks analysis is actually a variable and part of the solution for each increment, and is always relative to the current deformed configuration of the structure. If no converged solution is possible for an incremental increase in load, then the equilibrium solutions can be searched for with a decreased load. This is the mechanism that allows post-buckled shapes to be predicted and visualized. In extreme cases, no equilibrium solution can be reached, even with decreased load, and only load reversal is predicted to lead to stability. This is actually the case for the tanks analyzed here, where once collapse has initiated, displacements are predicted to be essentially unbounded for any amount of positive load and thus, surface pressure becomes negative before peak overpressure values are achieved.

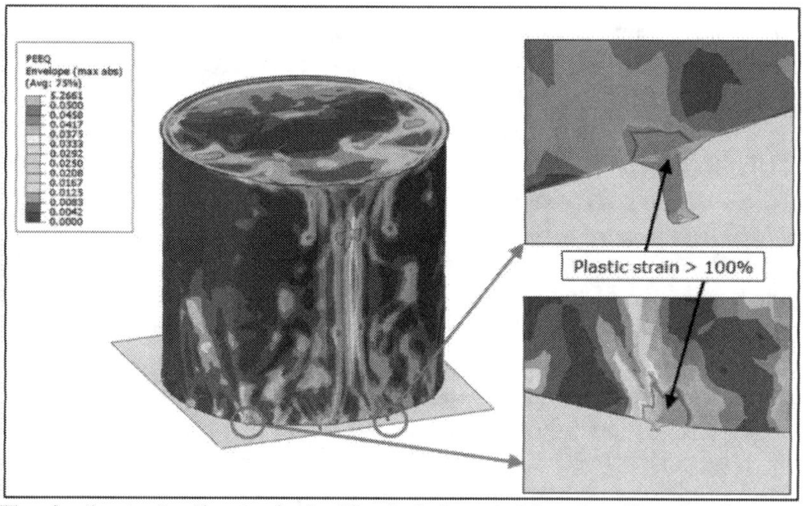

Figure 12: Equivalent plastic strain in Tank 1 due to blast loading (scale capped at 5%).

Based on the results discussed above, the three storage tanks subjected to the assumed postulated blast load are likely to fail via several possible failure modes. The first likely failure mode is plastic collapse of the tank walls and/or roofs. All three tanks see large deflections (the most severe occurring in Tank 1). Riks post-buckling analyses confirm that the magnitude of deflections seen in the tank walls would likely result in unstable collapse behavior before the peak blast overpressure is achieved. A second likely failure mode is extreme damage in the tanks at the hold-down bolts and/or regions of the tanks that are connected to the bolts. In all three tanks, equivalent plastic strains exceed 100% in these regions. Failure of multiple hold-down bolts or hold-down clips could cause a given tank to slide off its foundation and/or overturn. Furthermore, sloshing effects of the fluid inside the tanks are neglected in the simulations discussed herein, but would increase the likelihood of failure.

In short, the close proximity of all three tanks to the postulated blast source causes the corresponding overpressures acting on each tank to be relatively severe. Design changes to the tanks to supplement blast damage mitigation would likely have to be drastic in order to achieve a significant reduction in the probability of failure due to the postulated explosion considered in this study. Finally, in order to achieve practical simulation times, all of the explicit analyses discussed herein are conducted across multiple CPUs (often 16 or more) simultaneously. Utilizing parallel computing permits evaluation of complex three dimensional structures, such as the tanks discussed above, in reasonable amounts of time.

7. Conclusions

This study compares two different explicit computational methods for modeling blast loading on structures; incident wave loading and the CONWEP charge property. The underlying methodology for each method is discussed. CONWEP is well suited for conventional detonations and uses the equivalent TNT method, and it inherently calculates the appropriate blast wave parameters for a given explosive event. Additionally, CONWEP achieves realistic overpressure amplitudes, spatial decay, and reflection effects without having to model the fluid medium, as is the case for incident wave loading. Incident wave loading provides the user with the flexibility to define a time-varying overpressure amplitude. This is useful when performing sensitivity studies to blast loading parameters or when attempting to characterize the blast wave associated with a vapor cloud explosion that does not behave like a conventional detonation. The development and implementation of the CONWEP and incident wave blast loading models into commercial finite element programs makes accurately analyzing the response of complex three dimensional structures subjected to blast loading feasible.

Additionally, explicit, three dimensional finite element analyses of both single and double-walled storage tanks that carry an extremely high consequence of failure are discussed herein. Incident wave loading is used iteratively to determine the maximum permissible overpressure for a double-walled storage tank subject to a postulated vapor cloud explosion. The CONWEP blast loading model is used to analyze single and double-walled storage tanks subjected to a postulated chemical explosion (conventional detonation). The explicit computational methods, such as the ones examined herein, can provide valuable insight into the blast resistance of existing structures and can guide design decisions regarding how to mitigate blast damage for new structures in the petrochemical and related industries. Gaining a better understanding of the possible failure modes due to blast loading of critical process equipment is very beneficial in not only improving process safety, but also in enhancing safety protocol for catastrophic accidental explosions in industrial environments.

8. References

1. DOD, "Unified Facilities Criteria (UFC) 3-340-02: Structures to Resist the Effects of Accidental Explosions," *Department of Defense, United States of America*, 2008
2. Smith, P., and Hetherington, J., "Blast and Ballistic Loading of Structures, 1st ed," *Butterworth-Heinemann*, Oxford, 1994.
3. Kinney, G., and Graham, K., "Explosive Shocks in Air, 2nd ed," *Springer-Verlag*, Berlin, 1985.
4. Dusenberry, D., "Handbook for Blast-Resistant Design of Buildings, 1st ed," *John Wiley and Sons, Inc.*, Hoboken, New Jersey, 2010.
5. ASCE, "Design of Blast-Resistant Buildings in Petrochemical Facilities, 1st ed," Task Committee on Blast-Resistant Design, *American Society of Civil Engineers.*, Reston, Virginia, 1997.
6. Bulson, P., "Explosive Loading of Engineering Structures, 1st ed," *Taylor and Francis Group*, London, 1997.
7. Abaqus 6.10-1, DSS Simulia, "Analysis Users' Manual," 2010.
8. Hyde, D., "Users' Guide for Microcomputer Programs CONWEP and FUNPRO – Applications of TM 5-885-1," *U.S. Army Engineer Waterways Experimental Station*, Vicksburg, VA, 1988.
9. Huntington-Thresher, W., and Cullis, I., "TNT Blast Scaling for Small Charges," *19th International Symposium of Ballistics*, 2001.
10. Baker, W., "Design Considerations for Toxic Chemical and Explosives Facilities – Chapter 1, Blast Pressure Effects: An Overview," *American Chemical Society*, 1987.
11. Tang, M.J., and Baker, Q.A., "A New Set of Blast Curves from Vapor Cloud Explosion," *Process Safety Progress*. Vol. 18, No. 3, 1999.
12. American Institute of Chemical Engineers (AiChE), "Guidelines for Evaluating the Characteristics of Vapor Cloud Explosions, Flash Fires, and BLEVEs," *Center for Chemical Process Safety of the American Institute of Chemical Engineers.*, New York, 1994.
13. Lees, F.P., "Loss Prevention in the Process Industries," *Butterworth-Heinemann*, Oxford, 1996.
14. Mercx, W.P.M., van den Berg, A.C. and van Leeuwen, D., TNO report PML 1998-C53, "Application of correlations to quantify the source strength of vapor cloud explosions in realistic situations, Final report for the project: GAMES," *Netherlands Organization for Applied Scientific Research (TNO)*, 1998.
15. Van den Bosch, C.J.H. and R.A.P.M. Weterings, eds., "Methods for the calculation of physical effects: due to releases of hazardous materials (liquids and gases) ("TNO Yellow Book"), CPR 14E, Parts 1 and 2, 3rd Edition," Director-General for Social Affairs and Employment, *Committee for the Prevention of Disasters*. The Hague, 1997.

16. Woodward, J. L., "Estimating the Flammable Mass of a Vapor Cloud," *Center for Chemical Process Safety of the American Institute of Chemical Engineers*, New York, 1998.
17. Strehlow, R.A., Luckritz, R.T., Adamczyk, A.A., and Shimpi, S.A. "The Blast Wave Generated by Spherical Flames," *Combustion and Flame*, Vol. 35, 297-310, 1979.
18. Prueter, P.E., Dewees, D.J., and Brown, R.G. "Examining Computational Techniques to Simulate the Effects of External Chemical Explosions Using Explicit Finite Element Analysis," *In ASME Pressure Vessels and Piping Conference*. PVP2012-78566, 2012.
19. Maremonti, M., Russo, G., Salzano, E., and Tufano, V., "Post-Accident Analysis of Vapor Cloud Explosions in Fuel Storage Areas," *Trans IChemE*, 77(B), 1999.
20. Buncefield Investigation Board, "The Buncefield Incident 11 December 2005: The Final Report of the Major Incident Investigation Board," Volume 1, 2008.
21. Chang, J., and Lin, C., "A Study of Storage Tank Accidents," *Journal of Loss Prevention in the Process Industries*, 19(1) 51-59, 2006.

(44o) The Hidden Risks of Deposits, by-Products, and Residues On Exhaust and Waste Streams

Carlos A. Barrera and Abid Kemal, Thermal Sciences, Exponent, Menlo Park, CA

The safety risks associated with reactive compounds are usually captured and addressed during the conceptual design and pre-construction process safety reviews as, and for well-established chemistries these risks are well understood for the main processes and reactions. Lest time, if any, is devoted to the analysis of the reactive nature of the compounds present waste streams or exhaust lines. These lines are often overlooked during HAZOPs because of their secondary nature, or because they go to a scrubber or flare, which erroneously is considered a "catch-all" device that will "take care of everything." Waste streams and exhaust lines from certain physicochemical processes contain highly reactive compounds in the form of residues or by-products that in the right conditions can accumulate and form dangerous deposits. Some examples of such deposits are pyrophoric, shock-sensitive, and corrosive compounds. The purpose of this paper is to create awareness in the process safety community about the importance of a complete understanding of the chemistry of the reactive species present in these streams and the possible physical mechanisms than can lead to the formation of dangerous deposits, as well as the importance of including these streams in the process hazard analysis. Examples drawn from multiple processes will be presented.

(44q) Explosion Characteristics of Carbon Nano-Fibers

Jiaqi Zhang, Artie McFerrin Department of Chemical Engineering, Texas A&M Univerisity, Mary Kay O'Connor Process Safety Center, College Station, TX, Yi Liu, Mary Kay O'Connor Process Safety Center, Department of Chemical Engineering,Texas A&M University, College Station, TX and Dr. M Sam Mannan, Artie McFerrin Department of Chemical Engineering, Mary Kay O'Connor Process Safety Center, College Station, TX

Dust explosion is a hazard with serious consequences. While the main principle of dust explosion is still being exploring, the gap of dust explosion between particles of micro-size and those of nano-size also needs tremendous efforts. The explosion characteristics of CNFs are studied in Mary Kay O'Connor Process Safety Center. These experiments are conducted in a 36-L explosion vessel, which has been calibrated with a standard 20-L explosion vessel. Among the experiments, different kinds of CNFs of different grades and different conditions (milled and unmilled) are used. Compared with collected published data of graphic particles of micro-scale, the experimental data shows that the carbon nano-fibers create a higher Kst than the micro-scale ones. The results also show that explosion of the XT-PS grade (Pyrolytically Stripped) is much stronger than that of the XT-LHT grade (Low Temperature Heat-treated) and XT-HHT grade (High Temperature Heat-treated). Also, the explosion of milled CNFs is stronger than the unmilled ones. Further study to seek the causes is undergoing.

(44r) High-Fidelity Combined Dynamic Modeling of Depressurizing Vessels and Flare Networks: Accurately Assess the Effects of Process Changes On Mechanical Integrity and System Capacity

James Marriott, VP PSE Oil & Gas, Process Systems Enterprise Ltd, London, United Kingdom

The use of high-fidelity dynamic modeling for combined depressurization and flare network analysis can result in identification of additional capacity in flare networks without compromising process safety, presenting potentially significant opportunities for reducing or saving capital expenditure (CAPEX). The same analysis can simultaneously provide accurate assessment of temperatures in vessels and pipework, to identify areas where close approach to or violation of minimum metal temperatures may occur, thus compromising vessel and pipework mechanical integrity.

The flare networks for major plants represent a non-negligible part of the overall capital investment. Current industrial practice for their design is primarily based on steady-state analysis, as described in standards such as API 521 and supported by widely-used software tools. However, it is a widely recognized fact that the application of steady-steady considerations to what is fundamentally a dynamic system inevitably requires the use of conservative assumptions which often result in significant oversizing of flare headers and other components of the network. Another major contributor to capital cost is the use of special materials for the parts of the system that may be exposed to low-temperature fluids – typically the tail pipes attached to high-pressure process vessels – and which are therefore at risk of embrittlement. The key to being able to limit the additional capital expenditure is the accurate estimation of the length of piping that is subject to "abnormal" temperatures.

In addition to dynamic flare network analysis, the detailed dynamic modeling and simulation of the rapid depressurization ("blowdown") of high-pressure vessels is a key factor in determining both the load imposed on the pressure relief system (e.g. flare network) and also the temperatures of the vessel and downstream pipework walls. Current methods typically use 'lumped' equilibrium approaches that do not adequately predict these quantities; indeed, in contrast with the current over-conservative practice in flare network design, they may provide non-conservative solutions. By contrast, the high-fidelity approach incorporates a multi-phase non-equilibrium representation of the vessel fluid contents, as well as a 3-dimensional model of the metal walls that takes into account the transfer of heat between regions of the wall in contact with different phases. This allows a much more accurate estimation of the wall temperatures, and the direct computation of thermal stresses.

In the past it has been difficult to assess these important design criteria effectively because of the challenges in modeling such complex systems. This paper describes how the combination of high-fidelity depressurization capabilities and an advanced model-based system for flare system network design, coupled with evolving best practice in using these tools, represents a transformation for process safety management. The new techniques allow accurate assessment of the effects of process changes on mechanical integrity, thus enabling informed safety and CAPEX decision support.

An Engineering Method to Mitigate the Impact of Regulatory Focus on Relief System Installations by Prioritizing Risk

Dustin Smith
Smith & Burgess
5700 NW Central Dr. Ste 301 Houston, TX 77092
dustin.smith@smithburgess.com

Prepared for Presentation at
American Institute of Chemical Engineers
2013 Spring Meeting
9th Global Congress on Process Safety
San Antonio, Texas
April 28 – May 1, 2013

UNPUBLISHED

An Engineering Method to Mitigate the Impact of Regulatory Focus on Relief System Installations by Prioritizing Risk

Dustin Smith
Smith & Burgess
5700 NW Central Dr. Ste 301 Houston, TX 77092
dustin.smith@smithburgess.com

Keywords: Pressure Relief Valve Installation, API 3% Rule, Chatter, Relief Device Stability

Abstract

Since around 2005, regulators in the United States have put greater emphasis on relief device installations meeting the 3% rule. Spending large amounts of money to "fix" relief device installations that pose no safety risk decreases the plant's overall safety. To date, more incidents have occurred during facility construction / modifications than due to relief device chatter. This paper presents a method to assist engineers in determining if relief devices are susceptible to chatter. The methodology in this paper provides an engineering study to determine if existing installations are safe, which are allowed in the relevant engineering standards (API STD 520), and shows the research data by which it was validated. The model is used as a screening method that places the relief devices into two categories: (1) those installations that may chatter and (2) those installations that need no further review. The goal of any experimental comparison is that the model will error on the side of predicting chatter, but will be reliable enough to screen valves.

In addition to presenting the model, this paper will compare instances of known chatter to research conducted by API and work done by the Electric Power Research Institute in the 1980s. Thus far, based on research and acquired information, the method predicted all instances of chatter known to the authors. By providing a screening methodology that is supported by experimental data, plants can focus their spending on fixing real safety issues by identifying which relief installations are not expected to chatter. The paper will close by giving a brief explanation of the on-going research in relief valve stability.

1. Introduction

In November 2011, *Hydrocarbon Processing* published a paper that documented a method to determine if relief devices were susceptible to chatter. Other methods are being developed to determine the chances of chatter for a specific installation; however, the model discussed in the published paper is the only screening method that places the relief devices into two categories: (1) those installations that may chatter and (2) those installations that need no further review. The goal of any experimental comparison is that it will error on the side of predicting chatter, but will be reliable enough to screen valves. Since the publication of that article, the Oil & Gas industry has continued to struggle with the issue of relief device stability, so much so that API delayed issuance of *API STD 520 Part II Sizing, Selection, and Installation of Pressure-Relieving Devices in Refineries-Part II, Installation*. This paper compares instances of known chatter to research conducted by API, and uses this comparison to evaluate the model. Thus far, based on research and all acquired information, the method predicted all instances of chatter known to the authors.

This paper summarizes the analysis methods and validation for vapor service valves. The November 2011, *Hydrocarbon Processing* paper documents analysis methods for two-phase and liquid services. The focus on vapor / gas service valves is based on the validation data for these services analyzed herein.

2. Summary of the Stability Screening Method

The following is a summary of items used to confirm if chattering is likely to occur. Refer to the published paper for an extensive discussion of the methodology:

1. Confirm that the inlet line length is within the criteria to ensure that the pressure wave decay is not high enough to cause valve instability. The following equation (from Frommann & Friedel, Source 4) is used for this criteria:

$$L_i < 45{,}390\frac{d_i^2}{w_{\%O}}\left(\frac{P_S - P_{RC}}{P_S}\right)(P_S - P_B)t_o \qquad \textbf{Equation 1}$$

2. Confirm that the system's frictional and acoustic losses are less than the difference between the operating pressure and the valve's closing pressure. The formula for estimating the acoustical losses (from Singh, Source 5) is presented below:

$$\Delta P_{Acoustic} = \frac{Lw_{PSV}}{12.6d_i^2 t_O} + \frac{1}{10.5\rho}\left(\frac{w_{PSV}L}{cd_i t_O}\right)^2 \qquad \textbf{Equation 2}$$

3. Ensure that stability will not be compromised due to vortex shedding in the inlet line (this phenomenon occurs during normal operation and is generally not associated with relief conditions).

$$L_i < \frac{d_i c}{2.4U} \qquad \textbf{Equation 3}$$

4. Ensure that the relief rate or system size is at sufficient capacity to confirm that chattering will not occur from being oversized. For vapor service relief devices, this is assumed to not occur if the open-close cycle time for the system is greater than 1 second. This cycle is broken into two components: the time it takes for the system pressure to increase from the relief devices' blowdown/closing pressure to the set pressure and the time it takes for the system to depressure through the relief valve from the set pressure to the blowdown/closing pressure. This is represented by the following equation.

$$1 > t_{cycle} = t_{P_{RC} \to P_S} + t_{P_S \to P_{RC}} \qquad \textbf{Equation 4}$$

5. Ensure that the relief device is installed per the manufacturers' and/or general installation guidelines. These guidelines are shown in the catalogs, ASME B&PVC Section VIII, API STD 520, etc.

6. Review the valve's operational history to ensure that there is no known history of chatter.

For a complete discussion on the methodology of this analysis, the reader is referred to the work published by Smith Et. al, as the discussion is not further expanded in this paper.[1] Note that equation 17 does not follow the verbiage, and the *less than* sign should be replaced with a *greater than* sign.

3. Acceptance Criteria

This model is the only screening method to place the relief devices into two categories: (1) those installations that may chatter and (2) those installations that need no further review. The goal of any experimental comparison is that it will error on the side of predicting chatter, but will be reliable enough to screen valves. The author considers this model to be reliable as an engineering screening tool if the following criteria are met:

- *High Correlation to Experimental Data* – If the method predicts that valves will chatter and valves won't chatter most of the time, then it is a viable screening tool given the low cost of performing the analysis compared to making piping modifications.

- *No False Negatives* – The screening tool will be considered valid so long as it does not predict any installations as safe (no chatter) when in fact chatter was experimentally found.
- *Limited False Positives* – Given that the method is a conservative screening tool, some level of false positives are to be expected. This includes instances when the model would indicate the possibility of chatter when experimental results show valve stability.
- *Screens Valves as Acceptable* – The model will only be of value if the results of the model result in existing installations not requiring further modification. Based on the results presented in the referenced paper, 50% of the installations with inlet pressure losses greater than 3% were deemed acceptable as is.

The model, therefore, will be deemed conservative and reliable if the previous four criteria are met.

4. Comparisons / Validations

To confirm the validity of the procedure, the procedure has been reviewed against all cases of chatter known to the authors. The cases are divided into three different categories:

1. ***Known Installations*** – These installations are those that are known by the authors to have failed with a loss of containment and those installations that have sufficient information to perform the analysis. The information for some of the installations is not available in the public domain, as the confidentiality of these installations has been preserved.
2. ***ASME Studies*** – In the 80's, Zahorsky performed a set of research for the nuclear industry in which they experimentally set the blowdown as the minimum percentage needed for stable operation[2]. In order to meet the acceptance criteria, therefore, this method should determine that these installations are unstable. The results were re-checked with an increased blowdown of 2% above what the author listed to see what would happen.
3. ***API PERF Studies*** – In 2011 the results of the API PERF study were presented at the API 520 Committee meeting. In this work, the API subgroup tested 18 different valves with three different inlet line lengths to determine when chatter would occur. The results of the 54 trials were introduced in their presentation (some combinations were not tested).

5. Results

The following are the results of the comparisons:

Table 1 lists the results of the model comparison with the valves that have had a loss of containment where inlet line chatter was the cause or a contributor to the incident. The model predicted the potential for chatter for each of these installations.

Table 1: Summary of known installations that have chattered

Installation	Service	Chatter Predicted	Data Source
Refinery, North America	Liquid	Yes	Internal company incident investigation
Gas Plant, Middle East	Vapor	Yes	Internal company incident investigation
Refinery, North America	Liquid	Yes	Internal company incident investigation
Refinery, North America	Liquid	Yes	API Published Document [3]

Table 2 lists the blowdown at which the relief devices do not chatter as experimentally determined by Zahorsky. The table also lists the blowdown at which the model would predict stability. The difference between the two values can be taken as an estimate to which the model is conservative. Since the blowdown was experimentally set as the minimum percentage, the model is accurate, as it predicts false positives for these devices.

Table 2: Comparison of minimum predicted blowdown to the results experimentally derived by Zahorsky

Run Case	Experimentally Determined Blowdown	Predicted Acceptable Blowdown	(Δ Blowdown) Predicted – Expermintal)
1	3.9%	4%	0.1%
2	3.9%	5.6%	1.7%
3	5.6%	9.7%	4.1%
4	8.4%	16.7%	8.3%
5	8.3%	12.6%	4.3%
6	4.3	5.3	1.0%

Table 3 lists the comparison of the model results to the experimental PERF results. The model had agreement ~72% of the time if the cases that were not tested were excluded. If one assumes that the results for the untested cases would have chattered, the agreement is ~76%. Of the cases that the model did not agree with the experimental results, the model always predicted chatter on stable valves; thus, the model is accurately screening and is slightly conservative.

Table 3: Comparison of the model to the API PERF Study Results

Model Correlation	**PERF Results**	**Model Prediction**	**No. Of Cases**
Agreement	Chatter	Chatter	9
Agreement	Stable	Stable	25
False Negative	Chatter	Stable	0
False Positive	Stable	Chatter	15
Agreement[1]	Not Tested	Chatter	7

Note 1: There are a number of cases that were not tested, but were assumed to chatter as the reason for not being tested was not included.

6. Conclusion

The model results always predicted when valves may chatter (need modifications). There are no instances of valves that have chattered where the model predicted stability.

- *High Correlation to Experimental Data* – There is a 70% to 75% agreement with the API PERF Studies and all industry installations with known chatter where identified with the model. This model predicted if a valve would chatter accurately 75% of the time (see the next item on false negatives).
- *No False Negatives* – No cases of chatter were found when the model predicted stable operation.
- *Limited False Positives* – 25% of the PERF study valves that were stable indicated that chatter was possible. Also, the work by Zahorsky indicated that the model predicts stable operation with an inlet blowdown generally 2% to 4% greater than needed.

The model, therefore, is a reliable method to screen relief devices based on the comparison to 60+ experimental and industrial data points.

7. Notation

The equations presented in this paper are dimensional and the following listing explains the variables and unit set.

c = speed of sound (ft/s)
d = diameter (in)
L = length (ft)
P = pressure (psig)
t = time (s)
U = process fluid velocity as it passes the PSV nozzle (ft/s)
w = mass flow rate (lb/s)

<u>Greek Letters</u>

ρ = fluid density (lb/ft^3)

Subscripts
B = backpressure on relief device
i = inlet
o = opening
PSV = process safety valve
RC = valve reclosing pressure
S = relief device set pressure
%O = flow rate at the valves percent open

8. References

[1] Smith, Dustin P.E., John Burgess P.E., and Craig Powers PhD. "Relief Device Inlet Piping: Beyond the 3% Rule." *Hydrocarbon Processing.* 2011.

[2] Zahorsky, J.R. "Testing and Analysis of Safety/Relief Valve Performance." The ASME Winter Annual Meeting. November 15-19, 1982. Atlanta, GA

[3] Myers, DF, Conoco. "Denver Refinery Fire." *API.* 1986.

[4] Frommann, O. and Friedel, L. "Analysis of safety relief valve chatter induced by pressure waves in gas flow." *Journal of Loss Prevention in the Process Industries.* 1998.

[5] Singh, A. "On the Stability of a Coupled Safety Valve-Piping System." The ASME Winter Annual Meeting. November 15-19, 1982. Atlanta, GA.

9TH GLOBAL CONGRESS ON PROCESS SAFETY

Making Relief Load Estimates Match Reality

Nicholas N. Cristea
Smith & Burgess, LLC
5700 Northwest Central Dr. Ste 301 Houston, Texas 77092
Nicholas.Cristea@smithburgess.com

Dustin Smith
Smith & Burgess, LLC
5700 Northwest Central Dr. Ste 301 Houston, Texas 77092
Dustin.Smith@smithburgess.com

Prepared for Presentation at
American Institute of Chemical Engineers
2013 Spring Meeting
9th Global Congress on Process Safety
San Antonio, Texas
April 28 – May 1, 2013

UNPUBLISHED

Making Relief Load Estimates Match Reality

Nicholas N. Cristea
Smith & Burgess, LLC
5700 Northwest Central Dr. Ste 301
Houston, Texas 77092
Nicholas.Cristea@smithburgess.com

Dustin J. Smith, P.E.
Smith & Burgess, LLC

Keywords:
Relief load estimates, relief system analysis, process simulations, reduced duty, inlet control valve failure, column systems

Abstract

A pressure relief system designer must make some assumptions when looking at a relief systems analysis. Some assumptions may result in costly modifications of the system, so knowing when to apply certain assumptions is paramount with the ability to accurately size a relief valve. When looking at systems, industry standards should be consulted before taking certain credits based on the given scenario. For example, the designer may be able to reduce the relief load by looking at a reduced reboiler duty and a temperature pinch during an overpressure event. When evaluating an inlet control valve failure, some credits may be taken for the minimum turndown for the outflow. However, when looking at how a downstream system may respond to the overpressure event, credit for flow continuing may not be acceptable and flow may be assumed to be lost. The application of assumptions is common practice in relief systems analysis; however, an assumption that does not match the true operational limits may lead to problems with the relief system that do not actually exist. These invalid assumptions could cost millions of dollars to fix the problem.

1. Introduction

When doing a Relief Systems Documentation Analysis, the ability to accurately model a "worst case" relief scenario is paramount to ensuring the system is protected from the overpressure event. The engineer is undoubtedly limited to the availability of equipment data and accuracy of the process simulation developed to predict upset conditions. However, if some conservative yet unrealistic assumptions are used, the relief device can be labeled as providing inadequate protection in the event of the initiating event. [1] Physically modifying the system to fix the problem, as well as the loss of profit from having to shut down the unit to perform the update, can cost the facility large sums of money. However, performing a detailed analysis on the system may better predict the "worst case" scenario, avoiding unnecessary modifications. The ability to make relief load estimates match what may happen out in the field can still be accomplished, but knowing the difference between a realistic conservative assumption and an unrealistic conservative assumption is paramount. The pressure relief system designer should consult industry standards (e.g. API 521or NFPA) to see how to estimate relief loads for given overpressure scenarios and what assumptions for mitigating items/actions are acceptable. This paper will look at two cases where relief load estimates were reduced and one where by matching the physical constraints of the system to reality resulted in a larger relief load estimate.

2. When Realistic Assumptions Lower Relief Estimates

Oftentimes relief estimates are based on normal operating flow rates, heat exchange, and/or flow paths continuing during the upsets. With the amount of optimization and debottlenecking, these assumptions may not be realistic. Equipment capacities that are right on the design limit may not be able to produce the same flow rate or heat transfer under upset conditions. Flow paths that are available under normal operation may be isolated or otherwise not be available during upsets. The following are three examples that show these effects in a typical plant situation. The first two are examples of assumptions that tended to increase the relief rate while the last example illustrates an optimistic assumption that under-predicted the relief requirements.

2.1 Reduced Reboiler Duty

When looking at a column system, the primary relief loads tend to consist of a combination of continued heat input with the loss of cooling. The relief load estimates can be complex due to the very nature of the system. The ability to do a detailed analysis of the column may be able to better predict upset conditions. The authors have seen many cases where a detailed column analysis prevented the relief device modifications from being made based on inaccurate sizing basis. The ability to accurately model the relief scenario is paramount because an error could lead to events such as an unnecessary facility modification or a loss of containment from an undersized relief device. When a pressure relief system designer looks at the loss of overhead cooling on a column system, the scenario is a transient event. The case can start with the loss of overhead condensing with the reflux continuing for a short duration. After some time, the loss of condensing and continued overhead liquid product output results in the loss of level and reflux flow. To estimate this relief condition, the column system is modeled as two steady state cases to bracket the relief load. Under either of these conditions, the bottoms temperature may increase due to the increase in system pressure, but this phenomenon may be offset by the reduction in separation; therefore, a reduced duty calculation should be performed for both calculation sets to ensure that the reboiler is capable of transferring the same heat into the column system as during normal operating conditions. When looking at the loss of overhead cooling with the reflux continuing, the reboiler duty may not be able to be reduced due to increased light component traffic in the tower (as compared to the top tower reflux failure) when compared to the high temperatures in the reboiler and lower duties expected during the top-tower reflux failure scenario due to the loss of the light ends liquid traffic in the column. Taking credit for the reduced duty is still a reasonable assumption as an upper conservative limit when the assumption is based on the maximum duty possible for the installed reboiler and is limited by equipment capacity. The design equation for heat exchangers is given below [2]:

$$Q = UA\Delta T_{LM}$$

Reduced duty calculations, limited to process/process or process/utility exchangers (e.g. shell and tube exchangers), are not typically performed on fired heater reboilers. The exchanger data sheet, if up-to-date, is the best resource for determining the heat transfer characteristics of an exchanger. Most process simulators can simulate the duty of shell and tube exchangers by specifying the clean UA and, if applicable, the utility stream (e.g. 50# steam). The duty for the exchanger may change as reflux is lost to the system, which the upset may result in hotter bottoms temperature to the reboiler. The reduced duty may reduce the relief load estimate, as there is less vapor generation as compared to using the process duty under normal conditions. Figure 1 shows a screenshot from a process simulator that estimates the rate for a column system with continued feed and heat input, but a loss of cooling and reflux.

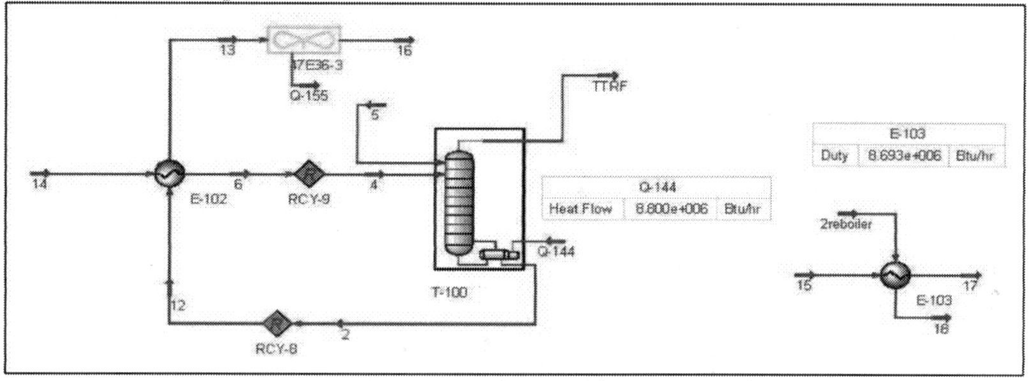

Figure 1: Example of a simulation for a column with loss of cooling and top-tower reflux failure with a reduced duty calculation performed.

2.2 *Using the correct composition and physical properties*

Generally the capacity of the reboiler (e.g. the reduced or LMTD pinched duty) will vary greatly if the calculation is based on the normal bottoms product as compared to the overhead reflux/liquid product. In the case of system upsets, the actual liquid inventory in the bottom of the tower may be lighter than normal. The pressure relief system designer is cautioned to use a composition in the reboiler to estimate the reduced duty that is consistent with the overpressure scenario being reviewed.

2.3 *Inlet Control Valve Failure Credits*

Another common overpressure scenario is the failure of automatic controls. Per API 521, this scenario is the complete failing open of a control valve (irrespective of its fail safe position).[3] If this upset condition results in overpressure, a relief device may be required to protect downstream equipment. When looking at an inlet control valve failure, the first common assumption pressure relief system designers make is to estimate the full capacity of the control valve and then verify if the relief valve can pass that fluid at the appropriate overpressure. The designers verify if the capacity of the relief device is greater than the control valve. If this estimation results in insufficient relief valve capacity, then some credits may need to be taken to make the relief load estimate more realistic.

> Per API 521§ 5.10.2,
> "in evaluating relieving requirements due to any cause, any automatic control valves that are not under consideration as causing a relieving requirement and that would tend to relieve the system should be assumed to remain in the position required for minimum normal processing flow. In other words, no credit should be taken for any favourable instrument response." [3]

For example, if the outlets from the system are expected to remain open, the calculation may assume that some of the fluid will flow through the normal path, see Figure 2. This credit is usually at the minimum turndown rate, as common industry practice is to assume that other control system remains in place and does not respond (unless it exasperates the single failure).[3] If the relief scenario is liquid, credit can be taken for the liquid outlet to reduce the fluid from the control valve that needs to be relieved by the pressure relief device by the minimum turndown rate. If the liquid outlet does not provide enough credit, the vapor outlet can be sized to take some relief as long as the vapor line is not lost by passing the liquid. A common example of this would be that a high-level alarm would trip a compressor; in this case, the relief device would have to pass both the liquid and the vapor feed to the vessel as the increased liquid level caused multiple failures. As long as the pressure relief system designer verified the liquid could flow through the vapor line and accumulate in the downstream system, credit can be taken through the vapor line. If the relief scenario is vapor, credit may be taken for flow through the vapor line at the minimum turndown rate (though the potential for credit for vapor flow through the normal liquid outlet is normally not considered). Consideration should be given for the difference in the relief fluid and the normal fluid and to adjust the credit based on the actual volume rate.

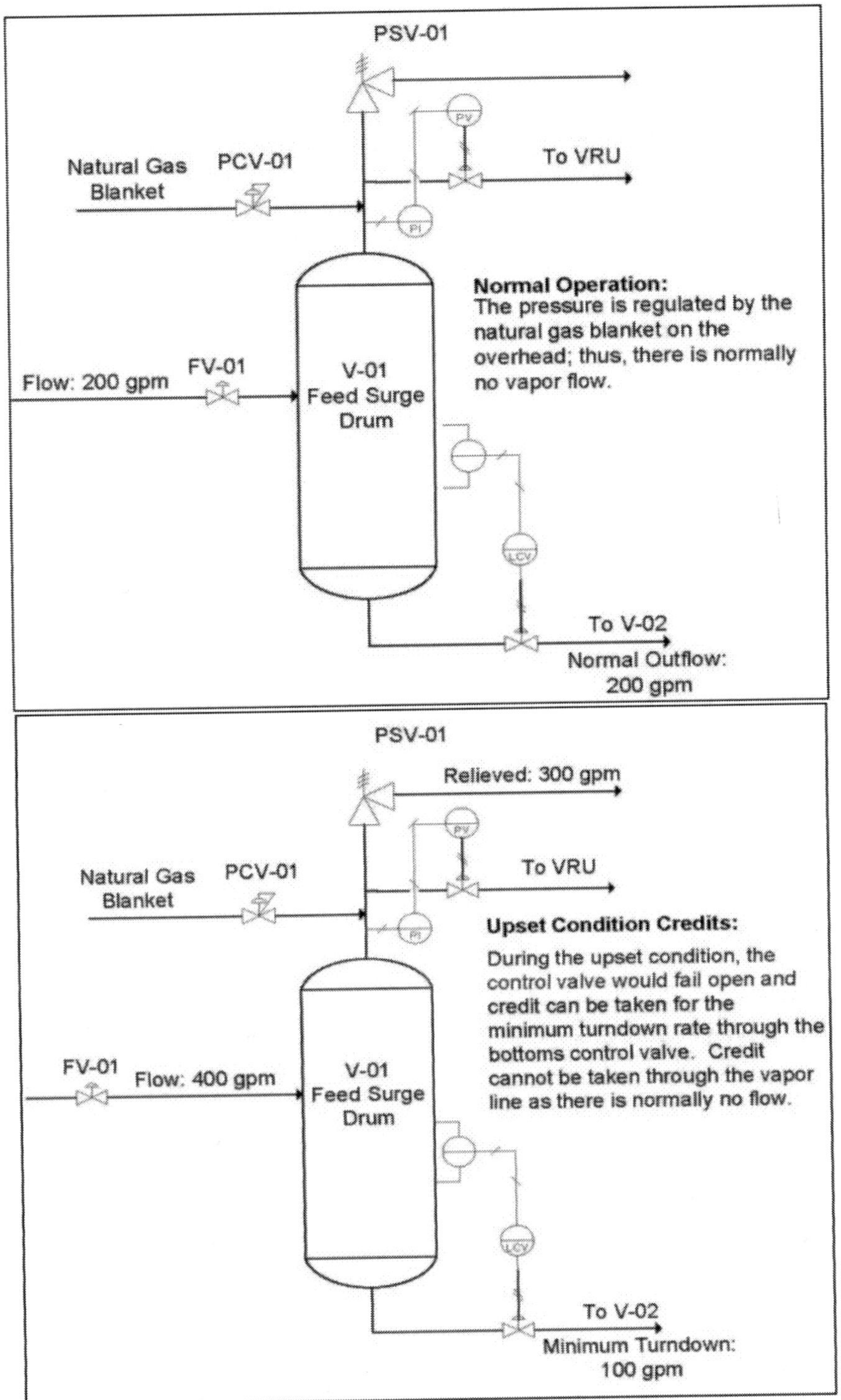

Figure 2: Looking at a feed surge drum during inlet control valve failure can lead to credits to be taken as long as they are reasonable for the system. Credit cannot be taken through the vapor line as there is normally no flow.

There is a special control valve failure case that pressure relief system designers need to be aware of. When a liquid control valve fails open, the vessel may de-inventory and gas may flow through the liquid control valve into the downstream vessel. This scenario, typically called a gas blow-by, should be reviewed for all control valve failures where there is liquid and gas in the upstream vessel. Figure 3 shows a typical system where gas blow-by may occur.

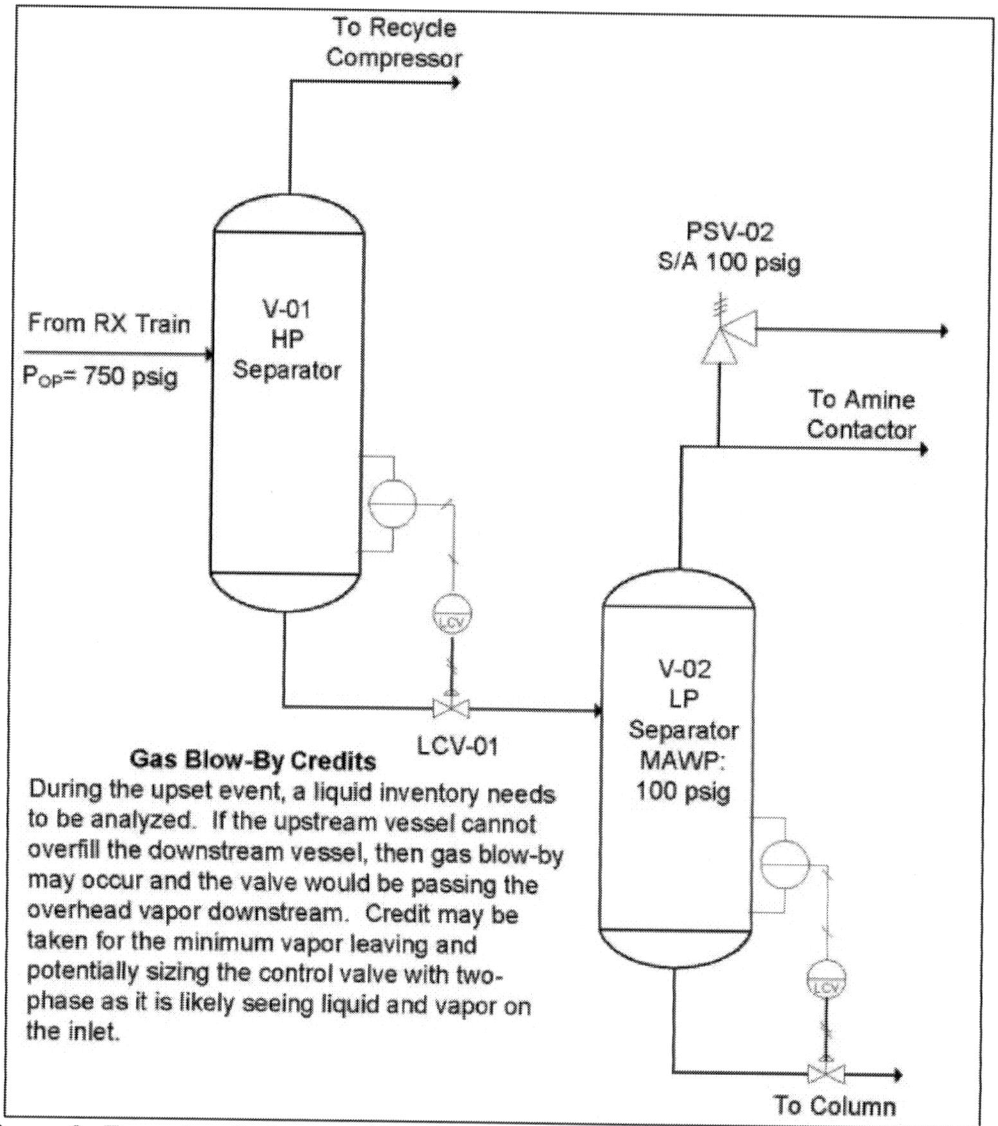

Figure 3: Example of when gas blow-by may occur. When looking at a low pressure separator, the sizing event should be vapor breakthrough the control valve from the upstream high pressure separator.

3. When Realistic Assumptions Make the Estimates Worse

3.1 Compressors Downstream of the System

In some instances, making relief load estimates match closer to reality may lead to a larger relief load estimate. In these cases, the pressure relief system designer must ensure that the relief system is accurately modeled and that the designer uses the larger rates and does not allow designs based on nonsensical assumptions to be used as the basis. There are many different assumptions that could cause this case to be true; the example that we are going to highlight is when the cascading effect of one failure causes the normal plant control system to take an action and increase the relief requirements for a pressure relief device. Oftentimes, a pressure relief system designer may want to take credit for some of the relief effluent to outflow from the system through a compressor. As stated above, this would not be a viable flow path for liquids, as almost all compressors have a high level trip on the inlet drum, but what about vapors? Depending on the scenario, the compressor may pass vapors at the upset conditions or the compressor could trip and result in a blocked outlet. Examples that may cause a compressor to trip under upset conditions include a higher inlet gas temperature trip, a high amp / power requirement trip (due to the increased suction pressure), high temperate trip (due to the potential loss of cooling to the compressor), vibration trips, and etc.[4,5] It is useful for the pressure relief system designer to discuss the scenario with personnel who are familiar with the operation of the compressor (usually either the operators or unit process engineer) to ensure that any assumptions made are consistent with operational history. Figure 4 shows a case where a compressor trip would not result in a larger relief load.

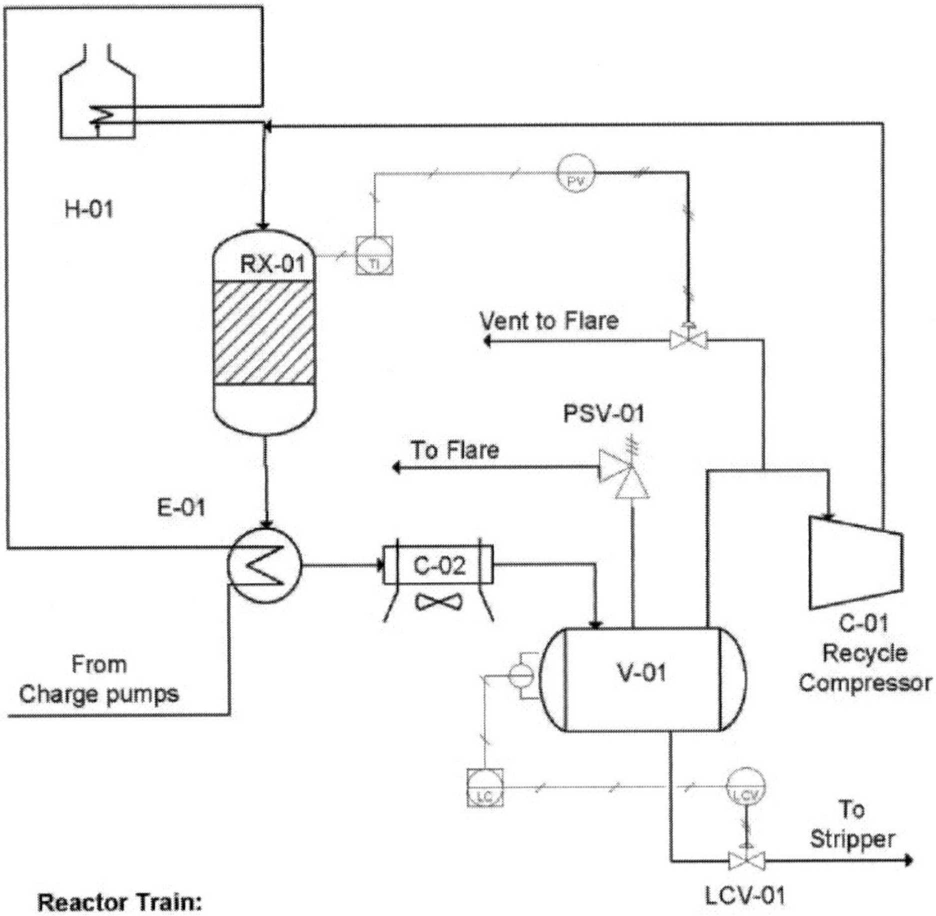

Reactor Train:

During an upset event due to the loss of the compressor, the relief load is not expected to be any larger as the recycle gas is lost. This source is not a feed source since it never enters or leaves the system. Also, the de-pressuring valve on the suction line to the compressor may open to vent to flare if a high temperature reading occurs.

Figure 4: An example of how compressors can affect the relief load. The loss of the compressor would not result in any excess relief as the recycle gas is lost.

3.2 Example of Compressors Shutdown Analysis

A larger estimated rate may occur in a scenario where there is a partial loss of cooling in a column system that uses a compressor to move the off gas from the overhead accumulator to the downstream system (e.g. a FCC Main Fractionator, Coker Bubble Tower, Atmospheric Tower, and etc.). After the simulation is developed to model the column operation during the partial loss of cooling, the modified characteristics of the off gas (e.g. temperature, density) are consistent with conditions that trip the compressor.

Additionally, the pressure relief system designer must ensure that the scenario itself (e.g. a unit wide power or cooling tower failure) could result in a compressor trip and result in this outlet being blocked. The relief load estimate would then need to include the cascading effects of the compressor shutting down and may be much larger than would be otherwise predicted (see Figure 5).

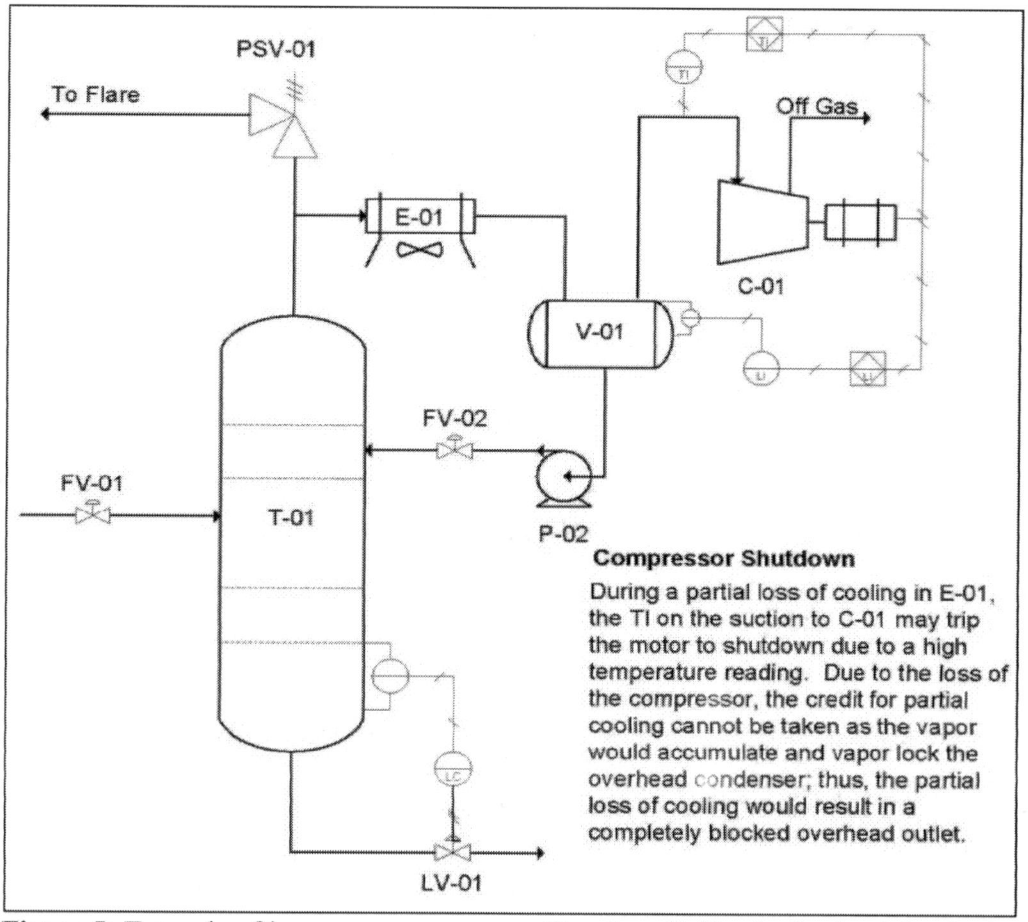

Figure 5: Example of how compressors can affect the relief load. On the right, the loss of the compressor would result in a larger relief as the vapor outlet is completely blocked.

4. Conclusion

To realistically represent the relief load estimate, the pressure relief system designer must understand the system, how it is operated, the limits and shutdowns of the associated equipment, and how these parameters interact during the specific overpressure event being reviewed. The pressure relief system designer must also ensure that all assumptions are consistent with industry, company, and regulatory requirements for the facility. There are many ways to check to see if the relief load estimate is based on realistic assumptions or on conservative/simplifying assumptions. In many cases, conservative/simplifying assumptions are acceptable as long as the assumptions don't result in deviations from standards or practices that result in spending more money. Oftentimes, conservative/simplifying assumptions do result in deviations, and the pressure relief system designer must then use more realistic assumptions to generate the relief load.

The pressure relief system designer must also be vigilant to not use simplifying assumptions that are not conservative. These assumptions tend to result in relief rates that may under-predict the true requirements in an overpressure event. Oftentimes industry, company, and regulatory requirements for the facility are set up to eliminate these scenarios, but the example in Section 3.2 illustrates the pressure relief system designer must check these assumptions (many of which can be implicit).

5. References

[1]. Smith, Dustin and John Burgess, "Relief Valve and flare action items: What plant engineers should know," *Hydrocarbon Processing*, Volume 91, NO.11, 2012

[2]. Middleman, Stanley, *An Introduction to Mass and Heat Transfer: Principles of Analysis and Design*, John Wiley and Sons, New York, NY. 1998

[3] ANSI/API Standard 521, "Pressure-relieving and Depressuring Systems," American Petroleum Institute, 5th ed., 2008.

[4]. GPSA, *Engineering Data Book: Volume I 10th Edition*, Gas Processors Suppliers Association, Tulsa, OK. 1987

[5]. Lieberman, Norman and Elizabeth Lieberman, *A Working Guide to Process Equipment 3rd Edition*, McGraw-Hill, New York, NY. 2008

9TH GLOBAL CONGRESS ON PROCESS SAFETY

Explosion incident simulation for Petrochemical plant

Zhao Xiangdi
State Key Laboratory of Safety and Control for Chemicals, SINOPEC Safety Engineering Institute
No.218, Yan'an San Road, Qingdao, P.R.China 266071
cheeryde@gmail.com

Yuan Jiwu
State Key Laboratory of Safety and Control for Chemicals, SINOPEC Safety Engineering Institute
No.218, Yan'an San Road, Qingdao, P.R.China 266071
Yuanjw.qday@sinopec.com

Jiang Chunming
State Key Laboratory of Safety and Control for Chemicals, SINOPEC Safety Engineering Institute
No.218, Yan'an San Road, Qingdao, P.R.China 266071
Jiangcm.qday@sinopec.com

Wang Zheng
State Key Laboratory of Safety and Control for Chemicals, SINOPEC Safety Engineering Institute
No.218, Yan'an San Road, Qingdao, P.R.China 266071
wz.qday@sinopec.com

Prepared for Presentation at
American Institute of Chemical Engineers
2013 Spring Meeting
9th Global Congress on Process Safety
San Antonio, Texas
April 28 – May 1, 2013

UNPUBLISHED

Explosion incident simulation for Petrochemical plant

Keywords: numerical simulation, explosion incident, TNT equivalent, Multi-Energy

Abstract

A new method of explosion simulation which is based on FLACS software was used for simulating and rebuilding an explosion incident which occurred in a coal gasification company. The other two commonly used methods which are based on TNT equivalent method and Multi-Energy method were considered to calculate the explosion overpressure also. Simulation results indicated that this new method can calculate explosion overpressure in three dimensional. It can make up for the shortfall of traditional assessment techniques. The result is more stable and reasonable when compared with the scene of the incident.

1. Introduction

Petrochemical plants may happen explosion incident easily than other facilities since they have many equipments and devices with lots of hazard material. Once the device exploded, it will cause large casualty and property loss. For example, Alon big spring refinery(2007) and Guangwei(2008) occurred fire and explosion incidents and caused many people dead. So it is important for our to calculate plant's capability against explosion. We could use the results to determine whether those plants should be firmed and changed or not. Researchers around the world found a series special assessment methods based on incident consequences or risk. Someone used those methods to design and renovate process safety of petrochemical plant[1][2]. However, researches on those technology were lags behind in China. We used traditional assessment technology which based on empirical formula. Those traditional methods did not consider the actual situation of factory. It couldn't show the those factor's effect on explosion incident such as the layer equipment, wind, fire protection system and other condition [3][4][5].

Nowadays the Computational Fluid Dynamics analysis technique is becoming more sophisticated. Someone in industry and academia studies explosion simulation used CFD. Herrmann, Hansen studied the consequences of gas cloud explosion using FLACS computational fluid dynamics software, then compared results with experiment, and found the results matched each other well[6][7]. Windhors, used AutoReaGas software to simulate the gas explosion and determine safe distance between petrochemical cracker and control room [8]. P.Hoorelbeke studied explosion risk assessment techniques commonly used [9]. However, there were only a little study on numerical simulation model for explosion incidents. Therefore, assessment on explosion incident by using FLACS were performed in the work.

2. Explosion incident and Simulate model

2.1 Explosion incident

23 December 2008 an explosion incident occurred in Hunan province in China. Gas released from one pipe in the carbon monoxide shift.

Then the gas cloud was ignited and caused explosion. The scenes after this incident with the insulation layer damaged was shown in figure 1.

(a)F2012， F2013, F2014 (b) T2101， R2103

(c) Office building with windows and glass damaged

Figure 1: The scene of the explosion incident with the insulation layer damaged

The gas which flowed in the pipeline was hydrogen. The loss rate of contain was 562kNm3/h in the release phase. The wind speed in-situ was about 3m/s with the direction was NE. The key equipment in this unit for simulation were shown in table 1.

Table 1. Information for the key equipment

Number	Name	Size(m)	Description
F2102	Gas-liquid separator	Φ4.2×13.4	40% of insulation damaged
F2103	Gas-liquid separator	Φ3.7×11.6	25% of insulation damaged
F2104	Gas-liquid separator	Φ3.7×15.6	Little effect
T2101	Regulate tower	Φ3.2×26.0	70% of insulation damaged

R2103	Converter	Φ4.9×13.8	50% of insulation damaged
R2104	Converter	Φ4.9×11.7	30% of insulation damaged
Control room		15.0×46.8×4	Some of the windows damaged; Lots of the windows broken
Office building		15×46.8×19.5	Some of the windows damaged; Lots of the windows broken

2.2 *Model for simulation*

3-D model which was based on the factory's actual layout was shown in figure 1. This model was built using the actual sizes of the carbon monoxide shift unit. The total size for simulation was 100m×80m×40m. Ignition position was in the center of carbon monoxide shift unit nearby leak point. The distance from ignition position to control room and office building was 65m. Shapes of equipment were modified for simulation. Monitor points were set for key equipments, as shown in figure 2.

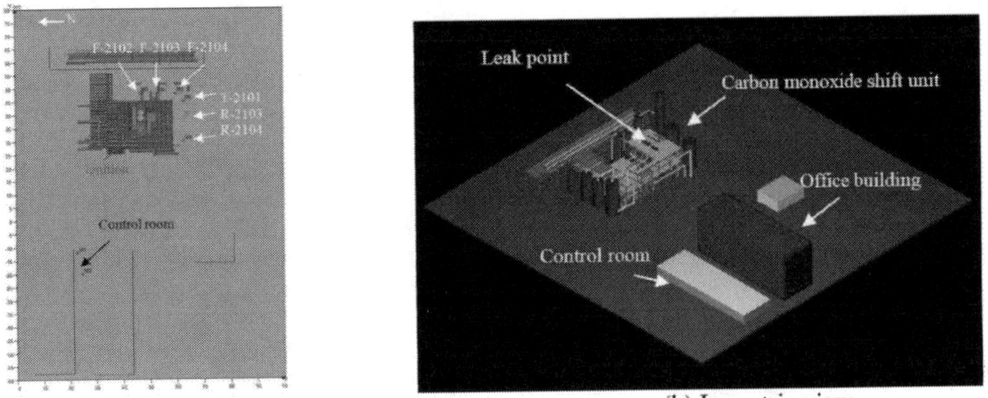

(a)Top view (b) Isometric view

Figure 2: Three dimensional model for simulation

2.3 *Results and discussion*

2.3.1 Leak and dispersion

The material in pipeline was hydrogen. Loss rate for contain was 562kNm^3/h. The time was 24 second between leak started and gas ignited. Wind speed was about 3m/s with direction was NE when the incident occurred. The new method which was based on FLACS was used to rebuilt dispersion process for hydrogen. Simulation results of fuel size after dispersion were shown in figure 3. The whole volume of the hydrogen was about 10x13x7 m^3 in a mixture concentration with equivalence ratio(ER).

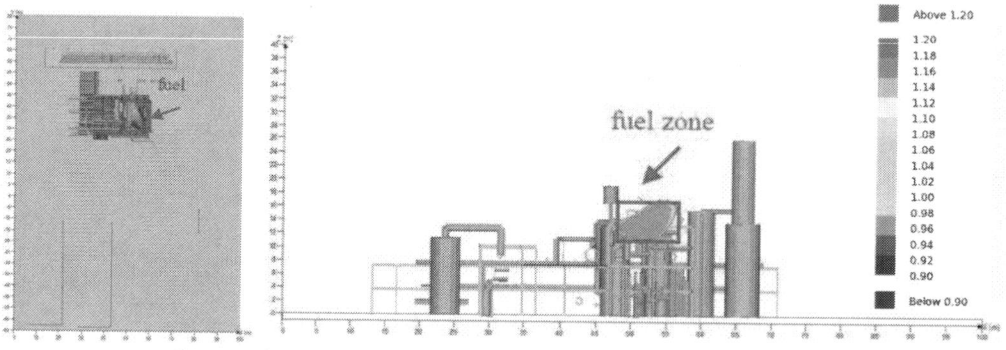

(a) top view (b)side view

Figure 3: The fuel size in this unit after dispersion (0.9~1.2 ER)

2.3.2 Explosion

Simulation assumes that hydrogen and air were in a mixture concentration with equivalence ratio. TNT equivalent method, Multi-Energy method and FLACS method were used to assess the process for gas explosion impact on the equipments and facilities. However, TNT equivalent method was simple and did not consider the effect of actual situation. Although the Multi-Energy method considered the explosion strength as the effect of actual situation, but it also did not consider reflection of complex equipment and did not predict the effect of risk mitigation system such as water spray.

Results of explosion overpressure using those three method were shown in table 2. It can be seen that TNT equivalent method and Multi-Energy method could only take into account explosion overpressure in two-dimensional. For example, overpressure calculated in the position 48m from center of explosion was 5.2 kPa using TNT equivalent method, however, the results increased to 7.9 kPa using Multi-Energy method. But, simulation results using FLACS were different. It could give overpressure in three-dimensional. The overpressure in position 48m from center of explosion was 6.7 kPa with 10m above ground, and 4.7 kPa with 19m above ground , as shows in table 2.

Table 2. Results of explosion overpressure using different methods.

		Overpressure in the different position from the ignition point (kPa)		
		48m (office building)	65m(control room)	74m
TNT equivalent method		5.2	3.48	2.95
Multi-Energy (blast strength 5~10)		7.9 ~ 24.7	5 ~ 16	5.1 ~ 13.4
FLACS method the height above ground(m)	1m	7.8	6.5	3.9
	10m	6.7	-	3.6
	19m	4.7	-	1.9

The overpressure in different monitor points(as shown in figure1) were shown in figure 4. The maximum explosion overpressure in carbon monoxide shift unit was about 62kPa, as well as, the overpressure was about 6.5kPa near the control room and office building.

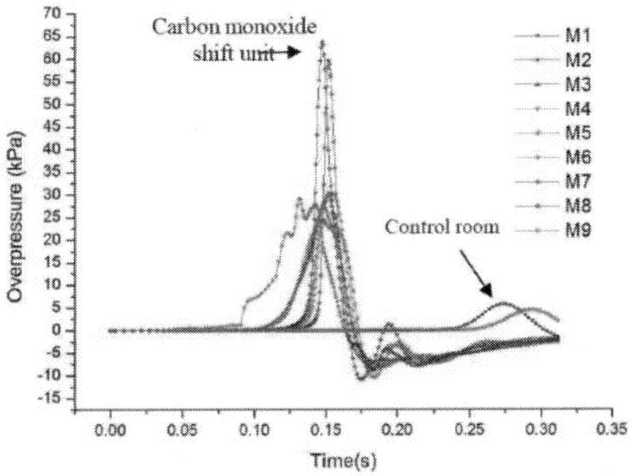

Figure 4: Overpressure of explosion in the different monitor points

The process for explosion overpressure wave affecting on equipments and facilities were shown in figure 5. As soon as explosion starting, a strong shock wave (a), causing major damage to the surrounding equipments, then it would gradually expand and weak (b-c), and subsequently affected office building and control room (d).

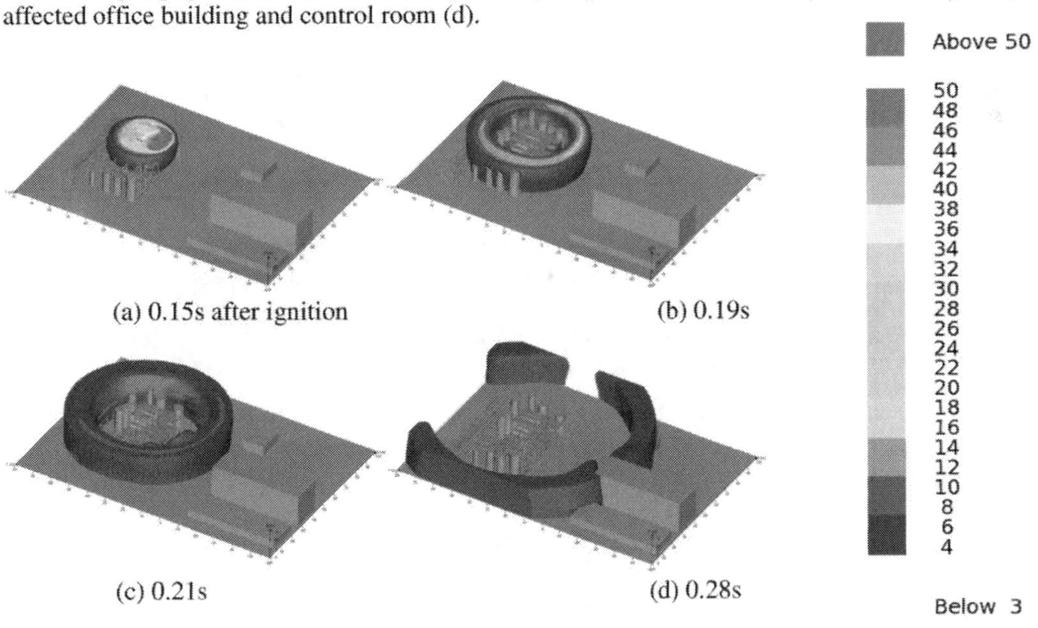

Figure 5: Explosion overpressure wave changed and expanded process (3-50 kPa)

3. Conclusion

The shortfall of traditional assessment techniques is that they can't predict the explosion effects with the layer equipment, wind, fire protection system and other condition's effect on the flame acceleration. A method of explosion simulation which is based on FLACS software for petrochemical plants was introduced in this paper. The results were as follows:

This method can calculate all parts of the space after explosion, and can make up for the shortfall of traditional assessment techniques. The overpressure simulation results using this method are between the results calculated by TNT equivalent method and Multi-Energy method. The result is more stable and reasonable when compared with the scene of the incident.

4. References

[1] D.Bjerketvedt, J.R.Bakke and K. Wingerden(1997), "Gas explosion handbook ,"*Journal of Hazardous Materials*, Volume 52, pp.1-150.

[2] S.B. Dorofeev(2007), "Evaluation of safety distances related to unconfined hydrogen explosions ," *International Journal of Hydrogen Energy*, Volume 32,p.2118-2124.

[3] B.H.Hjertager, I.Moen, J.H.S.Lee, R.k.Eckhoff, k.Fuhre, O. krest,(1981), "The influence of obstacles on Flame propagation and pressure development in a large vented tube ," *CMI Report No. 803403-2,* Chr. Michelsen Institute, Bergen, Norway.

[4] "Guidelines for chemical process quantitative risk analysis,"*American institute of chemical engineers*, NY,2000.

[5] A.C. van den Berg (1985), "The multi-energy method. A framework for vapour cloud explosion blast prediction ,"*Journal of Hazardous Materials*, Vol.12, pp.1-10.

[6] Herrmann, D.D. (1999), " Improved Vapor Cloud Explosion Predictions by Combining CFD Modeling with the Baker-Strehlow Method ," *International Conference and Workshop on Modeling the Consequence of Accidental Releases of Hazardous Materials.* CCPS/AIChE, Sep. 28 ,pp.479-494.

[7] Hansen, O.R., Talberg, O. and Bakke, J.R (1999), "CFD-Based Methodology for Quantitative Gas Explosion Risk Assessment in Congested Process Areas: Examples and Validation Status ," *International Conference and Workshop on Modeling the Consequence of Accidental Releases of Hazardous Materials.* CCPS/AIChE, Sep. 28 ,pp.457-477.

[8] Windhorst, J.C.A (1999), " Deriving Engineering Specifications from Release Dispersion and CFD Explosion," *International Conference and Workshop on Modeling the Consequence of Accidental Releases of Hazardous Materials.* CCPS/AIChE, Sep. 28 ,pp.495-514.

[9] P.Hoorelbeke, C.Izatt, J.R.Bakke, J.Renoult, R.W.Brewerton(2006), "Vapor cloud explosion analysis of onshore petrochemical facilities," *American Society of Safety Engineers Middle East Chapter 7th Professional Development Conference& Exhibition kingdom of Bahrain.* March 18-22,pp.1-15.

Using Reliability Centered Maintenance (RCM) to Reduce Fire Risk in High Temperature Oil Handling

Marc W. Yarlott, P.E.
Veolia Water North America
200 E Randolph St,
Chicago, IL 60601
marc.yarlott@veoliawaterna.com

Ryan Johnson
Veolia Water North America
200 E Randolph St,
Chicago, IL 60601
ryan.johnson@veoliawaterna.com

Steven Waters
Veolia Water North America
200 E Randolph St,
Chicago, IL 60601
steven.waters@veoliawaterna.com

Prepared for Presentation at
American Institute of Chemical Engineers
2013 Spring Meeting
9th Global Congress on Process Safety
San Antonio, Texas
April 28 – May 1, 2013

UNPUBLISHED

Using Reliability Centered Maintenance (RCM) to Reduce Fire Risk in High Temperature Oil Handling

Marc W. Yarlott, P.E.
Veolia Water North America
200 E Randolph St,
Chicago, IL 60601
marc.yarlott@veoliawaterna.com

Ryan Johnson
Veolia Water North America
Steven Waters
Veolia Water North America

Keywords: reliability, maintenance, RCM, critical spares, uptime

Abstract

Veolia Water North America contracted for the technically difficult Biosolids handling process Design, Build, Operate project with the Metropolitan Water Reclamation District of Greater Chicago at the Stickney Water Reclamation Plant. The plant design uses a high temperature oil to operate the Biosolids drying equipment. Several other plants using a similar high temperature oil system have had fires related to leakage of the oil. Having successfully employed the RCM analysis at many of Veolia's projects world wide, Veolia applied the RCM process to gain improvements in the overall reliability of the plant, and particularly the high temperature oil system, by optimizing maintenance strategies and identifying spare parts, procedures, and plans to maximize the uptime of the system. The RCM process resulted in identification of 425 spare parts at a cost of $525,000 which mitigated over $33 million dollars worth of potential risk. For a very technical process with significant risk of failures, the value of a complete RCM study will have a very high return on investment..

1. Introduction

Veolia Water North America (Veolia) operates a 150 dry ton per day biosolids pelletizer operation within the Stickney Water Reclamation plant in Chicago, Illinois. The facility is comprised of 4 independent pelletizer trains. The facility receives thickened solids (22-28%) and the solids are pumped to one of 4 large Seghers pelletizer units. The facility utilizes a gas fired boiler to heat the thermal oil that is used to transfer the heat needed to evaporate the moisture from the biosolids. The systems are very complex in nature and as with most solids processing facilities these systems are a challenge to operate and maintain.

As with any major start-up and commissioning, the operations staff gained a great deal of knowledge during the commissioning and start-up of the system. The process of converting the solids to a pellet is not just about heat transfer. Veolia's contract is a performance based one with plant uptime being the most significant driver for success. Veolia successfully completed its performance testing in late 2009. The management team at Veolia carefully evaluated each part of the performance test data and determined that it would be prudent to focus technical resources to build an uptime focused maintenance strategy.

In the Spring of 2010 Veolia initiated a comprehensive Reliability Centered Maintenance (RCM) review at the facility. This review was facilitated by internal Veolia employees and included the valuable input of the operations and maintenance staff at the facility. Veolia also brought in staff from other projects to help in analyzing specific functional areas. This RCM review incorporated a critical spares analysis that would be used to make cost/risk based storage decisions.

2. METHODOLOGY

2.1 Reliability Centered Maintenance

RCM is "a process used to determine the maintenance requirements of any physical asset in its operating context" [1]

Reliability Centered Maintenance has its roots in the civil aviation industry. In the late 1960s United Airlines invested a great deal of time and money in studying how and why equipment failed. The United Airlines management team understood that safety and reliability were of paramount importance to the operational mission. United Airlines mapped out a plan to better understand equipment failures, and focused the project on improving overall reliability of its aircraft. This analysis resulted in a startling conclusion that "there was no correlation of equipment age and failure" and that "equipment failures could not be prevented." In 1978, United Airlines published its findings in a document titled Reliability Centered Maintenance [2]. This publication was a significant document that challenged maintenance professionals to take a hard look at how and why maintenance tasks were performed.

The study concluded that equipment failure conformed to one of six failure patterns, see Figure 1 [1]. These curves plot the probability of failure on the y-axis and time on the x-axis. The failure curve analysis yielded very surprising results in that the statistics proved conclusively that equipment failures were not primarily dependent upon equipment age. The equipment failure distribution of these patterns is as follows:

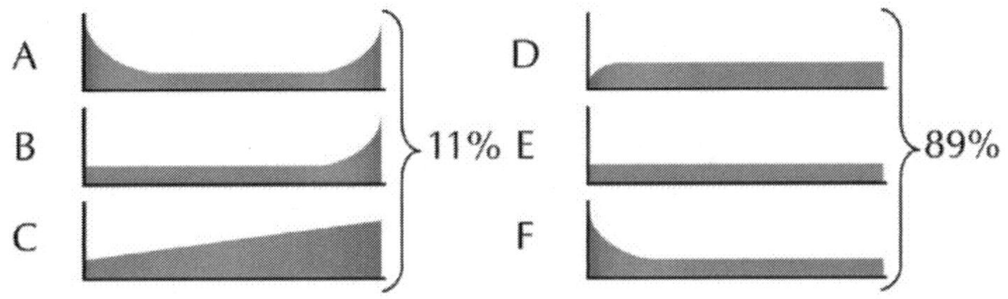

Pattern A	**4%**
Pattern B	**2%**
Pattern C	**5%**
Pattern D	**7%**
Pattern E	**14%**
Pattern F	**68%**

Figure 1, Six Patterns of Failure [1]

Based on this data, government regulators and the airlines decided to revamp the maintenance strategy of the civil aviation industry. In reviewing the failure patterns, one can quickly deduce that patterns A, B, and C have some dependency on equipment age, while patterns D, E, and F are definitely not age related.

It became quite obvious that performing calendar time based overhaul/intrusive maintenance was actually contributing to equipment failures rather than preventing them as planned. This realization resulted in drastic change to the airline maintenance programs and aircraft designs. This was also the beginning of the reliability-centered maintenance movement.

RCM will result in the correct maintenance mix of preventive maintenance (PM), condition assessment, repair (also called reactive maintenance) and proactive maintenance techniques to ensure an increase in the probability that a machine or component will function in the required manner over its design life cycle with a minimum of maintenance. RCM requires that maintenance decisions be based on maintenance requirements supported by sound technical and economic justification. As with any strategies, there are many ways to achieve the end goals.

It is important to note that RCM is not a "silver bullet" that can be applied in a lackadaisical manner. RCM requires discipline and attention to detail in the evaluation process. An RCM analysis is based on a detailed Failure Modes and Effects Analysis (FMEA) and includes probabilities of failure and system reliability calculations. The RCM process ensures that all the critical failure modes are being managed. The analysis is used to determine appropriate maintenance tasks to address each of the identified failure modes and their consequences. RCM is a proven approach that reduces risk and focuses limited maintenance resources on important assets.

In its simplest form the RCM process requires that a group knowledgeable in the operation and maintenance of the facility review systems in a methodical and organized manner. The goal is for the Operations and Maintenance (O&M) staff to learn from one another and freely exchange information related to the process, equipment, reliability, and operation. The RCM process requires that the O&M staff evaluate the following seven questions:

- **Functions:** What does this asset /system do (what do the users want this asset to do)?
- **Functional Failures:** In what ways can the asset/system fail to deliver the functions described above?
- **Failure Modes:** What causes the functional failures to occur?
- **Failure Effects:** What happens when the failure occurs?
- **Consequences:** Does this failure even matter?
- **Proactive Tasks:** Is there anything that can be done to predict or reduce the likelihood of the failure?
- **Default actions:** Otherwise, what other action is available to improve reliability?

Each failure mode is put through a decision tree (Figure 2) to determine its impact on environmental, safety, and financial risk. All of the critical failure modes are subsequently examined and assigned a proactive task (PM, PdM, or failure finding) to manage the risk.

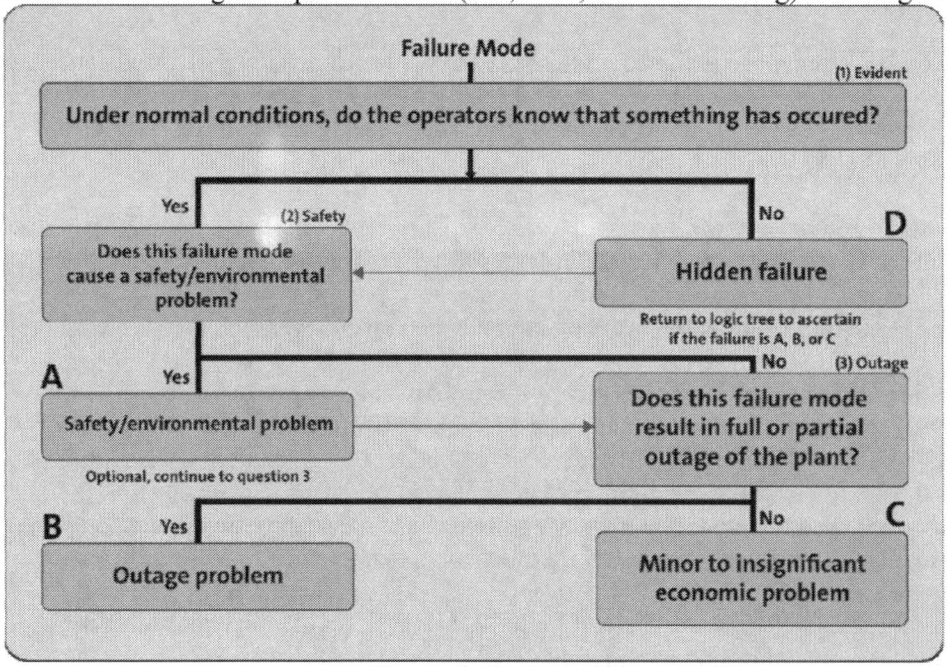

Figure 2, Decision Diagram [3]

3. RESULTS

3.1 Implementation of RCM

Prior to performing the RCM review, the project team was experiencing mechanical failures of system components that were resulting in some periods of downtime. The facility uptime was being negatively impacted by asset failures, and the need for rapid repairs resulted in little to no root cause analysis. The facility uptime had dropped off after performance testing and uptime was running on average of 50% June 2010. Staff worked out a plan that prioritized the RCM reviews based on the principle of analyzing problem systems first.

Initially the RCM project focused on one train of the processing facility. The train was further separated into 13 systems. Over a 16 week period, staff completed 13 classical RCM Reviews. In total over 1784 failure modes were analyzed in total.

One of the benefits of conducting the RCM review was that it also gave the project staff the opportunity to understand in more detail where the system choke points were, and why the failures were occurring. Many useful suggestions on improving uptime and reliability were recorded and noted in the RCM review as "items of interest". In total over 104 items were identified as recommended improvements to the existing installation. For example, one item of interest was related to the biosolids cake pumps. These pumps had some highly unusual failure modes and had become quite problematic from an O&M perspective. During the RCM review there were several failure modes that pointed the review team to questions if the hydraulic unit that was provided with these pumps was adequately sized. The hydraulic unit reservoir was running hot, and there were wide fluctuations and pressure swings in the system. After the RCM review the facility staff worked with a local engineering company to resize the drive unit. This change has significantly improved the uptime of the pumps and eliminated many of the "unusual" failure modes. Staff at the project are now working their way through the "items of interest" and implementing many of the group's recommendations.

A second benefit of the RCM review was a success directly related to the high temperature oil system. During the RCM Review the staff identified Failure Mode #61:

61	***Primary Oil***	Oil heater High Pressure alarm: Staff needs to confirm that there is indeed a High Pressure alarm on the oil heater when the PRV fails closed. This is a safety issue that can be mitigated with existing instrumentation

In following up on this item the plant staff confirmed that there was indeed a high pressure alarm; however, in checking the alarm, they discovered that the Pressure Relief Valves (PRV)'s were not the correct for this service. The Original Equipment Manufacturer (OEM) had supplied air service PRV's instead of liquid service. The process was shut down and immediately new PRVs were ordered and installed on the Heaters. The RCM review, allowed the staff to step through the features and failure modes of the equipment and identify and prevent a failure before it ever happened.

3.2 Critical Spare parts

There are many maintenance management practices that can help to properly manage risk. One area of maintenance management that must be included in a sound asset management program is the selection/identification of "critical spares." To achieve the desired availability of equipment, it is important to give proper consideration to having some spares parts/components available on-site. This is often implicitly considered by maintenance teams as an easy way to mitigate down time risks, thus leading to unnecessary amount of spares. One must be careful when evaluating critical spare parts. It is important that each spare part be properly vetted and justified through a proper risk based method.

Over the past 10 years equipment manufacturers have changed the way that they support their products. The days of having spares readily available for purchase have ended, in favor of the "just in time" manufacturing philosophy. In addition to less available stock, manufacturers have set some limitations on the timeframe that equipment is supported (planned obsolescence). These changes have created a need for operations and maintenance staff to better understand the challenges involved in obtaining needed parts.

Indeed, to be truly cost-effective the proper selection of critical spares has to be developed based on a methodical process. This process must take into consideration a cost-benefit/risk analysis that is aligned with the facility's goals and objectives. This analysis needs to include the following:

- Failure mode identification
- Consequences (safety, environmental, and economic) of not having the spare
- Lead time to procure
- Availability of spare (i.e. obsolete)
- Probability of failure
- Predictability of the failure
- Shelf life
- Storage considerations (space, environmental conditions, etc.)
- Cash flow

Each of these factors will now be considered in more detail.

3.3 Failure Mode Identification

A "failure mode" is an event that is reasonably likely to occur and that would cause the loss of an asset's functionality [1]. When developing a potential list of spare parts for an asset, an understanding of the equipment at the "failure mode" level is necessary.

The Chicago project identified 1784 failure modes in total. During the RCM review, staff decided which parts would be considered for spares. It is important that experienced operations and maintenance staff participate in the development of the failure modes. Special attention should be given to failure modes that are wear-related.

3.4 Consequences

Not all failure modes have the same impacts to the operations at a facility. Great care should be taken in evaluating the consequence of failure, utilizing a criticality review process (Figure 2). One should make an effort to monetize the risk of lacking a spare part when needed. For Chicago a project-specific matrix was developed and used to quantify lack of equipment availability into money. This allowed the staff to factor in items like cost of downtime, contract issues and safety into economic terms. This provided the framework upon which a proper spare part evaluation can be performed.

3.5 Lead Time

It is critically important that we understand the supply chain when it comes to spare parts. The availability and lead time for obtaining spare parts varies significantly based on the manufacturer/supplier and geographical location of the facility. When conducting a critical spares evaluation, one must always fully understand how long it will take to procure an item. Oftentimes excessive lead times may be reason enough to have a spare available at the facility. Initially a decision may be made not to stock a spare because a redundant asset is available. This decision may have to be reconsidered if one has to rely on solely the redundant system for a long period (while one waits for the spare to arrive). Many of the assets at the Chicago plant were custom built and manufactured overseas. Staff made contact with the vendors and determined the lead times on each of the items identified during the review.

3.6 Availability

Oftentimes equipment vendors discontinue product lines and support for products. Some of these changes are the result of market-driven forces, such as mergers and acquisitions within the manufacturing industry itself, while others are purely technologically-driven. Special consideration should be given to electronic components such as Variable Frequency Driver's (VFD) and I/O cards. In the Chicago review we discovered that some of the electronic components were no longer supported by manufacturers.

3.7 Probability of Failure

A proper critical spares evaluation needs to factor the probability of occurrence into the equation. This is usually done by experienced staff familiar with the common equipment in a similar operating context. There are published failure statistics that can also be used to benchmark this type of probability. Veolia has an enterprise Computerized Maintenance Management System (CMMS) that it uses to quantify the Mean Time Between Failures (MTBF). Understanding the failure rates will help in making a sound business decision as it relates to procuring spares.

3.8 Predictability of Failure

Some equipment/components will fail over time due to wear or age/fatigue. It may be possible to properly manage this type of failure by conducting periodic inspections to measure the overall condition of the asset. We identified components like chains, belts, sprockets, liners, and screws can be evaluated for wear and new parts can be ordered prior to loss of functionality.

Properly set up and monitored, condition assessment programs can minimize the amount of spares needed at a facility. Items like fuses, shear pins and electronics fail with little warning and should always be stocked if the asset is critical.

3.9 Shelf Life

Some components deteriorate with time (like probes, sensors, ball bearings, batteries and gaskets). These timing considerations should be factored into the decision-making process.

3.10 Storage Considerations

Prior to procuring any spare parts one must evaluate his/her location to determine if the facility has adequate storage facilities on site.

Consideration should be given to environmental conditions (humidity, temperature, sulfides, etc.) and the available space. Staff also needs to ensure that it has the proper systems in place to track the spares and to perform PM maintenance tasks on the inventory if needed (i.e., a spare pump may require that the shaft be rotated periodically to distribute lubrication and prevent shaft deflection). In Chicago we decided to store some of the spares on site and some in off site storage facilities. We had very good success in working with vendors in some cases to store items for us.

3.11 Critical Spares Methodology

Veolia Water utilized its internally developed critical spares methodology during the RCM review. The following is a small section of the spares analysis that was based on some contract terms and downtime estimates.

The contract is production-based and includes liquidated damages for downtime and loss of availability. What follows is a general description as to how the process was used to determine the need to stock a spare. For presentation purposes, only three scenarios are included.

3.11.1 Step One: "Monetize the Risk"

There are 4 process trains at this facility and three are needed to meet full capacity. The group determined that if all 4 trains were unavailable there would be a \$1,250/hour cost of downtime. The group then evaluated loss of capacity and assigned a lower dollar number for some unit operations (\$450/hr). After debating the liquidated damages clause of the contract and its impact on the business, the group settled on \$225/hr as being the cost of loss of redundancy. This process was extremely rigorous and debated by the review team. The assumptions were sent in a memorandum to the senior management team.

3.11.2 Step Two: Failure Mode Evaluation

	Failure Mode	Local	System	Plant
1	Bearing seizes due to lack of lubrication	Bearing wears, gets hot and eventually fails	Bearing wears, gets hot and eventually fails. Standby pump starts and carries load	Loss of redundancy
2	Temperature transmitter fails	Transmitter fails and thermal oil pumps stop pumping	Oil system shutdown and OLC initiates a system shutdown	Loss of one train
3	Heat trace thermostat fails	Heat trace fails, water line freezes, resulting in loss of flow	Biosolids pumps trip on loss of seal water.	Complete plant shutdown

3.11.3 Step Three: Match Critical Failure Mode to Spare

Failure Mode	Spare
Bearing seizes due to lack of lubrication	Bearing
Temperature transmitter fails	Transmitter
Heat trace thermostat fails	Thermostat

3.11.4 Step Four: Determine Spare Cost and Lead Time and Calculated Downtime Cost

Spare	Lead time (in operating hours)	Downtime cost/hour	Cost of downtime due to lack of spare
Bearing	48	$225	$10,800/event
Transmitter	24	$225	$5,400/event
Thermostat	24	$1250	$30,000/event

3.11.5 Step Five: Adjust Downtime Cost for Failure Probability

Spare	Downtime Cost (DC) due to lack of spare	Failure Probability (FP)**	Downtime Cost adjusted (DC * FP)	Cost of spare and storage
Bearing	$10,800/event	.2	$2,160/year	$1,200
Transmitter	$5,400/event	.1	$540/year	$250
Thermostat	$30,000	.1	$3,000/year	$600

** The Failure Probability (FP) is based on the number of failures anticipated in a ten-year period (e.g. a .1 signifies a 10% chance of at least one failure in a 10-year period). This can be based on past failure data or available MTBF data from other sources. One must consider contractual terms and residual contract years when making spare part decisions. For example, if the contract end date is within a two-year period and the risk of not carrying a spare is tolerable, one should not stock the item.

3.11.6 Step Six: Step Six: Evaluate Cost of Spare versus the Downtime Costs

Spare	Cost of spare and storage	Downtime cost adjusted	Spare required (yes or no)
Bearing	$1,200	$2,100	Yes
Transmitter	$250	$540	Yes
Thermostat	$600	$1250	Yes

As with any type of analysis it is important that the review group uses a bit of common sense when making spare part decisions. One must factor into the decision-making process contractual obligations, contract term, and the practicality of storing the item. Oftentimes the decision may come down to a very small risk (<$1,500 can be mitigated by spending about the same amount of money). One needs to use sound judgment before money is spent on spares.

In total, we analyzed 1784 failure modes for critical spare purposes. In total we decided to stock 425 spares parts at a cost of $525,000. These spares mitigated over $33 million dollars worth of potential risk.

4. CONCLUSIONS

Veolia Water has introduced a number of innovations to the standard framework for delivering RCM that have improved the business case for delivery. These innovations include techniques for prioritizing and simplifying the analysis, as well as more reliable means for estimating maintenance frequencies and estimating the true cost-benefit of critical spares. The use of RCM in the water/wastewater industry has been adapted from the original methodology used by United Airlines. The changes to the process allow for speedier delivery and execution of RCM. Implementation of RCM on 100% of the assets is not practical, nor cost-effective in our industry.

Veolia Water has been able to demonstrate a link between doing RCM well and delivering excellence in asset management. Overall, the benefit to the business and our customers that result from implementing the Veolia Water RCM process is as follows:

- Better identification of risks. Many new, and often significant, operational risks are identified with recommendations to mitigate them.

- Improved maintenance plan. New maintenance plans are developed to specifically manage the risks of asset failure. Frequencies and resource levels are optimized.

- Gains in operational efficiency. Opportunities for improvement are proposed for operating scenarios that offer risk and efficiency benefits.

- List of critical spares. Spares needs are determined, based on cost-benefit assessments, including financial estimates of risk.

- Prioritized capital improvements. Investigations are recommended for a number of potential capital works items designed to enhance contract performance and minimize risk.

- Effective knowledge capture. The RCM captures both documented information as well as the tacit knowledge held by staff about how the systems work.

- Tool for operations training. Participating staff are trained in the methodology and learn in detail how their systems work.

The project staff has been busy implementing the recommendations that came about from the RCM review. The identification of the wrong PRV's on the high temperature oil system was a very important safety related finding, but the benefits of the program continue to add up. The project has already seen drastic improvements in equipment availability and uptime. Each maintenance task, spare part, and equipment check has been developed to manage risk and improve uptime. The staff at the plant continues to learn from equipment failure and has validated that the time spent on the analysis and spares has tremendously improved the overall operation. The following, Figure 3, is a trend of uptime of one train since the RCM reviews were completed.

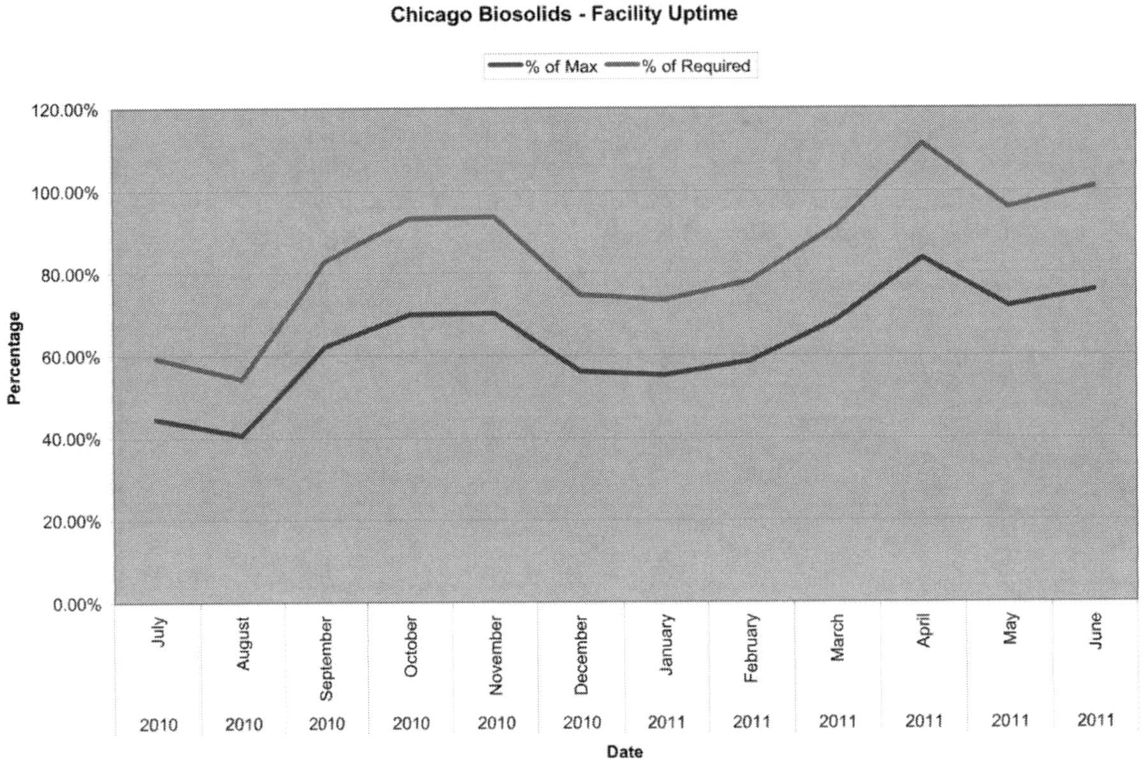

Figure 3, Uptime Performance Through June 2011

5. REFERENCES

[1] Moubray, John, *Reliability-centered Maintenance*, 2nd Edition, Industrial Press Inc., New York, New York, 1997.

[2] Nowlan, F. S.; Heap, H. F., *Reliability-centered Maintenance*, United Airlines and Dolby Access Press, sponsored and published by the office of Assistant Secretary of Defense, San Francisco, California, 1978.

[3] Smith, A.M.; Hinchcliffe, G.R., *RCM Gateway to World Class Maintenance*, Elsevier Butterworth Heinemann., Burlington, Massachusetts, 2004.

(44x) Characterization of the Emissive Power of Hybrid Methane-Air-Dust Flames

Brian Elias and Ali S. Rangwala, Fire Protection Engineering, Worcester Polytechnic Institute, Worcester, MA

Combustible dusts are commonplace in a wide variety of industrial settings ranging from coal mines to manufacturing facilities. The ignition of a dust cloud formed by the mixing of these combustible particles with the surrounding air may result in a deflagration or flash fire. The purpose of this study is to quantify the emissive power of the flame in order to better understand the combustion dynamics and the risk of burn injury associated with a dust flash fire.

Experimental tests are conducted by feeding combustible dust samples into a premixed dust-air burner apparatus to create a stabilized flame for analysis. The two dust types investigated in this study are Pittsburgh seam coal, and iron dust – both with sizes ranging from 70 to 90 um. In addition to the composition of the dust, other controlling parameters including dust concentration and turbulent intensity are also varied and their respective effect on the heat flux is reported. The total emissive power of the flame is calculated from the heat flux measurements in combination with flame surface area and an appropriate view factor.

Are You in Control of Process Safety?

Basis of Safety Assurance Can Provide the Answer

Graeme Ellis, Principal Lead Consultant
ABB Consulting
Daresbury Park, Warrington, WA4 4BT, UK
graeme.ellis@gb.abb.com

(Acknowledgement: Based on paper written by Phil Eames, ABB Consulting)

Prepared for Presentation at
American Institute of Chemical Engineers
2013 Spring Meeting
9th Global Congress on Process Safety
San Antonio, Texas
April 28 – May 1, 2013

UNPUBLISHED

Are You in Control of Process Safety?

Basis of Safety Assurance Can Provide the Answer

Graeme Ellis, Principal Lead Consultant
ABB Consulting
Daresbury Park, Warrington, WA4 4BT, UK
graeme.ellis@gb.abb.com

(Acknowledgement: Based on paper written by Phil Eames, ABB Consulting)

Keywords: Basis of Safety, Assurance, layers of protection, drill-down, vital signs, audit, safety critical elements

Abstract

Serious process safety accidents continue at an unacceptable frequency in the process industry, and senior managers are right to be concerned about whether all reasonable steps are being taken to control risks. Traditional methods for revalidation of Process Hazards Analysis and auditing of Process Safety Management systems can involve considerable time and resources. This paper will describe a 'Basis of Safety Assurance' approach that has been developed to provide a rapid view on the 'vital signs' related to the state of health of measures to prevent and mitigate major accident hazards. A representative set of scenarios for the site are selected and evidence gathered from documentation and site visits to confirm the robustness of the key layers of protection. Deficiencies related to plant, processes or people are identified resulting in a risk targeted set of improvement recommendations.

1. Introduction

Effective process safety management requires a thorough understanding of process risks so that measures to prevent and control them can be implemented and maintained. Structured techniques such as hazard identification (HAZID) and hazard and operability study (HAZOP) are used to identify major accident scenarios and decide how they will be prevented, controlled and mitigated. The description of major accident scenarios and the agreed layers of protection (or barriers) to reduce the risks to an acceptable level, is often termed the basis of safety.

The purpose of a process safety management system (PSMS) is to provide a framework of high level procedures (or risk control systems), in order to maintain protective equipment and procedures in a healthy state. Therefore many organizations, when they seek to assure themselves that process safety is being managed appropriately, base this assurance on detailed auditing of the PSMS.

However, process safety incidents are usually caused by failures of multiple risk control systems at a detailed level. The potential interactions between risk control systems are often not visible at the level of the PSMS, which is constructed of distinct generic elements (mechanical integrity, incident investigation, management of change etc.). Auditing of distinct elements may not identify the potential for an incident due to such interactions. Moreover, modern PSMS are tending to become more integrated with occupational safety management systems, environmental management systems and quality systems, making their role in the prevention of process safety incidents potentially less visible at system level.

Given that the prime concern in process safety management is to prevent process incidents from occurring, it makes sense to devote some assurance effort directly to the scenarios themselves and to their specific layers of protection. This is the essence of Basis of Safety Assurance. Whereas auditing addresses the component elements of the management system, Basis of Safety Assurance seeks to challenge the effectiveness of the process hazards analysis and robustness of the layers of protection for specific scenarios, thereby providing direct assurance that the risk reduction measures are in place and working.

This is not to suggest that auditing of the PSMS is not a necessary or effective tool. But focusing on an organization's key process incident scenarios can be a very direct and cost-effective way to provide assurance that process safety is being managed appropriately, complementing traditional audit-based assurance processes.

This paper describes the Basis of Safety Assurance technique and the experience of its use across a number of sectors in the process industry. It explains how the technique can be used to enhance an organization's process safety management audit and review program.

2. Limitations of Audit Based Assurance

Traditional process safety assurance or governance approaches are based on audits of the process safety management system (PSMS). However, in many organizations the components of the PSMS form part of a larger safety, health, environmental and possible quality management system, and may not be distinguished from components designed to control personal safety and health risks. This means that focus on process safety may diluted by or obscured by the larger management system.

Even if the PSMS is a stand-alone system, or is visible within a larger system, auditing focuses on the system as a whole and not on the control of specific risks. Auditing is often undertaken on an element-by-element basis, which is likely to miss the potential interactions between elements that characterize complex process safety events, typically represented by the "Swiss cheese" model. Moreover, even within a single element significant audit time is normally spent on understanding how the element is managed rather than on detailed field verification. Field verification on audits typically aims to confirm that the system is in compliance with requirements, for example that trip testing is being carried out to plan. It generally does not test that the system is effective in protecting against the risk of a specific scenario, for example the trip is identified as a layer of protection in HAZOP, there is a written test method, the trip test method is an "end to end", faults identified in previous tests have been corrected, etc.

3. Layers of protection and process safety "vital signs"

In his book on the management of complex process risks, James Reason [1] wrote of the need for proactive checking of the process or plant, or the 'system' as he describes it:

"*In the absence of bad outcomes, the best way - perhaps the only way - to sustain a state of intelligent and respectful wariness is to gather the right kinds of data. This means creating a safety information system that collects, analyses and disseminates information from incidents and near misses as well as from regular proactive checks on the system's vital signs. All of these activities can be said to make up an informed culture - one in which those who manage and operate the system have current knowledge about the human, technical, organisational and environmental factors that determine the safety of the system as a whole. In most important respects, an informed culture is a safety culture*".

The state of health of safety critical layers of protection is very much vital signs for the process in relation to major accident scenarios. If they are specified and designed correctly, maintained and tested, then the process is safe. The proactive checking that layers of protection are in place and healthy provides exactly the current knowledge of the state of the plant and the organization that Reason holds to be so important.

4. The 'Basis of Safety' Event

The testing of the state of health of layers of protection for specific major accident scenarios is the focus for a Basis of Safety Assurance event, a generic agenda for which is shown in Table 1.

Table 1. Basis of Safety Assurance Agenda

Phase	Day	Activity
Preparation	0	Arrange for immediate access to key site personnel and information during time on site
Introduction	1	Introduction to Process Safety on site. Review of major accident hazards Site tour Examination of Process Hazard Analyses Selection of 'Drill Down' scenarios
Drill Downs	2-N	Detailed drill downs Approx 4 scenarios per day including field visits
Feedback	Final	Preparation and delivery of feedback

The initial part of the event aims to understand the major accident hazards associated with the process by discussion, field visits and review of the process hazards analysis (HAZID and HAZOP documentation). This would typically take one day, during which time an assessment of the quality of the site's process hazards analysis and documentation of the basis of safety is also made. It culminates in the selection of specific major accident scenarios to be subjected to detailed assessment, sometimes termed a "drill-down" or "deep dive".

It is useful to select a range of different types of scenario where possible (fire, explosion, toxic release, environmental, etc.), so that as many different types of safeguard are subject to assessment. These then include for example; pressure relief systems, safety instrumented systems, passive and active fire protection systems, secondary and tertiary containment systems, etc.

The main part of the Basis of Safety event comprises the scenario assessments. These typically take around 2 hours per scenario, depending on the quality of the basis of safety documentation, the knowledge of the team and the speed of data retrieval, for example how fast plant records can be accessed to verify that maintenance and testing is taking place.

Findings from the scenario assessments provide the required assurance that layers of protection are healthy or lead to the generation of recommendations for improvement. Common findings across a number of scenario assessments can provide a useful indication of systemic issues, for example weaknesses in proof test procedures or compliance with critical equipment testing regimes. This part of the event complements conventional PSMS auditing.

5. The Scenario Assessment Technique

Most value can be gained from the scenario assessment technique if it is conducted as a team exercise. Experience has shown that the best results can be obtained by deploying two process safety specialists as assessors; one with a PSMS and operations background and the other with a plant engineering and asset integrity management background. This approach also helps from a leading and scribing standpoint since the event is recorded in a similar way to a process hazards analysis. In addition to the assessors, the involvement of process engineers, operating managers and maintenance engineers from the site is very important. This provides an understanding of the scenario and the location of information to verify that the identified layers of protection are in place and being maintained.

Since scenarios are normally selected from the site's process hazards analysis or basis of safety documentation, a scenario assessment normally starts with this documentation. The scenario is described and the relevant records are referenced, shown in the upper part of the recording form in Table 2.

It is important to note that the existing documentation may not fully describe the scenario, since the quality of HAZID and HAZOP documentation is heavily influenced by team composition, knowledge available, leader competence, availability of time and recording policy.

Table 2. Scenario Assessment Template (example)

Major Accident Hazard Drill Down Record
Scenario: Storage Tank V7706 (150m3) overfilled during (continuous) run-down and a large volume of solvent released into bund from the breather/emergency vent, subsequent ignition resulting in a flash fire/pool fire and potential for several fatalities.

PHA Reference: No specific scenario was identified for this material as incidents involving acetone were taken as the representative set. The acetone scenario for overfill was estimated to result in two fatalities.			
Safeguard	**Verification**	**Findings**	**Conclusions**
Monitoring of level transmitter by operators and dependent alarm function.	1. Calibration and maintenance of level transmitter and alarm LIA-7706. 2. SOP requiring level monitoring.	1. No regular maintenance is now performed on the level transmitter following the installation of the independent high level switches. Corrective maintenance is performed based on notifications raised by the operators. 2. There is no specific SOP relating to the operation of this solvent storage tank.	Consideration implementing routine level transmitter calibration on solvent storage tanks.
Independent high level switch with trip of run-down line.	1. Proof testing of high level trip LZA-7706H.	1. The trip is proof tested annually 2. The last proof test of this instrument was carried out on the 27/4/10. The proof test is an end- to-end test to a pre-defined test method including checklist. 3. No SIL assessment has been conducted.	Controls appear to be effective. However, conduct a SIL assessment for this scenario to verify that the proof test interval is adequate to provide the required level of protection.

It is therefore important that the assessors work with the team to identify all the relevant safeguards, which are recorded in the left hand column of the template shown in Table 2. Note that the term "safeguard" is used rather than "layer of protection" because the aim is to capture safeguards such as procedures and emergency plans as well as physical independent protection systems that are the focus of formalized Layers of Protection Analysis (LOPA) techniques.

Once safeguards have been listed the assessors ask - for each safeguard in turn - what can we check is in place? This is recorded in the "verification" column. Examples include procedure number and latest version, alarm test methods, maintenance routines and pressure safety valve test records.

Evidence for each verification is then sought and the findings recorded in the "findings" column of the template. It is important that documentary evidence is collected to ensure that the assessment is suitably probing and detailed.

Conclusions are then drawn for each safeguard. This may be a simple "Okay" if everything expected is in place, but may be a recommendation for improvement, a corrective action or a recommendation to consider whether a finding may be evidence of a systemic weakness. In viewing the robustness of the safeguards as a whole the team may determine that additional risk reduction is appropriate and make a recommendation to identify additional risk reduction measures. This would be required in a UK legislative context to demonstrate that risks have been reduced to "as low as reasonably practicable" or ALARP.

It is important to include a field visit for each scenario. This has two purposes, firstly to physically check that safeguards are installed in line with the design intent, for example that relief devices are in place and that passive fire protection and secondary containment are in good condition. Photographs of deficiencies provide high-impact evidence to site management. Secondly to talk to operators and maintenance technicians about their understanding of the process and of the major accident hazard that is being assessed. This gives an insight into the quality and experience of staff and the quality of process-specific training that they have received.

6. Experience with the Technique

There is now a significant body of experience of Basis of Safety Assurance across a number of process industry sectors. Companies have found it effective regardless of sector or scale of process hazard. Major hazards facilities are more familiar with the concepts but it is readily applicable to lower level process safety hazards, including those in the pharmaceutical and power generation sectors.

Some important advantages of using the scenario assessment "drill-down" technique are as follows:

- Drilling down into specific scenarios can identify significant risk exposures and weaknesses at a very detailed level, for example:
    - Trips tested only at commissioning;
    - Absence of detailed test methods and reliance on technicians' experience;
    - Trip testing not from sensor to final element (end-to-end);
    - Reliance on inexperienced 3rd parties for proof testing;
    - Demands on trips more frequent than the design basis.
- Examining multiple scenarios can provide evidence of whether there are systemic weaknesses.
- The technique is useful to identify weaknesses in process hazards analysis and in keeping records and basis of safety documentation up-to-date.
- The technique is a useful vehicle for providing training materials for specific process hazards.

7. Conclusion

Because it focuses on specific risks, Basis of Safety Assurance complements conventional PSM system audit activities. It helps to maintain a focus on major accident hazards and improve the understanding of site personnel and standard of documentation. It aligns with line management responsibilities for specific processes and assets, and it directly exposes risks of major accident hazards, thus providing real time assurance to senior management.

8. References

[1] Reason, J.T (1997), Managing the Risks of Organizational Accidents, Ashgate Publishing, Aldershot, UK.

A Methodology to Evaluate Flameless Venting Devices for Dust Explosions

Jenny Chao
FM Global, Research Division
1151 Boston-Providence Turnpike
Norwood, MA USA 02062
jenny.chao@fmglobal.com

Sergey B. Dorofeev
FM Global, Research Division
1151 Boston-Providence Turnpike
Norwood, MA USA 02062
sergey.dorofeev@fmglobal.com

Prepared for Presentation at
American Institute of Chemical Engineers
2013 Spring Meeting
9th Global Congress on Process Safety
San Antonio, Texas
April 28 – May 1, 2013

UNPUBLISHED

A Methodology to Evaluate Flameless Venting Devices for Dust Explosions

Jenny Chao
FM Global, Research Division
1151 Boston-Providence Turnpike
Norwood, MA USA 02062
jenny.chao@fmglobal.com

Sergey B. Dorofeev
FM Global, Research Division

Keywords: dust explosion, vented explosion, flameless venting, quench tubes

Abstract

A methodology that can be used to evaluate the effectiveness of explosion vents and of flameless venting devices is presented in this paper. This methodology is based on a series of explosion tests conducted in FM Global's 8 m^3 explosion vessel using cornstarch as an example dust. The overall performance of a flameless venting device is determined by its ability to successfully quench a vented explosion (while maintaining its structural integrity) and by its total efficiency, which is defined as the ratio of the effective vent area to the nominal vent area. The functional form of the total efficiency is taken as the product of an area efficiency (i.e., the ratio between the effective vent area of the device to that of a burst disc assembled with the device) and a burst efficiency of the burst disc itself (i.e., the ratio of the effective vent area of the burst disc to the nominal vent area). The effective vent areas are calculated from measured reduced overpressures using an empirical vent-sizing correlation.

1. Introduction

Venting is a very effective means to reduce potentially damaging overpressures that can develop inside an enclosure during an accidental explosion, but it can also generate several hazards outside of the enclosure such as an external explosion (of the vented unburned explosive mixture), a high-velocity flame jet, and blast wave effects. For dust explosions, a flameless venting device (which is essentially a box or a cylinder made up of layers of wire mesh) can be installed over an existing vent to minimize these external hazards. As a dust explosion is vented through the device, the layers of wire mesh retain a large portion of the burned and unburned dust and act as a heat sink, arresting the vented flame jet by reducing the reaction-zone temperature.

A flameless venting device, however, acts as an obstruction to the flow and essentially increases the reduced overpressure inside the enclosure (compared to venting without the device), which effectively decreases the vent area. Given an otherwise sufficiently-

Note: Do not add page numbers. Do not refer to page numbers when referencing different portions of the paper

designed vent, it is possible that the addition of a flameless venting device over that vent can result in reduced overpressures that exceed the design strength of the protected enclosure. Furthermore, the type and reactivity of the dust can strongly affect the performance of a flameless venting device. For example, fibrous and low melt-point dusts can severely clog the wire mesh, decreasing the effective vent area even further; in some cases, this can cause the device to fail catastrophically. It is, thus, important to evaluate the overall performance a flameless venting device in order to ensure its proper use. It is clear that this is determined by the ability of the device to successfully quench a vented explosion while maintaining its structural integrity; however, the decrease in the effective vent area should also be evaluated for different dust types and reactivities.

In the present investigation, a series of tests is conducted using cornstarch as a representative non-fibrous dust in an 8 m^3 vessel and is used to develop a methodology for determining the performance of a flameless venting device. The unvented effective reactivity of a fixed (optimal) concentration of cornstarch is first established in the vessel as a function of ignition delay time. By specifying the ignition delay time, vented tests are then conducted at the desired reactivity level, and the resulting overpressures from the various tests are used to calculate the total efficiency of the device (defined as the ratio between the effective vent area and the nominal vent area). In this paper, an overview of the methodology along with important results is presented.

2. Experimental Details

All experiments were performed in an 8 m^3 steel explosion vessel with a nominal diameter of 2 m and height of 2.9 m (shown in Figure 1). Three different sets of tests were conducted in order to evaluate the overall performance of a flameless venting device. These tests included unvented tests to establish the appropriate ignition delay times (between dust injection and ignition), vented tests with a burst disc only, and vented tests with a flameless venting device assembled with the same type of burst disc.

Figure 1. FM Global's 8 m^3 explosion vessel with the dust-dispersion system.

In all tests, 6 kg of cornstarch (for an optimal concentration of 750 g/m^3) was injected into the vessel at two different locations through 0.05-m dispersion nozzles using a Martin Engineering BB4-28-48 air cannon. Prior to any given experiment, the dust container (see Fig. 1) was filled with dried cornstarch and the vessel was partially evacuated to a pressure of about 0.85 bar. The air cannon was subsequently discharged, dispersing cornstarch and air into the vessel such that the initial pressure prior to ignition was 1 bar. Ignition of the cornstarch-air mixture was effected by a 10 kJ Sobbe pyrotechnic igniter placed in the centre of the vessel.

Venting occurred through the top flange. Either a burst disc with a nominal diameter of 0.76 m or flameless venting device/burst disc assembly (with the same nominal diameter) was mounted on the top flange. The static burst pressure of the burst disc was $\Delta p_{\text{burst}} = 0.1$ bar, and the actual vent area of the burst disc was 0.44 m^2.

A sketch of the locations of the different diagnostics is shown in Figure 2. For all tests, two Stellar Technologies piezoresistive pressure transducers (GT1600) were used to measure the pressure histories inside the vessel. The ranges of the pressure transducers were 0-10.3 bar and 0-3.4 bar for unvented and vented tests, respectively.

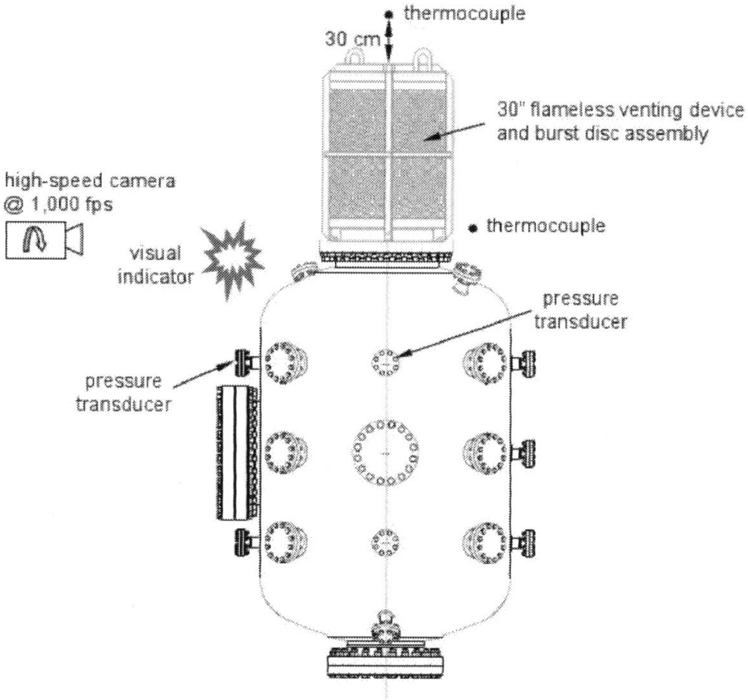

Figure 2. Location of the various diagnostics on the 8 m3 vessel.

For all vented tests, a burst indicator on the burst disc itself was used to determine the actual rupture time and pressure. For tests conducted with a burst disc only, the rupture time was also corroborated with a high-speed camera (at 1000 frames/s) and an electric match synchronized with ignition (as a visual indicator for the camera).

For tests using flameless venting devices, two K-type thermocouples (Omega 5SRTC-TT, 0.08 mm wire diameter) were used to measure the temperature of the vented combustion products. One thermocouple was located roughly 30 cm above the flameless venting device along the centerline of the vessel. The other thermocouple was placed next to the device (about 4 cm away from the surface of the device) where the initial rupture of the burst disc occurred. The high-speed camera was also used to monitor flame transmission through the venting device.

3. Efficiency of a Flameless Venting Device

Because a flameless venting device is always installed in conjunction with an impulsive bursting device (usually a burst disc), the total efficiency of the assembled device should account for the individual efficiencies of these two components.

3.1 Determining Area Efficiency (η_{area})

For a flameless venting device where the effective vent area, $A_{\text{v,eff}}$, is reduced compared to the vent size of the burst disc over which it is mounted, $A_{\text{v,burst}}$, an area efficiency can be defined such that:

$$\eta_{\text{area}} = \frac{A_{\text{v,eff}}}{A_{\text{v,burst}}}. \quad \text{[Eq. 1]}$$

In general, any vent-sizing correlation can be applied to Eq. 1 (see [1,2] for e.g.); however, in this case, FM Global's vent-sizing correlation (see [3] for the actual correlation) is manipulated to yield an expression for η_{area}:

$$\eta_{\text{area}} = \frac{A_{\text{v,eff}}}{A_{\text{v,burst}}} = \frac{K_{\text{eff,q}}}{K_{\text{eff,b}}}\frac{\Gamma_{\text{q}}}{\Gamma_{\text{b}}}\frac{\left(1+1.75\dfrac{p_{\text{burst,q}}-p_0}{p_0}\right)}{\left(1+1.75\dfrac{p_{\text{burst,b}}-p_0}{p_0}\right)}\times \frac{\left(1+3\left[1-\exp\left(-0.1\dfrac{a_{\text{cd}}A_{\text{v,eff}}\Delta p_{\max}}{V^{2/3}K_{\text{eff,q}}(H/D)^{-1/3}}\right)\right]\right)}{\left(1+3\left[1-\exp\left(-0.1\dfrac{a_{\text{cd}}A_{\text{v,burst}}\Delta p_{\max}}{V^{2/3}K_{\text{eff,b}}(H/D)^{-1/3}}\right)\right]\right)} \quad \text{[Eq. 2]}$$

where the subscripts q and b refer to the flameless venting device and the burst disc, respectively, and $H/D = 1.4$.

It can be seen that, in order to evaluate η_{area}, the performance of a flameless venting device (assembled with a burst disc) must be compared to the performance of the burst disc alone. This comparison must be based on reduced overpressures from tests. It should also be noted that actual burst pressures (rather than static nominal values) are used in Eq. 2; these values must be determined from tests as well.

3.2 Determining Burst Efficiency (η_{burst})

For a burst disc, the efficiency not only depends on the actual vent area but also on the actual vent-deployment pressure. In general, when the pressure at which the burst disc ruptures is greater than the nominal static pressure at which it is designed to release (i.e., $\Delta p_{burst} > \Delta p_{stat}$), the reduced pressure in the vessel increases. This increase in the reduced pressure corresponds to an equivalent decrease in the available vent size such that the effective vent area of the burst disc is smaller than the nominal vent area (i.e., $A_{v,burst} > A_{v,nominal}$). Therefore, the burst efficiency can be defined as:

$$\eta_{burst} = \frac{A_{v,burst}}{A_{v,nominal}} . \qquad \text{[Eq. 3]}$$

In order to evaluate Eq. 3, the effect of the increased vent-deployment pressure on the vent area for a given Δp_{red} should be properly described. Applying FM Global's vent-sizing correlation [3], the following expression can be used to determine η_{burst}:

$$\eta_{burst} = \frac{1 + 1.75\dfrac{\Delta p_{stat}}{p_0}}{1 + 1.75\dfrac{\Delta p_{burst}}{p_0}} \qquad \text{[Eq. 4]}$$

where Δp_{stat} is the nominal static burst pressure and Δp_{burst} is the actual burst pressure measured in the tests.

3.3 Determining Total Efficiency (η_{total})

The total efficiency of the entire assembled device is then determined by the product of the two efficiencies; i.e.,

$$\eta_{total} = \eta_{area}\eta_{burst} = \frac{A_{v,eff}}{A_{v,burst}}\frac{A_{v,burst}}{A_{v,nominal}} = \frac{A_{v,eff}}{A_{v,nominal}} . \qquad \text{[Eq. 5]}$$

This equation indicates that the effective vent area (instead of the nominal vent area) should be used to estimate the reduced pressure inside a vessel when a venting device is installed (i.e., $A_{v,eff} = \eta_{total} A_{v,nominal}$). Furthermore, for venting-device applications that have been designed using correlations for given nominal values of Δp_{stat} and Δp_{red}, the required vent area should be $1/\eta_{total}$ times larger than the designed vent area:

$$A_{v,required} = \frac{A_{v,design}}{\eta_{total}} . \qquad \text{[Eq. 6]}$$

4. Results and Discussion

4.1 Unvented Explosions

Unvented tests are first conducted in order to establish the required ignition delay times between dust injection and ignition. A summary of the effective unvented reactivity, (i.e., $K_{eff,unv} = V^{1/3}(dp/dt)_{max}$) as a function ignition delay time is shown in Fig. 3. The ranges of ST1 and ST2 dust classes are denoted by the shaded areas in the figure. Based on these results, ignition delay times of $\tau = 1.55$ and 1.44 s are used in the present study to simulate the behaviour of ST1 and ST2 dust classes using cornstarch.

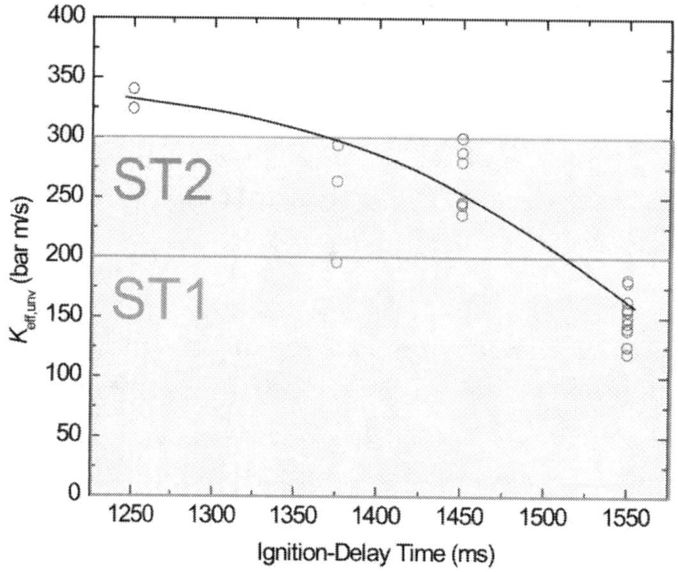

Figure 3. Variation of $K_{eff,unv}$ with ignition-delay time for 750 g/m^3 of cornstarch.

It should be noted that there exists fairly significant variability in the value of $K_{eff,unv}$ for a given ignition delay. In general, it was found that the effective reactivity of the dust is influenced by the initial conditions at the time of ignition, which results in variations in the growth of the flame ball. This seems to manifest itself in the initial portion of the pressure history; thus, the variation in pressure rise, Δp, at early times can be related to an effective flame velocity: $u_{eff} \propto \Delta p^{1/3}/t$ (see [4] for the simple analytical model that was used).

When the calculated values of u_{eff} for each test were compared to K_{eff} (for both unvented and vented tests), it was found by the authors that K_{eff} can, in fact, be characterized by u_{eff} and that the two quantities are linearly proportional (i.e., $K_{eff} = mu_{eff}$ where m is a proportionality constant that depends on the specific dust and apparatus used; see [5] for details). Therefore, test-to-test variability in K_{eff} can be for both unvented and vented tests can be accounted in Eq. 2 via u_{eff}.

4.2 Vented Explosions

A sequence of images from the high-speed camera are shown in Figs. 4 and 5 for tests conducted using a burst disc only and using a flameless venting device with a burst disc, respectively. In both cases, ignition occurred at $t = 0$ s.

In the first frame of Fig. 4 at $t = 0.116$ s, the burst disc ruptures and unburned cornstarch can be seen to begin to vent from the vessel. In the next frame at $t = 0.129$ s, a significant cloud of cornstarch is formed and is ignited when the expanding flame inside the vessel reaches the threshold of the vent. The resulting external explosion can be seen in the third frame at $t = 0.145$ s. In the last frame at $t = 0.173$ s, the remaining combustion products inside the vessel are then jetted through the vent and extend several meters above the vent exit.

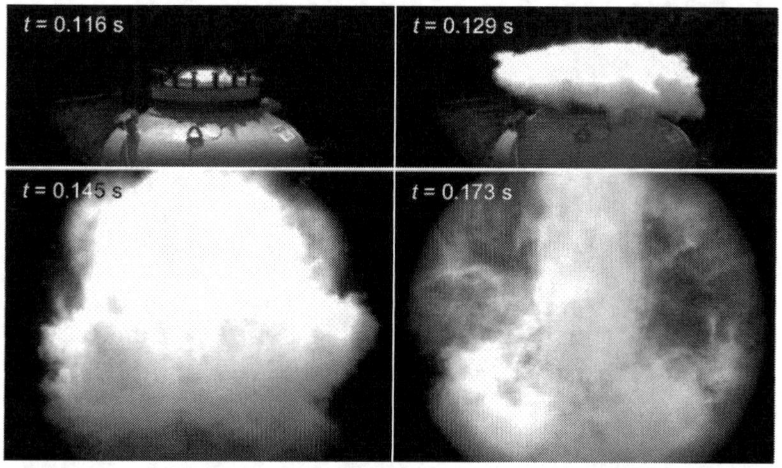

Figure 4. Images from a vented test using a burst disc only with an ignition-delay time of τ = 1.55 s (ST1 dust class).

In contrast, no visible activity can be observed in Fig. 5 in the first frame at t = 0.102 s when the burst disc ruptures (as indicated by the burst indicator). In the second frame at t = 0.121 s, unburned cornstarch can be seen venting through the device. In the third frame, a larger cloud of unburned cornstarch is formed outside of the flameless venting device but is not ignited as the combustion products are sufficiently cooled by the wire mesh. At no time during the course of the test were visible flames observed. In the last frame at t = 0.156 s, the pressure inside the vessel has already begun to decrease and only a cloud of unburned cornstarch and gaseous combustion products can be seen.

Examples of temperature measurements taken at two different locations (above the flameless venting device and to the side where the burst disc initially ruptures) are shown in Fig. 6a and b, respectively. The actual temperature measurements are shown in grey in the figure; smoothed temperature curves using a least-squares-fit method are shown in black. Above the flameless venting device, the peak temperature rise due to the venting of the combustion products is measured to be around 200°C. It can be seen that the duration of the temperature rise is very short (about 0.025 s). On the side of the flameless venting device where the burst disc initially ruptures, the temperature rise is not observed to increase over 100°C. From these temperature measurements, it appears that the vented combustion products were adequately cooled by the venting device.

Figure 5. Images from a vented test using a flameless venting device and a burst disc (τ = 1.55 s; ST1 dust class).

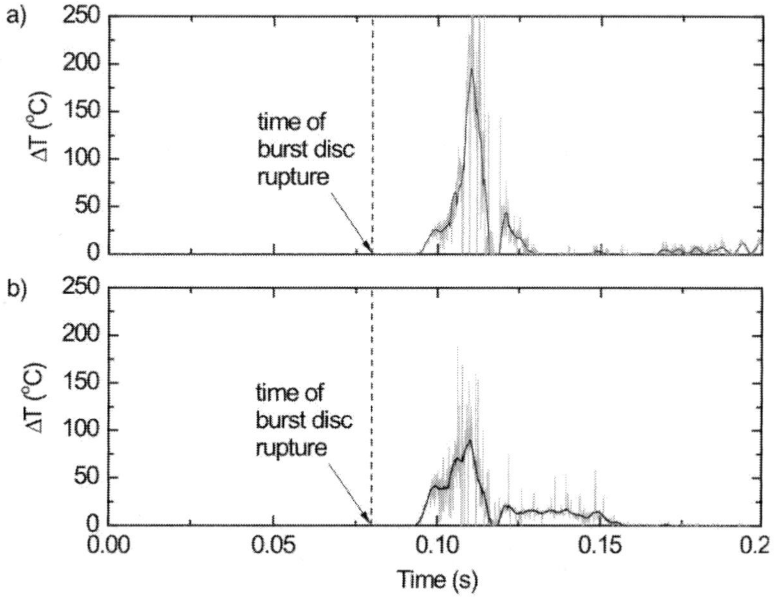

Figure 6. Temperature rise due to vented flow through a flameless venting device with $\tau = 1.45$ s at a) top and b) side where burst disc ruptures initially.

A typical overpressure history of a test using a flameless venting device and burst disc assembly is shown in Fig. 7 (similar trends are observed for tests using a burst disc only). Ignition occurs at $t = 0$ s. The rupture time of the burst disc is denoted by a vertical dashed line in the figure; the corresponding burst pressure is denoted by a horizontal dashed line.

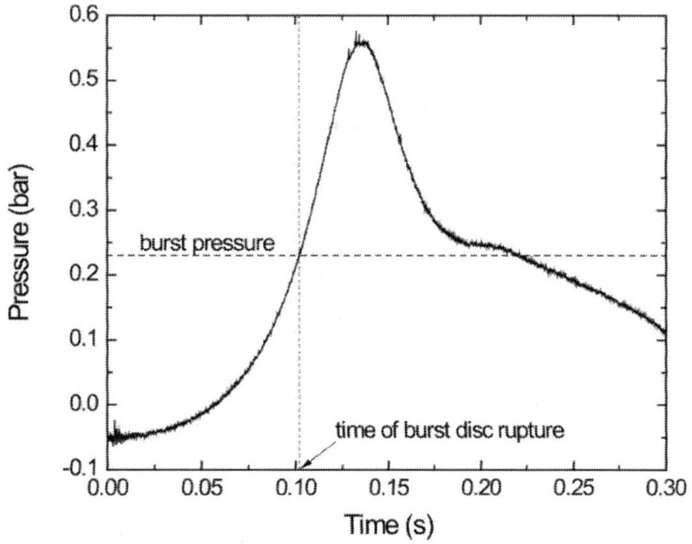

Figure 7. Overpressure history of a vented test using a flameless venting device and burst disc ($\tau = 1.55$ s; ST1 dust class).

It can be seen from the figure that the pressure inside the vessel increases gradually subsequent to ignition and during the deployment of the burst disc. The pressure inside the vessel eventually reaches a maximum reduced value of $\Delta p_{red,q} = 0.56$ bar. The actual pressure at which the burst disc ruptures ($\Delta p_{burst} = 0.23$ bar) is greater than the nominal static burst pressure of $\Delta p_{stat} = 0.1$ bar due to the inertia of the burst disc.

The results of vented tests conducted using a flameless venting device and using a burst disc only are summarized in Table 1. Each combination of a burst-disc/flameless-venting-device test and the corresponding area efficiency (calculated using Eq. 2) are shown. It can be seen in the table that the burst discs all ruptured at pressures greater than the nominal static burst pressure. It can also be seen that there exist test-to-test variability in the reduced overpressures; in some cases, the reduced overpressures using a flameless venting device are comparable to those when only a burst disc is used. This suggests that the reactivity of the dust is not constant from test to test; this variability is accounted for in the present methodology by relating K_{eff} to u_{eff}, which is calculated from the initial growth of the flame ball in each test.

Table 1. Summary of reduced pressures, actual burst pressures, outside temperatures, and corresponding η_{area} for all test combinations; A_v = 0.44 m^2 and $\Delta p_{stat} = 0.1$ bar.

Dust Class	burst disc			venting device					η_{area}
	Test #	Δp_{red} (bar)	Δp_{burst} (bar)	Test #	Δp_{red} (bar)	Δp_{burst} (bar)	ΔT_{top} (°C)	ΔT_{side} (°C)	
ST1	067	0.36	0.17	060	1.05	0.22	-	-	0.85
				081	0.56	0.23	475	225	0.90
	068	0.58	0.26	060	1.05	0.22	-	-	0.91
				081	0.56	0.23	475	225	0.97
	average±stdev								**0.91±0.05**
ST2	082	1.13	0.26	083	1.51	0.28	175	100	0.90
				085	1.56	0.22	125	225	0.84
	084	0.84	0.16	083	1.51	0.28	175	100	0.88
				085	1.56	0.22	125	225	0.94
	088	1.51	0.22	083	1.51	0.28	175	100	0.95
				085	1.56	0.22	125	225	0.89
	average±stdev								**0.90±0.04**

Burst efficiencies determined from Eq. 4 are shown in Table 2 for the five tests that were conducted using a burst disc only. The actual burst pressures that were determined from the experiments (see Table 1 for values) are used in the calculation.

Table 2. Summary of η_{burst} for all burst-disc-only tests.

Dust Class	Test #	η_{burst}
ST1	067	0.90
	068	0.80
	average±stdev	**0.85±0.08**
ST2	082	0.80
	084	0.90
	088	0.84
	average±stdev	**0.84±0.05**

In general, the results are fairly consistent for the two different dust classes (i.e., simulated ST1 and ST2 reactivities) and, thus, a single efficiency can be taken from the average values of η_{area} and η_{burst} to obtain $\eta_{area} = 0.90$ and $\eta_{burst} = 0.84$.

4.3 Determining Total Efficiency (η_{total})

Taking the product of the area and burst efficiencies, the total efficiency of the flameless venting device found in the present series of tests is found to be 76% (for non-fibrous ST1 and ST2 dusts). This implies that, when this device is installed over an existing vent, the effective vent area that should be used to estimate the reduced overpressure is 76% of the nominal vent area (without the device). Furthermore, the required vent area should be a factor of 1.3 (i.e., 1/0.76) greater than the designed vent area.

5. Concluding Remarks

It has been shown the methodology presented in this study can be used to evaluate the performance of a flameless venting device via a series of unvented and vented tests. It was also shown that the efficiency of a flameless venting device (defined by the ratio of the effective vent area to the nominal vent area) can be considered as the product of an area efficiency of the flameless venting device and a burst efficiency of the burst disc itself. The area efficiency is taken to be the ratio of effective vent area of the flameless venting device to that of the burst disc; the burst efficiency is defined as the effective vent area of the burst disc to the nominal vent area. Because test-to-test variability exists in the effective reactivity of the dust that is sufficiently large to influence the calculation for effective vent area, a correction factor must be included in the evaluation of the area efficiency. In the present study, this correction factor is applied to the effective reactivity, which is correlated to an effective velocity that describes the initial growth of flame ball.

6. References

[1] NFPA, "Standard on Explosion Protection by Deflagration Venting," National Fire Protection Association: Quincy, MA, 2007.

[2] VDI, "Pressure Venting of Dust Explosions," Verein Deutscher Ingenieure: Dusseldorf, Germany, 2002.

[3] Tamanini, F. "Scaling Parameters for Vented Gas and Dust Explosions," *Journal of Loss Prevention in the Process Industries*, Volume 14, No. 6, 2001.

[4] Nagy J., Conn J.W., Verakis H.C., "Explosion Development in a Spherical Vessel," U.S. Bureau of Mines, RI No. 7279, 1969.

[5] Chao J., Dorofeev S.B. "Evaluating the Decrease in Effective Vent Area of a Flameless Venting Device," manuscript prepared for *Journal of Loss Prevention in the Process Industries*.

Explosibility of Nano-Sized Metal Powders

Ashok Ghose Dastidar
Fauske and Associates LLC
16W070 83rd Street, Burr Ridge, Illinois, USA 60527
Dastidar@fauske.com

Simon Boilard
Dalhousie University, Halifax, Nova Scotia, Canada
sm416955@dal.ca

Paul R. Amyotte
Dalhousie University, Halifax, Nova Scotia, Canada
Paul.Amyotte@dal.ca

Leonid A. Turkevich
Division of Applied Research and Technology, NIOSH/CDC, Cincinnati, Ohio, USA 45226
LLT0@cdc.gov

Prepared for Presentation at
American Institute of Chemical Engineers
2013 Spring Meeting
9th Global Congress on Process Safety
San Antonio, Texas
April 28 – May 1, 2013

UNPUBLISHED

Explosibility of Nano-Sized Metal Powders

Ashok Ghose Dastidar
Fauske and Associates LLC
16W070 83rd Street, Burr Ridge, Illinois, USA 60527
Dastidar@fauske.com

Simon Boilard
Dalhousie University, Halifax, Nova Scotia, Canada
sm416955@dal.ca

Paul R. Amyotte
Dalhousie University, Halifax, Nova Scotia, Canada
Paul.Amyotte@dal.ca

Leonid Turkevich
Engineering and Physical Hazards Branch - Division of Applied Research and Technology, NIOSH-CDC, Cincinnati, Ohio, USA
llt0@cdc.gov

Keywords: dust explosions, nano-materials, metal powders

Abstract

The explosibility of micron- and nano-titanium was determined and compared according to explosion severity and likelihood using standard dust explosion equipment. ASTM methods were followed using commercially available testing equipment. The explosibility parameters investigated for the size ranges of titanium tested include explosion severity (maximum explosion pressure (P_{max}) and size-normalized maximum rate of pressure rise (K_{St})) and explosion likelihood (minimum explosible concentration (MEC), minimum ignition energy (MIE) and minimum ignition temperature (MIT)). The results indicate a significant increase in explosion severity as the particle size decreases from <150μm with an apparent plateau being reached at <45μm and ≤20 μm. Micron-size explosion severity could not be compared with that for nano-titanium due to pre-ignition of the nano-powder in the 20-L chamber. Explosibility screening tests were also performed on other nano-sized metal powders and similar results were observed for some materials.

The likelihood of an explosion increases significantly as the particle size decreases into the nano range. Nano-titanium is very sensitive and can self-ignite under the appropriate conditions. A similar phenomenon was observed for some of the other nano-metals. Safety precautions and procedures for the nano-metal handling are also discussed.

The findings and conclusions in this paper are those of the authors and do not necessarily represent the views of the National Institute for Occupational Safety and Health. Mention of product or company name does not constitute endorsement by the Centers for Disease Control and Prevention.

1. Introduction

Industrial accidents in the metal processing industries are not uncommon. Some of these accidents are in the form of dust explosions or flash fires. Two of the most cited incidents in North America for combustible metal dust explosions/flash-fires have been the Hayes Lemmerz accident in Huntington, Indiana in 2003 and, more recently, the Hoeganaes Corporation flash-fire in Gallatin Tennessee in 2011. It is important to note that combustible metals can be just as energetic and hazardous as organic fuels, and in some cases like aluminum, magnesium, titanium and zirconium even more energetic.

Dust explosions have been well documented for particles falling in the micron-range for different types of metals [1]. As the size is further decreased into the nano-range, new physical and chemical properties can occur changing the severity and likelihood of a dust explosion.

Interest in the production of nano-scale metal powders has developed in recent years. At the nano-scale, typically between 1 and 100 nm, properties of well-known substances change and they may exhibit unique physical and chemical changes. The bulk properties of the metal no longer hold and the surface properties of the material take precedence. As an example, bulk titanium and zinc are solid, opaque metals. However, at the nano-scale, the surface properties allow the metals to become transparent [2]. The small size of the nano-particles increases the reactive surface area within a bulk sample. This allows for the different physical and chemical properties of nano-particles. These properties are highly desirable for technological advancements but also come at a cost, with an increased explosion potential.

Dust explosion research involving nano-materials has been limited, but some research groups have performed tests with various materials to better understand the properties of nano-size powders. Holbrow et al. [3], with the UK Health and Safety Executive, performed dust explosion tests in a specially designed 2-L explosion chamber with different types of nano-materials including metals (aluminum, zinc, copper and iron). Explosions at a larger test scale using a 20-L sphere were also investigated by Vignes et al. [4] on carbon black, multi-walled carbon nanotubes and aluminum.

Using the nano-aluminum (100 nm and 200 nm) results from Vignes et al. [4], Dufaud et al. [5] compared the nano-aluminum to micron-size aluminum. The results for maximum explosion pressure, P_{max}, and maximum rate of pressure rise, $(dP/dt)_{max}$, for 100 nm samples were 8.2 barg and 1340 bar/s while the P_{max} and $(dP/dt)_{max}$ for 200 nm sample was 9.5 barg and 2480 bar/s. At the micron-scale, 3 μm and 7 μm aluminum gave a P_{max} of 9.8 barg and 9.1 barg and $(dP/dt)_{max}$ of 2090 bar/s and 1460 bar/s, respectively. Explosion severity in this case was limited by the size of the particles with the peak occurring around 1 μm [5].

Likewise, Wu et al. [6] performed nano-dust explosions with aluminum. Samples with an average particle size of 35 nm generated a P_{max} and $(dP/dt)_{max}$ value of 7.3 barg and 1286 bar/s. Samples having an average particle size of 100 nm, resulted in a P_{max} of 12.5 barg, and $(dP/dt)_{max}$ of 1090 bar/s, respectively. These results can be compared to larger micron-aluminum powder data obtained from Eckhoff [7]. For a mean particle size of 22 μm, P_{max} was 12.5 barg and $(dP/dt)_{max}$ was 1474 bar/s [7].

Wu et al. [8] tested the minimum ignition energy (MIE) of micron- and nano-titanium. As the particle size was decreased from 45 μm to 3 μm, the MIE dropped from 21.9 mJ to less than 1 mJ. For three nano-titanium sizes (35 nm, 75 nm and 100 nm), the MIE was lower than 1 mJ [8]. Nano-aluminum and carbon nanotubes also have similarly low MIE values.

These observations of the explosion hazard risk of micron and sub-micron sized metal powders have grave safety implications. The National Fire Protection Agency (NFPA) 484 document [9] is an occupancy standard which covers the mitigation of combustible metal dust explosion and fire hazards. While this document covers the unique hazards associated metal fine dusts and powders, it does not address the risks of metal dust combustibility at the nano-scales. Part of the reasoning for this may be due to the limited amount of experimental data in this field.

The current work is aimed at initiating an investigation into the explosion behavior of combustible metals by studying the explosion severity and likelihood of several nano-scale metal dusts and comparing the results to micron-scale combustibility data. Several metal powders including titanium, aluminum, chromium and copper were studied. The presentation here of the detailed titanium explosibility results and experimental challenges is drawn from Boilard et al. [10] and Amyotte et al. [11], with relevant excerpts.

2. MATERIALS

Various nano-metals were chosen for this study. Table 1 contains a list of the nano-metals on which dust cloud combustibility screening tests were conducted as per Section 13 of ASTM E1226 [12]. Detailed particle size analyses of the metals were not performed due to lack of sample availability. The manufacturer's specification was used for the particle size description.

Table 1 Nano-metals used for dust cloud combustibility screening tests.

Material Name	Manufacturer Reported Mean Particle Size [nm]
Tantalum Nanopowder	50-80
Chromium Nanopowder	35-45
Tungsten Nanopowder	40-60
Zinc Nanopowder	35-45
Molybdenum Nanopowder	45
Aluminum Nanoparticles 99.9% 40-60nm	40-60
Iron Nanoparticles 99.9% 40-60nm	40-60
Titanium Nanoparticles	40-60
Nickel Nanoparticles	40-60
Tin Nanoparticles	< 100
Copper Nanoparticles	25
Aluminum Nanoparticles 99.9%, 18nm	18
Cobalt Nanoparticles	25-30
Iron Nanoparticles 99.9% 25nm	60-80
Silicon Nanopowder	50-80
Iron Nanopowder – Carbon Coated	25
Nano Manganese	30-50

Nano-titanium was singled out for examination due to its reactivity and the availability of existing data for dust explosibility at the micron-scale. Six sample sizes were selected: <150 μm, <45 μm, ≤20 μm, 150 nm, 60-80 nm, and 40-60 nm.

Particle size distributions were determined to characterize the micron-size titanium. The powder manufacturer's literature states that the two smaller micron-titanium samples are nominally ≤20 μm and <45 μm – but are sized using different methodologies; a single-point BET surface area analysis is used to determine the size of the ≤20 μm titanium while a sieve analysis is used to analyze the <45 μm titanium. According to the manufacturer, the BET surface area analysis measures the average unagglomerated particle size. However, traditional sieve analysis measures what is essentially the particle size distribution of agglomerated particles. This can be seen in the sieve analyses performed in the current work and shown in Table 2. Particle sizes of the nano-titanium were taken to be as reported by the manufacturer and no further size analysis was performed.

Table 2 Sieve analysis of micron-size titanium powders.

		Wt. % Retained		
Sieve Size	**Micron Size**	**≤20 μm**	**<45 μm**	**<150 μm**
+325	>45 μm	0.98	0.00	91.88
-325+400	45-38 μm	10.55	6.45	5.06
-400+450	38-32 μm	5.96	4.50	0.38
-450+500	32-25 μm	23.54	23.03	1.72
-500	<25 μm	58.98	66.02	0.96

A scanning electron micrograph (SEM) of the ≤20 μm titanium sample is shown in Figure 1. Individual particles of this titanium are not spherical but have a granular shape. Agglomerates occur for this sample, as illustrated in Figure 1b. Two types of agglomerates are present; larger titanium particles are covered by smaller titanium "bits", and medium-sized particles are joined to form a larger particle. Figure 2 shows the 150 nm sample. While the individual particles are of course much smaller, agglomerates are still clearly present. The nano-agglomerates varied in composition between approximately 50 particles and thousands of particles.

a b

Figure 1 Scanning electron micrograph of ≤20 μm titanium powder.

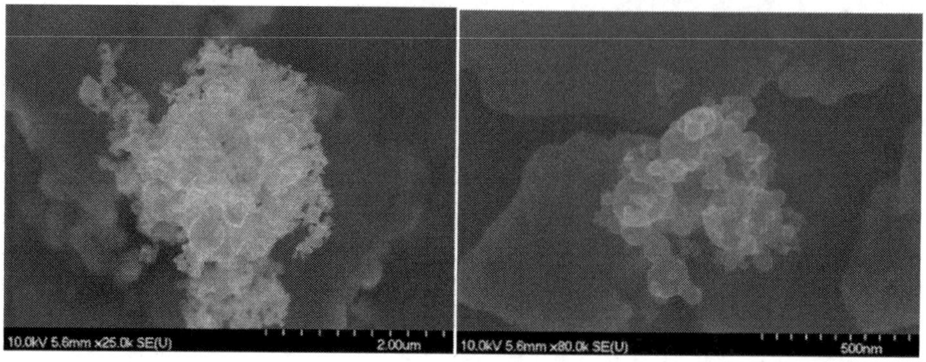

a b

Figure 2 Scanning electron micrograph of 150 nm titanium powder.

3. EQUIPMENT AND METHODOLOGY

The explosibility parameters investigated for the nano- and micron-size ranges of metal powders include dust cloud combustibility, maximum explosion pressure (P_{max}), size-normalized maximum rate of pressure rise (K_{St}), minimum explosible concentration (MEC), minimum ignition energy (MIE), and minimum ignition temperature (MIT). American Society for Testing and Materials International (ASTM) methods were followed using standard dust explosibility test equipment: Siwek 20-L explosion chamber, MIKE 3 apparatus and BAM oven. The applicable ASTM methods are ASTM E1226 [12], E1515 [13], E2019 [14], and E1491 [15], respectively.

Figure 3 shows a picture of the Siwek 20-L chamber with its corresponding components. The Siwek chamber was used to determine the dust cloud combustibility of the nano-metals listed in Table 1. These tests were conducted to establish a preliminary baseline on nano-metal dust explosibility from which a ranking, or prioritization, scheme could be established for further detailed study. The apparatus was operated with the standard dispersion methodology listed in the manufacturer's manual using air.

However, nitrogen was used in place of air as the dust dispersion medium for the detailed nano-titanium explosion severity testing. This was done to prevent pre-ignition of the powder in the dust storage chamber as observed by Wu et al. [6]. A lower-than-usual vacuum was therefore created in the 20-L chamber and oxygen was backfilled to create an elevated oxygen atmosphere (prior to dust dispersion). Once the dispersing nitrogen and the elevated oxygen level mixed, atmospheric conditions would be achieved with 79% nitrogen and 21% oxygen. CaRo11 (Kühner) dust was tested using this method to establish the procedural validity. The results from nitrogen dispersion and air dispersion were well-correlated. Therefore, all nano-titanium tests for explosion severity (P_{max} and $(dP/dt)_{max}$) were performed with nitrogen dispersal.

Figure 3 Siwek 20-L chamber.

4. RESULTS AND DISCUSSION

4.1 Dust Cloud Combustibility Screening Tests

Dust cloud combustibility screening tests were conducted per Section 13 of ASTM E1226 [12] for the metal nano-powders listed in Table 1. These screening tests were performed with the standard air dispersion methodology. The concentration of material tested and the subsequent explosion overpressure are reported in Table 3

Table 3 Dust Cloud Combustibility Screening Test Results

Material Name	Powder Concentration Tested [g/m^3]	Explosion Overpressure P_m [barg]
Tantalum Nanopowder	500	2.7
Chromium Nanopowder	500	Ignited on dispersion
Tungsten Nanopowder	500	0
	1000	0
Zinc Nanopowder	500	2.1
Molybdenum Nanopowder	500	0.8
	1000	4.0
Aluminum Nanoparticles 99.9% 40-60nm	500	5.1
Iron Nanoparticles 99.9% 40-60nm	500	2.1
Titanium Nanoparticles	500	Ignited on dispersion
	1000	Ignited on dispersion
Nickel Nanoparticles*	500	0.4
	1000	0.7**
Tin Nanoparticles	500	0.9
	1000	1.2
Copper Nanoparticles	Sample autoignited on removing from container	
Aluminum Nanoparticles 99.9%, 18nm	500	2.1
Cobalt Nanoparticles	500	0.0
	1000	0.0
	2500	0.8**

Table 3 Dust Cloud Combustibility Screening Test Results continued.

Material Name	Powder Concentration Tested [g/m³]	Explosion Overpressure P_m [barg]
Iron Nanoparticles 99.9% 25nm	500	0.0
	1000	0.9
	2500	1.3
Silicon Nanopowder	500	5.5
Iron Nanopowder – Carbon Coated	500	2.5
Nano Manganese	500	0.8
	1000	1.2

* sample started to get warm on exposure to air – no fire
** may be approaching combustibility at higher concentrations.

These preliminary explosibility screening test results indicate that not all nano-metals are explosible even though the metals do form oxides. The move from micron-scale to nano-scale did not make them explosible. Other nano-metals that do explode (i.e. aluminum, tantalum, tin etc) produce explosion overpressures comparable to micron-scale metal particles. It should be noted that only limited concentrations of these metals were tested and a full comparison of overpressures between nano-scale and micron-scale can only be accomplished with detailed explosion severity testing.

Of greater importance is that some nano-scale metals have highly increased ignition sensitivities; so much so that metals reacted on dispersion (e.g. titanium and chromium) or reacted on first exposure to air (e.g. copper and nickel). Any further study into nano-metal explosibility should focus on these materials. In the next sections we discuss a more detailed study into the explosion severity and ignition sensitivity of nano-titanium.

4.2 Explosion Severity

The explosion severity of the micron-size titanium samples can be seen in Table 4. P_{max} and K_{St} both increase significantly (40% and 400% increases, respectively) with decreasing size from <150 μm to <45 μm. As the particle size is further decreased to ≤20 μm, P_{max} decreases by approximately 10% and K_{St} remains essentially constant. The micron-size data therefore show the expected increase in P_{max} and K_{St} with decreasing particle size, but to a limit that may be imposed by particle agglomeration.

Table 4 Explosion severity of micron-size titanium powders.

Sample	P_{max} [bar(g)]	$(dP/dt)_{max}$ [bar/s]	K_{St} [bar·m/s]
<150 μm	5.5	84	23
<45 μm	7.7	436	118
≤20 μm	6.9	420	114

It is difficult to make a direct comparison of explosion severity between the micron- and nano-titanium due to pre-ignition of the nano-dust. Frictional or static sparking during the dispersal sequence ignited the dust before the chemical igniters were fired. (Dust ignition in the Siwek 20-L chamber is normally accomplished via one or more chemical igniters activated at a preset ignition delay time.) As previously described, nitrogen was used as the dispersing gas to prevent an explosion in the external dispersion reservoir. However, when the nano-titanium/nitrogen mixture encountered the elevated oxygen concentration in the 20-L chamber, immediate ignition occurred.

Consistent with the above discussion, Figure 4 shows a 40-60 nm titanium explosion at a dust concentration of 100 g/m^3. Here one sees an overlap of the dust dispersion and explosion steps, with the igniters firing after the dust has exploded. The pressure values in Figure 4 must be interpreted in light of the fact that the pressure transducers used are piezoelectric and hence measure only dynamic pressure changes, not static values.

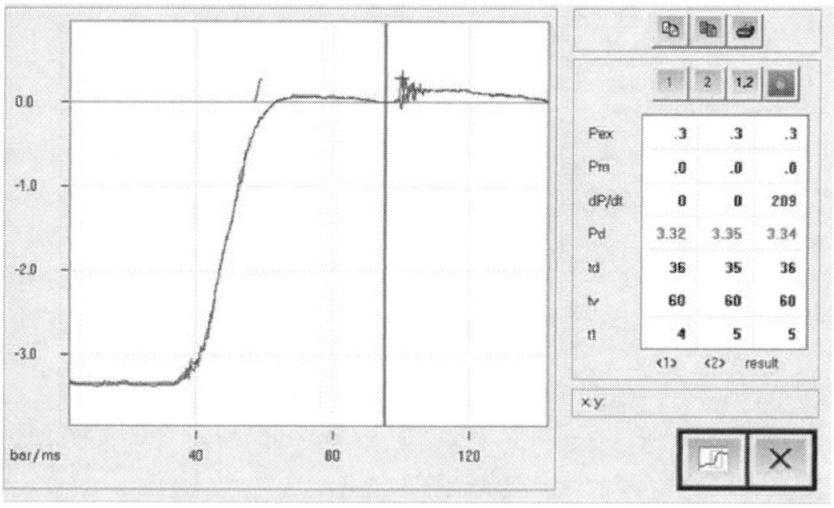

Figure 4 Sample pre-ignition pressure/time trace for nano-titanium.

The pre-ignition of the nano-titanium causes all the dust to be consumed and effectively prevents a typical evaluation of the maximum explosion pressure. The maximum rate of pressure rise can be estimated by a manual evaluation of the pressure/time trace. The explosion is very violent and causes complete combustion – again, even before the chemical igniters are activated.

All subsequent trials of this type involving nano-titanium were performed without igniters; results similar to those shown in Figure 4 were obtained. Because of the pre-ignition phenomenon, typical P_{max}/K_{St} testing was simply not possible with these materials. Therefore, estimates of the maximum explosion pressure and maximum rate of pressure rise were obtained from the pre-ignition phase of the pressure/time trace (the region to the left of the vertical line at 96 ms in Figure 4). It is particularly noteworthy that these explosibility parameters arise from an explosion scenario *with no external ignition source* (i.e., no chemical igniters as per typical Siwek 20-L chamber testing).

Figure 5 gives the maximum rate of pressure rise data acquired to date. For all nano-sizes, a quasi-plateau in the rate of pressure rise occurred at low concentrations. At approximately 125 g/m^3, the maximum rate of pressure rise for all three nano sizes was between 550 – 650 bar/s. It is important to note here that the rate of pressure rise data cannot be directly compared with K_{St} levels determined by international standards since the explosion is occurring in a more highly turbulent flow field as the powder jets into the test chamber from the reservoir. In the case of the

150 nm titanium, the rate of pressure rise displayed a decrease at concentrations higher than 500 g/m^3. Micron-size titanium did not reach its maximum rate of pressure rise until concentrations in the vicinity of 1500 g/m^3.

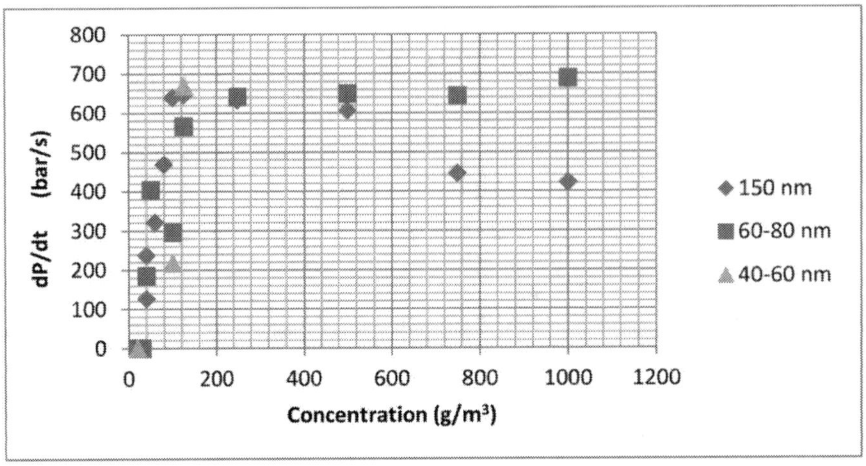

Figure 5 Maximum rate of pressure rise for nano-titanium dusts.

Surface area clearly plays a major role in the combustion of the different sizes of titanium studied in the current work. At the micron-scale, larger amounts of dust are needed to achieve the same effective surface area as with a smaller mass of nano-titanium. For complete combustion of all the oxygen in the 20-L chamber, high concentrations of micron-size titanium are needed. To react with the same amount of oxygen, much lower amounts of nano-titanium are required. There is thus a high reactivity at very low concentrations for nano-size titanium.

Maximum explosion pressure can be estimated for the nano-titanium samples via the P_d value given by the 20-L chamber control software. In the usual 20-L procedure, the chamber is partially evacuated to 0.4 bar(a) so the resulting chamber pressure upon dust dispersion and at the time of ignition (chemical igniter firing) is approximately 1 bara. The parameter P_d in the Kühner software reports the rise in pressure in the 20-L chamber due to dust dispersion. Acceptable values of P_d are in the range of 0.55 – 0.7 bar, meaning that when the value of P_d is added to the initial chamber pressure of 0.4 bar, the sum is approximately 1 bar.

Recalling that the piezoelectric pressure transducers measure only dynamic pressure changes (i.e., not static pressures), P_d values as shown in the pressure-time trace in Figure 4 are a combination of the pressure rise due to dust dispersion *and* the actual explosion itself occuring at sub-atmospheric pressures. The nano-size P_d results are therefore not directly comparable to the micon-size P_{max} results. They are, however, self-consistent as a data set as shown in Figure 6. Similar to the case for rate of pressure rise shown in Figure 5, nano-titanium reaches its highest values of "maximum explosion pressure" at lower concentrations than the micron-titanium, which again required dust concentrations around 1500 g/m^3.

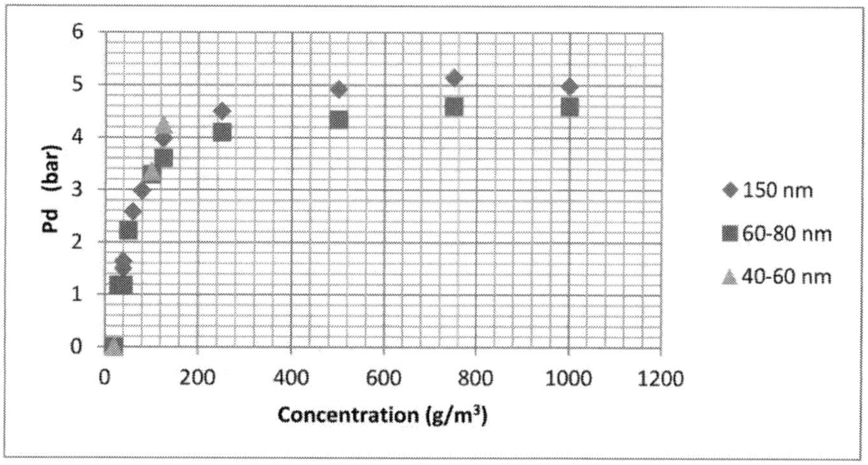

Figure 6 P_d values ("maximum explosion pressure") for nano-titanium dusts.

In these preliminary stages of studying explosion severity of nano-sized metals is appears that the explosion output is very similar between nano-sized and very small micron-sized metal powders, however the concentrations at which these maxima are reached are much lower. For the purposes of hazard mitigation where explosion protection schemes are devised for the worst-case explosion severity independent of fuel concentration, these results for nano-titanium may indicate that protection for small micron-size metals can be used for nano-sized metals as well.

4.3 Explosion Likelihood

The explosion likelihood of a dust can be assessed by measurement of three parameters: minimum ignition energy (MIE), minimum ignition temperature (MIT) and minimum explosible concentration (MEC). Results from the work reported here are given in Table 5.

Table 5 Explosion likelihood of titanium powders.

Sample	MIE [mJ] (no inductance)	MIT [°C]	MEC [g/m³]
<150 μm	1 – 3	>590	60
<45 μm	1 – 3	460	60
≤20 μm	<1	460	50
150 nm	<1	250	30 –40
60-80 nm	<1	240	Not determined yet
40-60 nm	<1	250	Not determined yet

Micron-size titanium has been previously shown to be ignitable at very low spark energies. Randeberg and Eckhoff [16] demonstrated that 3 μm and 9 μm titanium can ignite at energies as low as 0.012 mJ and 0.36 mJ, respectively. Low MIE values for titanium were also measured in the current work as shown in Table 5 (1 mJ being the lowest spark energy attainable with the MIKE 3 apparatus).

There is, however, a fundamental difference between the MIEs of the micron- and nano-size samples. As shown in Figure 7 for the ≤20 μm sample, ignition at 1 mJ occurred only at a high dust loading of 3000 mg. Figure 8, a MIKE 3 ignition graph for the 150 nm titanium, illustrates the significantly lower dust loadings are ignitable at 1 mJ (as low as 100 mg). Note that in Figures 7 and 8, dust quantity appears on the abscissa and spark energy on the ordinate; the open boxes indicate no ignition at the particular delay time and the solid boxes indicate ignition.

MIE tests with dust loadings greater than 900 mg were not attempted for the nano-titanium samples due to the risk of damage to glass components of the MIKE 3 apparatus and operator safety concerns.

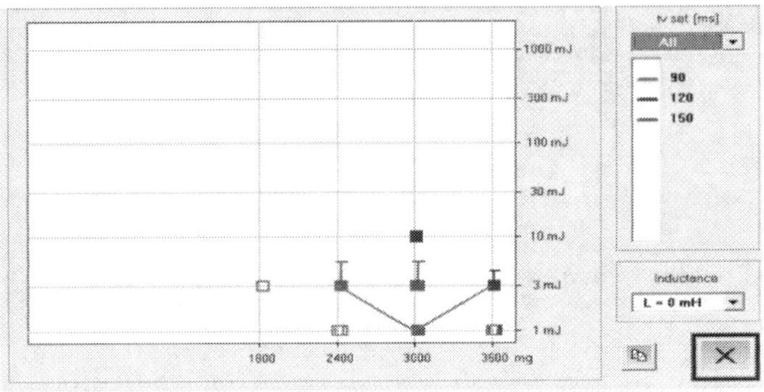

Figure 7 MIE data (no inductance) for ≤20 μm titanium.

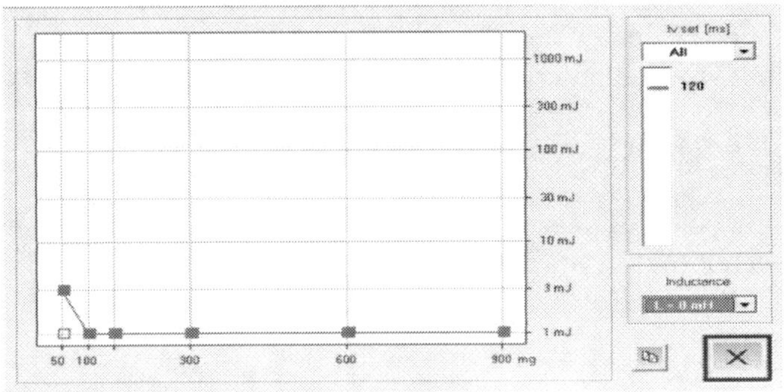

Figure 8 MIE data (no inductance) for 150 nm titanium.

A significant increase in explosion likelihood occurs as MIT decreases with decreasing particle size (Table 5). Large <150 μm particles could not be ignited in the BAM oven at 590 °C (which is the highest temperature attainable with our apparatus). Smaller micron-size particles at <45 μm and ≤20 μm ignited at a lower oven temperature of 460 °C. The minimum ignition temperature of the nano-titanium (all sizes) was in the range of 240 – 250 °C. These temperatures are obviously much lower than those required to ignite the micron-titanium, thus clearly demonstrating the enhanced potential for nano-titanium ignition by hot surfaces. This observed trend is also indicated from tests reported in Table 2, some nano-metals become pyrophoric and autoignite at room temperature once exposed to air (see copper and nickel).

It was in fact difficult to determine the minimum ignition temperature of the nano-sizes due to the nature of the powder. Titanium ignition produces flames that are very bright and fast, and test temperatures exceeding 350 °C would cause the nano-titanium to flash-off rapidly. In these tests, a 'hissing' sound could be heard as the powder entered the BAM oven upon dispersal. Visual observation of flames exiting the oven was more readily accomplished at temperatures below 300 °C.

While MEC determination for the micron-size samples in the Siwek 20-L chamber was relatively straightforward, the nano-size samples again posed an experimental challenge – this time due to the previously described pre-ignition of the powder during the dispersion sequence. With the 150 nm titanium, pre-ignition did not occur at low concentrations of 20 and 30 g/m^3; these concentrations could therefore be tested with nitrogen dispersion and the chemical igniter (5 kJ) in place. No explosions in the 20-L chamber were recorded at these conditions, meaning that the MEC of the 150 nm sample is greater than 30 g/m^3.

With a dust concentration of 40 g/m^3, and again nitrogen dispersion and a 5-kJ igniter in place, pre-ignition occured as evidenced by a P_d value of 1.64 bar. Given that approximately 0.6 bar of the P_d value would normally be associated with dust dispersal, and a 1 bar overpressure is the accepted explosion criterion for MEC testing, it can be argued that a P_d value greater than 1.6 bar indicates that an explosion occurred. The main difference from this procedure and usual MEC testing is that the nano-dust 'MEC explosions' occur with no external ignition source.

5. EXPERIMENTAL CHALLENGES

There are many experimental challenges that arise when performing dust explosion research with micron- and nano-titanium. These include challenges related to the experimental methods, material handling, waste disposal and laboratory safety. These challenges were especially apparent with the nano-materials.

5.1 Experimental Methods

Issues with titanium powder were first noticed during the 20-L P_{max}/K_{St} testing of the <45 μm and ≤20 μm samples. Ignition at high concentrations would cause explosion residue to be lodged beneath the dispersion valve of the Siwek chamber and disrupt subsequent trials. The dense titanium could not be completely dispersed, leaving a few grams beneath the dispersion valve. It seems that this remaining powder would then be ignited by the intense heat of the explosion occurring in the 20-L chamber and transferred through the valve assembly.

Figure 9 shows a 'normal' pressure/time trace for a test in the Siwek 20-L chamber. In this routine trace for one of the micron-size samples, one can see the pressure rise due to dust dispersion into the 20-L chamber followed by the igniters firing and then the dust itself exploding. This may be compared with the pre-ignition behavior displayed in Figure 4.

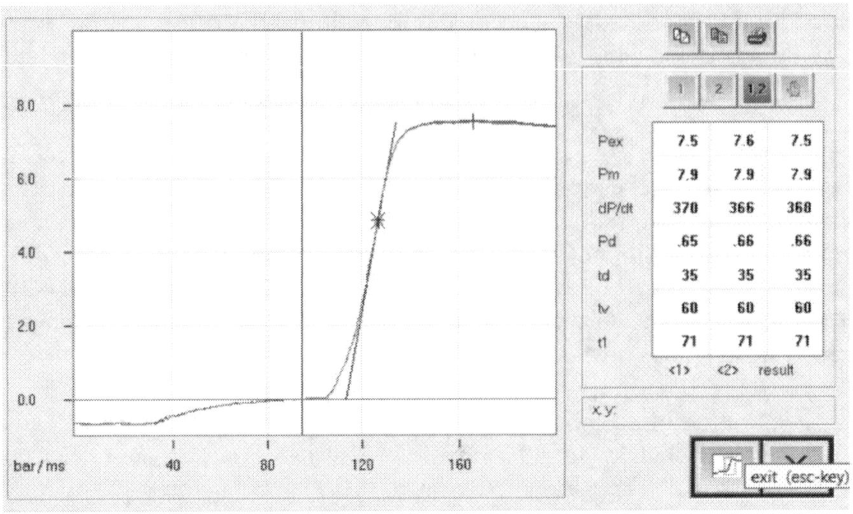

Figure 9 Sample pressure/time trace for micron-titanium.

Table 6 illustrates that no suitable combination of experimental conditions could be identified so as to disperse the nano-titanium into the 20-L chamber and achieve a pressure/time trace of the type shown in Figure 9. Numerous attempts were made by varying: (i) dust concentration, (ii) dispersion gas (nitrogen or air), (iii) dispersion gas pressure and ignition delay time (both of which affect dust cloud turbulence intensity), (iv) location of the dust prior to dispersion (placed in either the external dust storage container or the 20-L chamber itself), and (v) expected oxygen concentration in the 20-L chamber once dispersion was complete. Although no external ignition source (i.e., chemical igniters) was used in these tests, the result was an explosion similar to that shown in Figure 4 in every case.

Table 6 Trials with various experimental conditions for 150 nm titanium (no igniters).

Dust Concentration [g/m^3]	Dispersion Gas	Dispersion Pressure [barg]	Ignition Delay Time [ms]	Dust Placement	Final O_2 % in 20-L Chamber	Explosion During Dispersion?
125	N_2	20	60	Dispersion Reservoir	21.0	YES
125	N_2	20	60		19.5	YES
125	N_2	20	60		17.6	YES
125	N_2	20	60		15.0	YES
125	N_2	20	60		9.0	YES
500	Air	20	60	Under Rebound Nozzle	21.0	YES
500	Air	20	60	On Rebound Nozzle	21.0	YES
500	Air	10	60		21.0	YES
500	Air	10	120		21.0	YES
500	Air	5	120		21.0	YES
500	N_2	20	60		9.0	YES
250	N_2	10	60		9.0	YES

5.2 Material Handling

Titanium has been shown to react with water at elevated temperatures around 700 °C. The metal has a high affinity for oxygen, stripping the oxygen from water to form titanium dioxide and hydrogen gas [17]. Hydrogen explosions have occurred in the past when cooling water unintentionally comes in contact with molten titanium [18]. At ambient temperatures, micron-size titanium does not pose a hazard in this regard. Nano-size titanium, however, can have the same effect as molten titanium; it reacts with moisture to create hydrogen gas but this occurs at ambient temperatures. Hydrogen co-existing with easily ignitable nano-dust causes many concerns for handling the nano-titanium. Therefore, nano-titanium should have limited exposure to air or moisture and should be kept under dry nitrogen or argon. A nitrogen-filled glove bag was an essential feature of the safety precautions taken for the handling of nano-titanium in the current work. This reduced the potential nano-titanium contact with moisture and air. These precautions are also warranted for other nano-metals presented in Table 1 that autoignited and may be necessary for the other nano-metals so as to minimize self-oxidation prior to experimentation.

5.3 Waste Disposal

Disposal of unused nano-titanium powder was important to minimize storage of the material for extended periods of inactivity. The nano-titanium should be deactivated and stored in a fume hood. Nano-titanium was carefully mixed in small quantities with water or dilute nitric acid. Mixing the titanium with water creates stable titanium dioxide while forming hydrogen gas within a vented and controlled environment. Adding nitric acid to the mixture promotes deactivation of the nano-titanium.

5.4 Laboratory Safety

Before any experimentation began, the laboratory was assessed for proper ventilation equipment and personal protective equipment (PPE). An improved ventilation extraction arm was installed and the vacuum exhaust sent through the ventilation system. Appropriate PPE included: a filtered mask, safety glasses, face shield (when working with chemical igniters), lab coat, and double nitrile gloves.

Air quality measurements for airborne nano-particles were performed during normal operation of the 20-L chamber. Two types of particle counters were used: a FLUKE 983 particle counter for particles in the range of 300 nm to 10 μm, and a KANOMAX handheld CPC model 3800 for particles in the range of 15 nm to 1 μm. It was determined that no excess particles become airborne as a result of the use of nano-materials.

6. CONCLUSION

Preliminary results seem to indicate that explosion severity is not significantly different in the nano-scale than in the micron-scale however, the likelihood of an explosion increases significantly as the particle size decreases into the nano range. These nano-scale materials are very sensitive and can self-ignite under the appropriate conditions during laboratory testing and handling. As a result, safety precautions and procedures for the nano-scale metal dust combustibility testing should include laboratory personnel wearing appropriate PPE which should include, but not limited to, full face protection (face-shield or full-face respirator), breathing protection (half-face or full-face respirator), anti-static clothing, and fire-retardant clothing. Additionally, precautions should be taken to maintain an inert gas blanketing over unused samples.

The results obtained in the current detailed work on nano-titanium illustrate that it is much more sensitive to ignition than micron-titanium. Relatively low quantities and very low spark energies can cause nano-titanium to ignite yielding severe explosions. Nano-titanium is so reactive that traditional explosion test procedures using the Siwek 20-L chamber were not possible. Modifications to these procedures were attempted with limited success at preventing pre-ignition of the powder.

Based on this preliminary study, implications to explosion hazard mitigation are mixed. Given that the explosion overpressure of the various nano-metals, and the explosion severity data of the more detailed study of nano-titanium, are not that different than explosion severity of small micron-size metal powders; mitigation methods based upon overpressure containment, deflagration venting or suppression of micron-size nano-metals may be applicable to nano-sized metals as well.

The same is not true for mitigation methodologies involving ignition sensitivity. Here a new strategy must be developed factoring in lower autoignition temperatures, lower minimum ignition energies and lower concentrations of material at which combustion is possible. Such a discussion is beyond the scope of the current work.

ACKNOWLEDGEMENT

The authors gratefully acknowledge the financial support of the Natural Sciences and Engineering Research Council (NSERC) of Canada in the form of a strategic grant. Additionally, the authors would like to acknowledge partial financial support from the Nanotechnology Research Center (NTRC); part of the National Institute for Occupational Safety and Health (NIOSH) of the Centers for Disease Control and Prevention (CDC).

7. REFERENCES

[1] Cashdollar, K. L. and Zlochower I. A. (2007). Explosion Temperatures and Pressures of Metals and Other Elemental Dust Clouds. Journal of Loss Prevention in the Process Industries 20: 337–348.

[2] Pritchard, D.K. (2004): Literature Review – Explosion Hazards Associated with Nanopowders, UK Health and Safety Executive. EC/04/03.

[3] Holbrow, P., Wall, M., Sanderson, E., Bennett, D., Rattigan, W., Bettis, R. and Gregory, D. (2010): Fire and Explosion Properties of Nanopowders. UK Health and Safety Executive. RR782.

[4] Vignes, A., Traoré, M., Dufaud, O., Perrin, L., Bouillard, J. and Thomas, D. (2009): Assessing Explosion Severity of Nanopowders with the 20 L Sphere, 8th World Congress of Chemical Engineering, Montréal, Canada, 23-27 August, 2009.

[5] Dufaud, O., Vignes, A., Henry, F., Perrin, L. and Bouillard, J. (2011): Ignition and Explosion of Nanopowders: Something New Under the Dust. Journal of Physics: Conference Series 304: 012076.

[6] Wu, H.-C., Kuo, Y.-C., Wang, Y.-H., Wu, C.-W. and Hsiao, H.-C. (2010): Study on Safe Air Transporting Velocity of Nanograde Aluminum, Iron, and Titanium. Journal of Loss Prevention in the Process Industries 23: 308-311.

[7] Eckhoff, R.K. (2003): Dust Explosions in the Process Industries, 3rd edition, Gulf Professional Publishing/Elsevier: Boston, MA.

[8] Wu, H.-C., Chang, R.C. and Hsiao, H.-C. (2009): Research of Minimum Ignition Energy for Nano Titanium Powders and Nano Iron Powder. Journal of Loss Prevention in the Process Industries 22: 21-24.

[9] NFPA 484, Standard for Combustible Metals (National Fire Protection Association 2012).

[10] Boilard, S.P., Amyotte, P.R., Khan, F.I., Dastidar, A.G. and Eckhoff, R.K., "Explosibility of Micron- and Nano-Size Titanium Powders", Paper No. 009, Proceedings of Ninth International Symposium on Hazards, Prevention, and Mitigation of Industrial Explosions, Krakow, Poland (July 22-27, 2012).

[11] Amyotte, P., Khan, F., Boilard, S., Iarossi, I., Cloney, C., Dastidar, A., Eckhoff, R., Marmo, L. and Ripley, R., Explosibility of Nontraditional Dusts: Experimental and Modeling Challenges", Hazards XXIII, IChemE Symposium Series No. 158, Southport, UK (November 12-15, 2012), pp. 83-90.

[12] ASTM (2010): Standard Test Method for Explosibility of Dust Clouds, E1226. West Conshohocken, PA: ASTM International.

[13] ASTM (2007): Standard Test Method for Minimum Explosible Concentration of Combustible Dusts, E1515. West Conshohocken, PA: ASTM International.

[14] ASTM (2003): Standard Test Method for Minimum Ignition Energy of a Dust Cloud in Air, E2019. West Conshohocken, PA: ASTM International.

[15] ASTM (2006): Standard Test Method for Minimum Autoignition Temperature of Dust Clouds, E1491. West Conshohocken, PA: ASTM International.

[16] Randeberg, E. and Eckhoff, R.K. (2007): Measurement of Minimum Ignition Energies of Dust Clouds in the <1 mJ Region. Journal of Hazardous Materials 140: 237-244.

[17] Patnaik, P. (2003): Handbook of Inorganic Chemicals, McGraw-Hill: New York, NY.

[18] Poulsen, E. (2000): Safety-Related Problems in the Titanium Industry in the Last 50 Years. JOM 52: 13-17.

9TH GLOBAL CONGRESS ON PROCESS SAFETY

On-Going Developments In Addressing Combustible Dust Hazards

Walt Frank, P.E.
Frank Risk Solutions, Inc.
1110 Shallcross Ave.
Wilmington, DE 19806
wlf@frankrisk.com

Guy R. Colonna, P.E.
National Fire Protection Association
1 Batterymarch Park
Quincy, MA 02169
gcolonna@nfpa.org

Prepared for Presentation at
American Institute of Chemical Engineers
2013 Spring Meeting
9th Global Congress on Process Safety
San Antonio, Texas
April 28 – May 1, 2013

UNPUBLISHED

On-Going Developments In Addressing Combustible Dust Hazards

Walt Frank, P.E.
Frank Risk Solutions, Inc.
1110 Shallcross Ave.
Wilmington, DE 19806
wlf@frankrisk.com

Guy R. Colonna, P.E.
National Fire Protection Association
1 Batterymarch Park
Quincy, MA 02169
gcolonna@nfpa.org

Keywords: combustible dust, fire, explosion, standards, regulations, prevention, process safety

Abstract

The National Fire Protection Association (NFPA) continues to play an important role in creating guidance for identifying and controlling hazards associated with combustible dusts. Five separate standards address hazards associated with wood-related materials, metals, agricultural products and foodstuffs, sulfur, and all other combustible dusts (e.g., paper, plastics, chemicals, pharmaceuticals). This paper will address a number of developments in NFPA's initiatives addressing combustible dust hazards. Topics to be addressed include:

- The status of, and recent revisions to, NFPA's combustible dust standards;
- NFPA's plans for a new combustible dust standard, which will address the common issues associated with controlling combustible dust hazards, leaving the commodity-specific standards to address the issues unique to their scope;
- A new technical correlating committee, which will help ensure greater consistency between the approaches addressed in the various combustible dust standards;
- The increasing inclusion of process safety management principles in the combustible dust standards; and
- The increasing allowance provided to users of NFPA standards for the use of risk-based decision-making in the selection of combustible dust hazard control options.

Options exist for AIChE to provide technical support to a number of the above areas. The paper will also preview a new CCPS project for providing tools for addressing combustible dust hazards.

1. Introduction

Explosions and fires involving combustible dusts continue to cause injuries, property loss, and business interruptions all too-frequently. Increased awareness and attention within the public; the press; the insurance industry; and licensing, regulatory, and investigatory agencies has sharpened the focus on common-sense, practical methods for abating combustible dust hazards. This paper addresses a number of recent initiatives – with particular, but not sole, focus on NFPA's initiatives addressing combustible dust hazards.

2. Background on NFPA Combustible Dust Standards

NFPA continues to play an important role in creating guidance for identifying and controlling hazards associated with combustible dusts. There are currently five separate standards addressing the safe handling and processing of combustible dusts, as shown in Table 1.

Table 1: NFPA Combustible Dust Standards

NFPA Std.	Facility/Material Covered	Scope of Coverage
61	Agricultural and food processing facilities	• All facilities that receive, handle, process, dry, blend, use, mill, package, store, or ship dry agricultural bulk materials, their by-products, or dusts that include grains, oilseeds, agricultural seeds, legumes, sugar, flour, spices, feeds, and other related materials • All facilities designed for manufacturing and handling starch, including drying, grinding, conveying, processing, packaging, and storing dry or modified starch, and dry products and dusts generated from these processes • Those seed preparation and meal-handling systems of oilseed processing plants not covered by NFPA 36, *Standard for Solvent Extraction Plants*
484	Combustible metals	• The production, processing, finishing, handling, recycling, storage, and use of all metals and alloys that are in a form that is capable of combustion or explosion
655	Sulfur	• Size reduction of sulfur and the handling of sulfur in any form • Does not apply to the mining of sulfur, recovery of sulfur from process streams, or transportation of sulfur
664	Wood processing and woodworking facilities	• Industrial, commercial, or institutional facilities that process wood or manufacture wood products, using wood or other cellulosic fiber as a substitute for or additive to wood fiber, and that process wood, creating wood chips, particles, or dust
654	Any combustible particulate solid not covered by one of the above standards	• All phases of the manufacturing, processing, blending, conveying, repackaging, and handling of combustible particulate solids or hybrid mixtures, regardless of concentration or particle size, where the materials present a fire or explosion hazard

As Table 1 indicates, the first four standards are specific to particular commodities or industries. The fifth standard, NFPA 654, *Standard for the Prevention of Dust Explosions from the Manufacturing, Processing, and Handling of Combustible Particulate Solids* [1], serves as a "catch all" by covering facilities and materials not explicitly addressed by one of the first four standards. As such, it is currently the most broadly applicable, and most general, of the NFPA combustible dust standards. Materials commonly addressed by NFPA 654 would include, but not be limited to, plastics and rubber products, chemicals, pharmaceuticals, and paper and textile products.

All five standards have been recently revised. The next sections of this paper addresses some of the more significant revisions to these documents, and highlights some trends in the evolution of NFPA's combustible dust standards – including the gradual incorporation of selected process safety management principles.

3. The Evolution of the NFPA Combustible Dust Standards

The NFPA combustible dust hazard standards emerged in the early 1920's and today retain some of the fundamental principles for hazard identification and control introduced at that time. The first evidence of safety practices for combustible dusts came following the tremendous explosion in 1878 at the Washburn "A" flour mill in Minneapolis, which destroyed the mill and two other nearby mills killing fourteen Washburn employees and four other persons in nearby buildings. This incident triggered implementation of safety practices within mills nationwide, such as the addition of ventilation systems designed to limit the hazards from combustible dusts associated with flour and other agricultural processes. [2]

Then in 1924, NFPA provided owner/operators a standardized framework, or basis for safety, for protecting industrial operations involving solids production, including all aspects of storage, handling, and use. That framework stressed three primary steps to control the hazards. First, control the formation and release of the particulate materials (manage the formation of the fuel). Second, identify and control potential ignition sources, keeping in mind that ignition source control should not be the only layer of protection implemented. Lastly, recognize that if a combustible dust fueled fire or explosion occurs, it is essential to limit the spread of any incidents through protection measures such as building construction, equipment isolation, and housekeeping.

The earliest NFPA documents related to combustible dusts were developed for pulverized fuel systems and flour, feed, sugar, and cocoa in 1924. NBFU 60, *Regulations for the Installation of Pulverized Fuel Systems*, published by the National Board of Fire Underwriters (NBFU) and recommended by NFPA described the types of pulverized fuel systems in use at the time and noted that "...all involve the hazard incident to the creation and possible liberation of combustible dusts unless properly designed, constructed, and operated." [3] Even this earliest of NFPA documents used construction techniques involving detached, segregated, or separated building design that are common to today's hazardous materials building and fire codes. Pulverizing equipment, dust collection systems, dryers and other equipment that could contribute to the combustible dust hazard were recommended to be isolated if not located outside the facility. Though no specific thickness for tolerable dust layer accumulations within the facility was provided, the requirement for coal drying and pulverizing plants was that they "...be kept free from even small accumulations of coal dust." [4] To assist in ensuring that dust accumulations were easily detected, surfaces within the facility "...shall be smooth and of a color, preferably white, that contrasts with the dust." [5] Housekeeping requirements included intervals frequent enough to prevent accumulations, specifying the use of methods that did not scatter or distribute the dust. Nearly 90 years ago it was recognized that vacuuming or some similar pneumatic collection system was the preferred method for cleaning accumulated dust, rather than the use of compressed air blowdowns. These requirements remain the same in today's NFPA combustible dust standards.

That same year NBFU published NBFU 61 C, *Regulations for the Prevention of Dust Explosions in Flour and Feed Mills*, and NBFU 62 A and B, *Regulations for the Pulverizing Systems for Sugar and Cocoa*. [6 & 7] These documents served as the forerunners for today's NFPA 61, *Standard for the Prevention of Fires and Dust Explosions in Agricultural and Food Processing Facilities* [8], and included provisions quite similar to those highlighted above in NBFU 60. The Committee on Dust Explosion Hazards recognized in 1925 the importance of housekeeping, maintenance of mechanical equipment (such as screw conveyors and blowers), building construction, and dust collection when establishing control of the potential fire and explosion hazards associated with combustible dusts.

In 1948, NFPA approved *Fundamental Principles for the Prevention of Dust Explosions in Industrial Plants*, NFPA 63 (which was derived from NBFU 63 first published in 1938 under the same title) [9]. Today, there are currently 5 combustible dust standards; they address specific dust-type hazards associated with agricultural products and food processing, combustible metals, sulfur, wood working and wood processing, and non-dust-type specific combustible particulate solids.

NFPA 61, *Standard for the Prevention of Fires and Dust Explosions in Agricultural and Food Processing Facilities*, applies to facilities that handle, process, store, and ship dry agricultural bulk materials, their by-products, or agricultural dusts [8]. Also included are those facilities designed for manufacturing and handling starch as well as oilseed processing plants. In 1995, four existing standards dealing with starch processing, grain elevators, feed mills, and food processing were combined into the current single standard covering all applications for the agricultural and food processing facilities.

NFPA 484, *Standard for Combustible Metals*, applies to the production, processing, finishing, handling, recycling, storage, and use of all metals and alloys that are in a form that is capable of combustion or explosion [10]. This standard also completed a consolidation effort in 2002, whereby six metal-specific standards were combined into this single document. The current edition of the standard incorporates an applicability threshold based on the quantity or amount of metal present within a specific occupancy type. If the quantity of metal is exceeded (much like the Maximum Allowable Quantity (MAQ) concept in fire and building codes) then NFPA 484 applies.

NFPA 655, *Standard for Prevention of Sulfur Fires and Explosions*, applies to the crushing, grinding, or pulverizing of sulfur and to the handling of sulfur in any form [11].

NFPA 664, *Standard for the Prevention of Fires and Explosions in Wood Processing and Woodworking Facilities*, establishes the minimum requirements for fire and explosion prevention and protection of industrial, commercial, or institutional facilities that process wood or manufacture wood products [12]. Similar to NFPA 484, this standard defines an applicability threshold. The standard applies to woodworking operations that occupy areas of more than 465

m^2 (5000 ft^2) or where dust producing equipment requires an aggregate dust collection flow rate of more than 2549 m^3/hr (1500 ft^3/min).

The one remaining standard, NFPA 654, *Standard for the Prevention of Fire and Dust Explosions from the Manufacturing, Processing, and Handling of Combustible Particulate Solids*, applies to all phases of the manufacturing, processing, and handling of combustible particulate solids or hybrid mixtures where the materials present a fire or explosion hazard. [1] NFPA 654 is the single standard currently published by NFPA that is not dust-type specific. As such, it utilizes a more process safety-oriented philosophy for many of its core requirements.

During the past two years, each of the standards just described completed a revision – process safety elements formed the basis for many of the recent changes to the standards. Nearly all the standards include requirements that are predicated on the existence of a combustible dust fire or explosion hazard, but previously no clear guidance was given on how to determine whether such hazards existed. With the most recent revisions to NFPA 484, 654, and 664 criteria have been provided for determining whether a dust flash fire or dust explosion hazard exists. Where those conditions exist, additional safeguards are then required to protect the facility, equipment, and personnel.

One difference found in the NFPA standards over the past 20 years is the introduction of safety management system concepts. Safety and loss prevention experts recognize the role of people in the process and address that recognition by considering the influence of the worker on everyday safety. For that reason, NFPA 654 and NFPA 664 began in the 1990's to include process safety concepts within their requirements to highlight that, in addition to the explicit engineered requirements, it was essential also to have a system for managing the various elements of process safety. The elements that currently have been included in the standards are: process hazard analysis (or hazard analysis); management of change; operating procedures; emergency response; incident investigation; inspection, testing, and maintenance; and training. More will be provided on this topic later in the paper.

4. The Regulatory Environment

Following its investigation of three particularly severe combustible dust incidents in 2003, the U.S. Chemical Safety and Hazard Investigation Board (CSB) initiated a broad investigation of dust explosions in general industry. In its report, issued in November 2006, the CSB concluded that combustible dust explosions were a serious hazard in American industry, and that existing efforts inadequately addressed this hazard [13]. The CSB issued a recommendation to the Occupational Safety and Health Administration (OSHA) to develop a comprehensive combustible dust federal standard based, in part, on NFPA dust explosion standards, including NFPA 654 and NFPA 484. At that time, and to this day, the only existing industry specifically covered by a federal standard applicable to combustible dust processes is the grain handling industry. OSHA standard 29 CFR 1910.272 covers grain handling facilities and was issued in the late 1980s [14].

In addition to the recommendation to develop a standard, the CSB also recommended that OSHA establish a Special Emphasis Program (SEP) for combustible dusts. As part of the SEP, OSHA implemented a National Emphasis Program (NEP) focused on increasing its inspections for those industries identified as having the potential for combustible dust hazards. The NEP for combustible dusts was initiated in October 2007 and further expanded in March of 2008 following a catastrophic sugar refinery explosion earlier that year.

Then, in April 2009, the Secretary of Labor announced that OSHA would be initiating the rulemaking process for a comprehensive federal standard for combustible dust hazards in general industry. OSHA published the Advanced Notice of Proposed Rulemaking (ANPRM) in October 2009 and called for public comment to be submitted by January 19, 2010 [15]. In addition to soliciting public comments in response to the ANPRM, OSHA conducted three stakeholder meetings in 2009 and 2010 to provide additional opportunities to gather information on the hazards from combustible dusts, the steps various industry stakeholders were already taking to address those hazards, the status of consensus standards, and what was needed from an OSHA standard to protect workers from the fire and explosion hazards posed by combustible dusts. Information was also being gathered through the inspections being conducted as a result of the NEP.

One step in OSHA's regulatory process is the requirement to conduct a review of the potential impact that a new regulation might have on small employers. This review, required by the Small Business Regulatory Enforcement Fairness Act (SBREFA), has been scheduled and rescheduled several times since the end of 2010. While awaiting the SBREFA process to begin, OSHA convened an "expert panel" in May 2011 for the purposes of providing the agency staff with further information regarding combustible dust hazards and possible solutions that should be considered by OSHA in developing the proposed standard. The economic impact on small employers was one of the underlying questions for this special meeting. The SBREFA process remains the next step for OSHA prior to issuance of a Notice of Proposed Rulemaking (NPRM) which would finally provide stakeholders and other interested parties with the first look at specific regulatory text. The SBREFA review is now scheduled for October 2013 according to OSHA's current published regulatory agenda [16]. With that proposed schedule, it is not likely that the rulemaking process would be completed before 2014.

5. Tying It All Together – the Search for Common-Sense Consistency

During the Public Comment period for OSHA's ANPRM on combustible dusts, some commenters suggested that NFPA's multiple combustible dust standards for the various industries and dust types complicated use of those standards. It was suggested that those standards could be confusing for users and that inconsistent or possibly conflicting requirements existed within the documents. While progress with a comprehensive federal OSHA standard has been slow, and the outcome uncertain, NFPA has taken heed of the public comments and has initiated a restructuring of the current combustible dust committee projects. NFPA acted to establish a new committee structure by creating a Correlating Committee (CC) and a new Technical Committee (TC).

The CC will enable some of the core topics to be coordinated or correlated between the five current standards so that issues such as the definition of what constitutes a combustible dust and methods for determining that a hazardous condition exists can be reviewed, discussed, and treated in a consistent manner. Underlying NFPA's approach is the belief that fundamental or core requirements exist for all the industrial processes where combustible solids, powders, or dusts are handled or created and, thus, NFPA established the new TC on fundamentals. This new committee is charged with developing a standard on the fundamentals of combustible dust that will address those common considerations that apply across the five current standards – agricultural, combustible metals, general, sulfur, and wood processing/woodworking – and then direct the user to the dust-type specific standard for those requirements unique to that industry sector or dust type. The new standard, NFPA 652, *Standard on Combustible Dusts*, is currently in the first stage of development and is slated for completion in Fall 2014 [17]. Its scope is to provide the basic principles of, and requirements for, identifying and managing the fire and explosion hazards of combustible dusts and particulate solids.

6. Inclusion of Process Safety Management Principles

NFPA publications have long been recognized as a source of guidance for the design, construction, and maintenance of buildings; mechanical equipment; safety systems; and personal protective clothing and equipment related to mitigating the risks of fires and explosions. In addition to this "hardware" guidance, NFPA publications have also provided guidance on administrative requirements. Training requirements, for example, have been commonly addressed, where relevant, in NFPA publications. To further illustrate the point, it is noted that the 1982 edition of NFPA 654 incorporated requirements for operating procedures and emergency plans, and for training of personnel on these. In the chemical industry, we have come to aggregate such administrative requirements under the concept of a process safety management system.

Several years after the 1992 effective date of the OSHA process safety management (PSM) regulation, 29 CR 1910.119 [18], the NFPA technical committee responsible for NFPA 654 began the development of what would become the 1997 edition of the standard. On the committee at that time were a number of names and faces that would be recognized by (at least the older) attendees of the Loss Prevention Symposium. Collectively, we asked ourselves "Given that PSM is intended to prevent toxic releases and flammable gas explosions, and their unintended consequences, shouldn't there be aspects of PSM that could be equally effective in preventing combustible dust explosions?"

In answer to our own question, we strengthened the section on procedures and training in NFPA 654, including explicitly addressing maintenance procedures. We also introduced a requirement for management of change (MOC) of process materials, technology, equipment, procedures, and facilities. These initial ventures into PSM content were made with a concern that we did not want to be perceived as imposing a "chemical industry" solution on a variety of industries that were beyond the coverage boundaries of 29 CFR 1910.119. What we sought to add to NFPA 654 was a "PSM-lite" treatment of some of the more important aspects of process safety management, providing even less specificity than the "performance-based" OSHA PSM regulation.

The 2006 edition of NFPA 654 specified, along with other requirements, that the design of the fire and explosion safety provisions for a facility must be based on a process hazard analysis

(PHA) of the facility, the process, and the associated fire or explosion hazards.[1] The most recent revision to NFPA 654 (2013 edition) added requirements for incident investigation and the qualification and training of contractors [1]. In addition, the content on emergency procedures and inspection, testing, and preventive maintenance was strengthened.

Fortunately, there is a degree of overlap in the membership of the committees responsible for the five current combustible dust standards. This provides for cross-fertilization of ideas between the committees. As Table 2 illustrates, these process safety principles are being increasingly incorporated in the NFPA combustible dust standards – a trend that is anticipated to continue.

Table 2: Process Safety Elements Incorporated in NFPA Combustible Dust Standards

NFPA Standard	Process Safety Element						
	Hazard Analysis	Management of Change	Incident Investigation	Contractors	Emergency Procedures	Inspection, Testing, and Preventive Maintenance	Procedures /Training
61	N	N	N	N	Y	Y	Y
484	Y	Y	N	N	Y	Y	Y
654	Y	Y	Y	Y	Y	Y	Y
655	Y	Y	N	Y	Y	Y	Y
664	Y	Y	Y	Y	Y	Y	Y

7. Additional Combustible Dust Safety Guidance from NFPA

Several additional NFPA sources of information and guidance on addressing combustible dust hazards warrant mention. NFPA and the Fire Protection Research Foundation sponsored symposia addressing combustible dust hazards in 2009, 2010, and 2011. Proceedings from these symposia can be accessed on the NFPA web site. [19 - 21]

In 2012, NFPA published the *NFPA® Guide to Combustible Dusts*, which was co-authored by the authors of this paper and Sam Rodgers (Fig. 1) [22]. This 300-page reference provides numerous photographs and case studies of combustible dust incidents. Chapters address the following topics:

Figure 1: NFPA Guide to Combustible Dusts

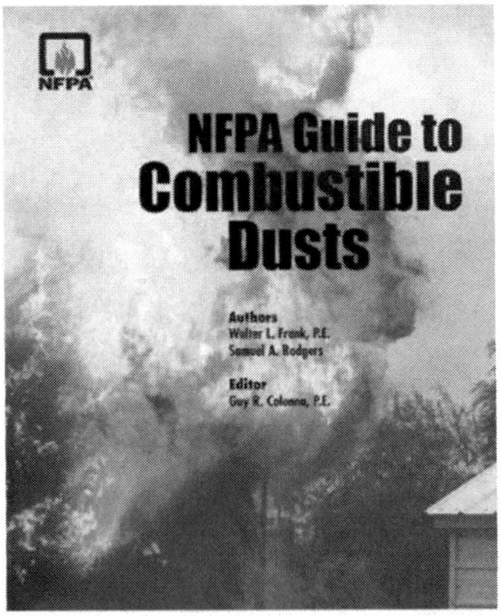

- Explosion fundamentals – extensive background on combustion principles, including discussion of the more commonly applied combustible dust hazard

[1] The 2000 edition of NFPA 654 was an "out-of-cycle" revision that focused on combining NFPA 650 with NFPA 654. No significant consideration of PSM content was added to this revision.

characterization tests
- Process hazard analysis (PHA) – use of PHA to identify, evaluate, and propose controls for combustible dust hazards
- Fire hazard control – focusing on some of the unique issues involved in combatting fires involving combustible dusts
- Explosion hazard control – discussion of the application of the various explosion prevention techniques recognized by NFPA
- Industrial processes – illustrates the application of basic combustible dust safety concepts to some commonly encountered processes and types of equipment
- Safety management systems – discussion of the application of selected safety management system elements to the control of combustible dust hazards.

In addition, the NFPA web site provides a number of investigation reports addressing significant combustible dust incidents.

8. Risk-Based Decision Making

No human endeavor, including the handling and processing of combustible dusts, can be made risk-free [23]. We can, however, seek to mitigate risks to a tolerable level. The NFPA standards described above are certainly sources of risk-mitigation guidance. The requirements in these standards are based upon decades-old proven industry practices, more recent technological advances, and the collective experience and judgment of the members of the technical committees responsible for the standards. Implicit in the standards-making process is the concept that conformance with the standard requirements should help the user attain a tolerable level of risk in operations.

NFPA standards have traditionally been compilations of very explicit, specific requirements – "in such-and-such a situation the user *shall* do this-and-that." [2] NFPA has recognized that the literal requirements of a standard may not always be attainable for all users. For this reason, the front-end "boiler-plate" of an NFPA standard contains an equivalency clause that states, in part:

> "Nothing in this standard is intended to prevent the use of systems, methods, or devices of equivalent or superior quality, strength, fire resistance, effectiveness, durability, and safety over those prescribed by this standard." [for example, 1]

In other words, the user is given the latitude to propose means, alternative to those explicitly established by the standard, to mitigate the risk of operations. Presumably, these alternate means should control the risk to a level no greater than that which would remain had the literal NFPA requirements been implemented. These alternate means are subject to approval by the authority having jurisdiction (AHJ). [3]

---

[2] NFPA standards and codes establish mandatory requirements and are written in the normative form, using "shall." Other NFPA documents, such as recommended practices, contain non-mandatory guidance.

[3] NFPA defines the authority having jurisdiction as "An organization, office, or individual responsible for enforcing the requirements of a code or standard, or for approving equipment, materials, an installation, or a procedure." [for example, 1]

About a decade ago, NFPA began gradually incorporating into its standards a new option for compliance. This new option, the performance-based option, allows the user to bypass certain of the traditional, prescriptive requirements and propose alternative approaches to achieving the objectives established for the standard; e.g., protecting life safety, ensuring structural integrity of the facility sufficient to allow for protection or evacuation of occupants, and ensuring continuity of operations to a degree satisfactory to the owner/operator. In application, the performance-based option applies to certain of the engineered controls, and not the administrative controls. For example, the housekeeping and training requirements remain governed by the prescriptive portions of the standard.

In effect, the performance-based option provides more detailed guidance for implementation of the equivalency clause described above. The performance-based option requires that specific fire and explosion scenarios be addressed in the design of the facility and systems, and that quantitative substantiation be provided to show how the performance objectives will be met. This requires documentation of all calculations, references, assumptions, and sources from which material characteristics and other data have been obtained. The AHJ is permitted to retain an independent third-party to review the proposed design.

More recently, some standards such as NFPA 654, have incorporated language such as:

> "A documented risk evaluation acceptable to the authority having jurisdiction shall be permitted to be conducted to determine the level of protection to be provided per this chapter."

The above is quoted from Chapter 7 of NFPA 654 [1], which addresses requirements for fire and explosion protection for specific types of equipment. For example, Chapter 7 might require explosion protection, such as deflagration venting, on a dust collector. The cited paragraph introduces the option for demonstrating, by risk assessment, that explosion protection is not necessary or warranted for a particular dust collector; e.g., because the inventory of combustible dust in the collector is so low that the risk of an explosion is low enough to be tolerable.

As will be discussed below, application of the risk-based option places a burden on the both the user and the AHJ to understand basic concepts related to risk assessment and risk-based decision making.

9. Areas to Address Going Forward

As NFPA and industry continue to advance practices to control combustible dust hazards, several topics remain in the forefront for research and discussion. Foremost on the list is the development of approaches for assessing whether a dust flash fire or dust explosion hazard condition exists. This is important since many requirements in the NFPA standards only apply to dust fire or explosion hazard areas. The 2013 edition of NFPA 654 includes several options for assessing whether a dust flash fire or dust explosion potential exists in a particular area [1]. These all are premised on scenarios involving the suspension of ambient dust layers to form combustible dust clouds. The Fire Protection Research Foundation sponsored a project, completed in 2011, to study entrainment of combustible dust from layer accumulations [24].

While much has been accomplished, additional work will likely improve our ability to define areas that require protection from dust flash fires and explosions.

The annex of the 2013 edition of NFPA 654 highlights an issue related to the results from the testing conducted for several combustible dust properties, such as maximum deflagration pressure and deflagration index. Chapter 2 of the NFPA® *Guide to Combustible Dust* describes the potential for obtaining false results – either false positives or false negatives – from screening tests used to determine whether dusts are in fact combustible and, thus, warrant coverage under the NFPA standards. One such test is the K_{st} test. It is often difficult to infer whether a low K_{st} value is indicative of a marginally combustible dust that still warrants a degree of explosion protection or, alternatively, whether it is a false positive produced by the nature of the test protocol. Additional guidance is needed on this topic.

OSHA has cited employers for failing to provide protective clothing for employees where a dust flash fire exposure potential exists. As a result of recent CSB investigations of the iron dust flash fires from incidents that occurred in early 2011 [25], greater focus has been placed on the use of fire-resistant clothing in combustible metal dust applications. More research on the effectiveness of fire-resistant clothing on various types of dusts may be warranted.

The current revision of NFPA 61, *Standard for the Prevention of Fires and Dust Explosions in Agricultural and Food Processing Facilities* [8], is seeking to address the design and installation of deflagration venting for silos. NFPA 68, *Standard on Explosion Protection by Deflagration Venting* [26], provides the requirements for determining the mandatory vent area for silos and includes calculations that account for the fact that some enclosures containing solids or powders may only be partially filled by combustible dust clouds. However, silos provide particular challenges to deflagration venting and further research on the placement of the vents on silos and limitations of the venting of certain silo designs may be warranted.

Prior definitions of combustible dust included a maximum particle size threshold; e.g., 420 microns. However, the most current definition used in NFPA 654 has removed the size specification and placed the discussion of the influence of particle size on the behavior of combustible dusts in the annex of the standard. During recent discussions within the various NFPA Technical Committees, alternative particle size thresholds have been suggested. For example, FM Global and operators in Europe are apparently using 500 microns rather than 420, while others suggest the upper limit could be somewhat greater than 500 microns. The concept of whether there is some particle size, above which combustible dust flash fire and explosion concerns could largely be discounted – and, if so, what that value might be – remain topics for debate.

The concept of risk-based decision making for determining prudent combustible dust hazard mitigation is now embodied in a number of NFPA standards. Additional guidance to owner/operators and AHJs will be required to facilitate the use of this option.

10. The Role for AIChE

Earlier, this paper described how "PSM-lite" management system concepts are gradually being incorporated into the NFPA combustible dust standards. The publications of the Center for Chemical Process Safety (CCPS) provide useful sources of guidance for those seeking to implement such "PSM-lite" provisions.

As noted above, additional guidance on risk assessment and risk-based decision making will be required to assist those seeking to use the flexibility recently incorporated in the NFPA standards. As a leader in the development of guidance on risk assessment for the process industries, CCPS is an obvious and likely source for additional guidance in this area.

The CCPS publication *Guidelines for Safe Handling of Powders and Bulk Solids* [27] should be noted as an existing source of valuable information on the mitigation of combustible dust hazards.

Finally, recognizing the need for additional technical guidance on the mitigation of combustible dust hazards, CCPS is initiating a project to develop and publish a concept book providing guidance on how to control and mitigate dust hazards in the workplace. The specific focus of this book will be defined over the next few months.

11. References

[1] NFPA 654, *Standard for the Prevention of Dust Explosions from the Manufacturing, Processing, and Handling of Combustible Particulate Solids,* 2013 ed., National Fire Protection Association, Quincy, MA.

[2] Minnesota Historical Society, http://www.mnhs.org/library/tips/history_topics/73washburn.html

[3] NBFU 60, *Regulations for the Installation of Pulverized Fuel Systems*, National Board of Fire Underwriters, 1924, page 1.

[4] *Ibid.*, page 4.

[5] *Ibid.*, page 4.

[6] NBFU 61 C, *Regulations for the Prevention of Dust Explosions in Flour and Feed Mills*, National Board of Fire Underwriters, 1924.

[7] NBFU 62 A and B, *Regulations for the Pulverizing Systems for Sugar and Cocoa*, National Board of Fire Underwriters, 1924.

[8] NFPA 61, *Standard for the Prevention of Fires and Dust Explosions in Agricultural and Food Processing Facilities,* 2013 ed., National Fire Protection Association, Quincy, MA.

[9] NFPA 63, *Fundamental Principles for the Prevention of Dust Explosions in Industrial Plants*, 1948 ed., National Fire Protection Association, Quincy, MA.

[10] NFPA 484, *Standard for Combustible Metals*, 2012 ed., National Fire Protection Association, Quincy, MA.

[11] NFPA 655, *Standard for Prevention of Sulfur Fires and Explosions*, 2012 ed., National Fire Protection Association, Quincy, MA.

[12] NFPA 664, *Standard for the Prevention of Fires and Explosions in Wood Processing and Woodworking Facilities*, 2012 ed., National Fire Protection Association, Quincy, MA.

[13] *Combustible Dust Hazard Study*, U.S. Chemical Safety and Hazard Investigation Board, Washington, DC, 2006.

[14] Grain handling facilities, 29 CFR 1910.272, OSHA: http://www.osha.gov/pls/oshaweb/owadisp.show_document?p_table=STANDARDS&p_id=9874

[15] Combustible Dust; Advance notice of proposed rulemaking, Docket No. OSHA-2009-0023, OSHA, https://www.osha.gov/pls/oshaweb/owadisp.show_document?p_table=FEDERAL_REGISTER&p_id=21152

[16] 2013 Regulatory Agenda, U.S. Department of Labor, accessible via: http://resources.regulations.gov/public/component/main?__dmfClientId=1360389595618&__dmfTzoff=300

[17] NFPA 652, *Standard on Combustible Dust,* National Fire Protection Association, Quincy, MA, currently under development, see: http://www.nfpa.org/aboutthecodes/AboutTheCodes.asp?DocNum=652

[18] Process safety management of highly hazardous chemicals, 29 CR 1910.119, OSHA: http://www.osha.gov/pls/oshaweb/owadisp.show_document?p_table=STANDARDS&p_id=9760

[19] Symposium on Dust Explosion Hazard Recognition and Control: New Strategies, Fire Protection Research Foundation, 2009: http://www.nfpa.org/itemDetail.asp?categoryID=993&itemID=44141&URL=Research/Fire%20Protection%20Research%20Foundation/Proceedings

[20] Symposium on Dust Explosion Hazard Recognition and Control: New Strategies, Fire Protection Research Foundation, 2010: http://www.nfpa.org/itemDetail.asp?categoryID=993&itemID=49551&URL=Research/Fire%20Protection%20Research%20Foundation/

[21] Symposium on Dust Explosion Hazard Recognition and Control: New Strategies, Fire Protection Research Foundation, 2011: http://www.nfpa.org/itemDetail.asp?categoryID=993&itemID=54153&URL=Research/Fire%20Protection%20Research%20Foundation/

[22] *NFPA® Guide to Combustible Dusts,* National Fire Protection Association, Quincy, MA, 2012.

[23] Frank, W. L., "Challenges in Developing and Implementing Safety Risk Tolerance Criteria," Process Safety Progress, Volume 30, Issue 3, September 2011.

[24] *Towards Estimating Entrainment Fraction For Dust Layers*, Fire Protection Research Foundation, 2011: http://www.nfpa.org/itemDetail.asp?categoryID=2481&itemID=55916&URL=Research/Fire%20Protection%20Research%20Foundation/Reports%20and%20proceedings/Hazardous%20materials/Other%20hazards/

[25] *Hoeganaes Corporation Fatal Flash Fires*, U.S. Chemical Safety and Hazard Investigation Board, Washington, DC, 2011.

[26] NFPA 68, *Standard on Explosion Protection by Deflagration Venting*, National Fire Protection Association, Quincy, MA, 2007.

[27] *Guidelines for Safe Handling of Powders and Bulk Solids,* American Institute of Chemical Engineers, Center for Chemical Process Safety, New York, NY, 2004.

9TH GLOBAL CONGRESS ON PROCESS SAFETY

Dust Flame Propagation in Industrial Scale Piping
Part 1: Empirical Study in a Conveying Vessel-Pipeline System

Thomas M. Farrell, M.Sc.
Fike Corporation, Industrial Protection Group
704 S. 10th St., Blue Springs, MO 64015
Tom.Farrell@fike.com

Jim Vingerhoets, M.Sc.
Fike Europe
Toekomstlaan 52, BE-2200 Herentals, Belgium
Jim.Vingerhoets@fike.com

Jef Snoeys, M.Sc.
Fike Europe
Toekomstlaan 52, BE-2200 Herentals, Belgium
Jef.Snoeys@fike.com

John E. Going, PhD.
Fike Corporation, Industrial Protection Group
704 S. 10th St., Blue Springs, MO 64015
John.Going@fike.com

Prepared for Presentation at
American Institute of Chemical Engineers
2013 Spring Meeting
9th Global Congress on Process Safety
San Antonio, Texas
April 28 – May 1, 2013

UNPUBLISHED

Dust Flame Propagation in Industrial Scale Piping

Part 1: Empirical Study in a Conveying Vessel-Pipeline System

Thomas M. Farrell, M.Sc.
Fike Corporation, Industrial Protection Group
704 S. 10th St., Blue Springs, MO 64015
Tom.Farrell@fike.com

Jim Vingerhoets, M.Sc.
Jef Snoeys, M.Sc.
Fike Europe

John E. Going, PhD.
Fike Corporation, Industrial Protection Group

Keywords: Dust Explosion, Flame Propagation, Vessel, Pipelines, Active Conveyance

Abstract

The use of pipes to connect vessels and transport hazardous materials, such as flammable gases or dusts, is a common practice in the process industry. The potential of fires and explosions has been recognized for decades and is addressed in standards NFPA 68 [1] and NFPA 69 [2]. The purposes of this paper are to present experimental evidence and to further develop a detailed understanding of combustion propagation in industrial scale piping. The experimental evidence on dust flame propagation is scarce, and specific attention was given to expand the knowledge in this field.

Earlier work [7] studying flame propagation in industrial scale piping focused primarily on the propagation of gaseous flames (propane, ethylene, hydrogen) and conditions for deflagration-to-detonation transition.

The current research program, as a part of an ongoing effort to study the complex phenomenon of flame propagation, was conducted with a focus on dust fuels. A 5 m^3 vented initiating vessel and interconnected pipeline which incorporates active conveyance capability was used for this program.

A systematic investigation of flame propagation speed and arrival time dependencies upon vessel vent area, fuel concentration, and conveyance velocity is presented. A discussion of propagation in the process flow direction is also included. This research also provides the empirical basis for a parametric study using computational fluid dynamics (CFD), which is presented in a separate paper [3].

1. Introduction

Several research programs have been undertaken in the past to study the behavior of flame propagation in industrial scale pipelines. Chatrathi et al. [7] investigated gas deflagrations and transition to detonation conditions in deflagrations originating on closed-ended pipelines. Pineau and Ronchail [8] studied dust deflagrations and transition to detonation conditions in standalone pipelines and also dust explosion propagation initiated in 1 m^3 and 0.1 m^3 closed vessels into small diameter pipelines. Vogl [4] focused on bidirectional dust flame propagation in high L/D conveying pipeline situations which do not involve an initiating vessel. Further work by Vogl [11] included a vented initiating vessel and clearly demonstrated that flame is capable of traveling both with and against the conveyance direction. Holbrow et al. [6] investigated propagation between interconnected vented vessels, focusing on the increased reduced pressures developed in a secondary vessel, but flame positions were not measured. Valiulis et al. [5] conducted work with an initiating vessel and propagation in the direction of the flow and focused on flame propagation distance as a function of vessel reduced pressure. Roser et al. [9] extracted empirical equations predicting the flame front position over time from a large dataset of dust explosion propagation tests initiated from different sized vented and contained vessels, connected to different sized pipelines, with and without conveyance. The value of the maximum reduced vessel explosion pressure is a highly influential factor in the equations.

These studies indicate that the speed and associated pressures with which an explosion propagates through pipelines is considerably affected by many application parameters and that it is not a trivial task to predict the phenomenon for a given set of application parameter values. Therefore, specific knowledge is of primary importance to determine the conditions under which explosion isolation devices can be used to stop flame propagation in industrial pipelines, as well as to enable trained factory personnel to design such protection systems.

In an effort to obtain additional industry-relevant data from and insights into explosion flame propagation through dust-air mixtures, Fike Corporation has decided to engage an ambitious and extensive research program dedicated to the effects of dust fuel characteristics, dust concentration profile, conveyance direction and speed, explosion venting parameters, ignition location and equipment geometry. This paper presents a subset of the current results of the ongoing empirical program. The accompanying paper [3] presents a part of the current results of a numerical simulation study.

This paper reports on a series of experiments that was performed in a rig which consisted of a 5 m^3 vented vessel connected to two 16-inch diameter pipelines with different lengths. Three different fuels were selected for the larger research program: wood flour, cornstarch, and a phenolic dust. These fuels were selected to represent the majority of market segments (agricultural, food, and chemical). Only the results for tests with wood flour will be presented in this paper. Results with other fuels are included in the accompanying paper [3]. Vessel reduced pressure was varied by using different vent areas and fuel concentrations. Two conveyance velocities, 15 m/s and 30 m/s, were used to investigate conveyance velocity influence upon upstream flame propagation.

A discussion regarding the technical challenges posed by testing in an actively conveyed system is presented, along with the solutions employed for the present work.

2. Test Equipment Setup

The test rig used in this research program consists of an inlet pipeline, a vessel, and an outlet pipeline, as shown below. The system is arranged with the inlet and outlet pipelines oriented 90° to each other, with the outlet pipeline aligned with the major axis of the vessel. All explosion venting was constrained to a single panel on the vessel, on the wall opposite the outlet pipeline interconnection. The test rig was arranged as a positive-pressure "push" system.

Figure 1 - Experimental System Layout

Fuel concentration was varied *systemically* (same concentration in vessel, inlet, and outlet pipelines test-to-test), as well as initiating vessel vent area and conveyance velocity to allow for the systematic investigation of flame arrival times and velocities.

Fuel concentration was also varied *zonally* (with different concentrations in vessel and pipelines) in order to determine how much of flame propagation behavior is driven by the initiating vessel pressure as opposed to a self-sustaining reaction within the pipelines. This simulates realistic aspiration conditions in pipelines with a higher concentration in primary vessel consistent with discharge events such as reverse pulse jet cleaning of filter cartridges or bags.

2.1 $5m^3$ Test Vessel

The vessel is constructed of two identical 2.55 m3 cube halves attached to each other without a partitioning wall in between. The cube is fitted with three dust injectors, which are comprised (each) of a 10-liter dust injection bottle, a pneumatically actuated ball valve, and a dust dispersion nozzle. The dust dispersion nozzles are scaled-up versions of the rebound nozzles commonly used in 20-liter explosibility testing. The injectors are mounted on the top of the vessel. Prior to test execution, each of the injector sample bottles were filled with fuel (divided equally) and pressurized to 20 bar (290 psig) with dry air.

An igniter probe was installed in the upper surface of the vessel to provide an ignition source within the vessel. This probe was affixed with two 5 kJ (10 kJ total energy) Sobbe chemical igniters, pointed downwards with a separation angle of approximately 90 degrees. The igniter probe was situated in the center of the upstream cube half in order to mitigate the effects of the migration of the initiating fireball into the downstream pipeline prior to significant vessel pressure generation.

The vessel was also fixed with eight Lexan (polycarbonate) viewing windows for visualizing the internal dust dispersion patterns, as well as to observe developing deflagrations, as shown below.

Figure 2 - $5m^3$ Vessel Lexan Windows for Deflagration Visualization

The forward panel of the assembly is fitted with four 0.42 m (16.5 in.) square removable panels, each capable of being fitted with either a venting membrane or a blind flange. Each panel contributes approximately 0.17 m^2 (1.9 ft^2) of vent area. The vent membranes used have a static activation pressure of approximately 100 mbarg (1.5 psig). This forward panel is shown below for reference.

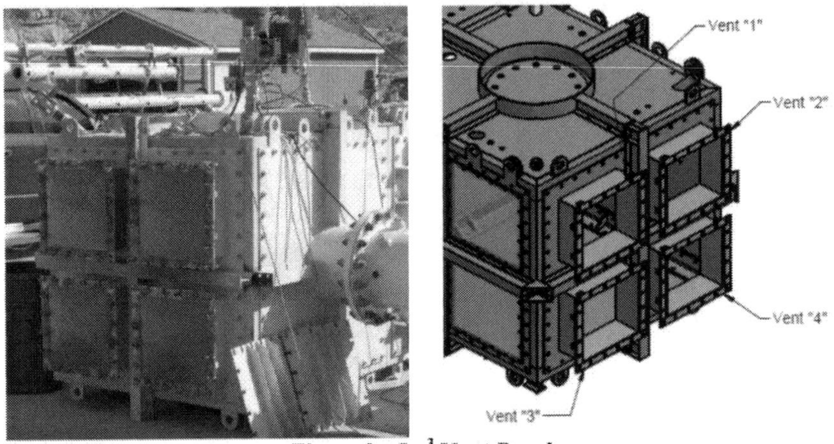

Figure 3 - 5m³ Vent Panels

2.2 Interconnected Pipelines

Two pipelines were connected to the 5 m^3 initiating vessel to facilitate flow establishment, as well as upstream and downstream flame propagation. The outlet (downstream) pipeline was 400 mm (16 in.) in diameter, and was approximately 17 m (57 ft) in length resulting in an inlet L/D ratio of 42.8. The inlet (upstream) pipeline was 400 mm (16 in.) in diameter, and was approximately 10 m (33 ft) in length resulting in an outlet L/D ratio of 24.8. Both pipelines were fitted with many ports for mounting flame detectors and pressure detectors.

2.3 Instrumentation and Control

Vessel pressure measurements were taken in two locations on the vessel through the use of Omegadyne PX409 pressure transducers with a range of 0-2 barg (0-30 psig).

Fike Optical Detectors were used as the primary measurement device for flame arrival in both pipelines. These detectors consist of an infra-red sensing element, amplifier, and associated signal conditioning circuitry. They have a full scale response time of less than 1 millisecond. The optical field-of-view of these detectors was mechanically restricted to a cone angle of approximately 10° in order to achieve a more accurate flame position measurement. Eight detectors were placed on the outlet pipeline, and six were placed on the inlet pipeline.

All signals (pressure and flame detector output voltages) were recorded by a National Instruments NI-PCI6229 data acquisition system controlled by the Putty.StripChart operating system, written by Fike Corporation in LabVIEW software. This system also controlled the digital outputs to sequence the injection and ignition processes. Both the acquisition and control cycles run synchronously on the same 1 kHz clock.

2.4 Injector Development and Conveyance Differences

In order to create a known fuel concentration in a flowing system, traditional injection methods required reassessment. In a traditionally injected static system, a known mass (kg) of fuel is injected into a known volume (m^3). In a flowing system, the volume is in a constant state of flux, with a moving timeframe. The volume, therefore, is most easily expressed as a volume flux rate (m^3/s) past a single point of injection; this is also expressed as the volumetric flow rate through the system. In order to control the fuel concentration in this shifting frame of reference, an injector is required to provide not a known mass of fuel, but a known mass flux rate (kg/s) which is consistent over a period of time. The direct division of the mass flux rate by the air volume flux rate yields the resultant fuel concentration (kg/m^3) flowing through the system past the point of injection.

Typical dust injection systems used in other research programs, such as hoper feeders and rotary valves, have shortcomings which can result in significant test variance. Hopper feeders, for example, lack the fuel feed rates required for testing of this nature, as well as the backpressure protection required to prevent flame flashing into the feeder. Rotary valves, while capable of withstanding and isolating limited explosion overpressure, create "pulses" of dust clouds, but do not independently provide consistent (constant) fuel delivery rates. The experimental difficulty of creating a homogeneous flowing dust cloud has challenged researchers, and has led to relatively few studies being published concerning dust deflagration propagation down a duct or pipe [10].

The Constant Mass Flow Rate (CMFR) injector was designed to deliver dust fuel to a flowing system in a known consistent feed rate, thereby allowing for the creation of a homogeneous concentration atmosphere.

At the core of the system is a pressurized vessel filled with fuel. The fuel flows up through a dip tube inserted at the neck of the system and reaching near the bottom of the container. A pneumatically actuated ball valve initiates the flow of fuel, which is initially dense-phase conveyance. An air bypass allows mixing of motive air with the dense phase conveyance to create a constant mass flow rate. The outlet of the bottle flows through a flexible tube and into the process connection. A process-mounted nozzle (simple elbow with a constriction nozzle to control flow rate) injects the fuel into the flowing system. This injector is pictured below for reference.

Figure 4 - CMFR Injector Bottle

This system is capable of wood flour mass delivery rates ranging from approximately 100 g/s up to 2000 g/s. In a 400 mm diameter pipeline flowing air at between 15 and 30 m/s, this can create fuel concentrations ranging from approximately 60 g/m^3 to 375 g/m^3. Doubling the number of CMFR injectors in a single location will locally double the fuel concentration. Flow is generally consistent from 200 ms after time of initiation until approximately 1500 ms after time of initiation, depending on the bulk density of the fuel.

Two CMFR injectors were used on the current test rig, as can be seen in the photo below. One CMFR (white bottle) was installed at the inlet of the upstream pipeline just downstream of the induction assembly, and the other CMFR (pink bottle) was installed at the inlet to the downstream pipeline within 1 pipe diameter of its interconnection with the initiating vessel.

Figure 5 - Injection System Components

In this configuration, concentration can be independently varied zonally, with different concentrations in the initiating vessel, upstream, and downstream pipelines. Numerous tests were conducted to determine the proper nozzle configurations and injection times to generate known concentration conditions.

2.5 Induction Assembly

An induction assembly was manufactured and installed between the flow motive fan and the active pipeline to provide overpressure protection for the fan. Explosion diverters are available for this purpose; however, in an attempt to keep the test rig "pure" and not introduce any equipment that would render the rig system-specific, it was decided not to use one. This system also allows for testing free from complications associated with filter fires, bag rupture, or dust collection system damage and maintenance, thereby allowing more tests in a short timeframe. Since several potentially significant variables are removed through the use of this assembly, a higher degree of repeatability can be achieved from test to test.

This induction assembly comprises a nozzle which directs a high pressure jet of air from the flow motive fan through a windband (which is open to the atmosphere), thereby creating an inductive flow of atmospheric air into the pipeline. This open interface also provides an escape path for combustion products expanding from the interconnected vessel, thus sparing the fan from damaging pressure waves and high temperature combustion gases. The elbow leading into the induction assembly is also equipped with an explosion vent to relieve reflective pressure waves which may transmit into the high pressure zone, providing further protection for the fan.

Figure 6 - Induction Assembly (Left) and Motive Fan (Right)

The use of such an inductive device also allows for the flame propagation research to be conducted with a very low flow resistance as viewed by the expanding gases flowing from the vessel. While a restrictive device such as an explosion diverter would see back pressure increases proportional to the square of the flow reversal velocity, the backpressure caused by the induction assembly remains relatively constant, influenced primarily by the jet stream within the nozzle. This system is therefore viewed from the flow resistance perspective of the combustion gases expanding in the upstream direction as an open-ended pipeline with a fixed (flow rate independent) backpressure.

2.6 Fuel Properties

The fuel used for the testing presented here is wood flour (AWF Grade 14020), with the following parameters as reported from standard 20-liter explosibility testing.

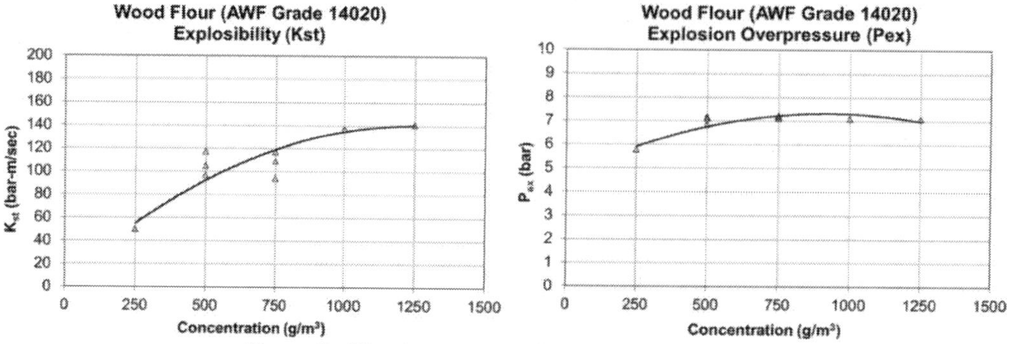

Figure 7 - Wood Flour Reactivity in 20-Liter Vessel

This fuel was also evaluated in the 1 m^3 vessel to determine its susceptibility to K_{st} enhancements through turbulence, as well as the minimum time delay that can be used for testing in an actively conveyed system. The minimum time delay must be chosen so as to ensure that all fuel is injected into the system, turbulence enhancement is minimized, and migration of the injected fuel cloud into the downstream pipeline through conveyance is minimized. This series of tests was conducted by gradually reducing the injection-ignition delay time from test to test, thereby increasing the turbulence at time of ignition.

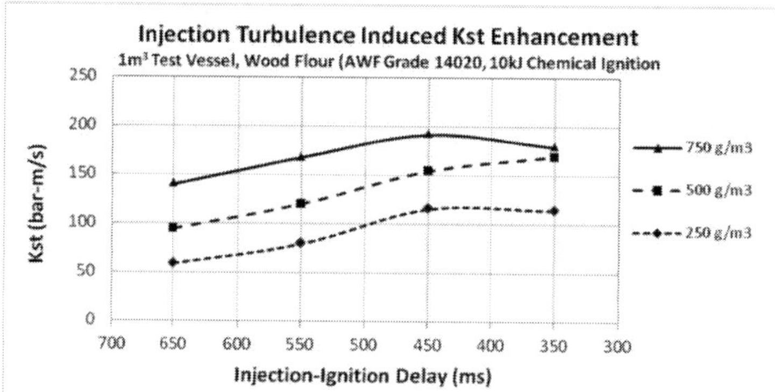

Figure 8 - Wood Flour Reactivity and Turbulence Enhancement in 1m³ Vessel

This fuel, while rated for a K_{st} of 140 bar-m/s from 20-liter testing, can be capable of K_{st} values ranging from 60 to 185 bar-m/s depending on delay time and concentration; the actively conveyed tests in this research program focus on a time delay of 400 milliseconds and concentrations of 250, 500, and 750 g/m^3, yielding K_{st} values of 115, 162, and 185 bar-m/s, respectively.

It is important to note that the maximum pressure rate of rise at 750 g/m^3 with enhanced turbulence is 54% greater than the values obtained for K_{st} in the 20-liter vessel (with standard turbulence). Understanding the behavior of dust deflagrations in realistic scenarios is of paramount concern in effective venting and isolation system design.

2.7 Combustible Atmosphere Setup

With the minimum time delay established at 400 milliseconds for the 5 m^3 test series, the injection timing for the two CMFR injectors was established. This required injection in each pipeline first, allowing for enough time for dust fuel to travel the full length of the upstream and downstream pipelines prior to ignition. The timing requirements for the CMFR injectors to flow fuel into their respective pipelines were dependent upon the conveyance velocity. 400 milliseconds before the fuel flowing in the pipelines reached their respective outlets (with the upstream pipeline discharging into the vessel), the static injectors on the primary vessel were initiated. For the next 400 milliseconds, the static injectors created the primary vessel dust cloud while the CMFR injectors completed establishing concentrations in the upstream and downstream pipelines. After this 400 millisecond time delay, the igniters were fired, and the deflagration began. Data was recorded for a minimum of 2 seconds after ignition in each test. A representative timing diagram, illustrating the relative execution order for different components on the system, is shown below for reference.

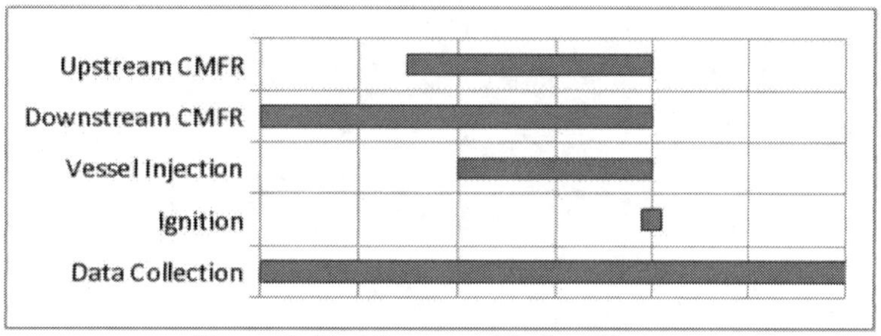

Figure 9 - Test Sequence Timing Diagram

3. Results and Discussion

Many tests were conducted to determine flame speeds and flame arrival times as a function of venting area, vessel fuel concentration, pipeline fuel concentration, and conveyance velocity. All tests were repeated a minimum of two times and the data presented here are averages of tests with similar results. Only results with wood flour are presented.

3.1 General Observations

In almost all experiments, the flame propagated with high speed to the extents of both pipelines, including the ejection of large fireballs, as shown in the figure below.

Figure 10 - Upstream and Downstream Propagation Example

The following figure gives an overview of the recorded average upstream and downstream flame speeds in the first ten meters of propagation, as well as the maximum vessel pressures that were reached. The systematic variance of fuel properties and vent system design show a strong correlation between vessel maximum reduced pressure and average interconnected pipeline flame speed over the first 10 meters of propagation.

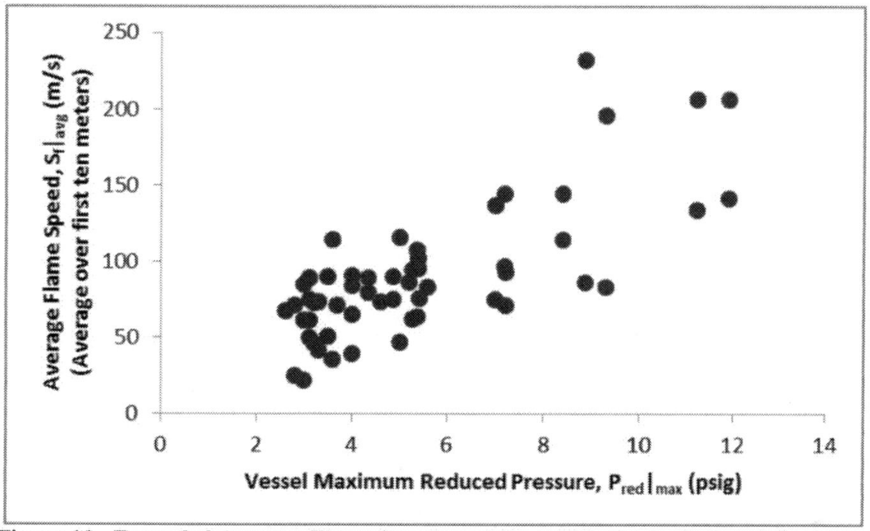

Figure 11 - Recorded Average Flame Speeds and Vessel Maximum Reduced Pressures

In the tests reported in this paper, the highest recorded flame speed was 232 m/s. This velocity is the maximum average flame speed over the first ten meters of propagation with highest conveyance velocity (30 m/s) and 500 g/m^3 of wood flour.

A correlation between vessel pressure and flame speed is noticed. This is due to the fact that high explosion pressures in the vessel lead to higher outflow of gases from the vessel through the pipelines. Indeed, the flame speed is the superposition of the expansion gas velocity and the turbulent burning velocity:

$$S_{f,pipe} = S_g + S_T.E$$

where $S_{f,pipe}$ is the flame speed in the pipeline, S_g is the pipeline gas flow induced by the explosion pressure in the vessel, S_T is the turbulent burning velocity, and E is the expansion factor due to turbulent burning in the pipeline.

3.2 Effect of Vent Area

The first series of tests to determine the effect of vessel vent area upon flame arrival times were conducted at a systemic (vessel and both pipelines) fuel concentration of 250 g/m^3 and a conveyance velocity of 15 m/s. The flame arrival times measured during testing are shown below for reference. Both the available vent area and vessel maximum reduced pressure are indicated. Downstream flame position locations (flame propagation in the direction of conveyance flow) are shown as positive values for X_f (flame position in pipeline), whereas upstream flame position locations (flame propagation *against* the direction of conveyance flow) are shown as negative values for X_f. In this series (1-3), P_{red} was controlled only by varying the available vent area.

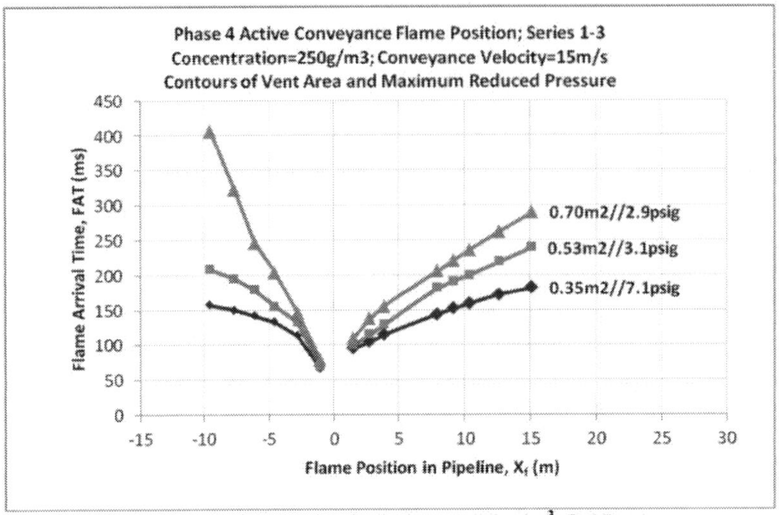

Figure 12 - Flame Arrival Times, 250 g/m^3 @ 15 m/s

As can be seen in this plot, as the venting area decreases and the vessel reduced pressure increases, the flame arrival times decrease. For high reduced pressures (bottom curves), the system is 'balanced' showing similar flame arrival times for up- and downstream propagation. For lower reduced pressures, the flame propagates faster downstream than upstream as the push fan has a more significant influence in the tested configuration.

As vent area is increased, it is clear that the duration of the deflagration overpressure is significantly decreased for tests with similar $P_{red}|_{max}$ values. This appears to have a significant influence upon flame speed, especially in the upstream pipeline. In the highlighted example tests below in Figure 13, the increase of vent area from 0.53 to 0.70 m^2 caused a reduction of peak overpressure duration (time above 1 psi) from 98 ms to 42 ms, respectively.

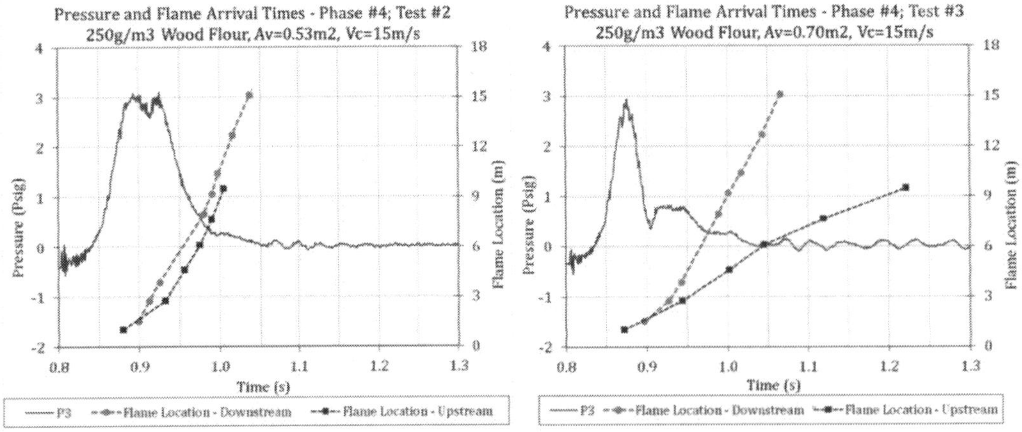

Figure 13 - Pressure Peak Duration and Flame Location

This loss of motivating pressure causes significant reductions in upstream flame propagation speed, as well as inciting deceleration of the flame front, as shown in the highlighted example above. The effect upon downstream propagation is far less pronounced.

This is consistent with the empirical findings of Valiulis et al. [5] relating to the influence of both $P_{red|max}$ and explosion overpressure duration upon flame propagation distance.

3.3 Effect of Concentration

With the system concentration (vessel and both pipelines) increased to 500 g/m^3, the flame arrival times decrease slightly for a given vent area, as shown below in the series 4-6 test results.

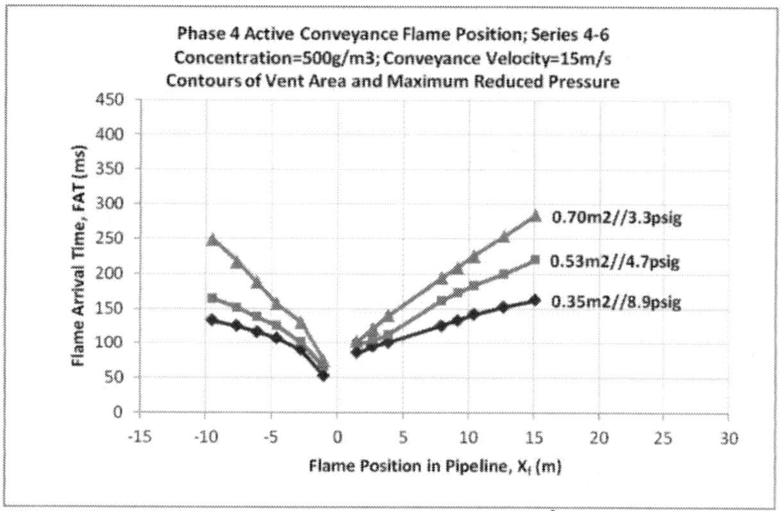

Figure 14 - Flame Arrival Times, 500 g/m^3 @ 15 m/s

The reduction in flame arrival times in series 4-6 compared to series 1-3 is attributed to the increase in vessel reduced pressure, which is in turn caused by the increase in K_{st} from 115 bar-m/s (for 250 g/m^3) to 162 bar-m/s (for 500 g/m^3) with the same available vent areas.

The test rig is still balanced under these test conditions, with similar flame arrival times in the upstream and downstream pipelines at 10 meters.

<u>3.3.1 Systemically Varied Concentration</u>

It was also observed that for tests with similar reduced pressures, the flame arrival times tend to remain relatively constant as system concentration is increased. The data plotted below represent simulated values (not actual test data) where results from tests with similar reduced pressures were interpolated to create simulated tests with varying fuel concentrations, but a fixed vessel reduced pressure of 5 psig. The flame arrival times do not differ greatly as concentration is varied systemically.

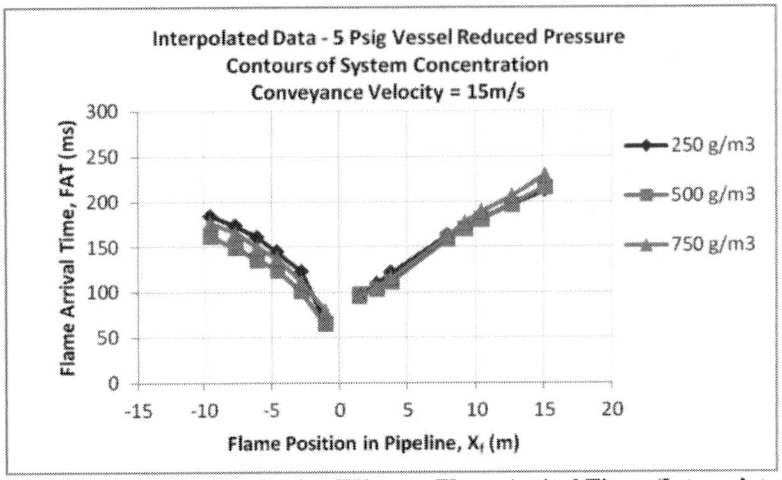

Figure 15 – System Concentration Effect on Flame Arrival Times (Interpolated)

The data suggest that for deflagration events involving ST1 dusts with reduced pressure above 350 mbarg (5 psig), flame speed in interconnected pipelines is more strongly a function of combustion gas expansion than fuel explosibility parameters alone. This finding is also limited to the current study L/D values, as the gas expansion is expected to be the governing factor in the current geometry.

3.3.2 Zonally Varied Concentration

The effect upon flame arrival times of fuel concentration within the inlet and outlet pipelines was tested by holding the vessel concentration constant at 500 g/m^3 and the vessel vent area constant at 0.53 m^2. The pipeline concentration was then varied in the range from 50 g/m^3 to 500 g/m^3 and flame arrival times were measured. In all tests completed in this series, the vessel reduced pressure was 5.3 psig (±10%). The results of this test series are shown below for reference.

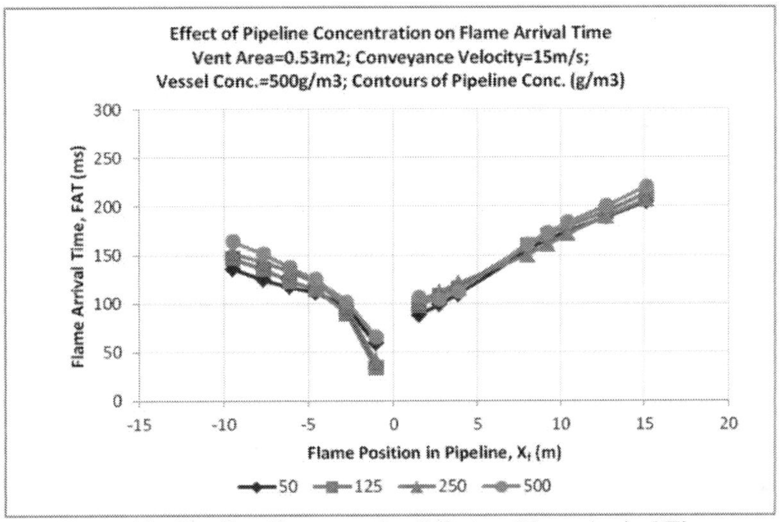

Figure 16 – Pipeline Concentration Effect on Flame Arrival Times

As can be seen in this data, the addition of fuel into the pipelines has very little effect upon flame arrival time, although it seems that highly fuel-laden lines tend to delay flame arrival, with the highest flame velocities occurring in near-MEC conditions of 50 g/m^3. This effect may be due to the added inertia introduced by the fuel particles themselves slightly reducing the expansion flow from the vessel into the pipelines. Vogl [4] found similar behavior. This concentration-coupled inertia effect is investigated through numerical simulations in the accompanying paper [3].

3.4 Effect of Conveyance Velocity

Systemically varied concentration tests were repeated with the conveyance velocity increased to 30 m/s, and similar trends in flame arrival times were observed as a function of vessel reduced pressure. Vessel reduced pressure was again controlled by varying the available vent area and fuel concentration. As vessel reduced pressure increases, the flame arrival times decrease.

With the conveyance velocity increased from 15 m/s to 30 m/s, the flame arrival times also appear skewed, favoring higher flame speeds in the downstream pipeline. This is more pronounced in the upstream pipeline than in the downstream pipeline, as can be seen in Figure 17 below.

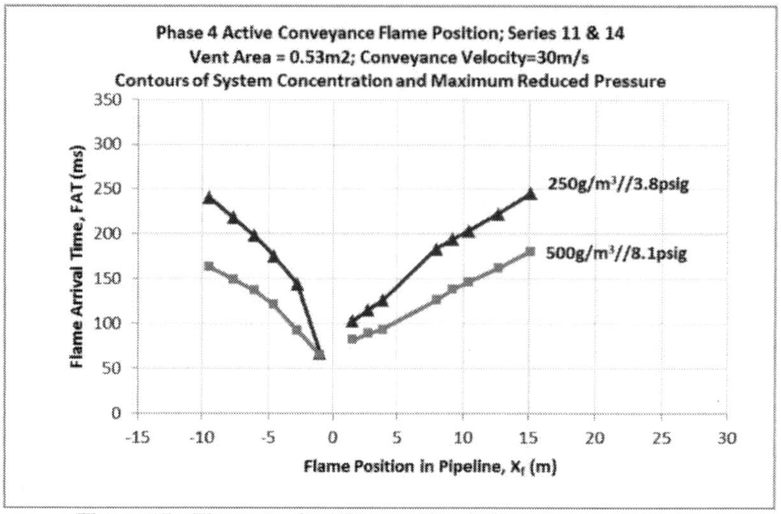

Figure 17 - Flame Arrival Times, 30 m/s Conveyance Velocity

By doubling the conveyance velocity, the dynamic pressure that must be overcome to facilitate upstream propagation quadruples per Bernoulli's equation ($\frac{1}{2}\rho V_c^2$; V_c is conveyance velocity and ρ *is gas density*), yielding a significant difference in flame arrival times at 10 meters. As vessel maximum reduced pressure increases, the skew effect becomes less significant.

4. Summary

This research has clearly demonstrated that flame can propagate upstream, even against relatively strong conveyance flows.

Flame speed shows a high degree of dependency upon vessel reduced pressure peak and duration ($\int p(t)dt$), whether generated by concentrations near optimum or by reduction in venting area.

Fuel concentration within pipelines does not appear have a strong influence upon flame speed during the initial phase of pipeline propagation. It is unclear from the current data whether pipeline fuel concentration plays a significant role in continued propagation violence; future research involving longer pipelines is planned. Deflagration flame fronts tend to propagate marginally faster through dust clouds near MEC than near optimum concentration.

Determination of the conditions under which the flame propagation is no longer driven by initiating vessel pressure, but instead transitions to a self-sustaining and accelerating reaction would be a valuable extension of the currently presented research. The effect of bends is not included in the current study, but is expected to have a significant effect on flame propagation, especially as it applies to determining conditions under which flame propagation decouples from vessel gas expansion and becomes self-sustaining and/or accelerating. Further research to investigate these trends will likely involve significantly higher L/D pipelines attached to the vessel inlet and outlet than were used in the current study.

An increase in conveyance velocity tends to skew flame arrival times, with propagation favoring the downstream direction. Further research will include negative pressure pull systems to determine if the conveyance type has a strong influence on flame arrival times and velocities.

5. Conclusion

Understanding the mechanisms of explosion flame propagation is the first step in designing optimum explosion protection solutions for interconnected process equipment that handle flammable substances. The present research has demonstrated that several application parameters may considerably influence the speed of dust flame propagation and consequently the required performance of protection methods. Not taking these influencing factors into account may result in under-dimensioning the isolation solution with potential failure to protect against the hazard, or over-dimensioning with significant room for improvement either in functionality or in cost savings.

This study further irrefutably demonstrates that the first step towards a safe design of explosion isolation is to manage the explosion pressure in the initiating vessel. The explosion protection engineer should therefore consider the explosion protection problem in its totality. Optimum protection systems will therefore consist of a combination of vessel explosion pressure protection and pipeline explosion propagation protection. Expert assistance for collaborative problem solving is available to the industry.

6. REFERENCES

[1] NFPA 68, Standard on Explosion Protection by Deflagration Venting, 2007

[2] NFPA 69, Standard on Explosion Prevention Systems, 2008

[3] Vingerhoets, J., et al., "Dust Flame Propagation in Industrial Scale Piping Part 2: CFD Study of a Conveying Vessel-Pipeline System", *Submitted for 9th Global Congress on Process Safety,* San Antonio, April 28 - May 1, 2013

[4] Vogl, A., "Flame propagation in pies of pneumatic conveying systems and exhaust equipment", *Process Safety Progress*, Vol. 15, No. 4, 1996

[5] Valiulis, J., et al., "Experiments on the Propagation of Vented Dust Explosions to Connected Equipment," *Process Safety Progress*, Vol. 18, No. 2, 1999

[6] Holbrow, P., et al., "Dust Explosions in Interconnected Vented Vessels," *Journal of Loss Prevention in the Process Industries* Vol. 9, No. 1, 1996

[7] Chatrathi, K., et al., "Flame Propagation in Industrial Scale Piping," *Process Safety Progress*, Vol. 20, No. 4, 2001

[8] Pineau, J.P., and G. Ronchail, "Propagation of Dust Explosions in Ducts," *Symposium on the Control and Prevention of Dust Explosions*, Bâle, France, Nov. 16-17, 1982

[9] Roser, M., et al., "Investigations of Flame Front Propagation Between Interconnected Process Vessels. Development of a New Flame Front Propagation Time Prediction Model," *Journal of Loss Prevention in the Process Industries,* No. 12, 1999

[10] Chatrathi, K., and J. Going, "Pipe and Duct Deflagrations Associated with Incinerators," *Process Safety Progress,* Vol.15, No.4, 1996.

[11] Vogl, A., "Explosionsübertragung aus Behältern in Rohrhleitungen pneumatischer Anlagen," VDI Berichte No. 1272, 1996.

Additional references not cited:

[12] Jian, W., et al., "Flame Propagation Through Potato Starch/Air Mixtures in Pipe of Interconnected Vessels," *8th International Conference on Measurement and Control of Granular Materials* (p376-379), 2009

[13] Lunn, G. A., "Dust Explosions in Totally Enclosed Interconnected Vessel Systems," *Journal of Loss Prevention in the Process Industries,* No. 9, 1996

[14] Lunn, G. A., "Flame Behavior in Pipelines and Linked Enclosures: A Review," *Safetynet Internet Seminar*, 2001

9TH GLOBAL CONGRESS ON PROCESS SAFETY

Assessing the Hazard of Marginally Explosible Dusts

Timothy J. Myers
Exponent, Inc.
9 Strathmore Road
Natick, MA 01760
dust@exponent.com

Alfonso F. Ibarreta
James Bucher
Kevin Marr
Exponent, Inc.
9 Strathmore Road
Natick, MA 01760

Prepared for Presentation at
American Institute of Chemical Engineers
2013 Spring Meeting
9th Global Congress on Process Safety
San Antonio, Texas
April 28 – May 1, 2013

UNPUBLISHED

Assessing the Hazard of Marginally Explosible Dusts

Timothy J. Myers
Exponent, Inc.
9 Strathmore Road
Natick, MA 01760
dust@exponent.com

Alfonso F. Ibarreta
James Bucher
Kevin Marr
Exponent, Inc.
9 Strathmore Road
Natick, MA 01760

Keywords: combustible dust, dust explosion, overdriving, marginally explosible dust

Abstract

In this study, we review existing literature comparing tests of marginally explosible dusts (MEDs) in both the 20-L and 1-m^3 chambers. We also present new test data of MEDs in both chambers for non-metal and metal dusts. We find that many of these dusts are explosible in the 20-L chamber but not the 1-m^3 chamber. However, we also find that some of these low K_{St} materials are explosible in both chambers, and may produce higher P_{max} and K_{St} values in the 1-m^3 chamber. For some dusts that are explosible in the 20-L chamber but not in the 1-m^3 chamber, we conduct additional testing with smaller energy pyrotechnic igniters to determine the igniter energy that matches data from the 1-m^3 chamber. Similar to previous studies, we find that the ignition energy required to match explosibility data in the two chambers varies for different dusts.

Hybrid explosion tests were performed using methane and an MED in the 20-L chamber with a spark igniter and lean mixtures of flammable gas to determine if the dust would contribute to the severity of a gas explosion. The MED that was tested had a small effect on the maximum overpressure and rate of pressure rise of the hybrid mixture relative to a lean gas explosion, and in some cases the MED decreased the explosion severity relative to methane alone. In contrast, similar hybrid explosions performed with a more explosible dust significantly increased the maximum overpressure and rate of pressure rise of the hybrid mixture relative to a lean gas explosion. The contribution of the MED to the hybrid explosion was most pronounced for low equivalence ratio (fuel lean) methane mixtures. We propose that this hybrid explosion method can be used to evaluate the potential of MEDs to increase the severity of gas explosions in industrial settings.

Finally, explosion and thermal hazard calculations are performed on MEDs in enclosures. The analysis shows that for some enclosures with existing openings, a deflagration of an MED with a low P_{max} and K_{St} value will not produce damaging overpressure, and no additional explosion venting may be required. However, the deflagration will still produce fireballs that can create a thermal hazard to personnel and property.

1. Introduction

Workplace regulatory agencies and the scientific community in general have greatly increased their focus on combustible dust in the last ten years due to a number of high profile incidents across various process industries. The industrial effort to categorize process dusts as either explosible or non-explosible has increased correspondingly. Increasingly, combustible dust standards use go/no-go explosibility screening tests, such as ASTM E1226, to determine if a material is a combustible dust and whether the standard applies to the material. Marginally explosible dusts (MEDs), defined herein as dusts with a K_{St} of roughly 50 bar-m/sec or less, have proven difficult to group into either classification, as test results in different apparatus for the same dust can vary. Others have classified similar MEDs or hard-to-ignite dusts based upon a low value of P_{max} or high ignition energy.

The two apparatus most commonly used to measure dust explosion severity are the 1-m^3 chamber, the use of which is outlined in ISO 6184-1 [1] and the 20-L chamber, the use of which is outlined in ASTM E1226 [2]. Both apparatuses typically use a pyrotechnic 10-kJ ignition source which is much stronger than the spark ignition energy source historically used in the smaller cylindrical chambers based on the Hartmann apparatus. Studies comparing explosion indices (P_{max}, K_{St}) in the 20-L and 1-m^3 chambers have found that in most cases they can be correlated provided the turbulence level or other parameters in the 20-L chamber are adjusted in order to reproduce the results of reference dusts in the 1-m^3 chamber. [3,4,5]. However, some studies have found that dusts near their explosibility limits (MEDs, dusts near their minimum explosible concentration, and dusts inerted with solids) have tested as explosible in the 20-L chamber and non-explosible in the 1-m^3 chamber when using similar energy igniters in both chambers [6,7,8,9].

Often cited as a reason for this difference is the much greater influence of the ignition source on the pre-ignition pressure and temperature rise in the 20-L chamber as compared to that in the 1-m^3 chamber. The pyrotechnic igniters used in testing cause preheating and pre-pressurization of the closed chamber immediately prior to ignition of the dust. This phenomenon has been labeled "overdriving" and is thought to give false positive results in the 20-L chamber. Possible explanations of the overdriving phenomenon in the 20-L chamber include preheating of air and/or dust by the igniter, or material effectively burning in the igniter that would not propagate a flame on its own. The increase in temperature at the time of ignition for the 20-L chamber may also promote volatilization of some dust immediately prior to its ignition. It has been debated whether the ignition conditions in the 20-L chamber when using typical energy igniters are realistic compared to likely ignition sources in industrial facilities.

The effect of chemical igniters on the chamber initial conditions will depend on the size of the test chamber. Table 1 shows the effect of the ignition energy on the starting pressure and temperature in the chamber. In this calculation, all of the igniter energy is assumed to be transferred to the gas, resulting in an increase in temperature and pressure. The measured values were obtained by recording the pressure rise inside a 20-L chamber when an igniter was fired at atmospheric pressure in air without dust. Measured values are lower than calculated values because a fraction of the heat is retained by the largely non-condensable reaction products of the igniters and potentially due to heat loss to the chamber walls. Applying the energy of the commonly used 10-kJ ignition source to 20-L of air in a closed volume causes a calculated pressure rise of approximately 2 bar and temperature rise of approximately 575°C. The same energy applied to 1-m^3 (1,000 L) of air only causes a calculated pressure rise of approximately 0.04 bar and a temperature rise of approximately 12°C. The ideal ignition source used in a test should be above the ignition energy of the material without significantly increasing the chamber pre-ignition pressure and temperature. As the data in **Table 1** shows, much larger ignition energies can be tolerated in larger chambers without causing significant preheating.

Table 1: Changes in 20-L chamber pressure and temperature with ignition energy.

Ignition Energy (J)	20-L Chamber				1-m^3 Chamber	
	ΔP Predicted (bar)	ΔP Measured (bar)	ΔT Predicted (°C)	ΔT Measured (°C)*	ΔP Predicted (bar)	ΔT Predicted (°C)
250	0.05	0.03	15	9	0.001	0.3
500	0.10	0.03	29	9	0.002	0.6
1,000	0.20	0.1	58	29	0.004	1.2
2,500	0.50	0.18	146	53	0.01	2.9
5,000	1.00	0.56	292	164	0.02	5.8
10,000	1.99	1.14	584	334	0.04	11.7

(*) Note: Measured temperature inferred from recorded pressure rise, assuming an ideal gas.

The ambiguity between explosibility results in the 20-L and 1-m^3 chamber can provide a source of confusion to industrial facilities that generate marginally explosible dust. If a given dust will not deflagrate where it accumulates, companies can spend significant resources attempting to mitigate a hazard that does not exist. However, worker injuries and fatalities have reportedly occurred due to the deflagration of marginally explosible dust in industrial settings. The incidents at a powder metal facility in Tennessee provide an example; iron dust with a reported K_{St} less than 20 bar-m/sec is reported to have contributed to one or more of those incidents. This dust was tested to be explosible by the U.S. Chemical Safety Board (CSB) according to the procedure in the 20-L chamber, but the result in the 1-m^3 chamber indicated the dust was non-explosible [10]. It is not clear if the sample tested in the 1-m^3 chamber was identical to the sample tested in the 20-L chamber or if the samples were representative of the material involved in the incidents.

The remainder of this paper examines trends identified in the scientific literature comparing explosion indices in the aforementioned apparatus. These trends are then compared to test results for a number of marginally explosible dusts in both chambers. Finally, the consequences of deflagrations involving marginally explosible dusts are evaluated.

2. Background

Richard Siwek conducted the first studies that correlated explosion indices in the smaller 20-L chamber with those in the 1-m^3 chamber [4,11]. The procedure that Siwek developed for the 20-L apparatus differed from that for the 1-m^3 chamber in two main ways: the 20-L chamber was evacuated to 0.4 bar of total pressure before dust injection, and the time delay between introduction of dust into the chamber and ignition was decreased by an order of magnitude. A 10-kJ ignition energy was used in both chambers. The 20-L apparatus that Siwek designed bears his name today.

Bartknecht also provided analysis based on Siwek's data comparing the explosion indices of dusts in both the 20-L and 1-m^3 explosion chambers [3]. He found good correlation for P_{max} between the two chambers from approximately 4 to 8 bar-g; at maximum explosion pressures higher than 8 bar-g, the values found in the 20-L chamber were lower than those found in the 1-m^3 chamber. The K_{St} values in the 20-L and 1-m^3 chambers correlated well for dusts up to 700 bar-m/sec, although the number of marginally explosible dusts tested was limited. For a subset of coating powders, Bartknecht again found good correlation between the two chambers for explosion indices in the range of 5 to 10 bar-g for P_{max} and 100 to 200 bar-m/sec for K_{St}. No marginally explosible dusts were tested within this subset.

More recently, Proust et al. performed a comparison between the 20-L and 1-m^3 chambers for P_{max} and K_{St} for a variety of non-metal dusts [6]. Four of these dusts were found to be non-explosible in the 1-m^3 chamber yet explosible in the 20-L chamber when tested with a 10-kJ ignition energy; these dusts had P_{max} ranging from 1.3 to 7.4 bar and K_{St} ranging from 9 to 62 bar-m/sec in the 20-L chamber. The dusts that were found to be explosible in both chambers had P_{max} ranging from 4.2 to 9.1 bar and K_{St} ranging from 45 to 226 bar-m/sec. This finding raises the question as to whether marginally explosible dusts were overdriven in the 20-L chamber in this study.

Based on their data, Proust et al. concluded that dusts with a K_{St} below 45 bar-m/sec measured in the 20-L chamber may not be explosible in the 1-m^3 chamber. Furthermore, they proposed that a correction factor of 30% be applied to the P_{max} of dusts with a K_{St} above 45 bar-m/sec measured in the 20-L chamber to correlate them to the results found in the 1-m^3 chamber.

The principal manufacturer of the Siwek 20-L chamber, Kühner AG, recommends corrections to the explosion pressures developed within the apparatus so that they are comparable to those measured in the 1-m^3 chamber. These corrections are automatically calculated in the control software for the Kühner 20-L chamber [12]. Explosion pressures above 5.5 bar-g are adjusted upward due to the reportedly "less favorable surface to volume ratio" in the 20-L chamber as compared to the 1-m^3 chamber. In the correction, $P_m = 0.775P_{ex}^{1.15}$, where P_{ex} is the explosion pressure in the 20-L chamber and P_m is the equivalent explosion pressure in the 1-m^3 chamber. Explosion pressures below 5.5 bar-g are adjusted downward to account for the greater influence of the chemical igniters on the explosion pressure. In this correction, $P_m = 5.5(P_{ex} - P_{ci})/(5.5 - P_{ci})$, where P_{ci} is a pressure factor proportional to the igniter energy. The P_{max} is then based on the corrected values. This value of P_{max} is widely reported in the literature for testing in 20-L chambers, and the remainder of this paper compares overpressure values based upon the Kühner corrections.

Other studies have investigated the difference between the two chambers for dusts near their explosibility limits when measuring parameters aside from P_{max} and K_{St}. Going et al. compared the minimum explosible concentration (MEC) of explosible dusts in the 20-L and 1-m^3 chambers; they found comparable results between the 20-L chamber using 2,500 J igniters and the 1-m^3 chamber using 10,000 J igniters [8]. At ignition energies greater than 2,500 J, the authors reported strong dependence of the MEC on ignition energy in the 20-L chamber, providing an indication that overdriving was occurring in the smaller chamber. Dastidar and Amyotte found that the concentration of inert dust used to prevent the deflagration of a combustible dust was higher in the 20-L chamber than in the 1-m^3 chamber when the same igniter strength was used [9]. The authors found good agreement for a variety of dusts for minimum inerting concentrations between the two chambers when an ignition energy of 500 to 1,000 J was used in the 20-L chamber. They hypothesized that the stronger influence of the igniters in the 20-L chamber increased the amount of diluent required to mitigate the explosivity of those dusts for constant ignition energy between the two chambers.

3. Comparison of P_{max} and K_{St} Values in the 20-L and 1-m^3 Chambers

3.1 Test Method

Twenty marginally explosible dusts, thirteen of them metallic and seven of them non-metallic, were tested to determine the maximum explosion overpressure, P_{max}, and the volume adjusted pressure rise rate, K_{St}. Testing in the 20-L chamber was conducted in the Siwek apparatus manufactured by Kühner AG according to the procedure outlined in ASTM E1226. Testing in the 1-m^3 chamber was conducted according to the procedure outlined in ISO 6184-1. The reader should refer to those standards for further details about the equipment and procedure.

Full K_{St}/P_{max} tests were run in both chambers at the same particle size for five of the twenty samples. For samples for which full K_{St}/P_{max} data was not available in one or both chambers, explosion overpressures and volume corrected rates of pressure rise were compared at the same dust cloud concentration. The testing for these samples was performed at or near concentrations at which the P_{max} and K_{St} would typically be observed (1000-3000 g/m^3 for metal dusts, 500-2000 g/m^3 for non-metal dusts). Therefore, the reported data approximate P_{max} and K_{St} and are reported generally as P and K values hereafter.

3.2 Results

In the following analysis, the explosion pressures in the 1-m^3 chamber are compared to the explosion pressures in the 20-L chamber corrected according to the earlier described Kühner algorithm [12]. Previously published values for P_{max} in the 20-L chamber are assumed to be based on the corrected explosion pressures and are compared to the corrected 20-L P values of this study. The values for dP/dt in the 20-L chamber are calculated according to the cube-root-law to obtain K_{St}.

The twenty dust samples of the present study were observed to be explosible in the 20-L chamber with K values less than 50 bar-m/sec with the exception of one sample with a slightly higher K value (63 bar-m/sec). Of the thirteen metal samples that were tested, only one tested as non-explosible in the 1-m^3 chamber; this sample had tested as very weakly explosible in the 20-L chamber. For the remaining twelve metallic dust samples the P and K values were higher in the 1-m^3 chamber than in the 20-L chamber.

Six of the seven non-metal samples were observed to be non-explosible in the 1-m^3 chamber. In the 20-L chamber these samples ranged from 2.4 to 5.6 bar with an average of 4.4 bar for explosion overpressures and from 2 to 39 bar-m/sec with an average of 18 bar-m/sec for K values. Figure 1 compares the K values observed in the two explosion test chambers; the top figure shows K values up to 250 bar-m/sec, whereas the bottom figure displays values up to 100 bar-m/sec to allow better visualization of the data for MEDs.

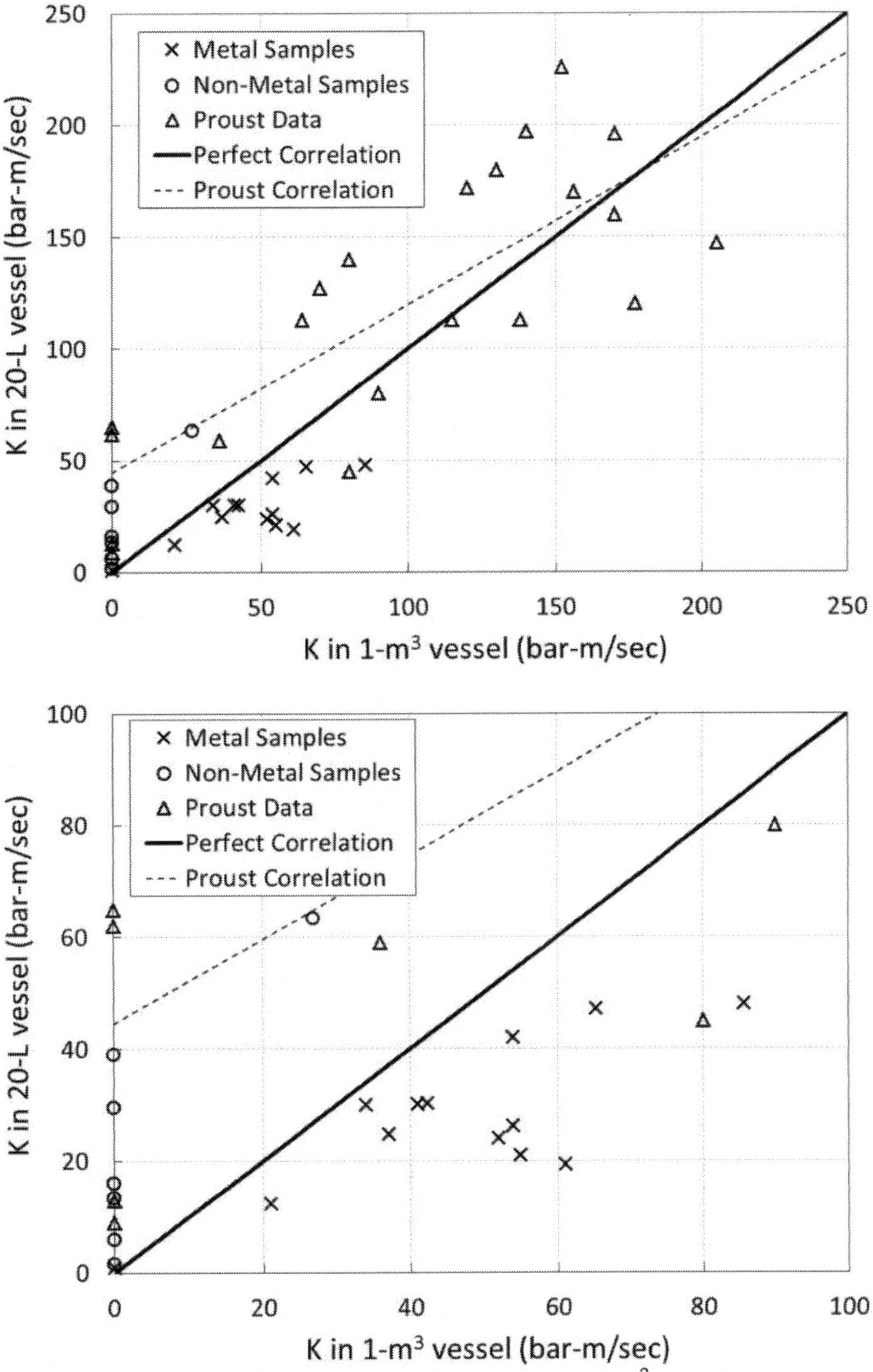

Figure 1: Comparison of K values between the 20-L and 1-m^3 chambers.

The data in Figure 1 are compared to a perfect cube-root-law correlation and the correlation proposed by Proust et al., who predicted that a dust with a K_{St} below 45 bar-m/sec in the 20-L chamber may not be explosible in the 1-m^3 chamber [6]. The K values observed for the marginally explosible metal samples of the present study correspond neither to the cube-root-law nor to the Proust correlation; the 20-L K values are lower than those predicted by the cube-root-law and higher than those predicted by Proust (whose correlation predicts a K_{St} of zero in the 1-m^3 chamber). For some marginally explosible metal samples, the K values are higher in the 1-m^3 chamber than in the 20-L chamber. However, metal samples in this and other studies with very low K values in the 20-L chamber appear to be non-explosible in the 1-m^3 meter chamber.

The data for the non-metal samples of the present study, while limited, correlate well with the findings of Proust. The six non-metal samples with K values from 2 to 39 bar-m/sec were observed to be non-explosible in the 1-m^3 chamber, while the one non-metal sample that was explosible in both chambers falls very close to the Proust correlation.

A comparison of P values between the 20-L and 1-m^3 chambers is shown in Figure 2; the top figure displays a broader scale to include the data of Proust, while the bottom figure is scaled to better visualize the MEDs of the current study. The data and correlation of Proust are shown alongside the data of the present study. Also depicted is a "perfect correlation" between the 20-L and 1-m^3 chambers, whereby the raw data from the 20-L chamber are corrected according to the Kühner algorithm.

For the metal samples, the trend fit by Proust correlates well to the explosion pressure data, indicating the trend holds for marginally explosible metal dusts. Neither Proust's P_{max} data nor the explosion pressure data for the metal samples of the present study fall on the perfect correlation line. The marginally explosible dusts of this study are in closer agreement with the correlation of Proust et al.

A significant number of the non-metal samples have positive P values ranging from 2.9 to 5.6 bar-g in the 20-L chamber and no explosion overpressure in the 1-m^3 chamber. However, this trend agrees with the prediction of Proust, as each of these samples has a K value below 45 bar-m/sec.

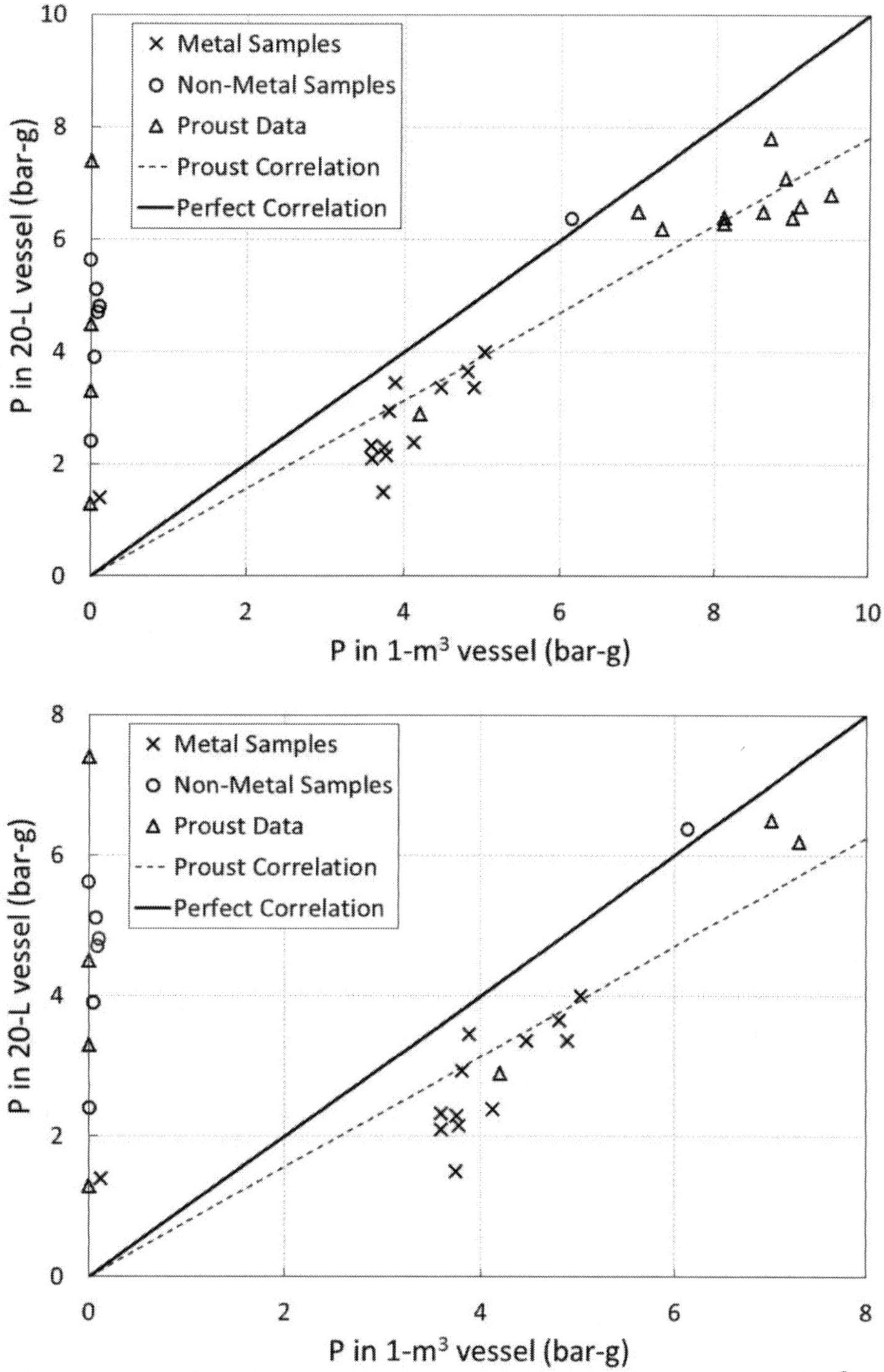

Figure 2: Comparison of explosion overpressures between the 20-L and 1-m^3 chambers.

3.3 Discussion

The question of overdriving in the 20-L chamber remains an issue in the hazard analysis of combustible dusts. Several studies prior to this work have attributed overdriving to pre-heating of the chamber air and/or dust by the igniters in the 20-L chamber. Proust et al. concluded that a combustible dust that was tested to have a K_{St} below 45 bar-m/sec may not be explosible in the 1-m^3 chamber.

The current study finds a distinct difference between marginally explosible dusts of metal and non-metal composition with regard to overdriving behavior. A significant majority of metal dusts that tested to be explosible in the 20-L chamber with K values below 50 bar-m/sec were found to have greater values of both P and K in the 1-m^3 chamber. This finding is in agreement with neither the cube-root-law nor the correlation of Proust et al. for K. However, the explosion overpressures of these metal dusts follow the general relationship that Proust et al. proposed between P_{max} in the 20-L and 1-m^3 chambers. The marginally explosible non-metal dusts in this study were found to be non-explosible in the 1-m^3 chamber. These results may indicate that such non-metal dusts are indeed overdriven in the 20-L chamber, consistent with the findings of Proust et al., while for metal dusts, overdriving occurs at lower K_{St} values than those for non-metal dusts.

Combustion of organic dusts typically involves pyrolysis or volatilization of the solid prior to combustion, whereas metal combustion is usually heterogeneous, occurring at the surface without volatilization of the metal. Preheating caused by igniters may have a more significant effect on volatilization of nonmetals, than on the surface combustion of metals.

4. Effect of Ignition Energy

To determine the effect of ignition energy on the results described in the previous section, the ignition energy in the 20-L and the 1-m^3 chambers was varied in an attempt to determine an ignition energy that would provide similar go/no-go results. This testing was performed using a marginally explosible carbonaceous dust, hereafter referred to as Sample A, with a minimum ignition energy (MIE) above 4,630 mJ (limit of test apparatus), a dust cloud minimum autoignition temperature (MAIT) of 654°C, and 3% volatile mater. Figure 3 depicts the explosion overpressure and volume normalized rate of pressure rise with ignition energy in both the 20-L and 1-m^3 chambers.

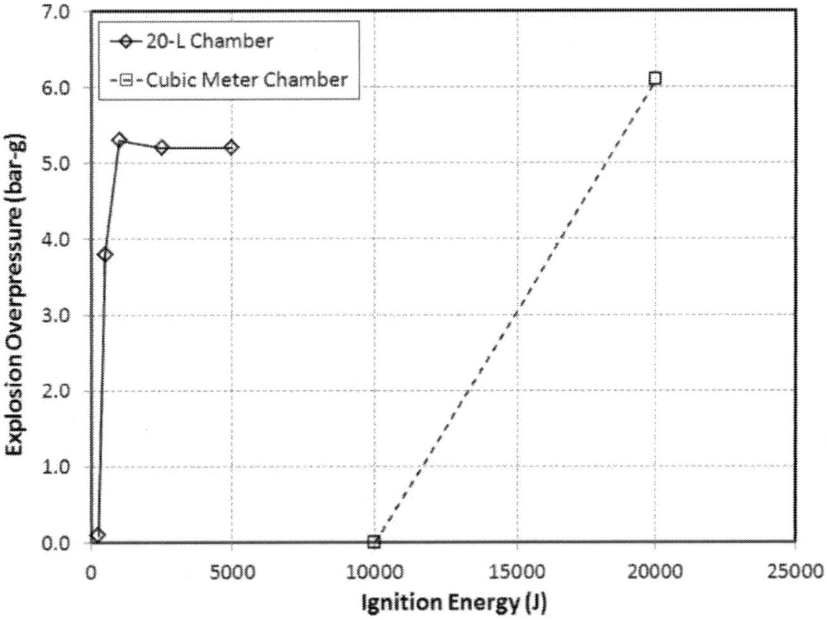

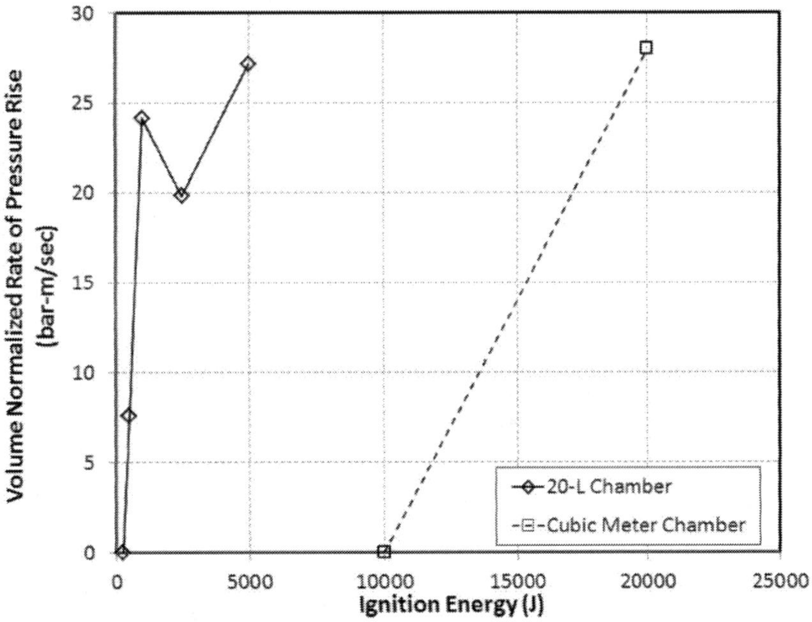

Figure 3: Explosion overpressure and volume normalized rate of pressure rise of an MED vs. ignition energy in the 20-L and 1-m^3 chambers.

The go/no-go threshold for this particular carbon dust occurred between 10 kJ and 20 kJ in the cubic meter chamber; in the 20-L chamber, this threshold was observed between 250 and 500 J, a factor of forty less than that observed in the cubic meter chamber. The observed explosion overpressure in the 20-L chamber increased with ignition energy to 1 kJ, after which it remained constant up to 5 kJ. These results are not consistent with the explosibility being dependent on ignition energy as the required ignition energy is expected to be independent of the chamber volume.

Figure 4 depicts the explosion overpressure and volume adjusted rate of pressure rise against ignition energy density (energy/volume) for the marginally explosible carbon dust in both explosion chambers. Once ignition energy is normalized to the size of the chamber, the go/no-go threshold between the two chambers appears to be very similar. For this particular MED, the ignition energy density that produces similar K values in both chambers is nearly directly proportional to chamber volume – a fiftyfold increase in chamber size necessitates a fortyfold increase in ignition energy to yield similar results. The fact that the material becomes explosible at a common energy density in both chambers suggests that the ignition may be due to the preheating effects of the igniters, which would be expected to be similar in both chambers at the same ignition energy density.

Table 2 presents results for two other marginally explosible carbon samples, labeled Carbon Samples B and C, that were tested in the 20-L and 1-m^3 chambers at varying ignition energies. Neither of these samples was explosible in the 1-m^3 chamber at an ignition energy of 20 kJ, but both samples produced explosions in the 20-L chamber at different ignition energies than Sample A. These results demonstrate that the ignition energy at which an MED is overdriven can vary; and it is difficult to recommend a single ignition energy for the 20-L chamber that will match data in the 1-m^3 chamber for all dusts.

Table 2: Go/No-Go test results for marginally explosible carbon dust samples at varying ignition energies.

MED Sample	IE in 20-L (kJ)	Result	IE in 1-m^3 (kJ)	Result
Carbon Sample A	0.25	Non-Explosible	10	Non-Explosible
	0.5	Explosible	20	Explosible
	1	Explosible		
	2.5	Explosible		
	5	Explosible		
Carbon Sample B	1	Non-Explosible	20	Non-Explosible
	2.5	Non-Explosible		
	5	Explosible		
	10	Explosible		
Carbon Sample C	0.5	Non-Explosible	20	Non-Explosible
	1	Explosible		
	2.5	Explosible		
	5	Explosible		

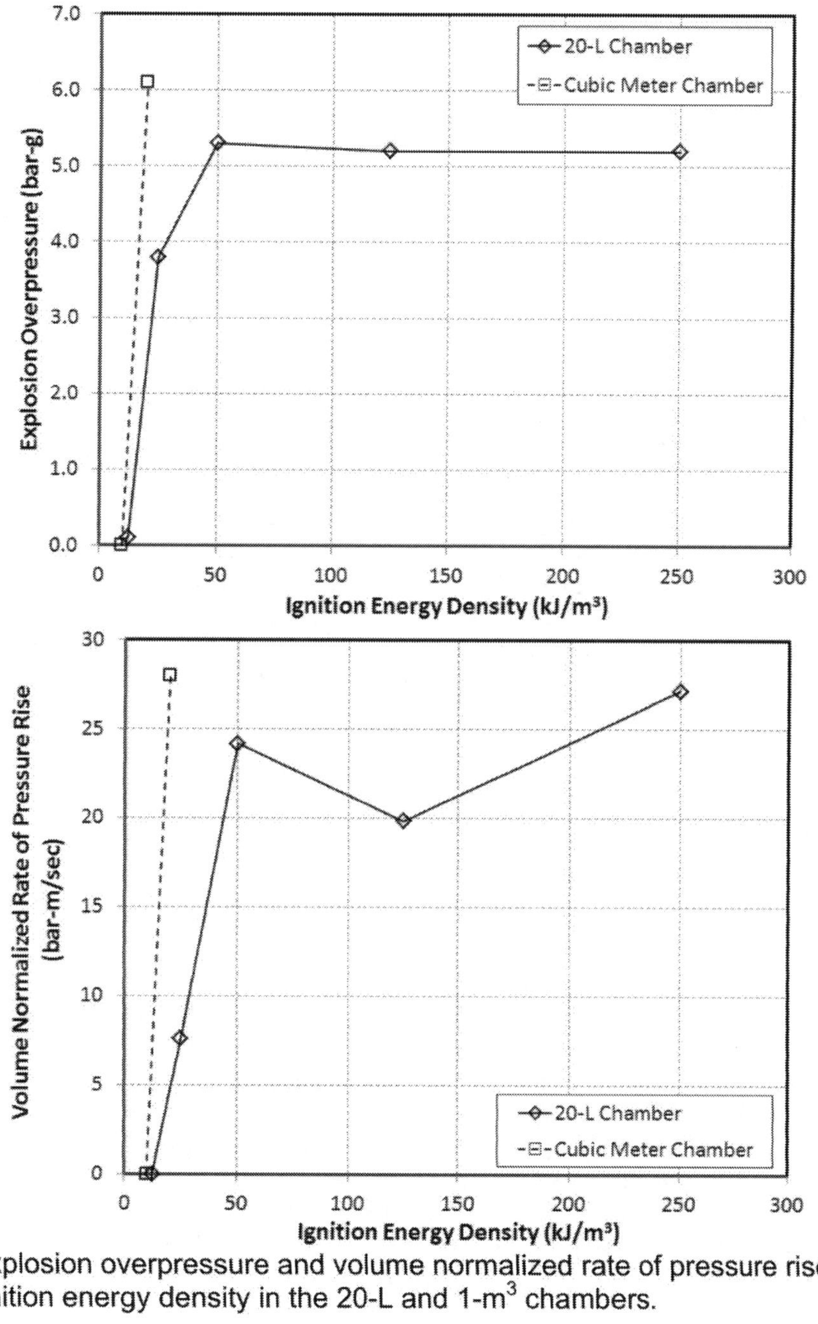

Figure 4: Explosion overpressure and volume normalized rate of pressure rise of an MED vs. ignition energy density in the 20-L and 1-m^3 chambers.

5. Hybrid Explosions

As shown in the present study and previous publications, a dust that is explosible in the 20-L chamber yet non-explosible in the 1-m^3 chamber may be overdriven due to the greater influence of the igniters on the combustibility of the material. However, ignition sources may be present in an industrial facility that are capable of overdriving a dispersed accumulation of an MED to the point at which it will combust in the flame of that ignition source. For example, the authors of this paper have heard arguments that a primary gas fire ball or explosion may be able to initiate a dust explosion with such an MED.

One way to examine this hypothesis is to perform hybrid dust explosions with a lean gas mixture and the dust. However, this hybrid scenario is different than an isolated gas fire ball igniting a dust cloud that does not contain gas. Testing was performed using lean methane mixtures at equivalence ratios (hereafter referred to as Φ) of 0.5 and 0.7 with three dusts: Pittsburgh Pulverized Coal (PPC), a marginally explosible carbon dust, and sand. PPC is typically reported to have approximately 37% volatile matter and a minimum ignition energy of 30 to 140 mJ. The MED consisted of a 3% volatile matter carbonaceous dust that was explosible in the 20-L chamber when tested with ignition energy greater than 500 J, and non explosible in the 1-m^3 chamber when tested with ignition energy less than 20 kJ and was used for the testing described in Section 4. A spark igniter was used with the standard injection pressures and ignition delays whether or not dust was present.

Figure 5 shows the overpressure (P_{hybrid}) and volume normalized rate of pressure rise (K_{hybrid}) as a function of dust concentration for the three dusts in hybrid explosions with methane at a Φ value of 0.7. The data in this figure are normalized by values measured with methane alone, with air injection into the chamber and the same ignition delay as used in dust tests. The PPC hybrid explosion produced a greater overpressure over the range of tested dust concentrations with a maximum increase of 20%. As was expected, the noncombustible sand reduced the overpressure over the tested dust concentration range due to the absorption of heat generated by the combustion of methane. The MED slightly raised the overpressure over most of the range of tested dust concentrations, with a maximum increase of about 12% at one low concentration

A non-monotonic increase in the rate of pressure rise was observed at low dust concentrations. The cause of this phenomenon is not known with certainty, but the authors hypothesize that small amounts of dust may increase the turbulence in methane relative to injection of air alone. The PPC significantly increased the rate of pressure rise over the tested range of dust concentrations, with a peak value approximately 70% higher than that of methane alone. As was expected, the inert sand decreased the rate of pressure rise over the entire range of tested dust concentrations. Other than an initial non-monotonic increase, the MED hybrid explosion produced a rate of pressure rise that was at or below that of methane alone across the tested range of dust concentrations.

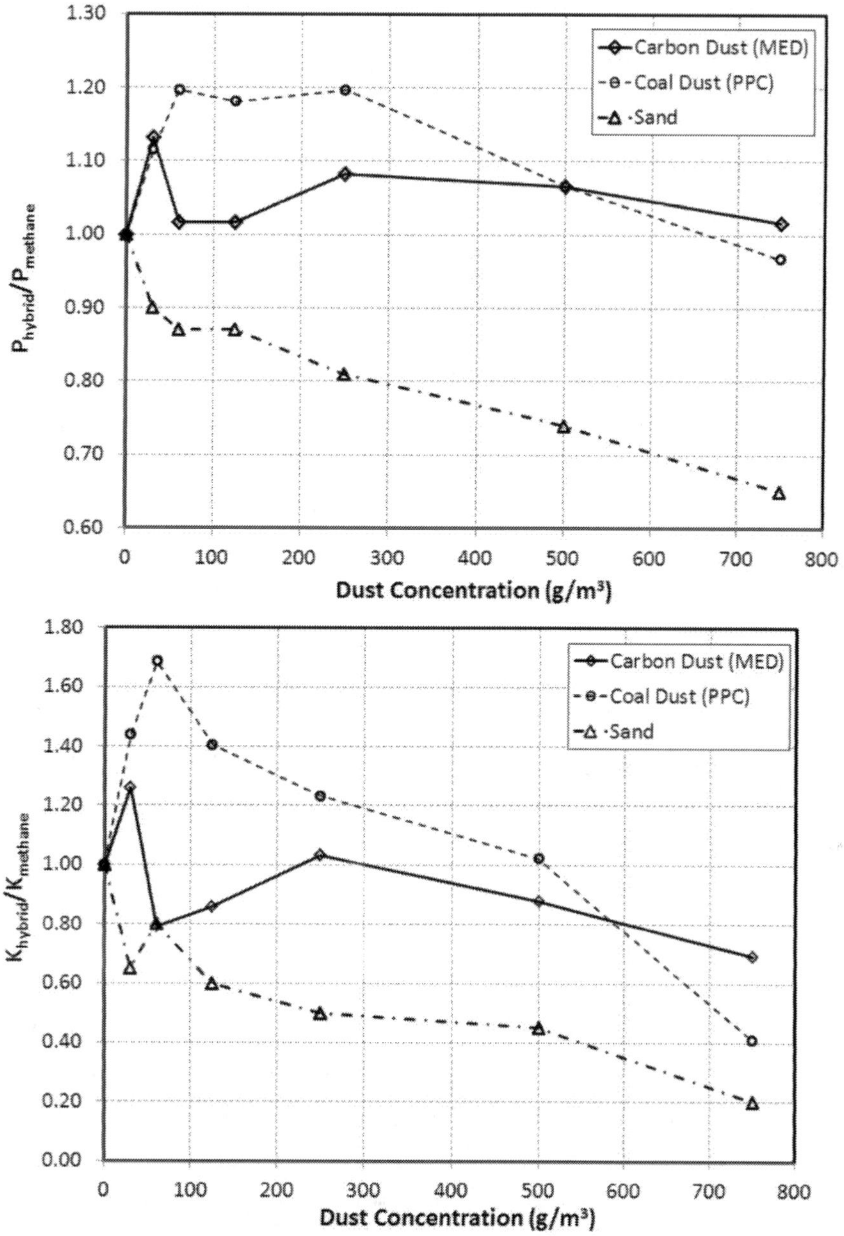

Figure 5: Ratio of overpressure and rate of pressure rise of hybrid explosion relative to methane alone for a methane equivalence ratio of 0.7.

Figure 6 depicts $P_{hybrid}/P_{methane}$ and $K_{hybrid}/K_{methane}$ against dust concentration at a methane Φ of 0.5 with PPC and the MED (sand was not tested at this Φ). Although the explosion severity ratios compared to methane underwent a non-monotonic increase at low dust concentrations at a methane Φ of 0.7, they initially decreased at a methane Φ of 0.5 before exceeding a value of 1.0 at higher dust concentrations. This initial decrease was observed for both PPC and the MED.

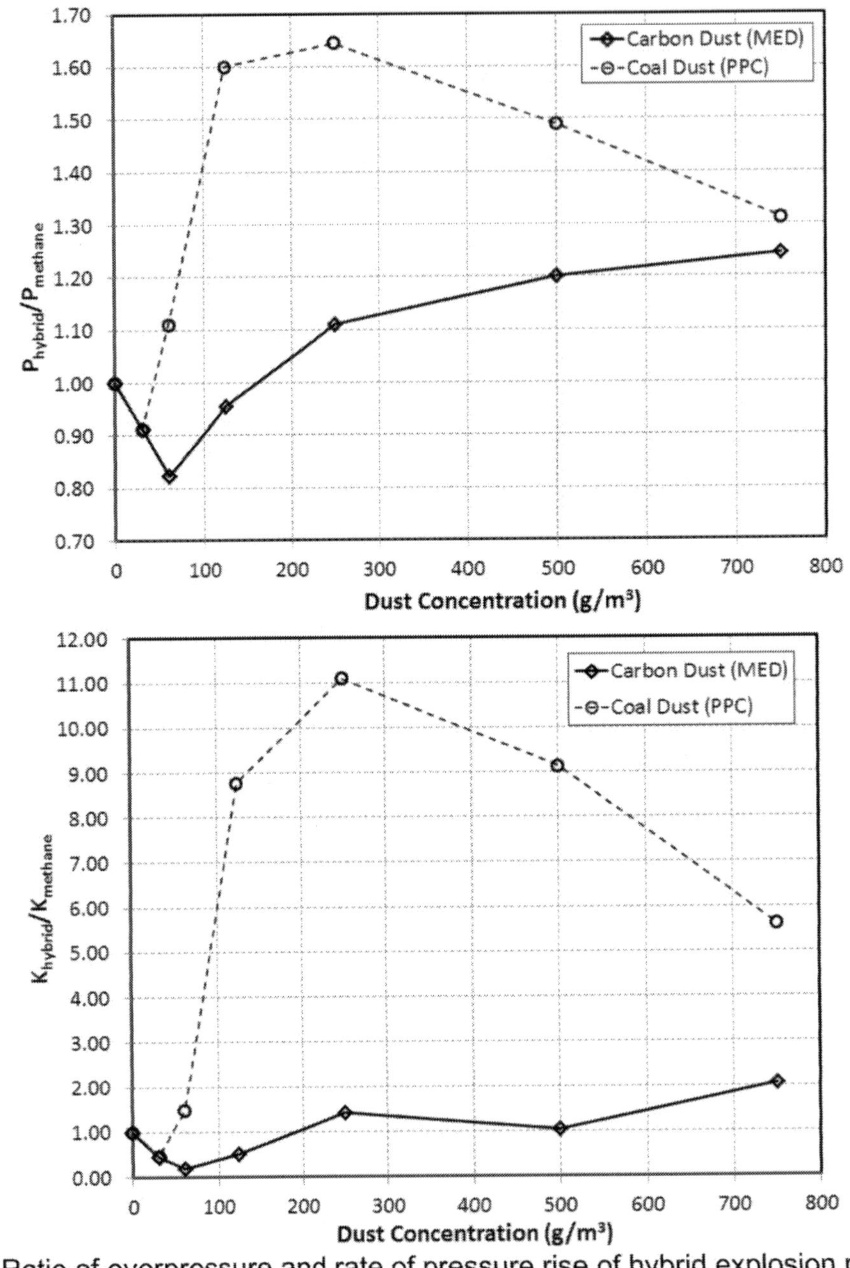

Figure 6: Ratio of overpressure and rate of pressure rise of hybrid explosion relative to methane alone for a methane equivalence ratio of 0.5.

At a methane Φ of 0.5, the explosion severity of PPC exceeded that of methane alone at a dust concentration of 60 g/m^3 and continued to increase until a concentration of 250 g/m^3, after which explosion severity declined. However, the explosion overpressure and rate of pressure rise of the MED relative to methane decreased from 30 to 60 g/m^3, after which these parameters increased relative to methane for the remainder of the tested dust concentration range. Whereas the highest values of P and K for PPC and methane occurred at a dust concentration of 250 g/m^3, this ratio was shifted to higher dust concentrations when the MED was tested at methane ER of 0.5.

These results indicate that an MED may not contribute to the severity of a hybrid explosion as significantly as a dust with higher P_{max} and K_{St} values. In some cases the MED may decrease the severity of a hybrid explosion relative to that of a methane explosion. The tested MED, was more likely to exceed the explosion severity of the gas explosion at low methane equivalence ratios and high dust concentrations. Additionally, the presence of MED in the combusting gas may increase the thermal radiation emitted by the fireball.

6. Hazard Analysis of Marginally Explosible Dusts

When dusts are found to be explosible but with low P_{max} and K_{St} values, facilities often question whether the same level of protection is necessary with those dusts as with dusts with higher P_{max} and K_{St} values. Most NFPA combustible dust standards allow a risk or hazard analysis to determine the appropriate level of protection. In some cases, this risk-based level of protection may be lower than the standard prescriptive requirements. An examination of properties of the dust and the specific process is typically needed. Many dusts with low P_{max} and K_{St} values also have high minimum ignition energies or high autoignition temperatures indicating that the material is difficult to ignite. Thus, the likelihood of ignition and consequence of ignition may be lower in some processes for marginally explosible dusts than dusts with a high P_{max} and K_{St}.

Historically, some practitioners have employed rules of thumb by which they did not use explosion protection such as venting on dust collectors handling material with a K_{St} lower than 50 or 100 bar-m/sec. NFPA 61 *Standard for the Prevention of Fires and Dust Explosions in Agricultural and Food Processing Facilities* does not require explosion protection on small cyclones used as process air-material separators provided certain criteria are met, including that the dust has a K_{St} less than 200 bar-m/s. To determine if there is technical justification to the argument that lower K_{St} materials may not require explosion protection on certain enclosures, it is useful to examine the vent area required on dust collectors using NFPA 68 *Standard on Explosion Protection by Deflagration Venting*. Figure 7 shows the calculated explosion vent area for a 2,500-cfm, 32-ft^3 dust collector and a 13,000-cfm, 200-ft^3 dust collector for a large number of dusts with different P_{max} and K_{St} values. Also plotted in the figure are the areas of the inlet and combined inlet and outlet ducts on the dust collectors.

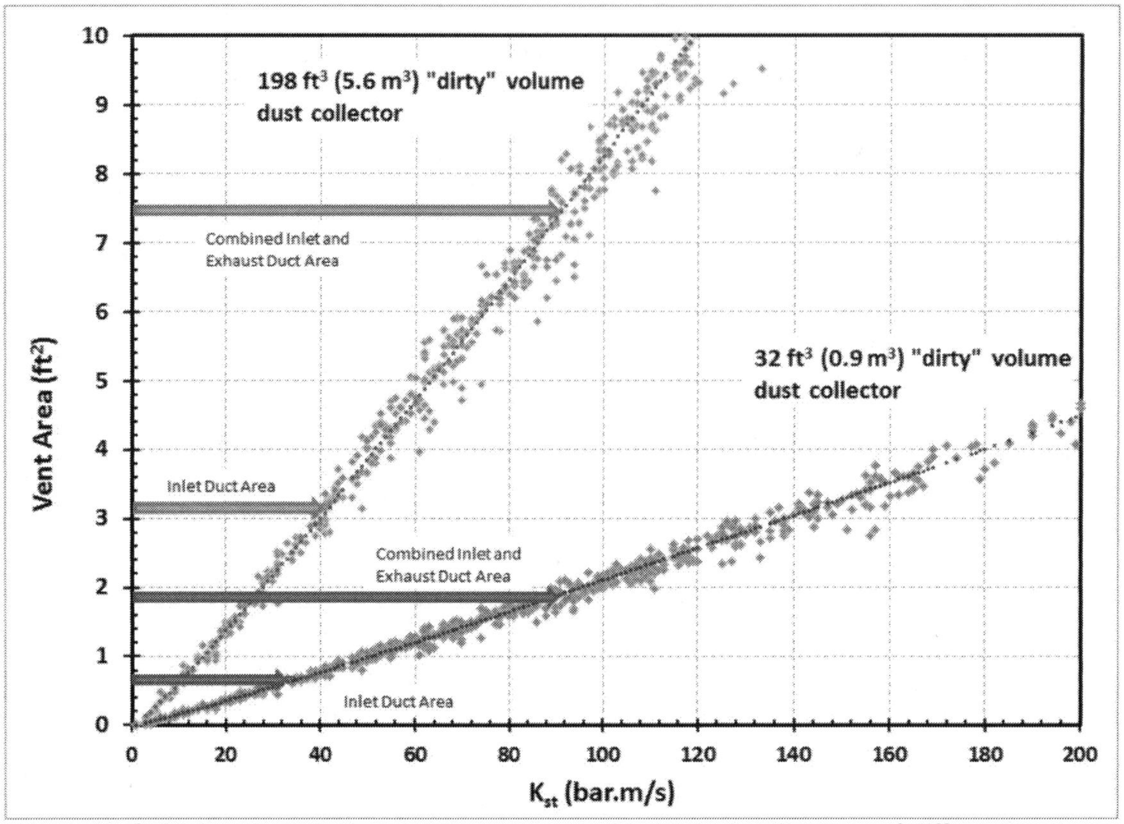

Figure 7: Calculated explosion vent area versus K_{St} in dust collectors of different volumes and flow rates.

The figure shows that the required vent area is less than the area of the inlet duct for dusts with a K_{St} below approximately 40 bar-m/s and less than the combined area of the inlet and exhaust ducts for dusts with a K_{St} below approximately 90 bar-m/s. Practically, this means that for a dust with a low K_{St} (e.g., 25 bar-m/s) and these dust collectors, a deflagration may be able to be vented by the inlet duct and to a lesser extent the outlet duct without bursting the dust collector, if the inlet or outlet ducts provide a low flow resistance path to the environment. A more formal analysis would be required for a specific dust collector to include the appropriate flow resistances in the inlet and outlet ducts and filter media to determine whether additional explosion venting would be necessary. Such an analysis may show that for some pieces of equipment, existing inlets and outlets can provide adequate explosion protection. However, it is important to note that even if the dust collector would not burst, flames could be ejected from the inlet and outlet of the dust collector and may require isolation or another form of protection.

Additionally, this analysis identifies that if pressure activated explosion protection or isolation is used on a dust collector with a low K_{St} dust, venting through inlet and outlet ducts may prevent the collector from reaching a sufficient overpressure to activate the system. For explosion suppression and active isolation systems, thermal or flame sensors should be considered in addition to pressure sensors for low K_{St} materials.

The minimum amount of combustible dust mass that represents a combustible dust explosion or flash fire hazard can be computed using the Mass Method B equations described in NFPA 654 *Standard for the Prevention of Fire and Dust Explosions from the Manufacturing, Processing, and Handling of Combustible Particulate Solids* (2013). These equations are dependent on the combustible dust P_{max} value, but that relationship can also be plotted as a function of K_{St} for a number of dusts where both P_{max} and K_{St} are known. Figure 8 shows the flash fire hazard amounts and dust explosion hazard amounts using a best-fit relation between P_{max} and K_{St} for a large number of combustible dusts. The calculations were performed for a square cross-section room of 10,000 ft^2 and 20-ft height walls, with an enclosure strength of 0.5 psi. The dusts were assumed to have a worst-case concentration (at which dP/dt is highest) of 1 kg/m^3. Additionally, an explosion hazard mass dust threshold was calculated for the same room with a 2.5% open area along the walls using the NFPA 68 compartment venting methodology, and making similar simplifying assumptions as the NFPA 654 analysis.

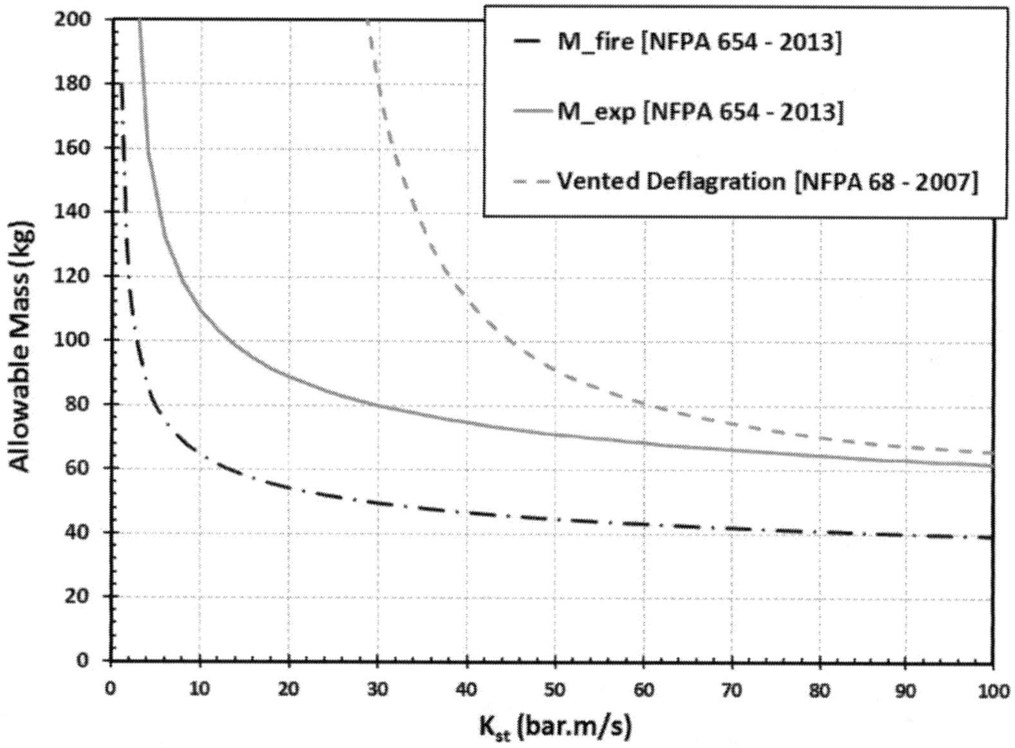

Figure 8: Minimum amount of combustible dust that represents a combustible dust hazard for a 10,000 ft^2 room, based upon NFPA 68 and 654 methodologies.

These results demonstrate that the flash fire scenario is more stringent for the analyzed room geometry, allowing less combustible dust within the enclosure than the NFPA 654 explosion hazard threshold. Which threshold is lower depends on the height and strength of the building. In the case of a completely closed volume, the allowable combustible mass only increases substantially for dusts with a K_{St} less than 20 bar.m/s, that have substantially lower P_{max} values. Venting of the enclosure significantly reduces the hazard posed by a combustible dust explosion for low K_{St} dusts. At low enough K_{St} values (11 bar.m/s in this case), there is no limit on the allowable dust accumulation because even the worst case concentration of dust in the room could be adequately vented. The effect of venting is lower or even negligible for higher K_{St} dusts.

These analyses show that that low K_{St} and corresponding P_{max} of a MED most significantly increase the combustible dust mass thresholds for vented deflagration scenarios. While, lower K_{St} and P_{max} values do increase the combustible mass thresholds for confined deflagration scenarios, the effect is not as pronounced.

7. Summary

The explosion hazard of marginally explosible dusts (MEDs) can be difficult to characterize in the commonly used 20-L explosion chamber due to potential overdriving by strong igniters. Similar to previous studies, we find that results from the 20-L chamber showing K_{St} values of non-metal dusts below approximately 50 bar-m/sec may be due to overdriving. In this study, the threshold K_{St} values below which overdriving may occur in the 20-L chamber with metal dusts appears to be lower than for nonmetal dusts. If 20-L chamber testing indicates the dust is marginally explosible, it is advisable to perform follow-up testing in the 1-m^3 chamber where strong igniters have a significantly lower effect on the pre-ignition temperature and pressure of the chamber. To date, no lower ignition energy has been identified that can universally be used in the 20-L chamber to match data obtained in the 1-m^3 for all MEDs. Where marginally explosible dust is found to be explosible in the 20-L but not the 1-m^3 chamber, it is prudent to conduct follow-up testing on multiple samples collected at different times to determine whether process variations could render the dust explosible in the 1-m^3 chamber.

This study showed that hybrid dust explosion testing can be used to evaluate the likelihood that an MED would contribute to the severity of a gas explosion. The contribution of the MED is most pronounced for low equivalence ratio (fuel lean) mixtures of gas.

With marginally explosible dusts, hazard analyses can be performed looking at the probability and consequences of a deflagration event. Such an analysis may determine that a lower level of protection can be justified relative to materials that produce higher values of P_{max} and K_{St}. In this study, for specific dust collectors and a building compartment, it was found that dusts with sufficiently low K_{St} and P_{max} values may not be able to produce damaging pressure given existing openings on the enclosures. However, even if the pressure of the deflagration of an MED does not damage an enclosure, fireballs can create a thermal hazard to personnel and property.

References

[1] ISO 6184-1: 1985, *Explosion Protection Systems – Part 1: Determination of Explosion Indices of Combustible Dusts in Air.*

[2] ASTM E1226-10, *Standard Test Method for Explosibility of Dust Clouds.*

[3] Bartknecht, W. Dust Explosions: Course, Prevention, Protection. Berlin: Springer, 1989.

[4] Siwek, R., *20-L Laborapparatur für die Bestimmung der Explosionskenngrößen Brenbarrer Stäube*, Ph.D. thesis from the Technical University of Winterthur, Switzerland.

[5] Britton, L.G., and Chippett, "Practical aspects of dust deflagration testing", *Journal of Loss Prevention in the Process Industries,* 2, 1989, 161-169.

[6] Proust, C., Accorsi, A., and Dupont, L., "Measuring the violence of dust explosions with the '20l sphere' and with the standard 'ISO 1m3 chamber': Systematic comparison and analysis of the discrepancies," *Journal of Loss Prevention in the Process Industries*, 20, 2007, 599-606.

[7] Rodgers, S.A.,Ural, E.A., "Practical issues with marginally explosible dusts – evaluating the real hazard", Process Safety Progress, 30(3), 2011, 266-279.

[8] Going, J., Chatrathi, K., and Cashdollar, K., "Flammability limit measurements for dusts in the 20-L and 1-m3 chambers," *Journal of Loss Prevention in the Process Industries* 13, 2000, 209-219.

[9] Dastidar, A. and Amyotte, P., "Determination of Minimum Inerting Concentrations for Combustible Dusts in a Laboratory-Scale Chamber," *Process Safety and Environmental Protection* 80, 2002, 287-297.

[10] *Case Study: Hoeganaes Corporation: Gallatin, TN – Metal Dust Flash Fires and Hydrogen Explosion*, published by CSB in December 2011.

[11] Siwek, R., "Experimental Methods for the Determination of Explosion Characteristics of Combustible Dusts," *Proceedings of the 3rd International Symposium on Loss Prevention in the Process Industries*, 1980, Vol. III, 839-850.

[12] Cesana, C. and Siwek, R., Manual for Kühner AG 20-L Apparatus, p. 6.

Influence of Liquid and Vapourized Solvents on Explosibility of Pharmaceutical Excipient Dusts

Paul Amyotte
Dalhousie University, Halifax, NS, Canada
paul.amyotte@dal.ca

Ashok Dastidar,[a] Faisal Khan,[b] Rolf Eckhoff,[c]
Md Nur Hossain,[d] Kalie Symington,[d] Vincent Boilard[d] & Meftah Abuswer[d]
[a]Fauske & Associates, Burr Ridge, IL, United States
[b]Memorial University, St. John's, NL, Canada
[c]University of Bergen, Bergen, Norway
[d]Dalhousie University, Halifax, NS, Canada

Prepared for Presentation at
American Institute of Chemical Engineers
2013 Spring Meeting
9th Global Congress on Process Safety
San Antonio, Texas
April 28 – May 1, 2013

UNPUBLISHED

Influence of Liquid and Vapourized Solvents on Explosibility of Pharmaceutical Excipient Dusts

Paul Amyotte
Dalhousie University, Halifax, NS, Canada
paul.amyotte@dal.ca

Ashok Dastidar,[a] Faisal Khan,[b] Rolf Eckhoff,[c]
Md Nur Hossain,[d] Kalie Symington,[d] Vincent Boilard[d] & Meftah Abuswer[d]
[a]Fauske & Associates, Burr Ridge, IL, United States
[b]Memorial University, St. John's, NL, Canada
[c]University of Bergen, Bergen, Norway
[d]Dalhousie University, Halifax, NS, Canada

Keywords: dust explosion, hybrid mixture, solvent admixture, pre-wetting, excipient dust, MCC, microcrystalline cellulose, lactose, explosion severity, explosion likelihood

Abstract

Hybrid mixtures of a combustible dust and flammable gas are found in many industrial processes. Such fuel systems are often encountered in the pharmaceutical industry when excipient (non-active ingredient) powders undergo transfer in either a dry or solvent pre-wetted state into an environment possibly containing a flammable gas.

The research described in this paper simulated the conditions of the above scenarios with microcrystalline cellulose (MCC) and lactose as excipients, and methanol, ethanol and isopropanol as solvents. Standardized dust explosibility test equipment (Siwek 20-L explosion chamber, MIKE 3 apparatus and BAM oven) and ASTM test protocols were used to determine the following explosibility parameters: maximum explosion pressure (P_{max}), size-normalized maximum rate of pressure rise (K_{St}), minimum explosible concentration (MEC), minimum ignition energy (MIE), and minimum ignition temperature (MIT).

Because the MIKE 3 apparatus and BAM oven are not closed systems, only baseline excipient-alone testing and excipient pre-wetted with solvent testing were possible for MIE and MIT determination. With the Siwek 20-L chamber (a closed system), it was feasible to conduct P_{max}, K_{St} and MEC testing for all three cases of the dust alone, pre-wetted with solvent, and with solvent admixed to the combustion atmosphere at 80 % of the lower flammability limit for each solvent prior to dust dispersal.

The experimental results demonstrate the significant enhancements in explosion likelihood and explosion severity brought about by solvent admixture in either mode. The extent of solvent influence was found to be specific to the given excipient and method of solvent addition. Solvent burning velocity considerations help to account for some of the experimental observations but for others, a more rigorous evaluation of solvent and excipient physical property data is needed.

1. INTRODUCTION

Hybrid mixtures consist of a flammable gas and a combustible dust, each of which may be present in an amount less than its lower flammable limit (LFL)/minimum explosible concentration (MEC), and still give rise to an explosible mixture [1]. The focus here is often on admixture of a flammable gas in concentrations below the LFL of the gas itself to an already explosible concentration of dust [1]. As described by Amyotte et al. [2], hybrid mixture research is typically conducted with three possible approaches: (i) gaseous solvent at room temperature existing in the combustion atmosphere prior to dust dispersal, (ii) liquid solvent at room temperature requiring flashing-off for admixture to the combustion atmosphere prior to dust dispersal, and (iii) liquid solvent at room temperature admixed as a liquid with the dust prior to dust dispersal.

Amyotte and Eckhoff [1] note that the influence of the co-presence of a flammable gas on the explosibility parameters of a fuel dust alone is well-established. These effects include higher values of explosion overpressure and rate of pressure rise, and lower values of minimum explosible concentration and minimum ignition energy [1]. There remains, however, a need for continued research on hybrid mixtures [3] given the range and diversity of industrial applications that can give rise to hybrid fuel systems as seen in recent studies [4-11].

The scope of the current work is the prevention and mitigation of explosions of hybrid mixtures consisting of a combustible dust and a flammable gas, or a combustible dust pre-wetted with a flammable solvent (i.e., the latter two scenarios described in the first paragraph above). This research is motivated by the occurrence of these scenarios in the pharmaceutical industry during transfer of dry or solvent-laden powders into a process unit that may contain a flammable gas [12]. The objective is the provision of explosion likelihood and explosion severity data acquired through best-practice testing methodologies using standardized apparatus. The specific hybrid fuel systems were selected to be representative of common pharmaceutical excipients (non-active ingredients) and solvents; the testing thus involved lactose and microcrystalline cellulose (MCC) dusts admixed with methanol (methyl alcohol), ethanol (ethyl alcohol) and isopropanol (isopropyl alcohol) solvents.

2. MATERIALS, APPARATUS AND PROCEDURES

All materials tested were pharmaceutical-grade in terms of composition and, in the case of the solids, particle size distribution (PSD) as-received from the suppliers. Tables 1 and 2 summarize the relevant material characteristics of the dusts (excipients) and solvents, respectively. Sieve analysis was used for the lactose PSD determination because of initial concerns about lactose solubility with Malvern light scattering analysis (which was performed for the MCC). The PSD results are consistent with the trend of the nominal mean diameters of 50 μm and 75 μm provided by the suppliers of the MCC and lactose samples, respectively.

Figures 1 and 2 show scanning electron micrographs (SEMs) of the MCC and lactose samples, respectively. MCC is observed to be fibrous or flocculent in nature, while lactose consists of irregularly-shaped, oblong particles.

Table 1 Material characterization of excipient powders.

Characteristic	MCC	Lactose
Supplier	Sigma-Aldrich	Hilmar Ingredients
Particle Size [weight %]	Malvern Analysis: 90 % < 56 μm 50 % < 27 μm 10 % < 9 μm	Sieve Analysis: 98 % < 250 μm 84 % < 150 μm 62 % < 89 μm 18 % < 75 μm
Moisture Content [weight %]	4.5	5.1

Table 2 Material characterization of solvents.

Characteristic	Methanol	Ethanol	Isopropanol
Formula[a]	CH_3OH	C_2H_5OH	C_3H_7OH
Molecular weight[a]	32	46	60
Lower flammability limit [volume %][b]	6.7	3.3	2.2
Laminar burning velocity [cm/s] (Methanol,[c] Ethanol,[d] Isopropanol[c])	56	42	41
Vapour pressure at 25 °C [mm Hg][a]	127	59	43
Specific heat capacity (liquid) at 25 °C [J/mol·K][e]	81	112	155
Boiling point at 1 atm [°C][a]	64.7	78.5	82.2
Heat of vapourization at boiling point and 1 atm [kJ/mol] (Methanol,[a] Ethanol,[a] Isopropanol[e])	35.2	38.5	39.9
Heat of combustion (liquid) at 25 °C and 1 atm with $H_2O(l)$ product [kJ/mol][a]	–726.6	–1366.9	–1986.6
Specific gravity [20°C/4°C][a]	0.792	0.789	0.785

[a]Felder, R.M. and Rousseau, R.W., "Elementary Principles of Chemical Processes (3rd edition", John Wiley & Sons, Inc., Hoboken, NJ (2005).
[b]Kuchta, J.M., "Investigation of Fire and Explosion Accidents in the Chemical, Mining, and Fuel-Related Industries: A Manual (Appendix A)", Bureau of Mines, US Department of the Interior, Avondale, MD (1985).
[c]NFPA, "NFPA 68: Standard on Explosion Protection by Deflagration Venting", National Fire Protection Association, Quincy, MA (2007).
[d]Lipzig, J.P.J., Nilsson, E.J.K., Goey, L.P.H. and Konnov, A.A., "Laminar Burning Velocities of n-Heptane, Iso-octane, Ethanol and their Binary Mixtures, Fuel, **90**, 2773-2781 (2011).
[e]Murphy, R.M., "Introduction to Chemical Processes. Principles, Analysis, Synthesis", McGraw-Hill, New York, NY (2007).

(a) (b)

Figure 1 Scanning electron micrograph of MCC powder: (a) 250 magnification, (b) 600 magnification.

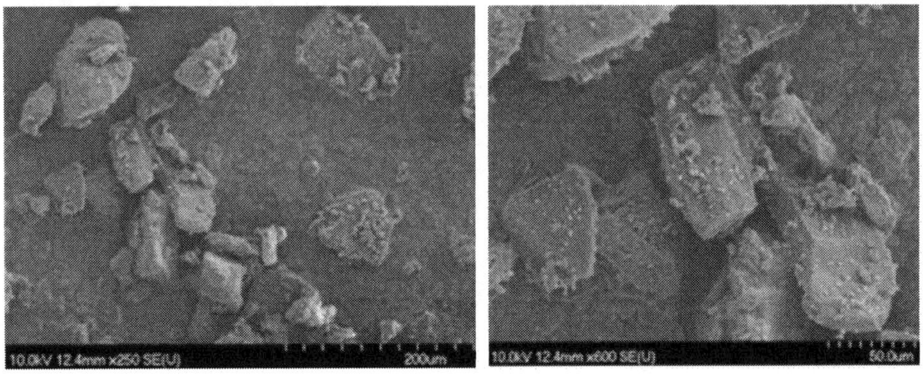

(a) (b)

Figure 2 Scanning electron micrograph of lactose powder: (a) 250 magnification, (b) 600 magnification.

Explosibility parameters investigated include maximum explosion pressure (P_{max}), size-normalized maximum rate of pressure rise (K_{St}), minimum explosible concentration (MEC), minimum ignition energy (MIE), and minimum ignition temperature (MIT). ASTM (American Society for Testing and Materials) protocols [13-16] were followed using standardized dust explosibility test equipment: (i) Siwek 20-L explosion chamber for P_{max}, K_{St} and MEC, (ii) MIKE 3 apparatus for MIE, and (iii) BAM oven for MIT. Apparatus and procedural descriptions can be found on the equipment manufacturer's web site (www.kuhner.com) [2].

All hybrid mixture testing involved a fixed solvent concentration of 80 % of the respective lower flammability limit (Table 2), calculated according to the volume of the particular test apparatus. For the pre-wetted (PW) tests, the required amount of liquid solvent – methanol (M), ethanol (E) or isopropnaol (IPA) – was mixed with the amount of dust corresponding to the dust concentration being tested. The pre-wetted dust was then dispersed into the specific apparatus (Siwek 20-L chamber, MIKE 3 apparatus or

BAM oven) via the usual procedure. For the atmospheric (ATM) tests (Siwek 20-L chamber only), the chamber was first evacuated to a pressure of 185 mm Hg (i.e., as close as possible to the respective solvent vapour pressures given in Table 2). The required amount of liquid solvent was then injected through a septum into the 20-L chamber, with the majority of the solvent flashing to vapour. The chamber was subsequently backfilled with air to a pre-dispersion pressure such that the chamber pressure at the time of dust ignition was approximately 1 bar. From a material balance perspective, any small amount of remaining liquid solvent would be vapourized by the shower of sparks originating from the chemical ignitors acting as the ignition source in the 20-L chamber [6].

3. RESULTS AND DISCUSSION

In this section we present the experimental results according to severity of explosion consequences (overpressure and rate of pressure rise) and likelihood of explosion occurrence (minimum explosible concentration, ignition energy and ignition temperature). Preliminary data analysis drawing on Amyotte et al. [2] is presented, with further data interpretation ongoing.

3.1 Explosion Severity

Figures 3 and 4 show the influence of dust concentration on explosion overpressure, P_m, and rate of pressure rise, $(dP/dt)_m$, for the excipient dusts by themselves. (All figures in this section give average values of P_m and $(dP/dt)_m$ at dust concentrations for which replicate testing was performed according to ASTM E-1226-10 [13].)

The data in Figures 3 and 4 display the expected trend of an increase in the measured explosibility parameter as dust concentration increases. Eventually, peak values of P_m and $(dP/dt)_m$ are attained, followed by a parameter decrease or leveling-off with further increases in dust concentration. The higher peak values and lower optimum concentrations for MCC over lactose are indicative of both compositional differences between the two materials and the smaller particle size of the MCC (Table 1).

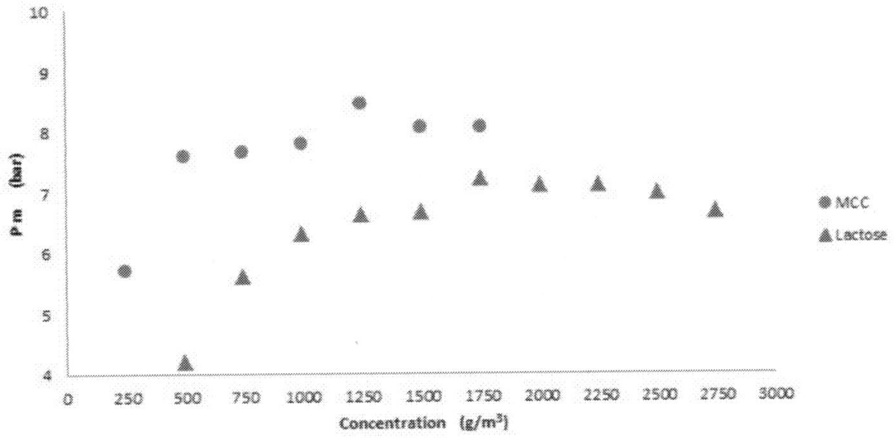

Figure 3 Influence of dust concentration on explosion overpressure of MCC and lactose (baseline excipient alone).

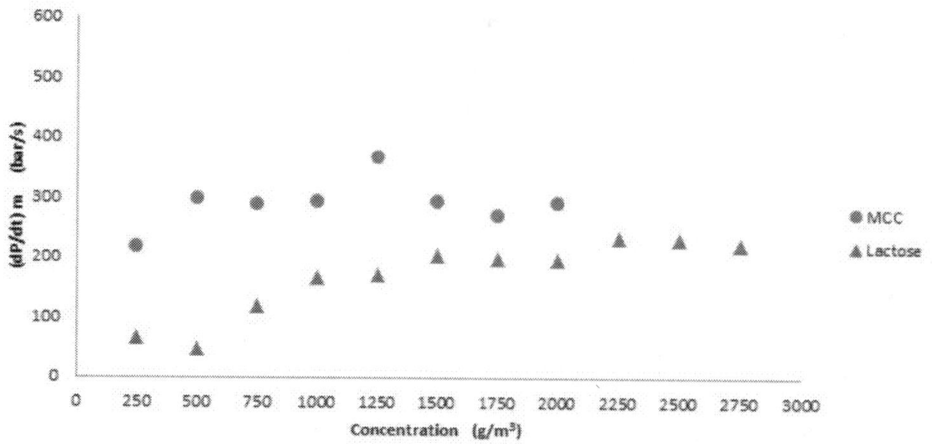

Figure 4 Influence of dust concentration on rate of pressure rise of MCC and lactose (baseline excipient alone).

Tables 3 and 4 give the complete P_{max} and K_{St} data sets for all test conditions.

Table 3 P_{max} and K_{St} data for MCC.

Material	**P_{max} [bar(g)]**	**K_{St} [bar·m/s]**
MCC	8.5	103
MCC + M (PW)	7.9	144
MCC + E (PW)	7.8	117
MCC + IPA (PW)	7.7	116
MCC + M (ATM)	7.9	168
MCC + E (ATM)	8.3	149
MCC + IPA (ATM)	8.4	172

Table 4 P_{max} and K_{St} data for lactose.

Material	**P_{max} [bar(g)]**	**K_{St} [bar·m/s]**
Lactose	7.1	65
Lactose + M (PW)	8.1	149
Lactose + E (PW)	8.4	148
Lactose + IPA (PW)	8.6	144
Lactose + M (ATM)	8.0	155
Lactose + E (ATM)	7.4	94
Lactose + IPA (ATM)	7.8	102

As preliminary commentary [2], we offer the following thoughts on the trends displayed by the data in Tables 3 and 4. In all cases, pre-wetting (PW) of MCC and lactose with solvent had a measurable impact on both P_{max} and K_{St}. As expected, the influence was generally an enhancement of each explosibility parameter; the lone exception was P_{max} for MCC which displayed a decrease of 0.6-0.8 bar(g) with solvent admixture by pre-wetting.

While the magnitude of the effect on K_{St} of solvent pre-wetting for MCC was generally distinguishable for the different solvents, this was not the case for lactose. Pre-wetting of lactose with each of the three solvents resulted in similar K_{St} values. This suggests an approximate correlation of K_{St} with burning velocity (Table 2) for pre-wetted MCC but not for pre-wetted lactose. Such a correlation was previously shown to hold in the atmospheric-type tests conducted by Amyotte et al. [6] for polyethylene admixed with various hydrocarbons.

The atmospheric (ATM) test data in Tables 3 and 4 show a reversal of the above trend for the two excipients. Here, the admixed solvent has generally the same effect on K_{St} of MCC regardless of the nature of the solvent. On the other hand, the lactose K_{St} values can be approximately ranked according to solvent burning velocity.

These observations are somewhat speculative, and it is likely that some of the other solvent physical properties shown in Table 3 (and perhaps others related to solubility) will be required to advance the phenomenological modeling of these data. What seems clear at present is that the influence of each solvent is specific to the particular excipient and the method of admixture (pre-wetting or atmospheric). This is clearly demonstrated by Figures 5 and 6 which display overpressure data for the lactose/ethanol and MCC/methanol systems, respectively.

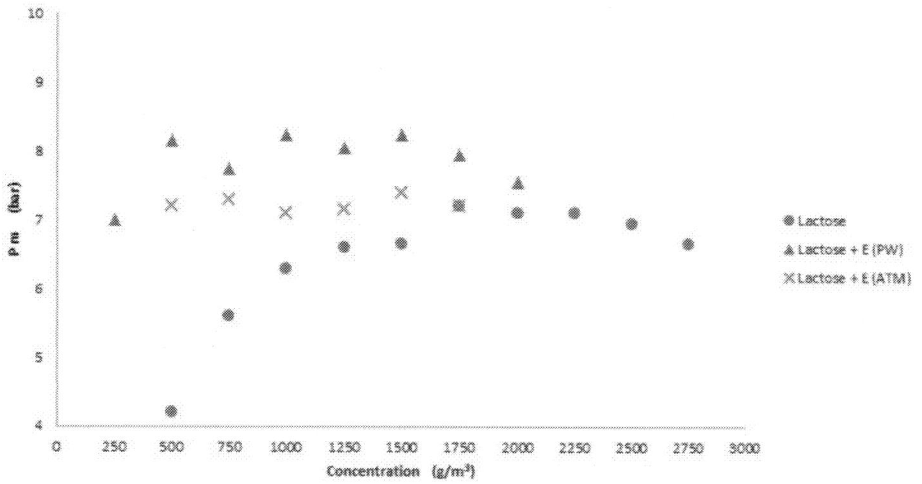

Figure 5 Influence of ethanol admixture on explosion overpressure of lactose.

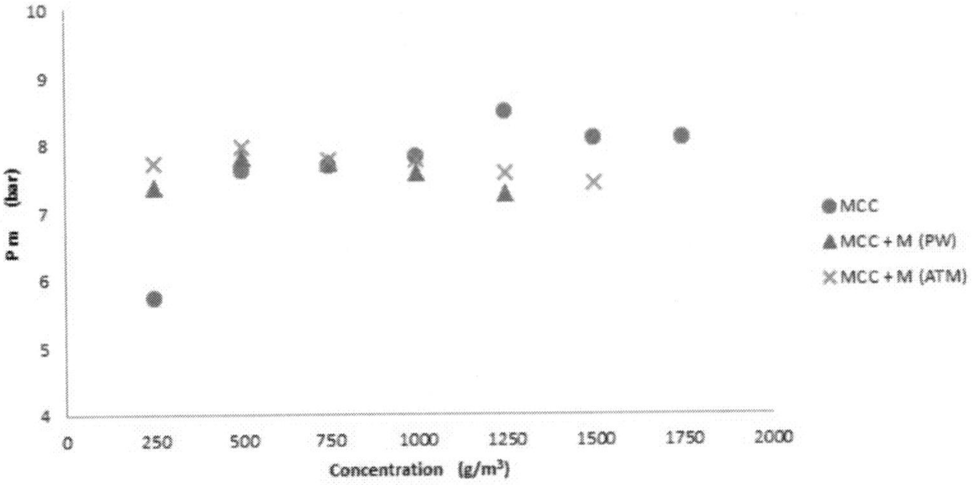

Figure 6 Influence of methanol admixture on explosion overpressure of MCC.

Similarly, Figures 7, 8 and 9 give rate of pressure rise data that further demonstrate the excipient- and admixture-specific nature of the influence of a given solvent. The systems shown are lactose/isopropanol, lactose/methanol and MCC/methanol, respectively.

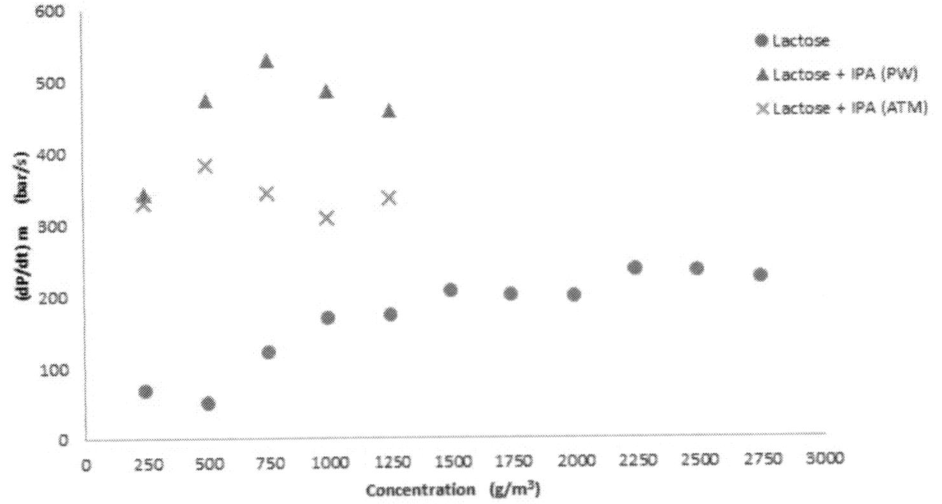

Figure 7 Influence of isopropanol admixture on rate of pressure rise of lactose.

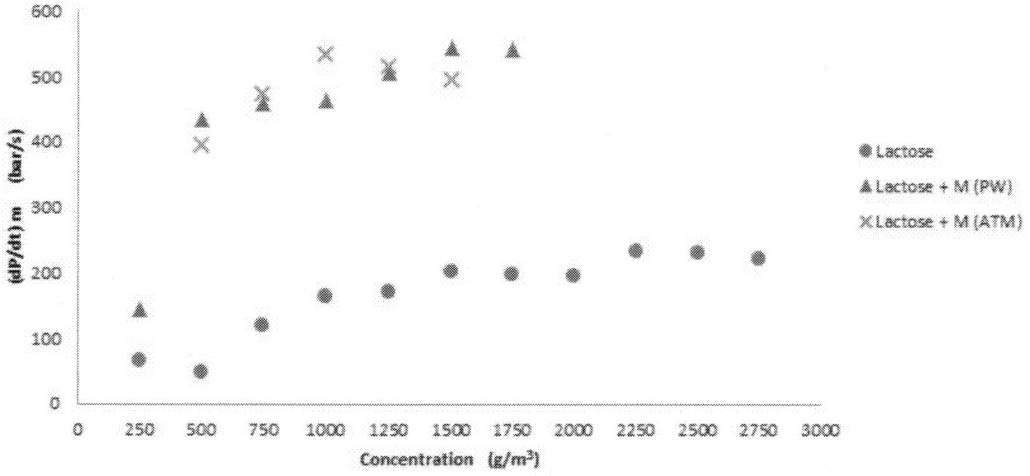

Figure 8 Influence of methanol admixture on rate of pressure rise of lactose.

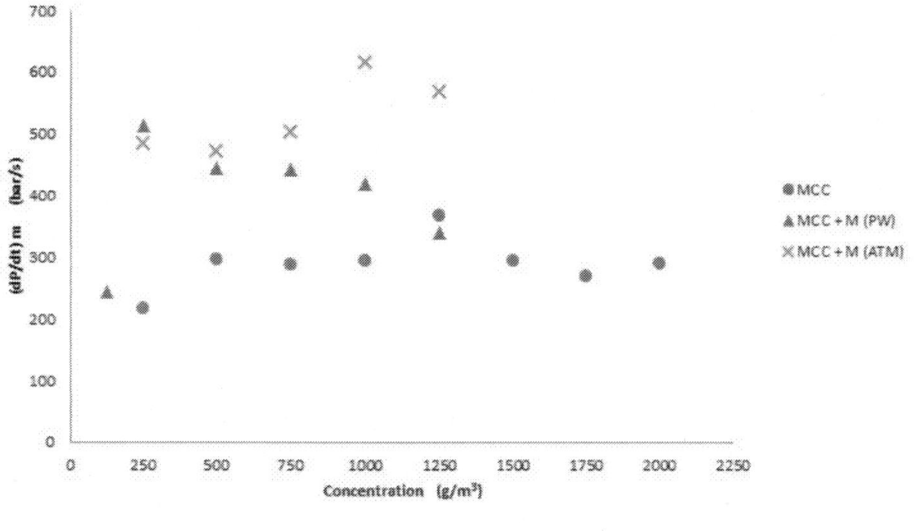

Figure 9 Influence of methanol admixture on rate of pressure rise of MCC.

3.2 Explosion Likelihood

Tables 5 and 6 give the complete MEC, MIE and MIT data sets for all test conditions. As with P_{max} and K_{St}, the influence of solvent admixture was generally an enhancement of these explosion likelihood (or ignition sensitivity) parameters – i.e., a reduction in MEC, MIE and MIT. Consistent with the pre-wetted lactose K_{St} values in Table 4, the pre-wetted lactose MIE values are all similar. The effect of inductance via the production of a protracted spark leading to lower MIEs is also seen in Tables 5 and 6.

Table 5 MEC, MIE and MIT data for MCC.

Material	**MEC [g/m³]**	**MIE [mJ] (Inductance)**	**MIE [mJ] (No Inductance)**	**MIT [°C]**
MCC	50	30–100[c] (74)[d]	300–1000 (540)	430
MCC + M (PW)	DL[a]	30–100 (55)	30–100 (67)	380
MCC + E (PW)	DL	10–30 (27)	300–1000 (380)	410
MCC + IPA (PW)	40	30–100 (42)	100–300 (180)	400
MCC + M (ATM)	< 10 (2.6 bar(g))[b]	ND[e]	ND	ND
MCC + E (ATM)	< 10 (1.5 bar(g))	ND	ND	ND
MCC + IPA (ATM)	< 10 (3.6 bar(g))	ND	ND	ND

[a]DL = Dispersion Limitation. The excipient dissolved in the admixed solvent to the extent that dust dispersion was not possible.
[b]Explosion overpressure at dust concentration of 10 g/m³. The explosion criterion is an overpressure of 1 bar(g).
[c]Range of ignition energies from lower value at which no ignition occurred to higher value at which ignition did occur.
[d]E_s (statistic energy) determined by manufacturer (Kuhner)-supplied software.
[e]ND = Not Determined. The MIKE 3 apparatus used for MIE measurement and the BAM oven used for MIT measurement are not closed systems (unlike the Siwek 20-L chamber used for determination of MEC).

Table 6 MEC, MIE and MIT data for lactose. (Same footnotes as Table 6.)

Material	**MEC [g/m³]**	**MIE [mJ] (Inductance)**	**MIE [mJ] (No Inductance)**	**MIT [°C]**
Lactose	70	30–100 (55)	100–300 (250)	420
Lactose + M (PW)	DL	10–30 (17)	100–300 (140)	350
Lactose + E (PW)	DL	10–30 (19)	100–300 (200)	380
Lactose + IPA (PW)	DL	10–30 (14)	100–300 (170)	400
Lactose + M (ATM)	< 10 (4.3 bar(g))	ND	ND	ND
Lactose + E (ATM)	< 10 (2.6 bar(g))	ND	ND	ND
Lactose + IPA (ATM)	< 10 (3.4 bar(g))	ND	ND	ND

4. CONCLUSION

The current work has provided an additional example of how common pharmaceutical solvents such as methanol, ethanol and isopropanol can significantly increase the explosion likelihood and explosion severity of common pharmaceutical excipients such as microcrystalline cellulose and lactose. These explosion enhancement effects have been demonstrated for two modes of solvent admixture – pre-wetting of the excipient powder and direct addition to the combustion atmosphere. The magnitude of the solvent influence on basic explosibility parameters (P_{max}, K_{St}, MEC, MIE and MIT) is dependent on both the nature of the excipient and the method of admixture. Further analysis is underway to facilitate better understanding of the observed phenomena.

5. ACKNOWLEDGEMENT

The authors gratefully acknowledge the financial support of the Natural Sciences and Engineering Research Council (NSERC) of Canada in the form of a strategic grant. We also appreciate the assistance of Jeff Dinyer of Patheon Inc. in helping us select the hybrid fuel systems for study.

6. REFERENCES

[1] Amyotte, P.R. and Eckhoff, R.K., "Dust Explosion Causation, Prevention and Mitigation: An Overview", Journal of Chemical Health and Safety, **17** (1), 15-28 (2010).

[2] Amyotte, P.Khan, F., Boilard, S., Iarossi, I., Cloney, C., Dastidar, A., Eckhoff, R., Marmo, L. and Ripley, R., "Explosibility of Nontradtional Dusts: Experimental and Modeling Challenges", Hazards XXIII, IChemE Symposium Series No. 158, Southport, UK (November 12-15, 2012), pp. 83-90.

[3] Worsfold, S.M., Amyotte, P.R., Khan, F.I., Dastidar, A.G., Eckhoff, R.K., "Review of the Explosibility of Nontraditional Dusts", Industrial & Engineering Chemistry Research, **51**, 7651-7655 (2012).

[4] Garcia-Agreda, A., Di Benedetto, A., Russo, P., Salzano, E. and Sanchirico, R., "Dust/Gas Mixtures Explosion Regimes", Powder Technology, **205**, 81-86 (2011).

[5] Sanchirico, R., Di Benedetto, A., Garcia-Agreda, A. and Russo, P., "Study of the Severity of Hybrid Mixture Explosions and Comparison to Pure Dust-Air and Vapour-Air Explosions", Journal of Loss Prevention in the Process Industries, **24**, 648-655 (2011).

[6] Amyotte, P., Lindsay, M., Domaratzki, R., Marchand, N., Di Benedetto, A., Russo, P., "Prevention and Mitigation of Dust and Hybrid Mixture Explosions", Process Safety Progress, **29**, 17– 21 (2010).

[7] Perry, A.J., Ozog, H., Murphy, M., Stickles, R.P., "Conducting Process Hazard Analyses for Dust-Handling Operations", Chemical Engineering Progress, 28-35 (February 2009).

[8] Dufaud, O., Perrin, L., Traore, M., Chazelet, S., Thomas, D., "Explosions of Vapour/Dust Hybrid Mixtures: A Particular Class", Powder Technology, **190**, 269-273 (2009).

[9] Dufaud, O., Perrin, L., Traoré, M., "Dust/Vapour Explosions: Hybrid Behaviours?", Journal of Loss Prevention in the Process Industries, **21**, 481-484 (2008).

[10] Pilao, R., Ramalho, E., Pinho, C., "Overall Characterization of Cork Dust Explosion", Journal of Hazardous Materials, **133**, 183-195 (2006).

[11] Pilao, R.; Ramalho, E.; Pinho, C., "Explosibility of Cork Dust in Methane/Air Mixtures", Journal of Loss Prevention in the Process Industries", **19**, 17-23 (2006).

[12] Glor, M., "A Synopsis of Explosion Hazards During the Transfer of Powders into Flammable Solvents and Explosion Prevention Measures", Pharmaceutical Engineering, **30** (1), 1-8 (2010).

[13] ASTM, "ASTM E1226-10 – Standard Test Method for Explosibility of Dust Clouds", American Society for Testing and Materials, West Conshohocken, PA (2011).

[14] ASTM, "ASTM E1515-07 – Standard Test Method for Minimum Explosible Concentration of Combustible Dusts", American Society for Testing and Materials, West Conshohocken, PA (2011).

[15] ASTM "ASTM E1491-06 – Standard Test Method for Minimum Autoignition Temperature of Dust Clouds", American Society for Testing and Materials, West Conshohocken, PA (2011).

[16] ASTM, "ASTM E2019-03 – Standard Test Method for Minimum Ignition Energy of a Dust Cloud in Air", American Society for Testing and Materials, West Conshohocken, PA (2011).

9TH GLOBAL CONGRESS ON PROCESS SAFETY

A Systematic Approach to Relief and Flare Systems Evaluation

Georges A. Melhem, Ph.D.
ioMosaic Corporation
93 Stiles Road | Salem, NH 03079
melhem@iomosaic.com

Prepared for Presentation at
American Institute of Chemical Engineers
2013 Spring Meeting
9th Global Congress on Process Safety
San Antonio, Texas
April 28 – May 1, 2013

UNPUBLISHED

A Systematic Approach to Relief and Flare Systems Evaluation

Georges A. Melhem, Ph.D.
ioMosaic Corporation
2401 Fountain View Drive, Houston, TX
melhem@iomosaic.com

Keywords: Relief Systems, Flare Systems, Emergency Relief Systems, Workflow, Reaction Systems, Depressuring Systems

Abstract

The process industries are primarily concerned with the reliability, availability, auditability and maintainability of relief and flare systems data. This data is a critical component of process safety information (PSI) and its lifecycle must be properly managed to ensure sound process safety management (PSM) and loss prevention programs.

For most large facilities, the process of managing the lifecycle of relief and flare systems data is complex and fraught with challenges and risks, whether the work is performed internally or contracted out. For existing large facilities the process of relief and flare systems evaluations require mechanical and process data collection, field verification, up to date heat and material balances, information about process safeguards, scenario identification, establishing relief requirements, identification and risk ranking of deficiencies, and managing the corrective actions process for addressing deficiencies where applicable.

Reliability is influenced by many technical and human factors including the quality of data, adequacy of tools used for analysis, the qualifications of the relief systems engineers performing the scenario identification, and relief and flare systems evaluations.

Availability primarily deals with how quickly can one access accurate and up to date relief and flare systems data. This is especially challenging since relief systems data is not all "structured" data and is interconnected with other engineering data systems.

Auditability involves version control and the management of revisions and/or modifications of relief and flare systems that typically result from plant/process modifications, process hazard analysis, incident investigations, etc.

Maintainability requires keeping the relief and flare systems data forever green and enabling efficient reviews and revisions.

This paper describes a systematic web-based workflow methodology for managing the lifecycle of relief and flare systems data for a single site or at a corporate level. The workflow methodology breaks the flare and relief systems data lifecycle into discrete components and activities, with built-in review, approval, quality management, and reporting. Built-in business and engineering rules ensure that all activities can only progress when specific quality criteria are met.

This system was developed based on our experience with the execution of many such large scale projects for refineries, chemical, and petrochemical facilities.

1. Introduction and Relief Systems Regulatory Requirements in the US

Any process dealing with relief systems design and/or evaluation has to comply with regulatory requirements and good engineering practice. In the US, the process safety information element of the OSHA PSM regulation requires a PSM covered facility to demonstrate that:

- Relief systems comply with recognized and generally accepted good engineering practice (RAGAGEP) 29 CFR 1910.119 (d)(3)(ii), and
- that the relief systems design and design basis information is properly documented 29 CFR 1910.119 (d)(3)(i)(D).

Under section 29 CFR 1910.119 (d)(3)(iii): "For existing equipment designed and constructed in accordance with codes, standards, or practices that are no longer in general use, the employer shall determine and document that the equipment is designed, maintained, inspected, tested, and operating in a safe manner". This is especially important for older facilities that were constructed well before the PSM regulation was enacted.

OSHA's definition of equipment is broad as defined in the mechanical integrity element and includes relief and vent systems and devices. This also creates additional requirements for relief and flare systems testing and inspection such as under (j)(4) as well as the requirement to address mechanical integrity deficiencies in a timely manner:

- CFR 1910.119 (j)(4)(ii): "Inspection and testing procedures shall follow recognized and generally accepted good engineering practices".

The PSM requirements needed to be satisfied by May 26, 1997.

Corporate standards, policies, best practices, and procedures qualify as in-house RAGAGEP for company facilities worldwide, especially where the corporate standard is more restrictive. Whether the facilities have their own procedures or follow corporate procedures or both, it is clear that OSHA expects that the facility will comply ("*do what they have committed to do*") with these procedures to satisfy PSM requirements regardless of how they are classified. More importantly, most corporate standards include by reference other standards and codes that they designate as RAGAGEP.

The term RAGAGEP encompasses a wide body of industry codes and standards. Moreover, RAGAGEP defines the standard of care expected of companies by regulatory agencies, government, and society in operating chemical manufacturing (and other) businesses [1].

RAGAGEPs will change with time as new standards are developed and existing standards are updated. This allows the PSM standard to stay current with RAGAGEPs without necessitating a change in the regulation. In many cases, the OSHA PSM standard has in fact resulted in many RAGAGEP s to be developed or updated.

Considerable effort is required to stay in compliance with a performance-oriented standard like the OSHA PSM regulation. Compliance requires (1) a good understanding of the RAGAGEPs that apply to the facility, (2) that the PSM program procedures incorporate these requirements, (3) constant vigilance to stay up to date on changes to RAGAGEPs since many industry standards are reissued every few years, (4) and most importantly an effective compliance audit program to verify that procedures include all applicable RAGAGEPs and that the procedures are being implemented as written.

On June 7, 2007 OSHA initiated a Petroleum Refinery Process Safety Management National Emphasis Program (NEP). The PSM covered Chemical facilities NEP program was effective on July 27, 2009. The NEP approach provided a particular set of requirements to be addressed during inspections including review of documents, interviews of employees, and verification of implementation for specific processes, equipment and procedures.

The NEP program most frequently cited items included mechanical integrity (MI), process safety information (PSI), operating procedures (OP), process hazard analysis (PHA) and management of change (MOC). The Petroleum Refinery NEP program focused heavily on relief and flare systems.

The OSHA NEP program citations statistics confirm that PSI compliance is lacking. Missing or incomplete PSI also cause deficiencies in MI and other PSM elements since PSI, MI, and MOC PSM compliance are highly interrelated. A significant component of PSI is relief and flare systems design and design basis documentation.

2. Relief Systems PSI Requirements

Relief systems are critical chemical process safeguards that protect from catastrophic vessel failures (bursts) due to overpressure and/or underpressure. They are often referred to as the "last line of defense" and are expected to function properly when all other safeguards fail to mitigate the scenarios causing overpressure or underpressure. Relief systems should not be the only line of defense or layer of safeguarding. Relief systems should be one of the protection layers, and preferably the last layer that will be called upon when all other layers fail to address deviations from normal operation.

A relief system includes both relief devices and vents containment / flare equipment. A relief system includes the relief device, the inlet line, the discharge line, the common header, the catch tank, etc.

A design basis is established by examining all possible scenarios (or contingencies) that can lead to overpressure and/or underpressure of the equipment protected by the relief system. The scenario that leads to the largest relief requirement is typically chosen as the "design basis". Proper scenario development and analysis for relief systems design requires substantial process safety information as shown by Table 1. Establishing the design basis for one vessel protected by one or more relief devices is a similar exercise to performing a PHA on that one vessel.

The documentation of "design and design basis" has to include much of the information provided in Table 1 in addition to the calculation details in order to comply with numerous published RAGAGEP requirements by NFPA, ASME, API, CCPS, AIChE/DIERS, ISA, ANSI, etc. Note that justification needs to be provided for why scenarios apply or for why they do not apply.

Table 1. Typical Process Safety Information Required for Establishing and Documenting Relief and Flare Systems Design and Design Basis [1]

Process Design and Description	Piping and Instrumentation Diagrams (P&ID) Heat and Material Balances (H&MB) Process Flow Diagrams (PFD) Process Safety Flow Diagrams (PSFD) Process Descriptions / Operating Procedures Plot Plans / Elevation Plans	
Utility and Piping Design	Utility operating conditions (electrical, instrument air, cooling water, steam, etc.) Electrical one-line diagrams Piping designations and ratings Insulation designations and ratings	
		Typical Data Source
Fluid and Mixture Properties	Thermo-Physical Properties	1. Properties Databases such as DIPPR and SuperChems 2. Company generated data 3. Estimates based on structure
	Chemical Reactivity and Reaction Kinetics	1. MSDS (starting point) 2. Client **adiabatic** calorimetry data 3. Open literature data 4. Externally generated adiabatic calorimetry data
Pressure Relief Devices	Manufacturer / model number Inlet / outlet / discharge area sizes Opening pressure and temperatures	Relief Device Information: 1. Maintenance records 2. Relief device specification sheets 3. Original design basis 4. P&ID 5. Valve Tag Inlet / Outlet Piping Details: 1. Existing isometric drawings 2. Field sketches
Fixed Process Equipment (General)	MAWP, MAWT, and vacuum rating Design conditions Equipment Dimensions	1. U-1A forms 2. Mechanical drawings 3. Equipment specification sheets 4. Operating Manuals 5. P&IDs 6. Nameplate
Vessels	Liquid levels	1. Operating procedures 2. P&IDs 3. Equipment design drawings 4. Level alarm set-points 5. Level-gauge tapping locations (from equipment design drawings)
	Elevation	1. Equipment elevation drawings 2. Equipment arrangements drawings
	Insulation type, thickness, fire proofing status	1. Maintenance records 2. Equipment design specification 3. P&IDs

Heat Exchangers	Design type Rated and normal duty Tube ID / length	1. U-1A forms 2. Heat exchanger specification sheets 3. P&ID 4. Nameplate
Heaters / Steam boilers	Tube design pressures Furnace design duty Boiler dimensions and design duty	1. Heater / Boiler specification sheets 2. U-1 Forms 3. P&ID 4. Nameplate
Rotating Process Equipment (General)	MAWP, MAWT Design conditions	1. Equipment specification sheets 2. P&ID 3. Equipment nameplate
Centrifugal Pumps	Pump capacity curve, rated capacity, and installed impeller size Suction Conditions	1. Performance curves 2. Pump specification sheets 3. Maintenance records (installed impeller and corresponding curve) 4. P&ID 5. Nameplate
Centrifugal Compressors	Compressor capacity curve and rated capacity Suction conditions Isentropic or polytropic efficiencies	1. Performance curves 2. Compressor specification sheet 3. Original design data 4. P&ID 5. Nameplate
Positive Displacement Pumps	Pump casing MAWP / MAWT, design conditions Rated capacity	1. Pump specification sheets 2. P&ID 3. Nameplate
Reciprocating Compressors	Compressor manufacturer/model Cylinder type (double acting, etc.), diameter Stroke length, Rod diameter, Piston displacement, Engine speed, Volumetric efficiency	1. Compressor specification sheets 2. Original design specification 3. P&ID 4. Nameplate
Turbines	Exhaust casing MAWP / MAWT, design conditions, Steam throughput	1. Turbine specification sheets 2. P&ID 3. Nameplate
Control Valves	Sizes (inlet / outlet / port) Manufacturer and model number Fail safe position	1. Control valve data sheets 2. Vendor data 3. Nameplate

3. Challenges Associated with Developing and Documenting Relief and Flare Systems Design and Design Basis

Significant resources are required in order to properly establish compliance with RAGAGEP and to document the design and design basis for relief and flare systems. Compliance requires up to date information for all that is listed under the OSHA PSI element (d). For example, the P&ID data has to be provided and field verified as well as vessel design ratings.

The level of effort required for performing an analysis to establish/verify/document the design and design basis for one relief device/system will depend on the nature of the system and the state of the existing information required to perform the analysis. One has to also account for missing PSI data development or mitigation costs that would be required if deficiencies are discovered. For example, some systems will require the testing and development of chemical reaction information for undesired chemistry.

For plants constructed prior to 1992, it expected that up to 40 % of relief device installations will not meet the OSHA requirements for compliance with RAGAGEP and/or for design and design basis documentation.

Experience indicates that most relief systems challenges are similar across a variety of industries. Typical issues encountered during relief and flare systems evaluation projects include but are not limited to:

- Outdated or non-existent design basis and supporting calculations including vent containment and flare system design basis
- Non-compliance with new API-521/ISO/OSHA documentation recommendations
- Missing or outdated process and mechanical information such as material and energy balances, piping isometrics, vessel design data, and control system specifications
- Relief devices that discharge hazardous materials to atmosphere, or the counterpart: overloaded flare systems
- Ignorance of chemical reactivity and multiphase flow in existing relief calculations
- Pressure relief valve installation inadequacies such as excessive inlet pressure drop and excessive backpressure
- Improperly designed depressuring systems, particularly with high-pressure operations – examples include cold temperatures downstream of PRV and wall temperatures/failure pressures under fire exposure

In addition, there are many gray areas in the current state of the art for relief systems evaluation that further exacerbate and complicate how compliance is achieved. Technical issues that are currently the subject of debate and some controversy amongst relief systems experts including but not limited to:

- Use of actual (best estimate) flow vs. required flow for (a) inlet pressure loss, (b) backpressure, (c) sub-header/ flare header hydraulics, and (d) effluent handling equipment (knockout drums, flare tips) design
- Use of 3 % inlet pressure loss requirement vs. a more relaxed requirement for existing installations such as blowdown minus 2 %
- Fire exposure and cold temperature development for depressuring systems, especially for gas filled vessels
- Correct usage of 2-phase discharge coefficient
- Estimation of 2-phase density with slip
- Use of fire flux for dynamic simulations. Decreasing wetted surface area for all gas flow as well as use of total vessel wetted surface area for 2-phase flow
- Level of documentation that is sufficient to meet the OSHA PSI requirements

In addition to the numerous technical complexities, the most significant challenge with relief and flare systems documentation is "Management of the Data Lifecycle" and how that integrates with overall process safety management (PSM). In particular, the connectivity of non-structured data becomes very important. A relief device design and design basis record might include spreadsheets, e-mails, sketches, articles and references, in addition to structured information that may already exist in a database.

Lack of PSI, uncertainties in the actual relief systems evaluation process dealing with some of the topics above as well as the availability of competing and inconsistent RAGAGEP, make this process very challenging.

Reliability, availability, auditability, and maintainability of this critical information continue to be a challenge for most companies. The ioXpress™ / SuperChems™ server-client platform was developed to easily address these four key requirements and to facilitate site-wide and large scale multi-site relief systems studies.

4. Relief and Flare Systems Information Reliability Requirements

The quality of relief systems evaluations is heavily influenced by many technical and human elements. The qualification of engineers performing the analysis should be verified. While drafting of isometrics and data entry can be adequately performed by a non-relief system expert, scenario identification, modeling of complex dynamics, and the assessment of corrective actions will require deep expertise in process safety and relief systems.

The quality of the end result of any process simulation or relief systems assessment will also depend on the quality of thermodynamic, transport, chemical reactivity, and kinetic data available for the calculations. Vapor-liquid equilibrium data in particular is very important, especially for reaction systems.

The evaluation process should use field verified equipment ratings, relief piping isometrics, and flare network data for existing systems. For new design, the initial evaluation should be revalidated after the construction drawings are issued and also audited for compliance after construction is complete and prior to startup. This is necessary as isometrics can be changed during construction without considerations of pressure loss, elevation changes, and other factors that can impact relief device stability and structural dynamics and integrity.

Scenario identification and development is best performed by experienced process safety and relief systems engineers. Once scenarios are developed, a formal scenario review with plant operations should be conducted in order to validate the scenarios. Operating personnel in facilities can provide a wealth of knowledge regarding scenarios that have occurred in the past or more importantly near misses as well as deep insight into the control systems.

Up to date and truly representative material and energy balance data is very important, especially for relief systems scenarios that are dependent on the flow capacity of a plant or a particular unit within a plant. Note that many older plants in the US have been debottlenecked and increases in capacities have pushed operating conditions much closer to the operating design limits.

Consideration for reaction forces and vibration risk developed during relief is often poorly addressed in many systems we have reviewed. In some unique cases poorly supported and/or designed relief piping can exhibit resonance with other systems components and fail catastrophically.

There are many unique systems that require special expertise such as high pressure ethylene systems, where the pressure can be as high as 1500 bars during normal operations and pressure relief can lead to very large dynamic reaction forces as well as cold cryogenic temperatures downstream of the flow restricting devices. In general, it is very difficult to support structures that are exposed to more than 25,000 lbf of thrust during relief.

Reaction and multi-phase flow systems also require special expertise. Unlike non-reaction systems, where small uncertainties in PVT data and/or other design variables such as fire flux can be tolerated, reaction systems are very unforgiving. Reactions rates will typically double every 10 degrees. An erroneous estimate of the actual temperature at the relief device set point pressure can yield much faster reaction rates and ultimate failure due to inadequate relief. The vast majority of reaction systems will cause multiphase flow. Unlike single-phase flow, two-phase flow estimates are more uncertain and the reliability of designs involving two-phase flow are best validated by an expert. [2]

Depressuring systems represent another area where special expertise is required and where poor designs have lead to accidents. For depressuring systems, we have to worry about both hot and cold temperatures during depressurization. Cold temperatures reached in vessel or downstream (especially if dew point is reached) can reach the embrittlement temperature of typical carbon steel. High superficial vapor velocities during depressuring that can cause liquid carryover and two-phase flow, excessive noise and vibration risk. [3, 4]

Condensing gas/two-phase flow downstream can lead to more liquid accumulation and cold temperatures. Condensation can lead to the formation of liquids at their associated dew point and can cause localized stress concentrations and embrittlement. Many offshore depressuring systems are suspect to hydrate formation as small amounts of water are present in the vessel hydrocarbon contents. Hydrate formation can cause plugging.

Even under fire exposure, large temperature differences at the vapor/liquid wall interface have been observed. This is where most failures are observed to initiate for two-phase vessels engulfed in a pool fire. A vessel under fire exposure that is rapidly depressurized can still be exposed to cryogenic temperatures because of the poor free convection heat transfer characteristics from the vessel wall to the vapor space during the first few minutes as the walls have not reached a sufficiently high temperature for radiation heat transfer to become dominant between the inner metal wall surface and the vapor/liquid contents. Localized flame jet heating on vessel walls (dry) and thermal heating from ignited PRV discharge are scenarios where protection by a relief device may not be possible or practical.

Software tools are a critical component of any relief systems design/evaluation. We often use a variety of tools including process simulators, spreadsheets, and commercially available specialized software tools such as SuperChems for DIERS (marketed by AIChE/DIERS). When selecting a software tool to evaluate relief and flare systems one needs to consider the pedigree of the software, how long it has been on the market, how many users are currently using the software, how the software is supported, how the software is updated and how backward compatibility is addressed, whether the software has undergone industry wide benchmarking and review, etc. These calculation engines typically do not address data integrity requirements such version control, access history, encryption, etc. The web-based workflow system described in this paper is based on the ioXpress/SuperChems server-client platform where the data management is centralized using ioXpress and the calculations are performed by clients using SuperChems Lite, for DIERS, or Expert, as well as a variety of other tools.

5. Relief and Flare Systems Information Availability Requirements

"Availability" deals primarily with how quickly and easily one can access ALL relevant relief and flare systems data. This is a very challenging requirement since relief systems data is not all "structured data". Data access is required by the OSHA PSM regulation 29 CFR 1910.119 (c)(3) and (p)(1).

Relief systems data is often interconnected or embedded with other plant data systems and covers more than just a "specification" sheet for a relief device. This important data must be available and easily accessible to audit, incident investigation, PHA and QRA teams.

The web-based workflow platform described later in this paper addresses this requirement by providing the user with the ability to access all the interconnected relief systems data using a single point of access web form or by drilling down from a PFD and/or a PID.

6. Relief and Flare Systems Information Auditability Requirements

The integrity of relief and flare systems information is enhanced by having a system with audit functions and features. "Auditability" deals primarily with version control and tracking of access, revisions, and modifications to all relief systems data and how modifications and updates influence and/or are influenced by management of change. Changes can result from field changes, overall process and equipment capacity changes, process changes, calculation revisions, incident investigations, revisions triggered by "MOC". Note that old versions should be maintained and accessible for the entire lifecycle of the plant.

The web-based platform described in this paper incorporates version control as well as the ability to reverse changes that are rejected by the workflow in order to preserve the integrity of the data.

7. Relief and Flare Systems Information Maintainability Requirements

"Maintainability" deals primarily with keeping the relief and flare systems data forever green. Many operating companies have requirements for automatic periodic review/revisions, once every five years for example. Revisions may also be needed because of changes in RAGAGEP, corporate policies, regulatory requirements, for example blowdown systems or facility siting, PHA, Incident Investigation, or MOC driven revisions.

These revisions should be easily handled by the system and should not necessitate a complete rework of the relief systems data.

The ioXpress/SuperChems platform addresses this requirement by providing a robust and easy to use action assignment, tracking, and resolution engine. This web-based action tracking system is integrated with the workflow engine.

8. The Need for Workflow

The execution and QA/QC of large, and complex site-wide flare relief and flare systems projects can be substantially enhanced and improved by using visual workflow to manage all aspects of a relief systems project:

- Visually represent complex engineering and/or business processes
- Associate one or more data forms with each workflow node
- Associate one or more business/engineering rules with each workflow node
- Associate sub-workflows with each workflow node
- Implement notifications, approvals, rejections, delegation of authority, etc.
- Implement tracking and reporting
- Implement alerts and subscriptions
- Management oversight

The workflow requirements described in the previous sections form the foundation of the automated ioXpress/SuperChems workflow process illustrated in Figure 1. Each workflow block controls the flow of data, approvals, and quality assurance.

A relief systems project in the workflow shown in Figure 1 can be initiated for a variety of reasons including:

- Establishing a baseline or a new design
- Revalidation of relief system design basis
- PHA review item
- MOC review item
- Corrective action
- Effluent handling evaluation
- Incident investigation
- Consequence analysis and/or QRA

Figure 1. The ioXpress/SuperChems Visual Workflow Process for Site-Wide Relief and Flare Systems Data

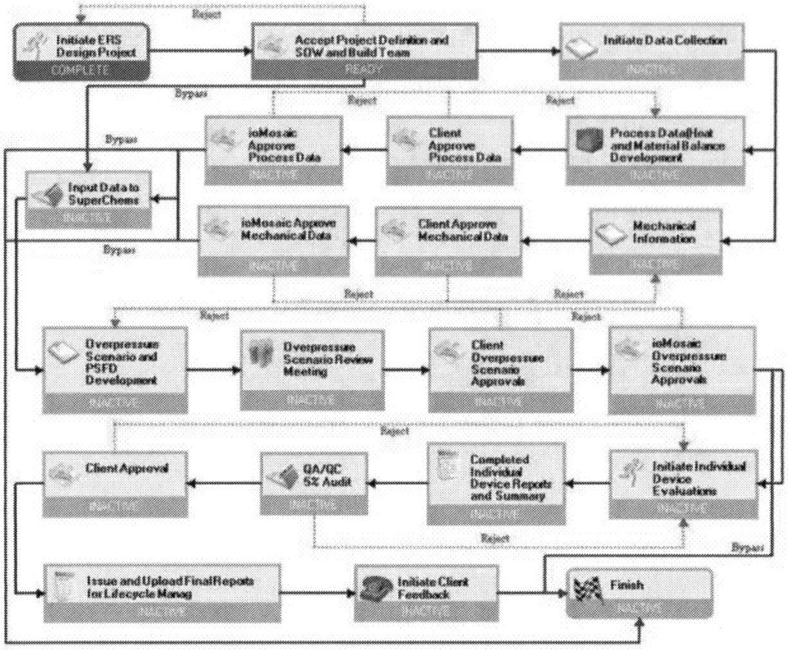

9. Overall Description of the Actual Work Process

A systematic work process for relief and flare systems evaluation was developed and implemented with the aid of electronic workflow in order the address the data quality and compliance challenges discussed above. Overall, the work process is divided into distinct phases:

Step I – Develop Overall Evaluation Guidelines and Quality Management

These guidelines are developed before any project work is started. They govern all aspects of performing the relief systems evaluation and data requirements. More importantly, they declare and summarize the applicable (RAGAGEP) to follow as well as what specific scenarios will be considered for different types of equipment and what the data hierarchy will be for performing the evaluations.

Step II – Data Collection

For existing facilities, the relief and flare systems evaluation begin with data collection, i.e. obtaining a copy of all the available data listed in Table 1. During this phase of the project, a team of engineers will go to the actual site to collect the data required for performing the relief systems evaluations. The team will field verify critical data and will draw the actual relief system isometrics for protected equipment. Prior to the data collection trip, a list of relief devices and their associated P&IDs are developed/reviewed and an understanding of the process is obtained from a process description. After the data collection is complete, data gaps are identified and missing data is developed from original or alternate/new sources.

Step III – Scenario Identification
Once all the data is obtained, vetted, and approved by all the stakeholders, the next project phase will involve scenario identification. This is a critical aspect of the relief systems evaluation. A team of senior engineers will evaluate scenarios that can cause temperature and pressure deviations for specific pieces of equipment. When the scenario development is complete, a formal scenario review is scheduled with the end user to go over the identified scenarios. The scenario list is finalized and agreed upon before the actual evaluation of relief requirements proceeds in the next step.

Step IV – Preliminary Evaluation of Relief Requirements
Using the guidelines, quality assurance (QA) procedures, site and plant data, and scenarios developed earlier, steady state and dynamic calculations of relief requirements are performed. Preliminary evaluation reports are developed. Where deficiencies exist, they are identified and recommendations for improvement are provided. These preliminary evaluations are discussed with the end user to agree upon the most practical solutions.

Step V – Final Evaluation of Relief Requirements
Final evaluations are provided based on step IV and these reports become the actual process safety management (PSM) relief systems design and design basis documentation.

Step VI – Evaluate Vent Containment and Flare Systems
If vent containment or flare systems are involved, specific scenarios considered and evaluated earlier are consolidated into global scenarios and the adequacy of the vent containment and /or flare system is evaluated. If the vent containment system is found to be deficient, recommendations for improvement are provided.

10. The Importance of Quality Assurance / Quality Control

The ioXpress/SuperChems platform described in this paper places strong emphasis on quality assurance and quality control as evident from the various levels of approval gates and requirements shown in the workflow in Figure 1.

For large scale pressure relief and effluent handling systems analyses the steps outlined below are executed using the ioXpress/SuperChems platform. Utilizing the electronic workflow requires the steps outlined below to be completed prior to advancing the workflow. This automatically ensures version and revision control in ioXpress.

1. **Dedicated Project Lead and Supervision:** Each project has a dedicated project lead to provide ongoing supervision and review during the execution of the project. The project lead has to sign off on all work products prior to submission for a final audit (step 6) after ensuring that the QA/QC procedures outlined below have been followed and implemented correctly.

2. **Formal Scenario Reviews:** Prior to commencement of any calculations or report generation, these reviews will be conducted with one or more operations process knowledgeable engineers to ensure that the correct equipment, piping, and process data is used and that all scenarios of concern have been selected and discussed [5].
3. **Automated Audit Alert SuperChems Scripts:** These scripts check the accuracy and consistency of a report against the requirements established earlier and are executed by the project design engineer to assist with validating the evaluation data prior to completing the checklist (step 4). These scripts look for missing comments, inconsistent discharge coefficients for selected devices and/or flow type, missing calculations, values being within expected normal limits, etc. The scripts ensure adherence to both client and industry standards and practices.
4. **Relief Systems Checklist:** This is a checklist that guides the design and evaluation of each relief system to ensure that no detail has been overlooked. The design engineer utilizes the checklist prior to submitting the evaluated relief device package for independent reviewer signoff.
5. **Independent Reviewer Signoff:** Each completed package will be checked and approved by a senior process safety or relief systems specialist confirming that the report is accurate and that it meets the requirements. This review focuses on completeness, accuracy and methodology.
6. **Independent Audit of a 5 % Sample:** For each project, a 5 % sample will be selected at random (but has to represent different types of equipment or operations) for detailed audit review. The process is intended to ensure that no systemic issues exist and is repeated until no further issues can be identified or until the entire pool of reports is exhausted.
7. **Electronic Version and Revision Control:** This is an automated feature of ioXpress that ensures version and revision control for all relief systems data and project files. It provides an auditable trail of all changes and revisions.
8. **Client Feedback:** Consistent with our internal quality control systems, we routinely seek and learn from client feedback. Results from the feedback are integrated in the next revisions of requirements as applicable in order to ensure continuous improvements are achieved.

11. General Recommendations for Relief Systems Engineers

Before attempting to perform relief and flare systems evaluation and design, one should have a thorough understanding of the process and all available layers of protection. Chemical systems involving reactions tend to be complex and require an assessment of both desired and undesired chemistries. Systems with supercritical fluids, depressuring systems, high pressure / low pressure interfaces, and fractionators and distillation towers will require more experience than other simpler systems. Review of operating procedures, batch records, PHA, risk assessment, and accident investigation reports can provide a lot of insight.

Develop and document all possible scenarios the can lead to overpressure, underpressure, and temperature excursions. Develop a "Protected Equipment Envelope" and identify the weakest system link. Indentify all available "Protection Systems" – Note that relief devices can fail too! It is very useful to draw a process safety flow diagram (PSFD) which shows all the available layers of protection including the relief system. Understand common mode and cascading failures and develop preferred and consistent methods for modeling specific scenarios.

Always perform a formal scenario review with plant operations and process knowledgeable engineers and/or chemists. Getting by-in from operations is always wise, especially if the evaluations reveal deficiencies that need to be corrected. Confirm potential operation deviations with plant operations before embarking on lengthy and expensive dynamic calculations. Confirm that the material and energy balance data is representative of current operating conditions. Establish and confirm all data required to model specific scenarios before modeling the scenarios.

Always field verify the data used for relief and flare systems evaluation for existing systems. This includes P&IDs, piping isometrics, equipment ratings, and relief device information.

Before attempting any calculations make sure you understand all relevant standards, laws, and RAGAGEP. Watch out for gray areas and develop philosophy documents that govern the overall evaluation efforts in order to ensure consistency. Develop standardized approaches and methods for modeling specific scenarios. If dealing with unique systems, obtain relief systems training from those who have expertise in such systems. Be prudent but not paranoid – There are always inherent risks in all kinds of technologies.

Establish relief requirements and actual flow capacities. Watch out for reaction systems, high pressure and depressuring systems. Look for computational tools that have a large installed user base and that are updated on a regular basis. Look for computational tools that can address both reactive and non-reactive chemicals and can address steady state and dynamic estimates as well as high pressure systems (as applicable). Do not use methods that are not compliant with RAGAGEP and always retain an archive copy of the actual software used to produce the evaluation.

If you identify deficiencies, risk rank the deficiencies and mitigate (where necessary) high risk contributors first. Always ensure that the final corrective action design matches field installations and /or modifications.

It is recommended that you split your documentation into three parts (see Figures 2 and 3) and connect the data using a system such as the ioXpress/SuperChems system:

- Part I – where most changes in documentation are likely to occur
- Part II – data used for part I including isometrics and that is not likely to change as often
- Part III – reference data, old calculations and all other data that is not going to change

Figure 2. Suggested General Content of Relief Systems Design and Design Basis Documentation

Part I – Automated Data and Calculations	Part II – Supporting Data	Part III – Reference Data
Approval Sheet	E-mail correspondence	MFD Legends
Specification Sheet(s)	QA/QC Checklists	Piping Specifications
Scenario Summaries	Other calculations (Excel, MathCAD)	PRD Vendor Information
Project Notes	Isometric Sketch	Company Guidance Documents
Equipment Data	Photos	Reaction Models Papers
Piping Isometrics	P&IDs, PFDs	E-mail Correspondence
Process Safety Diagram	H&MB	Copies of Reference Documents
Mixtures / Chemicals	Operating Procedures	Old PRD Calculations
Chemical Reactions	Process Information	International Standards
Streams	Equipment specs / fabrication and installation drawings / U-1A forms	RAGAGEP Papers
Scenario Applicability	Insulation specifications	
Scenario Calculations	Control Valve Data	
Scenario Notes	Pump Data	
Author Data	PRD Inspection Report	
	PRD field survey data	
	Prior PRD specification	

Figure 3. A Sample ioXpress/SuperChems Platform Dynamic Form Connecting Structured and Non-structured Relief Systems Data/Information

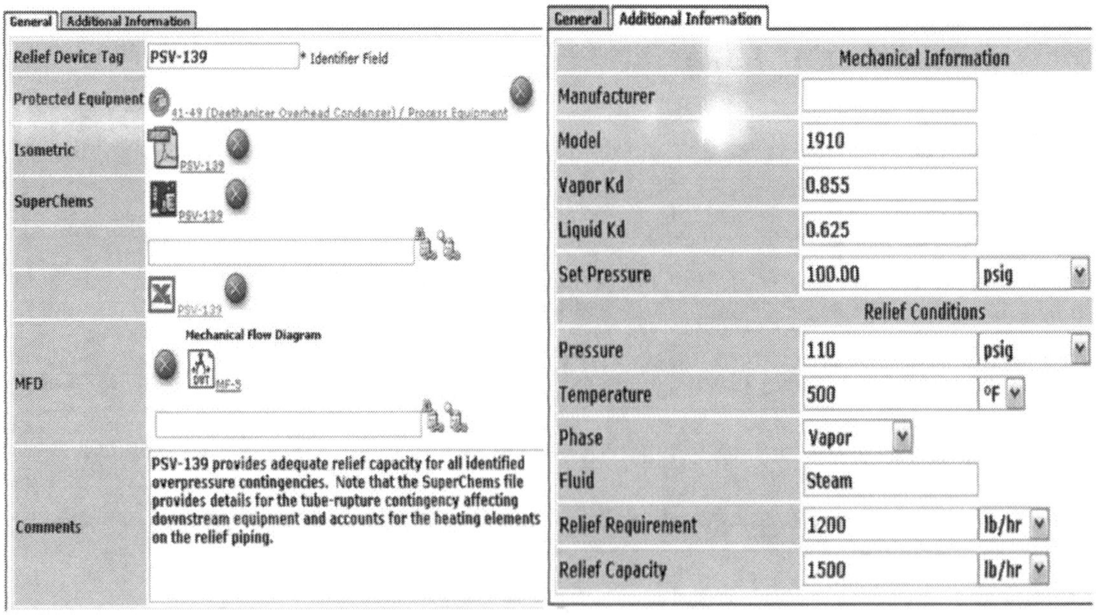

12. Conclusion

We provided in this paper a summary of a workflow system that we developed for the execution of large scale/site-wide relief systems studies as well as a general description of the work processes covered by the workflow. Benefits of this system include (a) compliance assurance and standardization, (b) enhanced change management, process operations, and loss avoidance, (c) easy integration of legacy/existing documentation and supporting calculations regardless of what calculation method, format, or tool is or was used, (d) easy, quick, and secure web-based access (internet or intranet) to the right information when needed, (e) single point access to different data formats, (f) cost efficient reviews, and most importantly (g) information life cycle management.

13. References

[1] Melhem, G. A., *Relief System's Last Line of Defense, Only Line of Defense?* Process Safety Progress, December 2006.

[2] Melhem, G. A., and P. Howell, "*Designing Emergency Relief Systems for Runaway Reactions*", Chemical Engineering Progress, Pages 23-28, September 2005.

[3] Melhem, G. A., "*Fire and Ice - How to Properly Model Depressuring Systems*", DIERS Users Group Meeting, Kansas City, May 2012.

[4] Melhem, G. A., "*Assess Piping Vibration Risk with Experience Based Methods*", DIERS Users Group Meeting, Kansas City, May 2012.

[5] Prophet, N. and G. Gorski, "*The Importance of a Team-Based Approach to Overpressure Scenario Development*", CCPS 9th Global Congress on Process Safety, San Antonio, April 2013.

9TH GLOBAL CONGRESS ON PROCESS SAFETY

Heat Integration and Relief Systems Design

Jason White, P.E.
Smith & Burgess
5700 NW Central Dr. Ste 301
Houston, Texas 77092
jason.white@smithburgess.com

Jason Spearow, P.E.
Smith & Burgess
5700 NW Central Dr. Ste 301
Houston, Texas 77092
jason.spearow@smithburgess.com

Prepared for Presentation at
American Institute of Chemical Engineers
2013 Spring Meeting
9th Global Congress on Process Safety
San Antonio, Texas
April 28 – May 1, 2013

UNPUBLISHED

Heat Integration and Relief Systems Design

Jason White, P.E.
Smith & Burgess
5700 NW Central Dr. Ste 301
Houston, Texas 77092
jason.white@smithburgess.com

Jason Spearow, P.E.
Smith & Burgess

Keywords: Heat Integration, Relief Systems Design, Relief load, Relief requirement, Flare Analysis, Flare hydraulic, Flare load

Abstract

Heat integration is commonly skipped when evaluating a process' relief systems design. Failing to evaluate the heat integration (or energy balance) for a process can result in overly conservative results and possibly lead to costly expenditures. Evaluating the effects of heat integration in relief systems design is a process that evaluates the physical limitations of the system and does not require taking credit for control system response. It is important to understand the effects that heat integration has on both individual relief devices, and in some cases, the significant affect it can have on flare system design. This becomes even more imperative in cases where relief devices that once discharged to the atmosphere are modified to discharge into a closed flare system. This paper covers two examples where evaluating the effects of heat integration on the relief systems prevented costly modifications to those systems.

1. Introduction

Heat integration is a common practice in process optimization, and it is becoming more and more important in the oil, gas, and petrochemical industry. This optimization often involves increases in process throughput. The effects these changes have on relief systems and disposal systems are often misunderstood. Understanding the physical limits of a process during an upset condition can be very important and ignoring these limits can be a costly mistake.

Heat integration is process of utilizing energy already present in a system to minimize the utility consumption. An example of heat integration can be seen in Figure 1 and Figure 2. The column system is shown with a feed preheater and a product cooler prior to heat integration and a feed/product exchanger after a heat integration project. Employing heat integration is a common plan in new facilities and a common cost savings upgrade for existing facilities. Heat integration can pose start-up/shutdown concerns, along with control system difficulties; however, these concerns are not covered in the scope of this paper.

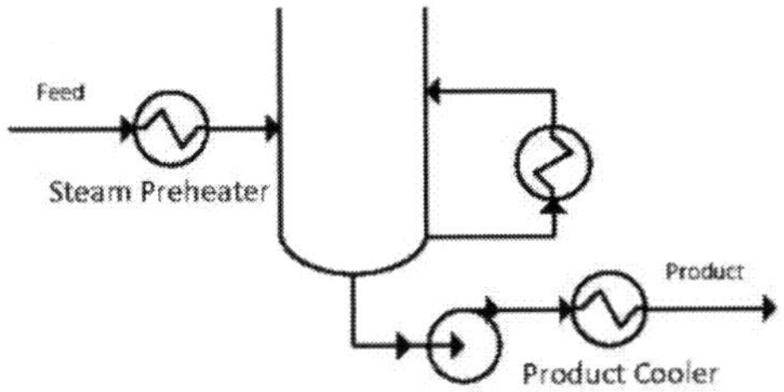

Figure 1 – Example of process prior to heat integration.

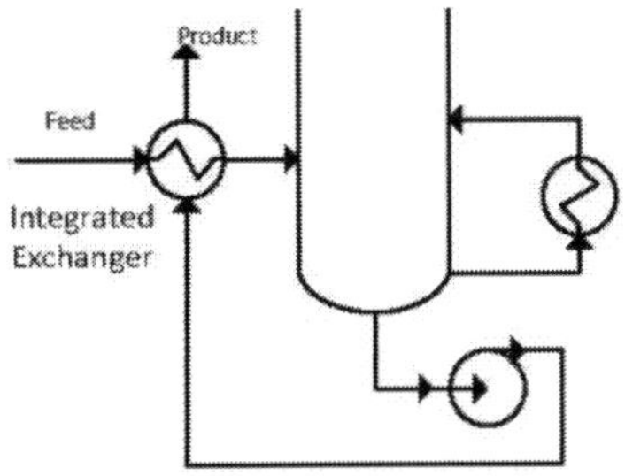

Figure 2 – Example of similar process after heat integration.

When evaluating relief systems design, either for existing facilities or for new construction, a conservative/consistent approach is taken. Due to the increased cost of the analysis, it is typical that the potential effects of heat integration are not recognized or considered. Many people also worry about what would qualify as taking credit for "positive control," which is not permitted by the regulating bodies. This paper discusses two projects in which the use of conservative methods would have resulted in millions of dollars in modifications and lost time. Sound engineering analysis was performed in a relatively short period of time, and both of these cases were resolved such that no modifications were required. This was not a risk assessment analysis or a risk ranking tool, but a practical application of understanding the situation at hand and assessing the true limits of the process during that situation.

2. Heat Integration Case Study 1

2.1 Problem Statement

A refinery was investigating implementing a heat integration project in their crude fractionation unit to increase the diesel fraction from the unit. The feasibility study identified that the existing relief devices may have inadequate capacity in the event of a partial power failure when many of the column's reflux/sidestream pumps are lost while the feed continues. The financial impact of installing additional relief capacity, which included modifications to the flare header, was high enough to warrant cancellation of the project. The effect of heat integration on the relief system was not fully considered during the analysis.

2.2 Analysis

In the partial power failure scenario, the previous calculation assumed the outlet temperature of the feed furnace would remain constant. The fuel gas to the burners is on temperature control and the fuel gas flow will increase in order to maintain a constant temperature. However, the majority of the heat input to the preheat train is lost during this power failure scenario, and the furnaces may not have sufficient heating capacity to maintain the normal outlet temperature. A simplified crude fractionator system with preheat train is shown in Figure 3 below. The pumps and exchangers shown in red are lost during this power failure scenario.

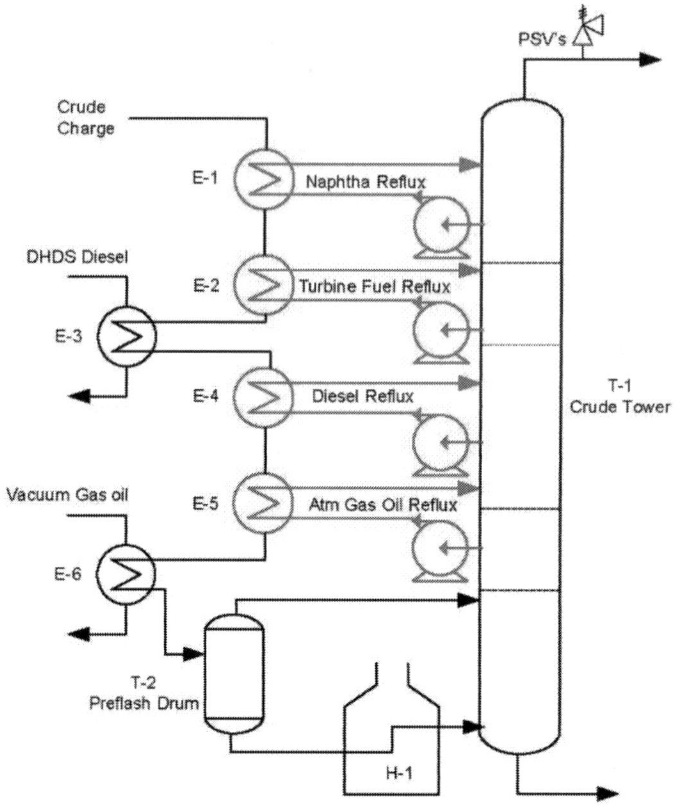

Figure 3 – Crude fractionator with preheat train

In order to estimate the temperature feeding the furnaces, the preheat train must first be modeled considering all exchangers that continue to provide heat input during the relief scenario. The exchangers in the preheat train should be modeled with their actual heat transfer coefficient (U) and surface area (A), along with the hot side fluid conditions, rather than merely modeled by applying the normal duty. During the relief scenario, there is a greater temperature difference between the hot and cold side of the exchangers in the preheat train; thus, the duties will be higher than normal. Applying the normal duty will underestimate the furnace feed temperature and is not conservative. The assumptions for the exchanger hot side flows and conditions will vary depending on the relief scenario. For example, it is important to consider factors such as any pumps in spare service and whether or not a column draw tray is expected to dry up. If the hot side of the preheat exchanger gets it's feed from the column being analyzed, then it is conservative to assume that draw is lost, resulting in additional cooling loss in the tower. There is typically a preflash drum at the end of the preheat train that flashes off the vapor and sends the liquid to the freed furnace. During a particular relief scenario, there may no longer be vapor present at the preflash drum; therefore, it is necessary to consider whether or not flow to the furnace will increase as the liquid level in the preflash drum rises.

Now that the new feed conditions to the furnace are determined, the maximum furnace duty is applied. The maximum furnace duty should be obtained from the manufacturer burner curves, with the furnace efficiencies applied. The crude tower can now be modeled with the new feed temperature and a required relief capacity can be calculated.

2.3 Results

Table 1 below shows the difference between the duties applied to the crude charge during normal conditions and during the power failure relief scenario. Notice that even though the furnace duty is significantly higher than normal, it is not enough to make up for the duty lost in the preheat train.

Table 1 - Comparison between normal and relief duties when accounting for heat integration (MMBTU/hr)

Case	E-1	E-2	E-3	E-4	E-5	E-6	H-1	Total
Normal	65.8	35.1	27.2	62.3	50.7	47.6	235.7	524.4
Relief	0	0	45.0	0	0	106.5	317.9	469.4
Difference	-65.8	-35.1	17.8	-62.3	-50.7	58.9	82.2	-55.0

After modeling the crude preheat train and applying the maximum feed furnace duty, a lower tower feed temperature and required relief rate was calculated as shown in Table 2.

Table 2 - Required relief with and without considering heat integration

Case	Tower Feed Temp (^oF)	Required Relief (lb/hr)
Without considering heat integration	680	719,900
Considering heat integration	610	622,800

The reduction in required relief rate was enough for the existing relief devices to provide adequate relief capacity; thus, there was no need for spending a substantial amount of money on installing additional relief capacity and making flare header modifications. The calculations did not involve dynamic or rigorous modeling, so the associated engineering was able to be done in a short time frame at minimal cost.

3. Heat Integration Case Study 2

3.1 Problem Statement

A refinery was working to resolve some concerns pertaining the radiation levels identified previously for the controlling flare scenario (total power failure). The system in question can be best described as a fully integrated fluidized catalytic cracking unit (FCCU) and gas condensates unit (Gas Con) where the pump-arounds providing cooling to the FCCU fractionator are integrated exchangers also acting as the reboilers to the distillation systems in the Gas Con (as shown in Figure 4).

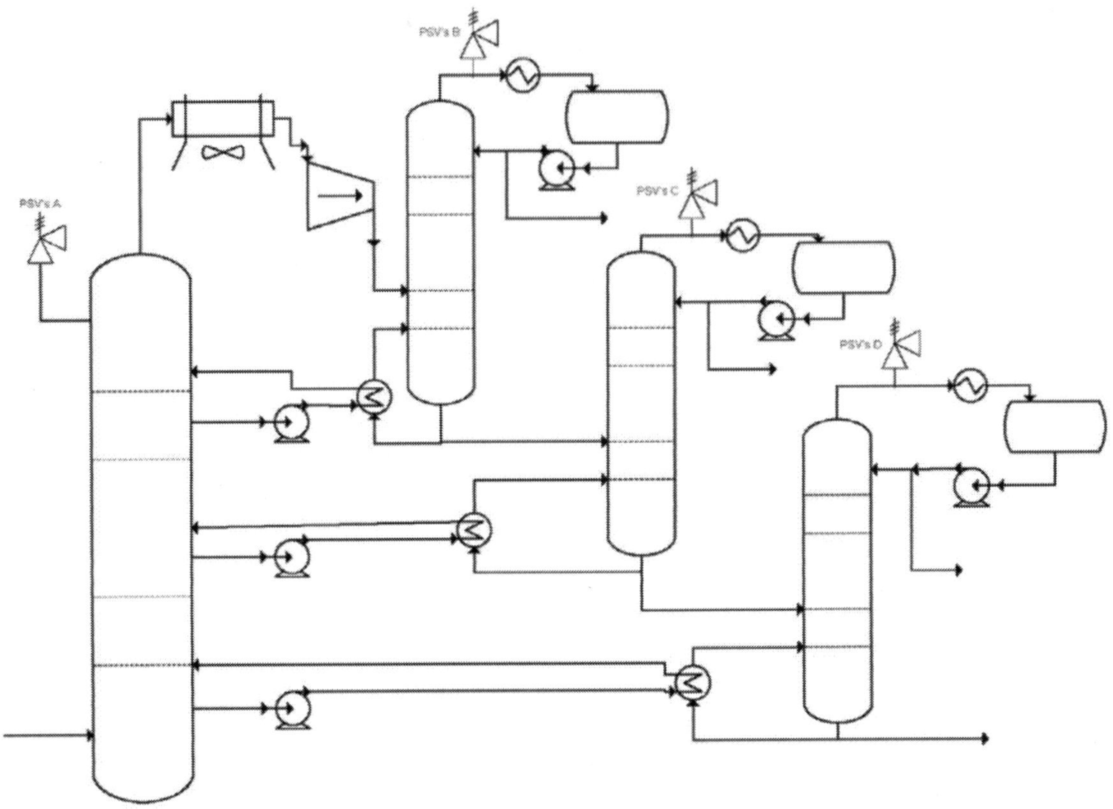

Figure 4 - System sketch of FCCU and Gas Con integrated exchangers.

The previous analysis used data generated from the individual relief device sizing reports that did not address the effects of heat integration. The loads from the relief systems (multiple valves used in most of the above cases) are represented in Table 3.

Table 3 - Load summary prior to heat integration analysis.

Valves	Location	Load (lb/hr)
PSV's A	FCCU Fractionator	454,123
PSV's B	Gas Con De-Propanizer	151,400
PSV's C	Gas Con De-Butanizer	433,700
PSV's D	Gas Con De-Isobutanizer	148,900
Total	FCCU and Gas Con	1,188,123

3.2 Analysis

The key to the analysis is to understand the scenario at hand. For any of the individual relief systems, there did not appear to be any errors. The pump-arounds that may or may not fail in the event of the power failure (due to steam driven spares) must be analyzed as the worst possible case for individual systems. When analyzing the fractionator, the pump-arounds provide cooling duty to the system and the failure of these provides the conservative sizing basis for the relief valves. When analyzing the tower systems in the Gas Con Unit, continued heat from the reboilers provides the conservative sizing basis for these relief valves and the pumps are assumed to remain in operation.

For the analysis of the flare, the conservative loads are not additive and it is overly conservative to do so. The pump-arounds cannot simultaneously fail and remain on, which the original analysis would indicate. For the purposes of the flare analysis, one or the other must be assumed. The secondary problem that comes from this is the inability to estimate which of the possible cases is actually the "worst case" for the flare. For this reason, both circumstances must be considered as shown in
Figure 5.

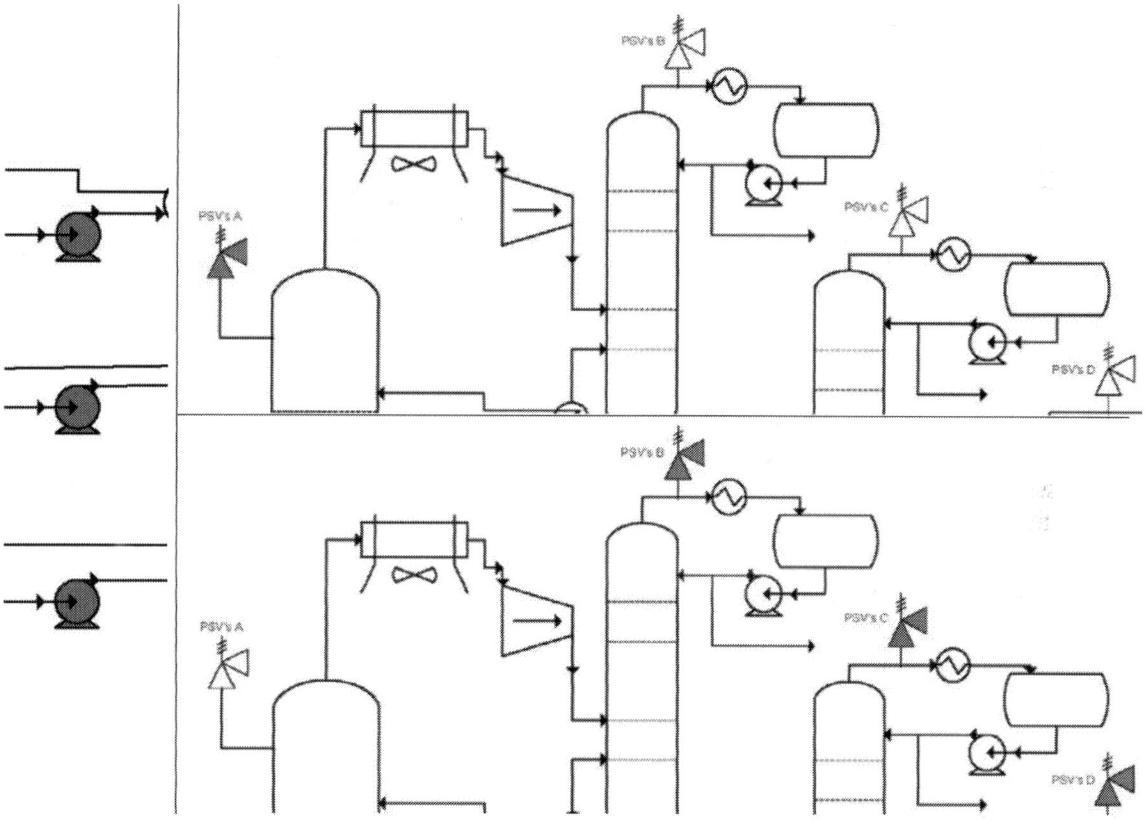

Figure 5 - Summary of the analysis clockwise from the left: the 3 pumps in question; the first case analysis with only the relief valve shown in red discharging; the second case analysis with the 3 relief valve discharging shown in red.

3.3 Results

By simply understanding the physical limits of the system, the flare loading was significantly reduced as shown in Table 4. To emphasize the importance of analyzing both cases, it should be noted that Case 1 provided more severe back-pressure issues (due to the lower set pressures in the fractionator system) while Case 2 provided the controlling scenario for flare radiation (due to the higher mass flow).

Table 4 - Load summary after the heat integration is accounted for.

Valves	Location	Case 1 Load (lb/hr)	Case 2 Load (lb/hr)
PSV's A	FCCU Fractionator	454,123	0
PSV's B	Gas Con De-Propanizer	0	151,400
PSV's C	Gas Con De-Butanizer	0	433,700
PSV's D	Gas Con De-Isobutanizer	0	148,900
Total	FCCU and Gas Con	454,123	734,000

4. Conclusion

Heat integration has become significantly more common in the oil, gas, and petrochemical industry to reduce utility costs. The basic premise of heat integration is that heat (typically from a product stream) is transferred to a different stream (typically a feed stream) through a process heat exchanger rather than using heating/cooling utilities to add/remove this energy. By understanding the effects it has on relief system evaluations, one can avoid over predicting required relief loads that may result in significant financial impacts. Understanding heat integration is also important so that required relief loads are not under predicted, which may result in unsafe design. It was demonstrated by the examples in this paper that examining the limitations of the system does not require significant time or rigorous modeling software. A relatively simple steady state analysis can be performed that captures the limitations of the systems and still provides a conservative result that does not take any credit for positive control. By simply understanding how heat is conserved in the process, numerous engineering hours and countless dollars can be saved.

API Standard 521 New Alternative Method to Evaluate Fire Relief for Pressure Relief Device Sizing and Depressuring System Design

Edward Zamejc
EZ Relief Systems Consulting, Inc.
4110 Pine Blossom Trail
Houston, TX, 77059
edzamejc@ezreliefsystems.com

Prepared for Presentation at
American Institute of Chemical Engineers
2013 Spring Meeting
9th Global Congress on Process Safety
San Antonio, Texas
April 28 – May 1, 2013

UNPUBLISHED

API Standard 521 New Alternative Method to Evaluate Fire Relief for Pressure Relief Device Sizing and Depressuring System Design

Edward Zamejc
EZ Relief Systems Consulting, Inc.
4110 Pine Blossom Trail
Houston, TX, 77059
edzamejc@ezreliefsystems.com

Keywords: pool fire exposure; fire relief sizing; fire heat input; depressuring systems; pool fire tests; over-temperature failure

Abstract

Since the 1950's, API Standards have provided guidance on determining relief loads for equipment exposed to pool fires. The API method is empirical based on tests performed in the 1940's. There is increasingly widespread interest in analytical methods based on heat transfer principles to model fire heat input. The API committee agreed to include an analytical method in the next edition of API Standard 521 to establish relief loads for pressure relief devices and to design depressuring systems for the fire scenario. The analytical method provides more flexibility than the empirical method but has limitations (e.g., too many permutations are possible leading to potential under-sizing of the pressure relief device).

This paper discusses the basis for the empirical method in API Standard 521 and provides comparisons of the empirical and analytical method with two more recent large-scale pool fire tests. This pool fire test data indicates that the empirical method will provide a conservative estimate of pool fire heat input for most applications and is still the method of choice when designing pressure relief systems. However, these recent tests indicate the empirical method needs to be modified when a vessel or equipment is partially confined by adjacent embankments or walls equal or greater than the vessel height. In such cases, the wetted area exponent should be 1.0 instead of 0.82.

The analytical method is useful in determining time-versus-temperature profiles for heating unwetted vessels of varying wall thicknesses and materials of construction. These profiles can be combined with tensile strength and stress-rupture data to specify a depressuring system's pressure-versus-time profile; this will minimize failure and/or mitigate the effects of failure due to overheating from a pool or jet fire exposure..

1. Introduction

American Petroleum Institute (API) Standard 521 "Pressure Relieving and Depressuring Systems" is an internationally recognized engineering standard used to design pressure relief systems, disposal systems (e.g., flares), and depressuring systems [1]. It is continually being reviewed, with new editions published in about 5-year intervals. A technical committee consisting of industry representatives, engineering contractors, and regulators recommend and integrate modifications into the Standard. These modifications involve lessons-learned from incidents or near-misses, advances in engineering methodologies, and new guidance based on shared experiences of the members or inspired by technical inquiries.

The upcoming API 521 6th edition (to-be-published) will include an analytical method to establish relief loads for pressure relief devices and to design depressuring systems for the fire scenario. The analytical method will complement, but not replace, the existing empirical method.

It is important to establish the scope of API 521 and differentiate it from API 2000 [2]. API 521 covers pressure vessels and processing equipment (e.g., vessels design in accordance with ASME Section VIII, Division 1 [3] and other pressure vessel design codes). In contrast, API 2000 covers low pressure, atmospheric, and refrigerated storage tanks designed in accordance with storage tank standards such as API 650 [4]. The current pool fire heat input equations in API 2000 are the same as those in NFPA 30 [4]. They were established in a 1963 meeting between API and NFPA and are based on a fire test and experience with storage tank fires. Because the origin/basis of API 521 and API 2000 fire equations are different and the scope of the equipment design codes are different, the fire exposure guidance API 521 and API 2000 can neither be interchanged nor compared (i.e., use API 521 for pressure vessels and API 2000 for storage tanks). The subsequent discussion relates only to API 521.

2. Fire Scenario - API Empirical Method

2.1 Basis of the API empirical method to evaluate pool fires

The fire scenario generates the most technical inquiries of any topic in API 521. The current method, shown in Equations 1 and 2, is an empirical method based on fire tests performed in the 1940's.

With adequate drainage and prompt firefighting:

$$Q = C_1 \cdot F \cdot A_{ws}^{0.82} \quad \text{[Eq. 1]}$$

Without adequate drainage and prompt firefighting:

$$Q = C_2 \cdot F \cdot A_{ws}^{0.82} \quad \text{[Eq. 2]}$$

Where:

Q is the total heat absorption (input) to the wetted surface, expressed in W (Btu/h);

C_1 is a constant [= 43,200 in SI units (21,000 in USC units)];

C_2 is a constant [= 70,900 in SI units (34,500 in USC units)].

F is an environment factor;

A_{ws} is the total wetted surface, expressed in square metres (square feet).

Note: the SI equation constants include a conversion factor for $(A_{ws})^{0.82}$.

The origin of the empirical method can be traced back to the 1950's, when the API Pressure Relief Systems (PRS) technical committee analyzed the available pool fire test data and developed empirical equations to determine the pool fire heat input to a vessel. This heat input could then be used to calculate the fire relief load by dividing by the heat of vaporization. These empirical equations include the:

- Maximum fire heat input (i.e., maximum heat flux absorbed by the vessel and its contents)
- Effect of wetted surface area of the vessel or equipment (i.e., area of the equipment in contact with liquid or below liquid level) on fire heat input
- Effect of drainage (i.e., whether the pool fire is under the vessel) on fire heat input

The maximum fire heat input into a vessel is sometimes confused with the flame surface heat flux (i.e., pool fire heat duty divided by flame surface area) and the incident heat flux at a vessel exposed to the fire. The incident heat flux excludes reduction in heat flux due to the absorptivity of the vessel and re-radiation from the vessel. Based on plotting hydrocarbon pool fire test data (see Figure 1), the maximum heat input into the vessel was determined by the API committee to be 34,500 BTU/h-ft^2 (see constant C_2 in Equation 2). This maximum heat input would occur when the wetted surfaces of the vessel are completely and continuously exposed to flame.

Figure 1. Pool fire heat input versus wetted area exposure

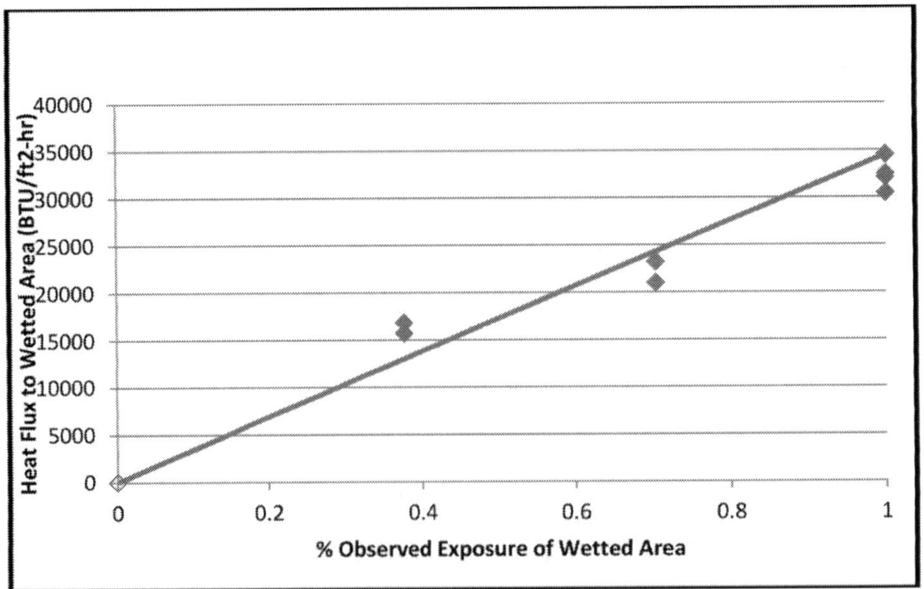

Test data sources: a) API project test No. 1, b) API project test No. 2, c) Rubber Reserve Corporation test No. 17, d) Standard Oil Company of California, e) Underwriters Laboratories, Inc. (See Reference [1] Table A-1).

Outdoor pool fires are easily influenced by even relatively calm wind conditions; wind causes flames to move around, thereby only intermittently exposing surfaces of larger vessels to the highest incident heat flux. To determine the effect of vessel wetted area, the API Committee plotted the total heat input versus the wetted area from several pool fire tests and an actual pool fire [Table A.1 in Ref.1]. The results, shown in Figure 2, indicate the heat input correlates with a 0.82 exponent on the wetted area (A_{ws}). It should be noted that, per convention, the heat input across only the wetted surfaces of vessels containing a liquid that can boil is considered when designing pressure relief systems for the fire scenario using the empirical method. The effect of heating unwetted surfaces and gas-filled vessels is discussed later in this article.

Figure 2. API 521 Table A.1 fire tests - Wetted area versus fire heat input

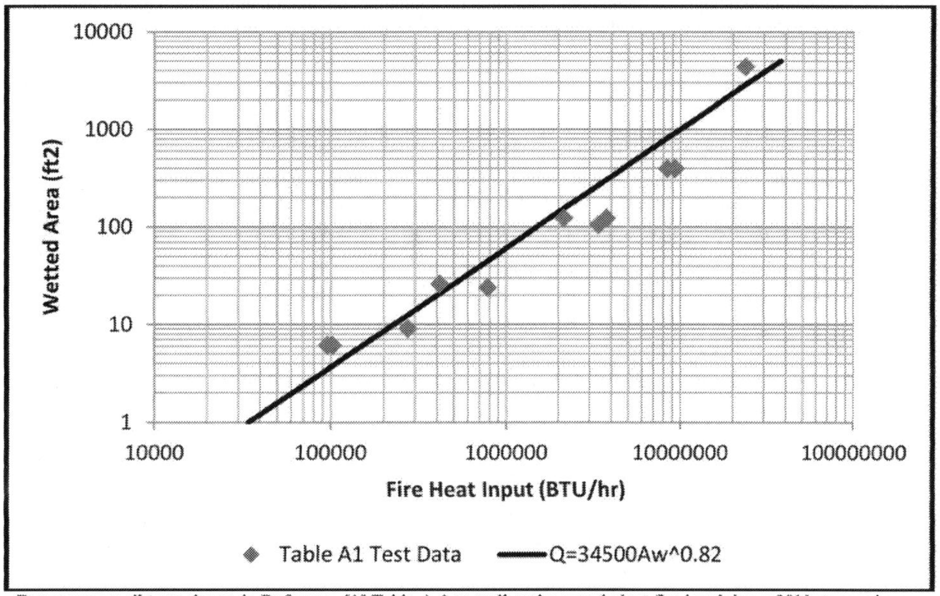

Data sources: all tests shown in Reference [1] Table A-1 as well as the actual plant fire involving a 38' butane sphere.

The effect of drainage was determined from Hottel's pool fire test data (see Reference 1, Table A.1, Test 1 and Test 2). Test 1 actually consists of the average of 31 tests without either drainage or firefighting. Test 2 consists of the average of 8 tests with drainage and 5 tests with both drainage and firefighting. The ratio for the pool fire heat input from Test 2 to Test 1 (no drainage nor firefighting) is 17,400 / 30,500 = 0.6. Hence, the maximum pool fire heat input to vessels with adequate drainage and firefighting is 34,500 * 0.6 = 21,000 BTU/h-ft^2 (see constant C1 in Equation 1). Hottel's test data from Reference 6, shown in Table 1, suggest that drainage alone (no firefighting) has a comparable reduction in heat input.

Table 1. Hottel pool fire test data showing effect of drainage with and without firefighting [6]			
Run	Drainage	Firefighting	Average heat input, BTU/ h-ft^2 (1)
10	Yes	None	22,862
11	Yes	None	30,081
12	Yes	None	9,819
13	Yes	None	19,517
14	Yes	None	21,707
15	Yes	None	26,953
17	Yes	None	2,166
19	Yes	None	16,436
18	Yes	Chemical foam for 30 sec. then mechanical foam	6,594
22	Yes	Chemical foam for 30 sec. then mechanical foam	9,337
23	Yes	Chemical foam for 30 sec. then mechanical foam	14,535
24	Yes	Chemical foam for 30 sec. then mechanical foam	25,028
25	Yes	Chemical foam for 30 sec. then mechanical foam	20,672
		Average for all tests	17,362
		Average for tests with drainage, but no firefighting	18,693
		Average for tests with both drainage and firefighting	15,233
(1) Heat input into wetted surface area			

The remaining parameter of the empirical method, designated "F", is the environment factor which credits for adequate fireproofing (See Reference 1). The 6th edition will have an expanded expression to account for multi-layer fireproofing.

2.2 Limitations of the API empirical method to evaluate fires

There are two types of fires relevant to pressure relief and depressuring system design – pool fire and jet fire. A pool fire is defined as a "burning pool of liquid." A jet fire is a "fire created when a leak from a pressurized system ignites and forms a burning jet." A pool fire can be classified as an open pool fire, a confined pool fire, or somewhere in between. A confined pool fire is defined as a "fire inside a building or a compact process module where the walls and/or surrounding equipment can reradiate and preheat the combustion air causing higher heat fluxes than an unconfined (i.e., open) fire." Generally only pool fires are considered when designing pressure relief systems while both pool fires and jet fires are often considered when designing depressuring systems.

Typical ranges for peak fire heat intensity (i.e., incident heat flux) are [7]:

- Open pool fire – 16,000 to 48,000 Btu/h· ft^2 (50 to 150 kW/m^2)
- Confined pool fire – 32,000 to 79,000 Btu/h· ft^2 (100 to 250 kW/m^2)
- Jet fire – 32,000 to 127,000 Btu/h· ft^2 (100 to 400 kW/m^2)

These peak fire intensities generally correspond to locations within the fire where the stoichiometric fuel-to-air ratio is equal to one. Because of the effects of ventilation (e.g., wind effects and confinement), fuel type, fuel-air stoichiometry and other factors, the peak fire heat intensity is generally only observed in localized parts of the flame volume. Most of the flame volume has significantly lower fire intensities than the peak.

Pressure relief systems designed for fire exposure require a total heat input into the relevant surfaces (e.g., wetted areas) using an averaged fire heat flux. In these cases, the fire heat intensity should be averaged across the entire flame volume. This average is designated as the "surface average heat flux". In contrast, when designing depressuring systems, the peak fire heat intensity (designated as the "local peat heat flux") is important because localized overheating in a small area can result in equipment failure due to overheating.

It should be noted that API 521 5th and prior editions provide design guidance for pressure relief and depressuring systems only for open pool fires. Also, the 34,500 Btu/h·ft^2 maximum fire heat input used in the API empirical method includes the vessel absorptivity, which can reduce the incident heat flux by 20-70%.

3. Analytical method to evaluate fires

Several authors recommended using an analytical method to model the fire scenario to overcome limitations with the empirical method and to provide more flexibility in modeling (e.g. [7] [8] [9] [10] [11] [12]). Because these methods are becoming more widely used, the API 521 committee agreed to incorporate the analytical method as an alternative to the empirical method and to include guidance on its application. The empirical method will still be recommended as the preferred method to evaluate most fire scenarios involving pressure relief system design with the analytical method preferred for special cases and fires outside the scope of the empirical method.

The analytical method to evaluate fires is a basic heat transfer as shown in Equation 3. The method determines the fire heat input to a vessel and conservatively ignores internal heat transfer limitations. It can be applied to all types of fires including open pool fires, confined pool fires, and jet fires. The proposed typical ranges in the parameters are given in Table 2 for the surface average heat flux for pool fires. Recommended values, which should be used where data or other resources are unavailable, will be given in the API 521 6th Edition. API 521 6th edition will also provide typical values of parameters for the local peak heat flux of pool fires and for the local peak and surface average heat fluxes for jet fires.

$$q_{absorbed} = \sigma \cdot \left(\alpha_{surface} \cdot \varepsilon_{fire} \cdot T_{fire}^4 - \varepsilon_{surface} \cdot T_{surface}^4\right) + h \cdot \left(T_{gas} - T_{surface}\right) \qquad \text{[Eq. 3]}$$

Where:

$q_{absorbed}$ is the absorbed heat flux from the fire, expressed in Btu/h·ft^2 (W/ m^2);

σ is the Stefan-Boltzmann's constant = 0.1713e-8 Btu/h·ft^2·R^4 (5.67e-8 W/ m^2·K^4);

$\alpha_{surface}$ is the equipment absorptivity, dimensionless;

ε_{fire} is the fire emissivity, dimensionless;

$\varepsilon_{surface}$ is the equipment emissivity, dimensionless;

T_{fire} is the fire temperature, expressed in °R (°K);

$T_{surface}$ is the equipment temperature, expressed in °R (°K);

T_{gas} is the temperature of air\fire in contact with the equipment surface, expressed in °R (°K);

h is the convection heat transfer coefficient of air\fire in contact the equipment, Btu/h·ft²·R (W/m²·K);

$\sigma \cdot \alpha_{surface} \cdot \varepsilon_{fire} \cdot T^4_{fire}$ is the radiative heat flux to the equipment;

$\sigma \cdot \varepsilon_{surface} \cdot T^4_{surface}$ is the re-radiation from the equipment;

$h \cdot (T_{gas} - T_{surface})$ is the convection heat transfer between the combustion gases and the equipment's surface

Caution must be taken when specifying the parameters because a wide range in fire heat inputs can result. When applying the analytical method to sizing pressure relief devices, the total heat input into the vessel shall use the wetted area to the 1.0 exponent, not the 0.82 exponent used in the API empirical method as shown in Equations 1 and 2.

Table 2 —Typical range in analytical method (Equation 3) parameters for an open pool fire "surface average heat flux"

Parameter	Description	Pool fire Surface Average Heat Flux Parameter Range	
ε_{fire}	Hydrocarbon flame emissivity	0.6 – 1.0	
$\varepsilon_{surface}$	Equipment emissivity	0.3 - 0.8	
$\alpha_{surface}$	Equipment absorptivity	0.3 - 0.8	
h	Convective heat transfer coefficient between equipment and surrounding air	1.76 - 5.28 Btu/h·ft²·R (10 - 30 W/m²·K)	
T_{gas}	Temperature of combustion gases flowing over the surface	1 392 – 2,112 °R 773 – 1,173 °K	(932 – 1,652 °F) (500 – 900 °C)
T_{fire}	Fire temperature	1,572 – 2,292 °R 873 – 1 273 °K	(1,112 – 1,832 °F) (600 – 1 000 °C)
$T_{surface}$	Equipment temperature	Increases as surface heats up	
σ	Boltzmann's constant	0.1713e-8 Btu/h·ft²·R⁴ (5.67e-8 W/m²·K⁴)	
q_{fire}	Fire heat flux – a wider range is possible	9,510 – 31,700 Btu/h·ft² (30 - 100 kW/m²)	
$q_{absorbed}$	Absorbed heat flux at start of the fire	7,925 – 23,775 Btu/h·ft² (25 - 75 kW/m²)	

4. Application of the analytical method to model recent pool fire tests

4.1 Use of the analytical method to model the Ballistics Research Laboratory (BRL) pool fire test wall temperature versus time

In 1974, a pool fire test was performed by the Ballistics Research Laboratory [13]. A 33,000 gallon (125 m^3) LPG tank car was exposed to a semi-enclosed pool fire where the pool fire and tank car were located in a pit, with embankments on all sides exceeding the tank car height (no roof). Fire and wall temperatures versus time at the top of the front and rear walls of the tank car are shown in Figures 3a and 3b, respectively.

The analytical method (Equation 3) was used in an attempt to reproduce the wall temperature versus time. Values for the variables were adjusted until the calculated time-temperature profile was approximated. The parameters selected to model tank car rear wall temperature versus time at several locations are:

ε_{fire}	= 0.62 (determined by BRL)
$\varepsilon_{surface}$	= 0.5
$\alpha_{surface}$	= 0.5
h	= 1.76 Btu/h·ft^2·R (10 W/m^2-K)
T_{gas}	= 2,112 ° R (1,652 °F) 1,173 °K (900 °C)
T_{fire}	= 2,112 ° R (1,652 °F) 1,173 °K (900 °C)
$T_{surface}$	= Initially @ 529 °R (69 °F) 294 °K (21 °C)
q_{fire}	= Maximum of 16,700 Btu/h·ft^2 (52.7 kW/m^2) (calculated)
$q_{absorbed}$	= Maximum of 9,560 Btu/h·ft^2 (30.2 kW/m^2) (calculated)

q_{fire} was calculated by setting $\varepsilon_{surface} = 0$, $\alpha_{surface} = 1$. The maximum absorbed heat flux is predicted by the analytical model to occur at the start of fire when the equipment is at ambient temperature. A similar set of values was used to predict the front wall temperature versus time with the exception that the fire and gas temperature was set at 1,472 °F (900 °C). Other combinations of values in the analytical model can be used that may provide an equal or better fit to the test data. Note that, because the vessel was engulfed in the pool fire, the gas temperature should be set equal to the fire temperature. The gas temperature will be lower than the fire temperature for non-engulfing pool fires. Also, the fire and gas temperatures were assumed to be constant throughout the pool fire.

Figure 3a. Ballistics Research Laboratory pool fire test data illustrating fire temperature versus time at the top of the front and rear walls of a rail tank car

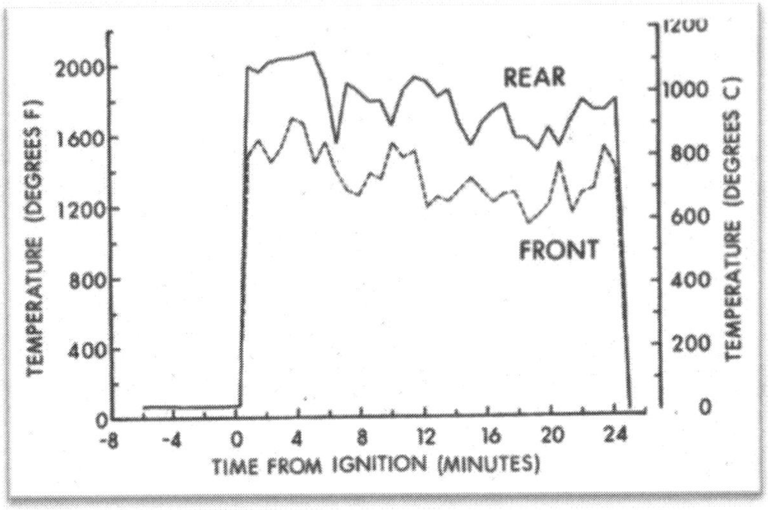

Figure 3b. Ballistics Research Laboratory pool fire test data illustrating rail tank car wall temperature versus time at the top of the front and rear walls

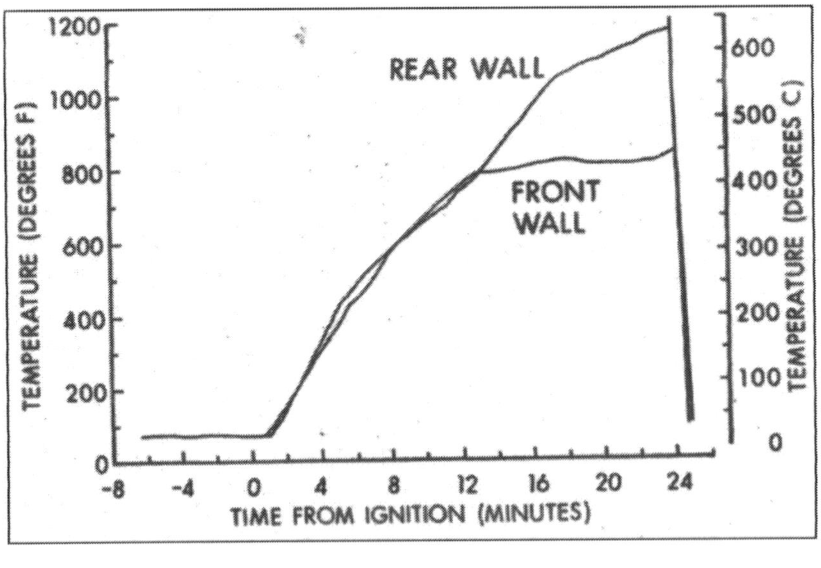

A comparison of the analytical model with the wall temperatures recorded during the test is shown in Figure 4. The analytical method provides a reasonable approximation to the observed rear wall temperature versus time. The leveling off of the front wall temperature at about 800 °F (425 °C) as observed in the test cannot be approximated with a single set of parameters, indicating that one or more parameters changed during the course of the fire. Once the temperature versus time profile is approximated by the analytical method, then the resultant vessel heat input (i.e., $q_{absorbed}$) can be determined. The analytical method will indicate the maximum heat input is at the start of the pool fire unless parameters change during the course of the fire.

Figure 4. Comparison of rail tank car wall temperature versus time between the analytical model and Ballistics Research Laboratory pool fire test data

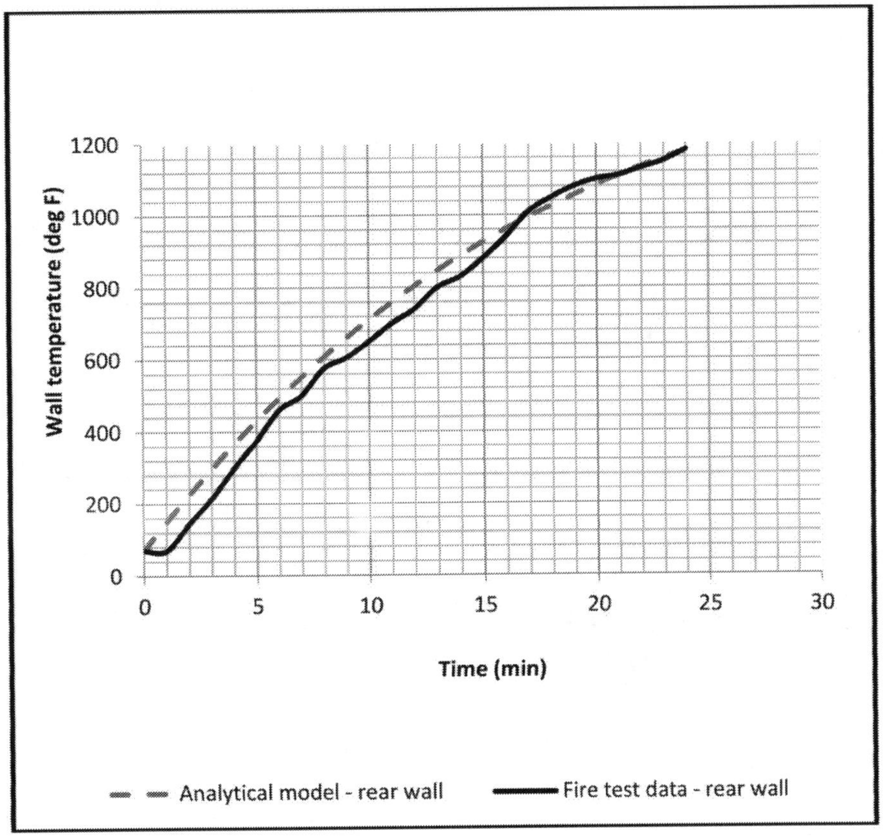

4.2 Use of the analytical method to model the Federal Institute for Materials Research and Testing (BAM) pool fire test [14][15] wall temperature versus time

In 1999, a full scale pool fire test was performed by the Federal Institute for Materials Research and Testing (BAM or Bundesanstalt für Materialforschung und -prüfung) in Germany [14] [15]. The test evaluated and compared fire exposure effects on a rail tank car containing propane and a Castor car used to transport radioactive material. The test setup is shown in Figures 5 and 6. Although the tank car was semi-confined by embankments on 3 sides, a light to calm northerly wind was still able to significantly affect the pool fire exposure of the tank car as shown in Figure 7. The tank car maximum pressure reached 25 bar (362 psig) about 15 minutes after the start of the pool fire at which time the tank car ruptured, resulting in a boiling liquid expanding vapor explosion (BLEVE). The BLEVE aftermath is shown in Figure 8. Pool fire flame/gas and tank car wall temperatures versus time at the various locations around the tank car, as shown in Figure 9, are illustrated in Figures 10 and 11, respectively. It is important to note that failure occurred before the pressure reached the pressure relief device opening pressure. A discussion of this failure as it relates to depressuring system design is given in Section 6.3.

Figure 5. BAM pool fire test setup involving propane rail tank car [14] [15]

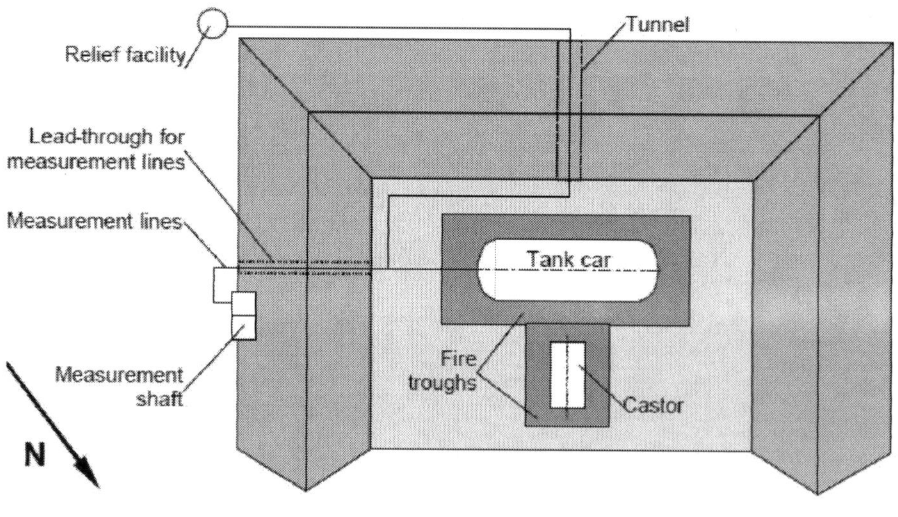

- Embankment dimensions: 60 x 50 x 6 meters (197 x 164 x 20 ft)
- Tank car capacity = 12,000 gallons (45.36 m^3) T-T length = 5.95 m; (19.5 ft) Diameter = 2.9 m (9.5 ft)
- Tank car test pressure = 28 bar (406 psia)
- Tank car contained liquid propane
- Fuel oil pool fire in troughs under the tank car and Castor container
- Castor container is used to store and transport radioactive materials and was tested along with the tank car

Figure 6. BAM pool fire test setup involving propane rail tank car [14] [15]

Figure 7. BAM pool fire test [14] [15]

Note: taken near end of test (calm wind speed).

Figure 8. BAM pool fire test – BLEVE aftermath [14] [15]

Figure 9. BAM pool fire test –Temperatures measurement locations [14] [15]

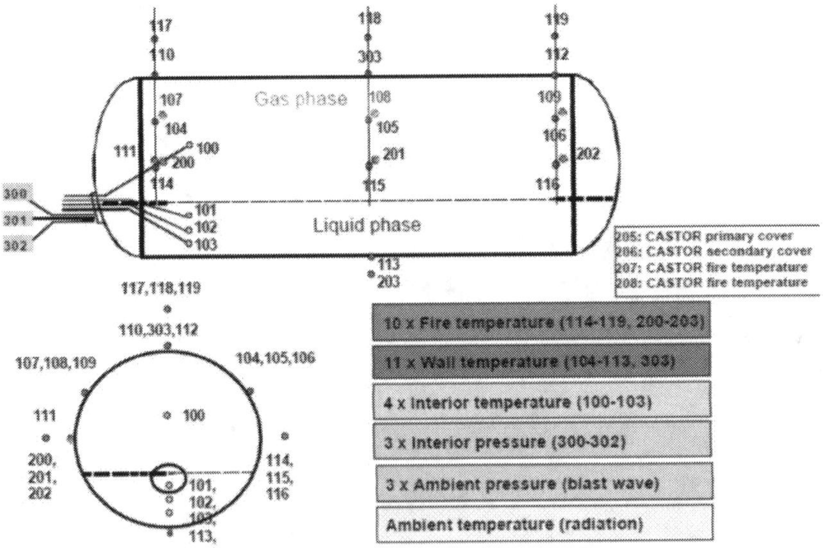

Figure 10. BAM pool fire test – Fire temperatures versus time [14] [15]

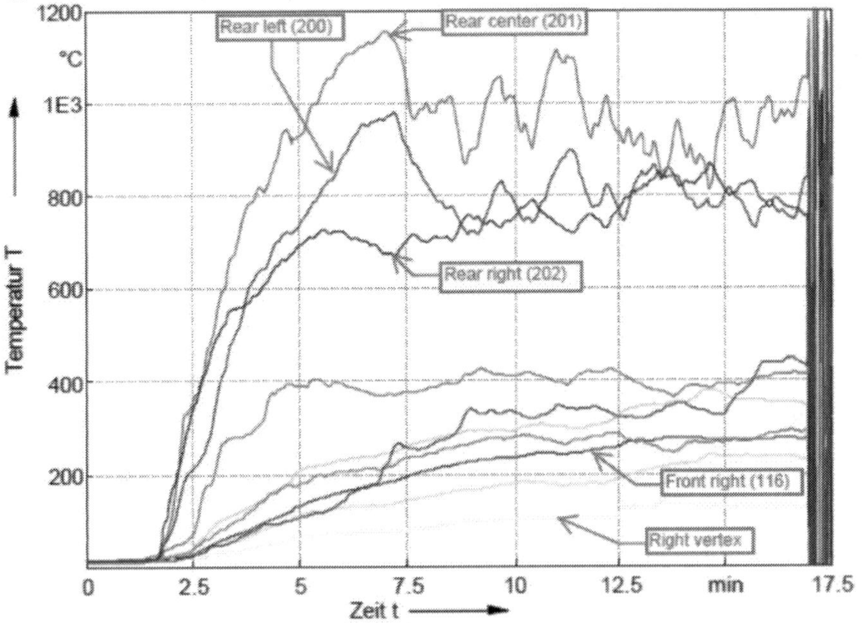

Note 1: Time = 0 is when gasoline starter fluid in a small plastic container was ignited. The main pool fire started about 100 seconds later when the plastic container failed and spilled burning gasoline into the fuel oil pool.
Note 2: The unmarked temperature curves were primarily in the front of the tank car (upwind location and without an adjacent embankment).

Figure 11. BAM pool fire test – Tank car wall temperatures versus time [14] [15]

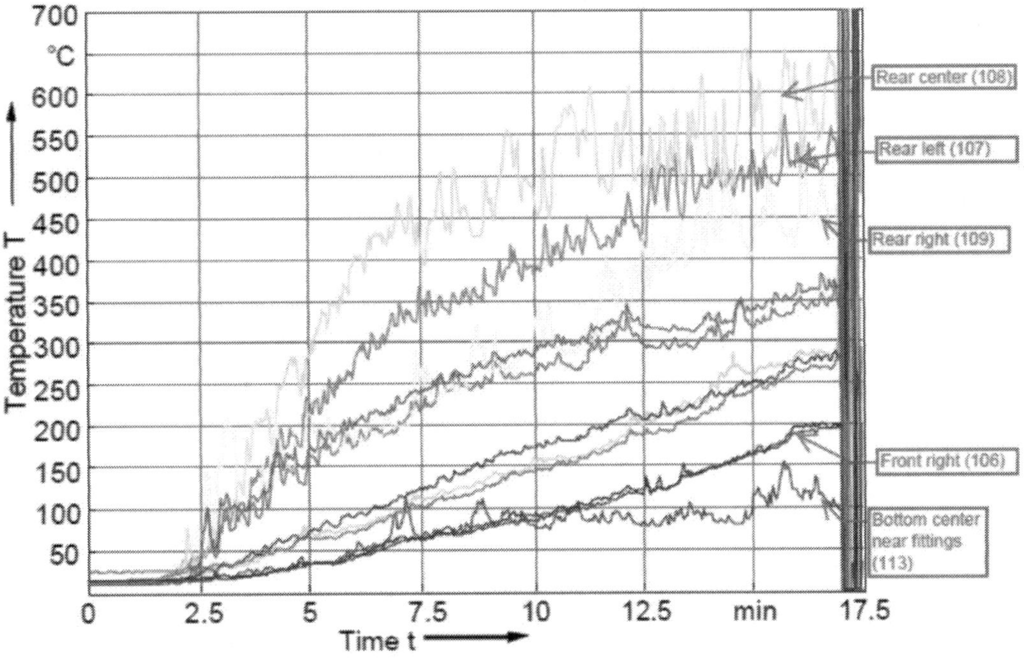

Because of the wide pool fire and wall temperature ranges shown in Figures 10 and 11, a single set of values for the parameters in the analytical method would not predict all variations. Table 3 illustrates parameters selected to model tank car wall temperature versus time at two locations. Other combinations of values in the analytical model can be used that may provide an equal or better fit to the test data. It should be noted that a transient approach to the analytical method, where the fire and gas temperatures were varied with time based on the test data shown in Figure 10, was evaluated; however, did not appear to significantly improve the fit with the test data.

A comparison of the analytical model with the wall temperatures recorded during the test, shown in Figure 12, indicates the analytical method can provide reasonable approximations to wall temperatures versus time.

Table 3. Analytical method parameters used to model tank car wall temperature versus time at two locations of the BAM pool fire test tank car

Parameter	Rear Center	Front Right
Fire temperature, °F (°C)	1832 (1000)	482 (250)
Gas temperature, °F (°C)	1832 (1000)	482 (250)
Convective heat transfer Coefficient, BTU/h-ft^2-R (w/M^2-K)	5.28 (20)	5.28 (20)
Fire emissivity	0.6	0.6
Metal emissivity	0.5	0.4
Metal absorptivity	0.5	0.4
The following were calculated using the values above:		
Calculated initial incident heat flux, BTU/h-ft^2 (kW/m^2)	34,570 (109.0)	2,260 (7.1)
Calculated maximum absorbed heat flux, BTU/h-ft^2 (kW/m^2)	20,321 (64.1)	1,721 (5.4)

Figure 12. Comparison of rail tank car wall temperature versus time between the analytical model and BAM test data

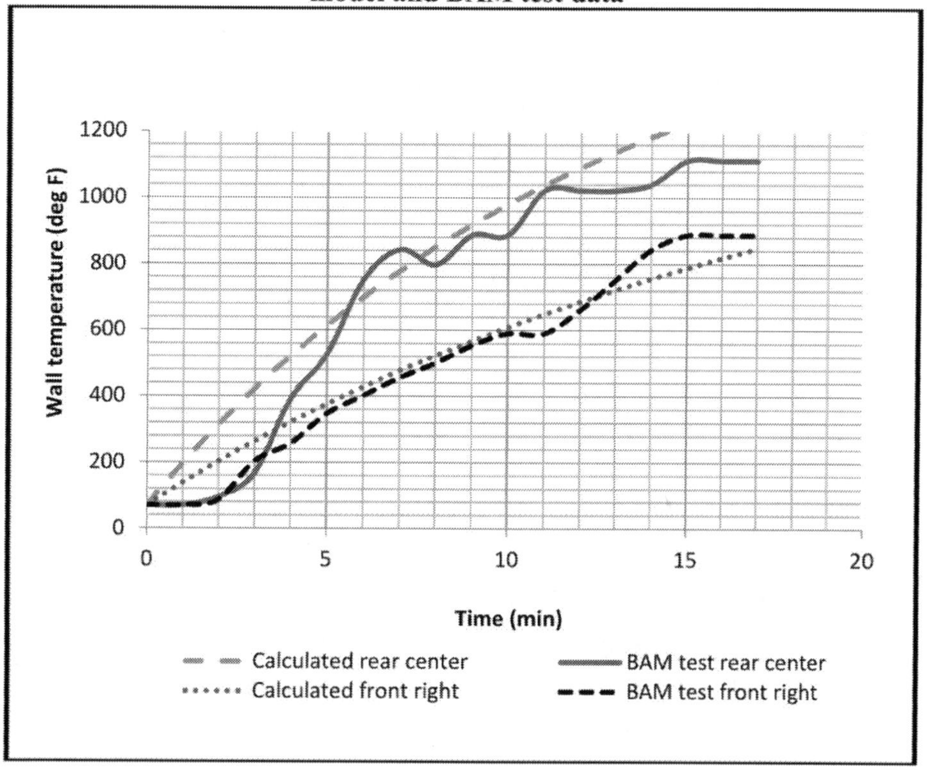

5. Comparison of the pool fire heat inputs between the empirical method, the analytical method and pool fire test data

5.1 Comparison with pool fire heat input based on BAM time-versus-temperature test data

The pool fire heat input determined by the empirical method and the analytical method can be compared using the BAM pool fire test data. The rail tank car was filled with about 2,650 gallons (10 m^3) of 95% liquid propane, resulting in an initial wetted surface area of about 249 ft^2 (23.16 m^2). The pool fire heat input for the empirical method without adequate drainage (see Equation 2), and the analytical method for several locations around the tank car are given in Table 4. There are two locations in the rear of the tank car where the analytical method indicated higher heat inputs than the empirical method. However, when averaged across the entire tank car, as one should do if sizing a pressure relief device, the empirical method resulted in about 30% more heat input than the analytical method.

Table 4. Pool fire heat inputs using the empirical and analytical method along with BAM fire test data

Location	Max Fire Heat Flux, BTU/ft^2-h (kW/m^2)	Max Absorbed Heat Flux, BTU/ft^2-h (kW/m^2)	Aw Factor	Total Heat Input, BTU/h (kW)	% of API
Analytical - Rear center	34,560 (109)	20,330 (64.1)	1	5.07x10^6 (1,485)	159%
Analytical - Front right	2,260 (7.1)	1,720 (5.4)	1	0.43 x10^6 (126)	14%
Analytical - Rear average (Note 1)	24,350 (76.79)	15,040 (47.41)	1	3.75 x10^6 (1,098)	118%
Analytical - Front average (Note 1)	6,650 (20.97)	4,620 (14.57)	1	1.15x10^6 (337)	36%
Analytical - Total average	15,500 (48.88)	9,830 (30.99)	1	2.45x10^6 (718)	77%
Empirical method	N/A	34,500 (70.9) (Note 2)	0.82	3.18x10^6 (933)	100%

Note 1: Average of left, center and right locations.
Note 2: The API maximum absorbed heat flux has units of BTU/hr/[(ft^2)$^{0.82}$] or kW/[(m^2)$^{0.82}$].

5.2 Comparison with pool fire heat input based on BAM fire test liquid sensible heating

Test data on the sensible heating of the propane liquid was obtained during the BAM test. This data can be used as an independent means to determine pool fire heat input during the BAM test. Note that the pressure did not reach the pressure relief device opening pressure prior to rail tank car failure during the BAM test. The test data indicated an average temperature rise of 7.2 °F/min (4.48 °C/min). Hence, the calculated total heat input due to sensible heating of the liquid is about 3.805x10^6 BTU/h (1,115 kW). For comparison, the empirical method (assuming inadequate drainage) predicted a total heat input of 3.18 x10^6 BTU/h (933 kW) per Table 4. This is roughly the same as the analytical approach using only the averaged tank car rear temperature data. A possible reason for these differences is discussed below.

Liquid swelling as the liquid heats up would increase wetted surface area; however, the temperature did not increase enough during the test for it to explain the difference between the test data and the empirical and analytical methods. A likely explanation is that the embankment on three sides of the tank car heated up during the fire and caused higher heat fluxes due to re-radiation, preheating of combustion air, and enhanced heat transfer. Indeed, the fire should be classified as semi-confined because the height of the embankment walls exceeded the height of the tank car. In such cases, the API 521 empirical method (Equations 1 and 2) does not directly apply. However, the equations can be modified by using a wetted surface area (Aw) exponent of 1.0 instead of 0.82. This would be appropriate in scenarios where the pool fire flames directly and continuously contact all of the wetted surfaces. Applying this to the BAM test rail tank car assuming 50% of the rail tank car is partially confined due to the embankments on three sides results in the Equation 4:

$$Q \text{ API modified empirical method} = 70{,}900 * [(\text{Aw confined})^{1.0} + (\text{Aw open})^{0.82}] \quad \text{[Eq. 4]}$$

$$Q \text{ API} = 70{,}900 * [(11.58)^{1.0} + (11.58)^{0.82}] = 1.349 \times 10^6 \text{ Watts} = 4.604 \times 10^6 \text{ BTU/h}$$

This is a conservative estimate of the total heat input as compared with the 3.805×10^6 BTU/h (1,115 kW) determined from liquid sensible heating. Based on the test data, the analytical method should use of the rear averaged heat input predicted by the analytical model (i.e., 3.75×10^6 BTU/h (1,098 kW)) to obtain a reasonable approximation. Where validating data is unavailable, the highest heat input obtained from the analytical model should be used.

The fire relief load can be determined by dividing the fire heat input by the heat of vaporization of the fluid at relieving pressure.

These results indicate that the API empirical method can be applied to some semi-confined configurations, where adjacent embankments exceed the vessel height, by using a wetted area exponent of 1.0 instead of 0.82 for the portion of the vessel adjacent to the embankment. This would not apply to completely confined situations (e.g., enclosed buildings or structures with a roof) which would require special modeling.

5.3 Comparison with pool fire heat input based on the BRL test

The BRL test obtained data on the relieving rate versus time, which was compared with that obtained with the empirical and analytical methods. A transient approach was used in these methods whereby the relief rate was varied with time to correspond to the decrease in wetted area as fluid is relieved. A comparison of the actual relief rate and that predicted by the empirical and analytical methods is given in Figure 13. Both the empirical and analytical methods predicted a decrease in relief rate with time because the wetted area is decreasing as fluid is relieved. However, the test indicated the relief rate actually increased with time.

One explanation that can increase the relief rate versus time is that there was an increase in heat flux with time due to heating of the surroundings. Adjustments to the analytical method were made to account for enhanced heat transfer due to heat-up of the surrounding embankments during the test. Figure 13 illustrates a modified analytical method where the convective heat transfer coefficient and the vessel absorptivity were increased by 20% every 2.5 minutes with a limit of 1.0 for the absorptivity. This should be considered an example of adjustments that can be made to adjust the model to fit test data, but they may not represent actual conditions nor be applicable to other fires. Note that the fire and gas temperature were not adjusted because the data showed the fire temperature to slightly decrease during the test.

Figure 13. Empirical and analytical method calculated relief rates versus BRL test data

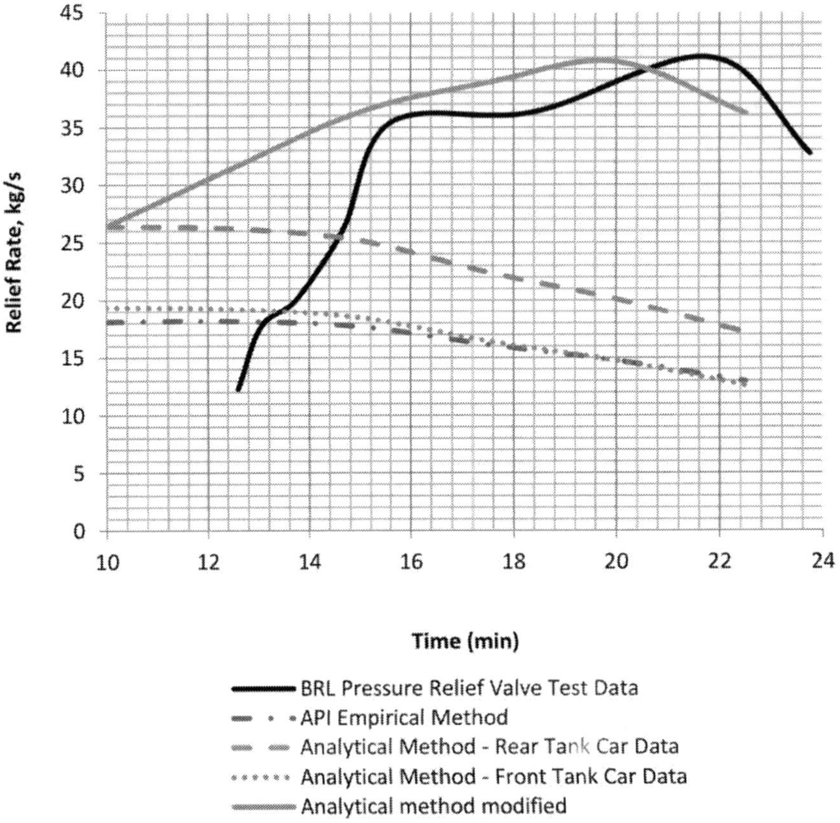

Instead of modifying the analytical method, the empirical method can be adjusted to account for the apparent increased heat input with time by increasing the exponent on the wetted area versus time. This effect can be illustrated by inserting the heat input determined from the actual relief rate and the wetted area in the empirical method for inadequate drainage; the equation is then solved for the wetted area exponent versus time. The results, shown in Figure 14, indicate the wetted area exponent approaches 1.0 toward the end of the test. This suggests that flame contact with the entire vessel surfaces increases with time. Using a wetted area exponent of 1.0 for the entire vessel (located in a pit with embankments exceeding the vessel height on all sides), with the empirical method, would provide a conservative pressure relief system design.

Figure 14. Empirical method wetted area exponent versus time using BRL test data

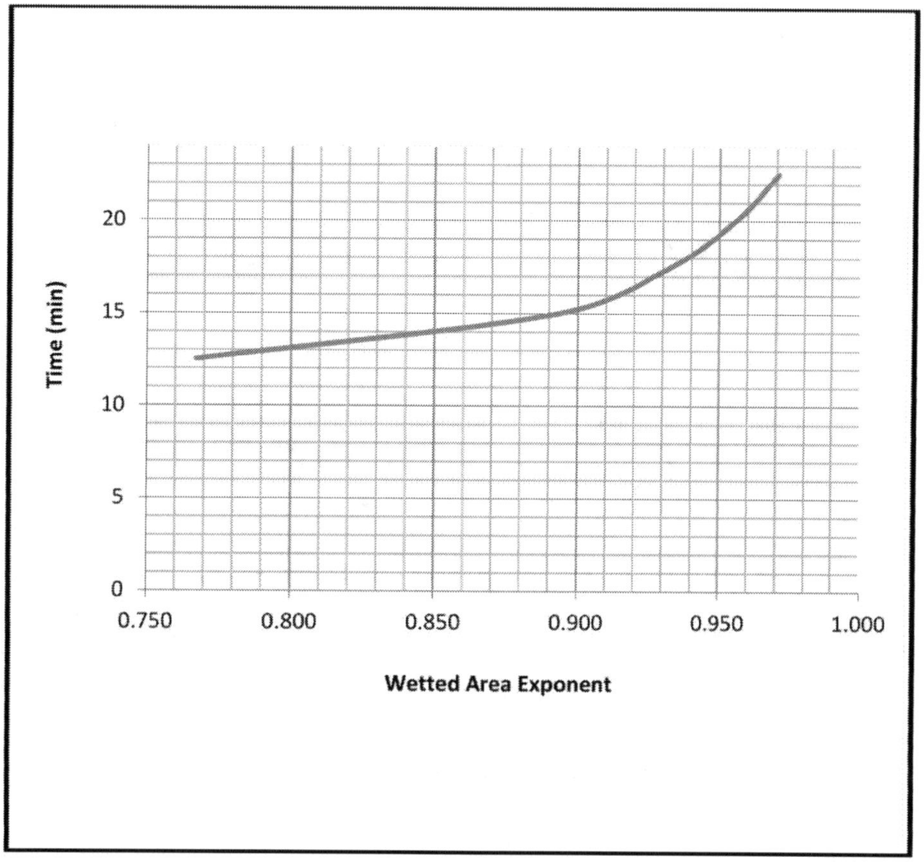

6. Application of the analytical method to depressuring system design

6.1 Effect of overheating unwetted metal plates

Unwetted metal surfaces are not cooled by boiling liquid inside the vessel. Hence, the metal temperature can get high enough such that metals such as carbon steel lose significant strength. Table 5 illustrates the effect of high temperatures on the tensile strength of carbon steel and 304 stainless steel. The loss of strength due to pool or jet fire exposure could exceed the safety factor used in the design of the vessels, thereby resulting in vessel rupture due to overheating, rather than overpressure.

Table 5. High temperature tensile strength of carbon steel and 18-8 stainless steel [16]

		18-8 Stainless Steel (304, 304L)		Carbon steel (SA-515, SA-516)	
Temperature °F	Temperature °C	Tensile Strength psi	Tensile Strength MPa	Tensile Strength psi	Tensile Strength MPa
900	482			45,500	313.7
1,000	538	53,000	365.4	36,500	251.7
1,100	593	48,500	334.4	27,200	187.5
1,200	649	43,000	296.5	20,000	137.9
1,300	704	35,000	241.3	13,500	93.1
1,400	760	27,000	186.2	9,025	62.2
1,500	816	20,500	141.3		
1,600	871	17,650	121.7		

The specific pressure vessel design code and material used will determine the appropriate safety factor to use in the vessel design. For example, the current edition of ASME Section VIII, Division 1 Pressure Vessel Design Code [3] includes a safety factor (now termed "design margin") of 3.5 between the tensile strength of the vessel and the allowable stress at room temperature for materials in which the tensile strength governs (e.g., carbon steel). For carbon steel, the safety factor implies the design pressure is a minimum of 3.5 times the burst pressure (assuming the weak link in the vessel is the wall plate, there are no imperfections in the wall, etc.).

ASME Section VIII, Division 1, UG-27 [13] provides equations that relate the allowable stress, vessel design pressure and wall thickness. In the case of circumferential stress for a cylindrical shell, the Equation [5] applies if P does not exceed 0.385SE:

$$P = S * E * t / [R + (0.6 * t)] \qquad \text{[Eq. 5]}$$

Where:

P = internal design pressure, psi
E = joint efficiency = 1.0 for full x-ray
S = maximum allowable stress value, psi
t = minimum thickness of the shell, inches
R = inside radius of the shell, inches

It should be noted that carbon steel vessels constructed to pre-1999 versions of the ASME Section VIII, Div. 1 code used a safety factor of 4.

For example, at room temperature, ASTM A515 Grade 70 carbon steel plate has a tensile strength of 70,000 psi [17;] therefore, the allowable stress will be S = 70,000 / 3.5 = 20,000 psi.

If a vessel fabricated from this material and designed to this allowable stress is heated to 1,200 °F, the tensile strength will decrease to 20,000 psi, as shown in Table 5. In other words, the material strength is reduced to the equivalent of a zero safety factor. Vessel rupture would be a certainty if the pressure then exceeded the design pressure because the loads on the vessel would exceed the tensile strength. Rupture would occur at even lower internal pressures if there are other coincidental loadings on the vessel (such as the weight of the vessel and attached equipment, temperature gradients, static head, internals, etc.) or defects in the vessel. In all these cases, a pressure relief valve would not provide protection because it is designed to reseat at its blowdown pressure and maintain pressure near the design pressure. Instead of a pressure relief valve, a depressuring system can be used to provide vessel protection, or at least mitigation of the effects of failure.

6.2 Depressuring criteria

In order to be effective, the depressuring system needs to depressure at a high enough rate to compensate for the loss of strength as the vessel heats up. The vessel heat up rate is dependent on the type of fire, materials of construction, and wall thickness.

API Std. 521 Figure 1 illustrates the heat-up of carbon steel plates of several thicknesses in an open pool fire [1]. One curve (Plate 2) was obtained from pool fire test data while the others were extrapolated based on the test data. Combining these temperature-versus-time curves along with the tensile strength data shown in Table 5 will allow determination of a minimum depressuring rate to keep the pressure below the tensile strength of the vessel. An appropriate safety factor should be considered given the uncertainties. Results, applying a 25% safety factor (i.e., Table 5 tensile strengths were multiplied by 0.75), are shown in Figure 15. The depressuring profile for a specific wall thickness needs to stay to the left of the specific curve shown in Figure 15. Failure will occur if the depressuring profile either intersects or is on the right side of the curve for the thickness in question. As noted in the previous section, failure can occur at even lower pressures, depending upon the amount of additional loads on the vessel.

An often used criteria for depressuring is to depressure to 50% of the design pressure in 15 minutes. As shown in Figure 15, this would be appropriate for vessels whose wall thickness is 1 inch or greater. A second criteria often used is to depressure to 100 psig in 15 minutes. This is more conservative and would provide protection for more high pressure equipment exposed to pool fires. The latter criteria would be preferred when protecting against jet fire exposure.

The user is cautioned that the design of a depressuring system to protect all equipment in a facility, regardless of size and/or design pressure, would be impractical in most cases. This is because the resultant large depressuring rates would likely require multiple large headers and multiple flares. The user needs to establish a list of large, high pressure equipment and determine which equipment from that list will be protected. A mitigating factor is that there are numerous small connections and piping sections in a unit which are relatively thin walled and likely to rupture in a large fire. Their rupture would provide additional depressuring capacity although there is no defensible way to quantitatively estimate it.

Figure 15. Reduction of carbon steel plate tensile strength versus time due to pool fire exposure

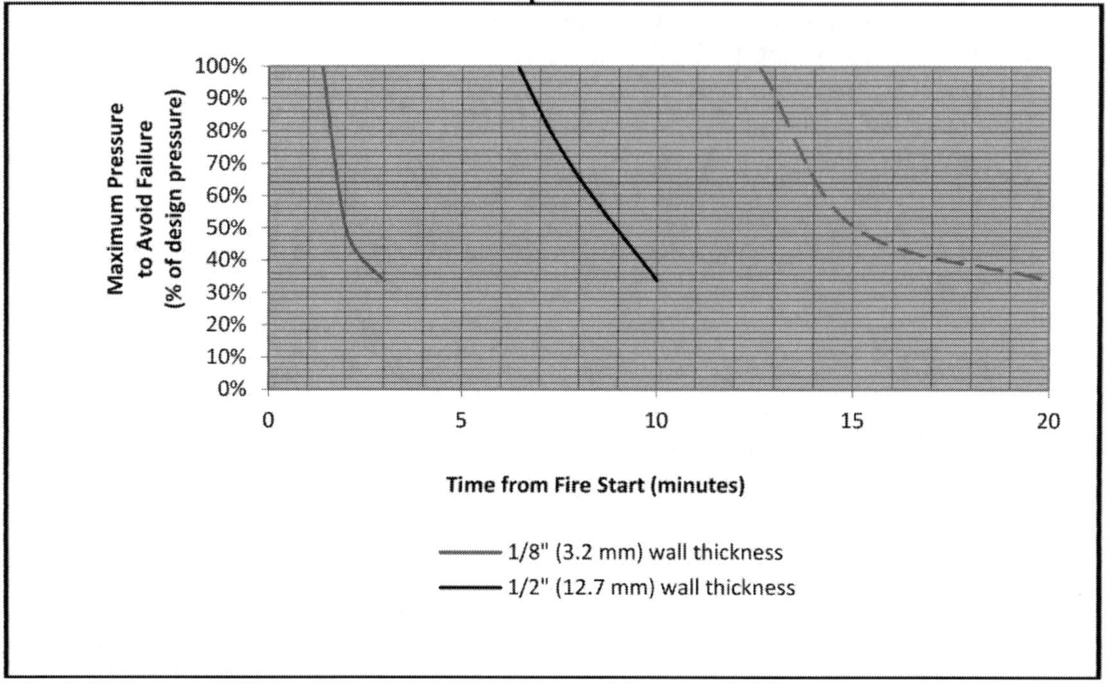

6.3 Application of the analytical method to depressuring system design

The analytical method can be used to extend the curves in Figure 15 to other wall thicknesses. The analytical method along with the parameters determined in section 6.2, for example, can be set up in a spreadsheet as a transient model in which the wall temperature change with time is calculated. At each time interval, the metal wall mass can be conservatively assumed to absorb all of the heat input, thereby increasing the wall temperature. The effect of wall thickness is accounted by the metal mass. This temperature-versus time profile is then combined with tensile strength data as in Section 6.2.

For example, the BAM pool fire test data indicated failure of the rail tank car occurred at rear center wall (in unwetted zone) [14] [15]. Test data further indicated the wall temperature ranged from 1,020 to 1,200°F (550 to 650°C), but it is possible that local temperatures got even higher because temperature recorded only at a few locations. Failure occurred 15 minutes after the start of the pool fire, or about 10 minutes after the fire temperature reached about 1,832°F (1,000°C). The rail tank car wall thickness was 0.59 inches (14.9 mm) and the material of construction was assumed to be comparable to carbon steel. The failure pressure of 362 psig (25 bar) was slightly lower than the test pressure of 406 psig (28 bar). The "Rear Center" parameters were used in the analytical model to predict the time-versus-temperature profile. This was combined with the tensile strength and stress rupture data to obtain the depressuring profile shown in Figure 16. In order to minimize the potential for rupture due to overheating, a depressuring system would need to stay to the left of the curve shown in Figure 16. Because the pressure at failure was slightly lower than the test pressure, Figure 16 predicts that failure would occur about 14 minutes after the start of the pool fire, which is a reasonable approximation as failure actually occurred about 15 minutes after the main pool fire started (see Figures 10 and 11). Note the first 2 minutes of the pool fire test is not considered because the fire was localized to a small igniter assembly that did not cause any significant increase in rail tank temperatures.

Figure 16. Depressuring profile to minimize failure potential of the rail tank car due to overheating in the BAM pool fire test

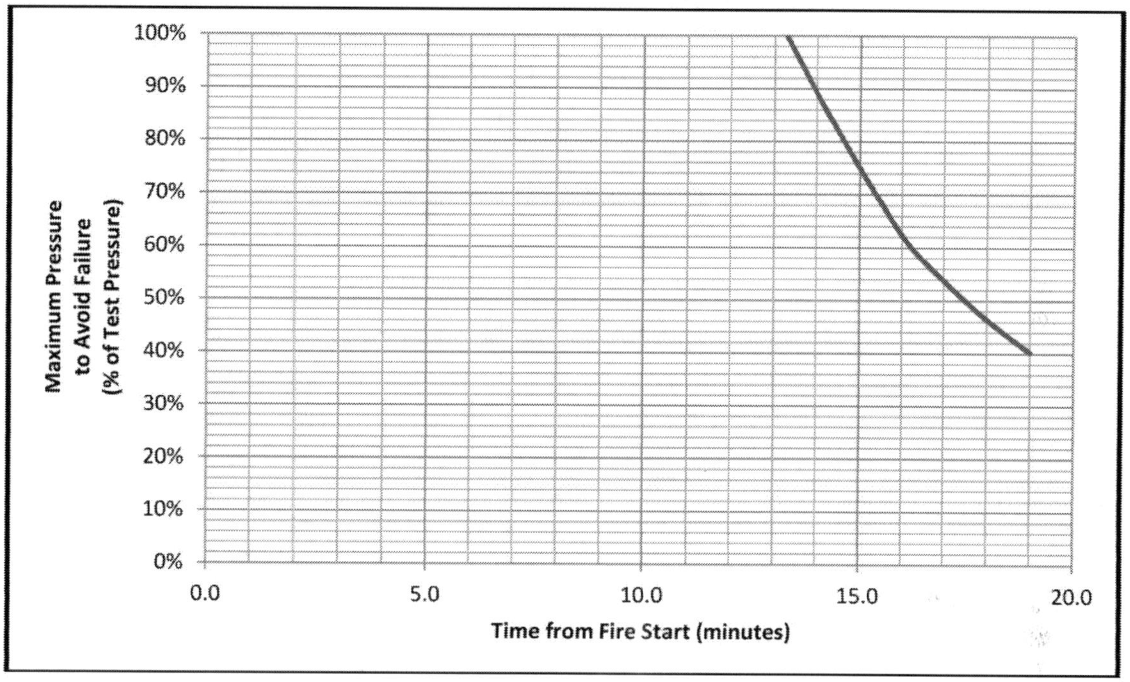

6.4 Effect of material of construction

The material of construction can significantly affect the depressuring requirements. The preceding sections discussed carbon steel vessels. As shown in Table 7, 304 stainless steel is superior to carbon steel regarding high temperature effects on tensile strength. A comparison of the depressuring profiles to minimize the potential for failure of a ½ inch wall thickness carbon steel vessel and a ½ inch wall thickness stainless steel vessel is illustrated in Figure 17. The depressuring system pressure-versus time profile would need to stay to the left of the applicable curve. These results indicate that the depressuring system for the stainless steel vessel would require a significantly lower depressuring rate than for the carbon steel vessel of comparable wall thickness. This method can be extended to other materials provided tensile strength data at high temperature is available.

Figure 17. Depressuring profiles to minimize failure potential of a 0.5 inch wall thickness carbon steel and stainless steel vessel due to overheating

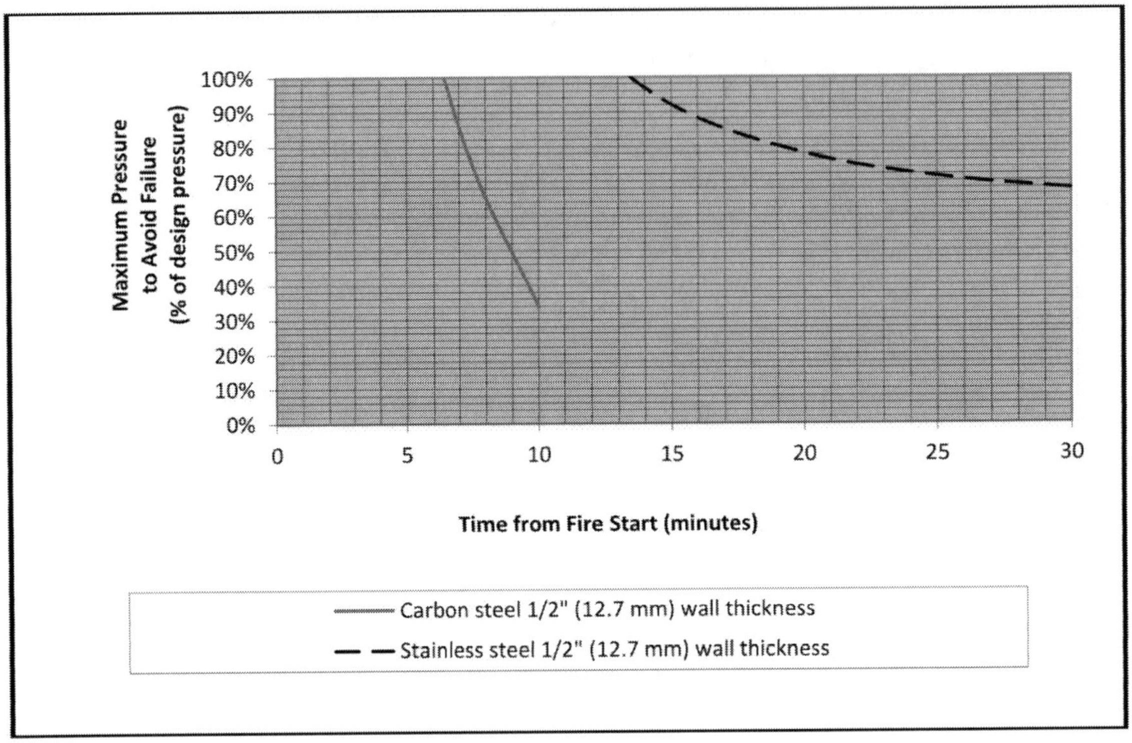

7. Conclusions

The increasingly widespread use of analytical methods to evaluate fire exposure of equipment prompted the API Std. 521 committee to include an analytical method in the upcoming 6th edition as an alternative to the existing empirical method. The analytical method provides more flexibility than the empirical method but has limitations (e.g., too many permutations are possible). More recent pool fire test data indicates the empirical method will provide a conservative estimate of pool fire heat input for most applications and is still the method of choice when designing pressure relief systems. However, these recent tests indicate the empirical method needs to be modified when a vessel or equipment is partially confined by adjacent embankments or walls equal or greater than the vessel height. In such cases, the wetted area exponent should be 1.0 instead of 0.82.

The analytical method is useful in determining time-versus-temperature profiles for heating unwetted vessels of varying wall thicknesses and materials of construction. These profiles can be combined with tensile strength and stress-rupture data to specify a depressuring system's pressure-versus-time profile to minimize failure and/or mitigate the effects of failure due to overheating from a pool or jet fire exposure.

8. References

[1] ANSI/API Standard 521, "Pressure-relieving and Depressuring systems", 5th Edition, Addendum, April 2008.

[2] ANSI/API Standard 2000, "Venting Atmospheric and Low-pressure Storage Tanks", 6TH Edition, November 2009.

[3] ASME Section VIII, Division 1, Pressure Vessel Code, 2007 with 2008a Addenda.

[4] API Standard 650, "Welded Tanks for Oil Storage", 11th Edition, October 2011.

[5] NFPA 30: Flammable and Combustible Liquids Code

[6] Personal correspondence from H.C. Hottel to L.W.T. Cummings December 12, 1950.

[7] Institute of Petroleum (Energy Institute, "Guidelines for the Design and Protection of Pressure Systems to Withstand Severe Fires", March 2003, ISBN 0 85293 279 0

[8] SCANDPOWER, "Guidelines for the Protection of Pressurised Systems Exposed to Fire", Report no. 27.207.291/R1 - Version 2, March 31, 2004

[9] Roberts, T. A.; Medonos, S.; Shirvill, L. C., "Review of the Response of Pressurised Process Vessels and Equipment to Fire Attack", OFFSHORE TECHNOLOGY REPORT, OTO 2000-051, June 2000

[10] Shirvill, L. C., "Heat Fluxes in Severe Fires", 2002 presentation to API Pressure Relief Systems Committee

[11] Salater, P., Overaa, S. J., Kjensjord, E., "Size Depressurization and Relief Devices for Pressurized Segments Exposed to Fire", Chemical Engineering Progress, September 2002, p38

[12] Salater, P., "Proposed changes to the next revision of API 521", 2006 presentation to API Pressure Relief Systems Committee

[13] Anderson, C., Townsend, W., Zook, J. and Cowgill, G., "The effects of a fire environment on a rail tank car filled with LPG", FRA-OR&D Report Number 75-31, PB-241358, September 1974.

[14] Ludwig J. and Heller, W., "Fire test with a propane tank car", BAM Test Report III.2/9907, 1999.

[15] Balke, C.; Heller, W., Konersmann, R., Ludwig, J. , "Study of the failure limits of a railway tank car filled with liquefied petroleum gas subjected to an open pool fire test", BAM Final Report September 13, 1999.

[16] Wharton, H. R., "Digest of Steels for High-Temperature Service", published by Timken Steel, 1946

[17] ASME Section II, Part D, Materials – Properties, 2007

E.I. DuPont de Nemours & Co., Inc. Flammable Vapor Explosion

Mark Wingard
U.S. Chemical Safety and Hazard Investigation Board
2175 K St NW
Suite 400
Washington D.C. 20037
mark.wingard@csb.gov

Prepared for Presentation at
American Institute of Chemical Engineers
2013 Spring Meeting
9th Global Congress on Process Safety
San Antonio, Texas
April 28 – May 1, 2013

UNPUBLISHED

E.I. DuPont de Nemours & Co., Inc. Flammable Vapor Explosion

Mark Wingard
U.S. Chemical Safety and Hazard Investigation Board
2175 K St NW
Suite 400
Washington D.C. 20037
mark.wingard@csb.gov

Keywords: hot work, explosion, incident, process safety

Abstract

This case study examines a November 9, 2010, explosion at an E.I. duPont de Nemours and Co. Inc., Yerkes chemical plant in Buffalo, New York which killed a contract welder and injured a foreman who were repairing the agitator support atop an atmospheric storage tank containing flammable vinyl fluoride. The U.S. Chemical Safety Board (CSB) determined that flammable vinyl fluoride (VF) vapor from interconnected, in-service process tanks flowed undetected into the tank and ignited when the welder was repairing the agitator support assembly.

1 Incident Summary

This case study examines a November 9, 2010, explosion at an E.I. duPont de Nemours and Co. Inc., Yerkes chemical plant in Buffalo, New York when a contract welder and foreman were repairing the agitator support atop an atmospheric storage tank containing flammable vinyl fluoride. The welder died instantly from blunt force trauma, and the foreman received first-degree burns and minor injuries. The explosion blew most of the top off the tank. The top and agitator assembly hung over the side of the tank supported only by a 2-foot section of the top (cover photo). The explosion caused minor overpressure damage in the tank farm area and the adjacent production building.

The U.S. Chemical Safety Board (CSB) determined that flammable vinyl fluoride (VF) vapor from interconnected, in-service process tanks flowed undetected into the tank and ignited when the welder was repairing the agitator support assembly. In February 2010, the CSB issued a "Hot Work Safety Bulletin" that summarized 11 similar fatal incidents. Like the incidents described in the bulletin, this was another example of improperly monitored hot work activities involving flammable conditions inside a container.

2 Background

2.1 *DuPont*

E.I. duPont de Nemours and Co., Inc (DuPont) is a Fortune 100 company which was founded in Wilmington, Delaware in 1802. The American chemical company operates in about 90 countries and employs more than 60,000 people. The company offers a broad range of products for industry and consumer use, including pesticides, electronics, apparel, and biomedical supplies. The company offers services as a safety resource for other corporations to evaluate and improve workplace safety.

DuPont has owned and operated the 100 acre facility in Buffalo, New York since 1921. The facility employs over 600 workers and manufactures Tedlar®, a polymer used as a film in photovoltaic panels and Corian®, used to make countertops and other durable surface products.

2.2 Maintenance Contractor

Mollenberg-Betz, founded in 1910, operates a large steel fabrication facility in Buffalo, New York, and provides piping system design, fabrication, field construction and maintenance services. It employs more than 75 professionals and skilled construction trade staff. Mollenberg-Betz had contractors working on site at the DuPont facility repairing a steam line. DuPont used Mollenberg-Betz to repair the heavily corroded steel on the agitator support structures atop tanks 1 and 2 due to their availability on site. At the time of the incident, the Mollenberg-Betz contractors had been working at the DuPont facility for two months. The welder and foreman assigned to the work had many years of piping system and tank welding experience.

2.3 Process Description

The Tedlar process converts VF into polyvinyl fluoride (PVF).VF is generated at another DuPont facility, shipped in on tanker trucks and stored at the Buffalo facility in high pressure storage tanks. VF is pumped from the storage tanks to a reactor and reacts to form PVF slurry in water and unreacted VF. After the reactor, the PVF water slurry passes through separators to remove the VF. The VF is pumped from the separators by compressors and recycles back to the reactor. After the separators, steam is injected into the PVF slurry to raise the temperature and vaporize any VF present in the slurry. The heated mixture passes through a small slurry flash tank where the residual VF is released to the atmosphere[1].

The non-combustible PVF slurry then passes to one of three insulated slurry holding tanks (numbered 1, 2, or 3) in the tank farm adjacent to the production building (Figure 1). Under normal operating conditions, slurry is pumped to slurry tank 3. Slurry tank 3 is newer than slurry tanks 2 and 1; it had been replaced in 2009 due to corrosion[2]. Slurry tanks 2 and 1 were used as overflow storage space in the event slurry tank 3 was filled and slurry tank 1 was generally kept empty of slurry.

If the slurry flash tank level is too high, hot slurry passes through the flash tank overflow line directly into slurry tank 2. A liquid trap (seal loop) on the end of the overflow line inside slurry tank 2 was designed to prevent VF vapor and steam from passing directly from the flash tank into the slurry tank. The PVF slurry is then pumped from the slurry tanks to the cpro\ production area for further processing. PVF is used in a variety of applications as a film or a surface protector due to its resistance to weathering and flammability-lowering properties.

---

[1] The Buffalo facility normally vents approximately 5 pounds per hour of VF to the atmosphere from the flash tank..

[2] Slurry tank 1 was removed after the 2010 incident.

2.3.1 Physical, Chemical, and Biological Properties of Vinyl Fluoride

VF is a colorless, highly flammable gas at ambient conditions with an ether-like odor.[3] VF vapors are heavier than air and will accumulate in low areas. VF is classified as a reasonably anticipated human carcinogen[4]. National Institute for Occupational Safety and Health recommends an exposure limit of 1 ppm as a time weighted average and a ceiling limit of 5 ppm for workers.

[3] The VF lower explosive limit is 2.6 percent and the upper explosive limit is 21.7 percent. The autoignition temperature is 725° °F (385 °C).

[4] According to the National Toxicology Program from the National Institute of Environmental Health Sciences in its 2011, 12th Edition of *Report on Carcinogen*

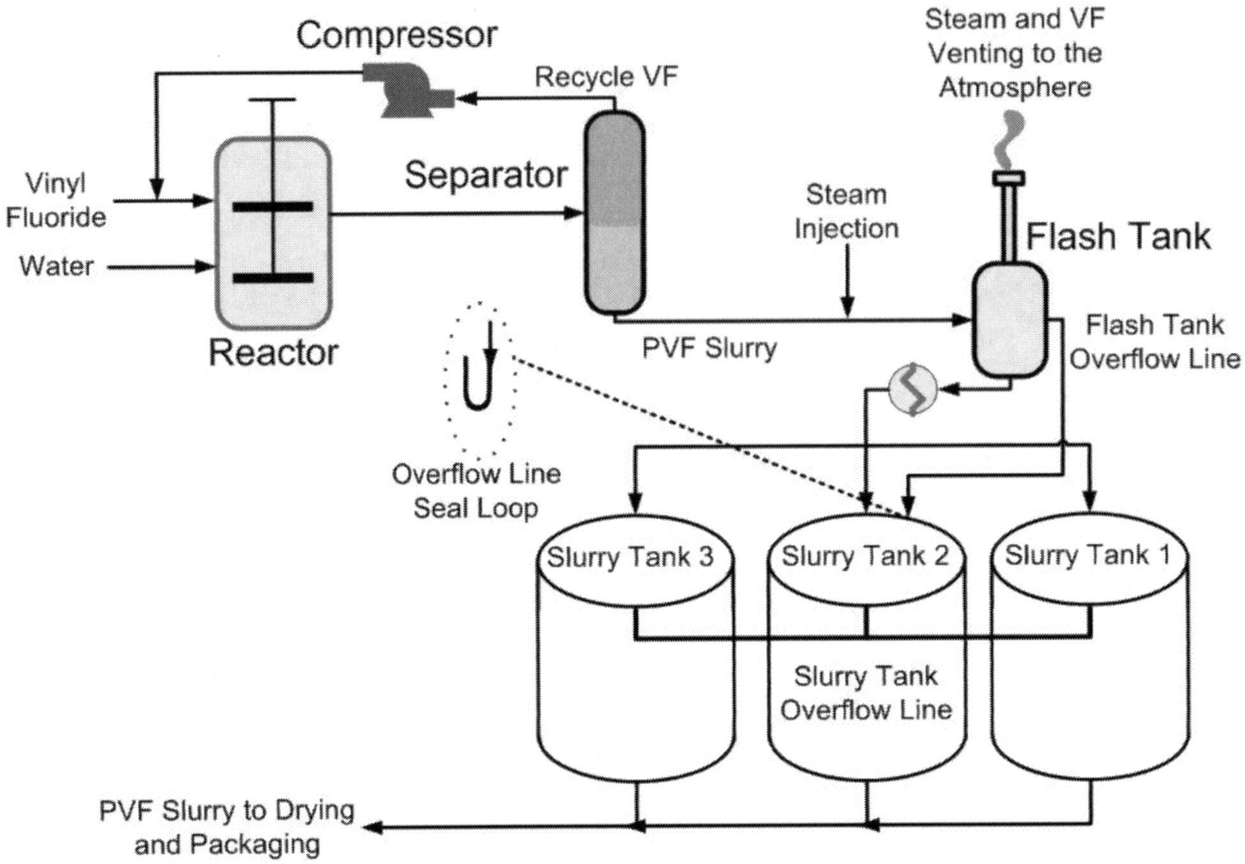

Figure 1. PVF process flow diagram

2.4 PVF Slurry Tank Farm

The flash tank and three slurry tanks are located in a tank farm adjacent to the Tedlar manufacturing building. The slurry tanks are 11 feet in diameter, 19 feet tall, and have a capacity of 10,800 gallons (Figure 2). The tank shell and top are one-quarter inch thick stainless steel. Each tank top has a hinged, unsealed steel cover on a 24-inch diameter manway and a large agitator motor drive and gearbox[5]. The agitator assembly is bolted to structural steel beams that are welded directly to the tank top.

The slurry tank feed and drain lines are equipped with isolation valves. The slurry tanks have a common overflow line attached near the top of the tanks. Blind flanges are installed on the overflow line and isolate one tank from another if one or two tanks are operating while the third is undergoing maintenance.

The 3-foot diameter flash tank vents small concentrations of flammable vapor directly into the ambient atmosphere, so the area is equipped with a flammable vapor air monitoring devices to notify personnel when explosion hazards are present. The continuous air monitors are located on the catwalk near slurry

[5] The PVF slurry tank farm process area was classified as a general purpose electrical classification area at the time of the incident at ground level, on top of the tanks, and on the tank farm elevated catwalk.

tank 3 and on the flash tank vent pipe (Figure 3). An alarm in the Tedlar control room activates if either device detects flammable vapor above the instrument set point.[6]

[6] The low alarm set point is 10 percent of the lower explosive limit (LEL) and the high alarm set point is 2 percent of the LEL.

Figure 2. PVF slurry tanks

Figure 3. Slurry flash tank (left). Agitator and open manway on tank. Second air monitor is located on catwalk near tank 3 (lower right).

3 Incident Description

3.1 Events prior to incident

The Tedlar process area conducted a planned shutdown from October 22 to November 6, 2010, ending 3 days before the fatal explosion. During this shutdown, the asbestos insulation on slurry tanks 1 and 2 was to be removed. New, non-asbestos insulation was to be installed and the slurry tanks would be externally inspected for corrosion similar to that previously found on slurry tank 3. The process was shut down on the evening of October 21 and all slurry was pumped out of slurry tanks 2 and 1.

On October 22, the slurry tanks were locked out[7] by DuPont maintenance personnel so the tanks could be cleaned and entered. On October 29, DuPont discovered damaged agitator support after the insulation was removed from the slurry tanks 1 and 2. A DuPont engineer wrote up a work order for the repair of the agitator supports and had the construction field engineer in charge of contractors begin writing a work

[7] OSHA defines lock out as "The placement of a lockout device on an energy isolating device, in accordance with an established procedure, ensuring that the energy isolating device and the equipment being controlled cannot be operated until the lockout device is removed."

scope and gather contractors and materials. DuPont hired Mollenberg-Betz to repair the agitator support based on the fact that they were already engaged in hot work on site and could be quickly made available.

The valves on the tank 1 fill and discharge lines were locked out November 1. Following the completion of DuPont site lockout and hot work safety procedures, the contractor performed the grinding and welding repairs atop tank 2 on November 1 (Figure 4), but had to delay the tank 1 work because repair materials were unavailable. DuPont personnel determined the slurry tank 1 job could be safely completed after the process restarted and scheduled the repairs for November 9, when the Mollenberg-Betz contractors would again be available.

During the tank 2 internal inspection on November 3, DuPont engineers discovered that the U-leg seal loop on the flash tank overflow line had a "fishmouth" split in the pipe (Figure 5), likely the result of the PVF water slurry freezing in the line. The engineers concluded that the slurry tank could be returned to service without repairing the split. DuPont engineers who saw the broken seal loop concluded its likely purpose was to limit steam in the flash tank from flowing into the slurry tanks. They further concluded that the operation was safe and that the broken seal loop presented no hazards. No acknowledgement was given to the fact that flammable VF gas had a pathway into the slurry tanks, an area classified as "low hazard" and not expected to contain any flammable materials. DuPont engineers scheduled a repair on this seal loop during the next planned unit outage. No management of change (Section 4.7.2) was carried out for the process to continue with this compromised seal loop.

A DuPont crew reconnected the tank 2 and 3 process piping and removed the valve locks to prepare for unit restart. Tank 1 remained out of service while awaiting the materials needed to repair the agitator support. The overflow line to all three tanks was never blinded[8], which provided a direct flow path from the vapor space in tanks 2 and 3 to the out-of-service tank 1.

---

[8] Blinding is the absolute closure of a pipe by fastening of a solid plate that completely isolates the interior of the pipe and is designed to withstand the maximum pressure generated within the pipe.

Figure 4. Repair weld on agitator support beam to the slurry tank 2 top plate

On November 6, DuPont restarted the Tedlar process with the valves aligned so that the PVF slurry flowed into slurry tank 3. The overflow line remained connected to all three tanks, and was not isolated or disconnected from tank 1 before the hot work was authorized. On November 8, a compressor within the Tedlar unit malfunctioned. Although unit operation without the ring compressor was listed in the Tedlar® operating procedures, and was acceptable as part of the site environmental emissions, restarting without the compressor, significantly increased the VF vapor present in the PVF slurry flowing into the slurry flash tank. The process operated in this mode until the incident.

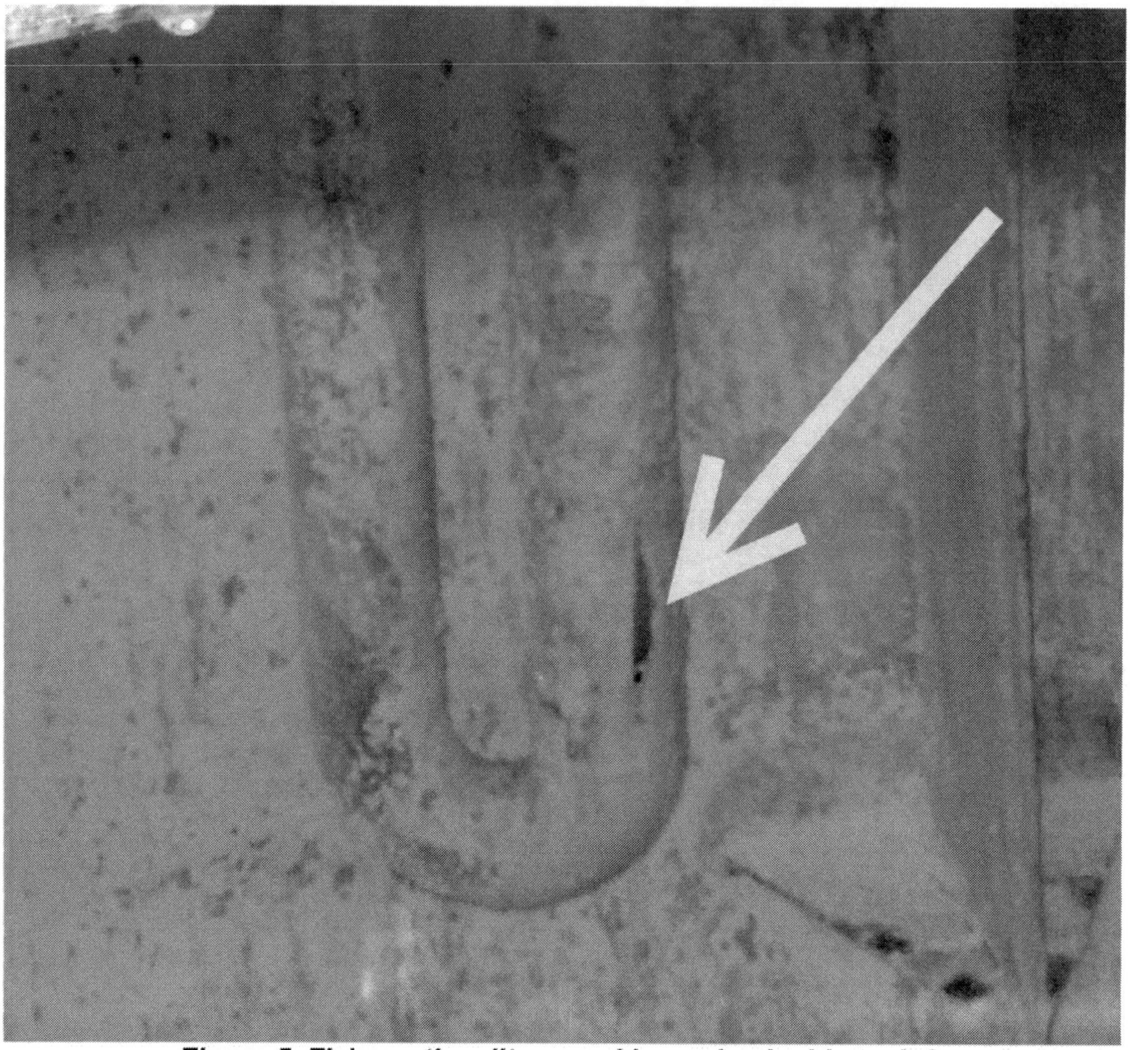

Figure 5. Fishmouth split on seal loop pipe inside tank 2

3.2 Permits in Place on Day of Incident

On November 7, 2010, a lockout card[9] completed for tank 1 work indicated that that all five valves leading to and from tank 1 and the agitator motor had been locked out. The card instructed workers to try to start the motor before beginning work, but did not reference the overflow line, which had no valves. On November 9, 2010, the contractors completed a hot work permit before starting work on slurry tank 1. The contractors did not check the lock out valves on slurry tank 1 with any DuPont employees prior to starting the hot work.

[9] A lockout card details what has been done to ensure an area has been effectively locked out for an area or a piece of equipment.

3.3 Incident Description

The morning of November 9, a DuPont lab technician tested the area around the top of the slurry tanks for flammable vapor concentration, including the region above tanks 2 and 3 and the immediate area above tank 1 where the repair work would be performed. The testing indicated there were no flammables present around the top of slurry tanks. Continuous air monitoring was also present near the top of tank 1, but the lab technician never tested the atmosphere inside the tank. The lab technician was not required to test the air inside tank 1 for flammable chemicals even though the repair work involved significant grinding and welding directly on the tank top. Nor did the DuPont construction field engineer or contractor crew ask the lab technician to test the air inside the tank.

Shortly after 9 am, the contractor crew went to the catwalk above tank 1 and started work. The contractor attached his safety harness lanyard to the agitator on tank 1, stepped out onto the tank top, and began the grinding and welding repair work. The foreman remained on the catwalk adjacent to the closed manway cover to supervise and act as fire watch. At the time of the incident the contractor was using an electric arc welder to weld metal support on to a C-channel beam on the top of slurry tank 1. The C-Channel beam was then to be welded to the agitator support.

At 11:04 am the tank 1 level transmitter in the control room recorded a sudden increase as a fire erupted inside the tank. The overpressure blew the steel manway cover off the tank, hurling it more than 100 feet. Nearly the entire 25-feet-long tank top-to-shell circumferential weld tore apart; only a segment about 24 inches long remained connected. This segment acted like a hinge as the top, agitator motor and drive assembly, and agitator shaft violently swung over the side of the tank (see cover photo). The welder died instantly in the explosion. The flash fire burned the foreman's arms and head, burst an eardrum, and scratched his eye. The fire quickly consumed the flammable vapor and self extinguished.

4 Incident Analysis

4.1 Fuel Source

The fuel source for the hot work explosion was flammable vinyl fluoride in slurry tank 1. DuPont technical engineers in process hazard analyses carried out before the incident assumed that any residual VF gas would vent safely to the atmosphere more than 10 feet above the work area through the flash tank vent. However, the defective seal loop provided a direct path for the flammable vapor and uncondensed steam to flow from the flash tank into tank 2. VF was also present in small amounts in the PVF slurry entering slurry tank 3. DuPont modeling published in July 2010 and based on data obtained from slurry samples on the Tedlar process calculated that 0.02 pounds per hour of VF would be present in the slurry going to the slurry tank. Prior to the incident, other DuPont models had stopped after the separators and made the assumption that any VF present would vent from the flash tank. DuPont did not see this as a hazard because this 0.02 pounds per hour is below the LEL for VF. DuPont did not consider that VF could accumulate in the large vapor space in the slurry tanks and reach flammable levels.

Prior to the incident DuPont calculated that the loss of the liquid ring compressor would more than double the amount of VF going to the flash tank and venting. The VF present in the flash tank had an open path to slurry tank 2 through the cracked seal loop. VF from the flash tank was the main source of flammables on the day of the incident.

The VF present in the slurry in tank 3 and from the defective seal loop in tank 2 had an unimpeded path to slurry tank 1 through the unblinded overflow line. VF, which is heavier than air accumulated to a flammable concentration in the bottom of the tank. As the steam and VF entered slurry tank 1, the steam condensed, due in part to the earlier removal of insulation from the slurry tank. The slurry tank insulation had been removed during the shutdown and not replaced and the condensing steam exacerbated the VF flow into slurry tank 1 at increased concentrations.

DuPont personnel sampled the air around the work area for flammable gas as required by the company hot work permit procedure; however, neither they nor the contractor crew recognized the possibility of VF vapor entering the tank. The sampling procedure did not require testing the atmosphere inside the tank even though the work required welding directly to the tank top. If air monitoring had been done inside slurry tank 1 prior to and during hot work, the flammable VF would have been detected and the incident would not have occurred.

4.2 Ignition Source

The ignition source came from the repair work activity of welding the agitator support on the tank top. The agitator shaft passed through an unsealed hole, approximately one-half inch, in the tank top, which provided a path for sparks to enter the tank or for flammable vapor to escape into the work area.

Welding the C-channel on the tank top using the arc welding technique increased the metal temperature above the steel melting point of 2800 °F (1538 °C). The surface of the metal would quickly get "red hot" as welding progressed, a temperature in excess of 1200 °F (650 °C), which was significantly above the VF vapor autoignition temperature of 725 °F (385 °C). The accumulated VF vapor inside the tank would ignite as soon as it contacted the extremely hot steel or if welding sparks entered the tank through the agitator shaft opening.

4.3 Hot Work Procedure

Hot work is any flame or spark producing operation including welding, grinding, and riveting. The DuPont corporate *Contractor Safety Handbook* states that a "Welding, Open Flame, and Sparking Equipment Permit" is required for flame or spark-producing activity at DuPont facilities. The two contractors who performed the hot work on slurry tank 1 completed such a hot work permit, titled the "Yerkes Work & Flame Permit," prior to starting work. The purpose of the permit was to ensure that, prior to any hot work, communication between the supervisor of the hot work and a proprietor of the area where the work was to be performed was appropriate, and to ensure the hot work was carried out safely. The permit indicated the scope of the work and listed needed safety precautions.

DuPont operating procedure "Yerkes Work & Flame Permit" provided guidance on completing the hot work permit of the same name at the Buffalo plant. The permit was required prior to the initiation of any hot work. This document was designed primarily for contractor use. As part of the hot work permitting procedure, contractors were required to complete a safety task assignment (STA) to address potential hazards in the hot work area. All employees doing the hot work were required to read this STA and sign that they had read it.

After the contractor completed the paperwork, the area proprietor[10] would sign the permit and notify the contractor of any process changes or safety information that might affect the job. This required the area proprietor to be knowledgeable about processes going on in the area. This procedure was in accordance with the guidelines set forth in the National Fire Protection standard NFPA 51B which states the individual authorizing the permit will, (Section 4.2.3)"ensure the protection of combustibles from ignition", and (Section 4.1.7) "advise all contractors about site-specific flammable materials, hazardous processes or conditions, or other potential fire hazards." The contractors had completed this same permit before doing hot work on slurry tank 2 the previous week.

In the hot work procedure, DuPont recognized that the contractor may be unfamiliar with process safety or activities in the area they would be working. DuPont determined that the construction field engineer[11] and the area proprietor would be responsible for helping the contractor understand potential hazardous conditions. On the day of the incident however, neither the construction field engineer nor the employee signing off as the area proprietor had understanding of the area, the process, or potential hazardous conditions.

4.4 DuPont Hot Work Permitting

On the morning of November 9, 2010, the contractor supervisor completed the required hot work permit for the repair work on slurry tank 1. The permit specified that the Mollenberg-Betz contractors would be burning, welding, and grinding work on the top of slurry tank 1. The requirement section of this permit asked if flammable material would be within 35 feet of the work. This section was not completed, even though the hot work was carried out within 35 feet of the slurry flash tank that was designed to vent VF to the atmosphere. The contractors were unfamiliar with the Tedlar process and the process equipment involved. The contractors did not know what the flash tank was or which chemicals were present inside it.

The safety precautions section of the permit included an item for "Lock, Tag, Try, Think" that was not checked as necessary, even though slurry tank 1 had been cleaned and locked out for the hot work. In another part of the permit the contractor supervisor indicated that lockout had been completed for this work.

The contractor supervisor also completed a safety task assignment with the construction field engineer as part of the permit. On this part of the permit the supervisor did not answer the questions "Are chemicals present? What chemicals?" and "Flammables/Combustibles present? Identify."

The "Yerkes Work & Flame Permit" requires signatures from two DuPont employees, an area proprietor and construction engineer, before contractors can begin work. DuPont's intent in structuring hot work procedures, from the corporate to the plant level, was to ensure that these individuals would understand the work and potential hazards.

In addition, the DuPont construction engineer for the slurry tank work had no working knowledge of the Tedlar process. He would not have known the effect of the loss of the compressor on increased VF in the system. On the morning of the incident he reviewed the work permit with the contractor supervisor. Since the contractor had performed hot work on slurry tank 2, the construction engineer expected the contractors

[10] An area proprietor is a DuPont employee who is knowledgeable about the area where any work would be done.
[11] The construction field engineer is the DuPont employee responsible for coordinating the contractor's work.

to obtain the proprietor signature from someone in the area where they were working, as required in the DuPont policies, and that the area proprietor would provide the contractors any plant-specific, process safety information necessary for the hot work.

After the construction engineer signed the permit, he obtained the proprietor signature from someone in the service department. The service department is responsible for mail, yard work, and related tasks. It was not located near nor was it related to the Tedlar process. DuPont policy states that the proprietor should be "knowledgeable" about the area they are signing off on and that the proprietor should "walk down" the area where the hot work is to be done. On the day of the incident, this was not done.

The contractors would not have known about any of the chemicals present in the Tedlar area or if they were flammable or combustible; neither did they have knowledge of the Tedlar process nor if flammable vapors might be present around the slurry tanks. Despite this they were allowed to complete the hot work permit and begin hot work without getting approval from any DuPont employee knowledgeable about the process.

Prior to this incident, the DuPont Buffalo facility allowed someone in the service department, rather than someone knowledgeable about the area, to sign off on contractor work permits. The individual in the service department who signed as area proprietor had no knowledge of the Tedlar area or any associated dangers and had been signing permits for the Mollenberg-Betz contractors for months while they did hot work in other parts of the Buffalo facility. Information gathered from CSB interviews indicated the DuPont service worker believed that he was simply releasing them to do their jobs and that the construction engineer had briefed them on the job and hazards. He was not aware that he needed to know about potential hazards or that he was not the appropriate DuPont employee to sign these permits. On the day of the incident the service department worker was still under the assumption that the Mollenberg-Betz contractors were working on the steam line and believed that was the purpose of the hot work permit.

DuPont's policies and procedures are structured to ensure that someone knowledgeable checks before hot work is performed. However, on the day of the incident, no individual with plant-specific knowledge of the Tedlar process reviewed the hot work permit. DuPont's practices had fallen short of its policies..

4.5 Lockout Procedure

DuPont's corporate *Contractor Safety Handbook* requires a lockout procedure to protect workers from injury caused by energized equipment, the opening of valves, or exposure to electrical sources. The DuPont Buffalo facility's *Yerkes Contractor Safety Handbook* states that any equipment locked for safety must be deactivated, tagged, cleared, and tried. Contractors working in a locked out area were responsible for verifying that the lockout procedure had been completed. The handbook also states that isolating piping systems in hazardous processes should include block valves or a blind flange. Piping systems were to be examined after lockout to ensure they were isolated.

DuPont Buffalo's procedure "Lockout Procedure Lock, Tag, Try, and Think" describes the lockout/tagout procedures at the Buffalo facility. This document states that "all sources of hazardous energy shall be removed or controlled prior to potential exposure to the hazards." The list of examples of hazardous energy sources does not include chemicals or fire hazards however the OSHA standard for lockout, Control of Hazardous Energy (29 CFR 1910.147) includes chemicals as energy sources for which a lockout procedure must address (Section 5.0). DuPont designated the area job representative, usually a DuPont field engineer, as responsible for securing required lockouts for contractor work. In addition, the field engineer and the contractor foreman were required to verify that the lockout procedure was done. The lockout card for slurry tank 1 did not include a requirement for blinding the overflow line; that is, tank 1 was considered completely "locked out" with the overflow open to the tank.

4.6 OSHA Citations

As a result of this incident, OSHA investigated and issued ten citations to DuPont, and eight to Mollenberg-Betz. All of the citations referenced either OSHA regulation 29 CFR 1910.147 or 29 CFR 1910.252 which regulate lockout/tagout and hot work, respectively. OSHA found several deficiencies in DuPont's lockout and hot work procedures. DuPont was fined for not including steps to isolate the slurry tank overflow line in their energy control procedure, and for not installing blinds in the overflow line.

OSHA also found that DuPont had not provided the Mollenberg-Betz contractors with specific control procedures for de-energizing slurry tank 1. OSHA fined Mollenberg-Betz because the contractors did not ensure that the lockout procedure had been completed and that the tank was not properly isolated. OSHA fined DuPont and Mollenberg-Betz for not informing their employees about the potential explosion hazards related to hot work on slurry tank 1.

4.7 Process Safety Management (PSM)

4.7.1 Process Hazard Analysis

In its process hazard analysis (PHA) for the Tedlar process, DuPont considered whether VF could ever enter the slurry tanks. In the PHA process description, DuPont stated that its separators would recover "all the VF in the slurry" and that the steam injected before the slurry flash tank "allows the remaining VF to be flashed off". DuPont told the CSB that "no tests were conducted prior to the incident to determine whether vinyl fluoride ("VF") was entering the slurry tanks".

The PHA also considered what would happen if the liquid ring compressor failed. DuPont reached the conclusion that "additional VF will leave through the flash tank vent". No consideration was given to VF making it past the flash tank to the slurry tanks.

Despite the assumption in this PHA that VF was not present in the process after the flash tank during normal operating conditions, DuPont evaluated the consequence of "more VF in slurry going to Slurry Tanks" if the steam injection failed. DuPont found that in this theoretical situation "VF left in slurry would be max 300 ppm [parts per million]" and would not pose a safety risk. In a 2009 PHA carried out on the downstream portion of the Tedlar process, DuPont considered the potential of VF in the slurry past the flash tank. DuPont determined that small amounts of VF may be present in the slurry "at a concentration below the LEL" (lower explosive limit). No recognition was given that VF below the LEL in a slurry tank with a large vapor space could accumulate until it reached explosive levels.

4.7.2 Management of Change

Management of change (MOC) is a written procedure for a company to safely regulate any changes in process chemicals, technology, equipment, and procedures. MOC requires considering impact of change on safety and health, modifications to operating procedures, and technical basis for the proposed change. MOC also requires that any employees involved in operating a process and maintenance and contract employees whose job will be affected by the change be informed of, and trained in, the change prior to start-up of the process or affected part of the process. MOC is mandated by OSHA's PSM program.

No MOC documentation or authorization was carried out for the damaged overflow line from the flash tank or for the failed liquid ring compressor. DuPont did not require a management of change if adverse safety impacts were not anticipated. DuPont engineers determined that any residual VF vapor would vent safely to the atmosphere from the flash tank above the slurry tanks. DuPont did not consider the possibility that flammable vapor could flow into, and accumulate in, the slurry tanks.

4.7.3 Mechanical Integrity

The OSHA PSM standard requires that the mechanical integrity of critical process equipment is maintained to ensure it is designed and installed correctly and operates properly. PSM mechanical integrity includes storage tanks, piping systems, and vent systems. The mechanical integrity section of PSM states:

> The employer shall correct deficiencies in equipment that are outside acceptable limits… before further use or in a safe and timely manner when necessary means are taken to assure safe operations

After DuPont personnel noticed the cracked seal loop, DuPont technical engineers decided not to repair the loop until the next turn around. If DuPont had included the slurry tanks and slurry flash tank overflow line in their PSM program they would have been required to fix the cracked seal loop before restarting the unit or ensure that the cracked seal loop posed no safety hazard.

4.7.4 DuPont PSM Classifications

The DuPont Tedlar process unit uses flammable VF in quantities exceeding 10,000 pounds; thus, the unit is covered by OSHA Process Safety Management (PSM) standard 29 CFR 1910.119. The PSM standard includes additional requirements to assess and reduce hazards associated with highly hazardous and flammable materials. DuPont included the VF storage vessels, reactors, separators, and flash tank in the PSM program but excluded the outdoor PVF slurry tank farm from the requirements of the PSM standard because DuPont concluded VF carryover into the flash tank and product storage tanks was unlikely and insignificant. DuPont further concluded that VF emissions from the flash tank would not accumulate in the tank farm area.

Prior to the incident, DuPont classified the slurry tank as a lower hazard operations (LHO) area, which is an area containing "any activity that exclusively manufactures, handles, stores, or uses any substances with low potential for death or irreversible human health effects." Post-incident, DuPont increased the slurry tank area rating to a higher hazard process (HHP), or an area containing "any activity manufacturing, handling, storing, or using hazardous substances that, when they are released, ignited, or intentionally combined, can result in death or irreversible human health effects." DuPont changed this because it modified the HHP definition to include any area that could contain more than 1 percent VF by weight or 0.65 percent by volume. DuPont did not conclude that OSHA PSM coverage extended to the slurry tanks.

The OSHA PSM standard does not apply to "Flammable liquids stored in atmospheric tanks or transferred which are kept below their normal boiling point without benefit of chilling or refrigeration." Although the slurry tanks were atmospheric vessels, they contained a flammable gas which was above its normal boiling point (-98°F, -72°C). The slurry tanks also served to agitate the slurry mixture, making them pieces of process equipment and not solely storage tanks.

In 1997, an OSHA memo about the chemical company Akzo Nobel provided guidance and serves as a letter of interpretation on how a facility should set the boundaries of PSM for a process:

> Aspects of the process which contain an HHC [Highly Hazardous Chemical] would be covered by all PSM elements, such as information, process hazard analysis and mechanical integrity. Aspects which do not contain HHC, but are interconnected or located nearby are part of the process. Such aspects may or may not be covered by the PSM standard based on whether the particular aspects could cause a HHC release or interfere with mitigating the consequences if there was [sic] a HHC release.

According to OSHA definitions, VF is an HHC. This incident clearly demonstrated that VF can enter the slurry tanks; additionally, the tanks are interconnected to the process that uses VF and are physically close to the flash tanks that remove VF. Despite DuPont's failure to correctly include the slurry tanks in the Tedlar PSM boundary limits, OSHA did not cite DuPont for failing to apply PSM to the slurry tanks.

4.8 DuPont Belle Incident

On September 20, 2011, the CSB released an investigation report[12] which detailed three incidents which occurred at a DuPont facility in Belle, West Virginia on January 22 and 23, 2010. The last of these incidents resulted in one fatality and one employee exposure to phosgene, a toxic gas.

The DuPont Belle and DuPont Buffalo incidents had a number of similarities. In the Belle report, the CSB found that DuPont practices at the Belle facility had deviated from their own policies and standards. It was determined that DuPont did not follow its own standards for the change-out of phosgene transfer hoses. This similarity was seen at the Buffalo facility where DuPont practice deviated from the hot work permitting process. The Belle report discussed DuPont Belle's PHA process and how some process hazards were not addressed. Similarly, the DuPont Buffalo PHA did not adequately consider the threat the potential threat that VF in the slurry tanks posed.

4.9 DuPont Corporate Audit

In November 2006, DuPont corporate performed a PSM audit of the DuPont Buffalo facility. DuPont awarded DuPont Buffalo a score of 99% on this audit, the highest score the auditor team had ever awarded a facility. The audit commended the Buffalo facility on their operating procedures and safe work practices and made no recommendations to improve these programs. No recommendations were made for the Buffalo facility's MOC procedure for subtle changes, stating that "all systems comply".

Another DuPont audit was carried out in October 2009. This document commended DuPont Buffalo's PHA process. It stated that the PHA's were "very well managed and executed" and are of "consistently

[12] Report available at www.csb.gov

high quality". These latest audits carried out by DuPont at the DuPont Buffalo facility missed many deficiencies that became apparent as a result of the November 9, 2010, incident.

5 Occupational Safety and Health Administration Standards

5.1 Lockout/Tagout

OSHA standards do not explicitly require testing the atmosphere inside a tank before performing hot work on the outside of a tank.

The OSHA Control of Hazardous Energy (lockout/tagout) standard (29 CFR 1910.147) requires positive control of hazardous energy including:

> Appropriate lockout devices or tagout devices to energy isolating devices, and to otherwise disable machines or equipment to prevent unexpected energization, start up or release of stored energy in order to prevent injury to employees.

This standard does not specifically address isolating tanks from hazards associated with flammable, combustible, or toxic material flow through pipes, but many elements of the lockout/tagout standard directly apply to assuring that tanks are properly isolated for safe confined space entry and hot work activities.

The lockout standard defines energy source as

> Any source of electrical, mechanical, hydraulic, pneumatic, chemical, thermal, or other energy.

5.2 Hot Work

The OSHA standard "Welding, Cutting and Brazing, General Requirements" (29 CFR 1910.252) includes requirements directly applicable to tank repair work. Specifically, §1910.252(a)(2)(vi)(C) prohibits cutting or welding

> in the presence of explosive atmospheres (mixtures of flammable gases, vapors, liquids, or dusts with air), or explosive atmospheres that may develop inside uncleaned or improperly prepared tanks or equipment which have previously contained such materials, or that may develop in areas with an accumulation of combustible dusts.

Furthermore, §1910.252(a)(3)(i) requires that "[a]ny pipe lines or connections to the drum or vessel…be disconnected or blanked" before welding or cutting. The overflow line, which provided the direct path for the flammable vapor to enter tank 1, was left unblinded and was not included on the lock out card for slurry tank 1.

6 Industry Standards

6.1 National Fire Protection Association

The National Fire Protection Association (NFPA) publishes consensus standards and guides about fire prevention and related subjects. Many of the standards are incorporated by reference in OSHA and industry consensus standards.

The standard "Fire Prevention During Welding, Cutting, and Other Hot Work" (NFPA 51B) provides a guidance for contractors and property managers who manage, supervise, and perform hot work. The standard is not intended to contain all necessary safety precautions and work practices involved in job-specific work, such as hot work on atmospheric storage tanks. Job-specific hot work prerequisites and practices are contained in other NFPA standards.

NFPA 51B appendices address hot work on atmospheric tanks by providing general guidance for a hot work permit program. The sample "Hot Work Permit" in Appendix A of NFPA 51B contains a checklist that includes two precautions directly applicable to the tank repair work:

- Enclosed equipment cleaned of all combustibles
- Containers purged of all flammable liquids/vapors

Appendix B summarizes more than 25 hot work fire and explosion incidents to "illustrate how such incidents occur and to emphasize the provisions of this standard." Three summarized incidents are similar to the DuPont incident in that each involved welding and other hot work on tanks and attached piping without adequate testing or other safety precautions.

Chapter 4 of the standard, "Safeguarding of Tanks and Containers for Entry, Cleaning, or Repair" (NFPA 326) contains basic precautions for performing hot work on storage tanks: "Extreme caution shall be used when work is performed on a tank or container that contains vapors related to the substances that are stored or were previously stored therein." Paragraph 4.1.6 adds, "Work on tanks or containers shall be permitted only after the characteristics…of the atmosphere within the tank or container have been determined." Isolating equipment prior to hot work is addressed in NFPA 326, paragraph 5.3.2: "The vents for the tank or container on which work is to be performed shall be isolated from the vents of other tanks or containers that might still be in service."

NFPA 326 6.1.1 states:

> To determine that an atmosphere is safe for the designated entry, cleaning, or repair work, tests for oxygen and for flammable, combustible, or other hazardous substance vapors, fumes, or dusts shall be made with an appropriate instrument as follows.
>
> (1) Before entry or re-entry
>
> (2) Before start of alterations or repairs
>
> (3) Before and during any hot work, cutting, welding, or heating operations
>
> (4) Continuously or periodically during the course of the work to be as determined by a qualified person

(5) After cleaning the interior of each tank or container to determine that the cleaning procedures have been effective

(6) After any process or activity has occurred that might change the atmosphere within the tank or container

The standard specifies that testing is needed before the start of alterations or repairs and before and during any hot work, cutting, welding, or heating operations. The OSHA hot work standard [29 CFR 1910.252] does not require air testing prior to the initiation of hot work.

6.2 American Welding Society

The American Welding Society (AWS) publishes consensus standards for welding, brazing, and thermal-cutting practices. The standard "Safety in Welding, Cutting, and Allied Processes" (ANSI Z49.1)[13] has the same purpose and scope as the NFPA 51B standard. It provides technique-specific safety criteria, detailed guidance on welder personal protective equipment such as arc-flash eye shield selection criteria, and welding fume hazard awareness. Explanatory information item E6.4, "Welding or Cutting Containers," specifically includes tanks and other large vessels in the definition of a container and warns

> All containers should be considered unsafe for welding or cutting unless they have been rendered safe…Even a water tank should be considered hazardous unless a qualified person has declared it safe to weld or cut.

"Recommended Safe Practices for the Preparation for Welding and Cutting of Containers and Piping" (AWS F4.1) discusses precautions for welding and thermal cutting on containers. The standard warns that "the container should be tested for hazardous gases, fumes, and vapors periodically to ensure that the container and work area are safe during welding." The standard is not intended for welding on "containers and confined spaces that can be entered by workers" but the DuPont tank repair work involved welding work on a tank that contained hazardous vapors; thus, the safety principles in the standard were applicable.

6.3 American Petroleum Institute

The American Petroleum Institute (API) publishes consensus standards for the petroleum industry. API Recommended Practice 2009 "Safe welding and cutting practices in refineries, gas plants, and petrochemical plants" discusses hot work in plants that deal with flammable vapors. API RP 2009 states that to effectively use a combustible gas detector, consider the characteristics of the gas (lighter or heavier than air) as well as the appropriate locations and frequency of test.

The lack of rigor and deviation from accepted practices in the permitting process and incorrect engineering assumptions resulted in VF vapor flowing into tank 1 and accumulating. When DuPont personnel discovered the damaged seal loop on the flash tank overflow line, they incorrectly concluded that it would still prevent VF vapor and steam from entering the slurry tank. In addition, they left tank 1 unblinded and still open to the overflow line when they restarted the Tedlar unit. Finally, DuPont authorized the hot work to begin without determining if the interior of tank 1 contained a hazardous atmosphere.

7 Other Hot Work Incidents

In February 2010, the CSB issued a hot work safety bulletin, "Seven Key Lessons to Prevent Worker Deaths During Hot Work In and Around Tanks", which discusses 11 incidents involving hot work activities where flammable vapor accumulated and ignited while repair work or process modifications were performed in or near storage tanks. Three of the incidents, which involved welding on a tank that contained flammable materials, closely parallel the incident at DuPont.

In June 2006, three contract workers were killed and one was seriously injured when a welding torch ignited flammable hydrocarbon vapors at a Partridge-Raleigh Oilfield in Raleigh, Mississippi. The contractors had been installing piping between two oil tanks. Like the DuPont incident, all of the tanks were interconnected by piping and the interconnected tanks contained flammable vapors which ignited.

In July 2008, three workers were killed and another seriously injured in an explosion at the Packaging Corporation of America fiberboard manufacturing facility in Tomahawk, Wisconsin. Workers were welding a temporary metal clamp on a damaged flange atop an 80-foot tall atmospheric storage tank. Company hot work procedures did not require sampling the atmosphere inside the tank prior to welding directly to the attached piping. The tank contained recycled water and fiber waste, which the company assumed did not pose a fire or explosion hazard. The CSB concluded that bacteria likely produced highly flammable hydrogen gas inside the tank. Sparks from the welding or the heated metal ignited the hydrogen. Two workers fell to their death and the third died after being hit by debris when the explosion blew off part of the tank top.

In February 2009, a welding contractor was killed while repairing a large crack in the bottom of a water clarifier tank at a ConAgra Foods facility in Boardman, Oregon. Water used to wash potatoes had leaked through the crack and accumulated inside the inaccessible space under the tank. Examination of a sample of the liquid indicated that bacterial decomposition of the organic matter likely produced flammable gas that accumulated inside the tank support skirt. The gas ignited when the contractor started welding the tank bottom.

This case study reiterates the key lessons of the CSB hot work bulletin to minimize fire and explosion hazards when performing hot work on atmospheric tanks, including the need to clearly understand and manage hot work activities, especially in terms of testing for a flammable atmosphere both outside and inside any tank or container where welding or grinding is done directly on the container surface.

Hot work incidents across the country continue to cause death, injury, and property damage with alarming frequency. The CSB is currently conducting a more comprehensive evaluation of hot work regulations and standards and expects to issue related safety recommendations in the near future.

8 Key Findings

1. DuPont PHAs made the incorrect assumption that VF in the Tedlar process could not reach flammable levels in the slurry tanks.

2. DuPont restarted the unit after incorrectly concluding that the defective seal loop did not increase the risk of VF vapor transfer into tank 2. Flammable VF vapor flowed directly from the flash tank into tank 2.
3. DuPont did not properly isolate and lockout tank 1 from in-service tanks 2 and 3 prior to authorizing hot work on tank 1. Consequently, flammable VF vapor passed directly from tank 2 into tank 1 through the overflow line and accumulated to a concentration above the lower explosive limit.
4. The DuPont hot work permit procedure did not require testing the atmosphere inside tank 1 for flammable vapor even though the work required welding directly to the tank top.
5. The individuals who signed off on the hot work permit were not knowledgeable in the operations and hazards of the Tedlar process.
6. The repair work created multiple ignition sources. Welding directly to the tank top increased the metal temperature inside the tank significantly above the VF vapor autoignition temperature. Welding and grinding generated hot sparks, which likely ignited the flammable vapor.
7. OSHA's hot work standard does not specifically require gas monitoring inside containers intended for hot work even though it is recommended by industry safe practice guidelines.[14]

9 Lessons Learned

1. Welding to the outside surface of a tank or container generates heat and sparks near the weld area. The extreme temperatures and sparks will likely ignite flammable materials present inside the equipment.
2. Before starting hot work activities, all process connections on tanks and similar containers should be completely isolated by closing valves, installing blanks, and disconnecting pipes to ensure that all possible and known sources of flammable materials cannot enter the container at any time.
3. The atmosphere inside any container previously containing flammables, regardless of size, should be tested before authorizing any hot work involving grinding, cutting, or welding on the outside surface and the atmosphere must be continuously monitored during the work.

10 Recommendations

[14] NFPA 326 and ANSI Z49.1

E.I. DuPont de Nemours and Co., Inc.

2011-01-I-NY R1

Develop and enforce a corporate quality control policy and procedures to require that all DuPont facilities audit their hot work permitting systems to ensure that:

- All potential explosion hazards associated with hot work activities are identified and mitigated before hot work is started
- All relevant forms required for permits are completed in accordance with corporate policies and industry standards (including NFPA 326 and NFPA 51B) before hot work is started
- Appropriate DuPont personnel officially approve hot work permits, by signature or equivalent, consistent with DuPont policies, before hot work is started

2011-01-I-NY R2 Revise corporate policies and procedures to require all process piping, vent piping, or similar connections to be positively isolated (using closed valves, blind flanges or pancake blanks) before authorizing any hot work.

2011-01-I-NY R3 Revise corporate policies and procedures to require that the atmosphere inside the container be monitored for flammable vapor prior to performing any welding, cutting, or grinding on the container surface.

2011-01-I-NY R4 Revise corporate policies and procedures to require air monitoring for flammable vapor inside the container for the duration of the hot work consistent with industry standards (NFPA 326, NFPA 51B). Create a policy for determining criteria for requiring continuous or periodical testing for the duration of hot work.

9TH GLOBAL CONGRESS ON PROCESS SAFETY

Can Fire Suppressants Promote Ignition? A Study of HFC-125 and HFC-227

Carmen H. Osorio[1], E. L. Petersen[2] and M.S. Mannan[1]

Corresponding author: carmenhelena@tamu.edu

[1]Mary Kay O'Connor Process Safety Center,
Artie McFerrin Department of Chemical Engineering,
Texas A&M University,
College Station, Texas 77843-3122, USA

[2]Department of Mechanical Engineering,
Texas A&M University,
College Station, Texas 77843-3123, USA

Prepared for Presentation at
American Institute of Chemical Engineers
2013 Spring Meeting
9th Global Congress on Process Safety
San Antonio, Texas
April 28 – May 1, 2013

UNPUBLISHED

Can Fire Suppressants Promote Ignition? A Study of HFC-125 and HFC-227

C. Osorio[1], E. L. Petersen[2] and M.S. Mannan[1]

[1]Mary Kay O'Connor Process Safety Center, Artie McFerrin Department of Chemical Engineering, Texas A&M University, College Station, Texas, USA
[2]Department of Mechanical Engineering, Texas A&M University, College Station, Texas, USA

Corresponding author: carmenhelena@tamu.edu

Abstract

C_2HF_5 (HFC-125,) and C_3HF_7 (HFC-227ea or FM-200) have gained considerable acceptance in the field of fire protection, especially after the restriction of several ozone-depleting substances under the Montreal Protocol. These two substitutes are currently included in the list of the Significant New Alternatives Policy (SNAP) Program of the U.S Environmental Protection Agency (EPA). Their desired properties such as low Ozone Depleting Potential (ODP), relatively low toxicity, electrical non-conductivity, high dispersion capabilities, and low flammability, make them two of the most-used total flooding fire suppressants at the moment. In terms of Minimum Extinguishing Concentration (MEC), C_3HF_7 and C_2HF_5, although less effective than bromine-containing compounds, are more efficient than other clean alternatives in the market. Nevertheless, the combustion properties of these agents are not well understood. In fact, previous numerical studies have shown that these compounds can act as combustion promoters under certain conditions, but there are scarce experimental data of HFCs that can be used to certify such predictions. This lack of data shows the necessity of having accurate measurements that can be used for model validation but also to better understand the role of different fire suppressants on the combustion initiation stage towards an optimal and safer application. To this end, this paper examines the effect of C_2HF_5 and C_3HF_7 on the high-temperature chemistry of methane and propane flames at different conditions. Ignition delay time measurements were obtained using a shock tube with reaction monitoring capability. Theoretical analysis was conducted using a detailed chemical kinetics mechanism that includes the hydrocarbon and fire suppressants set of reactions. Results indicate that C_2HF_5 and C_3HF_7 can actually act as ignition promoters under certain conditions. These results suggest that these fire suppressants may not be good alternatives to prevent flames, however they can still be used as fire extinguishers after the flame is well established. Finally, sensitivity analysis was carried out to identify the most significant reactions responsible for such behavior, and the results were compared with CF_3Br over a similar range of conditions.

1. Introduction

Despite that halogenated compounds have been shown to inhibit developed flames, some studies have demonstrated that they can also promote combustion at certain conditions [1-6]. Even the well-known CF_3Br (Halon 1301), as well as C_2HF_5 and C_3HF_7 (two of the most common fire suppressants alternatives), can act as both combustion inhibitors and promoters [2, 7-9]. Therefore, identification and understanding of such phenomena became extremely necessary for the safe and optimal application of such compounds. This analysis requires experimental data and validated models that help to predict the effect on different fire suppressants in a variety of systems and conditions. The chemical kinetics mechanism is also used to identify the key reactions, species and pathways, and therefore to better understand their fundamental properties.

Currently, there is an extensive research area that focuses on the HFC mechanism development; however, there are scarce experimental measurements that can be used for its validation. Considering this, the present work remedies this deficiency and focuses on providing accurate ignition delay time measurements for different fire suppressants ($C_3HF_7/C_2HF_5/CF_3Br$) and Fuels (CH_4/C_3H_8) systems, which can be compared directly with modeled data. The next sections describe the experimental and numerical methodology used, followed by the results, analysis and conclusions.

2. Background

To understand how fire suppressants work, it is important to have some background related to fire theory. Fire is the result of a rapid exothermic reaction between a fuel (i.e. methane) and an oxidizer (i.e. oxygen). For this process to take place, an ignition source should be present, thus the fuel can reach its ignition temperature. For the flame to be sustainable, it is necessary to have a continuous chemical chain reaction, where enough fuel and oxygen produce the required active radicals responsible for the major heat release. This combustion process can be represented through the fire tetrahedron (Fig. 1). Based on this concept, a fire can be extinguished by eliminating one of the four elements present in the process. In this work, we are focusing on understanding how the tested fire suppressants can affect the chemical chain reaction using chemical kinetics mechanisms.

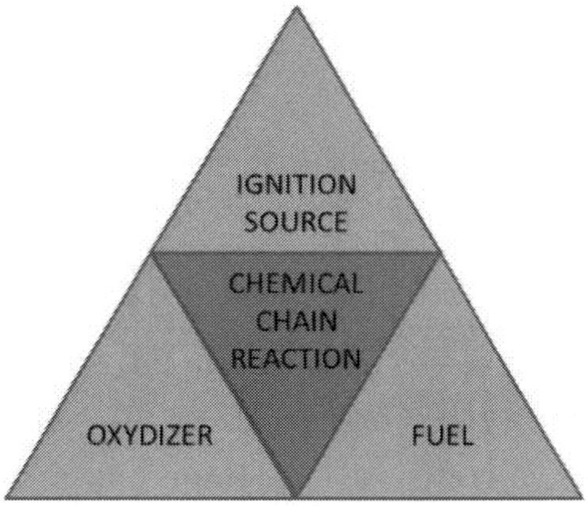

Figure 1. Fire tetrahedron for a flame.

Overall, the chemical process of combustion can be represented by individual steps called elementary reactions. To analyze the different properties of such a system, it is necessary to gather information on all of the elementary reactions that can be involved along with their kinetics parameters (A, n, Ea). This database is known as the chemical kinetic mechanism. Chemical processes are very dependent on highly reactive intermediates called free radicals such as H, OH, and HO_2, among others. In fact, elementary reactions can be categorized in four groups based on the production or consumption of the free radicals (Table 1). This classification can help to understand different phenomena. For example, generation and propagation of the flame can be attributed to branching reactions which are predominant over termination steps. It occurs when a large amount of radicals are produced due to a fast fuel decomposition.

Table 1. Classification of elementary reactions.

Initiation	One free radical is produced from stable molecules in the reactants
Propagation	One radical is consumed, and one radical-molecule is generated
Branching	Two free radicals are produced per single radical in the reactants
Termination	One radical-molecule is consumed to form stable species

3. Methodology

3.1 Modeling

Modeling is conducted using CHEMKIN 4.1 [10]. This program can solve the differential equations that result from the kinetics analysis taking into account the different kinetic parameters of all elementary reactions involved. For the ignition delay time modeling, the batch reactor routine was assumed. This model allows simulating the conditions behind the reflected shock wave in shock-tube experiments. Internal energy and volume are considered constant. Chemical kinetics mechanism consist of different sets of reactions including a recently updated H_2 and C_1-C_5 hydrocarbon mechanism [11], a widely used CF_3Br and HFC chemical kinetic mechanism, provided by Luo et al. [12-15], and a OH* chemistry mechanism [16, 17]. The final assembled chemical mechanism is validated by comparing the modeled and measured ignition delay times (τ_{ign}). To identify the most significant reactions that affect the combustion phenomena studied, sensitivity analysis is conducted using SENKIN [18].

3.2 Experimental

Ignition delay time measurements were conducted using the High Pressure Shock Tube at Texas A&M University. This apparatus has been designed to generate an incident and reflected shock wave that travels through the test mixture. This process allows the pressure and temperature to remain constant until ignition takes place. The time between the pressure rise after the reflective shock wave (time zero) and the rapid increase of OH* emission [19] is known as the ignition delay time (Figure 2). Pressure is monitored through equally spaced pressure transducers located alongside the tube and one at the endwall. Wave speed is calculated using the signal from the pressure transducers which give information of the time at which the shock wave travels from one transducer to the next. A window located at 1.6 cm from the endwall allows the light emitted after ignition to be captured. The optical signal is filtered at 310 nm (OH* wavelength) and then amplified by a photomultiplier tube detector. Pressure and emission signals are examined using GageScope software. More information about this facility is provided by Aul et al. [20].

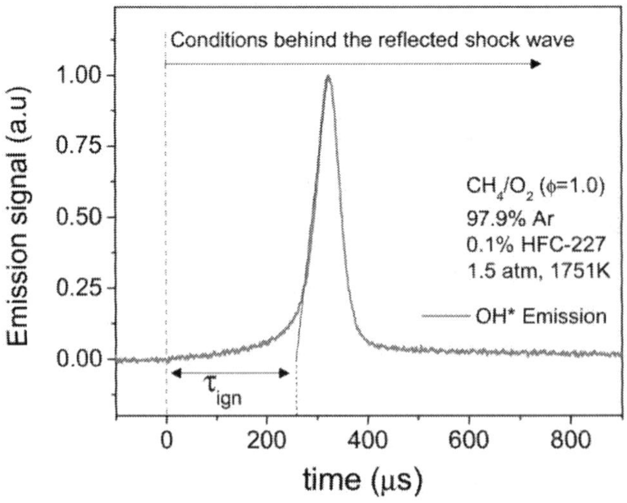

Figure 2. Definition of ignition delay time based on an OH* profile

4. Results and Discussion

4.1 Ignition delay time measurements

Different mixtures of fuel (CH_4 or C_3H_8), oxygen (O_2), and Argon (Ar) were analyzed with and without fire suppressants (i.e., CF_3Br, C_2HF_5, and C_3HF_7) at atmospheric pressure over a range of temperatures of 1350K-2200K. Figures 2 and 3 show the resulting experimental and modeled ignition delay times for the different mixtures evaluated. These results are very important because they help us to answer some general questions such as: Can a fire suppressant promote ignition? If this result is the case, at which conditions or for what fuels does it occur? Additionally, these results provide information regarding the chemical mechanism accuracy.

Experimental and modeled data (in Figure 3) show that the tested fire suppressants tend to decrease the ignition delay time in methane mixtures. Their promoting effect can be ranked as $CF_3Br < C_2HF_5 < C_3HF_7$. However, this trend is not general. Experimental data in Figure 4 show that the ignition of the propane mixtures can be slightly promoted by C_2HF_5, but retarded by the action of CF_3Br. C_3HF_7 does not show a significant effect on propane mixtures. Note that the range of practical application of the present experimental results remains strictly limited to the experimental conditions defined. This is when numerical approaches are very important since they allow predicting the combustion properties of different systems. However, such mechanism should be validated using accurate measurements, which recall one of the main objectives of this work.

Comparison between modeled and experimental data (see Figures 3 and 4) shows that the model very well predicts the effect of CF_3Br.on methane and propane; however, improvements on the HFC-chemistry are suggested for better prediction; especially for those including propane. The following section describes the results obtained from the sensitivity analysis which serves as the basis for future mechanism improvements.

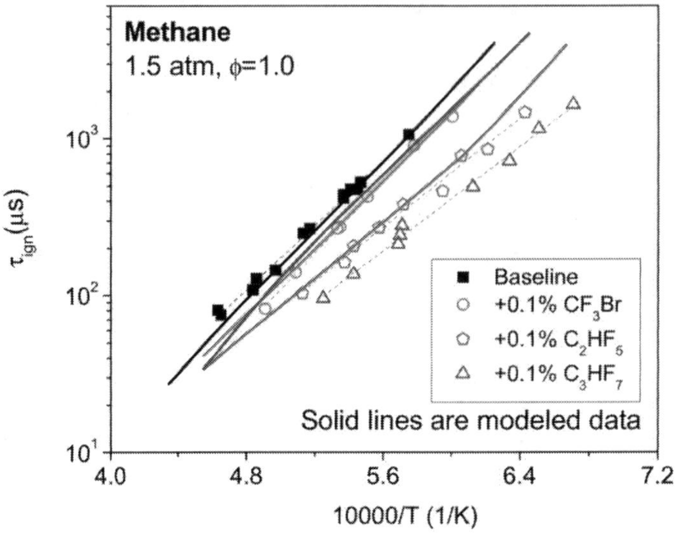

Figure 3. Experimental and modeled Ignition delay times (τ_{ign}) as a function of temperature for $CH_4/O_2/Ar$ Mixtures with and without CF_3Br, C_2HF_5, C_3HF_7.

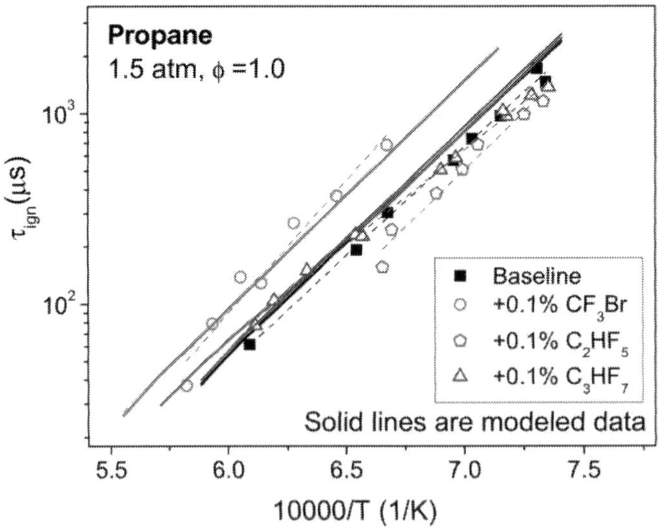

Figure 4. Experimental and modeled Ignition delay times (τ_{ign}) as a function of temperature for $C_3H_8/O_2/Ar$ Mixtures with and without CF_3Br, C_2HF_5, and C_3HF_7.

4.2 Sensitivity Analysis

Sensitivity analysis was carried out to find the key reactions responsible for the major effects of the tested agents on a specific reacting mixture. For this analysis, SENKIN [10] was used through the CHEMKIN [18] interface, which provides the reactions with the largest sensitivity coefficients. For convenience, the reaction numbering in this paper retains the same numbering used in the assembled mechanism. Tables 2 and 3 present a summary of such findings.

Results show that halogenated species can participate in promoting and inhibiting reactions that compete to give a net effect. The resulting effect depends on many factors such as the fire suppressants-fuel chemical interaction and the conditions of the system (i.e. temperature, pressure).

In terms of sensitivity coefficients, it can be seen that bromine-containing species mainly participate in strong inhibiting reactions. It can be explained due to the participation of bromine species in catalytic cycles where highly reactive radicals (i.e. H radicals) are removed from the flame reaction zone, as it was suggested by Westbrook [21]. On the other hand, fluorinated species may act as scavengers with lower radical removal efficiency. However, they can also participate in several ignition-promoting reactions. Despite all the research in this area, the flame inhibition mechanism of fluorinated species remains unclear, and further research may be conducted.

Current sensitivity analysis results can be used to improve the HFC chemistry. Particular attention is required to those reactions with significant influence, where the focus then is to find more-accurate Arrhenius kinetic parameters that lead to better predictions.

5. Summary and Conclusions

This paper examined the fundamentals of fire chemistry to understand the ignition effects of two of the most currently used HFC-clean fire suppressants (C_2HF_5 and C_3HF_7) on two common hydrocarbons. It also compared current results with the properties of the CF_3Br (Halon 1301), which is considered as the reference. This study was based on ignition delay times measurements and modeling of different, highly diluted mixtures of CH_4 and C_3H_8, with and without the fire suppressants. Our results suggest that the tested agents may not be good alternatives as ignition preventers; however, we do not deny that they can still be used as fire extinguishers. Considering the scarcity of HFC ignition delay time measurements in the literature, the current data set represents a valuable tool for model validation. Comparison between modeled and experimental data shows that the HFC-mechanism behaves well, however it can be improved. Sensitivity analysis was conducted to identify the most relevant reactions that participate at the ignition process. These results can be used as the basis for future HFC-chemical kinetics mechanism research and improvements.

Table 2. Summary of the sensitive analysis results for methane systems

Agent	Effect	Key Reactions	Net Effect
0.1% CF_3Br	Promoting	1281. $CF_3+O_2=CF_3O+O$ 57. $CH+O_2=CO+OH^*$ 1257. $CF_3+CH_4=CHF_3+CH_3$ 1737. $C_2HF+O=CFCO+H$ 1188. $CF_3Br+M=CF_3+ Br +M$ 1182. $CH_3+BR=CH_2+HBR$ 1169. $CH_2O+Br=HCO+HBr$	Promotion
	Inhibiting	1166. $CH_4+Br=CH_3+HBr$ 1208. $Br+HCO=HBR+CO$ 1143. $HBr+H=Br+H_2$	
0.1 % C_2HF_5	Promoting	57. $CH+O_2=CO+OH^*$ 1375. $CHF_2\text{-}CF_3=CF_2:CF_2+HF$ 1453. $CF_3+CHF_2=CHF_3+CF_2$ 1737. $C_2HF+O=CFCO+H$ 1305. $CF_2+O_2=CF_2:O+O$ 1760. $CH_3+CF_2=CH_2:CF_2+H$ 1281. $CF_3+O_2=CF3O+O$ 1280. $CHF_2+O_2=>CF_2:O+O+H$ 1328. $CF+O_2=CF:O+O$ 1226. $CHF_3+M=CF_2+HF+M$ 1320. $CF_2+H=CF+HF$	Promotion
	Inhibiting	1415. $CH_3+CF_3=CH_2:CF_2+HF$ 1424. $CHF_2+CF_3+M=CHF_2\text{-}CF_3+M$	
0.1% C_3HF_7	Promoting	1665. $CHF+CF_2=CHF:CF_2$ 1764. $CHF_2+CF_2=CHF:CF_2+F$ 1281. $CF_3+O_2=CF_3O+O$ 57. $CH+O_2=CO+OH^*$ 1280. $CHF_2+O_2=>CF_2:O+O+H$ 1305. $CF_2+O_2=CF_2:O+O$ 1375. $CHF_2\text{-}CF_3=CF_2:CF_2+HF$ 1453. $CF_3+CHF_2=CHF_3+CF_2$ 1737. $C_2HF+O=CFCO+H$ 1305. $CF_2+O_2=CF_2:O+O$ 1760. $CH_3+CF_2=CH_2:CF_2+H$ 1281. $CF_3+O_2=CF_3O+O$ 1280. $CHF_2+O_2=>CF_2:O+O+H$ 1305. $CF_2+O_2=CF_2:O+O$ 1320. $CF_2+H=CF+HF$ 1453. $CF_3+CHF_2=CHF_3+CF_2$ 1257. $CF_3+CH_4=CHF_3+CH_3$ 1328. $CF+O_2=CF:O+O$	Promotion
	Inhibiting	176. $CH+O_2=HCO+O$ 1412. $CH_3+CHF_2=CH_2:CHF+HF$ 1761. $CH_2F+CF_2=CHF:CF_2+H$ 1776. $CH_4+F=CH_3+HF$ 1415. $CH_3+CF_3=CH_2:CF_2+HF$	

Table 3. Summary of the sensitive analysis results for propane systems.

Agent	Effect	Key Reactions	Net Effect
0.1% CF_3Br	Promoting	1188. $CF_3BR+M=CF_3+BR+M$	Inhibition
	Inhibiting	1146.$Br+HO_2=HBr+O_2$ 1189. $CF_3Br+H=CF_3+HBr$ 1208. $Br+HCO=HBr+CO$ 1210. $C_2H_3+Br=C_2H_3Br$ 1143. $HBr+H=Br+H_2$	
0.1 % C_2HF_5	Promoting	-	Promotion
	Inhibiting	1286. $CHF_2+OH=CHF{:}O+HF$ 1776. $CH_4+F=CH_3+HF$	
0.1% C_3HF_7	Promoting	1358. $CO+F+M=CF{:}O+M$ 1328. $CF+O_2=CF{:}O+O$ R1849. $CF_3\text{-}CHF(+M)=CHF{:}CF_2+F(+M)$	Promotion
	Inhibiting	1339. $CF+C_2H_4=>C_2H_2+CH_2F$ 1397. $CF_3\text{-}CHF+H=CH_2F+CF_3$	

References

1. Shebeko, Y.N., et al., *The influence of fluorinated hydrocarbons on the combustion of gaseous mixtures in a closed vessel.* Combust. Flame, 2000. **121**(3): p. 542-547.
2. Suzuki, A., et al., *Effect of Bromotrifluoromethane on the Ignition in Methan and Ethane-Oxygen-Argon Mixtures behind Shock Waves.* Bull. Chem. Soc. Jpn., 1991. **64**: p. 3345-3354.
3. Saso, Y., et al., *Laminar Burning Velocities of Trifluoromethane–Methane Mixtures: Experiment and Numerical Simulation.* Combust. Flame, 1998. **114**(3–4): p. 457-468.
4. Linteris, G.T. and L. Truett, *Inhibition of premixed methane-air flames by fluoromethanes.* Combust. Flame, 1996. **105**(1-2): p. 15-27.
5. Babushok, V.I., et al., *Simulation Studies of the Influence of Fluorine and Bromine Containing Fire Suppressants on Ignition Behavior*, in *Halon Options Conf. Proc.*1994: Albuquerque, NM. p. 205-216.
6. Hamins, A. and P. Borthwick, *Suppression of ignition over a heated metal surface.* Combust. Flame, 1998. **112**(1–2): p. 161-170.
7. Osorio, C.H., E.L. Petersen, and M. S. *Kinetic Modeling and Experimental Study of CF_3Br Effect on Methane and Ethane Oxidation Behind Reflected Shock Waves.* in *7th US National Technical Meeting of the Combustion Institute.* 2011. Atlanta, GA.
8. Linteris, G.T., et al., *Stirred reactor calculations to understand unwanted combustion enhancement by potential halon replacements.* Combust. Flame, 2012. **159**(3): p. 1016-1025.
9. Babushok, V.I., G.T. Linteris, and O.C. Meier, *Combustion properties of halogenated fire suppressants.* Combust. Flame, 2012. **159**(12): p. 3569-3575.
10. Kee, R.J., et al., *CHEMKIN Release 4.1.1*, 2007: San Diego, CA.
11. Donato, N.S., et al., *Ignition and Oxidation of 50/50 Butane Isomer Blends.* J. Eng. Gas Turbines Power, 2010. **132**(5): p. 051502.

12. Luo, C., *Study on the Inhibition Mechanisms of $CBrF_3$, CF_3I and C_3F_7H in Methane Fuelled Premixed Flames*, in *Department of Chemical Engineering*2010, The university of Newcastle: Australia. p. 270.
13. Luo, C., B.Z. Dlugogorski, and E. Kennedy, *Influence Of CF_3I and $CBrF_3$ on Methanol – Air and Methane – Air Premixed Flames.* Fire Technol, 2008. **44**: p. 221-237.
14. Kennedy, E.M., B.Z. Dlugogorski, and C. Luo, *Burning Velocity And Flame Structure Of Methane Flames Inhibited With Heptafluoropropane (C_3HF_7)*, in *Proceedings of the Sixth Asia-Oceania Symposium on Fire Science and Technology*2004: Daegu, South Korea.
15. Luo, C., et al., *Effect of Bromine and Iodine on Methanol and Methane Flames.* , in *Suppression, Detection and Signaling Research and Applications - a Technical Working Conference (SUPDET 2010)*2010: Orlando, Florida.
16. Healy, D., et al., *n-Butane: Ignition delay measurements at high pressure and detailed chemical kinetic simulations.* Combust. Flame, 2010. **157**(8): p. 1526-1539.
17. Healy, D., et al., *Methane/n-Butane Ignition Delay Measurements at High Pressure and Detailed Chemical Kinetic Simulations.* Energy Fuels, 2010. **24**(3): p. 1617-1627.
18. Lutz, A.E., et al., *Senkin: A Fortran Program For Predicting Homogeneous Gas Phase Chemical Kinetics With Sensitivity Analysis. SAND87-8248*, 1988: Livermore, CA.
19. Petersen, E.L., *Interpreting Endwall and Sidewall Measurements in Shock-Tube Ignition Studies.* Combust. Sci. Technol., 2009. **181**(9): p. 1123-1144.
20. Aul, C.J., DeVries, J. and Petersen, E. L. . *New Shock Tube Facility for Studies in Chemical Kinetics at Engine Conditions.* in *Eastern States Fall Technical Meeting of the Combustion Institute.* 2007. Charlottesville, VA.
21. Westbrook, C.K., *Numerical Modeling of Flame Inhibition by CF3Br.* Combust. Sci. Technol., 1983. **34**: p. 201-225.

A Focus on Fire Fundamentals Including Emergency Response Training at the National Fire Agency in Taiwan

Ronald J. Willey
Department of Chemical Engineering
Northeastern University, Boston, MA 02115
r.willey@neu.edu

Jack, Huan-Chang Hsiao, Ph.D., Director, Training Center National Fire Agency, Ministry of the Interior, Executive Yuan, ROC; No. 100, Dagong St., ZhuShan Township, Taiwan 55795, ROC, **jack@nfa.gov.tw**
Roy E. Sanders, Chemical Process Safety Consultant, Lake Charles, LA, **sanders.roy@suddenlink.net**
Arcady Kossoy, Dr. Sc. R&D Deputy Director General, Cheminform, St. Petersburg (CISP) Ltd., PO Box 25, 193232, St. Petersburg, Russia, **kossoy@cisp.spb.ru**
Chi-Min Shu, Ph. D., Center for Industry-Academia Collaboration, National Yunlin University of Science and Technology, 123 University Road, Section 3, Douliu, Yunlin 64002, Taiwan, **shucm@yuntech.edu.tw**

Prepared for Presentation at
American Institute of Chemical Engineers
2013 Spring Meeting
9th Global Congress on Process Safety
San Antonio, Texas
April 28 – May 1, 2013

A portion of this article appeared in the March 2013 issue of AIChE PSP Vol 32, No. 1 pp 2-7

A Focus on Fire Fundamentals Including Emergency Response Training at the National Fire Agency in Taiwan

Ronald J. Willey
Department of Chemical Engineering
Northeastern University, Boston, MA 02115
r.willey@neu.edu

Jack, Huan-Chang Hsiao, Ph.D., Director, Training Center National Fire Agency, Ministry of the Interior, Executive Yuan, ROC; No. 100, Dagong St., ZhuShan Township, Taiwan 55795, ROC, **jack@nfa.gov.tw**
Roy E. Sanders, Chemical Process Safety Consultant, Lake Charles, LA, **sanders.roy@suddenlink.net**
Arcady Kossoy, Dr. Sc. R&D Deputy Director General, Cheminform, St. Petersburg (CISP) Ltd., PO Box 25, 193232, St. Petersburg, Russia, **kossoy@cisp.spb.ru**
Chi-Min Shu, Ph. D., Center for Industry-Academia Collaboration, National Yunlin University of Science and Technology, 123 University Road, Section 3, Douliu, Yunlin 64002, Taiwan, **shucm@yuntech.edu.tw**

Keywords: Emergency Response Training, Fire Prevention

1.0 ABSTRACT

This article reflects on process safety fundamentals of fires and explosions for the benefit of new professionals entering the field. It provides a brief review of fire fundamentals such as the fire triangle and the respect that must be paid to the unknown ignition source. Various control methods are reviewed at the elementary level, including the reminder that the ignition source is always free. After a review of fire fundamentals, an emergency response training exercise is presented for a burning propane fire witnessed at a training facility in Taiwan. The National Fire Agency (NFA) of Taiwan has built the third largest emergency response-training center in the world. The 270 acres facility is located in central Taiwan near Taichung, Taiwan. The site has 49 separate emergency response training scenarios. Scenario pads include training for fire attack on mock-ups of chemical storage facilities, tanker trucks, power transformers, passenger airliners, and shipping vessels. Overall, this paper serves as a reminder of the safe guards used in fire safety, and the importance of training beforehand in the event of an emergency.

2.0 INTRODUCTION

This article is intended for young professionals entering the field of process safety. We will use the scenario of "fire" as our example for process safety fundamentals, and emergency response training. Let's begin with the nature of fires and some elementary considerations by any professional performing hazard analysis of process areas for fire prevention.

3.0 FUNDAMENTALS ABOUT FIRES

3.1 The Nature of a Fire and Elementary Parameters Explained

Fire in its simplest terms is an exothermic reaction. It is a reaction between an oxidizer (e.g., air, oxygen, chlorine, fluorine, or hydrogen peroxide) and a fuel (e.g., methane, hydrogen, propane, gasoline, carbon disulfide, or ether). A "third" requirement is an external source of "energy" to initiate the reaction. Typically this is a called an ignition source. The combination of these, oxidizer, fuel, and ignition source form the sides to the equilateral "fire" triangle. Eliminate any one of the three, and a fire will not initiate. An outstanding primer on fires is the textbook by Crowl and Louvar, Chapter 6 [1]. Young engineers working in the field of process safety are encouraged to purchase this reference and retain it on their desks.

Two of us (Sanders and Willey) are disciples of Trevor Kletz – a pioneer in modern process safety methods (HAZOPS for example). In honor of him, we'd like to bring forth to a younger audience a portion of his simple, yet effective, teachings in the area of fire prevention and his wise words about ignition.

The following is extracted from Trevor's "What Went Wrong"[2]

The inevitability of ignition (paraphrased for the young engineer)

You must be familiar with the fire triangle as you begin a fire hazards analysis. Air, fuel, and a source of ignition are necessary for a fire or explosion to occur; remove one of the three, and an explosion is impossible. However, when flammable gases or vapors are handled on an industrial scale, this view, though theoretically true, is misleading. If flammable gases or vapors are mixed with air in flammable concentrations, experience shows that sources of ignition are likely to turn up as the amount of energy required to ignite a flammable mixture can be very small. As little as 0.2 mJ, or the energy produced when a one-cent coin falls 5 mm, can be enough energy to ignite a combustible mixture.

"THE IGNITION SOURCE IS ALWAYS FREE"

To further follow-up this statement, ignition sources can include the hot surfaces such as an automobile exhaust manifold located nearby the site, a hot bearing, static electricity discharge, an open flame, a discharged burning cigarette butt, lightning, or an electrical switch being turned on.

With the above "warning" written, it is prudent to include in a hazard analysis, management of change assessments, and general operation practice, the consideration of where ignition sources are present. When an external contractor arrives to a your facility, where should that pick-up truck be allowed? Are permits issued for vehicular equipment and tools capable of emitting sparks? If that contractor lights up a cutting torch, has full consideration been made to all of the consequences? Are there any potential for a fuel-air mixture? Several noteworthy fires can be traced to secondary tanks, connected but not isolated, to the tank undergoing "hot work" becoming a source of a flammable mixture. The U.S. Occupational Health and Safety Administration (OSHA) provides guidance and regulations for hot work activities that should serve as the basis for individual companies' safety programs [3].

Material safety data sheets (MSDS) or the newer label, safety data sheets (SDS), provide guidelines to help understand the nature of a flammable or combustible chemical or product. The auto ignition temperature (AIT) is a property of a material that if it is in the combustible range, will ignite upon contact with a surface at that temperature or above. The flammable limits or combustible ranges are also frequently presented as lower explosion limit (LEL) or upper explosion limit (UEL). These are the concentrations at ambient conditions in air in which the mixture is either flammable or combustible. Another set of terms is frequently used interchangeably with LEL and UEL. These are lower flammable limits (LFL) and upper flammable limits (UFL). There is a minor difference on the order of 0.1%, between the explosive and flammable limits. For example, propane has an AIT of 450°C, a LEL of 2.1% and a UEL of 9.5% [4]. Fire prevention begins by recognizing that if we have propane (fuel) plus air (oxidizer) between 2.1 to 9.5% propane, we have a potential for fire or explosion no matter what (see statement above). An additional reference source for flammability properties is a book by Yaws et al [5].

Another indicator of fire potential is the flash point [6]. This number is lowest temperature that a material can catch fire but not be self-sustaining. Generally, lower the flash point, the more flammable. It is related to vapor pressure of the material and to the method of testing (open cup or closed cup for example). Gasoline has a flash point below -43°C. Inside a closed container, at room temperature, the vapor pressure of gasoline brings it into the non-flammable range because the equilibrium mixture is well above its UFL. This is critical in maintaining safety within the world's transportation system. On the other hand, ethanol's flashpoint is 13°C and its fire point (point where a self-sustaining flame is possible) is 18°C. This presents a challenge related to storage of ethanol at ambient temperatures (20 to 30°C). The vapor-air mixture above the liquid is in the mist of the flammable limits for ethanol/air at ambient temperatures. Thus, an inert such as nitrogen must be used as a purging gas. A small flow or small positive head of nitrogen is used to keep the gas mixture within a storage tank and process vessels within the non-flammable range. This gas mixture must remain below the minimum oxygen concentration (MOC) by using nitrogen to displace the oxygen present in ambient air. Crowl and Louvar discuss methods of purging [1]. Large scale purging is one of the largest industrial applications of nitrogen either as a delivered gas/liquid or from on-site membrane separation processes.

3.2 Elementary Controls for Fire Mitigation

Fire fighting measures begin with the design of the process area. Indoor areas almost always require overhead fire sprinklers. There are two main types, wet pipe, where water is readily available at all times, and dry pipe (pressurized air is within – that trips a water valve when the pressure is released) for areas that may drop below freezing. The sprinkler head has a fusible link that typically breaks apart at 165°F (other temperature ranges are available for higher temperature areas). Water then flows out a ½" nozzle on the order of 30 gallons per minute and is dispersed by the cap of the head. Water damage is also a concern when a system goes off. Thus, access and ease of operation of control valves should also be considered in the layout of a system. Guidelines are provided by in NFPA 13 [7].

Outdoor areas, especially areas storing large amounts of flammables, may need fire monitors. These devices spray large amounts of water, several hundred gallons of minutes, into a process area. Some monitors are moveable, either manually or remotely, so that the water spray can be directed to the base of the fire. In the case of liquefied propane storage, the water spray onto the storage tank has an even more critical function of preventing one of the most severe petrochemical explosions, the boiling liquid expanded vapor explosion (BLEVE) [8].

3.3 Professional Guidelines to Consider in Fire Prevention and Design

A number of key guidelines for the control of fires exist for the young professional entering the process safety field. The Dow Fire and Explosion Index [9] provides guidance on spacing of processing units to achieve safer layouts and prevention of major catastrophes. Another Trevor Kletz axiom comes to mind. Paraphrased: "What you don't have can't leak, can't catch on fire, and can't explode" [10]. Proper hazard analysis in terms of fire and explosion prevention includes minimizing the amount of flammable and keeping inventory rotated, and away from personnel.

Fire prevention also includes reviewing materials from the National Fire Protection Association (NFPA) such as NFPA 30 [11] and NFPA 58 [12]. Is the space under assessment located inside a building, or confined space, or in an open-air situation? If inside a building or shelter, are fire sprinklers present overhead? If outside, are fire monitors required? Is there a need for flammable gas detectors? Is the primary flammable lighter than air such as methane or hydrogen? If yes, flammable gas detectors may need to be located at an elevation above ground level. If no, generally, closer to ground level is the appropriate placement as most flammable vapors are heavier than air.

Another elementary lesson in fire and explosion includes the understanding of the two major NFPA fire ratings. Unfortunately, the number system meaning is reversed and depends on which set of NFPA guidelines the user is following. The foremost is the NFPA diamond (NFPA 704) that contains four diamonds within (See Figures 1 and 2). The red diamond, upper center, contains a number from 0 to 4. Four is the most severe fire hazard. The relative ranking for flammables within the diamond is as follows [13]:

Table 1. NFPA 704 Red Diamond Flammability Ratings (adapted from [13]).

Number	Description	Example
Red - 0	Will not burn.	carbon dioxide
Red - 1	Must be heated in order to ignite. Flash point exceeds 90°C or 200°F **(NFPA 30 Class IIIB)**	mineral oil
Red - 2	Moderate heat or relatively high ambient temperature is required for ignition. Flash point between 38°C or 100°F and 93°C or 200°F **(NFPA 30 Class II and IIIA)**	diesel fuel
Red - 3	Liquids or solids that readily ignite at most ambient temperature conditions. Liquids have a flash point below 23°C (73°F) and boiling point at or above 38°C (100°F) or flash point between 23°C (73°F) and 38°C (100°F) **(NFPA 30 Class I B and I C)**	gasoline
Red - 4	Rapidly or completely vaporizes at normal temperature and pressure or readily disperses in air and readily burns. Flash point below 23°C (73°F) **(NFPA 30 Class I A)**	hydrogen, propane

The second set up NFPA rankings appear for flammable and combustible liquids and are used within NFPA 30 and NFPA 45 [14] guidelines. The boiling point is also used to segregate the hazards. These guidelines are used around the USA by fire inspectors accessing the amount of storage that flammable quantities may occupy within any enclosed room or process area. Here Class I flammable means highly flammable, and the amounts that can be stored are less than a Class II or Class III, which are considered combustibles. Table 2 presents a summary of the classifications.

Table 2. NFPA 30 Classification for Liquids [11].

Class	Description	Example
Class I A	Flash point is below 22.8°C (70°F) or ambient temperature and a boiling point below 37.8°C (100°F) **(NFPA 704 - Red 4)**	Butane
Class I B	Flash point is below 22.8°C (70°F) or ambient temperature and a boiling point above 37.8°C (100°F) **(NFPA 704 – Red 3)**	Hexane
Class I C	Flash point is between 22.8°C (70°F) to 37.8°C (100°F) and a boiling point above 100°F **(NFPA 704 – Red 3)**	Nonane
Class II	Flash point is between 37.8°C (100°F) to 60°C (140°F) **(NFPA 704 – Red 2)**	Decane
Class III A	Flash point is between 60°C (140°F) to 93.3°C (200°F) **(NFPA 704 – Red 2)**	Phenol
Class III B	Flash Point above 93.3°C (200°F) **(NFPA 704 – Red 1)**	Heat Transfer Fluid

4.0 EMERGENCY RESPONSE SHOULD A FIRE BREAK OUT

First and foremost, employees in any chemical industrial complex, must understand all of the hazardous properties of the raw materials, intermediates and finished products along with the hazards of all utilities involved in the emergency. When it comes to combustible and flammable liquids or flammable gases, they must understand the principles of flammable limits, flash points, auto-ignition temperatures and be ready to handle a potential fire. They must also understand the products of combustion of the burning materials and review for proper protection. When vinyl chloride burns the fire will emit HCL fumes. If carbon disulfide is ablaze combustion produces toxic gases and irritants including oxides of sulfur.

Emergency responders must know their personal protective equipment. NFPA 704 should be consulted for protective clothing [13]. Typically full protective clothing includes helmet, self-contained breathing equipment, bunker coast and pants, rubber boots and gloves.

Once a fire has erupted, what actions should be involved? If the fire is small enough for hand held or wheeled fire extinguishers, then what type of fire extinguisher should be used? There are three major categories, A, B, and C. "A" ratings are for trash or wood fire, "B" ratings are for extinguishing fires of combustible or flammable liquids, and "C" ratings are for electrical equipment fires [15]. Every individual within an organization should experience pulling the pin and putting out a small "B" type fire using a CO_2 extinguisher. The training should include the demarcation point as to when to "fight" or to "flee." It is okay to flee, and this decision has saved lives!

Another key assessment in fire prevention is the location of the emergency isolation valves to stop the flow of flammables from outside sources. In a hazard analysis ask: "Should emergency isolation valves be automated? If not, can they be accessed safely in an emergency?" There are numerous examples where a fire burned uncontrollably because the proper isolation valve could not be closed.

Your assessment should include the emergency response training expected of personnel. If that response includes a fire response by personnel working in the facility, then fire training is mandatory. If an in-house fire brigade is going to be set up, then they should be given the opportunity to attend a fire training session at a professional fire training facility. Numerous professional training facilities are willing to work with private and industrial groups on fire training. Often local fire departments are willing to conduct fire response training with local industries and fire brigades for training exercises. Recent *Process Safety Progress* articles about emergency response exercises and the need for emergency response are provided in references [16-19].

4.1 Consider an Emergency Response Practice with Local Fire Agencies

Successful emergency response includes working with the local fire agencies beforehand, practicing various levels of drills. At least annually, a practice run with local fire fighters is warranted, especially in rural locations where these fire fighters may be volunteers. One of the authors (Willey) witnessed an emergency response where a million dollar a day process was brought to a complete halt because of a fume from butane lighter brought near an outside flammable detector. An operator halted plant operations who had less than one year of experience. Because of training, she hit the emergency panic button when the flammable detector alarm sounded at her control panel. Within minutes the local fire department arrived and practiced applying foam to a butane storage tank that was not burning. The exercise was then fully reviewed with plant and fire-fighting personnel on how well the group responded to this practice event. During the past 30 years, this facility has had an excellent safety record. Part of the reason can be attributed to emergency response practice with local agencies.

4.2 Fire Response Training – Practice Beforehand – An Example from Taiwan

The National Fire Agency (NFA) of Taiwan is a key part of the Ministry of the Interior and has national responsibility for fire prevention and fire fighting as well as medical response and ambulance services. It was established on March 1, 1995 to improve emergency medical services, promote professional fire prevention and fire fighting service, enhance disaster rescue efficiency, and protect civilian life & property within Taiwan. The NFA Training Center is located in central Taiwan in Zhushan, Nantou County. It occupies 109 hectares (270 acres) and is now the 3rd largest such facility in the world. The main mission of this center is to provide training to young emergency response personnel working in Taiwan who select this as their career path. A typical trainee is a person between the ages of 20 to 25 years old who has selected emergency response as a career. They attend training at the facility for a period of 6 to 18 months depending upon which authority they are responsible to. The facility provides living quarters to trainees during their training period at the facility.

Full training includes practice on a wide variety of fire scenarios reflecting real life emergencies, earthquake responses, and water rescues. The fire responses include the extinguishment of a high intensity propane fire, the entry into a completely dark building with a concurrent smoke event, the entry into a major roadway tunnel with vehicle fire occurring, the entry into an airplane that may be on fire within, the entry into a ship, the approach to a burning tank truck fire, the approach to a chemical complex under a fire condition, and the rescue of stranded persons flowing in a very powerful body of flowing water such as a strong rapids.

The Taiwanese training facility uses propane as a fuel for some of its fire evolutions. Recall the fundamentals about fires above. One must have fuel, an oxidizer, and an ignition source. As shown in Figure 3, we see a scenario area for the training used on putting out a propane fire. Propane and air can flow out through the burners without any signs of a flame. Then, an ignition switch is thrown, and very large intimidating flames evolve and become the focus of the team's attention.

In this training session, the firefighting crew of nine is manning two 2 ½ inch fire hoses in approaching the threatening inferno. Water is the medium used as the responders inch their way forward. Their tactic is to begin with a very broad spray of water (Figure 4). This water spray protects them from the heat of the flame and allows them to move closer to the fire. During this particular training session the fire actually intensifies before the firefighters approach This intensification is by design and assists on protecting the trainees for their initial fire training exercises, allowing them to learn how to approach a large propane fire appropriately. Note that there are nine crew members on these two hoses. For the novice, one is not aware of the momentous forces that exist on any curved large diameter flexible hose. At 2 ½ inches, a large hose releasing 100 gallons per minute, if not supported properly, can develop enough momentive force to break a leg of a human. It is important that emergency responders have experienced this momentum before entering a major flame up. A minimum crew is also required to handle the hose safely.

Figures 5 through 7 show the crew extinguishing the flame. As it is approached, and the flame is brought under control, the water stream narrows, and focuses towards the base of the flame. This entire exercise occurred over an absolute time period of 2 minutes. Capable supervisors, in an on-site centrally located control tower, monitor the training via closed circuit television and can instantly shut down any fire exercise if there is any emergency.

Typically, fires from pipelines or large vessels should not be extinguished unless the fuel can be shutoff and isolated. If a fire is extinguished and the fuel continues to be emitted, a cloud can form and when it reignites an explosion can occur.

4.3 Additional Fire Training Scenarios at the Taiwan Training Facility

The facility has several other fire training scenarios available. Included is a chemical area training facility. There are fires in mock up apartment building, warehouses, theaters, a commercial kitchen fire, a subway car, a large commercial airliner mock up, and much more. There are approximately 50 "simulation points" where qualified instructors can create the conditions that would warrant live firefighting. It would take a visitor a couple of days to see all responders training in each of the available mockup areas.

4.4 Additional Emergency Response Training - Swift Water Rapids Training

Taiwan, due to its geographic location is subject to many natural disasters. It experiences earthquakes, typhoons, and other natural disasters in a much higher proportion than many other locations around the world. Taiwan became interested in such a disaster-training center after a devastating earthquake in 1999.

One natural disaster is the result of its mountainous central areas and the frequency of typhoons. Swift water flows are frequent occurrences. This facility has a large flume with flow rates of 100,000 gallons per minute for the training of emergency response personnel in the rescue of swimmers and others were stranded in a rapidly flowing stream. Figure 8 shows an example of a person moving down the 100,000 gpm flume in a practice for water rescue.

The agency is looking for partnerships with other countries and international corporations. It offers 1-week training courses in swift water, tunnel fire, and aircraft fire. For those working in the process industries, there is an 8, 32, or 56 hour HAZMAT training that follows NFPA guidelines, with division into awareness, operation, and technician levels. The facility is willing to share its training facilities for the preparation of emergency response personnel associated with these groups.

5.0 CONCLUSIONS

Emergency training before an event happens makes a difference. Lack of training often contributes to the problem rather than mitigate it. Further, this training should include working with outside groups with similar interests and resources. By working together both groups learn more effective methods to mitigate emergencies. The facility in Taiwan is willing to welcoming emergency response personal from public and private affiliations worldwide. For more information please write the director, an author of this paper, Dr. Jack, Huan-Chang Hsiao, at **jack@nfa.gov.tw**. We also note that the United States has several excellent fire fighter training schools with large fire scenarios. Places include the Texas A&M- College Station, Texas and Louisiana State University- Baton Rouge and the reader can contact these sources as well for potential fire and emergency response training.

Figure 1. An example of the NFPA Diamond. The center red diamond is related to fire hazard of the material within. Steam is rated "0" as it is not a fire hazard.

Figure 2. A flammable solvent storage tank with a flammability rating of 3.

Figure 3. Crew approaches a propane fire.

Figure 4 Crew walks towards the flame with a cone of streaming water.

Figure 5. Roughly 20 seconds into the training session, crew now move decisively towards the fire, water stream is starting to be concentrated.

Figure 6. Fire close to extinguishment.

Figure 7. Fire extinguished, total elapsed time – 2 minutes.

Figure 8. Swift water training facility during a training session.

6.0 LITERATURE CITED

1. Crowl, D.A.; Louvar, J.F. *Chemical Process Safety Fundamentals with Applications, 3rd Ed.* Prentice Hall PTR, 2011.
2. Kletz, T. " What Went Wrong?, 5th Edition, Case Histories of Process Plant Disasters and How They Could Have Been Avoided," Butterworth- Heinemann, 2009.
3. OSHA Regulations 29 CFR 1910.252, related to hot work, welding, cutting, and brazing, Retrieved December 22, 2012, from http://www.osha.gov/pls/oshaweb/owadisp.show_document?p_id=9853&p_table=standards
4. Propane MSDS, Retrieved December 22, 2012, from http://www.airgas.com/documents/pdf/001045.pdf.
5. Yaws, Carl L.; Braker, William; *Matheson Gas Data Book* Published by McGraw-Hill Professional, 2001.
6. Flash point. Wikipedia, Retrieved December 26, 2012, from http://en.wikipedia.org/wiki/Flash_point.
7. *NFPA 13: Automatic Sprinkler Systems Handbook*, Quincy, MA, 2013.
8. Boiling liquid expanding vapor explosion, Wikipedia, Retrieved December 28, 2012, from http://en.wikipedia.org/wiki/Boiling_liquid_expanding_vapor_explosion.
9. *Dow's Fire & Explosion Index Hazard Classification Guide, Seventh Edition, 2010,* Wiley On Line Books. Retrieved December 24, 2012, from http://onlinelibrary.wiley.com/book/10.1002/9780470938195.
10. Kletz, T. "*By Accident...a Life Preventing Them in Industry*," PFV Publications, London, 2000.
11. R. P. Benedetti and P. E. May, *NFPA 30 and NFPA 30A Flammable and Combustible Liquids Code Handbook 2012 edition,* National Fire Protection Association, Quincy, MA, 2012.
12. *NFPA 58, Liquefied Petroleum Gas Code, 2011 edition,* National Fire Protection Association, Quincy, MA, 2011.
13. NFPA 704 or the Fire Diamond, Retrieved December 24, 2012 from http://chemistry.about.com/od/safetysignsandsymbols/a/Nfpa-704-Or-The-Fire-Diamond.htm.
14. NFPA 45: Standard on Fire Protection for Laboratories Using Chemicals, 2011 Edition, National Fire Protection Association, Quincy, MA, 2011.
15. What do the A B C ratings mean on Fire Extinguishers? Retrieved December 24, 2012, from http://www.brkelectronics.com/faqs/oem/what-do-the-abc-ratings-mean-on-fire-extinguishers.
16. M. Idrees and S. Aslam, "Prevent major failures and accidents with well-trained and empowered operators," Process *Safety Progress,* Volume 29, Issue 2, June 2010, Pages: 174–178, DOI: 10.1002/prs.10348.
17. R. V. Kalnins, "Emergency preparedness and response," *Plant/Operations Progress*, 6 (1987) 6–10, DOI: 10.1002/prsb.720060106.
18. J. A. Klein. "Two centuries of process safety at DuPont," *Process Saf Prog*, 28 (2009) 114–122, DOI: 10.1002/prs.10309.

19. Q. Wang, T. Ma, J. Hanson and M. Larranaga, " Application of incident command system in emergency response," *Process Saf Prog,* 31(4) (2012) 402–406, DOI: 10.1002/prs.11538.
20. Kidde Fire Trainers, retrieved December 22, 2012 www.kiddeft.com/utcfs/ws-385/Assets/Taiwan%20Site%20Sheet.pdf

9TH GLOBAL CONGRESS ON PROCESS SAFETY

Practical Examples of Alternate Overpressure Protection Systems
An Owner's Perspective

Scott A. Tipler
The Dow Chemical Company
Midland, MI 48667
satipler@dow.com

Dow Co-Authors:
Don Eure [Plaquemine, LA]
Ken First [Midland, MI]
Robert Stack [Hillsdale, MI]
Tim Wagner [Midland, MI]

Prepared for Presentation at
American Institute of Chemical Engineers
2013 Spring Meeting
9th Global Congress on Process Safety
San Antonio, Texas
April 28 – May 1, 2013

UNPUBLISHED

Practical Examples of Alternate Overpressure Protection Systems An Owner's Perspective

Scott A. Tipler
The Dow Chemical Company
Midland, MI 48667
satipler@dow.com

Dow Co-Authors:
Don Eure [Plaquemine, LA]
Ken First [Midland, MI]
Robert Stack [Hillsdale, MI]
Tim Wagner [Midland, MI]

Keywords: High Integrity Protection Systems, HIPS, HIPPS, API-521 Annex E, UG-140

Abstract

In 2007, the American Petroleum Institute (API 521/ISO 23251) published guidance on the use of High Integrity Protective Systems (HIPS). In 2009, The Dow Chemical Company (Dow) updated its internal work process to apply Dow's risk based work processes to HIPS design, application and evaluation. The overall result is an efficient process that links together the existing work processes for conventional relief design, Layers of Protection Analysis (LOPA) and Safety Instrumented Systems (SIS). This produces a risk based method for applying and designing protection layers into HIPS. For additional details, refer to "High Integrity Protection System Design Using a Risk Based Approach" by Robert J. Stack, AICHE 2010 spring meeting 6th Global Congress on Process Safety [1].

A natural extension of HIPS is to use system design and operating discipline to mitigate overpressure scenarios where use of fully instrumented protection layers or conventional relief devices is neither practical nor effective. For example, fire protection systems may be used to mitigate the consequences associated with a fire and reaction scenario where conventional relief devices and instrumentation systems are not possible. In Dow, this approach is referred to as Alternate Overpressure Protection (AOP). HIPS and AOP uses system design to mitigate the overpressure scenario or reduce the size of the relief device.

System design (HIPS or AOP) are commonly used to mitigate overpressure scenarios when: 1) a conventional pressure relief device (PRD) is not practical or effective, 2) a conventional PRD will not be reliable or 3) a conventional PRD will work but is not cost effective. In addition, HIPS or AOP can also be used to reduce the required relief size by limiting the operating window of the process.

The purpose of this paper is to provide several practical examples on the use of HIPS and AOP and to describe some of the challenges and associated strategies to ensure successful implementation and sustained process safety performance.

1. INTRODUCTION

Conventional pressure relief devices provide highly reliable and cost-effective protection for most pressure containing equipment. However there are scenarios where conventional relief devices are not practical, not reliable, or cost effective. System design is then often used to mitigate the overpressure scenario or reduce the size of the relief device for the following situations:

<u>NOT PRACTICAL OR POSSIBLE:</u>
A relief device may not be practical or effective for:

- reactive chemical relief scenarios (e.g. runaway reactions)
- existing equipment that would require extensive modifications (e.g. new nozzle)
- relief scenarios protecting against excessive temperature, such as over-firing a cracker or heat transfer furnace

<u>NOT RELIABLE:</u>
A conventional pressure relief device may not be reliable when used in applications:

- that may cause plugging(e.g. monomer services, polymer solutions)
- that contain highly corrosive materials
- with fluids that tend to freeze, or cure in the relief system

<u>NOT COST EFFECTIVE:</u>
When relief valves are routed to a flare or scrubber system, the cost of the treatment system can greatly exceed the cost of the relief device and associated piping. Consider the use of system design to mitigate the overpressure and associated effluent scenarios when:

- the consequence of routing the effluent directly to atmosphere is not acceptable and the cost of a new effluent treatment system is not cost effective
- the incremental load from a relief device to an existing flare system or scrubber system exceeds the existing capacity, resulting in a costly capacity expansion project
- the initial cost for effluent treatment system can be greatly reduced by selectively using system design to eliminate high loads from a limited number of relief devices

For example, plant-wide loss of cooling is often the controlling scenario that determines the size of a flare system. For distillation columns, the pressure relief load due to loss of cooling is often much higher than that required by the other scenarios.

The American Society of Mechanical Engineers (ASME) Boiler and Pressure Vessel Code, Section VIII, Division 1, UG-140(b) [2], provides guidance on the protection of equipment from overpressure by system design or by a combination of system design and pressure relief devices.

High Integrity Protection Systems[1] (HIPS) or Alternate Overpressure Protection (AOP) can be used to mitigate the overpressure scenario or reduce the size of the relief device by system design. Use of either HIPS or AOP reduces the likelihood of a specific overpressure scenario to a tolerable frequency. Protection by system design (HIPS and AOP) can be used for both low pressure tanks and pressure vessels.

So what is "HIPS"? HIPS is a collection of protection layers that is used in lieu of or in combination with a relief device to mitigate a single overpressure scenario. In Dow, the HIPS solution must include a Safety Instrumented System (SIS) with a minimum Safety Integrity Level (SIL) 1. LOPA is used to identify all of the required protection layers and the SIL for any SIS protection layer(s). Fault Tree Analysis can also be used to determine the frequency of the initiating event or the probability of failure on demand (PFD) of the protections layers, especially for complicated systems (e.g. multiple sensors or final elements).

So what is "AOP"? AOP is a collection of protection layers that are used in lieu of, or in combination with, a relief device to mitigate a single overpressure scenario. As with HIPS, LOPA is used to identify all of the required protection layers. In contrast to HIPS, AOP does not include fully automated protection layers such as Safety Instrument Systems (SIS) and Basic Process Control System (BPCS) protective systems because a fully automated loop is neither practical nor effective. For AOP, the layers of protection include operator response to alarm and other Safety Related Protection Systems (SRPS). SRPS includes operating discipline (e.g. sampling and associated checklist to test for positive identification), fire protection systems (e.g. drainage to a remote containment area), and safety design features (e.g. unique connections). A human reliability analysis (HRA) may be required to confirm the reliability of operator response to alarms or operating discipline.

The use of HIPS is internationally accepted as a method for protecting pressure vessels from excessive pressure. In ASME jurisdictions, the authority to use HIPS is found in ASME Section VIII, Division 1, UG-140(b). However, ASME Section VIII is a high-level document which doesn't contain specific application guidance for the use of HIPS. Specific application guidance for complying with ASME Section VIII is found in the API 521 standard (5th edition – 2008 Addendum) [3], including detailed guidance for applying HIPS (API 521 Annex E).

In European countries which are covered by Pressure Equipment Directive (PED) regulation, the authority to use HIPS is found in PED Guideline 1/43, and the application standard for HIPS is EN 764-7, Pressure Equipment Part 7: Safety Systems for Unfired Pressure Equipment [4]. HIPS are called Safety Related Measurement, Control and Regulation Systems (SRMCR) in Europe.

In ASME jurisdictions, the authority to use AOP for pressure vessels is also found in ASME Section VIII, Division 1, UG-140(b). Specific application guidance is not available for AOP.

It is imperative that there are management systems in place to maintain the integrity of the protection layers for both HIPS and AOP. This can be accomplished by predictive preventative maintenance and testing, management of change, periodic audits, periodic reviews of the Process Hazard analysis, and periodic revalidation of the LOPA.

[1] Sometimes referred to as High Integrity Pressure Protection Systems (HIPPS).

See table 1 for a comparison of HIPS versus AOP.

Table 1. Comparison of HIPS versus AOP

Component	HIPS	AOP
Consequence/risk	Structured Hazard Analysis	Structured Hazard Analysis
Scenario likelihood	LOPA	LOPA
Layers of Protection	Operator Response to Alarm (OPR) Basic Process Control System (BPCS) Safety Instrumented System (SIS) Safety Related Protection Systems (SRPS)	OPR SRPS
Protection reliability and safety verification	Safety Requirements Specification (SRS) FTA	HRA FTA
Performance validation	Initial and periodic performance testing	Initial and periodic SRPS performance testing Procedure training
Documentation	HIPS form SRS Process and Instrument Diagrams (P&IDs)	AOP form P&IDs
Auditing	Periodic Internal Auditing Periodic PHA review Periodic LOPA revalidation	Periodic Internal Auditing Periodic PHA review Periodic LOPA revalidation

2. HIPS Examples

Below are examples of HIPS.

2.1 High Pressure Supply System – Basic Process Control Failure to limit high pressure to downstream equipment

Example 1: A high pressure supply system is connected to a lower pressure system via a control valve. Failure of the BPCS control loop (e.g. control valve sticks open or sensor fails low) can result in the overpressure of the lower pressure system. (See Figure 1)

HIPS Example: Propane Vaporizer Drum
Scenarios:

1. Fire Exposure (3"x4" Pressure Safety Valve (PSV) required)
2. BPCS Control Loop Failure (8"x10" PSV required)

Problem: Scenario #2 can create an additional hazard by dumping large amounts of liquid propane into the flare header, possibly overloading the flare system and creating other hazards

Solution:

- Size the PSV for the fire scenario
- Use HIPS to mitigate the BPCS control loop failure scenario

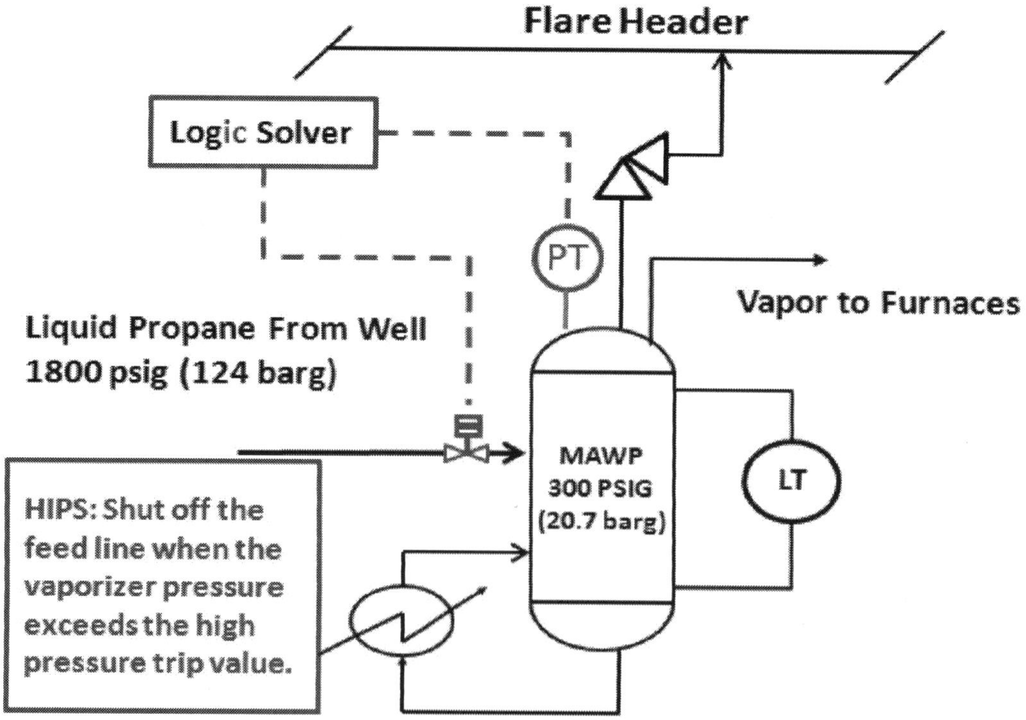

Figure 1: Liquid Overfill of propane vaporizer drum

In this example, a high pressure interlock is used to close an Emergency Block Valve (EBV) in the propane supply line from the well. The required safety integrity level (SIL) for the high pressure EBV interlock is determined by LOPA.

2.2 Fired Dowtherm™ Heater – Blocked In with Continued Heating

Example 2: If a heater coil is blocked in with continued heat applied, the coil temperature may exceed its maximum allowable working temperature (MAWT). As a result, the coil could rupture releasing the hot oil into the heater combustion chamber and potentially resulting in a process safety event (e.g. explosion). Fired oil heaters are typically not designed to contain a deflagration nor equipped with deflagration vent panels. (See Figure 2)

<u>HIPS Example:</u> Dowtherm™ Heater
Scenarios:

1. Liquid Thermal Expansion
2. Overheating and failure of the heater coil

Problem: For Scenario #2, the relief valve will not prevent the heater coil from overheating due to the low vapor pressure of the material. In addition, two phase venting will occur which is impractical to relieve due to the inlet losses.

Solution:

- Size the PSV for the liquid thermal expansion
- Use HIPS to mitigate the overheating and coil failure scenario

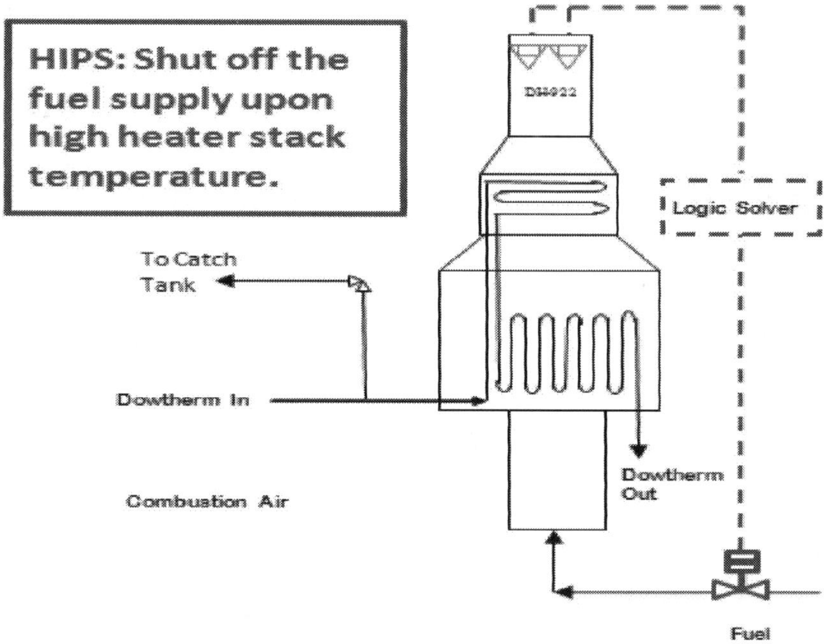

Figure 2: Dowtherm™ Heater Tube Failure

In this example, a high stack temperature interlock is used to close an EBV in the fuel supply. The required safety integrity level (SIL) for the high temperature EBV interlock is determined by LOPA.

2.3 Runaway reaction of tower bottoms – Loss of cooling or power

Example 3: The heating media supply temperature (e.g. steam) for distillation towers may be greater than the temperature of no return (TNR). The TNR is defined as the temperature when the heat losses equal the heat of reaction. In addition, heat neutral or endothermic reactions may produce significant amounts of gas at elevated temperature. Loss of power or cooling can result in a runaway reaction (e.g. decomposition) if the heating media supply valve should leak through causing the tower bottoms to exceed the TNR. Use of a relief device to protect against overpressure from the runaway reaction may be impractical if the relieving temperatures exceed the maximum allowable working temperature (MAWT) or if the decomposition gas generation rate is excessive. (See Figure 3)

<u>HIPS Example: Distillation Tower and Reboiler</u>
Scenarios:

1. Loss of Power
2. Loss of cooling (e.g. cooling pump failure)

Problem: Both scenarios will exceed the MAWT of the equipment and lead to additional decomposition reactions which are difficult to relieve.

Solution: Use HIPS to mitigate both scenarios

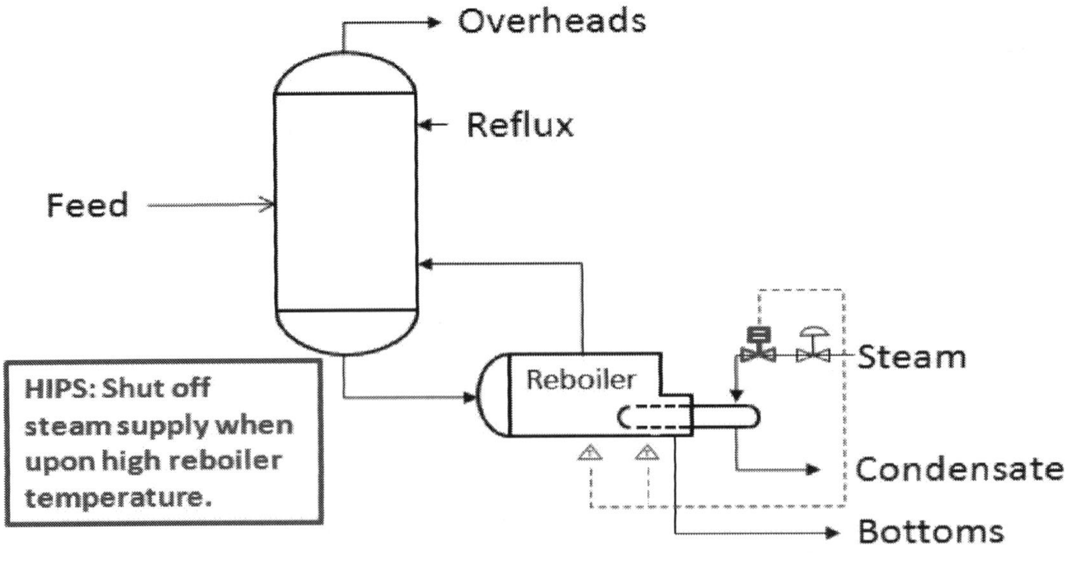

Figure 3: Distillation Tower Bottoms Runaway

In this example, a high temperature interlock is used to close an EBV in the steam supply line. In addition, the bottom recirculation pump may need to be shut off, especially if the system is insulated. In addition, a LOPA credit can also be taken for the basic process control action to close the steam control valve since the initiating event is loss of power or cooling and the valve is designed to fail closed. The required safety integrity level (SIL) for the high temperature EBV interlock is determined by LOPA.

2.4 Hydrocarbons Facility – Loss of Cooling

Example 4: Loss of cooling to a hydrocarbons facility due to a cooling pump failure or power loss can result in multiple relief systems relieving into a flare header at the same time. HIPS can be used to reduce the flare header load. This is especially beneficial when considering an expansion of existing facilities. (See Figures 4 and 5)

HIPS Example: Hydrocarbons facility flare system
Global Scenarios for flare header:

1. Fire Scenario
2. Loss of cooling

Problem: The loss of cooling load from the new process unit results in the flare and the flare header being overloaded.

Solution: Reduce the loss of cooling load by strategically applying HIPS. Apply the number of HIPS necessary to reduce the overall load to an acceptable value.

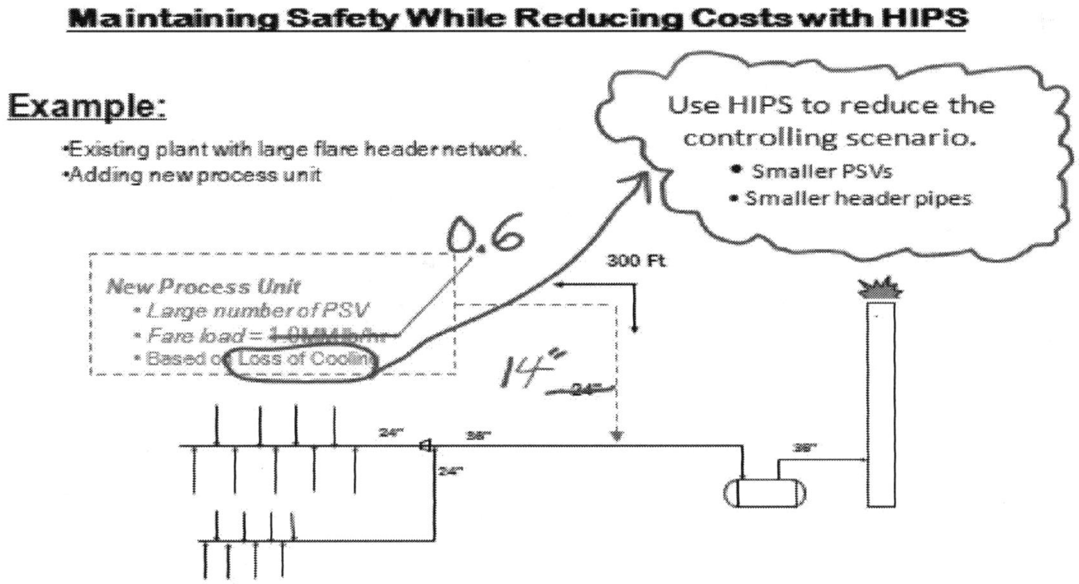

Figure 4: Reducing Flare Header Loads with HIPS

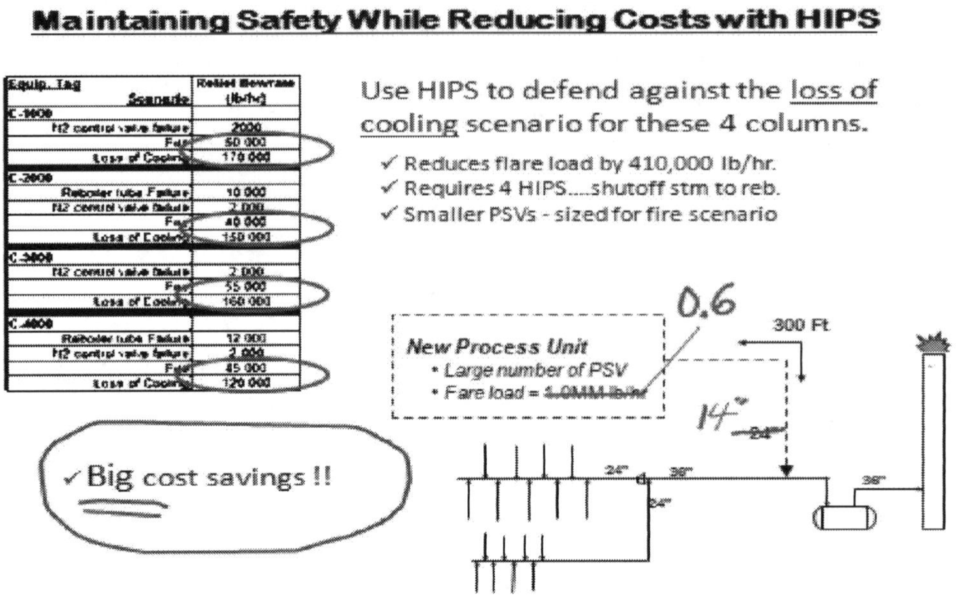

Figure 5: Reducing Flare Header Loads with HIPS

In this example, high pressure interlocks are used to close the EBVs in the steam supply lines and these are implemented as SIS. In addition, a LOPA credit can also be taken for the basic process control action to close the steam control valve since the initiating event is loss of power or cooling. The required safety integrity level (SIL) for each SIS is determined by LOPA.

3. AOP Examples

Below are examples of AOP.

3.1 Monomer Storage Tank – Fire and Reaction

Example 1: Fire results in an exothermic runaway polymerization and decomposition reaction in a nearby monomer storage tank.

AOP Example: Low Pressure Fixed Roof Monomer Storage Tank
Scenarios:

1. Fire with Reaction
2. Filling and Thermal Breathing

Problem: Required relief size for the fire with reaction scenario is impractical

Solution:

- Size the relief for the filling and thermal breathing scenario(no reaction)
- Use AOP to mitigate the fire with reaction scenario

In this example, AOP is provided by fire protection systems which mitigate the fire and reaction scenario. Fire protection systems can include drainage to a remote containment area, direct water spray, area water deluge, and foam suppression systems. Note that some fire protection systems can reduce the effectiveness of other fire protection systems. An alternative approach for providing overpressure protection may be to use a frangible roof. However, the advantage to using AOP is that it mitigates both the overpressure and the hazards associated with the effluent for the fire and reaction scenario (e.g. flash fire, vapor cloud explosion, and/or toxic exposure).

In order to take a LOPA credit for a remote containment system, it should have sufficient capacity to contain the largest spill and firewater [5]. If the capacity of the containment system is inadequate, flammable or combustible materials could back up into the process area and expose the other equipment to fire.

In order to take a LOPA credit for direct water spray, the system should havesufficient fire water nozzle coverage and density to limit the wall temperature below the TNR or the temperature where significant gas is generated [6]. Sprinkler systems where individual nozzles are activated do not qualify for the water spray credit. For single density spray, the vessel wall temperature can be limited to approximately 90 °C. For double density spray, the wall temperature can be limited approximately to 70 °C [7]. In addition the firewater supply system should have adequate capacity for uninterrupted flow for the duration of the fire or spill [7].

In order to take a LOPA credit for an area water deluge system, the material must be highly soluble in water (e.g. ethanol) or have a density greater than water such that the fire water will suppress the vaporization of the material. For soluble materials, the firewater supply system should have adequate capacity to reduce vapor pressure of the spilled quantity below the ignitable concentrations. For materials that are denser than water, the firewater supply system should have adequate capacity to maintain adequate water coverage for the entire diked areas without drainage. Slabs or areas that drain to a remote containment area should have adequate capacity for uninterrupted flow of tank contents and water for the duration of the fire or spill.

In order to take a LOPA credit for foam deluge, the foam is applied through an area water deluge system or by individual foam applicators such that the foam suppresses the vaporization of the material over the entire surface area [8]. The foam and firewater supply system should have adequate capacity to maintain proper coverage for the duration of the fire or spill [9].

To be effective, prompt detection and activation of the water or foam system is required. As a result, fully automated systems are recommended (e.g. deluge systems with heat activated detectors). For fully automated systems, a minimum time delay (e.g. 5 minutes) from the start of the fire till system activation is typically assumed. Spill detection by human observation or local sensor/alarm and foam or water addition that is tripped by plant personnel may qualify for LOPA credit depending on the time required to heat the vessel to the TNR. A human reliability analysis (HRA) may be required to confirm the reliability of a human activating a water or foam system that is not fully automated [10].

There are multiple initiating events and outcomes for the pool fire scenario. As a result, the pool fire frequency is typically determined outside of LOPA (e.g. Fault Tree Analysis (FTA)). The probability of ignition is considered in the pool fire frequency. If the maximum operating temperature is well below the flash point of the material, the probability of ignition is greatly reduced [11]. Once the consequence of the scenario is determined, the pool fire frequency is determined, and protection layers PFDs are determined, LOPA can be used to determine if the fire protection systems are adequate to mitigate the fire scenario (close the risk gap). A graphic representation is shown in the following bow tie diagram. (See Figure 6)

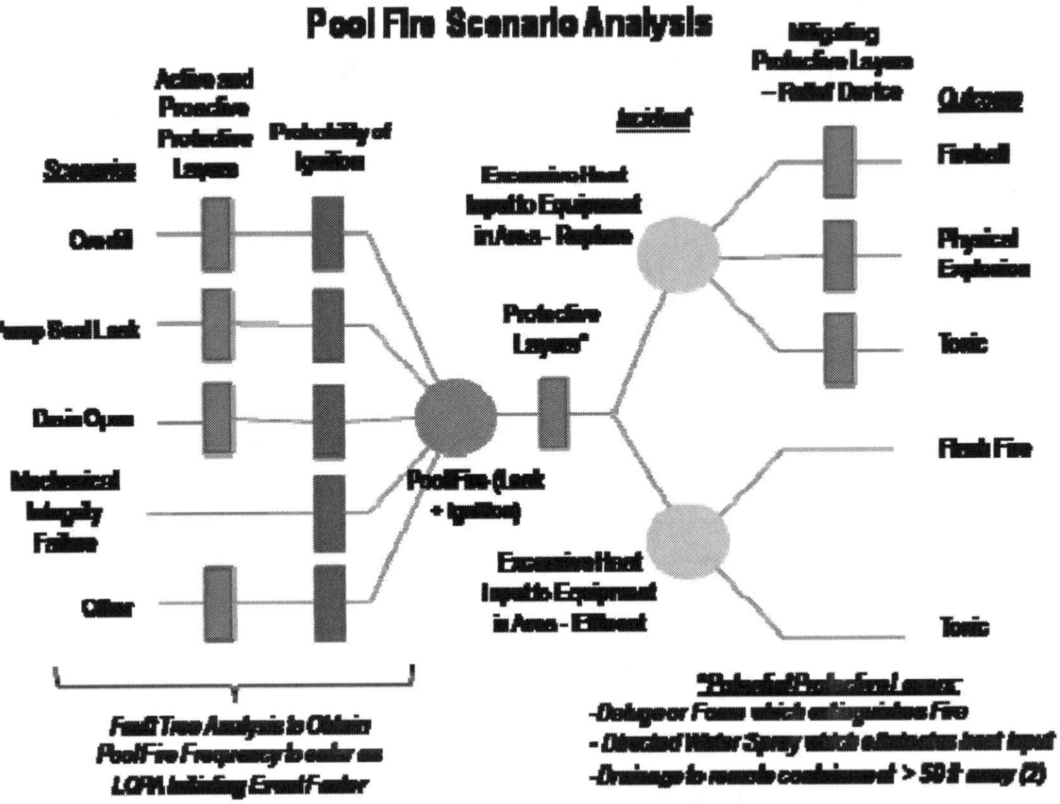

Figure 6: Pool Fire Bow Tie Analysis

Fire protection systems are typically considered as Safety Related Protection Systems (SRPS). However fully automated deluge systems (e.g. tripped by gas detection or temperature sensors) can be designed as Safety Instrumented Systems, in which case HIPS applies.

3.2 Hydrogen Peroxide Storage Tank – Contamination

Example 2: Contamination of hydrogen peroxide storage tank by unloading an incompatible material can result in an exothermic decomposition reaction. Hydrogen peroxide decomposition may occur if contaminated with iron, caustic, or an organic material [12]. Contamination may occur if the wrong material is added to storage tank for a variety of reasons including: 1) supplier sent the wrong material, 2) supplier mislabeled the container, or 3) logistics group located the wrong material at the unloading station.

AOP Example: Iron contamination of hydrogen peroxide storage tank

Scenarios:

1. Fire with decomposition
2. Contamination with reaction

Problem: Required relief size for the contamination and reaction scenario is impractical

Solution:

- Size the relief for the fire with decomposition scenario
- Use AOP to mitigate the contamination with reaction scenario

In this example, AOP is provided by Safety Related Protection Systems (SRPS) which mitigate the contamination with reaction scenario. SRPS may include confirmation of identity from a certificate of analysis, sampling with positive identification, dedicated unloading stations, unique unloading connections, dedicated or unique containers, and unloading procedures with associated checklist. A human reliability analysis (HRA) may be required to confirm the reliability of the unloading procedures. An alternative approach for providing overpressure protection may be to use a frangible roof. However, the advantage to using AOP, is that it mitigates both the overpressure and the hazards associated with the effluent for the contamination scenario (e.g. toxic exposure).

Use of technology may also afford protection against these types of scenarios. For example, plant personnel can verify the identity and enter the railcar or truck number and dome seal numbers into the process control system. The computer system ensures that the information matches the correct material before allowing the unloading EBV to open (permissive function). Likewise, bar coding can be used to confirm the identity of the material before the unloading valve would be allowed to open. This is especially useful for batch processes that have a number of manual additions. LOPA, with supporting analyses such as HRA, is then used to confirm that there are adequate lines of defense to mitigate the contamination scenario (close the risk gap).

4. Conclusion

System design (UG-140) can be used to mitigate overpressure scenarios when: 1) a conventional PRD is not practical or possible, 2) a conventional PRD will not be reliable or 3) a conventional PRD will work but is not cost effective. Within Dow, High Integrity Protection Systems (HIPS) or Alternate Overpressure Protection (AOP) are used to mitigate scenarios or reduce the size of the relief devices by system design.

HIPS is a collection of protection layers that is used in lieu of, or in combination with, a relief device to mitigate a single overpressure scenario. In Dow, the HIPS solution must include as a minimum a Safety Instrumented System (SIS) of Safety Integrity Level (SIL) 1. LOPA is used to define the required protection layers and the SIL requirements for the SIS protection layers.

An AOP system is also a collection of protection layers that is used in lieu of, or in combination with, a relief device to mitigate a single overpressure scenario. As with HIPS, LOPA is used to identify all of the required protection layers. In contrast to HIPS, AOP does not include fully automated protection layers such as Safety Instrument Systems (SIS) and Basic Process Control System (BPCS) Protective Systems because a fully automated loop is neither practical nor effective. For AOP, the layers of protection include operator response to alarm and other Safety Related Protection Systems (SRPS). SRPS includes operating discipline (e.g. sampling and associated checklist to test for positive identification), fire protection systems (e.g. drainage to a remote containment area), and safety design features (e.g. unique connections).

Implementing a HIPS or an AOP system is not an insignificant task. The analysis, specification, and documentation of a HIPS or an AOP system is often more complex, more time consuming, and more costly than designing and installing a conventional relief device. It is imperative that there are management systems in place to maintain the integrity of the protection layers for both HIPS and AOP. This can be accomplished by predictive preventative maintenance and testing, management of change, periodic audits, periodic reviews of the Process Hazard analysis, and periodic revalidation of the LOPA.

The examples and associated discussion above provide guidance on how to mitigate scenarios that are difficult to mitigate with conventional relief devices. This approach has many potential applications beyond the examples discussed in this paper.

5. References

[1] "High Integrity Protection System Design Using a Risk Base Approach" by Robert J. Stack, AICHE 2010 spring meeting 6th Global Congress on Process Safety.
[2] ASME Code UG-140, Pressure Vessels With Overpressure Protection By System Design, 2007 ASME Boiler and Pressure Vessel Code, Section VIII Division 1, 2009b Addenda, June 2009
[3] ANSI/API Standard 521 and ISO 23251, "Petroleum and natural gas industries - Pressure-relieving and depressuring systems", Fifth Edition, May 2008 Addendum
[4] EN 764-7, Pressure Equipment Part 7: Safety Systems for Unfired Pressure Equipment, 2006
[5] National Fire Protection Codes, Flammable and Combustible Liquids Code, NFPA-30, 2012 Edition
[6] National Fire Protection Codes, Standard for Water Spray Fixed Systems for Fire Protection, NFPA-15, 2012 Edition
[7] Vent Sizing for Fire Considerations: External Fire Duration, Jacketed Vessels and Heat Flux Variations Owing to Fuel Consumption, J. Hauser, et. al. Journal of Loss Prevention in the Process Industries, Volume 14, 2001, p. 403-412.
[8] National Fire Protection Codes, Standard for Low-, Medium-, and High-Expansion Foam, NFPA-11, 2010 Edition
[9] National Fire Protection Codes, Standard for the Installation of Foam-Water Sprinkler and Foam-Water Spray Systems, NFPA-16, 2011 Edition
[10] Center for Chemical Process Safety. (2006). *Human Factors Methods for Improving Performance in the Process Industries.* New York: American Institute of Chemical Engineers.
[11] Lee's Loss Prevention in the Process Industries, F. Lees and S. Mannan, 3rd edition, 2005, p. 147-149.
[12] Bulk Storage Guideline of Hydrogen Peroxide, CEFIC, March 2012

Additional references not cited:

[1] Anderson Greenwood Crosby Technical Seminar Manual - 4.0 Department Of Transportation Code Requirements - Gas Transmission And Distribution Piping Systems 4.2.4, 4.2.5 (2001)
[2] API Recommended Practice 520: Sizing, Selection, and Installation of Pressure-Relieving Devices in Refineries
[3] Sizing, Selection, and Installation of Pressure-Relieving Devices in Refineries," American Petroleum Institute (API), API Recommended Practice 520, July 1990
[4] International Electrotechnical Commission (IEC) - IEC 61511-1, Functional Safety: Safety Instrumented Systems for the Process Industry Sector, 2002
[5] Application of Safety Instrumented Systems for the Process Industries, ISA Standard, ISA-S84.01-1996, ISA TR-84.02 and S84.01-2004
[6] Guidelines for Pressure Relief and Effluent Handling Systems, Sec. 5.9, CCPS, 1998
[7] International Electrotechnical Commission (IEC) IEC 61508, Functional Safety of electrical/electronic/programmable electronic Safety Related Systems, 7 Parts, 1998 – 2000

[8] API Recommended Practice 553, Refinery Control Valves, First Edition (API, 1998)
[9] VDI/VDE 2180 Safety device of plants of process engineering with means of the process instrumentation (PLT) introduction, terms, conception (April 2007)
[10] A Study of Flare Load Reduction by a Safety Instrumented System based on a High Integrity Protective System (HIPS) Process Safety Progress (2013)

9TH GLOBAL CONGRESS ON PROCESS SAFETY

Quality Assurance and Quality Control of Pressure Relief Systems

Subhash S. Kale
Sr. Process Engineer
Lloyd's Register Energy Americas
1330 Enclave Parkway, Suite 200
Houston, TX 77077
subhash.kale@lr.org

Prepared for Presentation at
American Institute of Chemical Engineers
2013 Spring Meeting
9th Global Congress on Process Safety
San Antonio, Texas
April 28 – May 1, 2013

UNPUBLISHED

Quality Assurance and Quality Control of Pressure Relief Systems

Subhash S. Kale
Sr. Process Engineer
Lloyd's Register Energy Americas
1330 Enclave Parkway, Suite 200
Houston, TX 77077
subhash.kale@lr.org

Keywords: Quality Assurance, Quality Control, Pressure Relief

Abstract

Pressure relief systems are important for the safe operation of refineries and chemical plants. They are generally considered to be the last line of defense against abnormal operation which could lead to serious accidents, resulting in loss of life and property. To ensure that the system provides proper protection for the range of potential process deviations, it is important to properly size the pressure relief systems considering all the possible applicable overpressure scenarios. Quality Assurance (QA) and Quality Control (QC) are important work processes, which help to ensure that relief systems are properly designed and documented. This paper discusses best practices regarding Quality Assurance and Quality Control protocols for pressure relief systems.

Definitions are given for the terminology used for the QA/QC process of pressure relief system studies. A sample workflow of a typical pressure relief system analysis is also provided. Different steps of the pressure relief analysis process are briefly discussed, and sample work instructions are given for each step. An example set of documentation required for a pressure relief system study, as well as a recommended sequence of documents therein, are described. Work instructions are given for the QA/QC process in the different sections of the studies - system information, overpressure contingency analysis, relief rate calculation and relief device sizing, deficiency identification and final documentation. Lastly, some common mistakes which have been observed during the QA/QC of numerous studies of pressure relief systems from commercial refineries and chemical plants are reviewed.

1. Introduction:

QA/QC Definitions

It is important to start by defining "Quality Assurance" and "Quality Control" in general terms.

1.1 Quality Assurance

Quality Assurance is based on a process approach. It ensures that processes and systems are developed and adhered to in such a way that the deliverables are of good quality. This process is meant to produce defect-free goods or services, which means being "right the first time" with little to no rework.

Quality Assurance is done at the beginning of a project. The requirements of the customers are defined. Based on these requirements, processes and systems are established and documented. This is done to ensure that the requirements of the customers are met stringently. In summary, QA is a proactive process developed to avoid defects [1].

1.2 Quality Control

Quality Control is a product-based approach. It checks whether or not the deliverables satisfy the specifications as well as the quality requirements of the customers. QC is done after the product is produced. Based on the customer requirements developed during the QA process, the quality control personnel check whether or not the product meets the customer requirements. Thus, QC is a corrective process developed to identify defects in order to correct them [1].

The QC process checks each system evaluated for a given process or the portion of the process. A large process can have hundreds of such systems. There may be an additional QC step performed by the client or a third party to ensure that the system evaluation meets the client requirements and identifies any gaps in the analysis.

2. QA/QC of Pressure Relief Systems

A pressure relief system includes a pressure relief device (a general term used to refer to a relief valve, a rupture disk, or other device like a rupture pin used to relive overpressure). It is also comprises of inlet and discharge piping as well as effluent handling equipment.

To complete an evaluation of pressure relief systems, it is necessary to assemble different background information about the process and identify the pressure relief devices in the system. Numerous potential overpressure scenarios applicable to the system are analyzed, and calculations are performed to determine if the pressure relief devices and the associated piping mitigate any potential overpressure situation. The completed evaluation may include recommendations to correct any deficiencies found during the study. If a pressure relief system does not exist, a suitable relief system design is proposed for incorporation into the process.

The quality assurance process establishes the various steps required in the analysis of pressure relief systems and produces work instructions and checklists for each section based on customer requirements.

During the quality control process, each relief system study is checked using the checklists developed in the QA process. Any deficiencies are reported, and the deficiencies are reviewed for correction. A final check is made after the corrections are incorporated before sending the generated report of the system evaluation for client review.

2.1 Workflow

Figure 1 gives a schematic of the workflow in the QA/QC of pressure relief systems. Subsequent sections describe the various steps in the QA/QC process.

Figure 1: Acceptance Workflow of a Typical Relief Device Sizing and QA/QC Process

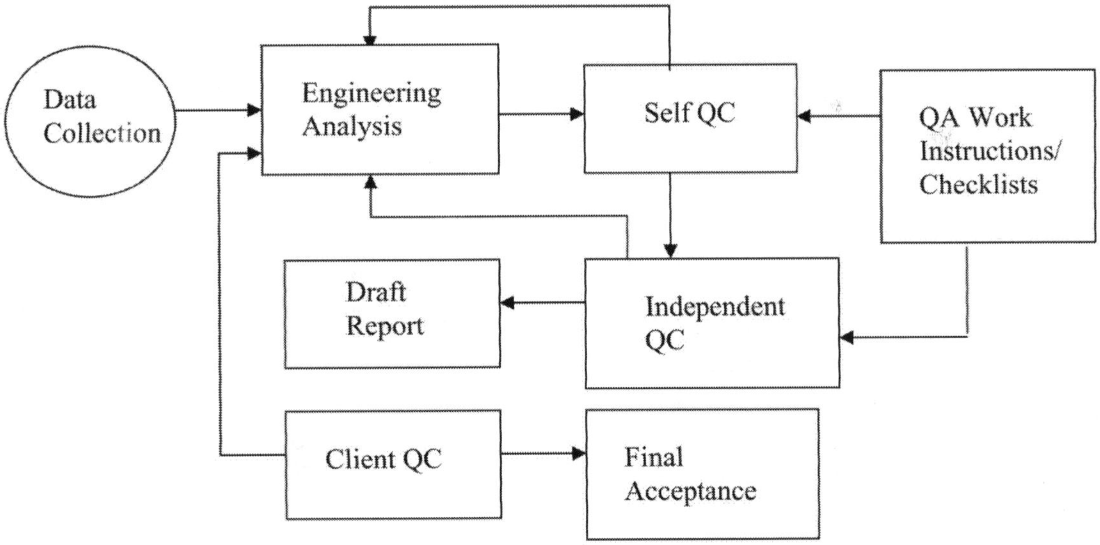

In the first step, the plant data is collected and the relief system analysis is completed using work instructions generated in QA process. The relief system designer does his own QC before submitting the study for an independent QC analysis. Any corrections are sent back to the relief system designer, who incorporates the corrections. After checking the corrections, the study is finalized and submitted to the client. The client QC may also indicate the need for additional changes, which are incorporated prior to finalizing the analysis, completing the documentation, and storing the study.

2.2 Work Instructions

Work instructions may be given by clients to be used as standard procedures for different sections of the relief device system folder. There could be work instructions internally generated within a company to provide deliverables of consistently high quality. Examples of key sets of work instructions are mentioned below:

- Work instructions providing guidance on relief system analysis scope
- Work instructions providing guidance on relief system folder contents
- Work instructions providing guidance on overpressure scenarios
- Work instructions regarding evaluation of equivalent length calculations
- Work instructions regarding evaluation of relief device sizing calculations

The following gives a brief description on what each of these sets of work instructions may contain. This is general information only and can vary from one company to another. Some companies have detailed, sometimes exhaustive work instructions for the aforementioned areas.

2.2.1 Work Instructions for Developing Relief System Analysis Scope:

These work instructions give guidance on what systems are to be included in the Pressure Relief System Analysis. It gives guidance on what documentation may be needed to conduct the analysis of system.

Definition of a System: A system is defined as a collection of pressurized equipment and/or piping that may require protection from various causes of overpressure. A simple system can have just one vessel. A more complex system may have a combination of different equipment; an example is a fractionation tower with a condenser, reboiler, pump around heat exchangers and the interconnecting piping etc. The system is separated from the rest of the process by block valves on the inlet and the outlet or equipment like pumps or compressors, etc. The equipment contained in the system requires an open flow path without any block valves that can be closed inadvertently. If there are any valves within the system, they must be chain locked open or car seal opened.

There may be single or multiple pressure sources in the system. Some examples of pressure sources are rotating equipment, other process equipment that feeds downstream equipment by pressure differential, or any utility being supplied to the system. There may be instrumentation associated with the different components of the system as well as valves (either automatic or manual), that may need to be considered in the analysis of the system.

There may be separate work instructions regarding the documentation needed for the relief system evaluation, which are discussed in a later section. The work instructions may give guidance on different types of pressure relief devices such as Pressure Safety Valves (PSV), Pressure Vacuum Relief Valves (PVRV), and Sentinel Valves (which give only an audible warning in case of overpressure). The QA/QC analysis should be restricted to the given portion of equipment under consideration. For example, a system considering the tube side of a heat exchanger does not consider its shell side and the shell side-associated equipment.

2.2.2 Work Instructions Regarding Relief System Folder Contents:

These instructions give guidance on what to include in the relief system folder. The instructions specify a cover sheet which provides an executive summary, an overpressure contingency analysis, relief device calculations and the supporting information. The supporting information includes inlet and outlet piping sketches, Process Flow Diagrams (PFDs), Piping and Instrumentation Diagrams (P&IDs), equipment data sheets, pump data sheets and pump curves, utility conditions, and any other supporting information used in the relief device calculations. The documentation for each of the relief devices in the system may be in their own folders. The relief system folder will typically have the following items:

- Relief valve (and any associated rupture disk) data sheet(s)
- Executive summary page
- Overpressure contingency summary (which gives a list of all the overpressure cases considered and also the results of the relief device calculations for each of the credible cases)
- Hydraulic calculations for the relief device inlet and the outlet piping
- Relief device sizing calculations
- Supporting information (like piping isometric drawings used for the relief device inlet and outlet equivalent length calculations)

The details of the documentation in the folder are discussed in Section 2.3.

2.2.3 Work Instructions for Evaluating Overpressure Scenarios:

Table 1 gives a list of some of the typical overpressure scenarios per American Petroleum Institute (API) 521 guidelines [2]. This list is only a general list. Each company may have its own customized list, which may have additional scenarios to consider. For atmospheric vents on low pressure tanks per API 2000 [3], there may be additional scenarios like inbreathing, outbreathing, or liquid draining. Each of these scenarios is to be analyzed for its credibility for the given system. If a scenario is credible, the reason why the scenario is credible is documented and the scenario is included in the relief device sizing calculations. If the scenario under consideration is similar to another scenario, a reference may be made to that scenario identification number. If a scenario is considered not to be credible, a remark is made to that effect and an explanation is given as to why it is not a credible scenario. If the scenario is out of scope of the current analysis, this is also noted. For example, some companies do not consider reactive systems or explosions to be within the scope of their studies.

Table 1 List of typical scenarios considered in relief system analysis[2]

#	**Scenario Description**
1	Blocked outlet
2	Failure of cooling systems
3	Overfilling
4	Failure of automatic controls
5	Abnormal heat or vapor input
6	Split exchanger tube
7	Hydraulic expansion
8	Exterior fire
9	Power failure (Steam, electric or other)
10	Instrument air failure
11	Inadvertent valve opening
12	Vacuum formation
13	Chemical reaction
14	Other

2.2.4 Work Instructions for Performing Equivalent Length Calculations:

The hydraulic calculation of equivalent lengths is important in determining the relief device inlet and outlet pressure drops. There are different methods for determining the equivalent lengths of piping. Piping isometric drawings are required for these calculations. If these types of drawings are not available from the client, a field assessment is necessary to obtain the piping isometric sketches. Straight pipe lengths, different types of fittings in the piping, or expanders or reducers are accounted for in these calculations. If there is a rupture disk in the relief valve inlet or outlet lines, that needs to be accounted for in the calculation as well. Also, the pipe entrance and exit losses are considered. Standard values of K factors are obtained from Crane Technical paper 410 [6] or other company specific information. The equivalent length of each section is calculated from the K values, the associated diameter of the piping and the friction factor calculated based on the flow through the piping.

2.2.5 Work Instructions for Relief Device Sizing Calculations:

These calculations consist of determining the relieving rate for each applicable scenario. API 521 [2] gives a description of calculation methods of relief rates which can be used for different overpressure scenarios. API 520 [3] gives methods to calculate the relief device size based on the relieving rate and various factors like discharge coefficients, backpressure correction factors, etc. Different companies also have their own proprietary tools for these calculations, such as Microsoft Excel spreadsheet templates, MathCAD templates, or database internal calculations. Since this paper focuses on the QA/QC analysis, details of the relief device calculations will not be discussed. API 520 and 521 can be referred to for additional details regarding relief device calculations.

2.3 Documentation

This section describes what is included in the documentation of a typical relief system folder. Different companies may have their own company-specific guidelines regarding the format of the documents and their sequence. The following gives an example of what is generally contained in the relief system documentation.

2.3.1 Sequence of Documents

Each company may have their own preferred order of the documents needed for pressure relief analysis. This example considers the sequence of documents in the following order:

- Executive summary
- Description of protected system
- System sketch
- Summary of overpressure scenarios
- Relief device calculations including inlet and outlet pressure drops
- Relief valve and/or any associated rupture disc datasheet
- Any background information

The following provides some information about the typical contents for each section.

2.3.2 Executive Summary

This document is intended to provide the overall results of the pressure relief analysis. Typical information shown in this document includes protected equipment and design conditions, relief device(s) including set pressure, summary of findings, and any deficiencies found during the analysis. This document may also contain a listing of credible overpressure scenarios and any assumptions made during the analysis.

2.3.3 Description of Protected System

This information may be detailed in the executive summary or can be shown in a separate document. If the description of the protected system is shown in a separate document, it is with the intent to provide more details than are shown in the executive summary. The equipment and relief device information from the executive summary may be repeated with additional information given. Typical information might include relevant control valves, listing of manual valves with administrative controls, relevant utilities to the system, or key rotating equipment.

2.3.4 System Sketch

Some companies include this section to provide a simplified schematic drawing of the protected system, its inlet and outlet streams, and their sources and destinations. It shows the relief devices in the system and their set pressures. This kind of a diagram helps to provide clarity for a complicated system with multiple equipment located in the protected system.

.2.3.5 Summary of Overpressure Scenarios

This document considers each case from a list of all the potential overpressure scenarios based on the applicable API standard and client specifications. It indicates if the particular scenario is applicable or not and includes a brief justification of why a given case is or is not a credible overpressure scenario. As indicated earlier, a credible scenario may refer to another similar credible scenario. This is done to avoid duplication of relief device calculations and reporting of the pertinent results. If a given case is not credible, it includes justification as to why the case is not considered credible. If a given case is out of the scope of a given study, this is also indicated in the description.

2.3.6 Relief Device Sizing and Pressure Drop Calculations

For each of the credible cases, the required relief rate, the required area, the capacity of the valve, and the inlet and outlet pressure drops are calculated. The required relief rate and required area are compared with the relief device capacity and available area to determine if the valve is adequate or not using the methods based on those given in the API 520 and API 521 standards [2, 3]. For low pressure tanks with design pressures less than 15 psig, API 2000 correlations are used [4].

The pressure drops in the relief device inlet and outlet are calculated based on the equivalent lengths of all piping and fittings in each segment and compared with API guidelines. For the relief device inlet, the pressure drop should be no greater than 3% of the relief device set pressure [7]. The pressure drop in the relief device outlet should be no greater than 10% of the set pressure for conventional relief valves. (Balanced bellows and pilot operated relief valves can have a higher outlet pressure drop; however, at higher values, relief device capacity may be affected.) API guidelines [2, 3] and vendor information indicate a conservative pressure drop value to be about 30% for bellows type relief valve and about 50% for pilot operated relief valves.

2.3.7 Relief Device Data Sheets

The relief valve or rupture disk data sheet summarizes key information about the relief device. This includes:

- Relief device tag
- Protected equipment and its service
- Fluid phase (liquid, vapor, or two phase fluid)
- Controlling overpressure scenario
- Relief device set pressure
- Equipment MAWP and MAWT and the operating conditions of pressure and temperature
- Relief device calculated and actual area
- Relief device manufacturer and the model number of the relief device
- Relief device material of construction

2.3.8 Background Information

Background information includes the process flow diagrams (PFD), piping and instrumentation diagrams (P&IDs), and equipment data sheets for the key equipment like pumps, compressors, heat exchangers, control valves, etc. For pumps and compressors in the system, the pump and compressor curves are provided. The relief device manufacturer supplied data may be included. The piping isometric drawings and field sketches of the relief device inlet and outlet piping and the hydraulic calculations for the relief device inlet and outlet equivalent length are also included. The results of any process simulations used to estimate thermodynamic properties of the relief fluid or required relief rates, and any other information supporting the relief device calculations, is provided in this section.

2.4 Work Instructions for QA/QC of Relief Device System

The purpose of quality assurance and quality control of pressure relief systems is to ensure that the analysis is done correctly and that the final report to the client is accurate and error-free. The QA/QC process is conducted after the analysis is complete and the documentation is assembled, as mentioned in the earlier section. QA/QC covers the following sections:

- System information
- Overpressure contingency analysis
- Relief rate calculation and relief device sizing
- Deficiency identification
- Final documentation

2.4.1 System Information

This section covers the protected equipment, pressure relief devices, pressure sources and any auxiliary equipment.

- **Protected equipment:** This section checks the following information:

    - The correct equipment is included for the system to ensure that no extra equipment is included in the system and no equipment is left out of the system.
    - The correct design pressure and design temperature are documented. The maximum allowable working pressure (MAWP) and maximum allowable working temperature (MAWT) are the important values to document since they indicate the conditions for which the equipment was designed.
    - Any P&ID numbers referenced for the listed equipment are correct. If P&ID revisions are documented, these should be checked as well.
    - The correct operating data are referenced in the relief device data sheet. This information may not always be available, and any assumptions for these values should be documented.

- **Pressure relief devices:** This section checks the following information:

  - The relief device manufacturer and model number match the device nameplate information or documentation provided by the client.
  - Any P&ID numbers referenced for the relief device(s) are correct. If P&ID revisions are documented, these should be checked as well.
  - The correct set pressure is referenced for each relief device, and it is ensured that the set pressure does not exceed the stated MAWP (or the equipment design pressure, if MAWP is not available).
  - The correct effluent location is mentioned along with corresponding backpressure.
  - The correct relief device inlet and outlet sizes are used.
  - The relief device orifice area and the sizing coefficients are selected correctly from the vendor-supplied values per the guidelines provided by the American Society of Mechanical Engineers (ASME) or those provided by American Petroleum Institute (API). API discharge coefficients should be used with standard API areas, while ASME discharge coefficients should be used with ASME areas. Where coefficients are not available from the vendor, default values may be used for the discharge coefficients. API discharge coefficients should not be used with ASME areas and vice versa to avoid errors in calculations of relief device sizes.
  - Also for liquid relief cases, use of liquid-certified valves should be confirmed since this affects the relief device sizing calculations. Some older relief devices, such as valves manufactured prior to 2005, may not have been certified for liquids. As a result, their capacity may need to be derated using the manufacturer's curves for use of the relief device with liquids.
  - The inlet and outlet piping is used correctly in the calculations. This can be determined by a review of the field sketches, or if not available, the piping isometric drawings. Also, the equivalent length calculations of RV inlet and outlet are correct and adequately documented in the supporting information.

- **Pressure sources:** This section checks the following information

  - The pressure sources are specified correctly and pump or compressor curves are provided in the supporting information.
  - Any pump or compressor in the system does not have the potential to be isolated from required overpressure protection equipment.

- **Additional equipment:** This section checks the following information:
  - The auxiliary equipment - control valves or key block valves - are mentioned correctly and the P&ID numbers on which they are located are identified.
  - The correct inlet and outlet sizing of the control valve is specified.
  - The correct upstream pressure is referenced and is greater than the protected system MAWP (to necessitate a credible case).

A final check is done to ensure the system is named correctly and all the associated process equipment, relief devices and additional equipment are included and documented.

2.4.2 Overpressure Contingency Analysis

This section checks each of the overpressure cases mentioned previously for the following:

- The correct applicability is designated for each case.
- The rationale explains the scenario accurately and the case correctly is considered or ruled out. (This activity is crucial; if cases which should be credible are ruled out, erroneous conclusions about the relief device adequacy can result.)
- Any P&ID numbers that are referenced in the scenario rationale are correct.
- All industry, client, and site specific guidelines are followed correctly.
- The description is correct and free of grammatical and spelling errors.

2.4.3 Relief Rate Calculation and Relief Device Sizing

This is one of the most important sections of the relief device study - and the one in which most mistakes are likely to happen. In order to minimize errors, some companies have standard tools like customized calculation templates or spreadsheet templates. Some of the key points to be checked for QA/QC of relief device calculations are as follows:

- The correct set pressures are used in the calculations.
- The correct overpressure percentages (expressed as percent of set pressure) are used and the relieving pressures are calculated correctly.
- The correct thermodynamic properties are calculated and used in the calculations.
- The correct overpressure percentages (expressed as percent of set pressure) are used and the relieving pressures are calculated correctly.
- The correct calculation methodology or calculation template is used.
- The correct relief device information is used. This includes orifice area, discharge coefficients, inlet and outlet pipe sizes, and equivalent lengths of piping for the inlet and outlet piping taken from the isometric drawing or field sketch.
- The correct units of measurement are used (English or metric).
- The correct heat of vaporization and equipment data are used for the relief device sizing for the fire exposure.
- The correct friction factor and pressure drop correlations are used in the relief device inlet and outlet pressure drop calculations.
- The correct backpressures are calculated in the relief device sizing. This could be an iterative calculation depending on the type of the valve used, and the iterations need to be converged properly.
- Any comments that provide additional information about the calculation should be clear and concise while providing more understanding for the details of the calculation.
- All the streams mentioned in the calculations should have correct references, and the corresponding supporting information about their rates and process conditions should be provided in the supporting documents. For example, key supporting data could include the relevant portions of the material balance, operating procedures or simulation outputs for process simulations used to calculate the stream properties and flow rates.

2.4.4 Deficiency Identification

In the QA/QC process of this section, any reported deficiencies are verified for their credibility and any missing deficiencies are identified. Any description of the deficiency should clearly explain the issue and may provide some recommendations to mitigate these issues.

2.4.5 Final Documentation

As mentioned earlier, the final documentation should consist of the following items:

- Executive summary
- Relief device overpressure summary
- Relief device specification sheet
- Overpressure contingency summary
- Required rate and relief device sizing calculations for all the credible cases
- Relief device inlet and outlet piping summary
- All the relevant supporting information, which may include:

  - Stream properties and supporting information of stream flow rates
  - P&IDs for the protected system and the upstream and downstream equipment which is connected to the protected system
  - PFDs for the protected system and the associated upstream and downstream equipment.
  - Heat and material balance for the protected system
  - Protected equipment specification sheets, if available from the client
  - Rotating equipment specification sheets and the pump and compressor curves
  - Relief device data from the vendor of the particular relief device or from ASME NB-18 book [8] information, if it was used in the relief device sizing calculations.
  - Operating procedures or any other information about the operating conditions for the protected system.

It is critical that the relief device calculations are correct and the results are correctly reported and consistent in all the relevant documents. The correct units of measurement should be used in the relief device sizing calculations and the specification sheets. Some companies use a database to automatically populate the relevant information consistently to avoid errors in manually inputting this information. If the calculations are changed for some reason, the information in the relevant documents should be automatically updated, saving a lot of rework and the associated costs.

3. Common Mistakes in Pressure Relief Analysis

Pressure relief analysis should not be done at the end of a project; it should be done in a timely manner during the design and periodically revalidated during the plant operation.

The QA/QC process is designed to minimize any errors in relief system analysis to ensure safe operation.
Based on experience in QA/QC analysis of multiple refineries and chemical plants, some common mistakes have been observed in the relief system studies.

3.1 Missing or Incorrect Data

It is important to have good data available from the plant for a proper pressure relief system study. Updated P&IDs, PFDs, pump and compressor curves, heat and material balances, equipment datasheets, control valve data sheets and relief device data should be available from the plants.

One of the most important pieces of information needed is the MAWP of the protected equipment. This can often be found on the equipment's U-1 form, which is the equipment manufacturer's data report on pressure vessels, per ASME Section VIII Boiler and Vessel code rules. In some cases, the vessel nameplate information can be used to obtain this information, if the U-1 form is not available. If this data is missing, the design pressure shown on the P&IDs can be used. Getting this information from the P&IDs, however, may not be reliable. The P&IDs may not be up-to-date, and there could be errors in the equipment information listed on them.

Control valve data is also important in relief device sizing calculations for failure of control cases. In the absence of this data, the control valves of the correct given size can be assumed to be similar to those from a specific manufacturer and of a specific model. Vendor data from these types of valves can be used to get the control valve information.

Incomplete or incorrect heat and material balances can require assumptions to be made about important stream information in order to complete the relief device study. Any assumptions made in the analysis should be stated clearly and should be approved by the client. These kinds of assumptions, if incorrect, can cause significant errors in relief device sizing.

3.2 Omitting or Incorrectly Ruling out Some Overpressure Scenarios

It is important that the relief system designer doing the relief system analysis have adequate experience and proper training in this kind of study. Following are some typical cases which can sometimes be overlooked but which require careful analysis.

- Reverse flow: These are sometimes hard to identify unless one has a good picture of the setup of the protected equipment and how it is interconnected with the upstream and the downstream equipment.
- Vacuum formation in atmospheric tanks: It is possible to consider only the pressure relief scenarios and miss the vacuum formation scenarios due to the liquid movement out of the tank or a drop in tank temperatures. Incorrectly sizing the relief device on the tank or not providing vacuum protection can cause tank ruptures.
- Gas breakthrough: One needs to consider the effect of liquid from upstream columns being drained into the downstream protected equipment, with subsequent vapor breakthrough to the protected equipment. The maximum pressure of the vapor needs to be considered in determining if the case is credible and in sizing the relief devices.
- Power failure: Failure of power in a complex system like a distillation column can affect multiple pieces of process equipment, the effects of which need to be considered properly. For example, for a column using a fired heater as a reboiler having both electric and steam turbine driven pumps, the steam turbine driven pump may continue circulation even though the feed and the reflux may be lost causing a boil-up situation, which can be overlooked.
- Tube rupture: For tube rupture in a heat exchanger, the effects of tube rupture on downstream equipment, which may be at lower design pressures than the heat exchanger being considered need to be analyzed.
- Inadvertent opening of valves: For control valves with a bypass valve, in some cases, flow through both the control valve and the bypass valve may need to be considered in sizing the relief device. The effect of some normally closed valves inadvertently opening and overpressuring the protected system needs to be properly considered.

3.3 Errors in Hydraulic Calculations

Relief device inlet and the outlet piping should be correctly accounted for using the field sketches or the piping isometric drawings. The correct type and number of fittings should be used in the calculations. Errors in assumptions regarding the relief device inlet or outlet piping will cause errors in pressure drop calculations; this can lead to incorrect conclusions about the relief device adequacy.

3.4 Errors in Relief Device Calculations

The correct relief device sizing equations should be used based on the applicable API standard. Correct values of vapor and liquid relief valve discharge coefficients should be obtained from the vendor information or from ASME supplied data [8]. For the failure of control case, the correct control valve sizing equations and characteristic constants should be used in determining the relief rate through the control valve. Two-phase flows at the inlet or the outlet of relief device should be correctly accounted for in the calculations if there is a phase change.

3.5 Errors in Documentation

It is important that the results of the relief device calculations be properly reported consistently in all the relevant documents, especially the relief device data sheet, which is used by the relief device vendors to supply the relief device. Any errors in this process can have a significant impact in the validity of the analysis. Errors in documentation can also result in fines in the event of an audit. It is critical that the relief designer be diligent in performing this work and that the QA/QC process be used effectively to ensure that the system is designed with the overpressure protection required for safe operation.

4. Conclusions

Pressure relief device analysis, and its quality assurance and control, are important since relief devices are generally used as a last line of defense against overpressure incidents for a plant. Proper sizing and documentation of the pressure relief devices is important to meet OSHA mandates and to prevent costly and tragic errors. It is important to understand the workflow for QA/QC of relief devices and to formulate a set of work instructions to support proper relief system analysis.

It is also necessary to have proper work instructions for the QA/QC analysis of the study to make sure the deliverables are error free and of high quality. Some examples of work instructions for conducting different sections of pressure relief study and for performing an efficient QA/QC analysis were given in this work. Some common mistakes observed in QA/QC of numerous pressure relief system studies from commercial refineries and chemical plants were also discussed, and care should be taken to avoid these costly mistakes.

References

1. Web download from http://www.totalqualityassuranceservices.com/quality-control-vs quality-assurance-the-difference-between-them, 11/17/2011
2. Pressure Relieving and Depressuring Systems- ANSI/API Standard 521, 5th edition, January 2007.
3. Sizing, Selection and Installation of Pressure Relief Devices in Refineries- Part 1: Downstream Segment, API Standard 520, 8th edition, December, 2008.
4. Venting Atmospheric and Low Pressure Tanks, ANSI/API Standard 2000, 6th edition, November 2009.
5. ASME Pressure Vessel Code, Section I, Rules for Construction of Power Boilers.
6. Crane Technical Paper 410, Crane Co. September, 2009.
7. Sizing, Selection and Installation of Pressure Relief Devices in Refineries- Part 2: Installation, API Recommended practice 520, 5th edition, August, 2003.
8. National Board NB-18 Relief Device Certifications NB-18, Aug, 2008.

The Influence of Hydrophilic Properties on the Venting of Foamy Three Phase Systems

Henrik Leimeister
Technische Universität Berlin
Straße des 17. Juni 135, 10623 Berlin, Germany
henrik.leimeister@tu-berlin.de

Jörg Steinbach
Technische Universität Berlin
Straße des 17. Juni 135, 10623 Berlin, Germany
joerg.steinbach@tu-berlin.de

Prepared for Presentation at
American Institute of Chemical Engineers
2013 Spring Meeting
9th Global Congress on Process Safety
San Antonio, Texas
April 28 – May 1, 2013

UNPUBLISHED

The Influence of Hydrophilic Properties on the Venting of Foamy Three Phase Systems

Henrik Leimeister
Technische Universität Berlin
Straße des 17. Juni 135, 10623 Berlin, Germany
henrik.leimeister@tu-berlin.de

Jörg Steinbach
Technische Universität Berlin

Keywords: venting, depressurization, foamy, three phase systems

Abstract

The correct design of a safety valve or rupture disk can only be done if predictions on the venting behavior can be made. To date, those predictions have limitations with regard to foaming or highly viscous systems. An experimental study is presented that describes the influence of hydrophilic particles on the venting of strongly and weakly foaming systems. It is shown that hydrophilic particles increase the mass discharge over time and reduce the pressure decrease over time by increasing the foaming of aqueous systems. These effects are not significantly influenced by the reactor characteristics. At higher additive concentrations the addition of particles is not as influential anymore. Thus models for foamy two-phase systems are applicable for the design of the described foamy three-phase systems as well.

1. Introduction

Excessive pressure is one of the major hazards in the chemical process industry. To protect vessels such as reactors from overpressure, rupture disks or safety valves are implemented as part of an overall process safety strategy.

A key parameter in the design of the right safety valve or rupture disk for a given setup is the minimum required diameter of the line. The minimum required diameter is highly influenced by various factors. The source of the excessive pressure in the vessel, the flow regime inside the venting device, the location of the venting device on the vessel, and the density of the liquid media are highly influential on the venting behavior. Each one of these factors drives the need for one specifically fitted model.

When only top-vented, slightly viscous vapor pressure systems are considered, the possibilities are narrowed down to three major scenarios. These scenarios are single-phase flow consisting of vapor, two-phase flow consisting of vapor and liquid, or three-phase flow with vapor, liquid, and particles present in the pressure relief device.

Single-phase flow can be modeled as vapor streaming through a jet nozzle. Thus, models for the determination of the minimum required diameter are well suited for this application. Two-phase flow is more difficult to model. Effects like slip, radial dispersion profiles, and thermo dynamical disequilibria have to be taken into consideration for precise results. Several models have been developed for this scenario. They range from conservative models that neglect slip, disequilibria and dispersion profiles like the Homogeneous Equilibrium Model (HEM) developed by the HSE [2006] to more detailed models like the HNE-DS model of the ISO 4126-10, which is presented in the work of Schmidt [2007]. For three-phase discharge, no influence of the particle on the system has been shown to this point. Several works like the ones from Poli et al [2009] and the HSE [2006] recommend to design as if the three-phase systems were two-phase flow.

However, all those models meet their limits if amphiphile components induce foaming during the venting process. To date foamy systems cannot be modeled precisely. Rather, the minimum required diameter is set based on over-conservative assumptions that can lead to significant cost increase and, in some cases, unsafe conditions. The direct influence of particles on the venting behavior of foamy systems is scarcely described. To determine the influence of particles, venting experiments of foaming systems with and without particles have been conducted and are presented in this work.

2. Experimental Setup

Experiments were carried out in a modified RC1e by Mettler Toledo. An additional pressure sensor was added as well as a vent line that led to an open 100 l vessel. The vent line was opened and closed electronically by a pneumatic ball valve. An aperture plate with a diameter of 2 mm was placed directly in front of the ball valve, setting the minimum diameter in the vent line. Thus, the ball valve and the aperture plate were used to represent a rupture disk. The experimental setup can be seen in Figure 1.

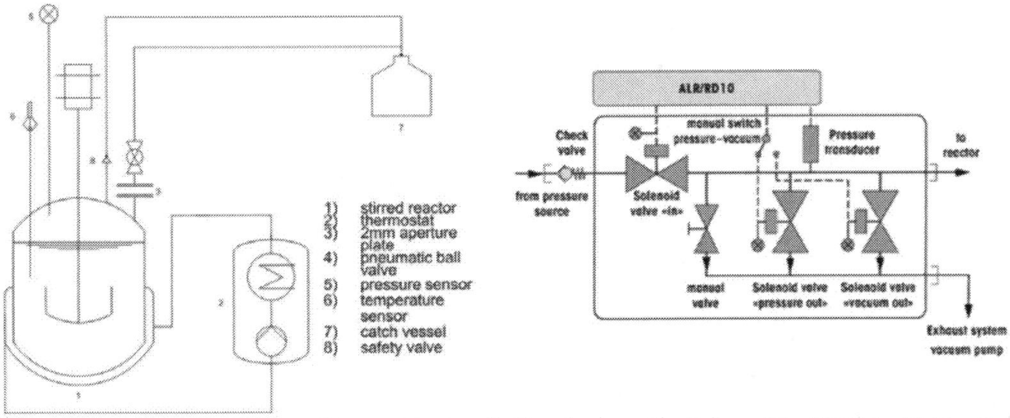

Figure 1: Modified RC1e for venting experiments (left) and schematic design of the RD10 by Mettler Toledo [2012] (right)

An RD10 pressure control loop by Mettler Toledo was also connected to the RC1e for experiments investigating the foaminess of the systems. A set of valves controlled the flow to and from the reactor, depending on the desired and actual pressure. Thus, pressure increase or decrease over time was controlled. The schematic design can be seen in Figure 1.

The results were than compared to experiments carried out in an Adiabatic Dewar Calorimeter II by Chilworth Technology. The ADCII has been used in past research on three-phase venting behavior. A detailed description of the setup can be found in the work of Poli et al [2009].

2.1 Systems and Parameters

The venting experiments were conducted in foaming aqueous non-reactive vapor systems. To induce the foaming, two surfactants were used. Falterol is an industrial cleaner consisting mostly of sodium dodecyl sulfate (SDS) that has been used in quite a few venting experiments of foamy systems. To reduce side effects by additives found in Falterol, pure SDS was also used in a different set of experiments. When working with SDS in aqueous systems, one has to take into account that SDS hydrolyses in water. An alcohol is formed that functions as co-surfactant and increases the foaminess of the systems. Thus all experiments were prepared with a solution of 1 g/l that had been aged a minimum of 2 days before use.

Silibeads were used as the solid phase. They were chosen since they have a perfectly round shape with a plane surface and are chemically inert. The full set of the experimental parameters can be found in Table 1.

Table 1: Experimental parameters for venting experiments

Reactor volume	Filling level	Mass fraction of particles	Set pressure	Additives	Particles
1.2 l (MP10), 1.1 l (ADCII)	80 %	20 %	6.0 bar	Falterol, sodium dodecyl sulfate	Silibeads 40-70 μm

2.2 Measurement of Foaminess

For the experiments on foaminess, the systems were heated until a pressure of 1.7 bar was established in the reactor. Now the pressure was reduced via the pressure control loop to 1.0 bar within 90 s. Thus, a thermodynamic disequilibrium was induced and the system started to boil. The bubbles created build foam on top of the liquid, and the maximum height of the foam column was measured. This method had the advantage that the same setup equipped with the same fixtures and materials could be used for measurements of foaminess and for venting experiments. Also, the liquid phase and the gassy phase were at similar conditions for both the foaming and the venting experiments.

3. Results

3.1 Foaminess

To show the effect of particles on the foaminess the system, SDS was chosen since it is a component in all systems. Figure 2 shows that the results indicate a strong influence of particles on the foaminess of foamy systems.

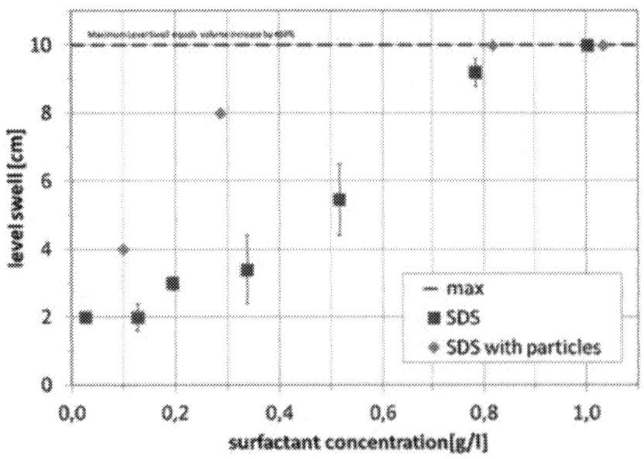

Figure 2: Foaminess of SDS solutions with and without particles

The level swell is given in centimeters. It was limited by the height of the reactor, indicated by the dashed line. The initial fill level was at 4 cm. Figure 2 shows that, already at low concentrations of SDS, the foaminess was significantly increased by the particles. The effect was at least twice as strong until a concentration of 0.3 g/l. Afterwards, precise interpretation was limited by the fact that the systems with particles reached the lid of the reactor before the end of the pressure decrease while the systems without particles do not. At 1.0 g/l, systems without particles also reached this limit, and further increase would not have led to additional results.

The reason for the increased foaminess is that there is a stronger bubbling in three-phase systems than in two-phase systems without particles. In two-phase systems, bubble nucleation and rise can only be observed on the metal fixtures of the reactor; however, in systems with particles, bubble nucleation is also observed on the particles themselves. Thus, a more turbulent regime arises that induces foaming over the cross section of the reactor.

3.2 Venting Experiments

The results of the venting experiments with Falterol followed the results of the experiments for foaminess. Adding particles leads to significant effects on the pressure-time curve and the mass discharge. An example for the pressure behavior and the overall mass discharge can be seen in Figure 3.

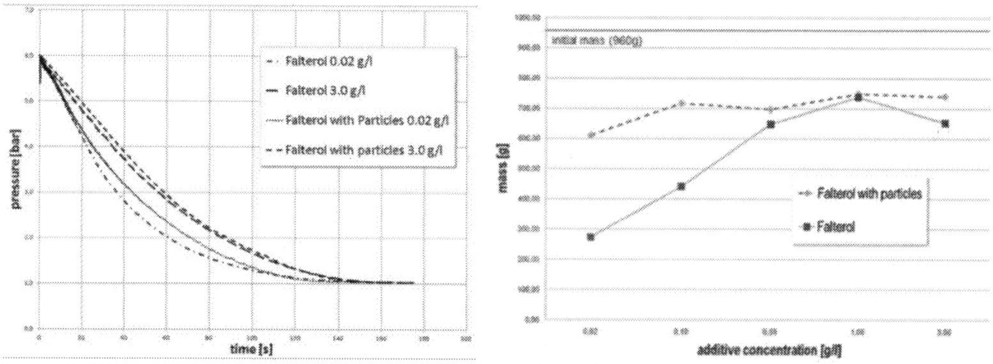

Figure 3: Pressure time curve (left) and mass discharge (left) of venting experiments with Falterol

Figure 3 clearly shows that adding particles slowed the pressure decrease over time. This was induced by the stronger foaming due to the addition of the particles. This effect was extremely pronounced at low concentrations. At higher concentrations of particles, the pressure curves for two- and three-phase systems were very close and almost overlapped.

The increased foaminess also led to an increased mass discharge. This effect was more pronounced at low Falterol concentrations. At 0.02 g/l of Falterol, mass discharge was more than doubled. Significant solid discharge was not observed, however. Both effects can be explained by a prolonged two-phase discharge. It appears that, due an increased foaminess, the level swell was higher and thus reached the vent line for a longer period.

At concentrations larger than 1.0 g/l of Falterol, the influence of the particles was less pronounced. Pressure decrease and mass discharge of the three-phase systems were similar to the ones of two-phase systems.

The results of the experiments with pure SDS (without the additives found in Falterol) differ in comparison. Although similar effects were seen in the mass discharge, the effects were less pronounced.

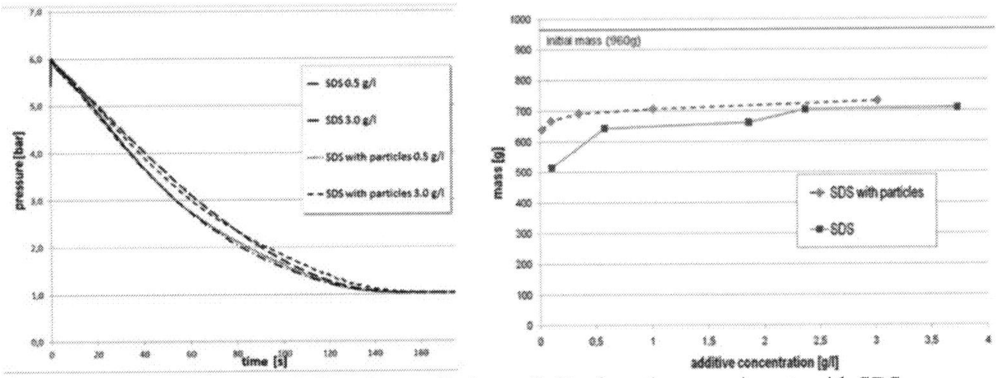

Figure 4: Pressure time curve (left) and mass discharge (left) of venting experiments with SDS

The mass discharge of three-phase systems was still larger than the mass discharge of two-phase systems. However, the mass discharge of three-phase systems didn't exceed the mass discharge of two-phase systems by more than 30%. As in the results of Falterol, the effects of SDS got smaller once higher concentrations were reached.

An effect of the particles on the pressure decrease over time was hardly observable. At the beginning of the venting, the three-phase system's pressure decreased faster. At later times, it slowed down and crossed the two-phase curve again. This might be a result of a longer two-phase discharge with a lower mass fraction of the liquid at the beginning of the vent line.

When comparing the results of the venting experiments conducted in the RC1e to those in the ADCII, the effects were comparable. Although only a few results can be presented at the current time, the delay of the pressure decrease in the presence of particles was again clear. The mass discharge behaved accordingly. For 0.005g/l SDS, the presence of particles increased the mass discharge from 180g to 280g, and for 0.01g/l SDS, the presence of particles increased the mass discharge from 300g to 480g. Again, no significant particle discharge was observed.

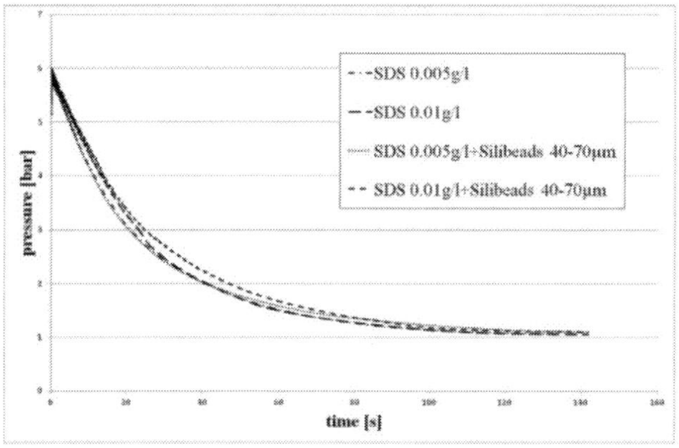

Figure 5: Pressure over time in venting experiments in the ADCII

To directly compare the effect of the foaminess on the venting, the mass discharge was plotted against the level swell for venting experiments conducted in the RC1e only. The comparison is seen in Figure 5. An obvious direct dependence between the level swell and the mass discharge can be seen. If the foaminess is increased, the mass discharge rises. At the beginning, this effect is quite pronounced. However, the impact is reduced when a relative mass discharge of 70% is reached. This mass discharge is characterized by the point at which a two-phase discharge is observed throughout the complete duration of the venting process. Afterwards, only a small change can be observed. However, the same trend is followed no matter what system is considered.

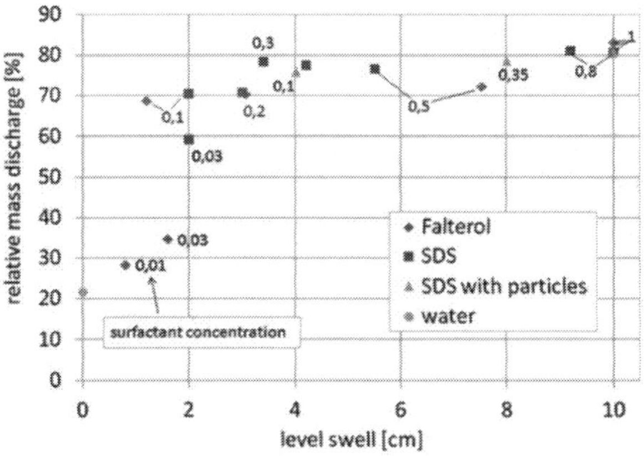

Figure 6: Mass discharge over foaminess for systems containing SDS

4. Conclusions

The presence of hydrophilic particles have a significant impact on the venting behavior of foamy systems. They increased the mass discharge and reduced the pressure decrease over time. Hardly any particles were vented. The reactor geometry and material did not show any major influence in these experiments. Thus, one can model foamy three-phase as foamy two-phase system under consideration of a prolonged two-phase discharge due to increased foaminess and use models recommended for foamy systems. One approach to model foamy systems can be found in the work of Schecker [2003].

Future experiments with different particles are planned. Current work involves experiments that are carried out using Silibeads which are coated with a thin hydrophobic layer. Also, weakly foaming systems with isobutonal as amphiphile component are under current investigation.

5. References

[1] Joshi, K.S., Baumann, A., Jeelani, S.A.K., Blickenstorfer, C., Naegeli, I., Windhab, E.J., "Mechanism of bubble coalescence induced by surfactant covered antifoam particles", *Journal of Colloid and Interface Science,* Volume 339, 446-453, 2009.

[2] Mettler Toledo, "RCPress10 Pressure Controller", <*www.us.mt.com*> accessed 15.01.2013.

[3] Poli, M., Imhof, H., Holst, N., Steinbach, J., "Venting of Foaming Three-Phase Systems", *Chem. Ing. Technol.,* Volume 32, NO. 2, 312-318, 2009.

[4] Ravi Kumar, N.V., Ramachandra Rao, N., Sudhakar, B., Gokhale, A.A., "Foaming experiments on LM25 alloy reinforced with SiC particulates", *Materials Science and Engineering*, Volume A 527, NO. 21-22, 6082-6090, 2010.

[5] Schecker, J., Friedel, L., "Untermodell für das Aufwallen von schäumenden Gemischen bei Druckentlastung", *Forschung im Ingenieurwesen,* Volume 69, 44-56, 2004 (German).

[6] Schmidt, J., "Sizing of nozzles, venturis, orifices, control and safety valves for initially sub-cooled gas/liquid two-phase flow – The HNE-DS method", *Forschung im. Ingeniuerwesen,* Volume 71, 47-58, 2007.

[7] Waldram, S., McIntosh, R., Etchells, J., "Reactor Pressure Relief of Fluids Containing Suspended Solids", *Process Safety Progress*, Volume 25, NO. 3, 214-226, 2006.

[8] Wasan, D., Nikolov, A., Shah, A., "Foaming-Antifoaming in Boiling Suspensions", *Ind. Eng. Chem. Res.,* Volume 43, 3812-381, 2004.

Lightning Related Ignition Mechanism and Associated Protection Techniques for Storage Applications

Mitchell Guthrie
Engineering Consultant
Blanch, NC USA
esecmg@embarqmail.com

Alain Rousseau
SEFTIM
Vincennes, FRANCE
alain.rousseau@seftim.fr

Prepared for Presentation at
American Institute of Chemical Engineers
2013 Spring Meeting
9th Global Congress on Process Safety
San Antonio, Texas
April 28 – May 1, 2013

UNPUBLISHED

Lightning Related Ignition Mechanism and Associated Protection Techniques for Storage Applications

Mitchell Guthrie
Engineering Consultant
Blanch, NC USA
esecmg@embarqmail.com

Alain Rousseau
SEFTIM
Vincennes, FRANCE
alain.rousseau@seftim.fr

Keywords: lightning, flammables storage, lightning ignition, grounding, impedance, potential equalization, strike termination

Abstract

Lightning can pose a myriad of threats to operations involving flammable vapors and liquids that give off flammable vapors. This paper focuses primarily on storage facilities and associated operations but the principles can be applied to any operation or installation where a flammable environment may exist. General principles of protection from a lightning threat are provided, along with a discussion of specific threats presented by operations in which flammable vapors may be present. The evolution of protection considerations is also discussed. Techniques for protection from direct attachment are provided to address those situations where an external flammable vapor could exist in the vicinity of a storage vessel. Threats produced by nearby lightning and strikes to associated piping and electrical conduits are also addressed. The elimination of both external and internal potential differences that could lead to dangerous arcing is discussed. This paper also provides a discussion of the difference in grounding system impedance versus static grounding techniques as they relate to the dissipation of lightning currents and explains their significance in minimizing the probability of arcing between structural components and systems. The principles discussed are summarized in the context of the development of a lightning protection plan.

1. Introduction

The hazard presented by lightning activity to operations associated with the presence of flammable vapors and liquids that give off flammable vapors is well recognized, but the specific threats associated with lightning activity and associated failure mechanisms may not be as well understood. At the request of the American Petroleum Institute (API), the National Fire Protection Association Committee on Lightning Protection (NFPA 780) has established a Task Group to review the requirements of their chapter on the protection of flammable vapors, flammable gases, or liquids that can give off flammable vapors [1]. It is important to note that these standards and other recommended practices are developed as "minimum acceptable" requirements. Ignition sensitivity of the contents and configuration of the containment structure may require additional considerations in protecting the contents from a lightning threat.

It is also noted that many incident investigations involving flammable materials which cite lightning as a cause go into little detail in explaining the specific failure mechanism. Those papers that do discuss failure mechanisms generally only speculate as to the cause of the event; which leaves little for those developing standards to determine how the event could have been prevented. This paper provides information that may be of use to investigators of lightning-related incidents in identifying possible ignition mechanisms that may be related to a specific incident. It also provides information that may be helpful in determining specific failure modes when lightning is suspected to be a contributor.

Finally, this paper identifies the basic elements of a lightning protection system for structures used to protect structures containing flammable vapors along with a description of their function in providing protection from lightning. It is intended that this may be useful to those specifying lightning protection and/or assessing the need for additional protection measures.

2. Lightning Threat Mechanisms

2.1 General

Lightning activity can create a variety of threats to operations involving flammable vapors or liquids that may give off flammable vapors. These threats can be manifested in the form of physical (mechanical) damage to structures and their appurtenances, thermal damage including the ignition of contents, and electrical damages due to overvoltages/overcurrents or electrical arcing. The net result of such threats can be loss of productivity, loss of production equipment, release of material and, in the worst-case, explosive rupture of containment vessels due to ignition of flammable vapors.

2.2 Physical (mechanical) threat

A direct strike to a storage vessel can lead to the threat of mechanical forces related to the shock wave created by the rapid expansion of air in the channel associated with the return stroke. Significant mechanical forces are also produced when a lightning impulse current is routed through sharp bends in conductors. These forces can be sufficient to separate conductors from fasteners as the electromagnetic forces straighten the bends and in some extreme cases, damage the conductor itself.

2.3 Thermal threat

The threat of thermal ignition of flammable vapors that may exist in the vicinity of a direct strike attachment point is difficult to mitigate. This is a consideration for both non-metallic and metallic roofs. The temperature in the plasma channel of a return stroke can reach 50,000° F. Ignition of a flammable atmosphere can result should the attachment point be in the vicinity of a faulty seal or gasket.

There can also be erosion of metal at the arc attachment point. Burn-through of the metal skin of a storage vessel is possible if the thickness of the shell is too thin. Such a penetration can allow leakage of material into and out of the tank if the hot spot does not ignite the internal contents. It is also important to consider that energies sufficient to burn through the metal may not be required to cause thermal ignition of the contents.

Kern [2] reported that laboratory testing has shown that temperatures at un-punctured interior surfaces opposite to the point of strike exceeded 100° C for 2 mm thick aluminum and stainless steel metal sheets when subjected to impulse currents. A temperature of 94° C was recorded for 2 mm thick soft steel from the same threat. Of greater importance is the thermal threat presented by strikes with long duration currents which Kern reported can lead to internal temperatures of more than 1000° C without burn-through for some sheets of unspecified metal thickness. Long duration current testing was also conducted on metal sheets of thicknesses specified to meet strike termination requirements. Temperatures recorded for 4 mm thick soft steel and stainless steel reached 950° C with values for 5 mm thick soft steel reaching 750° C and 570° C for stainless steel. The thermal time constant for the steels create concern because these high temperatures can remain on the interior surface for a long time.

2.4 Electrical threat

Direct or nearby lightning strikes can produce electrical overvoltages and overcurrents resulting in possible damages to electrical power distribution, instrumentation, and control circuits and associated hardware. The damages could result in electrical data upset, generate erroneous control and instrumentation signals, damage communication hardware, or cause permanent loss to electrical and electronic hardware.

Arcing due to voltage differences between metallic components has been suggested to be the most likely cause of tank fires due to lightning; especially in floating roof tanks. API/EI Research Report 545-A [3] reports that 52 of the 55 rim seal fires investigated as part of the 1997 LASTFIRE survey were caused by lightning. There are a number of scenarios which can produce a threat of electrical arcing. Some of the most common scenarios are discussed below.

A direct strike to an open-top metal floating roof tank can lead to electrical arcing across the floating roof seals regardless of whether the attachment point is to the tank shell or the floating roof itself. Static bonding cables can possibly limit the sparking but cannot prevent the arcing from such a threat [3].

When impulse currents resulting from direct or nearby lightning strikes are injected onto a grounding electrode, a substantial voltage could be developed. The magnitude of this voltage is a function of the value of peak current and the impedance to remote earth of the grounding system. As the voltage of the grounding system increases from 0 volts to hundreds of thousands or millions of volts, there is a probability of arcing between items connected to the grounding system and any ungrounded conductive items or independently grounded conductive items. This arcing can not only occur between the metal shell of a tank and isolated or insulated conductive items in or on the structure but also occur across flanges with isolating gaskets or piping which is not electrically continuous.

A charged cloud produces a vertical electric field onto the earth's surface. A resulting charge is distributed over the ground and objects on the ground. Horizontal flat surfaces will exhibit a uniform charge distribution. Structures in the vicinity of a charged cloud will concentrate the local electric field depending upon the geometry of the structure, with maximum charges appearing at the areas of maximum electric field concentration. Typical examples of locations on a storage vessel where maximum charge is likely to concentrate are the rim at the top of the vessel, handrails, lights, or other objects (such as conservation vents, etc.) on the top of the vessel. The slow movement of ground charge provides induction charging of a storage vessel. This charge generally is not significant for fixed roof metallic tanks and should not affect the charging of internal material due to the Faraday-type shield provided by the construction of the vessel. However, it can present a significant threat to a floating roof tank.

The charge that will be induced on the tank is neutralized almost instantaneously by a lightning strike that collapses the field. At that time, a heavy ground current flows to neutralize the ground charge. The voltage on the vessel shell is relaxed by the dissipation of the lightning current into the earth much faster than what is sometimes described in the petroleum industry as "bound" charge in the vapor space and along the oil surface for low conductivity liquids; creating a potential for internal arcing in the storage volume between the liquid surface or charge pockets in the vapor space and the tank shell.

3. Strike Termination

3.1 General

The component of a lightning protection system designed to intercept a downward leader and provide a path to earth of the resulting return stroke(s) is the strike termination system. Types of strike termination systems recommended for use for the protection of vessels containing liquids that may produce flammable vapors are Faraday-type and isolated strike termination systems.

3.2 Faraday-type strike termination

It is generally accepted that an electrically continuous metallic shell of a tank containing flammable vapors having a thickness equivalent to 3/16-inch (4.8 mm) steel or greater can serve as a lightning strike attachment point [4][5]. The international lightning protection standard indicates that a thickness of 5 mm steel or 7 mm aluminum requires no additional strike termination components provided that the temperature rise of the inner surface at the point of strike does not constitute a danger [6] (even though it allows for a 4 mm thickness where there is no danger of explosion).

The majority of flammables storage vessels often have protrusions that extend above the tank frame. These items would be required to be provided with strike termination devices if they fail to meet the metal thickness requirement. Typical examples are light fixtures, conservation vents, thin pipes (such as dip pipes), and thin electrical conduit. The following paragraphs offer methods for providing this supplemental protection.

3.3 Isolated strike termination

A number of technical papers that address direct strikes to structures containing flammable vapors discuss threats associated with high resistivity materials such as petroleum products, toluene, and xylene [7]-[13]. Buccella and Orlandi [12] indicates that a direct lightning strike to a grounded metal vessel filled with an insulating charged liquid (such as some hydrocarbons) can create a strong electric field in the air space above the oil and a high potential at the oil surface.

Britton has described through numerous publications [14][15][16] the process of charging of internal liquids having conductivities of 1pS/m or less and the resulting liquid surface potentials produced by the movement of such liquids. Extensive literature is available concerning the threat of electrostatic discharges associated with the charges that may exist along a liquid's surface and in charged vapors but little literature other than Buccella [13] is available to suggest the presence of internally charged vapors in metallic containers as a result of lightning activity.

Buccella [13] proposes a numerical method to compute the reduction of the voltage at the interior of a metallic tank provided by an external lightning protection system (LPS). The lightning protection system, its associated grounding system, and the shell of the metallic tank are modeled according to their equivalent circuit parameters and a SPICE analysis is conducted on the lumped circuit network to determine the electric voltage on the vessel boundary. A Finite Difference Time Domain method then uses the boundary conditions valid at the interior of the vessel defined by the computed voltage distribution from the SPICE analysis, to determine the probability of an internal voltage sufficient to initiate an internal spark of sufficient energy to ignite charged vapors that may appear in the vessel.

Buccella uses this method to provide an example of the magnitude by which the isolation of the lightning current from a tank can decrease the voltage on the internal shell of a metal tank. A 5 millimeter thick, 10 meters tall, 6.4 meter diameter steel tank filled to a height of 5 meters with a petroleum product having a conductivity of 1pS/m is considered with a 100 kA lightning impulse current having a rise time of 0.8 microseconds and time to half value of 80 microseconds (similar to those values given in IEC 62305-1 [17] as characteristic of a subsequent stroke). It is shown that a reduction of potential at the inner boundary of the container of about 90% is achieved by a LPS with 5 interconnecting conductive elements isolated from the container as little as 20 centimeters. It was also shown that increasing the spacing to 40 centimeters and/or increasing the number of parallel conductors to 10 provided little additional reduction of voltage potential on the interior of a tank.

Isolated lightning protection systems allowed by NFPA 780 [4] for the protection of structures containing flammable vapors utilize masts and/or overhead (or catenary) wires. IEC 62305-3 [6] requires that all parts of the strike termination devices and down conductors for structures containing hazardous areas shall be located a minimum of 1 m from a hazardous area, where possible.

3.4 Non-isolated or hybrid strike termination

Neither United States standards nor the international lightning protection standard specifically allow a lightning protection system with air terminals and down conductors installed directly on a structure containing flammable vapors. However, it is identified in 3.2 above that the metallic shell of a tank could serve as both the strike termination device and down conductor. Buccella [12][13] explains the significant benefit from even a minor isolation of the lightning current from the shell of a tank containing a flammable atmosphere. While the authors concur with Buccella's assessment that there is benefit to the isolation of lightning current from the shell of a flammables storage vessel that may contain charged liquids or vapors, it is unclear why a Faraday-type strike termination system is allowed and yet it may not be allowed to add an air terminal integral to the structure (properly bonded to the metal shell) to protect protrusions from the metal shell that may not meet the thickness requirements (such as lights, conservation vents, etc.). The limited use of air terminals as a supplement to a Faraday-type system to provide protection for such protrusions could be considered in future revisions of standards.

4. Grounding

The international lightning protection standard for structures with risk of explosion [6] requires that all tanks used for the storage of liquids that can produce flammable vapors or used to store flammable gases shall be grounded at a minimum of one point. It is recommended that this grounding component be a ground ring electrode external to the structure or a foundation earth electrode but allows that a steel tank bottom serve as the ring electrode as long as it maintains contact with the earth. For tanks in tank farms, the tanks shall be interconnected. This interconnection may be provided by electrically continuous piping. Isolated tanks (which are not interconnected with other tanks) require an additional grounding connection only when the diameter or maximum dimension exceeds 20 meters.

NFPA 780 [4] identifies that any of the following methods can provide a minimum acceptable grounding method for aboveground tanks:

(1) a grounded metallic piping system without insulated joints,
(2) vertical cylindrical metallic tank of at least 6 meters (20 feet) in diameter resting on earth or concrete, or at least 15 meters (50 feet) in diameter if resting on bituminous pavement,
(3) a minimum of two grounding electrodes at maximum 30 meter (100 feet) intervals along the perimeter of the tank.

For tanks installed where an insulating membrane isolates the tank from remote earth, NFPA 780 requires grounding using method (3) above. Numerous other standards and recommended practices, including API 2003 [5] and API 545 [18], accept the minimum grounding methods identified above although neither API recommended practice identifies a minimum contact dimension (diameter or contact surface area) as in NFPA 780. Unlike NFPA 780 and API 2003, API 545 does not suggest that supplemental grounding be required when there is an elastomeric liner in or under the tank bottom; even though it would be expected to have the effect of reducing the conductivity between the tank floor and the ground. The logic is that additional paths to earth will be provided by piping and cables attached to the tank.

API recommended practices suggest that tank grounding is not an important contributing factor in the prevention of arcing, rim-seal fires, or ignition of the contents in tanks. It is agreed that the grounding method used for storage tanks has little influence on the prevention of rim-seal fires or ignition of vapors in the vicinity of a direct strike but a low impedance grounding system could have a beneficial influence on the maximum level of ground potential rise and the peak voltage that will appear on the shell of the tank. Lowering the peak voltage on the shell of a tank will reduce the probability of arcing to ungrounded or remotely grounded objects as well as between the interior of the tank shell and any charged vapor-air mixtures or liquid surface potential that may exist in the interior of the storage vessel.

It is a common assumption that the low frequency value of grounding system resistance measurements is sufficient to characterize the response of the grounding system from a direct lightning strike threat. However, Figure 1 forwards a plot of resistance and impedance versus frequency derived from measurements made on a 6 meter diameter tank sitting on a concrete base which is immersed in a mixture of sand and water [19]. The blue (top) curve represents the overall impedance of the grounding system, the green (middle) curve represents the resistive component and the dotted red line represents the negative of the reactive component of the impedance (-X is used for clarity because it mirrors the impedance trace). The grounding system frequency response of this metal tank is similar in characteristics to the limited number of measurements of petroleum or chemical tanks made at locations worldwide (although the specific values change from site to site). As can be observed from the data provided in the figure, a very good grounding system value of approximately 1 ohm begins to increase in impedance at frequencies above 20 kHz due to inductive reactance and exceeds 40 ohms at the highest frequency of 1 MHz. The actual value of voltage that would appear during the onset of a lightning current pulse would be significantly greater than one would assume based on a calculation using only the low frequency value of grounding system resistance.

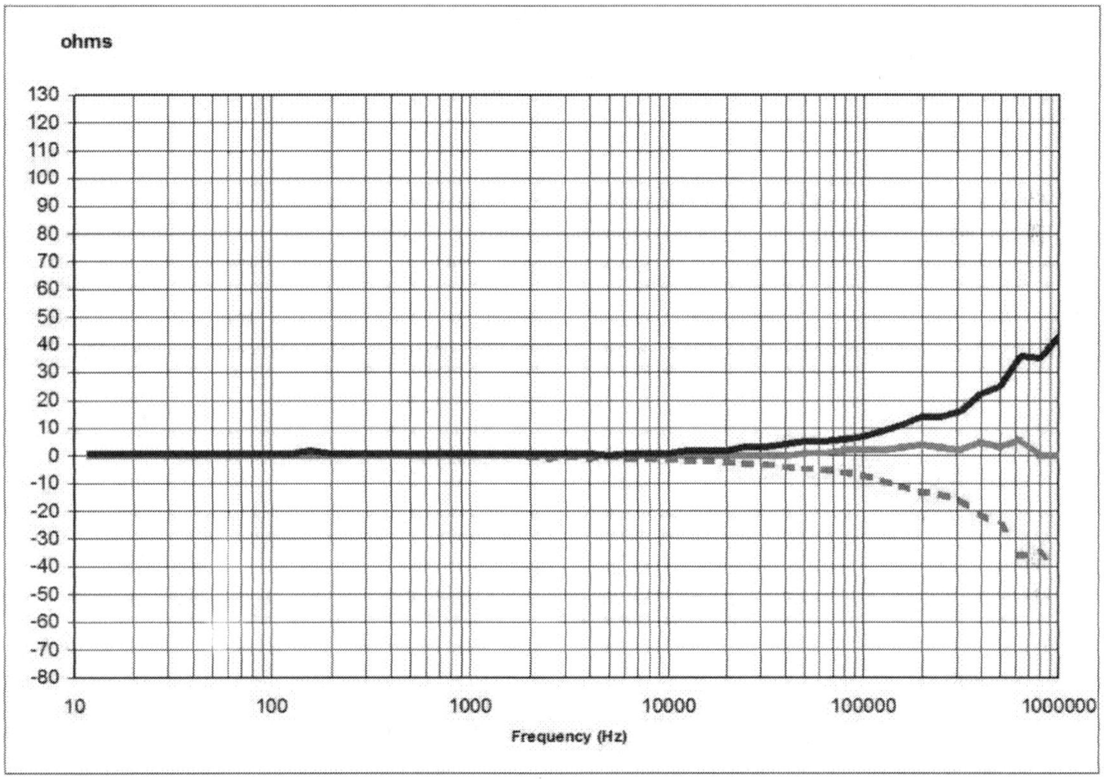

Figure 1. Impedance versus Frequency for 6 meter diameter metal tank

Grounding system impedance analysis has been reported for several applications [20][21]. However, there is not enough grounding impedance data on metallic tanks sitting on concrete pads to draw clear conclusions as to efficacy of this grounding technique. The data available does not appear to suggest that the capacitive effect noted from the impedance measurements reflects that which would be expected from analytical calculations. Additional study is recommended to determine whether this discrepancy is due to the limited amount of data currently analyzed or whether there are other factors such as interconnection of reinforcing steel in the pad, age of the concrete pad, moisture in the concrete, location of insulating membranes, or depth of concrete between the metal tank bottom and reinforcing steel in the concrete pad.

Much literature is available discussing the reduction of the overall resistance of a grounding system due to the ionization of the local earth in the proximity of grounding electrodes conducting significant impulse currents into the earth. The absolute value of the effect of such ionization varies so much by the specifics of each installation that the authors suggest the phenomena be acknowledged but not considered in the design or analysis of the response of the lightning protection grounding system.

5. Potential Equalization

5.1 General

It is generally agreed that the most common cause of ignition of flammable atmospheres in and around structures housing materials that may produce flammable vapors is electrical arcing due to voltage differences between conductive objects. The primary method to reduce such electrical arcing is to equalize the potential between the grounding system for such structures and any ungrounded or independently (remotely) grounded metal objects that may be installed in or on the structure.

5.2 General bonding

The general bonding requirements for lightning protection systems are applicable as the minimum acceptable potential equalization requirements for structures housing materials which may produce flammable vapors. The installation of a ground loop conductor is recommended for these applications as the primary method for interconnection of incoming conductors. In the case of metallic storage vessels, national and international standards allow metallic shell to serve as the ground loop conductor.

All incoming conductors, including piping and electrical conduit, must be interconnected with this common ground bus regardless of whether it enters the structure above or below earth. Electrically-continuous structural components such as structural steel or significant metallic ladders may be used as bonding points to transition to the equipotential ground bus where allowed by code. Permanent internal metallic components should be welded to the metal frame of a steel storage vessel where applicable to decrease the probability of insulation due to contamination from by-products such as wax or corrosive material.

Finally, attention should be paid to the length of conductor used to provide potential equalization. The inductive reactance associated with the length and routing of a conductor could result in a potential difference sufficient to cause arcing at the time of maximum current rate of rise even though it provides an effective bond at lower frequencies associated with significant current flow.

5.3 Floating roof tanks

Floating roof tanks provide significant potential equalization challenges. The construction techniques required to create open top floating roof tanks present potential ignition hazards due to arcing across the non-continuous metallic surfaces. In the case of floating-roof tanks, the floating-roof must be effectively bonded to the main tank shell. This bonding is provided through the installation of shunts bonding the floating roof to the tank shell or through a metallic ladder bonded to the shell.

The design of seals and shunts and their relative locations needs to be carefully considered but is outside the scope of this paper. Additional information on the design and testing of shunts for floating roof petroleum applications can be found in API/EI Research Report 545-A [22].

Culham Laboratory conducted lightning current testing on shunt/shell interfaces and identified that flexible bonding conductors of 35 mm width and minimum thickness of 3 mm should be applied between the tank shell and the floating roof at about 1.5 m intervals around the roof periphery [22]. The bonding conductors must be arranged so that they cannot form a re-entrant loop. They also found that some sparking at the shunts cannot be avoided. Culham results indicated that even clean steel shunt/shell interfaces sparked. When these sparks occur in the vicinity of damaged or leaky seals, ignition of flammable vapors will likely occur. For such installations, maintenance of the seals is critical to reducing the susceptibility to direct and nearby lightning strikes. Immersion of the shunts in the liquid is also an option. The immersion will reduce the oxygen to levels below that which would support ignition. However, Culham reported their testing revealed that currents in immersed shunts tended to cause an eruption of fluid, due to the arc pressure. It is unclear as to whether this eruption of fluid would provide sufficient oxygen to support ignition.

6. Lightning Protection Plan

6.1 General

It is recommended that a lightning protection plan be developed for structures containing flammable vapors to identify the need for lightning protection, identify maintenance and inspection procedures, and establish procedures for termination of any necessary operations at the approach of a thunderstorm. Lightning risk assessments are available in both the IEC and NFPA lightning protection standards to support in determining the need for protection of structures by specific application and can also be used to determine any requirements for the termination of operations. Maintenance and Inspection recommendations are discussed below.

6.2 Maintenance and inspection

IEC 62305-3 requires that lightning protection and grounding systems installed to protect structures with the risk of explosion (including flammables storage tanks) be inspected at regular periodic intervals. This inspection shall include electrical testing every 12 to 14 months as determined by the authority having jurisdiction. NFPA 780 also provides recommendations for the maintenance and inspection of lightning protection systems but is not specific to flammables applications.

Maintenance and inspection plans for flammables storage areas should be tailored to the specific application and physical limitations of the site. For example, in older plants there may not be available real estate to perform earth resistance testing of grounding systems; even when individual electrodes can be isolated. In those cases, it may be more effective to conduct point-to-point bonding testing. The use of visual inspections to verify the efficacy of bonding connections and critical installations can be very effective in many applications.

Periodic inspections for flammables storage areas should also address the inspection and maintenance of seals for floating roof tanks to minimize the probability of a flammable vapor in the vicinity where sparking may appear between the tank shell and floating roof. This is also critical for all tanks in those locations where direct attachment is most probable.

6.3 Termination of operations

There are some operations that may be susceptible to lightning activity or that may result in the generation of charged vapors. The termination of such operations may be beneficial in reducing the risk of ignition of a flammable atmosphere. For example, the loading operations of materials with conductivities of 1 pS/m or less could create an internal charge cloud that is susceptible to internal arcing. A direct or nearby lightning strike could create a collapsing electric field that could lead to an internal arc between the charged vapor and the internal wall of the metal tank, causing ignition of the contents. The use of lightning warning systems in the lightning protection plan can provide advance warning to allow the termination of any necessary operations.

7 Conclusion

Some incident investigations citing lightning as a cause of ignition of flammable vapors are unable to identify the specific ignition mechanism because of the damage to the storage vessel.

Some, such as Oland [10], cite the possibility that a flanged connection for a level alarm, manway, or PV valve which provided enough of a gap to produce a spark. It is critical in the analysis to not only consider the possibility of the creation of an arc but also consider whether the energy in the arc is sufficient to ignite the atmosphere.

Others cite the possibility of arcing between the tank shell and "bound space charge" inside the tank even though there is no identified charging mechanism to create such a "bound charge." While the generation of charged liquid surfaces and vapor clouds is acknowledged, the application will produce a relaxation time for such charges. For materials with conductivities of 50 pS/m or more, the relaxation time is less than one second.

It is acknowledged that there are applications where a "minimum acceptable" lightning protection system for a structure containing flammable vapors could consist of nothing more than a 5 mm or more thick steel continuous metallic container sitting on earth, with or without a release prevention barrier membrane. The 5 mm thick shell should be sufficient to prevent burn through and a tank diameter of 6 m or more should provide enough surface area contact with earth to dissipate currents into the earth regardless of whether surface arcing is produced. The majority of these vessels will also contain numerous connections to piping and grounded electrical conduit that will provide additional paths to earth that will act as supplemental grounding electrode. However, it must also be noted that additional considerations may be required depending upon the conductivity and ignition sensitivity of the contents.

The most common failure mode resulting in ignition of flammable atmospheres is generally agreed to be arcing in the vicinity of a flammable vapor. This is most often caused by a difference in potential between ungrounded or independently grounded conductors located close enough that the potential difference exceeds the breakdown voltage of the atmosphere. This can be resolved by either increased bonding or separation of the independently grounded conductors. For floating roof tanks, the design and location of the bonds are critical.

An additional threat that must also be assessed is the thermal threat that can result from a direct strike. The thermal effect of lightning current flow is not thought to be an issue but the actual lightning attachment point may be. If the material thickness is not sufficient, burn through can occur or a hot spot sufficient to ignite the contained vapors can appear on the inner surface of the metal. This is best addressed by sufficient thickness of the shell of a tank if a Faraday-type protection system is used or to provide strike termination devices that would terminate the strike prior to attachment to the structure.

It can be shown that isolation of lightning currents from the shell of a metallic tank can reduce the voltage that can appear on the inner shell of a metallic tank. The use of isolated lightning protection systems such as masts or groups of masts interconnected by overhead wires can be particularly useful when the acceptable risk of an event is very low and the contents of the structure contains a liquid of conductivity of 1 pS/m or less. The limited use of air terminals as a supplement to a Faraday-type system to provide protection for such protrusions could be considered in future revisions of standards.

Finally, future study is suggested to determine the efficacy of the grounding electrode represented by a metallic tank sitting on a concrete pad. It is also suggested that standards committees consider the possibility of the use of a hybrid lightning protection system to allow air terminals to be installed on metal tanks to protect protrusions that may not meet the thickness requirements of a Faraday-type system.

8 References

[1] "NFPA 780 Report on Comments," National Fire Protection Association, www.nfpa.org/Assets/files/AboutTheCodes/780/780_A2013 Report on Comments_(ROC).pdf , 1 February 2013.

[2] Kern, Alexander, "Simulation and measurement of melting effects on metal sheets caused by direct lightning strikes," *Proceedings of International Aerospace and Ground Conference on Lightning and Static Electricity*, Cocoa Beach, FL, 16-19 April 1991.

[3] CUL/LT-0234, "Review of Lightning Phenomena and the Interaction with Above Ground Storage Tanks," API/EI Research Report 545-A, "Verification of lightning protection requirements for above ground hydrocarbon storage tanks," First Edition, American Petroleum Institute and Energy Institute London, ISBN 978 0 85293 3763, Colchester, UK, October 2009.

[4] NFPA 780, "Standard for the Installation of Lightning Protection Systems," 2011 Edition, National Fire Protection Association, Quincy, MA, 21 June 2010.

[5] API 2003, "Protections Against Ignitions Arising Out of Static, Lightning, and Stray Currents," API Recommended Practice, 7th Edition, American Petroleum Institute, Washington, DC, January 2008.

[6] IEC 62305-3, "Protection against lightning; Part 3: Physical damage to structures and life hazard, Annex D: Additional information for LPS in the case of structures with a risk of explosion," Edition 2, International Electrotechnical Commission, Geneva, pp. 52-58, December 2010.

[7] H. Kramer, H. and K. Asano, "Incendivity of sparks from surface of electrostatically charged liquids", *Journal of Electrostatics*, Volume 6, pp. 361-371, 1974.

[8] Asano, K. "Relaxation time of space charge in partly filled tank," *IEE Proceedings on Physical Science, Measurement and Instrumentation*, Volume 137, Issue 2, pp. 111-116, March 1990.

[9] Matsubata, Y. and T. B. Jones, "Numerical modeling for potential relaxation of charged oil containers" *IEEE Transactions on Industrial Applications*, Vol. 29, No. 2, pp. 295-299,1993.

[10] Oland, Jeffrey, "Tank 15 – A Perfect Storm," *43rd Annual Loss Prevention Symposium*, Paper 6F, American Institute of Chemical Engineers, Tampa, FL, April 2009.

[11] HU Hai-yan and LIU Quan-zhen, "Research on Lightning Sparks Discharge and Protection Measures of Large Floating Roof Tank", *International Conference on Lightning Protection*, Paper No. 61, Vienna, Austria, 2012.

[12] Buccella, C and A. Orlandi, "An Efficient Technique for the Evaluation of Lightning-Induced Voltage in a Cylindrical Vessel Containing Charged Oil," *IEEE Transactions on Industry Applications*, Volume 39, No. 2, pp.368 –373, Mar/Apr 2003.

[13] Buccella, Concettina, "An Accurate Numerical Method to Compute the Effects of the Protections Against Direct Lightning on Charged Liquid Tanks," *IEEE Industry Applications Conference 2002*, Conference Record of the 37th IAS Annual Meeting, Volume 2, pp. 1006 – 1011, 13 – 18 October 2002.

[14] Britton, Laurence G. and Jack A. Smith, "Static Hazards of the VAST," Journal of Loss Prevention in the Process Industries, Volume 25, pp. 309-328, 2012.

[15] Britton, Laurence G., Avoiding static ignition hazards in chemical operations, AIChE-CCPS Concept Book, Center for Chemical Process Safety, New York, 1999.

[16] Britton, L. G., & T.J. Williams, "Some characteristics of liquid-to-metal discharges involving a charged "low risk" oil," *Journal of Electrostatics*, Volume 13, pp. 185-207, 1982.

[17] IEC 62305-1, ""Protection against lightning – Part 1: General principles," 2nd Edition, International Electrotechnical Commission, Geneva, December 2010.

[18] API 545, "Recommended Practice for Lightning Protection of Aboveground Storage Tanks for Flammable or Combustible Liquids," First Edition, American Petroleum Institute, Washington, DC, October 2009.

[19] Rousseau, Alain and Pierre Gruet, "Measurement of a lightning earthing system", *SIPDA* 2005.

[20] Rousseau, Alain, Mitchell Guthrie, and Vladimir Rakov, "High Frequency Earthing Impedance Measurements at Camp Blanding, Florida," *30th International Conference on Lightning Protection*, Cagliari, Italy, September 2010.

[21] Guthrie, Mitchell, John Tobias, and Alain Rousseau, "Grounding System Design and Measurements for Critical Installations," presented at *33rd Department of Defense Explosives Safety Seminar*, Palm Springs, CA, August 2008.

[22] API/EI Research Report 545-A, "Verification of lightning protection requirements for above ground hydrocarbon storage tanks," First Edition, American Petroleum Institute and Energy Institute London, ISBN 978 0 85293 3763, Colchester, UK, October 2009.

Additional references not cited:

[23] Diego, Arturo Galvan, "Lightning protection of structures with risk of fire and explosion," Lightning Protection edited by Vernon Cooray, Chapter 17, pp. 821-841.

[24] Grcev, Leonid, "High-frequency grounding," Lightning Protection edited by Vernon Cooray, Chapter 10, pp. 503-529.

[25] Golde, R.H., Lightning Protection, Chemical Publishing Co., London, 1973.

[26] Rakov, Vladamir and Martin Uman, Lightning, Physics and Effects, Cambridge University Press, ISBN 0521 583276, Cambridge, UK, 2003.

Guidelines for Evaluating the Likelihood of Ignition of Released Flammable Masses

Bob Stack
Dow Chemical
165 Uran St. Hillsdale, MI 49242
stackrj@dow.com

Adrian Sepeda
CCPS Staff Consultant
PO Box 250183, Plano, TX 75025
Adrian_L@swbell.net

Mike Moosemiller
Baker Engineering and Risk Consultants, Inc.
8430 West Bryn Mawr, Suite 720, Chicago, IL 60631
mmoosemiller@BakerRisk.com

Prepared for Presentation at
American Institute of Chemical Engineers
2013 Spring Meeting
9th Global Congress on Process Safety
San Antonio, Texas
April 28 – May 1, 2013

UNPUBLISHED

Guidelines for Evaluating the Likelihood of Ignition of Released Flammable Masses

Bob Stack
Dow Chemical
165 Uran St. Hillsdale, MI 49242
stackrj@dow.com

Adrian Sepeda
CCPS Staff Consultant
PO Box 250183, Plano, TX 75025
Adrian_L@swbell.net

Mike Moosemiller
Baker Engineering and Risk Consultants, Inc.
8430 West Bryn Mawr, Suite 720, Chicago, IL 60631
mmoosemiller@BakerRisk.com

Keywords: Ignition, risk, risk analysis, ignition probability, LOPA

Abstract

Almost since the beginning of time, and certainly since the beginning of structured risk analyses, a core question in studies involving flammable releases has been, "Will the release ignite?" Technology has been developed to predict consequences of ignition with varing degrees of precision. However, the "Will it actually ignite" question has received much less attention and technical development. We know that when a flammable material is released from process equipment, the result may be a fire, an explosion, or the release may simply dissipate with no apparent effect other than a minor environmental impact. Depending on the circumstances, the probability (likelihood) of ignition can range from 0 to 1. Most of the current methods used for estimating the likelihood of ignition are rather crude, and in many or most cases are not based on actual process industry data. For anyone performing a risk-based analysis of any type (QRAs, LOPAs, risk-based facility siting studies, or even the application of a PHA risk matrix), this results in overly conservative or overly optimistic expectations of ignition. In other words, the current methods are often not much better than a guess and can be misleading, with potentially catastrophic results.

It is not difficult to find situations in which the probability of ignition for a given set of conditions varies widely from source to source. There are also some variables that are known to be important to the likelihood of ignition that have not yet been rigorously quantified. This results in uncertainty in risk-based studies, limiting the ability of the risk manager to justify spending for appropriate projects that would reduce the likelihood of ignition as a way of managing risk.

For this reason, CCPS commissioned a new book for release in early 2013, that consolidates the available information on this subject and proposes ignition probability algorithms that advance the state of the art. Purchasing the book also allows access to software that codifies (and automates some of) the methods discussed in the book. The intended audiences for the book are Process Safety Subject Matter Experts (SMEs) involved in risk assessments and hazard evaluations. The timing of the new book's release is after the release of the revision to the CCPS book, *"Guidelines for Evaluating Process Plant Buildings for External Fires, Explosions, and Toxic Releases,"* and near the release time of another CCPS book, *"Guidelines for Conditional Modifiers and Enabling Conditions"* in order to support the risk-based analyses described in those publications.

The algorithms proposed in this book represent what we believe to be the state of the art, but the reader must recognize that the "art" is still evolving. It is anticipated that software updates will occur periodically based on user feedback from the first release. A communication method will be developed and employed so that people who purchase the book, complete with the Calculation Tool, will benefit from those software updates. Therefore, the goal of this project is not only to provide a 'snapshot' of the current technology in an effort to manage risk, but to promote its continued improvement via user involvement and feedback.

Note: - To avoid confusion, definitions of the following terms as used for the purposes of this paper are provided below:

Frequency – The rate at which a certain event occurs, expressed in terms of events per unit time (e.g. "explosions per year").
Probability – The odds or expectations that a certain event will occur, given a stated starting point (e.g. the probability that I roll a "3" on a single die is 1/6). The units of probability are dimensionless.
Likelihood – A term that can be used in place of either "frequency" or "probability," and generally both in the context of this paper.

1. Introduction

1.1 Project Background

The hazards of flammable liquids and vapors have been evaluated for centuries, and in recent decades, has become progressively more quantitative. The Center for Chemical Process Safety (CCPS), an AIChE Technological Alliance, has addressed this issue by generating literature regarding "siting" (protection of people in buildings) and a large range of other process safety issues. The *"Guidelines for Evaluating Process Plant Buildings for External Explosions, Fires, and Toxic Releases, 2nd edition"* was released by CCPS in the spring of 2012. Its purpose is to provide the "how to do" for the recommendations of API 752-3rd Edition on building siting. With the technology and literature now available, industry has the tools necessary to make technologically-based decisions regarding the consequences of flammable releases, and specifically regarding the consequences on building occupants should those releases ignite (or explode). This information has validated that buildings themselves are often a significant hazard should an explosion occur. Therefore, the belief that seeking refuge in a building during or immediately after an ignigatable release is often exactly the wrong thing to do.

Another recently published book and a valuable resource on a related subject is the CCPS *"Guidelines for Conditional Modifiers,"* a follow-on to another CCPS book on the subject of Layers of Protection Analysis.

In reality, not all releases of flammable materials ignite or explode when released to the atmosphere in an operating chemical or petrochemical facility! It is often simply a matter of the very specific circumstances and atmospheric conditions existing at that particular moment in time. This is important because industry, while charged with conducting safe and environmentally sound operations, is not able to eliminate all risk associated with manufacturing, just as an individual is not able to eliminate all risk associated with living their daily lives. Industry is, however, expected to manage the associated risk and stay within a tolerable level. That involves making risk-based decisions that are defendable and repeatable regarding siting and/or any other hazards in the workplace. Risk-based decision making involves considering, calculating, and/or determining data for the two components of risk - likelihood that an event will occur and the expected consequences if it does. The Risk equation is often written as:

$$\text{Risk} = \text{Frequency} \times \text{Consequence}$$

and read as "Risk is a measure of human injury, environmental damage, or economic loss in terms of both the incident likelihood and the magnitude of the loss or injury." (CCPS 2009).

Consequences are predictable and quantifiable if enough data exists. In the manufacturing sector that handles volatile chemicals, there is technology in place to make those calculations with acceptable precision. However, there is a large technological void when it comes to calculating the probability or likelihood that the released flammable material (gas, liquid, or vaporizing liquid) will ignite! History has shown that sometimes it does and sometime it does not.

The determination of ignition likelihood (of a released flammable mass) is the objective of this book. When both components of risk are able to be expressed in quantitative terms, then risk can be mathematically, and more accurately, determined and compared to other risks and/or other scenarios. The result is that this book and the technology herein, when combined with the second edition of the CCPS book "Guidelines for Evaluating Process Plant Buildings for External Explosions, Fires, and Toxic Releases" and/or the "Guidelines for Conditional Modifiers" provides the methodology and technical basis for making risk-based siting decisions. While industry has been making risk-based decisions for many years, this is the first time that they can make these decisions with a complete set of technologies for all branches of a scenario event tree. This allows defensible, technologically-based, risk-based decisions for siting—protection of people in buildings.

Additionally, the methodology and data provided by this new book will provide the platform for future risk reduction efforts and the setting of priorities involving releases. For instance, from a post-event reaction standpoint, the information gained from the risk analysis could be used to develop better and more appropriate emergency response plans—on site and off site. From a risk reduction standpoint, the analysis would identify the major contributors to the risk, allowing industry to prioritize and focus on those areas or events so that available resources are used wisely, and are based on technology, not emotions, fear or misconceptions

1.2 Need for This Book

What distinguishes this effort from similar efforts in the past is recognition that the currently available data in the literature suffers from three significant shortcomings:

1. Data are limited because there is no practical means to conduct controlled experiments of many situations – we are unable to go into a variety of 'live' operating units, release quantities of flammables, and record how often the clouds result in a fire, explosion or non-ignition.

2. The data that are available are either from situations that are different from a normal process plant (e.g. offshore blowouts) or are likely to be skewed (e.g. events resulting in a fire or an explosion will be in the public record much more often than releases that did not ignite, resulting in an overestimate of the probability of ignition).

3. There are variables that we can expect will be important in whether a release is ignited, or that the ignited release is a fire or an explosion, etc. but for which data do not exist or are extremely limited or biased at best. Examples include the impact of the rate of ventilation in indoor releases, the effect of different electrical classification types in the area into which the material is released or the operating conditions of the equipment relating to expected operational temperature, the inate characteristics of failure mode(s) and even simple things such as how hot pump shaft seals may get during normal opereations and during failure.

The net effect of these and other shortcomings is not only a lack of information. It also means that risk managers have a greatly reduced set of risk mitigation options in their toolkit. For example, if a risk manager strongly feels that prohibiting vehicular traffic near a unit will reduce the risk, his or her opinion is all well and good. But given the available literature, there are no hard data to support this conjecture. In the absence of justification from this project, the risk manager will be unable to defend an inconvenient or costly change submitted to management. Thus, there is a very real danger that prudent risk reduction measures would not be undertaken due to a lack of data to justify it.

1.3 Scope

The scope of this new book project was limited to the ignition of flammable liquids and vapors in external environments when exposed to an ignition source. It did not include ignitions for other reasons (e.g. static buildup during top tank filling), and it did not include dust ignition. Also, the intended application is for a land-based operation. While many of the principles developed here could be applied to operations such as offshore or shipping operations, the book and tools were not specifically designed with those applications in mind.

1.4 Approach to Project and Deliverables

The project has two primary deliverables:

- The book—which describes the approach and methodologies available for various degrees of precision;
- Software which allows users to predict ignition probabilities using their own source inputs and risk tolerance evaluation methodology.

2. Literature

The project was initiated by collecting the relevant prior work in the field, much or most of which was not cited in previous related publications. In all, useful information from over 60 citations is provided. Some reviewers of the book have commented that this is the most valuable content in the book. Following are examples of some of the information gleaned from the literature:

2.1 Factors Relevant to Ignition Probability

When developing algorithms for ignition probability prediction, it is important to make the resulting tools as effortless to use as possible; otherwise the tools will be underutilized. It is therefore important to capture those factors that are important, but at the same time try to distill them to the minimum number necessary to get the job done.

2.2 Factors Related to the Release Source

There are a significant number of variables that might be considered important in determining the ultimate outcome of a release of flammables. Primary among these are the physical and chemical properties of the materials being released, such as:

- Autoignition temperature (AIT) of the chemical
- Flash point
- Normal boiling point
- Vapor pressure
- Minimum ignition energy (MIE) of the chemical
- Flame speed
- Electrical properties of the chemical, and potential presence of contaminants

Other important factors relate to the release point itself:

- Source pressure
- Source temperature
- Release rate (hole size)
- Release duration
- Release location (open or congested area)
- Location of ignition sources

In at least one case (AIT), the approach would seem to be straightforward – if the material is above the AIT, it ignites; if not, it doesn't. Unfortunately, the real world situation is not so straightforward. The measured AIT has a number of shortcomings related to predicting the actual spontaneous ignition of a chemical, including variability in AIT test apparatuses, as well as the effects of variability in the surfaces encountered when a release occurs in a real plant environment.

Therefore, since the measured AIT is an imperfect measure of a release in the real world, it is not treated as a discrete value above which ignition is 100% certain and below which it is not. Rather, it is assumed that there is a range of temperatures above and below the reported AIT at which autoignition can actually occur. This and other complexities were addressed in this effort.

It was noted earlier that one of the goals of any work such as this is to minimize user effort. This includes consolidating the number of inputs where possible. For example, if a particular chemical property can be reasonably estimated using another more widely-available property, then only one entry is required instead of two. This cannot always be done – for example, in the case of AIT, one can see that there is no relationship between AIT and boiling point:

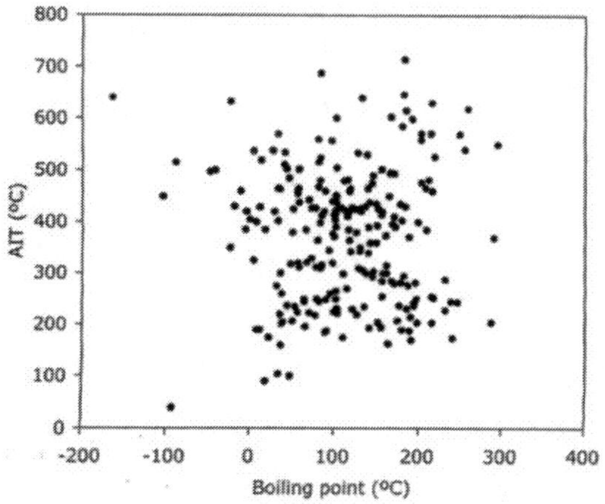

Figure 1. Non-Relationship Between AIT and Boiling Point (from Babrauskas 2003)

There is more of a correlation with carbon chain length:

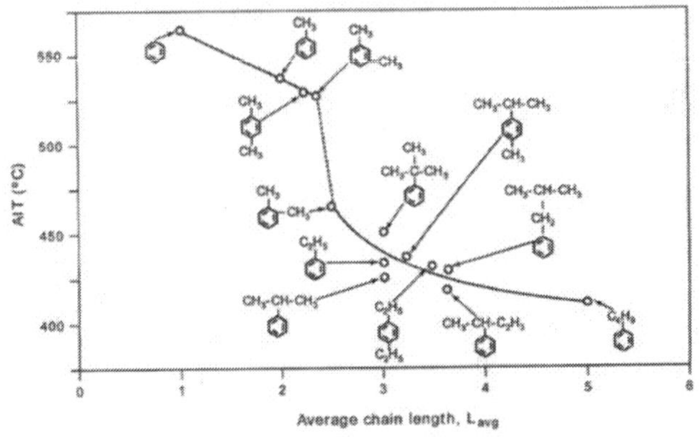

Figure 2. AIT as a Function of Chain Length (from Babrauskas 2003)

However, that relationship likely cannot be extended for all flammables, and if used, there would be no net reduction in the number of inputs required.

2.3 Factors Related to the Environment into Which Material is Released

There are additional variables that can impact releases that are not ignited at the source. These include:

- Transport behavior of the released material (e.g. heavier/lighter-than-air).
- Presence of ignition sources such as running equipment, vehicles, fired heaters, hot surfaces, etc.
- Electrical classification of the area.
- (If indoors) Indoor ventilation rate.
- Area congestion.

For example, Britton (1999) and Pratt (2000) describe the factors that are involved in electrostatic ignition, which could occur at the source or in the open environment. Britton (Figure 3) describes the ability of some of these to ignite various chemicals:

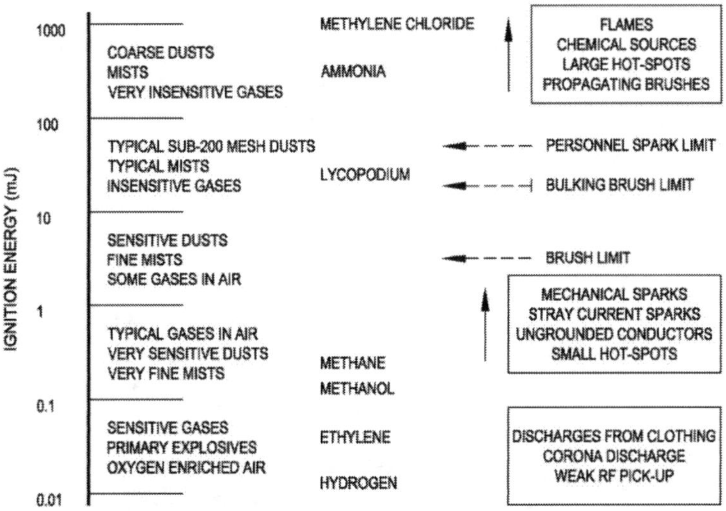

Figure 3. Ignition energies of various materials and types of ignition source that may ignite them (updated from Britton 1999)

These books provide schematic and visual representations of electrostatic phenomena, e.g.:

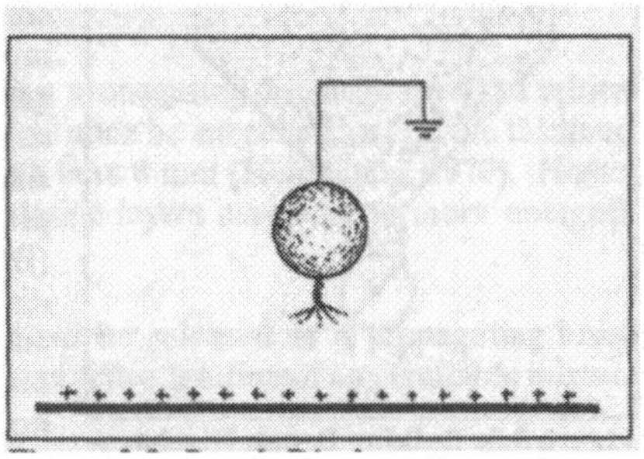

Figure 4. Brush Discharge (from Pratt 2000)

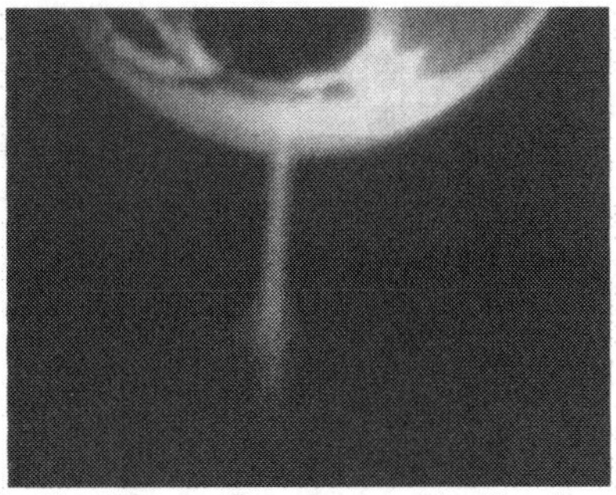

Figure 5. Positive brush discharge from negatively charged plastic to grounded sphere (from Britton 1999)

Other references discuss some of the complexities alluded to earlier. For example, multiple authors describe how ignition by a hot surface generally occurs at a temperature substantially higher than the reported AIT. This is not only because of the uncertainties in the laboratory value itself, but also because of flow patterns around the hot surface, as depicted in the following figure:

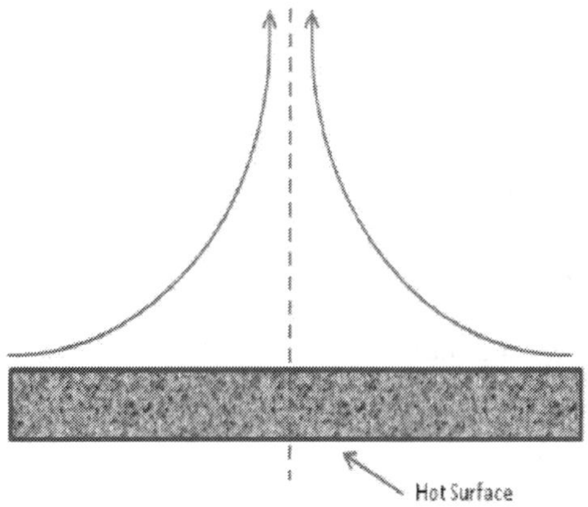

Figure 6. Flow Patterns of a Horizontal Upward Facing Heated Surface (from Duarte et al. 1998)

It was also recognized during the study that there are measures that may be implemented that affect these probabilities but are so situation-specific that they require the user to estimate their effectiveness in preventing ignition and/or explosion. These include gas detection systems, deluge/Halon, explosion vent panels, and the like. The software tool provided with this book will provide these input options so that the user can incorporate them into the prediction but will not provide methods to develop the input values themselves.

These and other ignition controls and phenomena are discussed in Chapter 1 of the book.

2.4 Data Sources (the basis for Chapter 3)

There are a number of investigators and regulators who have looked at the POI issue before, subject to the shortcomings of data described earlier (although generally not acknowledged). Some of this work resulted directly in algorithms for POI prediction, such as multiple studies conducted for the UK Health and Safety Executive (Table 1):

Table 1. Ignition Source Parameters for typical plant (from Daycock and Rew 2004)

Land-use type	Ignition sources	Base case, or 'typical', ignition source parameters						
		p	ta	ti	a	λ	μ	Loc.
1. Car park	'Rush hour' vehicles	0.2	6	474	0.0125	0.0021	160	Out.
	'Other' vehicles	0.2	6	54	0.1	0.0167	3	Out.
	Smoking	1	10	470	0.021	0.0021	8	Out.
2. Road area	'Rush hour' vehicles	0.1	6	474	0.0125	0.0021	160	Out.
	'Other' vehicles	0.1	6	54	0.1	0.0167	3	Out.
	Delivery vehicles	0.1	6	24	0.2	0.0333	20	Out.
	Traffic control	1	0	15	0	0.0667	20	Out.
3. Controlled roads	Delivery vehicles	0.2	6	24	0.2	0.0333	20	Out.
4. Waste ground	None	0	-	-	0	0	0	Out.
5. Boiler house	Boiler	1	120	360	0.25	0.0021	200	In.
6-11. Flames	Continuous (indoors)	1	-	0	1	0	200	In.
	Continuous (outdoors)	1	-	0	1	0	200	Out.
	Infrequent (indoors)	1	60	420	0.125	0.0021	200	In.
	Infrequent (outdoors)	1	60	420	0.125	0.0021	200	Out.
	Intermittent (indoors)	1	5	55	0.0833	0.0167	200	In.
	Intermittent (outdoors)	1	5	55	0.0833	0.0167	200	Out.
12. Kitchen facilities	Smoking	1	5	115	0.042	0.0083	200	In.
	Cooking equipment	0.25	5	25	0.167	0.0333	100	In.
13-15. Process areas	'Heavy' equipment levels	0.5	-	-	1	0.028	50	In.
	'Medium' equipment	0.25	-	-	1	0.035	50	In.
	'Light' equipment levels	0.1	-	-	1	0.056	50	In.
16. Classified	None	0	-	-	0	0	0	In.
17. Classified (Ex.)	Material handling	0.05	5	25	0.167	0.0333	10	Out.
18. Storage (Ex.)	Material handling	0.1	10	20	0.333	0.0333	10	Out.
19. Office	'Light' equipment levels	0.05	-	-	1	0.056	20	In.

In this case, the algorithms take the form of the following equation:

$$Q_A = Q_{A1}Q_{A2}\cdots Q_{AJ} = \prod_{j=1}^{J} Q_{Aj} = \prod_{j=1}^{J}\left\{\exp\left\{\mu_j A\left[\left(1-a_j p_j\right)e^{-\lambda_j p_j t}-1\right]\right\}\right\}$$

Some other sources do not provide actual POI data, but are nonetheless useful in developing the background for ignition predictions. For example, work performed for the American Petroleum Institute describes the effect of environmental conditions on ignition of various materials by a hot surface:

Table 2. Open Air Auto-Ignition Tests Under Normal Wind and Convection Current Conditions (from API 2003)

	Published Ignition Temperatures (Approximate at Time of Test)		Hot Surface Temperature Without Ignition Occurring	
Hydrocarbon	°C	°F	°C	°F
Gasoline	280 – 425	540 – 800	540 – 725	1000 – 1335
Turbine Oil	370	700	650	1200
Light Naphtha	330	625	650	1200
Ethyl Ether	160	320	565	1050

Yet other sources provide historical data that can be useful in describing the relative importance of various ignition source types, e.g.

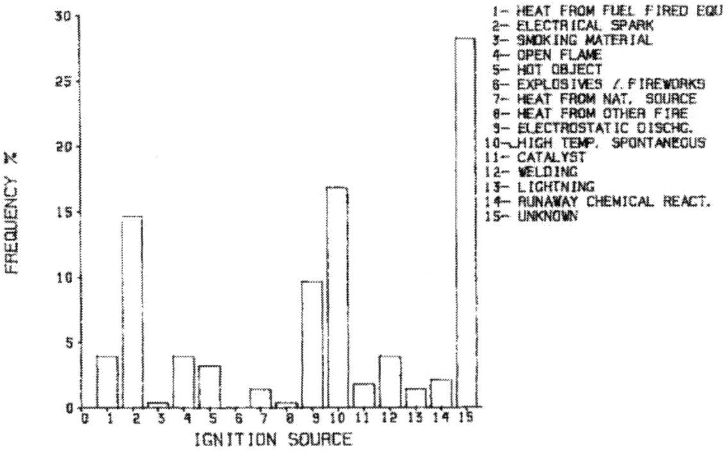

Figure 7. Hydrogen Incident Statistics – 1961 to 1977 – Distribution by Ignition Source (from Zalosh et al. 1978)

Some references help describe other useful properties such as potential mixing rules for multi-component flammable streams:

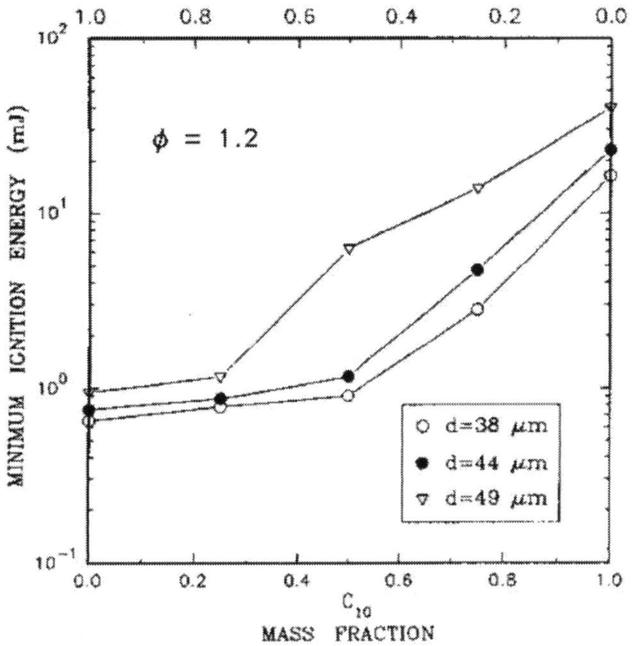

Figure 8. Minimum Ignition Energies Versus Component Mass Fraction for Bicomponent Fuel Sprays (from Lee et al. 1996)

These references and many others were considered in the development of the algorithms used for predicting ignition and explosion probabilities.

2.5 Subject Matter Expert Survey

A survey of subject matter experts was taken to assess their opinions on the probability of ignition/explosion in a variety of situations. The survey included consideration of a range of scenarios with hundreds of combinations of variables that had previously been determined to be significant. Interestingly, the survey results revealed many of the same inconsistencies (differences in experiences) that were the reason for the project in the first place, and 'validated' the variance in experimental and field test results for hydrogen, for example.

3. Project Deliverables

3.1 Overview

The book and algorithms were released earlier this year. The algorithms take the following forms, in their most complex manifestation:

- Probability of Immediate Ignition = f (AIT vs. release temperature, release pressure, chemical MIE, etc.)
- Probability of Delayed Ignition = f (temperature, release pressure, MIE, etc.)

There is also an appendix discussion of explosion probability. It was originally intended that this discussion would be in the main part of the book, but it was recognized that it should be treated differently because there are key variables (e.g. congestion and confinement of the space in which the ignition takes place) that are outside the scope of this effort. For this reason, the explosion discussion is framed in terms of the probability of explosion *given the requisite environmental conditions necessary for an explosion to take place*.

Probability of Explosion = f (chemical, degree of congestion/confinement in release area, presence/effective of explosion prevention measures, etc.)

Simpler versions requiring lesser degrees of input effort are also provided for applications such as PHA team risk matrix use and LOPAs. However, chemical properties for over 200 chemicals are built into the software (and book), reducing the level of effort required from the user.

Some of the more interesting or novel relationships adopted or developed include the following:

Probability of Immediate Ignition = $POII_{ai} + POII_{static}$,
where $POII_{ai}$ = f (T and AIT) and $POII_{static}\ \alpha\ P^{1/3}$ and $MIE^{-0.6}$

Probability of Delayed Ignition = $PODI_{S/D} \times M_{MAG} \times M_{MAT}$,
where $PODI_{S/D}$ = f (ignition source strength and duration of exposure), M_{MAG} = f (hole $size^{const.}$) and M_{MAT} = f (log (MIE))

Other relationships are being developed to account for the special issues associated with indoor releases. The Level 3 analyses will also provide modifications based on refinements to some of the inputs (e.g. MIE as f (T) and MIE as f (droplet size)).

3.2 Book Contents

The next step was to compile and analyze the information from the literature and convert it into book form. The contents of the book include the following:

Chapter 1: Introduction – This chapter provides an overview of the reasons for the book, its scope, and an overview of ignition concepts.

Chapter 2: Estimation Methods – This chapter shows the development and result of the ignition probability algorithms. An illustration of the method is also provided.

Chapter 3: Technical Background and Data Sources – This chapter compiles the available literature (discovered during the research and developmental phases of the complete project) on ignition probabilities and illustrates the highlights of that data.

Chapter 4: Additional Examples – This chapter discusses the limitations of ignition likelihood models and other important conceptual information. Twenty case studies that illustrate the algorithms in both typical, atypical and "out of scope" situations are included in this chapter. It also shows how the methods can be used for risk mitigation.

Chapter 5: Software Illustration – This chapter gives step-by-step instructions, with software screen captures to help the user navigate the software and use it effectively and efficiently.

3.3 Software

The development of ignition probability algorithms took place in parallel with the development of Chapters 2, 4 and 5 of the book. The book and software provide ignition probability algorithms at three levels of detail:

Basic Level – Default values for probability of immediate ignition (POII), probability of delayed ignition (PODI), and probability of explosion given delayed ignition (POE). These will be independent of process or environmental variables, and so are primarily intended for screening-level purposes only.

Second Level – Algorithms for POII, PODI and POE that incorporate the widely accepted variables affecting these probabilities, and for which published information exists (e.g. chemical type, release temperature vs. autoignition temperature, release rate, 'strength' of ignition sources and duration of exposure).

Third Level – Algorithms for POII, PODI and POE that incorporate additional variables that are considered or assumed to be important, but for which data is extremely limited or non-existent. We expect up to five such additional variables to be included. Examples include: (1) electrical classification area type, (2) indoor vs. outdoor operation, (3) indoor ventilation rates, (4) administrative controls (e.g. prohibitions to vehicular traffic), and (5) physical controls (e.g. deluge systems). Note that by definition, these algorithms will be expert-based as opposed to data-based. As such, straightforward models will be developed that will be acknowledged as being "prototypes" for each variable.

4. Worked Example

It is not possible to fully explain the basis of the technology in this short paper, but the appendix to this paper provides an example from Chapter 4 of the book that illustrates the workings of the technology. When reading through the example, note that simply adding in a factor for the wind rose-based probability that the wind would blow the vapor cloud toward the truck (probability that the vapor cloud reaches the ignition source) would lead to close agreement between the Level 1 and Level 2 results.

5. Summary

CCPS published the findings of this ignition probability review and analysis in book and software form. The timing of the book/software was intended to complement the release of two revised CCPS books: *"Guidelines for Evaluating Process Plant Buildings for External Fires, Explosions, and Toxic Releases"* and *"Guidelines for Conditional Modifiers."* The new book supports risk-based analyses that are described in those references. However, the potential uses of this information are much wider, and will likely include LOPA, QRA, event tree, PHA risk matrix applications, and more. The result will be a series of algorithms organized across three levels of sophistication (and input requirements) that allow the effort put into the analysis to be commensurate with the level of refinement needed by the user.

In contrast to previous efforts to predict ignition probabilities, this effort focuses more effort on the physical and chemical properties of the material being released and the immediate release environment. It also considers relevant variables not quantified in previous efforts. The result is an approach which incorporates more variables than previous efforts. The corollary benefit to risk managers is that the treatment of the added variables will provide a basis for quantifying the benefits of risk reduction measures that were heretofore not quantified in publically-available references. Thus, the toolkit of the risk manager will be significantly broadened. And, in the end, more realistic risk estimations can be made.

As is, it may not be the complete "missing link," but with your help and feedback we can over time enhance its precison, reliability and scope. With that thought in mind, CCPS will develop and maintain a website for users to give feedback, suggestions, etc. all in the effort to enhance the value of this new tool and, ultimately, better understand and manage risk.

REFERENCES

API (American Petroleum Institute), "Risk-Based Inspection Base Resource Document," API Publication 581, 1st Ed., May 2000.

API (American Petroleum Institute), "Ignition Risk of Hydrocarbon Liquids and Vapors by Hot Surfaces in the Open Air," API Recommended Practice 2216, 3rd Ed., December 2003.

Babrauskas, V., *Ignition Handbook*, Fire Science Publishers/SFPE, 2003.

Britton, L.G., Avoiding Static Ignition Hazards in Chemical Operations, Center for Chemical Process Safety, American Institute of Chemical Engineers, 1999.

CCPS, "Guidelines for Developing Quantitative Safety Risk Criteria", Center for Chemical Process Safety/American Institute of Chemical Engineers, New York, 2009.

Daycock, J.H., and P.J. Rew, "Development of a method for the determination of on-site ignition probabilities," Health & Safety Executive Research Report 226, 2004.

Duarte, D., Rohalgi, J. and R. Judice, "The Influence of the Geometry of the Hot Surfaces on the Autoignition of Vapor/Air Mixtures: Some Experimental and Theoretical Results," Process Safety Progress, Vol. 17, Spring 1998.

Klinkenberg, A., van der Minne, J., eds., "Electrostatics in the Petroleum Industry – The Prevention of Explosion Hazards," Elsevier Publishing Co., 1958.

Lee, K-P, Wang, S-H and S-C Wong, "Spark Ignition Characteristics of Monodisperse Multicomponent Fuel Sprays," Combustion Science and Technology, Vol. 113-4, 1996.

Moosemiller, M.D., "Development of Algorithms for Predicting Ignition Probabilities and Explosion Frequencies", 2009 AIChE Spring National Meeting, 43rd Annual Loss Prevention Symposium, Tampa, 2009.

Pratt, T.H., Electrostatic Ignitions of Fires and Explosions, Center for Chemical Process Safety, American Institute of Chemical Engineers, 2000.

Zalosh, R., Short, T., Marlin, P. and D. Coughlin, "Comparative Analysis of Hydrogen Fire and Explosion Incidents, Progress Report No. 3," for Division of Operational and Environmental Safety, United States Department of Energy, Contract No. EE-77-C-02-4442, July 1978.

Appendix – Worked Example From Book

4.3.3 "Spill from Methyl Ethyl Ketone Tank"

4.3.3.1 The Event Description

Methyl Ethyl Ketone (MEK) is stored at 20 °C under a slight nitrogen pad in a 7500-gallon vertical storage tank. The scenario is a tank leak or discharge line leak/failure that would empty the tank contents to the dike and spill pit over a short period (10 minutes to a few hours). Because of the leak location, it is not expected that it could be safely stopped.

There is a good chance that a tank truck would be unloading in the adjacent area about 50 feet from the tank, and the dike drainage would bring a draining pool of MEK liquid within 25 feet of the tank truck. The tank truck is not running during unloading; however, it would be running during positioning and the exhaust would be hot for about 10 minutes after tank truck positioning. Thus, a high temperature ignition source could exist 10% of the time during unloading, with tank truck unloading occurring 25% of the time during the day. The wind, if from the west (P=0.3), would blow the MEK vapor cloud over the tank truck area with the plume distance to the LFL being 30 feet and the plume being 20 feet wide for a release rate of 15 kg/sec (the lowest release rate case of interest). For a 150 kg/sec (~ "catastrophic") release rate, the LFL cloud is still only 40 feet long and 40 feet wide. There could be other trucks or cars in the area but they would be more than 50 feet from the dike. The key question is, what would the POI be, given that an LFL cloud blows toward the hot engine exhaust?

Figure 4.6. Spill from MEK tank.

4.3.3.2 Potential Ignition Sources

There are some aspects of the problem that can be solved using the tools in this book, and some that cannot. For the time being, it will be assumed that the cloud can reach the ignition source (that the wind direction and truck presence permit this). The conditional probabilities for these will be discussed later.

There is just one ignition source – the hot truck engine. The calculations will be performed at both Level 1 and Level 2.

4.3.3.3 Ignition Probability Calculations

Immediate Ignition Probability Calculations

Level 1 – As per Section 2.7.1, the value of POII is set to 0.05, since there is no contribution from autoignition at the ambient storage temperature.

Level 2 – As per Section 2.8.1.1, the contributions to immediate ignition include a static contribution that is dependent on the tank pressure and the effective MIE of the material being released:

$$POII_{static} = 0.003 \times P^{1/2} \times \{MIE_{ingested} \times (10{,}000/P_{liquid})^{0.25} \times \exp[0.0044(60\text{-}T)]\}^{-0.6}$$

P (= P_{liquid}) is considered to be 5 psig, to account for liquid head in the tank. The MIE for MEK is 0.53 mJ. T is 68 °F. Therefore:

$$POII_{static} = 0.003 \times (5)^{1/2} \times \{(0.53) \times (10{,}000/5)^{0.25} \times \exp[0.0044(60\text{-}68)]\}^{-0.6}$$

$$= 0.00245$$

Section 2.8.1.2 describes the contribution of autoignition. Since the temperature of this mixture is well below its AIT, there is no contribution of autoignition to the POII. The predicted overall POII is therefore 0.00245.

Delayed Ignition Probability Calculations

Level 1 – As per Section 2.7.2, since the chemical is on the standard pick list, the PODI is calculated automatically as per Eqn 2-17:

$$PODI = 0.15 - 0.25 \log MIE$$

$$PODI = 0.15 - 0.25 \log(0.53) = 0.22$$

Level 2 – The strength "S" can be calculated one of two ways, either by assuming that the truck is a "motor vehicle" as per Table 2.2, or by treating the truck as a "hot surface" as per the discussion following Table 2.2. "S" for each case is:

$$S_{motor\ vehicle} = 0.3$$

$$S_{hot\ surface} = 0.5 + 0.0025\,[1000 - 941 - 100(1)] = 0.40$$

[In the calculation above, the temperature of the engine is assumed to be 1000 °F, the velocity of the gas flowing over the engine is assumed to be 1 meter/second, and the autoignition temperature of MEK is 941 °F.] Select an average value of 0.35.

The baseline PODI is calculated as per Section 2.8.2.2, given the original statement that the truck engine could remain hot for 10 minutes:

$$PODI_{S/D} = 1 - [(1\text{-}S^2) \times e^{-Qt}]$$

$$PODI_{S/D} = 1 - [(1\text{-}0.35^2) \times e^{-0.25\times10}] = 0.973$$

The first factor to be applied to the baseline PODI is the "Magnitude of Release" multiplier. This can estimated using the amount released for the two release cases described earlier, then applying the weight-based relationship in Section 2.8.2.3:

Amount released @ 15 kg/sec = 15 kg/sec (2.2 lb/kg)(600 sec) = 19,800 lbs, and $M_{MAG\_Amount\ Released\ (liquid)}$ = (Amount Released/5,000)$^{0.3}$ = 1.51

Amount released @ 150 kg/sec = 150 kg/sec (2.2 lb/kg)(600 sec) = 198,000 lbs. This amount is greater than the tank capacity of 7500 gallons (~ 50,000 lbs), so it is assumed that the release amount is limited to 50,000 lbs, and the event duration is limited to 2.5 minutes. This in turn changes the earlier baseline calculation for PODI to the following:

$$PODI_{S/D(150kg/s)} = 1 - [(1\text{-}0.35^2) \times e^{-0.25\times2.5}] = 0.634$$

$$M_{MAG\_Amount\ Released\ (liquid)} = (50{,}000/5{,}000)^{0.3} = 2.00$$

The second multiplier to the PODI is based on the MIE of the material being released. As per Section 2.8.2.4:

$$M_{MAT} = 0.5 - 1.7 \log(\text{MIE}), \text{ or}$$

$$M_{MAT} = 0.5 - 1.7 \log(0.53) = 0.97$$

The third factor to be applied is based on the release temperature vs. normal boiling point, as described in Section 2.8.2.5:

$$M_T = 1 - (\text{NBP} - T)/230, \text{ or}$$

$$M_T = 1 - (175 - 68)/230 = 0.535$$

The last multiplier is for indoor vs. outdoor releases, which is set to 1 for this outdoor release. Then,

$$PODI_{Level\ 2} = PODI_{S/D} \times M_{MAG} \times M_{MAT} \times M_T \times M_{IN/OUT}$$

$$PODI_{Level\ 2,\ 15kg/s} = 0.973 \times 1.51 \times 0.97 \times 0.535 \times 1 = 0.762$$

$$PODI_{Level\ 2,\ 150kg/s} = 0.634 \times 2.00 \times 0.97 \times 0.535 \times 1 = 0.658$$

Final Ignition Probability Calculations

Level 1 – POII and PODI were estimated above to be 0.05 and 0.22, respectively. Since immediate ignition precludes the possibility of a delayed ignition, the probability of delayed ignition is:

$PODI_{final} = PODI_{calc}$ (1-POII) = 0.22 (1-0.05) = 0.21

This PODI includes both fire and explosion outcomes.

Level 2 – POII was estimated above to be 0.00245. PODI was estimated to be 0.762 and 0.658, respectively, for the 15 kg/sec and 150 kg/sec release cases. Since immediate ignition precludes the possibility of a delayed ignition, the probability of delayed ignition is:

$$PODI_{final,\ 15kg/sec} = PODI_{calc}\ (1\text{-}POII) = 0.762\ (1\text{-}0.00245) = 0.76$$

$$PODI_{final,\ 150kg/sec} = PODI_{calc}\ (1\text{-}POII) = 0.658\ (1\text{-}0.00245) = 0.66$$

This PODI includes both fire and explosion outcomes.

Discussion of Results

Here is a case where the Level 2 analysis gives higher delayed ignition probabilities than the Level 1 analysis, thus illustrating the points made in Section 4.2.1 about conservatism in probability calculations.

One factor that could be incorporated into the analysis outside of the algorithms in this book is the probability of the wind blowing in the unfavorable wind direction (toward the hot engine). If the wind blows toward the hot engine 25% of the time, the PODI would be multiplied by 0.25, giving a PODI of 0.165 (close to Level 1 results).

Development of a Steel Component Combustion Model for Fires Involving Pure Oxygen Systems

Delmar "Trey" Morrison III, Ph.D., P.E.
Exponent, Inc.
4580 Weaver Parkway, Suite 100, Warrenville, Illinois 60555
tmorrison@exponent.com

Ryan J. Hart, Ph.D., Sean J. Dee, and Brenton Cox, Ph.D.
Exponent, Inc., Warrenville, Illinois

Prepared for Presentation at
American Institute of Chemical Engineers
2013 Spring Meeting
9th Global Congress on Process Safety
San Antonio, Texas
April 28 – May 1, 2013

UNPUBLISHED

Development of a Steel Component Combustion Model for Fires Involving Pure Oxygen Systems

Delmar "Trey" Morrison III, Ph.D., P.E.
Exponent, Inc.
4580 Weaver Parkway, Suite 100, Warrenville, Illinois 60555
tmorrison@exponent.com

Ryan J. Hart, Ph.D., Sean J. Dee, and Brenton Cox, Ph.D.
Exponent, Inc., Warrenville, Illinois

Keywords: oxygen, steel, pure oxygen systems, process design, loss of containment, flammability

Abstract

Pure oxygen systems pose significant fire and explosion risk to equipment, operations, and personnel if a loss of oxygen containment occurs. In industrial oxygen systems, combustion of metal can be initiated by a variety of ignition sources. The combustion of process vessel walls, components, or piping can result in loss of containment of pressurized oxygen. Automatic shutdown and isolation systems are designed to detect loss of containment, incipient loss of containment due to process parameters (e.g., pressure drop), or internal fire conditions. The effectiveness of these isolation systems is dependent upon the speed with which they activate after the initiating event.

The prediction of failure modes and time scales can be an important tool in the design, operation, and maintenance of industrial high purity oxygen systems. Modeling of metal consumption in oxygen fueled fires requires consideration of multiple ignition mechanisms. Previously published experimentation has been motivated by the need to compare the relative burn resistance of metals and nonmetals under a variety of operating conditions, but has not produced a tool to inform the potential for loss of containment in oxygen systems based on the different ignition mechanism failure modes.

This work aims to develop a series of theoretical models using experimental results and kinetic relationships, to predict loss of containment in piping for industrial systems for combinations of different failure modes. Using this type of analysis, process designers can make informed decisions on the design of high purity oxygen systems, the timescale required by emergency shutdown systems, and schedules for routine maintenance and inspection.

1 Introduction

Many industrial systems rely upon pressurized oxygen (O_2) supplies as part of their design. Pressurized O_2 may be found at small scale in health care applications, but our focus here is on large scale industrial systems that may rely upon O_2 at high pressures and continuous supplies.

Such O_2 distribution systems can be found in space rocket engines, oxygen-enhanced combustion systems, and chemical oxidation process units, among others.

Purity and concentration of oxygen are two important factors to consider when designing a safe oxygen handling system. Specifically, gaseous oxygen (enriched) systems are defined as containing at least 23.5% O_2 by volume with the remaining volume being inert [1]. Oxygen systems can be further classified into one of three categories according to their purity [1]:

- Low purity oxygen – contains less than 35% O_2 by volume
- Standard purity oxygen - equal or greater than 99.5% O_2 by volume
- Ultra high purity oxygen - greater than 99.999% O_2 by volume

The safe handling and use of oxygen gas in industrial applications requires special considerations for fire and explosion safety. Industrial standards and the underlying research efforts by industrial oxygen users have promulgated standards including ASTM G94-05, ASTM G63-99, ASTM G88-05, CGA G-4.4, NFPA 50, and NFPA 53, which provide safety and design guidance for oxygen distribution systems.[1-6] The ignition of both metallic and nonmetallic components is a concern in enriched O_2 atmospheres (i.e., greater than 23.5% O_2). System components that would not normally burn in ambient air (i.e., 21% O_2) may burn vigorously in an enriched O_2 environment. This can lead to loss of containment of high purity oxygen and rapid escalation of fire and explosion incidents [7, 8].

The fire hazard in oxygen systems is significant because O_2 is normally present at high concentration. High oxygen concentration combined with a potential fuel and ignition source can result in rapid loss of containment of pressurized oxygen, rapid fire growth, and rapid fire extension to other areas.

1.1 Modes of Ignition

Initial ignition of a component or particle within an oxygen system may lead to a kindling chain mechanism of fire growth where the energy released from an initial, smaller, easily ignitable fuel source causes ignition and combustion of a more ignition-resistant material [3, Section 5.2.14]. The kindling chain may quickly lead to fire growth within a system and overcome (i.e., ignition and continued combustion) the burn resistance of the components in a piping system leading to loss of O_2 containment. Ignition and burn resistance of materials are related to the material composition, thickness, configuration, and environment (e.g., temperature, pressure, and velocity). Several potential ignition mechanisms for metal alloys have been identified in the literature [1]:

- Particle impact
- Mechanical friction[1]

---

1 While frictional heating may occur from gaseous flow against a pipe wall or through a small opening, flow frictional heating causing ignition of metals is generally only supported by "unverifiable anecdotes" in high pressure oxygen systems (e.g., greater than 500 psi) [ASTM G88-05, 5.2.12 – 5.2.12.1]. Hence, frictional heating is more appropriately discussed from a standpoint of surface friction induced by internal moving components in valves, pumps, or other devices with moving parts.

- Adiabatic compression
- Promoted ignition
- Electrical arcing (e.g., static electric discharge, lightning strike, electrical system fault)

The collision of high velocity particles with stationary surfaces can generate localized heating of the particles due to the conversion of kinetic energy to heat at the impact point. If the heat generated is sufficient, ignition and combustion of the particle may occur which could then ignite the impacted material or other nearby fuels (e.g., polymeric gaskets). Both impinging (impact face-on in the direction of flow), and non-impinging (impact obliquely or parallel to the direction of flow) impact with internal piping system surfaces may result in ignition of either the particle or the surface material [3, Section 5.2.6]. Particles may be introduced from the attrition of surface scale deposits from corrosion in piping, mechanical wear of system internals, or other sources of particle contamination. The nature of the particle along with the velocity and other system conditions will affect the likelihood of ignition of the particle. For example, silica particles are non-ignitable, but rust particles are ignitable.

Mechanical frictional heating may result from metal-to-metal rubbing, which can be experienced in moving valve internals. This rubbing mechanism can wear protective coatings from surfaces as well as generate small particles [3, Section 5.2.8.1]. Adiabatic compression of oxygen can occur when a valve is opened to a dead end space in a pipe with O_2 at a lower pressure. The resulting mechanical compression of the gas in the dead space as it is pressurized to the pipeline pressure can result in heating sufficient to ignite non-metallic valve seats or gaskets [3, Section 5.2.7]. These could, in turn, ignite metallic components leading to a system fire.

Several design steps can be taken to limit the potential for ignition, or to stop the kindling chain before the system is breached. The use of filters can reduce the presence of particles, but it is infeasible to remove all particles for an oxygen system [3, Section 5.2.6.1]. The likelihood of ignition from both frictional heating and particle impact can be limited by selecting combustion-resistant materials of construction [3, Sections 7.6.3 and 7.8.3]. The likelihood of ignition by adiabatic compression can be limited by system design and throttling controls, for example to limit the rate of valve opening/closure or ignition resistance of materials in the adiabatic compression zone.

1.2 Steel Alloys in Oxygen Service

Metal alloys for use in O_2 service are classified as either combustion-resistant or exempt, based upon the alloy, configuration, temperature, pressure, and gas flow velocity. A combustion resistant material at the given conditions may undergo ignition, but the combustion process will quench with limited propagation away from the point of ignition. An exempt material has no thickness or pressure limitation for use in O_2 systems, based upon the standards. Although the alloys listed are referred to as "exempt" from velocity limitations, this does not mean that the metals are immune to oxygen-enhanced combustion. Sufficient evidence exists in the literature and from practical experience to indicate that damage can occur to these materials from repeated non-propagating combustion events. Ignition and self-sustained combustion of materials can be evaluated based upon the energy balance.

Any constitutive equation for mass consumption must be solved in parallel with the conservation of energy equation by solving simultaneously for the reaction temperature and rate of propagation. The energy balance is summarized as:

$$Q_b - Q_s - Q_r - Q_c - Q_h = 0 \quad \text{[Eq. 5]}$$

where Q_b is the heat generated by oxidation in the molten mass, Q_s is the heat transferred to the solid metal across the metal/melt interface, Q_r is the radiative heat loss, Q_c is the convective heat loss, and Q_h represents the heat that is used to warm the molten metal from melting temperature to the burning temperature. Q_b is the product of the heat of combustion of the metal times the combustion rate, but the other terms are geometry dependent.

Metals are more difficult to ignite than non-metals due to their higher ignition temperature, higher thermal conductivity, and ability to generate ignition-resistant oxide coatings [2, Section 5.4.2]. If sufficient energy can be added to initiate combustion (i.e., ignition), then a given material may start to burn. However, if the oxygen concentration, density, and velocity are not within certain bounds, the combustion event will extinguish. In practical terms, repeated particle impact may be able to eventually cause a hole in a non-exempt material part.

Examples of metallic alloys used in O_2 distribution systems are provided in Table 1 [1, Appendix C and D]. The table lists some common stainless steel alloys and nickel alloys. Carbon steel alloys are unsuitable for use in oxygen systems at concentrations greater than 35% by volume due to their reduced resistance to combustion [1, Section 4.2.3.2]. The exemption status of a given metal is determined based upon a distance of burning greater than 30 mm in the standardized tests. For a material in a thickness less than that tested, the standards indicate that it should be treated as flammable at pressure and velocity above the exemption curves as displayed below.

Table 1. Example O_2 oxygen system alloy exemption pressures and minimum thicknesses.

Material	Nominal Composition	Min. Thickness	Exemption Pressure
Monel 400	67Ni, 32Cu, 1Fe	None Specified	21 MPa (3000 psig)
Monel K-500	66.5Ni, 30Cu, 3(Al, Ti)	None Specified	21 MPa (3000 psig)
Hastelloy C-276	56Ni, 12Cr, 13.5Mo, 4Fe, 3W, 2.5Co	None Specified	5.3 MPa (750 psig)
Inconel X-750	74Ni, 15.5Cr, 7Fe, 2.5Ti, 1Al	None Specified	6.9 MPa (1000 psig)
Stainless Steel, cast, CF-8M[2]	19Cr, 11Ni, 3Mo, 1.5Si, 1.5Mn, Bal. Fe	3.18 mm (0.125 in) 6.35 mm (0.250 in)	1.4 MPa (200 psig) 2.0 MPa (290 psig)
Stainless Steel, wrought, 316	17Cr, 12Ni, 2Mn, 3Mo, 1Si, Bal. Fe	3.18 mm (0.125 in) 6.35 mm (0.250 in)	1.4 MPa (200 psig) 2.0 MPa (290 psig)
Gray Cast Iron	3C, 2Si, 0.8Mn, Bal. Fe	3.18 mm (0.125 in)	0.27 MPa (25 psig)

[2] Cast analog of 316L/316 stainless steels.

For non-exempt materials (i.e., combustion-resistant), velocity limitations for impingement and non-impingement oxygen flow are shown in Figure 1 and Figure 2, respectively reproduced from pgs. 12-13 in reference [1]. For each graph, systems containing oxygen-contact stainless steel alloy components should be designed to keep the pressure and velocity below and to the left of the curve to limit particle kinetic energy [3, Section 7.6.2]. The velocity limit as a function of pressure is lower for the impingement curve than the non-impingement curve because a theoretical particle impact face-on to a surface will generate more energy than an oblique impact.

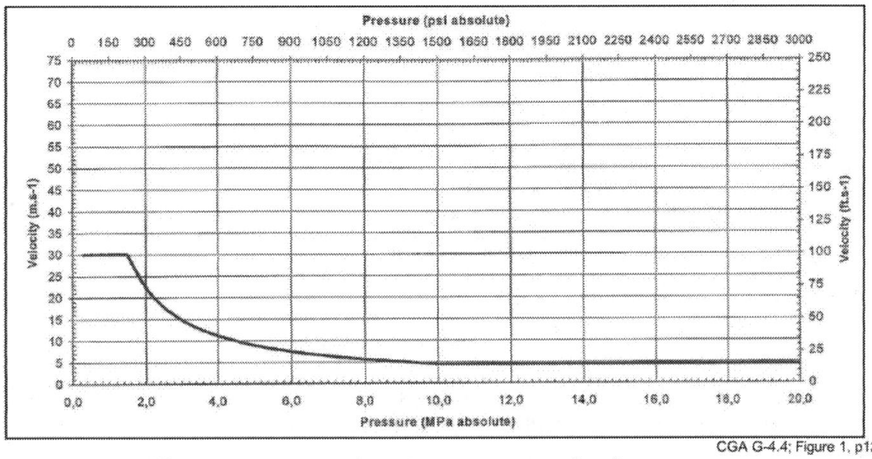

CGA G-4.4; Figure 1, p12

Figure 1. Impingement velocity curve.

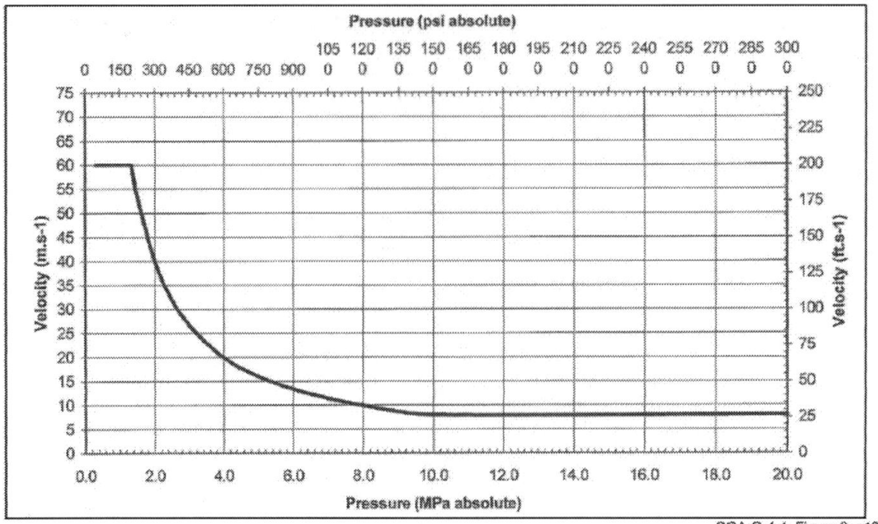

Figure 2. Non-impingement velocity curve.

Stainless steels are widely used materials of construction in industrial oxygen systems, despite their well-established flammability. At high oxygen pressure and flow velocity, stainless steel components are susceptible to ignition due to various mechanisms, including particle impact and frictional heating. Even when isolated via emergency shutdown (ESD) systems, releases of pressurized O_2 have occurred and resulted in significant damage to process systems.

Additional information relevant to past incidents and for the design and maintenance of oxygen systems can be found in the cited references.

2 Steel Combustion Models[3]

System designers have several resources to use in the prevention or mitigation of potential ignition sources in oxygen systems, but ignition incidents may still occur. In order to prevent potentially catastrophic releases of pressurized oxygen, it may be helpful to quantify the speed at which an ESD system must detect and respond to an impending loss-of-containment. Because steel components may be susceptible to ignition and combustion, an initially small hole can quickly grow to a large size. The resulting release of oxygen in the area of the leak may therefore accelerate with time and lead to a rapidly escalating fire or explosion event involving nearby equipment or structures. In this section, models of increasing sophistication are presented to describe steel consumption due to combustion. First, propagation rates directly measured from experiments are compiled. Second, a semi-empirical model for oxidation rate constructed from propagation rate data and Arrhenius-type kinetics is considered. Then a model derived from transport equations and process geometry is discussed.

[3] For sake of brevity, the term "steel" will be applied to all grades of stainless steel.

Combustion of steel involves the oxidation of iron (Fe) in the steel along with other elemental components. The carbon in carbon steel alloys may be partially oxidized to carbon monoxide (CO) or fully oxidized to carbon dioxide (CO2). However, the oxidation of Fe is the primary reaction involved in the burning of steel. Based upon the literature, the oxidation reaction occurs between molten Fe and gaseous oxygen. Molten Fe is oxidized to molten iron oxide (FeO), which then falls or is blown away from the combusting metal surface [9]. The balanced combustion reaction is given in Eq. 1.

$$Fe_{(molten)} + \frac{1}{2} O_{2\,(gas)} \rightarrow FeO_{(liquid)} \quad \Delta H^{o}_{298} = -63.8 \text{ kcal/mole} \quad \text{[Eq. 1]}$$

Thermal conductivity and heat transfer are very important in metal combustion due to the high melting temperature. If the heat input is insufficient to melt the metal given the boundary conditions, then propagating combustion will not occur.

While a sophisticated model describing these processes may seem to be optimal, the available process and material data required to employ such a model may necessitate that a less sophisticated model be employed for a desired level of confidence. The availability of experimental data required to compute model parameters and the difficulty to employ the model are also discussed.

2.1 Propagation Rate Approach

The simplest means of estimating the consumption rate of steel components in a pure oxygen environment is the direct application of propagation rate $\tilde{V}$ (cm/s) measured in a comparable scenario. As discussed by Benz, et al., measurements of propagation rate in metal rods dates back to the late 1950s, and a breadth of experimental data is available [11]. In particular, Benz et al. cites German studies by L. Kirschfeld that summarized the results from a variety of experiments in which he measured average propagation rates of solid metal rods. Unfortunately, the original data from this article was not available to us. Here, we summarize Benz's results for 316 and 304 stainless steel rods with diameters of 1/8 inch (0.32 cm) and lengths of 5 inches (12.7 cm) (see Figure 3 (a)). Propagation rate, $\tilde{V}$ (cm s^{-1}) was measured as a length-wise rate of consumption (see Figure 3 (b)). Propagation rate data for other metals from this and other sources can also be found in ASTM G94-05 [2].

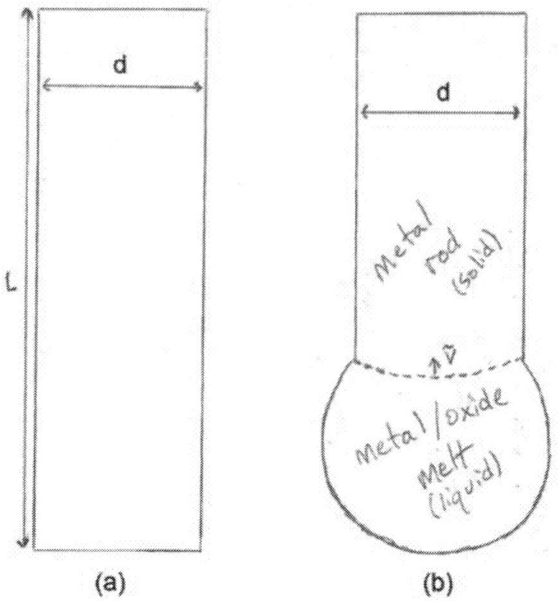

Figure 3. Rod combustion.

2.1.1 Iron Rods, 1/8 Inch (320 mm) Diameter

Steinberg et al. burned pure iron rods of 320 mm diameter in 100% oxygen at 1000 psig (6.9 MPa) [13]. The rods were oriented vertically and ignited at the bottom to allow upward propagation of the burning front. The authors recorded a melting rate of 1.14 ± 0.025 cm s^{-1}.

2.1.2 316 Stainless Steel, 1/8 Inch Diameter

Benz reported data at five different pressures between 1,000 psig and 10,000 psig for 316 stainless steel, as summarized in Table 2. While experiments were performed at 500 psig, the fire extinguished after consuming less than 40% of the rod, so no propagation rate was reported.

Table 2. Data for 316 Stainless Steel [10]

Pressure (psig)	Pressure (MPa)	Propagation rate (cm s^{-1})	Propagation rate (in s^{-1})
500	3.45	< 40% consumed	< 40% consumed
1,000	6.89	1.12	0.441
3,000	20.67	1.22	0.480
4,000	27.57	1.24	0.488
7,000	48.24	1.44	0.567
10,000	68.91	1.58	0.622

2.1.3 304 Stainless Steel, 1/8 Inch Diameter

Benz reported data at 3,000 psig and 5,000 psig for 304 stainless steel, as summarized in

Table 3. Again, experiments were performed at 500 psig, but the fire extinguished after consuming less than 40% of the rod, and so no propagation rate was reported.

Table 3. Data for 304 Stainless Steel [10].

Pressure (psig)	Pressure (MPa)	Propagation rate (cm/s)	Propagation rate (in/s)
500	3.45	< 40% consumed	< 40% consumed
3,000	20.67	1.19	0.469
5,000	34.46	1.30	0.512

A more recent study by Slockers, et al. investigated the burn behavior of 304 stainless steel rods at 1000 °F, of 1/8 in. diameter and 4.4 in. length, at 85 psig (0.7 MPa) [12]. The authors reported a burn length of around 0.28 in. (0.7 cm), consistent with the finding that at lower pressures, the combustion of steel is not likely to be self-propagating.

Given the similarity in the data, the most simplistic model would be to use the linear burning rate (e.g., 1 cm/s) to calculate the rate of hole size growth. If one assumes that the hole grows in a flat sheet as an approximation of a pipe or component wall, then the rate of hole size growth is shown in Equation 2, where R_{hole} is the rate of hole size increase in units of length^2/time during the time interval Δt from initiation of the hole. This approach can be used for an average propagation rate or a value from the literature representing the conditions of concern. This equation can be applied with more complex models for the linear rate of combustion as described below.

$$R_{hole} = \pi \tilde{V}^2 \Delta t \quad \text{[Eq. 2]}$$

2.2 Semi-empirical Kinetic Approach

A semi-empirical equation was employed by Slockers and Robles-Culbreth to capture the dependence of propagation rate ($\tilde{V}$) on oxygen pressure (P_{ox}) and an Arrhenius-type dependence on substrate temperature (T_s) and activation energy (E_a). [11] The general form for this correlation is shown in Equation 3,

$$\tilde{V} = C\, P_{ox}^{\,l}\, T_s^{\,n} \exp(-E_a/RT_s) \quad \text{[Eq. 3]}$$

where C is an unknown proportionality constant, l and n are exponential quantities established experimentally, and R is the universal gas constant. Values for C, n and E_a were not reported, but a value of l was reported for mild steel and 316 stainless steel. Benz et al., also cited Kirschfeld's observation that $\tilde{V}$ was inversely proportional to the diameter of the rod. Thus, for a cylindrical geometry, comparison to measured values of $\tilde{V}$ and P_{ox} can be achieved by the Equation 4:

$$\tilde{V} = \tilde{V}_0 (d_0/d)(P_{ox}/P_{ox,0})^l \quad \text{[Eq. 4]}$$

For mild (carbon) steel, a value l=0.4 +/- 0.5 was reported for P_{ox} between 44 and 1450 psig [11].For Type 316 stainless steel, a value of l = 0.075 was reported for P_{ox} between 1000 and 4000 psig and l=0.26 for P_{ox} greater than 4000 psig. For example, Eq. 3 is employed using the measured values of $\tilde{V}_0$=1.12 cm/s at $P_{ox,0}$=1000 psig for P_{ox} less than 4000 psig, and $\tilde{V}_0$=1.24 cm/s at $P_{ox,0}$=4000 for P_{ox} 4000 psig and greater. The projection compares favorably to subsequently measured values, as shown in Figure 4(note: this is not surprising, as these values were used to find l).

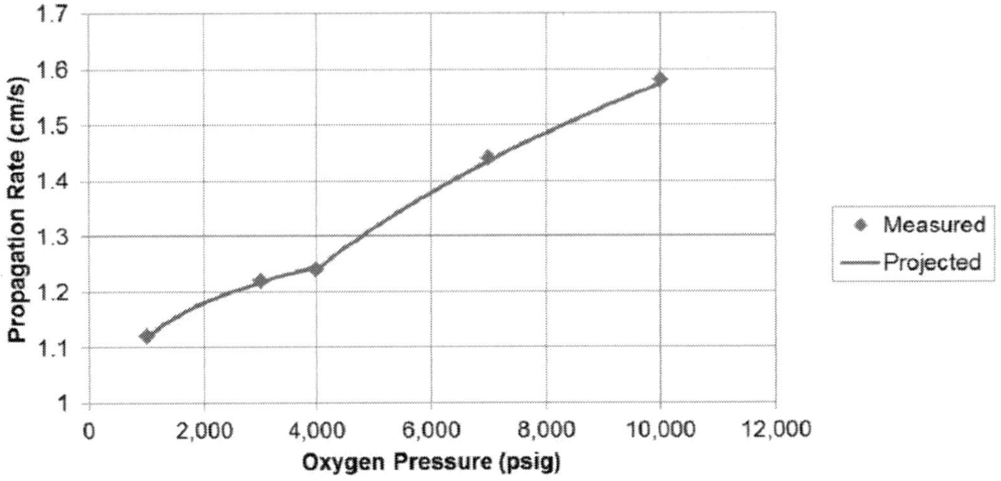

Figure 4. Propagation rate versus oxygen pressure using Eq. 4.

2.3 Conservation of Mass and Energy Approach

A model derived from more fundamental mass and energy balances was presented by Sato, et al., for the specific scenario of a burning metal rod, in which it was assumed that the oxidation reaction occurs in a mass of molten metal which forms from the solid [12]. In this model, the mass balance is calculated from the rate at which oxygen is absorbed into the molten metal, which results in an equation functionally similar to Eq. 3. Specifically,

$$\tilde{V} = j_{m/o} S\, C\, P_{ox}^{\,l} exp(-E_a/RT_b) * 1/(\beta A_s \rho_m) \quad \text{[Eq. 5]}$$

where, in addition to previously defined quantities, $j_{m/o}$ is the weight ratio of the metal oxide and oxygen, S is the surface area of the molten mass, T_b is the molten mass temperature, ρ_m is the density of the metal, C is a proportionality constant, and l is the exponent discussed in Section 2.2. Additionally, A_s is the cross-sectional area of the rod, and β is the product of A_c and h_c, a parameter corresponding to the configuration of the solid/liquid interface and the heat transfer coefficient across that interface, respectively. Note that in their study, Sato, et al. assign β a value of 1, which assumes a flat interface and neglects the effects of convection.

While models of this type allow for greater adaptability, the mass balance requires two more parameters than Eq. 3 (S, A_c and h_c, in exchange for eliminating the temperature exponential n). Additionally, the energy balance requires several more circumstance-specific parameters in the form of heat transfer coefficients and the emissivity of the molten mass. It is unlikely that this type of model would provide substantial improvement to the accuracy of propagation rates in most industrial applications.

3 Conclusions

Three approaches for calculating the flame propagation rate in steel components were presented. These approaches were presented in order of complication. The first method employed direct experimental evidence at varying pressures as a look-up table. The second method was based upon a semi-empirical equation but resulted in an empirical fit, by which propagation rates could be predicted from experimental values. The third method, based upon the conservation of mass and energy, was outlined.

4 References

1. CGA G-4.4, "Oxygen Pipeline and Piping Systems," Compressed Gas Association, 2012.
2. ASTM G94-05, "Standard Guide for Evaluating Metals for Oxygen Service," ASTM International, October 2005.
3. ASTM G88-05, "Standard Guide for Designing Systems for Oxygen Service," ASTM International, November 2005.
4. ASTM G63-99, "Standard Guide for Evaluating Nonmetallic Materials for Oxygen Service," ASTM International, May 2007.
5. NFPA 50, "Standard for Bulk Oxygen Systems at Consumer Sites," National Fire Protection Association, 2001.
6. NFPA 53, "Recommended Practice on Materials, Equipment and Systems Used in Oxygen-Enriched Atmospheres," National Fire Protection Association, 2011.
7. Kletz, T. A., "Fires and Explosions of Hydrocarbon Oxidation Plants," Plant/Operations Progress, Volume 7, NO. 4, 1988, pp. 226-230.
8. Christianson, R., and Stoltzfus, J. M., "Selecting Materials for Use in Oxygen-Enriched Atmospheres," ASME Winter Annual Meeting, 1993.
9. Lanyi, M. D., "Discussion on Steel Burning in Oxygen (from a Steelmaking Metallurgist's Perspective)," Flammability and Sensitivity of Materials in Oxygen-Enriched Atmospheres: Ninth Volume, ASTM STP 1395, T. A. Steinberg, H. D. Beeson and B. E. Newton, Eds., American Society for Testing and Materials, West Conshohocken, PA, 2000.
10. Benz, F. J., Shaw, R. C., and Homa, J. M., "Burn Propagation Rates of Metals and Alloys in Gaseous Oxygen," Flammability and Sensitivity of Materials in Oxygen-Enriched Atmospheres: Second Volume. ASTM STP 910. M. A. Bennings, ED., American Society of Testing and Materials, Philadelphia, 1986, pp. 135-152.

11. Slockers, M. J., and Robles-Culbreth, R, "Ignition of Metals at High Temperatures in Oxygen," Journal of ASTM International, Volume 3, NO. 9, 2006.

12. Sato, J. and Hirano, T., "Fire Spread Limits Along Metal Pieces in Oxygen," Flammability and Sensitivity of Materials in Oxygen-Enriched Atmospheres: Third Volume, ASTM STP 986, D. W. SchroU, Ed., American Society for Testing and Materials, Philadelphia, 1988, pp. 158-171.

13. Steinberg, T. A., Mulholland, G. P., Wilson, D. B., and Benz, F. J., "The Combustion of Iron in High-Pressure Oxygen," *Combustion and Flame*, Vol. 89, 1992, pp. 221-228.

5 Other Useful References

A. ASTM G63 – 99 (Reapproved 2007), Standard Guide for Evaluating Nonmetallic Materials for Oxygen Service

B. ASTM G72 Test Method for Autogenous Ignition Temperature of Liquids and Solids in a High-Pressure Oxygen-Enriched Environment

C. ASTM G86 Test Method for Determining Ignition Sensitivity of Materials to Mechanical Impact in Ambient Liquid Oxygen and Pressurized Liquid and Gaseous Oxygen Environments

D. G88 Guide for Designing Systems for Oxygen Service

E. G93 Practice for Cleaning Methods and Cleanliness Levels for Material and Equipment Used in Oxygen-Enriched Environments

F. G124 Test Method for Determining the Combustion Behavior of Metallic Materials in Oxygen-Enriched Atmospheres

G. G126 Terminology Relating to the Compatibility and Sensitivity of Materials in Oxygen Enriched Atmospheres

H. G128 Guide for Control of Hazards and Risks in Oxygen Enriched Systems

I. Rosales, K., Shoffstall, M., and Stoltzfus, J., "Oxygen Compatibility Assessment on Components and Systems," Journal of ASTM International, Volume 3, NO. 7, 2006.

Prevention and Control of Ignition Sources in the Explosives Industry – BOS (Basis of Safety)

Dr. Martin Held
Austin International, Inc.
25800 Science Park Drive
Cleveland, OH 44122
martin.held@austinpowder.com

Prepared for Presentation at
American Institute of Chemical Engineers
2013 Spring Meeting
9th Global Congress on Process Safety
San Antonio, Texas
April 28 – May 1, 2013

UNPUBLISHED

Prevention and Control of Ignition Sources in the Explosives Industry – BOS (Basis of Safety)

Dr. Martin Held
Austin International, Inc.
25800 Science Park Drive
Cleveland, OH 44122
martin.held@austinpowder.com

Keywords: Control of Ignition Sources, Explosives, Basis of Safety

Abstract

Prevention and control of ignition sources is the key element in manufacturing and processing of explosives. The fundamental approach of BOS (Basis of Safety) and its application principles will be described.

Examples will be given to demonstrate how safe operation principles based on BOS are implemented in inherent plant design for manufacturing of explosives of varying sensitivity and also describe trends in modern plant processes.

1. Introduction

As with other industries, the explosives manufacturing industry has gotten safety policies and procedures, regular safety audits and tries and usually does comply with any available safety standard. However, despite all this paperwork, time and attention the explosives industry continues to carry a fair number of safety incidents, the vast majority of which could have been anticipated. Conclusively, it can be said that for a whole variety of reasons there is in many cases a lack of attention to basic explosive safety, or so called BOS (Basis of Safety) in the explosives industry. In simple words, BOS is a set of simple established principles to do just two things:

1) prevent unintentional explosions (by preventing ignition) and

2) manage the consequences of a potential explosion.

The consequent application of BOS principles ensures that all (process) controls are appropriate and in place.

With general principles on prevention and control of ignition sources from the chemical process industry in place as applicable, this paper is putting a focus on principles for materials and substances with an immanent explosion hazard that does not require explosive atmospheric conditions as reactive oxygen and fuel are chemically bound in molecules.

2. How does an unintended explosion occur?

2.1 Explosive definition

In the environment of explosives manufacturing an explosive or a substance with explosive properties is defined as a substance/mixture that **under the conditions in which it is present** can react violently to produce an explosive effect. These substances include (non-exhaustive)

- Detonating explosives – lead azide, lead styphnate, PETN (pentaerythrite tetranitrate), nitroglycol, nitroglycerine, emulsion and watergel explosives
- Pyrotechnics – delay composition, fuse head dips (in detonators)
- Reactive mixtures – wastes, raw materials
- Ammonium nitrate and ammonium nitrate emulsion matrix

Explosives are designed to rapidly liberate vast amounts of energy. This energy is released by a shock wave which can travel up to 8,000 meters per second through the explosive material.

To avoid an explosion and with this the release of the stored energy, we need to understand what is happening and required to lead to it.

Application of BOS principles is not a separate exercise, but does complement existing safety management and QC systems.

2.2 Sequence of events - BOS hierarchy

To characterize and be able to control ignition sources and with this an explosion hazard, we need to understand what is happening. These are the conditions for an ignition or the sequence of events.

2.2.1 Presence of explosive material

All explosives have the oxygen and the fuel necessary for a reaction contained either on a molecular level or in a chemical mixture. An explosion once initiated will travel through all the material present leaving the products of explosion and a destructive and expanding high pressure wave or air blast.

Thus, looking at the sequence of events, this means, if there is no explosive material present then there won't be an explosion. So, if these materials can be prevented at early stages in a production process (technology) or be kept to a minimum (inventory, batch size, reactor size) it should be done. From a plant design perspective, this could be a change from a batch to a continuous process.

2.2.2 Sources of ignition

In the sequence of events, adding any energy to explosive materials can potentially trigger an initiation leading to ignition. Preventing of ignition is the best place to intervene to stop the chain of events leading to an explosion. Ignition sources in the explosives industry can be summarized as F.I.S.H. which are Friction (between surfaces), Impact (between impacting or colliding surfaces, increase of pressure and temperature), Static (electrostatic discharge) and Heat (fire, decomposition, adiabatic compression, radiant heat, prolonged friction) as illustrated in Figure 1. Most explosive substances are well characterized regarding their sensitivity towards ignition sources.

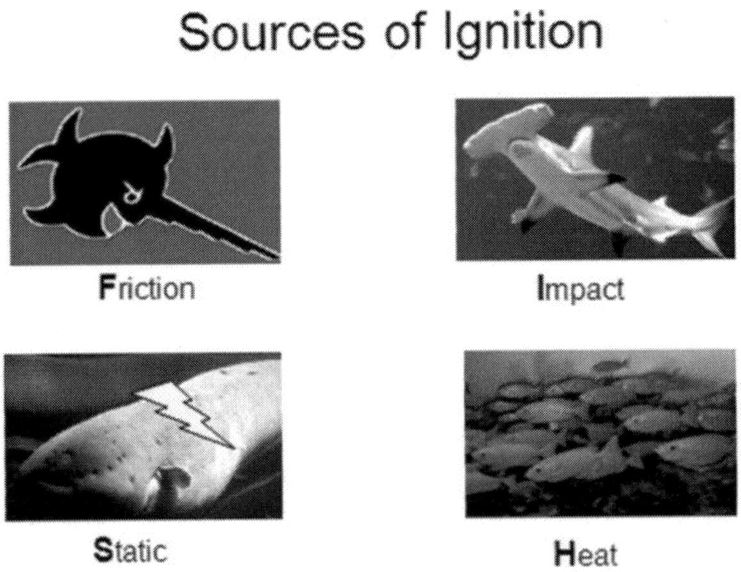

Figure 1: Sources of Ignition (**F**riction, **I**mpact, **S**tatic, **H**eat)

However, they can behave differently under slightly changing process or operating conditions and may become unpredictable. Therefore, the process design has to take a very conservative approach.

2.2.3 Decomposition

Once an ignition triggered by one of the above mentioned ignition sources has started, the decomposition of an explosive material may be local and only consume a small quantity of material.

But it could propagate and spread rapidly (this may be immediately for highly sensitive explosives up to minutes for less sensitive material). Eventually, this may lead to deflagration or burning or at worst case to an explosion.

The actual behavior of an explosive material after an ignition depends on several factors that include the chemical properties of the material, the temperature and pressure, the confinement, the amount of solvents, the shape (solid material), impurities etc.

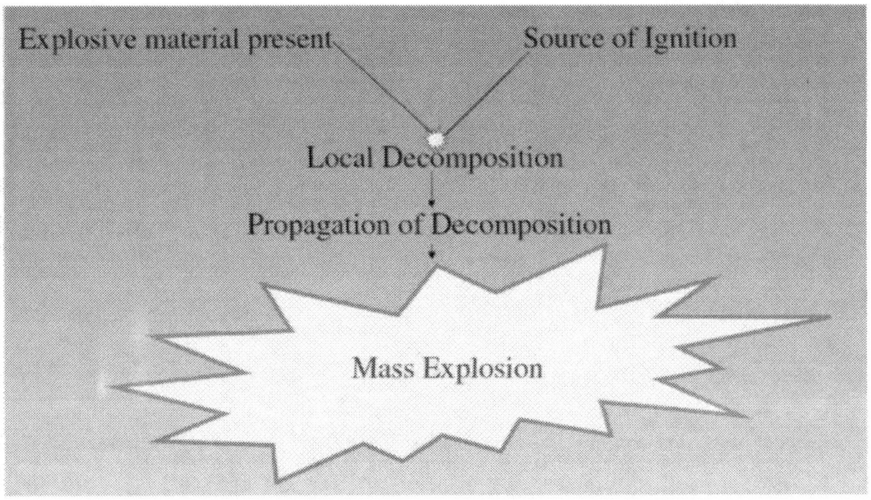

Figure 2: Sequence of events leading to an explosion [1]

3. Hazards with explosives

Explosives are designed to explode when used in a specific application such as breaking rocks in a mining operation. Most of the explosives in today's world are very well characterized in their physical and chemical properties, thermo-chemistry, stability, minimum ignition sources, and compatibilities. However, it needs to be kept in mind that the properties have been determined under exactly defined conditions that in general are not entirely transferable to manufacturing process conditions.

The actual processing environment from design of equipment, control systems integrity over maintenance conditions, raw material quality towards people at any given time is not characterized as quite as exactly and much smaller deviations (lower threshold levels) as compared to the non-explosives chemical process industry can trigger a step change in system entropy and make it impossible to regain containment and control (normal operation) at a very early stage.

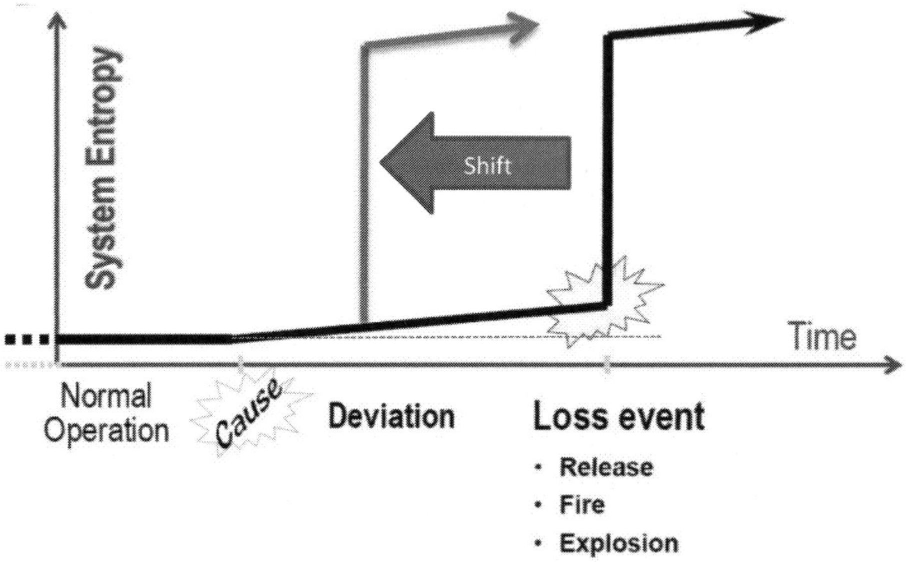

Figure 3: Step change in system entropy [2]

The explosive's behavior under even small deviations may become unpredictable. Therefore, the safety systems applied have to be conservative.

As a consequence, process design and safeguards as applied in a wide area of the chemical process industry may not be sufficient for the explosives manufacturing industry. Consequences of a loss event (explosion) are massive very often.

Picture 1: Explosion aftermath

4. BOS principles

BOS is based on the principles of using lines of defences to block hazards from turning into incidents.

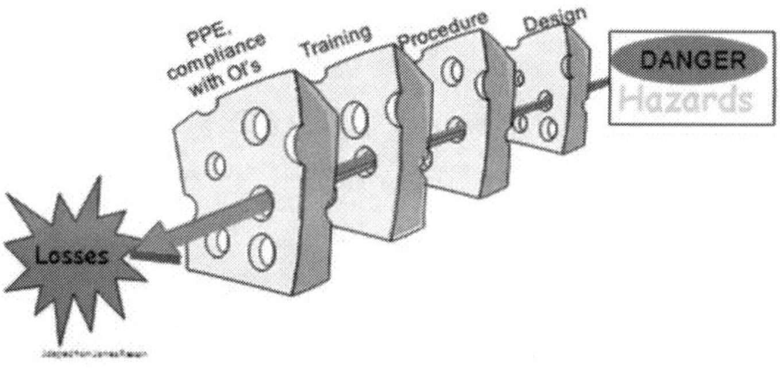

Figure 4: Line of defences [3]

These lines could be:

- Learning from the past

-Inherent Design (e.g. remote operation, process equipment), plant layout (e.g. inventory, separation of processes, safety barriers), Hazard Studies

- Workplace controls

- Individual operator controls

- Organizational controls

BOS principles are intended to assist all personnel involved in designing explosives manufacturing operations, working in these and handling explosives and sensitive (raw) materials to understand:

- The hazards associated with those materials

- How hazardous conditions can arise under normal and abnormal situations/process conditions

- How to avoid those hazardous conditions arising (prevention of ignition sources)

- How to minimize the consequences in the event of an incident (not covered in this paper)

4.1 Principles for safe operation

The principles for safe explosives operation to warrant a manufacturing process without sources of ignition can be summarized as follows:

- Slow
- Soft
- Light
- Low
- Gentle
- Cool
- Grounded
- In the correct place
- Under control

4.2 Inherent design

4.2.1 Quantity/Distance (Q/D) approach

From a historical perspective, one of the first principles applied in the explosives manufacturing industry is the Q/D approach, i.e. quantity and distance.

Basically, the required distance to separation one explosives operation from another or towards non-explosives operations and to the public (roads, railways, inhabited buildings) is calculated by

$$D = k * M^{1/3} \qquad \text{[Eq. 1]}$$

With D = distance (m)

M = net quantity of explosives (kg)

k = constant, number depending on target of an explosion

Calculating overpressure effects and the distribution of debris after simple physical equations lead to a design to physically separate hazardous operations at comparatively large distances to prevent the transmission of an explosion from one manufacturing process to the others, such as separating mixing/chemical batch reaction from cartridging and packaging.

Figure 5: Q/D violated due to managing of operations (explosion chain reaction, inventory control) [1]

This is also valid to separate explosives manufacturing operations from public life activities.

The Q/D principle was considered for the European chemical industry after the explosion of a large quantity of ammonium nitrate at the Tolouse/F incident in 2001 and a couple of major releases of hazardous substances at other occasions.

The explosives industry had observed a shift to apply less stringent Q/D principles inside an explosives operation after the introduction of 'safer' industrial explosives such as watergels and emulsions while keeping those principles to protect third parties and the public. However, incidents in the past recent years have changed this trend to become more conservative and/or to apply the approach of QRA (quantified risk assessment) by applying mathematical methods.

4.2.2 Remotely controlled operations

There are operational processes where despite of all measures taken rare events of explosions are unavoidable depending on ambient conditions (e.g. detonator manufacturing). In this case, processes have been designed to be operated remotely, i.e. the operator is protected by blast walls that can withstand the effect of an explosion and, in addition, the amounts capable to detonate are limited to one piece (detonator) or to a small amount of explosives.

Another reason for remote or robotic controlled operations is to avoid human failure in mechanical operations. As an example, operating a drying oven with trays carrying pyrotechnic mixtures sensitive to impact and friction to be moved in and out under robotic control would prevent an incident if an operator slipped and dropped a tray with the consequence of the mixture to ignite when the tray hits the floor.

4.2.3 Type of process, conveying by gravity and desensitizing prior to transfer

Batch nitration processes in manufacturing of nitroglycol and nitroglycerin (NG), where operators had to attend the nitration process and monitor temperature, have been replaced by continuous nitration processes under remote control.

Figure 6: Batch nitrator for nitroglycerine [4]

Another principle is to not apply mechanical energy to convey sensitive explosive material, but to apply gravity. This today is always the case in the manufacturing of NG where flow induced by gravity is used to convey the material and density effect are used to separate the product from the reaction mixture. This is the reason why dynamite plants have always been placed in a region with mountains or hills and the process following the slope of a valley.

Furthermore, in today's manufacturing processes NG is transferred as an emulsion mixed with water which is much less sensitive to effects of friction and impact. The NG/water emulsion is then separated for further processing of the NG. In addition, mixing processes for dynamite down the line to cartridging the final product are more and more transferred to remote operations.

4.2.4 Inventory control, confinement, pump selection, sealings and bearings, stage of process where explosives are present, detonation traps

Emulsion explosives (water-in-oil type emulsion containing an oversaturated, highly concentrated (ammonium) nitrate solution dispersed in a fuel phase consisting of oil/paraffin and an emulsifier)) are insensitive to static, friction and impact during normal process conditions compared to 'classical' explosives. Emulsion explosives are sensitive to heat (overheating, fire, thermal effect from prolonged friction), especially under confinement.

The reduced sensitivity in manufacturing processes lead to the introduction of more standard equipment as being used in the chemical process industry (pumps, mixers etc.), higher inventories and different process steps taking place in one building. However, still having a material with explosives properties in the process was sometimes neglected and the explosives industry learnt this lesson after a number of severe incidents.

After these incidents until recently, much more attention has been paid towards plant design (separation of processes) and the selection of equipment from the chemical process industry.

New emulsion plant designs are almost all continuous processes where the amount of hazardous material present is kept to a minimum (e.g. size of hoppers, volume of emulsion mixers, explosives limits in packaging areas) for inventory control. A similarity from the chemical process industry would be minimizing the amount of hazardous process intermediates such as the very toxic methyl isocyanate (MIC) in the production of carbamate pestizides.

Even though an ammonium nitrate based emulsion is a hazardous material (class 5.1, UN 3375), it is not an explosive until it is sensitized (this is done by introducing tiny gas bubbles either by chemical reaction or with glass or plastic microspheres). The stage of sensitization can be taken to the very end of material confinement in pipework during processing just before cartridging (into plastic film) or filling into containers.

Another option is to build in detonation traps (these could be reducing the pipe diameter into an array of (non-metal) pipes of much smaller diameter below the so-called 'critical diameter' for detonation (a detonation would not be sustainable below this diameter and stop) or to convey explosives cartridges on belts at high speed to generate a large distance between those cartridges) to prevent the transmission of a potential detonation from the main process area to further process steps. An analogy in the chemical process industry would be flame arrestors.

Very important is the selection and the control of pumps to convey emulsion and emulsion explosives. Progressive cavity pumps (PC pumps) are very common because of an almost pulsation free flow. However, in a dry running or dead heading (blocked flow in line) PC pump heat induced to continuous input of energy by friction can lead to high temperatures with the emulsion matrix starting to decompose with the potential of a pump explosion. There have been a number of incidents in the past with unattended PC pumps continuing to operate and eventually exploding. Depending on the application the safety devices for PC pumps comprise pressure and temperature trips and alarms with automatic shutoff, rupture discs, no flow detection and timeout devices (dead man handling).

Figure 7: Progressive cavity pump (operating principle)

For transfer only application where no accuracy of flow is required, diaphragm are in use very often as the maximum pressure is limited to the air pressure applied to the pump and dry running conditions with heat generated by friction as with a PC pump do not exist.

In today's world, packing glands for pump and mixer shaft sealing are almost entirely out of use and have been replaced with lip seals. The contact surface between the rotating shaft and the seal is under permanent friction. With a lip seal worn out, the emulsion matrix would leak to the outside and the seal would be replaced. In case of packing glands that are often retightened within maintenance intervals, the emulsion starts to migrate into the sealing material and to crystallize between the shaft and the packing glands. The crystals start to groove the shaft and with the enhanced friction the temperature in that contact area starts to rise and eventually may lead to a decomposition of the ammonium nitrate with the risk of fire and explosion.

In addition, it is good practice and very often standard to keep a certain distance between sealing and bearing that explosive material cannot migrate into the bearing and mix with grease and be exposed to friction. Also, the space between sealing and bearing should not be entirely encapsuled to allow visual control for leaking material and to avoid it to accumulate and reach the bearing.

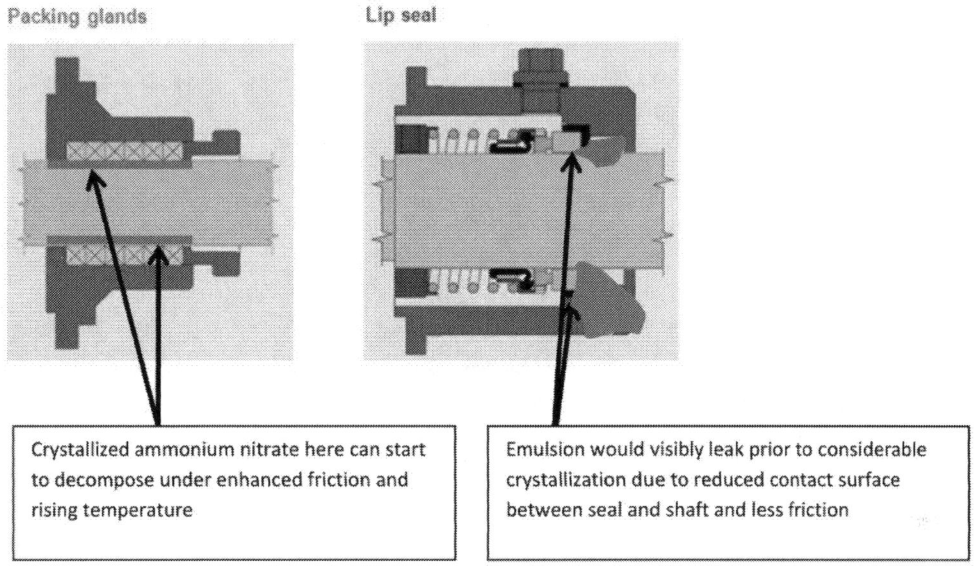

Figure 8: Lip seal and packing glands

It is also common to limit the number of revolutions (rpm) when using a PC type pump and not to operate the pump towards its limit to avoid the entry of too much energy into friction.

4.2.5 Other controls of ignition sources

Static mixers are common in continuous nitration processes and in mixing and sensitizing of water-based explosives (emulsions and watergels) versus the application of motor driven mixers that are transferring dynamic energy into the system.

When agitators and impellers are used in batch mixers or crystallizers, it has to be ensured that a safe clearance between the moving equipment and the inner walls of the tank/kettle is warranted to avoid metal/metal contact and with this friction and impact. Precautions have also to be taken to prevent distortion and the agitator or parts of it to become loose and drop into the tank when explosive material sensitive to friction and impact is present (primary explosives such as lead azide (PbN_3) are very sensitive to any kind of friction and impact).

Ball valves are a source for friction when the valve is actuated and this can be can ignition source when processing dissolved explosives. Pinch valves in this case are a safer alternative, but it has to be taken care that explosive material cannot migrate into fissures aging over time and becoming more sensitive than the original substance. Friction/impact effects when closing a valve could then also trigger decomposition that propagates into an explosion.

Screws, nuts and bolts becoming loose with the potential to fall into openings of equipment where explosive material is processed are a major concern when explosive material sensitive to friction and impact is present. Therefore, drilling and lacing of screws and bolts, gluing or self-securing nuts are very suitable in prevention of ignition sources and this can be taken care of at the design phase (definition of zones for securing).

Explosives dust and sublimate accumulate in threads (e.g. sublimed TNT in cast booster melting) over time even when good housekeeping is applied. When loosening a thread by unscrewing, friction is applied to entrapped material and could start an ignition. Control devices plugged into a lid rather than screwed in or caps on nuts after equipment has been fixed are good explosives practice.

4.3 Safe operation principles

The use of shadow boards to identify and control permitted articles and lanyards to prevent tools from falling into the process environment are very common in the explosives manufacturing industry. The correct use of permitted tools only (e.g. spatula from soft material instead of metal, brass vs stainless steel to prevent sparks, conductive shovels) is a main factor to prevent ignition sources from manual operation.

Picture 2: Shadow board for control of permitted articles

Consequent sealing of wall joints and cracks prevents explosives material to become entrapped and accumulate and also to migrate behind light wall constructions. This eases and prevents hidden material to be ignited especially during maintenance work and dismantling and when hot work is applied.

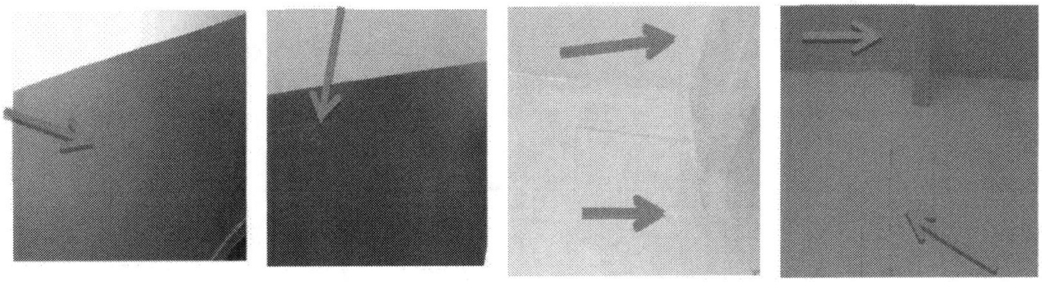

Picture 3: Sealing of cracks and wall joints

Clean floor policies, wet mats in entry areas and soft floors prevent spilt explosives material and dust to mix with grit and the risk of ignition by friction when personnel is walking over floors.

Soft bumpers on trolleys and around corners at room entries significantly reduce the effect from impact (shock) when material is transported on walk-/driveways from one location to another.

Picture 4: Soft bumpers on trolleys and containers

5. References

[1] Begg A.H., "Principles of BOS and GEP", Lecture, XVII SAFEX Congress 2011, Istanbul/Turkey.

[2] Johnson R.W., "The Anatomy of Process Safety Incidents", Unit 1, Lecture 2, AIChE eLearning Course ELS 104 (accessed Aug. 23, 2012).

[3] Swiss Cheese Model by James Reason published in 2000 ("Human error models and management", *British Medical Journal*, **320**, pp. 768 – 770).

[4] Picture from ICI.

Hazards of Performing Hot Work in Gasoline Storage Tanks

Scott G. Davis
GexCon US
4833 Rugby Avenue, Suite 100
Bethesda, MD 20814
sgdavis@gexcon.com

Peter C. Hinze, Kees VanWingerden
GexCon US, GexCon AS
Fantoftvegen 38
NO-5072 Bergen, Norway
pchinze@gexcon.com, kees@gexcon.com

Prepared for Presentation at
American Institute of Chemical Engineers
2013 Spring Meeting
9th Global Congress on Process Safety
San Antonio, Texas
April 28 – May 1, 2013

UNPUBLISHED

Hazards of Performing Hot Work in Gasoline Storage Tanks

Scott G. Davis
GexCon US
4833 Rugby Avenue, Suite 100
Bethesda, MD 20814
sgdavis@gexcon.com

Peter C. Hinze, Kees VanWingerden
GexCon US, GexCon AS

Keywords: CFD, computational fluid dynamics, gasoline, storage tanks, gas dispersion, explosion, accident investigation

Abstract

This paper presents a recent investigation into explosion that occurred in an unleaded gasoline storage tank during hot work operations. The storage tank had a floating roof design that was covered by a geodesic dome. The floating roof was subdivided into many isolated pontoons. Sometime after 12:00 pm, an explosion occurred during hot work operations in the tank that was cleaned and cleared for confined space entry. The explosion blew the geodesic dome roof off of the tank where it fractured into multiple sections, split the top plate of the floating roof from the bottom and propelled it up and out of the tank, deformed the bottom section of the floating roof towards the floor and northern section of the tank, deformed/collapsed the sidewalls of the tank, and killed three welding employees, who were on the floating roof at the time of the explosion and were blown out of the tank.

GexCon performed an advanced CFD explosion analysis into the cause of the accident, which could reconcile the observed evidence in the case. The study concluded that residual gasoline vapors were present within an exterior pontoon in the floating roof, which subsequently ignited when the welders began cutting into the floating roof. In addition, inspection revealed that this particular pontoon had a small hole on its sidewall near the liquid level of the tank. This hole allowed gasoline to seep within the pontoon resulting in residual vapors on the day of the incident. This study will present the lessons learned regarding isolation of the tank, cleaning the tank for hot work operations, confined space entry and continual atmosphere monitoring.

1. Introduction

Welding contractors were hired to install a gauge pole into an existing gasoline storage tank at a gasoline product terminal. The process was to be implemented as follows: (1) isolate the tank from operations; (2) clean the tank such that it was ready for hot work; and (3) install the gauge pole. The actual installation of the gauge pole involved cutting a hole in the plates of the floating roof, welding a sleeve into the floating roof, then slotting, grinding, and welding the gauge pole. Figure 1 shows an example of a gasoline storage tank with a floating roof and a geodesic dome.

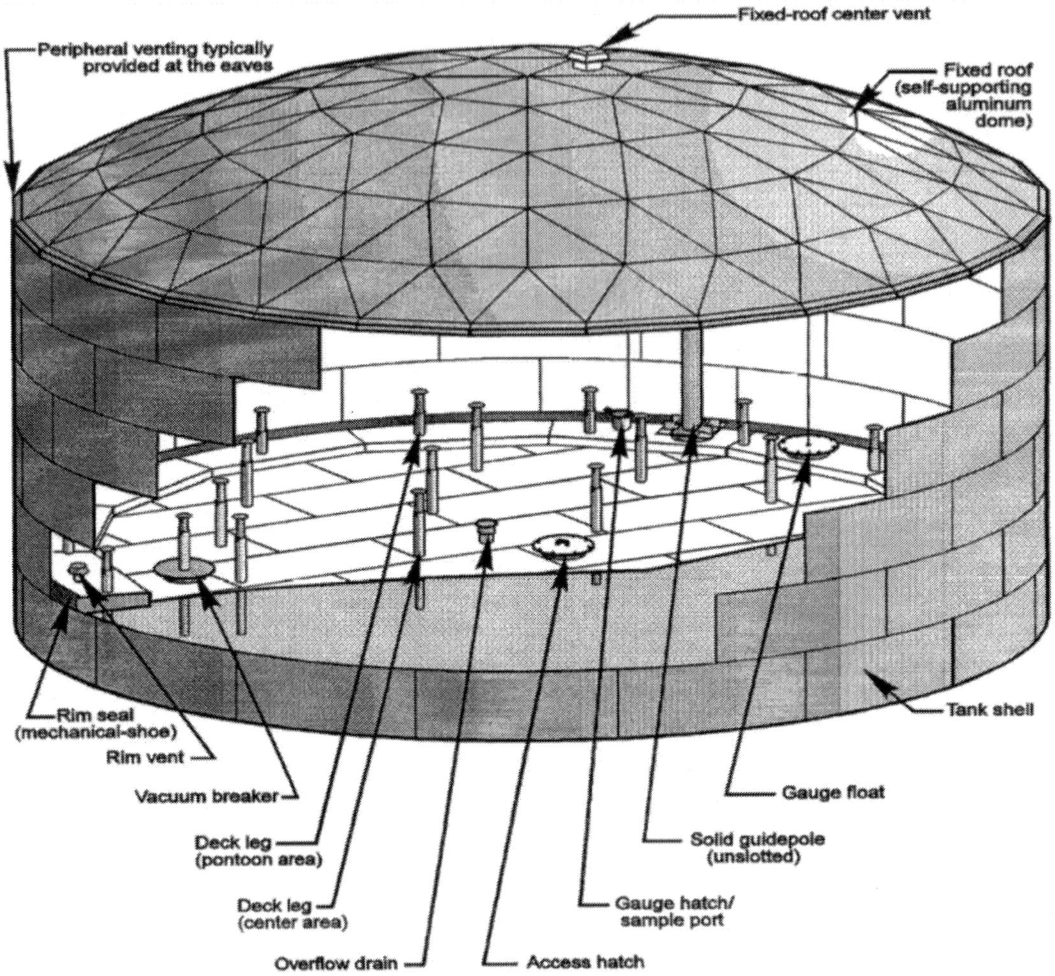

Figure 1: External floating roof tank with a geodesic dome.

The storage tank had previously held unleaded fuel and had been drained completely about a week before the incident occurred. The tank was isolated from operations a week and a half before the accident. Isolation consisted of stripping the lines feeding the tank, installing skillets (a type of blank to block the pipe), and removing manway covers. To facilitate the removal of gasoline vapors and to dry any liquid gasoline, a fan was installed in one of the manways on the side of the tank. The fan was operational for a total of five days. A cleaning company was contracted to clean the tank, a procedure which lasted for three days. Two days after the cleaning was completed, the welding contractors began their work.

The floating roof consisted of two circular plates of steel welded to a cylindrical steel plate approximately three feet in length. Two more concentric cylinders were welded inside the roof at radii of 1/3 and 2/3 of the roof radius, and the entire floating roof was divided with four plates that ran through its diameter. A manway was also present at the center of the floating roof. Figure 2 shows an image of the floating roof used in the computer simulations.

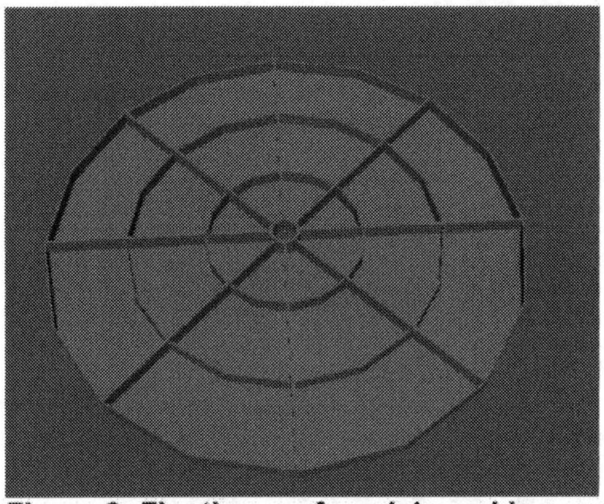

Figure 2: Floating roof model used in computer simulations. Note that the top plate has been removed to show the internal structure.

Several of the welding contractors were in the tank when an explosion occurred. The explosion tore the floating roof in half—separating the top plate from the bottom plate—and ejected the top plate of the floating roof up and out of the geodesic dome that covered the tank. The workers who were in the tank at the time of the explosion were all killed.

GexCon undertook an investigation of what happened to determine the cause of the accident. The goal of this work was to determine two things: whether gasoline product could remain in the pontoon(s) for the period of time that the workers were ventilating the tank, and whether one pontoon full of gasoline could cause the damage observed.

2. Description of the Incident

On the day of the explosion, the management company of the facility reported taking lower explosive limit (LEL) measurements inside the tank and that they detected no gasoline vapors and an oxygen level of 20.8%. A confined space entry work permit and a hot work permit were issued to allow entry into the tank. The welding company monitored the atmosphere inside the tank during their work.

In the early afternoon, the explosion occurred, blowing the geodesic dome roof off of the tank where it fractured into multiple sections (see Figure 3 and Figure 4), splitting the top plate of the floating roof from the bottom and propelling it up and out of the tank (see also Figure 3), deforming the bottom section of the floating roof towards the floor and northern section of the tank (see Figure 4), deforming/collapsing the sidewalls of the tank (see Figure 3 and Figure 5), and killing three welding company employees. These men were on the floating roof at the time of the explosion and were blown out of the tank.

Autopsy reports for the workers indicated that these men died of multiple or extensive blunt force injuries and concluded the cause of death to be accidental. There were no reported burn injuries to the decedents.

Figure 3: Incident tank after the explosion. The top of the floating roof is indicated. It has flipped over, so the outward facing side (painted white) is now on the ground, and the underside is facing upwards.

Figure 4: Bottom section of the floating roof, still in the tank.

Figure 5: Exterior photograph of the tank showing the buckling of the tank shell.

An overview photograph of the tank after the explosion is shown in Figure 5. The photograph shows that the failure mode of the tank was buckling due to a negative pressure event inside the tank (see Figure 6). Furthermore, the top of the floating roof deformed a section of the tank shell in the area where it was blown out of the tank. In addition, the bottom portion of the floating roof was crushed up against the north side of the tank shell and there were markings on the wall opposite where the bottom of the floating roof was found (Figure 6).

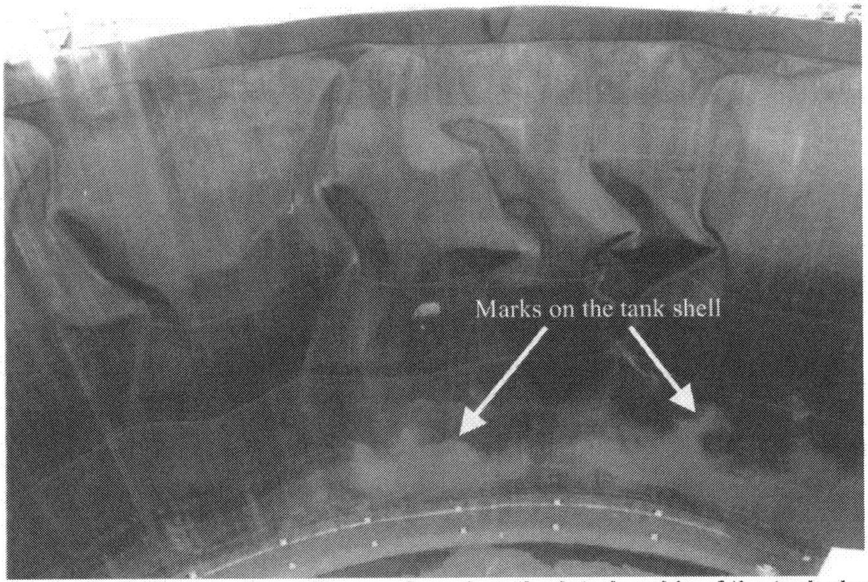

Figure 6: Evidence of buckling as viewed on the interior side of the tank shell on the south side

Regarding the floating roof, the explosion caused the top of the floating roof to separate from the bottom section and subsequently blew it out of the tank. The top of the floating roof was severely deformed but remained largely in one piece (see Figure 7). However, there was an area of the top floating roof near the exterior circumference (e.g., outer pontoon region) which had been separated into several fragments as a result of the explosion. Figure 8 shows an example of one such fragment, which had been labeled with an "X", and there was also evidence of grinding the exterior paint on another fragment.

Figure 7: Top portion of floating roof as it was found. It was flipped over in the explosion.

Figure 8: Separated fragment – top of floating roof. Note the "X" mark on this fragment.

The bottom section of the floating roof was significantly deformed and had been blown in the direction of the tank shell where the top plate had been blown out (see Figure 4). The explosion had caused the outer most pontoon section of the floating roof bottom (in the southern region – lower part of the image) to deform downward, followed by a general deformation and displacement towards the north. In the same southern region, the explosion caused the side wall of the floating roof to completely separate from the bottom plate (see Figure 9).

Figure 9: Side wall of the floating roof separated from the bottom plate.

A hole was found on the side wall of a pontoon. Moreover, this hole was in the region where the side wall had separated from the bottom of the floating roof (see Figure 10). Evidence of a failed weld was also found on the bottom of the floating roof in the areas where the side wall had separated from the bottom of the floating roof. The fragments from the top of the floating roof that had grinding marks and the hole with the "X" were in the vicinity of the hole in the side wall and where the failed weld was found.

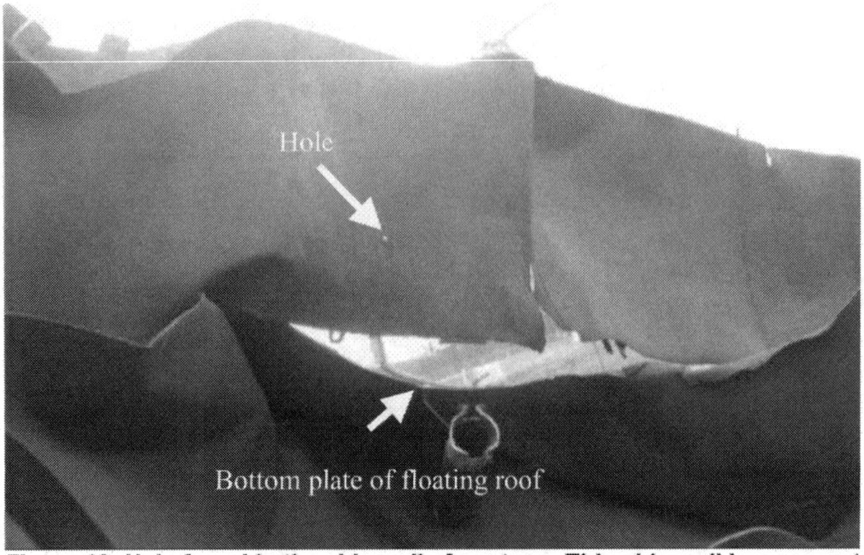

Figure 10: Hole found in the side wall of pontoon. This side wall has separated from the bottom surface of the floating roof.

The work for which the welding company was hired required cutting a hole in the floating roof using an oxyacetylene torch. The work also included grinding the paint off the surface of the floating roof, which produces sparks that are capable of igniting vapors. Though it is not known precisely what they were doing at the time of the explosion, there is evidence that they may have initiated grinding activities or cutting with the oxyacetylene torch.

Grinding wheels were found inside the tank and grind marks were noted on various surfaces of the floating roof, suggesting it was used just prior to or at the time of the explosion. The acetylene torch handle was blown out of the tank after the incident. The control valves on the acetylene torch handle and tank were found to be open. This indicates that the torch was in use when the explosion occurred and is consistent with the separated fragment of the floating roof with the "X" marked on it (see Figure 8) indicating the location where cutting was to occur.

2. Incident Investigation

The explosion was caused by an ignition of a vapor/air mixture within the tank. There were three general areas within the tank where the vapors could have migrated or accumulated (see Figure 11 for general locations): (1) above the floating roof; (2) below the floating roof; and (3) within the pontoons of the floating roof. The first two scenarios evaluate the likelihood of having vapor/air mixtures above or below the floating roof due to residual gasoline product remaining in the tank (e.g., in the drain) or from a leak of acetylene gas, while the third scenario evaluates the vapor clouds caused by residual gasoline liquid or vapor remaining within the pontoons of the floating roof. This analysis will evaluate these three scenarios and determine which one successfully reconciles the damage observed on the tank.

The resulting explosion damage observed on the tank included: (1) the geodesic dome failed into numerous fragments, most of which were found blown out of the tank; (2) the top of the floating roof separated from the bottom section and was blown out of the northern side of the tank causing deformation to the tank shell; (3) the outer most pontoon section of the floating roof bottom (in the southern region) was deformed downward, followed by a general deformation and displacement towards the northern side of the tank; (4) the side-wall of the floating roof in the southern region of the tank was completely separated from the bottom plate; (5) discoloration markings were observed on the southern tank shell at the approximately height of the floating roof; and (6) the tank shell buckled due to a negative pressure event during the explosion.

Inspection of the evidence revealed that the damage observed on the gasoline storage tank was caused by an explosion that was ignited within the external pontoon/pontoons on the southern side of the floating roof (see schematic in Figure 12). The explosion event was large enough to separate the top plate of the floating roof from the bottom section, blow the top plate off the floating roof and out of the tank, fail the geodesic dome, and partially collapse and deform the bottom section of the floating roof (see schematic in Figure 13 and Figure 14). In addition, the resulting negative pressure observed during the explosion was sufficient to cause buckling of the tank shell.

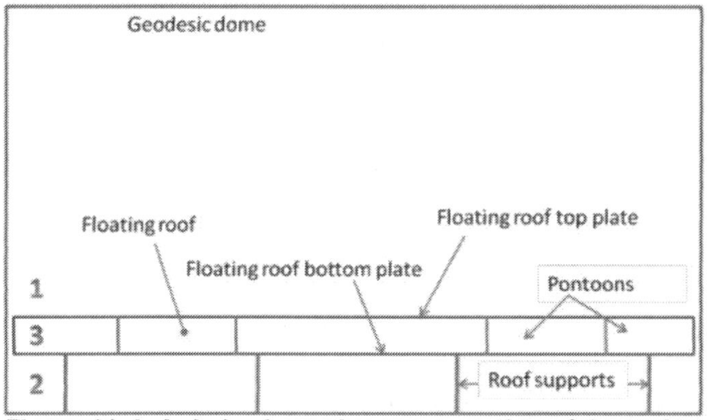

Figure 11: Labeled schematic representation of the tank and floating roof.

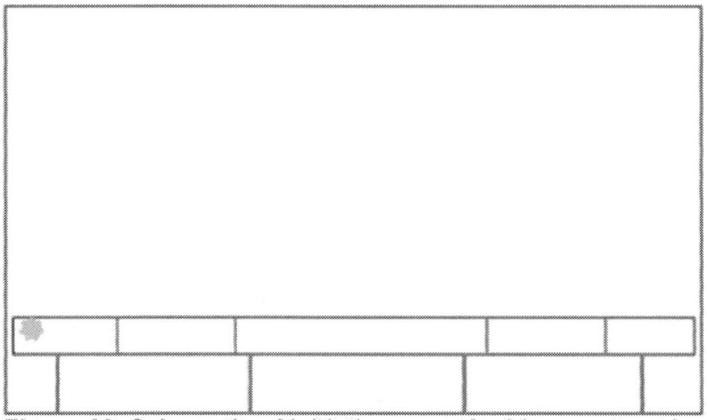
Figure 12: Schematic of initiating event inside an external pontoon/pontoons.

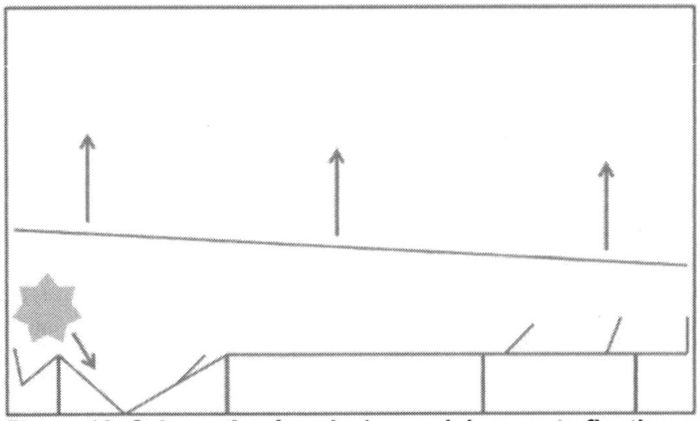

Figure 13: Schematic of explosion and damage to floating roof.

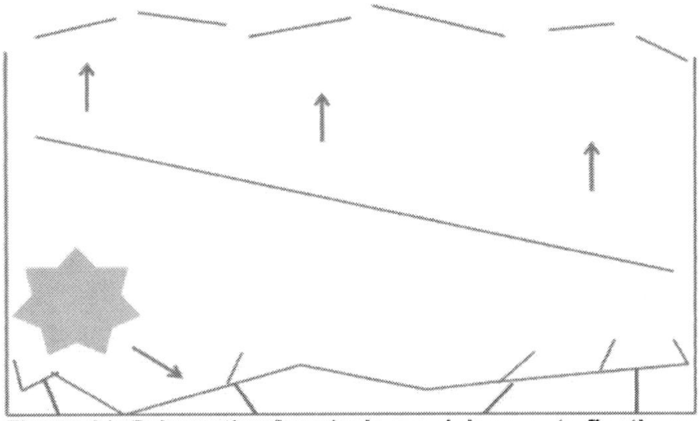

Figure 14: Schematic of explosion and damage to floating roof and geodesic dome.

Tanks such as this are designed to be able to withstand wind loading. An estimate for the design pressure can be found in API 650 *Welded Tanks for Oil Storage.* API 750 indicates that the tank shell shall be checked for stability against buckling from the design wind speed in accordance with formulas that are intended to cover tanks with either open tops or closed tops. The formulas are based on 3-second gust design wind speed = 190 km/h (120 mph) at 10 m (33 ft) above ground. The resulting pressure on the tanks is reported as 17.2 mbar (36 lbf/ft^2, 0.25 lbf/in^2), which includes a 2.4 mbar (5 lbf/ft^2, 0.035 lbf/in^2) negative pressure within the tank. Therefore, the explosion scenario would need to be capable of producing a vacuum inside the tank on the order of 17.2 mbar.

Despite the damage to the tank demonstrating that the explosion likely initiated within the pontoons, an extensive explosion modeling analysis was next conducted to evaluate the explosion dynamics of vapors clouds ignited at various locations within the tank. This analysis tested the hypotheses of fuel/air mixtures ignited above, below and within the pontoons of the floating roof. This analysis was performed to investigate the dynamics of the explosion, and it was not exhaustive in its scope.

3. CFD Modeling

GexCon undertook CFD simulations to determine scenarios that may have occurred to cause this explosion. There were four sets of simulations that were performed: dispersion scenarios to determine whether gasoline vapor could have been in the pontoons given the ventilation procedures undertaken by the workers, ignition of a vapor cloud on top of the floating roof, ignition of a cloud underneath the floating roof, and ignition of a cloud inside the pontoon(s) of the floating roof. The goal of the simulations was not to identically reproduce the exact failure scenario, but to identify which scenario produces the governing explosion dynamics that can explain the observed explosion damage.

To understand which mechanisms could lead to the accident and observed damage, a number of dispersion and explosion simulations were performed using the computational fluid dynamics (CFD) model FLACS. FLACS is a leading CFD tool for the simulation of gas dispersion and vapor cloud explosion scenarios, and is extensively used for safety studies in the petrochemical industry. FLACS is capable of modeling gas and aerosol releases, dispersion of vapors, ventilation in structures, and ignition of flammable fuel-air mixtures to evaluate the flame front progression and the overpressures due to explosions. Sensor points can be placed within the computational domain to track the time history of relevant variables such as static or dynamic pressure, velocities, fuel concentration, etc.

3.1 Dispersion Modeling

These simulations modeled the possible dispersion of pentane and heptane from a modified pontoon. Pentane and heptane are both components of gasoline with vapor pressures that are representative of different components of gasoline. Because the semicircular geometry of the pontoon setup is more complex to model and does not in any way affect the phenomena that are being investigated, the dispersion simulations were performed in a rectangular geometry that had the same volume and aspect ratio as one of the outermost pontoons. A two centimeter hole was included in one of the sides because a hole was observed in a pontoon after the explosion (see Figure 10).

These simulations were allowed to run for 21 days of simulation time, to account for the time that was claimed to have passed between the draining of the tank and the incident. Allegedly the tank had been drained a week or two before the welding company started the hot work, so three weeks was chosen as a conservative scenario. The initial conditions were set such that there was a concentration of the fuel at its vapor pressure for 15°C (59°F) in the pontoon; the equilibrium mass fraction of pentane at 15°C is approximately 0.68, and the equilibrium mass fraction of heptane at 15°C is approximately 0.12. Equilibrium mass fraction is defined as the mass fraction of the gas that is the fuel vapor at its vapor pressure for a specific temperature.

The dispersion simulation was run for 21 days, and the concentration was observed. Figure 15 shows the results for these simulations after three weeks. As this figure shows, the mass fractions for both heptane and pentane remain very close to their equilibrium mass fractions. This indicates that very little of the fuel vapors will escape from a 2 cm hole in the side of the pontoon.

The next simulations that GexCon performed were to estimate what would happen if the cover to the pontoon were removed for a few hours, as it was alleged that the welding contractors had removed the covers to the pontoons several hours before the explosion. The results of these simulations are shown in Figure 16. As the figure shows, the concentration of fuel vapor decreases with the top of the pontoon removed; however, there remains a flammable concentration after three hours. The LEL for both heptane and pentane is about 0.04 by mass fraction.

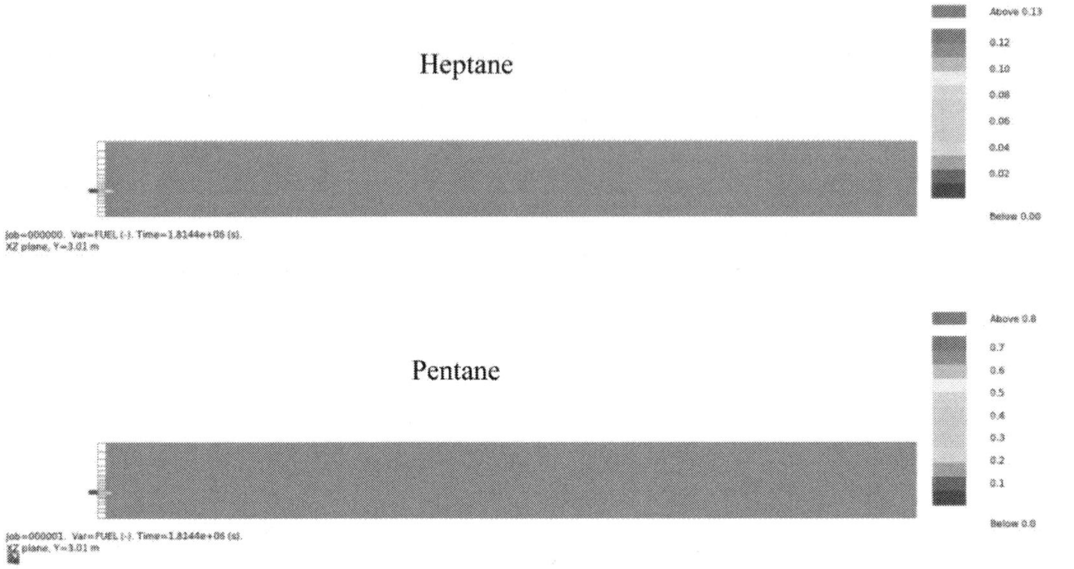

Figure 15: Mass fraction for heptane (top) and pentane (bottom) after three weeks of simulation time.

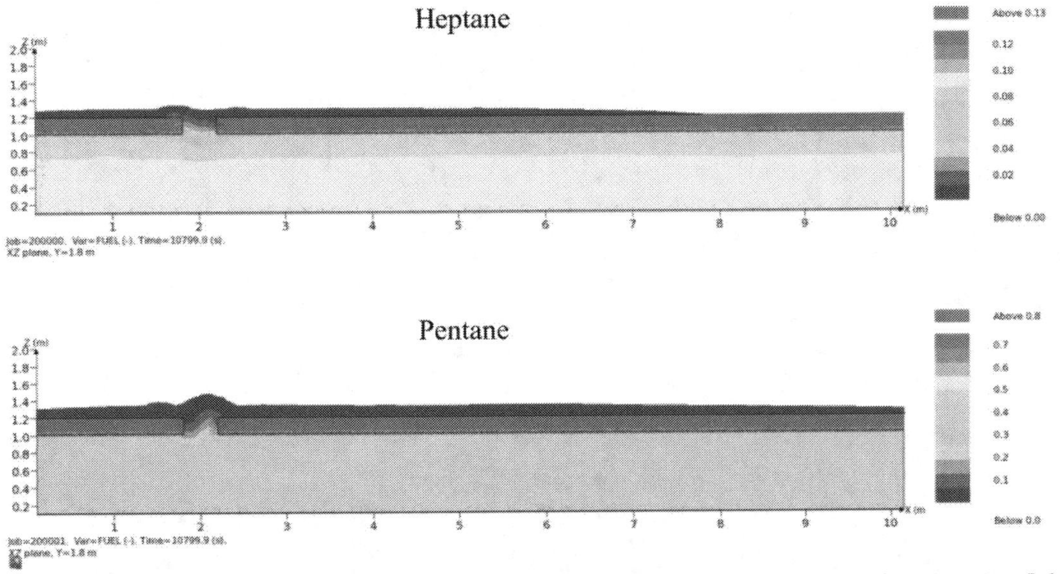

Figure 16: Mass fraction for heptane (top) and pentane (bottom) three hours after the top of the pontoon is removed.

These results show that, despite the actions performed by the workers to prepare the tank for hot work, it was possible for flammable vapors to have remained in one or more pontoons up until the welding contractors started their hot work. It seems likely that liquid gasoline could have been introduced to the pontoon via the hole, and vapor would have remained in the pontoon for the few weeks after the tank had been drained.

3.2 Explosion Simulations

The tank geometry for the explosion simulations is shown in Figure 17

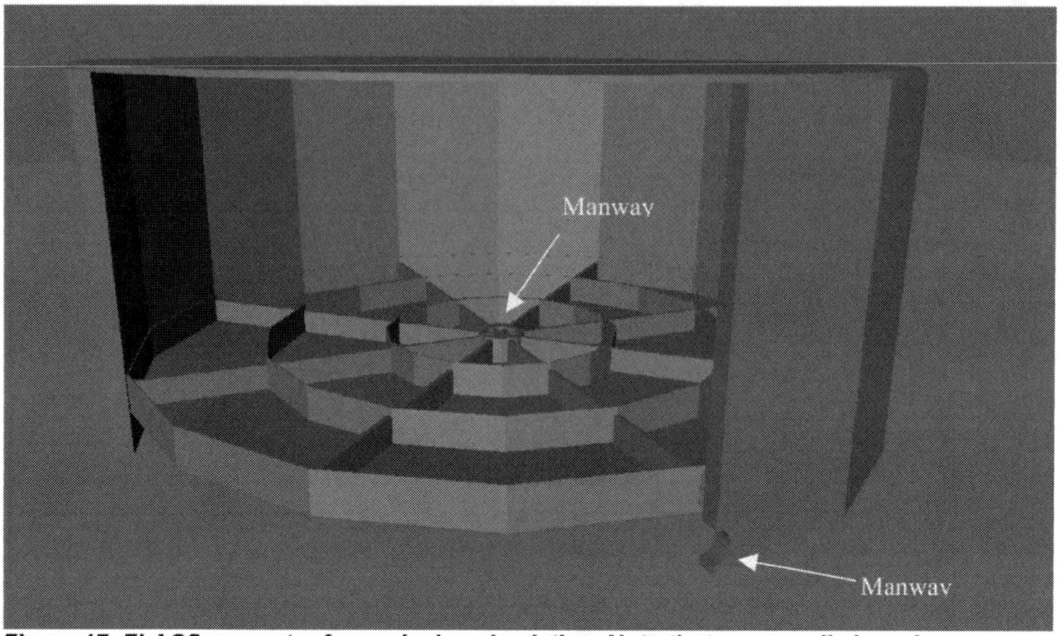

Figure 17: FLACS geometry for explosion simulation. Note that some walls have been removed to show the interior of the floating roof.

Three explosion scenarios were investigated: explosions of a vapor cloud underneath the floating roof, a vapor cloud above the floating roof, and a cloud inside two pontoons of the floating roof. As the pontoons in the floating roof were not hermetically sealed from each other, the explosion study evaluated a vapor cloud with a volume of 3885 ft^3 (110 m^3), which is approximately equal to the volume of two external pontoons.

While no vapors were detected within the upper and lower levels of the tank, the vapor clouds that were located underneath/above the floating roof were chosen to be larger than those within the pontoons (e.g., 5720 ft^3 or 162 m^3). To limit the amount of venting between the upper and lower levels, the clouds underneath/above the floating roof were located to one side of the manway in the floating roof. As the hot work was being conducted in the location of the external most pontoons, the cloud within the pontoons was largely located in two neighboring external pontoons.

As an initial estimate, the top and bottom plates of the floating roof were allowed to give way at pressures that were adequate to lift the individual plates. The entire floating roof was estimated to be approximately 153,400 lbs., or 76.7 tons, which would result in a lifting pressure of approximately 0.14 psig or 9.4 mbar. For these simulations, the failures of the top and bottom plates of the floating roof were considered individually. That is to say, each plate will fail individually, rather than lifting the entire floating roof. They were each estimated to be 6025 lbs., or 30.13 tons, resulting in a failure pressure of 0.05 psi or 3.7 mbar. For the explosions with the cloud beneath the floating roof, this assumption may be adequate; however, for clouds above the roof, the failure pressure may be larger due to the structural support of the baffles and legs of the roof. For clouds within the pontoon, this assumption will be slightly low for the top plate as it does not account for the failed weld, and it is also low for the lower plate due to the structural support of the roof legs. While the exact failure pressure of the top/bottom may not be known, the goal of the simulations was to evaluate the dynamic loading of the various surfaces and compare that to the known response of the failed components.

To measure the forces on the top and bottom plates of the floating roof, a tool available in FLACS was used. This tool is called Pressure Relief Panels, which are simulation elements that measure a pressure difference across an infinitely thin surface. These panels may be set to fail at a particular pressure, causing an area that is blocked to be cleared of obstruction. Four panels were placed on the top and bottom plates of the floating roof. The failure pressure for the geodesic dome was estimated to be 0.7 mbar, based on the force needed to lift the aluminum structure. As residual gasoline was the likely fuel source; heptane was used for the vapor in the explosion simulations.

Results for these simulations are shown below. Figure 18 shows the results of the flame propagation during the deflagration for the cloud ignited underneath the floating roof, while Figure 19 shows the flame propagation during the deflagration for the cloud ignited on top of the floating roof. Figure 20 shows the flame propagating beneath and through the failed top plate during the deflagration for the cloud ignited within the floating roof, inside one of the pontoons.

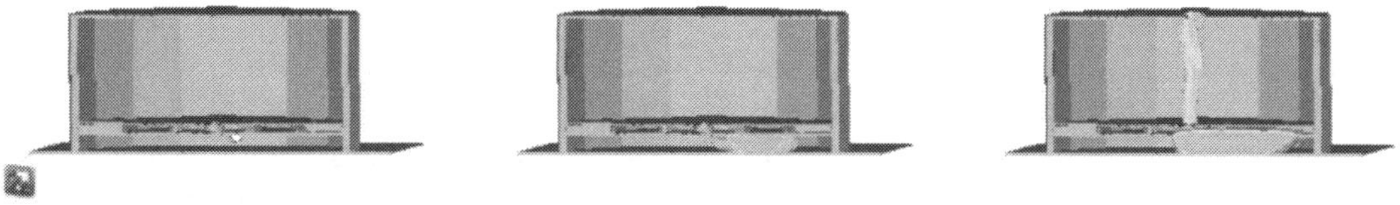

Figure 18: Flame propagation at 0.1 seconds (left), 0.5 seconds (middle), and 1 second (right) for ignition of cloud underneath the floating roof.

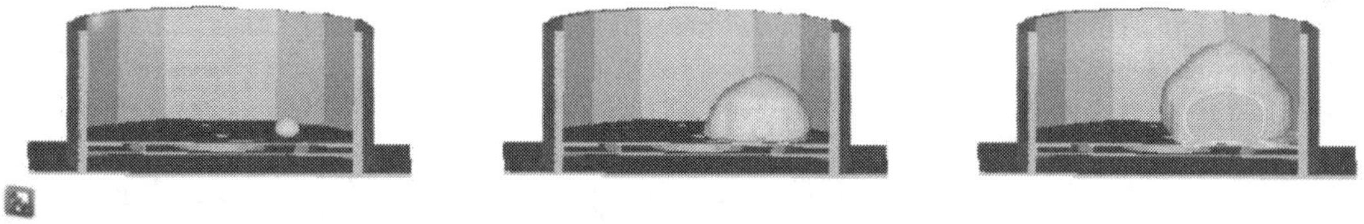

Figure 19: Flame propagation at 0.1 seconds (left), at 0.5 seconds (middle), and 1 second (right) after ignition for cloud above the floating roof.

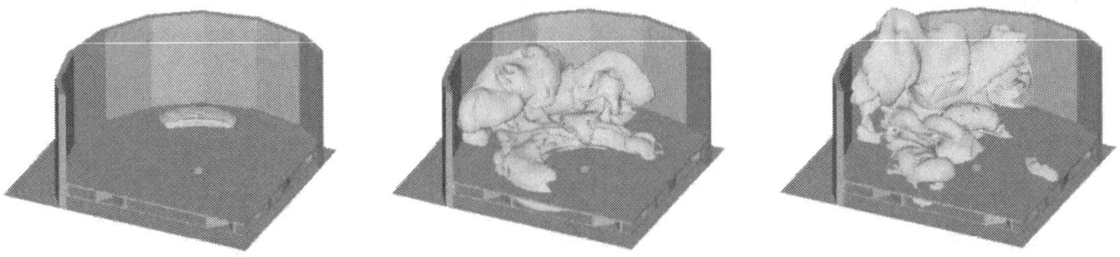

Figure 20: Flame propagation below and through the failed top plate of the floating roof after 0.2 seconds (left), 0.5 seconds (middle), and 1 second (right) for the cloud ignited inside external pontoon of the floating roof.

Generally, when a vapor cloud is ignited within a confined tank or vessel, the flame propagates through the flammable mixture causing the pressure to increase within the vessel due to the formation of hot combustion products. If the pressure within the vessel exceeds the strength of certain components, the explosion will vent through the openings. The momentum of the venting gases can reach a point where the tank becomes "over vented", creating a vacuum or negative pressure within the tank. Since it is challenging to output the absolute minimum negative pressures in FLACS, they will only be approximated in this study. That is to say, the minimum pressures presented will represent a time step in the simulation where the largest number of cells is at their minimum pressure.

Figure 21 shows maximum observed pressures for the simulation with the cloud underneath the floating roof. Figure 22 shows the same plot for the cloud above the floating roof, and Figure 23 shows this for the cloud inside a pontoon of the floating roof. For the vapor clouds ignited above and below the floating roof, the maximum pressures do not exceed 0.2 barg (2.9 psig) within the tank due to the relatively large, unobstructed volume, which readily vents the hot combustion gases through the open manways and failed geodesic dome. The maximum pressure in the pontoon cloud is over 0.4 barg inside the pontoon, which would be more than adequate to cause the damage observed in the accident (i.e., the shearing of the section of the pontoon where the explosion occurred from the rest of the top plate). Lastly, the explosion within the pontoon is the only scenario that could explain the observations of the top plate being displaced upward, but the bottom plate was forced downwards. Ignitions below the floating roof would displace the entire roof upward, while ignitions above the floating roof would displace the entire roof downward.

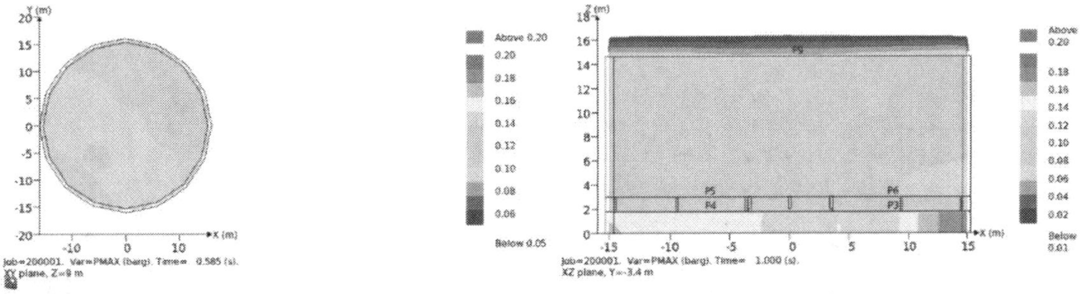

Figure 21: Maximum pressure in for the cloud underneath the floating roof, shown in two cut-planes.

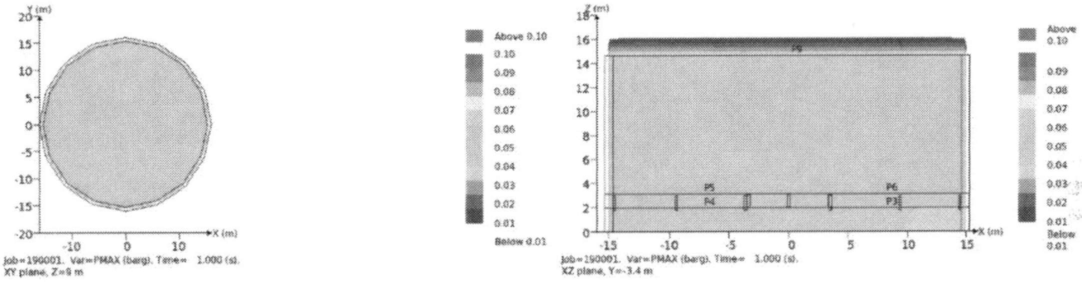

Figure 22: Maximum pressure in for the cloud above the floating roof, shown in two cut-planes.

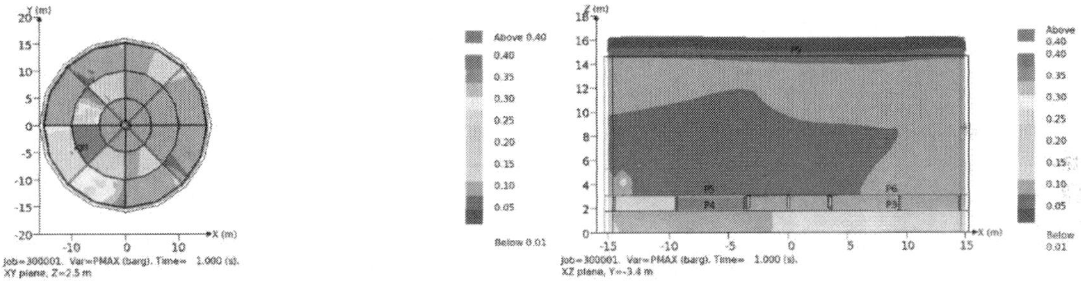

Figure 23: Maximum pressure in for the cloud inside a pontoon of the floating roof, shown in two cut-planes.

The minimum pressures observed in the tank are shown in Figure 24, Figure 25, and Figure 26. For all three cases, the negative pressure inside the tank is below -0.0172 barg, and thus all could have caused the collapse of the tank. It is worthy of note that the lowest pressures occur for the cloud located inside the pontoon, with pressures below –0.05 barg.

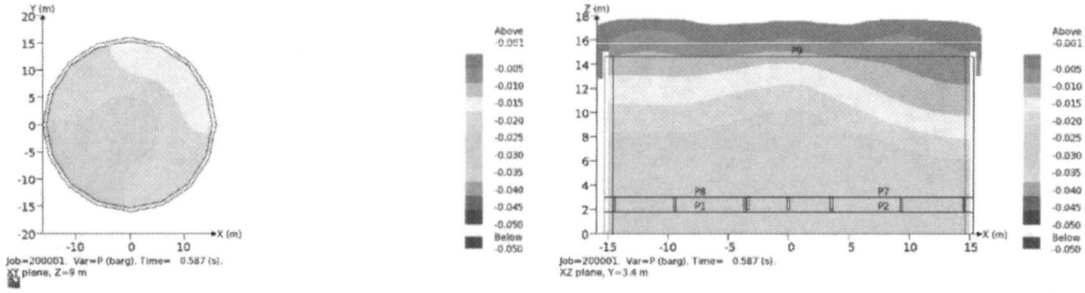

Figure 24: Minimum pressure for simulation with cloud underneath floating roof, shown in two cut-planes, top view (left) and side view (right).

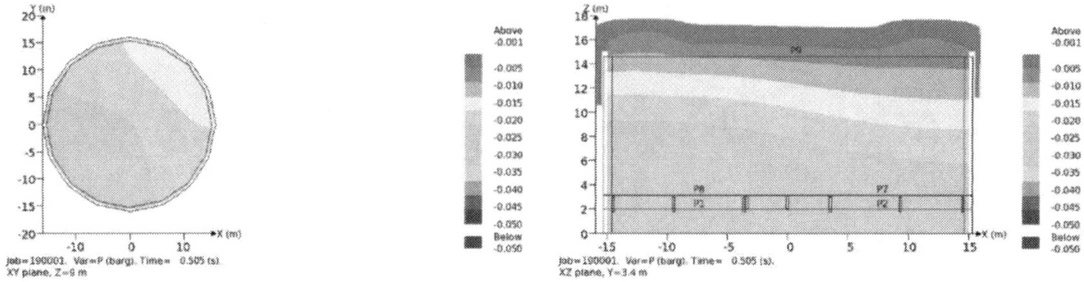

Figure 25: Minimum pressure for simulation with cloud above floating roof, shown in two cut-planes, top view (left) and side view (right).

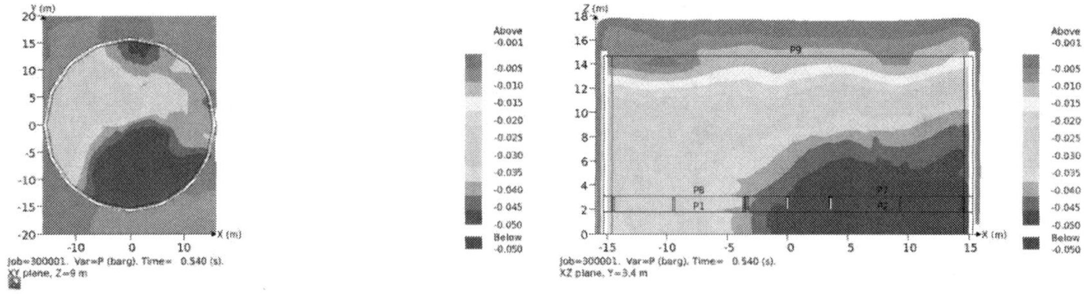

Figure 26: Minimum pressure for simulation with cloud inside pontoon of floating roof, shown in two cut-planes, top view (left) and side view (right).

CFD simulations confirmed observations of the resulting blast damage: a vapor cloud within the pontoons of the floating roof was ignited. This most likely scenario that occurred on the day of the incident was: (1) gasoline product or vapors seeped into one or more of the external pontoons; (2) the residual gasoline product or vapors were not removed from the pontoons prior to the hot work operations; (3) the welding company began hot work on the pontoon(s) with a flammable gasoline/air mixture inside; and (4) the resulting deflagration and explosion tore open the pontoon and split the top surface of the floating roof from the bottom, hurling the bottom of the floating roof away from the pontoon where the explosion initiated.

An oxyacetylene torch has a significantly higher likelihood of igniting fuel-air mixtures than mechanical sparks, because it is a much stronger ignition source. As the control valves for the oxyacetylene torch were found in the "on" position, the residual vapors in the pontoon were likely ignited by the cutting torch. However, sparks from the grinding wheel cannot be completely ruled out as a potential ignition source.

4. Conclusions

An accident occurred while a welding company was performing hot work inside a gasoline storage tank. Since it had been cleaned, the gasoline tank was classified as a non-permit required confined space under OSHA 1910.146. Hence, there was no requirement for an attendant at the manway or for a rescue crew to be available. However, atmospheric monitoring should have been conducted continuously. After an initial reading on the morning of the accident, no other values were recorded for the LEL and the oxygen concentration.

While it seemed that all precautions had been taken for a safe work environment, there were some major failures in the preparation for hot work. Most notably, the LEL was measured inside the tank, but not inside of the pontoons of the floating roof. Since most components of gasoline are significantly heavier than air, the chances of measuring any flammable gases, even when the covers to the pontoons have been removed, are severely impaired. LEL readings should be taken in all areas where vapors may accumulate. The tank had been inadequately cleaned, as the inside of the pontoons should have been scrubbed, or, at the very least, ventilated for longer than the few hours that the welding crew had allowed.

9TH GLOBAL CONGRESS ON PROCESS SAFETY

Explosion from a Smoldering Silo Fire

Russell A. Ogle,
Scott E. Dillon, and Mark Fecke
Exponent, Inc.
4580 Weaver Parkway, Warrenville, IL 60555
rogle@exponent.com, sdillon@exponent.com, and mfecke@exponent.com

Prepared for Presentation at
American Institute of Chemical Engineers
2013 Spring Meeting
9th Global Congress on Process Safety
San Antonio, Texas
April 28 – May 1, 2013

UNPUBLISHED

Explosion from a Smoldering Silo Fire

Russell A. Ogle,
Scott E. Dillon, and Mark Fecke
Exponent, Inc.
4580 Weaver Parkway, Warrenville, IL 60555

Keywords: smoldering fire, agricultural silo, carbon monoxide, self-heating, explosion

Abstract

Many organic granular materials are susceptible to self-heating when stored in silos. Under the right environmental conditions self-heating can lead to spontaneous ignition and smoldering combustion. This is a case study of an explosion that occurred in a grain storage silo with a smoldering fire. Although the hazards of combustible dust are relatively well recognized, the explosion hazard presented by a smoldering fire is less well known. There were two explosions in this incident: a primary explosion involving carbon monoxide and smoke generated by the smoldering fire, and a secondary explosion fueled by combustible dust. The explosion caused both injuries to personnel and significant property damage. This paper discusses the causal factors for the explosion and the lessons learned for effectively monitoring and responding to a smoldering fire in a silo. The potential explosion hazard of the carbon monoxide and smoke generated by a smoldering fire is emphasized.

1. Introduction

Many organic granular materials are susceptible to self-heating when stored in silos. Under the right environmental conditions self-heating can lead to spontaneous ignition and smoldering combustion. This is a case study of an explosion that occurred in a grain storage silo with a smoldering fire. Although the hazards of combustible dust are relatively well recognized, the explosion hazard presented by a smoldering fire is less well known. There were two explosions in this incident: a primary explosion involving carbon monoxide and smoke generated by the smoldering fire, and a secondary explosion fueled by combustible dust. The explosion caused several injuries to personnel and significant property damage.

The subject facility stored agricultural grain products in a concrete grain elevator comprised of several individual conventional silos. Material was moved into the grain elevator silos from railcars, trucks or barges using unloading stations, conveyors and a bucket elevator. Below the grain elevator was an access tunnel with discharge chutes to a conveyor. The silos were provided with aeration fans that could be used to provide a flow of air through the material to cool and dry hot, moist processed material.

One of the operators discovered the fire by detecting a distinctive burnt odor in the access tunnel. Several silos were examined before it was determined which silo was the source of the odor.

Subsequent investigation the next day revealed several unmistakable signs of combustion: burnt odor, light smoke, elevated temperatures measured by an optical pyrometer, some charred and discolored grain product material sampled from the silo, decreased oxygen concentration and elevated levels of carbon monoxide inside the silo. That evening one of the operators opened the roof hatch to measure the temperature of the silo contents and noted glowing coals on the surface of the grain product. Unsure of how to proceed, the facility owner attempted to suffocate the fire by preventing the intrusion of air into the silo. No attempt was made to consistently and effectively monitor the progress of this action.

Eventually, after several weeks, a grain salvage company was hired to remove the grain product from the silo. This was a process that was to take several days. During this time the rooftop hatch and the tunnel chute were often both open creating a chimney effect which accelerated product combustion. Water was frequently sprayed on smoking material from both the top of the silo and from the tunnel. On the day of the explosion, small flames were observed at an aeration vent. The tunnel hatch was closed but the rooftop hatch remained open. The explosion occurred half an hour later. Almost two months had elapsed after the discovery of the smoldering fire.

This paper discusses the causal factors for the explosion and the lessons learned for effectively monitoring and responding to a smoldering fire in a silo. The potential explosion hazard of the carbon monoxide and smoke generated by a smoldering fire is emphasized in the context of the special hazards of silo fires. Finally, strategies for prevention and suppression of silo fires are presented.

2. Accident Summary

The subject grain facility is located on the bank of a river and is adjacent to a railroad line. The facility stores various agricultural grain products. The facility consists of an office and a grain elevator. The exact age of the facility is unknown, but some construction drawings date back to the early 1960's. The elevator is a concrete structure with several conventional silos (not oxygen-limiting). The facility can receive or ship grain via truck, rail, or barge. The grains are conveyed to the rooftop of the elevator where they are distributed to the various silos via a drag conveyor. The silos have a nominal diameter of 25 feet and are approximately 80 feet in height. Two separate dust collections systems are used throughout the facility. The first system handles the barge conveyor and truck pit conveyor. The second system handles the drag conveyors on the rooftop, the tunnel and the elevator as well as the silos. Schematics of the silos can be seen in Figure 1 and Figure 2.

Facility personnel first documented evidence of a fire in Silo 1 early in the year. This evidence consisted of smoke, an unusual odor, an elevated temperature measurement, discolored grain and glowing embers. Low oxygen levels and high carbon monoxide levels were also measured in the headspace of the Silo 1. After taking these measurements, facility personnel intermittently monitored the temperature of the silo for three weeks. Several facility employees observed sporadic smoke and odor. Within a week or two of taking the initial temperature measurements, the smoke and temperatures declined. Facility employees then noticed an increase in both temperature and smoke emission several weeks after the initial observations.

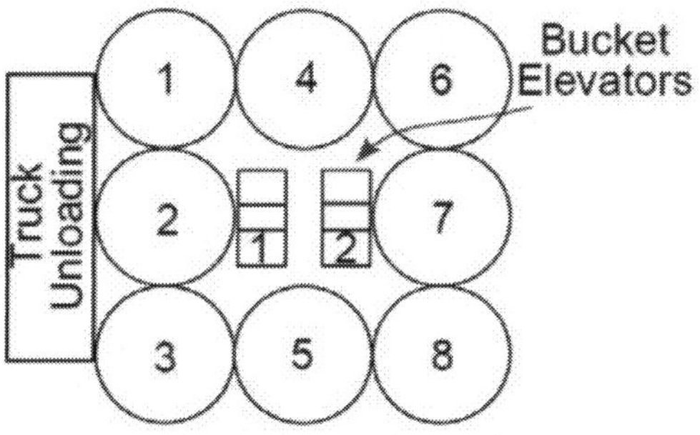

Figure 1. Top View of Grain Elevator and Silos

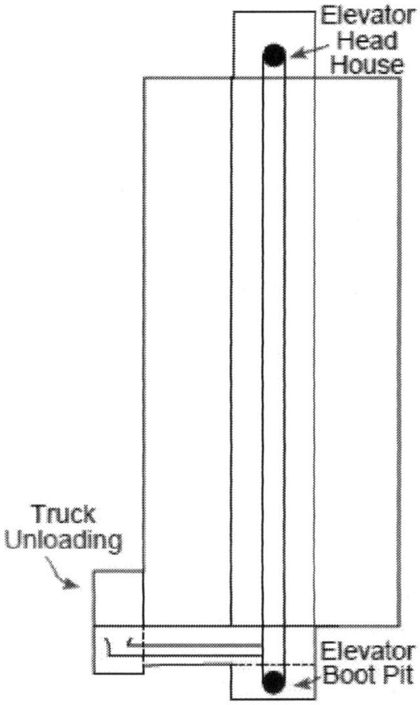

Figure 2. Side View of Grain Elevator and Silos

The facility then contracted a salvage company to remove the material from the silo as well as agglomerated grain from the walls of the silo. Material was being removed from the bottom of the silo through a hatch in the access tunnel. As the crew for the salvage company began removing grain from the silo, they noticed both an unpleasant odor as the grain was being removed in large chunks. During the grain removal process, water was frequently sprayed on smoking material from both the top of the silo and from the tunnel. Within several day of the contractor beginning to remove the material, the crew noticed an increase in temperature and odor. The crew was equipped with a 4-gas monitor that did alarm on carbon monoxide (CO) when the monitor was placed inside the hatch in the access tunnel from which they were removing the grain. This hatch was internal to the concrete elevator structure.

Within a week after the salvage company started removing grain from the silo, the removal process slowed due to the presence of hardened material. After returning from a break one day, the salvage contractor employees noticed that the smoke emission from Silo 1 significantly increased. The contractor employees then opened a manway on the side of the silo, near the aeration fan, and attempted to dislodge the hardened material inside the silo while smoke continued to pour out of the top of the silo. The salvage company then applied water to the grain inside the silo from a hatch on the roof using a garden hose. A timeline of the key events is seen below.

Table 1. Timeline of Key Events

Date	Event
Early May	Facility personnel noticed grain leaving Silo 1 was coming out discolored
Middle of May	Smoke, an unusual odor and elevated temperature measurement were coming from Silo 1
Middle of May	Low oxygen levels and high carbon monoxide levels were measured in the headspace of Silo 1
Middle of May - Early June	Facility personnel intermittently monitored the temperature of the silo. Water was frequently sprayed on smoking material from both the top of the silo and from the tunnel. Sporadic smoke and declining temperatures were noted throughout this period
Early June	Facility personnel noticed an increase in both Silo 1 temperature and smoke emissions for several weeks
Middle of June	Facility contracted a salvage company to remove material from the Silo
End of June	Salvage company personnel 4-gas meter alarmed on CO when placed inside the silo. Large chucks of grain were being removed.
Early July	Smoke emissions significantly increase. Salvage company opened manway near aeration fan trying to dislodge material while smoke poured out of the top of the silo. A garden hose was used to apply water to the silo from a hatch on the roof.

Just prior to the explosion, the salvage company was instructed to remove all the tools from the interior of the elevator access tunnel. To accomplish this, workers entered the access tunnel to remove the tools. During this process, two explosions occurred inside the grain elevator injuring several people, including the workers in the access tunnel. The first explosion occurred in Silo 1 and a secondary explosion propagated through one of the two bucket elevator legs and was released through the elevator boot into the access tunnel where the workers were located. Upon arrival to the scene by the Fire Department, they noted smoke venting from the northwest silo and roof structure.

3. Fire and Explosion Dynamics Analysis

Combustible Dust Fires and Explosions versus Smoldering Fires

Combustible grain dust fires and explosions in silos have been a significant concern for over a hundred years. NFPA 61 is the standard for the prevention of fires and dust explosions in agricultural and food processing facilities and was developed in 1923 to prevent dust explosions in grain terminals and flour mills. [1] NFPA 61 provides numerous requirements for preventing combustible dust explosions. However, NFPA 61 does not provide specific guidance on the control and extinguishment of smoldering fires.

Occupational Safety and Health Administration (OSHA) regulation 29 CFR 1910.272 covers grain handling facilities and provides requirements for the control of grain dust fires and explosions. While the CFR regulation provides requirements for issues such as providing training to employees regarding grain silo fire hazards, hot work requirements, and housekeeping, the CFR does not provide guidance on fire-fighting or how to suppress or control a fire in a grain silo, including a smoldering fire.

Furthermore, most of the standards and literature related to silo fires and explosions focus on the combustible dust hazards and provide limited guidance or requirements on dealing with the explosion hazards that are generated by a smoldering fire. Extensive guidance is available on limiting combustible dust emissions and accumulations, limiting ignition sources, and protecting against the results of a deflagration by isolation, suppression and/or venting. However, the unique hazards that are present as a result of the gases generated during a smoldering fire receive far less attention in the standards and literature. Some case studies of smoldering gas explosions are reported by Eckhoff. [2]

The Smoldering Fire

The two most common causes of smoldering fires in silos are (1) self-heating that results in spontaneous combustion and (2) the transport of a burning ember generated during material processing. For a material to undergo self-heating it must be capable of undergoing an exothermic reaction and for solid materials it usually needs to be in a porous or granular form. Most agricultural products and organic materials can typically undergo some form of self-heating when they are in a granular form. Material self-heating resulting in a smoldering silo fire requires the following: [3]

- *The material must exhibit self heating,*
- *Self-heating must reach a runaway, or critical, condition, that is, the reaction must rapidly accelerate so that a high temperature is reached,*
- *The runaway self-heating must start sustained smoldering,*
- *The sustained smoldering must reach the outside surface of the material and possibly erupt into flaming.*

The development of a self-heating reaction that transitions into a fire is presented in Figure 3. Whether or not a self-heating material can reach a thermal runaway condition is dependent on several factors including the overall size of the material mass, the material and ambient temperatures, the boundary conditions (the walls), ventilation and availability of oxygen, size of the particles, moisture content of the material as well as other factors and is beyond the scope of this discussion. According to Krause, et al, "The characteristic range of temperature for a self-ignition process turning into a smoldering fire is from 150 to 400°C. A range of temperature from 250 to 500°C indicates a propagating smoldering fire with the temperature being the higher the more oxygen is available for combustion." [3] Once initiated, a smoldering fire within a silo of material can go unnoticed for weeks or even months.

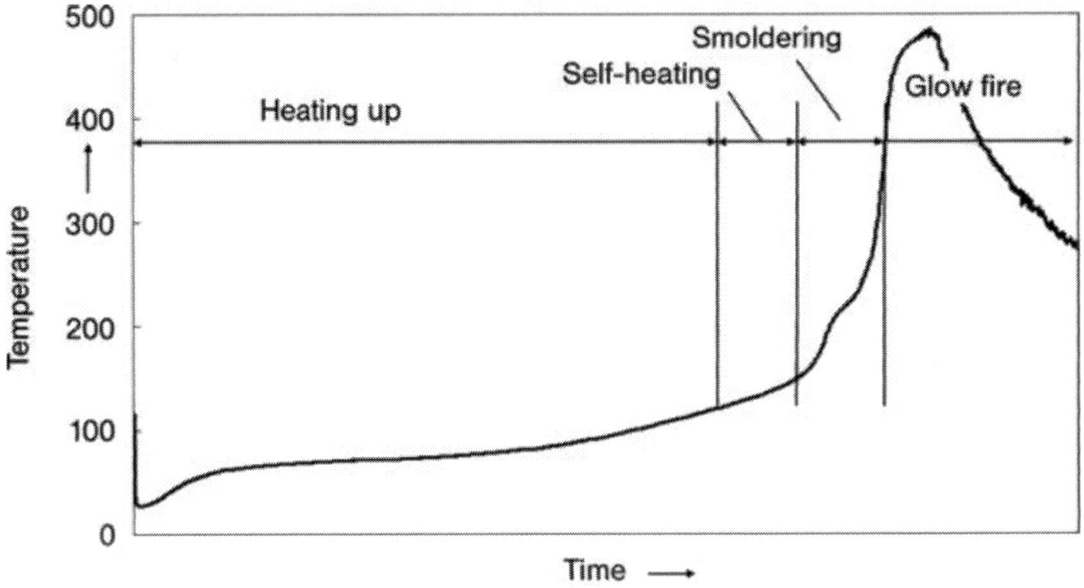

Figure 3. Stages of a fire in bulk material triggered by self-ignition (from Krause, et al) [3]

Burning embers entering the silo can typically be prevented through the use of fast-acting spark/ember detection and extinguishing systems that extinguish the hot ember before they can be introduced into the stored material.

Self-heating must be prevented or controlled through appropriate material processing, transport and storage to control the material temperature and moisture content. Detection of smoldering fires can be achieved through a combination of CO or hydrocarbon vapor detection within the silo, interior material temperature measurements, and infrared monitoring of the silo/collector wall temperature.

In the case study presented, there was a fire in the silo for several weeks leading up to the explosion. The evidence, observed by the facility and contract personnel, consisted of the following: smoke, elevated temperature measurements of the material bed, discolored material (brown and black as shown in Figure 4) accompanied by a burnt odor, a reduced oxygen concentration in the silo, and an elevated carbon monoxide (CO) concentration in the silo. In general, these observations are indicators of fire as defined by both the National Fire Protection Association (NFPA) and American Society of Testing and Materials (ASTM). [4, 5] More specifically, these observations (plus the detection of an unusual or burnt odor) are specific indicators of smoldering combustion in a bed of agricultural grain product. [3, 6]

Smoldering will typically produce glowing embers. Glowing will occur when the smoldering reaction reaches the outer surface of the material or when the material is penetrated to the location where the reaction front is located. [3] Therefore, glowing embers may not be visible if the smoldering wave is below the surface of the fuel and the material has voids or channels present due to material agglomeration ("rat holes"). However, glowing embers, which are indisputable evidence of smoldering combustion, were observed by facility personnel on at least one occasion prior to the explosion. An example of smoldering material and glowing embers in a silo after a silo explosion are shown in Figure 5. In the weeks before the explosion, communications between facility personnel acknowledged that a smoldering fire was present in the silo. These conditions were observed for several weeks prior to the explosion.

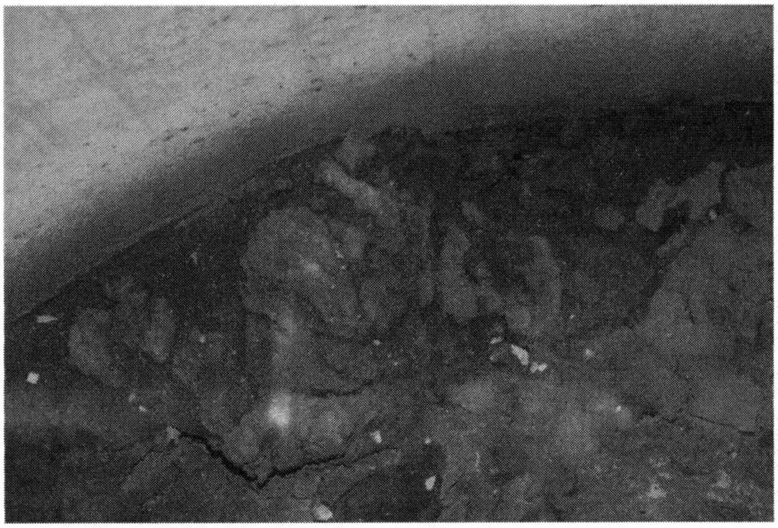

Figure 4. Smoldered material in the silo.

Conditions in the silo described in this case study were conducive for the development of self-heating and a smoldering fire. One important factor was the caking of agglomerated material to the walls of the silos. The material was sticking to the walls of the silos and "clumping" (agglomerating). This was a common occurrence in the silos and was attributed to the elevated moisture content and temperature of the material during production and transfer of the material to the silos.

Clumping is caused by loading the material into storage at too high of a temperature and with too much moisture. The high temperature and moisture conditions of the material not only increase their propensity for agglomeration, they also promote the growth of mold. For many organic materials such as agricultural grains, the initial self-heating step is microbiological growth. In one investigation it was found that pelletized wheat caked on the side of a silo contained on the order of at least 10 to 1,000 times the amount of mold species as the uncaked pellets. [7] Mold growth leads to biological heating which is the first step towards a silo fire. [8] Controlling temperature and moisture of materials during production, transfer, and storage can be essential to reducing the risk of smoldering fires in silos. The process of self-heating can only occur if the material were allowed to remain idle in the silo for an extended period of time, on the order of several weeks, which was what occurred in this case study.

Figure 5. Smoldering material and glowing embers after a silo explosion.

Conditions for an Explosion

There were essentially two explosions in this accident: the primary explosion in the silo and a secondary explosion that propagated through one of the two elevator legs and was released through the elevator boot into the access tunnel. In evaluating the source of the primary explosion three factors had to be considered: the fuel, the ignition source, and ventilation (the source of air/oxygen). Together these three factors comprise the fire triangle.

For the secondary explosion two additional factors had to be considered: dispersion of the fuel and confinement. These additional factors combined with the factors comprising the fire triangle make up the combustible dust pentagon.

Fuel

There were four potential fuels involved in the primary explosion: carbon monoxide (CO), smoke, "water gas" and combustible dust from the stored material. Efficient burning that has the proper ratio of fuel to oxygen (air) to allow a stoichiometric reaction typically produces carbon dioxide (CO_2) and water (H_2O) as the primary products of combustion. On the other hand, smoldering combustion tends to manifest itself as an under-ventilated fire that does not have sufficient oxygen available for efficient combustion. Under-ventilated fires generate large quantities of carbon monoxide (CO). Carbon monoxide can form a flammable atmosphere over a wide concentration range: between 12.5% and 74% by volume. In the incident presented above, the presence of carbon monoxide was demonstrated by measurements of the atmosphere within the silo. The measurement indicated that the CO concentration was significantly above the normal atmospheric levels and was increasing. This smoldering combustion process was also evident as an oxygen-deficient atmosphere – the oxygen concentration in the silo was measured to be approximately 30 – 40% below normal atmospheric concentrations.

The generation of smoke, the second potential fuel, was evident on several occasions. Smoke is a product of combustion, and smoke from incomplete combustion can contain carbon particles and unburned fuel gases that can serve as fuel in a deflagration. The existence of smoke within the silo was detected by plant personnel, both visually and by smell, for several weeks leading up to the explosion.

An additional potential fuel was "water gas." The term "water gas" refers to the generation of carbon monoxide (CO) and hydrogen (H_2) due to the application of water to a smoldering fire in a silo. According to the National Grain and Feed Association (NGFA) Emergency Preplanning and Firefighting Manual for grain elevators: [9]

> *Although the most common approach to a fire is to apply water, there is some hazard in using water, particularly in bin fires. A danger of using water on a hot grain fire in a confined space is the possibility of the water reacting with carbon, a product of combustion, to form a potentially explosive concentration of carbon monoxide and hydrogen known as "water gas."*

NFPA 61 provides the following explanation of the water gas reaction:

> *Application of small amounts of water on glowing grain in a partially confined space, such as a grain silo, and in the presence of air can generate a water gas reaction. The glowing grain must be at temperatures of at least 700°C to 800°C (1290°F to 1470°F), and initial water contact may not cool the mass of glowing grain below 600°C (1110°F).*
>
> *The partial oxidation reduction between carbon and water forms carbon monoxide and hydrogen as follows:*

$$C + H_2O \rightarrow +CO + H_2$$

In the presence of oxygen (air), the carbon monoxide and hydrogen burn, immediately releasing heat as follows:

$$C + H_2O + O_2 \rightarrow +CO_2 + H_2O + 19{,}440\ J/g\ (8350\ Btu/lb)$$

In a partially confined space, the combustion energy will rapidly pressurize the space beyond what the silo walls or tops can withstand, causing destruction of the silo.

The fourth potential fuel present in the silo was combustible grain dust within the silo. Agricultural products such as that in the incident described above have the potential to generate small particles as they are processed, handled and transferred. An "agricultural dust" is defined by NFPA 61 as, "Any finely divided solid agricultural material 420 microns [0.42 µm] or smaller in diameter (material passing a U.S. No. 40 Standard Sieve) that presents a fire or explosion hazard when dispersed and ignited in air." [1] A more general definition for a combustible dust, as presented in NFPA 654, is "A finely divided combustible particulate solid that presents a flash fire hazard or explosion hazard when suspended in air or the process-specific oxidizing medium over a range of concentrations." [10] To be an explosion hazard, combustible dusts need to be in a dispersed form, which can occur during the pneumatic conveying and discharging of material into a silo. However, any dust generated would tend to settle to the lower depths of the product mass. To participate as a fuel in an explosion, the dust must first be suspended before it is ignited. Just prior to the explosion described in this case study, there were no reports of actions or plant activities that could lead to dust suspension within the silo. Thus, although dust may have played a role in the secondary explosion within the bucket elevator, it is unlikely that it played a role in the primary explosion.

It is concluded that the fuel for the primary explosion was some combination of carbon monoxide, smoke, and hydrogen (in the form of water gas) forming a flammable atmosphere due to the smoldering material within the silo. It is typical for accidents of this nature to be attributed to combustible dust, and therefore the hazards of a smoldering fire, especially one that persists for some time, are often overlooked. For example, a method for minimizing dust cloud formation is to wet the surface of the material, as was done in this case study. However, the application of a relatively small amount of water to a smoldering organic material will often not cool the heated mass and can cause the material to undergo a biological degradation that can generate flammable gasses.

Ignition Source

As a result of the incident investigation it was determined that there were few competent ignition sources inside the silo. There were no moving parts that could produce mechanical friction, no electrical components that could malfunction, no hot work activities, and no signs of anyone using smoking materials in the vicinity. The only plausible ignition source was the self-heating and spontaneous ignition of the stored material. The self-heating process begins with biological heating caused by respiration of molds in the material. As the temperature of the bulk mass increases, chemical oxidation begins and generates heat at a faster rate. Eventually, if the temperature continues to rise, a smoldering combustion wave is formed in the product mass.

Smoldering produces glowing embers with a typical temperature between 600 to 750°C (roughly 1100 to 1400°F). [11] Glowing embers can become exposed to the atmosphere within the silo if the smoldering wave approaches the surface of the pellet mass or due to the removal of material covering the glowing mass. The glowing embers may persist or they may transition to flaming combustion. Once the smoldering combustion wave reaches the fuel surface, transition to flaming combustion can occur.

Ventilation

The progress of the smoldering wave in the stored material was influenced by the ventilation conditions in the silo in the weeks, days and hours leading up to the explosion. It is possible to quench a smoldering reaction by starving it of oxygen. In this incident, the facility took steps to try and seal the silo, but the construction of the silo was not gas-tight and there was interconnection between the silos – the silos were connected to one another by vents at the junction between the top of the silo and the concrete roof. Therefore, some infiltration of fresh air and exfiltration of combustion products (observed as smoke and odor) occurred. Furthermore, during the multi-week duration of the fire, the facility had run several pieces of process equipment and pneumatically conveyed product to the adjacent silos, thereby introducing additional air into the silo structure. The process also involved the running of a dust collection system connected to the silos, including the subject silo that would have pulled fresh air into the silos. Although the dust collection pipe to the subject silo had been blocked off, the operation of the dust collector would have increased the ventilation through the adjacent silos due to the interconnection of the silos.

Furthermore, some of the silos, including the subject silo, were provided with an aeration system to reduce the moisture content of the material. The system consisted of a perforated pipe that extended onto the lower portion of the silos which was connected to an external fan and air inlet. The investigation revealed that the aeration system was rarely used to dry the materials. As part of the attempts to seal the silo, the air inlet was covered with a sheet of plastic, however a nearby auxiliary air inlet had not been covered. During the weeks leading up to the explosion the plastic sheeting had been observed to melt and the seal had become compromised allowing fresh air into the silo and for the smoldering reaction in the material to continue.

Finally, just prior to the explosion, the lower manway on the bottom of the silo was opened in order to dislodge agglomerated material. The opening of the manway increased the natural draft (chimney effect) of air into the silo and accelerated the fire.

Although the facility had attempted to seal the silo, various circumstances allowed air to infiltrate the silo and provide oxygen to the smoldering material. These circumstances are attributed to continuing to operate the plant and convey material to nearby silos, limited knowledge of the silo construction, and an ineffective fire suppression strategy, which is addressed further in the Suppression Strategies section.

Explosion Dynamics

The primary explosion occurred when the flammable atmosphere inside the silo ignited. The resulting deflagration damaged the concrete roof of the grain elevator and the concrete structure of the bucket elevator and caused a secondary explosion in the elevator boot. The most probable cause of the primary explosion within the silo was the ignition of a flammable atmosphere of carbon monoxide, hydrogen and smoke ignited by either glowing embers or flames at the material bed surface. The concrete roof of the entire silo structure was damaged at the edges and concrete was cracked and spalled at locations of piping penetrations. This damage indicated that the roof was at least partially lifted by the impulse of the primary explosion and then dropped as the combustion product gases vented from the silo. The mass of the concrete roof was estimated to be between 80,000 and 120,000 pounds. The deflagration pressure within the silo could easily achieve a peak pressure between 80 and 150 pounds per square inch gauge (psig). Neglecting the effects of venting, this translates into a peak lifting force of between 1.5 and 2.8 million pounds force – more than sufficient potential to lift the entire roof of the silo.

The primary explosion propagated a pressure wave along two different paths through connected vent ducts. One path was the ducting for the silo dust collector. The pressure wave caused the explosion relief panel on the dust collector to function. The second path entered the vent ducts for the bucket elevator, fueled by combustible duct in the ducts and in the elevator itself dispersed by the primary explosion. The pressure wave triggered a secondary explosion in the boot of one of the bucket elevators resulting in a deflagration overpressure that ruptured the elevator boot allowing flames and hot gasses to enter the access tunnel, where the injured workers were standing. The overpressure was caused by the secondary explosion in the elevator leg.

Secondary explosions are a common feature of explosion events in a combustible dust environment. Grain elevator legs are especially susceptible to grain dust explosions due to the potential for combustible dust accumulation in the elevator boot. Despite the clear evidence of the overpressure in the elevator boot, no evidence of heat damage to the rubber belt or any of the plastic buckets was observed indicating an extremely rapid flame front velocity that did not have time to damage the plastic components. Similarly, no explosion damage was evident in the tunnel.

Following the explosion, the material in the silo continued to smolder and burn as is evidenced by the continued emission of smoke for several hours. Firefighters were finally able to suppress the fire by removing the material, spraying the exposed material with water and spraying water through the open manway at the bottom of the silo.

4. Prevention and Monitoring

The first step to preventing a smoldering fire gas explosion is to prevent a smoldering fire from occurring in the silo. As mentioned above, the two most common causes of smoldering fires in silos are self-heating that results in spontaneous combustion and the transport of a burning ember generated during material processing. Burning embers entering the silo can typically be prevented through the use of fast-acting spark/ember detection and extinguishing systems that extinguish the hot ember before they can be introduced into the material.

Self-heating must be prevented or controlled through silo cleaning, material turnover (i.e. removal of old material), as well as appropriate material processing, transport and storage to control the material temperature and moisture content. Detection of smoldering fires can be achieved through a combination of CO or hydrocarbon vapor detection within the silo, interior material temperature measurements, and infrared monitoring of the silo/collector wall temperature.

Early detection of a smoldering fire is desirable, but difficult to achieve using conventional methods. By the time that smoke or elevated concentrations of (for example) carbon monoxide are detected at the top of the silo, the pyrolysis process has probably been going on for several days. Once a smoldering fire starts in a silo, special suppression strategies need to be implemented, as traditional fire-fighting strategies have been shown to be ineffective and can increase the hazard of an explosion.

5. Suppression Strategies for Silo Fires

While fires and explosions are somewhat common in agricultural silos, the bulk of the literature regarding the hazards tends to focus on the combustible dust hazard. The smoldering fire hazards are less commonly addressed. This section provides a summary of effective strategies and good practices for dealing with smoldering fires in silos as well as a discussion of ineffective practices. Some of these practices, while similar in basic concept, can produce significantly different results in the way in which they are implemented.

Due to the unique nature of a smoldering fire in a silo, special fire suppression strategies need to be utilized. In addition, some traditional fire suppression methods have been determined to have limited effectiveness or can exacerbate the hazardous condition within the silo, and should be avoided. Education, training and a robust emergency response plan with regard to smoldering silo fires is extremely important to the safe mitigation of the hazards.

Water

A traditional fire suppression strategy employed by fire departments and fire brigades is to flood a fire with large volumes of water. The application of water to a flaming fire involving normal combustible materials (wood, paper, grains, etc.) is extremely effective in extinguishing the fire as the heat from the fire is removed by converting the liquid water to huge volumes of steam in a ratio of approximately 1,700 gallons of steam for each gallon of water. The steam also acts to displace oxygen from the fire to assist in extinguishment. Because of its abundance, simple application and tremendous effectiveness, water has been the fire suppression media of choice for most fires, especially those involving ordinary (e.g. Class A) combustible materials.

However, the application of small amounts of water and water spray systems such as sprinklers are completely ineffective against a smoldering fire in a silo. Dousing a silo fire with water is only effective if the water reaches the fire, as with a fire near the surface of the material that can be extinguished with a simple hose stream. For a deep-seated smoldering fire in a silo the application of water, even huge volumes of water, can be ineffective or even cause a more hazardous situation. In addition, the deeper the fire is in the material, the more difficult it will be to suppress with water. Penetration of water into the material is weak. For water that does penetrate the material, the water may evaporate due to the heated material as well as the flow of hot gases, smoke and combustion products without ever reaching the source of the fire. The application of water may wet the top surface of the material; however, the water can find routes and channels through the material to the bottom of the silo without ever reaching the burning material. Continued application of water will fill the silo with water and can lead to the structural integrity of the silo being compromised. The application of water to smoldering material or glowing embers inside a silo can also increase the fermentation process or generate flammable carbon monoxide and hydrogen gas (water gas) as described above.

The use of water on silo fires can have additional negative effects:

- The water spray can disturb and disperse combustible dust, leading to a potential combustible dust flash fire or explosion,
- Swelling of the goods stored in the silo which in turn are pressing open the cell walls,
- The fire extinguishing water does not reach the fire source because the goods stored in the silo are sticking together,
- The injection of an inerting agent at a later date is hindered or even made impossible, and
- The silo cells become statically overloaded due to hydrostatic pressure from the accumulated water.

Foam

There is no evidence to support using foam suppression agents to extinguish a fire in a silo, particularly a smoldering fire. While there is experimental data that indicates that foam can be effective, practical applications during actual fire scenarios have shown a variety of results. The purpose of foam is to provide a barrier that blocks the flow of air into the material from above. However, there may be other routes of air infiltration into the silo and/or there may be an extensive amount of air within the material itself. As with the application of water, foam will most likely not extinguish the fire and could generate an atmosphere within the silo that causes an explosion.

Sealing the Silo

Because of the dangers associated with the application of water on a hot grain fire in a confined space, some silo fires can be more effectively controlled and suppressed by the reduction of the oxygen supply. One method of eliminating the oxygen is to seal the silo. Sealing a silo containing a smoldering fire can limit the amount of air that can reach the smoldering reaction.

In the weeks leading up to the explosion, the facility attempted to seal the smoldering silo to extinguish the fire by consuming the oxygen in the silo while preventing the inflow of fresh air.

They attempted to seal the aeration fan inlet, exhaust and duct collector vents and sealed other places where smoke was exiting the silo or where air could enter. However, the design of the silos prevented the achievement of a completely air-tight seal because each silo was vented to its adjacent silos at the junction between the silo top and the roof. Thus, air could flow between adjacent silos. Spread of smoke and carbon monoxide between the adjacent silos was evident in smoke staining observed in the upper portions of the silos as shown in Figure 6. There is a soot line at a fairly uniform elevation within each of the silos. This is consistent with smoke flowing from the smoldering silo into the remaining silos. Smoke was transported from the silo and accumulated in the other silos via the hydraulic connection. Because the warm smoke was more buoyant (less dense) than the ambient air, the smoke filled each silo from the top down.

Figure 6. Smoke staining on the upper portions of the interconnected silos.

To determine the efficacy of a silo sealing effort, the facility needs to monitor the silo atmosphere. Successful sealing of the silo would be demonstrated by the depletion of oxygen without an increase in carbon monoxide or temperature. In this case study, the facility relied on temperature monitoring instead of atmospheric monitoring to check on the condition of the silo. However, the temperature monitoring strategy that was implemented was unable to determine that the silo sealing effort failed to extinguish the fire.

Oxygen Displacement / Inerting with Gaseous Agents

The principal of inerting a silo is similar to that of sealing the silo, and is designed to dilute and displace the available oxygen within the silo. The smoldering reaction consumes the oxygen in the air filling the voids between the particles. The reaction progresses by fresh air being introduced to the smoldering reaction zone. Sources for the fresh air are from within the material (away from the reaction zone) or from outside the material (i.e. from the empty space above the material, through leaks in the silo wall, through unsealed vents, etc.). The flow of air through the material is typically by diffusion or convection.

The concept of inerting a smoldering silo fire is to replace and/or reduce the available oxygen with a gaseous agent that is incapable of undergoing an oxidation reaction and will not react with the stored material. The most common gaseous agents used for inerting a silo fire are carbon dioxide (CO_2) and nitrogen (N_2), mostly due their availability and cost. The primary difference between the two gases is the density. CO_2 has a density greater than air while N_2 has a density much closer to air, which is comprised of approximately 79 percent nitrogen. Because of the density differences, CO_2 is best applied from the top of the silo where it can sink down through the material while N_2 can be applied from the top or the bottom. The two gases can also be applied simultaneously.

The most important factors for the application of a gaseous agent into the silo is sealing of openings, reduction of oxygen within the entire silo and continuous application of the gas to control and/or suppress or the fire prior to removal of the material. Some references have indicated that it is safe to begin emptying the silo only when the oxygen concentration is reduced to below 8 percent; however, no scientific basis for this claim has been established. [12] Intermediate-scale experiments involving the extinguishment of smoldering fires in silos conducted by Tuomisaari et al. concluded the following: [12]

- Extinguishing smoldering fires in silos is an extremely complicated issue primarily due to the inhomogeneity of the material,
- The application of water should not be used to extinguish silo fires,
- When suppressing smoldering silo fires with an inerting gas, monitoring the suppression progress inside real silos is extremely difficult because the oxygen concentration in the space above the material is not necessarily representative of the concentration within the material,
- Emptying a silo is only safe after the fire has been extinguished, and failed attempts to extinguish silo fires by inerting may be due to the silo being opened too early,
- Other inerting gases, or combinations of gases such as commercially available suppression agents, can be used to suppress silo fires; however, the use of CO_2 and N_2 are the most practical,
- CO_2 was found to be a more effective agent than N_2 due to the higher gas density which causes CO_2 to leak from the silo more slowly and remain in the silo longer than N_2,
- The most effective way of injecting a gaseous agent is from the bottom of a silo due to the upward buoyant flow in the silo,
- Inerting with a gaseous agent is more effective during the early stages of the fire due to the increased buoyant flow that is created by a larger fire, and
- No quantitative guidelines can be recommended to verify the extinguishment of a smoldering fire.

To enable effective inerting, silos should be provided with ports where an inerting gas can be introduced into the silo easily. In addition, means to completely seal the silo as tightly as possible need to be provided to prevent leakage of the inerting gas and to prevent the inflow of fresh air. Plates and/or doors can be used to cover any openings at the bottom of the silo. A source for large volumes of carbon dioxide and/or nitrogen should be identified before a fire occurs. Some facilities have contractors identified that can quickly respond with tankers of inert gas and can begin the introduction of gas into the silo quickly. Controlled emptying of the silo can be performed when temperature measurements and gas analyses indicate that the smoldering reaction has substantially reduced. As mentioned above, there is no quantitative guidance to determine exactly when a smoldering fire in a silo has been completely extinguished.

Examination of historical examples of smoldering fires in silos have demonstrated that inerting the material using a gaseous agent such as CO_2 and/or N_2 is the most effective means of controlling and suppressing a smoldering fire in a silo. The best method of extinguishment is to smother the fire with inert gas (in the gas phase) delivered as close to the bottom of the silo as possible. [2, 3] Unfortunately the use of inerting gases is not accepted by all emergency responders. A 2010 article featured on the Fire Chief magazine website states the following: [13]

> *Using carbon dioxide or nitrogen gas to extinguish conventional silo fires is a questionable practice. This is because the structure is open to atmosphere and oxygen usually can still reach the fire. Some departments have had success with these gases, but more experimentation is needed before a recommendation can be made regarding their use.*

Monitoring

Accurate measurements of the concentration of carbon monoxide and oxygen within the storage silo can provide information regarding the smoldering fuel and the effectiveness of suppression activities. Smoldering generates carbon monoxide and smoke, both of which can be fuels in gas explosions. In the case study presented, the facility attempted to institute a monitoring program for the silo temperature. Properly done, temperature measurements using a temperature probe could be an effective way to probe the location and extent of smoldering combustion. [14] By allowing the smoldering fire to continue to burn and by failing to control and monitor the ventilation into the silo, the facility allowed carbon monoxide and smoke to accumulate and achieve a flammable concentration in the silo headspace. If the formation of a flammable atmosphere within the silo had been prevented, there would have been no explosion. Additionally, if the silo had been properly sealed, the fire would have consumed all of the oxygen in the silo and the fire would have self-extinguished. During the weeks leading up to the explosion the facility attempted to seal the smoldering silo to starve the fire of oxygen. However, facility personnel did not adequately monitor the progress of the fire to evaluate the effectiveness of these actions. The most effective method to judge the progress of smoldering combustion in a confined space is to perform atmospheric monitoring within the silo. Instead, the facility monitored the silo temperature using inadequate instrumentation.

The temperature of the material in the silo can also provide information regarding the extent of the smoldering mass of material. However, temperature measurements can be extremely difficult, especially if there is a significant amount of material in the silo. According to the NFGA emergency preplanning and firefighting manual: [14]

> *A grain or grain product fire in a concrete bin or silo is one of the most difficult to extinguish. Although heat or smoke may be present in the grain over space, rarely will there be significant flame present. Thus, the fire may not be located*

easily within the bin. Since grain is a good insulator, temperatures extremes also may not always be detected by temperature cables.

Two effective methods for measuring the temperature of the material in the silo and locating hot spots are through the use of thermographic photography and a thermometer probe. [14]

The simplest means for obtaining temperature measurements is through the use of a noncontact thermometer, but this type of measurement can only measure surface temperatures (the outer silo wall, the top surface of the material, the external surfaces of piping, etc.) at a single point. This type of device is typically not suitable for temperature measurements in the material interior (the region that is smoldering). Furthermore, surface temperatures measured with this type of device may be inaccurate because smoke can interfere with the device's optics, resulting in the device measuring the average smoke temperature, thus reporting a lower temperature than the true surface temperature of the material. A noncontact thermometer cannot measure interior temperatures deep inside the material which is where smoldering combustion typically occurs, as in the case study described. Therefore, the device could not "see" the smoldering combustion. The surface temperature would have been much cooler than the interior temperature where smoldering is occurring.

Infrared (IR) thermal imaging cameras can be used to measure larger surfaces of the silo including the outer walls, ducts, manyways, etc. and may provide better indications of where hot spots are located and/or the progression of the fire. Unfortunately, like the noncontact thermometer, thermal imaging cameras only provide surface temperatures. To obtain temperatures within the material, thermometer probes can be inserted into the mass of material. A thermometer probe is a long, hollow probe made with pipe with a sharp tip that can be inserted into the material to locate hot spots caused by smoldering or burning material in a storage silo. Some of the drawbacks to a thermometer probe is that it has to be physically inserted into the material (which has inherent hazards), the probe must be inserted at the correct location to measure the temperature of the hot spot, and the level of the material may not allow the insertion of such a probe from an accessible location.

Rapid Response

When a smoldering fire is detected in a silo, a rapid response becomes very important. A delayed response will allow the fire to grow and spread through the material making the fire more difficult to suppress. As a smoldering fire grows, the temperature increases causing the buoyancy forces within the material to increase. An increased buoyant flow will carry inerting gases away from the smoldering material at a greater rate. A delayed response also allows flammable gases and smoke to build up within the silo.

The delay in material removal during the presented case study increased the hazard of the situation in two ways: 1) the time delay allowed the smoldering combustion region to grow in size, making it more difficult to remove the material, and 2) during the time delay, the facility periodically increased the ventilation to the silo, thus prolonging the life of the fire and increasing the generation rates of carbon monoxide and smoke containing flammable unburnt gases and particulates.

The agglomerated pellets would have been difficult to remove without the complication of smoldering. Smoldering complicates the problem because it hardens the agglomerated material by charring it. This has been observed in other silo fires where smoldering leads to bridging of the particulate material.

The facility had the salvage contractor dump the material recovered from the silo into the tunnel because if the material was dumped outdoors onto the ground it would interfere with the operation of the truck scale and essentially shut down the shipping and receiving function of the facility. Removal of the material through the manway onto the ground outside the silo would have allowed the material to be removed in a quicker, safer way. In addition to the delay in beginning the removal of the material, the slow removal of the material through the tunnel increased the overall hazard of the situation. Dumping the pellets into the cramped confines of the tunnel put the salvage workers at greater risk to personal injury.

Guidance Documents

Section 12.6 of NFPA 61 provides requirements for fire-fighting operations. Among these requirements, NFPA 61 states that, "Fires, when discovered, shall be reported promptly to facility management and emergency responders, including the fire department." NFPA 61 also recommends that, "If possible, incipient fires shall be manually extinguished or burning materials removed." For fires that cannot be controlled promptly in their incipient stage, NFPA 61 requires that the endangered structure(s) be evacuated. For the case study presented above, the facility failed to follow these requirements.

To fight incipient fires in a concrete bin or silo, NFPA 61 recommends the following:

- Concrete bin or silo fires should be extinguished by removing burning material from the bin or silo directly to the outside after wetting the top surface of the material with gentle application of water at a low flow rate directly to the burning materials.
- Water fog should be applied to walls and to the underside of the roof to reduce airborne dust.
- Fire should be located by thermometer probes, thermographic photography, or feeling heat on bin or silo surfaces.
- Openings to the bin or silo should be sealed to limit oxygen entry.
- Material flow to and from the bin or silo should be stopped.
- Fire-fighting operations should be done from outside the bin or silo.

Furthermore, NFPA 61 cites the NGFA Emergency Preplanning and Firefighting Manual for guidance on firefighting actions at grain elevator facilities. The NGFA manual indicates that, "Procedures for extinguishing bin fires vary with each fire." [9] The document discusses three options for fighting fires in concrete bins: (1) starve the fire of oxygen, (2) flood the bin with inert gas to displace the oxygen in the bin, or (3) apply water gently to the walls, grain surface, and hotspots. The document also recommends consulting with a structural engineer to get advice on how to cut access holes into the bin if the hotspots cannot be reached through the bin's existing openings. The recommended practices for fighting a concrete bin or silo fire as presented in the NGFA guidance document are (emphasis provided by NGFA):

- Stop all equipment operating in the area, particularly the system to the troubled bin.
- Seal all openings into the bin, particularly at the bottom and sides to reduce the supply of oxygen to the fire.
- Locate and determine the extent of the fire. **Never try to fight a fire from inside the bin.**
- Monitor temperature in adjoining bins.
- Consider displacing oxygen by injecting carbon dioxide or nitrogen into the bin overspace.
- **Gently** wet down interior exposed walls at the bin top and the grain surface.
- If water is used, apply it to hot spots **slowly** with a probe or through access holes cut in the bin walls.

If the smoldering material cannot be extinguished using this procedure the NGFA guidance document gives the following instruction for removing the smoldering material: [9]

- Wet down the exposed bin bottom in the tunnel area of the facility (or side doors if applicable) and all surrounding equipment to inhibit ignition if smoldering grain or flames erupt from the bin.
- Unload the material directly outside through manhole ports or loadout spouts by auger or other means. **Never run hot or burning grain from a troubled bin through an elevator leg or into another bin.**
- If necessary, slowly allow grain to exit the bottom onto the floor (in the tunnel), wetting all material exiting the bin. This can be accomplished by:
    - Using a temporary chute to direct grain flow onto the floor;
    - Cutting or otherwise removing the belt to allow the grain to flow directly onto the floor;
    - If it is not possible to have the grain flow directly onto the floor, allow it to flow onto a stationary belt, wetting the grain as it exits the bin and shoveling it off the belt.

Note that the NGFA manual recommends removing the material directly to the outside when possible. Examples of burning material being removed from two silos are shown in Figure 7 and Figure 8. In the case study presented above, the material was not removed directly to the outside, due to the use of the truck scale. When removing material into an enclosed tunnel, as was done in the case study, all of the material exiting the bin should be wetted. This was not the procedure that was followed.

Figure 7. Removal of stored material through silo wall to expose smoldering material.

Figure 8. Removal of burning material.

6. Conclusions

The storage in silos of organic granular materials that are prone to self-heating must be managed carefully. It is important to understand the self-heating capability of the granular material in question, and to define safe (subcritical) storage conditions. The storage conditions within the silo should be controlled to avoid supercritical storage temperatures or excessive humidity. Since the propensity for self-heating often increases with material age, other useful safety measures include inventory control (frequent material turnover) and inspection of silo contents to detect caked solids on the silo wall or granular flow anomalies within the silo.

If evidence of self-heating or smoldering is observed in a silo, it merits immediate, serious attention. The hazardous situation should be monitored with appropriate technology while a strategy for suppressing the smoldering fire is developed. Suppressing a smoldering fire requires specialized skills and should not be attempted by inexperienced personnel. The most effective suppression strategy is to flood the silo with an inert gas that can extinguish the fire, but there may be site-specific circumstances that make this strategy infeasible. If inert gas flooding is not feasible, then the only other options are to either let the fire burn itself out in the silo (a process with may take months) or to rapidly remove the entire contents of the silo so that the burning material can be extinguished outside the silo. The latter option may require substantial structural damage to the silo, and should, therefore, be implemented under the guidance of a structural engineer.

7. References

1. NFPA 61, "Standard for the Prevention of Fires and Dust Explosions in Agricultural and Food Processing Facilities," 2008 edition, National Fire Protection Association, Quincy, Massachusetts.
2. Eckhoff, R., Dust Explosions in the Process Industries, Gulf Professional Publishing, Elsevier Science (2003).
3. Krause, U. (editor), *Fires in Silos*, Wiley-VCH, Berlin, (2009).
4. NFPA Glossary of Terms, National Fire Protection Association, Quincy, MA (2008).
5. ASTM E176-10a Standard Terminology of Fire Standards, American Society of Testing and Materials, West Conshohocken, PA, (2011).
6. Mills, J.T., et al., Spoilage and Heating of Stored Agricultural Products: Prevention, Detection, and Control, Agriculture Canada (1989)
7. Blasi, D.A., et al., Wheat Middlings Composition, Feeding Value, and Storage Guidelines, Kansas State University Agricultural Experimental Station, Kansas State University, Manhattan, KS, (1998).
8. Bowes,P.C., Self-Heating: Evaluating and Controlling the Hazards, Elsevier, Amsterdam, (1984).

9. Goforth, K.J., et al., Emergency Preplanning and Firefighting Manual: a Guide for Grain Elevator Operators and Fire Department Officials, National Grain and Feed Association, Washington, DC (1987).

10. NFPA 654, "Standard for the Prevention of Fire and Dust Explosions from the Manufacturing, Processing, and Handling of Combustible Particulate Solids", 2013 ed., NFPA, Quincy, MA.

11. Drysdale, D., Introduction to Fire Dynamics, Second Edition, Wiley, Chichester, (1999).

12. Tuomissaari, M., Baroudi, D., and Latva, R., "Extinguishing smouldering fires in silos," BRANDFORSK project 745-961, VTT Building Technology, Technical Research Centre of Finland, Espoo, (1998).

13. Halpin, Ted, "A Different Animal: Agricultural silo fires present unique challenges and dangers to firefighters," Fire Chief, www.firechief.com, October 1, 2010.

14. Goforth, K.J., et al., Emergency Preplanning and Firefighting Manual: a Guide for Grain Elevator Operators and Fire Department Officials, National Grain and Feed Association, Washington, DC (1987).

Lessons Learned from Fire in Air Supply Ducts

Al Waller
FMC Corporation
Philadelphia, PA
alvin.waller@fmc.com

Michelle Brown, CSP
FMC Corporation
Philadelphia, PA
michelle.brown@fmc.com

Prepared for Presentation at
American Institute of Chemical Engineers
2013 Spring Meeting
9th Global Congress on Process Safety
San Antonio, Texas
April 28 – May 1, 2013

UNPUBLISHED

Lessons Learned from Fire in Air Supply Ducts

Al Waller
FMC Corporation
Philadelphia, PA
alvin.waller@fmc.com

Michelle Brown, CSP
FMC Corporation
Philadelphia, PA
michelle.brown@fmc.com

Keywords: Lessons learned, fire, HEPA filter, air filter, thermal inertia

Abstract

In 2012, loss of air flow caused activation of interlocks to shut off steam and electrical heat inputs to an air heating system that was part of a drying/milling operation. Shortly thereafter, upstream filters and air ducts ignited causing more than US$500K in damage and business interruption. Investigations concluded that the interlocks functioned as designed on loss-of-airflow but were not sufficient to prevent overheating the inlet air filters (including HEPA filters) by the residual heat in the air heaters. The two main lessons learned from this incident apply to a broad range of operations in the CPI:

1) Mundane unit operations such as air filters that have combustible components must be evaluated as thoroughly as the rest of the chemical process.
2) Designers and PHA leaders must recognize the potential of thermal inertia of heaters (and coolers) to heat (or cool) a stagnant volume of fluid when flow is interrupted. Gas streams are particularly susceptible due to the low heat capacity relative to the heat capacity of a metallic heat exchanger. Interlocking the supplies (to steam and electrical heaters in this incident) is often necessary but may not be sufficient.

Other lessons "re-learned" include the importance of learning from a previous near miss, the importance of limiting escalation aspects of this fire, and the importance of preventative maintenance.

1. Introduction

A drying and milling operation experienced a fire inside its air supply ducts in 2012. Total losses exceeded US$500k, including replacement of air duct components and business interruption. An internal investigation was conducted with assistance from a consulting company by first focusing on the source of the fuel and then on the ignition mechanism.

Figure 1 below is a simplified diagram of a portion of the system; instrumentation and controls are omitted for clarity. Ambient air is pulled into the system by negative draft where it is filtered through pre-filters then HEPA filters. This filtered air is then heated via a steam heat exchanger then an electrical resistance heater to approximately 200C. This hot air then contacts damp combustible powder to convey it through a drying zone and into a cyclone. The distance between the HEPA filters and the steam heater is less than 2 meters while the distance from the electrical heater to the damp solids entry point is approximately 25 meters. Section 3.2 will more completely describe the air duct system and the process mills.

FIGURE 1 – Sketch of Air Duct for Dryer System

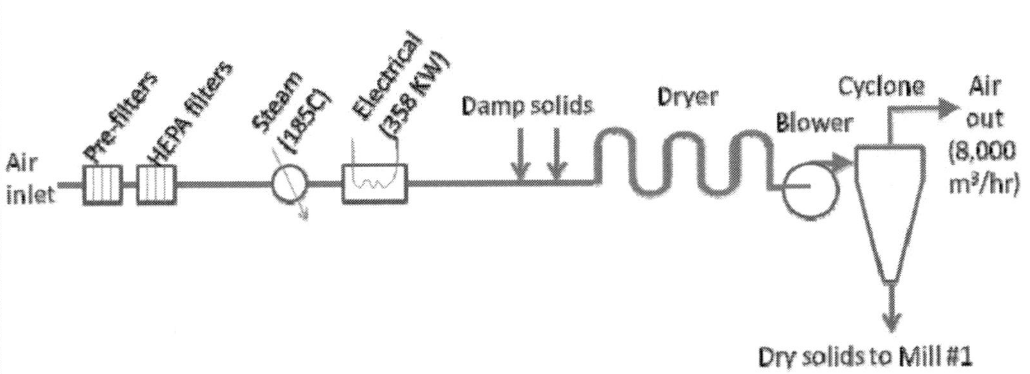

FIGURE 2 – Fire In Progress

The blower drive belts failed in September 2012 resulting in rapid loss of airflow so interlocks shut off steam and power to the heaters. Approximately 17 minutes later a portion of the air duct internally ignited.

2. "New" Lessons Learned

Forensic evidence pointed to the air filters as the source of fuel for the fire. Damage was limited near the air filters in Figure 1 plus there were no process solids or process ash upstream of the entry points for damp solids. The steam heater and electric heater were confirmed as tripped off. Calculations indicate that the residual heat (thermal inertia) was sufficient to ignite the air filters.

2.1 Filters As Fuel

The specific pre-filters and HEPA filters used at the incident site were rated for maximum continuous operation at 70C and 80C, respectively. Tables 1 and 2 list the key aspects of combustion testing of new pre-filters and HEPA filters. All components were susceptible to decomposition at 200C or less with the glass fiber and polyester component of the HEPA filter supporting a self-propagating flame in the oven test. Estimated energy release from combustion of the air filters was consistent with the damage observed from the incident.

Table 1. Stability of Air Filters

Unit	Component	Stability in 200C Oven Test	Stability in 500C Oven Test
Pre-filter	Glass fiber & polyester	Exothermic self-heating at 200C (to 600C). 47% weight loss	Exothermic self-heating (to 600C), 56% weight loss
HEPA Filter	Glass fiber & polyester	Exothermic self-heating at 160C (to 600C), self-propagating flame. 23% weight loss	Exothermic self-heating (to 600C), self-propagating flame. 39% weight loss
	Polyurethane sealant	38% weight loss	97% weight loss

Table 2. Estimated Air Filters Combustion Contributions

Unit	Component	Mass (kg)	Combustible Mass, from 500C Oven Test (kg)	Available Energy Release	
				(MJ)	(% of Total)
Pre-filter	Glass fiber & polyester	6.0	3.3	88	5%
HEPA Filter	Glass fiber & polyester	52.8	20.5	539	31%
	Polyurethane sealant	52.8	51.5	1089	63%

2.2 Residual Heat As Ignition Source

Recorded temperatures from the incident were not helpful in quantifying the residual heat remaining in the heaters after heating sources were tripped off. Recorded temperatures from a prior normal shutdown with the blower operating at 8000 cubic meters per hour were useful to calculate residual heats in the heaters as summarized in Figure 3 and Figure 4. The steam heater retains approximately 48 MJ while the electrical heater retains approximately 15 MJ.

FIGURE 3 - Steam Heater Residual Heat

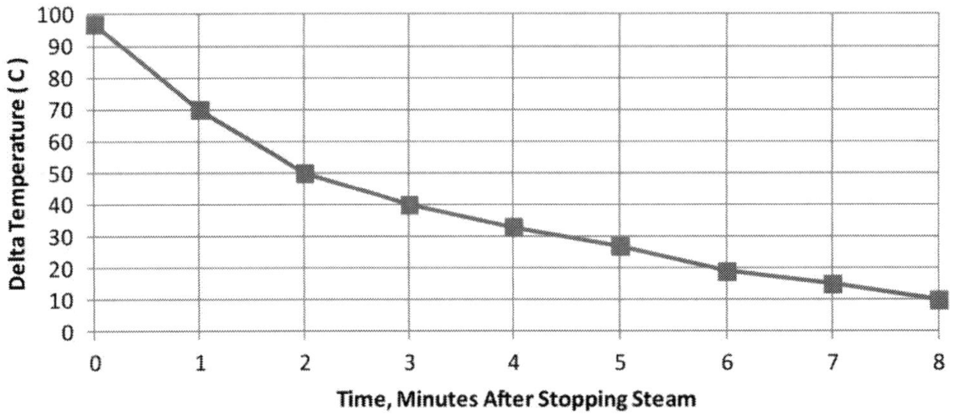

FIGURE 4 - Electrical Heater Residual Heat

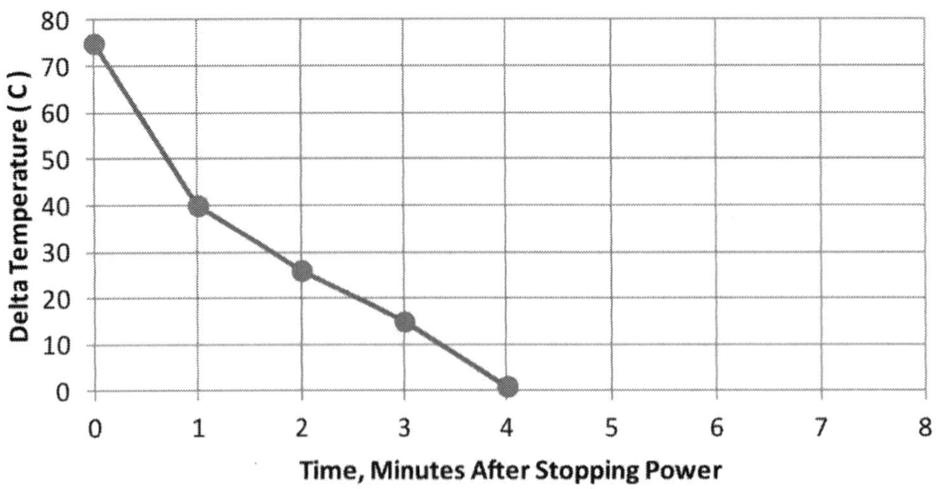

For the incident initiated by blower failure, this combined residual heat of 63 MJ can increase the temperature of stagnant air in the heater portion of the duct to well over 200C resulting in expansion driving the hot air backwards into the air filters. Alternatively, for this incident this combined residual heat could establish a convection current of dense cool air flowing forward at the bottom of the air filter duct, rising in the hot heaters zone, and hot air flowing backwards at the top of the air filter duct to the air inlet. Either incident mechanism results in excessively hot air in air filters that are rated for 70C or 80C maximum temperatures. The Standard Operating Procedures (SOP) for startup recognized the need to start the blower before heating and the SOP for normal shutdown recognized the need to keep the blower running until the air cooled down. Interlock designs recognized the necessity of tripping heating sources on loss-of-airflow. However tripping heating sources on loss-of-airflow is not sufficient to prevent the hazard of hot air reverse flow; therefore, an additional safeguard is needed. For this system, a new vertical vent duct with actuated valve was added between the heaters to open on loss-of-airflow as the primary additional safeguard to provide chimney-effect cooling of the heaters.

3. "Old" Lessons Re-Learned

In addition, lessons from this air duct fire have been refreshed multiple times and are widely known in the CPI.

3.1 Need to Investigate Near Misses and Minor Incidents

Prior to the September 2012 incident, a separate minor incident occurred during a normal shutdown. Some charring of the HEPA filters was attributed to a leaking steam heater isolation valve. The response was to repair the leaking steam valve; thus, the opportunity to investigate this precursor incident was missed.

3.2 Need to Limit Potential Escalation

The fire damage from the September 2012 incident extended beyond the Dryer air duct system (shown in Figure 1 above). The damage was much more severe to the other air filters (pre-filters and HEPA filters), other heat exchangers, and other ductwork interior surfaces as shown in Figure 5. The 3 other blowers in Figure 5 stayed on-line following the Dryer airflow failure and trip of the Dryer air heaters. Hot fumes of pyrolysis/combustion products from the Dryer air filters were pulled into the connected ductwork by these other blowers resulting in direct escalation of the fire or perhaps the Mill #2 electrical heater ignited fumes which backflashed to the air filters. While the Dryer air filters likely experienced restricted oxygen availability, the other air filters had ample oxygen to support combustion. The common air intake duct reduced the initial capital cost by a very small amount compared to separate intake ducts yet added unnecessary complications that led to major escalation. The system was rebuilt with separate air intakes.

FIGURE 5 – Sketch of Air Ducts for Complete System

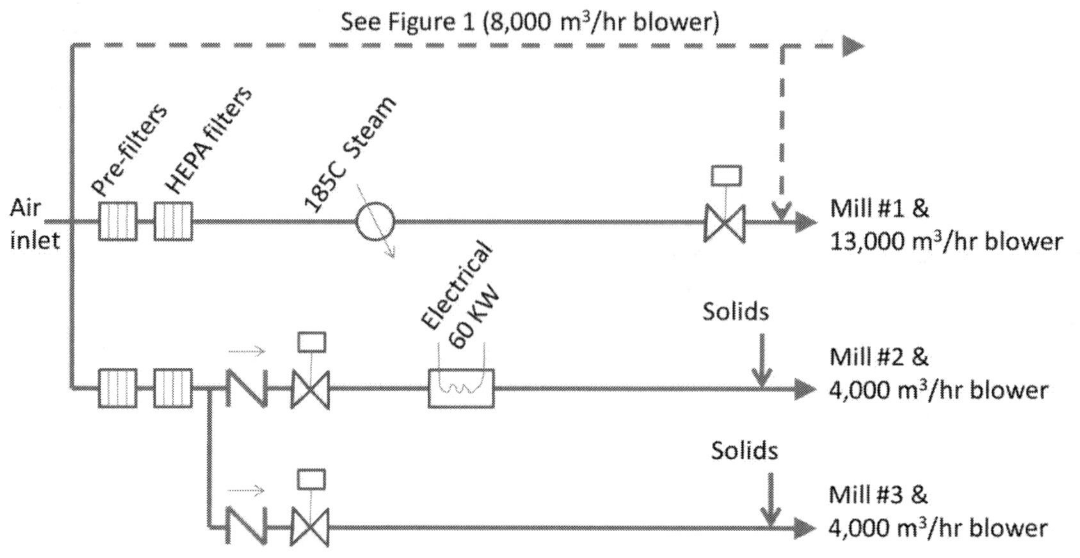

Modest external damage was also visible from smoldering combustible dust on the surfaces of some ducts due to accumulation. Housekeeping practices need to limit combustible dust accumulations below 3 mm (1/8-inch).

3.3 Need to Maintain Equipment Properly

Finally, the blowers did not have a specific installation procedure to ensure belt alignment plus they were not on a preventative maintenance program.

4. Conclusion

Those who forget the past are doomed to repeat. The main lessons learned from this incident apply to a broad range of operations in the CPI:

1) Mundane unit operations such as air filters that have combustible components must be evaluated as thoroughly as the rest of the chemical process.
2) Designers and HAZOP leaders must recognize the potential of thermal inertia of heaters (and coolers) to heat (or cool) a stagnant volume of fluid when flow is interrupted. Gas streams are particularly susceptible due to the low heat capacity relative to the heat capacity of a metallic heat exchanger. Interlocking the supplies (to steam and electrical heaters in this incident) is often necessary but may not be sufficient.
3) Near misses and minor incidents should be thoroughly investigated; an opportunity was missed at this site to prevent the major incident.

4) Keep it simple and keep areas clean. Significant escalation occurred due to a combined air duct that added unnecessary interactions between the systems. Minor escalation occurred from accumulated combustible dusts on exterior surfaces.
5) Maintain equipment properly. Rigorous installation procedures (such as belt alignment) and preventative maintenance improve both safety performance and on-stream reliability.

5. References

Additional references not cited:

[1] Kletz, Trevor. "What went Wrong? - Case Histories of Process Plant Disasters and How They Could Have Been Avoided" (5th Edition, 2009), Elsevier.

[2] DOE (U.S. Department of Energy), "Nuclear Air Cleaning Handbook" Chapter 10 – Fire Protection, DOE-HDBK-1169-2003 (2003).

Hydrogen Sulfide Exposure from Molten Sulfur – A Forgotten Hazard?

Mark J. Viz,
Sean J. Dee, Russell A. Ogle, and Brenton L. Cox
Exponent, Inc.
4580 Weaver Parkway, Warrenville, Illinois 60555
mviz@exponent.com, sdee@exponent.com, rogle@exponent.com,
and bcox@exponent.com

Prepared for Presentation at
American Institute of Chemical Engineers
2013 Spring Meeting
9th Global Congress on Process Safety
San Antonio, Texas
April 28 – May 1, 2013

UNPUBLISHED

Hydrogen Sulfide Exposure from Molten Sulfur – A Forgotten Hazard?

Mark J. Viz,
Sean J. Dee, Russell A. Ogle, and Brenton L. Cox
Exponent, Inc.
4580 Weaver Parkway, Warrenville, Illinois 60555

Keywords: molten sulfur; hydrogen sulfide; railcar; tank car; tank trailer; process safety management; PSM; hazardous materials; hazmat; PHMSA; OSHA; NA2448

Abstract

Molten sulfur is one of the most frequently transported hazardous materials (hazmat) in North America. Molten sulfur is typically transported at an elevated temperature by rail tank cars or over-the-road tank trailers as a Class 9 "Miscellaneous Hazardous Material" under the identification number NA2448. Perhaps the most prominent hazard of molten sulfur is that it is shipped at an elevated temperature, which can cause thermal burn injuries. A less obvious hazard is the potential release of hydrogen sulfide (H_2S), a highly toxic gas.

In this paper we first summarize the key features of the H_2S-molten sulfur equilibrium that are relevant to this hazard. Then we address the relative risk of H_2S exposure from an analysis of molten sulfur incident release statistics. Next, we discuss a few case studies in which vented H_2S gas from molten sulfur led to unexpected consequences for handling personnel. We close our discussion with suggestions for how to possibly prevent H_2S exposure through a combination of engineering controls, operating procedures for both loading and unloading operations, air monitoring, and personal protection equipment.

1. Introduction

Molten sulfur is one of the most widely used global feed stocks. In 2010, the global demand for sulfur reached 78.3 million tons, with elemental sulfur accounting for 66% of global production.[1] With growing demand in the Middle East and Asia, continued use of sulfuric acid, and a growing need for nickel, sulfur production is projected to grow steadily over the next decade.[2] In 2011, elemental sulfur was a $1.6 billion dollar industry in the United States, with 8.1 million tons of production.[3] Additionally, over the last thirty years there has been a significant shift in the production of molten sulfur, moving from mining of geologic sulfur deposits (Frasch process) to environmental byproduct sulfur (Claus process). This technology shift has increased the H_2S

[1] Messick, D. L., World Sulfur Outlook, The Sulfur Institute, 2010.

[2] Kitto, M., Sulfur Supply Demand Balance: The Outlook to 2015, British Sulfur Consultants, 2005.

[3] Apodaca, L. E., Sulfur, U.S. Geological Survey, Mineral Commodity Summaries, January 2012.

hazard in molten sulfur. Given the livelihood of the global sulfur market and the change in the source of the material, the need to communicate and control the hazards associated with the handling of molten sulfur is more critical than ever.

The hazard of dissolved H_2S in molten sulfur became widely recognized in 1963 with the disappearance of the molten sulfur tanker, the Marine Sulfur Queen.[4] This tanker disappeared without a distress signal off the coast of Florida with all 39 crew members. The U.S. Coast Guard searched for the vessel but was only able to recover a few life preservers and some floating debris. Reports came to light of frequent fires on the sulfur tanker and the frequent odor of rotten eggs.[5,6] Investigators from various countries performed headspace measurements in molten sulfur transportation vessels and found H_2S concentrations approaching the minimum flammable level. Recommendations were published that the dissolved H_2S content of molten sulfur should be maintained at 10 parts per million (ppm).[7]

As early as 1970, the U.S. Department of Transportation (USDOT) issued a rule under docket HM-70, that even though it was intended to address requirements for rail tank cars and tank trailers used to ship H_2S as its own commodity, clearly emphasized the hazards associated with exposure to H_2S. The comments to the federal rule state that H_2S is a "severe irritant and is highly toxic"; that exposure leads to "olfactory fatigue"; that H_2S is "heavier than air" and "tends to 'pool' rather than to dissipate into the atmosphere."[8]

Later, in the mid-1980s, the USDOT issued a Final Rule under docket HM-198 that clearly identified two significant hazards associated with the transportation of molten sulfur: the presence of unknown but potentially significant levels of H_2S and that the elevated temperature of molten sulfur could cause burns.[9] As reported in the Final Rule: "Several incidents involving the transport of molten sulfur, including three in the state of California since 1985, have emphasized the need for regulation of this material. Accidents involving spills near Barstow County and Culver City in 1986, and another in Benecia in 1985, demonstrated a lack of hazard communication. In each of these incidents, lack of swift and accurate identification of the material led to delays in handling the incidents and, in at least two instances, contributed to injuries sustained by persons on the scene."

A review of available Material Safety Data Sheets (MSDSs) for molten sulfur suggests that the range of H_2S concentrations and corresponding safe loading, unloading, and handling procedures for molten sulfur vary considerably among producers and shippers. As a consequence, the handling processes may be performed in a manner ignorant of the possible high concentrations of H_2S. According to several MSDSs, the primary hazards associated with molten sulfur are

---

[4] King, F.W., "Taking H_2S from liquid sulfur-why and how," Oil and Gas Journal, pp. 164-165, June 1974.

[5] Anon., "Investigations: The Queen with the Weak Back," TIME, March 8, 1963.

[6] U.S. Coast Guard, Marine Board of Investigations, "Disapearance of the SS Marine Sulphur Queen at sea on or about 4 February 1963 with the presumed loss of all persons on board," March, 1964.

[7] Lagas, J.A., "Stop emissions from liquid sulfur," Hydrocarbon Processing, pp. 85-89, October 1982.

[8] 35 FR 18919 et seq., December 12, 1970, docket no. HM-70, "Hydrogen Sulfide Gas in Cargo Tank Trucks and Tank Cars."

[9] 53 FR 17158 et seq., October 21, 1986, docket no. HM-198, "Molten Sulfur and Molten Materials Generally."

thermal burns, sulfur-fueled fires, dust explosions, exposure to H_2S, and, in the event of a sulfur fire, inhalation of sulfur dioxide.[10,11,12,13,14] The hazard of H_2S is not prominently communicated in many of these current MSDSs.

Elemental sulfur is typically transported in a molten state in railcars or tank trailers. Commonly, the H_2S hazard is reported as a potential risk, because the concentration of H_2S within the vapor space of the railcar, tank trailer or stationary storage or process vessel is specified as a range or threshold value. The actual concentration of H_2S present in the vessel headspace will likely change due to temperature variations and agitation (as occurs in railcars and tank trailers during movement), and the potential exposure will be dependent on the producer, shipper or consignee's handling procedures and best practices. The authors' own field experiences have also shown that particular to the unloading of molten sulfur railcars, unloading personnel can have a false sense of safety when opening the top fittings or manway cover to prepare a tank car for steaming. Unloading personnel occasionally believe that the solidified "crust" of molten sulfur that is usually present at the surface, especially in cold weather, will somehow prevent any significant release of H_2S. Our own investigatory experience has shown that this is not always true.

Occupational exposure to H_2S during handling can generally be categorized as an example of a low probability, high consequence risk. This is supported by a review of the Pipeline and Hazardous Materials Safety Administration (PHMSA) and Occupational Safety and Health Administration (OSHA) incident report databases, which confirms that over the last 20 years, there have been six OSHA-reported fatalities from H_2S exposure associated with molten sulfur containers and over the last 24 years, there have been no PHMSA-reported fatalities from molten sulfur releases (while the commodity is still under the jurisdiction of the federal hazmat transportation regulations).[15,16] Our collective experience based on our investigations of numerous hazmat releases is that the tendency is for successful management of low probability, high consequence hazards to lead to complacency and the occasional elimination of seemingly superfluous safeguards.

---

[10] Hovensa, Sulfur MSDS No. 1794, 2006.

[11] Tesoro, Safety Data Sheet: Sulfur, Molten, 2012.

[12] Keyera, Sulfur, Molten MSDS, 2011.

[13] Devon, Material Safety Data Sheet: Sulfur (Liquid), 2010.

[14] Marathon, Material Safety Data Sheet: Marathon Sulfur, 2010.

[15] Note that there can be overlap between the release incident reporting required by OSHA and PHMSA based on where the incident occurred and what function was being performed when the release occurred. So, for example, an accidental release of molten sulfur involving an injury or fatality that occurred while a tank car was connected to plant apparatus during unloading might be reported to OSHA and not to PHMSA even though releases during unloading can be covered by PHMSA regulations as well. Efforts to harmonize the regulations and reporting requirements promulgated by a variety of federal agencies has been the subject of federal rulemaking in dockets such as HM-223, "Applicability of the Hazardous Materials Regulations to Loading, Unloading and Storage," Docket No. PHMSA-98-4952 (HM-223), under which PHMSA issued a Final Rule on June 1, 2005.

[16] Note that at times, exposure to H_2S can result in a prolonged hospitalization that results in fatality at a date after the original paperwork related to regulatory agencies is submitted.

According to the USDOT PHMSA "Incident Reports Database," there have been 428 reported releases of molten sulfur in transportation between the years 1989 to 2012, inclusive. For all 428 reported releases, no fatalities and 19 injuries were reported. A portion of these injuries were related to exposure to fumes and odors that resembled "rotten eggs", a description commonly used for H_2S. Additionally, incident reports indicated that H_2S gas was released at various times while the commodity was in transit, being loaded or unloaded, or while being stored in transit. Although there may be some overlap between exposure incidents reported in PHMSA's Incident Reports Database and the analogous Accident Investigation Summaries from OSHA (OSHA Form 170), which are maintained in OHSA's Integrated Management Information System (IMIS), the OSHA database includes six reported fatalities from H_2S exposure associated with the handling of molten sulfur over the last twenty years.

Our review of H_2S exposure incidents is intended to act as a reminder of the importance of employing a hierarchy of safeguards when managing high consequence hazards like H_2S.

2. Technology Background

A major technological shift in the production of sulfur has resulted in an increased potential for exposure to H_2S. For most of the 20th century, the Frasch process was used to recover sulfur from underground, elemental deposits.[17] In the Frasch process, superheated water is used to melt solid deposits of sulfur which are driven to the surface with compressed air for recovery. However, with the growth of the refining industry, the supply of elemental sulfur from the desulfurizing of sour crude oil and natural gas overtook production from the Frasch process in 1970. The Claus process, which can produce elemental sulfur from feed stocks containing up to 20% H_2S, is now the preeminent production process for producing sulfur.

To convert H_2S to elemental sulfur, the Claus process uses a two-step process, first utilizing a high temperature thermal step, followed by a lower temperature catalytic step. In the thermal section, air is added to H_2S at temperatures in excess of 1800°F, to covert one-third of the H_2S to sulfur dioxide (SO_2) as shown in Equation 1.

$$H_2S + 3/2\ O_2 \rightarrow SO_2 + H_2O \qquad \text{[Eq. 1]}$$

The resulting SO_2 then reacts with the remaining H_2S in an endothermic equilibrium reaction to form elemental sulfur vapor as shown in Equation 2.

$$2H_2S + SO_2 \leftrightarrow 3/2\ S_2 + 2\ H_2O \qquad \text{[Eq. 2]}$$

To drive conversion in the endothermic equilibrium reaction shown in Equation 2, S_2 vapor is removed through condensation. As the vapor cools, the S_2 equilibrates with the S_6, and S_8 forms through multiple equilibrium reactions.

If one-third of the sulfur is consumed in Reaction 1, then the remaining two-thirds is consumed in Reaction 2 resulting in stoichiometric consumption of H_2S. However, several side reactions usually lower the conversion to SO_2 in Reaction 1, subsequently lowering the overall H_2S

[17] Greenwood, N. N. and Earnshaw, A., Chemistry of the Elements, 2nd ed., 1997.

conversion. This highlights both the importance of the air to H_2S feed ratio as a process control variable, and the need for a secondary catalytic reactor to further convert the H_2S that is not consumed in the thermal section.

The catalytic section of the Claus process is used to convert H_2S as well as side products including COS and CS_2 that are formed from side reaction in the thermal section. The catalytic section runs at significantly lower temperatures than the thermal section (inlet 400°F - 450°F, outlet 600°F - 630°F), but uses a Clause catalyst (usually aluminum (III) or titanium (IV) oxide) to covert H_2S to elemental sulfur as shown in Equations 3 and 4.[18,19]

$$2\ H_2S + SO_2 \rightarrow 3/6\ S_6 + 2\ H_2O \qquad [\text{Eq. 3}]$$

$$2\ H_2S + SO_2 \rightarrow 3/8\ S_8 + 2\ H_2O \qquad [\text{Eq. 4}]$$

The effluent of the catalytic section can be cycled through additional catalytic beds, or condensed to a final product stream at temperatures between 265°F - 350°F.

The one common feature to all of these chemical reactions is the constituent H_2S. While the Claus process can achieve 96% - 97.5% conversion with one thermal reactor, and three catalytic reactors in series, the solubility of H_2S in molten sulfur adds an additional processing constraint. The solubility of H_2S in molten sulfur actually increases with temperature up to approximately 425°F. Solubility remains approximately constant from 425°F until 630°F, where it begins to drop with further increases in temperature.[20] Thus, in the catalytic reactors and condensers, there is still an appreciable amount of H_2S dissolved in the molten sulfur. If the hot sulfur is then stored in a tank, as the tank cools, the dissolved H_2S will release into the headspace, increasing the concentration to potentially dangerous levels. If the sulfur continues to cool, then a crust will form as the surface of the product reaches temperatures below 240°F. This solid crust can serve as a physical barrier protecting personnel from potential H_2S exposure. In cases where a crust is observed, special consideration should be noted to avoid agitation that could disrupt the solid interface, resulting in a potential H_2S vapor exposure. However, it should be noted, that formation of a protective crust should not be regarded as a reliable safety indicator, since crust formation is not correlated to the concentration of H_2S in the vapor space.

The headspace concentration of H_2S that can be achieved with molten sulfur in a transportation or storage container is influenced by four primary factors: the dissolved H_2S concentration, the liquid temperature, agitation of the liquid, and the presence of any organic contaminants. To maintain an H_2S concentration below its lower flammable limit (LFL), typical specifications for the dissolved H_2S concentration range from 5 to 15 ppm. At the temperature range at which molten sulfur is typically handled, the LFL corresponds to a headspace concentration of 3.4%, or

---

[18] Goar, B. G. and Fenderson, S., Fundamentals of Sulfur Recovery by the Claus Process, presented at The Laurence Reid Gas Conditioning Conference, Norman, Oklahoma, March 3, 1996.

[19] Sulfur and Hydrogen Sulfide Recovery, Kirk-Othmer Encyclopedia of Chemical Technology, John Wiley & Sons.

[20] Fanelli, R., Solubility of Hydrogen Sulfide in Sulfur, Industrial and Engineering Chemistry, pp. 2031-2033, September 1949.

34,000 ppm (at room temperature the LFL for H_2S is 4.3%).[21] From the viewpoint of respiratory exposure, this is a potentially instantaneous lethal dose.

3. Hazard Communication

The MSDSs for molten sulfur identify several hazards. Common features were noted in the 19 MSDSs reviewed for this study. First, nearly all of the surveyed MSDSs noted the potential dangers associated with H_2S and listed H_2S as a material present in molten sulfur. The amount of H_2S present was usually noted as a small amount; descriptions included "trace", "< 0.1%", "< 1%", "< 500 ppm", and "< 1% (variable)". It was also common for MSDSs to include additional information about the potential hazards of H_2S exposure, the acceptable exposure limits (which even today are still a subject of debate), and to identify the need for proper ventilation during storage and handling.

In the MSDS review, there is significant confusion and room for improvement regarding the description of PPE and unloading procedures. In addition to H_2S hazards, respiratory protection also included consideration for dust hazards since molten sulfur can solidify to a yellow solid when cooled. In most MSDSs, it is noted that respiratory protection should be used, but only if the concentrations of H_2S are unknown/above regulated levels, or other engineering controls are not sufficient to control the exposure. One MSDS stated that respiratory protection was not commonly required when handling molten sulfur. While several MSDSs noted that unloading/opening of tanks should be done in well ventilated areas, it was not noted if proper ventilation is a sufficient engineering control to protect workers from potential exposure to H_2S. One MSDS included a picture of a dust mask in the recommended PPE, which is appropriate for the potential dust hazard, but not for H_2S exposure.

In addition to not specifying when a respirator should be worn, very few MSDSs had detailed recommendations for unloading procedures related to the H_2S hazard in molten sulfur. As previously discussed, most of the MSDSs recognized several key components to safe unloading: proper ventilation, respiratory protection if the H_2S concentration was unknown or above acceptable limits, and confined space warnings. Air monitoring was not commonly listed as a recommended engineering control in the MSDSs. Additionally, recommendations regarding venting of tank headspace were not usually recommended before sampling or unloading. This omission, however, could be the result of the inclusion of opening the tank car or tank trailer before steaming as part of the unloading procedure. Details such as these regarding unloading might not typically be specifically addressed in the MSDS. Other MSDSs specified venting, only if the product was excessively heated or exposed to elevated temperatures. The potential evolution of H_2S into the vapor space can be driven both by temperature increases in the presence of water or by temperature decreases which drive dissolved H_2S out of the molten sulfur. Including specific directions for when respiratory protection is required and directions for proper unloading safety controls before handling/unloading could improve the hazard communication for H_2S in molten sulfur.

---

[21] NFPA 655, Standard for Prevention of Sulfur Fires and Explosions, National Fire Protection Association, Quincy, MA, Section A.5.1, p. 17, 2007.

4. Review of PHMSA and OSHA Incident Data

Two meaningful sources of incident data regarding H_2S exposure from molten sulfur handling are the PHMSA Incident Reports Database and the Accident Investigation Summaries from OHSA Form 170. Although there may be some overlap in incident reporting between these two databases, in general the PHMSA database contains mandatory reports that involve the release of a regulated hazardous commodity while in transit, while being loaded or unloaded, or while in storage during transit. The OSHA database contains mandatory reports that involve incidents or accidents that have resulted in the injury or fatality of an employee. The differences between the type of incidents reported in each database and the corresponding limitations in the reports should be understood. For example, personnel may have been briefly exposed to H_2S shortly after a tank car has been opened for steaming or unloading, but if there was no release of molten sulfur, the incident would not necessarily need to be reported in the PHMSA database. In a similar manner, if a release of molten sulfur occurred after a transport vessel had been connected to plant apparatus but it did not involve the unloading equipment, this incident might only be reported in the OSHA database, and then only if a reportable injury or fatality occurred as a result. In this manner, then, the review and subsequent use of the information provided in these reports should be considered with care and properly qualified when attempting to estimate occurrence probabilities and severities.

Bulk shipments of molten sulfur in North America are typically identified by the number "NA2448." Using the PHMSA Incident Reports Database, we performed a search of reportable incidents involving NA2448 by all modes of transportation (air, water, highway and rail) over the entire range of dates for which electronic records are available. From 1989 to 2012, the database contained 428 reports of unintended releases of molten sulfur, all of which occurred either by rail or highway. In these 428 reports, no fatalities were reported and 19 injuries were reported. An interesting finding is that of the 428 reports, 43 of them were described (based on the responder's description provided in Part VI of the Form DOT F 5800.1) as being initially discovered based on the detection of a strong odor. In addition, the 43 reports that explicitly identified "odor" in the description all involved rail tank cars. Of these 43, three of them further described the odor as being that of "rotten eggs," which is the typical way that people refer to the unique smell of H_2S.

It is compelling that roughly 10% of the molten sulfur releases reported from 1989 to 2012 indicate that they were initially detected by the "odor" of sulfur, and in a few cases, the more telling odor of "rotten eggs." Many of these instances occurred when a rail tank car was technically "in transit" but was not "in movement," such as when a railcar is in a classification yard. The irony involved in the discovery of these situations is that while the inhalation hazard of the H_2S is a high consequence hazard, had it not been for personnel noticing the odor coming from these tank cars, detection of the lading loss might not have occurred.

For releases that occurred in rail transit, the most common causes were a corroded or otherwise breached rupture disk, an improperly secured manway cover (usually involving a number of loose or missing bolts) or an improperly seated gasket. These causes create concern for railcar handling personnel because unless they can detect a potential exposure by smell or see the physical indicators of lost lading (such as crusted sulfur on the outside of the tank car), they would likely approach the railcar assuming that it was completely sealed. This might lead to personnel being placed in danger if they approach an unsealed railcar that is venting high concentrations of H_2S that are not detected quickly enough for personnel to detect and identify the hazard and approach the railcar using more appropriate PPE.

The majority of releases from tank trailers involved human error on the part of the driver who failed to properly monitor the loading operation and allowed the tank trailer to be overfilled. Reports involving overfilling do not refer to sulfur or H_2S odors, likely because the loss of lading is immediately apparent by visual observation.

A review of the Accident Investigation Summaries from OSHA revealed five cases that noted an H_2S release from molten sulfur tanks during unloading.[22] A total of 16 employees were involved in the five incidents, with six total fatalities. There are several common features related to each of these incidents. First, all five incidents occurred in the fall or winter season in locations where the ambient temperatures are low. Second, none of the five incident reports makes mention of the exposed employees wearing proper respiratory protection, and two investigations noted that employees were not wearing respirators. In two of the incidents, the exposure to H_2S resulted in loss of consciousness which caused a fall for the employees who were not wearing fall protection.

The most recent OSHA report of an accident related to H_2S exposure from molten sulfur occurred in 1995. An employee was taking samples from a molten sulfur tank truck. After opening the manway cover, he reached approximately 15 inches into the tank to collect the sample. A respirator was not provided, and even though he was on a nearly nine-foot tall platform, he was not wearing fall protection. The employee was found unconscious on the gravel covered ground approximately 10 minutes after taking the sample. An autopsy listed blunt force injuries from the fall as the cause of death, but testing revealed H_2S levels up to 800 ppm using similar unloading procedures on site. OSHA maintains that H_2S was a causal factor related to the fall.

In addition to potential fatalities due to falls, H_2S can also cause poisoning or asphyxia. In 1993, three employees were unloading a molten sulfur railcar. Two employees were on a platform on top of the railcar, while the third employee was on the ground hooking up steam lines. The third employee found his two coworkers unconscious on the platform shortly after they had opened the manway cover on the railcar. The employee who opened the manway cover was found dead on arrival, the other employee on the platform was hospitalized for several days, while the third employee was examined and released.

5. Safeguards to Prevent or Control H_2S Exposures

H_2S exposure during the handling of molten sulfur can be generally categorized as a low probability, high consequence event based on an assessment of the data available from PHMSA and OSHA. However, these incidents were of high consequence due to the fatalities that have been reported. Managing this or any low probability, high consequence risk may benefit from the use of multiple safeguards for a variety of reasons.

Suggested safeguards to consider, most of which can be implemented independently or together, are:

- Closed-loop venting and treatment of H_2S vapor (e.g., adsorption systems, flares);
- Administrative controls (e.g., standard procedures and training for unloading, buddy system);
- External ventilation of unloading areas;
- Air monitoring systems with alarms;
- Respiratory protection (PPE) sufficient to the hazard level or that ensures the safety of unloading personnel when the hazard level is unknown.

In fact, some states have mandated safety procedures associated with the handling of H_2S, including specifications for personal protective equipment, venting requirements, and equipment specifications.[23] As always, the best system of safeguards will be site-specific and tailored to the capabilities of site personnel. It is also important to consider the consequences of any changes prior to their implementation, even the addition of safeguards, lest any new hazards be introduced.

6. Recommendations

The safe handling of molten sulfur requires understanding and planning regarding the potential release of H_2S. Herein, we have discussed the technical reasons for H_2S hazards in molten sulfur, reviewed accident statistics, and relevant safety documentation. The following recommendations might be considered when handling molten sulfur.

- Sulfur supply contracts, which typically include requirements for a dry weight analysis of the sulfur based on minimum or maximum amounts of ash, carbon, arsenic, selenium and possibly other elements, should also address maximum levels of H_2S. In addition, these requirements should address when and how the product sample should be taken and then analyzed, especially if it is determined that the H_2S content is strongly dependent on sample temperature and storage time.
- Documentation regarding the shipping, receipt, and unloading of molten sulfur should be maintained per regulatory and "best practices" requirements. These documents, as well as sulfur supply contracts, might include warnings regarding the potential increase in H_2S concentration during shipment.

[23] For example, California Title 8, Section 4414 and Washington Administrative Code 296-79-250.

- Development of site-specific loading and unloading procedures should be written and followed given due consideration to the assessment and management of the hazards associated with H_2S in molten sulfur. These procedures might include:
    - Forced exhaust ventilation requirements;
    - Proper temperature, humidity, and air monitoring regarding H_2S concentrations at tank inlets, outlets, and bulk concentrations within the unloading area;
    - Proper PPE for burn protection and H_2S exposure, possibly including a requirement for a SCBA respirator and an escape bottle;
    - Controlled tank headspace venting before sampling or unloading;
    - Consider fall protection requirements knowing that non-lethal exposure to H_2S can induce a loss of consciousness and potential injuries related to falls.

Finally, it is emphasized that there is typically no fixed set of safeguards that will work at every facility. A well-designed system of safeguards to protect workers from H_2S exposure should likely consider the personnel, equipment, and safety management system at that particular facility.

9TH GLOBAL CONGRESS ON PROCESS SAFETY

Calcium Stearate Deflagrations – A series of unfortunate events

Albert Ness
ABS Consulting
10301 Technology Drive
Knoxville, TN 37932
aness@absconsulting.com

Konanur Manjunath
Dow Chemical Co.
Freeport, Tx
KManjunath@Dow.com

Prepared for Presentation at
American Institute of Chemical Engineers
2013 Spring Meeting
9th Global Congress on Process Safety
San Antonio, Texas
April 28 – May 1, 2013

UNPUBLISHED

Calcium Stearate Deflagrations – A series of unfortunate events

Albert Ness
ABS Consulting
10301 Technology Drive
Knoxville, TN 37932
anness@absconsulting.com

Konanur Manjunath
Dow Chemical Co.
Freeport, Tx
KManjunath@Dow.com

Keywords: Calcium Stearate Solids, Combustible Dust, Bulking Brush Discharge, Particle Size Distribution, Resistivity, Bonding, FIBC, Inertion

Abstract

A series of deflagrations occurred in a plant unloading Calcium Stearate from bulk bags. Each one had a different root cause and illustrates the importance of management systems for ignition control and the need to evaluate a system for many types of ignition sources. The first event was caused by static discharge from an ungrounded tote. When upgrading the system Type C bags were specified for the product and a second deflagration occurred due to static discharge from and ungrounded bag. In the third deflagrations cause was most likely a conical pile discharge. The presence of fines and coarse material plus the use of a larger cubic vessel probably enhanced the likelihood of the brush discharge.

.

1. Introduction

This paper describes three deflagrations that occurred in a bulk bag unloading system that had previously run for over six years without incident.

1.1 Product data for Calcium stearate

Table 1 lists data for the Calcium Stearate that was being used in the facility.

Table 1 Calcium Stearate Properties

Parameter	Value
D_{50}	500 microns
Minimum Ignition Energy (MIE) (< 75 μ fraction)	8 mJ
Volumetric resistivity	$14\ \ 10^{14}$ Ω-m

2. Unloading station description

The unloading station consisted of a bulk bag of Calcium Stearate (CaSt) hung u above a feed hopper. An Iris valve was used to control the flow rate of the CaSt into the hopper. A flexible feeding chute was attached from the hopper outlet to the lid of a 2500 L tote. Local exhaust ventilation was provided at the feed hopper and tote.

Figure 1: Unloading station for calcium stearate bulk bags

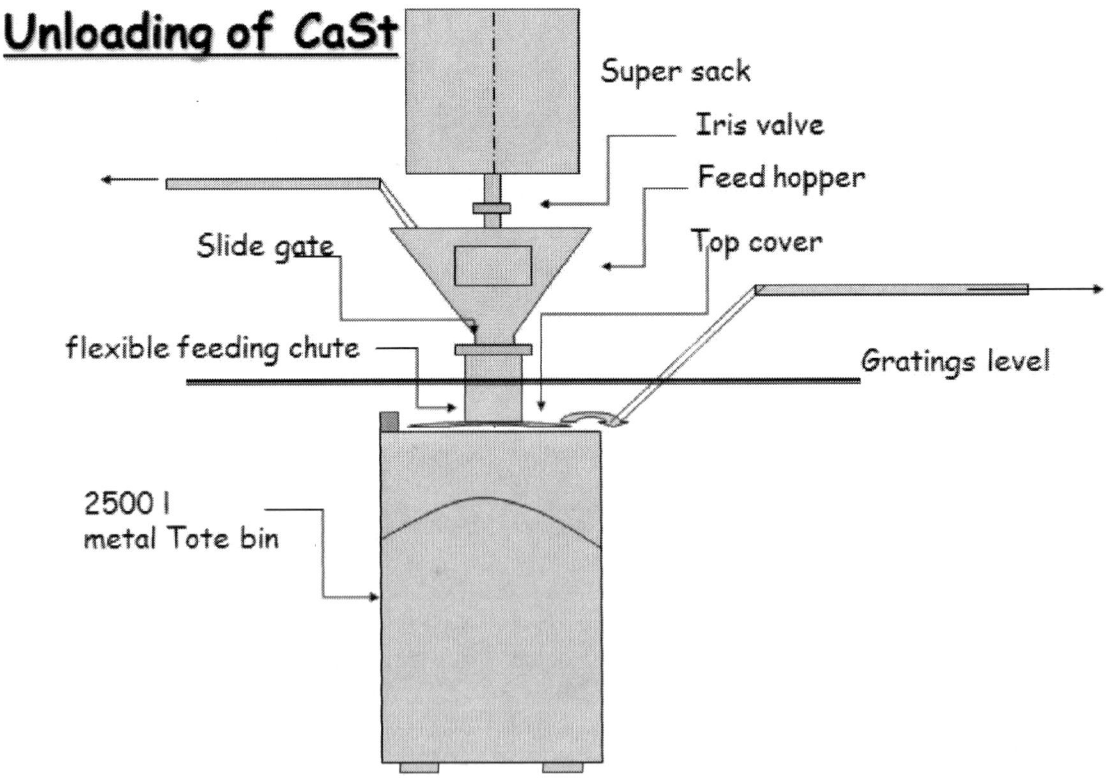

2.1 Event #1 while unloading

At the end of the loading cycle a deflagration occurred in the head space of the tote. The fire was quickly extinguished and little actual damage occurred. Investigation showed that the Feed hopper and the lid of the tote were not grounded and that the flexible chute was not static dissipative. Additionally, it was discovered that the tote vacuum system was not used. Figure 2 shows where the damage occurred.

Figure2: Unloading station at the time of event 1

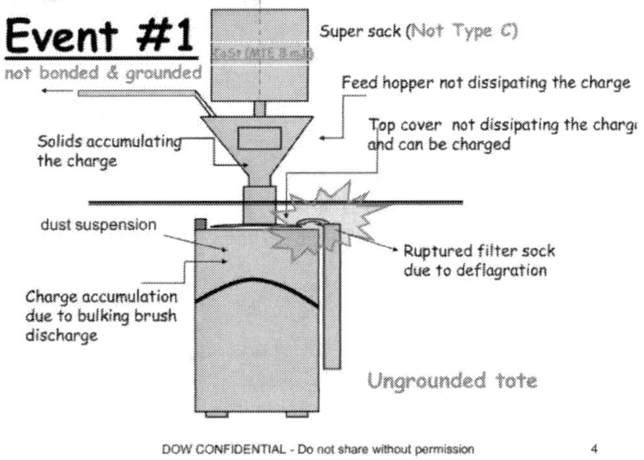

With the vacuum system inoperative, fines were able to stay suspended in the head space of the tote. A calculation of the capacitance of the lid of the lid showed that it had a capacitance of 180 pico Farads. Figure 3 shows that at that level, a static discharge energy of about 30-40 mJ for a charge of 20 EV, well above the 8 mJ MIE of the fraction under 75 microns, was possible. Therefore it was concluded that the most likely cause of the deflagration was a spark between the lid and tote ignition suspended fines in the tote.

Figure 3: Calculated electrostatic build up on process equipment

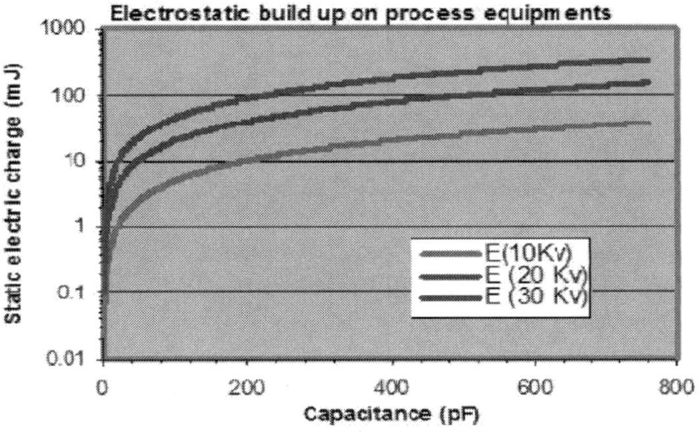

2.2 Event #2

After the first event, the metal equipment was grounded and the tote vacuum system was put back into operation. Also, it was decided to specify use of Type C bags for the bulk bags. Previously, the bulks bags were type B, that is, an all plastic bag that cannot be grounded, but has a breakdown voltage of 4kV. This allows for discharge of energy to the air. A Type C Bag has continuous conductive threads are sown into these bags, hence continuity is provided if the bag is grounded.

After starting up the improved system, a second deflagration occurred. Upon investigation, it was found that; 1) the bulk bag had not been grounded and 2) the Iris valve was not grounded due to its Teflon coating and the powder buildup on it. The failure to ground the new bulk bag is a reminder of the expression "No good deed goes unpunished". It is also a reminder of the need for a good Management of Change System. In the case of this change, the operator's training should include instruction on the hazard of NOT grounding the Type C bag.

2.3 Event #3 while unloading

Prior to restarting the unit, a ground verification system was installed to ensure the bulk bag was grounded before charging. Figure 4 shows where the damage occurred.

Figure 4: Unloading station at the time of event 3

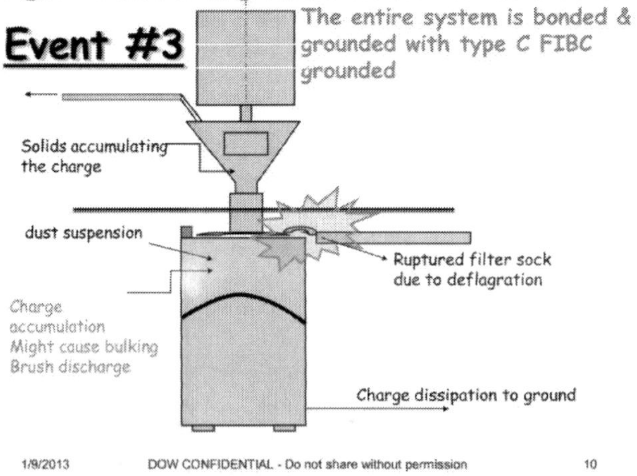

At this point, there was no obvious source of ignition. Several experts were called in and a more thorough investigation was done. Findings were:

1. The FIBC was examined for continuity of grounding. Except for the area of the bag that was damaged by the fire, the continuity was acceptable. Testing of FIBCs at the supplier found all the bags were conductive.
2. The CaSt in the FIBC from the last event was somewhat smaller than in the initial events. Table 2 shows the difference in size between the CaSt particles from the first and third events. The material from the third event was not out of spec, it was considered

acceptable material by the plant. The key point here is that normal variation in the raw material can affect the degree of hazard.

Table 2: Size Comparison

Cumulative Vol. %	Event 1	Event 3
90	800	800
50	500	425
10	200	150
% < 75 μ	2-3	~5

3. Fines collected from the top of a different tote had a D50 of approximately 10 μ and were found to have an MIE of 3 mJ.
4. The dust extraction system was again found to have been off.

2.4 Bulking or Conical Brush Discharges

Whenever solids are filled into containers, bulk bags and so on, bulking brush discharges are developed. It is the type of discharge observed on the conical pile of a heap of powder; thus it is often called conical pile discharge. Bulking brush discharges have an apparent maximum effective energy of 10-25 mJ (with respect to dust ignition), and are believed to be responsible for dust explosions in grounded silos [5, 17, 18].

Britton [17] states that bulking brush discharges cannot always be practically eliminated in equipment above about 1 meter diameter when non-conductive powders are transferred.

The following scenarios can lead to bulking brush discharges:

1. Fines in the suspension that coat the conductive wall rendering it less conductive or nonconductive. Now we have a situation of a system that is not grounded in practice. As an example the fines may be all fraction < 10 microns and the MIE for such a fraction will be much lower than the tested sample (< 74 microns). Even though this is a hypothetical situation, one needs to consider if the attrition is a factor in product processing.
2. Fines are in suspension in the free head room of the vessel and the filling continues with the mixture containing coarser particles. As the particle size increases, the equivalent energy increases. See Figure 3. This combination with fines to super fines and coarser particles is a concern as this can result in an incident.

Scenario 2 perfectly describes the situation that occurred in the third event.

2.4 Causes

The root cause of the third events is the high fines content of the CaSt in the FIBCs. The lack of dust extraction may have led to a suspension of the fines in the head space of the metal tote bin. Ignition could have been caused by either a defective Type C FIBC, or a conical brush discharge.

Testing of the existing bag was inconclusive due to the damage it sustained. Experts in static electricity and combustible solids believed that the conical pile discharge could not be ruled out.

Potential mitigation for conical brush discharges can include:

- Flow rate reduction
- Use of active neutralizers for relatively small silos.
- To minimize the probability of ignition, the first consideration
- Minimize the sub-200 mesh fines concentration to increase the MIE to greater than 20 mJ. Inertion, especially if process design or powder properties make deflagrating venting and suppression undesirable alternatives. Inerting is often considered where the powder MIE falls below the 3-10 mJ range.

In this application, inerting was chosen as the fire and explosion prevention technique

3. *Concluding remarks*

This series of events illustrates many points.

Not fully understanding the source of hazards can lead to jumping to obvious, but not comprehensive, conclusions in an investigation. It pays to consult the appropriate subject matter experts even for seemingly simple events.

No good deed goes unpunished. Changes made to improve the safety of a system can introduce new hazards. In this case the switch to Type C bags introduced a new potential ignition source. Training of operating personnel should include not only what the change is and how the procedures may change, but also on new hazards introduced.

Even on a simple system, it pays to consult with subject matter experts on potential hazards. The effect of variations of particle size on MIE and the concept of conical pile discharges, may very well not be known to process designers.

8. REFERENCES

1. Bartknecht, W., **Dust Explosions**, Berlin: Springer-Verlag, 1989.
2. Glor, M., **Electrostatic Hazards in Powder Handling**, Letchworth UK: Research Studies Press Ltd., 1988.
3. "**Recommended Practice on Static Electricity**", NFPA 77, 2003.
4. "**Recommended Practice for the Classification of Flammable Liquids, Gases, or Vapors and of Hazardous (Classified) Locations for Electrical Installations in Chemical Process Areas**" NFPA 497, 2004
5. "Recommended Practice for the Classification of Combustible Dusts and of Hazardous (Classified) Locations for Electrical Installations in Chemical Process Areas" NFPA 499, 2004
6. "Standard for the Prevention of Fire and Dust Explosions from the Manufacturing, Processing, and Handling of Combustible Particulate Solids, NFPA 654, 2006

7. Eckhoff, R.K., **Dust Explosions in the Process Industries**, London UK: Butterworth-Heinemann, 2003.
8. Britton, L.G., **Avoiding Static Ignitions Chemical Operations**, AIChE, CCPS Concept Book, 1999, Publication G-67
9. Jaeger, Norbert and Siwek, Richard; "**Prevent Explosions of Combustible Dusts**", Chemical Engineering Progress, June 1999
10. T.H. Pratt **Electrostatics ignitions of fires and explosions**. Center for Chemical Process safety. AIchE Industry technology Alliance, USA, 1997.
11. L. Britton Private communications, November , 2003
12. M. Glor **Ignition hazard due to static electricity in particulate processes**, Powder Technology 135–136 (2003) 223– 233

Effectiveness of improvised gas absorption techniques for emergency responders at releases of toxic gases

Hannes Kern
Montanuniversität Leoben
Franz-Josef Strasse 18, 8700 Leoben
hannes.kern@unileoben.ac.at

Matthias Taferner, Harald Raupenstrauch
Montanuniversität Leoben
Franz-Josef Strasse 18, 8700 Leoben
matthias.taferner@stud.unileoben.ac.at, harald.raupenstrauch@unileoben.ac.at

Prepared for Presentation at
American Institute of Chemical Engineers
2013 Spring Meeting
9th Global Congress on Process Safety
San Antonio, Texas
April 28 – May 1, 2013

UNPUBLISHED

Effectiveness of improvised gas absorption techniques for emergency responders at releases of toxic gases

Hannes Kern
Montanuniversität Leoben
Franz-Josef Strasse 18, 8700 Leoben
hannes.kern@unileoben.ac.at

Matthias Taferner, Harald Raupenstrauch
Montanuniversität Leoben
Franz-Josef Strasse 18, 8700 Leoben
matthias.taferner@stud.unileoben.ac.at, harald.raupenstrauch@unileoben.ac.at

Keywords: toxic gas release, emergency response, absorption techniques

Abstract

In September 2011 a spill of nitric acid at a galvanic company in Trieben, Austria caused the release of large amounts of nitrous fumes. An IBC tank with 1000 liter of nitric acid (53%) was flipped over by a forklift and cracked open in the top area. The content was fully released and lead to serious corrosion effects on storage racks combined with the production of critical amounts of nitrous fumes. 27 workers had to undergo medical treatment and a storage and chemical handling area of several hundred square meters was filled by nitrous fumes with concentrations of up to 100 ppm.
Emergency responders used an improvised gas absorption system to clear the area of toxic fumes. Outside the sealed area concentrations of nitrous fumes well below the AEGL-2 (4h) value for nitrous dioxide where measured. In the course of the analyses of the response the effectiveness of improvised gas absorption techniques was investigated.
For the investigations chlorine and ammonia in combination with a mobile exhauster, (10.000 m^3/h) commonly used by emergency responders in Austria were used. Different methods of gas absorption where investigated and tested upon their practical relevance.

1. Introduction

Toxic gases are commonly used in industrial and public areas, both as base products for chemical production and as technical gases for various purposes. In Europe gases like chlorine or ammonia are often used as disinfecting agents or coolant in cold storage warehouses. In February 2012 a release of several hundred kilograms of ammonia from a cold storage warehouse in Vienna (Austria) injured 62 Persons, 32 firefighters and 4 emergency medical technicians right in the busy city center of Vienna. Firefighters were able to contain the released ammonia within the warehouse and tried to absorb and disperse it with water by using water curtains and water shields. Half a year earlier, on September 5th 2011 the spillage of 1000 liter nitric acid 53% in a metallurgical production site Trieben (Austria) caused the occurrence of significant amounts of nitrous fumes in a large storage area. The nitrous fumes were produced due to the reaction of the acid with the material of storage racks and metal materials in the surrounding area. The nitrous fumes were contained in the storage area, but ventilation procedures were very critical due to the close surrounding neighborhood. At the incident in Trieben 27 workers were injured. Due to a lack of knowledge regarding improvised gas absorption techniques experiments with ammonia, chlorine and different absorption agents were carried out. The experiments were supported by the Hazmat Team of the Graz Fire Brigade.

2. Application and testing of absorption techniques

2.1 General overview on techniques for gas defense

In the field of gas defense and control different techniques are applied by emergency responders. Depending on the strategy followed by the responding crews and the situational demands there are three main categories of technical and strategic measures that can be used during a response situation dealing with toxic or hazardous gases. These methods are described in the following sections.

2.1.1 Active methods

As "active methods" all response techniques can be considered that aim to directly fight and disperse gas clouds or change their direction of spread. Most of the active response methods are based on technical devices that are able to generate large volume flows of water or/and air that are directed against the spreading gas cloud. Depending on the size of the gas release water nozzles, water monitors or different sized ventilators are used for the active response approach. Appliances like the LUF 60 (firefighting support vehicle) combine a high volume rate ventilator (90.000 m^3/h) with a circular water dispersion system and provide therefore an aerosol cloud that shows better kinetic effects than the sole air stream. For water soluble gases also absorption effects can be achieved by this method.

2.1.1 Passive methods

"Passive methods" aim to contain the released gas cloud or try to protect areas from gas by a restriction of spread. An example for such systems are, water shields normally used to protect from radiative heat during firefighting operations. Also known are water curtains mainly consisting of fire hoses with different numbers of upward directed spray nozzles. Water curtains are more flexible in application. Water shields and water curtains are available with adjustable flow rates and for different operating pressures. The effectiveness largely depends on the spilled gas type and the nozzle and flow configuration of the chosen appliance.

2.1.1 Absorption techniques

Absorption techniques for emergency responders are in general not well established. For different purposes the effects of water on water soluble gases is known and well understood. The application of knowledge from industrial processes and their applicability for emergency responders is scientifically not well reviewed. Commercial portable gas absorbers are established in some industrial emergency response crews but mostly have very low volume rates. For water soluble gases some absorption techniques concerning the use of water aerosols or spray streams are known but often improvised. For chlorine spills the effects of sodium thiosulfate were investigated and have proved to be very effective [1,2], Other chemicals for absorption purposes regarding emergency response have not been studied very well. Effluents from absorption processes could cause hazardous situations too, but absorption in general offers the chance to reduce the three dimensional problem of a hazardous gas release to a two dimensional problem of a hazardous liquid.

2.2 Response at Trieben (Austria)

On September 5th 2011 a forklift at a metallurgical production site in Trieben (Austria) by accident dropped an IBC with 1000 liter 53% of nitric acid in the chemical storage area. The local fire brigade was called at about 11:00 am and arrived about five minutes later. The incident commander noticed during size up operations orange/brown fumes inside the cellar of the building. Most of the cellar was affected, about 200m^2 were directly contaminated with nitric acid. It is assumed that before the fire brigade was called, workers tried to dilute the nitric acid with up to 3000 liter of water, which caused severe a deterioration of the situation. About 30 minutes later the hazmat team arrived and started measurements with detector tubes. Values of around 100 ppm for nitrous fumes were measured in all areas of the cellar. The nitrous fumes occurred from the reaction of nitric acid with the material of the storage racks in the storage area. The lower parts of the storage racks were principally protected by a coating material but at some points the coating was damaged. Emergency responders over nearly 24 hours tried to contain the spill with chemical binding agents and chemical vacuum cleaners. Due to the contamination with nitrous fumes gastight suits were used which lead to huge logistical demands. The normal working time for the cleaning teams was around 45minutes to an hour, they were supplied by air hoses from a mobile compressor for respiratory air. Due to the close vicinity of the surrounding neighborhood it was not possible to vent the nitrous fumes directly in the atmosphere. A mobile exhauster was placed a spray nozzle was fixed near the exhaust opening of the ventilator. The flow rate of the water stream was around 200 l/min at 6 bar. The exhauster had an output of around 20.000 m^3/h. During the exhaust process the concentration of nitrous fumes was constantly monitored and well below 8,2 ppm (AEGL 2, 4h for NO_2) at the direct vicinity of the exhauster. The effect might have not only occurred through absorption effects but also because of a dilution effect of excess air at the input side of the exhauster.

2.3 Absorption tests with the MSA Auer COMVEX Ventilator

To investigate the potential of improvised gas absorption techniques the applicability of the MSA Auer COMVEX Ventilator was chosen as an example. The COMVEX Ventilator is used by many Austrian fire departments and offers the possibility to mount a water spray system on the ventilator that is normally used for the production of high expansion foam.

2.3.1 MSA Auer COMVEX [3]

The MSA Auer COMVEX Ventilator mainly consists of an air fan and tubing for suction and exhaust purposes. The fan has an air consumption of 10.000 m³/h and a power of 2,6 kW. Additional to the fan, a top piece for the production of high expansion foam is available (Pic.1). The foam generating kit consists of short tube with five spraying nozzles and a mixer to be able to ad foaming agents. At an operating pressure of 4-8 bar the water consumption of the foam generating kit is around 100 l/min. Mixing rates of 1-3% are possible.

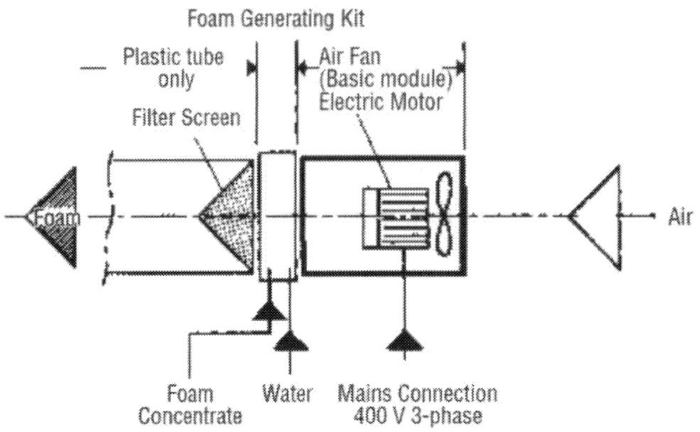

Pic. 1: Principal layout of the COMVEX Ventilator system [3]

2.3.2 Sodium thiosulfate as absorption material

Many absorption materials for technical purposes are well known. For emergency response purposes it is crucial to have absorption materials that present a low hazardous potential and have a good availability in large quantities. Logistically it is not possible to store large amounts of different absorption chemicals for specific tasks. For high risk chemicals (concerning toxicity and used quantities) it may be reasonable to foresee special measures. In case of chlorine that has a high toxic potential and is commonly used for chemical production processes and disinfection purposes, sodium thiosulfate gives good results as absorption agent in combination with water [2].
Sodium thiosulfate in an aqueous solution reduces elementary chlorine to chlorine anions. The reaction is shown in Equation 1:

$$2Na^+ + S_2O_3^{2-} + 4Cl_2 + 13H_2O \rightarrow 2Na^+ + 2HSO_4^- + 8\ H_3O^+ + 8Cl^- \quad [Eq.\ 1]$$

One mol of sodium thiosulfate is able to reduce 4 mol of elementary chlorine and has therefore a high potential to be used as absorption agent. Sodium thiosulfate pentahydrate has a solubility of around 680 g/l at 20 °C which equals a concentration of about 2.7 mol/l.

2.3.2 The MSA Auer COMVEX as improvised absorber

The focus of the investigations was the absorption of ammonia and chlorine at concentrations around 200-300 ppm. The ventilator was equipped with a suction tube (10 m) and an equally built exhaust tube. Tests with ammonia where carried out with inlet concentrations of 250 and 290 ppm. The ammonia feed stream was provided by a bottle of pressurized ammonia, the concentration was monitored by detector tubes. At the start of the experiment a constant ammonia stream was adjusted and measured at the exhaust tube while the ventilator was operating. When constant concentration conditions were reached, the injection of water started and the gas concentration was measured again. At the first test the ammonia concentration dropped from 250 ppm to 100 ppm (33.3%). The second run lead to a concentration of 150 ppm at a starting concentration of 290 ppm (48.3%).

Pic. 2: Improvised absorption system during experiments with chlorine

Due to the good solubility of ammonia in water, the tests were only carried out with water. The second series of experiments was carried out with chlorine in the same way. The first test with chlorine was started at a concentration of 490 ppm, the injection of water caused no measurable reduction of the concentration. In the second experiment with chlorine, a 2,7 mol/l solution of sodium thiosulfate was added to the water stream by the mixer of the foam generating kit (Pic.2). A mixing ratio of 3% was chosen. The use of sodium thiosulfate resulted in a reduction of the chlorine concentration from 260 ppm to 100 ppm (61.5%).

3. Conclusion

The effectiveness of gas absorption techniques is not very well known by emergency responders in Austria. Examples from real incidents show that the availability of gas absorption techniques may improve the effectiveness of emergency response at toxic gas releases. It is clearly not possible to have ready gas absorption appliances and agents for a wide range of specific problems. Scientific research on available improvised techniques and the development of technical improvements may enhance their effectiveness. Experiments with sodium thiosulfate in cases of chlorine spills show an effectiveness of over 60% at low chlorine concentrations. Keeping in mind that the investigated ventilator/nozzle system was not designed for gas absorption in general, the results may still leave a lot of room for improvements. Future research will focus on the improvement of the effectiveness of different methods for the control of toxic gas releases.

4. References

[1] Engelhardt, F., "Chlorabsorption an fallenden Tropfen und Übertragung auf Wasservorhänge," *Bergische Universität –Gesamthochschule Wuppertal*, Dissertation Juni 2002

[2] Bauch, M., Peine, T., "Versuche zur Erhöhung der Effizienz beim Niederschlagen von Chlor," *Brandschutz - Deutsche Feuerwehr Zeitung*, NO. 1, 2010

[3] "MSA AUER COMVEX Portable Ventilator and High Expansion Foam Generator - Instruction Manual," MSA AUER

Case Studies in Process Safety: Lessons Learned from Software-Related Accidents

Terry L. Hardy
Director, Safety & Risk Management
Great Circle Analytics, LLC
1238 Race Street
Denver, CO, 80206
thardy@gcirc.com

Prepared for Presentation at
American Institute of Chemical Engineers
2013 Spring Meeting
9th Global Congress on Process Safety
San Antonio, Texas
April 28 – May 1, 2013

Case Studies in Process Safety: Lessons Learned from Software-Related Accidents

Terry L. Hardy
Director, Safety & Risk Management
Great Circle Analytics, LLC
1238 Race Street
Denver, CO, 80206
thardy@gcirc.com

Keywords: Safety Analysis, Software Safety, Computing Systems, Process Safety Management

Abstract

Software and automation can play a significant role with respect to safety in the chemical process and energy production industries. Because computing systems are increasingly used to control critical functions, software may directly contribute to an accident. On the other hand, software can also be used as part of the hazard control strategy to reduce risks, and computing systems can provide valuable information to help make safety decisions. The importance of including software as part of an organization's efforts to analyze and manage hazards and risks seems clear, but for many organizations software is not effectively incorporated into process safety efforts. This paper reviews lessons learned from accidents and incidents to illustrate the potential for a software-related accident even when process safety management tools and techniques are used. This discussion is intended to provide insights to help improve process safety and software safety efforts.

1. Introduction

Process Safety Management generally refers to the application of management principles and systems to the identification, understanding, and control of process hazards to protect employees, facility assets, and the environment. Process Safety Management is focused on prevention of, preparedness for, mitigation of, response to, and restoration from catastrophic releases of chemicals or energy from a process associated with a facility. In Process Safety Management, a process is defined as any use, storage, manufacturing, handling, or the on-site movement of highly hazardous chemicals, or combination of these activities. Process Safety Management is regulated by OSHA through the Process Safety Management (PSM) Standard.

Software and computing systems are becoming essential to the operation of most industries, often through the automation of tasks, and as such can have a major impact on process safety. Software and computing systems offer tremendous advantages that encourage the use of automation. Software can allow complex operations to be performed, and can perform complicated calculations. Computing systems are reliable in the sense that there are a small number of components to perform a large number of actions. Computing systems are also flexible and can be modified to fit the application. From a safety perspective, automation can reduce the risks to operators by separating them from dangerous environments such as those in hazardous chemical processing facilities. In addition, automation can help by performing tasks that might pose a threat to humans, such as lifting large objects or operating high-speed cutting tools.

Software, computing systems, and automation have disadvantages as well. A major disadvantage to computing systems is that they often introduce increased complexity to systems, making them harder to understand, operate, maintain, and test. This complexity can also introduce new failure modes and interactions. Complex systems have a greater number of requirements, more interfaces, more subsystems, and more logical operations, increasing the potential for error. Computing systems can create a situation where no single operator can immediately foresee the consequences of a given action in the system. Computing systems can also increase coupling. Tight coupling occurs when processes are highly integrated and intrinsically time-dependent; for example, once a process has been set in motion, it must be completed within a certain period of time. Tightly coupled systems allow little time for operators to take action to prevent adverse effects. In addition, automation may require that operators think about their tasks in new ways, and may change the fundamental role of the operator from one who physically performed tasks to someone who monitors and communicates information. Finally, software will only do what a programmer has told it to do. Therefore, unlike humans, software cannot react to an abnormal situation unless specifically told to.

In spite of the fact that software is such an important part of complex systems, the analysis of hazards and risks from software has been inconsistent in the process industries. Safety analyses have historically been hardware-focused. Therefore, many analysts may not understand how to incorporate software into their system hazard analyses, and evaluators of those analyses may not understand what should be included and assessed. Organizations may be focused on compliance to regulations, which often do not address software, and as a result organizations may not properly assess or mitigate software risks. Organizations need to increase the attention given to addressing and analyzing the potential for hazards related to software and computing systems.

Software includes computer programs, procedures, scripts, rules, and associated documentation and data pertaining to the development and operation of a computer. Software can be developed by the organization implementing the system, by an outside software developer, or may be purchased as Commercial Off-The-Shelf (COTS) software. Software safety encompasses not just the software but also the computing system. A computing system includes the software and supporting hardware, sensors, effectors, humans who interact with the system, and data necessary for successful operation. Examples of computing systems include Programmable Logic Controllers (PLC) and Supervisory Control and Data Acquisition (SCADA) systems.

Through the review of hundreds of software-related accidents and incidents, and the author's personal experience in this discipline, common themes and lessons have been found.[1] This paper will use case studies to illustrate those themes and lessons learned. The discussion will center on software safety as part of a broader process safety effort and will include recommendations that can be used to improve the safety of software-driven systems.

2. Lessons Learned: Software and Computing System Related Accidents

This section discusses a number of lessons learned related to accidents and incidents where software and computing systems were contributors. These accidents are taken from detailed reports and investigative summaries from multiple industries and organizations, and are included to broadly illustrate hazards and risks in software and computing systems. Note that in discussing these accidents, this paper does not intend to oversimplify the events and conditions that led to the accidents, or place blame on any individuals or organizations. There is rarely a single identifiable cause leading to an accident. Accidents are usually the result of complex factors that include hardware, software, human interactions, environments, and procedures. The descriptions are meant to provide examples of where the analysis process failed in some way. Readers are encouraged to review the complete accident and mishap investigation reports referenced here to understand the often complex conditions and chain of events that led to each accident discussed here.

2.1 Decisions made in the acquisition and planning phases of development can profoundly affect safety

Planning typically involves trade-offs between many different facets of the program, including cost, schedule, performance, and safety. Poor planning can lead to unexpected safety consequences, and many safety decisions are actually made in the planning and acquisition phase. For example, inadequate resources may be allocated to the software safety effort. This can result in a failure to perform hazard analyses and identify safety requirements early in the program when these activities have the most impact. Therefore, safety personnel should be included in early phases of a program, especially when developing contractual requirements, and software safety considerations should be included in schedule and resource discussions.

On August 11, 1985, an Institute, West Virginia chemical production facility leaked methylene chloride and aldicarb oxime, toxic chemicals used to manufacture the pesticide Temik. Six workers were injured, and more than a hundred residents living near the facility were sent to the hospital. The toxic chemical leak occurred when a tank overheated after a steam valve failed. Due to the failure, steam leaked into the heating jacket of a storage tank containing a mixture of methylene chloride and aldicarb oxime. The methylene chloride vaporized from the heating, leaving concentrated aldicarb oxime to settle on the bottom of the tank. Operators later pumped the material from the storage tank for processing. They thought they had emptied the tank, but concentrated aldicarb oxime remained in the storage tank because a level gauge was broken and the operators performed no additional checks. The concentrated aldicarb oxime continued to be heated over several days by the steam in the heating jacket, ultimately resulting in a runaway chemical reaction days later. The high pressure from the reaction led to release of aldicarb oxime through a flare, and large amounts of the chemical were also released to the atmosphere when a burst disc failed.

The root causes of the leak were found to be the use of a tank that was not designed to hold aldicarb oxime, faulty level indicators, defective safety valves, mistaken transmission of steam to a tank, and failure of control room operators to notice critical pressure and temperature changes. In addition, a computerized monitoring system contributed to the severity of the accident. Earlier in the year, the company had installed a new air-monitoring system that could automatically identify and detect releases of hazardous gases. That computerized system could also predict whether the gases would be contained within the release area or would migrate and potentially harm the public. This information was important because it could be used to mitigate risks through evacuation or other measures if a vapor cloud appeared to be spreading. Company officials initially stated that the computer system had failed and therefore they were not informed of the lack of containment of the aldicarb oxime gas cloud. However, they later admitted that they had not purchased a version of the software programmed to detect aldicarb oxime. Had the company invested in the more expensive version of the software, they could have predicted that a cloud of aldicarb oxime would not have been contained in the production area. However, the company had only ordered the basic software model that did not contain aldicarb oxime properties.[2, 3]

2.2 Hazard causes may not include software and computing systems

Many organizations are improving their safety analyses with respect to software. However, software safety still lacks sufficient numbers of qualified practitioners, and the methods for performing comprehensive analyses are not universally known. In many cases, software causes may not be included in hazard analyses. Or if they are, the software causes may be stated in general terms in hazard analyses, such as including a generic hazard cause called "software error," instead of defining specifically what software functionality can lead to an undesirable outcome. In addition, the software hazard analyses may not pay enough attention to those cases where the software works exactly as intended, but the implemented functionality was unsafe.

On October 18, 2005, an explosion occurred at a natural gas decompression/recompression facility near Empress, Alberta. Extensive damage occurred to the facility, but no injuries were reported. The facility decompressed and conditioned high-pressure gas from a pipeline to achieve pressures and temperatures consistent with what was needed in the Natural Gas Liquids (NGL) stripping facilities. At the time of the accident the facility's "A" PLC lost communication with the NGL facility control room. Transportation Safety Board of Canada (TSB) investigators found that a processor card had failed, shutting down the PLC. When the PLC failed its outputs were de-energized. However, some of the microprocessor functions required an energized signal to initiate physical shutdown of some functions. These functions included main and auxiliary lube oil pumps, purge/damper valves, and suction and discharge valves. As a result of this condition, the pumps did not shut down and valves did not close in a safe position when the PLC failed. Operators in the NGL control room received an alarm indicating that the communications were lost. They investigated the problem and believed that the failure in communication had triggered an emergency shutdown of the facility. The operators believed that the emergency shutdown meant that all components were in a safe state. However, because the recompressor motor did not shut down, a surge event occurred in the pipe. This surge caused a pipe to break, leading to a natural gas leak. The leaking natural gas ignited, resulting in an explosion. The TSB noted that the loss of a PLC had not been identified as a safety concern in the hazard analyses prior to the accident.[4]

2.3 Risks may be underestimated or optimistically evaluated

The risk assessment process helps decision makers in their risk decisions (e.g., accept the risk without change, reduce the risk to an acceptable level through hazard elimination and mitigation, transfer the risk, or forego the activity), assists in justifying the acceptance of residual risk, and allows communication of risks throughout the organization. A risk assessment can either be qualitative or quantitative, although the emphasis in process safety is often on qualitative risk assessment. Evaluating software and computing system risk can be difficult, especially if an extensive history does not exist with the automated equipment. Risk must be carefully evaluated based on known technical factors, such as design complexity and maturity, degree of system testing, use of unproven technologies, potential for unexpected human intervention, and so on. Accidents have illustrated unfounded optimism in the use of software to assure safety.

On April 7, 2000, a pipeline failure occurred at the Chalk Point Generating Station in Maryland. The pipeline released approximately 140,400 gallons of fuel oil into surrounding wetlands, Swanson Creek, and the Patuxent River. The cost of the environmental response and cleanup exceeded $71 million. The U.S. National Transportation Safety Board (NTSB) determined that the probable cause of the accident was a fracture in a buckle in the pipe. The buckle went undiscovered because data from an in-line inspection tool were interpreted incorrectly. The NTSB stated that a contributor to the severity of the accident was inadequate operating procedures and practices for monitoring the magnitude of flow through the pipeline. When the leak occurred the 12-inch-diameter pipeline was being cleaned. An automated pipeline monitoring system existed, but this automated system could not adequately monitor pipeline conditions during this operation because the locations of the meters, pressure-sensing points, and temperature-sensing points had not been configured to be in the direct liquid flow path. Therefore, the automated system did not provide alarms or other information to operations personnel. Field personnel manually recorded tank level measurements at the upstream and downstream points, but did not use this information to evaluate whether any fuel oil had been lost. The manual monitoring procedures and lack of alarms from the automated system led to a 7-hour delay in responding to the leak. The field personnel finally realized there was a problem when the fuel pump began to cavitate. Using hand calculations the crew then realized that they had not received over 3000 barrels of fuel oil and they shut the pipeline down. Following the accident the company installed a SCADA system with software-based leak detection and radar tank gauges. In this case the operators underestimated the risks associated with a leak and overestimated their ability to detect a failure.[5]

2.4 Hazard controls may rely only on good software processes and testing

Hazard controls are devices and approaches to mitigate risk presented by a hazard by reducing either the severity of the hazard or the probability of its occurrence. Strong hazard control design follows the design order of precedence, where the first approach is to try to design out the hazard or minimize the risks through design selection. Many hazard reports will focus on the implementation of good software processes, or extensive unit testing, to prove that the design is safe. However, the software processes and testing will not help if the software design is flawed with respect to safe system operations (including hardware, software, processes, humans, and environments). In these cases the software may operate exactly as designed, but the operation may be unsafe.

On February 11, 2003, an employee of a manufacturer in Gonzales, Texas was fatally injured while performing maintenance on a reaction tank. The U.S. Mine Safety and Health Administration (MSHA) determined that the cause of the accident was a failure to close and secure a manual gate valve for a steam line and a failure to place the batch PLC in the stop mode. The company was a surface clay mill that purchased clay and blended, refined, milled and processed the material into products used in paints, inks, and grease. On the day of the accident the employee had been informed that there had been a product change in one of the batch processing systems. The employee was assigned to perform cleanup duties on a reactor tank. Two valves controlled steam entry into the tank: a manual gate valve and a butterfly valve with an automatic pneumatic actuator. The PLC controlled the functioning of the batch system based on sensors that monitored material flow. At the time of the accident the PLC was in "slurry hold" mode. In this mode the system was programmed to actuate the steam valve when the clay slurry level reached 5.5 feet. An aluminum extension ladder used by the employee caused the level sensor to falsely sense that slurry was in the reactor, which resulted in the PLC sending a command to open the steam valve. Because the manual valve had been left open, steam at 350° F then entered the tank, fatally burning the employee.[6]

2.5 Organizations may fail to consider the possibility of hazard controls not working

Hazard controls are never 100 percent reliable. Therefore, it is important for the analyst to consider what could happen if the controls fail. Consideration should be given to unexpected hardware, software, and human interactions, and a "defense in depth" approach should be used where multiple, independent methods are used to prevent a hazard from becoming an accident. Controls should be carefully evaluated to identify what could happen if they do not work as expected, especially if those controls rely heavily on software and personnel actions.

On February 18, 2009, an employee was fatally injured at a coal preparation plant in the Hunter Valley region of New South Wales, Australia. The accident occurred when 10 tons of waste rock were inadvertently released from the reject bin and fell onto the cabin of the employee's truck. At the plant, raw coal was extracted from the mine and usable coal was separated from waste rock. The waste rock was transferred approximately 2 kilometers on conveyers to the reject bin. The waste rock was then loaded from the reject bin onto trucks and hauled away. The process of loading the trucks with waste rock was controlled by a PLC system. The PLC system included truck detection sensors, traffic lights, bin capacity sensing, and remote control, hand-held transmitters used by the truck drivers. On the day of the accident the truck driver drove his truck under the reject bin delivery chute. A signal was sent from the handheld remote control to command the chute to open. The accident report stated that it was not clear whether the signal was sent inadvertently or intentionally. Opening the chute required that two of three lines of truck detection sensors be blocked in addition to a command from the remote control to assure that the truck was in the correct location. Each sensor line contained three sensors, and all three sensors had to be blocked for the entire line to be considered as blocked. At the time of the accident the truck was obscuring one line of sensors, and a second line of sensors was obscured by dirt on the lenses and therefore was not working correctly. Because two of the sensor lines were blocked and the remote control signal had been sent, the PLC automatically opened the reject bin chute door and dropped 10 tons of material on the truck cab before the driver had safely cleared the chute, resulting in the fatal injury.[7]

2.6 Testing may not provide information on subsystem and component interactions or hazard control operations

The safety verification and validation process is intended to determine that the design solution has met all the safety requirements (verification) and that the correct system is built (validation). The verification and validation process, if performed correctly, will provide evidence that risk has been reduced. Testing is an important part of the verification and validation process, and comprehensive software testing should be conducted throughout the development cycle. Testing should include not only nominal conditions based on requirements but also abnormal conditions such as improper inputs, inputs outside expected conditions, inputs exactly at and near the boundary conditions, and inputs stuck at some value. Safety-critical software must include full system integration testing of end-to-end events. That testing must include stressing of the software, and should include interactions of the software with hardware, humans, and environments. A number of accidents have occurred when no component failed in the conventional sense, but the interaction of components caused a system failure. Testing must include verifications of hazard controls, and it should assure that redundancy works when needed.

On October 24, 2002, a grinder exploded at a limestone processing plant in Foreman, Arkansas. An operator was killed when flammable waste fuel covered him and ignited. The operator had started the pump for solid waste fuel processing when the accident occurred. The MSHA stated in its report that the cause of the accident was that the safety monitoring system designed to shut off the waste fuel system pump had not been maintained so that it functioned properly. Kilns were used in processing activities at the plant, and these kilns were heated by burning coal, natural gas, and liquid waste fuel. The liquid waste fuel was delivered by truck or railcar and pumped into large storage tanks. From the storage area it was pumped through a grinder to reduce the particle size of the solids in the fuel. Two independent systems monitored and controlled the waste fuel delivery. A Foxboro Intelligent Automation Distribution Control System monitored and recorded normal operating parameters. The Foxboro also issued audible and visual alarms that were available at the plant control room. A PLC provided basic startup and shutdown of the system and responded to commands from the Foxboro. On the day of the accident the Foxboro sensed that the fuel delivery pressure was low, apparently due to blockage in the line. As designed, the Foxboro sent a command to the PLC to shut down the pumps. However, the PLC failed to respond and the pumps kept running. Three months prior to the accident this PLC had been installed; this was supposed to be a simple replacement of an older PLC of similar capability. However, the Foxboro had not been connected to the newer PLC, and the connections remained to the older non-functioning PLC. The complete system had never been tested with the new PLC. A system test had been scheduled three days prior to the accident but had been aborted when a pump failed during the test; the test had never been rescheduled. The accident report stated that the blockage may have broken free just prior to the accident. With the pumps running, the pressure elevated significantly and a "water hammer" effect caused overpressurization in the system at the grinder. The grinder was torn loose from its base, spraying fuel and pulling loose a 480-volt cable that ultimately served as an ignition source.[8]

2.7 Software change management and hazard analyses processes may not be integrated

Engineering by its very nature is an activity that requires change. Changes can come about for a variety of reasons, including the discovery of problems during development, changes in requirements, or routine upgrades of software or hardware. As the development cycle proceeds and as an organization gains more operational experience, new hazards are often uncovered and some hazards may no longer be relevant. If the hazard analysis is not updated to reflect these changes then resources may be expended on previously identified hazards that may no longer be relevant, and new hazards may not be discovered as the design matures. In addition, routine changes and modifications to software and computing systems must be analyzed for potential hazards as part of the broader hazard analysis process.

On June 10, 1999, a 16-inch-diameter pipe carrying gasoline ruptured in Bellingham, Washington. The ruptured pipeline released approximately 237,000 gallons of gasoline into a nearby creek, according to the NTSB. That gasoline then ignited, burning approximately 1½ miles along the creek. Three people died in the fire. One home and the Bellingham water treatment plant were also damaged in the accident.

The NTSB investigated the accident and determined that the probable cause of the rupture was damage to the pipe during a modification project performed in 1994. This damage weakened the pipeline, making it susceptible to rupture under increased pressure in the pipe. The NTSB also stated that inspections of the pipeline during the project were inadequate and the company did not identify and repair damage. The report noted that in-line pipeline inspection data should have prompted the company to excavate and examine that section of pipeline, but the company failed to perform such work after reported anomalies. The SCADA system computers also played a role in the accident. The SCADA system was used for operation of the pipeline, for example to open and close valves remotely as required or to operate pumps as needed. Just prior to the accident the operator was preparing to initiate delivery of gasoline to a terminal in Seattle, diverting delivery from another facility. During the process of switching delivery destinations, the pressure in the pipeline began to increase, which is a normal condition but one that required the operator to start a pump to reduce pressure. However, when the operator tried to start that pump, the SCADA system failed to execute the start command issued by the operator. The operator soon found that the SCADA system was unresponsive to any commands, something that had never happened before. The report stated that, "Had the controller been able to start the pump at Woodinville, it is probable that the pressure backup would have been alleviated and the pipeline operated routinely for the balance of the fuel delivery." Instead, the pressure in the pipe increased, and the increased pressure likely caused the damaged pipe to rupture.

The cause of computer system failure was likely a change made to the system database just prior to the accident. The NTSB accident report stated that the SCADA system administrator entered new records into the live database at the time of the accident. The system administrator however did not check the records or test the system software to see if those changes introduced any problems. The computing system problem could not be replicated after the accident and therefore the cause of the anomaly could not be definitively identified. The report stated, "The Safety Board concludes that, had the SCADA database revisions that were performed shortly before the accident been performed and thoroughly tested on an off-line system instead of the primary on-line SCADA system, errors resulting from those revisions may have been identified and repaired before they could affect the operation of the pipeline."[9]

2.8 Human-software interactions have significant safety implications that are often underestimated

Humans interact with hardware and software in a variety of ways. A number of accidents have occurred where these interactions have been a significant contributing cause. Of special concern is when a new computing system is implemented that changes what is expected of the operator in normal and emergency situations.

On August 28, 2008, an explosion occurred at a pesticide manufacturing plant in Institute, West Virginia. Two workers were killed in the explosion and eight others were injured. The U.S. Chemical Safety and Hazard Investigation Board (CSB) found that the explosion was the result of a runaway chemical reaction. The company was starting up a methomyl unit for the first time after several months of down time for maintenance to install a new computer control system and reactor vessel. Normal operations called for dissolved methomyl and waste chemicals to be fed into a preheated residue treater vessel partially filled with solvent. The treater allowed the methomyl to decompose safely, after which it was mixed with other waste chemicals and used to fuel facility boilers. On the day of the accident a methomyl-solvent mixture was added prematurely to the residue treater vessel before solvent had been added to the tank. This event occurred when operators were troubleshooting equipment problems during the startup. This mixture was supposed to be added to the vessel only after that vessel was filled with clean solvent and heated to a minimum safe temperature. An interlock system existed as part of the automatic feed control system to prevent inadvertent introduction of methomyl. This interlock was password-protected to prevent inadvertent override, but operators intentionally overrode the interlock. Bypassing the interlocks had apparently been a practice condoned by management in the past as a workaround for operational problems. Once the methomyl decomposition reaction started it could not be stopped, and the pressure rapidly rose in the vessel due to gas from the reaction, leading to the explosion. Uncrystallized methomyl existed in the tank due to equipment problems, and this material greatly increased the methomyl concentration in the residue treater, contributing to the runaway reaction.

The CSB stated that the company initiated this process startup before the company had completed critical checks, including valve lineups, equipment checkouts, and computer calibrations. The CSB said that the operating procedure for the startup was complex, but this procedure had not been reviewed or approved. In addition, the company had not performed training on the new computer system put in place as part of the maintenance. This new computer system offered significant improvements to help automatically control the operation. For example, the control system included graphical display screens that simulated the process flow. But the system was also complex and the modifications changed the way operators interacted with the system. The operators thought the screens were difficult to navigate, and responding to troubleshooting alarms was difficult. The accident report stated that, had the operators received adequate training on the new computer system, they may have been able to recognize problems in operation before the explosion.

The report faulted the company for not performing an adequate pre-startup safety review. System operators did not participate in the safety review, and review checklists showed items as completed when they were not. The CSB stated that the company had also failed to perform a thorough Process Hazard Analysis. The report stated that the Process Hazard Analysis was performed quickly because management had not allotted sufficient time for analysis. In addition, the CSB stated that the Process Hazard Analysis included invalid assumptions, and said that the team did not apply the tools properly, resulting in unmitigated accident scenarios.[10]

2.9 Support software, including models and simulations, may be as critical to safety as control software

While the focus of software safety efforts is usually on software directly controlling an application, support software may contribute to an accident or to the effectiveness of the emergency response. Examples of support software include databases used for maintenance activities, software to estimate load stability, models that provide design calculations or assurance information, and so on. A failure to analyze the hazards associated with this support software can result in unforeseen system failures.

On November 12, 2008, a 2 million gallon liquid fertilizer tank at in Chesapeake, Virginia collapsed. Two workers performing welding operations at the site were seriously injured and an adjacent neighborhood was partially flooded as a result of the accident. The CSB found that the company had not assured that welds to replace vertical joints met accepted industry standards, and the CSB faulted the company for its failure to perform inspections of the welds. The company was also faulted for not having proper procedures in place for filling the tanks following major facility modifications. In its report, the CSB also noted that the contractor hired by the company to calculate the maximum fill height had used some faulty assumptions. The maximum liquid level was supposed to be calculated in part based on the minimum measured shell thicknesses and the extent of the weld inspection (full, spot, or no radiography). The contractor used the maximum (not minimum) measured thickness, and assumed full inspection of the welds. These assumptions led to an overestimation of the allowable liquid level. The tank failed at a fill level of 26.74 feet, below the calculated maximum of 27.01 feet. The CSB also noted a number of previous overfilling accidents. The CSB found 16 other tank failures at nine facilities in other states between 1995 and 2008. These 16 failures resulted in one death, four hospitalizations, one community evacuation, and two releases to waterways. Eleven occurred due to defective welding.[11]

3. Recommendations for the Software System Safety Process

There is a great deal of information in the literature on how to improve software development processes. However, the above themes and lessons learned indicate that improving software and computing system safety is more than just providing additional training or improved software development techniques or tools. Improving software safety requires a change in the way process safety practitioners think. Software safety must be fully integrated into the process safety analysis process. And process safety practitioners must begin to ask critical questions about their software safety efforts. Some examples of these questions include the following:

- Do plans reflect how business is really done? Are plans reviewed? Do plans have unrealistic schedules or resource allocations? Are software and computing systems considering in the planning and acquisition processes? Poor or unrealistic plans may reflect an organization that does not truly place a priority on safety activities.
- Is there a convincing story that the safety analysis is complete? Is there a sufficient description of the system to support an analysis, including the software and computing system? Are computing systems treated as "black boxes"? Have software causes been thoroughly evaluated using a systematic approach, or are the software causes listed as "software error" with no further explanation? Do the hazard analyses include support software, such as design models or databases? Failure to show that the problem is being looked at systematically could indicate that there are holes in the analysis, with potentially significant problems overlooked.
- Are the hazard reports detailed enough? Are causes descriptive? Are the recommendations clear and sufficiently detailed? Does the logic make sense and is it complete? Do controls match up with the causes, showing a one-to-one or many-to-one relation? Lack of detail could be an indication of insufficient knowledge of the system, or lack of information on the system.
- Do the risk assessments have a logical basis? Have the risk assessments considered software complexity, maturity, testing, use of unproven technologies, etc.? Are assessments overly optimistic based on what is known about the computing system? A failure to provide a basis for your risk assessments may result in unrealistic assessments.
- Are the hazard controls primarily procedural rather than design changes, safety features, or devices? Is there an overreliance on alarms and signage? Do software controls only rely on good software processes? Is there an overreliance on humans and software to "save the day"? Overreliance on operational controls may indicate a weak safety design.
- Are hazard control recommendations being implemented? Can the selected hazard control strategy actually be implemented and verified? Is the control strategy so complex that it will be impossible to determine whether it will work when needed? Complex controls, overlapping control strategies, or inadequate implementation may be an indication of a weak safety design.
- Has the risk assessment truly considered the worst case? What is the basis for the likelihood levels? Is the risk analyzed only for steady-state operation, or have startups and shutdowns also been considered? Failure to provide good answers to these questions indicates a potential misunderstanding of the risk.
- Are problems found in test and design included in the hazard reports and factored into the design? Have problems and incidents been fully investigated? Does a process exist for tracking action items, including whether recommendations have been implemented? Are changes to software and computing systems properly factored into the hazard analysis? Failure to incorporate problems, changes, and corrective actions is an indication of the potential to miss serious design flaws.

These questions help to identify whether the hazard analysis process is robust. However, we must also ask questions specifically related to the use of software and computing systems in complex systems. The best questions come from real-world examples of accidents where software has been a contributor. Some examples of questions are as follows, and others can be found in the literature.[12]

- Have safety-critical software, commands, and data been identified?
- Do hazard controls for software-related causes combine good practices and specific safeguards?
- Is software and system testing adequate, and do tests include sufficient off-nominal conditions?
- Is the computing system design overly complex?
- Is the design based on unproven technologies?
- What happens if the software locks up?
- Are the sensors used for software decisions fault tolerant?
- Has software mode transition been considered?
- Has consideration been given to the order of commands and potential out of sequence inputs?
- Will the software and system start up and shut down in a known, safe state?
- Are checks performed before initiating hazardous operations?
- Will the software properly handle spurious signals and power outages?

These are by no means all the questions a decision maker should ask, and positive answers to these questions provide no assurance that an accident will be prevented. These questions should encourage critical thinking about safety and generate additional questions to provide further insight on system risk. A failure to ask these questions could mean that the potential for an accident is higher than we had assumed.

It is up to all stakeholders to look for those conditions that could lead to an accident and to recognize that the worst can happen. This means we should all express concerns about software safety management and engineering when necessary based on our knowledge, experience, and judgment, including lessons learned from accidents. We must ask questions to understand the potential for harm, to understand the steps taken to assure that the risks have been reduced, and to assure that there is proof that hazard controls are effective. We do not have to be software experts to ask questions, and in fact a lack of expertise could be an advantage when trying to understand how a system works. Most importantly, when thinking of software and computing system safety, we must think of the system and its interactions between software, hardware, humans, processes, and environments.

4. Conclusion

Process Safety Management can provide immense benefits in reducing operational risks in the chemical process and energy production industries. By proactively identifying hazards, assessing and characterizing risks, and taking actions to reduce those risks, organizations can prevent accidents and reduce the potential for death, injury, property damage, and environmental impacts. Given the importance of software and computing systems in the design and operation of many systems, software must be included as part of the broader process safety effort. However, the accidents provided in this paper, along with many others, illustrate that software is often not systematically considered as part of process safety activities. For example, software may not be identified as a hazard cause, risks may be underestimated, software-related hazard controls may be oversimplified, testing may not include sufficient safety scenarios, and system changes may not be factored into the design or hazard analysis. Process safety practitioners must take advantage of available software and computing system lessons learned, going beyond examples presented here, to improve their own safety efforts and, most importantly, to prevent accidents.

5. References

[1] Hardy, T.L., *Software and System Safety: Accidents, Incidents, and Lessons Learned,* AuthorHouse, 2012.

[2] Pool, B., "System Wrongfully Blamed in Union Carbide Leak: SAFER Fights Off Dark Cloud of Bad Publicity," *Los Angeles Times,* August 20, 1985.

[3] Schlager, N., *Breakdown: Deadly Technological Disasters,* Visible Ink Press, 1995.

[4] Transportation Safety Board of Canada, "Programmable Logic Controller Failure, Foothills Pipe Lines Ltd. Decompression/Recompression Facility, BP Canada Energy Company Empress Natural Gas Liquids Facility Near Empress, Alberta 18 October 2005," Report Number P05H0061, July 12, 2006.

[5] U.S. National Transportation Safety Board, "Rupture of Piney Point Oil Pipeline and Release of Fuel Oil Near Chalk Point, Maryland, April 7, 2000," Pipeline Accident Report NTSB/PAR-02/01, July 23, 2002.

[6] U.S. Mine Safety and Health Administration, "Report of Investigation: Fatal Other Accident (Steam Burns), February 11, 2003, Southern Clay Plants & Pits Southern Clay Prod. Inc., Gonzales, Gonzales County, Texas," Mine I.D. No. 41-00298, 2003.

[7] State of New South Wales (Australia), Department of Industry and Investment, "Fatality involving David Hurst Oldknow Ravensworth Underground Mine Coal Preparation Plant Reject bin 802 18 February 2009," May 2010.

[8] U.S. Mine Safety and Health Administration, "Report of Accident : Exploding Vessels Under Pressure Accident, October 24, 2002, Foreman Quarry and Plant, Ash Grove Cement Company, Foreman, Little River County, Arkansas," Mine I.D. No. 03-00256, 2003.

[9] U.S. National Transportation Safety Board, "Pipeline Rupture and Subsequent Fire in Bellingham, Washington, June 10, 1999," Pipeline Accident Report NTSB/PAR-02/02, October 8, 2002.

[10] U.S. Chemical Safety and Hazard Investigation Board, "Pesticide Chemical Runaway Reaction Pressure Vessel Explosion, Bayer CropScience, Institute, West Virginia, August 28, 2008," Report No. 2008-08-I-WV, January, 2011.

[11] U.S. Chemical Safety and Hazard Investigation Board, "Investigation Report: Allied Terminals, Inc. – Catastrophic Tank Collapse, Allied Terminals, Inc., Chesapeake, Virginia, November 12, 2008," Report No. 2009-03-I-VA, May 2009.

[12] Hardy, T.L., *Essential Questions in System Safety: A Guide for Safety Decision Makers,* AuthorHouse, 2011.

Technical Report on Chevron 2012 Pipe Rupture and Fire

Steve Cutchen, Lauren Grim
United States Chemical Safety and Hazard Investigation Board
1 Denver Federal Center Way
Building 20, Room A1419
Denver, CO 80225

Roger Evans, Don Holmstrom, Amanda Johnson, Dan Tillema, Kelly Wilson, Mark Wingard
United States Chemical Safety and Hazard Investigation Board

The CSB is an independent federal agency charged with investigating serious chemical incidents. The agency's board members are appointed by the President and confirmed by the Senate. CSB investigations look into all aspects of chemical incidents, including physical causes such as equipment failure as well as inadequacies in regulations, industry standards, and safety management systems. Please visit the CSB's website at www.csb.gov. For further information contact Communications Director Hillary Cohen at (202) 446-8094 or Sandy Gilmour at (202) 251-5496

Prepared for Presentation at
American Institute of Chemical Engineers
2013 Spring Meeting
9th Global Congress on Process Safety
San Antonio, Texas
April 28 – May 1, 2013

UNPUBLISHED

Technical Report on Chevron 2012 Pipe Rupture and Fire

Steve Cutchen, Lauren Grim
United States Chemical Safety and Hazard Investigation Board
1 Denver Federal Center Way
Building 20, Room A1419
Denver, CO 80225

Roger Evans, Don Holmstrom, Amanda Johnson, Dan Tillema, Kelly Wilson, Mark Wingard
United States Chemical Safety and Hazard Investigation Board

Keywords: corrosion, sulfidation corrosion

Abstract

The U.S. Chemical Safety and Hazard Investigation Board (CSB) and the California Division of Occupational Safety and Health (Cal/OSHA) released a technical evaluation report on piping samples taken from the Chevron Refinery in Richmond, California, where a hydrocarbon release and massive fire occurred on August 6, 2012. The report, prepared by Anamet, Inc., concluded the piping ruptured due to severe sulfidation corrosion, and that tested pipe samples showed a very low concentration of corrosion-inhibiting silicon.

1. Background

On August 6, 2012, a hydrocarbon release and resulting fire occurred at the Chevron U.S.A. Inc. Richmond Refinery in Richmond, California The incident occurred when a combustible hydrocarbon liquid known as "gas oil" leaked from an 8-inch pipe connected to an atmospheric crude oil distillation column in the refinery's crude unit. Workers initially noted the leak and were in the process of attempting to diagnose the source of the leak in the still-operating crude unit when the pipe ruptured catastrophically. Due to the high temperature, in excess of 600 degrees Fahrenheit, and physical properties of the material in the equipment, the gas-oil immediately formed a large hydrocarbon vapor cloud.

The CSB determined that nineteen Chevron employees were engulfed in a vapor cloud formed by the hydrocarbon release. Eighteen employees escaped before the fire started and one employee escaped without injury after the fire ensued. The fire burned for approximately 5 hours, and water was applied to the rupture site for 23 days. The incident resulted in six minor injuries. Production at the crude unit was suspended for over six months after the incident. More than 15,000 residents in the surrounding area sought treatment at area medical facilities as a result of the release and fire. Post incident on-site visual examination by Chevron, Cal/OSHA, and the CSB revealed extensive thinning of the pipe wall at the location of the rupture.

2. Sample Testing and Report

The evaluation, performed by Anamet, Inc. of Hayward, California was designed and overseen by the CSB, the California Division of Occupational Safety and Health Administration (Cal/OSHA), the United Steelworkers union, and Chevron. The evaluation was funded by the CSB. The report concludes that the 8-inch steel pipe, from a section designated as 4-sidecut which was installed in 1976, ruptured due to severe sulfidation corrosion, and that tested pipe samples showed a very low concentration of corrosion-inhibiting silicon.

The report cites wall thinning due to sulfidation corrosion as the cause of the piping failure. In crude oil distillation, the report notes, naturally occurring sulfur and sulfur compounds are available to react with steel components, particularly plain carbon steels. Corrosion rates vary according to the sulfur content of the oil being processed, temperature, and other factors, including silicon content and other materials in steel pipe.

The report concludes, "The reported service conditions of the #4 Crude Unit 4-sidecut piping, the composition of the carbon steel, presence of thick sulfide scale on the inside surfaces of the pipe, and generally uniform wall thinning indicate that sulfidation corrosion was the cause of the 8-inch 4-sidecut rupture."

The report noted that the failed pipe section was subjected to a higher corrosion rate than neighboring components due to low silicon content in the pipe. The report states that "Experience has shown that silicon in carbon steel is known to inhibit sulfidation" when the concentration is above a threshold value. Chemical analysis of the Chevron pipe showed the silicon concentration of the ruptured section to be far below this value, resulting in a significantly higher rate of sulfidation corrosion.

The report, which contains numerous photographs of the corrosion and pipe rupture, also notes the presence of an inward deformation toward the inside surface of the pipe. The CSB has determined this deformation was likely caused by a fire pike used during the plant's emergency response. A photograph depicts a tight fit between the fire pike tip and this inward deformation of the failed pipe.

3. Report Findings

The Anamet report provides the following conclusions:

1. Rupture of the 8-inch 4-sidecut resulted from wall thinning caused by sulfidation corrosion.
2. Post rupture corrosion destroyed fracture morphology that could have indicated the rupture initiation site. However, rupture likely originated in the thinnest region and initially followed a longitudinal path driven by the hoop stress. Consequently, an area of likely rupture initiation was identified.

3. The wall thickness of the ruptured section was less than all the other sections of the 8-inch 4-sidecut that were evaluated.

4. Chemical analysis showed the silicon concentration of the ruptured section was 0.01-wt%.

5. Six of twelve specimens (50%) from unique sections of the 8-inch 4-sidecut, 12-inch 4-sidecut, and 2011 12-inch samples were found by chemical analysis to have a silicon concentration of less than 0.1-wt%.

6. Deformation of a small region of the 8-inch 4-sidecut rupture edge was consistent with mechanical force applied from the outside surface of the pipe, possibly from a fire pike, such as samples E-099 and E-082-2, or other pointed object. This deformed region was located within the area of likely rupture initiation.

4. CSB Future Plans

The CSB investigation to determine the root causes of the incident is ongoing. The CSB will release a report detailing its findings and recommendations to key stakeholders later this year. The CSB, Cal/OSHA, the USW, and Chevron are cooperating under an agreement to test 4-sidecut carbon steel piping from Chevron's refinery in El Segundo, California, south of Los Angeles.

AUTHOR INDEX

AUTHOR INDEX